CBS Engineering Series

Numerical Methods in
Science and Engineering

CBS Engineering Series

Numerical Methods in
Science and Engineering

Useful for

- Undergraduate and Post Graduate students of Engineering, MCA, B.Sc. and M.Sc.

- GATE and All entrance examination for admission in M.Sc., M.Phil. and Ph.D. Courses

Dr. SUDHIR KUMAR PUNDIR

M.Sc., M.Phil, NET, Ph.D.
Associate Professor
Department of Mathematics
S.D. (P.G.) College,
Muzaffarnagar (U.P.)

CBS Publishers & Distributors Pvt. Ltd.

New Delhi • Bengaluru • Chennai • Kochi • Kolkata • Mumbai
Hyderabad • Nagpur • Patna • Pune • Vijayawada

NUMERICAL METHODS
in Science and Engineering

ISBN: 978-93-86217-64-6

First Edition: 2017

Published by:
Satish Kumar Jain for CBS Publishers & Distributors Pvt. Ltd.,
4819/XI Prahlad Street, 24 Ansari Road, Daryaganj, New Delhi - 110002
delhi@cbspd.com, cbspubs@airtelmail.in • www.cbspd.com
Ph.: 23289259, 23266861, 23266867 • Fax: 011-23243014
Corporate Office: 204 FIE, Industrial Area, Patparganj, Delhi - 110 092
Ph: 49344934 • Fax: 011-49344935
E-mail: publishing@cbspd.com • publicity@cbspd.com
Branches:
• *Bengaluru:* 2975, 17th Cross, K.R. Road, Bansankari 2nd Stage,
  Bengaluru - 70 • Ph: +91-80-26771678/79 • Fax: +91-80-26771680
  E-mail: cbsbng@gmail.com, bangalore@cbspd.com
• *Chennai:* No. 7, Subbaraya Street, Shenoy Nagar, Chennai - 600030
  Ph: +91-44-26681266, 26680620 • Fax: +91-44-42032115
  E-mail: chennai@cbspd.com
• *Kochi:* Ashana House, 39/1904, A.M. Thomas Road, Valanjambalam,
  Ernakulum, Kochi • Ph: +91-484-4059061-65
  Fax: +91-484-4059065 • E-mail: cochin@cbspd.com
• *Kolkata:* 6-B, Ground Floor, Rameshwar Shaw Road, Kolkata - 700014
  Ph: +91-33-22891126/7/8 • E-mail: kolkata@cbspd.com
• *Mumbai:* 83-C, Dr. E. Moses Road, Worli, Mumbai - 400018
  Ph: +91-9833017933, 022-24902340/41 • E-mail: mumbai@cbspd.com

Representatives:
• Hyderabad: 0-9885175004   • Nagpur: 0-9021734563
• Patna: 0-9334159340   • Pune: 0-9623451994
• Vijayawada: 0-9000660880

Printed at:
India Binding House, Noida (U.P.)

Preface

The book entitled "NUMERICAL METHODS IN SCIENCE AND ENGINEERING" meet the needs of engineering and science students of UG and PG levels. Besides, it will also be very useful for students preparing for various competitive examinations.

The contents of this book are derived from the curricula offered by various universities across the country. This book consists of eleven chapters. In each chapter an ample amount of theory is given which is supported by solved examples followed by exercises along with their answers. Numerical techniques using C-language have been explained in each chapter. Each of the program has been successfully run on a computer and tested. A list of objective questions are given at the end of each chapter.

I express my gratitude to the authors and publishers of various books I consulted during the preparation of the book.

I wish to sincerely thank **Sh S.K. Jain**, Managing Director, CBS Publishers and Distributors, New Delhi for his encouragement and help in bringing out this publication in a present nice form.

My special thanks to Sh. B.M. Singh, Sh. Sunil Dutt, Sh. Puneet Verma and entire team of CBS Publishers and Distributors, New Delhi whose encouragement and unstinted support enabled me to complete my book. Mr. Peeyush Goel, M/s Dreamshapers also deserve special mention for nice type setting.

I must also record my appreciation due to my wife Dr. Rimple, daughter Rijuta and son Shrish for their understanding and love during the long period that I have taken to complete this book.

Above all I am thankful to The Almighty God, without whose grace nothing is possible for any one.

Readers are welcomed to point out errors, if any and send their valuable suggestions for improving the quality of the book.

Dr. Sudhir Kumar Pundir
email : skpundir05@yahoo.co.in

Contents

1. Programming In C   1-92

1.1	Introduction	1
1.2	'C' and its development	1
1.3	Important features of C	1
1.4	Structured programming concept of modular programming	2
1.5	Working with 'C' program	2
1.6	Contents of C-program	3
1.7	Program writing in C	4
1.8	Operators in a C-program	8
1.9	Hierarchy of operators	11
1.10	Header files	12
1.11	Instructions of a C-program	13
1.12	Data output with printf () function	16
1.13	Use of backslash characters	17
1.14	Data input with scanf () function	18
1.15	Input-output manipulatiors	22
1.16	Decision control in the C-program	23
1.17	Repetition control statements	41
1.17	Working with arrays	60
1.18	String–the array of characters	74
1.19	Functions	83

2. Error and Approximations   93-154

2.1	Introduction	93
2.2	Accuracy of numbers	93
2.3	Errors and their analysis	96
2.4	Enherent errors	103
2.5	Rounding off error	104
2.6	Truncation error	104
2.7	The general formula for errors	106
2.8	Floating point arithmetic and errors	109
2.9	Computer storage	109
2.10	Concept of normalized floating point	111
2.11	Pitfalls of floating point representation	111
2.12	Error in a series approximation	116
2.13	Error in determinants	117
2.14	Application of error formula to the fundamental operations of arithmetics	117
2.15	Order of approximations	125
2.16	Propagation of error	126
2.17	Blunders	129
2.18	Numerical instability	130
2.19	Machine computations	130
2.20	Computer software	133
2.21	Number system	133
2.22	Base conversion	134
2.23	Binary arithmetic	143
	Computational technique lab	147

3. Numerical Solution of Algebraic and Transcendental Equations   155-250

3.1	Introduction	155
3.2	Properties of the equations and its roots	155

3.3   Methods of solution   156
3.4   Bisection method   156
3.5   Secant method   167
3.6   Iteration method   170
3.7   Iterative method for the system of non-linear equations   177
3.8   Regula-Falsi method (or method of false position)   179
3.9   Newton-Raphson's method   191
3.10   Complex roots   205
3.11   Newton's method for complex roots   206
3.12   Muller's method   208
3.13   Lin-Bairstow's method   213
3.14   Graeffe's root square method   221
       Computational technique lab   234

4.   Solution of Simultaneous Linear Algebraic Equations   251-298

4.1   Introduction   251
4.2   Linear equations   251
4.3   Existence of solution   252
4.4   Gauss elimination method   254
4.5   Lu decomposition method or method of factorization   260
4.6   Jordan's method   266
4.7   Crout's method   267
4.8   Iterative methods   273
       Computational technique lab   290

5.   Finite Differences and Interpolation   299-422

5.1   Introduction   299
5.2   Difference schemes   299
5.3   Relation between the operators   302
5.4   Factorial notation   303
5.5   To express a given polynomial into factorial notation   304
5.6   Fundamental theorem of difference calculus   307
5.7   Differences of zero   321
5.8   Divided difference   321
5.9   Difference between divided difference and ordinary difference   323
5.10   Interpolation   332
5.11   Methods of interpolation   332
5.12   Finite difference calculus   333
5.13   Newton's formulae for interpolation   333
5.14   Hermite's interpolation formula   351
5.15   Newton's divided difference formula   355
5.16   Lagrange's interpolation formula   363
5.17   Central differences interpolation formulae   379
5.18   Stirling's difference formula   380
5.19   Choice to select the suitable interpolation formula   381
5.20   Bessel's difference formula   382
5.21   Everett's difference formula   383
       Computational technique lab   403

6.   Numerical Differentiation and Integrations   423-526

6.1   Introduction   423
6.2   Derivative using Newton's forward interpolation formula   423
6.3   Derivatives using Newton's backward difference formula   425
6.4   Derivatives using Stirling's formula   426
6.5   Derivative using Newton's divided difference formula   426
6.6   Maxima and minima of tabulated function   427
6.7   Error analysis in numerical differentiation   442
6.8   Error in higher order derivatives   442
6.9   Numerical quadrature   449
6.10   Quadrature formula for equally spaced arguments   449

6.11	The Trapezoidal rule	450
6.12	Simpson's 1/3 rule	451
6.13	Simpson's 3/8 rule	453
6.14	Weddle's rule	455
6.16	Newton-Cote's formula	458
6.17	Properties of Cote's numbers	459
6.18	Deductions from Newton-Cote's formula	460
6.19	Gauss's quadrature formula	463
6.20	Chebychev's formula	466
6.21	Higher order rules	490
6.22	Numerical evaluation of the singular integral	491
	Computational technique lab	**513**

7. Matrix Inversion and Eigen Values Problems   527-630

7.1	Introduction	527
7.2	Gauss elimination method	527
7.3	Gauss-Jordan method	527
7.4	Triangularisation method	533
7.5	Choleski's method	541
7.6	Escalater method	547
7.7	Inversion of complex matrices	550
7.8	Eigen values and eigen vectors	553
7.9	Relation between eigenvalues and eigenvectors	554
7.10	Eigenvalues of special type of matrices	556
7.11	The Cayley-Hamilton theorem	576
7.12	Power method	588
7.13	Inverse power method	597
7.14	Rutishauser method	600
7.15	Jacobi's method	602
7.16	Given's method	608
7.17	House holder's method	613
	Computational technique lab	623

8. Difference Equations   631-668

8.1	Introduction	631
8.2	Difference equation as a relation among the value of y_x	631
8.3	Order of difference equation	632
8.4	Degree of difference equation	632
8.5	Solution of difference equation	632
8.6	Linear difference equation	634
8.7	Existence and uniqueness theorem	635
8.8	Solution of the equation $y_x + 1 = ay_x + b$	636
8.9	Solution as sequences	641
8.10	Linear homogeneous equation with constant coefficients	644
8.11	Linearly independent solution or fundamental set of solutions	646
8.12	General solution of second order homogeneous difference equation	650
8.13	General solution of the homogeneous difference equation of order n	652
8.14	Particular solution of the complete difference equation	655
8.15	Solution of simultaneous difference equations	662

9. Numerical Solution of Ordinary Differential Equations   669-742

9.1	Introduction	669
9.2	Existence and uniqueness of solution of differential equation	669
9.3	Euler's method	671
9.4	Euler's modified method	674
9.5	Solution by Taylor series	680
9.6	Picard's method of successive approximations	685
9.7	Runge-Kutta method	693
9.8	Simultaneous differential equations	703

9.9	Solution of second order differential equations	705
9.10	Milne's methods	708
9.11	Error analysis	722
9.12	Convergence of a method	724
9.13	Stability analysis	724
	Computational technique lab	730

10. Numerical Solution of Partial Differential Equations — 743-794

10.1	Introduction	743
10.2	Difference quotients	743
10.3	Classification of partial differential equations	745
10.4	Elliptic equations	745
10.5	Solution of Laplace equations by Libermann's iteration process	746
10.6	Parabolic equations	759
10.7	Solution by forward difference method	759
10.8	Solution by Bender-Schmidt's method	760
10.9	Crank-Nicholson method	761
10.10	Dufort and Frankel's method	762
10.11	Iterative method	763
10.12	Solution of two-dimensional heat equation: ADE method	769
10.13	Hyperbolic equation	771
	Computational technique lab	785

11. Method of Least Squares, Curve Fitting and Approximations — 795-884

11.1	Introduction	795
11.2	Continuous frequency distribution	796
11.3	Graphical representation	796
11.4	Curve fitting	798
11.5	Method of least squares	799
11.6	Method of curve-fitting	802
11.7	Fitting of some special curves	803
11.8	Curve fitting by sum of exponentials	817
11.9	Spline function	821
11.10	Regression analysis	827
11.11	Properties of regression coefficients	829
11.12	Angle between two lines of regression	830
11.13	Fitting a polynomial function	838
11.14	Non-linear regression	840
11.15	Simplified determination of regression analysis	843
11.16	Regression analysis of grouped data	844
11.17	Multiple linear regression	844
11.18	Data approximation of functions	848
11.19	Types of approximations	849
11.20	Use of orthogonal functions	853
11.21	Gram-Schmidt orthogonalizing process	854
11.22	Legendre and Chebyshev polynomials	856
11.23	Uniform Approximation	857
11.24	Chebyshev polynomial approximations	859
11.25	Lanczos economization of power series for a general function	860
11.26	Rational approximation	860
	Computational technique lab	871

■ **Glossary**	**885-888**
■ **Bibliography**	**889**
■ **Index**	**890-892**

CHAPTER 1

Programming in C

1.1 INTRODUCTION

When we work with the computers, we use the software programs for different purposes. These software programs are the set of instructions written in the form of coding. The instruction code writing is possible with the programming languages, because the language is the best medium of interaction between two objects. In case of computer, the programming language facilitates the communication between computer-user and computer. By using the instructions of the programming language, the software programs are developed to instruct the computer to work accordingly.

1.2 'C' AND ITS DEVELOPMENT

With the development phase of computers, different programming languages are developed from time to time, such as Machine Language, Assembly Language and High Level Language. The Machine Language was the first language developed to instruct the computer. After that, the Assembly Language was developed. But the work with Machine Language and Assembly language was very typical due to typical code writing. On the other hand, the instructions of High Level programming language are very easy due to its resemblance with the simple English language.

'C' language is also a such type of programming language, which was assumed to be categorized as High Level Language. But it was not classified as pure High Level Language, because, it lies in between Low Level Languages (Machine Language and Assembly language) and High Level Languages (BASIC, FORTRAN etc). It is kept in between these two categories because of having *"better machine efficiency, faster program execution and better program development"* and so it is classified as the Middle Level Language.

1.3 IMPORTANT FEATURES OF C

'C' language became very much popular due to its some specific features, which make it different from Low Level and High Level languages. Some of the important and useful features of 'C' language are as follows :

(1) 'C' is a Middle Level language and the middle level language provides the programmer a set of controls and data manipulation statements. These controls and statements are used to define the working instructions in a software program.

(2) The programming code of 'C' language is portable. It means the software for one type of computer system (such as Apple II Plus) can be used by another type of computer system (such as IBM PC).

(3) 'C' language was first language used for System programming that means the programming related to the operating system. For example: UNIX is the operating system which was written by using C-language programming code.

(4) The 'C' language is the Structured Programming Language that means the C-programs are made up of blocks.

(5) Blocks are the 'set of commands' that are logically connected. These are called "Routines" or "Functions". These are the Building Blocks of C-language.

(6) 'C' language has its own grammar, so has the specific syntaxes.

(7) 'C' language is a case sensitive language, because it uses the ASCII (American Standard Code for Information Interchange) standards for its grammar.

(8) One very important feature of C-language is that the execution of C-program produces an (.exe) executable file. The executable file is independent of the programming language or platform used for programming.

(9) The program of C-language provides facility to extend itself, because a C-program is made up of functions and each function is associated with a specific task.

(10) C-language has a proper collection of functions in standard form to be used in several C-programs as Standard Library of C Language.

(11) The structured programming concept and modular structure of the C-program facilitates easy debugging, testing and maintenance of the program.

(12) The C-language requires compiler for the C-program compilation, so the C-program execution is fast.

1.4 STRUCTURED PROGRAMMING CONCEPT OF MODULAR PROGRAMMING

A programming problem may require several lines of codes for making a program. This complete programming problem can be split into a series of modules. A Module is the block of instructions related logically. The module should be easy to read and understand. The modular programming provides Top-Down approach for program design by breaking a program into modules. The modular approach of the program design is useful to well organize the program. The module should not be more than half page long in general and has only one entry - one exit point. The program control cannot jump from one module to another using GOTO statement. The longer modules should be split into submodules if required.

1.5 WORKING WITH 'C' PROGRAM

Programming in C-language involves several steps as follows:

  (i) Program Writing : The programmer should take care of the C-language grammar for program writing, because the C-language has specific grammar and syntax rules for creating program instructions. This program is called Source program or Source Code.

 (ii) Program Compiling : After the complete program writing, the programmer requires to compile the C-program (Source Code).

(iii) Program Linking : Program linking refers to the linking of programming code to that of predefined functions provided by C-language.

(iv) Program Execution : Program execution is the final step in the program life cycle. This stage includes the processing of instructions stored in the object code of C-program and finally produces output of the program.

1.6 CONTENTS OF C-PROGRAM

A C-program is the collection of one or more functions. Every function is a collection of instructions or statements related logically to perform some specific task. The general structure of a C-program is as follows :

```
/*Comment Entry*/
#include<header file>
main()
{
statement 1;
statement 2;
------------3;
------------4;
}
```

Fig. 1

The components of a C-program are:

(i) Comment Entry : Comments are the remark statements of a program. These remark statements are used to describe the objectives of program, function or statement. But we can not put the comment inside another comment.

For example: /*This is a sample/*C-program*/ is invalid

(ii) Include Directive : The 'include directive' is used to make the inclusion of any header file available to the C-program. The Header file is the collection of functions and routines required frequently by the C-programmers in different C-programs. These header files make the Standard Library of C-language. Each header file has the extension name .h. The include directive is used in beginning of C-program with the header file as follows :

```
# include <header file name with extension>
```

Sometime file inclusion differs in C-program as follows –

```
#include "header files name with extension"
```

The commonly used header files are stdio.h, conio.h, math.h, dos.h etc.

(iii) Main() Function : It is the necessary part of a C-program. The use of main() indicates the compiler about the start-point of the program. The compiler starts compilation of a C-program always from the main() function in a sequential manner started from the first statement of main() functions to the last statement.

▶ **REMARK**

▸ Compiler is a software program, which translates the high-level language program to machine language code.

Every C-program must contain at least and only one main() function. The use of more than one main() function in a C-program may cause confusion for the compiler to start compilation. The set of statements belonging to a function is always kept enclosed within the pair of the curly brace {....}. The opening brace '{' indicated the beginning of main() function and the closing brace '}'indicated the end of function. The statements within the curly braces comprise the Body of main() function. Generally, the statement of a C-program is written in small case letters, because 'C' is a case sensitive language. Every statement in a C-program must end with a semicolon (;). In a statement of a C-program, there may exist space between two words.

1.7 PROGRAM WRITING IN C

The program of C-language is the collection of several instructions. The writing of instruction must follow some grammar rules. The C-language provides syntax-rules for program writing. Every C-program instruction consists of various components such as keyword, constants, variables, operators, special symbols. These are called C-tokens. The 'C-tokens' are the basic components of each and every C-program.

1.7.1 CHARACTER SET OF C-LANGUAGE

In a C-program every instruction consists of several words which are further consisted of characters. C-language has its own character set, which is used to create the words useful for instruction writing. Following are the characters provided by C-language.

Alphabets :

> Upper Case Alphabets A-Z (26 letters)
> Lower Case Alphabets a-z (26 letters)

Digits :

> Decimal Digits : 0, 1, 2, 3, 4, 5, 6, 7, 8, 9.

White Spaces :

> Blank space, Horizontal Tab.

Special Character symbols :

> Operators   = + – * / % < > ! & ? :
> Punctuation marks , .; ' "
> Symbols   # $ ^ ( ) { } [ ] ~ @

1.7.2 KEYWORDS

Keywords are the Reserve Words which have their fixed meaning. Their meaning is already known to compiler of C-language. The programmer cannot change the meaning of these keywords in the program. The keywords cannot be used as the variables or identifiers. C-language provides following keywords:

Auto	double	if	static	volatile
Break	else	int	struct	pascal
case	enum	long	switch	entry
char	extern	near	typedef	huge
const	float	register	union	continue
far	return	unsighed	default	for
short	void	do	goto	sighed
while				

Fig. 2

Some compiler may use other additional keywords. C-language has the syntax rules regarding the use of keywords, that means, all keywords must be written in lower case letters.

1.7.3 CONSTANTS IN A C-PROGRAM

A constant is a data value or quantity, which is not changed. The constants are used for processing by a program. The constants are not changed for a specific task or processing.

These can be directly used for processing or these can be used for processing after storing in the memory locations of computer system.

Basically the constants are classified as two types - Numeric constants and Character constants.

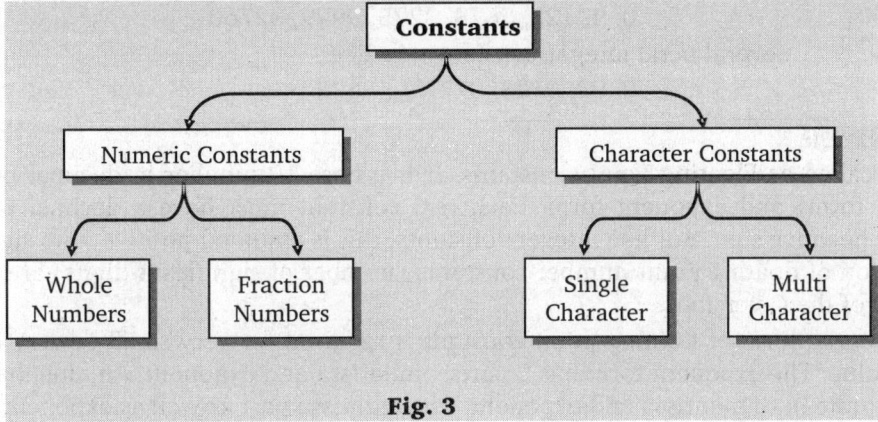

Fig. 3

But, these constants are broadly classified into two categories as – Primary constants and Secondary constants.

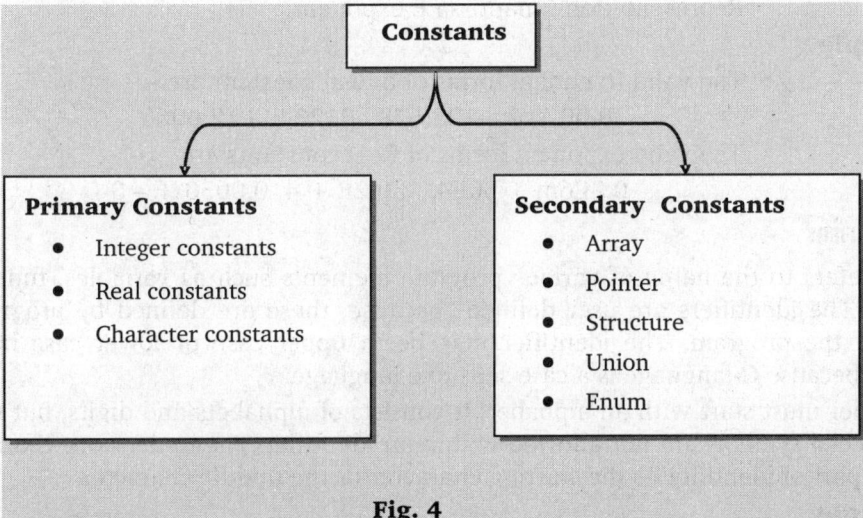

Fig. 4

The Primary constants are the basic constants to be used in a C-program as the Character Constants, Integer Constants and Real Constants.

The Secondary constants are made up from the primary constants. These are Array, Pointer, Structure, Union, Enum, etc.

(I) CHARACTER CONSTANTS

The Character constants refer to the alphanumeric characters. A character constant may be any single alphabet (lower case or upper case), any single digit, any special symbol or a space enclose in single quote (') always pointing to left.

For example :   'A', '7', '$', '+'

(II) INTEGER CONSTANTS

The integer constant is a numeric data value which can be used for calculation. It consists of the digits without decimal point. The integer constant may be either positive or negative. An integer constant can be made by the combination of digits from 0 to 9.

For example : Several valid decimal integer constants are like –

0, 2, 123, 5678, 3275, 9999, 32760

Several octal integer constants are like -

0, 02, 0123, 0777

(III) REAL CONSTANTS

These are called as Floating Point Constants. It is a base-10 number in decimal point from (fractional form) and exponent form. Each real constant must have a decimal point with positive or negative sign. But like integer constants, this is assumed positive if no sign is used. The precision of floating point number constants (number of significant digits) are generally six in most of the C-versions.

The exponential form of floating point constant is required for very small or very large real constant value. The exponent form has 2 parts – mantissa and exponent. An alphabet 'e' or 'E' should separate both mantissa and exponent. The mantissa part as well as exponent part may be positive or negative. The mantissa part may contain the decimal point but the exponent part can never.

Representation : mantissa E exponent.

For example :

The valid fractional forms of a Real constant are –

30.00, 0.2, –123.678, .01234, +12.0066

The valid exponent forms of Real constants are –

0.3E6m + 30E4, –30.2E + 4, 0.00502E – 3

1.7.4 IDENTIFIERS

Identifier refers to the name of various program elements such as variables, functions and arrays etc. The identifiers are user defined because, these are defined by programmer as required in the program. The identifier may be in upper case or lower case but treated differently because C-language is a case sensitive language.

The identifier must start with an alphabet. It consists of alphabets and digits, but the spaces and the special symbols are not allowed within an identifier. The underscore character (_) can be the part of identifier as the starting character or the middle character.

For example:

The valid Identifiers are like

n, num, name, num1, ltem_rate, CITY, _price

1.7.5 VARIABLES AND DATA TYPES

Variable is an Identifier, which is used to store a data value for processing. Sometimes we can refer it as the memory space having specific name and storing data value. A variable can have different data values at different times during program execution. Due to this behaviour, it is named Variable. The naming conventions for variable are similar to that of identifiers. Generally, the variable name consists of 1 to 8 characters. The characters may be alphabets, numbers or underscores, but may not be space or symbols. Some C-compilers permit the length of variable name up to 40 characters. The first character of the variable name should be an alphabet strictly. Any keyword can never be used as the variable name.

For example: simple_ints, age, work_hs, etc. are valid variable names.

Basically, the fundamental data types supported by most of the C-compiles are dependent on the constants of C-language. These are the "Primary Data Types" of C-language. These are described as follows:

    (i) Integer data type   **(ii)** Character data type   **(iii)** Float data type

(I) INTEGER DATA TYPE

Integers are the whole numbers having different storage range supported by particular machine. Integer data types are used to store these whole numbers using different memory spaces or variables. The general integer data type uses 2 bytes of memory space and has the storage range between -32768 and +32767. These 2 bytes (16 bits) of memory store the sign of integer constant in the last bit (16th bit), as 1 for negative and 0 for positive sign.

The short integer data types (both signed and unsigned) use 2 bytes of memory. The short signed integers have the storage range from -32768 to +32767. These are most commonly used and default integer types for a C-program. The short unsigned integer data types stores only positive integer constants in between the range of 0 and 65535.

The long integer data types (both signed and unsigned) occupies 4 bytes of memory. The storage range of long signed Integer ranges from -2147483648 to +2147483647. The long unsigned integer data type stores positive data in between the range of 0 and 4294967295. The default long integer is assumed to be signed long integer.

(II) CHARACTER DATA TYPE

The character data types are used to store the alphanumeric characters that means the character constants. One character data type uses 1 byte (8 bits) of memory to store single character. These are classified as signed characters and unsigned characters. The storage range of sign character data type is -128 to +127 and that of unsigned character data type is 0 to 255. By default, we use character data type which is signed character data type.

(III) FLOAT DATA TYPE

The real constants are stored in the memory space of 4 bytes or 8 bytes. Each floating point constant has 6 digits of precisions by default. There are different data types provided in C-language for storing the real constants of different range as float, double and long double.

Data Types, Keywords for Data Types, Size and Range

Data Type	Key word for Data Type	Size (Bytes)	Storage Range
Character	char or signed char	1	–128 to +127
Unsigned Character	unsigned char	1	0 to 255
Short Signed Integer	short signed int	2	–32768 to +32767
Short Unsigned Integer	short unsigned int	2	0 to 65535
Long Signed Integer	long signed int	4	–2147483648 to +2147483647
Long unsigned Integer	long unsigned int	4	0 to 4294967295
Float	float	4	–3.4e38 to +3.4e38
Double	double	8	–1.7e308 to 1.7e308
Long Double	long double	10	–1.7e4932 to +1.7e4932

1.8 OPERATORS IN A C-PROGRAM

Operator is a symbol associated with an action or operation on data such as arithmetic calculation, assignment or logical operations. C-language has a rich set of operators, which can be categorized as two types : based on number of operands and based on their operations.

1.8.1 OPERATORS BASED ON NUMBER OF OPERANDS

Operands are the variables or constants on which the specified operator performs some operation. Each operator requires at least one operand to perform the operation. The following types of operators are classified on the basis of required number of operands –

(a) Unary Operators : The unary operators require only operand to perform the specific operation. The operator is written to the left or right side of the operand. C-language provides many operators to be used as unary like increment operator (++), decrement operator (– –), not operator (!). These operators are used differently for different purpose.

For example :

++ variable/constant,   variable/constant ++

– – variable/constant,   variable/constant – –

! variable/condition/constant

(b) Binary Operators : The binary operators work on two operands. The operator is placed in between the left and right operands. The operands may be the variable or constant.

There are many binary operators provided by C-language which we can use for most of the arithmetic and logical operations. These are =, +, –, *, /, %, <, >, && :: and many others like <=, >=, !=, these operations are used for several purpose. These are to be used in a programming instruction as following –

| operand 1   operator operand 2 |

For example : Variable = constant/variable

Variable = variable/constant + variable/constant

(c) Ternary Operators : Ternary operators operate on three operands. The conditional operators (? :) are the ternary operators provided by C-language. The ternary operators provided by C-language works for decision making follows –

| (condition expression ? Statement   1 : statement 2) |

1.8.2 OPERATORS BASED ON THE OPERATIONS

The Operators operate on the data value and produce the results based on the operations. The operations performed by the operators are of many types such as arithmetic operations, logical operation etc. Following are the operators to be performed or operands–

(a) Assignment Operator : The 'equals to' (=) operator works as the assignment operator in C-language. The operator is used to assign/store a data value in a variable. There should be used the variable name on the left side of the assignment operator.

But the right side of assignment operator may be any variable or any constant. The assignment operator stoers the data value of the right side operand to the left side operand. This operator is used in following manner in a C-program-

| Variablename = variable/constant |

For example : Number = 10;

Rate = 6.50;

X = 'A';

(b) Arithmetic operators : The arithmetic operator are used to perform the arithmetic calculations on data value either variable or constant. C-language provides operator for performing arithmetic calculations:-

Operator	Operation
+	Addition
-	Subtraction
*	Multiplication
/	Division
%	Modulus (Remainder)

The arithmetic operators are the binary operators requiring two operands for processing. The operators is placed in the middle of left and right operands. There is no operator for performing exponential operations in the C-language. The operators provided for arithmetic calculation can be used as follows –

> operand 1   arithmetic   operand 2   operator

Most of the time, the arithmetic operators are used with the assignment operator, so that the arithmetic calculation result could be stored in a variable.

> Variable = variable/constants arithmetic-operator variable/constant

For example :

$$X = a + b$$
$$X = 10 * 5;$$

(c) Comparison Operators : In a C-program, we may require some comparison of data values for decision making process these are also known as Relational operators. For this work C-language provides many comparison operators described below with their function :

Operator	Operates on
= =	Equals to
<	Lesser than
>	Greater than
<=	Lesser than or equals to
>=	Greater than or equals to
!=	Not equals to

These comparison operators require two operands for performing operations. These operators return a Boolean value either True or False after the evaluation of comparison of two data values.

(d) Logical Operators : These are called "Boolean operators", because the comparison between values or operands is reduced to either 'True' or 'False'. The values returned by logical operators are zero or one. Zero indicates to False value and One to True. These operators never change the value of operands. Following logical operators are provided by C-language –

Operator	Action
&& (AND)	Evaluates first operand and then second and so on and works for all
| | (OR)	Evaluation both all the operands and works for any one
! (NOT)	Evaluates of negation of the operand

(e) Increment and Decrement Operator : These both are unary operators. These operators specify that the operands is either increased or decreased by one only.

Operator	Usage
++ (Increment operator)	Increment by one
– – (Decrement operator)	Decrement by one

These operators are used in two manners as – prefix and postfix. If the operators (increment/decrement) are placed to the left side or operands, then it is called prefix, such as – (operator operand) in following manner –

$$++\text{Variable/Constant}$$
$$--\text{Variable/Constant}$$

If these operators are used to the right side of the operand, then this is called postfix method of using the operators, such as – (operands operators) in following manner –

$$\text{Variable/Constant} ++$$
$$\text{Variable/Constant} --$$

For example : Assume, the variable x has initial value 10.

$$X = 10$$

  (i) Output 'x' at this time is also 10. (Show x)
  (ii) Output 'x++' also shows 10 (show x++) [Postfix operation]
 (iii) Output '++x' shows 11. (show ++x) [prefix operation]

In this example, (ii) step shows that the increment in the value of x is performed after display because, it is postfix operation. Thus, in postfix operation the increment or decrement is performed after assignment.

The step (iii), states that the increment is performed prior to other operation, because the operators is used as profix.

(f) Compound Assignment Operators : These operators are the combination of arithmetic operator and the assignment operator. These operators first perform the arithmetic operation and then assignment operation.

Operator	Usage
+= (Add Assignment)	operands += operand2
– = (Substract Assignment)	operands -= operand2
*= (Times Assignment)	operands *= operand2
/= (divide Assignment)	operands /= operand2
%= (Modulus Assignment)	operands %= operand2

These operators perform the operations similar to simple arithmetic calculation as follows :

        (+ =)   Operates like (Var1 += Var 2)
                 It is similar to (Var1 = Var1 + Var2)

For example :   X = 10;
                  Y = 20;
                  X = x + y; or x + y;
                  Both results in 30 as the value of x.

1.9 HIERARCHY OF OPERATORS

In a C-program, the various statements are written to perform some tasks. There may be various types of operation to perform like arithmetic or logical. For this purpose, many operators may be used in an instruction or statement. If a statement contains only one operator, then it is easy to perform that operation. But the problem regarding operations arises when the statement contains two or more operators. Then the execution of operations become typical.

In general, the expression (statement) is evaluated from left to right. But some rules of priority are to be followed during operations on multiple operators. These rules are called **Precedence Rules.** According to these rules, each operators has a fix priority to be operated. This priority or precedence is referred to as the Hierarchy of operators. So, the hierarchy of operators can be described as the evaluation sequence of the operators in a statement. The arithmetic operations are very much dependent on the hierarchy of arithmetic operators.

The following priority is given to the arithmetic operators needed for different arithmetic calculations –

Operator	Priority	Description
$\begin{bmatrix} ++ \\ -- \end{bmatrix}$	First priority	Increment, Decrement
$\begin{bmatrix} * \\ / \\ \% \end{bmatrix}$	Second priority	Multiply, Division, Modulus
$\begin{bmatrix} + \\ - \end{bmatrix}$	Third priority	Addition, Subtraction
[=]	Fourth priority	Assignment

This hierarchy shows that, the operations to be performed first in an arithmetic expression are increment or decrement operation. Then the multiplication, division or modulus takes place and then addition or subtraction and at last, the assignment operator is evaluated.

For example :

$$X = 5 + 2 * 4$$

This expression will be evaluated as follows –

$$X = 5 + 2 * 4$$
$$X = 5 + 8 \qquad \text{[First operation } *]$$
$$X = 13 \qquad \text{[Second operation } +]$$
$$X = 13 \qquad \text{[Third operation } =]$$

But, the evaluation problem again arises when a statement contains two or more operations of some priority. In such case, the operators of same priority are operated in the sequence of their presence in the statement, *i.e.*, first come first operated. But this technique is not applicable for the operators of different priority.

For example :

$$a = 12 - 5 * 2 + 12 / 4$$

This expression contains many operators, out of which some have similar priority (such as */, + –), it will be evaluated as follows –

a = 12 – 5 * 2 + 12/4	[First Priority * /]
a = 12 – 10 +12/14	[First presence *]
a = 12 – 10 – 3	[Second presence /]
a = 2 + 3	[Second priority first presence –]
a = 5	[Second priority second presence +]
a = 5	[Last priority =]

Sometimes, the arithmetic calculation becomes very large. Then we use parentheses to separate the group of operations. The evaluation of such arithmetic statement always starts from inner most parenthesis using all hierarchy rules.

For example :

$$B = 2 * 3 + (40 – 2 * 8) + 4/2$$

This expression has different operators of different hierarchy and also the parentheses. Its evaluation will be performed as follows –

B = 2 * 3 + (40 - (2 * 8) + 4)/2	[Innermost parenthesis (2 * 8)]
B = 2 * 3 + (40 – 16 + 4)/2	[Subtraction 40 – 16]
B = 2 * 3 + (24 + 4)/2	[Addition 20 + 4]
B = 2 * 3 + 28/2	[Multiplication 2 * 3]
B = 6 + 28/2	[Division 28/2]
B = 6 + 14	[Addition 6 + 14]
B = 20	[Assignment b = 20]
B = 20	

1.10 HEADER FILES

The C-language has a large set of header set of header files as the standard library. The standard C-library is full of library functions. In the standard library, several header files are present. Each header files has the extension name '.h'. Each header files has the collection of several functions which can perform many tasks such as input/output operations, arithmetic operations. These functions are useful for repeatedly use of some operations. The functions which are related in operations are kept in one header file. In any C-program, to use a function from the standard library, the related header file must be included in the program. Some commonly used header files and their functions are discussed below :

(2) strings.h

The "string.h" header file contains the string manipulation functions. The string is the collection of characters. Several functions of "string.h" provide the facility to manipulate the string.

(3) math.h

This header file has the collection of many functions which help in mathematical calculations.

(4) stdio.h

The header files named "stdio.h" has the collection of input/out put functions. These functions are associated with keyboard input/file input and screen output/file output.

(5) stdlib.h :

The common function of this header file are as follows :

Return Type	Functions	Description
double	atof(char *string)	Converts string to the double precision numbers (float)
int	atoi (char *string)	Convert string to an integer.
long	atol(char *string)	Convert string to a long integer.
	exit(int *n*)	Closes all files and buffer and terminate the program. It requires zero as, argument for successful termination of the program.
char	*iota(int var)	Converts the given integer to string type.
char	*ltoa(long var)	Converts given long data value to the string type.
int	system(char *command)	Pass the command to the operating system and returns zero for successful execution of the command.
int	rand( )	Return the random positive numbers.

(6) dos.h

The function of "dos.h" operates only where the dos environment is used as out put window for C-language.

1.11 INSTRUCTIONS OF A C-PROGRAM

In a C-program, we use several variables, constants, keywords and identifiers in different manner to get the desired solution of a programming problem. The C-program is also called as the collection of several instructions in sequential manner. These instructions are related with each other to solve a specific problem. The instructions of a C-program are also called **Programming Statements.** Thus, a programming statement can be defined as "a single line instructions to carry out some action". These instructions are of following type depending on their action —

- Declaration Instructions
- Algebraic expressions
- Input-Output Instructions
- Control Instructions

It is not necessary to use all type of the instructions in a C-program can hold any number of C-instructions to describe the operations.

1.11.1 DECLARATION INSTRUCTIONS

The Declaration instructions are used to specify the data type of variables, which need to be used within the program.

The syntax used for declaration instruction is as follow –

datatype   variablename;

Each declaration statement must end with a semicolon (;)

For example :

    Int   num1;
    Int   num;

Float   rate;

Char   sname;

Sometime, we need several variables of same data type, then it is not required to declare all such variables as separate declaration statements. But we can declare all such variables in a single declaration instruction by separating the variable name with comma sign (,).

For example:

int num1 num2;   [Variables num1 and num2 are integer type]

float rate, salary, discount;   [Variables rate, salary and discount are float]

char sname , cname;   [Variables sname name are of character type]

1.11.2 ALGEBRAIC EXPRESSIONS

The algebraic expressions are often called Arithmetic Instruction. Generally an algebraic expression contains at least one operator with operands.

▶ REMARK

▷ The arithmetic instructions use the arithmetic operators with the operands. The assignment operator also constitutes an important part of any arithmetic instruction. If we use the constants in an arithmetic instruction, then the constants are always placed to the right side of assignment operator. The arithmetic operators need to be used strictly in arithmetic instructions.

Example:

    i)   X = a + b;   (ii)   C = a * 10;   (iii)   Z = 20;

   v)   T = x;

   v)   Num = a (b + c)   it is invalid because there is no operator use between the operands 'a' and '(a + b)'.

In the instruction (i) we have used only variables with addition and assignment operators. So the data values stored in the variables 'a' and 'b' are added and the result is stored in the variable 'X'.

In the instruction (ii), the constant and variables both are used. The multiply operation is performed on the variable 'a' and constant value 10 and result is stored to the variable 'C'.

The instruction (iii), we have used only assignment operator, which store the constant data value on right side to the operator in the variable 'Z' to the left side of assignment operator.

The instruction (iv) also describe the assignment of data value from one variable to another variable. But point to remember is that the data value from right side of assignment operator (either constant or variable) is always stored to the operand on left side of the assignment operators (always one variable).

▶ REMARK

▷ Sometimes, many individual expressions are collected and enclosed within the pair of braces {}. Then the block of braces is referred to as the compound statement. The individual expression may be arithmetic instructions or control instructions or any other compound statement.

For example :

```
{
x = 10;
y = x * 2;
z = x + y';
}
```

An individual arithmetic instruction may contain any type of variables or constants as the operands. Based on the type of operands used, an arithmetic instructions may be of following type —

(a) **Integer arithmetic instructions:** These arithmetic instruction contain only integer variables or constants.

For Example :

$$Int\ x = 10,$$
$$y = 20,\ z;$$
$$z = x + y;$$

(b) **Real arithmetic instructions:** The arithmetic instruction contains only Real variables or constants.

For example :

$$float\ price,\ discount;$$
$$price = 15.60;$$
$$discount = (price*7.5)$$

(c) **Mixed arithmetic instructions:** The arithmetic instruction contains the operands of different type such as integer and float both.

For example :

$$float\ ta,\ da;$$
$$Int\ x,\ y,\ salary;$$
$$X = 10;$$
$$Y = 20;$$
$$Salary = 5000;$$
$$Ta = (salary * 15.4)/100;$$

The expressions can also be logical representing some logical conditions and returns the result either True (1) or False (0). The logical expressions use comparison operator like ==, <,>,<=, etc.

For example :

$X = y$	$a = 10$
$X <= y$	$b > = 50$
$X != y$	$c != 100$

1.11.3 INPUT-OUTPUT INSTRUCTIONS

The input instructions are associated with the taking of input form input devices (keyboard etc) and supplying them to the program for processing. For this purpose, the standard library of C-language provides several functions (like scanf() function).

The output instructions are used to obtain the output result from the program after processing. The output result is send to the output device (such as monitor screen) by several standard library functions (like printf() function).

1.11.4 CONTROL INSTRUCTIONS

The control statements are used to control the flow of various instructions in a C-program. It means, the sequence of statements execution is controlled by the control instructions. The instructions are of following types –

(a) **Sequence Control Instructions:** These instructions take care of the execution sequence of instructions. The programming instructions should execute in the order of their appearance in the program.

(b) **Decision control Instructions:** These instruction allow the program to make the execution of some instructions decision based.

(c) **Repetition Control Instructions :** These instruction control the execution of a group of statements repeatedly.

1.12 DATA OUTPUT WITH printf() FUNCTION

Printf() is an output function present in standard library header file stdio.h. This function is used to print the output result on the output screen (computer screen). The output result may involve the variables, constants as required by the user. The printf() function can be used in several formats in a C-program to produce different output on the screen. The different formats of printf() function are as follows –

      printf("Output Message");

      printf("<Format Specifiers>", <list of variables>);

      printf("Output Message <Format Specifiers">, <list of variables.);

The format specifiers are the code-representation of the data type for variables. The format specifiers of different data type are as follows –

Data Type	Format Specifier
signed char, unsigned char	%c
short signed int	%d
short unsigned int	%u
long signed int	%ld
long unsigned	%lu
float	%f
double	%lf
long double	%Lf

Besides these basic format specifiers for basic data types, some other format specifiers are used for other type of data as follows –

Data Type	Format Specifier
String	%s
Floating point number	%e, %g, %E
Short integer	%h
Octal integer	%o
Hexadecimal integer	%x or %X
Decimal, Octal or Hexadecimal integer	%i

PROGRAM-1

A sample C-program showing use of printf()

```
/* Sample C-Program */
#include <stdio.h>
main()
{
int x,  y;
x  =  10;
y  =  20;
printf("This is my First C  program")
printf(The value of x = % d", x);
printf("y = %d", y);
printf("End of Program !");
}
```

Output:

This is my First C program The value of x = 10, y = 20. End of program!

In this program, we have used all different format of printf() function. There are four output instructions written with printf() to those four output on the computer screen. But, the output of the program shows all the output statements in a single line. This is not the appropriate output of the program, because the output is not shown in separate lines. The output to next line is enforced by the new line character ('\n'), Which is lying under the special category of characters constants as "Backslash characters".

1.13 USE OF BACKSLASH CHARACTERS

C-language provides some special backslash character constants. These character constants are used to make the output more batter. So these are always to be used with output functions. Each backslash character constant consists of one backslash and one character enclosed in single quotes. But each one represents only one character because the presence of backslash is escaped. So these characters are also known as Escape Sequences. The list of backslash character constants is as follows

Backslash Character Constant	Works for
'\a'	Beep sound (audible bell)
'\b'	Back space
'\f'	Form feed
'\n'	New line
'\r'	Carriage return
't\'	Horizontal tab
'v\'	Vertical tab
'\"	Single quotes
'\"'	Double quotes
'\?'	Question Mark
'\\'	Back slash character
'\0'	Null character

PROGRAM-2

We can modify the output of the program by using the backslash character such as with the program of example 9 as follows;

```
/*Modify Sample Program */
#include<stdio.h>
main()
{
int x, y;
x = 10;
y = 20;
printf(/nThis is my First C Program,");
printf("/nThe value of x = /1%d", x);
printf("/n/ty = /t%d", y);
}
```

Output :
This is my First C Program.
The Value of X = 10
Y = 20
End of Program !

PROGRAM-3

Calculation of simple interest for the given amount of Rs. 4500, rate of interest 4.5% and duration 5 years.

```
/* Program for Simple interest Calculation */
# include<stdio.h>
main()
{
int amount = 4500, year = 5;
float int_rate = 4.5, simple_int;
simple_int = (amount*year*int_rate)/100;
printf("\n Amount %d Rs.", amount);
printf("\n Rate of Interest = %f", int_rate);
printf("\n Duration = %d years", year);
printf("\n The Simple interest calculated is = Rs. %f", simple_int);
}
```

Output :
Amount = 4500 Rs
Rate of Interest = 4.5
Duration = 5 years
The simple Interest calculated is = Rs. 1012.450000

1.14 DATA INPUT WITH scanf () FUNCTION

In a C-program the variables require data value for processing. So the value assignment should be done before processing. The value assignment of the variables is done by using assignment operator in the beginning of program or at run time of the program. When the program requires some input at run time, the input data is provided from the standard input device (such as key-board) and this data value is sent to the variables by a standard library function of C named scanf() function.

The scanf() function can be used to take different types of input like numerical values, single character or string etc. The general format of scanf() function is as follows

scanf("<Format specifies>", <Address of variable, ...>);

PROGRAM-4

Calculate simple interest by input data.

```
/* Simple interest Calculation */
#include<stdio.h>
main()
int amount, year;
float int_rate, simple_int;
printf("/n Enter amount, rate of interest and duration:");
scanf("%d%f%d", & amount, & int_rate, & year);
simple_int = (amount" int_rate"year)/100;
print ("/n Simple interest Calculated = %f", simple_int);
}
```

Output :

Enter amount, rate of interest and duration :   5000
                                                 4.5
                                                 5

Simple Interest calculated = 1125.000000

In the above example, we have used a special symbol (&) with the variable name in the input instruction with scanf() function. The symbol (&) is known as the Address Operator in C-programming language. This address operator is used to sent the input data value to the specified variables at specific memory locations.

PROGRAM-5

Calculate compound interest on the given amount for given duration.

```
/* Compound interest Calculation */
#include <stdio.h>
#include<math.h>
main()
{
float p, r, y, I, a;
/* input data */
print("/n Enter the principal amount (p):");
scanf("%f", &p);
Printf("/n Enter interest Rate (r):");
Scant("%f".&r);
Print("/n Enter number of years (y):");
Scant("%f", &y);
/* Calculate compound interest */
I = r/100;
A = p* pow ((1+i),y)'
/* writing output */
print ("/n the final amount after interest = f", a);
}
```

Output :

Enter the principal amount (p) : 1000

Enter Interest Rate (r) : 10

Enter number of Years (y) : 2

The final amount after interest = 1210.000000

In this example, we have used a special function pow () and for this, the header file "math.h" is included in the beginning of program. This function is used to calculate the result of raising given power to the given number.

1.14.1 POW() FUNCTION

This function is present in the standard library of C-language named "math.h". This function facilitates the calculation of x" without any extra instruction. The syntax used for solution is:

$$pow (x, n);$$

For example :   pow (2, 3) evaluates 8.

PROGRAM-6

Calculate net salary of the employee, whose basic salary is input through keyboard. TA = 15% and DA = 41% of basic salary is to be calculated.

```
#include <stdio.h>
#include <conio.h>
/* Program for salary Calculation */
main()
{
float bs, ta, da, ns;
clrscr();
print("/n Enter Basic Salary of the Employee (bs):");
scanf("%f", & bs);
/* Calculations */
ta = (bs* 15)/100;
da = (bs*41)/100;
ns = bs+ta+da;
print("/n T.A. = %f /n D.A. = %f", ta, da);
printt("/n Net Salary Calculated is = %f", ns);
printt("/n Press any key to continue …..");
getch();
}
```

Output :

Enter Basic Salary of Employee (bs): 1000

T.A. = 150.00000

D.A. = 410.000000

Net salary Calculated is = 1560.000000

Press any to key to continue ....

This program has used two new function – clrscr() and getch(). We have not discussed about these function so far. These function are used to make the program output better, user friendly and interactive. Both the function are provided in the standard library header file named conio.h (console Input/Output).

1.14.2 CLRSCR() FUNCTION

This function is used to clear the output screen during program execution. It activates the operating system to clear the output screen. This function can be used anywhere in the program as required for the interactive output.

1.14.3 GETCH() FUNCTION

This function gets a character from the keyboard and so directs the program to wait for a key press. Its use is to stop the program output on the output screen for a while. The program execution resumes after pressing a key. We can use this function anywhere in the program where wait is required.

▶ **REMARK**

▶ It is not necessary to use these two function in a C-program. But their presence affects the output of program. When we execute a C-program, the control immediately transfers from the output screen to program screen. Then, getch() function is useful for seeing the output properly because execution waits for a key press by keyboard.

PROGRAM-7

Calculate Real Roots of Quadratic Equation

```
/* Real Roots of Quadratic Equation */
#include <stdio.h>
#include <conio.h>
#include <math..h>
main()
{
float a,b,c,d,x1,x2;
clrscr();
/* data input /*
printf("/n Enter the values of a,b,c,/n");
scanf("%f%f%f". & a, & b, &c);
/* Calculations */
d = sqrt (b+b-4*c):
x1 = (-b+d)/(2+a);
x2 = (-b-d)/(2+a);
/* Result Display */
printf(/* First Root x1 = %e /n Second Root X2 = %e", x1,
x2);
getch();
}
```

Output :

Enter the values of a,b,c :

    3   10   3

First Root X1 = − 33

Second Root X2 = − 3

1.15 INPUT-OUTPUT MANIPULATIONS

In a program of C-language, we take the input data from printf() function and send the output to output screen by scanf() function generally. There are different format specifiers used for data input and output.

The integer number requires the format specifier as "%". This format is used to input/output a numeric value but the number of digits in the number depends on the range of integer But we can manipulate the data input or output as follows:

Integer Format	-	%wd
Float Format	-	%wf ofr % w.pf
Character Format	-	%wc
String Format	-	%s for %ws

PROGRAM-8

An example of string manipulations.

```
/* Input-Output Manipulations */
#include<stdio.h>
#include<conio.h>
main()
{
int a,b;
float c, d;
char x;
clrscr();
a = 1 2 3 4;
c = 16.4358;
x = 'A';
printf("\n Enter a number : ");
scanf("%2d", &b);
printf("\n Enter a floating point number : ");
scanf("%f", &d);
printf("\nA=%2d    \nB=%2s", a, b);
printf("\nC=%f    \nD=%f", c, d);
printf("\nC=%2f    \nD=%2f", c, d);
printf("\n%c    \n%3c    \n%5c",   x, x, x);
printf("\n%3c    \n%c", x, x);
getch();
}
```

Output :

Enter a number : 6536

Enter a floating point number : 63.43

A = 12

B = 65

C = 16.435800

D = 63.430000

C = 16.43

D = 63.43

A

A

A

A

A

1.16 DECISION CONTROL IN THE C-PROGRAM

In most of the C-program, we have used the instructions to be executed in the sequence of their appearance. But in some problem solving processes, we have to execute the instructions based on certain conditions,. This kind of situation is handled by Decision Making in C-programming, For this purpose C-language provides Decision Control Instructions. These decision control instructions works for a logical condition. The execution of statements is based on True/False evaluation of the logical condition. There are following types of decision control instruction supported by C-language-

(i) The if statement

(ii) The switch statement

(iii) The conditional operator

(iv) The go to statement

1.16.1 IF STATEMENT

The use of if statement in a C-program allows the programmer to perform several actions differently depending on the outcome of a logical test. This process is also called as Branching because the logical test is evaluated in two values either True or False. It means every logical test has at lest two branches which carry out the operations.

The logical test is also called as condition. The condition is a Boolean function of operands and operators, which results either in True or False. The operands used may be variables, constants or any other identifier. The condition is to be expressed as follows-

(operand1   operator   operand2)

Every condition must be specified by enclosing within the parentheses (). The conditional expression statement need not to be followed by semicolon (;). The operators used for conditional expression are known as Relational Operators or Comparison operators. These operators compare the data value of two operands and return true of False. If the condition is evaluated True, the related statement or set of statements is executed. If the condition is evaluated Not True of False, then the statement or group of statement is not executed. The following table shows the conditional expression and their evaluation-

Condition expression	Format Specifier
op1 = = op2	True if op1 is equals to op2.
op1<op2	True when op1 is lesser than op2.
op1>op2	True when op1 is greater than op2.
op1<op2	True when op1 is smaller than or equals to op2.
op1>op2	True when op1 is greater than or equals to op2.
op1!op2	True when op1 is not equals to op2.

(op1 → operand 1, op2 → operand 2).

The if statement is used in a variety of ways in the C-program to represent conditional expressions. The various formats of if- statement are described below-

  (i) if statement   (ii) if ..... else statement

  (iii) if ..... else. If .... ladder format   (iv) nested if .... Else statement

(I) IF STATEMENT

This format of if-statement works in a situation where a single instruction or multiple instructions are to be executed for a specified condition, the syntax used will be-

Syntax :   (For single statement within if)

                    if (condition/Boolean expression)

                        statement 1;

The statement will execute only and only if the condition specified returns True value. But there is no instruction to execute for the False value of the condition. So this type of if- format is not better for user interactor or user friendliness, because user will not have any output message for the False value of the condition.

The use of this format is illustrated with the help of following example-

Sometimes, we need to process a group of instructions for a specific condition. Then the group of instructions has to be enclosed within the pair of curly braces {}. The syntax used for multiple lines within if is as follows-

Syntax :

                    if (condition)

                    {

                    statement 1;

                    statement 2;

                    _____;

                    _____;

                    }

If the condition returns True on evaluation, then the set of instruction enclosed with in the pair of curly braces is executed, otherwise no instruction is to be processed.

PROGRAM-9

```
/* Single statement within if */
#include<stdio.h>
#include<conio.h>
main()
{
int x;
clrscr();
print("/n Enter a number in the range of 1 and 10:");
getch();
}
```

Output :

  (i) Enter a number in the range of 1 and 10: 7

      You have entered Right Number.

  (ii) Enter a number in the range of 1 and 10: 17.

      There is no output for the second value (17).

PROGRAM-10

```
/* Multiple statements within if /*
#include<stdio.h>
#include<conio.h>
main()
{
int age;
clrscr();
print("/n Enter the age of candidate (25 or above); ");
scant("%d", & age);
if (age>= 25)
{
print("/n The candidate is selected.");
print("/n Congratulations 1");
}
getch();
{
```

Output :

(i) Enter the age of candidate (25 or above): 23

The candidate is selected.

Congratulations !

(ii) Enter the age of candidate (25 or above): 18

No output message is produced because the condition is false for age = 18.

(II) IF....... ELSE STATEMENT

The if statement is responsible for executing a single statement or group of statements in case the evaluation of condition results in True value. But there is a problem with False value of Boolean expression, because if statement does not deal with the false value of the condition. Then the implementation of if......else statement has solved this problem. By using if...... else statement, we can execute one group of instructions for True value of condition and another group of instructions for False value for the condition. The syntax used for if......else statement is as follows-

Syntax :

(a) if ....else for single instruction

if (condition)

Statement 1;

Else

Statement 2;

(b) if ....else for multiple instructions

if (condition)

{

statement 1:

..................;

...............;

```
    }
    else
    {
    statement 2;
    ...............;
    ...............;
    }
```

When we use if...else statement in a C-program then the if...else statement consists of two blocks – if block and else block. The group of statements after "if condition" and before "else statement" comprises "if block". The statement after "else statement" from the "else block". There can never be any "else" without an "if". The use of "else" statement is optional for an "if" statement; but the use of "else" statement depends of "if" statement. We cannot specify the condition with "else " statement, because the condition is always to be specified with the "if" statement.

PROGRAM-11

Find largest and smallest of two numbers.

```
/* Largest and Smallest */
#include<stdio.h>
#include<conio.h>
main()
{
int x, y;
clrscr();
printf("\n Enter  the two numbers: ");
scanf("%d %d", &n, &y);
if (x>y)
printf("\n Largest = %d \n  smallest = %d", x, y);
else
printf("\n Largest = %d \n Smallest = d%",  y, x );
getch();
}
```

Output :

   (i) Enter the two numbers : 27 17

          Largest = 27

        Smallest = 17

   (ii) Enter the two numbers : 17 27

           Largest = 27

         Smallest = 27

(III) IF....ELSE IF....... LADDER FORMAT

Sometimes we have many different statements to perform in different situation. It means the program structure require the evaluation of several conditions individually. Then the statements are processed according to the evaluation of condition. First of all, the first condition is evaluated if it is found true then the respective statement is processed otherwise the next condition is evaluated. This process continues for several conditions the syntax used for " if ladder" is as follows –

Syntax :

  (a) For Single instruction

```
if (Condition 1)
        Statement 1;
else if (condition 2)
        Statement 2;
        _____
        _____
else if (condition n)
        statement n;
else
        statement;
```

  (b) For multiple instructions –

```
if (condition 1)
{
        statement 1;
        _____;
        _____;
}
else if (condition 2)
{
        statement 2;
        _____;
        _____;
}
        _____
        _____
else if (condition n)
{
        statement n;
        _____;
        _____;
}
else
{
        statement else
        _____;
        _____;
}
```

In case of " if – ladder", if a condition is evaluated as True, then the respective statements are processed and other conditions are not evaluated. But the False evaluation of a condition results in the evaluation of next condition. If all the specified conditions result False on evaluation, then the statements of last " else block" are executed.

PROGRAM-12

The marks for 5 subjects (each 100 max) of a student are input through keyboard. Calculate total, percentage and assign grade by following condition –

(1) Grade 'A' – Percentage 80 or above

(2) Grade 'B' – Percentage 70 to 79

(3) Grade 'C' – Percentage 60 to 69

(4) Grade 'D' – Percentage 50 to 59

(5) Grade 'F' – Percentage Below 50.

```
/* Grade Assignment */
#include<stdio.h>
#include<conio.h>
main()
{
int m1, m2, m3, m4, m5, total;
float perc;
clrscr();
printf("\n Enter marks in 5 subjects:  ");
scanf("%d %d %d %d %d , &m1, &m2, &m3, &m4, &m5);
total = m1+m2+m3+m4+m5
perc = total/5;
printf("\n Total Marks = %d", total);
printf("\n Percentage = %.2f", perc);
if (perc >= 80)
printf("\n Grade A");
else if (perc >= 70)
printf("\n Grade B");
else if (perc >= 60)
printf("\n Grade C");
else if (perc >= 50)
printf("\n Grade D")
else if (perc <= 50)
printf("\n Grade F");
getch();
}
```

Output :

```
Enter marks in 5 subjects : 50   80   40   60   70
Total Marks  = 30
Percentage = 60.00
Grade C
```

(IV) NESTED IF...ELSE STATEMENTS

In a C-program, sometimes the statements related to a condition may consist of another conditional statement. It means a condition on satisfying again test for the another condition and so on. For that situation, the "if-else" construct is to be written with in the body of "if-block" or "else-block". This type of program structure is called as "Nesting of if...else" or

"Nested if...else". There are several formats available for nested if...else such as

(a) Nested if...else for single instruction –

    (i) if (condition 1)
```
    {
    if (condition  2)
            statement 1;
    else
            statement 2;
    }
    else
            statement 3;
```
  (ii) if (condition 1)
```
            statement 1;
    else
    {
    if (condition 2)
            statement 2;
    else
            statement 3;
    }
```
 (iii) if (condition 1)
```
    {
    if (condition 2)
            statement 1;
    else
            statement 2;
    }
    if (condition 3)
            statement 3;
    else
            statement 4;
    }
```

(b) Nested if...else for multiple instructions –

    (i) if condition 1)
```
    {
    if (condition 2)
    {
            Statement 1;
            _ _ _;
            _ _ _;
    }
    else
    {
            Statement 2;
            _ _ _;
            _ _ _;
    }
    {
    else
```

```
        {
                Statement 3;
                _ _ _;
                _ _ _;
        }
(ii)  if (condition 1)
        {
                Statement 1;
                _ _ _;
                _ _ _;
        }
        else
        {
        if (condition 2)
        {
                Statement 2;
                _ _ _;
                _ _ _;
        }
        else
        {
                Statement 3;
                _ _ _;
                _ _ _;
        }
        }
(iii) if (condition 1)
        {
        if (condition 2)
        {
                statement 1;
                _ _ _
        }
        else
        {
                statement 2;
                _ _ _;
        }
        }
        else
        {
        if(condition 3)
        {
                statement 3;
                _ _ _;
        }
        else
        {
                statement 4;
```

```
    }
  }
```

we can have the nesting of "if-else" construct up to any level. It means one "if-block" or "else-block" can have many nested if...else being one within another.

PROGRAM-13

Find the largest number out of any three given numbers.

```
/*Largest of three numbers */
#include <stdio.h>
#include <conio.h>
main()
{
int x, y, z;
clrscr();
printf("\nEnter any three numbers : ");
scanf("%d  %d %d &x, &y, &z);
if (x>y)
{
if (x>z)
printf("\n Largest = %d", x);
else
{
if (z>y)
printf("\n Largest = %d", z);
}
}
else
{
if (y>z)
printf("\n Largest = %d", y);
else
printf("\n Largest = %d",z);
}
getch();
}
```

Output :
  (i) Enter any three numbers : 2 3 4
       Largest = 4
 (ii) Enter any three numbers : 2   4   3
       Largest = 4
(iii) Enter any three numbers : 4   3 5
       Largest = 5
 (iv) Enter any three numbers : 4   2   3
       Largest = 4

Using Logical Operators with the if...else Construct :

C-language facilities the following logical operators to operate to operate on boolean function –

(1)	&& Operator (AND)	Test two or many given condition simultaneously and process for True value of all condition.
(2)	|| Operator (OR)	Test for one condition out of many given conditions and process for true value of any condition.
(3)	! Operator (NOT)	Make a True expression False and vice versa. So it reverse evaluation result of any condition and works for negation.

The use of logical operators in a C-program leads to decrement in the writing of "if statements" that means decreases the size of program. Their use also solve the problem arised due to nested if...else for keeping track of corresponding if...else and pair of curly braces. The logical operators can be used in several format like –

(a) if (condition 1 && condition 2 && ....condition N)
```
        {
            statement 1;
            ___ ___ ___;
            ___ ___ ___;
        }
        else
        {
            statement 2;
            ___ ___ ___;
            ___ ___ ___;
        }
```

(b) if condition 1 II condition 2 II .....condition N)
```
        {
            statement 1;
            ___ ___ ___;
            ___ ___ ___;
        }
        else
        {
            statement 2;
            ___ ___ ___;
            ___ ___ ___;
        }
```

(c) if (! condition 1))
```
        {
            statement 1;
            ___ ___ ___;
            ___ ___ ___
        }
```

```
    else
    {
        statement 2;
        ___ ___ ___;
        ___ ___ ___;
    }
```

We can also use the two or all logical operators in one if-statement as required.

PROGRAM-14

The electricity power distribution company distribute the electricity to domestic users as per following charges –

Units Consumed	Charges per Unit
0 - 100 Units	Rs. 1.80 per unit
101 - 200 units	Rs. 2.40 per unit exceeding 100 units.
201 - 300 units	Rs. 3.20 per unit exceeding 200 units.
Above 300 units	Rs. 4.50 per unit exceeding 300 units.

```c
/*Calculate total amount paid by electricity consumer */
#include <stdio.h>
#include <conio.h>
main()
{
int units;
float charge;
clrscr();
printf("\n Enter units consumed : ");
scanf("%d", &units);
if (units>=0 &&  units)
charge = units*1.80;
else if  (units  = 101 &&  units <=200)
        charge = (100*1.80)+(units- 100)*2.40);
else if  (units  = 201 &&  units <=300)
        charge  =  (100*1.80)+(units- 100*2.40) + ((units -
200)*3.20);
else if  (units >= 201 && )
        charge = (100*1.80)+(100*2.40) + (100*3.20) + ((units -
300)*4.50);
else
printf("\n Units consumed cannot be negative");
printf("\n Total amount to be paid for units = %d is Rs. %.2f",
units, charges).
getch();
}
```

Output :

(i) Enter units consumed : 50

Total amount to be paid for units = 50 is Rs. 90.00

Decision Control in the C-program

(ii) Enter units consumed : 110

Total amount to be paid for units = 110 is Rs. 204.00

1.16.2 SWITCH STATEMENT

Sometimes we have to solve a programming problem which is related to the selection of choice out of a number of alternatives. The selection of one choice out of many alternatives is more complicated rather than selection between two alternatives. In fact " if-statement" can easily be used for such situation. But sometimes it becomes complicated with "if-statement" to write such instructions. C-language provides a special control statement which allows a programmer to make decision from the number of alternatives. This is called "switch" statement or "switch..case..default" statement. The "switch" statement has following program structure format –

```
switch (variable/expression)
{
case value 1:
        Statement 1;
        — — —;
        — — —;
        break;
case value 2:
        Statement 2;
        — — —;
        — — —;
        break;
        |          |          |
        |          |          |
        |          |          |
case value N:
        Statement N;
        — — —;
        — — —;
        break;
        default;
        statement default;
        — — —;
}
```

this switch-statement requires an expression or variable for testing. The value 1, value 2 ..... value N are the constant for which the test is to be done. The test by switch-statement is similar to the process of following if......else structure.

```
switch (variable)
{
```

```
if (variable ==value 1)                                    case value 1 :
{ statement 1;                                                |    |    |    |
_ _ _;                                                        |    |    |    |
_ _ _;                                                        |    |    |    |
}
else if (variable ==value 1)                               case value 2:
{statement 2;                                                 |    |    |    |
_ _ _;                                                        |    |    |    |
_ _ _;                                                        |    |    |    |
}
_____
else if (variable ==value N)                               case value N :
{statement n;                                                 |    |    |    |
_ _ _;                                                        |    |    |    |
_ _ _;                                                        |    |    |    |
}
else
{                                                          default :
statement default;                                           |    |    |    |
_ _ _;                                                        |    |    |    |
}                                              }
```

It means the switch-statement test, an expression only for equality (==). Any other type of comparison like lesser than, greater than, etc. are not supported by switch-statement.

During the process of switch statement, the variable or expression is evaluated. Its value is matched with the constant values given with "case statements" one by one. When a constant value of case-statement" matches with the variable/expression value, the statements following that case are executed. If no case-value matches with the value of variable/expression, then the statements following the "default-statement" are executed.

We have used a special statement "break" which indicates the end of a "case-block". Every "case statement" has a constant value and is followed by a colon sign (:). The "break" statement causes the immediate exit from the block, where it is used and transfer the program control to the statement just following that block.

In 'switch..case..default" statement, the use of "break" statement is used to exit from switch after successful execution of a case-block. The "break" statement transfers the program control immediately outside switch-block.

PROGRAM-15

```
/* Program to show the purpose of switch statement */
#include <stdio.h>
#include <conio.h>
main()
{
int n;
```

```
printf("\n Enter a number between 1 and 5:");
scanf("%d", &n);
switch(n)
{
case 1:
        printf("\n ONE");
        break;
case 2:
        printf("\n TWO");
        break;
case 3:
        printf("\n THREE");
        break;
case 4:
        printf("\n FOUR");
        break;
case 5:
        printf("\n FIVE");
        break;
default:
        print("\n Number is out of range (1-5)");
        }
        getch();
        }
```

Output:

   (i) Enter a number between 1 and 5 : 2

       Two

  (ii) Enter a number between 1 and 5 : 7

       Number is out of range (1-5).

Here, we see that the condition checking is done only for one case. When the variable value is matched with any case value, the corresponding statement is executed and break statement is executed immediately to stop further case match. If no case value is matched, the default case statement is executed. See that there is no break statement used in default case. This is because that after the default case, there is no other case statement from which the control is to exit. If the default case statement is executed, the program control automatically transfer to the outside of switch block because of sequential execution.

If we avoid the use of break statement in "switch-case" then the program flow is not appropriate as the user requires. The output of the program will not be like the user's requirement.

But we can skip the break statement and processing statements for several cases, when many case statements have one common output statements. Its use can be described by following example –

PROGRAM-16

Check the input character is vowel or consonant.

```
/* vowel or consonant */
```

```
#include <stdio.h>
#include<conio.h>
#include<ctype.h>
main()
{
char x;
clrscr();
print("\n Enter any alphabet:");
x = getchar();
if (isalpha(x)!=0)
{
switch(tolower(x))
{
case 'a':
case 'i':
case 'o':
case 'u':
printf("\n vowel.");
          break;
default :
          printf("\n constant.");
}
}
else
          printf("\n not an alphabet.");
getch();
}
```

Output :

  (i) Enter any alphabet : A

    Vowel.

 (ii) Enter any alphabet :T

    Consonant.

(iii) Enter any alphabet :5

    Not an alphabet.

▶ **REMARK**

▸ Using "switch-case" statement, we can perform the process by evaluation the character variable value or integer variable value. But for the checking of float constants, there may be nesting of switch that means one "switch-case" with in another "switch-case" statement. The use of "default" statement is optional for "switch" statement. But its use makes the program interactive. The "switch" statement is very useful for menu driven program. The menu-driven programs perform different process by selecting one choice out of given list (menu) of working options.

PROGRAM 17

Input 2 numbers and create a menu driven program having following options.

1. Addition
2. Difference
3. Multiplication
4. Exit

```c
/* Menu driven program with switch statement */
#include <stdio.h>
#include <conio.h>
#include <stdlib.h>
main()
{
int a,b,c, choice
printf("\n Enter 2 numbers :");
scanf("%d%d, &a, &b);
printf("\n The options are  :\n");
printf("\n1.Addition \n2.Difference  \n3.Multiplication);
printf("n4.Division \n5.Exit");
printf("\n Enter your choice (1,2,3,4,5)  :");
scanf(%d", &choice);
        switch (choice)
        {
        case 1 :
        c=a+b
        printf("\'n' sum of %d and %d = %d.", a,b,c);
        break;
        case 2 :
        if (a>b)
                c=a-b;
        else if (b>a)
                c= b-a;
        printf("\'n' Difference of %d and %d = %d.", a,b,c);
        break;
        case 3:
                c=a*b;
        printf("\'n' Multiplication Result of %d and %d = %d.",
                        a,b,c);
        break;
        case 4:
                c=a/b;
        printf("\'n' Division  Result of %d and %d = %d.", a,b,c);
        case 5 :
        exit(0);
```

```
                    default :
                    printf("\n Invalid choice ~");
                    {
                    getch();
                    }
```

Output :

Enter 2 Numbers : 15 12

The options are :
  1. Addition
  2. Difference
  3. Multiplication
  4. Division
  5. Exit

Enter your choice (1,2,3,4,5) : 2

Difference of 15 and 12 = 3

In this program, we have used a new function named exit(). It is standard library function present in the header file "stdlib.h". this function causes the immediate termination of the program. This function requires an argument. This argument value acts as the status value for program termination. This status value is returned to the operating system after program termination. The value zero indicates the normal termination and other values indicates the errors.

1.16.3 THE CONDITIONAL OPERATOR

The C-language provides the special operators used decision making. These operators are the combination of two characters? And :. These are classified as Ternary operator because it requires 3 operands. These work similarly "if-else statement, so called a "conditional operators". These are useful to make a program short in length because these operators work for shorting the if...else statement. The general syntax of using the conditions operators is as follows –

> Boolean expression ? expression 1 : expression 2

The "Boolean expression" is the logical expression which is to be tested and evaluated before processing if statements if the condition is evaluated as TRUE, then the statements given as "expression 1" are executed; otherwise the statement of "expression 2" are executed.

The conditional operators are the conversion of following "if...else" statement –

> If (Boolean expression)
>> expression 1;
>
> else
>> expression 2;

PROGRAM-18

Check the given number is even or odd using conditional operators.

```
/* Example of conditional operators */
#include<stdio.h>
#include<conio.h>
main()
```

```
{
int n;
clrscr();
printf("\n Enter a number  :");
scanf("%d", &n);
(n%2=0 ? printf(\n EVEN")  : printf("\n ODD"));
getch();
}
```

Output :

    Enter a number : 17

    ODD

So, we have seen that condition are easy way to shorten the length of program. But we can have only single line statement for the "expression 1" or "expression 2". This is the only limitation with the use of conditional operators. This limitation is useful because multiple statements can make the conditional operator statement complicated to write and understand.

1.16.4 THE GOTO STATEMENT

In a C-program we have used the sequential program structure. It means, the program execution starts from the first statement of the program and continues to the last statement of the program. Sometimes, the statements are not executed because of conditional evaluation.

C-language supports a special statement named "goto" watch causes unconditionally jumping of program control from one statement to another in the program. The "goto" statement transfer the program control to the backward statements or forward statements as the program requires.

The use of "goto statement" requires the label name, which identifies the statement where the program control is to be transferred, because C-language does not support line numbers. So the "label" acts as line-name through which the goto statement identifies a statement. The "label can be used anywhere in the program.

The label name is followed by a colon sign (:) when applied to any statement. But it is followed by semicolon (;) when used with "goto" statement.

In a C-program, we can use any number of "label" statement as per the requirements. But the use of "label" should be avoided in a program because use of "label" and "goto" statement s break the statement the sequential flow of program execution and effects the program performance by transferring program control here-and-there in the program.

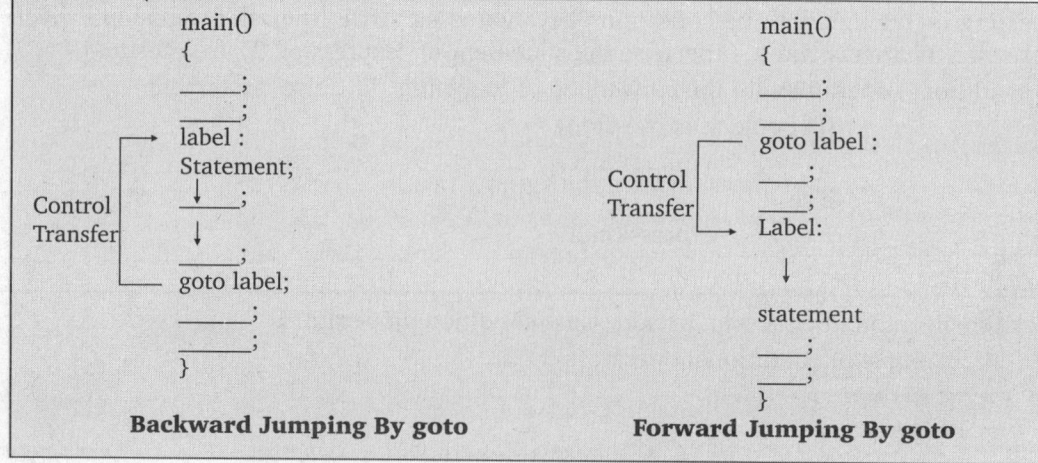

Backward Jumping By goto   **Forward Jumping By goto**

PROGRAM-19

Program to check a number even/odd with goto.

```c
/*Use of goto statement */
#include<stdio.h>
#include<conio.h>
main()
{
int n;
clrscr();
printf("\n enter the number  :");
scanf("%d", &n);
if (n%2 == 0)
        goto one;                   /*[EVEN]*/
else
        goto two;                   /*[ODD]*/
one:
        printf("\n  number is even  !");
        goto out;
two:
        printf("\n  number is odd  !");
        goto out;
out :
        printf("\n*End of program *");
        getch();
}
```

Output :

  (i) Enter the Number : 17
     Number is odd !
     *End of program *

 (ii) Enter the Number : 22
     Number is even!
     *End of program *

1.17 REPETITION CONTROL STATEMENTS

In a C-program, we have generally used the sequential flow of statement execution. Sometimes the statements are skipped to execute by using "Decision Control Statements". Sometimes, we need to execute some statements repeatedly. According to general program flow, we have to write such instructions many times in the program. This may lead to the increasing program length and make the program tedious. For solving this type of problem, the concept of "Repetition Control Statement" is applied to the programming language.

The "Repetition Control Statements" provided by C-language are "loop-structure". The repetition of some statements in a program make a loop like structure. So the "Repetition-Control Statements" are also called as "Loop Control Statements". The loop control Statements save the problem of rewriting of the repeatedly executing statements. In a general sequential flowing program, we have to write such statements many times in the program, such as, if we want to display a single line statements five times on the output screen, then we have to write that output instruction five times in the program as follows –

```
/* Repetition without loop */
# include <stdio.h>
#include <conio.h>
main( )
{
printf(" /n print This Line Five Times.");
printf("/n print This Line Five Times.")
printf("/n print This Line Five Times.");
printf("/n print This Line Five Times .");
printf("/n print This Line Five Times.");
getch();
}
```

Fig. 5

Sometimes, we can use the "goto statement" to repeating the statement execution. The "goto statement" in "Backward Jump Format" can only work for repetition of some statements. But we know, that the use of "goto-statements" should be avoided in the C- program. If we use "goto-statements" for the above program, then the length of program decreases as follows-

```
              /*Repetition with goto */
              # include <stdio.h.>
              #include <conio.h>
              main()
              {
              int ctr =1;
              start :
              printf( "/n Print T his line Five Times .");
                 ┌─→ if (ctr <5)
       loop   │     {      ↓
              │          ctr = ctr + 1;
              │              ↓
              │          goto start;
              │     }
              └──  getch( );
              }
```

Fig. 6

Here we see that a loop like structure is created by "label statement" and "goto statement" and the statements in between are repeatedly executed till the condition remains satisfied. A variable is required to control the repetition and also a condition is required. The control is transferred back if the condition is satisfied otherwise the program control flows forward in sequential manner. This concept is applied to "loop-structure" in the C-programming. It means, loop-structure can be said the improved format of "goto-statement". By using "loop -structure" we have no problem and is easy to use.

With the help of loop structure, a single statement or sequence of statements are to be executed till the termination of a condition. The repeated execution continues till the looping condition is satisfied. So, we need such a condition which can control the repetition of statements, and this condition is called "control-condition" for the loop. The condition is the boolean function made by the combination of variables and constants. By this fact, the loop-structure can be defined as the combination of two parts — One is the "control statement" consisting of control variable and control condition; another is the "body of loop" consisting of the statements to execute repeatedly.

Depending on the evaluation of control condition, the loop structure are classified as two types —

(i) Entry Controlled Loop

This loop evaluates the control condition first before the execution of loop-body. If the condition evaluation results in TRUE value, the program control enters the body of the loop otherwise not, so program-control entry in the loop body is restricted by the control condition and so it is called as "Entry controlled loop". C-language provides two types of loop-structure of this category as :

- while loop
- for loop

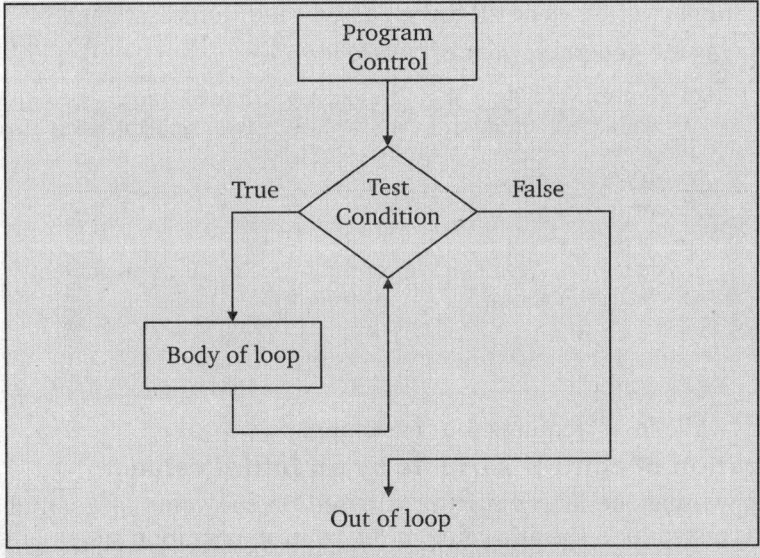

Fig. 7

(ii) Exit Controlled Loop

This loop first of all executes the statements of loop-body. Then the control condition is evaluated. If the condition is evaluated as TRUE, then the repetition continues otherwise the program-control exits the loop-body. So the exit of program control depends on evaluation of control condition but entry does not. This is because it called as "Exit controlled loop". C-language provides one type of loop of this category as :

- do...while loop

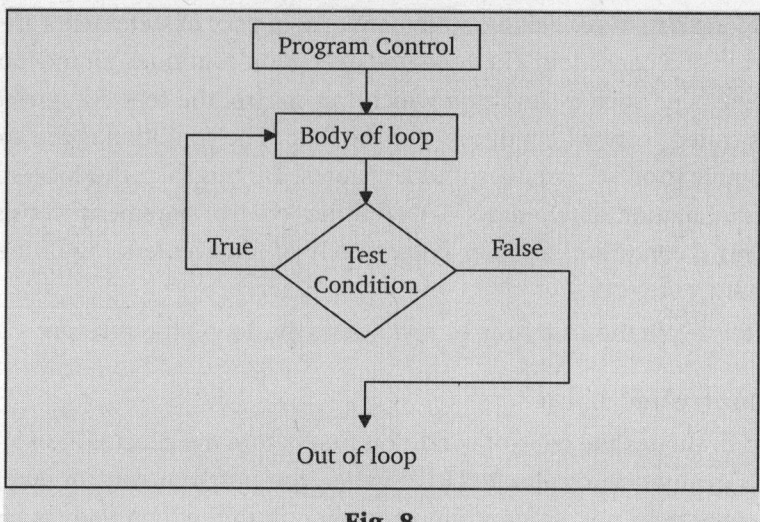

Fig. 8

It means , in "Exit controlled loop", the body of the loop execute first time unconditionally.

(iii) while Loop

The "while-loop" is the simplest loop structure available in C-language. It is an "entry controlled-loop". It involves the repetition of some programming expressions for a specific number of times or till the specified condition is satisfied.

Following is the general format of "while loop"—

Control variable = initial value;]→ Initialization
while (test condition)   ]→ Boolean expression
{
  Body ⌠ -----------
  of | -----------
  loop ⌡ -----------
Increment/Decrement in control variable;] Incrementation
or Determination
}

Fig. 9

The format of "while-loop" requires four basic parts –

(a) Initialization of control variable by an initial value

As we know, that the loop structure is based on the value of a variable. It means a variable is tested for a specific value in the control- condition statement. That variable is referred to as the "control variable" because the repetition of statement is controlled by that variable. It acts like the counter for repetition. It may be character or integer or float or any other datatype variable.

(b) Boolean expression for specifying the test condition

Every loop is controlled by a "Control Condition" which is used to control the repetition of loop-body. The "while-loop" evaluates the Boolean expression and executes loop-body if the condition evaluates "TRUE", otherwise the loop terminates. If the loop-body is executed, then the "Test condition" is again evaluated by the program control for further processing.

If the "Test condition" is again evaluated "TRUE", then the body of loop executes again. This process repeatedly continues until the "Test condition" evaluates "FALSE", because the False-Value of Test-condition transfers the program control out of the loop and starts execution of further programming statement.

(c) Body of loop consisting statements to execute

The "Body of the loop" consists of the statements which we have to execute again and again repeatedly. There may be single statement or many statements within the "Body of loop". The body of loop starts just after the "control-condition statement". The statements of two loop-body need to be enclosed within the pair of curly braces { }, if their number is two or more than two. But the use of curly braces can be omitted if the statement within loop-body is only one in number.

(d) Increment/Decrement in the value of control variables

The repetition of loop body depends on the control-condition. This control-condition sometimes be used for specifying the time limitation for loop execution. For that purpose, the control variable has to reach to the value specified with condition. Then we have to increase or decrease the value of control variable. In most of the loop-structures, the incrementation or decrementation statement is applied as the last statement of loop. But it is not necessary for all loop structure. If we avoid the use of Increment/Decrement statement, the loop execution may become "infinite" and body of the loop is executed again and again.

With the help of "while-loop", we can modify the program described in Fig. 6 as follows :

PROGRAM-20

Sample Program for while loop.

```
/* Example while loop */
#include<stdio.h>
#include<conio.h>
main()
{
int ctr;
ctr=1;
while(ctr<=5)
{
printf("'n Print This Line Five Times,");
ctr=ctr+1
}
getch();
}
```

Output :

Print This Line Five Times.
Print This Line Five Times.
Print This Line Five Times.
Print This Line Five Times.
Print This Line Five Times.

The Increment or Decrement in the control variable can be performed by using different operators as follows:

(i) Using Increment (++)/Decrement (– –) operator

$$\text{Increment} \binom{\text{Variable} + +}{+ + \text{Variable}} \binom{\text{Variable} - -}{- - \text{Variable}} \text{Decrement} \atop \text{operator}$$

The Increment operator (++) increases the value of variable always by one only.

Same is the case with Decrement operator (– –) which decreases variable value by one.

(ii) Using Compound Assignment operators

$$\text{Increment} \binom{\text{Variable} + = \text{Constant}}{\text{Variable}^* = \text{Constant}} \binom{\text{Variable} - = \text{Constant}}{\text{Variable}/ = \text{Constant}} \text{Decrement} \atop \text{value}$$

These operators perform arithmetic calculations as well as assignment of result to variable simultaneously. We can increase or decrease the control variable by any value.

For example: We have a variable 'X' with an arithmetic value '0'. Then ¬

X+=2	yields	12 in X	because it means X = X+2
X+=1	yields	11 in X	because it means X = X+1
X*=3	yields	30 in X	because it means X = X*3
X–=8	yields	2 in X	because it means X = X – 8
X/=2	yields	5 in X	because it means X = X/2

(iii) Using Arithmetic Operators

variable = variable + constant value

variable = variable – constant value

variable = variable * constant value

variable = variable / constant value

variable = variable % constant value

These operations allow the increment or decrement in the value of control variable by any numeric constant.

For example: We assume a variable 'X' with an initial value 10 and another variable 'Y' to store result of increment or decrement Then –

Y = X + 2	Results	12 in Y
Y = X – 2	Results	8 in Y
Y = X * 3	Results	20 in Y
Y=X/2	Results	5 in Y
Y = X%2	Results	5 in Y
X = X+2	Results	12 in X

So we can use any method for increment or decrement. Their work may be similar in some cases like –

$$X++, \ X+=1, \ X=X+1 \text{ [Increase by one in all cases.]}$$

But their work may be different sometimes like ¬

$$X=X+2 \text{ [Increase by two X++]} \quad \text{[Increase by one]}$$

$$X+=2$$

$$X+=4 \text{ [Increase by 4]}$$

Sometimes, we can use the loop-control variable in the other statements of loop body. The "while-loop" structure is clearly described in the following examples -

PROGRAM-21

Print counting from 1 to 10.

```
/*Counting From 1 to 10 with while loop */
#include<stdio.h>
#include<conio.h>
main()
{
int count=1;
clrscr();
while (count<=10)
{
print("\n %d", count);
count++;
}
getch();
}
```

Output :

1
2
3
4
5
6
7
8
9
10

PROGRAM-22

Calculate sum of digits of a given number.

```
/* Sum of digits like 123 = 1 + 2 + 3 = 6 */
#include<stdio.h>
#include,<conio.h>
main()
{
int num, sum, digit;
clrscr();
printf("\n Enter a number : ");
scanf("%d", &num);
sum = 0;
while (num>0)
```

```
                    {
                    digit=num%10;              /* Detacting last digit say 3 */
                    sum += digit;
                    num/=10;                   /* converting num by skipping */
                    }                          /* last digit and become say 12*/
                    printf("\n Sum of digits : %d', sum);
                    getch();
                    }
```

Output :

Enter a number : 567

Sum of digits : 18

1.16.5 FOR LOOP

It is also called as "Entry Controlled loop" which provides the short form of while-loop-structure. As we know that the loop-structure is generally consisting of four parts – initialisation, condition test, body of loop and Increment/Decrement operation. For this work, the while loop structure requires at least four statements written in four different lines. But using "for-loop-structure", we can decrease this length of loop-structure to only two line structure. So the "for-loop-structure" is very useful for shorting the program length. The general format of "for-loop-structure" is as follows :

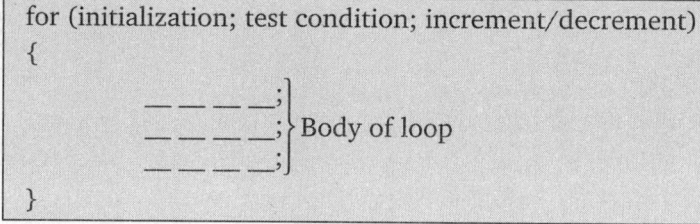

Fig. 10

The "for-loop-structure" holds the initialization, Test Condition and Increment/Decrement statements as single statement enclosing with in the pair of parenthesis (---), but separated by semicolon (;). After the "for statement", the body of the loop consist of only the processing statements. The curly braces are used only when the body of loop contains two or more statements.

The execution of "for-loop- structure" begins with the initialization of control variable in "for-statement". Then the control is transferred to the control-condition statement for testing the specified condition. If the condition is evaluated "TRUE", then the program control is transferred to the loop-body to execute its statements, otherwise the loop terminates and the program control is transferred to the statement just following the loop body. If the body of loop is executed, the program control is again transferred back to the "for-statement" to increase or decrease the control variable. After increment or decrement in the value of control variable, the program control again test the condition and the process goes on until the specified condition evaluates "FALSE". The use of increment or decrement operator is just like with the "while loop structure".

PROGRAM- 23

Print the following series of n numbers.

```
2, 5, 10, 17, 26, .... ...
/* series of 2, 5, 10, 17, ... ... */
#include <stdio.h>
#include <conio.h>
main()
{
int a, b, c, n, I;
clrscr();
printf("\n How many numbers you want in series :");
scanf("d", &n);
a=1;
b=1;
for(i=1; i<=n; i++)
{
c=a+b;
printf("\n\nt %d", c);
a=c;
b=b+2;
}
getch();
}
```

Output :

How many numbers you want in series : 7

    2
    5
    10
    17
    26
    37
    50

SPECIAL FEATURES OF for LOOP

(1) We can use the multiple initialisation or multiple test condition or multiple increment/decrement in the "for statement" of a loop.

PROGRAM-24

There are two series like –

    1, 2 ,3, 4, -- --- -- -- --- - 9, 10   Series 1
    10, 9, 8, 7 -- --- -- -- - -- 2, 1   Series 2

Print the series generated by subtraction of series 2 elements for corresponding series 1 elements.

```
/* subtraction series (1, 2, 3, -- 9, 10) - (10, 9, 8 -- --2, 1) */
#include <stdio.h>
#include<conio.h>
main()
{
int x, y,z;
clrscr();
printf("\n the Resultant (1,2,3 --, 9, 10) - (10, 9,8, -- 2, 1) series :\n");
for(x=1, y=10; x<=10, x++, y--)
{
z=x-y;
printf("\nn\t%d", z);
}
getch();
}
```

Output :

The resultant (1,2,3, ---, 9, 10) – (10,9,8, -- 2,1) series :

        −9
        −7
        −5
        −3
        1
        3
        5
        7
        9

In the above example program, we see that the multiple initialisation test condition and increment/decrement are separated by semicolon. But the two initialisations (x=1 and y =10) are separated by a comma sign and similarly the conditional statements and increment/ decrement statements.

(2) we can see the logical operator to join the multiple conditional statements as follows :

        for (--,--; x<=10 && y>1; --,--)
        {---;
        ---;
        ---;
        }

(3) But it is not necessary that the test conditions are to be made up of all control variable. It means for multiple initialization, only single control variable can also be used for specifying the control-condition, like –

        for (--, --; x<=10; --,--)
        {----------------------;}

(4) The 'initialization statement" of the loop may also contain the arithmetic expression, such as –

        for (y=10, n=y/10; --,--; --,--)
        {---------------------------------;}

(5) We can omit the part of "for statement" sometimes if required such as –

```
main()
{
int n;
for (n=1; n<=10;}
{
printf("\n%d", n);
n++;
}
}
```

Fig. 11

(6) we have seen that the omission of "Increment statement" from for statement causes no error but there is required the semicolon for that statement. Same is the case with "Initialization statement" omission.

```
main()
{
int x;
x=1
for (; n<10; n++)
{
printf("\n%d", x);
}
}
```

Fig. 12

(7) Some times we can omit both initialization and increment /decrement statement, but the omission of "control condition" could to the error in loop-execution.

```
main()
{
int x;
x=1;
for (; n<10;)
{
printf("\n%d", x);
}
}
```

Fig. 13

(8) Some programs use the "for-loop-structure" without body statement. These loops are used for delay-purpose and called as "time delay loops". These loops have the termination of "for-statement" with the semicolon sign (;) because there is no statement in the body of loop. The semicolon sign only indicate in the loop body is called "NULL statement".

```
for (i=1; i< 1000; i=i++);
```

This loop executes 1000 times but the execution does not produce any result because of "NULL statement" within the body. But the control variable value continuously increases from 1 to 1000 and does only time delay for execution of further statements.

1.16.6 THE DO...WHILE LOOP

This is an "Exit Controlled Loop", it means the entry within loop-body is not controlled by "control condition" but this controls the exit of program control from the body of loop. In this loop-structure, first of all the statements within "body or loop" are executed and then the "control-condition" is evaluated at the end. If the condition is evaluated "TRUE", then the program continues to execute the body of loop again until the "control-condition" evaluates "FALSE". The loop will terminate only when the condition becomes false. The "control-condition" is evaluated after one execution of loop body, so it is called "Exit Controlled Loop". This loop is helpful in the situation, when we have to execute the loop-body at the first attempt without testing the condition. The general format of do...while loop is as follows¬

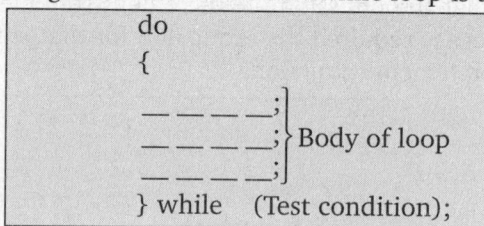

Fig. 14

Generally, the use of "do...while-loop" is preferred with menu-driven programs because we have to show the menu-list at least once.

PROGRAM-25

Input two numbers and perform the following operations as per user's requirement–

1. Addition
2. Subtraction
3. Exit

```c
/* Menu driven program with do-while loop */
#include<stdio.h>
#include<conio.h>
#include<stdlib.h>
main()
{
int x, y. choice. result;
char ans='y';
printf("\n enter two numbers: ");
scanf("%d%d", &X, &y);
do
{
printf("\n The options are: \n1.Adddition \n2.Subtraction \n3. Exit");
printf("\n enter your choice: ");
scanf("%d". &choice);
switch (choice)
{
case 1 :
        result=x+y;
```

```
                    printf("\n Sum of %d and %d = %d". x,y,result);
                    break;
        case 2 :
                    result=x-y;
                    printf("\n subtraction of %d and %d = %d.
                    x,y,result);
                    break;
        case 3 :
                    printf("\n program terminates !");
                    exit(0);
        }
        printf("\n Do you wish to continue the operations (y/n):");
        scanf("%C". &ans);
        } while (ans='y'  ||  ans == 'Y');
        getch();
        }
```

Output :

Enter two numbers: 12   8

The options are :

1. Addition

2. Subtraction

3. Exit

Enter your choice: 1

Sum of 12 and 8 = 20

Do you wish to continue the operations (y/n) : y

The options are :

1. Addition

2. Subtraction

3. Exit

Enter your choice: 2

Subtraction of 12 and 8 = 4

Do you wish to continue the operations (yln) : n

1.16.7 Difference Between "while-loop" and "do...while loop"

We have seen, that both "while loop" and "do...while loop" appear to be same because both are used to repeat a statement or group of statements for a specific time controlled by a "control condition", But both loops very much differ in their working.

In "while loop", the condition is evaluated before the execution of "loop-body". But in "do...while loop" the conditions evaluated after one execution of the "loop-body".

In "do...while loop", the body of the loop executes at least once always, whether the control-condition evaluates "TRUE" or "FALSE", But in "while-loop", it is not necessary that the body of loop will be executed. There may be some situations, when the body of loop is not executed at all in case of "while loop".

In "while-loop", the first execution as well as repetition of loop-body always depends on the control-condtion. But in "do-while loop", the first execution of the loop body is not dependent on control-condition but the repeated execution is always dependent on the condition.

We can easily illustrate the difference between "while-loop" and "do...while loop" by following example –

/* while loop */
main()
{
int n;
n=10;
while(n<10)
{
printf("\n Loop is executed");
n=n + 1;
}
getch();
}
	/* do...while loop */
	main()
	{
	int n;
	n=10;
	do
	{
	printf("\n Loop is executed");
	n=n + 1;
	} while (n<10);
	getch();
	}

No output is produced by this program using while loop	**Output:** Loop is executed
	This loop will execute one time.

Fig. 15

1.16.7 Breaking of Loop Execution

The loop-structure executes the body of loop many times. The execution is stopped only when the control-condition is evaluated "FALSE".' It means the loop termination is possible only with the control-condition-evaluation and the loop-execution cannot be stopped in the middle. But sometimes, we may require to S\OP or skip the loop-execution in the middle of normal execution. Such as the search of a number out of 10 number should terminate the loop as soon as search is over.

For this situation, C-language supports two operations¬

  (i) Stop the loop-execution (Jump out of loop)

  (ii) Skip a part of loop-execution

(i) Jump out of loop

  "Jump out of loop" means stop the loop execution before the loop-condition. It means exit from the loop execution before the complete execution cycle. This work is accompanied by using the "break" statement or the "goto" statement.

  The "break" statement is used to stop the execution of the block in which it is used and transfer the program control to the next statement following that block. If we use the "break" statement with loop-body, it will cause the termination of loop execution. But the use of "break" statement should be controlled by a condition. As soon as the given condition is satisfied, the termination of loop occurs, and the program control jumps out of the body of loop to the statement just following the loop structure. We can use the "break" statement within body of the loop in following formats¬

(a) Jump Out of while-loop

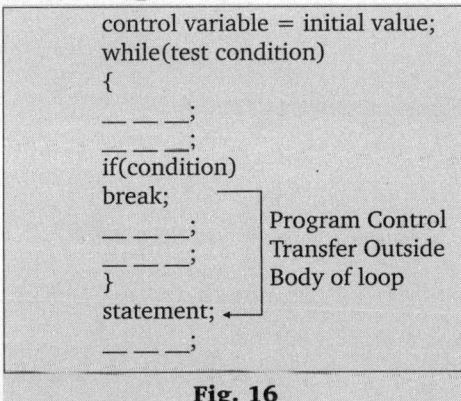

Fig. 16

(b) Jump Out of for-loop

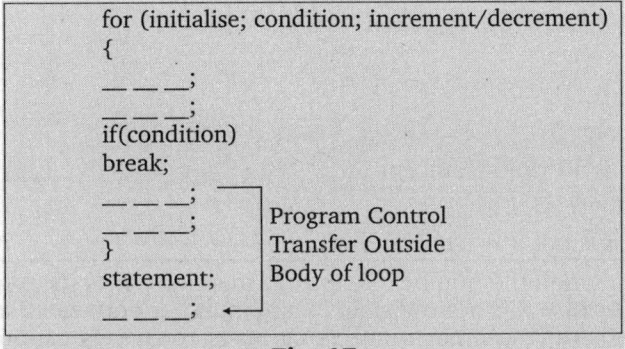

Fig. 17

(c) Jump out of do...while loop

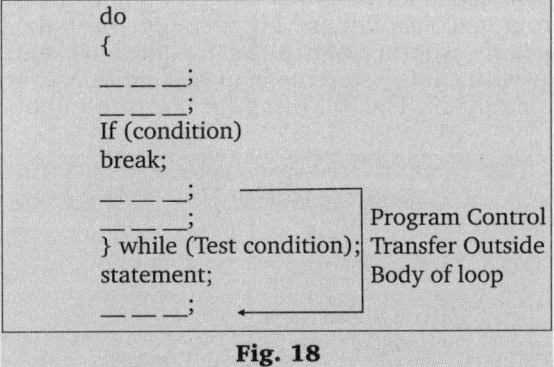

Fig. 18

The use of "break" statement with the loop structure can be described easily by example:

PROGRAM-26

Input an integer number and check it is prime or not

```
/* Prime number checking using break */
#include<stdio.h>
#include<conio.h>
```

```
main()
{
int num, i;
clrscr();
printf("\n Enter the number: ");
scanf("%d", &num);
for (i=2; knum; i++)
{
if (num%i==0)
{
printf("\n The number %d is not a Prime Number !"',num);
break;
}
}
if (i==num)
printf("\n The number %d is a Prime Number !", num);
getch();
}
```

Output :

  (i) Enter the number: 7

     The number 7 is a Prime Number!

  (ii) Enter the number: 25

     The number 25 is not a Prime Number!

In the above program, when the number input is 7, the loop starts its execution from $i=2$ and continues upto $i=6$ because 7 is not divisible by any number between 2 and 6 and loop is not broken in the middle. But, the input number 25 leads to the execution of break statement because 25 is divisible by $i=5$ and so the control jumps out to the if statement but value of i is 5 and num is 25.

The "goto" statement is also useful for jumping out of loop body because the "goto" statement is used to transfer the program control to any other statement in the program having a "label name". For exiting a loop body, we can assign a "label name" to a statement outside loop body and use that label name with "goto" statement in a condition contained within loop-body using "Forward Jumping method". The following formats are to be applied for the "Jumping out of loop with goto" –

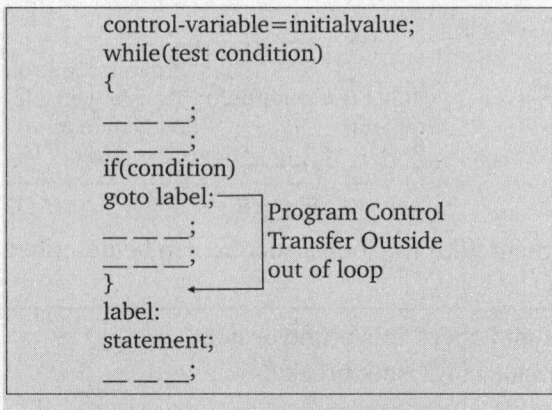

Fig. 19 (a) goto with while-loop

```
for (initialise; condition; increment/decrement)
{
__ __ __;
__ __ __;
if(condition)
goto label; ──
__ __ __;
__ __ __;
}
label:  ◄──
statement;
__ __ __;
```

Program Control
Transfer Outside
out of loop

Fig. 19 (b) goto with for-loop

```
do
{
__ __ __;
__ __ __;
if(condition)
goto label;
__ __ __;
} while (test condition);
label:
statement;
__ __ __;
```

Program Control
Transfer Outside
out of loop

Fig. 19 (b) goto with do...while-loop

The following shows, how the forward jump with goto-statement is useful to determine the loop execution.

PROGRAM-27
_____

Input numbers from the input device until the negative number is found. Count the total input positive numbers.

```c
/* Count total input numbers using goto-jumping */
#include<stdio.h>
#include<conio.h>
main()
{
int num, count, n, i;
clrscr();
count = 0;
printf("\n Enter how many numbers you input :   ");
scanf("%d", &n);
for (i=1; i<=n; i++)
{
printf("\n Enter a number: ");
scanf("%d", &num);
```

```
if (num>=0)
        count++;
else
{
printf("\n The number is negative !");
        goto out;
}
out:
printf("\n Total positive input numbers are:%d",count);
getch();
}
```

Output :

Enter how many numbers you input: 5

Enter a number: 15

Enter a number: 0

Enter a number: 7

Enter a number: 2

The number is negative !

Total positive input numbers are : 3

In the above example, we have seen that the loop terminates before n=5 due to the input of negative numbers, The input of negative number lead to the transfer of control outside the loop body immediately.

(ii) Skip a part of loop execution

Under certain condition, we may require to skip the statements of loop to execute for a specific iteration (control variable's value) but want to execute the loop statements for next iteration. The C-language has a special processing statement for this purpose and it is the "continue" statement. It causes the skipping of all statements following "continue" statement within the body of loop and transfer the program control to the next iteration execution. So it is called "continue" statement.

We can use the "continue" statement with any loop-structure as follows –

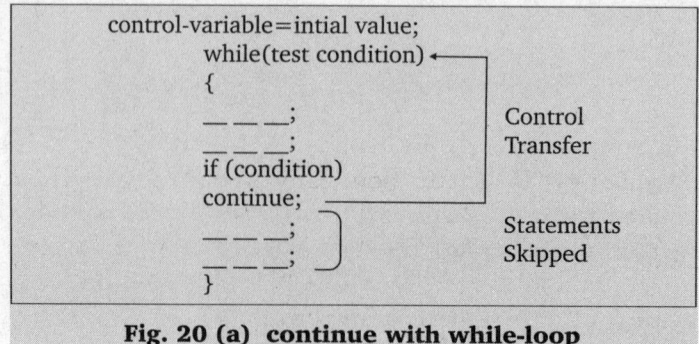

Fig. 20 (a) continue with while-loop

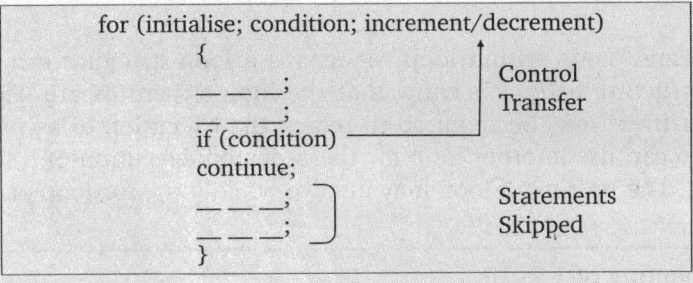

Fig. 20 (b) continue with for-loop

The processing of complete loop-body is not affected, but the statements after "continue" statement are only affected by the use of "continue" statement. The following example is useful to explain the use of "continue" statement¬

PROGRAM-28

Input 5 characters and count the total alphabets entered.

```
/* Count alphabets using continue */
#include<stdio.h>
#include<conio.h>
#include<ctype.h>
main()
int i, count;
char x;
clrscr();
count=0;
for(i=1; k=5; i++)
{
printf("Enter a character: ");
scanf("%c", &n);
if (isaalpha(x)==0)
continue;
count++;
}
printf("\n Total alphabets = %d", count);
getch();
}
```

Output :
Enter a character: A
Enter a character: $
Enter a charaacter : B
Enter a character : #
Enter a character: C
Total alphabets = 3

NESTED LOOP

"Nested Loop" means "Loop within loop". It means a loop-structure can be made the part of another loop-structure-body. We know that the loop structures are used to repeat a set of statements. But these may be required to repeat the execution of a complete loop in the program. Then we can use another loop for the repeated execution of a loop. This is called "Nesting of loops". The nesting of loop may involve while-loop, for-loop or do-while loop.

PROGRAM-29

Print five times counting of 1 to 10.

```
/*Counting 1-10 Five Times */
#include<stdio.h>
#include<conio.h>
main()
{int num, i;
clrscr();
for(i=1; i<=5; i++)
{
printf("\n");
        for (num=1; num <=10; num++)
                printf("\t%d", num);
}
getch();
}
```

Output :

1	2	3	4	5	6	7	8	9	10
1	2	3	4	5	6	7	8	9	10
1	2	3	4	5	6	7	8	9	10
1	2	3	4	5	6	7	8	9	10
1	2	3	4	5	6	7	8	9	10

1.17 WORKING WITH ARRAYS

In a computer program we use data for processing sometimes we require a large amount of data for processing. Then it is a little bit difficult to declare so many variable in a program for example: storage of 100 number require declaration for 100 integer variables.

One solution of such problem is the use of loop. Suppose loop start from 1 and continue to 100. Only one variable is used in this case for loop. But the problem arises due to variable nature. Whenever a new value is given to the variable the previous value is removed from the memory. After the complete execution of loop in the variable hold only the last value. so it is not possible to see all 100 number stored by loop in the same variable.

The array is the useful data type for this purpose. The array is group of various data items of same type sharing same name and present in continuous memory locations. The array can hold all the values for reuse because array contains many variables of same name and same data type. The complete set of values is referred to as an array and the individual value as array element.

The array can be of any primary data type as integer character or float. The array of character is called string and of float and integer as array. The use of array for storing large amount of data one of type makes the program easy and deficient and also solves the problem of many variable declarations. We can use the values stored as array-element again and again for different manipulations till the completion of programs.

The arrays are of following their types based on the dimensions for data storing:

   (i) One dimensional arrays   (ii) Two dimensional arrays

  (iii) Multi dimensional arrays

1.17.1 ONE-DIMENSIONAL ARRAYS

The list of data items have only variable name but differentiated by a subscripts number because we can not have multiple variable of same name in a c-program. Single subscripts number differentiates the subscript number acts as the identification number for multiple data items and so the one dimensional array variable are called as single subscripts variable. The subscript number of array-data-items is started from zero and x and be a positive integer number. For example when we have can integer array named X for 10 integer variable these variable are named as – X [0], X [1] ... X [9].

(i) Declaration of Array

      Array is also a variable. So it is to be declared in the program before its use. Every array has a data type and number of that array for its declaration. The general format for array decoration is as follows.

               | datatype arrayname [arraysize]; |

The data refers to the data type required for array such as char, int or float etc. the array name is the variable name to be given to all data items of the array. The array size indicates the number of data items required for the array. This size is to be mentioned within brackets []. All the data items of array are declared continuously in memory location horizontal or vertically.

For example:

In this declaration an array of integer type is declared with 10 data items having their name as num. Their subscript number starts from 0 and ranges upto 9. This array can be assumed to declared in memory as follows

	0	1	2	3	4	5	6	7	8	9
int num										
	num [0]	num [1]	num [2]	num [3]	num [4]	num [5]	num [6]	num [7]	num [8]	num [9]

This array can store maximum 10 integer constants in the variable named num [0], num [1], num [2]...num[9]. The item of array occupies the space in memory as the data type of array needs. In the above "num" is declared as int type. So every array element of the array occupies 2bytes of memory. The complete array require the memory space used by one variable of the multiple by total number of elements in the array such as 20 bytes in the above example. We can also get the memory address of all array elements if the memory address of first array elements is known. Suppose the address of the first array elements in above example is 2020. Then the memory address of another array elements will be as follows.

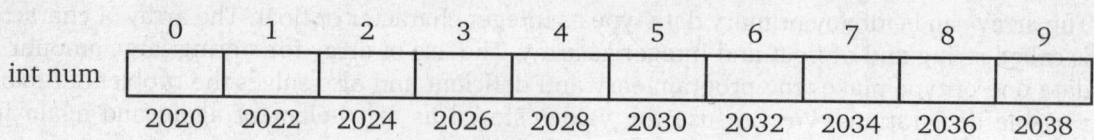

	0	1	2	3	4	5	6	7	8	9
int num										
	2020	2022	2024	2026	2028	2030	2032	2034	2036	2038

We can declare the of any other data type also like-

char name [25];
float rate [12];

REMARK

- C-language does not use any bound checking for data value to be stored in the array with reference to the pre-specified array size. It means we can store the data value in a subscripts-item of array beyond the array size also and this data value is stored in the memory by overwriting the another data value may be any important data. This may produce the unpredictable resulting data values. But the C-language does not produce any error message as warning for going beyond the array size. So the programmer should take care of the array size during data entry for getting the right output.

(ii) Initalization of Array

The array elements can be initialized by constant data values in a C-program, if required, there several formats for assigning the initial data values to an array.

For example: int num[5]={2, 4, 6, 8, 10};
int num [ ]= {2, 4, 6, 8, 10};

Both the initialization states that integer array num is declared and some data values are put in these variables using list of values, the number of values must be equal to the array size specified. But we can skip the array size specification because initialization at the time of declaration specify the array size. If we do not store any data value in the array elements by initialization, then these array elements are assumed to contain the garbage data value.

Here, we have used two types of array initialization. We see that the size of array is optional if we initialize the array with declaration.

The another method of array initialization is the subscript method and involves the assignment of values to the array elements individually as shown below-

```
int num[5];
num[0] = 2;
num[1]=4;
num[2]= 6;
num[3]= 8;
num[4]=10;
```

These all methods for array initialization differ in their format only because in memory the values are stored as follows-

	0	1	2	3	4
int num	2	4	16	8	10

But the subscript method is also useful for changing the value of any array element in the program after initialization. **For example,** we want to change the value of third array element to '15. Then we can use the following statement-

num[2] = 15;

After this value assignment, the elements of array num will become-

	0	1	2	3	4
int num	2	4	15	8	10

In this value assignment, we have used the subscript number 2 with the array name for referring to for referring to the third array element , because the subscript number of array starts form zero. The subscript method is the most appropriate method for referring to an individual array element also. The subscript number of the required array element is specified number of the required array element is specified within brackets [ ] with the array name.

(iii) Data Input and Data-Output with Arrays

In an array, the data-input is performed by scanf() function and data-output is produced by print() function similar to other variables.

Let us take an example to describe data-input and output with array. Suppose we want to store 5 integer number by using array and display them as output after storage. The following example program ode will describe this problem.

PROGRAM-30

Store 5 integer numbers in array and show them as output.

```
/* Input-Output with Array */
#include<stdio.h>
#include<conioh.>
main()
{
int x[5];
printf('\n Enter five numbers:');
scanf('%d%d%d%d%d", &x[0], &x[1], &x[2], &x[3], &x[4]);
printf("\n the numbers stored in Array :\n");
printf("\n%d\n%d\n%d\n%d\n%d\n%d", x[0], x[1], x[2], x[3], x[4]);
getch();
}
```

Output :

Enter five numbers : 12 13 14 15 16

The numbers stored in Array :

12

13

14

15

16

In this example, we have used the functions scanf() and printf() for input-output with array respectively. We have seen that scanf0 uses five format specifiers for taking input of 5 integer values as array-data elements. But this type of input-output is easy for an array of small size. It will be very difficult to input or output the array of size 100 or other using this method. The solution of this problem is possible with the use of loop-structure. The scanf() statement will have to be repeated 100 times. So we can use loop-structure for this purpose. The problem

arises with the subscript number because the subscript number should change for every scanf()
statement. Then the control variable to represent the subscript numbers from 0 upto the size
of array by using the loop-structure, the program shown in example 1 can be modified and the
input-output in array becomes easy and the program length decreases.

PROGRAM-31

Input-Output with Array using loop structure.

```
/* Input-Output Array using loop */
#include<stdio.h>
#include<conioh.>
main()
{
int x[10], i;
clrscr();
printf('\n Enter 10 array elements :\n");
for(i=0; i<=9; i++)
{
printf("\n Enter the number :   ");
scanf("%d", &x[i]);
}
printf("\n The stored array elements are :\n");
for(i=0; i<=9; i++)
printf("n x[%d]=%d", i, x[i]);
getch();
}
```

Output :
Enter array elements :
Enter the number : 12
14
16
18
20
The stored array elements are :
x[0] = 12
x[1] = 14
x[2] = 16
x[3] = 18
x[4] = 20

PROGRAM-32

Print sum of 10 input numbers.

```
/* Sum of any 10 numbers */
#include<stdio.h>
#include<conio.h>
```

```
                main()
                {
                int arr[10], i, sum;
                clrscr();
                sum=0;
                printf('\n Enter 10 numbers :\n");
                for(i=0; i<=9; i++);
                scanf("%d", &arr[[i]);
                for(i=0; i<=9; i++)
                sum=sum+x[i];
                printf("\n Sum = %d", sum);
                getch();
                }
```

Output :

Enter 10 numbers :

2

4

6

8

1

3

5

7

9

4

Sum = 49

(iv) Searching Within Array

Searching means find a specific datavalue within the array elements. Several searching methods are provided by C-language like linear search, binary search.

Linear search allows to search an elements in linearly that means starting from zero to the last array element. But this type of search takes a large amount of time.

Binary search is useful for saving because maximum number of time taken is lesser than linear search.

PROGRAM-33

Input n number in an array and search a number using linear search method .

```
                /* Linear search */
                #include <stdio.h>
                #include<conio.h>
                main()
                {
                int a[20], n, num, i, flag;
```

```
clrscr();
flag=0;
start :
printf("\enter the size required for array:");
scan f ("%",&n);
if (n>20)
{
printf("\n size required for array size ");
printf("\n enter the size from 1 to 20 ! );
go to start;
}
printf("\n enter %d number for array… \n" ,n);
for (i=0; i<=n-1; i++)
scanf("%d", &a[i]);
print f( "\n enter the number to search : ");
scanf("%d",& num);
printf("\n now searching starts …");
for(i=0; i<=n-1; i++);
{
if(a[i]==num)
{
printf("\n%d is found at %d position !",num, i+1);
flag=1;
break;
}
}
if(f==0)
printf("\n% d is not in entire array !);
getch();
}
```

Output :

Enter the size required for array	:	35

Size required is out of array size.

Enter the size required for array	:	5

Enter 5 number for array...

15

25

30

35

Enter the number to search .	:	25

Now searching starts...

25 is found at 3 position !

In this program, we have used goto statement for bound-checking the array size during the input. Now the numbers entered in this program will not exceed the size of 20 numbers. The flag is a variable used to evaluate true or false. Firstly, its value is taken as zero. When the search is successful the value of flag variable is turned to one. This value of flag variable is evaluated for unsuccessful or successful search.

PROGRAM-34

Program for binary search in an array

```
/* Binary search */
#include <stdio.h>
#include <conio.h>
main()
{
int arr[2], n, i, beg, end, mid, num;
clrscr();
start;
printf("\n enter array size required : m");
scanf("%d'.&n);
if (n>20)
{
printf("\n  enter %d number in ascending order..." ,n);
printf("\n");
for(I=0; i<=1; i++)
scanf("%d", &arr[i]);
prinf("\n enter number to search in array : ");
scanf("%d", & num);
beg =0
end =n - 1
mid=(beg + end)/2;
{
while(num<arr[mid] && beg <=end)
{
if (num<arr[mid]}
end=mid - 1;
else
beg=mid + 1;
mid=(beg+end)/2;
}
printf("\n %d is found at %d!' num, mid+1);
else
printf("\n%d is not found in entire array !" num);
getch();
}
```

Output :

Enter array size required   : 5

Enter 5 numbers in ascending order

12

15

27

35

42

Enter number to search in array :15

15 is found at 2 !

In this program, the search of a number is controlled by a loop. The loop continues till the search-number is not found at mid position or the array element terminated. The number to be searched is given as 15. The search starts from middle array element which is 27. The search is unsuccessful. Now the search number (15) is smaller than middle number (27). So search is continued to the left of middle array element. Now new mid element is 15 and the search element is also 15, so the search is successful. This process continues further is the search becomes unsuccessful.

(v) Sorting the Array

Sorting is the process to organize the data either in ascending order or descending order. The array can hold many data items of similar type at a time so we can arrange these data items either in ascending order or descending order like a sorted list. There are several sorting processes developed to sort a list of detail-items like selection sort, bubble sort etc. we will discuss the process of selection sort here for sorting a list of numbers.

SELECTION SORT PROCESS

The selection sort process uses the technology to select the minimum number from a list and place to the beginning of list ascending or descending order sorting respectively. We can describe the actual process taken place in selection sort ascending order by following example :

Let us take an array of 10 numbers as follows :

int x[10]	22	56	33	80	78	46	25	65	35	12

	0	1	2	3	4	5	6	7	8	9
int x	22	56	33	80	78	46	25	65	35	12

After Step 1	12	56	33	80	78	46	25	65	35	22
After Step 2	12	22	33	80	78	46	25	65	35	56
After Step 3	12	22	33	80	78	46	25	65	35	56
After Step 4	12	22	25	33	78	46	80	65	35	56

After Step 5	12	22	25	33	35	46	80	65	78	56
After Step 6	12	22	25	33	35	46	80	65	78	56
After Step 7	12	22	25	33	35	46	56	65	78	80
After Step 8	12	22	25	33	35	46	56	65	78	80
After Step 9	12	22	25	33	35	46	56	65	78	80
After Step 10	12	22	25	33	35	46	56	65	78	80

In the selection sort process for ascending order sorting, the step-1 shows the selection of minimum number (12) from entire array and its replacement with the zero subscript array element (22). In step-2, the minimum number is selected out from the array except zero subscript element and its replacement is performed with subscript one array element x[1]. This process continues till the complete sorting of array. The minimum numbers are push down side to the array.

PROGRAM-35

Sort the list of number in ascending order by using selection sort method.

```
/* Selection sort process */
# include <stdio. h>
# include <conio. h>
main()
{
int x[20], n, min, p, temp;
clrscr();
start:
printf("\n Enter Array Size Required:");
if (n>20)
{
printf("\n Out of Range. \n Enter from 1 to 20 ! \n ");
goto start;
}
printf("\ Enter %d Number to sort :\n");
for(i=0; i<=n-1; i++)
scanf(" %".&x[i]);
printf("\n Now sort process starts ...\n");
for(i=0; i<=1; i++)
{
pos=i;
```

```
min=x[i];
for(p=i+1;p<=n-1; p++)
{
if(xp[p])
{
min=x[p];
pos=p;
}
}
temp=x[i];
x[i]=x[pos];
x[pos]=temp;
}
printf("\n Sorted Array in Ascending Order : \n");
for(i=0; i<=n-1; i++);
printf("\n%d",x[i]);
getch();
}
```

Output :

Enter array size required: 10

Enter 10 numbers to sort :

22

56

33

80

78

46

25

65

35

12

Now sort process starts-

Sorted Array in Ascending Order :

12

22

25

33

35

46

56

65

78

80

TWO DIMENSIONAL ARRAYS

We know that the array is used to store the data items in the form of a list by using the one-dimensional array. But in some situations, we have to store the tabular data values. Then the one-dimensional array is not able to do this kind of work. The C-language provides the facility to work with tabular data using two-dimensional array.

The two dimensional array can hold the data values in the form of table containing rows and columns. These are also useful to solve the matrices problem of mathematics, so these are also called "Matrices". The data is stored in two dimension horizontal as well as vertical logically.

(i) Declaration of Two-dimensional Array

The declaration of two-dimensional array requires specification of two dimensions as row-size and columnsize with array name and datatype. The general syntax used for two dimensional array declaration is

> datatype   arrayname   [Rowsize]   [Columnsize];

For example

    int n[2] [3];

By this declaration a two dimensional array of 2 rows and 3 columns is created as follows-

	Col. 0	Col. 1	Col. 2
Row 0			
Row 1			

Fig. 21

An individual data-element of any two dimensional array requires two subscripts for its identification such as the third element of first row in this array is referred to as x[0][2x], because the subscript number of first array is zero and of third column is two. The subscript number for row is used first and then column. Then the data elements of the array specified above are identified individually as –

	Col. 0	Col. 1	Col. 2
Row 0	x[0] [0]	x[0] [1]	x[0] [2]
Row 1	x[1] [0]	x[1] [1]	x[1] [2]

Fig. 22

But this tabular representation of two-dimensional array is only logical. These are not stored in this form in the memory because memory does not allow the arrow declaration in rows and columns. In memory, the array data elements are always stored in continuous memory location collectively, whether they are one-dimensional, two-dimensional or multi-dimensional.

Let us consider the above declared array, if the array element starts storage from 2024 memory address, then its memory representation will be as follows-

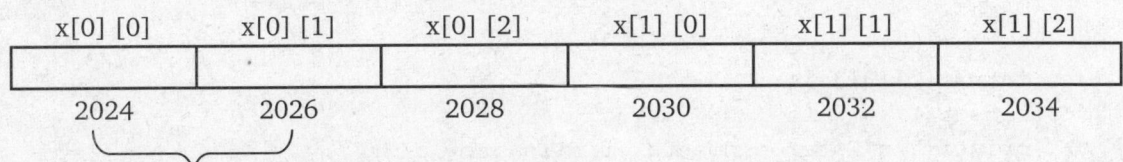

Difference of Two Bytes due to Integer Type

Fig. 23 : Row-Major Form

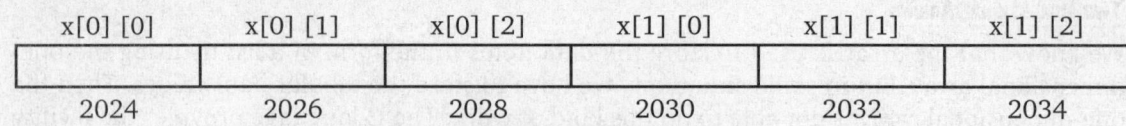

x[0] [0]	x[0] [1]	x[0] [2]	x[1] [0]	x[1] [1]	x[1] [2]
2024	2026	2028	2030	2032	2034

Fig. 24 : Column-Major Form

In Row-Major form the elements are stored row-wise in memory. In Column-Major Form, the elements are stored column wise in memory.

(ii) Initialization of Two-Dimensional Array

We can initialize the two-dimensional array using different formats. Consider the following examples –

```
int num [2] [3] = {
                  {2,4,6},
                  {8.10.12.}
                  };
int num [2] [3] = {2,4,6,8,10,12};
int num [ ] [3] = {2,4,6,8,10,12};
```

All the above initialisation will do the same work and creates the array with date as follows–

	Col. 0	Col. 1	Col. 2
Row 0	2 x[0] [0]	4 x[0] [1]	6 x[0] [2]
Row 1	8 x [1] [0]	10 x [1] [1]	12 x [1] [2]

Fig. 25

(iii) Input-Output in Two-Dimensional Array

We have seen the one-dimensional array (one row) requires at least one loop structure for array input or output. Now, the two-dimensional array has more than one rows. So it is difficult to enter date for more than one rows in two-dimensional array. Consider the following example.

PROGRAM-36

Program to describe the Input-Output in Two-Dimensional Array.

```
/* Input-Output a matrix */
#include<stdio.h>
#include<conio.h>
main()
{
int x [2][3], i;
clrscr();
printf("/n Enter elements of First now : /n");
for (i=0; i<=2; i++)
printf("%d" &x {0}(i));
print ("/n Enter elements of Second row : /n");
```

```
for (i=0; i<=2; i++)
scanf("%d", & n (i)(i));
}
printf("the matrix entered:/n");
for(i=0, i<=2;i++)
printf(n%d/t", x[0] [i];
print("/n");
for(i=0; i<=2; i+i)
printf("%d/t", x[1](i));
getch();
}
```

We have seen, that a loop structure is required to repeat for each now entry again and again. So we can modify this program using Nested loop-structure in the following program.

PROGRAM-37

Program describing Input-Output in Matrix using Nested loops.

```
/* Matrix */
#include<stdio.h>
#include<conio.h>
main()
{
int x[2][3], i, j;
clrscr();
printf(" Enter 6 matrix elements : /n");
for (i=0; i<=1; i++)
scanf("%d", &x [i][i]);
}
printf"/n The matrix entered is :/n");
for (i=0; i<=1; i++)
for (i=0; i<=2; i++)
printf("%d/t", x[i][i];
printf("n");
}
getch();
}
```

Output :

Enter 6 matrix elements:

1

2

3

4

5

6

The matrix entered is :

1	2	3
4	5	6

Then two-dimensional array is very useful for solving the matrix problem like matrix addition, matrix-multiplication, etc., easily. We can also maintain the tabular data also using two-dimensional array like student-details, employee details, etc.

1.18 STRING – THE ARRAY OF CHARACTERS

The string is the Array of Characters which is used to work on a group of characters. It is called "String" because string means line of characters. These play a very important role in the programming with C-language. In many programs, we may need to store the words of lines of characters like name of a person, address, cityname, etc. These groups of characters are handled by strings. The string is the constant represental in between double quotation marks.

For example :   "C is an interesting language".

One string is the continuous collection of many character variables. The character strings are very useful for simple manipulations on words or lines like read, write, copy, compare, etc.

(i) Declaration and Initialization

The declaration of string is similar to other datatype array. The string name should be any valid variable name. The general declaration format of character string is as follows-

char arrayname[size];

For example: char str[10];

This array declaration creates an array of 10 character variables with the name str.

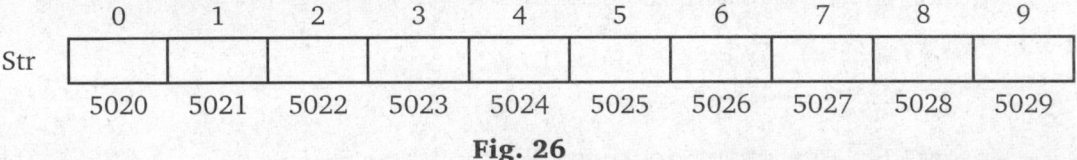

0	1	2	3	4	5	6	7	8	9
5020	5021	5022	5023	5024	5025	5026	5027	5028	5029

Str

Fig. 26

All these character variables are present in continuous memory location and each variable occupies one byte (8 bits) of memory. Let us assume the start element str[0] of the array shown in Fig. 25 is present at address 5020. Then the next variable str[1] is present at the address of 5021.

We can initialize the character array same as that of integer array at the time of declaration. The following formats are used for initializing the character array :

char str[6] = {'H', 'E', 'L', 'L', 'O', '\0'};

char str[5] = {"HELLO"};

char str[5] = "HELLO";

char str [ ] = {'W', 'O', 'R', 'L', 'D', '\0'};

char str [ ] = "WORLD"

These all initializations store each character in individual character variable.

0	1	2	3	4	5
H	E	L	L	O	\0

Fig. 27

In a string, "\0' is always stored at the end of string. This character is used as "string terminator". It is called as "NULL Character" and is different from zero because its ASCII value is different from zero. The string stores the characters using ASCII value. The ASCII value of "0' is zero. But the ASCII value of zero is 48. The role of "10' is very important in handling the string.

The work with integer array depends on the size of array or number of data elements stored. When we manipulate the integer arrays, the end limit of these is known to us. But the character arrays are not similar in working. In case of character strings, we are not aware of the character to be input in the string, because the strings are generally used to store words, lines like name, address etc. While entering this kind of data in the string, we do not know the accurate number of characters. It is not an easy task to input the data in string after counting the characters. If it is not possible then the string manipulation becomes difficult. For making the string manipulation easy without knowing the total characters in the string, the concept of "0' is used. So the "\0" is used as "End of String". Let us see the following example to show the use of "0' in a string

PROGRAM-38

String manipulation without use of "0'.

```
/* Without '\0' */
#include<stdio.h>
main()
{
char str[5] = "HELLO";
int i;
i=0;
while(i)
{
printf("%c", str[i]);
i++;
}
}
```

Output : HELLO

But the manipulation is not easy. So this program can be manipulated as follows –

PROGRAM-39

```
/* With '\0' */
#include <stdio.h>
main()
{
char str [  ] = "HELLO";
int i;
for (i=0; str[i]!='\0'; i++)
printf("%c", str[i]);
}
```

Output : HELLO

In both the above examples, the importance of "0' in string handling is shown very clearly. The program in Example 2 above shows that the string terminator "0' is useful in the strings of unknown number of characters. It means, the number of character in the string to manipulate are not known. The loop for manipulating the string uses the '\0' as the control condition and so the number of characters are not important to know.

(ii) Input/Output of Strings

The initialization of string is not sufficient for string manipulations differently. Like other variables, the data value for character array is also required. In a character array, there are present many character variables with same name and different subscript number. So we can input/output the string data value by splitting it into individual character, and using the function scanf() and prinf() respectively.

PROGRAM-40

Input a character string one by one character using loop structure.

```
#include<stdio.h>
main()
{
char str[5]
int I;
printf("\n Enter Five character string :");
for(I=0; I<5; I++)
scanf("%c",&str[i]);
printf("\n Entered string is :");
for(i=0; i<5; i++)
printf("%c", str[i]);
}
```

Output :

Enter five character string : HELLO

Enter string is : HELLO

This type of string input-output is not convenient. This case requires the knowledge of number of characters in the string, and the input-output is performed like the array of other type (integer). The "for-loop" is used to input-output the characters individually using the control-variable of loop as the subscript number variable. But this type of input-output is very typical with string, because the string input or string output should not be dependent on the total number of characters. So the input-output in string is manipulated by using a special type of format specifier –"%s" provided by C-language.

The use of "%s" for string input-output does not require the knowledge about number of characters. Using his format, we can store any number of character in the string without loop; within the specified array size or beyond it. As we know, that the data storage beyond array size may overwrite some important data. This fact is also applicable for character arrays (string). The following examples will described the use of scanf() and printf() for input-output of string –

PROGRAM-41

Use of printf() to print the string on output screen

```
/* string output with printf()*/
#include<stdio.h>
#include<conio.h>
main()
{
char str[ ]="PASSION";
clrscr();
printf("\n The string stored is : %", str);
getch();
}
```

Output :

The string stored is: PASSION

The format "%s" is used within prinft() function a string which is initially assigned to a string variable in the beginning of program. Note that the string variable in Example (7.4) is not given any array size. So the character array is created depending on the number of characters assigned during initialization. According to this example, the character array will be created as follows –

	0	1	2	3	4	5	6	7
char str	P	A	S	S	I	O	N	\0

Fig. 28

This figure shows that the last character is always string terminator '\0' and stored automatically. So, it is assumed that "%s" format is able to read the string up to the string terminator '\0'.

In the same manner, we can take the input from the keyword also by using "%s" with scanf() function.

PROGRAM-42

Program to show input of a string.

```
/*Input string */
#included<stdio.h>
#include<conio.h>
main()
{
char str[20];
clrscr();
printf("\n Enter Your Name :");
scanf("%s",&str);
printf("\n The entered name is : %s", str);
getch();
}
```

Output :

Enter Your Name : John

      The entered name is: John

Enter Your Name : John Smith

      The entered name is: John

According to the example 5, the character "str" is declared for 20 character size. The scanf() function reads the characters from keyboard and stores into the array one by one continuously till the space key or enter key is pressed. When the space key or enter key is pressed, the string terminator '\0' is put in the array. It means the scanf() function terminates the input of characters on finding the space, tab space, new line character, carriage returns or from feed characters. This point is well described in the example 5 (Output (i) &(ii)). The first output (i) the input string was only 1 word. So the array stored the complete word. But in the output (ii) the input is given in the form of two word (john Smith) using one space. The result shows only one word (John). This result describes that the entry of space in the input string terminates the string and the string terminator '\0' is put in the string. It means, the use of scanf() function with '%s' format is not sufficient for proper text entry. The text may be a character, word or line. So the use of another function gets() is proposed. This function is used to input a string with space, or say, multiword string. The input of string is not terminated by space entry, but the pressing of enter key puts the '\0' in the last of string. The following format should be used for taking string input with gets() function-

> gets (string variable name);

Another function puts() acts like the counterpart function of gets() function. The puts() function prints one line of text at a time on screen and always in new line. It means the use of puts() function for printing a line does not require '\n' for new line. The use of both functions – gets() and puts() is described below in the example 7.6-

PROGRAM-43

Program to read and write a line of text with space .

```
/* use of gets() & puts () */
#include <stdio.h>
#include <conio.h>
main()
{
char str[50];
clrscr();
printf("\n  Enter your name :");
gets (str);
puts("welcome ......!");
puts(str);
puts ("in you are in Heaven  !!");
getch();
}
```

Output :

Enter your name : Arundhati Mishra

Welcome ....!
Arundhati Mishra
You are in heaven !!

(iii) Output Manipulations on String

The string output is generally performed by using "%s" format in printf() function or puts() function . we can manipulate string output by using width specification with the format "%s". the width specification for string requires field width and number of characters.

For example, the format "%5.3s" will print three characters of string in the field width of 5 columns and it will be right justified. The format "%-5.3s" will print the three characters of string in the field width of 5 columns in left justification. If the field width is lesser than the length (number of characters) of string given, then the width specification is not followed during output. But the complete string is printed. The zero number of characters for printing will result in no print output. The following example illustrates the use of width manipulators during string output-

PROGRAM-44

Program to show the output manipulation on string.

```
/* string output manipulation */
main()
{
char city[ ] = "Muzaffar Nagar!";
clrscr();
printf("\n%16s", city);
printf("\n%6s", city);
printf("\n%16.8s", city);
printf("\n%-16.8s", city);
printf("\n%10.0s", city);
printf("\n%8.0s", city);
printf("\n%s", city);
getch( );
}
```

Output :

    Muzaffar Nagar !
    Muzaffar Nagar !
    Muzaffar !
    Muzaffar
    Muzaffar
    Muzaffar Nagar

(iv) Read the Entire String

The Passing through the entire string depends on the number of characters in that string. Passing through the string means reach to each and every character of string. Generally, we are not aware of the number of characters in the string. So the string terminator '\0' is used for passing through entire array.

PROGRAM-45

Input a string and print it character by character.

```
/* Pass through String */
#include <stdio.h>
#include<conio.h>
main()
{
char str[20];
int  I;
clrscr();
print ("\n Enter Any word :   ");
scanf( "%s", str);
printf("\n  the string entered is  :\n");
for (1=0; str[I]!='\0',; i++)
printf("%c", str[i]);
printf("\n Character by character in different lines :");
for(i=0; str[I]!='\0'; i++)
printf("\n%c", str[i]);
getch();
}
```

Output :

Enter any word : HELLO

The string entered is :

HELLO

character by character in different lines :

H

E

L

L

O

According to the above example, we can pass through the entire array using "%c" as the format for processing each character uniquely. In this example, a string of character variables is created to store a word. Suppose the string entered is "HELLO". Then the string terminator '\0' is put as the last character in the array as shown below

char str	0	1	2	3	4	5
	H	E	L	L	O	\0
	str [0]	str [1]	str [2]	str [3]	str [4]	str [5]

Fig. 29

In the program (Example 8), variable is used to control the subscript number of array elements starting from zero, while manipulating the array elements within loop structure. The loop structure is controlled by the presence of string terminator within the string. The manipulation loop finds the '\0' as a character and stops the loop repetition. This type of processing does not require the knowledge of total number of characters in the string. So this type of processing plays an important role in string manipulation

PROGRAM-46

Input a string and count total number of characters in it.

```
/*count number of characters*/
# include<stdio.h>
# include<conio.h>
main()
{
char str [20];
int I,  count;
clrscr();
printf("\n Enter a string :");
gets(str);
count=0;
for (I=0; str[I]='\0'; i++)
count++;
printf("\n Total characters  :%d", count);
getch();
}
```

Output :

(i) Enter a string : MY INDIA
   Total characters :8

(ii) Enter a string : Computer-World
   Total characters :14

This program counts total number of characters in the string. The number of characters in the string make the length of string.

PROGRAM-47

Input two string and compare them .

```
/* Comparison of two string */
#include <stdio.h>
#Include<conio.h>
main()
{
char str1[20], str2[20];
```

```
int ctr1, ctrt2, I, flag;
clrscr();
printf("\ n Enter first string : ");
gets (str1);
printf("\ n Enter second string : ");
gets (str2);
for (ctr1=0; str1[ctr1]!='\0';  ctr1++);
for (ctr2=0; str1[ctr2]!='\0';  ctr2++);
if(ctr1!=ctr2)
{
printf
("\n Both strings are not equal  !");
goto out;
}
flag=0;
for  (I=0;  I<ctr1 II  I<ctr2;  I++)
{
if (str1[I]! = str2[I])
flag=1;
}
if(flag==0)
printf("nBoth strings are equal  !");
else
printf("\n Both strings are un equal !");
out :
getch();
}
```

Output :

   (i) Enter first string : Hello

       Enter second string : World

       Both string are unequal !

  (ii) Enter first string : Hello

       Enter second string : Welcome

       Both string are not equal !

 (iii) Enter first string : Internet

       Enter second string : Internet

       Both string are equal !

(v) String Handler Functions

We have performed so many types of manipulation the strings using various logical concepts. There are several types of processing which can be performed on string. The standard library of C-language also provides some functions for handling the string manipulation. These are function already stored in a header file named "string.h". Some most commonly used string-handler functions are as follows.

strlen () function

This function is used to find out the length of a string. The length of a string refers to total numbers of characters in the string including space. The result returned by strlen () function is an integer value being number of characters. The counting of character starts form subscript zero and continues upto string terminator '\0'. The general format used for strlen() function is as follows-

> integer var = strlen(string/string var);

strcmp() function

This function is used to compare two given strings and find whether both strings are similar or not. The comparison of strings starts from zero subscripts and compare character by character till the mismatch character is not found or any one string terminates. The string comparison function has the following formats :

> integer value = strcmp (str1, str2);

the integer var may have any one of the following return value –

$$
\text{integer} \longrightarrow
\begin{cases}
(+)\text{ve} & (\text{string1} > \text{string2}) \\
0 & (\text{string1} = \text{string2}) \\
(-)\text{ve} & (\text{string1} < \text{string2})
\end{cases}
$$

1.19 FUNCTIONS

It is already discussed in the beginning of C-language features that the functions are the building blocks of C-language. C-language is said to be the combination of functions. "A function is a self-contained block of statements which are kept together to perform a specific task in related manner".

Every C-program is required to use at least one function. We have used several functions in different C-programs till now for different purposes. If all the processing step of a program are contained in one block, then this block is termed as the main() function. So the main() function is the necessary part of each and every C-program. Besides main() function, we have used several other function in our C-program, which are present in the header files as the standard library of C-language, such as printf(), getch() function etc.

1.19.1 USE OF FUNCTIONS

The functions make the C-language. A C-program consists of one or more functions. The main() function is the necessity of each C-program. So it is must for a C-program to have a least one function as main() function, because the program execution requires it as recognization point for starting.

In general, we use main() function to hold all the programming code for solving a specific problem. But this type of program writing leads to many problems. The large programs are difficult to understand, test and debug. So the large program code is splitted into small pieces of program • Sometimes a group of programming statements are needed to use more than once in a program • So these programming step are contained in a block as the function of the function of standard library.

The main advantage of using the function is in easy design. The program is designer using the modular approach of the program design. The modular program design is based of working with modules. The module is the another name function. A module is easy to understand because of small length. Generally we assume that the module should not be large than on page.

The use of function also facilitates reusability of programming code in the same program different program, because the code to reuse is referred to as function.

The programming of functions enables the programmer to easily expand a program by adding a new function to existing program. A new person can also identify the working of module/function easily.

We see that the main() function is the basic part of a C-program. We can use many other functions in a C-program to work with main() function. So we can assume the main() function as the Header person of an organization. The other functions can be assumed as the subordinates of the main() function, so that the work of the main() function can be divided. Because of this reason, the function are also called as subprograms.

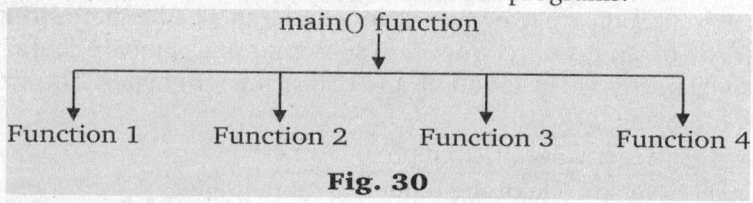

Fig. 30

1.19.2 CLASSIFICATION OF FUNCTIONS

Basically, the functions in a C-program can be of following two types –

- System defined functions
- User defined functions

This classification of functions is based on the function definition of function refers to as writing the steps to be performed by the function. The programming steps are written only once for a function. Both types of functions classified are discussed below.

(i) System Defined Functions

The system defined functions are the library function of C-language. The language has a large collection of function in the header files. These are called system-defined because these are already defined within the header files and we can simply use these functions in our C-program my where. We have no need to write the processing step of these function in our C-program such as the function for mathematical processing are kept in the header file "math.h". Similarly be functions related to string manipulation are present in the header file "string.h". We have used so many functions of different header files in previous C-programs. The use of printf() and scanf() function is also one of example of system-defined functions because we have used these function in various C-programs for input-out. It is necessary to include the related header file in the beginning of C-program for using any system defined function.

(ii) User Defined Function

The library functions has predefined work. So these are useful only for specific purpose and defined operations. But, most of the time user write the program for several different purpose because the requirement for program writing is different. Every C-program contains at least one. These type of function are generally used to contain the statements needed to use many time in a program.

PROGRAM-48

A program to find factorial of a number using without arguments with return function

```
                /* with arguments without return */
                #include<stdio.h>
                #include<conio.h>
                border(); /*Function declaration*/
                fact();
                main()
                {
                clrscr();
                printf("\n This is main function!");
                printf("\n Creating border :");
                border();
                fact();
                border();
                pritf("\n Back to main function!");
                getch();
                }
                border() /* creating border line*/
                {
                int c;
                printf("\n");
                for(c=21; c<=40; c++)
                printf("-");
                }
                fact() /*Finding factorial */
                {
                long int f;
                int n, m;
                printf("\n Enter the number to find factorial :");
                scanf("%d &n);
                for(m; n; m>=1; m--)
                f=f*m;
                border();
                printf("\n Factorial of %d=%d", n, f);
                }
```

Output :

This is main function!

Creating border:

Enter the number to find factorial: 5

Factorial of 5 = 120

Back to main function

In the above example, we have created 2 functions for separate work. The border() function is written to print a line of 40 characters required wherever required in the output of program. We see that there is no restriction on the function calling. We have called the border() function by both main() and fact() function as needed. Both the functions – fact() and border() perform their operations independently and there is data transfer among main(), border() and fact() function during function call.

(a) Function without Arguments and with Return Values

These functions does not receive the data values or variables as argument from the caller function. But these functions are able to send back the result to caller function. So there way data-transfer from the called-function to caller function only. The following figure shows the layout of such functions –

```
          returntype Function();  ───────►  Function declaration
          Main()
          {
          ---;
          ----;
          ----;
      ┌──► Variable = Function();  ───────►  Function calling
      │   ----;
      │   ----;
      │   ----;
      │   }
      │   returntype Function1()  ───────►  Function definition
      │   {
      │   ---;
      │   ---;
      └── return (variable/datavalue);  ──►  Return statement
          /* Result Back */}
```

Fig. 31

The functions declared with a returntype, which specify the type of datavalue to be returned by a function to caller function. The caller function captures the data value returned by the called function and can use for further processing by storing in a variable of same data type as the return type. A function can return only one data value to the caller function. The return statement is generally the least statement of a function-body but it can be used anywhere in the function body as the user requires. But the program control is transferred back immediately by the return statement to the caller function. The datatype of the datavalue/variable to be returned must be similar to the function. Otherwise we may not get the appropriate result.

PROGRAM-49

Find factorial of a number using without argument and with return functions.

```
/*Without argument with return*/
#include<stdio.h>
#include<conio.h>
long int fact(); /*Function declaration*/
main()
```

```
{
long int r;
clrscr();
printf("\n This is main function !")
r = fact();
printf('\n Back to main function !");
printf("\n Factorial is %!d", r);
getch();
}
long int fact()
{
int n;
long int f=1;
printf("\n This is fact function to find factorial!");
printf("\n Enter the number to find factorial :");
scanf("%d; &n);
for (; n>=1; n--)
f =f*n;
return(f);
}
```

Output :

This is main function!

This is Fact function to find factorial!

Enter the number to find factorial:

Back to main function!

Factorial is 120!

In the example 2, we have seen the transfer of program control and data value between main() and fact() functions. The fact() function return back the datavalue of variable "f" and this returned value is captured by main() function through the variable "r". In this case the name of both the variables may be same or may be different because there is direct copy of values from the variable of a function to another function.

(b) Functions with Argument and Return Values

These function ask for some datavalue to be given by the caller function gives the data to the called-function during function call but the called-function does not send the result back to caller function. These functions are very useful because several different function can take the same data value for different processing. The figure (8.4) describes the function layout

Function(arguments type list ------);

main()

{

---;

---;

function(actual argument list ---);

```
---;  |   value  transferred from caller
---;  ↓   to called function
```

function(formal arguments list ---)
```
{
---;
---;
}
```

The function declaration requires the datatypes of all the variable/values given as argument separately. The function call step requires the variable name or values to be used as argument. This step is also known as "argument passing". The function definition uses the declaration of formal argument list before use in the function body.

The argument given by the caller function to called function in function-call statements are known as Actual argument. The argument in which the function receives value from caller function in its definition are known as Formal arguments. There should be similarity in the number, datatype and sequence of actual arguments and formal arguments with the argument type list in function declaration. The values are copied from the actual arguments to formal arguments in the specified sequence. If the number of datatype of actual and formal arguments do not match, we may get the garbage result without any error message. The actual arguments may be variables, constant or expression. But the formal arguments should always be variables only.

PROGRAM-50

Program to find sum of two numbers using with arguments and without return function

```c
/*With arguments without return*/
#include<stdio.h>
#include<conio.h>
sum(int, int);

main()

{

int x,y;

clrscr();

printf("\n This is main function!");

printf("\nEnter Two Number:");

scanf("%d%d; &n, &y);

sum(x,y);

getch();   }

sum(int a, int b)

{

int z;

z=a+b;

printf("\n This is sum function!");

printf("\n Result %d", z);

}
```

Output :

This is main function!

Enter Two Number: 12 15

This is Sum Function!

Result = 27

In the example 3, we have created the formal arguments within the parenthesis () with function name (sum) in its definition. This is "modern method" of declaring the arguments with function definition in one line.

The previous method was "classic method" which the function argument were declared in separate line after the definition of function name, as follow-

> sum(a, b) ⟶ Function header
>
> int a,b;
>
> {
>
> ---;
>
> ---;
>
> ---;
>
> }

Now-a-day the modern compilers support both "classic-method" as well as "modern-method" for declaration of the formal argument.

(c) Functions with Argument and With Return Values

These functions are most valuable type of the functions. These functions receive the data from caller function for processing and then send the result back to caller function. So the result received by caller function can be used by any other function for further processing. In a large program, the result of one calculation can be used by any other calculations. These functions have two-way data transfer from caller function to called function and vice-versa. The following figure shows the layout of such type of function

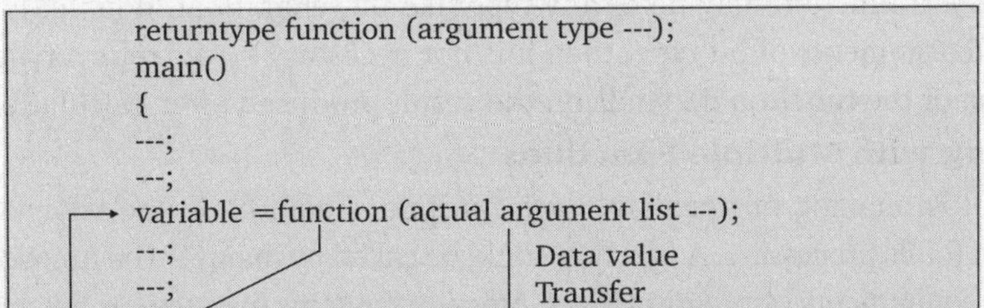

```
        returntype function (argument type ---);
        main()
        {
        ---;
        ---;
        variable =function (actual argument list ---);
        ---;                        Data value
        ---;                        Transfer
        }
    Call    returntype Function (formal argument list ---)
            {
            ---;
            ---;
    Back    return (datavalue/variable);
            }
```

Fig. 32

PROGRAM-51

Program to find sum of Two numbers using with argument and with return function.

```
/* With Argument and with return */
#include<stdio.h>
#include<conio.h>
int sum <int,  int);
main()
{
int x, y, z;
clrscr();
printf("enter two numbers  :")
scanf( "%d%d", &x, &y);
z-sum(x, y);
printf("\n sum of %d & %d=%d', x, y, z);
getch();
}
int sum(int a, int b)
{
int r;
r = a + b
return(r)
}
```

The argument passed to a function may be of some datatype as shown in the example(4) or these may be of different datatypes, such as – function(inta,floatb,charc,intd). Similarly, the returntype of a function may be of any type like int, char, float, double, long int etc. If a function has all arguments of int type, then it is not necessary for its return type to be integer. The return type of the function depends on the result produced after calculation.

(d) Working with Multiple Functions

In the programming of functions, each function should be in the knowledge of main() function for its processing. A function could be called by main(). The function call depends on the requirements of program. Each function contains instructions for the processing of specific operation. The function is called in the program wherever its operation is required in the program.

C-language calling of one function by another, then another and so on up to main() function. This is called "Nesting of Function" up to any level. Only one fact should be considered that the chain of function call is to be ended in main() function. Consider the following example program.

PROGRAM-52

Program to check a number is armstrong or not.

Armstrong Number

```
153 = (1 x 1 x 1) + (5 x 5 x 5) + (3 x 3 x 3) = 153
/*Nesting of function*/
#include<stdio.h>
#include<conio.h>
int arm (int);
int cube(int);
main()
{
int n, s;
clrscr();
printf("\n enter a number   :");
scanf("%d", &n);
s=arm(n);
if(s==n)
printf("\n Number %d is Armstrong!");
else
printf("\n Number %d is not Armstrong !");
getch();
}
int arm(nt n)
{
int sum=0, d;
while (n>0)
{
d=n%10;
sum=sum+cube(d);
n=n/10;
}
}
int cube (int d)
{
int c;
c= dxdxd;
return (c);
}
```

EXERCISE 1.1

Write the C program for following problems :

1. Input the basic salary of an employee. Provide bonus 15% of basic salary for the employee. Provide bonus 15% of basic salary for the employee having basic salary upto Rs. 8000 otherwise 18%.

2. Input an integer number of four digits. Check whether its last two digits and first two digits are perfect square or not.

   (**Hint :** 3625, 36 is a perfect square of 6, 25 of 5).

3. Input three numbers. Print them in ascending order.

4. Input a character. Check it is alphabet or digit.

5. Input an year number. Check it is leap year or not.

6. Write a C program to print even numbers in the range n1 – n2 inclusive.

7. Print all perfect square numbers in the range n1 – n2 inclusive.

   (**Hint :** n1 = 1, n2 = 10, perfect square = 1, 4, 9).

8. Input a number. Reverse it and check whether the reverse number is equal to original number or not.

9. Input a number (x) and power (n) for raising to x. Calculate x^n using C program.

10. Print sum of all numbers from 1 to 10 except 5.

11. Input n numbers and display sum of all even, all odd elements individually.

12. Input an array of 10 numbers. Reverse it and print the resulting array.

13. Input an array of 10 numbers and print product and sum of them.

14. Accept a decimal number as input data. Convert it to binary number and print.

15. Find the maximum number in an array of 10 numbers and print it.

16. Count number of characters in a given string.

17. Convert the lowercase letters to uppercase and uppercase to lowercase in a string.

18. Input two numbers and create the functions for sum, subtraction, multiplication and division to perform these operations as per user's choice.

19. Input two numbers and write functions for multiplication and division of these two numbers without the use of * and / operators.

20. Input a number and make a function FACT to calculate the factorial of given number.

⌘⌘⌘⌘⌘⌘

CHAPTER 2 — Error and Approximations

2.1 INTRODUCTION

Approximations and errors are in integral part of our life. These are exist everywhere, and sometime are unavoidable. A number of different types of errors arise during the process of numerical computing. These errors contribute to the total error in the final result.

Also the numerical data used in solving the problems are usually not exact, and the numbers expressing such data are therefore not exact. They are merely approximations, two to three, four or more figures. Not only are the data of practical problems usually result is to be obtained are also approximate. Therefore, an approximate calculation is one which involves approximate data, or approximate methods or both. Therefore, it is evident that the error in a computed result may be due to one or both sources, *i.e.*,

    (i) error in data   and   **(ii)** error in calculation.

The first type of error can not be decrease, but the second type can be made as small as we please, by taking the number to as many figure as we desired. Therefore, we can assume that the calculations are always carried out in such a manner as to make the errors of calculation negligible. In this chapter, we examine the sources of various types of computational errors and their subsequent propagation.

2.2 ACCURACY OF NUMBERS

  (i) Exact numbers : The numbers in which, there is no uncertainity and no approximation, it said to be exact numbers. **For example:** $5, 6, 7, \dfrac{8}{2}, \dfrac{1}{5}, \ldots$ are exact numbers.

  (ii) Approximate numbers : These are numbers which are not exact.
    For example: 1.41421 .... 3.141592.... are not exact numbers, since they contains infinitely many digits, are called approximate numbers.

▶ REMARKS
- The approximate number is a number which can not be expressed by a finite number of digits.
- Although, the numbers $\pi, \sqrt{2}$, etc. are exact numbers, they can not be expressed exactly by a finite number of digits. But when we expressed these numbers in digital form 3.141592, 1.41421, etc. such numbers are therefore only approximation to the true values and in such cases are called approximate numbers.
- Some authors always insist that one must say "approximate value" of a number in place of approximate number.
- Here, we used the symbol $\simeq$ for approximately equal to.
- Such numbers which represents the given numbers to a certain degree of accuracy are called approximate numbers.

(iii) Rounding-off a Number : If we divide 22 by 7 we get $\frac{22}{7} = 3.142857143...$ a quotient which a non-terminating decimal fraction. For use this type of number in practical computation, it is to be cut-off to a manageable size such as 3.14, 3.143,... . *The process of cutting-off superfluous digits and retaining as many digits as desired is known as rounding off a number.*

REMARK

- To round off a number is to retain a certain number of digits, counted from the left and dropped the others. Thus, to round off π to three, four or five and six figures respectively, we have 3.14, 3.142, 3.1416, 3.14159.

ALGORITHM

To rounding off a number or digit to n significant figures, discard all digits to the right of the nth place using the following concepts.

Step 1.	If this number is less than half a unit in the nth place, leave the nth digits as it is.
Step 2.	If the discarded number is greater than half a unit in the nth place, add 1 to the nth digit.
Step 3.	If the discarded number is exactly half a unit in the nth place, leave the nth digit unchanged.

For Example : The following numbers are rounded off correctly to four significant figures

(i) 38.63243 becomes 38.63 **(ii)** 91.8773 becomes 91.88

(iii) 21.64489 becomes 21.64 **(iv)** 87.495 becomes 87.50.

ALGORITHM

The old rule of rounding off the number says that when a 5 is dropped the preceding digit should always be increased by 1. It is not a good exercise and give inaccuracy in computations. Since, it is obvious that when a 5 is cut off, the preceding digit should be increased by one in only half the cases and should be left unchanged in the other half.

REMARK

- The numbers rounded off to n significant figures are said to be correct to n significant figures.

(iv) Significant Figures : Here, we have that all the digits 1, 2 ... upto 9 are significant figures and 0 is a significant figure except when it is used to fix the decimal point or to fill the places of unknown digits, *i.e.*, 0 may or may not be a significant figure. It depends upon the position in which zero has been used. As discussed earlier when zero is used to fixup the decimal point or to fill up the places of discarded digits, it is not a significant figure.

For example: Consider the numbers 0.00086 and 5800, correct to two significant figures. Then all zeros, which are used are insignificant. On the other hand, zero used in 430, correct to three significant figures, is a significant figure.

REMARKS

- The zeroes used between two non-zero digits are always significant figure e.g. 408.
- To round off a number or figure to r significant digits, discard all the digits or replace by zeros to the right of rth digit according as the number to be rounded off is a decimal fraction or whole number. Then r^{th} digit to be increased by 1 or to be left unaltered, according as the portion to be discarded or replaced by zeroes as greater than or less than half of the unit at the rth places (counted from the left). In case the discarded portion is exactly half of the rth unit, then the rth unit is to be increased by 1, if it is odd, otherwise it is left unchanged.

ALGORITHM

Step 1.	Significant digits are counted from left to right starting with the left most non-digits.

Step 2.	The significant figure in a number in positional notations consists of
	(a) all non-zero digits
	(b) zero digits which
	  – lie between significant digits
	  – lie to the right of decimal points and at the same time, to the right of a non-zero digit.
	  – are specifically indicated to be significant

Step 3.	The significant figure in a number written in scientific notation e.g. $M \times 10k$ consists of all the digits explicitly in M.

For Example

(1) The number 8.3678235, when rounded to three places of decimal, we get it as 8.368. Because, we leave the portion 0.0008235 which is more than half of 0.001.

(2) The number 83988235, when rounded to five significant digits, we get as 83988. Because the portion left out is 235, which is less than half of 1000.

(3) The number 8.6325 when rounded to three decimal places, we get 8.632 as the rounded number.

(4) 83675, rounded to four significant figures as obtained as 83680. Here the fourth place, when we counted from the left is 7 which is odd and the portion left out is exactly half of the unit at this place. Therefore we increase 7 by one.

SOLVED EXAMPLES

EXAMPLE 1. *Round-off the following numbers correct to four significant figures*
         68.3643, 878.367, 8.7265, 56.395

SOLUTION. Here, we have to retain first four significant figures. Therefore
  (i) 68.3643 becomes 68.36
  (ii) 878.367 becomes 878.4
  (iii) 8.7265 becomes 8.726 (Because the digit in the fourth place is even).
  (iv) 56.395 becomes 56.40 (Because the digit in fourth place is odd).

EXAMPLE 2. *Find the sum of the following approximate numbers, each being correct to its last figures*   396.56, 657.2, 758.9826, 3.052

SOLUTION. Since the number 657.2 is correct to one decimal place. Therefore, it is not worth while to retain digits beyond two decimal places. Hence, we rounded off the given numbers to two decimal places, and then found the sum.
Therefore, the required sum
$$= 396.56 + 657.20 + 758.98 + 3.05 = 1815.79 \approx 1815.8$$

▶ **REMARKS**

▸ When we deal with the approximate numbers of unequal accuracies, retain one more significant figure is more accurate numbers then are contained in the least accurate number as it being done in above example. In the end the sum has been rounded to one decimal place.

▸ The concept of accuracy and precision are closely related to the significant digits, as follows:
  (a) Accuracy refers to the number of significant digits in a value. For example, the number 86.498 is accurate to five significant digits.
  (b) Precision refers to the number of decimal positions, *i.e.*, the order of magnitude of the last digit in a value. Here the number 86.498 has a precision of 0.001 or 10^{-3}.

2.3 ERRORS AND THEIR ANALYSIS

Definition : *The quantity, True value – Approximate value is called the error.*

2.3.1 SOURCES OF ERRORS

Following are some sources of error in numerical computations.

(i) Input Errors: The input information is rarely exact. It comes from the experiments and any experiment can give results of any limited accuracy.

(ii) Algorithmic Errors: Sometimes, the direct algorithms based on a finite sequence of operations are used. Errors due to limited steps don't amplify the existing errors. Since the application of some formula is not possible for a infinite number of times, algorithm has to be stopped after a finite number of steps. Hence, the obtained results are not exact.

(iii) Computational Errors: Sometimes, when we performing elementary operations, the number of digits increases greatly. Therefore, the result can not be held fully in a register available in the given system.

2.3.2 TYPES OF ERROR

(i) Absolute error: If x^A is the approximate value of exact number x^T, then the absolute error denoted by E_a is defined by

$$E_a = \Delta x = |x^T - x^A|$$

$$\Rightarrow \quad E_a = |x^T - x^A|$$

REMARK

- In error analysis, the magnitude of the error is not important, not the sign of error. Therefore, we consider the absolute error generally.

(ii) Relative Error: In many cases, absolute error may not reflect its influence correctly as it does not take into account the order of magnitude of the value under consideration. **For example-** An error of 1 gram is much more significant in the weight of 10 grams Gold, that in the weight of a bag of sugar. Due to this reason the concept of relative error is introduced.

The relative error is the absolute error divided by the true value of the given quantity. It is denoted by E_r and defined as

$$E_r = \left|\frac{x^T - x^A}{x^T}\right| = \frac{\text{Absolute error}}{\text{True value}}$$

(iii) Percentage Error: The percentage error in x^A, which is the approximate value of x^T is

$$E_p = 100 \times E_r = 100 \times \left|\frac{x^T - x^A}{x^T}\right|$$

REMARKS

- The relative error is also known as normalized absolute error.
- If $\bar{x}$ be a number such that $\left|x^T - x^A\right| \le \bar{x}$, then $\bar{x}$ is said to be an upper limit on the magnitude of absolute error and measures the absolute accuracy.
- The relative and percentage errors are independent of the units of measurement, while absolute errors are expressed in terms of unit used.
- If a number is correct to n significant figures then its absolute error can not be greater than half a unit in a nth places.

▸ If a number is correct to n decimal places then the error $= \frac{1}{2} \cdot 10^{-n}$.

For example: If the number 8.869 correct to three decimal points its absolute error is not greater than $0.001 \times \frac{1}{2} = \frac{1}{2} \times 10^{-3} = 0.0005$.

SOLVED EXAMPLES

EXAMPLE 1. *Find the sum of* 392, 780.56, 64320, 72300, 23657 *assuming that the number 72300 is known to only three significant figures.*

SOLUTION. Since we have, that the number 72300 is known to hundred places.

Therefore, we round off other numbers correct to tens places and then find the sum, *i.e.*,

$$\text{Sum } S = 390 + 780 + 64320 + 72300 + 23660$$
$$= 161450 \simeq 161400$$

Here, we observe that, the last significant digit (counting from left) is 4 which is uncertain by one unit of this place.

THEOREM 1. *If the first significant figure of a number is r and the number is correct to n significant figures, then the relative error is less than* $\dfrac{1}{r \times 10^{n-1}}$.

PROOF. Let us suppose that N be any given exact number which contains n significant figures and m denotes the number of correct decimal places.

Then, there are following three cases :

Case (i): If $m < n$

In this case the number of digits in the integral part of N is given by $(n - m)$. Let us denote the first significant figure of N by r. Then, we have

Absolute error $\qquad E_a \leq \dfrac{1}{10^m} \times \dfrac{1}{2}$

and $\qquad N \geq r \times 10^{n-m-1} - \dfrac{1}{10^m} \times \dfrac{1}{2}$

which gives $\qquad E_r \leq \dfrac{\dfrac{1}{10^m} \times \dfrac{1}{2}}{r \times 10^{n-m-1} - \dfrac{1}{10^m} \times \dfrac{1}{2}}$

$$E_r = \dfrac{10^{-m}}{2r \times 10^{n-1} \times 10^{-m} - 10^{-m}}$$

$$= \dfrac{1}{2r \times 10^{n-1} - 1} = \dfrac{1}{2\left(r \times 10^{n-1} - \dfrac{1}{2}\right)}$$

Now, since n is any positive integer and r stands for any digits 0, 1, ..., 9. Then we have $2r \times 10^{n-1} > r \times 10^{n-1}$ in all cases except $r = 1$ and $n = 1$. (We can ignore this case, because it is a trivial case when $N = 1$, 0.001, 0.0001 etc., *i.e.*, the case in which N contains only one digit different from zero, which would not occur in common practice). Therefore, we may assume that

$$2r \times 10^{n-1} - 1 > r \times 10^{n-1} \text{ for all cases}$$

Then, the relative error $E_r < \dfrac{1}{r \times 10^{n-1}}$

Case (II): If $m = n$

Here we have N is a decimal and r is the first decimal figure, then we have

the absolute error $\qquad E_a \le \dfrac{1}{10^m} \times \dfrac{1}{2}$

and $\qquad\qquad N \ge r \times 10^{-1} - \dfrac{1}{10^m} \times \dfrac{1}{2}$

$\Rightarrow \qquad\qquad E_r \le \dfrac{10^{-m} \times \dfrac{1}{2}}{r \times 10^{-1} - 10^{-m} \times \dfrac{1}{2}}$

$$= \dfrac{10^{-m}}{2r \times 10^{-1} - 10^{-m}} = \dfrac{1}{2r \times 10^{m-1} - 1}$$

$$= \dfrac{1}{2r \times 10^{m-1} - 1} < \dfrac{1}{r \times 10^{m-1}}$$

Case (III): If $m > n$

Here we have $m > n$, therefore, r occupies the $(m - n + 1)^{\text{th}}$ decimal place.

$\Rightarrow \qquad\qquad N \ge r \times 10^{-(m-n+1)} - \dfrac{1}{10^m} \times \dfrac{1}{2}$ and $E_a \le \dfrac{1}{10^m} \times \dfrac{1}{2}$

Therefore, $\qquad\qquad E_r \le \dfrac{10^{-m} \times \dfrac{1}{2}}{r \times 10^{-m} \times 10^{n-1} - 10^{-m} \times \dfrac{1}{2}}$

$$= \dfrac{10^{-m}}{2r \times 10^{-m} \times 10^{n-1} - 10^{-m}}$$

$$= \dfrac{1}{2r \times 10^{n-1} - 1} < \dfrac{1}{r \times 10^{n-1}}$$

Here, we can say that the theorem is true in all the three possible cases.

▶ **REMARKS**

▸ Except in the case of approximate numbers of the form $r(1.000...) \times 10^k$, in which r is the only digit from zero, the relative error is less than $\dfrac{1}{2r \times 10^{n-1}}$.

▸ If $r \ge 5$ then the given approximate number is not of the form $r(1.000...) \times 10^k$, then $E_r < \dfrac{1}{10^n}$; for in the case $2r \ge 10$ and therefore $2r \times 10^{n-1} \ge 10^n$.

THEOREM 2. _If the relative error in an approximate number is less than_ $\left[\dfrac{1}{(r+1) \times 10^{n-1}}\right]$, _the number is correct to n significant figures or at least is in error by less than a unit in the nth significant figures._

PROOF.   Let us assume

N = The given number, *i.e.*, the exact value,

n = number of correct significant figure in N,

r = first significant figure in N,

k = number of digits in the integral part of N.

Then, we have

$n - k$ = number of decimal in N,

Also, given $N \le (r+1) \times 10^{k-1}$

Now, let the relative error

$$E_r < \frac{1}{(r+1) \times 10^{n-1}}$$

Then, we have the absolute error

$$E_a < (r+1) \times 10^{k-1} \times \frac{1}{(r+1) \times 10^{n-1}} = \frac{1}{10^{n-k}}$$

Now, $\dfrac{1}{10^{n-k}}$ is one unit in $(n-k)^{\text{th}}$ decimal places or in the nth significant figure.

Therefore, the absolute error E_a is less than a unit in the nth significant figure.

Now, let us suppose that the given number is pure decimal number. Also let k = number of zero between the decimal point and the first significant figure. Then $(n + k)$ is equal to the number of decimals in N.

and $N \le \dfrac{(r+1)}{10^{k+1}}$

Therefore, if $E_r < \dfrac{1}{(r+1) \times 10^{n-1}}$ then, we have

$$E_a < \frac{(r+1)}{10^{k+1}} \times \frac{1}{(r+1) \times 10^{n-1}} = \frac{1}{10^{n+k}}$$

Now, $\dfrac{1}{10^{n+k}}$ is one unit in $(n+k)^{\text{th}}$ decimal places or in the nth significant figure.

Hence the absolute error E_a is less than a unit in the nth significant figure.

◣ REMARKS

▸ If $E_r < \dfrac{1}{[2(r+1) \times 10^{n-1}]}$, then E_a is less than half a unit in the nth significant figures and the given number is correct to n^{th} significant figures.

▸ If the relative error of any number is not greater than $\dfrac{1}{(2 \times 10^n)}$, the number is certainly correct to n significant figures.

▸ The absolute error is always connected with the number of decimal places, whereas the relative error is connected with the number of significant figures.

SOLVED EXAMPLES

EXAMPLE 1.   *Verify the theorem (1) for the number 875.32 correct to five significant figures.*

SOLUTION.   The given number $N = 875.32$

We observe that $r = 8$ and $n = 5$

Since, we have the absolute error $E_a \ngtr 0.01 \times \dfrac{1}{2} = 0.005$

Therefore, the relative error $\le \dfrac{0.005}{875.32} = \dfrac{5}{875320}$

$$= \dfrac{1}{2 \times 87532} < \dfrac{1}{2 \times 80000} = \dfrac{1}{2 \times 8 \times 10^4}$$

$$< \dfrac{1}{8 \times 10^4} = \left(\dfrac{1}{r \times 10^{n-1}} \right)$$

Hence, the theorem is verified.

EXAMPLE 2. *Round off the numbers 865250 and 37.46235 to four significant figures and compute E_a, E_r and E_p.*

SOLUTION. Here, the given numbers are (i) 865250 and (ii) 37.46235

(i) 865250

If we rounded off the given number to four significant figures, then we get 865200.

Therefore, the absolute error

$$E_a = \left| x^T - x^A \right| = |865250 - 865200| = 50$$

Now, the relative error

$$E_r = \dfrac{E_a}{x^T} = \dfrac{50}{865250} = 6.71 \times 10^{-5}$$

Also, the percentage error

$$E_p = E_r \times 100 = 6.71 \times 10^{-5} \times 100 = 6.71 \times 10^{-3}.$$

(ii) 37.46235

If we rounded off the given number to four significant figures, then we get 37.46.

Then $\qquad E_a = |37.46235 - 37.46| = 0.00235$

$$E_r = \dfrac{E_a}{x^T} = \dfrac{0.00235}{37.46235} = 6.27 \times 10^{-5}$$

and $\qquad E_p = E_r \times 100 = 6.27 \times 10^{-3}$

EXAMPLE 3. *If 0.333 is the approximate value of $\dfrac{1}{3}$, find the absolute, relative and percentage errors.*

SOLUTION. Here, we have

$$x^T = \dfrac{1}{3}, x^A = 0.333$$

Therefore,

(i) Absolute error

$$E_a = \left| x^T - x^A \right| = \left| \dfrac{1}{3} - 0.333 \right| = \left| \dfrac{1}{3} - \dfrac{333}{1000} \right| = \dfrac{1}{3000} = 0.00033$$

(ii) Relative Error

$$E_r = \dfrac{E_a}{x^T} = \dfrac{0.00033}{1/3} = 0.00099$$

(iii) Percentage error

$$E_p = 100 \times E_r = 100 \times 0.00099 = 0.099$$

EXAMPLE 4. *Let x = 0.005998. Find the relative error if x is truncated to three decimal digits.*

(UPTU MCA–2006; UPTU B.TECH.–2004)

SOLUTION. Given that $\quad x = 0.005998 = 0.5998 \times 10^{-2}$.

Now, $\qquad x_a = 0.599 \times 10^{-2}$ (after truncating to three decimal places)

$$\text{Relative error} = \left|\frac{x - x_a}{x}\right| = \left|\frac{0.5998 \times 10^{-2} - 0.599 \times 10^{-2}}{0.5998 \times 10^{-2}}\right|$$

$$= 0.00333 = 0.333 \times 10^{-2}.$$

EXAMPLE 5. *If 1.414 is used as an approximation to $\sqrt{2}$. Find the absolute and relative errors.*

SOLUTION. We have

True value = $\sqrt{2}$ = 1.41421356

and approximate value = 1.414

Therefore,   Error = True value – Approximate value

$$= \sqrt{2} - 1.414 = 1.41421356 - 1.414 = 0.00021356$$

Then, absolute error = $|0.00021356|$ = 0.21356×10^{-3}

Finally, the relative error = $\dfrac{\text{Absolute error}}{\text{True value}} = \dfrac{0.21356 \times 10^{-3}}{\sqrt{2}} = 0.151 \times 10^{-3}.$

EXAMPLE 6. *Find the sum $S = \sqrt{3} + \sqrt{5} + \sqrt{7}$ to 4 significant digits and find its absolute and relative errors.*

SOLUTION. It is known that

$$\sqrt{3} = 1.732, \sqrt{5} = 2.236, \sqrt{7} = 2.646$$

$$\therefore \qquad S = 1.732 + 2.236 + 2.646 = 6.614$$

Now, absolute error E_a = 0.0005 + 0.0005 + 0.0005 = 0.0015

The total absolute error shows that the sum is correct to 3 significant figures only.

Thus, we take $S = 6.61$

Then, we have relative error = $\dfrac{0.0015}{6.61} = 0.0002$

EXAMPLE 7. *It is required to obtain the roots of $X^2 - 2X + \log_{10}2$ to four decimal places. To what accuracy should $\log_{10}2$ be given?*

SOLUTION. The roots of the given equation are

$$X = \frac{2 \pm \sqrt{4 - 4\log_{10}2}}{2} = 1 \pm \sqrt{1 - \log_{10}2}$$

Then $\left|\Delta X\right| = \dfrac{1}{2}\dfrac{\Delta(\log 2)}{\sqrt{1 - \log_{10}2}} < 0.5 \times 10^{-4}$

$$= \Delta(\log 2) < 2 \times 0.5 \times 10^{-4}(1 - \log 2)^{1/2} < 0.83604 \times 10^{-4}$$

$$= 8.3604 \times 10^{-5}$$

EXAMPLE 8. *If $a = 10.00 \pm 0.05$, $b = 0.0356 \pm 0.0002$, $c = 15300 \pm 100$, $d = 62000 \pm 500$.*
Find the maximum value of absolute error in $a + b + c + d$.   [MDU(BE)–2005]

SOLUTION. We have

Absolute error in a = $|\pm 0.05|$ = 0.05

Absolute error in b = $|\pm 0.0002|$ = 0.0002

Absolute error in c = $|\pm 100|$ = 100

Absolute error in d = $|\pm 500|$ = 500

Hence, the maximum absolute error in $a + b + c + d$

$$= 0.05 + 0.0002 + 100 + 500 = 600.0502$$

EXAMPLE 9.   *Three approximated values of number* $\frac{1}{3}$ *are given as 0.30, 0.33 and 0.34. Which of these three is the best approximation?*

SOLUTION.   We know that the best approximation will be the one which has the least absolute error.

Here,   true value $= \frac{1}{3} = 0.33333$

Case I. Approximate value = 0.30

∴   Absolute error = |True value – Approximate value| = |0.33333 – 0.30|
$$= 0.03333$$

Case II. Approximate value = 0.33

∴   Absolute error = |True value – Approximate value| = |0.33333 – 0.33|
$$= 0.00333$$

Case III. Approximate value = 0.34

∴   Absolute error = |True value – Approximate value| = |0.33333 – 0.34|
$$= |-0.00667| = 0.00667$$

We observe that, absolute error is least in case II. Hence, 0.33 is the best approximation.

EXAMPLE 10.   *Given the solution of a problem as $x_A = 35.25$ with the relative error in the solution atmost 2%. Find, to four decimal digits, the range of values within which the exact value of the solution must lie.*   (UPTU MCA–2002)

SOLUTION.   It is given that

(i)   Maximum relative error in the solution = 2% = 0.02

(ii) Approximate value of the solution is $x_A = 35.25$.

Let x be the exact value of the solution, then as per given, we have

$$\left| \frac{x - x_A}{x} \right| < 0.02 \text{ , i.e., } \left| 1 - \frac{x_A}{x} \right| < 0.02$$

$$\Rightarrow \qquad -0.02 < \left(1 - \frac{x_A}{x} \right) < 0.02$$

If $\left(1 - \frac{x_A}{x} \right) > -0.02$ then

$$-\frac{x_A}{x} > -1 - 0.02 \qquad \Rightarrow \qquad -\frac{x_A}{x} > -1.02$$

$$\Rightarrow \qquad \frac{x_A}{x} < 1.02 \qquad \Rightarrow \qquad x_A < 1.02x \text{ .}$$

$$\Rightarrow \qquad x > \frac{x_A}{1.02} = \frac{35.25}{1.02} = 34.558823594$$

Also, if $\left(1 - \frac{x_A}{x} \right) < 0.02$, then we have

$$-\frac{x_A}{x} < -1 + 0.02 \qquad \Rightarrow \qquad -\frac{x_A}{x} > -0.98$$

$$\Rightarrow \qquad \frac{x_A}{x} > 0.98 \qquad \Rightarrow \qquad x_A > 0.98x$$

$$\Rightarrow \qquad x < \frac{x_A}{0.98} = \frac{35.25}{0.98} = 35.9693877551$$

Thus, we have

$$34.558823594 < x < 35.9693877551$$

Hence, the range of values within which the exact value of the solution lies, correct to four decimal places is given by

$$34.5588 < x < 35.9694.$$

EXERCISE 2.1

1. Round off the following numbers correct to four significant figures :
   (i) 58.3643   (ii) 979.267
   (iii) 7.7265   (iv) 0.065738
   (v) 3.26425   (vi) 35.46735
   (vii) 7326583000   (viii) 18.265101

2. Find the relative error if 2/3 is approximated to 0.667.

3. If the number r is correct to 3 significant digits, what will be the maximum relative error.

4. A carpenter measures a 10-foot beam to the nearest eighth of an inch and a mechanist measures a $\frac{1}{2}$ inch bolt to the nearest thousandth of an inch. Which measurement is more correct ?

5. The following numbers are all approximate and are correct as far as their last digit only. Find their sum 136.421, 28.3, 321, 68.243, 17.482.

6. If the number p is correct to three significant digits, what will be the maximum relative error ?

7. The height of an observation tower was estimated to be 47 m whereas it's actual height was 45 m. Find the percentage relative error in the measurement.

8. If true value $= \frac{10}{3}$, approximate value $= 3.33$. Then, find absolute and relative errors.

9. Round off the number 75462 to four significant digits and then calculate the absolute error and percentage error.
   (UPTU–2004)

10. Find the relative error in taking $\pi = 3.141593$ as 22/7.
    (VTU–2007)

11. Suppose that you have a task of measuring the lengths of a bridge and a rivet, and come up with 9999 and 9 am, respectively. If the true values are 10,000 and 10 cm. respectively, compute the percentage relative error in each case.
    (Pune–2004)

12. Given $a = 9.00 \pm 0.05$, $b = 0.0356 \pm 0.0002$, $c = 15300 \pm 100$, $d = 62000 \pm 500$. Find the maximum value of absolute error in $a + b + c + d$.
    (PTU–2001)

13. Find the absolute error and the relative error in the product of 432.8 and 0.12584 using four digit mantissa.
    (Kerala–2003)

Answers

1. (i) 58.36   (ii) 979.3   (iii) 7.726   (iv) 0.06574   (v) 3.264
   (vi) 35.45   (vii) 7327×10^6 (viii) 18.26   **2.** 0.0005   **3.** 0.0005
   4. Beam measurement   **5.** 571   **6.** 0.0005   **7.** 4.44%
   8. 0.003333, 0.000999   **9.** $0.7546; -0.0002 \times 10^5; 0.00265$ **10.** -0.0004
   11. 0.01%; 10%   **12.** 600.0002   **13.** 0.17312; 0.0003178

2.4 INHERENT ERRORS

The errors which are already present in the statement of a problem before its solution are called Inherent errors. These types of errors arise either due to the given data being approximated or due to limitations of the mathematical measurements.

The inherent error contains two components :

(i) **Data errors:** The data error arises when data are obtained by some experimental methods with limited accuracy and precision. This may be due to some special limitations in instrument or in reading.

(ii) Conversion errors: The conversion error arise due to the limitations of the computer to store the data exactly. Generally, it occurs in the floating- point representation which retains only a specified number of digits. The digits which are not retain gives the round off error.

▶ REMARKS
- ▸ The inherent errors is also known as input errors.
- ▸ Data errors is also known as empirical errors.
- ▸ Conversion errors are also known as representation errors.

2.5 ROUNDING OFF ERROR

It occurs from the process of rounding off the numbers during the computations, *i.e.*, it occur when a fixed number of digits are used to represent exact numbers. Such types of errors are unavoidable in most of the calculations due to the limitations of the computing aids.

If a number x has the floating point representation of the form

$$x = d_1 d_2 \dots d_t d_{t+1} \dots \times B^e \qquad \qquad \dots(1)$$

where d_1, d_2 ,..., d_t ... are integers and satisfies $0 \leq d_i \leq B$ and e is the exponent. Then Rounding a number can be done by the following two ways :

    (i) Chopping: Here, we neglect d_{t+1}, d_{t+2} ... in (1) and obtain the number

$$= d_1 d_2 \dots d_t \times B^e$$

    (ii) Symmetric rounding: Here the fractional part in (1) is written as

$$d_1 d_2 \dots d_t d_{t+1} + \frac{1}{2} B$$

    and the first t digits are taken to write the floating point number.

For Example- Find the sum of 0.223×10^3 and 0.556×10^2 and write the result in three digit mantissa.

Solution. Here, the number of the smaller magnitude is adjusted so that its exponent is same as that of the number of larger magnitude. We have

$$0.2230 \times 10^3$$
$$0.0556 \times 10^3$$
$$\overline{0.2786 \times 10^3}$$

$$\Rightarrow \qquad \begin{cases} 0.278 \times 10^3, & \text{for chopping} \\ 0.279 \times 10^3, & \text{for rounding} \end{cases}$$

▶ REMARKS
- ▸ In chopping, the extra digits are dropped, which is called truncating the number.
- ▸ In symmetric round off method, the last retained significant digit is rounded up by 1 if the first discarded digit is larger or equal to 5, otherwise the last retained digits is unchanged.
  - **For example:** The numbers 83.8893 becomes 83.89 and the number 86.6431 would become 86.64.
- ▸ The rounded off error can be reduced by retaining at least one more significant figure at each step than that given in the data and rounded off at the last step.

2.6 TRUNCATION ERROR

The truncation errors arises by using some approximations in place of an exact mathematical procedure.

For example- When we calculate the sine of an angle using the following series

$$\sin x = x - \frac{x^3}{3!} + \frac{x^5}{5!} - \frac{x^7}{7!} + \dots$$

Then, we can not use the infinite terms of above series. After a certain number of terms, we terminate the process. Then, an error which is introduced here, is called truncation error.

▶ REMARKS

▸ Truncation error is a type of algorithm error.

▸ In numerical computing, we used many iterative procedures, which are infinite. Therefore, a knowledge of the truncation error is very much important.

▸ This error can be reduced by using a better numerical model which increases the number of arithmetic operations.

▸ When we use a number of discrete steps in the solution of a differential equation, then the error which is introduced here, is called discretisation error.

SOLVED EXAMPLES

EXAMPLE 1. *Obtain a second degree polynomial approximation to*
$$f(x) = (1 + x)^{1/2}, x \in [0, 0.1]$$
Using the Taylor series expansion about $x = 0$. Use the expansion to approximate $f(0.05)$ and found the truncation error.

SOLUTION. Here, the given function is
$$f(x) = (1 + x)^{1/2}$$

Then, we get
$$f(x) = (1 + x)^{1/2} \quad \Rightarrow \quad f(0) = 1$$
$$f'(x) = \frac{1}{2}(1+x)^{-1/2} \quad \Rightarrow \quad f'(0) = \frac{1}{2}$$
$$f''(x) = -\frac{1}{4}(1+x)^{-3/2} \quad \Rightarrow \quad f''(0) = -\frac{1}{4}$$
$$f'''(x) = \frac{3}{8}(1+x)^{-5/2} \quad \Rightarrow \quad f'''(0) = \frac{3}{8}$$

Now, using the Taylor series expansion, we get
$$(1+x)^{1/2} = 1 + \frac{x}{2} - \frac{x^2}{8} + R_n$$

where R_n is the remainder term and given by
$$R_n = \frac{1}{16} \cdot \frac{x^3}{[(1+\theta)^{1/2}]^5}, 0 < \theta < 0.01$$

Then the truncation error is given by
$$T = (1+x)^{1/2} - \left(1 + \frac{x}{2} - \frac{x^2}{8}\right) = \frac{1}{16} \cdot \frac{x^3}{[(1+\theta)^{1/2}]^5}$$

Now,
$$f(0.05) = 1 + \frac{0.05}{2} - \frac{(0.05)^2}{8} = 0.10246875 \times 10^1$$

Then, the bound of the truncation error for $x \in [0, 1]$ is given by
$$|T| \le \frac{(0.1)^3}{16[(1+8)^{1/2}]^5} \le \frac{(0.1)^3}{16} = 0.625 \times 10^{-4}$$

EXAMPLE 2.   *Find the truncation error in the result of the following functions for $x = \dfrac{1}{5}$ when we use*

(a) *First three terms*   (b) *First four terms*

$$e^x = 1 + x + \frac{x^2}{2!} + \frac{x^3}{3!} + \frac{x^4}{4!} + \frac{x^5}{5!} + \frac{x^6}{6!}$$

SOLUTION.   (a) Let T denote the truncation error. If we add first three terms then

$$T = \left(1 + x + \frac{x^2}{2!} + ... + \frac{x^6}{6!}\right) - \left(1 + x + \frac{x^2}{2!}\right) = \frac{x^3}{3!} + \frac{x^4}{4!} + \frac{x^5}{5!} + \frac{x^6}{6!}$$

Now, T at $x = \dfrac{1}{5} = \dfrac{(0.2)^3}{6} + \dfrac{(0.2)^4}{24} + \dfrac{(0.2)^5}{120} + \dfrac{(0.2)^6}{720} = 0.1402755 \times 10^{-2}$

(b) Now, we find the truncation error, when first four terms are added

$$T = \left(1 + x + \frac{x^2}{2!} + ... + \frac{x^6}{6!}\right) - \left(1 + x + \frac{x^2}{2!} + \frac{x^3}{3!}\right) = \frac{x^4}{4!} + \frac{x^5}{5!} + \frac{x^6}{6!}$$

Now, T at $x = \dfrac{1}{5} = \dfrac{(0.2)^4}{24} + \dfrac{(0.2)^5}{120} + \dfrac{(0.2)^6}{720} = 0.694222 \times 10^{-4}$

2.7 THE GENERAL FORMULA FOR ERRORS

Let $Y = f(x_1, x_2, ..., x_n)$ be a function of n variables $x_1, x_2, ..., x_n$. Suppose, ΔY is the error in Y due to the errors $\Delta x_1, \Delta x_2, ..., \Delta x_n$ in $x_1, x_2, ..., x_n$ respectively.

Then we have

$$Y + \Delta Y = f(x_1 + \Delta x_1, x_2 + \Delta x_2, ..., x_n + \Delta x_n) \qquad ...(1)$$

Expanding by Taylor series, we get

$$Y + \Delta Y = f(x_1, x_2, ..., x_n) + \left(\Delta x_1 \frac{\partial Y}{\partial x_1} + \Delta x_2 \frac{\partial Y}{\partial x_2} + ... + \Delta x_n \frac{\partial Y}{\partial x_n}\right)$$

$$+ \frac{1}{2}\left[(\Delta x_1)^2 \frac{\partial^2 Y}{\partial x_1^2} + (\Delta x_2)^2 \frac{\partial^2 Y}{\partial x_2^2} + ... + (\Delta x_n)^2 \frac{\partial^2 Y}{\partial x_n^2} + 2\Delta x_1 \Delta x_2 \frac{\partial^2 Y}{\partial x_1 \partial x_2} + ...\right] + ...$$

$$...(2)$$

Now, since the errors $\Delta x_1, \Delta x_2, ..., \Delta x_n$ all are very small. So, that we can neglect $(\Delta x_i)^2$ and higher order terms of Δx_i.

Then, we have

$$Y + \Delta Y = f(x_1, x_2, ..., x_n) + \left(\Delta x_1 \frac{\partial Y}{\partial x_1} + \Delta x_2 \frac{\partial Y}{\partial x_2} + ... + \Delta x_n \frac{\partial Y}{\partial x_n}\right) \qquad ...(3)$$

$$\Rightarrow \qquad \Delta Y = \Delta x_1 \frac{\partial Y}{\partial x_1} + \Delta x_2 \frac{\partial Y}{\partial x_2} + ... + \Delta x_n \frac{\partial Y}{\partial x_n} \qquad ...(4)$$

$$[\because Y = f(x_1, x_2, ..., x_n)]$$

Now, divide the equation (4) by Y, we get the relative error is

$$\frac{\Delta Y}{Y} = \frac{\Delta x_1}{Y} \cdot \frac{\partial Y}{\partial x_1} + \frac{\Delta x_2}{Y} \cdot \frac{\partial Y}{\partial x_2} + ... + \frac{\Delta x_n}{Y} \cdot \frac{\partial Y}{\partial x_n} \qquad ...(5)$$

Now, taking the modulus of (4) and (5), the maximum absolute error and relative error are given by

$$|\Delta Y| \le \left|\Delta x_1 \frac{\partial Y}{\partial x_1}\right| + \left|\Delta x_2 \frac{\partial Y}{\partial x_2}\right| + \dots + \left|\Delta x_n \frac{\partial Y}{\partial x_n}\right|$$

and

$$\left|\frac{\Delta Y}{Y}\right| \le \left|\frac{\partial x_1}{Y} \cdot \frac{\partial Y}{\partial x_1}\right| + \left|\frac{\partial x_2}{Y} \cdot \frac{\partial Y}{\partial x_2}\right| + \dots + \left|\frac{\partial x_n}{Y} \cdot \frac{\partial Y}{\partial x_n}\right|$$

SOLVED EXAMPLES

EXAMPLE 1. *In a $\triangle ABC$, $a = 6$ cm , $c = 15$ cm and $\angle B = 90°$. Find the possible error in the computed value of A, if the errors in the measurement of a and c are 1 mm and 2 mm respectively.*

SOLUTION. Here, we have $a = 6$ cm

$$c = 15 \text{ cm}$$
$$\angle B = 90°$$

Then, we have the triangle given by fig. 1.

From figure 1, we have $A = \tan^{-1}\dfrac{a}{c}$

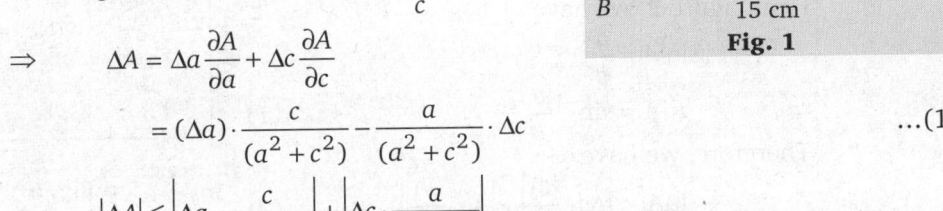

Fig. 1

$$\Rightarrow \qquad \Delta A = \Delta a \frac{\partial A}{\partial a} + \Delta c \frac{\partial A}{\partial c}$$

$$= (\Delta a) \cdot \frac{c}{(a^2 + c^2)} - \frac{a}{(a^2 + c^2)} \cdot \Delta c \qquad\qquad \dots(1)$$

or $\qquad \left|\Delta A\right| \le \left|\Delta a \cdot \dfrac{c}{a^2 + c^2}\right| + \left|\Delta c \cdot \dfrac{a}{a^2 + c^2}\right|$

Given that $\Delta a = 1$ mm $= 0.1$ cm, $\Delta c = 2$ mm $= 0.2$ cm, $a = 6$ cm and $c = 15$ cm. Putting all these values in equation (1), we get

$$|\Delta A| \le \left|\frac{0.1 \times 15}{(6)^2 + (15)^2}\right| + \left|\frac{0.2 \times 6}{(6)^2 + (15)^2}\right| = \frac{1.5 + 1.2}{261} = \frac{2.7}{261} = 0.0103 \text{ Radians}$$

$$\Rightarrow \quad |\Delta A| \le 0.0103 \text{ radians}$$

or $\quad |\Delta A| \le 35'25''$

EXAMPLE 2. *If $u = \dfrac{4x^2 y^3}{z^4}$ and $\Delta x = \Delta y = \Delta z = 0.001$, compute the relative maximum error in u when $x = y = z = 1$.*

SOLUTION. Here, we have $a = 6$ cm

$$u = \frac{4x^2 y^3}{z^4} \qquad\qquad \dots(1)$$

From eq. (1), we have

$$\frac{\partial u}{\partial x} = \frac{8xy^3}{z^4}, \frac{\partial u}{\partial y} = \frac{12x^2 y^2}{z^4} \text{ and } \frac{\partial u}{\partial z} = -\frac{16x^2 y^3}{z^5}$$

Now, we have

$$\Delta u = \frac{\partial u}{\partial x} \Delta x + \frac{\partial u}{\partial y} \Delta y + \frac{\partial u}{\partial z} \Delta z \qquad\qquad \dots(2)$$

Now, putting the values of $\dfrac{\partial u}{\partial x}, \dfrac{\partial u}{\partial y}$ and $\dfrac{\partial u}{\partial z}$ in eq. (2), we get

$$\Delta u = \frac{8xy^3}{z^4}\Delta x + \frac{12x^2 y^2}{z^4}\Delta y - \frac{16x^2 y^3}{z^5}\Delta z$$

Now, $(\Delta u)_{max} = \left|\dfrac{8xy^3}{z^4}\Delta x\right| + \left|\dfrac{12x^2 y^2}{z^4}\Delta y\right| + \left|\dfrac{16x^2 y^3}{z^5}\Delta z\right|$

$$= 8(0.001) + 12(0.001) + 16(0.001) = 0.036$$

Therefore, the maximum relative error is

$$= \frac{(\Delta u)_{max}}{(u)_{at\ x=y=z=1}} = \frac{0.036}{4} = 0.009$$

EXAMPLE 3. *In a $\triangle ABC$, $a = 30$ cm, $b = 80$ cm, $\angle B = 90°$. Find the maximum error in the computed value of A, if possible errors in a and b are $\dfrac{1}{3}\%$ and $\dfrac{1}{4}\%$ respectively.*

SOLUTION. Here, we have

In $\triangle ABC$, $a = 30$ cm, $b = 80$ cm, $\angle B = 90°$

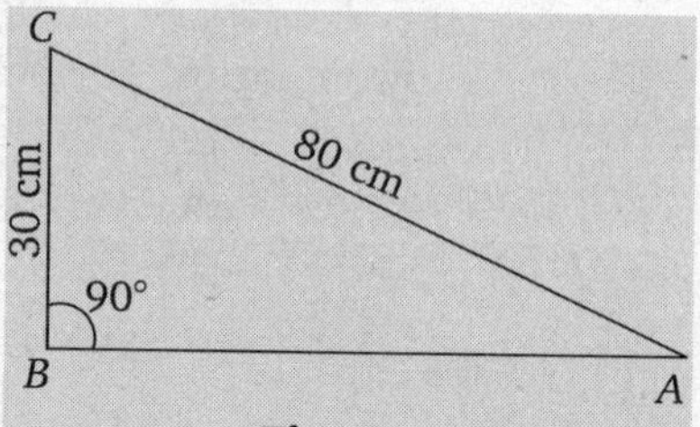

From figure 2, we have

$$\sin A = \frac{a}{b}$$

$\Rightarrow \qquad A = \sin^{-1}\dfrac{a}{b}$ ...(1)

Therefore, we have

$$|\Delta A| < \left|\Delta a \cdot \frac{\partial A}{\partial a}\right| + \left|\Delta b \cdot \frac{\partial A}{\partial b}\right| \qquad ...(2)$$

Fig. 2

Now, we have the possible errors in a and b are 1/3% and 1/4% respectively, then

$$\frac{\Delta a}{a}\times 100 = \frac{1}{3} \qquad \Rightarrow \qquad \Delta a = 0.1$$

and $\dfrac{\Delta b}{b}\times 100 = \dfrac{1}{4} \qquad \Rightarrow \qquad \Delta b = 0.2$

Also, from equation (1)

$$\frac{\partial A}{\partial a} = \frac{1}{\sqrt{b^2 - a^2}} \quad \text{and} \quad \frac{\partial A}{\partial b} = \frac{a}{b\sqrt{b^2 - a^2}}.$$

Putting all these values in equation (2), we get

$$|\Delta A| < |0.00135 + 0.00100| = 0.00235 \text{ radians}$$

$\Rightarrow \qquad \Delta A < 8'5''$

EXAMPLE 4. *Find the relative error in the function $y = ax_1^{m_1} x_2^{m_2}...x_n^{m_n}$*

SOLUTION. Here, we have

$$y = ax_1^{m_1} x_2^{m_2}...x_n^{m_n} \qquad ...(1)$$

Taking log of both sides, we get

$$\log y = \log a + m_1 \log x_1 + m_2 \log x_2 + ... + m_n \log x_n \qquad ...(2)$$

Now, differentiating eq.(2), we get

$$\frac{1}{y}\cdot\frac{\partial y}{\partial x_1} = \frac{m_1}{x_1}$$

$$\frac{1}{y} \cdot \frac{\partial y}{\partial x_2} = \frac{m_2}{x_2}, \ldots \frac{1}{y} \cdot \frac{\partial y}{\partial x_n} = \frac{m_n}{x_n}$$

Therefore, the error

$$E_r = \frac{\partial y}{\partial x_1} \cdot \frac{\Delta x_1}{y} + \frac{\partial y}{\partial x_2} \cdot \frac{\Delta x_2}{y} + \ldots + \frac{\partial y}{\partial x_n} \cdot \frac{\Delta x_n}{y}$$

$$= m_1 \frac{\Delta x_1}{x_1} + m_2 \frac{\Delta x_2}{x_2} + \ldots + m_n \frac{\Delta x_n}{x_n}$$

Hence, $(E_r)_{max} \leq m_1 \left| \frac{\Delta x_1}{x_1} \right| + m_2 \left| \frac{\Delta x_2}{x_2} \right| + \ldots + m_n \left| \frac{\Delta x_n}{x_n} \right|$

▶ REMARK

▶ The relative error of a product of n numbers is approximately equal to the algebraic sum of their relative errors. This result can be verified easily by taking $a = 1, m_1 = m_2 = \ldots = m_n = 1$, then

$$E_r = \frac{\Delta x_1}{x_1} + \frac{\Delta x_2}{x_2} + \ldots \frac{\Delta x_n}{x_n}.$$

2.8 FLOATING POINT ARITHMETIC AND ERRORS

Generally, there are two types of numbers, which we used in calculations

(i) Integers : $0, \pm 1, \pm 2, \pm 3, \ldots$.

(ii) Real numbers : Such as numbers with decimal.

Since, we used finite digit arithmetic in computers, therefore all the integers can be represented easily with finite digits. On the other hand, all real numbers can not be represented as a finite

digits numbers like $\left(\frac{2}{3} \right) = 0.666\ldots$ Hence, we use floating point representation.

(iii) Floating Point Numbers:

An n digit floating point number β has the form

$$x = \pm (d_1 d_2 \ldots d_n)_\beta \cdot \beta^e, \quad 0 \leq d_i < \beta, \quad m \leq e \leq M$$

where $(d_1 d_2 \ldots d_n)_\beta$ is a β fraction called mantissa and its value is given by

$$(d_1 d_2 \ldots d_n)_\beta = d_1 \times \frac{1}{\beta} + d_2 \times \frac{1}{\beta^2} + \ldots + d_n \times \frac{1}{\beta^n}$$

Also e is called the exponent.

▶ REMARKS

▶ A floating point number is said to be normalised if $d_1 \neq 0$ or else $d_1 = d_2 = \ldots = d_n = 0$.

▶ The precision or length n of floating-point numbers on any computer is usually determined by the word length of the computer. **For example:** IBM 1130, in single precision 6 decimal digits and inextend precision, *i.e.*, double precision, nine decimal digits are used.

▶ Calculation in double precision usually doubles the storage requirements and running time as compared with single precision.

▶ The exponent e is also limited to range $m < e < M$, where m and M are integers varying from computer to computer.

2.9 COMPUTER STORAGE

Computer storage has its own limitations. Storage is provided into locations. Each location or word has a storage capacity which means a finite number of digits. The limitation causes errors and concept of floating point becomes more important. To discuss it, we must keep in mind the constants of number of digits that can be stored in one word or location *i.e.*, it would be very difficult to store a number as 1, 2, 3, 4, ..., 10.

The solution to this problem to some extent can be used of floating point, *i.e.*, representation of this number to same digits of accuracy and with power of 10. For example, say representing this number to 4 digits of accuracy as 1.234×10^9.

Although, these two are not same, yet second option will be significantly accurate for most application purpose.

To convert to floating point, the major concern is number of digits of accuracy to return.

To discuss this concept let us assume that each location can store 6 digits:

•	•	•	•	•	•

Location or word

Initially we can assume, first 3 digits represents integer portion of a fractional number and last 3 as fractional part. **For example:** to store 123.456

1	2	3	4	5	6

↑assumed decimal position

Decimal point is assumed in middle and this sign does not exist physically. In this system range is very limited. Tracking of decimal point will be more difficult in this system as we perform mathematical operations like +, –, *, /.

Range is ±999.999 to 000.001.

To improve this range concept, most usual representation is to use 4 digits for integers and 2 for floating, *i.e.*, 1234.56 is stored as

1	2	3	4	5	6

↑assumed decimal position

Range is increased from 9999.99 to 0000.01 still is very inadequate for most of computations. To remove this problem we use concept of floating point in power notation form.

For example : 1234.56 is *represented as* 0.1234×10^7 *and written as* 1234 E07 *is i.e.,*

1	2	3	4	07

↑

Clearly range is increased

$$0.9999 \times 10^{99} \text{ to } 0.1000 \times 10^{-99}$$

This is much larger. Problem still arise as sign is not a available. If sign bit is used then representation of negative numbers will be reduced to 10^{-9} only as one bit will be consumed as sign bit. To avoid this a concept of Excess method is used. This is a split range of exponent with 50 as base from 00 to 99.

50 is centre so all exponent > 50 are positive and < 50 are negative. Range will be from –50 to 49.

Excess –50 Method says add 50 to exponent.

For example: 0.123456×10^3 *will be stored as*

1	2	3	4	5	3

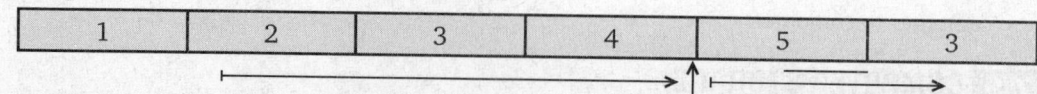

And say 0.123456×10^{-3} will be stored as

For example: 0.123456×10^3 *will be stored as*

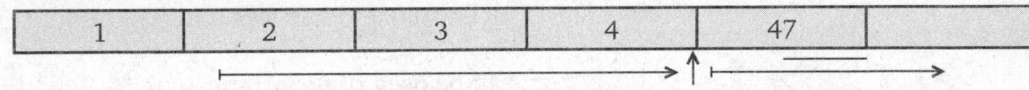

1	2	3	4	47	

Range is !0.9999 × 10^{49} to 0.1000 × 10^{-50}.

2.10 CONCEPT OF NORMALIZED FLOATING POINT

Consider a number 0.001234 × 10^{-5}, which is to be stored. It will be stored as

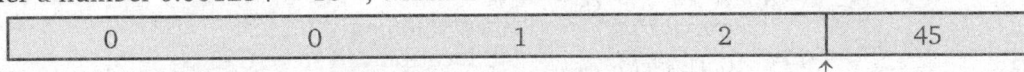

0	0	1	2	45

We loose 2 significant digits. If we represent this number as 0.1234 × 10^{-7}, the storage will be which is much reliable representation.

1	2	3	4	43

So removing zeroes in beginning is termed as normalized floating. In normalized floating range is further increased.

2.11 PITFALLS OF FLOATING POINT REPRESENTATION

We know that mantissa have to be truncated to four digits in order to fit into the normalized floating-point format of the hypothetical system.

For example.

$$4x = x + x + x + x \qquad \qquad ...(1)$$

When arithmetic is performed using normalized floating point representation, equation (1) may not hold true.

SOLVED EXAMPLES

EXAMPLE 1. (i) *Add* 0.1234 × 10^{-3} *and* 0.5678 × 10^{-3} *using concept of normalized floating point.*

SOLUTION. We have 0.1234 × 10^{-3} + 0.5678 × 10^{-3}

$\Rightarrow$ 0.1234 E3
+ 0.5678 E3
0.6912 E3 $\Rightarrow$ 0.6912 × 10^{-3}

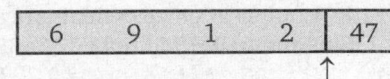

6	9	1	2	47

(ii) *Add* 0.2315 × 10^{2} + 0.9443 × 10^{2}

$\Rightarrow$ 0.2315 E02
+ 0.9443 E02
1.1758 E02 $\Rightarrow$ 0.1175 × 10^{3}

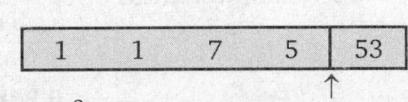

1	1	7	5	53

(iii) *For different base* 0.1234 × 10^{3} + 0.4567 × 10^{2}

$\Rightarrow$ 0.1234 E3
+ 0.4567 E2
0.5801 E3   Make base as same

$$\begin{array}{r} \Rightarrow 0.1234\ E3 \\ +\ 0.0456\ E3 \\ \hline 0.1690\ E3 \end{array} \Rightarrow 0.1690 \times 10^3$$

1	6	9	0	53

↑

EXAMPLE 2. *Subtract the following :*

(i) $0.4567 \times 10^8 - 0.1234 \times 10^8$

$$\begin{array}{l} 0.4567\ E8 \\ 0.1234\ E8 \\ \hline 0.3333\ E8 \end{array} \Rightarrow 0.3333 \times 10^8$$

(ii) Different base $0.4567 \times 10^8 - 0.1234 \times 10^7$

$$0.4567\ E8$$

$$0.1234\ E7 \Rightarrow \begin{array}{l} 0.4567\ E8 \\ 0.0123\ E8 \\ \hline 0.4444\ E8 \end{array} \Rightarrow 0.4444 \times 10^8$$

(iii) Normalized answer $0.4567 \times 10^8 - 0.4566 \times 10^8$

$$\begin{array}{l} 0.4567\ E8 \\ 0.4566\ E8 \\ \hline 0.0001\ E8 \end{array} \Rightarrow 0.1 \times 10^5$$

(iv) Condition of overflow :

$$0.4568 \times 10^{49}$$

$$0.7767 \times 10^{49}$$

$$\begin{array}{l} 0.4568\ \ E49 \\ 0.7767\ \ E49 \\ \hline 0.12335\ \ E49 \end{array} \Rightarrow 0.1233 \times 10^{50} \text{ over flow}$$

(v) Condition of underflow:

$$\begin{array}{l} 0.4567\ E52 \\ 0.4500\ E52 \\ \hline 0.0067\ E52 \end{array} \Rightarrow 0.67 \times 10^{-52}$$

↓ under flow

REMARKS

- In multiplication, exponents are added and mantissa multiplied. If added expanded >99 overflow

For example: Multiply 0.55432 * 0.4111 E7

= 0.22787273*$E9$

decreased

= 0.22789 E9

- In division exponents are subtracted

For example: Divide 0.9380 E5 by 0.3500 E2

$$= \frac{0.9380\ E5}{0.3500\ E5}$$

= 0.2680 E3

EXAMPLE 3. *Apply the procedure of multiplication of two floating point numbers for the following multiplications :*

$$(0.5334 \times 10^9) \times (0.1132 \times 10^{25})$$

and $$(0.1111 \times 10^{74}) \times (0.2000 \times 10^{80})$$

indicate if the result is overflow or underflow.

SOLUTION. The procedure for multiplication of two floating point numbers is

(i) multiply the mantissas of the two normalized floating point numbers.

(ii) and their exponents.

(iii) Resultant mantissa is normalized.

Therefore, $(0.5334 \times 10^9) \times (0.1132 \times 10^{-25})$

$$= (0.5334) \times (0.1132) \times (10^9 \times 10^{-25})$$

$$= 0.06038038 \times 10^{-16}$$

$$= (0.6038\ E -17)$$

and $(0.1111 \times 10^{71}) \times (0.20000 \times 10^{80})$

$$= (0.1111) \times 9.20000 \times (10^{74} \times 10^{80})$$

$$= (0.02222) \times 10^{154}$$

$$= (0.2222\ E\ 153)$$

Since exponent is greater than 99, therefore, the result is "overflow".

EXAMPLE 4. *In normalized floating point mode, carry out the following mathematical operations*

(*i*) $(0.4546\ E3) + (0.5454\ E8)$

(*ii*) $(0.9432\ E - 4) - (0.6353\ E -5)$

SOLUTION. We have

(i) $0.5454\ E8$

   $+0.0000\ E8$

   $\overline{\quad 0.5454\ E8\quad}$ $(\because 4546\ E\ 3 = 0.0000\ E8)$

(ii) $0.9432\ E - 4$

   $-0.0635\ E\ -4$

   $\overline{\quad 0.8797\ E\ -4\quad}$ $(\because 6353\ E -5 = 0.0635\ E -4)$

EXAMPLE 5. *Multiplying the following floating point number 0.1111 E10 and 0.1234 E 15.*

SOLUTION. We have $0.1111\ E\ 10 \times 0.1234\ E\ 15 = 0.1370\ E\ 24$.

EXAMPLE 6. *For e = 2.7183 calculate the value of e^x when x = 0.5250 E1, where*

$$e^x = 1 + x + \frac{x^2}{2!} + \frac{x^3}{3!} + \dots$$

SOLUTION. We have $e^{0.5250\ E1} = e^5 \times e^{0.25}$

Now $e^5 = (0.2718\,E1) \times (0.2718\,E1) \times (0.2718\,E1) \times (0.2718\,E1) \times (0.2718\,E1)$

$$= 0.1484\ E3.$$

Also, $e^{0.25} = 1 + (0.25) + \frac{(0.25)^2}{2!} + \frac{(0.25)^3}{3!}$

$$= 1.25 + 0.03125 + 0.002604 = 0.1284\ E1$$

Therefore,

$$e^{0.5250\,E1} = (0.1484\,E3) \times (0.1284\,E1) = (0.1905\,E3)$$

EXAMPLE 7. *Find the smallest root of equation $x^2 - 400x + 1 = 0$ using four digit arithmetic.*

SOLUTION. It is known that, roots of equation $ax^2 - bx + c$ are

$$\frac{b - \sqrt{b^2 + 4ac}}{2a} \text{ and } \frac{b - \sqrt{b^2 - 4ac}}{2a}$$

Also, product of roots are $\dfrac{c}{a}$.

$\therefore$ smaller root is

$$\frac{c/a}{\left(\dfrac{b+\sqrt{b^2-4ac}}{2a}\right)} = \frac{2c}{b+\sqrt{b^2-4ac}}$$

Here, $a = 1 = 0.1000\,E1, b = 400 = 0.4000\,E3, c = 1 = 0.1000\,E1$

Now, $b^2 - 4ac = 0.1600\,E6 - 0.4000\,E1 = 0.1600\,E6$

$\Rightarrow \sqrt{b^2 - 4ac} = 0.4000\,E3$

Hence, smaller root $= \dfrac{2 \times (0.1000\,E1)}{0.4000\,E3 + 0.4000\,E3} = \dfrac{0.2000\,E1}{0.8000\,E3} = 0.25\,E - 2 = 0.0025$

EXAMPLE 8.   *Determine the number of terms of the exponential series.*

$$e^x = 1 + x + \frac{x^2}{2!} + \frac{x^3}{3!} + \dots + \frac{x^n}{n!} + \dots$$

Such that their gives the values of e^x correct to six decimal places for $0 \le x \le 1$.

SOLUTION.   Given that $e^x = 1 + x + \dfrac{x^2}{2!} + \dfrac{x^3}{3!} + \dots + \dfrac{x^n}{(n-1)!} + R_n(x)$

Where $R_n(x) = \dfrac{x^n}{n!}e^\theta, 0 < \theta < x$

Max, absolute error (at $\theta = x$) $= \dfrac{x^n}{n!}e^x$

and the maximum relative error $= \dfrac{x^n}{n!}$

Hence $(E_r)_{\max}$ at $x = 1 = \dfrac{1}{n!}$

For a six decimal accuracy at $x = 1$ we have

$$\frac{1}{n!} < \frac{1}{2} \times 10^{-6} \text{ or } n! > 2 \times 10^6$$

Which gives n = 10

EXAMPLE 9.   *In case of normalized floating point representations, associative and distributive laws are not always valid. Give example to prove the statement.*

Or

If the normalization on floating point is carried out at each stage, prove the following

   (i) $a(b - c) + ab - ac$*, where* $a = 0.5555\,E1$*,* $b = 0.4545\,E1$*,* $c = 0.4535\,E1$*.*

   (ii) $(a + b) - c \ne (a - c) + b$*, where* $a = 0.5565\,E1$*,* $b = 0.5556\,E1$*,* $c = 0.5644\,E1$*.*

SOLUTION.   In normalized floating point representations, the associative and the distributive laws of arithmetic are not always valid.

Consider the following examples:

Non-distributivity of Arithmetic

Since $a = 0.5555\,E1$, $b = 0.4545\,E1$, $c = 0.4535\,E1$

$\therefore$ $(b - c) = 0.0010\,E1 = 0.1000\,E{-1}$

$\Rightarrow$ $a(b - c) = (0.5555\,E1) \times (0.1000\,E{-1})$

                  $= (0.0555\,E0) = 0.5550\,E{-1}$

Also, $ab = (0.5555\,E1) \times (0.4545\,E1) = 0.2524\,E2$

$$ac = (0.555\ E1) \times (0.4535\ E1) = 0.2519\ E_2$$
$$\Rightarrow \qquad a(b - c) \neq ab - ac$$

Non-Associativity of Arithmetic

Let $a = 0.5665\ E1$, $b = 0.5556\ E{-}1$, $c = 0.5644\ E1$

Therefore, $\quad (a + b) = 0.5665\ E1 + 0.5556\ E{-}1$

$$= 0.5665\ E1 + 0.0055\ E1 = 0.572\ E1$$

$\therefore \qquad (a + b) - c = 0.5720\ E1 - 0.5644\ E1 = 0.0076\ E1 = 0.7600\ E{-}1$

$$(a - c) = 0.5665\ E1 - 0.5644\ E1 = 0.0021\ E1 = 0.2100\ E{-}1$$

$\Rightarrow \qquad (a - c) + b = 0.2100\ E{-}1 + 0.5556\ E{-}1 = 0.7656\ E{-}1$

$\Rightarrow \qquad (a + b) - c \neq (a - c) + b$

EXAMPLE 10. *Calculate the value of polynomial* $x^3 - 4x^2 + 0.1x - 0.5$ *for* $x = 4.011$, *using floating point arithmetic with 4 digit mantissa in two different ways. Find the relative errors in the two methods.*

SOLUTION. We have $\qquad x = 4.011$

Value of x in floating point representation is

$$x = 0.4011\ E1$$

Now value of given polynomial in real arithmetic is

$$x^3 - 4x^2 + 0.1x - 0.5 = (4.011)^3 - 4(4.011)^2 + 0.1(4.011) - 0.5$$
$$= 64.529453 - 4(16.088121) + (0.4011) - 0.5$$
$$= 0.0780693 \qquad \qquad \text{...(i)}$$

Now, in normalised floating point

$$x^3 - 4x^2 + 0.1x - 0.5 = x \cdot x \cdot x - 4 \cdot x \cdot x + 0.1x - 0.5$$
$$= (0.4011\ E1)(0.4011\ E1)(0.4011\ E1) - 4(0.4011\ E1)$$
$$(0.4011\ E1) + 0.1(0.4011\ E1) - 0.5000\ E0$$
$$= 0.6452\ E2 - 0.6435\ E2 + 0.4011\ E0 - 0.5000\ E0$$
$$= 0.0017\ E2 - 0.0989\ E0$$
$$= 0.1700\ E0 - 0.989\ E0$$
$$= 0.0611\ E0 \qquad \qquad \text{...(2)}$$

Now relative error in two methods

$$= (1) - (2) = 0.0780 - 0.0611 = 0.0179$$

EXAMPLE 11. *For* $e = 2.7183$, *calculate the value of* e^x *when* $x = 0.5250\ E1$. $\qquad$ (UPTU–2001)

SOLUTION. Here, $\qquad e^{0.5250\ E1} = e^5 \cdot e^{0.25}$

Now, $\qquad e^5 = (0.2718\ E1) \times (0.2718\ E1) \times (0.2718\ E1)$
$$\times (0.2718\ E1) \times (0.2718\ E1)$$
$$= 0.1484\ E3$$

and $\qquad e^{0.25} = 1 + (0.25) + \dfrac{(0.25)^2}{2!} + \dfrac{(0.25)^3}{3!}$

$$= 1.25 + 0.3125 + 0.002604 = 0.1284\ E1$$

Hence, $\qquad e^{0.5250\ E1} = (0.1484\ E3) \times (0.1284\ E1) = 0.1905\ E3$

EXAMPLE 12. *Add the following floating point numbers* 0.4546 E5 *and* 0.5433 E7. $\qquad$ (UPTU–2001)

SOLUTION. Clearly, the exponent are not equal.

So, $\qquad$ 0.5433 E7

$\underline{+\ 0.0045\ E7} \qquad\qquad |0.4546\ E5 = 0.0045\ E7$

$\quad\ $ 0.5478 E7

2.12 ERROR IN A SERIES APPROXIMATION

The Taylor's series for $f(x)$ at $x = a$ is

$$f(x) = f(a) + (x-a)f'(a) + \frac{(x-a)^2}{2!}f''(a) + ... + \frac{(x-a)^{n-1}}{(n-1)!}f^{n-1}(a) + R_n(x)$$

where $R_n(x)$ is the remainder term and given by

$$R_n(x) = \frac{(x-a)^n}{n!}f^n(\theta), a < \theta < x$$

Here, we have that, if the series is convergent, $R_n(x) \to 0$ as $R \to \infty$. Now, if $f(x)$ is approximated by the first n terms of this series, then the maximum error will be given by the $R_n(x)$. Also if the accuracy required in a series approximation is preassigned, then we can find the number of terms which gives the desired accuracy.

2.12.1 SERIES WITH REMAINDER TERMS

(1) The Binomial series

$$(1+x)^m = 1 + m \cdot x + \frac{m(m-1)}{2!}x^2 + \frac{m(m-1)(m-2)}{3!}x^3 + ...$$

$$+ \frac{m(m-1)...(m-n+2)}{(n-1)!}x^{n-1} + R_n$$

where

(a) $R_n = \dfrac{m(m-1)(m-2)...(m-n+1)}{n!}x^n(1+\theta x)^{m-n}, 0 < \theta < 1$

(b) If $x > 0$ then $R_n < \left| \dfrac{m(m-1)...(m-n+1)}{n!} \cdot x^n \right|$

(c) If $x < 0$ and $n > m$ then $R_n < \left| \dfrac{m(m-1)(m-2)...(m-n+1)}{n!} \cdot \dfrac{x^n}{(1+x)^{n-m}} \right|$

(2) Exponential Series

(a) $e^x = 1 + x + \dfrac{x^2}{2!} + \dfrac{x^3}{3!} + ... + \dfrac{x^{n-1}}{(n-1)!} + R_n$ with $R_n = \dfrac{x^n}{n!}e^{\theta x}$ **[MDU(BE)2005]**

In general $e < 3$ and $\theta \le 1$

$$\Rightarrow \qquad R_n < \frac{3}{n!}$$

(3) Logarithmic Series

$$\log_e(m+1) = \log_e m + 2\left(\frac{1}{2m+1} + \frac{1}{3(2m+1)^3} + \frac{1}{5(2m+1)^5} + ... \right.$$

$$\left. + \frac{1}{(2n-1)(2m+1)^{2n-1}} \right) + R_n$$

where $\qquad R_n = 2\left[\dfrac{1}{(2n+1)(2m+1)^{2n+1}} + \dfrac{1}{(2n+3)(2m+1)^{2n+3}} + ... \right]$

Also, we have $R_n < \dfrac{1}{2} \cdot \dfrac{1}{m(m+1)(2n+1)(2m+1)^{2n-1}}$

(4) Series a^x

$$a^x = 1 + x\log a + \frac{(x\log a)^2}{2!} + ... + \frac{(x\log a)^{n-1}}{(n-1)} + R_n \quad \text{where} \quad R_n = \frac{(x\log a)^n}{n!}a^{\theta x}$$

2.13 ERROR IN DETERMINANTS

If the elements of a determinant are not exact due to rounding or otherwise, then the value of the determinant may be seriously affected, due to the loss of some important significant figures. The amount of such type of losses can not be determined in advance. Here we determine the upper limit of the error in a determinant as follows:

Let us define a determinant as

$$D = \begin{vmatrix} x_1 & x_2 & x_3 \\ y_1 & y_2 & y_3 \\ z_1 & z_2 & z_3 \end{vmatrix} \qquad \qquad ...(1)$$

Now, let Δx_i, Δy_i and Δz_i are the errors in x_i, y_i and z_i respectively and ΔD as the error in D, then we have

$$D + \Delta D = \begin{vmatrix} x_1 + \Delta x_1 & x_2 + \Delta x_2 & x_3 + \Delta x_3 \\ y_1 + \Delta y_1 & y_2 + \Delta y_2 & y_3 + \Delta y_3 \\ z_1 + \Delta z_1 & z_2 + \Delta z_2 & z_3 + \Delta z_3 \end{vmatrix} \qquad \qquad ...(2)$$

From eq.(1), we have

$$dD = \begin{vmatrix} dx_1 & x_2 & x_3 \\ dy_1 & y_2 & y_3 \\ dz_1 & z_2 & z_3 \end{vmatrix} + \begin{vmatrix} x_1 & dx_2 & x_3 \\ y_1 & dy_2 & y_3 \\ z_1 & dz_2 & z_3 \end{vmatrix} + \begin{vmatrix} x_1 & x_2 & dx_3 \\ y_1 & y_2 & dy_3 \\ z_1 & z_2 & dz_3 \end{vmatrix}$$

$$\Rightarrow \qquad dD = (y_2 z_3 - y_3 z_2)dx_1 - (x_2 z_3 - x_3 z_2)dy_1 + (x_2 y_3 - x_3 y_2)dz_1$$
$$- (y_1 z_3 - y_3 z_1)dx_2 + (x_1 z_3 - x_3 z_1)dy_2 - (x_1 y_3 - x_3 y_1)dz_2$$
$$+ (y_1 z_2 - y_2 z_1)dx_3 - (x_1 z_2 - x_2 z_1)dy_3 + (x_1 y_2 - x_2 y_1)dz_3 \qquad ...(3)$$

Here, we observe that, the maximum possible error would occur when the signs of the elements and the signs of the errors are such that all the eighteen terms in equation (3) are of the same sign.

Now, equation (3) shows that the error in a determinant composed of non-exact elements may be anything from zero upto a number of sufficient magnitude.

2.14 APPLICATION OF ERROR FORMULA TO THE FUNDAMENTAL OPERATIONS OF ARITHMETICS

(i) Error in Addition of Numbers:

Let $y = x_1 + x_2 + ... x_n$ be a function.

Let us suppose Δx_i to denote the error in x_i. Then we have

$$y + \Delta y = (x_1 + \Delta x_1) + (x_2 + \Delta x_2) + ... + (x_n + \Delta x_n)$$
$$= (x_1 + x_2 + ... + x_n) + (\Delta x_1 + \Delta x_2 + ... + \Delta x_n)$$
$$\therefore \qquad \Delta y = \Delta x_1 + \Delta x_2 + ... + \Delta x_n$$

Now, dividing by y, we get

$$\frac{\Delta y}{y} = \frac{\Delta x_1}{y} + \frac{\Delta x_2}{y} + ... + \frac{\Delta x_n}{y}$$

$$\left| \frac{\Delta y}{y} \right| \leq \left| \frac{\Delta x_1}{y} \right| + \left| \frac{\Delta x_2}{y} \right| + ... + \left| \frac{\Delta x_n}{y} \right|$$

Then, the absolute error is obtained by the relation given by

$$\Delta y = \left| \frac{\Delta y}{y} \right| \cdot y = \text{Product of Relative error and the number } y.$$

(ii) Error in subtraction of Numbers:

Let $y = x_1 - x_2$ be given.

Let us suppose Δy, Δx_1 and Δx_2 denote the errors in y, x_1 and x_2 respectively.

Then, we have

$$y + \Delta y = (x_1 + \Delta x_1) - (x_2 + \Delta x_2) = (x_1 - x_2) + (\Delta x_1 - \Delta x_2)$$

$$\Rightarrow \qquad \Delta y = \Delta x_1 - \Delta x_2 \qquad\qquad (\because y = x_1 - x_2)$$

$$\Rightarrow \qquad \frac{\Delta y}{y} = \frac{\Delta x_1}{y} - \frac{\Delta x_2}{y}$$

But, we have

$$|\Delta y| \le |\Delta x_1| + |\Delta x_2| \Rightarrow \left| \frac{\Delta y}{y} \right| \le \left| \frac{\Delta x_1}{y} \right| \le + \left| \frac{\Delta x_2}{y} \right|$$

Therefore, the relative error and absolute errors are given by

$$\text{Relative error} = \left| \frac{\Delta y}{y} \right| \le \left| \frac{\Delta x_1}{y} \right| \le + \left| \frac{\Delta x_2}{y} \right|$$

and Absolute error $= |\Delta y| \le |\Delta x_1| + |\Delta x_2|$

(iii) Error in Product of Numbers:

Let $\qquad y = x_1 x_2 \dots x_n$

Now, suppose that Δy, Δx_1, Δx_2, ..., Δx_n denote the errors in y, x_1, x_2, ..., x_n respectively.

Then, we have

$$\frac{\Delta y}{y} = \frac{\Delta x_1}{y} \cdot \frac{\partial y}{\partial x_1} + \frac{\Delta x_2}{y} \cdot \frac{\partial y}{\partial x_2} + \dots + \frac{\Delta x_n}{y} \cdot \frac{\partial y}{\partial x_n}$$

Now $\qquad \dfrac{1}{y} \cdot \dfrac{\partial y}{\partial x_1} = \dfrac{x_2 x_3 \dots x_n}{x_1 x_2 x_3 \dots x_n} = \dfrac{1}{x_1}$

$$\frac{1}{y} \cdot \frac{\partial y}{\partial x_2} = \frac{x_1 x_3 \dots x_n}{x_1 x_2 x_3 \dots x_n} = \frac{1}{x_2}$$

$$\dots\dots\dots\dots\dots\dots\dots\dots\dots\dots\dots\dots\dots$$

$$\dots\dots\dots\dots\dots\dots\dots\dots\dots\dots\dots\dots\dots$$

$$\frac{1}{y} \cdot \frac{\partial y}{\partial x_2} = \frac{x_1 x_2 \dots x_{n-1}}{x_1 x_2 \dots x_n} = \frac{1}{x_n}$$

$$\therefore \qquad \frac{\Delta y}{y} = \frac{\Delta x_1}{x_1} + \frac{\Delta x_2}{x_2} + \dots + \frac{\Delta x_n}{x_n}$$

Therefore, the Relative error and absolute error are given by

$$\text{Relative error} = \left| \frac{\Delta y}{y} \right| \le \left| \frac{\Delta x_1}{x_1} \right| + \left| \frac{\Delta x_2}{x_2} \right| + \dots + \left| \frac{\Delta x_n}{x_n} \right|$$

$$\text{Absolute error} = \left| \frac{\Delta y}{y} \right| \cdot y = \left| \frac{\Delta y}{y} \right| \cdot (x_1 x_2 \dots x_n)$$

(iv) Error in Division of Two Numbers:

Let $y = \dfrac{x_1}{x_2}$

Since, we have

$$\frac{\Delta y}{y} = \frac{\Delta x_1}{y} \cdot \frac{\partial y}{\partial x_1} + \frac{\Delta x_2}{y} \cdot \frac{\partial y}{\partial x_2} = \frac{\Delta x_1}{x_1/x_2} \times \frac{1}{x_2} + \frac{\Delta x_2}{x_1/x_2}\left(\frac{-x_1}{x_2^2}\right) = \frac{\Delta x_1}{x_1} - \frac{\Delta x_2}{x_2}$$

$$\therefore \qquad \left|\frac{\Delta y}{y}\right| \le \left|\frac{\Delta x_1}{x_1}\right| + \left|\frac{\Delta x_2}{x_2}\right|$$

Thus, the relative error is given by

$$\text{Relative Error} \le \left|\frac{\Delta x_1}{x_1}\right| + \left|\frac{\Delta x_2}{x_2}\right|$$

(v) Error in Evaluating x^k :

Let $y = x^k$, where k is any integer or a fraction. Then, we have the relative error

$$= \left|\frac{\Delta y}{y}\right| < \frac{\Delta x}{y} \cdot \frac{dy}{dx}$$

i.e.,
$$\left|\frac{\Delta y}{y}\right| < \frac{\Delta x}{x^k} \cdot k \cdot x^{k-1} = k \cdot \frac{\Delta x}{x}$$

Thus, relative error in evaluating $x^k = k \cdot \left|\dfrac{\Delta x}{x}\right|$

(vi) Inverse Problem:

Let $y = f(x_1, x_2, ..., x_n)$ be a function, which have a desired accuracy, i.e., if Δy is error in y. Then we have to determine errors $\Delta x_1, \Delta x_2, ..., \Delta x_n$ in $x_1, x_2, ..., x_n$.

Since, we have

$$\Delta y = \Delta x_1 \cdot \frac{\partial y}{\partial x_1} + \Delta x_2 \cdot \frac{\partial y}{\partial x_2} + ... + \Delta x_n \cdot \frac{\partial y}{\partial x_n}$$

Now using the principal of equal effects, we have

$$\Delta x_1 \cdot \frac{\partial y}{\partial x_1} = \Delta x_2 \cdot \frac{\partial y}{\partial x_2} = ... = \Delta x_n \cdot \frac{\partial y}{\partial x_n}$$

$$\Delta y = \Delta x_1 \cdot \frac{\partial y}{\partial x_1} + \Delta x_1 \cdot \frac{\partial y}{\partial x_1} + ... + \Delta x_1 \cdot \frac{\partial y}{\partial x_1} = n \Delta x_1 \cdot \frac{\partial y}{\partial x_1}$$

$$\therefore \qquad \Delta x_1 = \frac{\Delta y}{n \dfrac{\partial y}{\partial x_1}}$$

Similarly
$$\Delta x_2 = \frac{\Delta y}{n \dfrac{\partial y}{\partial x_2}} ... \quad \Delta x_n = \frac{\Delta y}{n \dfrac{\partial y}{\partial x_n}}$$

Thus
$$\Delta x_1 = \frac{\partial y}{n \dfrac{\partial y}{\partial x_1}}, \Delta x_2 = \frac{\partial y}{n \dfrac{\partial y}{\partial x_2}}, ..., \Delta x_n = \frac{\partial y}{n \dfrac{\partial y}{\partial x_n}}$$

SOLVED EXAMPLES

EXAMPLE 1. *Find the possible relative error and absolute error in the sum of 0.1429 and 0.0909, where 0.1429 and 0.0909 are the approximate values of 1/7 and 1/11, correct to four decimal places.*

SOLUTION.   Since, we consider the approximation in four decimal places, therefore in each case, the maximum error is

$$\frac{1}{2} \times 0.0001 = 0.00005$$

Now

(i) The relative error $= \left|\frac{\Delta y}{y}\right| < \left|\frac{0.00005}{0.2338}\right| + \left|\frac{0.00005}{0.2338}\right|$

$$(\because y = x_1 + x_2 = 0.1429 + 0.0909 = 0.2338)$$

$$\therefore \qquad \left|\frac{\Delta y}{y}\right| < \frac{0.00001}{0.2338} = 0.00043$$

(ii) The absolute error $= \left|\frac{\Delta y}{y}\right| y = \frac{0.00001}{0.2338} \times 0.2338 = 0.0001$

EXAMPLE 2.   *Find the relative error in the difference of following two numbers, given by $\sqrt{5.5} \approx 2.345$ and $\sqrt{6.1} \approx 2.470$, correct to four significant figures.*

SOLUTION.   Here we have $\qquad \Delta x_1 = \Delta x_2 = \frac{1}{2}(0.001) = 0.0005$

$$(\because \text{ we consider the approximation into four significant figures})$$

$$\therefore \qquad \text{The relative error} < \left|\frac{\Delta x_1}{y}\right| + \left|\frac{\Delta x_2}{y}\right|$$

$$= 2\left|\frac{\Delta x_1}{y}\right| = 2\left|\frac{0.0005}{2.470 - 2.345}\right| \qquad (\because y = x_1 - x_2)$$

$$= 2\left|\frac{0.0005}{0.125}\right| = \frac{0.001}{0.125} = 0.0008$$

Hence, the possible maximum error is $= 0.0008$.

EXAMPLE 3.   *Find the product of 346.1 and 865.2 and state how many figures of the results are trustworthy, given that the numbers are correct to four significant figures.*

SOLUTION.   Since we consider the approximation in one decimal place, therefore

$$\Delta x_1 = \frac{1}{2}(0.1) = \Delta x_2 = 0.05$$

and $y = 346.1 \times 865.2 = 299446$
which is correct to six significant figures.

Then, the relative error $\leq \left|\frac{\Delta x_1}{x_1}\right| + \left|\frac{\Delta x_2}{x_2}\right| = \left|\frac{0.05}{346.1}\right| + \left|\frac{0.05}{865.2}\right|$

$$= 0.000144 + 0.000058 = 0.000202$$

Therefore, the absolute error = Relative error $\leq 0.000202 + 299446 \approx 60$

The true value of the product of the numbers gives lies between

$$299446 - 60 = 299386 \text{ and } 299446 + 60 = 299506$$

Now, the mean of these values is $\dfrac{299386 + 299506}{2} = 299446$ which can be

written as 299.4×10^2 correct to four significant figures.

EXAMPLE 4.   *Find the number of trustworthy figures in $(0.491)^3$ assuming that the number is 0.491 is correct to last figure.*

SOLUTION.   Since, we know that the Relative error $E_r = \frac{\Delta y}{y} < k \frac{\Delta x}{x}$

Since we consider the approximation of given number up to three decimal places

$$\therefore \qquad \Delta x = \frac{1}{2}(0.001) = 0.0005$$

Also, here $k = 3$

$$\Rightarrow \qquad k\frac{\Delta x}{x} = \frac{3 \times 0.0005}{(0.491)^3} = \frac{3 \times 0.0005}{0.118371} = 0.01267$$

$\therefore$ The absolute error $= E_r \cdot y$

$$< 0.01267 \times (0.491)^3$$
$$= 0.1267 \times 0.118371 = 0.0015$$

Since the error affects the third decimal places, therefore, $(0.491)^3 = 0.1183$ is correct to second decimal places.

EXAMPLE 5. *The error in the measurement of the area of circle is not allowed to exceed 0.1%. How accurately should the diameter be measured?*

SOLUTION. Let d be the diameter of the circle.

Then area $\qquad A = \dfrac{\pi d^2}{4}$

$$\Rightarrow \qquad \frac{\partial A}{\partial d} = \frac{\pi d}{2}$$

$$\Delta A = \Delta d \cdot \frac{\partial A}{\partial d}, \qquad \therefore \quad \Delta d = \frac{\Delta A}{\dfrac{\partial A}{\partial d}}$$

Now   percentage error in $A = \dfrac{\Delta A}{A} \times 100 = 0.1$

$$\therefore \qquad \Delta A = \frac{0.1 \times A}{100} = 0.001 \times A = \frac{0.001 \times \pi d^2}{4}$$

$\therefore$ The percentage error in $d = \dfrac{\Delta d}{d} \times 100 = \dfrac{100}{d} \times \dfrac{\Delta A}{\partial A / \partial d}$

$$= \frac{100}{d}\left(\frac{0.001 \times \pi d^2}{4}\right)\frac{\pi d}{2} = \frac{0.1\pi d^2}{4d} \times \frac{2}{\pi d} = \frac{0.1}{2} = 0.05$$

EXAMPLE 6. *The percentage error in R, which is given by $R = \dfrac{r^2}{2h} + \dfrac{h}{2}$, is not allowed to exceed 0.2%. Find allowable error in r and h when r = 4.5 cm and h = 5.5 cm.*

SOLUTION. The percentage error in R

$$= \frac{\Delta R}{R} \times 100 = 0.2$$

$$\Delta R = \frac{0.2}{100} \times R = \frac{0.2}{100} \times \left[\frac{(4.5)^2}{2 \times 5.5} + \frac{5.5}{2}\right] \qquad \left(\because R = \frac{r^2}{2h} + \frac{h}{2}\right)$$

$$= \frac{0.2}{100} \times \frac{50.5}{11} = \frac{0.002 \times 50.5}{11} \qquad\qquad \dots(i)$$

(i) Percentage error in $r = \dfrac{\Delta r}{r} \times 100$

$$= \frac{100}{r}\left(\frac{\Delta R}{\dfrac{2\partial R}{\partial r}}\right) \qquad\qquad \left(\because \Delta r = \frac{\Delta R}{\dfrac{2\partial R}{\partial r}}\right)$$

$$= \frac{100}{r} \times \frac{\Delta R}{2\left(\dfrac{r}{h}\right)} = \frac{100(\Delta R) \cdot h}{2r^2} \qquad \qquad ...(ii)$$

Put $r = 4.5$ and value of ΔR from equation (1), in equation (2), we get

$$\text{Percentage error} = \frac{100}{2 \times (4.5)^2} \times \frac{0.002 \times 50.5}{11} \times h$$

$$= \frac{0.1 \times 50.50 \times 5.5}{11 \times 20.25} = 0.12$$

(ii) Percentage error in $h = \dfrac{\Delta h}{h} \times 100$

$$= \frac{100}{h} \times \frac{\Delta R}{2\dfrac{\partial R}{\partial h}} = \frac{100}{h} \cdot \frac{\Delta R}{2\left(\dfrac{-r^2}{2h^2} + \dfrac{1}{2}\right)}$$

$$= \frac{100\Delta R}{\left(\dfrac{-r^2}{h^2} + h\right)} = \frac{100}{20/11} \times \frac{50.5 \times 0.002}{11} = 0.505$$

EXAMPLE 7. *Use the Series* $\log_e\left(\dfrac{1+x}{1-x}\right) = 2\left(x + \dfrac{x^3}{3} + \dfrac{x^5}{5} + ...\right)$

to compute the value of log (1.2) *correct to seven decimal place and find the number of terms retained.*

SOLUTION. Let $\dfrac{1+x}{1-x} = 1.2 \quad \Rightarrow \quad x = \dfrac{1}{11}$

If we retains n terms, then $(n+1)^{\text{th}}$ term $= \dfrac{x^{2n+1}}{2n+1} = \dfrac{\left(\dfrac{1}{11}\right)^{2n+1}}{2n+1}$

For seven decimal accuracy, we have

$$\frac{1}{2n+1} \cdot \left(\frac{1}{11}\right)^{2n+1} < \frac{1}{2} \times 10^{-7} \Rightarrow (2n+1)(11)^{2n+1} > 2 \times 10^7$$

$$\Rightarrow \qquad n \geq 3$$

Hence, retaining the first three terms of the given series, we get

$$\log_e(1.2) = 2\left(x + \frac{x^3}{3} + \frac{x^5}{5}\right)_{\text{at } x = \frac{1}{11}} = 0.1823215.$$

EXAMPLE 8. *For $x = 0.4845$ and $y = 0.4800$. Calculate the value of* $\dfrac{x^2 - y^2}{x+y}$ *by using normalized floating point arithmetic. Compare with the value of $(x - y)$ indicate error in the former.*

SOLUTION. Given that $x = 0.4845, y = 0.4800$

Now, $(x^2 - y^2) = (0.4845\ E0 \times 0.4845\ E0) - (0.4800\ E0 \times 0.4800\ E0)$

$$= (0.0043\ E0) = (0.4300\ E -2)$$

$$(x + y) = (0.4845\ E0 + 0.4800\ E0) = (0.9645\ E0)$$

So, $\dfrac{(x^2 - y^2)}{(x+y)} = \dfrac{(0.4300\ E - 2)}{(0.9645\ E0)}$

$$x - y = (0.4845 \ E0) - 0.4800 \ E0$$
$$= (0.0045 \ E0) = (0.4500 \ E - 2)$$

Hence in normalized floating point arithmetic, the value of $\dfrac{(x^2 - y^2)}{x + y} \neq x - y$

The error is $(0.4500 \ E - 2) - (0.4458 \ E - 2) = (0.0042 \ E - 2) = (0.4200 \ E - 4)$

EXAMPLE 9. *Compare the percentage error in the time period* $T = 2\pi\sqrt{\dfrac{l}{g}}$ *for* $l = 1$ *m if the error in measurement of l is 0.01.*

SOLUTION. We have $\qquad T = 2\pi\sqrt{\dfrac{l}{g}}$

Taking log both the sides, we get

$$\log T = \log 2\pi + \frac{1}{2}\log l - \frac{1}{2}\log g$$

$$\Rightarrow \qquad \frac{1}{T}\delta T = \frac{1}{2}\cdot\frac{\delta l}{l}$$

$$\Rightarrow \qquad \frac{\delta T}{T}\times 100 = \frac{\delta l}{2l}\times 100 = \frac{0.01}{2\times 1}\times 100 = 5\%$$

EXAMPLE 10. *The discharge Q over a notch for head H is calculated by the formula* $Q = kH^{5/2}$, *where k is a given constant. If the head is 75 cm and an error of 0.15 cm is possible in its measurement, estimate the percentage error in computing the discharge.*

SOLUTION. Here, we have $Q = kH^{5/2}$

Taking log of both the sides, we get

$$\log Q = \log k + \frac{5}{2}\log H$$

On differentiating, we get

$$\frac{\delta Q}{Q} = \frac{5}{2}\cdot\frac{\delta H}{H}$$

$$\therefore \qquad \frac{\delta Q}{Q}\times 100 = \frac{5}{2}\times\frac{0.15}{75}\times 100 = \frac{1}{2} = 0.05$$

EXAMPLE 11. *If* $r = 3h(h^6 - 2)$. *Find the percentage error in r at h = 1 if the percentage error in h is 5.*

SOLUTION. We have $\delta r = \dfrac{\partial r}{\partial h}\cdot\delta h = (21h^6 - 6)\delta h$

$$\therefore \quad \frac{\delta r}{r}\times 100 = \left(\frac{21h^6 - 6}{3h^7 - 6h}\right)\delta h \times 100 = \left(\frac{21 - 6}{3 - 6}\right)\left(\frac{\delta h}{h}\times 100\right) = \frac{15}{-3}\cdot 5\% = -25\%$$

Now, percentage error is $\quad = \left|\dfrac{\delta r}{r}\times 100\right| = 25\%$

EXAMPLE 12. *If* $\sqrt{29} = 5.385$ *and* $\sqrt{\pi} = 3.317$ *correct to four significant figures, find the relative error in their sum and differences.*

SOLUTION. The numbers 5.385 and 3.317 are correct to four significant figures. Therefore. Maximum error in each case is

$$\frac{1}{2}\times 10^{-3} = 0.0005$$

$$\therefore \qquad \Delta x_1 = \Delta x_2 = 0.0005$$

Now, relative error in their sum is

$$\left|\frac{\Delta X}{X}\right| \le \left|\frac{\Delta x_1}{x}\right| + \left|\frac{\Delta x_2}{x}\right| \qquad\qquad (\because X = x_1 - x_2 = 8.702)$$

$$\le \left|\frac{0.0005}{8.702}\right| + \left|\frac{0.0005}{8.702}\right| < 1.149 \times 10^{-4}$$

Also, relative error in their difference is

$$\left|\frac{\Delta X}{X}\right| \le \left|\frac{\Delta x_1}{x}\right| + \left|\frac{\Delta x_2}{x}\right| \text{ where } X = x_1 + x_2 = 2.068$$

$$\le \left|\frac{0.0005}{2.068}\right| + \left|\frac{0.0005}{2.068}\right| < 4.835 \times 10^{-4}$$

EXERCISE 2.2

1. If $\sqrt{29} = 5.385$ and $\sqrt{\pi} = 3.317$ correct to four significant figures. Find the relative errors in their sum and differences.

2. Find the number of terms of the exponential series such that their sum gives e^x correct to six decimal places at $x = 1$.

3. If $R = \dfrac{4xy^2}{z^3}$ and errors in x, y, z be 0.001. Show that the maximum relative error at $x = y = z = 1$ is 0.006.

4. If $R = \dfrac{1}{2}\left(\dfrac{r^2}{h} + h\right)$ and error in R is at the most 0.4%. Find the percentage error allowable in r and h when $r = 5.1$ cm and $h = 5.8$ cm.

5. Determine the number of terms required in the series for $\log(1 + x)$ to evaluate $\log 1.2$ correct to six decimal places.

6. Find the relative error in calculation of $\dfrac{7.342}{0.241}$, where the number 7.342 and 0.241 are correct to three decimal places. Determine the smallest interval in which true result lies.

7. Find the number of trustworthy figures in $(367)^{1/5}$ where 367 is correct to three significant figures.

8. How accurately, the length and time of vibration of a pendulum should be measured in order that the computed value of g be correct to 0.01%.

9. Let n_0 be the approximate cube root of n and

let $x = \dfrac{n}{n_0^3} - 1$, show that cube root of n is given by

$$n^{1/3} = n_0\left[1 + \frac{x}{3} - \frac{x^2}{9} + \frac{5x^3}{81} - \frac{10x^4}{243} + \dots\right]$$

Hence, find the value of $(6)^{1/3}$ correct to four significant figures.

10. If n_0 is the approximate value of the square root of n and $x = \dfrac{n}{n_0^2} - 1$, show that the square root of n is given by

$$n^{1/2} = n_0\left[1 + \frac{x}{2} - \frac{x^2}{8} + \frac{x^3}{16} - \frac{5x^4}{128} + \dots\right]$$

Hence, find the square root of 5 correct to three decimal places.

11. Write a short note on 'Error in Numerical computations'.

12. Let x^* approximate x correct upto n significant figures if e^x is evaluated for x, $-8 \le x \le 9$. Then, what should be the relative error.

13. If $R = 4x^2y^3z^{-4}$, find the maximum absolute error and maximum relative error in R when errors in $x = 1$, $y = 2$, $z = 3$ respectively are equal to 0.001, 0.002, 0.003. (UPTU–2003)

14. Represent 44.85×10^6 in normalized floating point mode. (UPTU–2004)

15. If $r = h(4h^5 - 5)$, find the percentage error in r at $h = 1$, if the error in h is 0.04. (WBTU–2005)

Answers

1. 1.149×10^{-4}, 4.836×10^{-4} **2.** $n = 10$ **4.** 0.23, 0.14 **5.** $n = 10$

6. 0.0021, (30.4647 – 0.0639) **7.** 3.26, correct to three figures

8. (i) Percentage error in length = 0.005 (ii) Percentage error in time = 0.0025 **9.** 1.817

10. 2.236 **13.** 0.00355, 0.0089 **14.** 0.4485 $E8$ **15.** 76

2.15 ORDER OF APPROXIMATIONS

Let us suppose $f(h)$ be a function with approximation $g(h)$ and the error bound is known to be $\mu(h^n)$ where n is a positive integer so that

$$|f(h) - g(h)| \leq \mu|h^n|$$

where h is sufficiently small.

Then, we say that $g(h)$ approximate the function $f(h)$ with order of approximation $O(h^n)$ and write

$$f(h) = g(h) + O(h^n)$$

For example: (i) Consider $(1 - h)^{-1} = 1 + h + h^2 + h^3 + h^4 + \dots$

is written as $\qquad (1 - h)^{-1} = 1 + h + h^2 + h^3 + O(h^4)$

to the fourth order approximations.

Similarly $\qquad \cosh = 1 - \dfrac{h^2}{2!} + \dfrac{h^4}{4!} - \dfrac{h^6}{6!} + \dots \quad = 1 - \dfrac{h^2}{2!} + \dfrac{h^4}{4!} + O(h^6)$

2.15.1 ORDER OF APPROXIMATION FOR SUM AND PRODUCT

(i) Approximation for Sum: Consider, from the previous example

$$(1 - h)^{-1} = 1 + h + h^2 + h^3 + O(h^4) \qquad \dots(1)$$

and $\qquad \cosh = 1 - \dfrac{h^2}{2!} + \dfrac{h^4}{4!} + O(h^6) \qquad \dots(2)$

Then, for the approximation of sum of eq. (1) and (2), we get

$$[(1 + h)^{-1} + \cosh] = 2 + h + \frac{h^2}{2!} + h^3 + O(h^4) + \frac{h^4}{4!} + O(h^6) \qquad \dots(3)$$

Now since $\qquad O(h^4) + \dfrac{h^4}{4!} = O(h^4)$

and $\qquad O(h^4) + O(h^6) = O(h^4)$

Therefore, from eq. (3), we get

$$[(1 + h)^{-1} + \cosh] = 2 + h + \frac{h^2}{2!} + h^3 + O(h^4)$$

a fourth order approximation.

(ii) Approximation for Product:

For the approximation of product of (1) and (2), we get

$$[(1 + h)^{-1} \cosh] = (1 + h + h^2 + h^3)\left[1 - \frac{h^2}{2!} + \frac{h^4}{4!}\right] + (1 + h + h^2 + h^3)O(h^6)$$

$$+ \left(1 - \frac{h^2}{2!} + \frac{h^4}{4!}\right)O(h^4) + O(h^4)O(h^6)$$

$$= 1 + h + \frac{h^2}{2} + \frac{h^3}{2} - \frac{11h^4}{24} + \frac{11h^5}{24} + \frac{h^6}{24} + \frac{h^7}{24} + O(h^4)$$

$$+ O(h^6) + O(h^4)O(h^6) \qquad \dots(4)$$

Now since

$$O(h^4)O(h^6) = O(h^{10})$$

$$\Rightarrow \quad -\frac{11h^4}{24} + \frac{11}{24}h^5 + \frac{h^6}{24} + \frac{h^7}{24} + O(h^4) + O(h^6) + O(h^{10}) = O(h^4)$$

Therefore, from eq. (4), We get

$$[(1-h)^{-1}\cosh] = 1 + h + \frac{h^2}{2} + \frac{h^3}{2} + O(h^4)$$

which is of the first order approximation.

2.16 PROPAGATION OF ERROR

Let us suppose $g(n)$ represents the growth of error after n steps of a computation process. Then, we have the following observations

   (i) If $|g(n)| \sim n\varepsilon$ then, the growth of error is linear.

   (ii) If $|g(n)| \sim \delta^n \varepsilon$ then, the growth of the error is exponential.

   (iii) If $\delta > 1$ then the exponential will grow indefinitely as $n \to \infty$ and

   (iv) If $0 < \delta < 1$ then exponential error decrease to zero as $n \to \infty$

2.16.1 SOME IMPORTANT OBSERVATIONS ON ERRORS

- If C_1 and C_2 are the first significant figures of two numbers which are each correct to n significant figures and if neither number is of the form $C(1.00...) \times 10^p$, then their product or quotient is correct to :

   (a) $(n-1)$ significant figures if $C_1 \geq 2$ and $C_2 \geq 2$.

   (b) $(n-2)$ significant figures if either $C_1 = 1$ or $C_2 = 1$.

- If C is the first significant figure of a number which is correct to n significant figures, and if this number contains more one digits different from zero, then its pth power is correct to:

   (a) $(n-1)$ significant figures if $p \leq C$

   (b) $(n-2)$ significant figures if $p \leq 10C$.

   and its rth root is correct to

   (a) n significant figures if $rC \leq 10$.

   (b) $(n-1)$ significant figures if $rC \leq 10$.

- If C is the first significant figures of a number which is correct to n significant figures and if this number contains more than one digit different from zero, then for the absolute error in its common logarithms we have

$$E_a < \frac{1}{4C \times 10^{n-1}}$$

- If a logarithm (base 10) is not in error by more than two units in the mth decimal places, the antilog is certainly correct to $(m-1)$ significant figures.

2.16.2 PROPAGATED ERROR

In any numerical problem, the true value of numbers may not be used exactly, *i.e.*, in place of true values of the numbers, some approximate values like floating point numbers are used initially. The error arising in the problem due to those inexact/approximate values is called propagated error.

Let x^A, y^A be approximation to x and y respectively and w be arithmetic operation.

Then,   the propagated error $= xwy^A - x^A wy^A$

and   the relative propagated error $= \dfrac{xwy - xw^A y^A}{xwy}$

$$\text{Total relative error} = \frac{xwy - x^A w^A y^A}{xwy}$$

$$= \frac{xwy - x^A wy^A}{xwy} + \frac{x^A wy^A - x^A w^A y^A}{xwy}$$

▶ **REMARK**

▸ For the first approximation.

Total relative error = relative propagated error + relative generated error.

2.16.2 PROPAGATION OF ERROR IN FUNCTION EVALUATION OF A SINGLE VARIABLE

Let $f(x)$ be evaluated and x^A be an approximation to x. Then, the absolute error in evaluation of $f(x)$ is $f(x) - f(x^A)$ and relative error is

$$\gamma_{f(x)} = \frac{f(x) - f(x^A)}{f(x)}$$

Let us suppose $\qquad x = x^A + \rho_x$

Then, by Taylor's series expansion, we get

$$f(x) = f(x^A) + \rho_x f'(x^A) + \dots$$

$$\Rightarrow \qquad \gamma_{f(x)} = \frac{\rho_x f'(x^A)}{f(x)} \qquad\qquad \text{(By neglecting the higher order terms)}$$

$$= \frac{\rho_x}{f(x)} \approx \frac{\gamma f'(x^A)}{f(x)} = \gamma_x \frac{x f'(x^A)}{f(x)}$$

$$\left|\gamma_{f(x)}\right| = \left|\gamma_x\right| \left|\frac{x f'(x^A)}{f(x)}\right|$$

▶ **REMARKS**

▸ For evaluation of $f(x)$ in denominator of R.H.S. after simplification, $f(x)$ must be replaced by $f(x^A)$ in some cases so

$$\left|\gamma_{f(x)}\right| = \left|\gamma_x\right| \left|\frac{x f'(x^A)}{f(x)}\right|$$

The expression $\left|\dfrac{x f'(x^A)}{f(x)}\right|$ is called condition number $f(x)$ at x.

▸ If the condition number is very large, then function is said to be more ill-conditioned.

SOLVED EXAMPLES

EXAMPLE 1. Let $f(x) = x^{1/10}$ and x^A approximates x correct to n significant decimal digit. Show that $f(x^A)$ approximates $f(x)$ correct to $(n + 1)$ significant decimal digits.

SOLUTION. We have

$$\gamma_{f(x)} = \gamma_x \cdot \frac{x f'(x^A)}{f(x)}$$

$$= \gamma_x \cdot \frac{x \cdot \dfrac{1}{10} x^A{}^{\frac{9}{10}}}{x^{1/10}} = \left(\frac{1}{10}\right)\gamma_x$$

$$\therefore \qquad \left|\gamma_{f(x)}\right| = \left(\frac{1}{10}\right)\left|\gamma_x\right| \le \frac{1}{10} \cdot \frac{1}{2} \cdot 10^{1-n} = \frac{1}{2} \cdot 10^{1-(n+1)}$$

$\Rightarrow \quad f(x^A)$ approximates $f(x)$ correct to $(n+1)$ significant digits.

EXAMPLE 2. *The function $f(x) = \cos(x)$ can be explained as*

$$\cos x = 1 - \frac{x^2}{2} + \frac{x^4}{4!} - \frac{x^6}{6!} + \dots$$

compute the number of terms requires to estimate $\cos\left(\dfrac{\pi}{4}\right)$ *so that the result is correct to least two significant digits.*

SOLUTION. We know that the pre-specified tolerance e_s can be obtained by using
$$e_s = (0.5 \times 10^{2-n})\%$$
Therefore, we have
$$e_s = 0.5 \times 10^{-m} = 0.5 \times 10^{-2}$$
The remainder term R_n is given by $R_n = \dfrac{x^{2n}}{(2n)!}\cos\xi$

Then, maximum relative error $= \dfrac{(\pi/4)^{2n}}{(2n)!}$

Therefore,

$$0.5 \times 10^{-2} \ge \frac{(\pi/4)^{2n}}{2n!}$$

i.e., $\dfrac{1}{0.5 \times 10^{-2}} \le \dfrac{2n!}{(\pi/4)^{2n}}$

$\Rightarrow \qquad 200 \le \dfrac{2n!}{(\pi/4)^{2n}}$

n	$\dfrac{(2n)!}{(\pi/4)^{2n}}$
1	3.24
2	63.074
3	3067.561

Thus $\qquad n = 3$

EXAMPLE 3. *The function $f(x) = \tan^{-1}x$ can be expanded as follows:*

$$\tan^{-1}x = x - \frac{x^3}{3} + \frac{x^5}{5} - \dots + (-1)^{n-1}\frac{x^{2n-1}}{(2n-1)} + \dots$$

Compute number of terms n such that the series determines $\tan^{-1}1$ correct to eight significant figures.   [MDU(BE)–2006]

SOLUTION. Proceed same as above, we get
$$e_s = 0.5 \times 10^{-m} = 0.5 \times 10^{-8}$$
Also, the remainder term after n terms is given by
$$R_n = \frac{x^{2n+1}}{2n+1}\tan^{-1}\xi, 0 < \xi < x$$
Therefore, the maximum relative error is given by
$$\left(\frac{x^{2n+1}}{2n+1}\right)_{x=1} = \frac{1}{2n+1}$$
Since, the error must be less than e_s, therefore
$$0.5 \times 10^{-8} \ge \frac{1}{2n+1}$$

$$\Rightarrow \qquad \frac{1}{0.5 \times 10^{-8}} \le 2n+1$$

$$\Rightarrow \qquad 2 \times 10^{8} \le 2n+1$$

Therefore $\qquad n = 10^{8}+1.$

EXAMPLE 4. *Determine the number of terms of the exponential series*

$$e^{x} = 1 + x + \frac{x^{2}}{2!} + \frac{x^{3}}{3!} + \ldots + \frac{x^{n}}{n!} + \ldots$$

such that their sum gives the values of e^{x} correct to six decimal places for $0 \le x \le 1$.

[UPTU(MCA)–2002]

SOLUTION. Here $\qquad e^{x} = 1 + x + \frac{x^{2}}{2!} + \frac{x^{3}}{3!} + \ldots + \frac{x^{n-1}}{(n-1)!} + R_{n}(x) \qquad \ldots(1)$

Where $\quad R_{n}(x) = \frac{x^{n}}{n!}e^{\theta}, 0 < \theta < x$

Max, absolute error (at $\theta = x$) $= \frac{x^{n}}{n!}e^{x}$ and the max. relative error $= \frac{x^{n}}{n!}$

Hence, $(E_{r})_{\max}$ at $x = 1 = 1/n!$

For a six decimal accuracy at $x = 1$ we have $\frac{1}{n!} < \frac{1}{2} \times 10^{-6}$ or $n! > 2 \times 10^{6}$,

which gives $n = 10$.

EXERCISE 2.3

1. Obtain polynomial approximation to $f(x) = (1-x)^{1/2}$ over [0, 1] by means of Taylor series about $x = 0$. Find the number of terms required in the expansion of obtain results correct to 5×10^{-1} for $0 \le x \le 1/2$.

2. Obtain a second degree polynomial approximation to $f(x) = (1+x)^{1/2}, x \in [0, 0.1]$ using Taylor series expansions about $x = 0$. Use the expansions to approximate $f(0.5)$ and found to truncation error.

Answers

2. Truncation error $= 0.625 \times 10^{-4}$.

2.17 BLUNDERS

Blunders are errors which arises due to human imperfection. Since these errors are due to human mistakes, it should be possible to avoid them. These types of errors can occur at any stage of the numerical processing due to the

   (i) lack of understanding of the problem

   (ii) wrong assumptions

   (iii) selecting a wrong method

   (iv) wrong guessing the initial values.

The solution have its care, coupled with a careful examination of the results for reasonableness. Sometimes a test run with known results is worthwhile, but it is no guarantee of freedom from foolish error. When hand computation was more common check sums were usually computed. They were designed to reveal the mistake and permits its correction.

2.18 NUMERICAL INSTABILITY

We know that every arithmetic operation performed during computations, gives some errors, which may grow or decay in subsequent calculations. In some cases errors may grow so large as it make the computed result totally redundant. Such a procedure is called numerically unstable.

On the other hand, in some cases it can be avoided by changing the calculation procedure, which avoids subtractions of nearly equal numbers or division by a small number or by retaining more digits in the mantissa.

There are the following types of instability:

    (i) Inherent instability: This instability may arise due to the ill-condition ness of the problem. Here, we can not avoid the inherent instability by changing the method of solution. It is the property of the problem itself. We can avoid this instability by suitable reformulation of the problem.

    (ii) Induced instability: The induced instability may arise due to the wrong choice of the method of solution. Although, the problem is well conditioned in this case. Induced instability can be avoided by a suitable modification or change of the method of solution.

2.18.1 SENSITIVITY ANALYSIS

Investigation to see how small changes (or perturbations or disturbances) in input parameters influence the output are termed as sensitivity analysis, when problem is sensitive to small changes in its parameter, it is impossible to make a numerically stable method for its solution.

2.19 MACHINE COMPUTATIONS

When we solve any problem using computers, then to obtain meaningful results, we have the following phases:

    (i) Choice of a method: A method is defined by a mathematical formula for finding the solution of the given equation. In some cases, there may be more than one methods are available to solve the same problem. Choose the method which suits the given problem best. The assumptions and limitations of the method must be studied carefully.

    (ii) Designing the Algorithm: Since, we know that the computer do not solve problem rather they are used to implement the solution to problems.

The logical and concise list of procedure for solving a problem is called an algorithm. It describes the steps that lead to required results in a finite number of operations. Here, it may be noted that the computer is concerned with the algorithm and not with the method. The algorithm tells the computer where to start, what information to use, what operation to be carried out and in which order, what information to be printed and when to stop. An algorithm should also include steps to identify and abnormal data or results and take corrective measures. In case of large problem we use the modular approach. A module is a program unit or entity that is responsible to a single task. It is also known as sub-programs.

An algorithm has five important properties:

(i) Algorithm should be completed after a finite number of steps.

(ii) Every step of algorithm should be well defined.

(iii) Algorithm should clearly specify which quantities are to be read.

(iv) Algorithm should clearly specify which quantities are to be displayed.

(v) In an algorithm, all operations should be executable.

Algorithm to find the root of a quadratic Equation: If we design an algorithm to find the real roots of the equation.

$$ax^2 + bx + c = 0 \qquad a, b, c \in \mathbf{R}$$

for 10 set of values, using the usual method $x = \dfrac{-b \pm \sqrt{b^2 - 4ac}}{2a}$

Then, we have the following computational steps:

1. Set I = I
2. Read a, b, c
3. Check is $a = 0$? If yes print wrong data and go to step 9
4. Calculated $d = b^2 - 4ac$
5. Check : is $d < 0$? If yes, print, roots and complex and goto step 9
6. Calculate $e = \sqrt{d}$
7. Calculate x_1 and x_2 using the usual method
8. Print x_1 and x_2
9. $I = I + 1$
10. Check : is $I \leq 10$? If yes, goto step 2, otherwise goto step (11)
11. Stop

(iii) **Flowchart:** A flowchart is a graphical representation of a specific number of sequences of steps (algorithm) to be followed by the computer to produce the solution of a given problem. It maks use of flowchart symbols to represent the basic operations to be carried out and the arrow indicate the flow of information and processing.

Flow chart symbols:

	Symbols	Meaning		Symbols	Meaning
1.	⬭	Start or End	4.	◇	Decision making and branching
2.	▭	Computational steps	5.	◯	Connector
3.	▱	Input or output	6.	⟶	Flow of control

FLOW CHART FOR FINDING REAL ROOTS OF THE QUADRATIC EQUATION

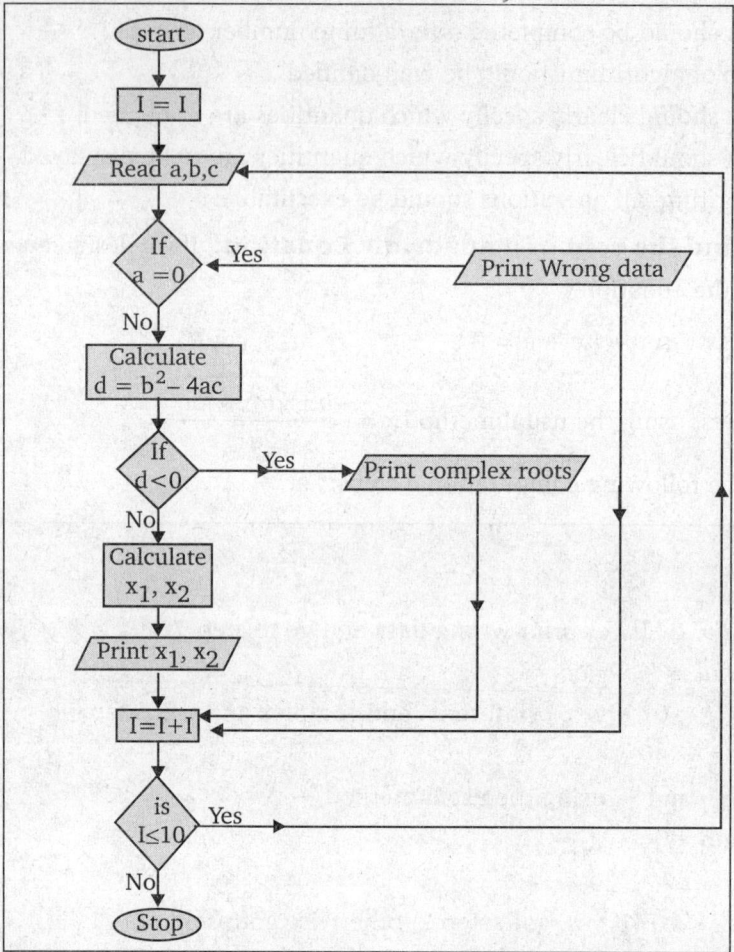

▶ REMARKS

▶ Flowchart provide a graphic representation of the problem so it is easy to understand the plan of the solution.

▶ If helps in reviewing and correcting the program.

▶ It provides a convenient aid to writing computer instructions.

(iv) Programming: In this phase we write the program into any computer language.

Program for the roots of a Quadratic equation in 'C' Language

```
# include <math.h>
main()
{
float a, b, c d;
float root 1, root 2;
printf ("Input the values a, b, c\n")
else
```

```
            }
      root 1= (-b + sqrt (d))/(2.0*a);
      root 2= (-b - sqrt (d))/(2.0*a);
printf ("/n/nRoot = % f/n/nRoot 2 = % f\n" root 1, root 2);
            }

      }
```

(v) Computer Execution : After writing the program of instruction for the computer in a suitable computer language, check the errors in program and remove. After that, prepare the data in the required form. Then, the computations are performed by the computer and the results are given out.

2.20 COMPUTER SOFTWARE

The computer software provide a useful computational tool for users. The writing of a computer software requires a good understanding of the programming. A good computer software must contain some criteria of:

   (i) Self starting   (ii) Accuracy and reliability
  (iii) Minimum number of levels   (iv) Good documentation
   (v) Criteria of portability

  (i) Self starting: A good computer software should be self starting as far as possible. Since, any numerical method involves some parameters, whose values are to be determined. The program will be more acceptable, if it can be made automatic in the sense that the program will select the initial approximation itself rather than requiring the user to specify them.

  (ii) Accuracy and reliability: Accuracy and reliability are measures of the performance of an algorithm on all similar problem. Fixed the error criteria and get the solutions of all similar problems to that accuracy. The program must be able to prevent most of the exceptional conditions.

  (iii) Minimum number of levels: A good software must contain the minimum number of levels, because if the number of levels are increased, then there is a wasted of time due to the interlinking and transfer of parameters.

  (iv) Good documentation: A good documentation should clarify what kind of problems can be solved using the software, what parameters are to be supplied, what accuracy can be achieved, which method has been used and other useful details. It should be noted that the program must have some comments lines at various places giving more explanation about the method and steps.

  (v) Criteria of portability: The software should be made independent of the computer being used as far as possible. Therefore we have that the software must be machine independent, *i.e.*, the same program should be able to run on any machine with minimum modifications.

2.21 NUMBER SYSTEM

It is a mechanical language which provides a facility to make the numbers. We may define a number system as a system which consists of,

- a set of symbol used for formation of numbers.

- a set of rules which may be used to form numbers from these symbols and assign values to them.
- a set of rules performing common arithmetic operation on this system.

There are many types of number system. Some important number systems are as follows:

(i) Decimal number system   **(ii)** Binary number system.

(iii) Octal number system   **(iv)** Hexadecimal number system

(i) Decimal number system: This number system has a base of 10, *i.e.*, 0, 1,2,3,4,5,6,7,8,9. Number of digits needed to represent a number are changed after every 10^n intervals, where n is an integer. A number can be written in expended notation form by breaking every digit according to its place value. **For examples**

1. The number 456 can be written as $4 \times 10^2 + 5 \times 10^1 + 6 \times 10^0$

2. The number 6428.31 can be written as

$$6 \times 10^3 + 4 \times 10^2 + 2 \times 10^1 + 8 \times 10^0 + 3 \times 10^{-1} + 1 \times 10^{-2}$$

(ii) Binary number system: In binary number system, numbers can be represented using 2 digits only so the base of binary numbers system is 2. The two digits that are used in binary number system are 0 and 1. A binary numbers can be written in expanded notation form by breaking the number into digits according to their place value.

e.g., $1010 = (1 \times 2^3) + (0 \times 2^2) + (1 \times 2^1) + (0 \times 2^0)$

$$= 1 \times 8 + 0 \times 4 + 1 \times 2 + 0 \times 1 = 8 + 2 = 10$$

This means $(1010)_2 = (10)_{10}$.

(iii) Octal number system: Octal number system is the number system with base 8. This means in this number system, there are 8 symbols or digits which are used for formation of the numbers. These symbols are 0, 1, 2, 3, 4, 6 and 7. The place value in octal number system are the power of 8. Consider, a number $(156)_8$. This can be written in the expanded form as

$$156_8 = 6 \times 8^0 + 5 \times 8^1 + 1 \times 8^2$$
$$= 6 \times 1 + 5 \times 8 + 1 \times 64$$
$$= 6 + 40 + 64$$

The means, $(156)_8 = (110)_{10}$

(iv) Hexadecimal number system: This number system is number system with base 16. Using the symbols 0, 1, 2, 3, 4, 5, 6, 7, 8, 9, A, B, C, D, E and F. In number system, in addition to decimal digits 0 to 9, this symbols A, B, C, D, E and F are used to represent the numbers 10, 11, 12 ,13, 14 and 15 respectively.

Consider a number $(13BD)_{16}$. This number can be written in expanded form as

$$(13BD)_{16} = 1 \times 16^3 + 3 \times 16^2 + B \times 16^1 + D \times 16^0$$
$$= 4096 + 768 + 11 \times 16 + 13 \times 1$$
$$= 4096 + 768 + 176 + 13 = (5053)_{10}.$$

2.22 BASE CONVERSION

(i) Decimal to Binary (To convert the Integer part): To convert the number in decimal number system to the number in binary number system. We apply the method of repeated division. The division is done by 2.

ALGORITHM

To rounding off a number or digit to n significant figures, discard all digits to the right of the nth place using the following concepts.

Step 1.	Divide the given number by 2.
Step 2.	Note the quotient and remainder. Remainder will be either 0 or 1.
Step 3.	If quotient is not 0, then divide the quotient by 2 and go to step 2.
Step 4.	If quotient is 0, then stop the process of division.
Step 5.	The process of first remainder is called least significant digit (LSD) and last remainder is called most significant digit (MSD).
Step 6.	Arrange all the remainders from MSD to LSD in a sequence from left to right.

Then the combination of 0 and 1 thus obtained is the required binary equivalent of given number.

For example: *Convert* $(45)_{10}$ *into binary number system.*

Solution: Performing repetitive division by 2.

2	45	remainder	
2	22	1	LSD
2	11	0	
2	5	1	
2	2	1	
2	1	0	
	0	1	MSD

Thus $(45)_{10} = (101101)_2$

To convert the fractional part: For converting a fractional decimal number in binary, we use the method of repeated multiplication. The multiplier is 2.

ALGORITHM

Step 1.	Multiply the given number by 2 and separate the integral part.
Step 2.	Multiply the fractional part again by 2 and separate the integral part.
Step 3.	Continue this process, till the fractional part reduces to zero.
Step 4.	Write the integral parts and prefix the binary point.

This will be the desired binary fraction.

SOLVED EXAMPLES

EXAMPLE 1. *Convert* $(0.8176)_{10}$ *to binary number system.*

SOLUTION.

		0	0.8176×2
MSD		1	0.6352×2
		1	0.2704×2
		0	0.5408×2
LSD		1	0.0816×2
		0	0.1632×2

$$(08176)_{10} = (0.11010\ ...)_2$$

EXAMPLE 2.   *Convert* $(67.25)_{10}$ *to binary number system.*
SOLUTION.   First we convert the integral part into binary equivalent.

2	67	remainders
2	33	1
2	16	1
2	8	0
2	4	0
2	2	0
2	1	0
	0	1

Now we convert the decimal part

MSD	0	0.25×2
	0	0.50×2
LSD	1	0.00×2

Thus $(67.25)_{10} = (1000011.01)_2$

(iii) Binary to Decimal: To convert the binary number to decimal number.

ALGORITHM

Step 1. Multiply the digit of whole binary number with powers of 2. The power for integral part of number are positive and negative for fractional part of number.

Step 2. Add the total result which are obtained by multiplying the power of digits. We obtain the final result after addition.

For example: *Convert the following binary numbers to decimal number*

  (i) $(1100111)_2$ **(ii)** $(11001101.01)_2$

Solution. (i) $(1100111)_2$

$$= 1 \times 2^6 + 1 \times 2^5 + 0 \times 2^4 + 0 \times 2^3 + 1 \times 2^2 + 1 \times 2^1 + 1 \times 2^0$$
$$= 64 + 32 + 0 + 0 + 4 + 2 + 1 = (103)_{10}$$

    (ii) $(11001101.01)_2 = 1 \times 2^7 + 1 \times 2^6 + 0 \times 2^5 + 0 \times 2^4 + 1 \times 2^3 + 1 \times 2^2$
$$+ 0 \times 2^1 + 1 \times 2^0 + 0 \times 2^{-1} + 1 \times 2^{-2}$$
$$= 128 + 64 + 0 + 0 + 8 + 4 + 0 + 1 + 0 + 0.25$$
$$= (205.25)_{10}$$

    (iii) Binary to Octal : To convert a binary number into octal number system.

ALGORITHM

Step 1. Firstly we convert binary number to decimal and then decimal to octal. We make the groups of three digits. We start the grouping from right to left.

Step 2. Now each group of three digits converts the decimal number system. After that written the decimal numbers combinedly.

The group of three binary digits from an octal number as shown the table given below:

0	1	2	3	4	5	6	7
00	001	010	011	100	101	110	111

For example: *Convert the following binary number to octal*

   (i) $(101011101)_2$ **(ii)** $(111100011)_2$

   (iii) $(10011011101010)_2$

Solution. **(i)**   Grouping these into three bits each we get

101	011	101	Group of three bits from
III	II	I	right octal equivalent.
5	3	5	

Thus $(101011101)_2 = (535)_8$

   (ii)

111	100	011	Group of three bits from
III	II	I	right octal equivalent.
7	4	3	

$\Rightarrow \quad (111100011)_2 = (743)_8$

   (iii)

010	011	011	101	010	Group of three bits from
V	IV	III	II	I	right octal equivalent.
2	3	3	5	2	

$\Rightarrow \quad (10011011101010)_2 = (23352)_8$

(iv) Binary to Hexadecimal: To convert an integer:

ALGORITHM

Step 1.	For this conversion we divide all binary digit of the number to be converted in the groups of four bits each and start the grouping from right to left.
Step 2.	Now each of these groups of four bit each will be converted to decimal number system and written below the groups.

A group of four binary digits forms one hexadecimal as shown in the table below:

Hexadecimal digit	Binary equivalent
0	0000
1	0001
2	0010
3	0011
4	0100
5	0101
6	0110
7	1111
8	1000
9	1001
10 or A	1010
11 or B	1011
12 or C	1100
13 or D	1101
14 or E	1110
15 or F	1111

For example: *Convert* $(1110101101)_2$ *to hexadecimal equivalent.*

Solution.   Grouping these into four bits each we each

11	1010	1101

Here, we see that 11 is alone so we have written two zero's to its lefts.

Now we have four groups as

0011	1010	1101
III	II	I
3	10 or A	13 or D

Thus $(1110101101)_2 = (3AD)_{16}$

To convert a fraction:

ALGORITHM

Step 1.	For this conversion we divide all binary digit of the fraction part to be converted in the groups of four bits each. Start the grouping from left to right.
Step 2.	Now each of these groups of four bits each will be converted to decimal number system. After that these numbers written in groups.

For example: *Convert* $(100011.01)_2$ *to hexadecimal equivalent.*

Solution.   After grouping of 100011.01, we get

0100	0011	0100
III	II	I
4	3	4

Thus $(100011.01)_2 = (43.4)_{16}$

(v) Decimal to Octal : To convert the integer: For converting the decimal number to octal we apply the following process step by step as,

ALGORITHM

Step 1.	Divide the number by 8.
Step 2.	Note down the quotient and remainder. Remainder will be any digit from 0 to 7.
Step 3.	If quotient is not 0, then divide the quotient again by 8 and go to step 2.
Step 4.	If quotient is 0, then stop the process of division.
Step 5.	Write all remainder from left to right.

The combination of digit 0 to 7 thus obtained is the required octal equivalent of number.

For example: *Convert* $(8765)_{10}$ *to octal number system.*

Solution.

8		8765	remainders
8		1095	5
8		136	7
8		17	0
8		2	1
		0	2

Thus $(8765)_{10} = (21075)_8$

To convert the fraction: To convert a fractional decimal number is octal, use the method of repeated multiplication. The multiplier is 8.

ALGORITHM

Step 1.	Multiply the number by 8.
Step 2.	Note down the integer part and fractional part of the result separately.
Step 3.	If the fractional part of the result satisfies any two conditions, stop the process of multiplication. Conditions are: (i) fractional part is 0. (ii) fractional part achieved has already appeared before that position.
Step 4.	If the resultant fraction does not satisfy any of the above conditions, then go to step 9.

Write all carries from left to right. The combination of digit 0 to 7 thus obtained is the required result.

SOLVED EXAMPLES

EXAMPLE 1. *Convert* $(0.1015625)_{10}$ *to octal number system.*

SOLUTION. Multiply repeated by 8.

MSD	0	0.1015625×8
	0	0.8125000×8
	6	0.5000000×8
LSD	4	0.0000000×8

Thus $(0.1015625)_{10} = (0.064...)_8$.

EXAMPLE 2. *Convert* $(1093.21875)_{10}$ *to octal number system.*

SOLUTION. Converting both integral part and fractional part separately

8	1093	remainders		0	0.21875×8
8	136	5		1	0.75000×8
8	17	0		6	0.00000×8
8	2	1			
8	0	2			

$\Rightarrow \quad (1093)_{10} = (2105)_8 \qquad \Rightarrow \qquad (0.21875)_{10} = (0.16)_8$

Thus $(1093.21875)_{10} = (2105.16)_8$.

(vi) Decimal to hexadecimal

To convert an integer: For converting the number in decimal number system to the number in hexadecimal number system, use the method of repeated division.

ALGORITHM

Step 1.	Divide the number by 16.
Step 2.	Note down the quotient and remainder. Remainder will be from 0 to 9 or A to F.
Step 3.	If quotient is not 0, then divide the quotient by 16, and go the step 2.
Step 4.	If quotient is 0 or any digit or symbol less than 16 then stop the process of division.
Step 5.	Write all remainder from left to right. The combination obtained is the desired Hexadecimal number.

For example: *Convert* $(198275)_{10}$ *to hexadecimal equivalent.*
Solution.

16	198275	remainders
16	12392	3
16	774	8
16	48	6
16	3	0
	0	3

Thus $(198275)_{10} = (30683)_{16}$.

To convert a fraction: To convert a fraction decimal number in hexadecimal, use the method of repeated multiplication.

ALGORITHM

Step 1.	Multiply the fraction part by 16.
Step 2.	Note down the integer part (carry) and fractional part of the result separately.
Step 3.	If the fractional part is 0 or achieved has already appeared before that position, stop the process of multiplication.
Step 4.	If the resultant fraction, does not satisfy the condition of step 3, then go to step 1.

After this process we write first carry to last carry in the sequence. This sequence obtained is the required result.

For example: *Convert 0.6875875 to hexadecimal number system.*
Solution.

0	0.6875875×16
11	0.00110000×16
0	0.0176×16
0	0.2816×16
4	0.5056×16
8	0.896×16

Thus $(0.68756875)_{10} = (0.110049)_{16} = (B0049)_{16}$

(vii) Octal to decimal: For the conversion of octal number to decimal number, multiply the whole octal number with power of 8. These powers are positive for integral part of number and negative for fractional part of number.

For example: *Convert* $(1727)_8$ *to decimal equivalent.*
Solution. $(1727)^8 = 1 \times 8^3 + 7 \times 8^2 + 2 \times 8^1 + 7 \times 8^0$
$$= 512 + 448 + 16 + 7 = (983)_{10}$$

Example: *Convert* $(3027.105)_8$ *to decimal equivalent.*
Solution. $(3027.105)8 = 3 \times 8^3 + 0 \times 8^2 + 7 \times 8^1 + 2 \times 8^0 + 1 \times 8^{-1} + 0 \times 8^{-2} + 5 \times 8^{-3}$
$$= (1559.124765625)_{10}$$

(viii) Octal to Binary: The conversion octal to binary is very easy. Every digit of the number which is to be converted from octal to binary, is individually converted to the 3-bit binary equivalent. The combination of 0 and 1 is our desired result.

For example: *Convert* $(103.2)_8$ *to binary equivalent.*
Solution.

$$(103.2)_8 = \quad 1 \quad\quad 0 \quad\quad 3 \quad\quad 2$$
$$\quad\quad 001 \quad 000 \quad 011 \quad 010 \quad\quad \text{Binary equivalent}$$

Thus $(103.2)_8 = (001000011.010)_2$.

(ix) Octal to hexadecimal: For converting an octal number to hexadecimal number.

ALGORITHM

| **Step 1.** | Convert the octal number to binary equivalent. |
| **Step 2.** | Now convert this binary equivalent to hexadecimal number system. The number obtained is the required result. |

For example: *Convert* $(72232321)_8$ *to hexadecimal equivalent.*

Solution. Firstly we convert the given octal number to Binary equivalent.

$$(72232321)_8 = 7 \rightarrow 111$$
$$2 \rightarrow 010$$
$$2 \rightarrow 010$$
$$3 \rightarrow 011$$
$$2 \rightarrow 010$$
$$3 \rightarrow 011$$
$$2 \rightarrow 010$$
$$1 \rightarrow 001$$

Thus, $(72232321)_8 = (111010010011010011010001)_2$

Now we convert this number into hexadecimal equivalent. Grouping these into four bits each we get

1110	1001	0011	0100	1101	0001
14 or E	9	3	4	13 or D	1

Thus $(111010010011010011010001)_2 = (E934D1)_{16}$.

(x) Hexadecimal to Binary: For converting an hexadecimal number to binary equivalent, we individually convert to the 4-bit binary equivalent. Then the combination of 0 and 1 thus obtained the desired result.

For example: *Convert* $(A92)_{16}$ *to Binary equivalent.*

Solution. $(A92)_{16} = A \times 16^2 + 9 \times 16^1 + 2 \times 16^0$

$$= 10 \times 256 + 9 \times 16 + 2 \times 1 = 2560 + 144 + 2 = (2706)_{10}$$

Now

2	2706	Remainder
2	1353	0
2	676	1
2	338	0
2	169	0
2	84	1
2	42	0
2	21	0
2	10	1
2	5	0
2	2	1
2	1	0
	0	1

Thus $(2706)_{10} = (101010010010)_2$

Hence $(A92)_{16} = (101010010010)_2$

Alternate Method:

A	9	2	
10	9	2	
1010	1001	0010	Binary equivalent

$\Rightarrow \quad (A92)_{16} = (101010010010)_2$.

(xi) **Hexadecimal to Decimal:** For converting hexadecimal to decimal equivalent. We individually separate the number and multiply the whole number with power of 16. After this process add the total resultant numbers, which will be desired Decimal number.

SOLVED EXAMPLES

EXAMPLE 1. *Convert* $(5009B)_{16}$ *to Decimal equivalent.*

SOLUTION.
$$(5009B)_{16} = 5 \times 16^4 + 0 \times 16^3 + 0 \times 16^2 + 9 \times 16^1 + B \times 16^0$$
$$= (327680 + 0 + 0 + 144 + 11) = (327835)_{10}$$

Thus $(5009B)_{16} = (327835)_{10}$

EXAMPLE 2. *Convert* $(BCD)_{16}$ *to Decimal equivalent.*

SOLUTION.
$$(BCD)_{16} = B \times 16^2 + C \times 16^1 + D \times 16^0$$
$$= B \times 256 + C \times 16 + D$$
$$= 11 \times 256 + 12 \times 16 + 13$$
$$= 2816 + 192 + 13$$
$$= (3021)_{10}$$

(xii) **Hexadecimal to Octal:** For converting the hexadecimal number to octal number system, firstly convert the hexadecimal number to binary equivalent. After this process, convert this binary equivalent to octal number system. The number obtained is the direct result.

For example: *Convert* $(E934D1)_{16}$ *to hexadecimal number system.*

SOLUTION.

$(E934D1)_{16}$	=	E	9	3	4	D	1
		1110	1001	0011	0100	1101	0001

$\Rightarrow \quad (E934D1)_{16} = (111010010011010011010001)_2$

Now we convert this binary number to octal equivalent.

Grouping these into three bits each, we get

111	010	010	011	010	011	010	001
7	2	2	3	2	3	2	1

Therefore $(111010010011010011010001)_2 = (72232321)_8$

This implies $(E934D)_{16} = (72232321)_8$

A conversion table between decimal, hexadecimal octal and binary relation is given below:

Decimal O_{10}	Hexadecimal O_{16}	Octal O_8	Binary O_2
0	0	00	0000
1	1	01	0001
2	2	02	0010
3	3	03	0011
4	4	04	0100
5	5	05	0101
6	6	06	0110
7	7	07	0111
8	8	10	1000
9	9	11	1001
10	A	12	1010
11	B	13	1011
12	C	14	1100
13	D	15	1101
14	E	16	1110
15	F	17	1111

2.23 BINARY ARITHMETIC

Arithmetic operations additions, subtraction, multiplication and division on binary numbers constitute binary arithmetic.

(i) Binary Addition : The rules of binary addition are

$$0 + 0 = 0$$
$$0 + 1 = 1$$
$$1 + 0 = 1$$
$$1 + 1 = 10 \text{ Sum 0 with carry 1.}$$

Like in decimal system when the sum of two digits exceed the highest digit, 1 is carried to the next higher bit position in binary system when the sum exceeds 1 a one is carried to the next higher bit position.

SOLVED EXAMPLES

EXAMPLE 1. *Add the binary numbers* $(10110)_2$ *and* $(1101)_2$.

SOLUTION.

```
_ _ _ 1 _  ← carry
    10110
   +1101
   _____
   100011
```

EXAMPLE 2. *Add the binary numbers* $(11001)_2$ *and* $(10011)_2$.

SOLUTION.

```
    11001
   +10011
   _____
   101100
```

(ii) Binary subtraction: The rules for binary subtraction are

$$0 - 0 = 0$$
$$1 - 0 = 1$$
$$0 - 1 = 1$$
$$1 - 1 = 0$$

with borrow or 1 from the next column to the left.

If we need to borrow from a digit which is 0, then two or more borrows must be made toward the left. We borrow from the first non zero digit to the left and each. intervening 0 becomes 1 in the process.

EXAMPLE 3.   *Subtract* $(1111)_2$ *from* 1110101.

SOLUTION.
$$
\begin{array}{r}
1110101 \\
-1111 \ i.e., (1100110)_2 \\
\hline
1100110
\end{array}
$$

EXAMPLE 4.   *Subtract* $(101111)_2$ *from* $(110101)_2$.

SOLUTION.
$$
\begin{array}{r}
110101 \\
-101111 \ i.e., (110)_2 0. \\
\hline
000110
\end{array}
$$

EXERCISE 1.4

1. Convert the following numbers binary to decimal equivalent:
  (i) 110111   **(ii)** 0.101
  (iii) 11010111.1101

2. Convert the following number decimal to binary:
  (i) 5233   **(ii)** 0.8125
  (iii) 9342.982

3. Convert the following number into octal system:
  (i) $(9786)_{10}$ **(ii)** $(8765.27)_{10}$
  (iii) $(100000000)_2$ **(iv)** $(1110111011)_2$

4. Convert the following number into hexadecimal:
  (i) $(19)_{10}$ **(ii)** $(286)_{10}$
  (iii) $(100110101111)_2$
  (iv) $(360.13)_8$

5. Convert the following number into octal:
  (i) $(1011101)_2$ **(ii)** $(A985B)_{16}$
  (iii) $(5834E.B93)_{16}$

6. Fill in the blanks:
  (i) $(FA9B)_{16} = ($ _____ $)_{10}$

(ii) $(217)_{10} = ($ _____ $)_8$
(iii) $(1046.25)_{10} = ($ _____ $)_{16}$
(iv) $(A92)_{16} = ($ _____ $)_{10}$
(v) $(1100110)_2 = ($ _____ $)_{10}$
(vi) $(42.25)_{10} = ($ _____ $)_2$

7. What is the decimal equivalent to the hexadecimal number $(BCDE)_{16}$?

8. Find the sum of following binary numbers:
  (i) 1001, 101010
  (ii) 10110, 1101
  (iii) 110101, 101111
  (iv) 111011, 10111000
  (v) 1001011, 1101001

9. Find the difference of following binary numbers:
  (i) 1000 – 1   **(ii)** 11010 – 101
  (iii) 1110001 – 100110
  (iv) 11011 – 1101100 **(v)** 110.110 – 1.1011

10. Calculate the following:
  (i) $(100111)_2 - (111010)_2$
  (ii) $(111111)_2 + (10101)_2 + (11011)_2$

Answers

1. (i) $(55)_{10}$ (ii) $(0.625)_{10}$ (iii) $(215.15)_{10}$
2. (i) 1010001110001   (ii) 0.1101   (iii) 10011001110010.11111011
3. (i) $(23072)_8$ (ii) $(21075.212...)_8$ (iii) $(400)_8$ (iv) $(733)_8$
4. (i) $(13)_{16}$ (ii) $(AF9)_{16}$ (iii) $(9AF)_{16}$
5. (i) $(135)_8$ (ii) $(2514133)_8$ (iii) $(3701516.5623)_8$
6. (i) $(64155)_{10}$ (ii) $(330)_8$ (iii) $(416.4)_{16}$ (iv) $(2706)_{10}$
6. (v) $(102)_{10}$ (vi) $(101010.01)_2$ **7.** $(3021.875)10$
8. (i) 110011   (ii) 100011   (iii) 1100100   (iv) 11110011 (v) 10110100
9. (i) 111   (ii) 10101   (iii) 1001011   (iv) 1010001 (v) 101.0001
10. (i) 1101   (ii) 1011111

Objective Evaluations

FILL IN THE BLANKS

1. The numbers in which, there is no uncertainty and approximation is said to be _____ numbers.

2. The numbers which are not exact is called _____ numbers.

3. If x^A is the approximate value of exact number x^T, then the absolute error denoted by E_a is

defined by _____ .

4. If 1.414 is used as an approximation to $\sqrt{2}$. Then the absolute error is _____.

5. If $a = 10.00 \pm 0.05$, $b = 0.0356 \pm 0.0002$, $c = 15300 \pm 100$, then $a+b+c+d =$ _____.

6. The smallest root of equation ax^2-bx+c is _____ .

TRUE/FALSE

Write 'T' for True and 'F' False statement.

1. The approximate number is a number which can be expressed by a finite number of digits. **(T/F)**

2. The number rounded off to n significant figures are said to be correct to n significant figures. **(T/F)**

3. The percentage error in x^A, which is the approximate value of x^T is

$$E_p = 100 \times E_r = 100 \times \left| \frac{x^T - x^A}{x^T} \right|$$ **(T/F)**

4. The relative error is also known as normalized absolute error. **(T/F)**

5. Truncation error is a type of algorithm error .

6. The number of terms required in the series for $\log(1+x)$ to evaluate by $\log 1.2$ correct to six decimal places is 9. **(T/F)**

7. If $r = h(4h^5 - 5)$ then the percentage error in r at $h = 1$, if the error in h is 0.04 is 76. **(T/F)**

8. Every equation of odd degree has atleast one real root. **(T/F)**

MULTIPLE CHOICE QUESTIONS

Choose the most appropriate one.

1. The number in which, there is no uncertainty and no approximation, is said to be:
   (a) exact number
   (b) approximate number
   (c) both (a) and (b) are true
   (d) none of these

2. The error in x^A, which is the approximate value of x^T is

$$E_p = 100 \times E_r = 100 \times \left| \frac{x^T - x^A}{x^T} \right|$$ is called :

   (a) absolute error
   (b) relative error
   (c) percentage error
   (d) none of these

3. The error $E_r = \left| \dfrac{x^T - x^A}{x^T} \right|$ is known as :
   (a) absolute error
   (b) relative error
   (c) percentage error
   (d) none of these

4. The normalized absolute error is known as :
   (a) relative error   (b) relative error
   (b) percentage   (d) none of these

5. Inherent error is also known as :
   (a) input error
   (b) empirical error
   (c) representation error
   (d) none of these

6. Conversion error is also known as :

(a) input error

(b) empirical error

(c) representation error

(d) none of these

7. Addition of binary numbers $(10110)_2$ and $(1101)_2$ is :

(a) 110010

(b) 100011

(c) 101011

(d) 110011

Answers

FILL IN THE BLANKS

1. exact   **2.** approximation   **3.** $E_a = |x^T - x^A|$ **4.** 0.21356×10^{-3} **5.** 600.0502   **6.** 0.0025.

TRUE/ FALSE

1. F   **2.** T   **3.** T   **4.** T   **5.** T   **6.** F   **7.** T   **8.** T

MULTIPLE CHOICE QUESTIONS

1. (a)   **2.** (c)   **3.** (b)   **4.** (a)   **5.** (a)   **6.** (c)   **7.** (b)

COMPUTATIONAL TECHNIQUE LAB

1. ERRORS CALCULATION

SYMBOLS USED

TV = True Value
AV = Approximate Value
EA = Absolute Error
ER = Relative Error
EP = Percentage Error

ALGORITHM : CALCULATION OF ERRORS

Step 1. START
Step 2. Input TV, AV
Step 3. EA = |TV – AV|
Step 4. ER = EA/TA
Step 5. EP = ER * 100
Step 6. Print EA, ER, EP
Step 7. STOP

FLOW CHART : ERRORS CALCULATION

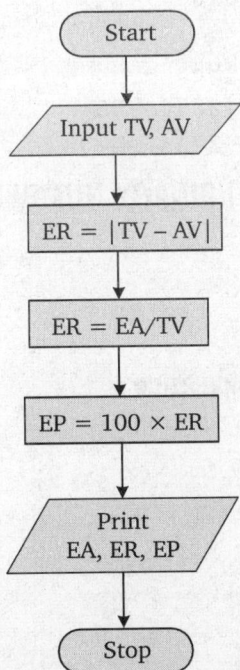

PROGRAM : *Write a program to calculate the errors.*

//error calculation

```
#include<stdio.h>
#include<conio.h>
#include<math.h>
void main()
{
    float tv,av,er,ep,ea;
    clrscr();
    printf("\n enter true value \n");
    scanf( "%f",&tv);
    printf("\n enter approximate value\n");
    scanf( "%f",&av);
    ea=fabs(tv-av);
    er=ea/tv;
    ep=100*er;
    printf("\n\n absolute error= %e ",ea);
    printf("\n\n relative error= %e ",er);
    printf("\n\n percentage error= %e ",ep);
    getch();
}
```

Output: ERROR CALCULATION

```
enter true value
37.46235
enter approximate value
37.46
absolute error= 2.349854e-03
relative error= 6.272574e-05
percentage error= 6.272574e-03
```

2. CONVERSION OF DECIMAL TO BINARY NUMBER

SYMBOLS USED

bn = binary number
dn = decimal number
r = remainder

ALGORITHM : DECIMAL TO BINARY CONVERSION

Step 1. START
Step 2. Input dn
Step 3. row = 4, col = 30
Step 4. Perform steps 5 to 8 while (dn > 0)
Step 5. r = dn%2
Step 6. dn = dn/2
Step 7. Print r 'at' col, row
Step 8. col = col – 1
Step 9. STOP

Flow Chart : Decimal to Binary Conversion

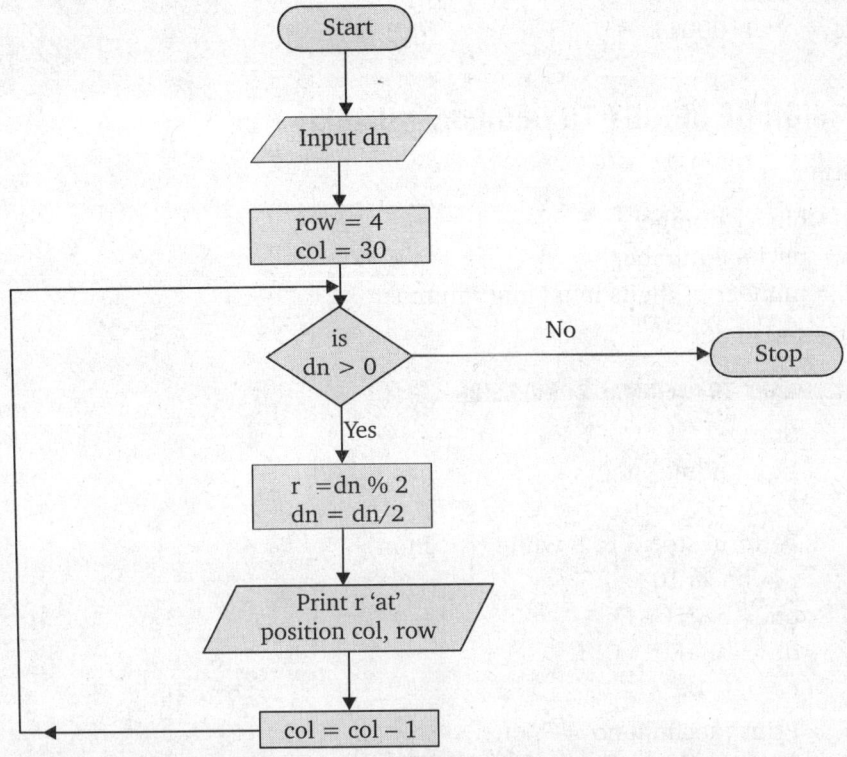

Program. *Write a C program to convert a given decimal number into binary number.*

```c
// Decimal to binary conversion
#include<stdio.h>
#include<conio.h>
void main()
 {
   int dn,r,row=4,col=30;
   clrscr();
  printf("\n enter  decimal number   ");
  scanf("%d",&dn);
  printf("\n binary no. is = ");
  while(dn>0)
      {
      r=dn%2;
      dn=dn/2;
      gotoxy(col,row);
      printf("%d",r);
      col- -;
      }
      getch();
  }
```

Output: DECIMAL TO BINARY CONVERSION

enter decimal number 99

binary no. is =   1100011

3. CONVERSION OF BINARY TO DECIMAL NUMBER

SYMBOLS USED

bn = binary number

dn = decimal number

num = number of digits in a binary number

r = remainder

ALGORITHM : BINARY TO DECIMAL CONVERSION

Step 1.	Start
Step 2.	Input num, bn
Step 3.	i = 0
Step 4.	Perform step 5 to 8 while (i < num)
Step 5.	r = bn % 10
Step 6.	bn = bn/10
Step 7.	dn = dn + r * (2^i)
Step 8.	i = i + 1
Step 9.	Print "decimal no = ", dn
Step 10.	Stop

FLOW CHART : BINARY TO DECIMAL number

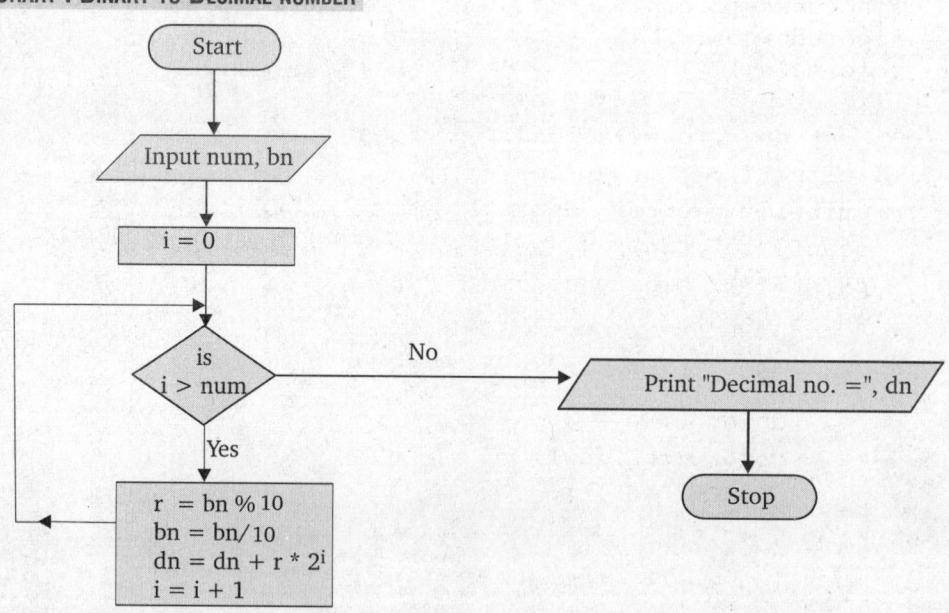

PROGRAM. *Write a C-program to convert a given binary number to decimal number.*

```c
// Binary to decimal conversion
#include<stdio.h>
#include<conio.h>
#include<math.h>
void main()
 {
    int r,i,num;
    long int bn,dn=0;
    clrscr();
    printf("\n enter  number of digits\n ");
    scanf("%d",&num);
    printf("\n enter binary number    ");
    scanf("%ld",&bn);
    printf("\n decimal no. is = ");
    for(i=0;i<num;i++)
       {
          r=bn%10;
          bn=bn/10;
          dn=dn +r*pow(2,i);
       }
   printf("%ld",dn);
   getch();
   }
```

Output : BINARY TO DECIMAL CONVERSION
enter number of digits
7
enter binary number   1100011
decimal no. is = 99

3. CONVERSION OF DECIMAL NUMBER TO OCTAL NUMBER

SYMBOLS USED

dn = decimal number
on = octal number

ALGORITHM : CONVERT DECIMAL NUMBER INTO EQUIVALENT OCTAL NUMBER

Step 1.	Start
Step 2.	Input dn
Step 3.	on = 0, i = 1

Step 4.   Perform steps 5 to 8 while (dn > 0)

Step 5.   r = dn % 8

Step 6.   dn = dn/8

Step 7.   on = on + r* i

Step 8.   i = i* 10

Step 9.   Print "Equivalent octal number" on,

Step 10.   Stop

FLOW CHART : CONVERSION OF DECIMAL NUMBER TO OCTAL NUMBER

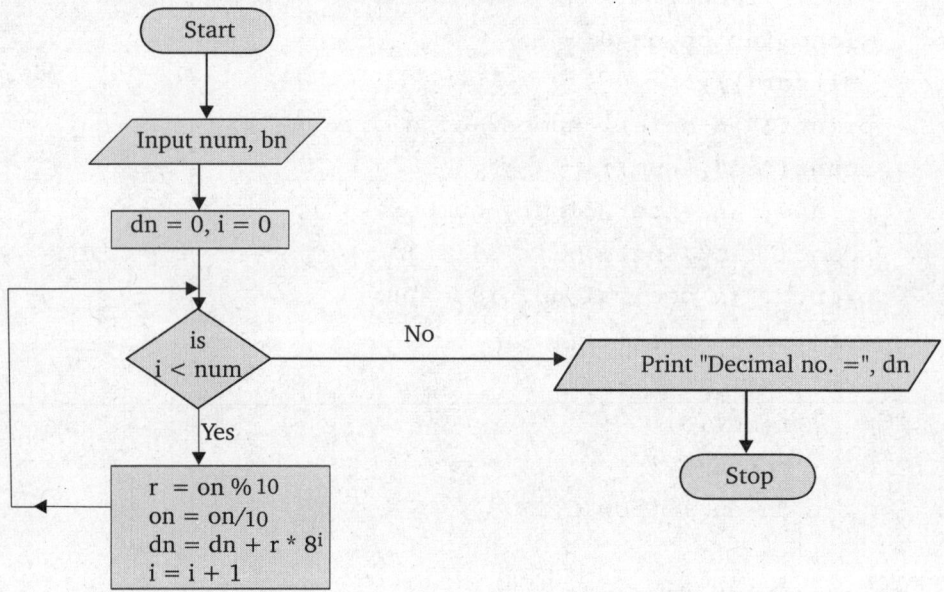

PROGRAM : *Write a C program to convert a decimal number into octal number.*

```
// Decimal  to octal conversion
#include<stdio.h>
#include<conio.h>
void main()
{
int dn,r,on=0,i=1;
clrscr();
printf("\n enter decimal number   ");
scanf("%d",&dn);
printf("\n octal no. is = ");
while(dn>0)
{
        r=dn%8;
        dn=dn/8;
        on=on+r*i;
        i=i*10;
}
printf("%d",on);
```

```
getch();
}
```

Output : DECIMAL TO OCTAL CONVERSION

enter decimal number 9876

Octal no. is = 23072

4. CONVERSION OF GIVEN OCTAL NUMBER INTO AN EQUIVALENT DECIMAL NUMBER

SYMBOLS USED

dn = decimal number

on = octal number

num = number of digits in Octal number

ALGORITHM : CONVERT OCTAL NUMBER INTO DECIMAL NUMBER

Step 1. START

Step 2. Input "Enter number of digits", num

Step 3. Input "Enter Octal number", on

Step 4. i = 0; dn = 0

Step 5. Perform steps 6 to 8 while (i < num)

Step 6. r = on % 10

Step 7. on = on/10

Step 8. dn = dn + r * 8^i

Step 9. Print "decimal number = ", dn

Step 10. STOP

FLOW CHART : CONVERSION OF OCTAL NUMBER INTO DECIMAL NUMBER

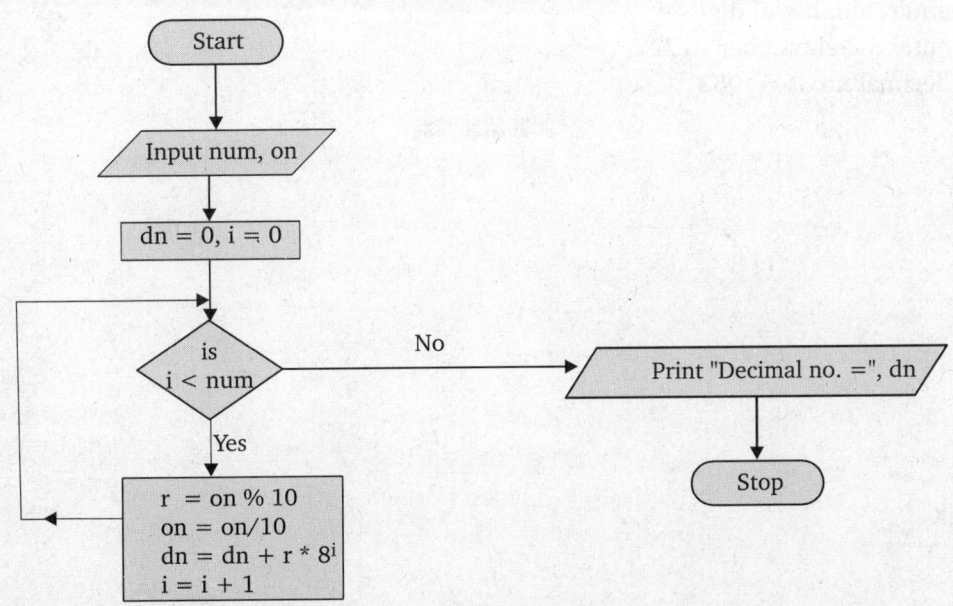

PROGRAM : *Write a C program to convert a given octal number into a decimal number.*

// Octal to decimal conversion

```c
#include<stdio.h>
#include<conio.h>
#include<math.h>
void main()
 {
int r,i,num;
int on,dn=0;
clrscr();
printf("\n enter  number of digits\n ");
scanf("%d",&num);
printf("\n enter octal number   ");
scanf("%d",&on);
printf("\n decimal no. is = ");
for(i=0;i<num;i++)
{
        r=on%10;
        on=on/10;
        dn=dn+r*pow(8,i);
}
printf("%d",dn);
getch();
 }
```

Output : OCTAL TO DECIMAL CONVERSION

enter number of digits 4

enter octal number   1727

decimal no. is = 983

⌘⌘⌘⌘⌘⌘

Numerical Solution of Algebraic and Transcendental Equations

CHAPTER 3

3.1 INTRODUCTION

The solution of the equation of the form $f(x) = 0$ is used in science and engineering. In this chapter we shall discuss various useful methods for evaluating the root of any equation having numerical coefficients. To find the roots of an equation $f(x) = 0$ we start with a known approximate solution and apply any method discussed in this chapter.

Definition 1. *An expression of the form* $f(x) = a_0x^n + a_1x^{n-1} + ... + a_{n-1}.x + c_n$

where all a_i's are constants, provided $a_n \neq 0$ and n is a positive integer called a polynomial in x of degree n.

Definition 2. *An equation of the form $y = f(x)$ is said to be algebraic if it can be expressed in the form* $f_ny_n + f_{n-1}y_{n-1} + ... + f_1y_1 + f_0 = 0$ *where f_i is an ith order polynomial in x.*

Definition 3. *A non-algebraic equation is called a transcendental equation i.e., $f(x)$ is an expression involving some other functions such as trigonometric, logarithmic, exponential etc.*

For example :

   (i) $4\sin x - e^x = 0$ (ii) $\log x^3 - 5\tan x = 0$

> ▼ REMARK
> ▶ A transcendental equation may have a finite or an infinite number of real roots or may not have real root at all.

Definition 4. *The process of finding the roots of an equation is called the solution of that equation, and the value of x for which $f(x) = 0$ is satisfied is called its root.*

Geometrically, we can say that a root of an equation $f(x) = 0$ is that value of x, where the graph of $y = f(x)$ cuts the x-axis.

3.2 PROPERTIES OF THE EQUATIONS AND ITS ROOTS

(1) If $f(x)$ is exactly divisible by $(x - a)$ then a is the root of $f(x) = 0$.

(2) Every algebraic equation of degree n has only n roots real as well as imaginary.

(3) If $f(x)$ is continuous in the interval (a, b) and $f(a)$, $f(b)$ have opposite signs, then the equation $f(x) = 0$ has atleast one root between $x = a$ and $x = b$.

(4) The complex roots of an equation always occur in pairs.

(5) Every equation of odd degree has atleast one real root.

(6) If a polynomial of degree n vanishes for more than n values of x, it must be identically zero.

(7) If an equation of nth degree has at the most p positive roots and at the most of negative roots, then it follows that the equation has atleast $\{n - (p + q)\}$ imaginary roots.

(8) In any algebraic equation $f(x) = 0$, the number of positive roots cannot exceed the number of changes of sign from positive to negative and from negative to positive, the number of negative roots cannot exceed the number of changes of sign in $f(-x)$. **(Descarte's Rule of signs).**

3.3 METHODS OF SOLUTION

The methods of finding the solution of an equation includes:

    (i) Direct Methods   (ii) Graphical Methods

    (iii) Trial and Error Methods   (iv) Iterative Methods.

3.3.1 DIRECT METHOD

In some cases, roots can be found by using direct analytical methods. For example, for a quadratic equation $ax^2 + bx + c = 0$, the roots of the equation, obtained by

$$x = \frac{-b \pm \sqrt{b^2 - 4ac}}{2a}$$

3.3.2 GRAPHICAL METHOD

This method involves plotting the given function and finding the points where it crosses the x-axis. These points represent approximate values of the roots of the function.

3.3.3 THE TRIAL AND ERROR METHODS

Involves a series of guesses for x, each time evaluating the function to see whether it is close to zero. The value of x that causes the function value closer to zero is one of the approximate roots of an equation.

3.3.4 AN ITERATIVE TECHNIQUE

Begins with an approximate value of the root (which is known as initial guess), which is then successively improved by iteration. The process of iteration stops when the desired level of accuracy is obtained.

In this chapter we shall discuss few iterative methods such as Bisection, Regula-Falsi, Newton-Raphoson, Muller's methods, etc. that are commonly used.

3.4 BISECTION METHOD

Bisection method of solving the equation $f(x) = 0$ is based on the following theorem:

Statement. *Let $f(x)$ be a function, which is continuous in the closed interval $[a, b]$ and $f(a)$, $f(b)$ are of opposite signs, then there exist at least one value c, (say) of x where $c \in]a, b[$ satisfy $f(c) = 0$.*

Proof. Let $f(a)$ be negative and $f(b)$ be positive in the interval $]a, b[$ then, at least one root of the equation $f(x) = 0$ lies in the interval $]a, b[$. Let the rounded

value of this root be $c = \dfrac{1}{2}(a + b)$, which is obtained by

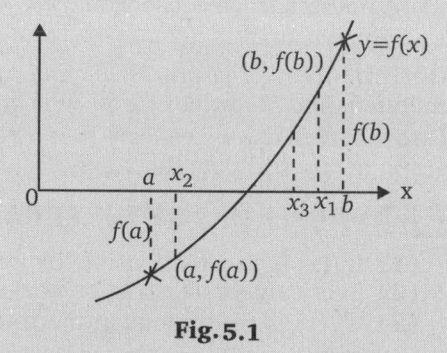

Fig. 5.1

dividing the distance between the points a and b into two equal parts. If $f(c) = 0$, then c is the required root of the equation $f(x) = 0$, otherwise root will lie in the interval $]a, c[$ or $]c, b[$ which will depend on whether the value of $f(c)$ is positive or negative.

Let $f(c)$ be positive then as before, we divide the interval $]a, c[$ into two equal parts and let $d = \dfrac{1}{2}(a + c)$. The iterative cycle is terminated when the search interval becomes smaller than the prescribed tolerance or the value of the function nearly vanishes at the new x-value.

ALGORITHM

Step 1.	Take initial approximation for x_1 and x_2.
Step 2.	Compute $f_1 = f(x_1)$ and $f_2 = f(x_2)$
Step 3.	If $f_1.f_2 > 0$, x_1 and x_2 do not bracket any root
Step 4.	Compute $x_0 = \dfrac{x_1 + x_2}{2}$; $f_0 = f(x_0)$
Step 5.	If $f_1.f_0 < 0$, then set $x_2 = x_0$ else set $x_1 = x_0, f_1 = f_0$
Step 6.	Root is given by $(x_1 + x_2)/2$

▶ **REMARKS**

▸ If ε is the prescribed tolerance in the required root then the iterative cycle terminates when the absolute error becomes less than or equal to ε, *i.e.*, $|x_1 - x_0| < \varepsilon$.

▸ For prescribed tolerance ε, the approximate number of iterations required in bisection method, can be determined by the relation.

$$\frac{x_1 - x_0}{2^n} < \varepsilon.$$

3.4.1 PROPERTIES OF BISECTION METHOD

(1) If $f(x)$ is not continuous in the closed interval $[a, b]$ then bisection method is failed.

(2) If $f(x)$ is continuous in a closed interval and does not cut the x-axis, then $f(x)$ does not have a real root.

(3) Bisection method gives the real root of $f(x) = 0$.

(4) Bisection method is also named as Bolzano method or interval halving method.

(5) Bisection method is always converges.

3.4.2 ORDER OF CONVERGENCE OF BISECTION METHOD

In Bisection method, at each iteration, the interval in which the root lies is divided into half interval. If we consider the middle points of successive intervals to be the approximation to the root, one half of the current interval (the interval in which the root lies) is the upper bound to the error. Therefore, in bisection method

$$e_{n+1} = \frac{1}{2} e_n \qquad \text{so} \qquad e_{n+1} \propto e_n.$$

Hence, the bisection method is linearly convergent.

THEOREM 1. *Bisection method is always convergent.*

PROOF. Let $[p_n, q_n]$ be the interval at n^{th} iteration of Bisection method in which a root of the equation $f(x) = 0$ lies. Also, let x_n be the nth approximation to the root.

Initially, assume that $p_1 = a, q_1 = b$

Then $\qquad\qquad x_1 = \text{First approximation} = \dfrac{p_1 + q_1}{2}$

$\Rightarrow \qquad\qquad p_1 < x_1 < q_1$

Now, suppose that either the root lies in $[p_1, x_1]$ or $[x_1, q_1]$

Therefore, either $\qquad [p_2, q_2] = [p_1, x_1] \qquad$ or $\qquad [p_2, q_2] = [x_1, q_1]$

i.e., either $\qquad p_2 = p_1, q_2 = x_1$ or $\qquad\qquad p_2 = x_1, q_2 = q_1$

$\Rightarrow \qquad\qquad p_1 \leq p_2, q_2 \leq q_1$

Further let x_2 be the second approximation

$$\therefore \qquad x_2 = \frac{p_2 + q_2}{2} \qquad \Rightarrow \qquad p_2 < x_2 < q_2$$

Continuing in this way, we get

$$x_n = \frac{p_n + q_n}{2} \qquad \Rightarrow \qquad p_n < x_n < q_n$$

Therefore, here we get following two sequences

$$p_1 \leq p_2 \leq p_3 \leq \ldots \leq p_n \qquad \text{and} \qquad q_1 \geq q_2 \geq q_3 \geq \ldots \geq q_n$$

The sequence $\langle p_n \rangle$ is non-decreasing bounded by b and the sequences $\langle q_n \rangle$ is non-increasing bounded by a. Thus, both these sequence converge.

Let $\lim_{n \to \infty} p_n = p$ and $\lim_{n \to \infty} q_n = q$

We observe that length of interval decreases of each iteration, therefore

$$\lim_{n \to \infty} (q_n - p_n) = 0 \qquad \Rightarrow \qquad \lim_{n \to \infty} q_n - \lim_{n \to \infty} p_n = 0$$

($\because$ Limit of the difference of two functions is equal to the differences of their limits.)

$$\Rightarrow \qquad q - p = 0 \qquad \Rightarrow \qquad q = p$$

But

$$p_n < x_n < q_n \qquad \Rightarrow \qquad \lim_{n \to \infty} p_n < \lim_{n \to \infty} x_n < \lim_{n \to \infty} q_n$$

$$\Rightarrow \qquad p_n < \lim_{n \to \infty} x_n < q_n \qquad \Rightarrow \qquad \lim_{n \to \infty} x_n = p = q \ (\because p = q) \ldots(1)$$

Since a root lies in (p_n, q_n), therefore, we have

$$f(p_n)f(q_n) < 0 \qquad \Rightarrow \qquad \lim_{n \to \infty} \{f(p_n) + f(q_n)\} \leq 0$$

$$\Rightarrow \qquad f(p).f(q) \leq 0 \qquad \Rightarrow \qquad (f(p))^2 \leq 0$$

$$\Rightarrow \qquad f(p) = 0$$

$$(\because p = q)$$

($\because$ Square of any real number cannot be negative)

$$\Rightarrow \qquad p \text{ is a root of } f(x) = 0 \qquad \qquad \ldots(2)$$

Hence, from (1) and (2) we conclude that the sequence $\langle x_n \rangle$ converges necessarily to a root of the equation $f(x) = 0$.

SOLVED EXAMPLES

EXAMPLE 1. *Find the root of the equation $x^3 - x - 1 = 0$ lying between 1 and 2 by bisection method.*

(UPTU-2004)

SOLUTION. Let $f(x) = x^3 - x - 1 = 0$

since $f(1) = 1^3 - 1 - 1 = -1$, which is negative

and $f(2) = 2^3 - 2 - 1 = 5$, which is positive

Therefore, $f(1)$ is negative and $f(2)$ is positive, so at least one real root will lie between 1 and 2.

First approximation: Now using Bisection method, we can take first approximation

$$x_1 = \frac{1+2}{2} = \frac{3}{2} = 1.5$$

Then, $f(1.5) = (1.5)^3 - 1.5 - 1 = 3.375 - 1.5 - 1 = 0.875$

$\therefore \qquad f(1.5) > 0$, that is, positive.

Second approximation

$$x_2 = \frac{1+1.5}{2} = 1.25$$

Then $\qquad f(1.25) = (1.25)^3 - 1.25 - 1 = 1.956 - 2.25 = -0.297 < 0$

$\therefore f(1.25)$ is negative.

Therefore, $f(1.5)$ is positive and $f(1.25)$ is negative, so that root will lie between 1.25 and 1.5.

Third approximation: The third approximation is given by

$$x_3 = \frac{1.25+1.5}{2} = 1.375$$

Now, $\qquad f(1.375) = (1.375)^3 - 1.375 - 1 = 0.2246$

$\therefore f(1.375)$ is positive

$\therefore$ The required root lies between 1.25 and 1.375.

Fourth approximation: The fourth approximation is given by

$$x_4 = \frac{1.25+1.375}{2} = 1.313$$

Now, $\qquad f(1.313) = (1.313)^3 - 1.313 - 1 = -0.0494$

Therefore, $f(1.313)$ is negative and $f(1.375)$ is positive. Thus, root lies between 1.313 and 1.375.

Fifth approximation: The fifth approximation is given by

$$x_5 = \frac{1.313+1.375}{2} = 1.344$$

$\therefore \qquad f(1.344) = (1.344)^3 - 1.344 - 1 = 0.0837$

$\therefore \qquad f(1.344) > 0$

$\therefore f(1.313)$ is negative and $f(1.344)$ is positive, so root lies between 1.313 and 1.344.

Sixth approximation: The sixth approximation is given by

$$x_6 = \frac{1.313+1.344}{2} = 1.329$$

$\therefore \qquad f(1.329) = (1.329)^3 - 1.329 - 1 = 0.0183$

$\therefore \qquad f(1.329) > 0$

$\therefore \quad f(1.313)$ is negative and $f(1.329)$ is positive, so that the required root lies between 1.313 and 1.329.

Seventh approximation: The seventh approximation is given by

$$x_7 = \frac{1.313+1.329}{2} = 1.321$$

$\therefore \qquad f(1.321) = (1.321)^3 - 1.321 - 1 = -0.0158$

$\therefore f(1.321)$ is negative and $f(1.329)$ is positive, the required root lies between 1.321 and 1.329.

Eighth approximation: The eighth approximation is given by

$$x_8 = \frac{1.321+1.329}{2} = 1.325$$

From above iterations, the root of $f(x) = x^3 - x - 1 = 0$ upto two places of decimals is 1.32, which is of desired accuracy.

EXAMPLE 2.   *Find a root of the equation $x^3 - x - 4 = 0$ between 1 and 2 to three places of decimal by bisection method.*

SOLUTION.   Given $f(x) = x^3 - x - 4 = 0$

We want to find the root lie between 1 and 2.

At $x_0 = 1$

$\Rightarrow$ $f(x_0) = (1)^3 - 1 - 4 = -4$ (Negative)

At $x_1 = 2$

$\Rightarrow$ $f(x_1) = (2)^3 - 2 - 4 = 2$ (Positive)

which implies that root lies between 1 and 2.

First approximation : Here,

$$x_0 = 1, x_1 = 2, x_2 = \frac{x_0 + x_1}{2} = \frac{1+2}{2} = 1.5$$

Now,

$$f(x_0) = -4, f(x_1) = 2,$$

and

$$f(x_2) = (1.5)^3 - 1.5 - 4 = -2.125$$

Since $f(1.5)$ is negative and $f(2)$ is positive.

Thus the root lies in the interval $]1.5, 2[$.

Second approximation: Here,

$$x_0 = 1.5, x_1 = 2, x_2 = \frac{x_0 + x_1}{2} = \frac{1.5+2}{2} = 1.75$$

Also,

$$f(x_0) = -2.125, f(x_1) = 2,$$
$$f(x_2) = (1.75)^3 - 1.75 - 4 = -0.39062$$

Since $f(1.75)$ is negative and $f(x)$ is positive, therefore the root lies between 1.75 and 2.

Third approximation: Here,

$$x_0 = 1.75, x_1 = 2, x_2 = \frac{1.75+2}{2} = 1.875$$

$$f(x_0) = -0.39062, f(x_1) = 2,$$

$$f(x_2) = (1.875)^3 - 1.875 - 4 = 0.71679$$

Since $f(1.75)$ is negative and $f(1.875)$ is positive, therefore root lies between 1.75 and 1.875.

Fourth approximation: Here,

$$x_0 = 1.75, x_1 = 1.875,$$

so

$$x_2 = \frac{1.75+1.875}{2} = 1.8125$$

Also

$$f(x_0) = -0.39062, f(x_1) = 0.71679,$$

and

$$f(x_2) = (1.8125)^3 - 1.8125 - 4 = 0.14184$$

Since $f(1.75)$ is negative and $f(1.8125)$ is positive therefore root lies between 1.75 and 1.8125.

Fifth approximation: Here,

$$x_0 = 1.75, x_1 = 1.8175,$$

$$x_2 = \frac{1.75+1.812}{2} = 1.78125$$

Also
$$f(x_0) = -0.39062, f(x_1) = 0.14184,$$
$$f(x_2) = (1.78125)^2 - 1.78125 - 4 = -0.12960$$

Since $f(1.78125)$ is negative and $f(1.8125)$ is positive, therefore root lies between 1.78125 and 1.8125.

Repeating the process, the successive approximation are

$x_6 = 1.79687$	$x_7 = 1.78906$	$x_8 = 1.79296$	$x_9 = 1.79491$
$x_{10} = 1.79589$	$x_{11} = 1.79638$	$x_{12} = 1.79613$	

From the above discussion, the value of the root to three decimal places is 1.796.

EXAMPLE 3. *Find a root of the equation $f(x) = x^3 - 4x - 9 = 0$, using the bisection method in four stages.* (Mumbai–2003, JNTU–2009)

SOLUTION. Given $f(x) = x^3 - 4x - 9 = 0$ ...(1)

Then $f(2) = (2)^3 - 4(2) - 9 = -9$ and $f(3) = (3)^3 - 4(3) - 9 = 6$

Therefore, $f(2)$ is negative and $f(3)$ is positive, a root lies between 2 and 3.

First approximation: First approximation to the root is given by
$$x_1 = \frac{2+3}{2} = 2.5$$

Thus $f(2.5) = (2.5)^3 - 4(2.5) - 9$ [Using (1)]
$$= 15.625 - 19 = -3.375$$

Therefore, the root lies between 2.5 and 3.

Second approximation: The second approximation to the root is given by
$$x_2 = \frac{2.5+3}{2} = 2.75$$

Then $f(2.75) = (2.75)^3 - 4(2.75) - 9 = 20.797 - 20 = 0.797$

∴ $f(2.75)$ is positive and $f(2.5)$ is negative. Thus root lies between 2.5 and 2.75.

Third approximation: The third approximation to the root is given by
$$x_3 = \frac{2.5+2.75}{2} = 2.625$$

Now, $f(2.625) = (2.625)^3 - 4(2.625) - 9 = 18.088 - 19.5 = -1.412$

∴ $f(2.625)$ is negative while $f(2.75)$ is positive. Then the root lies between 2.625 to 2.75

Fourth approximation: The fourth approximation to the root is given by
$$x_4 = \frac{2.625+2.75}{2} = 2.6875$$

Hence, after the four steps the root is 2.6875 approximately.

EXAMPLE 4. *Using bisection method determine a real root of the equation $f(x) = 8x^3 - 2x - 1 = 0$.*

SOLUTION. It is given that
$$f(x) = 8x^3 - 2x - 1 \qquad \text{...(1)}$$
Then $f(0) = 8(0)^3 - 2(0) - 1 = -1$
and $f(1) = 8(1)^3 - 2(1) - 1 = 5$

Therefore, $f(0)$ is negative and $f(1)$ is positive so that a root lies between 0 and 1.

First approximation: First approximation to the root is given by

$$x_1 = \frac{0+1}{2} = 0.5$$

$$f(0.5) = 8(0.5)^3 - 2(0.5) - 1 \hspace{2cm} \text{[Using (1)]}$$
$$= 1 - 1 - 1 = -1, \text{ which is negative}$$

Thus $f(0.5)$ is negative and $f(1)$ is positive. Then root lies between 0.5 and 1.

Second approximation: The second approximation to the root is given by

$$x_2 = \frac{0.5+1}{2} = 0.75$$

Now $\hspace{1.5cm} f(0.75) = 8(0.75)^3 - 2(0.75) - 1$
$$= 3.375 - 2.5 = 0.875, \text{ which is positive}$$

Since $f(0.5)$ is negative, while $f(0.75)$ is obtained positive. Therefore, the root lies between 0.5 and 0.7.

Third approximation: The third approximation value of the root is given by

$$x_3 = \frac{0.5+0.75}{2} = 0.625$$

Now $\hspace{1cm} f(0.625) = 8(0.625)^3 - 2(0.625) - 1 = 1.953 - 2.25 = -0.297$

Therefore, $f(0.75)$ is positive, while $f(0.625)$ is obtained negative. Thus the root lies between 0.625 and 0.75.

Fourth approximation: The fourth approximation to the root is

$$x_4 = \frac{0.625+0.75}{2} = 0.688$$

Next, $\hspace{1cm} f(0.688) = 8(0.688)^3 - 2(0.688) - 1 = 2.605 - 2.376 = 0.229$

Thus $f(0.688)$ is obtained positive and $f(0.625)$ is negative.

Then the root lies between 0.625 and 0.688.

Fifth approximation: The fifth approximation to root is given by

$$x_5 = \frac{0.625+0.688}{2} = 0.657$$

Now, $\hspace{1cm} f(0.657) = 8(0.657)^3 - 2(0.657) - 1$
$$= 2.269 - 2.314 = -0.045$$

Therefore $f(0.657)$ is negative and $f(0.688)$ is positive so that root lies between 0.657 and 0.688.

Sixth approximation: The sixth approximation to the root is given by

$$x_6 = \frac{0.657+0.688}{2} = 0.673$$

Now, $\hspace{1cm} f(0.673) = 8(0.673)^3 - 2(0.673) - 1$
$$= 2.439 - 2.346 = 1.093$$

Therefore, $f(0.673)$ is positive and $f(0.657)$ is negative so the root lies between 0.657 and 0.673.

Seventh approximation: The seventh approximation to the root is given by

$$x_7 = \frac{0.657+0.673}{2} = 0.665$$

Now, $\hspace{1cm} f(0.665) = 8(0.665)^3 - 2(0.665) - 1$
$$= 2.353 - 2.33 = 0.023$$

Therefore, $f(0.665)$ is positive and $f(0.657)$ is negative so the root lies between 0.657 and 0.673.

Eighth approximation: The eighth approximation to the root is given by

$$x_8 = \frac{0.657 + 0.665}{2} = 0.661$$

From last two approximations, i.e., $x_7 = 0.665$ and $x_8 = 0.661$, it is observed that the approximate value of the root of $f(x) = 0$ upto two decimal place is 0.66.

Computational table

Iteration No.	x_1	x_2	$x_2 = \dfrac{x_0 + x_1}{2}$	$f(x_0)$	$f(x_1)$	$f(x_2)$
1	0	1	0.5	−1	5	−1
2	0.5	1	0.75	−1	5	0.875
3	0.5	0.75	0.625	−1	0.875	−0.297
4	0.625	0.75	0.688	−0.297	0.85	0.229
5	0.625	0.688	0.657	−0.297	0.229	−0.045
6	0.657	0.688	0.673	−0.045	0.229	1.093
7	0.657	0.673	0.665	−0.045	1.093	0.023

From the above table, approximate value of the root to two decimal places is 0.66.

EXAMPLE 5. *Using bisection method, find a real root of the equation $f(x) = 3x - \sqrt{1 + \sin x} = 0$.*

SOLUTION. The given equation $f(x) = 3x - \sqrt{1 + \sin x} = 0$ is a transcendental equation.

Given

$$f(x) = 3x - \sqrt{1 + \sin x} = 0$$

Then

$$f(0) = 3(1) - \sqrt{1 + \sin 1} = 3 - \sqrt{1.8414}$$

$$= 3 - 1.3570 = 1.643 > 0$$

Thus $f(0)$ is negative and $f(1)$ is positive, therefore, a root lies between 0 and 1.

First approximation: The first approximation to the root is given by

$$x_1 = \frac{0 + 1}{2} = 0.5$$

Now,

$$f(0.5) = 3(0.5) - \sqrt{1 + \sin(0.5)} \qquad \text{[Using (1)]}$$

$$= 1.5 - \sqrt{1.4794} = 1.5 - 1.2163$$

$$= 0.2837 > 0$$

Thus, $f(0.5)$ is positive, while $f(0)$ is negative, therefore a root lies between 0 and 0.5.

Second approximation: The second approximation to the root is given by

$$x_2 = \frac{0 + 0.5}{2} = 0.25$$

Again

$$f(0.25) = 3(0.25) - \sqrt{1 + \sin(0.5)}$$

$$= 0.75 - \sqrt{1.2474} = 0.75 - 1.1169 = -0.3669 < 0$$

Thus, $f(0.25)$ is obtained to the negative and $f(0.5)$ is positive, therefore a root lies between 0.25 and 0.5.

Third approximation: The third approximation to the root is given by

$$x_3 = \frac{0.25 + 0.5}{2} = 0.375$$

Now, $f(0.375) = 3(0.375) - \sqrt{1 + \sin(0.375)}$

$$= 1.125 - \sqrt{1 - 3663} = 1.125 - 1.1689 = 0.0439 < 0$$

Thus, $f(0.375)$ is negative and $f(0.5)$ is positive, therefore, a root lies between 0.375 and 0.5.

Fourth approximation: The fourth approximation to the root is

$$x_4 = \frac{0.375 + 0.5}{2} = 0.4375$$

Again, $f(0.4375) = 3(0.4375) - \sqrt{1 + \sin(0.4375)}$

$$= 1.3125 - \sqrt{1.4237} = 1.3125 - 1.932 = 0.1193 > 0$$

Thus, it is observed that $f(0.4375)$ is positive and $f(0.375)$ is negative therefore, the root lies between 0.375 and 0.4375.

Fifth approximation: The fifth approximation to the root is given by

$$x_5 = \frac{0.375 + 0.4375}{2} = 0.4063$$

Again, $f(0.4063) = 3(0.4063) - \sqrt{1 + \sin(0.4063)}$

$$= 1.2189 - \sqrt{1.3952} = 1.2189 - 1.1812 = 0.0377 > 0$$

Thus, $f(0.4063)$ is positive and $f(0.375)$ is negative, therefore, the root lies between 0.375 and 0.4063.

Sixth approximation: The sixth approximation to the root is given by

$$x_6 = \frac{0.375 + 0.4063}{2} = 0.3907$$

Again, $f(0.3907) = 3(0.3907) - \sqrt{1 + \sin(0.3907)}$

$$= 1.1721 - \sqrt{1.3808}$$

$$= 1.1721 - 1.1751 = -0.003 < 0$$

Thus, $f(0.3907)$ is negative and $f(0.4063)$ is positive, therefore the root lies between 0.3907 and 0.4063.

Seventh approximation: The seventh approximation to the root is given by

$$x_7 = \frac{0.3907 + 0.4063}{2} = 0.3985$$

From the last two observations, that is, $x_6 = 0.3907$ and $x_7 = 0.3985$, the approximate value of the root upto two places of decimal is given by 0.39. Hence, the root is 0.39 approximately.

EXAMPLE 6. *Find a real root of the equation $x \log_{10} x = 1.2$ by bisection method.*

SOLUTION. Let $f(x) = x \log_{10} x - 1.2 = 0$

so that $f(1) = 1 \log_{10} 1 - 1.2 = -1.2 < 0$

and $f(2) = 2 \log_{10} 2 - 1.2$ [Using (1)]

$$= 0.602 - 1.2 = -0.598 < 0$$

$$f(3) = 3\log_{10} 3 - 1.2$$
$$= 3(0.4771) - 1.2 = 1.4313 - 1.2 = 0.2313 > 0$$

Thus $f(2)$ is negative and $f(3)$ is positive, therefore, the root will lie between 2 and 3.

First approximation: The first approximation to the root is

$$x_1 = \frac{2+3}{2} = 2.5$$

Again, $\qquad f(2.5) = 2.5\log_{10} 2.5 - 1.2 = 2.5(0.3979) - 1.2$
$$= 0.9948 - 1.2 = -0.2052 > 0$$

Thus, $f(2.5)$ is negative and $f(3)$ is positive, therefore, the root lies between 2.5 and 3.

Second approximation: The second approximation to the root is

$$x_2 = \frac{2.5+3}{2} = 2.75$$

Now, $\qquad f(2.75) = 2.75\log_{10} 2.75 - 1.2 = 2.75(0.4393) - 1.2$
$$= 1.2081 - 1.2 = 0.0081 > 0$$

It is found that $f(2.75)$ is positive and $f(2.5)$ is negative, so the root will lie between 2.5 and 2.75.

Third approximation: The third approximation is given by

$$x_3 = \frac{2.5+2.75}{2} = 2.625$$

Again, $\qquad f(2.625) = 2.625\log_{10} 2.625 - 1.2 = 2.625(0.4191) - 1.2$
$$= 1.1001 - 1.2 = -0.0999 < 0$$

Thus, $f(2.625)$ is found to be negative and $f(2.75)$ is positive, therefore, the root lies between 2.625 and 2.75.

Fourth approximation: The fourth approximation to the root is

$$x_4 = \frac{2.625+2.75}{2} = 2.6875$$

Further, $\qquad f(2.6875) = 2.6875\log_{10} 2.6875 - 1.2 = 2.6875(0.4293) - 1.2$
$$= 1.1537 - 1.2 = -0.0463 < 0$$

Thus, $f(2.6875)$ is negative and $f(2.75)$ is positive, therefore, the root lies between 2.6875 and 2.75.

Fifth approximation : The fifth approximation to the root is given by

$$x_5 = \frac{2.6875+2.75}{2} = 2.7188$$

Now, $\qquad f(2.7188) = 2.7188\log_{10} 2.7188 - 1.2 = 2.7188(0.4344) - 1.2$
$$= 1.1810 - 1.2 = -0.019 < 0$$

Thus, $f(2.7188)$ is positive and $f(2.75)$ is negative, therefore, the root lies between 2.734 and 2.75.

Sixth approximation: The sixth approximation to the root is given by

$$x_6 = \frac{2.7188+2.75}{2} = 2.7344$$

Again, $f(2.734) = 2.734 \log_{10} 2.734 - 1.2 = 2.734(0.4368) - 1.2$

$$= 1.1942 - 1.2 = -0.0058 < 0$$

Thus, $f(2.734)$ is negative and $f(2.75)$ is positive, therefore, the root lies between 2.734 and 2.75.

Seventh approximation: The seventh approximation to the root is given by

$$x_7 = \frac{2.734 + 2.75}{2} = 2.742$$

Again, $f(2.742) = 2.742 \log_{10} 2.742 - 1.2 = 2.742(0.4381) - 1.2$

$$= 1.2012 - 1.2 = 0.0012 > 0$$

Thus, $f(2.742)$ is positive and $f(2.734)$ is negative, therefore, the root lies between 2.734 and 2.742.

Eighth approximation: The eighth approximation to the root is

$$x_8 = \frac{2.734 + 2.742}{2} = 2.738$$

Hence, from the approximate value of the root x_7 and x_8 we observed that, upto two places of decimal, the root is 2.74 approximately.

EXAMPLE 7.   *Perform five iterations of bisection method to obtain the smallest positive root of equation*

$$f(x) = x^3 - 5x + 1 = 0$$

(Agra(B.E.)–2002)

SOLUTION.   We have $f(2.1) = (-)\text{ve}, f(2.15) = (+)\text{ve}$

Therefore the root lies between 2.1 and 2.15.

First approximation to the root is

$$x_1 = \frac{2.1 + 2.15}{2} = 2.125$$

Now, $f(2.125) = (-)\text{ve}$

Therefore the root lies between 2.125 and 2.15

Second approximation to the root is

$$x_2 = \frac{2.125 + 2.15}{2} = 2.1375$$

Now, $f(2.1375) = (+)\text{ ve}$

Therefore the root lies between 2.125 and 2.1375.

Third approximation to the root is

$$x_3 = \frac{2.125 + 2.1375}{2} = 2.13125$$

Now, $f(2.13125) = (+)\text{ve}$

Therefore the root lies between 2.125 and 2.13125.

Fourth approximation to the root is

$$x_4 = \frac{2.125 + 2.13125}{2} = 2.1281$$

Now, $f(2.1281) = (-)\text{ve}$

Therefore the root lies between 2.125 and 2.1281.

Fifth approximation to the root is

$$x_5 = \frac{2.125 + 2.13125}{2} = 2.129$$

Hence, the required root is 2.129.

EXERCISE 3.1

1. Find a real root of $e^x = 3x$ by bisection method.

2. If $f(x) = 2^x - x - 3$. Find $f(x)$ for $x = -4, -3, -2, -1, 0, 1, 2, 3, 4$ and compute between which integers values roots are lying.

3. Find the root of $\tan x + x = 0$ upto two decimal places which lies between 2 and 2.1 using bisection method

4. Find a root, correct to three decimal places for each of the equations using bisection method.

   (i) $x^3 + x^2 + x + 7 = 0$ (ii) $x^3 - 5x + 3 = 0$

5. Compute the root of $f(x) = \sin 10x + \cos 3x$ by taking initial approximations 4 and 5.

6. Find a positive root of the equation $x^3 + 3x - 1 = 0$ by bisection method.

7. Find the approximate root of the equation $3x - \sqrt{1 - \sin x} = 0$ by bisection method.

8. Compute the real root of the equation $3x + \sin x - e^x = 0$.

9. Find the positive real root of $x - \cos x = 0$ by bisection method upto four places of decimals between 0 and 1.   (Mumbai-2004; UPTU 2002)

10. Using bisection method, find the negative root of the equation $x^3 - 4x + 9 = 0$.   (JNTU–2009)

11. Find a root of the equation $x^3 - x - 11 = 0$, using the bisection method correct to three decimal places.   (Kerala–2003)

Answers

1. 1.5121375

2.
x	−4	−3	−2	−1	0	1	2	3	4
$f(x)$	1.0625	0.125	−0.75	−1.5	−2	−2	−1	2	9

Root lies between (−3, −2) and (2, 3)

3. 2.02875625   4.(i) 0.0552   (ii) 2.279   5. 4.712389   6. 0.322

7. 10.39188   8. 0.3604   9. 0.7391   10.−2.7064   11. 1.3437

3.5 SECANT METHOD

In this method the graph of the function $y = f(x)$ in approximated by a secant line at each iteration. Secant line is nothing but the chord joining the initial limits of the interval. Taking x_0, x_1 as the initial limits of the interval, we write the equation of the chord joining the points $(x_0, f(x_0))$ and $(x_1, f(x_1))$ is given by

$$y - f(x_1) = \frac{f(x_1) - f(x_0)}{x_1 - x_0}(x - x_1)$$

This secant line cuts the x-axis, $i.e., y = 0$ then the abscissa of the point is given by

$$x_2 = x_1 - \frac{x_1 - x_0}{f(x_1) - f(x_0)} f(x_1)$$

which is an approximation to the root. In the general way, the successive approximations is given by

$$x_{n+1} = x_n - \frac{x_n - x_{n-1}}{f(x_n) - f(x_{n-1})} f(x_n), n \geq 1$$

▶ REMARKS
▸ In this method, $f(x_0)$ and $f(x_1)$ are not necessarily of opposite signs.
▸ In this method, it is not necessary that the interval (x_0, x_1) must contain the root.
▸ If the secant method converges, its rate of convergence is faster than that of the method of false position.
▸ The order of convergence of secant method is 1.62.

SOLVED EXAMPLES

EXAMPLE 1. *Determine the root of the equation* $f(x) = \cos x - xe^x = 0$ *using the secant method upto four decimal places.* (Mumbai B.Tech-2004, UPTU-2009)

SOLUTION. Since $f(x) = \cos x - xe^x = 0$ ...(1)

Taking the initial approximations $x_0 = 0$, $x_1 = 1$

So that $f(x_0) = f(0) = \cos 0 - 0(e^0) = 1$

and $f(x_1) = f(1) = \cos 1 - 1(e) = -2.1780$.

Then by secant method, we get

$$x_2 = x_1 - \frac{x_1 - x_0}{f(x_1) - f(x_0)} f(x_1)$$

$$= 1 - \frac{1-0}{-2.1780-1}(-2.1780) = 1 - \frac{2.1780}{3.1780} = 1 - 0.6853 = 0.3147$$

Now $f(x_2) = f(0.3147) = \cos(0.3147) - (0.3147)e^{0.3147}$

$$= 0.9509 - 0.4311 = 0.5198 \qquad \text{[From (1)]}$$

Thus the second approximation to the root is

$$x_3 = x_2 - \frac{x_2 - x_1}{f(x_2) - f(x_1)} f(x_2) = 0.3147 - \frac{0.3147-1}{0.5198+2.1780}(0.5198)$$

$$= 0.3147 + \frac{0.3562}{2.6978} = 0.4467$$

Now, $f(x_3) = f(0.4467)$

$$= \cos(0.4467) - (0.4467).e^{0.4467}$$

$$= 0.9019 - 0.6983 = 0.2036 \qquad \text{[From (1)]}$$

Then, the third approximation to the root is

$$x_4 = x_3 - \frac{x_3 - x_2}{f(x_3) - f(x_2)} f(x_3)$$

$$= 0.4467 - \frac{0.4467-0.3147}{0.2036-0.5198}(0.2036) = 0.4447 + \frac{0.0269}{0.3162}$$

$$x_4 = 0.5318$$

Now, $f(x_4) = f(0.5318)$

$$= \cos(0.5318) - (0.5318).e^{0.5318}$$

$$= 0.8619 - 0.9051 = -0.0432 \qquad \text{[From (1)]}$$

Then, the fourth approximation to the root is

$$x_5 = x_4 - \frac{x_4 - x_3}{f(x_4) - f(x_3)} f(x_4)$$

$$= 0.5318 - \frac{0.5318-0.4467}{-0.0432-0.2036}(-0.0432)$$

$$= 0.5318 - \frac{0.0037}{0.2468} = 0.5318 - 0.0150 = 0.5168$$

Now, $f(x_5) = f(0.5168) = \cos(0.5168) - (0.5168).e^{0.5168}$ [From (1)]

$$= 0.8694 - 0.8665 = 0.0029$$

Then, the fifth approximation to the root is

$$x_6 = x_5 - \frac{x_5 - x_4}{f(x_5) - f(x_4)} f(x_5) = 0.5168 - \frac{0.5168 - 0.5318}{0.0029 + 0.0432}(0.0029)$$

$$= 0.5168 + \frac{0.0000435}{0.0461} = 0.5177$$

Now, $f(x_6) = f(0.5177) = \cos(0.5177) - (0.5177).e^{0.5177}$ [From (1)]

$$= 0.8690 - 0.8688 = 0.0002$$

Thus, the sixth approximation to the root is

$$x_7 = x_6 - \frac{x_6 - x_5}{f(x_6) - f(x_5)} f(x_6)$$

$$= 0.1577 - \frac{0.5177 - 0.5168}{0.0002 - 0.0029} \times 0.0002 = 0.5177 + 0.0006 = 0.517763$$

Hence the root of the equation $f(x) = \cos x - xe^x$ is 0.5177 correct to four decimal places.

EXAMPLE 2. *Find a root of the equation $x - e^{-x} = 0$ correct to three decimal placed by the secant method.*

SOLUTION. Let $f(x) = x - e^{-x} = 0$ ...(1)

so that $f(0) = -1$ and $f(1) = 1 - e^{-1} = 0.6321$

Taking $x_0 = 0,\, x_1 = 1.$

Now applying secant method.

$$x_2 = x_1 - \frac{x_1 - x_0}{f(x_1) - f(x_0)} \times f(x_1)$$

$$= 1 - \frac{1 - 0}{0.6321 + 1} \times 0.6321 = 1 - \frac{0.6321}{1.6321} = 0.6127$$

Now, $f(x_2) = f(0.6127) = (0.6127) - e^{-0.6127}$ [From (1)]

$$= 0.6127 - 0.5419 = 0.0708$$

Then, the second approximation to the root is given by

$$x_3 = x_2 - \frac{x_2 - x_1}{f(x_2) - f(x_1)} \times f(x_2)$$

$$= 0.6127 - \frac{0.6127 - 1}{0.0708 - 0.6321} \times 0.0708$$

$$= 0.6127 - \frac{0.0274}{0.5613} = 0.6127 - 0.0488 = 0.5639$$

Now, $f(x_3) = f(0.5639) = 0.5639 - e^{-0.5639}$ [From (1)]

$$= 0.5639 - 0.5690 = -0.0051$$

Then, the third approximation to the root is

$$x_4 = x_3 - \frac{x_3 - x_2}{f(x_3) - f(x_2)} \times f(x_3)$$

$$= 0.5639 - \frac{0.5639 - 0.6127}{-0.0051 - 0.0708} \times (-0.0051)$$

$$= 0.5639 + \frac{0.00025}{0.0759} = 0.5639 + 0.0033 = 0.5672$$

Now, $f(x_4) = f(0.5672) = 0.5672 - e^{-0.5672}$ [From (1)]

$$= 0.5672 - 0.5671 = 0.0001$$

Then, the fourth approximation to the root is

$$x_5 = x_4 - \frac{x_4 - x_3}{f(x_4) - f(x_3)} \times f(x_4)$$

$$= 0.5672 - \frac{0.5672 - 0.5639}{0.0001 - 0.0051} \times 0.0001 = 0.5672 - 0.00063 = 0.5670$$

Hence, the approximate root is 0.567 correct to three decimal places.

EXAMPLE 3. *Compute root of the equation $x^2 e^{-x/2} = 1$ in the interval [0, 2] using secant method. The root should be correct to three decimal places.*   (UPTU(MCA)–2005)

SOLUTION. Taking initial approximation $x_0 = 0, x_1 = 2$

We obtain for the secant method

$$f(x) = e^{-x/2}.x^2 - 1 \Rightarrow f(0) = -1, f(2) = 0.4760$$

Apply secant method

$$x_2 = x_1 - \left| \frac{x_1 - x_0}{f_1 - f_0} \right| f_1 = 2 - \left| \frac{2 - 0}{0.4760 + 1} \right| \times 0.4760 = 1.3550$$

$$\Rightarrow \qquad f(x_2) = -0.0665$$

Similarly

$$x_3 = x_2 - \left| \frac{x_2 - x_1}{f_2 - f_1} \right| \times f_2 = 1.3550 - \left| \frac{1.355 - 2}{-0.0665 - 1} \right| \times (-0.0665) = 1.3148$$

3.6 ITERATION METHOD

This method is used when we find the value of the root $x = \alpha$ in terms of a function x, *i.e.*,

$$x = f(x) \text{ from } f(x_0) = 0 \qquad ...(1)$$

In this method, we first obtain the approximate value of the root, say x_0 of the given equation and then substitute it in $f(x)$ and get second approximation as

$$x_1 = f(x_0) \qquad ...(2)$$

Now putting $x = x_1$ in $f(x)$ and get the third approximation

$$x_2 = f(x_1) \qquad ...(3)$$

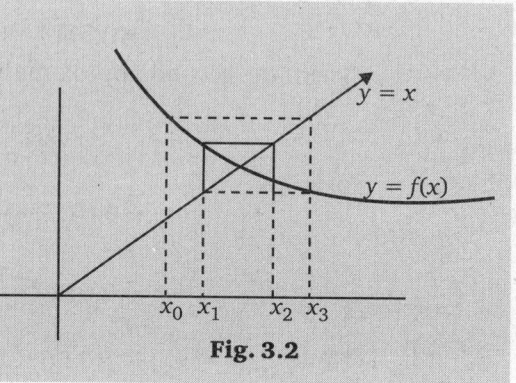

Fig. 3.2

Continuing this process, we may get

$$x_n = f(x_{n-1}) \quad \text{or} \quad x_{n+1} = f(x_n)$$

▶ Remarks

▶ If $f(x)$ be a root of $F(x) = 0$ and let I be the interval containing the point $x = a$. Let $f(x)$ and $f'(x)$ be continuous in I when $f(x)$ is given by the equation $x = f(x)$ which is equivalent to $F(x) = 0$. Then if $|f'(x)| < 1 \forall x \in I$, the sequence of approximations $x_0, x_1, ..., x_n$ defined by $x_{n+1} = f(x_n)$ converges to the root a, provided that the initial approximations x_0 is chosen in I.

▶ The advantages of iterative methods are :
(i) the use of the same set of instructions will save space in memory unit of the computer.
(ii) the round off errors are minimised where as these errors are compounded by other method.

▶ The iterative method $x = g(x)$ is convergent if $|g'(x)| < 1$
if $|g'(x)| = 1$ and $|g'(x)| < 1$, the number of iterations would be large. But if $g'(x)$ is very small, the number of iteration will be lessor.

▶ When $|g'(x)| > 1 \Rightarrow g'(x) > 1$ or $g'(x) < -1$, the iterative process is divergent.

3.6.1 Rate of Convergence of Iteration Method

Let $f(x) = 0$ be the equation which is being expressed as $x = g(x)$. The iterative formula for solving the equation is
$$x_{i+1} = g(x_i)$$
If α is the root of the equation $x = g(x)$ lying in the interval $]a, b[, \alpha = g(\alpha)$.

The iterative formula may also be written as $x_{i+1} = g(x + \overline{x_i - \alpha})$

Then, by mean value theorem
$$x_{i+1} = g(\alpha) + (x_i - \alpha)g'(c_i) \text{ where } a < c_i < b$$
But
$$g(\alpha) = \alpha \qquad \Rightarrow \qquad x_{i+1} = \alpha + (x_i - \alpha)g'(c_i)$$
$$\Rightarrow \qquad x_{i+1} - \alpha = (x_i - \alpha)g'(c_i) \qquad \qquad ...(1)$$

Now, if e_{i+1}, e_i are the error for the approximations x_{i+1} and x_i.
$$\therefore \qquad e_{i+1} = x_{i+1} - \alpha, e_i = x_i - \alpha$$
Using this in (1), we get $\qquad e_{i+1} = e_i g'(c_i)$
Here $g(x)$ is a continuous function, therefore, it is bounded
$$\therefore \qquad |g'(c_i)| \leq k, \text{ where } k \in]a, b[\text{ is a constant.}$$
$$\therefore \qquad e_{i+1} \leq e_i k$$
Since, the index of e_i being 1, the rate of convergence of the iterative method is $x_{i+1} = g(x_i)$ is linear.

SOLVED EXAMPLES

EXAMPLE 1. *Solve $x = 0.21 \sin(0.5 + x)$ by iteration method starting with $x = 0.12$.* (UPTU-2003)

SOLUTION. Here, we have
$$x = 0.21 \sin(0.5 + x) \Rightarrow f(x) = 0.21 \sin(0.5 + x)$$
Here, we observe that $|f(x)| < 1$

$\Rightarrow$ Method of iteration can be applied.
Now first approximation of x is given by
$$x^{(1)} = 0.21 \sin(0.5 + 0.12) = 0.21 \sin(0.62)$$
$$= 0.21(0.58104) = 0.1220$$

The second approximation is given by
$$x^{(2)} = 0.21\sin(0.5 + 0.122) = 0.21\sin(0.622)$$
$$= 0.21(0.58267) = 0.1224$$
The third approximation of x is given by
$$x^{(3)} = 0.21\sin(0.5 + 0.1224) = 0.21\sin(0.622)$$
$$= 0.21(0.58299) = 0.12243$$
The fourth approximation of x is given by
$$x^{(4)} = 0.21\sin(0.5 + 0.12243) = 0.21\sin(0.62243)$$
$$= 0.21(0.58301) = 0.12243$$
Here, we observe that $x^{(3)} = x^{(4)}$

Hence, the required root is given by
$$x = 0.12243.$$

EXAMPLE 2.   *Find a real root of the equation*
$$\cos x = 3x - 1$$

Correct to three decimal places, using Iteration method.   (JNTU-2009, UPTU-2003)

SOLUTION.   Here, we have
$$f(x) = \cos x - 3x + 1 \qquad \qquad \qquad ...(1)$$
We observe that $f(0) = 2 = +$ve

and $$f(\pi/2) = -3\left(\frac{\pi}{2}\right) + 1 = -\text{ve}$$

Roots lies between 0 and $\dfrac{\pi}{2}$.

Now, rewriting the given equation as follows
$$x = \frac{1}{3}(\cos x + 1) = g(x) \quad \text{(Say)}$$
Then, we have

$$g'(x) = -\frac{\sin x}{3} = |g'(x)| < 1 \text{ in } \left]0, \frac{\pi}{2}\right[$$

$= $ Iteration method can be applied.

Take the first approximation, $x_0 = 0$
Then we can find the successive approximation as follows.

$$x_1 = g(x_0) = \frac{1}{3}[\cos 0 + 1] = 0.667$$

$$x_2 = g(x_1) = \frac{1}{3}[\cos(0.667) + 1] = 0.5953$$

$$x_3 = g(x_2) = \frac{1}{3}[\cos(0.5953) + 1] = 0.6093$$

$$x_4 = g(x_3) = \frac{1}{3}[\cos(0.6093) + 1] = 0.6067$$

$$x_5 = g(x_4) = \frac{1}{3}[\cos(0.6067) + 1] = 0.6072$$

$$x_6 = g(x_5) = \frac{1}{3}[\cos(0.6072) + 1] = 0.6071$$

Now x_5 and x_6 being almost same.

Hence, the required root is given by 0.607.

EXAMPLE 3. *Find the reciprocal of 41 correct to 4 decimal places by iterative formula.*

$$x_{i+1} = x_i = (2 - 41x_i).$$

SOLUTION. Iterative formula is given by

$$x_{i+1} = x_i = (2 - 41x_i) \qquad \qquad \ldots(1)$$

Putting $i = 0$ in equation (1), we get

$$x_1 = x_0(2 - 41x_0)$$

Let $x_0 = 0.02$

$$x_1 = (0.02)(2 - 0.82) = 0.024$$

Put $i = 1$ in equation (1), we get

$$x_2 = (0.024)[2 - (41 \times 0.024)] = 0.0244$$

Again putting $i = 2$ in equation (1), we get

$$x_3 = 0.02439$$

Hence, reciprocal of 41 is 0.0244.

EXAMPLE 4. *Find the square root of 20 correct to 3 decimal places by using recursion formula*

$$x_{i+1} = \frac{1}{2}\left(x_i + \frac{20}{x_i}\right)$$

SOLUTION. Here, we have

$$x_{i+1} = \frac{1}{2}\left(x_i + \frac{20}{x_i}\right) \qquad \qquad \ldots(1)$$

Putting $i = 0$ in equation (1), we get

$$x_1 = \frac{1}{2}\left(x_0 + \frac{20}{x_0}\right)$$

Let $x_0 = 4.5$

$$x_1 = \frac{1}{2}\left(4.5 + \frac{20}{4.5}\right) = 4.47$$

Again putting $i = 1$ and $x_1 = 4.47$ in equation (1), we get

$$x_2 = \frac{1}{2}\left(4.47 + \frac{20}{4.47}\right) = 4.472$$

Hence, $\sqrt{20} \approx 4.472$ correct to three decimal places.

EXAMPLE 5. *Using the method of iteration, find a positive root between 0 and 1 of the equation* $xe^x = 1.$

SOLUTION. The given equation can also be written as

$$x = e^{-x}$$

Then

$$g(x) = e^{-x} \Rightarrow g'(x) = -e^{-x}$$

$$\Rightarrow \qquad |g'(x)| < 1 \quad \text{for } x < 1$$

$\Rightarrow$ Iteration is convergent.

Now, starting with $x_0 = 1$, we find the successive iterations such that

$$x_1 = \frac{1}{e} = 0.3678974$$

$$\Rightarrow \qquad x = e^{-0.3678974} = 0.6922006$$

$$.............$$
$$.............$$
$$.............$$

$$x_{20} = 0.5671477 \text{ and so on.}$$

EXAMPLE 6. *Find a real root of $2x - \log_{10} x = 7$ correct to four decimal places using iteration method.*

(UPTU-2004)

SOLUTION. Here, we have $f(x) = 2x - \log_{10} x - 7$

We observe that

$$f(3) = 6 - \log 3 - 7$$
$$= 6 - 0.4771 - 7 = -1.4471$$

and $\qquad f(4) = 0.398$

$\Rightarrow$ A root of $f(x)$ lies between 3 and 4.

The given equation can also be written as

$$x = \frac{1}{2}(\log_{10} x + 7) = g(x)$$

Then $\qquad g'(x) = \frac{1}{2}\left(\frac{1}{x}\log_{10} e\right)$

$$\Rightarrow \qquad |g'(x)| < 1 \quad \text{when } 3 < x < 4$$

$$(\because \log_{10} e = 0.4343)$$

Also, since $|f(4)| < |f(3)|$, the root is near to 4.

Taking $x_0 = 3.6$, the successive approximations are given by

$$x_1 = g(x_0) = \frac{1}{2}(\log_{10} 3.6 + 7) = 3.77815$$

$$x_2 = g(x_1) = \frac{1}{2}(\log_{10} 3.77815 + 7) = 3.78863$$

$$x_3 = g(x_2) = 3.78924$$

$$x_4 = g(x_3) = 3.78927$$

Since, x_3 and x_4 are almost equal, hence the root is 3.7892 correct to four decimal places.

EXAMPLE 7. *Show that the following rearrangement of equation $x^3 + 6x^2 + 10x - 20 = 0$ does not yield a convergent sequence of successive approximation by iteration method near $x = 1$.*

$$x = \frac{20 - 6x^2 + x^3}{10}$$

SOLUTION. We have

$$x = \frac{20 - 6x^2 + x^3}{10} = g(x)$$

$$\Rightarrow \qquad g'(x) = \frac{12x - 3x^2}{10}$$

We observe that $g'(x) < -1$ in the neighbourhood of $x = 1$. Thus $|g'(x)| > 1$.

Hence, the sequence $\langle x_n \rangle$ of approximations does not converge.

EXAMPLE 8.   *Find the smallest root of the equation*

$$1 - x + \frac{x^2}{(2!)^2} - \frac{x^3}{(3!)^2} + \frac{x^4}{(4!)^2} - \frac{x^5}{(5!)^2} + \ldots = 0$$

SOLUTION.   The given equation can also be written as

$$x = 1 + \frac{x^2}{(2!)^2} - \frac{x^3}{(3!)^2} + \frac{x^4}{(4!)^2} - \frac{x^5}{(5!)^2} + \ldots = g(x)$$

Omitting x^2 and higher powers of x, we get $x = 1$ (approximately)

Taking $x_0 = 1$ we get

$$x_1 = g(x_0)$$

$$= 1 + \frac{1}{(2!)^2} - \frac{1}{(3!)^2} + \frac{1}{(4!)^2} - \frac{1}{(5!)^2} + \ldots = 1.2239$$

$$x_2 = g(x_1)$$

$$= 1 + \frac{(1.2239)^2}{(2!)^2} - \frac{(1.2239)^3}{(3!)^2} + \frac{(1.2239)^4}{(4!)^2} - \frac{(1.2239)^5}{(5!)^2} + \ldots$$

$$= 1.3263$$

Similarly, we may get

$$x_3 = g(x_2) = 1.38,$$

$$x_4 = 1.409, x_5 = 1.425, x_6 = 1.434$$

$$x_7 = 1.439, x_8 = 1.442$$

Since, the values of x_7 and x_8 are almost equal. Hence, the required root is 1.44 correct to two decimal places.

EXAMPLE 9.   *Suggest a value c so that the iteration formula $x = x + c(x^2 - 3)$ may converge at a good rate. Given that $x = \sqrt{3}$ is a root.*

SOLUTION.   Let $f(x) = x + c(x^2 - 3)$

It will converge if

$$|f'(x)| < 1 \quad \Rightarrow \quad -1 < f'(x) < 1$$

i.e., if $-1 < 1 + 2cx < 1$

Also, the convergence will be rapid if $f'(a) = 0$

i.e., if $1 + 2ca \approx 0$ *i.e.*, if $1 + 2c\sqrt{3} \approx 0$

$$\Rightarrow \qquad c = -\frac{1}{2\sqrt{3}} < -\frac{1}{4}$$

Hence, if we may take $c = -\frac{1}{4}$, then the given iterative formula may converge at a good rate.

EXAMPLE 10.   *If α, β are the root of $x^2 + ax + b = 0$, show that the iteration $x_{n+1} = -\left(\dfrac{ax_n + b}{x_n}\right)$ will converge near $x = \alpha$ if $|\alpha| > |\beta|$ and the iteration $x_{n+1} = -\dfrac{b}{x_n + a}$ will converge near $x = \alpha$ if $|\alpha| < |\beta|$.*

(UPTU(MCA)–2005)

SOLUTION.   It is given that α, β are the roots of $x^2 + ax + b = 0$. Then we have $\alpha + \beta = -a$ and $\alpha\beta = b$

Also, the formula $x_{n+1} = -\left(\dfrac{ax_n + b}{x_n}\right)$ which is of the form $x_{n+1} = f(x_n)$ will converge to $x = \alpha$ if

$$\left|\frac{d}{dx}\left\{\frac{-(ax+b)}{b}\right\}_{x=x_n}\right| < 1$$

$\Rightarrow \qquad \left|\dfrac{b}{x_n^2}\right| < 1 \qquad$ or $\qquad x_n^2 > |b|$

$\Rightarrow \qquad \left|x_n^2\right| > |b|$

or $\qquad |\alpha|^2 > |b| \qquad$ as $\qquad x_n \to \alpha$

$\Rightarrow \qquad |\alpha|^2 > |\alpha|.|\beta| \qquad\qquad\qquad\qquad (\because \alpha\beta = b)$

$\Rightarrow \qquad |\alpha| > |\beta|$

In a similar way $x_{n+1} = -\dfrac{b}{x_n + a}$ will converge to $x = \alpha$ if

$$\left|\frac{d}{dx}\left(\frac{-b}{x+a}\right)_{x=x_n}\right| < 1 \qquad \text{or} \qquad \left|\frac{b}{(x_n + a)^2}\right| < 1$$

$\Rightarrow \qquad (x_n + a)^2 > |b| \qquad \Rightarrow \qquad (\alpha + a)^2 > |b| \qquad (\text{as } x_n \to a)$

$\Rightarrow \qquad \beta^2 > |b| \qquad \Rightarrow \qquad |\beta|^2 > |\alpha|.|\beta|$

$\Rightarrow \qquad |\beta| > |\alpha| \quad \text{or} \qquad |\alpha| < |\beta|$

EXAMPLE 11. *The equation* $\sin x = 5x - 2$ *can be put as* $x = \sin^{-1}(5x - 2)$ *and also as* $x = \dfrac{1}{2}(\sin x + 3)$ *suggesting two iterating procedures for its solutions. Which of these, if any would succeed and which would fall to give root in neighbourhood of 0.5 ?*

SOLUTION. **Case I:** When $g(x) = \sin^{-1}(5x - 2)$

Therefore, the method would not give a convergent sequence.

Case II: $\qquad g(x) = \dfrac{1}{5}(\sin x + 2)$

$\Rightarrow \qquad g'(x) = \dfrac{1}{5}\cos x$

$\Rightarrow \qquad |g'(x)| \le \dfrac{1}{5}$ for all x because $|\cos(x)| \le 1$

$\Rightarrow g(x)$ will succeed.

Therefore, taking $x = g(x) = \dfrac{1}{5}(\sin x + 2)$ and initial values $x_0 = 0.5$.

We get, the first approximation is given by

$$x_1 = \frac{1}{5}(\sin 0.5 + 2) = 0.4017$$

Similarly, $x_2 = g(x_2)$

$$= \frac{1}{5}(\sin x_1 + 2) = \frac{1}{5}(\sin(0.40175) + 2)$$

$$= \frac{1}{5}(2.0070112) = 0.40140$$

Also, $x_3 = g(x_2) = \frac{1}{5}[\sin(0.40140) + 2]$

$$= \frac{1}{5}(2.00700) = 0.4014$$

Hence, the required root upto four decimal places is 0.4014.

EXAMPLE 12. *Determine p, q and r so that order of the iterative method given by*

$$x_{n+1} = px_n + q.\frac{a}{x_n^2} + r.\frac{a^2}{x_n^5}$$

For computing $a^{1/3}$ becomes as high as possible. (UPTU(MCA)–2007)

SOLUTION. Let $x = a^{1/3}$

$$f(x) = x^3 - a = 0$$
$$f'(x) = 3x^2$$

By Newton Raphson method

$$x_{n+1} = x_n - \frac{f(x_n)}{f'(x_0)} = x_n - \frac{x_n^3 - a}{3x_n^2} = \frac{3x_n^3 - x_n^3 + a}{3x_n^2} = \frac{2x_n^3 + a}{3x_n^2}$$

$\therefore$
$$x_{n+1} = \frac{2}{3}x_n + \frac{a}{3x_n^2}$$

Comparing the given equation

$$x_{n+1} = px_n + q.\frac{a}{x_n^2} + r.\frac{a^2}{x_n^5}$$

Hence, $p = 2/3, q = 1/3$ and $r = 0$.

3.7 ITERATIVE METHOD FOR THE SYSTEM OF NON-LINEAR EQUATIONS

Consider non-linear equations
$$\left.\begin{array}{l} f(x,y) = 0 \\ g(x,y) = 0 \end{array}\right] \qquad \qquad ...(1)$$

whose real root are to be required within the given degree of accuracy.

Let us take

$$x = F(x,y) \qquad \text{and} \qquad y = G(x,y) \qquad ...(2)$$

provided $\left|\frac{\partial F}{\partial x}\right| + \left|\frac{\partial F}{\partial y}\right| < 1$ and $\left|\frac{\partial G}{\partial x}\right| + \left|\frac{\partial G}{\partial y}\right| < 1$ in the neighbourhood of the root.

Let (α, β) be the exact root of (1) and let (x_0, y_0) be the initial approximations.
Then from (2), the successive approximations are given by

$$\begin{array}{ll} x_1 = F(x_0, y_0); & y_1 = G(x_0, y_0) \\ x_2 = F(x_1, y_1); & y_2 = G(x_1, y_1) \\ x_3 = F(x_2, y_2); & y_3 = G(x_2, y_2) \end{array}$$

and so on, we get following two iterative formulae.

$$x_{n+1} = F(x_n, y_n) \qquad \text{and} \qquad y_{n+1} = G(x_n, y_n)$$

If these formulae converges, then in the limit; $\alpha = F(\alpha, \beta)$ and $\beta = G(\alpha, \beta)$
Hence, (α, β) gives the root of system (1).

SOLVED EXAMPLE

EXAMPLE. *Find a real root of the system of the equations by iterative method*

$$x = 0.2x^2 + 0.8; y = 0.3xy^2 + 0.7$$

SOLUTION. Let us assume that

$$F(x, y) = 0.2x^2 + 0.8$$

and

$$G(x, y) = 0.3xy^2 + 0.7$$

Then

$$\frac{\partial F}{\partial x} = 0.4x, \frac{\partial F}{\partial y} = 0, \frac{\partial G}{\partial x} = 0.3y^2, \frac{\partial G}{\partial y} = 0.6xy$$

Let us choose initially $x_0 = \dfrac{1}{2}, y_0 = \dfrac{1}{2}$. Then, we have

$$\left|\frac{\partial F}{\partial x}\right|_{(x_0, y_0)} + \left|\frac{\partial F}{\partial y}\right|_{(x_0, y_0)} = 0.2 < 1$$

and

$$\left|\frac{\partial G}{\partial x}\right|_{(x_0, y_0)} + \left|\frac{\partial G}{\partial y}\right|_{(x_0, y_0)}$$

$$= \frac{0.3}{4} + \frac{0.6}{4} = \frac{0.9}{4} < 1$$

Hence, the successive approximations are given by

$$x_1 = F(x_0, y_0) = (0.2)\left(\frac{1}{2}\right)^2 + 0.8 = 0.85$$

$$y_1 = G(x_0, y_0) = (0.3)\left(\frac{1}{2}\right)\left(\frac{1}{2}\right)^2 + 0.7 = 0.74$$

$$x_2 = F(x_1, y_1) = 0.2x_1^2 + 0.8 = 0.2(0.85)^2 + 0.8 = 0.9445$$

$$y_2 = G(x_1, y_1) = 0.3x_1y_1^2 + 0.7$$
$$= 0.3(0.85)(0.74)^2 + 0.7 = 0.8396$$

Also

$$x_3 = F(x_2, y_2) = 0.2x_2^2 + 0.8$$
$$= 0.2(0.9445)^2 + 0.8 = 0.9784$$

$$y_3 = G(x_2, y_2) = 0.3x_2y_2^2 + 0.7$$
$$= 0.3(0.9445)(0.8396)^2 + 0.7 = 0.8997$$

From these three approximations, we conclude that the root converges to (1, 1). Also from (1), we get

$$1 = F(1, 1) \text{ and } 1 = G(1, 1)$$

EXERCISE 3.2

1. By iterative method, find $\sqrt{30}$.
2. By iterative method, find the real root of the equation $3x - \log_{10} x = 6$ correct to four significant figures.
3. Find the cube root of 15 correct to four significant figures by iterative method.
4. Find a real root of each of the following equations, using the iterative method.

   (i) $e^x = \cot x$ (ii) $1 + x - x^3 = 0$

(iii) $\sin x = 10(x - 1)$ (iv) $x^2 - 1 = \sin^2 x$

5. If $F(x)$ is sufficiently differentiable and the iteration $x_{n+1} = F(x_n)$ converges, show that the order of convergence is a positive integer.
6. Find a real root of the equation $x^3 + x^2 - 1 = 0$ in (0, 1) with an accuracy of 10^{-4} by iteration method.   (UPTU-2006)
7. Use the iteration method to find a root of the equation $x^3 - 9x + 1$ to four decimal places.   (Madras–2006)

Answers

1. 5.477225575   **2.** 2.108   **3.** 2.466   **4.** (i) 0.5314   (ii) 1.466   (iii) 1.088   (iv) 1.404

7. 2.9428.

3.8 REGULA-FALSI METHOD (OR METHOD OF FALSE POSITION)

This method also known as method of false position is the oldest method of finding the real roots of the equation $f(x) = 0$ and it is somewhat similar to the bisection method.

A graphical description of the false position method is shown in the adjoining fig.(3.3).

Consider the equation $f(x) = 0$. Let x_0 and $x_1(x_0 < x_1)$ be two values of x such that $f(x_0)$ and $f(x_1)$ are of opposite signs. Then the graph of $y = f(x)$ crosses the x-axis at some point between x_0 and x_1.

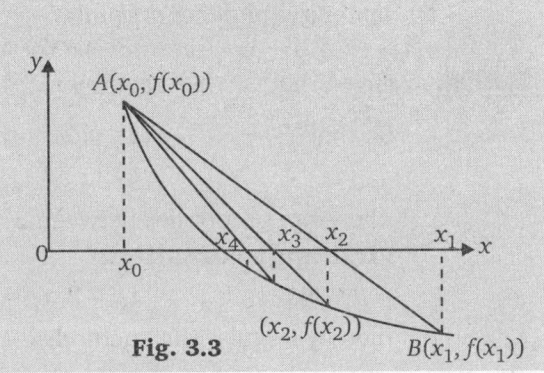

Fig. 3.3

∴   The equation of the chord joining two points $A(x_0, f(x_0))$ and $B(x_1, f(x_1))$ is

$$y - f(x_0) = \frac{f(x_1) - f(x_0)}{x_1 - x_0}(x - x_0) \qquad \ldots(1)$$

Since equation (1) intersects the x-axis between the points, where $y = 0$. Thus, we get

$$-f(x_0) = \frac{f(x_1) - f(x_0)}{x_1 - x_0}(x - x_0) \quad \Rightarrow x = \frac{x_0 f(x_1) - x_1 f(x_0)}{f(x_1) - f(x_0)}$$

Therefore, the first approximation of the root is given by

$$x_2 = \frac{x_0 f(x_1) - x_1 f(x_0)}{f(x_1) - f(x_0)} = x_0 - \frac{f(x_0)(x_1 - x_0)}{f(x_1) - f(x_0)} \qquad \ldots(2)$$

In general $$x_{n+1} = x_{n-1} - \frac{f(x_{n-1})(x_n - x_{n-1})}{f(x_n) - f(x_{n-1})} \qquad \ldots(3)$$

ALGORITHM

Step 1. If $f(x_2)$ and $f(x_0)$ are of opposite sign, then the root lies in between x_0 and x_2, therefore we replace x_1 and x_2 in equation (2) and get the next approximation x_3. On the other hand, if $f(x_2)$ and $f(x_0)$ are of the same sign then $f(x_2)$ and $f(x_1)$ will be of opposite signs and therefore, the root lies between x_2 and x_1. Hence, we replace x_0 by x_2 in equation (2) and get the next approximation x_3. This process is to be repeated, until we get the root to desired accuracy.

▰ **REMARKS**

▸ In this method, we choose such points x_0 and x_1 for which $f(x_0)f(x_1) < 0$.

▸ We choose the points x_0 and x_1 such that they form sufficiently small interval, then in this interval the curve is considered as a straight line. Thus, it is called false position method.

▸ The rate of convergence of false position method is faster than that of the bisection method.

▸ It is also called linear interpolation method.

SOLVED EXAMPLES

EXAMPLE 1. *Find a real root of the equation* $f(x) = x^3 - 2x - 5 = 0$ *by the method of false position upto three places of decimals.*

(Rohtak–2006, Manipal–2005, Tamilnadu(MCA)–2005, 08, IAS–2002)

SOLUTION. Given that $\qquad f(x) = x^3 - 2x - 5 = 0$

So that $\qquad f(2) = (2)^3 - 2(2) - 5 = -1$

and $\qquad f(3) = (3)^3 - 2(3) - 5 = 16$

Therefore, a root lies between 2 and 3.

First approximation:

∴ Taking $\qquad x_0 = 2, x_1 = 3, f(x_0) = -1, f(x_1) = 16,$

then by Regula-Falsi method,

We get $\qquad x_2 = x_0 - \dfrac{x_1 - x_0}{f(x_1) - f(x_0)} \cdot f(x_0)$

$$= 2 - \frac{3 - 2}{16 + 1} \cdot (-1) = 2 + \frac{1}{17} = 2.0588$$

Now,

$$f(x_2) = f(2.0588)$$

$$= (2.0588)^3 - 2(2.0588) - 5 = -0.3911$$

[Using (1)]

Therefore, root lies between 2.0588 and 3.

Second approximation:

Now, taking $x_0 = 2.0588$

and $\qquad x_1 = 3, f(x_0) = -0.3911, f(x_1) = 16$

Then we get

$$x_3 = x_0 - \frac{x_1 - x_0}{f(x_1) - f(x_0)} \cdot f(x_0)$$

$$= 2.0588 - \frac{3 - 2.0588}{16 + 0.3911} \times (-0.3911)$$

$$= 2.0588 + 0.0225 = 2.0813$$

Now,

$$f(x_3) = f(2.0813)$$
$$= (2.0813)^3 - 2(2.0813) - 5$$
$$= 9.0158 - 9.1626 = -0.1468 \qquad \text{[Using (1)]}$$

∴ Root lies between 2.0813 and 3.

Third approximation : Taking $x_0 = 2.0813$ and $x_1 = 3, f(x_0) = -0.1468$, $f(x_1) = 16$. Then by false position method, we get

$$x_4 = x_0 - \frac{x_1 - x_0}{f(x_1) - f(x_0)} . f(x_0)$$
$$= 2.0813 - \frac{3 - 2.0813}{16 + 0.1468} \times (-0.1468)$$
$$= 2.0813 + 0.0084 = 2.0897$$

Now,

$$f(x_4) = f(2.0897)$$
$$= (2.0897)^3 - 2(2.0897) - 5$$
$$= 9.1254 - 9.1794 = -0.054 \qquad \text{[Using (1)]}$$

Fourth approximation: The root lies between 2.0897 and 3. Therefore, taking

$$x_0 = 2.0897, x_1 = 3, f(x_0) = -0.054$$

and

$$f(x_1) = 16.$$

Then by false position method, we have

$$x_5 = x_0 - \frac{x_1 - x_0}{f(x_1) - f(x_0)} . f(x_0)$$
$$= 2.0897 - \frac{3 - 2.0897}{16 + 0.054} \times (-0.054)$$
$$= 2.0897 + 0.0031 = 2.0928$$

Now,

$$f(x_5) = f(2.0928) = (2.0928)^3 - 2(2.0928) - 5$$
$$= 9.1661 - 9.1856 = -0.0195$$

Fifth approximation: The root lies between 2.0928 and 3 so taking $x_0 = 2.0928, x_1 = 3, f(x_0) = -0.0195$ and $f(x_1) = 16$. Then

$$x_6 = x_0 - \frac{x_1 - x_0}{f(x_1) - f(x_0)} . f(x_0)$$
$$= 2.0928 - \frac{3 - 2.0928}{16 + 0.0195} \times (-0.0195)$$
$$= 2.0928 + 0.0011 = 2.0939$$

Now,

$$f(x_6) = f(2.0939) = (2.0939)^3 - 2(2.0939) - 5$$
$$= 9.1805 - 9.1879 = -0.0074$$

Thus, the root lies between 2.0939 and 3.

Sixth approximation:

Taking $x_0 = 2.0939, x_1 = 3, f(x_0) = -0.074$

and $f(x_1) = 16$. Then, we get

$$x_7 = x_0 - \frac{x_1 - x_0}{f(x_1) - f(x_0)} \cdot f(x_0)$$

$$= 2.0939 - \frac{3 - 2.0939}{16 + 0.0074} \times (-0.0074)$$

$$= 2.0939 + 0.00042 = 2.0943$$

Now

$$f(x_7) = f(2.0943) = (2.0943)^3 - 2(2.0943) - 5$$

$$= 9.1858 - 9.1886 = -0.028$$

Seventh approximation: The root lies between 2.0943 and 3, so taking $x_0 = 2.0943$, $x_1 = 3$, $f(x_0) = -0.0028$ and $f(x_1) = 16$. Then

$$x_8 = x_0 - \frac{x_1 - x_0}{f(x_1) - f(x_0)} \cdot f(x_0)$$

$$= 2.0943 - \frac{3 - 3.0943}{16 + 0.0028} (-0.0028)$$

$$= 2.0943 + 0.00016 = 2.0945$$

Hence, the root is 2.094 correct to three decimal places.

Hence, approximate value of the root to four places of decimal is 2.0945.

EXAMPLE 2.   *Find the real root of the equation $x^3 - 9x + 1 = 0$ by Regula-Falsi method.*

SOLUTION.   Let $\qquad\qquad f(x) = x^3 - 9x + 1 = 0$ ...(1)

So that

$$f(2) = (2)^3 - 9(2) + 1 = 8 - 18 + 1 = -9$$

$$f(3) = (3)^3 - 9(3) + 1 = 27 - 27 + 1 = 1$$

Since $f(2)$ and $f(3)$ are of opposite signs, therefore, the root lies between 2 and 3, so taking $x_0 = 2$, $x_1 = 3$, $f(x_0) = -9$, $f(x_1) = 1$, then by Regula-falsi method, we get

First approximation:

$$x_2 = x_0 - \frac{x_1 - x_0}{f(x_1) - f(x_0)} \cdot f(x_0)$$

$$= 2 - \frac{3 - 2}{1 + 9} \times (-9) = 2 + \frac{9}{10} = 2.9$$

Now,

$$f(x_2) = f(2.9)$$

$$= (2.9)^3 - 9(2.9) + 1$$

$$= 24.389 - 25.1 = -0.711$$

Second approximation: The root lies between 2.9 and 3. Therefore, taking $x_0 = 2.9$, $x_1 = 3$, $f(x_0) = -0.711$, $f(x_1) = 1$.

Then

$$x_3 = x_0 - \frac{x_1 - x_0}{f(x_1) - f(x_0)} \cdot f(x_0)$$

$$= 2.9 - \frac{3 - 2.9}{1 + 0.711} (-0.771)$$

$$= 2.9 + 0.0416 = 2.9416$$

Now, $f(x_3) = f(2.9416)$

$$= (2.9416)^3 - 9(2.9416) + 1$$
$$= 25.4537 - 25.4744 = -0.0207$$

Third approximation: The root lies between 2.9416 and 3. Therefore, taking $x_0 = 2.9416, x_1 = 3$, $f(x_0) = -0.0207$ and $f(x_1) = 1$. Then, we get

$$x_4 = x_0 - \frac{x_1 - x_0}{f(x_1) - f(x_0)} . f(x_0)$$

$$= (2.9416) - \frac{3 - 2.9416}{1 + 0.0207} . (-0.0207)$$
$$= 2.9416 + 0.0012 = 2.9428$$

Now, $f(x_4) = f(2.9428)$

$$= (2.9428)^3 - 9(2.9428) + 1$$
$$= 25.4849 - 25.4852 = -0.0003$$

Fourth approximation: The root lies between 2.9428 and 3.

Therefore, taking $x_0 = 2.9428$, $x_1 = 3$, $f(x_0) = -0.0003$ and $f(x_1) = 1$. Then by false position method, we have .

$$x_5 = x_0 - \frac{x_1 - x_0}{f(x_1) - f(x_0)} . f(x_0)$$

$$= 2.9428 - \frac{3 - 2.9428}{1 + 0.0003} (-0.0003)$$
$$= 2.9428 + 0.000017 = 2.942817$$

Hence, the root is 2.9428 correct to four places of decimal.

EXAMPLE 3. *Find a root between 1 and 2 of the equation $f(x) = x^3 - x - 4$ to four places of decimal using Regula-Falsi method.*

SOLUTION. Here, by usual procedure, we have the following computational table.

Computational table

Iteration No.	x_0	x_1	x_2	$f(x_0)$	$f(x_1)$	$f(x_2)$
1	1.00000	2.00000	1.66666	–4.0000	2.00000	–1.3709
2	1.66666	2.00000	1.78048	–1.03709	2.00000	–0.13616
3	1.78048	2.00000	1.79447	–0.13616	2.00000	–0.01606
4	1.79447	2.00000	1.79610	–0.01606	2.00000	0.00193
5	1.79610	2.00000	1.79629	–0.00193	2.00000	–0.00028
6	1.79629	2.00000	1.79631	0.00028	2.00000	–0.00024
7	1.79631	2.00000	1.79633	0.00024	2.00000	–0.00007

Hence, the approximate value of the root to four places of decimal is 1.7963.

EXAMPLE 4. *Find a real root of the equation $x^3 - x^2 - 2 = 0$ by Regula-Falsi method.*

SOLUTION. Let $f(x) = x^3 - x^2 - 2 = 0$

Then $f(0) = -2, f(1) = 1^3 - 1^2 - 2 = -2$

and $f(2) = 2^3 - 2^2 - 2 = 2$

Thus, the root lies between 1 and 2.

First approximation : Taking $x_0 = 1, x_1 = 2, f(x_0) = -2, f(x_1) = 2$.

Then by Regula-Falsi method, an approximation to the root is given by

$$x_2 = x_0 - \frac{x_1 - x_0}{f(x_1) - f(x_0)} \cdot f(x_0)$$

$$= 1 - \frac{2-1}{2+2}(-2) = 1 + \frac{1}{2} = 1.5$$

Now, $f(x_2) = f(1.5)$

$$= (1.5)^3 - (1.5)^2 - 2 \text{ [Using (1)]}$$

$$= 3.375 - 4.25 = -0.875$$

Thus the root lies between 1.5 and 2.

Second approximation: Taking $x_0 = 1.5$, $x_1 = 2$, $f(x_0) = -0.875$, $f(x_1) = 2$. Then the next approximation to the root is

$$x_3 = x_0 - \frac{x_1 - x_0}{f(x_1) - f(x_0)} \cdot f(x_0)$$

$$= 1.5 - \frac{2 - 1.5}{2 + 0.875}(-0.875)$$

$$= 1.5 + 0.1522 = 1.6522$$

Now, $f(x_3) = f(1.6522)$,

$$= (1.6522)^3 - (1.6522)^2 - 2 \text{ [Using (1)]}$$

$$= 4.5101 - 4.7298 = -0.2197$$

Thus, the root lies between 1.6522 and 2.

Therefore, taking $x_0 = 1.6522$, $x_1 = 2$, $f(x_0) = -0.2197$ and $f(x_1) = 2$.

Third approximation: The next approximation to the root is

$$x_4 = x_0 - \frac{x_1 - x_0}{f(x_1) - f(x_0)} f(x_0)$$

$$= 1.6522 - \frac{2 - 1.6522}{2 + 0.2197}(-0.2197) \quad = 1.6866$$

Next, $f(x_4) = f(1.6866)$

$$= (1.6866)^3 - (1.6866)^2 - 2 \quad = -0.0469$$

Thus, the root lies between 1.6866 and 2

Fourth approximation: Taking $x_0 = 1.6866$, $x_1 = 2$, $f(x_0) = -0.046$ and $f(x_1) = 2$, Then we have

$$x_5 = x_0 - \frac{x_1 - x_0}{f(x_1) - f(x_0)} f(x_0)$$

$$= 1.6866 - \frac{2 - 1.6866}{2 + 0.0469}(-0.0469) \quad = 1.6938$$

Now, $f(x_5) = f(1.6938)$

$$= (1.6938)^3 - (1.6938)^2 - 2 = -0.0096$$

Fifth approximation: The root lies between 1.6938 and 2. Therefore, taking $x_0 = 1.6938$, $x_1 = 2$, $f(x_0) = -0.0096$ and $f(x_1) = 2$. Then next approximation to the root is

$$x_6 = x_0 - \frac{x_1 - x_0}{f(x_1) - f(x_0)} \cdot f(x_0)$$

$$= 1.6938 - \frac{2-1.6938}{2+0.0096}(-0.0096) = 1.6953$$

Now, $f(x_6) = f(1.6953)$

$$= (1.6953)^3 - (1.6953)^2 - 2$$

$$= 4.8724 - 4.8740 = -0.0016$$

Therefore, the root lies between 1.6953 and 2. Therefore, taking $x_0 = 1.6953$, $x_1 = 2$, $f(x_0) = -0.0016$, $f(x_1) = 2$.

Sixth approximation: The next approximation to the root is

$$x_7 = x_0 - \frac{x_1 - x_0}{f(x_1) - f(x_0)} \cdot f(x_0)$$

$$= 1.6953 - \frac{2-1.6953}{2+0.0016}(-0.0016)$$

$$= 1.6953 + 0.0002 = 1.6955$$

Hence, the root is 1.695 correct to the three places of decimal.

EXAMPLE 5. *Find a real root of the equation $xe^x - 3 = 0$, using Regula-Falsi method correct to three decimal places.*

SOLUTION. We have $f(x) = xe^x - 3 = 0$.

Then $f(1) = e - 3 = -0.2817$ ...(1)

and $f(1.5) = (1.5)e^{1.5} - 3 = (3.7225)$

∴ The root lies between 1 and 1.5. Therefore, taking $x_0 = 1$, $x_1 = 1.5$, $f(x_0) = -0.2817$ and $f(x_1) = 3.7225$. Then approximation to the root is

First approximation:

$$x_2 = x_0 - \frac{x_1 - x_0}{f(x_1) - f(x_0)} \cdot f(x_0)$$

$$= 1 - \frac{1.5-1}{3.7225 - 0.2817}(-0.2817) = 1 + \frac{0.14085}{4.0042} = 1.0352$$

Now, $f(x_2) = f(1.0352)$

$$= (1.0352)e^{1.0352} - 3$$

$$= 2.9148 - 3 = -0.0852$$

Thus, the root lies between 1.0352 and 1.5. Then taking $x_0 = 1.0352$, $x_1 = 1.5$, $f(x_0) = -0.0852$ and $f(x_1) = 3.7225$.

Second approximation: The next approximation to the root is

$$x_3 = x_0 - \frac{x_1 - x_0}{f(x_1) - f(x_0)} \cdot f(x_0)$$

$$= 1.0352 - \frac{1.5-1.0352}{3.7225 + 0.0852}(0.0852)$$

$$= 1.0352 + 0.0104 = 1.0456$$

Next, $f(x_3) = f(1.0456)$

$$= (1.0456) e^{1.0456} - 3 = -0.0252$$

Thus, the root lies between 1.0456 and 1.5.

Therefore, taking $x_0 = 1.0456$, $x_1 = 1.5$, $f(x_0) = -0.0252$ and $f(x_1) = 3.7225$.

Third approximation: The next approximation to the root is

$$x_4 = x_0 - \frac{x_1 - x_0}{f(x_1) - f(x_0)} \cdot f(x_0)$$

$$= 1.0456 - \frac{1.5 - 1.0456}{3.7225 + 0.0252}(-0.0252) = 1.0487$$

Now, $\quad f(x_4) = f(1.0487)$

$$= (1.0487) e^{1.0487} - 3 = 2.9929 - 3 = -0.0071$$

The root lies between 1.0487 and 1.5.

Fourth approximation: The next approximation to the root is

$$x_5 = x_0 - \frac{x_1 - x_0}{f(x_1) - f(x_0)} \cdot f(x_0)$$

$$= 1.0487 - \frac{1.5 - 1.0487}{3.7225 + 0.0071}(-0.0071) = 1.0496$$

$$= 1.0487 + \frac{0.0032}{3.7296} = 1.0496$$

Now, $\quad f(x_5) = f(1.0496) = (1.0496)e^{1.0496} - 3$

$$= 2.9929 - 3 = -0.0018$$

The root lies between 1.0496 and 1.5.

Fifth approximation: Taking $x0 = 1.0496$, $x1 = 1.5$, $f(x_0) = -0.0018$ and $f(x_1) = 3.7225$. Then the next approximation to the root is given by

$$x_6 = x_0 - \frac{x_1 - x_0}{f(x_1) - f(x_0)} \cdot f(x_0)$$

$$= 1.0496 - \frac{1.5 - 1.0496}{3.7225 + 0.0018}(-0.0018)$$

$$= 1.0496 + \frac{0.00081}{3.7243} = 1.0498$$

Hence, the root is approximately 1.0498 correct to three decimal places.

EXAMPLE 6.   *Find a real root of the equation* $3x + \sin x - e^x = 0$ *by false position method.*

                                                     (UPTU(MCA)–2007)

SOLUTION.   We have $\qquad f(x) = 3x + \sin x - e^x = 0$

$$f(0.3) = -0.154 \; (- ve)$$

$$f(0.4) = 0.975 \; (+ ve)$$

$\Rightarrow$ Root lies between 0.3 and 0.4.

First Approximation : Using Regula-Falsi method, we have

$$x_2 = x_0 - \frac{x_1 - x_0}{f(x_1) - f(x_0)} f(x_0)$$

$$= 0.3 - \frac{0.4 - 0.3}{(0.0975) - (-0.154)}(-0.154)$$

$$= 0.3 + \left(\frac{0.1 \times 0.154}{0.2515}\right) = 0.3612$$

Also, $\qquad f(x_2) = f(0.3612) = 0.0019 \; (+ ve)$

$\Rightarrow$ Root lies between 0.3 and 0.3612.

Second Approximation: We have

$$x_3 = x_0 - \frac{x_2 - x_0}{f(x_2) - f(x_0)} f(x_0)$$

$$= 0.3 - \left\{ \frac{0.3612 - 0.3}{0.0019 - (-0.154)} \right\} (-0.154)$$

$$= 0.3 + \left(\frac{0.0612}{0.1559} \right) (0.154) = 0.3604$$

Also, $\quad f(x_3) = f(0.3604) = -0.00005 \; (-\text{ve})$

$\Rightarrow$ Root lies between 0.3604 and 0.3612.

Third Approximation: We have

$$x_4 = x_3 - \frac{x_2 - x_3}{f(x_2) - f(x_3)} f(x_3)$$

$$= 0.3604 - \left\{ \frac{0.3612 - 0.3604}{0.0019 - (-0.00005)} \right\} (-0.00005)$$

$$= 0.3604 - \left(\frac{0.0008}{0.00195} \right) (0.00005) = 0.36042$$

We observe that x_3 and x_4 are approximately the same. Hence, the required real root is 0.3604 correct to four decimal places.

EXAMPLE 7. *Find the root of the equation $xe^x = \cos x$ in the interval $(0, 1)$ using Regula-Falsi method correct to four decimal places.* (Bhopal–2009, UPTU-2004)

SOLUTION. We have $\quad f(x) = \cos x - xe^x = 0$

$\therefore \qquad f(0) = 1, f(1) = \cos 1 - e = -2.17798$

$\Rightarrow$ Root lies between 0 and 1.

Then, by Regula-Falsi method.

$$x_2 = x_0 - \frac{(x_1 - x_0)}{f(x_1) - f(x_0)} f(x_0) = 0 - \frac{1 - 0}{-3.17798} \cdot 1 = 0.31467$$

Also, $\quad f(x_2) = f(0.31467) = 0.51987$

$\Rightarrow$ Root lies between 0.31487 and 1.

Thus, $\quad x_3 = 0.31487 - \frac{(1 - 0.31487)}{(-2.17798 - 0.51987)} (0.51987) = 0.44673$

Also, $\quad f(x_3) = 0.20356$

Repeating this process upto a required number of times, we get the required root is 0.5177 upto 4 decimal places.

EXAMPLE 8. *Using the method of false position, find the root of the equation $x^6 - x^4 - x^3 - 1 = 0$ upto four decimal places.* (Nagarjuna–2001; UPTU-2003)

SOLUTION. We have $\quad f(x) = x^6 - x^4 - x^3 - 1$

Then $\quad f(1.4) = -0.056$ and $f(1.41) = 0.102$

$\Rightarrow$ Root lies between 1.4 and 1.41

First Approximation : By Regula-Falsi method

$$x_2 = x_0 - \frac{x_1 - x_0}{f(x_1) - f(x_0)} f(x_0)$$

$$= 1.4 - \left(\frac{1.41 - 1.4}{0.102 + 0.056} \right)(-0.056) = 1.4035$$

Now, $f(x_2) = -0.0016$ which is negative.

$\Rightarrow$ Root lies between 1.4035 and 1.41

Second Approximation : Again by Regula-Falsi method

$$x_3 = x_2 - \frac{x_1 - x_2}{f(x_1) - f(x_2)} f(x_2)$$

$$= 1.4035 - \left(\frac{1.41 - 1.4035}{0.102 + 0.0016} \right)(-0.0016) = 1.4036$$

Now, $f(x_3) = -0.0003$, which is negative.

$\Rightarrow$ Root lies between 1.4036 and 1.41.

Third Approximation : By Regula-Falsi method

$$x_4 = x_3 - \left\{ \frac{x_1 - x_3}{f(x_1) - f(x_3)} \right\} f(x_3)$$

$$= 1.4036 - \left(\frac{1.41 - 1.4036}{0.102 + 0.00003} \right)(-0.00003) = 1.4036$$

Since x_3 and x_4 are approximately equal. Hence, the required root is given by 1.4036.

EXAMPLE 9.   *Find a real root of the equation* $x \log_{10} x = 1.2$ *by Regula-Falsi method correct to four decimal places.*   [UPTU(MCA)–2004, Agra–2006, VTU–2003, 10, JNTU–2008,

Kottayam–2005, Mumbai–2004, Burdwan–2003]

SOLUTION.   We have $f(x) = x \log_{10} x - 1.2$

$\therefore$ $f(2.74) = -0.0005634$

and $f(2.741) = -0.0003087$

First Approximation : By Regula-Falsi method, we have

$$x_2 = x_0 - \left\{ \frac{x_1 - x_0}{f(x_1) - f(x_0)} \right\} . f(x_0)$$

$$= 2.74$$

$$- \left\{ \frac{2.741 - 2.74}{0.0003087 - (-0.0005634)} \right\}$$

$$(-0.0005634)$$

$$= 2.74$$

$$- \left\{ \frac{0.001}{0.0008721} \right\} (0.0005634)$$

$$= 2.740646027$$

Also, $f(x_2) = -0.00000006016$, *i.e.,* negative

$\Rightarrow$ Root lies between 2.740646027 and 2.741.

Second Approximation : We have

$$x_3 = x_2 - \left\{\frac{x_1 - x_2}{f(x_1) - f(x_2)}\right\} f(x_2)$$

$$= 2.740646027$$

$$-\left(\frac{2.741 - 2.740646027}{0.0003087 + 0.000000060616}\right)$$

$$-(0.000000060616) = 2.740646096$$

We observe that x_2 and x_3 agree upto seven decimal places. Hence, the required root correct to four decimal places is 2.7406.

EXAMPLE 10. *Find the smallest positive root of the equation* $x - e^{-x} = 0,$ *using false position method.* [UPTU(MCA)–2003]

SOLUTION. We have $\qquad f(x) = x - e^{-x}$

$\therefore \qquad\qquad f(0.56) = -0.01121$ and $f(0.58) = 0.201$

$\Rightarrow$ Root lies between 0.56 and 0.58.

Let $x_0 = 0.56$ and $x_4 = 0.58$

First Approximation: We have

$$x_2 = x_0 - \left\{\frac{x_1 - x_0}{f(x_1) - f(x_0)}\right\} . f(x_0)$$

$$= 0.56 - \left(\frac{0.58 - 0.56}{0.0201 + 0.01121}\right)(-0.01121) = 0.56716$$

Also, $\qquad f(x_2) = 0.00002619,$ *i.e.,* positive

$\Rightarrow$ Root lies between 0.56 and 0.56716.

Second Approximation: We have

$$x_3 = x_0 - \left\{\frac{x_2 - x_0}{f(x_2) - f(x_0)}\right\} . f(x_0)$$

$$= 0.56 - \left(\frac{0.56716 - 0.56}{0.00002619 + 0.01121}\right)(-0.01121) = 0.567143$$

We observe that x_2 and x_3 agree upto four decimal places. Hence, the required root correct to three decimal places is 0.567.

EXAMPLE 11. *The equation* $x^6 - x^4 - x^3 - 1 = 0$ *has one real root between* 1.4 *and* 1.5 *using false position method find the root* (*three iterations*) (UPTU(MCA)–2001)

SOLUTION. $\qquad f(x) = x^6 - x^4 - x^3 - 1$

$$f(1.4) = -2.20736 \text{ and } f(1.5) = 1.953125$$

Hence, root lies between 1.4 and 1.5.

First Approximation :

Taking $x_0 = 1.4, x_1 = 1.5, f(x_0) = -2.20736$ in the formula

$$x_2 = x_0 - \frac{(x_1 - x_0)}{f(x_1) - f(x_0)} f(x_0) \qquad\qquad ...(1)$$

$$= 1.4 - \frac{(1.5-1.4)(-2.20736)}{1.953125+2.20736} = 1.453055353$$

$$f(x_2) = 0.86419171$$

i.e., the root lies between 1.4 and 1.453055353

Second Approximation :

Taking $\quad x_0 = 1.40, x_1 = 1.453055353,$
$$f(x_0) = -2.20736$$

$$f(x_1) = 0.8641917 \text{ and putting in (1)}$$

$$x_3 = 1.4$$
$$- \frac{(1.453055353-1.4)}{0.86419171+2.29736}(-2.20736)$$

$$= 1.4 + \frac{(0.053055353)}{3.07155171}(2.20736) = 1.438128046$$

$$f(x_3) = 0.594926086$$

Hence, root lies between 1.4 and 1.438128046.

Third Approximation: Taking

$$x_0 = 1.4, x_1 = 1.438128046, f(x_0) = -2.20736,$$

$$f(x_1) = 0.594926086 \text{ becomes (1)}$$

$$x_4 = 1.4 - \frac{0.038128}{2.802286086}(-2.20736)$$

$$= 1.4 + 0.30033415 = 1.430033415$$

$$f(x_4) = -0.445778722.$$

Hence, root lies between 1.430033415 and 1.438128046.

Hence, approximate root is 1.430.

EXERCISE 3.3

1. Find a positive root if $xe^x = 2$ by the method of false position. (SVTU–2007, UPTU-2003)
2. Find real cube root of 18 by Regula-Falsi method.
3. Find the real root of $f(x) = x^3 - 98$ by Regula-Falsi method within $E_8 = 0.01\%$.
4. Find the real root of the following equations:
   (i) $x^4 - x^3 - 2x^2 - 6x - 4 = 0$
   (ii) $x = \tan x$
   (iii) $(5-x)e^x = 5$ near $x = 5$
   (iv) $x^4 + x^3 - 7x^2 - x + 5 = 0$ lying between 2 and 3.

(v) $x^3 - 3x + 4 = 0$ between –2 and –3
(vi) $x^2 - 9 = 0$
5. Find a real root of the following equations correct to three decimal places, by the method of false position:
   (i) $x^3 + x - 1 = 0$ (Ranchi–2000)
   (ii) $x^3 - 4x - 9 = 0$ (VTU–2007)
6. Locate the root of $f(x) = x^{10} - 1 = 0$, between 0 and 1.3 using bisection method and method of false position comment on which method is preferable. (Pune–2004)

Answers

1. 0.852605　2. 2.620741394　3. 4.6104　4.(i) 2.7320506　(ii) 4.4934
(iii)4.9651142　(iv) 2.0608526　(v)–2.195823345　(vi) 3　5. (i) 0.686　(ii) 2.7065
6. (i) 0.99976　(ii) 0.99931

3.9 NEWTON-RAPHSON'S METHOD

One of the most widely used methods of solving equations is Newton's method. Like the previous ones, this method is also based on a linear approximation of the function, but does so using a tangent to the curve.

3.9.1 DERIVATION OF NEWTON-RAPHSON METHOD USING TAYLOR'S SERIES EXPANSION

Let x_n be an estimate of a root of the function $f(x)$. Also, let h be a small interval such that

$$h = x_{n+1} - x_n \qquad \qquad ...(1)$$

Now by Taylor's series expansion, we have

$$f(x_{n+1}) = f(x_n + h) = f(x_n) + hf'(x_n) + \frac{h^2}{2!} f''(x_n) + ...$$

Since h is small we can neglect second, third and higher degree terms in h. Therefore, we get

$$f(x_{n+1}) = f(x_n) + hf'(x_n)$$

If x_{n+1} is a root of $f(x)$, then $\qquad f(x_{n+1}) = 0$

which implies $\quad f(x_n) + hf'(x_n) = 0 \qquad \qquad \Rightarrow \qquad h = -\dfrac{f(x_n)}{f'(x_n)}$

Now using (1), we get $\qquad h = x_{n+1} - x_n = -\dfrac{f(x_n)}{f'(x_n)} \qquad \Rightarrow \qquad x_{n+1} = x_n - \dfrac{f(x_n)}{f'(x_n)}$

ALGORITHM

Step 1.	Take initial approximation say x_0.
Step 2.	Find $f(x_0)$ and $f'(x_0)$.
Step 3.	Compute next approximation using $x_1 = x_0 - \dfrac{f(x_0)}{f'(x_0)}$.
Step 4.	Check the accuracy of latest approximation.
Step 5.	Using above procedure, find other successive approximation x_{i+1}.

▶ **REMARKS**

▸ If $f'(x_0)$ is large, i.e., the graph of $y = f(x)$ is nearly vertical to the x-axis, while crossing it, then Newton's method is very useful.

▸ This method can also be used when the roots are complex.

▸ The convergence of Newton's method for double root is linear.

▸ Newton's method is conditionally convergent.

▸ Newton's method converges if $\left| f(x) f''(x) \right| < \left| f'(x) \right|^2$.

3.9.2 GEOMETRICAL INTERPRETATION OF NEWTON'S METHOD

Let x_0 be an approximation very close to the root a of the equation $f(x) = 0$. Let $A(x_0, f(x_0))$ be the point on curve $y = f(x)$. Then the equation of the tangent at $A(x_0, f(x_0))$ is given by

$$y - f(x_0) = f'(x_0)(x - x_0)$$

The line cuts the x-axis where $y = 0$ then

$$x_1 = x_0 - \frac{f(x_0)}{f'(x_0)}$$

The point x_1 is a first approximation to the root a.

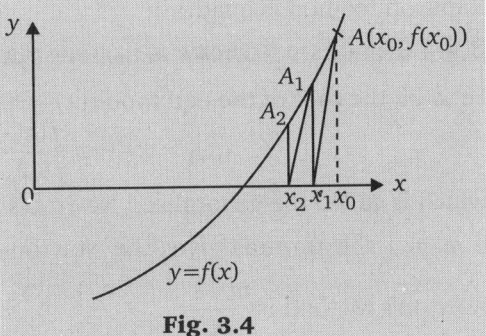

Fig. 3.4

Let $A_1(x_1, f(x_1))$, be a point corresponding to x_1, then the tangent at A_2 cuts the x-axis at x_2 which is second approximation to the root. Continue this process, we get better approximation to the root a.

3.9.3 RATE OF CONVERGENCE OF NEWTON'S METHOD

(UPTU–2002, 07, UPTU(MCA)–02, 03, 06, 08)

Let x_n be different from the root a of $f(x) = 0$ by a small quantity ϵ_n so that

and
$$\left. \begin{array}{l} x_n = a + \epsilon_n \\ x_{n+1} = a + \epsilon_{n+1} \end{array} \right\} \qquad \ldots(1)$$

Since Newton's method gives

$$x_{n+1} = x_n - \frac{f(x_n)}{f'(x_n)} \qquad \ldots(2)$$

From (1) and (2) we get,

$$a + \epsilon_{n+1} = a + \epsilon_n - \frac{f(a + \epsilon_n)}{f'(a + \epsilon_n)} \quad \Rightarrow \quad \epsilon_{n+1} = \epsilon_n - \frac{f(a + \epsilon_n)}{f'(a + \epsilon_n)}$$

Using Taylor's theorem

$$\Rightarrow \epsilon_{n+1} = \epsilon_n - \frac{f(a) + \epsilon_n \ f'(a) + \dfrac{\epsilon_n^2}{2!} f''(a) + \ldots}{f'(a) + \epsilon_n \ f''(a) + \dfrac{\epsilon_n^2}{2!} f'''(a) + \ldots} = \epsilon_n - \frac{\epsilon_n \ f'(a) + \dfrac{\epsilon_n^2}{2!} f''(a) + \ldots}{f'(a) + \epsilon_n \ f''(a) + \dfrac{\epsilon_n^2}{2!} f'''(a) + \ldots}$$

$$(\because f(a) = 0)$$

$$= \frac{\dfrac{\epsilon_n^2}{2!} f''(a)}{f'(a) - \epsilon_n \ f''(a)} \qquad \text{(Neglecting third and higher power of } \epsilon_n)$$

$$= \frac{\epsilon_n^2 \ f''(a)}{2 f'(a)} \left[1 + \frac{\epsilon_n \ f''(a)}{f'(a)} \right]^{-1}$$

$$\therefore \qquad \epsilon_{n+1} = \frac{\epsilon_n^2 \ f''(a)}{2 f'(a)}$$

(Expand by Binomial theorem and neglecting ϵ_n^3 and higher powers of ϵ_n)

$$\therefore \qquad \epsilon_{n+1} \propto \epsilon_n^2$$

This shows that subsequent error is proportional to the square of the previous error. Hence, the convergence is of second order or a quadratic. Hence, rate of convergence of Newton-Raphson method is quadratic.

3.9.4 GENERALISED NEWTON'S METHOD FOR MULTIPLE ROOTS

(UPTU(MCA)–2003)

Let a be the root of the equation $f(x) = 0$ of order m. Then

$$x_{n+1} = x_n - m \frac{f(x_n)}{f'(x_n)}$$

which is called the generalised Newton's Method.

If $m = 1$ this formula gives the Newton-Raphson's method. In this case, the convergence of Newton's Method is $\dfrac{m-1}{m}$.

3.9.5 LIMITATIONS OF NEWTON-REPHSON'S METHOD

Newton-Raphson method is not applicable in the following situations:

1. Division by zero may occur if $f'(x_i)$ is zero.
2. If the initial guess is too far away from the required root, the process may converge to some other root.
3. A particular value in the iteration sequence may repeat, resulting in an infinite loop.

▶ REMARK

▸ The Newton-Rephson method approximate the curves of $f(x)$ by tangents. Some complication will arise if the derivatives $f(x_n)$ is zero. In such cases a new initial value for x must be chosen to continue the procedure.

SOLVED EXAMPLES

EXAMPLE 1. *Find a real root of the equation $3x = \cos x + 1$ by Newton's method.*

(UPTU(MCA)–2004, Rohtak–2005; VTU-2009; UPTU-2004)
...(1)

SOLUTION. Let $\qquad f(x) = 3x - \cos x - 1$

So that $\qquad f(0) = -2$

and $\qquad f(1) = 3 - \cos 1 - 1 = 3 - 0.5403 - 1 = 1.4597$

So a root of the equation $f(x) = 0$ lies between 0 and 1.

Let us take $\quad x_0 = 0.6$

From (1) $\quad f'(x) = 3 + \sin x$ ...(2)

Therefore, the Newton's Method gives

$$x_{n+1} = x_n - \frac{f(x_n)}{f'(x_n)}$$

$$= x_n - \frac{3x_n - \cos x_n - 1}{3 + \sin x_n} \qquad \text{[From (1) and (2)]}$$

$$\therefore \qquad x_{n+1} = \frac{x_n \sin x_n + \cos x_n + 1}{3 + \sin x_n} \qquad ...(3)$$

First approximation: Putting $n = 0$, we get first approximation

$$x_1 = \frac{x_0 \sin x_0 + \cos x_0 + 1}{3 + \sin x_0}$$

$$= \frac{0.6 \sin(0.6) + \cos(0.6) + 1}{3 + \sin(0.6)} = \frac{0.6(0.5646) + 0.8253 + 1}{3 + 0.5646}$$

$$= \frac{2.16406}{3.5646} = 0.6071$$

Second approximation: Putting $n = 1$ into (3), we get second approximation

$$x_2 = \frac{x_1 \sin x_1 + \cos x_1 + 1}{3 + \sin x_1}$$

$$= \frac{0.6071 \sin(0.6071) + \cos(0.6071) + 1}{3 + \sin(0.6071)}$$

$$= \frac{0.6071(0.5705) + 0.8213 + 1}{3 + 0.5705} = \frac{2.1677}{3.5705} = 0.6071$$

Hence $x_1 = x_2$. Then root is 0.6071 correct to four decimal places.

EXAMPLE 2. *Find an iterative formula to find $\sqrt{N}$, where N is a positive number and hence, find $\sqrt{12}$ correct to four decimal places.*

SOLUTION. Let
$$x = \sqrt{N} \Rightarrow x^2 = N$$
i.e., $\quad x^2 - N = 0$

Let $\quad\quad f(x) = x^2 - N$

By Newton-Raphson method, we have
$$x_{n+1} = x_n - \frac{f(x_n)}{f'(x_n)} = x_n - \frac{x_n^2 - N}{2x_n}$$

$$\therefore \quad\quad x_{n+1} = \frac{1}{2}\left[x_n + \frac{N}{x_n} \right] \quad\quad\quad\quad ...(1)$$

which is the required iteration formula.

Now putting $N = 12$ we have
$$f(x) = x^2 - 12$$

Clearly $f(3) < 0$ and $f(4) > 0$

Therefore, root lies between 3 and 4.

Let the initial approximation x_0 by 3.1. Then

First approximation:
$$x_1 = \frac{1}{2}\left[x_0 + \frac{12}{x_0} \right] = \frac{1}{2}\left[3.1 + \frac{12}{3.1} \right]$$
$$= 3.485839$$

Second approximation:
$$x_2 = \frac{1}{2}\left[x_1 + \frac{12}{x_1} \right] = \frac{1}{2}\left[3.4854839 + \frac{12}{3.4854839} \right]$$
$$= 3.4641672$$

Third approximation:
$$x_3 = \frac{1}{2}\left[3.4641672 + \frac{12}{3.4641672} \right]$$
$$= 3.4641016$$

Hence, the value of $\sqrt{12}$ correct to four decimals is 3.4641.

EXAMPLE 3. *Find the real root of the equation $x\log_{10} x = 1.2$ by Newton-Raphson's Method.*

(UPTU(Ag.)–2004, Rohtak–2004, 06, Mumbai-2004)

SOLUTION. Let $\quad\quad\quad f(x) = x\log_{10} x - 1.2 = 0$

Then $\quad\quad\quad f(1) = -1.2$
$$f(2) = 2\log_{10}2 - 1.2 = -0.5979$$
and $\quad\quad\quad f(3) = 3\log_{10}3 - 1.2 = 0.2314$

$\therefore$ A root of $f(x) = 0$ lies between 2 and 3.

Let us take $x_0 = 2$

From (1), we have
$$f'(x) = \log_{10} x + \frac{1}{x}.x\log_{10} e$$
$$= \log_{10} x + 0.4343 \quad\quad\quad\quad ...(2)$$

Now by Newton's Method we have

$$x_{n+1} = x_n - \frac{f(x_n)}{f'(x_n)}$$

$$x_{n+1} = x_n - \frac{x_n \log_{10} x_n - 1.2}{\log_{10} x_n + 0.4343}$$

[From (1) and (2)]

$$\Rightarrow \qquad x_{n+1} = \frac{0.4343 x_n + 1.2}{\log_{10} x_n + 0.4343}, n = 0, 1, 2, \ldots$$

...(3)

Putting $n = 0$, we get the first approximation.

First approximation:

$$x_1 = \frac{0.4343 x_0 + 1.2}{\log_{10} x_n + 0.4343}$$

$$= \frac{0.4343(2) + 1.2}{\log_{10}(2) + 0.4343} = \frac{2.0686}{0.7353} = 2.8133$$

Putting $n = 1$ into (3), we get second approximation.

Second approximation:

$$x_2 = \frac{0.4343 x_1 + 1.2}{\log_{10} x_1 + 0.4343}$$

$$= \frac{0.4343(2.8133) + 1.2}{\log_{10} 2.8133 + 0.4343} = \frac{2.4128}{0.8835} = 2.7411$$

Putting $n = 2$ into (3), we get third approximation.

Third approximation:

$$x_3 = \frac{0.4343 x_2 + 1.2}{\log_{10} x_2 + 0.4343}$$

$$= \frac{0.4343(2.7411) + 1.2}{\log_{10} 2.7411 + 0.4343} = \frac{2.3905}{0.8722} = 2.7408$$

Putting $n = 3$ into (3), we get fourth approximation.

Fourth approximation:

$$x_4 = \frac{0.4343 x_3 + 1.2}{\log_{10} x_3 + 0.4343}$$

$$= \frac{0.4343(2.7408) + 1.2}{\log_{10} 2.7408 + 0.4343} = \frac{2.3903}{0.8721} = 2.7408$$

Since, $x_3 = x_4$. Hence, the root of $f(x) = 0$ is 2.7408 correct to four decimal places.

EXAMPLE 4. *Find the real root of the equation $x^2 - 5x + 2 = 0$ by Newton-Raphson's Method.*

SOLUTION. Let $\qquad f(x) \equiv x^2 - 5x + 2 = 0$ ...(1)

Now $\qquad f(4) = 4^2 - 5(4) + 2 = -2$

and $\qquad f(5) = 5^2 - 5(5) + 2 = 2$

Therefore, the root lies between 4 and 5.

From (1), we get $f'(x) = 2x - 5$ ... (2)

Now, Newton-Raphson's Method becomes

$$x_{n+1} = x_n - \frac{f(x_n)}{f'(x_n)}$$

$$\Rightarrow \qquad x_{n+1} = x_n - \frac{x_n^2 - 5x_n + 2}{2x_n - 5}$$

[Using (1) and (2)]

$$\Rightarrow \qquad x_{n+1} = \frac{x_n^2 - 2}{2x_n - 5}, n = 0, 1, 2, \dots \qquad \dots (3)$$

Let us take $x_0 = 4$

Putting $n = 0$ into (3), we get first approximation to the root.

First approximation:

$$x_1 = \frac{x_0^2 - 2}{2x_0 - 5} = \frac{4^2 - 2}{2(4) - 5} = \frac{14}{3} = 4.6667$$

Putting $n = 1$ into (3), we get second approximation,

Second approximation:

$$x_2 = \frac{x_1^2 - 2}{2x_1 - 5} = \frac{(4.6667)^2 - 2}{2(4.6667) - 5}$$

$$= \frac{19.7781}{4.3334} = 4.5641$$

Putting $n = 2$ into (3), we get third approximation.

Third approximation:

$$x_3 = \frac{x_2^2 - 2}{2x_2 - 5} = \frac{(4.5641)^2 - 2}{2(4.5641) - 5}$$

$$= \frac{18.8310}{4.1282} = 4.5616$$

Putting $n = 3$ into (3), we get fourth approximation.

Fourth approximation:

$$x_4 = \frac{x_3^2 - 2}{2x_3 - 5} = \frac{(4.5616)^2 - 2}{2(4.5616) - 5} = \frac{18.8082}{4.1232} = 4.5616$$

Since $x_3 = x_4$. Hence, the root of the equation is 4.5616 correct to four decimal places.

EXAMPLE 5. *Solve* $x^3 + 2x^2 + 10x - 20 = 0$ *by Newton-Raphson method.*

SOLUTION. Let $\qquad f(x) = x^3 + 2x^2 + 10x - 20$

$$\Rightarrow \qquad f'(x) = 3x^2 + 4x + 10$$

Now, by Newton-Raphson method, we have

$$x_{n+1} = x_n - \frac{f(x_n)}{f'(x_n)}$$

$$= x_n - \left[\frac{x_n^3 + 2x_n^2 + 10x_n - 20}{3x_n^2 + 4x_n + 10} \right]$$

$$= \frac{2(x_n^3 + x_n^2 + 10)}{3x_n^2 + 4x_n + 10} \qquad \dots (1)$$

Clearly $\qquad f(1) = -7 < 0$

and $\qquad f(2) = 16 > 0$

$\therefore$ Root lies between 1 and 2.

Let $x_0 = 1.2$ be the initial approximation

Then $[\because f(1.2) < 0]$

First approximation:

$$x_1 = \frac{2(x_0^3 + x_0^2 + 10)}{3x_0^2 + 4x_0 + 10} = \frac{2[(1.2)^3 + (1.2)^2 + 10]}{3(1.2)^2 + 4(1.2) + 10} = \frac{26.336}{19.12}$$

$$= 1.3774059$$

Second approximation:

$$x_2 = \frac{2(x_1^3 + x_1^2 + 10)}{3x_1^2 + 4x_1 + 10}$$

$$= \frac{2[(1.3774059)^3 + (1.3774059)^2 + 10]}{3(1.3774059)^2 + 4(1.3774059) + 10}$$

$$= \frac{29.021052}{22.01364} = 1.3688295$$

Third approximation:

$$x_3 = \frac{2(x_2^3 + x_2^2 + 10)}{3x_2^2 + 4x_2 + 10}$$

$$= \frac{2[(1.3688295)^3 + (1.3688295)^2 + 10]}{3(1.3688295)^2 + 4(1.3688295) + 10}$$

$$= \frac{28.876924}{21.0964} = 1.3688081$$

Fourth approximation:

$$x_4 = \frac{2(x_3^3 + x_3^2 + 10)}{3x_3^2 + 4x_3 + 10}$$

$$= \frac{2[(1.3688081)^3 + (1.3688081)^2 + 10]}{3(1.3688081)^2 + 4(1.3688081) + 10} = 1.3688081$$

Hence, the required root is 1.3688081.

EXAMPLE 6. *Find the real of root of the equation* $\log x - \cos x = 0$ *correct to the three places of decimal by Newton-Raphson's Method.*

SOLUTION.   Let $f(x) = \log x - \cos x = 0$ ...(1)

So that $f(1) = -0.5403$

$$f(2) = \log 2 - \cos 2$$
$$= 0.6931 + 0.4161 = 1.1092$$

$\therefore$ The root lies between 1 and 2.

Also, $f(1.1) = \log 1.1 - \cos 1.1 = 0.0953 - 0.4535 = -0.3582$

$$f(1.2) = \log 1.2 - \cos 1.2 = -0.18$$

$$f(1.3) = \log 1.3 - \cos 1.3 = 0.2623 - 0.2674 = -0.0051$$

and $f(1.4) = \log 1.4 - \cos 1.4 = 0.3364 - 0.1699 = 0.1665$

Thus the root will lies between 1.3 and 1.4.

Now, from (1) $f'(x) = \dfrac{1}{x} + \sin x$ ...(2)

Then by Newton-Raphson's Method, we get

$$x_{n+1} = x_n - \frac{f(x_n)}{f'(x_n)} \Rightarrow x_{n+1} = x_n - \frac{\log x_n - \cos x_n}{\dfrac{1}{x_n} + \sin x_n}$$

$$\text{[Using (1) and (2)]}$$

$$x_{n+1} = \frac{x_n + x_n^2 \sin x_n + x_n \log x_n + x_n \cos x_n}{1 + x_n \sin x_n} \qquad ...(3)$$

Let us take $x_0 = 1.3$

Now putting $n = 0$ into (3), we get first approximation

First approximation:

$$x_1 = \frac{x_0 + x_0^2 \sin x_0 - x_0 \log x_0 + x_0 \cos x_0}{1 + x_0 \sin x_0}$$

$$= \frac{1.3 + (1.3)^2 \sin(1.3) - 1.3 \log(1.3) + 1.3 \cos(1.3)}{1 + 1.3 \sin(1.3)}$$

$$= \frac{1.3(1 + 1.2526 - 0.2623 + 0.2674)}{1 + 1.2526} = \frac{2.93501}{2.2526} = 1.3029$$

Putting $n = 1$ into (3), we get second approximation

Second approximation:

$$x_2 = \frac{x_1(1 + x_1 \sin x_1 + \log x_1 + \cos x_1)}{1 + x_1 \sin x_1}$$

$$= \frac{1.3029(1 + 1.3029 \sin 1.3029 + \log 1.3029 + \cos 1.3029)}{1 + 1.3029 \sin 1.3029}$$

$$= \frac{1.3029(1 + 1.2526 - 0.2645 + 0.2647)}{1 + 1.2564} = \frac{2.9401}{2.2564} = 1.3030$$

Hence, the required root is 1.303 correct to three decimal places.

EXAMPLE 7.   *Apply Newton's formula to prove that the recurrence formula for finding the nth roots of a is*

$$x_{i+1} = \frac{(n-1)x_i^n + a}{n(x_i^{n-1})}$$

Hence, evaluate $(240)^{1/5}$.

SOLUTION.   Let $x = a^{1/n}$ $\Rightarrow$ $x^n = a$ or $x^n - a = 0$

Let $f(x) = x^n - a = 0 \Rightarrow f'(x) = nx^{n-1}$

Now, by Newton-Raphson Method, we have

$$x_{i+1} = x_i - \frac{f(x_i)}{f'(x_i)} = x_i - \frac{x_i^n - a}{nx_i^{n-1}}$$

$$\Rightarrow \qquad x_{i+1} = \frac{nx_i^n - x_i^n + a}{nx_i^{n-1}} = \frac{(n-1)x_i^n + a}{nx_i^{n-1}} \qquad\qquad ...(1)$$

Now to find the value of $(240)^{1/5}$.

We know that
$$(243)^{1/5} = (3^5)^{1/5} = 3$$
Take $a = 240$ and $n = 5$ in (1), we get
$$x_{i+1} = \frac{4x_i^5 + 240}{5x_i^4} \qquad\qquad ...(2)$$

First approximation: Let $i = 0, x_i = x_0 = 2.9$ (say), then from (2), we get

$$x_i = \frac{4x_0^5 + 240}{5x_0^4} = \frac{4(2.9)^5 + 240}{5(2.9)^4}$$

$$= \frac{4(205.111) + 240}{5 \times 70.7281} = \frac{1060.444}{353.6405} = 2.99$$

Second approximation: Let $i = 1, x_i = x_1 = 2.99$ (say), then from (2), we get

$$x_2 = \frac{4x_i^5 + 240}{5x_i^4} = \frac{4(2.99)^5 + 240}{5(2.99)^4}$$

$$= \frac{4(238.977) + 240}{399.627} = 2.9925$$

Hence, the required value of $(240)^{1/5}$ correct to three places of decimal is 2.993.

EXAMPLE 8. *Show that the modified Newton's-Raphson method*

$$x_{n+1} = x_n - \frac{2f(x_n)}{f'(x_n)} \qquad\qquad ...(1)$$

gives a quadratic convergence when $f(x) = 0$ has a pair of double roots in neighbourhood of $x = x_n$.

SOLUTION. Let ε_n be the error in x_n.

Then, we have $x_n = x_0 + \varepsilon_n$.

Therefore (1) gives

$$\varepsilon_{n+1} = \varepsilon_n - \frac{2f(a + \varepsilon_n)}{f'(a + \varepsilon_n)}$$

Now, expanding in powers of ε_n and using $f(a) = 0, f'(a) = 0$, since $x = a$ is a double root near $x = x_n$.

we get

$$\varepsilon_{n+1} = \varepsilon_n - \frac{\left[\dfrac{2\varepsilon_n^2}{2!} f''(a) + ...\right]}{\left[\varepsilon_n.f''(a) + \dfrac{\varepsilon_n^2}{2!} f'''(a) + ...\right]}$$

$$= \varepsilon_n - \frac{2\varepsilon_n^2 \left[\frac{1}{2!} \{f''(a)\} + \frac{1}{3!} f'''(a) + \dots \right]}{\varepsilon_n \left[f''(a) + \frac{\varepsilon_n}{2!} f'''(a) + \dots \right]} = \varepsilon_n - \frac{2\varepsilon_n \left[\frac{1}{2!} f''(a) + \frac{1}{3!} f'''(a) \right]}{f''(a) + \frac{\varepsilon_n}{2!} f'''(a)}$$

$$\Rightarrow \quad \varepsilon_{n+1} = \frac{1}{6} \varepsilon_n^2 \frac{f'''(a)}{\left[f''(a) + \frac{\varepsilon_n}{2!} f'''(a) \right]}$$

$$\Rightarrow \quad \varepsilon_{n+1} \approx \frac{1}{6} \varepsilon_n^2 \frac{f'''(a)}{f''(a)} \quad \Rightarrow \quad \varepsilon_{n+1} \propto \varepsilon_n^2$$

Hence, rate of convergence of given modified Newton-Raphson method is quadratic.

EXAMPLE 9. *Prove the Chebyshev formula*

$$x_1 = x_0 - \frac{f(x_0)}{f'(x_0)} - \frac{1}{2} \cdot \frac{[f(x_0)]^2 \cdot f''(x_0)}{[f'(x_0)]^3}$$

for the roots of the equation $f(x) = 0$.

SOLUTION. Let $f(x) = 0$ be the given equation, therefore

$$f(x) = f(x + x_0 - x_0) = 0$$

$$= f(x_0) + (x - x_0) f'(x_0) \dots = 0$$

(On expanding by Taylor's series)

$$\Rightarrow \quad x_1 = x_0 - \frac{f(x_0)}{f'(x_0)}, \text{ which is the first approximation to the root.}$$

Again by Taylor's series. We have

$$f(x) = f(x_0) + (x - x_0) f'(x_0) + \frac{(x - x_0)^2}{2!} f''(x_0)$$

$$\Rightarrow \quad f(x_1) = f(x_0) + (x - x_0) f'(x_0) + \frac{(x - x_0)^2}{2!} f''(x_0)$$

But $f(x_1) = 0$ as x_1 is an approximation to the root. Therefore,

$$f(x_0) + (x_1 - x_0) f'(x_0) + \frac{1}{2!} (x_1 - x_0)^2 f''(x_0) = 0$$

$$\Rightarrow \quad x_1 = x_0 - \frac{f(x_0)}{f'(x_0)} - \frac{1}{2} \cdot \frac{[f(x_0)]^2 f''(x_0)}{[f'(x_0)]^3}$$

$$\Rightarrow \quad x_1 = x_0 - \frac{f(x_0)}{f'(x_0)} - \frac{1}{2} \cdot \frac{\{f(x_0)\}^2 f''(x_0)}{\{f'(x_0)\}^3}$$

▶ REMARKS
▸ The above formula is also known as Newton-Raphson's extended formula.
▸ It can be used interactively.

EXAMPLE 10. *By using Newton-Raphson method, find the root of $x^4 - x - 10 = 0$, which is nearer to $x = 2$ correct to three places of decimals.*

(JNTU-2008)

SOLUTION. Here $f(x) = x^4 - x - 10$

$$f'(x) = 4x^3 - 1$$

$$\therefore \qquad x_{n+1} = x_n - \frac{f(x_n)}{f'(x_n)} = x_n - \frac{x_n^4 - x_n - 10}{4x_n^3 - 1} = \frac{3x_n^4 + 10}{4x_n^3 - 1}$$

The approximate value of the root is given to be 2. Taking $x_0 = 2$ we get

$$x_1 = \frac{3x_0^4 + 10}{4x_0^3 - 1} = \frac{3.2^4 + 10}{4.2^3 - 1} = 1.871$$

$$x_2 = \frac{3x_1^4 + 10}{4x_1^3 - 1} = \frac{3.(1.871)^4 + 10}{4.(1.871)^3 - 1} = 1.856$$

$$x_3 = \frac{3x_2^4 + 10}{4x_2^3 - 1} = \frac{3.(1.856)^4 + 10}{4.(1.856)^3 - 1} = 1.856$$

EXAMPLE 11. *Find a real root of the equation* $x = e^{-x}$ *by Newton-Raphson method.*

[UPTU(MCA)–2003, 08]

SOLUTION. Here, we have

$$f(x) = xe^x - 1 \quad \Rightarrow \quad f'(x) = (1 + x)e^x$$

Let $x_0 = 1$

Then $\qquad x_1 = 1 - \left(\dfrac{e-1}{2e} \right) = \dfrac{1}{2}\left(1 + \dfrac{1}{e} \right) = 0.6839397$

Now, $\qquad f(x_1) = 0.3553424$

and $\qquad f'(x_1) = 3.337012$

So that, $\qquad x_2 = 0.6839397 - \dfrac{0.3553424}{3.337012} = 0.5774545$

Proceeding in the same way, we get

$$x_3 = 0.5672297 \text{ and } x_4 = 0.5671433$$

Hence, the required root is 0.5671 correct to four decimal places.

EXAMPLE 12. *Find the smallest root of the equation* $e^{-x} = \sin x$, *upto four places of decimal.*

SOLUTION. We have $\qquad f(x) = e^{-x} - \sin x = 0$

so that $\qquad x_{n+1} = x_n + \dfrac{e^{-x_n} - \sin x_n}{e^{-x_n} + \cos x_n}$

Taking $x_0 = 0.6$, we have

$$x_1 = 0.58848, x_2 = 0.588559$$

Hence, the required value of the root is 0.5885.

EXAMPLE 13. *Show that the following two sequence, both have convergence of the second order with the same limit* $\sqrt{a}$.

$$x_{n+1} = \frac{1}{2} x_n \left(1 + \frac{a}{x_n^2} \right) \text{ and } x_{n+1} = \frac{1}{2} x_n \left(3 - \frac{x_n^2}{a} \right).$$

SOLUTION. We have

$$x_{n+1} = \frac{1}{2} x_n \left(1 + \frac{a}{x_n^2} \right)$$

$$\Rightarrow \quad x_{n+1} - \sqrt{a} = \frac{1}{2} x_n \left(1 - \frac{a}{x_n^2} \right) - \sqrt{a} = \frac{1}{2} \left(x_n + \frac{a}{x_n} - 2\sqrt{a} \right)$$

$$= \frac{1}{2} \left(\sqrt{x_n} - \frac{\sqrt{a}}{\sqrt{x_n}} \right)^2 = \frac{1}{2x_n} \left(x_n - \sqrt{a} \right)^2$$

$$\Rightarrow \quad e_{n+1} = \frac{1}{2x_n} e_n^2$$

which shows the convergence of second order.
Similarly,

$$x_{n+1} - \sqrt{a} = \frac{1}{2} x_n \left(3 - \frac{x_n^2}{a} \right) - \sqrt{a} = \frac{1}{2} x_n \left(1 - \frac{x_n^2}{a} \right) + (x_n - \sqrt{a})$$

$$= \frac{x_n}{2a} (a - x_n^2) + (x_n - \sqrt{a}) = (x_n - \sqrt{a}) \left[1 - \frac{x_n}{2a} (x_n + \sqrt{a}) \right]$$

$$\Rightarrow \quad e_{n+1} = \frac{x_n - \sqrt{a}}{2a} [2a - x_n^2 - x_n \sqrt{a}]$$

$$= -\left(\frac{x_n - \sqrt{a}}{2a} \right) (x_n - \sqrt{a})(x_n + 2\sqrt{a}) = -\frac{(x_n + 2\sqrt{a})}{2a} e_n^2$$

which also shows the quadratic convergence.

EXAMPLE 14. *Find a positive value of $(17)^{1/3}$ correct to four decimal places by Newton-Raphson method.* (UPTU-2003)

Solution. The iterative formula is given by

$$x_{n+1} = \frac{1}{3} \left(2x_n + \frac{a}{x_n^2} \right) \qquad \qquad \dots(1)$$

Here, we have $a = 17$
Taking initial approximation $x_0 = 2.5$
Putting $n = 0$ in (1), we get

$$x_1 = \frac{1}{3} \left(2x_0 + \frac{17}{x_0^2} \right) = \frac{1}{3} \left(5 + \frac{17}{6.25} \right) = 2.5733$$

Putting $n = 1$ in (1), we get

$$x_2 = \frac{1}{3} \left(2x_1 + \frac{17}{x_1^2} \right) = \frac{1}{3} \left(5.1466 + \frac{17}{6.6220} \right)$$

$$= 2.5713$$

Putting $n = 2$ in (1), we get

$$x_3 = \frac{1}{3} \left(2x_2 + \frac{17}{x_2^2} \right) = \frac{1}{3} \left(5.1426 + \frac{17}{6.61158} \right)$$

$$= 2.57128$$

Putting $n = 3$ in (1), we get

$$x_4 = \frac{1}{3}\left(2x_3 + \frac{17}{x_3^2}\right) = \frac{1}{3}\left(5.14256 + \frac{17}{6.61158}\right)$$

$$= 2.57128$$

We observe that x_3 and x_4 agree to four places of decimal. Hence, the required root is 2.5713 correct to four places of decimals.

EXAMPLE 15. *Solve* $x^4 - 5x^3 + 20x^2 - 40x + 60 = 0$ *by Newton-Raphson method when all the roots of the given equations are complex.* (UPTU-2003)

SOLUTION. We have

$$f(x) = x^4 - 5x^3 + 20x^2 - 40x + 60 = 0$$

$\Rightarrow$ $\quad f'(x) = 4x^3 - 15x^2 + 40x - 40$

Thus, Newton-Raphson method gives

$$x_{n+1} = x_n - \frac{f(x_n)}{f'(x_n)} = x_n - \frac{x_n^4 - 5x_n^3 + 20x_n^2 - 40x_n + 60}{4x_n^3 - 15x_n^2 + 40x_n - 40}$$

Putting $n = 0$, and taking $n_0 = 2(1 + i)$ by trial, we get

$\quad x = 1.92(1 + i)$

Further $\quad x_2 = 1.915 + 1.908i$

We know that imaginary roots occur in conjugate pair, therefore, roots are 1.915 $\pm$ 1.908i upto three places of decimals.

If $\alpha \pm i\beta$ be the other pair of roots, then

sum of the roots $= (\alpha + i\beta + \alpha - i\beta + 1.915 + 1.908i + 1.915 - 1.908i)$

$$= 2\alpha + 3.38 = 5$$

$\Rightarrow$ $\quad \alpha = 0.585$

Again, product of the roots

$$= (\alpha^2 + \beta^2)[(1.915)^2 + (1.908)^2] = 60$$

$\quad \beta = 2.805$

Hence, other two roots are $0.585 \pm 2.805i$.

EXAMPLE 16. *The graph of* $y = 2\sin x$ *and* $y = \log x + c$ *touch each other in the neighbourhood of point* $x = 8$. *Find c and the coordinates of the point of contact.*

SOLUTION. We know that graphs will touch each other if values of $\dfrac{dy}{dx}$ at their point of contact is same.

For $\quad y = 2\sin x, \dfrac{dy}{dx} = 2\cos x$

For $\quad y = \log x + c, \dfrac{dy}{dx} = \dfrac{1}{x}$

Thus, $\quad 2\cos x = \dfrac{1}{x} \quad \Rightarrow \quad x\cos x - 0.5 = 0$

Let $\quad f(x) = x\cos x - 0.5 \quad \Rightarrow \quad f'(x) = \cos x - x\sin x$

Then, Newton's formula is given by

$$x_{n+1} = x_n - \frac{x_n\cos x_n - 0.5}{\cos x_n - x_n\sin x_n}$$

For $n = 0$, $x_0 = 8$, the first approximation is given by $x_1 = 7.793$

Also, second approximation

$$x_2 = 7.789 \approx 7.79$$

Now, $y = 2 \sin 7.79 = 1.19960$

Thus, the point of contact is (7.79, 1.996)

Now $y = \log x + c \Rightarrow 1.996 = \log 7.79 + c$

Hence, $c = -0.054$

EXAMPLE 17. *Find the value of p and q so that the rate of convergence of the iterative method is two.*

$$x_{n+1} = px_n + q.\frac{N}{x_n^2} \text{ is 3.}$$ [UPTU(MCA)–2003]

SOLUTION.   Here, we have

$$x^3 = N$$

Let $f(x) = x^3 - N$

If α be the exact root, we have

$$a^3 = N$$

Putting $x_n = \alpha + e_n, x_{n+1} = \alpha + e_{n+1}, N = \alpha^3$ in $x_{n+1} = px_n + q.\dfrac{N}{x_n^2}$

we get $\alpha + e_{n+1} = p(\alpha + e_n) + q.\dfrac{\alpha^3}{(\alpha + e_n)^2}$

$$= p(\alpha + e_n) + q.\frac{\alpha^3}{\alpha^2\left(1 + \dfrac{e_n}{\alpha}\right)^2} = p(\alpha + e_n) + q\alpha.\left(1 + \frac{e_n}{\alpha}\right)^{-2}$$

$$= p(\alpha + e_n) + q\alpha.\left[1 - 2\frac{e_n}{\alpha} + 3\left(\frac{e_n}{\alpha}\right)^2 + ...\right]$$

$$= p(\alpha + e_n) + q\alpha - 2qe_n + 3q\frac{e_n^2}{\alpha} - ...$$

$$= (p + q - 1)\alpha + (p - 2q)e_n + O(e_n^2) + ...$$

For the method to become of order as high as possible, *i.e.*, of order 2 we must have

$$p + q = 1 \quad \text{and} \quad p - 2q = 0$$

Hence $p = \dfrac{2}{3} \quad \text{and} \quad q = \dfrac{1}{3}.$

EXERCISE 3.4

1. Using Newton's iterative formula, find the real root of $x \sin x + \cos x = 0$ which is near to $x = \pi$ correct to 3 decimal places.

2. Using Newton's method, find the roots of the following equations.

   (i) $2x - \log_{10} x = 7$ (ii) $x^2 - 25 = 0$

   (iii) $4(x - \sin x) = 1$ (iv) $x^4 + x^2 - 80 = 0$

   (v) $\log x = \cos x$

   (vi) $x^2 + 4 \sin x = 0$ (Hazaribagh–2009)

3. Show that

$$x_{i+1} = \frac{1}{3}\left(2x_i + \frac{N}{x_i^2}\right).$$

4. Show that the square root can be obtained by

the recursion formula $x_{i+1} = x_i \left(1 - \dfrac{x_i^2 - N}{2N} \right)$.

5. Show that the iterative formula for finding the reciprocals of n is $x_{i+1} = x_i(2 - nx_n)$.

6. Show that the double root of the equation $x^3 - x^2 - x + 1 = 0$ is 1.0001.

7. Show that the equation $f(x) = 1 - xe^{1-x} = 0$ has a double root at $x = 1$. The root is obtained by using the modified Newton-Raphson method with $m = 2$ starting with $x_0 = 0$.

8. Find the Newton-Raphson method, a root of the following equations correct to three decimal places:

   (i) $x^3 - 3x + 1 = 0$ (Bhopal–2009)

   (ii) $x^3 - 2x - 5 = 0$ (PTU–2005)

   (iii) $x^3 - 5x + 3 = 0$ (Mumbai–2004)

   (iv) $3x^3 - 9x^2 + 8 = 0$ (Madras–2003)

   (v) $x^2 + 4\sin x = 0$ (Hazaribagh–2009)

   (vi) $e^x = x^3 + \cos 25\, x$ which is near 4.5   (VTU–2007)

   (vii) $x \log_{10} x = 12.34$, start with $x_0 = 10$ (Anna–2004)

   (viii) $\cos x = xe^x$ (JNTU–2009)

9. Develop an algorithm using N.R. method, to find the fourth root of a positive number N, and hence find

   (i) $4\sqrt{32}$ (WBTU–2005)

   (ii) $\sqrt{5}$ (Anna–2007)

   (iii) $3\sqrt{24}$ (Madras–2003)

10. Evaluate the following (correct to 3 decimal places) by using the Newton-Raphson method:

   (i) $1/18$ (JNTU–2004)

   (ii) $1/\sqrt{15}$

Answers

1. 2.798   2. (i) 3.7892   (ii) 5   (iii) 1.171   (iv) 2.908   (v) 1.303   (vi) –1.934

8. (i) 1.532 (ii) 2.095 (iii) 1.834 (iv) 1.226 (v) –1.9338 (vi) 4.545 (vii) 0.052 (viii) 0.518

9. $x_{n+1} = \dfrac{1}{4}\left(3x_n + \dfrac{N}{x_n^3} \right)$ (i) 2.3784 (ii) 2.2361 (iii) 2.8845   10. (i) 0.0555   (ii) 0.2582

3.10 COMPLEX ROOTS

If the given equation $f_n(x) = 0$ of degree n has at most M positive roots and almost N negative roots, then the equation $f_n(x) = 0$ has at least $n - (M + N)$ imaginary roots. Here it should be noticed that the imaginary roots always occur in pair.

SOLVED EXAMPLE

EXAMPLE. *Solve the equation* $3x^3 - 4x^2 + x + 88 = 0$, *if one root is* $2 + i\sqrt{7}$.

SOLUTION. Here, the given equation is

$$3x^3 - 4x^2 + x + 88 = 0 \qquad \ldots(1)$$

Since, one root is $2 + i\sqrt{7}$ then the other root will be $2 - i\sqrt{7}$. Now we find the equation having these two roots

$$(x - 2 - i\sqrt{7})(x - 2 + i\sqrt{7}) = 0$$

$\Rightarrow \qquad (x - 2)^2 + 7 = 0 \Rightarrow x^2 - 4x + 11 = 0$

Now divide the equation (1) by $x^2 - 4x + 11 = 0$, we get

$$3x + 8 = 0 \Rightarrow x = -8/3.$$

Hence, all the roots are $(2 + i\sqrt{7})$ and $-\dfrac{8}{3}$.

3.11 NEWTON'S METHOD FOR COMPLEX ROOTS

Let $f(z)$ be a given equation such that $f(z) = 0$ where $z = x + iy$ ...(1)

Then, (1) can be written as $f(z) = u(x, y) + iv(x, y)$ ...(2)

Where, $u(x, y)$ and $v(x, y)$ are the real and imaginary parts of $f(z)$ respectively.

Thus the problem of finding the roots of the single equation (1) is equivalent to determining the roots of two simultaneous equations

$$\left.\begin{array}{c} u(x, y) = 0 \\ v(x, y) = 0 \end{array}\right] \qquad ...(3)$$

Now, assume that (x_i, y_i) is an initial approximation to a solution of (3) and the exact solution is $(x_i + \Delta x, y_i + \Delta y)$

Then, we have

$$\left.\begin{array}{c} u(x_i + \Delta x, y_i + \Delta y) = 0 \\ v(x_i + \Delta x, y_i + \Delta y) = 0 \end{array}\right] \qquad ...(4)$$

Expanding (4) by Taylor's series expansion about (x_i, y_i), we get

and

$$\left.\begin{array}{l} u(x_i + \Delta x, y_i + \Delta y) = u(x_i, y_i) + [\Delta x u_x(x_i, y_i)] + [\Delta y u_y(x_i, y_i)] + ... = 0 \\ v(x_i + \Delta x, y_i + \Delta y) = v(x_i, y_i) + [\Delta x v_x(x_i, y_i)] + [\Delta y v_y(x_i, y_i)] + ... = 0 \end{array}\right] \qquad ...(5)$$

where suffixes with respect to x and y represent partial differentiation.

Neglecting the higher powers of Δx and Δy we get

$$\left.\begin{array}{l} u(x_i, y_i) + \Delta x u_x(x_i, y_i) + \Delta y u_y(x_i, y_i) = 0 \\ v(x_i, y_i) + \Delta x v_x(x_i, y_i) + \Delta y v_y(x_i, y_i) = 0 \end{array}\right] \qquad ...(6)$$

Now, solving (6) for Δx and Δy we get

$$\Delta x = -\left[\frac{u(x_i, y_i)v_y(x_i, y_i) - v(x_i, y_i)u_y(x_i, y_i)}{J}\right]$$

and

$$\Delta y = -\left[\frac{u_x(x_i, y_i)v(x_i, y_i) - u(x_i, y_i)v_x(x_i, y_i)}{J}\right]$$

where J is the Jacobian of the functions u and v at (x_i, y_i) and given by

$$J = \begin{vmatrix} u_x & u_y \\ v_x & v_y \end{vmatrix}$$

Thus, an improved approximation (x_{i+1}, j_{i+1}) to the exact solution is given by

$$x_{i+1} = x_i - J^{-1}[u(x_i, y_i)v_y(x_i, y_i) - v(x_i, y_i)u_y(x_i, y_i)]$$

and

$$y_{i+1} = y_i - J^{-1}[u_x(x_i, y_i)v(x_i, y_i) - u(x_i, y_i)v_x(x_i, y_i)] \quad ...(7)$$

$$\text{where } i = 0, 1, 2, ...$$

This is the Newton-Raphson method for two variables and has quadratic convergence.

Let us assume that $f(z)$ is an analytic function of z. Then the function $u(x, y)$ and $v(x, y)$ satisfy the Cauchy-Riemann equation

$$u_x = v_y, u_y = -v_x \Big] \qquad ...(8)$$

Then, the equation (7) becomes

$$x_{i+1} = y_i - \left[\frac{u(x_i, y_i)v_y(x_i, y_i) - v(x_i, y_i)u_y(x_i, y_i)}{u_x^2(x_i, y_i) + v_x^2(x_i, y_i)}\right]$$

and
$$y_{i+1} = y_i - \left[\frac{v(x_i, y_i)u_x(x_i, y_i) - u(x_i, y_i)v_x(x_i, y_i)}{u_x^2(x_i, y_i) + v_x^2(x_i, y_i)}\right] \quad ...(9)$$

where $i = 0, 1, 2, ...$

Generally, the Newton-Raphson method for (1) may be defined as

$$z_{k+1} = z_k - \frac{f(z_k)}{f'(z_k)}, \quad k = 0, 1, 2, ... \quad ...(10)$$

Then, it is easy to verify that (1) is equivalent to (9) when $f(z)$ is analytic function of z. The initial approximation z_0 in (10) must be a complex number.

SOLVED EXAMPLES

EXAMPLE 1. *Find the complex root of the equation $f(z) = z^3 + 1$ correct to eight decimal places.*

SOLUTION. Here, we have
$$f(z) = z^3 + 1 = 0 \quad ...(1)$$

Let $z = x + iy$ and $f(z) = u(x, y) + iv(x, y)$

Then (1), reduces to

$$u(x, y) + iv(x, y) = (x + iy)^3 + 1$$

$$\Rightarrow \quad u(x, y) + iv(x, y) = (x^3 - 3xy^2) + i(2x^2y - xy - y^3)$$

Equating real and imaginary part of both the sides, we get

$$u(x, y) = (x^3 - 3xy^2) \quad \text{and} \quad v(x, y) = 2x^2y + xy - y^3$$

Now using the Newton-Raphson method for complex numbers with the initial approximation $z_0 = (0.25, 0.25)$, we get the following successive iteration.

k	z_k	z_{k+1}	$f(z_{k+1})$
0	(0.25, 0.25)	(0.16666667, 2.83333333)	(0.9687, 0.3125(–1))
1	(0.16666667, 2.83333333)	(0.15220505, 1.89374026)	(–0.3009(1), –0.2251(2))
2	(0.15220505, 1.89374026)	(0.19263553, 1.27724322)	(–0.6340, –0.6660(1))
3	(0.19263553, 1.27724322)	(0.31932197, 0.91041889)	(0.6438(–1), –0.1941(1))
4	(0.31932197, 0.91041889)	(0.49252896, 0.83063199)	(0.2385, 0.4761)
5	(0.49252896, 0.83063199)	(0.49983161, 0.86738607)	(0.1000, 0.3140(–1))
6	(0.49983161, 0.86738607)	(0.49999870, 0.86602675)	(–0.3284(–2), –0.2484(–2))
7	(0.49999870, 0.866602675)	(0.50000000, 0.86602540)	(–0.1548(–5), 0.5414(–5))

Which gives that, the second root is

$$(-0.5, -0.86602540).$$

EXAMPLE 2. *Find the roots of* $3x^3 - 7x^2 + 10x - 20 = 0$. *By taking* $x = 3 + i$.

SOLUTION.

$$f(3+i) = 3(3+i)^3 - 7(3+i)^2 + 10(3+i) - 20$$

$$= 8 + 46i$$

$$f'(x) = 9x^2 - 14x + 10$$

$$\Rightarrow \quad f'(3+i) = 9(3+i)^2 - 14(3+i) + 10$$

$$= 9(8 + 6i) - 42 - 14i + 10$$

$$= 40 + 40i$$

Now, $$x_1 = x_0 - \frac{f(x_0)}{f'(x_0)}$$

$$\Rightarrow \quad x_1 = (3+i) - \frac{8+46i}{40+40i} = \frac{72+114i}{40+40i}$$

$$= 2.33 + 0.52i$$

Proceed further in the same manner we can get $x_2, x_3, \ldots$

3.12 MULLER'S METHOD

In this method $y = f(x)$ is approximated by a second degree curve passing through three points in the nbd of root. Thus the root of the quadratic curve are then assumed to be the approximation to the roots of the equation $f(x) = 0$.

Let us suppose x_{i-2}, x_{i-1} and x_i be the three approximations to the root r of the given equation

$$y = f(x) = 0 \qquad \ldots(1)$$

Suppose y_{i-2}, y_{i-1} and y_i be the corresponding value of the function $f(x)$ corresponding to x_{i-2}, x_{i-1} and x_i respectively.

Now, consider the equation of parabola

$$y = ax^2 + bx + c$$

The equation of the parabola for

$(x_{i-2}, y_{i-2}), (x_{i-1}, y_{i-1})$ and (x_i, y_i) becomes

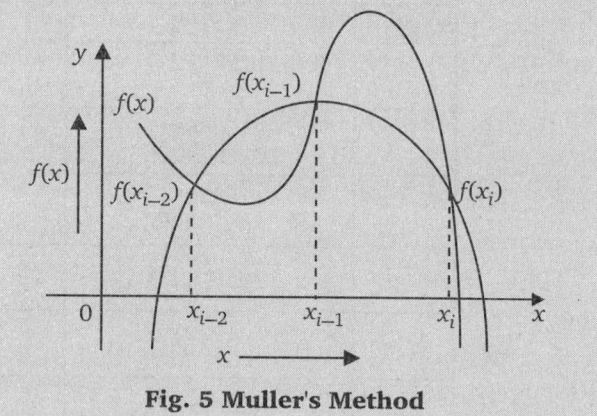

Fig. 5 Muller's Method

and

$$\left. \begin{array}{c} y_{i-2} = ax_{i-2}^2 + bx_{i-2} + c \\ y_{i-1} = ax_{i-1}^2 + bx_{i-1} + c \\ y_i = ax_i^2 + bx_i + c \end{array} \right] \ldots(3)$$

On eliminating a, b and c from (2) and (3), we get

$$\begin{vmatrix} y & x & x & 1 \\ y_{i-2} & x_{i-2}^2 & x_{i-2} & 1 \\ y_{i-1} & x_{i-1}^2 & x_{i-1} & 1 \\ y_i & x_i^2 & x_i & 1 \end{vmatrix} = 0$$

$$\Rightarrow \quad y = \frac{(x-x_{i-1})(x-x_i)}{(x_{i-2}-x_{i-1})(x_{i-2}-x_i)} y_{i-2} + \frac{(x-x_{i-2})(x-x_i)}{(x_{i-1}-x_{i-2})(x_{i-1}-x_i)} y_{i-1} + \frac{(x-x_{i-2})(x-x_{i-1})}{(x_i-x_{i-2})(x_i-x_{i-1})} y_i$$

$$\ldots(4)$$

Let us define

$$\lambda = \frac{x - x_i}{x_i - x_{i-1}}, \lambda_i = \frac{x_i - x_{i-1}}{x_{i-1} - x_{i-2}} \text{ and } \delta_i = \frac{x_i - x_{i-2}}{x_{i-1} - x_{i-2}} y_i \qquad \ldots(5)$$

Then (4) becomes

$$y = \frac{(y_{i-2}\lambda + y_{i-1}\lambda_i + y_i)\delta_i\lambda^2}{\delta_i} + \frac{y_{i-2}\lambda_i^2 - y_{i-1}\delta_i^2 + y_i(\lambda_i + \delta_i)}{\delta_i} + \lambda + y_i \qquad \ldots(6)$$

Also, from (5), we get

$$x = x_i + \lambda(x_i - x_{i-1}) \qquad \ldots(7)$$

To determine λ put $y = 0$ in (6), we get

$$(y_{i-2}\lambda_i - y_{i-1}\delta_i + y_i)\lambda_i\lambda^2 + \mu_i\lambda + \delta_i y_i = 0 \qquad \ldots(8)$$

where

$$\mu_i = y_{i-2}\lambda_i^2 - y_{i-1}\delta_i^2 + y_i(\lambda_i + \delta_i)$$

Solving (8) for $\dfrac{1}{\lambda}$ (after dividing throughout by $\lambda_i\lambda^2$), we get

$$\frac{1}{\lambda} = \frac{-\mu_i \pm \sqrt{[\mu_i^2 - 4y_i\delta_i\lambda_i(y_{i-2}\lambda_i - y_{i-1}\delta_i + y_i)]}}{2y_i\delta_i} \qquad \ldots(9)$$

Since λ must be small in magnitude therefore choose the sign as it make the numerator of (9), largest in magnitude.

▶ REMARKS

- ▶ If we solve (8) for λ then we get inaccurate results. Due to this reason, we solve (8) for $1/\lambda$.
- ▶ It is an extension of secant method.
- ▶ In this method the given equation $y = f(x)$ is approximated by a second degree curve passing through the points (x_{i-2}, y_{i-2}), (x_{i-1}, y_{i-1}) and (x_i, y_i) in the neighbourhood of the root. Then a root of this quadratic is taken as the next approximation.
- ▶ The Muller's method is iterative and converges for almost all initial approximations quadratically. In case no better approximations are known, we take $x_{i-2} = -1$, $x_{i-1} = 0$ and $x_i = 1$.

ALITER

Let the quadratic curve passing through the point $(x_1, f(x_1))$, $(x_2, f(x_2))$ and $(x_3, f(x_3))$ to estimate a root of $f(x)$. Let one of the root of quadratic polynomial $p(x)$ is taken as an approximate value of the root of $f(x)$. Let x_4 is one of the root of $p(x)$ is assumed as the next approximation for the root of $f(x)$.

Then $p(x)$ can be written as

$$p(x) = a_0 + a_1(x - c) + a_2(x - c)_4^2 \qquad \ldots(1)$$

Choose $c = x_3$, then (1) becomes

$$p(x) = a_0 + a_1(x - x_3) + a_2(x - x_3)^2 \qquad \ldots(2)$$

Now, since x_4 is a root of $p(x) \Rightarrow \quad 1 \, p(x_4) = 0$

then (2) becomes

$$a_2(x_4 - x_3)^2 + a_1(x_4 - x_3) + a_0 = 0$$

Solving this equation for $(x_4 - x_3)$, we get

$$x_4 - x_3 = \frac{-2a_0}{a_1 \pm \sqrt{a_1^2 - 4a_0a_2}}$$

where a_0, a_1 and a_2 are constants and have to be determined by the following equations. At $x = x_1, x_2$ and x_3 we have

$$a_2(x_1 - x_3)^2 + a_1(x_1 - x_3) + a_0 = p(x_1) = f(x_1)$$

$$a_2(x_2 - x_3)^2 + a_1(x_2 - x_3) + a_0 = p(x_2) = f(x_2)$$

$$a_2(x_3 - x_3)^2 + a_1(x_3 - x_3) + a_0 = p(x_3) = f(x_3)$$

Now, letting $h_1 = x_1 - x_3, h_2 = x_2 - x_3$ and $h_3 = x_3 - x_3$ denoting $f_i = f(x_i)$, we obtain

$$\left.\begin{array}{r}a_2 h_1^2 + a_1 h_1 + a_0 = f_1 \\ a_2 h_2^2 + a_1 h_2 + a_0 = f_2 \\ a_0 = f_3\end{array}\right] \qquad ...(1)$$

Solving the system (1), we obtain $a_0 = f_3$

$$a_1 = \frac{d_2 h_1^2 - d_1 h_2^2}{h_1 h_2 (h_1 - h_2)}$$

and $a_2 = \dfrac{d_1 h_2 - d_2 h_1}{h_1 h_2 (h_1 - h_2)}$ where $d_1 = f_1 - f_3$ and $d_2 = f_2 - f_3$.

Repeat this process, using x_2, x_3 and x_4 as the initial values to get the next approximation x_5.

$$x_5 = x_4 + h_5.$$

ALGORITHM

Step 1.	Take the initial approximation x_1, x_2, x_3.
Step 2.	Compute $f_1 = f(x_1), f_2 = f(x_2), f_3 = f(x_3)$.
Step 3.	Compute h_i and d_i using $h_i = x_i - x_3$, $d_i = f_i - f_3$
Step 4.	Compute a_0, a_1, a_2 by $a_0 = f_3, a_1 = \dfrac{d_2 h_1^2 - d_1 h_2^2}{h_1 h_2 (h_1 - h_2)}, a_2 = \dfrac{d_1 h_2 - d_2 h_1}{h_1 h_2 (h_1 - h_2)}$
Step 5.	Compute h by $h = \dfrac{-2a_0}{a_1 \pm \sqrt{a_1^2 - 4a_0 a_2}}$ (Choose the sign in the denominator that it makes it largest)
Step 6.	Compute $x_4 = x_3 + h$
Step 7.	Compute $f_4 = f(x_4)$
Step 8.	If $f(x_4)$ satisfies the given criteria, the root is obtained go to step 10.
Step 9.	Otherwise, set $x_1 = x_2, x_2 = x_3, x_3 = x_4$ and $f_1 = f_2, f_2 = f_3, f_3 = f_4$ go to Step 3.
Step 10.	Write the value of the root (x_4).

SOLVED EXAMPLES

EXAMPLE 1. *Apply Muller's method to obtain the root of the equation* $\cos x - xe^x = 0$ *which lies between 0 and 1.*

(Rohtak–2004, 07)

SOLUTION. The given equation can be written as

$$y = \cos x - xe^x$$

Take the initial approximation as

$$x_{i-2} = -1, x_{i-1} = 0 \text{ and } x_i = 1$$

we get
$$y_{i-2} = \cos 1 + e^{-1}$$
$$y_{i-1} = 1, y_i = \cos 1 - e$$
$$\lambda = x - 1, \lambda_i = 1, \delta_i = 2$$

and
$$\mu_i = (\cos x + e^{-1}) - 4 + 3(\cos 1 - e)$$

Now using

$$\frac{1}{\lambda} = [\frac{-\mu_i \pm \sqrt{\begin{matrix} \mu_i^2 - 4y_i\delta_i\lambda_i(y_{i-2}\lambda_i - y_i \\ - y_{i-1}\delta_i + y_i) \end{matrix}}}{2y_i\delta_i}]$$

Then we get two value of $\frac{1}{\lambda}$, i.e., $\frac{1}{\lambda} = \pm\frac{1}{0.5585}$ choose negative sign (To make the numerator is largest)

$\Rightarrow \qquad \lambda = -0.5585$

Then, the next approximation is
$$x_{i+1} = x_i + \lambda(x_i - x_{i-1}) = 1 - 0.5585 = 0.4415$$

Repeating this process, we get
$$x_{i+2} = 0.5125, x_{i+3} = 0.5177, x_{i+4} = 0.5177$$

$\Rightarrow$ The root is 0.518, correct to three decimal places.

EXAMPLE 2. *Solve the equation*

$$f(x) = x^3 + 2x^2 + 10x - 20 = 0$$

by Muller's method. (UPTU-2003)

SOLUTION. Here the given equation is

$$f(x) = x^3 + 2x^2 + 10x - 20$$

Let us assume the initial approximation as $x_1 = 0, x_2 = 1, x_3 = 2$

Then, we get $f_1 = -20, f_2 = -7$ and $f_3 = 16$

also
$$h_1 = x_1 - x_3 = -2, h_2 = x_2 - x_3 = -1,$$
$$d_1 = f_1 - f_3 = -36, d_2 = f_2 - f_3 = -23$$
$$D = h_1h_2(h_1 - h_2) = 2(-2 + 1) = -2$$

$\Rightarrow \qquad a_1 = \dfrac{(-23)(-2)^2 - (-36)(-1)^2}{-2} = 28$

$a_2 = \dfrac{(-36)(-1) - (-23)(-2)}{-2} = 5$

and $\qquad h = \dfrac{-2 \times 16}{28 \pm \sqrt{28^2 - 4.516}} = -\dfrac{32}{49.5 + 0.659} = -0.645934$

(By taking positive sign)

Now $\qquad x_4 = x_3 + h = 2 - 0.645934 = 1.3540659$

For second iteration.

Take the approximation, $x_1 = 1, x_2 = 2$

and $\qquad x_3 = 1.3540659$

Then, proceeding as above, we get
$$x_4 = 1.3686472$$

Continuing this process, to get better accuracy, we get the required root = 1.368808107.

EXAMPLE 3. *Using Muller's method, find the root of the equation* $y = f(x) = x^3 - 2x - 5 = 0$ *which lies between 2 and 3.*

(UPTU-2003)

SOLUTION. We have $\qquad y = f(x) = x^3 - 2x - 5$

Clearly, root lies between 2 and 3, so we take

$$x_{n-2} = 1.9, x_{n-1} = 2, x_n = 2.1$$

Then (1) gives

$$y_{n-2} = x_{n-2}^3 - 2x_{n-2} - 5 = -1.941$$

$$y_{n-1} = x_{n-1}^3 - 2x_{n-1} - 5 = -1$$

$$y_n = x_n^3 - 2x_{n-1} - 5 = 0.061$$

Now $\qquad \lambda = \dfrac{x - x_n}{x_n - x_{n-1}} = \dfrac{x - 2.1}{0.1} = 10(x - 2.1)$

$$\lambda_n = \frac{x_n - x_{n-1}}{x_{n-1} - x_{n-2}} = \frac{0.1}{0.1} = 1$$

$$\delta_n = \frac{x_n - x_{n-2}}{x_{n-1} - x_{n-2}} = \frac{0.2}{0.1} = 2$$

Since $\qquad \mu_n = y_{n-2}\lambda_n^2 - y_{n-1}\delta_n^2 + y_n(\lambda_n + \delta_n)$

$\Rightarrow \qquad \mu_n = (-1.941)(1)_n^2 - (-1)(2)^2 + (0.061)(1 + 2)$

$$= 2.242$$

Further, $\qquad \dfrac{1}{\lambda} = \dfrac{-\mu_n \pm [\mu_n^2 - 4y_n\delta_n\lambda_n(y_{n-2}\lambda_n - y_{n-1}\delta_n + y_n)]^{1/2}}{2y_n\delta_n}$

$$= \frac{-2.242 \pm [5.026564 - 8(0.061)(0.12)]^{1/2}}{4(0.061)}$$

$$= \frac{-2.242 \pm 2.229}{0.244}$$

$\Rightarrow \qquad \lambda = \dfrac{0.244}{-2.242 \pm 2.229}$

Taking negative sign, we get

$$\lambda = -\frac{0.244}{4.471} = -0.05457$$

The next approximation to the root is

$$x_{n+1} = x_n + \lambda(x_n - x_{n-1})$$

$$= 2.1 - 0.05457(2.1 - 2) = 2.1 - 0.005457 = 2.094543$$

This procedure can now be repeated with the three approximations as 1.9, 2 and 2.094543.

Let $\qquad x_{n-2} = 1.9, x_{n-1} = 2, x_n = 2.094534$

Then from (1), we get

$$y_{n-2} = -1.941, y_{n-1} = -1, y_n = -0.000947$$

Then $\qquad \lambda = \dfrac{x - x_n}{x_n - x_{n-1}} = \dfrac{x - 2.094543}{0.094543}$

$$\lambda_n = \frac{x_n - x_{n-1}}{x_{n-1} - x_{n-2}} = \frac{0.094543}{0.1} = 0.94543$$

$$\delta_n = \frac{x_n - x_{n-2}}{x_{n-1} - x_{n-2}} = \frac{0.194543}{0.1} = 1.94543$$

and

$$\mu_n = y_{n-2}\lambda_n^2 - y_{n-1}\delta_n^2 + y_n(\lambda_n + \delta_n)$$

$$= (-1.941)(0.94543)^2 - (-1)(1.94543)^2$$
$$+ (-0.0000947)(2.891)$$
$$= 2.049$$

Then

$$\frac{1}{\lambda} = \frac{-2.049 \pm [4.198401 - 4(-0.0000947)}{2(-0.0000947)(1.94543)}$$
$$\frac{(1.94543)(0.94543)(0.11026)]^{1/2}}{}$$

$$\Rightarrow \qquad \lambda = \frac{-0.0003685}{-2.049 \pm 2.04901} = 0.0000891$$

(By taking negative sign)

The next approximation to the root is
$$x_{n+1} = x_n + \lambda(x_n - x_{n-1}) = 2.094543$$
$$+ (0.0000899)(0.094545)$$

$$= 2.094552$$

Hence, the required root is 2.095 correct to three decimal places.

3.13 LIN-BAIRSTOW'S METHOD

This method is used for finding the complex roots of a polynomial equation $P_n(x) = 0$ with real coefficients, where

$$P_n(x) = a_0 x^n + a_1 x^{n-1} + a_2 x^{n-2} + \ldots + a_{n-1}x + a_n = 0$$

Consider a quadratic polynomial equation

$$P_2(x) = a_0 x^2 + a_1 x + a_2 = 0$$

Clearly, the roots of $P_2(x) = 0$ are given by

$$\frac{-a_1 + \sqrt{a_1^2 - 4a_0 a_2}}{2a_0} \quad \text{and} \quad \frac{-a_1 - \sqrt{a_1^2 - 4a_0 a_2}}{2a_0}$$

Hence, a second degree equation can be solved completely.

3.13.1 TO FIND THE COMPLEX ROOTS OF $P_n(x) = 0, n > 2$

We know that complex roots always occur in pair. If $\alpha + i\beta$ and $\alpha - i\beta$ are two roots of $P_n(x) = 0$, then $(x - \alpha - i\beta)$ and $(x - \alpha + i\beta)$ will be two linear factors of $P_n(x) = 0$.

Therefore, $(x - \alpha - i\beta)(x - \alpha + i\beta)$ or $(x^2 - 2ax + \alpha^2 + \beta^2)$ will be a quadratic factor of $P_n(x)$ which shows that the equation $P_n(x) = 0$ has a pair of complex roots which are obtained from a quadratic factor with real coefficient in polynomial $P_n(x)$.

Since we have to find the complex roots of $P_n(x) = 0$ so if $P_n(x) = 0$ has a real root, then the linear factor corresponding to that real root can be removed by using the method of synthetic division and the remaining polynomial has no real root.

In this method we start with an approximate quadratic factor $x^2 + ax + b$ and find such values of a and b, after applying an iterative process, such that $x^2 + ax + b$ becomes a factor of $P_n(x)$

and then the roots of $x^2 + ax + b$ will be the roots of $P_n(x) = 0$.

If $P_n(x)$ can be written as

$$P_n(x) = (x^2 + ax + b)P_{n-2}(x) + Ax + B \qquad \qquad \text{...(1)}$$

where $P_{n-2}(x)$ is a polynomial of degree $n - 2$, then $Ax + B$ represents the remainder when $x^2 + ax + b$ is a factor of $P_n(x)$, i.e., when $P_n(x)$ is divided by $x^2 + ax + b$.

From (1) we find that $x^2 + ax + b$ will be a factor of $P_n(x)$ if

$$Ax + B = 0 \Rightarrow A = 0 \text{ and } B = 0 \qquad \qquad \text{...(2)}$$

But both A and B depend on a and b, then the equation (2) can be taken as

$$A(a, b) = 0 \quad \text{and} \quad B(a, b) = 0 \qquad \qquad \text{...(3)}$$

Let $a + \Delta a$ and $b + \Delta b$ be the actual values of a and b that satisfy (3), then

and

$$\left. \begin{aligned} A(a + \Delta a, b + \Delta b) &= 0 \\ B(a + \Delta a, b + \Delta b) &= 0 \end{aligned} \right\} \qquad \qquad \text{...(4)}$$

Now expanding (4) by Taylor's series and neglecting the second and higher order term of Δa and Δb, we get

and

$$\left. \begin{aligned} A(a,b) + \frac{\partial A}{\partial a}\Delta a + \frac{\partial A}{\partial b}\Delta b &= 0 \\ B(a,b) + \frac{\partial B}{\partial a}\Delta a + \frac{\partial B}{\partial b}\Delta b &= 0 \end{aligned} \right\} \qquad \qquad \text{...(5)}$$

Solving (5) to get the values of Δa and Δb, the values of Δa and Δb thus obtained involving $\frac{\partial A}{\partial a}, \frac{\partial A}{\partial b}, \frac{\partial B}{\partial b}$ and $\frac{\partial B}{\partial b}$.

But A and B are not known explicitly as functions of a and b, so we cannot obtain the values of $\frac{\partial A}{\partial a}, \frac{\partial A}{\partial b}, \frac{\partial B}{\partial a}$ and $\frac{\partial B}{\partial b}$.

Bairstow had developed a method for obtaining the numerical values of $\frac{\partial A}{\partial a}, \frac{\partial A}{\partial b}, \frac{\partial B}{\partial a}$ and $\frac{\partial B}{\partial b}$ which is given as follows :

Since we have $P_n(x) = a_0 x^n + a_1 x^{n-1} + a_2 x^{n-2} + ... + a_{n-1}x + a_n$.

If $P_{n-2}(x) = b_0 x^{x-2} + b_1 x^{n-3} + b_2 x^{n-3} + ... + b_{n-3}x + b_{n-2}$ then equation (1) becomes

$$a_0 x^n + a_1 x^{n-1} + a_2 x^{n-2} + ... + a_{n-1}x + a_n$$

$$= (x^2 + ax + b)(b_0 x^{n-2} + b_1 x^{n-3} + b_2 x^{n-3} + ... + b_{n-3}x + b_{n-2}) + Ax + B$$

Now comparing the coefficients of like powers of x on both sides, we get

and

$$\left. \begin{aligned} a_0 &= b_0 \\ a_1 &= b_1 + ab_0 \\ a_2 &= b_2 + ab_1 + bb_0 \\ &\cdots\cdots\cdots\cdots\cdots\cdots\cdots \\ a_r &= b_r + ab_{r-1} + ba_{r-2} \\ &\cdots\cdots\cdots\cdots\cdots\cdots\cdots \\ a_{n-2} &= b_{n-2} + ab_{n-3} + bb_{n-4} \\ a_{n-1} &= A + ab_{n-2} + bb_{n-3} \\ a_n &= B + b_{n-2} \end{aligned} \right\} \qquad \qquad \text{...(6)}$$

Let us setting $b_{-1} = 0 = b_{-2}, b_{n-1} = A$ and $b_n = B - ab_{n-1}$ then equation (5) can be written as

$$a_r = b_r + ab_{r-1} + bb_{r-2}, r = 0, 1, 2, ..., n$$

or
$$b_r = a_r - ab_{r-1} + bb_{r-2}, r = 0,1,2,...,n \qquad ...(7)$$

Now differentiating (7) partially w.r.t. a and b, we get

$$\frac{\partial b_r}{\partial a} = -b_{r-1} - a\frac{\partial b_{r-1}}{\partial a} - b\frac{\partial b_{r-2}}{\partial a} \qquad ...(8)$$

and
$$\frac{\partial b_r}{\partial a} = -b_{r-2} - a\frac{\partial b_{r-1}}{\partial b} - b\frac{\partial b_{r-2}}{\partial b} \qquad ...(9)$$

Also $b_0 = a_0, b_{-1} = 0, b_{-2} = 0$ so that

$$\frac{\partial b_0}{\partial a} = 0 = \frac{\partial b_0}{\partial b}, \frac{\partial b_{-1}}{\partial a} = 0 = \frac{\partial b_{-1}}{\partial b}, \frac{\partial b_{-2}}{\partial a} = 0 = \frac{\partial b_{-2}}{\partial b} \qquad ...(10)$$

Putting $r = 1$ in (8) and (9), we get

$$\frac{\partial b_1}{\partial a} = -b_0 - a\frac{\partial b_0}{\partial a} - b\frac{\partial b_{-1}}{\partial b} = -b_0 \qquad \text{[Using (10)]... (11)}$$

$$\frac{\partial b_1}{\partial a} = -b_{-1} - a\frac{\partial b_0}{\partial b} - b\frac{\partial b_{-1}}{\partial b} = 0 \qquad \begin{array}{l}\text{[Using (10)]}\\ ... (12)\end{array}$$

Again put $r = 2$ in (8) and (9), we get

$$\frac{\partial b_2}{\partial a} = -b_1 - a\frac{\partial b_1}{\partial b} - b\frac{\partial b_0}{\partial a} = -b_1 + ab_0 \qquad \begin{array}{l}\text{[Using (10), (11)]}\\ ... (13)\end{array}$$

$$\frac{\partial b_2}{\partial b} = -b_0 - a\frac{\partial b_1}{\partial b} - b\frac{\partial b_0}{\partial b} = -b_0 \quad \text{[Using (10), (12)]} \quad ... (14)$$

Thus, from (11) and (13), we get

$$\frac{\partial b_2}{\partial b} = \frac{\partial b_1}{\partial a} \qquad ... (15)$$

Again, put $r = 3$ in (9), we get

$$\frac{\partial b_3}{\partial b} = -b_1 - a\frac{\partial b_2}{\partial b} - b\frac{\partial b_1}{\partial b}$$

or
$$\frac{\partial b_3}{\partial b} = -b_1 + ab_0 \qquad \text{[Using (13), (14)]} \quad ... (16)$$

Thus, from (13) and (16), we get

$$\frac{\partial b_3}{\partial b} = \frac{\partial b_2}{\partial a}$$

Hence, from (15) and (17), we conclude that

$$\frac{\partial b_{r+1}}{\partial b} = \frac{\partial b_r}{\partial a} \text{ for } r = 0, 1, 2, ... n, ...$$

∴
$$\frac{\partial b_{r+1}}{\partial b} = \frac{\partial b_r}{\partial a} \text{ for all } r. \qquad ... (18)$$

Now, if we again set $\dfrac{\partial b_{r+1}}{\partial b} = \dfrac{\partial b_r}{\partial a} = -c_{r-1}$ for $r = 0, 1, 2, ... n-1, ...$, then from (8) and (9),

we get $\qquad c_{r-1} = b_{r-1} - ac_{r-2} - bc_{r-3}$ and $\quad c_{r-2} = b_{r-2} - ac_{r-3} - bc_{r-4}$.

These two relations can be written as a single relation.

$$c_r = b_r - ac_{r-1} - bc_{r-2} \qquad ...(19)$$

where $r = 1,2,3,...,(n-1)$ and $\quad c_{-1} = -\dfrac{\partial b_0}{\partial a} = 0, c_0 - \dfrac{\partial b_1}{\partial a} = b_0$.

Next, we shall find the derivatives in terms of c_r's.

Since, we have

$$A = b_{n-1} \quad \text{and} \quad B = b_n + ab_{n-1}$$

Then, we have

$$\frac{\partial A}{\partial a} = \frac{\partial b_{n-1}}{\partial a} = -c_{n-2}, \frac{\partial A}{\partial b} = \frac{\partial b_{n-1}}{\partial b} = -c_{n-3}$$

$$\frac{\partial B}{\partial a} = \frac{\partial b_n}{\partial a} + b_{n-1} + a\frac{\partial b_{n-1}}{\partial b} = -c_{n-1} + b_{n-1} - ac_{n-2}$$

and

$$\frac{\partial B}{\partial b} = \frac{\partial b_n}{\partial b} + a\frac{\partial b_{n-1}}{\partial b} = -c_{n-2} - ac_{n-3}$$

Now putting the values of $\dfrac{\partial A}{\partial a}, \dfrac{\partial A}{\partial b}, \dfrac{\partial B}{\partial a}$ and $\dfrac{\partial B}{\partial b}$ in (5), we get.

$$b_{n-1} - c_{n-2}\Delta a - c_{n-3}\Delta b = 0$$

or

$$c_{n-2}\Delta a - c_{n-3}\Delta b = b_{n-1} \qquad \qquad \dots (20)$$

and $\quad b_n + ab_{n-1} + (-c_{n-1} + b_{n-1} - ac_{n-2})\Delta a + (-c_{n-2} - ac_{n-3})\Delta b = 0$

or $\quad b_n - (c_{n-1} - b_{n-1})\Delta a - c_{n-2}\Delta b + a(b_{n-1} - c_{n-2}\Delta a - c_{n-3}\Delta b) = 0$

or $\qquad \qquad \qquad b_n - (c_{n-1} - b_{n-1})\Delta a - c_{n-2}\Delta b = 0$

or $\qquad \qquad \qquad (c_{n-1} - b_{n-1})\Delta a + c_{n-2}\Delta b = 0 \qquad \qquad \dots (21)$

Substituting the values of c_r's and b_r's obtained from (19) and (7) in (20) and (21) and solving them to get the approximate values of Δa and Δb denoted by Δa^* and Δb^* respectively.

Now we shall take the new values $a + \Delta a^*$ and $b + \Delta b^*$ as the initial values and applying the above process again to get the better values of a and b.

In order to find the values of b_r's and c_r's we use the following (synthetic division) scheme :

$a_0 (= 1)$	a_1	a_2	a_3	...	a_{n-2}	a_{n-1}	a_n	
	$-ab_0$	$-ab_1$	$-ab_2$	...	$-ab_{n-3}$	$-ab_{n-2}$	$-ab_{n-1}$	$-a$
		$-bb_0$	$-bb_1$	...	$-bb_{n-1}$	$-bb_{n-3}$	$-bb_{n-2}$	$-b$
$b_0 (= 1)$	b_1	b_2	b_3	...	b_{n-2}	b_{n-1}	b_{n-1}	
	$-ac_0$	$-ac_1$	$-ac_2$	...	$-ac_{n-3}$	$-ac_{n-2}$		$-a$
		$-bc_0$	$-bc_1$	...	$-ac_{n-4}$	$-ac_{n-3}$		$-b$
$c_0 (= 1)$	c_1	c_2	c_3	...	c_{n-2}	c_{n-1}		

SOLVED EXAMPLES

EXAMPLE 1. *Find a quadratic factor of*

$$f(x) = x^4 - 3x^3 + 20x^2 + 44x + 54 = 0$$

SOLUTION.   Here, tile quadratic factor will close to $x^2 + 2x + 2$ so we shall start the method with $a = 2, b = 2$.

Now we find $b_r's$ and $c_r's$ by following scheme:

1	−3	20	44	54	
	−2	10	−56	4	−2
		−2	10	−56	−2
1	−5	28	$-2(b_{n-1})$	$2(b_n)$	
	−2	14	−80		−2
		−2	14		−2
1	$-7(c_{n-3})$	$40(c_{n-2})$	$-68(c_{n-1})$		

If Δa and Δb are the corrections in a and b then we have

$$c_{n-2}\Delta a' + c_{n-3}\Delta b = b_{n-1}$$

$$(c_{n-1} - b_{n-1})\Delta a + c_{n-3}\Delta b = b_n$$

Putting the values of $b_n, b_{n-1}, c_{n-1}, c_{n-2}$ and c_{n-3} from table, we get

$$40\Delta a - 7\Delta b = -2 \qquad \qquad \text{...(1)}$$

$$-66\Delta a + 40\Delta b = 2 \qquad \qquad \text{...(2)}$$

Solving (1) and (2), we get

$$\Delta a = -0.058, \Delta b = -0.0457$$

The new values of a and b are given by

$$a' = a + \Delta a = 2 - 0.058 = 1.942$$

$$b' = b + \Delta b = 2 - 0.0457 = 1.9543$$

Now repeating the same process by taking a' and b'.

1	−3	20	44	54	
	−1.942	9.5974	−53.683	0.0482	−1.942
		−1.9543	9.6582	−54.023	−1.9543
1	−4.942	27.6431	−0.0248	0.0252	
			(b_{n-1})	(b_n)	
	−1.942	13.3687	−75.849		−1.942
		−1.9543	13.4534		−1.9543
1	−6.884	39.0575	−62.4204		
	(c_{n-3})	(c_{n-2})	(c_{n-1})		

If $\Delta a'$ and $\Delta b'$ are the corrections in a' and b', then we have

$$c_{n-2}\Delta a' + c_{n-3}\Delta b = b_{n-1}$$

$$(c_{n-1} - b_{n-1})\Delta a' + c_{n-2}\Delta b' = b_n$$

Putting the values of $b_n, b_{n-1}, c_{n-1}, c_{n-2}, c_{n-3}$ from above table, we get

$$39.0575\Delta a' + 6.884\Delta b' = -0.0248 \qquad \qquad \text{... (3)}$$

$$-62.3956\Delta a' + 39.0575\Delta b' = 0.0252 \qquad \qquad \text{... (4)}$$

Solving (3) and (4), we get

$$\Delta a' = -0.00073, \Delta b' = -0.00054$$

The new values of a and b are given by

$$a'' = a' + \Delta a' = 1.942 - 0.00073 = 1.94127$$
$$b'' = b' + \Delta b' = 1.9543 - 0.00054 = 1.95376$$

Hence, $x^2 + a''x + b''$ i.e., $x^2 + 1.94127x + 1.95376$ is the required factor of $f(x)$.

EXAMPLE 2. *Using Lin-Bairstow's method, find the quadratic factor of the polynomial given by*

$$f(x) = x^3 - 2x^2 + x - 2.$$

SOLUTION. Let $x^2 + ax + b$ with $a = 1, b = 1$ be the quadratic factor of $f(x) = x^3 - 2x^2 + x - 2$.

Now we find b_r's and c_r's by the following synthetic scheme:

1	−2	1	−2	
	−1	3	−3	−1
		−1	3	−1
1	−3	$3(=b_{n-1})$	$-2(b_n)$	
	−1	4		−1
		−1		−1
1	−4	6		
$(=c_{n-3})$	$(=c_{n-2})$	$(=c_{n-1})$		

If Δa and Δb are the corrections in a and b, then we have

$$c_{n-2}\Delta a + c_{n-3}\Delta b = b_{n-1}$$
$$(c_{n-1} - b_{n-1})\Delta a + c_{n-2}\Delta b = b_n$$

Putting the values of $b_n, b_{n-1}, c_{n-1}, c_{n-2}$ and c_{n-3} we get

$$-4\Delta a + \Delta b = 3 \qquad \qquad \text{... (1)}$$

and $$(6 - 3)\Delta a - 4\Delta b = -2 \qquad \qquad \text{... (2)}$$

or $$3\Delta a - 4\Delta b = -2$$

Solving (1) and (2), we get

$$\Delta a = -0.769, \Delta b = -0.0769$$

Therefore, we get the new values of a and b, given by

$$a' = a + \Delta a = 1 - 0.769 = 0.231$$
$$b' = b + \Delta b = 1 - 0.0769 = 0.9231$$

Now repeating the same process by taking a' and b':

1	−2	1	−2	
	−0.231	0.5154	−0.1368	−0.231
		−0.9231	2.0594	−0.9231
1	−2.231	0.5923	0.0774	
		(b_{n-1})	(b_n)	
	−0.231	0.05687		−0.231
		0.9231		−0.9231
1	−2.462	0.2379		
(c_{n-3})	(c_{n-2})	(c_{n-1})		

If $\Delta a'$ and $\Delta b'$ are the corrections in a' and b', then, we have

$$c_{n-2}\Delta a' + c_{n-3}\Delta b' = b_{n-1}$$

$$(c_{n-1} - b_{n-1})\Delta a' + c_{n-2}\Delta b' = b_n$$

Putting the values of $b_n, b_{n-1}, c_{n-1}, c_{n-2}, c_{n-3}$ we get

$$-2.462\Delta a' - \Delta b' = 0.5923 \qquad \text{... (3)}$$

$$-0.3544\Delta a' - 2.462\Delta b' = 0.0774 \qquad \text{... (4)}$$

Solving (3) and (4), we get

$$\Delta a = -0.2394, \Delta b' = 0.00289$$

The new values of a' and b' are given by

$$a'' = a' + \Delta a' = 0.231 - 0.2394 = -0.0084 = 0$$

$$b'' = b' + \Delta b' = 0.9231 + .00289 = 0.9256 = 1$$

Hence $x^2 + a''x + b''$, i.e., $x^2 + 1$ is the required quadratic factor of

$$f(x) = x^3 - 2x^2 + x - 2.$$

EXAMPLE 3. Solve $x^4 - 5x^3 + 20x^2 - 40x + 60 = 0$ (given that all the root of this equation are complex), by using Lin-Bairstow method.

SOLUTION. Let $x^2 + ax + b$ be one of the quadratic factor, with $a = -4, b = 8$
Now we find b_r's and c_r's by the following scheme :

1	−5	20	−40	60	
	4	−4	32	0	4
		−8	8	−64	−8
1	−1	8	0	−4(b_n)	
			(b_{n-1})		
	4	12	48		4
		−8	−24		−8
1	3	12	24		
	(c_{n-3})	(c_{n-2})	(c_{n-1})		

If Δa and Δb are the corrections in a and b, then we have

$$c_{n-2}\Delta a + c_{n-3}\Delta b = b_{n-1}$$

$$c_{n-1} - b_{n-1}\Delta a + c_{n-2}\Delta b = b_n$$

Putting the values of $b_n, b_{n-1}, c_{n-1}, c_{n-2}, c_{n-3}$ from above table, we get

$$12\Delta a + 3\Delta b = 0 \qquad \text{.... (1)}$$

$$24\Delta a + 12\Delta b = -4$$

Solving (1) and (2), we get

$$\Delta a = 0.16712, \Delta b = -0.667$$

The new values of a and b are given by

$$a' = a + \Delta a = -4 + 0.167 = -3.833$$

$$b' = b + \Delta b = 8 - 0.667 = 7.333$$

Now repeating the same process by taking a' and b':

1	−5	20	−40	60		
	3.833	−4.4731	31.4072	−0.1349	3.833	
		−7.333	8.5576	−60.086	−7.333	
1	−1.167	8.1939	−0.0352	−0.2209		
			(b_{n-1})	(b_n)		
	3.833	10.2188	42.4685		3.833	
		−7.333	−19.5498		−7.333	
1	2.666	11.0797	22.8835			
	(c_{n-3})	(c_{n-2})	(c_{n-1})			

If $\Delta a'$ and $\Delta b'$ are the corrections in a' and b', then we have

$$c_{n-2}\Delta a' + c_{n-3}\Delta b' = b_{n-1}$$
$$(c_{n-1} - b_{n-1})\Delta a' + c_{n-2}\Delta b' = b_n$$

Putting the values of $b_n, b_{n-1}, c_{n-1}, c_{n-2}$ and c_{n-3} from above table, we get

$$11.797\Delta a' + 2.666\Delta b' = -0.0352 \qquad \ldots(3)$$
$$22.9187\Delta a' + 11.0797\Delta b' = -0.2209 \qquad \ldots(4)$$

Solving (3) and (4), we get

$$\Delta a' = 0.00323, \Delta b' = -0.0266$$

The new values of a' and b' are given by

$$a'' = a' + \Delta a' = -3.833 + 0.00323 = -3.82977$$
$$b'' = b' + \Delta b' = 7.333 - 0.0266 = 7.3064$$

Therefore, one of the quadratic factor of the given equation is

$x^2 + a''x + b''$, *i.e.*, $x^2 - 3.8298x + 7.3064$

If $p \pm iq$ be the root of $x^2 - 3.8298x + 7.3064 = 0$, then $2p = 3.8298$ and $p^2 + q^2 = 7.3064$.

$$\Rightarrow \qquad p = 1.9149, q = 1.9077$$

Hence, a pair of roots is $1.9149 \pm 1.9077i$.

Next to find remaining two roots of the given equations we divide the given equation by $x^2 - 3.8298x + 7.3064$ as follows :

1	−5	20	−40	60		
	3.8298	−4.4816	31.4503	0	3.8298	
		−7.3064	8.5499	−60.0002	−7.3064	
1	−1.1702	8.212	0.0029	−0.0002		
			≈ 0	≈ 0		

$\therefore$ The other quadratic factor is

$$x^2 - 1.1702x + 8.212$$

If $\alpha \pm i\beta$ be its roots, then

$$2\alpha = 1.1702 \text{ and } \alpha^2 + \beta^2 = 8.212$$

$$\Rightarrow \qquad \alpha = 0.5851, \beta = 2.8053$$

Hence, the other pair of the roots is $0.5851 \pm 2.8053i$.

3.14 GRAEFFE'S ROOT SQUARE METHOD

Sometimes we come across with some algebraic equations which have complex roots. In such cases the root-squaring method of Graeffe is the best to use. This method gives all the roots at once both real and complex.

Graeffe's method is mainly based upon the process in which the given equation is transformed into another whose roots are higher powers of those of the original equation. The roots of the transformed equation are widely separated and because of this fact can easily be found.

3.14.1 THE ROOT-SQUARING PROCESS

In Graeffe's method the given equation is transformed to another by repeated application of a root squaring process. The first step of this process is to separate the even and odd powers of x and then making square of both sides to get second equation whose roots are the squares of the original equation. In the same way the second equation is then transformed into a third equation whose roots are the squares of the roots of second equation. This roots squaring process is continued in this manner until the roots of the last transformed equation are completely separated.

Consider the polynomial equation with real coefficients

$$x^n + a_1 x^{n-1} + a_2 x^{n-2} + a_3 x^{n-3} + \ldots + a_{n-1}x + a_n = 0 \qquad \ldots (1)$$

Separating the even and odd powers of x and squaring, we get

$$(x^n + a_2 x^{n-2} + a_4 x^{n-4} + \ldots)^2 = (a_1 x^{n-1} + a_3 x^{n-3} + \ldots)^2.$$

Putting $x^2 = y$ after simplifying above equation, we get

$$y^n + b_1 y^{n-1} + b_2 y^{n-2} + \ldots + b_{n-1}y + b_n = 0 \qquad \ldots (2)$$

where,
$$\left. \begin{array}{l} b_1 = -a_1^2 + 2a_2 \\ b_2 = a_2^2 - 2a_1 a_3 + 2a_4 \\ \ldots\ldots\ldots\ldots\ldots\ldots\ldots\ldots\ldots \\ \ldots\ldots\ldots\ldots\ldots\ldots\ldots\ldots\ldots \\ b_n = (-1)^n a_n^2 \end{array} \right\}$$

If $\alpha_1, \alpha_2, \alpha_3, \ldots, \alpha_n$ be the roots of equation (1), then the roots of equation (2) are $\alpha_1^2, \alpha_2^2, \alpha_3^2, \ldots, \alpha_n^2,$

Repeating above process m times, we get the new transformed equation as

$$z^n + c_1 z^{n-1} + c_2 z^{n-2} + \ldots c_{n-1}z + c_n = 0.$$

The roots of (4) are $\alpha_1^{2m}, \alpha_2^{2m}, \alpha_3^{2m} \ldots, \alpha_n^{nm}$ if γ_i are the root of (4), then $\gamma_i = \alpha_i^{2m}$ for

$i = 1, 2, 3, \ldots, n$ and $\gamma_i > 0$.

Let us assuming that $\qquad |\alpha_1| > |\alpha_2| > |\alpha_3| > \ldots |\alpha_n|$

then $\qquad |\gamma_1| >> |\gamma_2| >> |\gamma_3| >> \ldots >> |\gamma_n|$

where $>>$ indicates for much greater than.

Therefore, $\qquad \dfrac{|\gamma_2|}{|\gamma_1|} = \dfrac{\gamma_2}{\gamma_1}, \dfrac{|\gamma_3|}{|\gamma_2|} = \dfrac{\gamma_3}{\gamma_2}, \ldots \dfrac{|\gamma_n|}{|\gamma_{n-1}|} = \dfrac{\gamma_n}{\gamma_{n-1}},$

are negligible as compared to unity.

Now from equation (4), we have

$$\sum_{i=1}^{n} \gamma_i = -c_i$$

or

$$c_i = -(\gamma_1 + \gamma_2 + \gamma_3 + ... + \gamma_n) = -\gamma_1\left(1 + \frac{\gamma_2}{\gamma_1} + \frac{\gamma_3}{\gamma_1} + ...\right)$$

$$\sum_{i \neq 1}^{n} \gamma_i\gamma_j = \sum \gamma_1\gamma_2 = c_2 \quad \text{or} \quad c_2 = \gamma_1\gamma_2\left(1 + \frac{\gamma_3}{\gamma_1} + ...\right)$$

$$\sum \gamma_1\gamma_2\gamma_3 = -c_3 \quad \text{or} \quad c_3 = (-1)^n \gamma_1\gamma_2...\gamma_n$$

....

$$\gamma_1\gamma_2\gamma_3...\gamma_n = (-1)^n \gamma_1\gamma_2...\gamma_n.$$

Since $\dfrac{\gamma_2}{\gamma_1}, \dfrac{\gamma_4}{\gamma_3}$ ..., etc. are negligible as compared to unity, then above equations become

$$\gamma_1 \approx -c_1$$

$$\left.\begin{array}{l} c_1 \approx -\gamma_1 \\ c_2 \approx \gamma_1\gamma_2 \\ c_3 \approx -\gamma_1\gamma_2\gamma_3 \\ .... \\ c_n \approx (-1)^n \gamma_1\gamma_2...\gamma_n \end{array}\right\} \Rightarrow \begin{array}{l} \gamma_2 \approx -\dfrac{c_2}{c_1} \\[2mm] \gamma_3 \approx -\dfrac{c_3}{c_2} \\[2mm] .... \\[1mm] \gamma_n \approx -\dfrac{c_n}{c_{n-1}} \end{array}$$

But $\gamma_i = \alpha_i^{2m}$ for $i = 1, 2, 3, ..., n$, then

$$\alpha_i = (\gamma_i)^{1/2m}$$

$$|\alpha_i| = (|\gamma_i|)^{1/2m} = \left|\frac{c_i}{c_i - 1}\right|^{1/2m}$$

Hence, we can determine $\alpha_1, \alpha_2, ..., \alpha_n$ the roots of the given equation.

3.14.2 ROOTS ARE REAL AND NUMERICALLY EQUAL

If the magnitude of c_i is half the square of the magnitude of the corresponding coefficient in the preceding equation, then α_i will the double (equal) root of the given equation. We find the double root as follows :

$$\gamma_i = \frac{c_i}{c_i - 1} \quad \text{and} \quad \gamma_{i+1} = -\frac{c_{i+1}}{c_i}$$

$$\therefore \quad \gamma_i\gamma_{i+1} \approx \gamma_i^2 \approx \left|\frac{c_{i+1}}{c_{i-1}}\right| \Rightarrow \alpha_i^{2m} \approx \left|\frac{c_{i+1}}{c_{i-1}}\right| \Rightarrow |\alpha_i| \approx \left|\frac{c_{i+1}}{c_{i-1}}\right|^{1/2m}$$

Thus, $|\alpha_i|$ gives the magnitudes of the double root and on substituting it in the given equation we find its sign.

3.14.3 FOR COMPLEX ROOTS

When the given equation has only one pair of complex roots and others are real, we first find all real roots.

Let α_r and α_{r+1} be the complex roots, and let $\alpha_r = \xi + i\eta$, and $\alpha_{r+1} = \xi - i\eta$ then we have
$$\alpha_1 + \alpha_2 + \ldots + \alpha_{r-1} + 2\xi + \alpha_{r+2} + \ldots + \alpha_n = -a_1. \qquad \ldots(6)$$

From this equation we find the value of ξ.

If α_r and α_{r+1} form a complex pair $\rho_r e^{\pm i\phi_r}$, then for m (Number of squaring) sufficiently large, we obtain $\rho_r \phi_r$ as follows:

and

$$\left. \begin{array}{c} \rho_r^{2^{(2^m)}} \approx \left| \dfrac{c_{r+1}}{c_{r-1}} \right| \\[4mm] 2\rho_r^m \cos m\phi_r \approx \dfrac{c_{r+1}}{c_{r-1}} \end{array} \right\} \qquad \ldots(7)$$

Now η is given by $\eta = \sqrt{\rho_r^2 - \xi^2}$.

SOLVED EXAMPLES

EXAMPLE 1. *Solve $x^3 - 8x^2 + 17x - 10 = 0$ by Graeffe's root squaring method.*   (Madras-1996)

SOLUTION. Let
$$f(x) = x^3 - 8x^2 + 17x - 10 = 2 \qquad \ldots(1)$$
Now
$$f(x) = x^3 - 8x^2 + 17x - 10$$
$$ + - + -$$

Clearly, $f(x)$ has three changes of sign, then by Descarte's rule of sign, $f(x) = 0$ has all the three roots positive and real.

Rewriting (1), we get
$$x^3 + 17x = 8x^2 + 10$$
Squaring both sides, we get
$$(x^3 + 17x)^2 = (8x^2 + 10)^2$$
$$x^6 + 289x^2 + 34x^4 = 64x^4 + 100 + 160x^2$$
or
$$x^6 - 30x^4 + 129x^2 - 100 = 0$$
Putting $x^2 = y$, we get
$$y^3 - 30y^2 + 129y - 100 = 0 \qquad \ldots(2)$$
Again (2) can be written as
$$y^3 + 129y = 30y^2 + 100$$
Now, squaring both sides, we get
$$y^6 + 16641y^2 + 258y^4 = 900y^4 + 10000 + 6000y^2$$
or
$$y^6 - 642y^4 + 10641y^2 - 10000 = 0$$
Putting $y^2 = z$ we get
$$z^3 - 642z^2 + 10641z - 10000 = 0 \qquad \ldots(3)$$
rewriting (3), we get
$$z^3 + 10641z = 642z^2 + 10000$$
Squaring both sides, we get
$$(z^3 + 10641z)^2 = (642z^2 + 10000)^2$$
$$\Rightarrow \qquad z^6 + 113230881z^2 + 21282z^4$$
$$= 412164z^4 + 100000000 + 12840000z^2$$
or
$$z^6 - 390882z^4 + 100390881z^2 - 100000000 = 0$$

Putting $z^2 = u$, we get

$$u^3 - 390882u^2 + 100390881u - 100000000 = 0 \qquad \ldots (4)$$

If α_1, α_2 and α_3 are the roots of (1), then

$$\alpha_1^8 = 390882$$

$$\Rightarrow \qquad |\alpha_1| = (390882)^{1/8} = 5.00041 \approx 5$$

$$\alpha_2^8 = \frac{100390881}{390882}$$

$$\Rightarrow \qquad |\alpha_2| = \left(\frac{100390881}{390882}\right)^{1/8} = 2.00081 \approx 2$$

$$\alpha_3^8 = \frac{100000000}{100390881}$$

$$\Rightarrow \qquad |\alpha_3| = \left(\frac{100000000}{100390881}\right)^{1/8} = 0.999512 \approx 1$$

Hence, the required root are 5, 2 and 1.

EXAMPLE 2. *Find all roots of the equation*

$$x^3 - 2x^2 - 5x + 6 = 0$$

by Graeffe's method, squaring thrice. (Meerut 2014)

SOLUTION. Let $\qquad f(x) = x^3 - 2x^2 - 5x + 6 = 0 \qquad \ldots(1)$

$$\therefore \qquad f(x) = x^3 - 2x^2 - 5x + 6 = 0$$
$$ + - - +$$

Clearly $f(x)$ has two changes of sign, so by Descarte's rule of sign, $f(x) = 0$ will have two positive roots. Also

Clearly, $f(-x)$ has one change of sign, so by Descarte's rule of sign, $f(x) = 0$ will have one negative root. Hence, $f(x) = 0$ has all real roots. Rewriting the equation (1) as

$$x^3 - 5x = 2x^2 - 6$$

Squaring both sides, we get

$$(x^3 - 5x)^2 = (2x^2 - 6)^2$$

or $\qquad x^6 + 25x^2 - 10x^2 = 4x^4 + 36 - 26x^2$

or $\quad x^6 - 14x^4 + 49x^2 - 36 = 0$

Putting $x^2 = y$ we get

$$y^3 - 14y^2 + 49y - 36 = 0 \qquad \ldots(2)$$

Again rewriting (2)

$$y^3 - 49y = 14y^2 + 36$$

Squaring both sides, we get

$$(y^3 + 49y)^2 = (14y^2 + 36)^2$$

or $\quad y^6 + 2401y^2 + 98y^4 = 196y^4 + 1296 + 1008y^2$

or $\qquad y^6 - 98y^4 + 1393y^2 - 1296 = 0 \qquad \ldots(3)$

Again putting $y^2 = z$, we get

$$z^3 - 98z^2 + 1393z - 1296 = 0 \qquad \ldots(4)$$

Rewriting (4) as

$$z^3 + 1393z = 98z^2 + 1296$$

Squaring both sides, we get

$$(z^3 + 1393z)^2 = (98z^2 + 1296)^2$$

or $z^6 + 1940449z^2 + 2786z^4 + 1686433z^2 - 1679616 = 0$

Putting $z^2 = u$, we get

$$u^3 - 6818u^2 + 1686433u - 1679616 = 0 \qquad \dots(5)$$

If $\alpha_1, \alpha_2, \alpha_3$ be the roots of the equation (1), then we have

$$\alpha_1^8 = 6818$$

$$\Rightarrow \qquad |\alpha_1| = (6818)^{1/8} = 3.011443 \approx 3$$

$$\alpha_2^8 = \frac{1686433}{6818}$$

$$\Rightarrow \qquad |\alpha_2| = \left(\frac{1686433}{6818}\right)^{1/8} = 1.991425 \approx 2$$

$$\alpha_3^8 = \frac{1679616}{1686433}$$

$$\Rightarrow \qquad |\alpha_3| = \left(\frac{1679616}{1686433}\right)^{1/8} = 0.999499 \approx 1$$

Now, $f(3) = 0, f(-2) = 0, f(1) = 0$

Hence, the required roots are 3, –2, 1.

EXAMPLE 3. *Apply Graeffe's roots square method to find all the roots of the equation* $x^4 - 3x + 1 = 0$. (Madras-1995)

SOLUTION. Let $\qquad f(x) = x^4 - 3x + 1 = 0 \qquad \dots(1)$

$$\Rightarrow \qquad f(x) = x^4 \underset{+}{} - 3x \underset{-}{} + 1 \underset{+}{} = 0$$

Clearly, $f(x)$ has two changes of sign so by Descarte's rule of sign, $f(x) = 0$ has two positive real roots.

Also, $\qquad f(-x) = x^4 \underset{+}{} + 3x \underset{+}{} + 1 \underset{+}{}$

Clearly, $f(x)$ has no change of sign so $f(x)$ has no negative real root.

But the degree of $f(x)$ is four so $f(x)$ has a pair of complex root.

Rewriting the equation (1), we get

$$x^4 + 1 = 3x$$

Squaring both sides, we get

$$(x^4 + 1)^2 = 9x^2 \qquad \text{or} \qquad x^8 + 1 + 2x^4 = 9x^2$$

or $x^8 + 2x^4 - 9x^2 + 1 = 0$

Putting $x^2 = y$, we get

$$y^4 + 2y^2 - 9y + 1 = 0 \qquad \dots(2)$$

Again rewriting (2), we get

$$(y^4 + 2y^2 + 1)^2 = (9y)^2 = 81y^2 \quad \text{or} \quad y^8 + 4y^4 + 1 + 4y^6 + 2y^4 + 4y^2 = 81y^2$$

or $\qquad y^8 + 4y^6 + 6y^4 - 77y^2 + 1 = 0$

Putting $y^2 = z$, we get

$$z^4 + 4z^3 + 6z^2 - 77z + 1 = 0 \qquad \qquad ...(3)$$

Rewriting (3), we get

$$z^4 + 6z^2 + 1 = -4z^3 + 77z$$

Squaring both sides, we get

$$(z^4 + 6z^2 + 1)^2 = (-4z^3 + 77z)^2$$

or $\quad z^8 + 36z^4 + 1 + 12z^6 + 2z^4 + 12z^2 = 16z^6 + 5929z^2 - 616z^4$

or $\qquad z^8 - 4z^6 + 654z^4 - 5917z^2 + 1 = 0$

Putting $z^2 = u$, we get

$$u^4 - 4u^3 + 65u^2 - 5917u + 1 = 0$$

If α_1, α_2, α_3 and α_4 be the roots of equation (1), then

$$\alpha_1^8 = 4 \qquad \Rightarrow \qquad |\alpha_1| = (4)^{1/8} = 1.1892$$

$$\alpha_2^8 = \frac{654}{4} \qquad \Rightarrow \qquad |\alpha_2| = \left(\frac{654}{4}\right)^{1/8} = 1.891$$

$$\alpha_3^8 = \frac{5917}{654} \qquad \Rightarrow \qquad |\alpha_3| = \left(\frac{5917}{654}\right)^{1/8} = 1.3169$$

$$\alpha_4^8 = \frac{1}{5917} \qquad \Rightarrow \qquad |\alpha_4| = \left(\frac{1}{5917}\right)^{1/8} = 0.3379$$

From (3) and (4) we observe that the magnitudes of the coefficients c_1 and c_4 have become constant

i.e., from (4) $\qquad c_1 = -4, c_4 = 1$

and from (3), $\qquad b_1 = -4, b_4 = 1$

which indicates that α_1 and α_4 are the real roots whereas α_2 and α_3 form a pair of complex roots.

Let $\alpha_2 = \xi + i\eta$ and $\quad \alpha_3 = \xi - i\eta$.

Now we find the complex roots of the form

$$\rho_2 e^{\pm i\phi_2} = \xi + i\eta$$

From (4), we have

$$\rho_2^{2^{(2^3)}} \approx \left|\frac{c_2 + 1}{c_2 - 1}\right| \quad \Rightarrow \quad \rho_2^{16} \approx \left|\frac{c_3}{c_1}\right| = \frac{5917}{4} = 1479.25$$

$$\Rightarrow \qquad \qquad \rho_2 \approx (1479.25)^{1/16} = 1.5781$$

From equation (1), we have

$$\alpha_1 + \alpha_2 + \alpha_3 + \alpha_4 = 0 \qquad \Rightarrow \qquad \alpha_1 + 2\xi + \alpha_4 = 0$$

$$\Rightarrow \qquad \xi = -\frac{1}{2}(\alpha_1 + \alpha_2) = -\frac{1}{2}(1.1892 + 0.3379)$$

$$\Rightarrow \qquad \qquad \xi = -0.7636$$

and $\qquad \qquad \eta = \sqrt{\rho_2^2 - \xi^2} = \sqrt{(1.5781)^2 - (-0.7636)^2}$

$$= 1.381$$

$\therefore$ The complex roots are $-0.7636 \pm 1.381i$.
Hence, all the four roots are

$$1.1892, 0.3379, -0.7636 \pm 1.381i.$$

EXERCISE 3.5

1. Using Muller's method, find the root of the following equations
   (i) $x^3 - x^2 - x - 1 = 0$
   (ii) $x^3 - x - 1 = 0$
   (iii) $x^3 - 3x - 5 = 0$
   (iv) $x^3 + 2x^2 + 10x - 20 = 0$ by taking $x_0 = 0, x_1 = 1$ and $x_2 = 2$.
   (v) $x^3 - 2x - 1 = 0$ (UPTU-2006)

2. Apply Lin-Bairstow method to find a quadratic factor of the equation $x^4 + 5x^3 + 3x^2 - 5x - 9 = 0$ close to $x^2 + 3x - 5$.   (Madras-1997; UPTU-2003)

3. Find the roots of the equation $x^4 + 9x^3 + 36x^2 + 51x + 27 = 0$ to three decimal places using Lin-Bairstow method.

4. Find the quadratic factors of the equation $x^4 - 8x^3 + 39x^2 - 62x + 50 = 0$ by using Lin-Bairstow method (upto third iteration) starting with $a = 0, b = 0$.

5. By Lin-Bairstow method, find a quadratic factor of the equation $5x^6 + 4x^5 + 7x^4 + 3x^3 + 20x^2 + 30x + 1 = 0$.

6. Solve $x^3 - 5x^2 - 17x + 20 = 0$ by Graeffe's method.

7. Find the roots of $5x^3 + 2x^2 - 15x - 6 = 0$ by Graeffe's method, squaring two times.

8. Apply Graeffe's method to find all the roots of the equation $x^3 - 6x^2 + 11x - 6 = 0$.
                    (Madras-1997)

9. Find all the roots of the equation $x^3 - 4x^2 + 5x - 2 = 0$ by Graeffe's method, squaring three times.   (Madras-1998; Madras-2001)

10. Solve $x^3 - 9x^2 + 18x - 6 = 0$ by Graeffe's method.

Answers

1. (i) 1.839287   (ii) 1.324718   (iii) 2.2596   (iv) 1.368808   (v) 2.26
2. $x^2 + 2.9026x - 4.9176$ 3. $-0.759, -1.42, -3.411 \pm 2.903i$ 4. $(x^2 - 2x + 2)(x^2 - 6x + 25)$
6. 7.018, -2.974, 0.958   8. 3, 2, 1   9. 2, 1, 1   10. 6.3, 2.3, 0.4

MISCELLANEOUS EXERCISE

1. By using bisection method, find an approximate root of the equation $\sin x = 1/x$ that lies between $x = 1$ and $x = 1.5$ (measured in radians). Carry out computations upto the 7th stage.   (VTU-2003S)

2. Find the positive root of $x^4 - x = 10$ correct to three decimal places, using Newton-Raphson method.   (JNTU-2008, Madras-2006)

3. Find the Newton's method, the real root of the equation $3x = \cos x + 1$.   (VTU-2009, SVTU-2007)

4. Find a root of the following equation, using the bisection method correct to three decimal places.
   $$x^3 - x^2 - 1 = 0   \text{(JNTU-2009)}$$

5. Using Newton-Raphson method, find a root of the following equations correct to three decimal places.
   $x \tan x + 1 = 0$ which is near $x = \pi$.   (JNTU-2006, VTU-2006)

6. Develop a recurrence formula for finding $\sqrt{N}$, using Newton-Raphson method and hence compute to three decimal places.
   (i) $\sqrt{13}$ (UPTU-2008)
   (ii) $\sqrt{10}$ (JNTU-2008)

Answers

1. 1.11328   2. 1.856   3. 0.6071   4. 1.46   5. 2.7985
6. $x_{n+1} = 1/2(x_n + N/x_n)$;   (i) 3.605   (ii) 3.162

Objective Evaluations

FILL IN THE BLANKS

1. Every equation of the odd degree has atleast _____ .

2. If a polynomial of degree n vanishes for more than n values of x, then it must be _____ .

3. Every equation of the nth degree has only _____ roots.

4. If $\alpha + i\beta$ is a root of a polynomial of degree two, then its other root is _____ .

5. They type of the equation $e^x + \sin x = 0$ is _____ .

6. If ε be the error, then $\varepsilon_i = \alpha - x_i = x_{i+1} - x_i$, where α is a root of $f(x) = 0$ and if $\dfrac{\varepsilon_{i+1}}{\varepsilon_i}$ is almost constant, then convergence is _____ .

7. In question no. 6, if $\dfrac{\varepsilon_{i+1}}{\varepsilon_i^2}$ is nearly constant, then convergence is _____ .

8. If a root of $f(x) = 0$ lies between a and b, then $f(a).f(b)$ _____ .

9. In a secant method it does not required the condition that $f(a).f(b)$ _____ .

10. Newton's formula will convergence if $|f(x)/f''(x)|$ is less than _____ .

11. Newton's method has a convergence of order _____ .

12. The convergence of secant method is of order _____ .

13. The convergence of Newton-Raphson method for double root is _____ .

14. If a root of $f(x) = 0$ lies between a and b, then in Regula-falsi method the part of the curve $y = f(x)$ between a and b is replaced by _____ .

TRUE / FALSE

Write 'T' for True and 'F' for False statement.

1. If the secant method once convergence, its rate of convergence is faster than that of the method of false position. **(T/F)**

2. Newton-Raphson method has quadratic rate of convergence. **(T/F)**

3. If $f(x) = 0$ has a root α of order two, then $f'(\alpha) \neq 0$. **(T/F)**

4. The order of convergence of secant method is 1.62. **(T/F)**

5. Bisection method never converges. **(T/F)**

6. If $f(x) = x^2 + \log x$, then $f(x)$ is an algebraic equation. **(T/F)**

7. In bisection method, if $f(a).f(b) < 0$, then first approximation to the root of $f(x) = 0$ is $\dfrac{a+b}{2}$. **(T/F)**

8. Newton-Raphson method is not applicable when $f(x)$ has complex roots. **(T/F)**

9. Newton's method is conditionally convergent while Regula-falsi is surely convergent. **(T/F)**

10. Newton's method is failed if $f(x_0)/f'(x_0)$ is large. **(T/F)**

11. If $f(x) = 0$ has a root of order 2, then the convergence of Newton's method is given by $\varepsilon_{n+1} = \dfrac{1}{2}\varepsilon_n$. **(T/F)**

12. The secant method is failed if $f(x_{n-1}) = f(x_n)$. **(T/F)**

MULTIPLE CHOICE QUESTIONS

Choose the most appropriate one

1. An equation which can be expressed in the form of polynomial is called

(a) algebraic equation

(b) transcendental equation

(c) both (a) and (b) are true

(d) none of these

2. A non-algebraic equation is called

(a) transcendental equation

(b) polynomial

(c) both (a) and (b) are true

(d) none of these

3. Transcendental equation are

(a) trigonometric

(b) logarithmic

(c) exponential

(d) all are three

4. A transcendental equation may have

(a) a finite or infinite number of real root

(b) not have real root at all

(c) both (a) and (b) are true

(d) none of these

5. The process of finding the roots of the equation is called the

(a) rootification   (b) solution

(c) simplification   (d) none of these

6. If $f(x)$ is exactly divisible by $(x - a)$ then a is the root of

(a) $f(x) = 0$ (b) $f(x) \neq 0$

(c) $f(x) > 0$ (d) none of these

7. Every algebraic equation of degree n has

(a) n roots   (b) atleast n roots

(c) atmost n roots   (d) none of these

8. Every equation of odd degree has at least one

(a) real root   (b) complex root

(c) can't say   (d) none of these

9. If a polynomial of degree n vanishes for more than n values of x, it must be

(a) equal to 0   (b) identically zero

(c) not equal to 0   (d) none of these

10. If $f(x)$ is a function which is continuous in a closed interval $[a, b]$ and $f(a)$, $f(b)$ are of opposite signs then $\exists$ at least one value of $C \in \,] \, a, b \, [$ such that

(a) $f(c) = 0$ (b) $f(c) \neq 0$

(c) $f(c) > 0$ (d) none of these

11. If $f(x)$ is not continuous in the closed interval, than Bisection method :

(a) can be applied   (b) failed

(c) can't say   (d) none of these

12. If $f(x)$ is continuous in a closed interval and does not cut the x-axis then $f(x)$:

(a) has a real root

(b) does not have a real root

(c) may have a real root

(d) none of these

13. Bisection method is also known as :

(a) bolzano metod

(b) interval halving method

(c) both (a) and (b) are true

(d) none of these

14. Bisection method gives the :

(a) real roots only

(b) complex roots only

(c) both real or complex

(d) none of these

15. The rate of convergence of bisection method is :

(a) linear   (b) quadratic

(c) cubic   (d) none of these

16. If we have to find the value of the root in terms of the function, then we use:

(a) Bisection method

(b) Iterative method

(c) Picard's method

(d) none of these

17. The rate of convergence of iterative method is :

(a) linear   (b) quadratic

(c) cubic   (d) none of these

18. The formula

$$x_{n+1} = x_{n-1} - \frac{f(x_{n-1})(x_n - x_{n-1})}{f(x_n) - f(x_{n-1})}$$

is called:

(a) Regula-falsi method

(b) Secant method

(c) Newton-Raphson method

(d) None of these

19. Regula-Falsi method is also called

(a) Interval halving method

(b) Bolzano method

(c) Linear interpolation method

(d) none of these

20. The rate of convergence of Regula-Falsi method is :

    (a) linear   (b) quadratic

    (c) cubic   (d) none of these

21. The formula $x_{n+1} = x_n - \dfrac{f(x_n)}{f'(x_n)}$ is called :

    (a) Newton-Raphson method

    (b) Regula Falsi method

    (c) Bisection method

    (d) none of these

22. The rate of convergence of Newton-Raphson method is :

    (a) linear   (b) quadratic

    (c) cubic   (d) none of these

23. The convergence of Newton's method for double roots is :

    (a) linear   (b) quadratic

    (c) cubic   (d) none of these

24. Newton's method is :

    (a) always convergent

    (b) conditionally convergent

    (c) divergent

    (d) none of these

25. Newton's method converges if :

    (a) $|f(x)f''(x)| < |f'(x)|$

    (b) $|f(x)f''(x)| < |f'(x)|^2$

    (c) $|f(x)f'(x)| < |f''(x)|^2$

    (d) none of these

26. Newton-Raphson method is suitable for :

                   (Rohilkhand–2011)

    (a) $f'(x)$ is large   (b) $f'(x)$ is small

    (c) $f'(x) < 0$ (d) none of these

27. The method which is not applicable for finding the roots?

    (a) Secant method

    (b) Lagrange's method

    (c) Newton's method

    (d) none of these

28. The real root of the equation $f(x) = 0$ are the abscissae of the points where the graph crosses the :

(a) *y*-axis   (b) line $y = x$

(c) *x*-axis   (d) none of these

29. Newton's formula for finding the square root of *a* is :

    (a) $x_{n+1} = \dfrac{1}{2}\left(x_n - \dfrac{a}{x_n}\right)$

    (b) $x_{n+1} = \dfrac{1}{2}\left(x_n + \dfrac{a}{x_n}\right)$

    (c) $x_{n+1} = 2x_n$

    (d) none of these

30. If x_0 be an approximate value of the desired root of the equation $f(x) = 0$ then by Newton Raphson method the improved value of the root is $x_0 + h$ where $h =$

    (a) $-\dfrac{f(x_0)}{f'(x_0)}$ (b) $-\dfrac{f'(x_0)}{f(x_0)}$

    (c) $\dfrac{f(x_0)}{f'(x_0)}$ (d) none of these

31. If the equation $f(x) = 0$ be expressed in the form $x = \phi(x)$. If x_0 is the initial approximation to the solution $x = \phi(x)$, then by iteration method, $x_{n+1} =$

    (a) $\phi(x_{n+1})$ (b) $\phi(x_n)$

    (c) $\phi(x_{n-1})$ (d) none of these

32. In Regula-Falsi method, if a root of $f(x) = 0$ lies between x_1 and x_2 then the approximate value of the desired root is x_{1+h} where $h =$

    (a) $\dfrac{(x_2 - x_1)|y_1|}{|y_1| + |y_2|}$ (b) $\dfrac{(x_1 - x_2)|y_1|}{|y_1| + |y_2|}$

    (c) $\dfrac{x_1 - x_2}{y_1 + y_2}$ (d) none of these

33. If one root of the equation $f(x) = 0$ is near to x_0, then the first approximation of this root as calculated by Newton-Raphson method is the absicssa of the point where the following straight line intersects the *x*-axis :

    (a) tangent to the curve $y = f(x)$ at the point $(x_0, f(x_0))$

    (b) normal to the curve $y = f(x)$ at the point $(x_0, f(x_0))$

    (c) both (a) and (b) are true

    (d) none of these

34. By false positioning, the second approximation of a root of equation $f(x) = 0$ is :

(a) $\dfrac{x_0 f(x_0) - x_1 f(x_1)}{f(x_0) - f(x_1)}$

(b) $\dfrac{x_0 f(x_1) - x_1 f(x_0)}{f(x_1) - f(x_0)}$

(c) both (a) and (b) are true

(d) none of these

35. The condition for convergence of Newton-Raphson method to a root α is :

(a) $\dfrac{f'(\alpha)}{f''(\alpha)} < 1$ (b) $\dfrac{f'(\alpha)}{f''(\alpha)} > 1$

(c) $\dfrac{1}{2} \dfrac{f'(\alpha)}{f''(\alpha)} > 1$ (d) none of these

36. Newton-Raphson method is applicable only when :

(a) $f(x) \neq 0$ in the neighbourhood of actual root $x = a$

(b) $f'(x) \neq 0$ in the neighbourhood of actual root $x = a$

(c) both (a) and (b) are true

(d) none of these

37. Starting with $x_0 = 1$, the next approximation x_1 to $(2)^{1/3}$ by Newton's method is given by:

(a) 4/3   (b) 5/3

(c) 2/3   (d) none of these

38. If α be the exact root, then Newton-Raphson method converges fast if $f'(\alpha)$ is :

(a) very large   (b) 0

(c) < 0 (d) none of these

39. After second iteration of Newton-Raphson method, the positive root of the equation $x^2 = 3 \left(x_0 = \dfrac{3}{2} \right)$ is:

(a) 5/4   (b) 97/56

(c) 4/3   (d) none of these

40. The method which always converges to root of equation $f(x) = 0$ is:

(a) Secant method

(b) Regula Falsi method

(c) Newton Raphson method

(d) none of these

41. To find the positive square root of $a > 0$ by solving $x^2 - a = 0$ by Newton-Raphson

method, if x_n denote the nth iteration with $x_0 > 0$, $x_0 \neq \sqrt{a}$ then the sequence $< x_n >$ is:

(a) strictly increasing

(b) strictly decreasing

(c) monotonic

(d) none of these

42. Newton-Raphson method for finding p^{th} root of a number N is :

(a) $\dfrac{1}{p} \left\{ (p-1)x_n - \dfrac{N}{x_n^{p-1}} \right\}$

(b) $\left\{ (p-1)x_n - \dfrac{N}{x_n^{p-1}} \right\}$

(c) both (a) and (b) are true

(d) none of these

43. Given that one root of the equation $x^3 - 10x^2 + 31x - 30 = 0$ is 5, the other two roots are :   (GATE (CE)–2007)

(a) 2 and 4   (b) 2 and 3

(c) 3 and 4   (d) none of these

44. The Newton Raphson algorithm for the function will be: (GATE (CE)–2005)

(a) $x_{k+1} = 2x_k - ax_k^2$

(b) $x_{k+1} = x_k - \dfrac{a}{2}x_k^2$

(c) $x_{k+1} = \dfrac{1}{2}\left(x_k + \dfrac{a}{x_k} \right)$

(d) none of these

45. The following equation needs to be numerically solved using the Newton-Raphson method.

$$x^3 + 4x - 9 = 0$$

The iterative equation for this purpose is (k indicates the iteration level).   (GATE (CE)–2007)

(a) $x_{k+1} = x_k - 3x_k^2 + 4$

(b) $x_{k+1} = \dfrac{3x_k^2 + 4}{2x_k^2 + 9}$

(c) $x_{k+1} = \dfrac{2x_k^3 + 9}{3x_k^2 + 4}$

(d) none of these

46. For $a = 7$ and starting with $x_0 = 0.2$, the first two iterations will be :   (GATE (CE)–2005)

(a) 0.11, 0.1299   (b) 0.13, 0.1428

(c) 0.12, 0.1416   (d) 0.12, 0.1392

47. Starting from $x_0 = 1$, one step of Newton-Raphson method in solving the equation $x^3 + 3x - 7 = 0$ gives the next value (x_1) as :

(GATE (ME)–2005)

(a) $x_1 = 1.5$ (b) $x_1 = 0.5$

(c) $x_1 = 2$ (d) none of these

48. The square root of a number N is to be obtained by applying the Newton Raphson iterations to the equation $x^2 - N = 0$. If i denotes the iteration index, the correct iterative scheme will be: (GATE (CE)–2011)

(a) $x_{i+1} = \dfrac{1}{2}\left(x_i + \dfrac{N^2}{x_i}\right)$

(b) $x_{i+1} = \dfrac{1}{2}\left(x_i - \dfrac{N}{x_i}\right)$

(c) $x_{i+1} = \dfrac{1}{2}\left(x_i + \dfrac{N}{x_i}\right)$ (d) none of these

49. The equation $e^X - 1 = 0$ is required to be solved using Newton's method with an initial guess $x_0 = -1$. Then, after one step of Newton's method, estimate x_1 of the solution will be given by :   (GATE (EE)–2008)

(a) 0.36784   (b) 0.71828

(c) 0.0000   (d) none of these

50. When the Newton-Raphson method is applied to solve the equation $f(x) = x^3 + 2x - 1 = 0$ the solution at the end of the first iteration with the initial guess value as $x_0 = 1.2$ is:

(GATE(EE)–2013)

(a) –0.82   (b) 1.69

(c) 0.49   (d) 0.705

51. Roots of the algebraic equation $x^3 + x^2 + x + 1 = 0$ are: (GATE (EE)–2011)

(a) $(-1, +j, -j)$ (b) $(+1, -1, +1)$

(c) $(0, 0, 0)$ (d) none of these

52. Let $x^2 + 17 = 0$. The iterative steps for the solution using Newton-Raphson's method is given by: (GATE (EE)–2009)

(a) $x_{k+1} = x_k - \dfrac{117}{x_k}$

(b) $x_{k+1} = \dfrac{1}{2}\left(x_k + \dfrac{117}{x_k}\right)$

(c) $x_{k+1} = x_k - \dfrac{X_k}{117}$ (d) none of these

53. Solution of the variables x_1 and x_2 for the following equations is to be obtained by employing the Newton-Raphson iterative method equation: (GATE (EE)–2011)

(i) $10x_2 \sin x_1 - 0.8 = 0$

(ii) $10x_2^2 - 10x_2 \cos x_1 - 0.6 = 0$

Assuming the initial values $x_1 = 0.0$ and $x_2 = 1.0$, the Jacobian matrix is:

(a) $\begin{bmatrix} 10 & 0 \\ 0 & 10 \end{bmatrix}$ (b) $\begin{bmatrix} 10 & 0 \\ 0 & -10 \end{bmatrix}$

(c) $\begin{bmatrix} 10 & -0.8 \\ 0 & -0.6 \end{bmatrix}$ (d) none of these

54. The recursion relation to solve $x = e^{-x}$ using Newton-Raphson method is   (GATE (EC)–2008)

(a) $x_{n+1} = e^{-x_n}$

(b) $x_{n+1} = (1 + x_n)\dfrac{e^{-x_n}}{1 + e^{-x_n}}$

(c) $x_{n+1} = x_n - e^{-x_n}$

(d) none of these

55. The equation $x^3 - x^2 + 4x - 4 = 0$ is to be solved using the Newton-Raphson method. If $x = 2$ is taken as the initial approximation of the solution, then the next approximation using this method will be:(GATE (EC) – 2007)

(a)   (b) $\dfrac{2}{3}$

(c) 1   (d) 0

56. A numerical solution of the equation $f(x) = x + \sqrt{x} - 3 = 0$ can be obtained using Newton-Raphson method. If the starting value is $x = 2$ for the iteration, the value of x that is to be used in the next step is:

(GATE (EC)–2011)

(a) 0.739   (b) 1.694

(c) 0.306   (d) none of these

57. Consider the series $x_{n+1} = \dfrac{x_n}{2} + \dfrac{9}{8x_n}, x_0 = 0.5$ obtained from the Newton-Raphson method. The series converges to: (GATE (CS)–2007)

(a) $\sqrt{2}$ (b) 1.5

c) 1.4   (d) 0

58. Newton-Raphson method is used to compute a root of the equation $x^2 - 13 = 0$ with 3.5 as the initial value. The approximation after one iteration is: (GATE (CS)–2010)

(a) 3.575   (b) 3.677

(c) 3.607   (d) none of these

59. The Newton-Raphson iteration
$$x_{n+1} = \frac{1}{2}\left(x_n + \frac{R}{x_n}\right)$$
can be used to compute the : (GATE (CS) –2008)

(a) square root of R (b) square of R

(c) reciprocal of R (d) none of these

60. The bisection method is applied to compute a zero of the function $f(x) = x^4 - x^3 - x^2 - 4$ in the interval [1, 9]. The method converges to a solution after ____ iterations.

(GATE (CS)–2012)

(a) 1   (b) 5

(c) 7   (d) 3

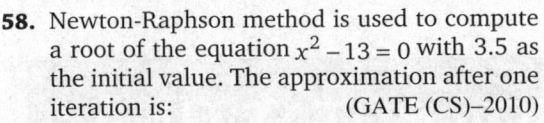

Answers

FILL IN THE BLANKS

  1. one real root   **2.** Identically zero   **3.** n **4.** $\alpha - i\beta$ **5.** transcendental

  6. linear   **7.** quadratic   **8.** less than zero   **9.** less than zero

  10. $|f'(x)|^2$ **11.** 2   **12.** 1.62   **13.** 1   **14.** a variable chord

TRUE/FALSE

  1. T   **2.** T   **3.** F   **4.** T   **5.** F   **6.** F   **7.** T   **8.** F   **9.** T

  10. T   **11.** T   **12.** T

MULTIPLE CHOICE QUESTIONS

  1. (a)   **2.** (a)   **3.** (d)   **4.** (c)   **5.** (b)   **6.** (a)   **7.** (a)   **8.** (a)   **9.** (b)

  10. (a)   **11.** (b)   **12.** (b)   **13.** (c)   **14.** (a)   **15.** (a)   **16.** (b)   **17.** (a)   **18.** (a)

  19. (c)   **20.** (a)   **21.** (a)   **22.** (b)   **23.** (a)   **24.** (b)   **25.** (b)   **26.** (a)   **27.** (b)

  28. (c)   **29.** (b)   **30.** (a)   **31.** (b)   **32.** (a)   **33.** (a)   **34.** (b)   **35.** (c)   **36.** (b)

  37. (a)   **38.** (a)   **39.** (b)   **40.** (b)   **41.** (a)   **42.** (a)   **43.** (b)   **44.** (a)   **45.** (c)

  46. (d)   **47.** (a)   **48.** (c)   **49.** (b)   **50.** (d)   **51.** (a)   **52.** (b)   **53.** (a)   **54.** (b)

  55. (a)   **56.** (b)   **57.** (a)   **58.** (c)   **59.** (a)   **60.** (d)

COMPUTATIONAL TECHNIQUE LAB

1. BISECTION METHOD

SYMBOL USED

x_1, x_2 = Initial approximations in which root lies.

err = allow error

x_3 = New approximation of the root in each iteration

itr = a counter which keeps track of the no. of iterations performed.

ALGORITHM : BISECTION METHOD

Step 1. Start

Step 2. Define $f(x)$

Step 3. Input x_1, x_2, err

Step 4. If $(f(x_1) * (f(x_2)) < 0$

          print "Initial approximations are correct"

    else

          print "Initial approximations are not correct"

          go to step 3

Step 5. itr = 1

Step 6. Perform steps 7 to 10 while $(|x_2 - x_1| > \text{err})$

Step 7. $x_3 = (x_1 + x_2)/2$

Step 8. if $f(x_3) = 0$

    Print "Solution converges in iteration", i

    Print "Root =", x_3

    go to step 13

Step 9. If $(f(x_1) * (f(x_3)) < 0$

        $x_2 = x_3$

    else

        $x_1 = x_3$

Step 10. itr = itr +1

Step 11. Print "Solution converges in iteration", i

Step 12. Print "Root=", x_3

Step 13. STOP

Flow Chart : Bisection Method

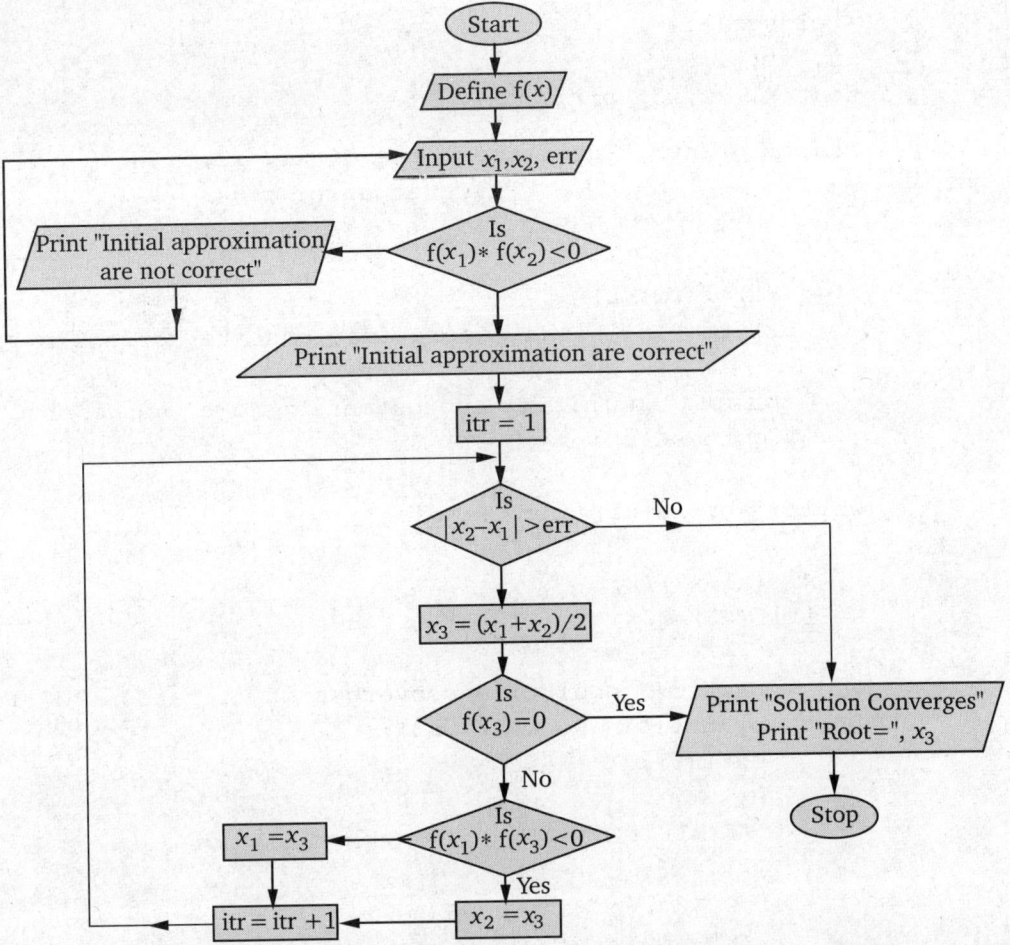

PROGRAM . *Following is a C program to find the root of the equation $f(x)=x^3-4*x-9$ by Bisection method.*

(UPTU B.Tech. 2003)

// Bisection Method

/*Bisection Method to find root of $x*x*x-4*x-9=0$ */

```c
# include<stdio.h>
# include<conio.h>
# include<math.h>
# include<process.h>

float f(float x)
  {
       return(x*x*x-4*x-9 );
  }
```

```c
  void  main()
{
   clrscr();
int itr=1;
float x1,x2,x3, err;

start: printf("Enter the value of x1,  x2,"
                    "allowed error\n");
        scanf("%f%f%f", &x1,&x2,&err);

if(f(x1)*f(x2)<0)
    printf("\n initial approximations are correct\n");
else
  { printf("\n initial approximations are  not correct\n");
     goto start;
     }
while(fabs(x2-x1)>err)
{
   x3=(x1+x2)/2;
   if(f(x3)==0)
      {
      printf("\nsolution converges in iteration=%d",itr);
      printf("\n Root=%f",x3);
      exit(0);
      }
    if(f(x1)*f(x3)<0)
        x2=x3;
    else
        x1=x3;
   itr++;
  }
    printf("\n solution converges in iteration=%d",itr);
    printf("\n Root=%f",x3);
    getch();
}
```

Output: BISECTION METHOD TO FIND ROOT OF x*x*x-4*x-9=0

```
Enter the value of x1,  x2, allowed error
3    2       .0005
initial approximations are correct
solution converges in iteration =12
Root=2.7065
```

Lab Assignment : Bisection Method

1. Write a C program to find the root of the equation by using Bisection method correct to two places of decimal.

$$f(x) = x^3 - x - 11 = 0$$

Hint: Define a function $f(x) = x * x * x - x - 11$

  Input: $x_1 = 2, x_2 = 3$, err $= 0.5 \times 10^{-2} = 0.005$

  Output: Root = 2.38

2. Write C program to find root for the given function by Bisection method correct to two decimal places.

$$f(x) = 3x - \sqrt{1 + \sin x}$$

 Hint: Define $f(x) = 3 * x - \text{sqrt}(1 + \sin(x))$

  Input : $x_1 = 0,\ x_2 = 1$ err $= 0.005$

  Output: Root = 0.39

3. Write a C program to find root using Bisection method correct to two decimal places for following function.

$$f(x) = x^2 - 4x - 10 = 0$$

 Hint: Define $f(x) = x * x - 4 * x - 10$

  Input $x_1 = 5, x_2 = 6$ and err $= 0.005$

  Output Root = 5.738

4. Write a C program to find root using Bisection method correct to two decimal places for following function.

$$x\ \log_{10} x = 1.2$$

 Hint: Define $f(x) = x * \log_{10}(x) - 1.2$

  Input $x_1 = 2, x_2 = 3$, err $= 0.005$

  Output Root = 2.74.

2. ITERATION METHOD

Symbols Used

  $x_1 =$ Initial approximations of the root

  $x_2 =$ New approximation of root in each iteration

  $x_0 =$ Used to store the previous value of x_1

Algorithm : Iteration Method

Step 1. Start

Step 2. For given function $f(x)$ defined $g(x)$

Step 3. Input x_1, err

Step 4. itr $= 0$

Step 5. Perform steps 6 to 9 while ($|x_2 - x_0| >$ err)

Step 6. $x_2 = g(x_1)$

Step 7. $x_0 = x_1$

Step 8. $x_1 = x_2$

Step 9.   itr = itr + 1

Step 10.   Print "Solution converges in iteration", i

Step 11.   Print "Root=", x_2

Step 12.   STOP

FLOW CHART : ITERATION METHOD

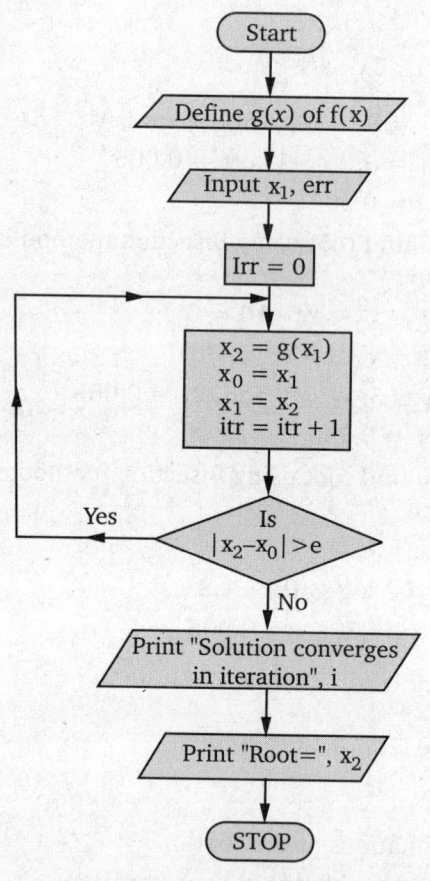

PROGRAM.   *Following is a C program to find the root of the equation $f(x) = 2x - \log_{10} x - 7 = 0$ by using iteration method correct to 4 decimal places.*

We have $f(x) = 2x - \log_{10} x - 7 = 0$

It can be written as

$$x = \frac{1}{2}(\log_{10} x + 7)$$

Define $g(x) = (\log_{10}(x) + 7)/2$

Program for iteration method

/*Iteration Method to find root of $2x-\log_{10}x-7=0$ */

```c
# include<stdio.h>
# include<conio.h>
#include<math.h>
#include<process.h>

float g(float x)
{
        return((log10(x)+7)/2);
}

void  main()
{
  clrscr();
int itr=0;
float x1,x2,xo, err;

        printf("Enter the value of x1, allowed error\n");
        scanf("%f%f", &x1,&err);
    do
        {
        x2=g(x1);
        xo=x1;
        x1=x2;
        itr++;
        } while(fabs((x2-xo)>err));

        printf("\n solution converges in iteration =%d",itr);
        printf("\n Root=%f",x2);
        getch();
}
```

Output: ITERATION METHOD TO FIND ROOT OF 2x-log10x-7=0

```
Enter the value of x1, allowed error
3      .00005

solution converges in iteration = 5

Root= 3.789278
```

LAB ASSIGNMENTS : ITERATION METHOD

1. Write a C program to find root of the equation $f(x) = x^3 - 2x + 1 = 0$ using iteration method correct to 3 decimal places.

Hint : $\qquad f(x) = x^3 - 2x + 1 = 0$

$$g(x) = x = \frac{x^3 + 1}{2}$$

Define $g(x) = (x*x*x+1)/2$

Input $x_1 = 0$,   err $= 0.0005$

Output   Root $= 0.6175$

2. Write C program to write real root of the equation cos x = 3x −1 correct to three decimal places using iteration method.

     Hint : $f(x) = \cos x = 3x - 1$

$$x = \frac{\cos x + 1}{3} = g(x)$$

     Define $g(x) = (\cos (x)+1)/3$

     Input : $x_1 = 0$,   err $= 0.0005$

     Output :   Root $= 0.607$

3. Write C program to find real root of the equation by iteration method upto 4 decimal places

$$f(x) = x^3 + x^2 - 1 = 0$$

     Hint : $x = \dfrac{1}{\sqrt{1+x}}$

     Define $x) = (1/sqrt (1+x))$

     Input $x_1 = 0.5$, err $= 0.00005$

     Output   Root $= 0.7548$

3. REGULA-FALSI METHOD

SYMBOLS USED

         $x_1, x_2 =$ Initial approximations near to the root

            $x_3 =$ New approximation in each iteration

           err $=$ allowed error

      max itr $=$ Maximum no. of iterations

         itr $=$ iteration counter

ALGORITHM : REGULA-FALSI METHOD

Step 1.   Start

Step 2.   Define function $f(x)$

Step 3.   Input x_1, x_2, err, maxitr

Step 4.   If $(f(x_1) * f(x_2)) \leq 0$

              Print "Initial approximation are correct"

       else

              Print "Initial approximation are not correct".

       goto step 3

Step 5.   itr $= 1$

Step 6.   Repeat steps 7 to 9 while (ite $<=$ maxitr)

Step 7. $x_3 = (x_1 * f(x_2) - x_2 * f(x_1))/f(x_2) - f(x_1))$

> if $(|f(x_3) < \text{err}|)$
>> Print "Solution converges in iteration", itr
>> Print "Root=", x_3
>> goto step 11

Step 8.   if $(f(x_1) * f(x_3)) < 0$
>> $x_2 = x_3$

> else
>> $x_1 = x_3$

Step 9.   itr $=$ itr $+ 1$

Step 10.   Print "Solution does not converges, iterations not sufficient"

Step 11.   STOP

FLOW CHART : REGULA-FALSI METHOD

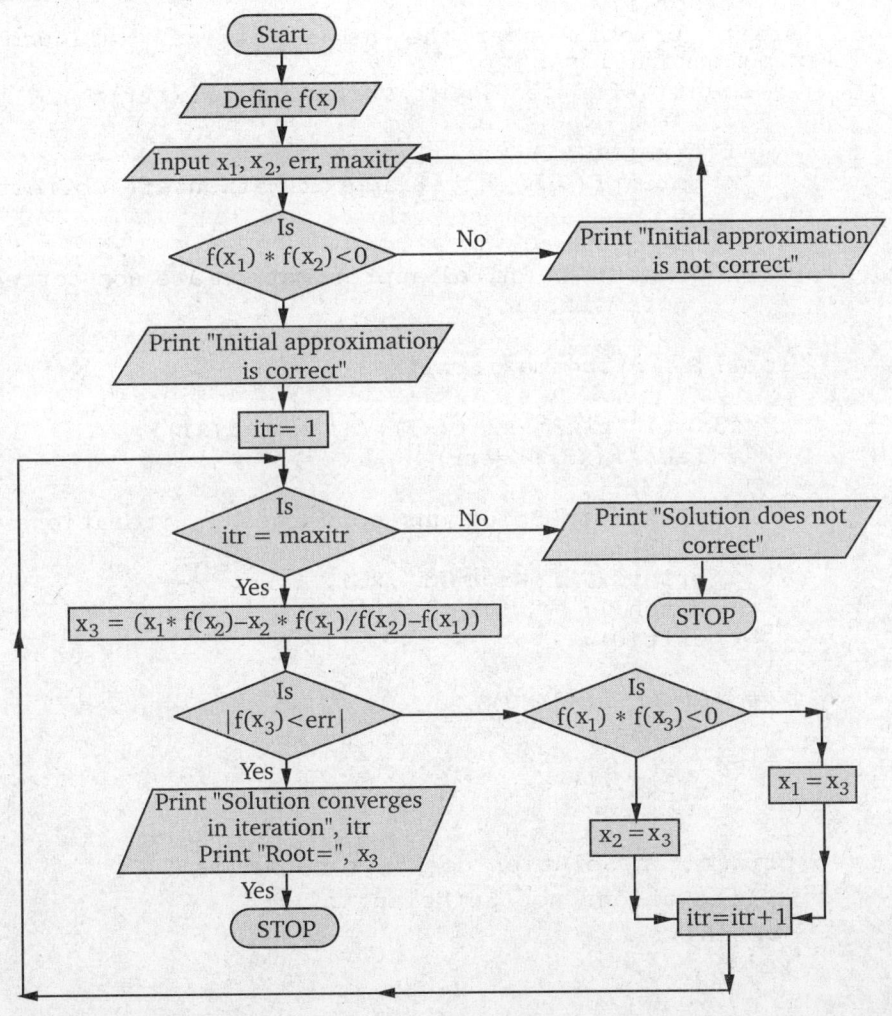

PROGRAM. *Following is a C program to find the root of the equation $f(x) = \cos x - xe^x$ by using Regula-Falsi method.*

                                                                     (UPTU B.Tech. 2002)

REGULA FALSI METHOD

```c
/* REGULA FALSI METHOD FOR cos(x)-xe^x */
#include<stdio.h>
#include<conio.h>
#include<math.h>
float f(float x)
{
return cos(x)-x*exp(x);
}
void main()
{
    int itr, maxitr;
    float x1,x2,x3,err;
    clrscr();
  start: printf("enter the value of x1,x2, allowed error,
maximum iteration\n");
    scanf("%f%f%f%d", &x1,&x2,&err, &maxitr);

    if(f(x1)*f(x2)<0)
        printf("\n inital approximation are correct");
     else
       {
        printf("\n inital approximation are not correct\n");
         goto start;
        }
  for(itr=1;itr<=maxitr;itr++)
   {
     x3= (x1*f(x2)-x2*f(x1))/(f(x2)-f(x1));
     if(fabs(f(x3))<=err)
     {
       printf("\n Solutions converges in iterations =%d",
         itr);
       printf("\n Root=%f",x3);
       getch();
       exit(0);
     }
   if (f(x1)*f(x3)<0)
       x2=x3;
   else
       x1=x3;
   }
  printf("\n Solution does not converge,"
    "iterations not Sufficient\n");
   getch();
  }
```

Output: REGULA FALSI METHOD FOR COS(X)-X*eX

```
enter the value of x1,x2, allowed error, maximum iteration

0    1      .00005   20

Solutions converges in iterations = 10

Root = 0.517748
```

LAB ASSIGNMENTS

1. Write C program to find root of the equation $x^3 - 9x + 1 = 0$ by Regula-Falsi method.

   Hint : Define $f(x) = x * x * x - 9 * x + 1$
   Input $x_1 = 2,\ x_2 = 3,\ err = 0.00005,\ maxitr = 8$
   Output Root = 2.9428

2. Write C program to find root of the equation $xe^x - 3 = 0$ by Regula-Falsi method correct to three decimal places.

   Input : $x_1 = 1,\ x_2 = 1.5,\ err = 0.0005,\ maxitr = 10$
   Output : Root = 1.049

3. Write C program to find real root of the equation by Regula-Falsi method
$$f(x) = x^2 - \log_e x - 12$$

   Hint : Define $f(x) = x * x - \log(x) - 12$
   Input : $x_1 = 3,\ x_2 = 4,\ err = 0.0005,\ maxitr = 8$

4. NEWTON-RAPHSON METHOD

SYMBOLS USED

x_1 = Initial approximation of the root
x_2 = New approximation of the root in each iteration
itr = iteration counter
err = allowed error
x_0 = old value of x_1 and x_1 is changed in each iteration
df(x) = first derivative of $f(x)$

ALGORITHM : NEWTON-RAPHSON METHOD

Step 1. Start
Step 2. Define function $f(x)$ and $df(x)$
Step 3. Input x_1, err
Step 4. itr = 0
Step 5. Repeat steps 6 to 10 until ($|x_2 - x_0| < e$)

Step 6.   if $df(x_1) = 0$

               Print "Initial approximation is not correct"

               goto step 3

Step 7. $x_2 = x_1 - (f(x_1)/ df(x_1))$

Step 8. $x_0 = x_1$

Step 9. $x_1 = x_2$

Step 10.   itr = itr + 1

Step 11.   Print "Solution converges in iterations", itr

Step 12.   Print "Root=", x_2

Step 13.   STOP

FLOW CHART : NEWTON-RAPHSON METHOD

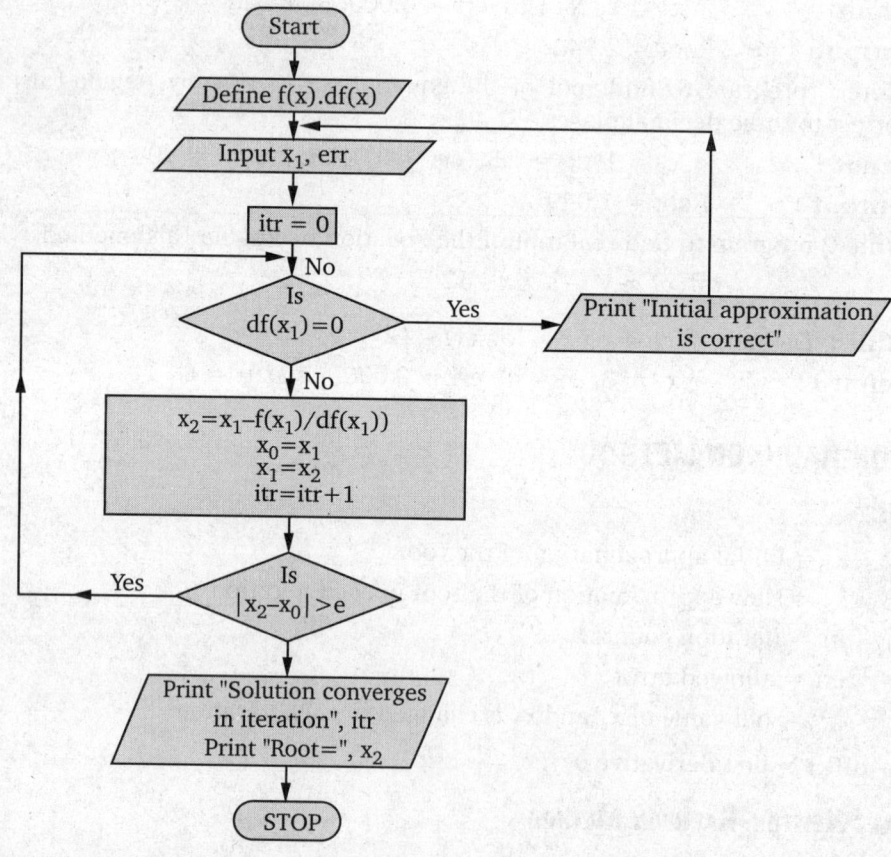

PROGRAM.   *Following is a C program to find the root of the equation $f(x) = x^4 - x - 10$ by using Newton-Raphson method.*

We have $f(x) = x^4 - x - 10$

$\Rightarrow$ $df(x) = 4x^3 - 1$

NEWTON RAPHSON METHOD

/*NEWTON RAPHSON METHOD TO FIND ROOT OF x^4-x-10 */

```c
#include<stdio.h>
#include<conio.h>
#include<math.h>

float f(float x)
{
     return  x*x*x*x-x-10;
}
float df (float x)
{
     return 4*x*x*x-1;
}
void main ()
{
     int itr=0;
     float x1,x2,err,xo;
     clrscr();

 start:      printf("\n Enter the value of x1,allowed error\
n\n");
     scanf("%f%f",&x1,&err);

     do
      {
        if(df(x1)==0)
           {
           printf("\n Initial approximation is not correct");
             goto start;
           }
        x2=x1-f(x1)/df(x1);
        xo=x1;
        x1=x2;
        itr++ ;
      } while(fabs(x2-xo)>err);
```

```
printf("\n Solution converges in iteration = %d\n",itr);
printf("\nRoot =%f",x2);
getch();
}
```

Output : NEWTON RAPHSON METHOD TO FIND ROOT OF x4-x-10

```
Enter the value of x1,allowed error
1.6   .0005
Solution converges in iteration = 4

Root=1.855585
```

LAB ASSIGNMENTS : NEWTON-RAPHSON METHOD

1. Write C program to find root of the equation $f(x) = x^3 - 2x - 5$ correct to 3 decimal places by Newton-Raphson method.

   Hint : Define $f(x) = x * x * x - 2 * x - 5$

   and $f(x) = 3 * x * x - 2$

   Input $x_1 = 2$, err = 0.0005

   Output   Root = 2.09455

2. Write C program to find root of the equation $x \log_{10} x = 1.2$ by Newton-Raphson method correct to four decimal places.

   Hint : Define $f(x) = x * \log_{10}(x) - 1.2$

                    $df(x) = \log_{10}(x) + 0.4343$

   Input : $x_1 = 2$, err = 0.00005

   Output :   Root = 2.7408

3. Write C program to find square root of 12 correct to three decimal places by Newton-Raphson method.

   Hint : Define $f(x) = x^2 - 12 = 0$

                    $f(x) = x * x - 12$

                    $df(x) = 2 * x$

   Input : $x_1 = 3.4$, err = 0.0005

   Output :   Root = 3.464

5. MULLER'S METHOD

Symbols Used

x_1, x_2, x_3 = Initial approximation

x_4 = New iteration in each step

err = allowed error

itr = iteration counter

Algorithm : Muller's Method

Step 1. Start

Step 2. Define $f(x)$

Step 3. Input x_1, x_2, x_3, err

Step 4. $f_1 = f(x_1), f_2 = f(x_2), f_3 = f(x_3)$, itr = 0

Step 5. Repeat steps 6 to 17 until $(|h| < e)$

Step 6. itr = itr + 1

Step 7. $h_1 = x_1 - x_3$, $h_2 = x_2 - x_3$

Step 8. $d_1 = f_1 - f_3$, $d_2 = f_2 - f_3$

Step 9. $a_1 = f_1$

Step 10. $d = h_1 * h_2 * (h_1 - h_2)$

Step 11. $a_2 = (d_2 * h_1^2 - d_1 * h_2^2)/d$

Step 12. $a_3 = (d_1 * h_2 - d_2 * h_1)/d$

Step 13. if $(a_2 > 0)$

$$h = -2 * a_1 / (a_2 + \sqrt{a_2^2 - 4 * a_1 * a_3})$$

else

$$h = -2 * a_1 / (a_2 - \sqrt{a_2^2 - 4 * a_1 * a_3})$$

Step 14. $x_4 = x_3 + h$

Step 15. $f_4 = f(x_4)$

Step 16. $x_1 = x_2, x_2 = x_3, x_3 = x_4$

Step 17. $f_1 = f_2, f_2 = f_3, f_3 = f_4$

Step 18. Print "Solution converges in iteration", itr

Step 19. Print "Root=", x_4

Step 20. STOP

FLOW CHART : MULLER METHOD

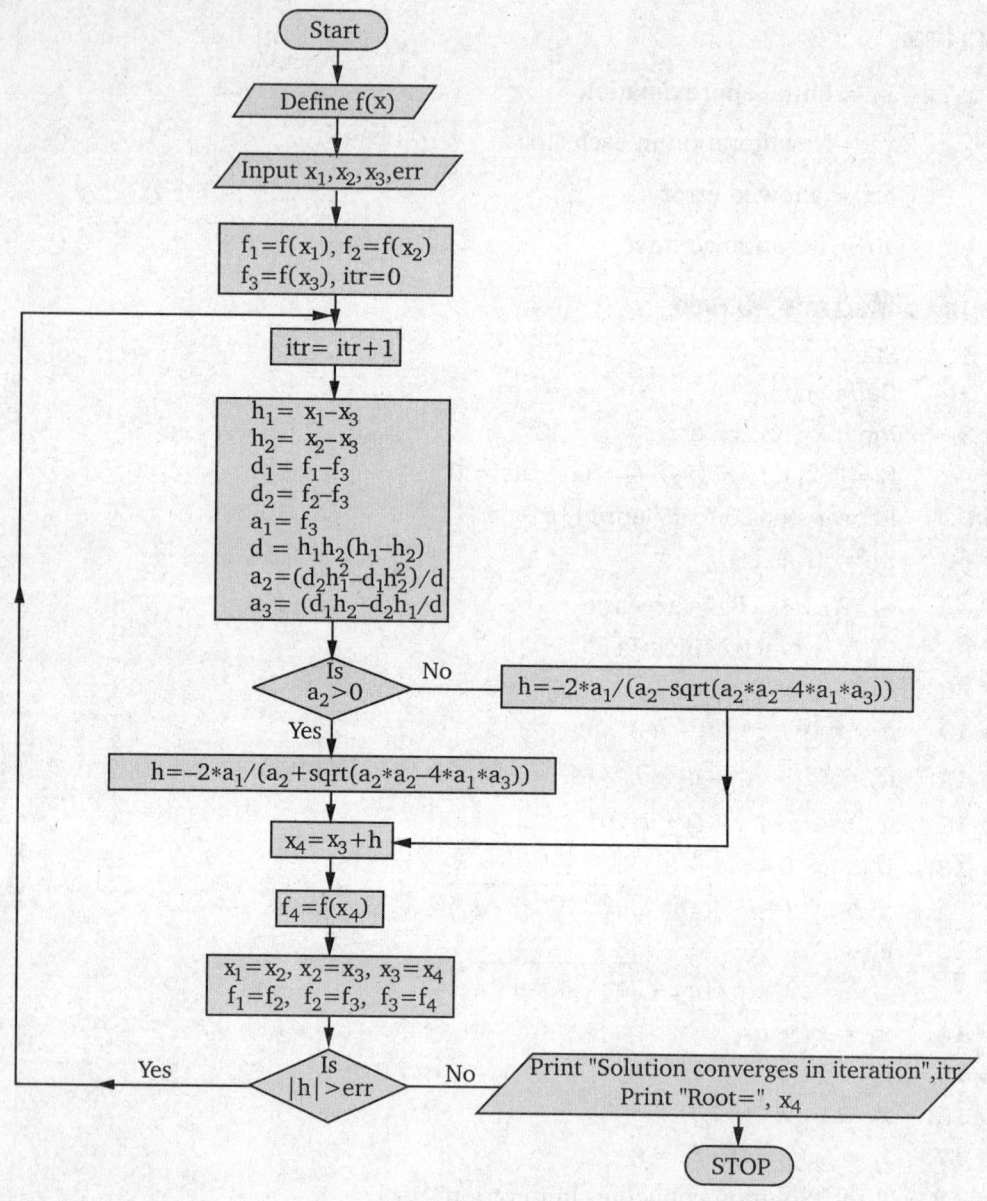

PROGRAM. *Following is a C program to find the root of the equation* $f(x) = \cos x - xe^x$ *by Muller's method.*

MULLER'S METHOD

```
/* MULLER'S METHOD cos(x)-x*ex */

#include<stdio.h>
#include<conio.h>
```

```c
#include<math.h>

float f(float x)
{
return cos(x)-x*exp(x);
}
void main()
  {
    int  itr=0;
    float x1,x2,x3,x4,err,d,d1,d2,h1,h2,a1,a2,a3,h;
    float f1,f2,f3,f4;
    clrscr();

  printf("\n Enter initial approximation x1,x2,x3 \n");
  scanf("%f%f%f",&x1,&x2,&x3);
  printf("\n  Enter the allowed error\n");
  scanf("%f", &err);
  f1=f(x1),  f2=f(x2),  f3=f(x3);
   do
    {
      itr++ ;
      h1=x1-x3;
      h2=x2-x3;
      d1=f1-f3;
      d2=f2-f3;
      a1=f3;
      d=h1*h2*(h1-h2);
      a2=(d2*h1*h1-d1*h2*h2)/d;
      a3=(d1*h2-d2*h1)/d;
      if(a2>0)
            h=-2*a1/(a2+sqrt(a2*a2-4*a1*a3));
      else
            h=-2*a1/(a2-sqrt(a2*a2-4*a1*a3));
      x4=x3+h;
      f4=f(x4);
      x1=x2,   x2=x3,   x3=x4 ;
      f1=f2,   f2=f3,   f3=f4 ;
    } while(fabs(h)>err);

    printf("\n  Solution converges in iterations=%d",itr);
    printf("\n\n Root =%f",x4);
    getch();
    }
```

Output: MULLER'S METHOD FOR cos(x)-x*ex

```
Enter initial approximation x1,x2,x3
-1   0      1

Enter the allowed error
.0005

Solution converges in iterations =4

Root = 0.517757
```

LAB ASSIGNMENTS : MULLER'S METHOD

1. Write C program to find the root of the equation $x^3 - x - 4 = 0$ by Muller method correct to 4 decimal places.

Hint : Define $f(x) = x*x*x - x - 4$

Input : $x_1 = 0, x_2 = 1, x_3 = 2$

err = 0.00005

Output Root = 1.7963

2. Write a C program to find root of the equation $3x + \sin x - e^x$ by Muller's method.

Hint : Define $f(x) = 3*x - \sin(x) - \exp(x)$

Input : $x_1 = 0.5, x_2 = 1.0, x_3 = 0.0$

Output : Root = 0.36042

3. Write C program to find the root of the equation

$$f(x) = x^3 + 2x^2 + 10x - 20 = 0 \text{ by Muller method.}$$

Hint : Define $f(x) = x*x*x + 2*x* + 10*x - 20$

Input : $x_1 = 0, x_2 = 1, x_3 = 2$, err = 0.0005

Output : Root = 1.3688

⌘⌘⌘⌘⌘⌘

Solution of Simultaneous Linear Algebraic Equations

4.1 INTRODUCTION

Simultaneous linear equations have great importance in the field of engineering and Science.

In the field of science, the analysis of electronic circuits having a number of invariant elements, analysis of a network under sinusoidal steady-state conditions, determination of the output of a chemical plant and finding the cost of reactions are some of the problem which depend on the solution of system of linear algebraic equations. We shall discuss a system of 'm' linear equations with n unknowns. If $m > n$, as a rule the equations cannot be satisfied. If $m < n$, the system of linear equations usually has an infinite number of solutions. In this chapter we will discuss the system of linear equation, if $m = n$.

4.2 LINEAR EQUATIONS

Let us consider n equation in m variables:

$$\left.\begin{array}{l} a_{11}x_1 + a_{12}x_2 + \ldots + a_{1m}x_m = b_1 \\ a_{21}x_1 + a_{22}x_2 + \ldots + a_{2m}x_m = b_2 \\ a_{31}x_1 + a_{32}x_2 + \ldots + a_{3m}x_m = b_3 \\ \ldots\ldots\ldots\ldots\ldots\ldots\ldots\ldots\ldots\ldots\ldots\ldots \\ \ldots\ldots\ldots\ldots\ldots\ldots\ldots\ldots\ldots\ldots\ldots\ldots \\ a_{n1}x_1 + a_{n2}x_2 + \ldots + a_{nm}x_m = b_m \end{array}\right\}$$

These equations can be written in a matrix form as follows

$$\begin{bmatrix} a_{11} & a_{12} & \cdots & a_{1m} \\ a_{21} & a_{22} & \cdots & a_{2m} \\ a_{31} & a_{32} & \cdots & a_{3m} \\ \cdots & \cdots & \cdots & \cdots \\ \cdots & \cdots & \cdots & \cdots \\ a_{n1} & a_{n2} & \cdots & a_{nm} \end{bmatrix} \begin{bmatrix} x_1 \\ x_2 \\ x_3 \\ \cdots \\ \cdots \\ x_m \end{bmatrix} = \begin{bmatrix} b_1 \\ b_2 \\ b_3 \\ \cdots \\ \cdots \\ b_m \end{bmatrix}$$

or $$\mathbf{AX = B}$$

Where $$A = [a_{ij}]_{m \times n}$$

$$X = \begin{bmatrix} x_1 \\ x_2 \\ x_3 \\ \cdots \\ \cdots \\ x_m \end{bmatrix}_{m \times 1} \quad \text{and} \quad B = \begin{bmatrix} b_1 \\ b_2 \\ b_3 \\ \cdots \\ \cdots \\ b_m \end{bmatrix}_{m \times 1}$$

If all the $b_i (i = 1, 2, ..., n)$ are zero, then the system of linear equations are called homogeneous. Otherwise it is called non-homogeneous. The solution of such equations can be obtained by direct or iterative methods. Amongst the direct methods, we will describe Gauss elimination, LU decomposition, Jordan's and Crout's methods. About the iterative methods, we will describe the Jocobi, Gauss-Seidel and Relaxation method.

4.3 EXISTENCE OF SOLUTION

Given an arbitrary system of equations, it is difficult to say whether the system has a solution or not. Sometimes there may be a solution but it may not be unique. There are six possibilities:

1. System has a unique solution.
2. System has no solution.
3. System has a solution but not unique (it has infinite solutions).
4. System is ill-conditioned.
5. System is under determined.
6. System is over determined.

(1) Unique Solution

Consider the system

$$2x_1 - x_2 = 2$$
$$x_1 + x_2 = 5$$

The system has a solution

$$x_1 = \frac{7}{3} \text{ and } x_2 = \frac{8}{2}$$

Since no other pair of values of x and y would satisfy the equation, the solution is said to be unique. The system is illustrated in Fig. 4(1).

(2) No Solution

The equations

$$2x_1 - x_2 = 2$$
$$2x_1 - x_2 = 0$$

have no solution. These two lines are parallel as shown in Fig. 4(2) and therefore, they never meet. Such equations are called inconsistent equations.

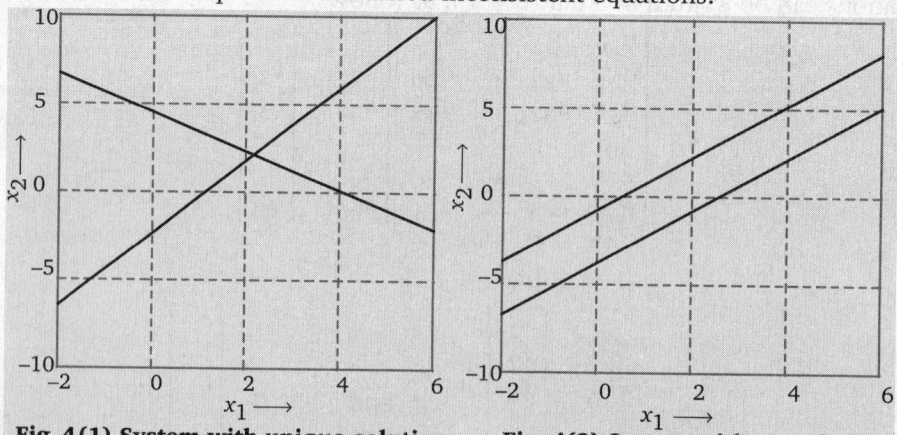

Fig. 4(1) System with unique solution   **Fig. 4(2) System with no solution**

(3) No Unique Solution

The System
$$2x_1 - x_2 = 2$$
$$-4x_1 + 2x_2 = -4$$

has many different solutions. We can see that these are two different forms of the same equation and, therefore, they represent the same line (Fig. 4(3)). Such equations are called dependent equations.

The systems represented in Figures (c) and (3) are said to be singular systems.

(4) ILL Conditioned Systems

There may be a situation where the system has a solution but it is very close to being singular. For example, the system
$$x_1 - 2x_2 = -2$$
$$0.45x_1 - 0.91x_2 = -1$$

has a solution but it is very difficult to identify the exact point at which the lines intersect (Fig. 4(4)). Such systems are said to be ill-conditioned. Ill-conditioned systems are very sensitive to round off errors and, therefore, may pose problems during computations of the solution.

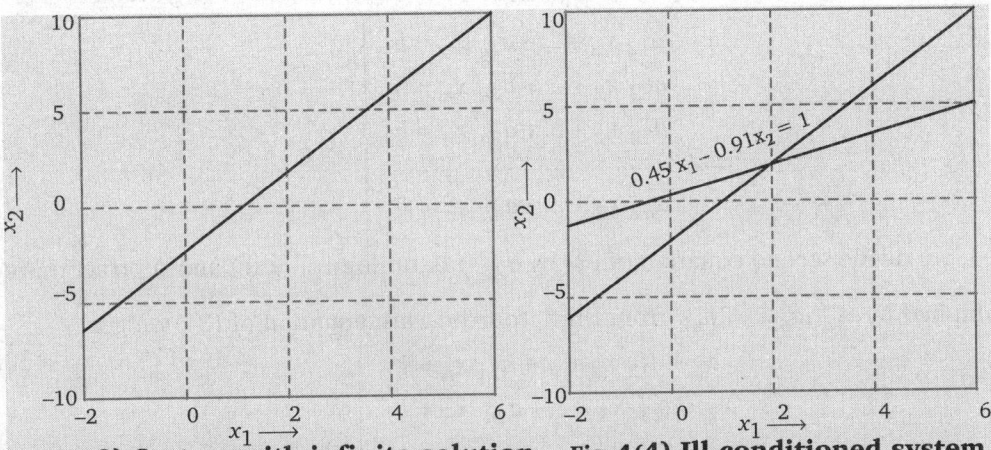

Fig. 4(3) System with infinite solution   **Fig. 4(4) Ill-conditioned system**

(5) Under determined System

If number of equations is less than number of unknown variables $(m < n)$, the system is said to undetermined and a unique solution for all unknowns is not possible e.g.,

$$3x_1 - 4x_2 + 5x_3 = -2$$
$$x_1 - 5x_2 + 3x_3 = 5$$

(6) Over determined System

If the number of equations is greater than number of unknowns $(m > n)$ then, the system is said to be over-determined, i.e., there are m independent equations, none of them can be derived from others.

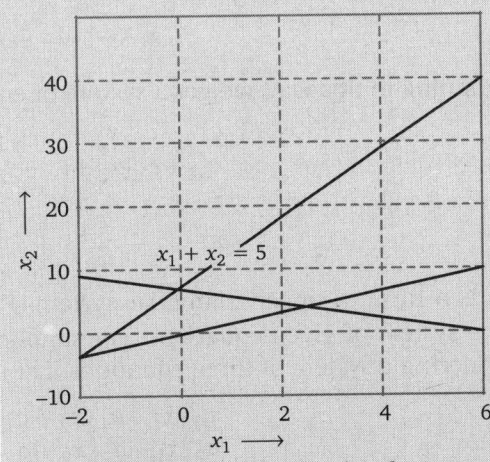

Fig. 4(5) Over determined system

4.4 GAUSS ELIMINATION METHOD

In this method, the variables from the system of linear equations are eliminated successively and the system of equations is therefore reduced to an upper triangular system from which the variable are determined by back substitution. This method is described as follows: Let us consider a system of linear equation

$$AX = B \qquad \qquad ...(1)$$

assuming $\det A \neq 0$. Equation (1) has the following form:

$$\left.\begin{aligned}
a_{11}x_1 + a_{12}x_2 + ... + a_{1n}x_n &= b_1 \\
a_{21}x_1 + a_{22}x_2 + ... + a_{2n}x_n &= b_2 \\
a_{31}x_1 + a_{32}x_2 + ... + a_{3n}x_n &= b_3 \\
\cdots\cdots\cdots\cdots\cdots\cdots\cdots\cdots\cdots \\
\cdots\cdots\cdots\cdots\cdots\cdots\cdots\cdots\cdots \\
a_{n1}x_1 + a_{n2}x_2 + ... + a_{nn}x_n &= b_n
\end{aligned}\right\} \qquad ...(2)$$

Assuming $a_{11} \neq 0$ and divide the first equation by a_{11} and then we subtract this equation multiplied by $a_{21}, a_{31}, ..., a_{n1}$ from second, third ... nth equation of (2), we get

$$\left.\begin{aligned}
x_1 + a_{12}'x_2 + ... + a_{1n}'x_n &= b_1' \\
a_{22}'x_2 + ... + a_{2n}'x_n &= b_2' \\
a_{32}'x_2 + ... + a_{3n}'x_n &= b_3' \\
\cdots\cdots\cdots\cdots\cdots\cdots\cdots\cdots \\
a_{n2}'x_2 + ... + a_{nn}'x_n &= b_n'
\end{aligned}\right\} \qquad ...(3)$$

Next, we divide second equation of (3) by a_{22}' (assuming $a_{22}' \neq 0$) and subtract this equation multiplied by $a_{32}', a_{42}', ..., a_{n2}'$ from third, fourth ... nth equation of (3), we get

$$\left.\begin{aligned}
x_1 + a_{12}'x_2 + ... + a_{1n}'x_n &= b_1' \\
x_2 + a_{23}''x_3 + ... + a_{2n}''x_n &= b_2'' \\
a_{33}''x_3 + ... + a_{3n}''x_n &= b_3'' \\
\cdots\cdots\cdots\cdots\cdots\cdots\cdots\cdots \\
a_{n3}''x_3 + ... + a_{nn}''x_n &= b_n''
\end{aligned}\right\} \qquad ...(4)$$

Continuing in this way, we get a system of equation as follows:

$$\left.\begin{aligned}
x_1 + c_{12}x_2 + c_{13}x_3 + ... + c_{1n}x_n &= d_1 \\
x_2 + c_{23}x_3 + ... + c_{2n}x_n &= d_2 \\
\vdots \\
\vdots \\
c_{nn}x_n &= d_n
\end{aligned}\right\} \qquad ...(5)$$

This is a form of upper triangular system. From back substitution we can find the solution of the system of given equations. For simplicity and clarity, we shall explain this method by considering a system of three equations. Let us consider these equations.

$$\left.\begin{aligned}
a_{11}x_1 + a_{12}x_2 + a_{13}x_3 &= b_1 \\
a_{21}x_1 + a_{22}x_2 + a_{23}x_3 &= b_2 \\
a_{31}x_1 + a_{32}x_2 + a_{33}x_3 &= b_3
\end{aligned}\right\} \qquad ...(6)$$

Step 1. First, eliminate x_1 from second and third equations. Assuming $a_{11} \neq 0$, now dividing first equation by a_{11} and then subtract from second and third after multiplied by a_{21} and a_{31} respectively, we get.

$$\left.\begin{array}{l} x_1 + a_{12}' x_2 + a_{13}' x_3 = b_1' \\ a_{22}' x_2 + a_{23}' x_3 = b_2' \\ a_{32}' x_2 + a_{33}' x_3 = b_3' \end{array}\right\} \qquad \ldots(7)$$

where

$$a_{12}' = \frac{a_{12}}{a_{11}}, a_{13}' = \frac{a_{13}}{a_{11}}, a_{22}' = a_{22} - a_{21}a_{12}', a_{23}' = a_{23} - a_{21}a_{13}',$$

$$a_{32}' = a_{32} - a_{31}a_{12}', a_{33}' = a_{33} - a_{31}a_{13}',$$

$$b_1' = \frac{b_1}{a_{11}}, b_2' = b_2 - a_{21}b_1', b_3' = b_3 - a_{31}b_1'.$$

Step 2. Now eliminating x_2 from third equation in (7).

Again assuming $a_{22}' \neq 0$. Dividing second equation in (7) by a_{22}' and then subtract from third equation after multiplied by a_{32}' we get

$$\left.\begin{array}{l} x_1 + a_{12}' x_2 + a_{13}' x_3 = b_1' \\ x_2 + a_{23}'' x_3 = b_2'' \\ a_{33}'' x_3 = b_3'' \end{array}\right\} \qquad \ldots(8)$$

where $a_{23}'' = \dfrac{a_{23}'}{a_{22}'}, a_{33}'' = a_{33}' - a_{32}'a_{23}'', b_2'' = \dfrac{b_2'}{a_{22}'}, b_3'' = b_3' - a_{32}'b_2''.$

Step 3. Evaluating x_1, x_2 and x_3 from (8) by back substitution.

▼ **REMARKS**

▶ The coefficient a_{11}, a_{22}' and a_{33}'' are called pivots.

▶ This method will fail if any one of the pivots a_{11}, a_{22}' and a_{33}'' becomes zero. In such cases, we rewrite the equations in a different order so that the pivots are non-zero.

▶ From each of the procedure, the largest coefficient of x is chosen as pivot element.

▶ This method proposes a systematic astrology for reducing the system of equations to the upper triangular form using the forward elimination approach and then for obtaining values of unknowns using back substitution process.

SOLVED EXAMPLES

EXAMPLE 1. *Solve the following equations by Gauss's elimination method*

$$6x + 3y + 2z = 6$$
$$6x + 4y + 3z = 0$$
$$20x + 15y + 12z = 0.$$

SOLUTION. Here pivot element is 6. Now Divide first equation by 6, we get

$$x + \frac{1}{2}y + \frac{1}{3}z = 1 \qquad \ldots(1)$$

Now eliminating x from second and third equation with the help of (1). Subtract

(1) multiplied by 6 and 20 from second and third equation, respectively we get

$$y + z = -6 \qquad \qquad ...(2)$$

$$5y + \frac{16}{3}z = -20 \qquad \qquad ...(3)$$

Now eliminating y from (3) with the help of (2), we get

$$\left(\frac{16}{3} - 5\right)z = -20 + 30$$

$$\frac{1}{3}z = 10 \quad \Rightarrow z = 30$$

Substitute the value of z into (2), we get

$$y = -6 - 30 = -36$$

and again substitute the values of y and z into (1), we get

$$x + \frac{1}{2}(-36) + \frac{1}{3}(30) = 1$$

$$x - 18 + 10 = 1 \quad \Rightarrow x = 9$$

Hence, the solution of the equations are

$$x = 9, y = -36, z = 30.$$

EXAMPLE 2.   *By Gauss's elimination method, solve the following equations*

$$5x - y - 2z = 142$$
$$x - 3y - z = -30$$
$$2x - y - 3z = -5$$

SOLUTION.   The largest coefficient in first equation is 5, which is pivot element. So divide first equation by 5, we get

$$x - \frac{1}{5}y - \frac{2}{5}z = \frac{142}{5} \qquad \qquad ...(1)$$

Now eliminating x from second and third equation with help of (1) we get

$$-\frac{14}{5}y - \frac{3}{5}z = -\frac{292}{5} \qquad \qquad ...(2)$$

$$-\frac{3}{5}y - \frac{11}{5}z = -\frac{309}{5} \qquad \qquad ...(3)$$

Eliminating y from (2) and (3), we get

$$-\frac{145}{5}z = -\frac{3450}{5} \quad \text{or} \quad z = \frac{3450}{145} = 23.79$$

Substitute the value of z into (3) we get

$$-\frac{3}{5}y - \frac{11}{5}(23.79) = -\frac{309}{5}$$

$$-\frac{3}{5}y = -\frac{309}{5} + \frac{11(23.79)}{5} \quad \Rightarrow \quad y = 15.77$$

Substitute the values of y and z into (1), we get

$$x - \frac{1}{5}(15.77) - \frac{2}{5}(23.79) = \frac{142}{5}$$

$$x = \frac{142}{5} + \frac{15.77}{5} - \frac{2(23.79)}{5} = \frac{205.35}{5}$$

or $$x = 41.07$$

Hence, the solution are given by

$$x = 41.07, y = 15.77, z = 23.79.$$

EXAMPLE 3.   *Using Gauss's elimination method solve*

$$2x_1 + 4x_2 + x_3 = 3$$
$$3x_1 + 2x_2 - 2x_3 = 2$$
$$x_1 - x_2 + x_3 = 6$$

(Marathwada-2008)

SOLUTION.   Dividing first equation by 2, we get

$$x_1 + 2x_2 + \frac{1}{2}x_3 = \frac{3}{2} \qquad \qquad \text{...(1)}$$

Multiplying (1) by 3 and subtract from second and also subtract (1) from third equation, we get

$$4x_2 + \frac{7}{2}x_3 = \frac{5}{2} \qquad \qquad \text{...(2)}$$

$$-3x_2 + \frac{1}{2}x_3 = \frac{9}{2} \qquad \qquad \text{...(3)}$$

Now dividing (2) by 4 and subtract after multiplies by -3 from (3), we get

$$25x_3 = 51$$

or

$$x_3 = \frac{51}{25} = 2.04$$

Substitute the value of x_3 into (2), we get

$$4x_2 + \frac{7}{2}(2.04) = \frac{5}{2}$$

$$4x_2 = \frac{5}{2} - \frac{7(2.04)}{2} = \frac{5 - 14.28}{2}$$

$$\Rightarrow \qquad x_2 = -\frac{9.28}{8} = -1.16$$

Now substitute the value of x_2 and x_3 into (1), we get

$$x_1 + 2(-1.16) + \frac{1}{2}(2.04) = \frac{3}{2}$$

$$x_1 = \frac{3}{2} + 2(1.16) - \frac{1}{2}(2.04)$$

$$= \frac{3 + 4.64 - 2.04}{2} = \frac{5.6}{2} = 2.8$$

Hence, the solutions are given by

$$x_1 = 2.8, x_2 = -1.16, x_3 = 2.04.$$

EXAMPLE 4.   *Solve by the Gauss-elimination method.*

$$2x + y + 4z = 12$$
$$8x - 3y + 2z = 23$$
$$4x + 11y - z = 33$$

(Meerut-2015)

SOLUTION.   Dividing first equation by 2, we get

$$x + \frac{1}{2}y + 2z = 6 \qquad \qquad \text{...(1)}$$

Now subtract (1) after multiplied by 8 and 4 respectively from second and third equation, we get

$$-7y - 14z = -45 \qquad \qquad \text{...(2)}$$

$$9y - 9z = 9 \qquad \qquad \text{...(3)}$$

Now multiplying (4) by 9 and subtract from (3), we get

$$-27z = 9 - \frac{405}{7}$$

or
$$-27z = -\frac{342}{7} \qquad \text{...(5)}$$

Hence, the system of equations reduces to upper triangular form as follows:

$$\left.\begin{aligned} x + \frac{1}{2}y + 2z &= 6 \\ y + 2z &= \frac{45}{7} \\ -27z &= -\frac{342}{7} \end{aligned}\right\} \qquad \text{...(6)}$$

By back substitution, we get

$$z = \frac{342}{189} = 1.81 \quad \text{and} \quad y + 2(1.81) = \frac{45}{7}$$

$$\Rightarrow \qquad y = \frac{45}{7} - 2(1.81) = 6.43 - 3.62 = 2.81$$

and
$$x + \frac{1}{2}(2.81) + 2(1.81) = 6$$

$$\therefore \qquad x = 6 - \frac{1}{2}(2.81) - 2(1.81) = 0.975$$

Hence, the solution is $x = 0.975$, $y = 2.81$, $z = 1.81$.

EXAMPLE 5.   *Apply Gauss elimination method to solve the equations*

$$x + 4y - z = -5$$
$$x + y - 6z = -12$$
$$3x - y - z = 4$$

(Mumbai–2009)

SOLUTION.   Elimination x from second and third equation with the help of first equation. Subtract first equation from second and after multiplied by 3 from third respectively, we get the system of equations as follows:

$$x + 4y - z = -5 \qquad \text{...(1)}$$
$$-3y - 5z = -7 \qquad \text{...(2)}$$
$$-13y + 2z = 19 \qquad \text{...(3)}$$

Eliminating y from (3) with help of (2). Divide (2) by –3 and then this equation is subtracted after multiplies by –13 from (3), get

$$\left.\begin{aligned} x + 4y - z &= -5 \\ y + \frac{5}{3}z &= \frac{7}{3} \\ \frac{71}{3}z &= \frac{148}{3} \end{aligned}\right\} \qquad \text{...(4)}$$

By back substitution from (4), we get

$$z = \frac{148}{71} \quad \text{and} \quad y = \frac{7}{3} - \frac{5}{3}z = \frac{7}{3} - \frac{5}{3}\left(\frac{148}{71}\right)$$

$$\Rightarrow \qquad y = -\frac{81}{71}$$

and
$$x = -5 - 4y + z = -5 - 4\left(-\frac{81}{71}\right) + \frac{148}{71}$$

$$= -5 + \frac{472}{71} = \frac{117}{71}$$

Hence the solution are
$$x = \frac{117}{71}, y = -\frac{81}{71}, z = \frac{148}{71}.$$

EXAMPLE 6. *Solve the following system by Gauss elimination method.*
$$2x + y + z = 10; 3x + 2y + 3z = 18; x + 4y + 9z = 16 \qquad \text{(PTU-2003, 05)}$$

SOLUTION. We have
$$2x + y + z = 10 \qquad \qquad \text{...(1)}$$
$$3x + 2y + 3z = 18 \qquad \qquad \text{...(2)}$$
$$x + 4y + 9z = 16 \qquad \qquad \text{...(3)}$$

Divide (1) and (2) and subtract after multiplied by 3 from (2) and subtract from (3), we get

$$x + \frac{1}{2}y + \frac{1}{2}z = 5 \qquad \qquad \text{...(4)}$$

$$\frac{1}{2}y + \frac{3}{2}z = 3 \qquad \qquad \text{...(5)}$$

$$\frac{7}{2}y + \frac{17}{2}z = 11 \qquad \qquad \text{...(6)}$$

Now divide (5) by $\frac{1}{2}$ and then subtract after multiplied by $\frac{7}{2}$ from (6), we get

$$x + \frac{1}{2}y + \frac{1}{2}z = 5$$

$$y + 3z = 6$$

$$-2z = -10$$

From back substitution from (9), (8) and (7), we get
$$z = 5$$

and
$$y + 3z = 6$$
$$y + 3(5) = 6$$
$$y = 6 - 15, y = -9$$

and $x + \frac{1}{2}(-9) + \frac{1}{2}(5) = 5$

$$x = 5 + \frac{9}{2} - \frac{5}{2}$$

Hence, the solution is $x = 7, y = -9, z = 5$.

EXAMPLE 7. *Solve by Gauss elimination method*
$$x + 2y + z = 3; 2x + 3y + 3z = 10; 3x - y + 2z = 13$$

SOLUTION. Given equations are
$$x + 2y + z = 3 \qquad \qquad \text{...(1)}$$
$$2x + 3y + 3z = 10 \qquad \qquad \text{...(2)}$$
$$3x - y + 2z = 13 \qquad \qquad \text{...(3)}$$

Here pivot element of (1) is 1 so no need for dividing (1). Now eliminating x from (2) and (3) by subtracting (1) after multiplied 2 and 3 respectively from (2) and (3), we get

$$x + 2y + z = 3 \qquad \qquad ...(4)$$
$$-y + z = 4 \qquad \qquad ...(5)$$
$$-7y - z = 4 \qquad \qquad ...(6)$$

Now eliminating y from (6) with the help of (5) by subtracting (5) after multiplied by -7 from (6), we get

$$x + 2y + z = 3 \qquad \qquad ...(7)$$
$$-y + z = 4 \qquad \qquad ...(8)$$
$$6z = 32 \qquad \qquad ...(9)$$

By back substitution from (7), (8) and (9), we get

From (9)
$$z = \frac{32}{6} = \frac{16}{3}$$

From (8)
$$-y = 4 - z$$

$$-y = 4 - \frac{32}{6} = -\frac{8}{6} \implies y = \frac{8}{6} = \frac{4}{3}$$

From (7) $\quad x + 2y + z = 3$

$$x + 2\left(\frac{8}{6}\right) + \frac{32}{6} = 3$$

$$\implies \qquad x = 3 - \frac{32}{6} - \frac{16}{6} = -\frac{30}{6}$$

$$x = -5$$

Hence, the solution is

$$x = -5, \ y = \frac{4}{3}, \ z = \frac{16}{3}$$

4.5 LU DECOMPOSITION METHOD OR METHOD OF FACTORIZATION

This method is based on the fact that every square matrix A can be expressed as the form of LU where L is a unit lower triangular matrix while U is upper triangular matrix and provided all the principal minors of A are non-singular.

That is, If $A = [a_{ij}]_{n \times n}$, then

$$a_{11} \neq 0, \begin{vmatrix} a_{11} & a_{12} \\ a_{21} & a_{22} \end{vmatrix} \neq 0, \begin{vmatrix} a_{11} & a_{12} & a_{13} \\ a_{21} & a_{22} & a_{23} \\ a_{31} & a_{32} & a_{33} \end{vmatrix} \neq 0, \text{ and so on.}$$

For simplicity and understanding this method, let us consider a system of three equations

$$a_{11}x_1 + a_{12}x_2 + a_{13}x_3 = b_1$$
$$a_{21}x_1 + a_{22}x_2 + a_{23}x_3 = b_2$$
$$a_{31}x_1 + a_{32}x_2 + a_{33}x_3 = b_3$$

These equations can be written in matrix form as follows:

$$\boldsymbol{AX = B}$$

Where
$$A = \begin{bmatrix} a_{11} & a_{12} & a_{13} \\ a_{21} & a_{22} & a_{23} \\ a_{31} & a_{32} & a_{33} \end{bmatrix}, X = \begin{bmatrix} x_1 \\ x_2 \\ x_3 \end{bmatrix}, B = \begin{bmatrix} b_1 \\ b_2 \\ b_3 \end{bmatrix} \qquad ...(1)$$

Let
$$A = LU$$

where
$$L = \begin{bmatrix} 1 & 0 & 0 \\ l_{21} & 1 & 0 \\ l_{31} & l_{32} & 1 \end{bmatrix}, U = \begin{bmatrix} u_{11} & u_{12} & u_{13} \\ 0 & u_{22} & u_{23} \\ 0 & 0 & u_{33} \end{bmatrix} \qquad \dots(2)$$

Then from (2)
$$\begin{bmatrix} a_{11} & a_{12} & a_{13} \\ a_{21} & a_{22} & a_{23} \\ a_{31} & a_{32} & a_{33} \end{bmatrix} = \begin{bmatrix} 1 & 0 & 0 \\ l_{21} & 1 & 0 \\ l_{31} & l_{32} & 1 \end{bmatrix} \begin{bmatrix} u_{11} & u_{12} & u_{13} \\ 0 & u_{22} & u_{23} \\ 0 & 0 & u_{33} \end{bmatrix}$$

or
$$\begin{bmatrix} a_{11} & a_{12} & a_{13} \\ a_{21} & a_{22} & a_{23} \\ a_{31} & a_{32} & a_{33} \end{bmatrix} = \begin{bmatrix} u_{11} & u_{12} & u_{13} \\ l_{21}u_{11} & l_{21}u_{12}+u_{22} & l_{21}u_{13}+u_{23} \\ l_{31}u_{11} & l_{31}u_{12}+l_{32}u_{22} & l_{31}u_{13}+l_{32}u_{23}+u_{33} \end{bmatrix}$$

Comparing the two matrices, we get

(i) $\quad u_{11} = a_{11}, u_{12} = a_{12}, u_{13} = a_{13}$ $\qquad$ (ii) $\quad l_{21}u_{11} = a_{21}$ $\quad$ or $\quad l_{21} = \dfrac{a_{21}}{u_{11}} = \dfrac{a_{21}}{a_{11}}$

(iii) $\quad l_{31}u_{11} = a_{31}$ $\qquad$ or $\qquad l_{31} = \dfrac{a_{31}}{u_{11}} = \dfrac{a_{31}}{a_{11}}$

(iv) $\quad l_{21}u_{12} + u_{22} = a_{22}$ $\qquad$ or $\qquad l_{21} = \dfrac{a_{22} - u_{22}}{u_{12}}$

$\qquad$ or $\qquad u_{22} = a_{22} - l_{21}u_{12} = a_{22} - \dfrac{a_{21}}{a_{11}}.a_{12}$

(v) $\quad l_{21}u_{13} + u_{23} = a_{23}$ $\qquad \Rightarrow \qquad u_{23} = a_{23} - l_{21}u_{13}$

$\qquad \qquad \qquad \qquad \qquad \Rightarrow \qquad u_{23} = a_{23} - \dfrac{a_{21}}{a_{11}}a_{13}.$

(vi) $\quad l_{31}u_{12} + l_{32}u_{22} = a_{32}$

$\qquad$ or $\qquad l_{32} = \dfrac{1}{u_{22}}\left[a_{32} - \dfrac{a_{31}}{a_{11}}a_{12} \right] = \dfrac{\left[a_{32} - \dfrac{a_{31}a_{12}}{a_{11}} \right]}{\left[a_{22} - \dfrac{a_{21}a_{12}}{a_{11}} \right]}$

(vii) $\quad l_{31}u_{13} + l_{32}u_{23} + u_{33} = a_{33}$

$\qquad$ or $\qquad u_{33} = a_{33} - l_{31}u_{13} - l_{32}u_{23}$

$$= a_{33} - \dfrac{a_{31}}{a_{11}}a_{13} - \dfrac{1}{u_{22}}\left[a_{32} - \dfrac{a_{31}}{a_{11}}a_{12} \right]\left[a_{23} - \dfrac{a_{21}}{a_{11}}a_{13} \right]$$

$$= a_{33} - \dfrac{a_{31}a_{13}}{a_{11}} - \dfrac{\left[a_{32} - \dfrac{a_{31}a_{12}}{a_{11}} \right]\left[a_{23} - \dfrac{a_{21}}{a_{11}}a_{13} \right]}{\left[a_{22} - \dfrac{a_{21}a_{12}}{a_{11}} \right]}$$

Thus from above we can find the elements of L and U. Now L and U are therefore obtained. Replacing A by LU in (1), we get

$$LUX = B \qquad \dots(3)$$

Now let
$$UX = Y \qquad \dots(4)$$

where
$$Y = \begin{bmatrix} y_1 \\ y_2 \\ y_3 \end{bmatrix}$$

From (3) and (4), we get
$$LY = B$$

or
$$\begin{bmatrix} 1 & 0 & 0 \\ l_{21} & 1 & 0 \\ l_{31} & l_{32} & 1 \end{bmatrix} \begin{bmatrix} y_1 \\ y_2 \\ y_3 \end{bmatrix} = \begin{bmatrix} b_1 \\ b_2 \\ b_3 \end{bmatrix}.$$

From this equation, we get

$$y_1 = b_1 ; y_2 = b_2 - l_{21} y_1 ; y_3 = b_3 - l_{31} b_1 - l_{32} b_2 \qquad \qquad ...(5)$$

Now from (4)
$$UX = Y$$

$$\begin{bmatrix} u_{11} & u_{12} & u_{13} \\ 0 & u_{22} & u_{23} \\ 0 & 0 & u_{33} \end{bmatrix} \begin{bmatrix} x_1 \\ x_2 \\ x_3 \end{bmatrix} = \begin{bmatrix} y_1 \\ y_2 \\ y_3 \end{bmatrix}$$

From this equation, we get

$$x_3 = \frac{y_3}{u_{33}}, x_2 = \frac{y_2 - u_{23} x_3}{u_{22}}, x_1 = \frac{y_1 - u_{12} x_2 - u_{13} x_3}{u_{11}}$$

With the help of L and U, x_1, x_2, x_3 can be calculated.

▶ **REMARKS**

▸ LU decomposition method is superior to Gauss elimination method.

▸ This method is applicable if the coefficient matrix can be expressed as the product of lower and upper triangular matrix.

▸ This method can also be renamed as Method of factorization.

SOLVED EXAMPLES

EXAMPLE 1.   *Solve the following equations by LU decomposition method.*

$$2x_1 + x_2 + x_3 = 2$$
$$x_1 + 3x_2 + 2x_3 = 2$$
$$3x_1 + x_2 + 2x_3 = 2$$

SOLUTION.   Given equations are

$$2x_1 + x_2 + x_3 = 2 \qquad \qquad ...(1)$$

$$x_1 + 3x_2 + 2x_3 = 2 \qquad \qquad ...(2)$$

$$3x_1 + x_2 + 2x_3 = 2 \qquad \qquad ...(3)$$

Here, the coefficient matrix A is given by

$$A = \begin{bmatrix} 2 & 1 & 1 \\ 1 & 3 & 2 \\ 3 & 1 & 2 \end{bmatrix} \text{ and } X = \begin{bmatrix} x_1 \\ x_2 \\ x_3 \end{bmatrix}, B = \begin{bmatrix} 2 \\ 2 \\ 2 \end{bmatrix}$$

$$\therefore \qquad AX = B$$

Now
$$A = LU$$

$$\begin{bmatrix} 2 & 1 & 1 \\ 1 & 3 & 2 \\ 3 & 1 & 2 \end{bmatrix} = \begin{bmatrix} 1 & 0 & 0 \\ l_{21} & 1 & 0 \\ l_{31} & l_{32} & 1 \end{bmatrix} \begin{bmatrix} u_{11} & u_{12} & u_{13} \\ 0 & u_{22} & u_{23} \\ 0 & 0 & u_{33} \end{bmatrix} \qquad \qquad ...(4)$$

$$= \begin{bmatrix} u_{11} & u_{12} & u_{13} \\ l_{21}u_{11} & l_{21}u_{12}+u_{22} & l_{21}u_{13}+u_{23} \\ l_{31}u_{11} & l_{31}u_{12}+l_{32}u_{22} & l_{31}u_{13}+l_{32}u_{23}+u_{33} \end{bmatrix}$$

On comparing two matrices, we get

$$u_{11} = 2, u_{12} = 1, u_{13} = 1$$

$$l_{21}u_{11} = 1 \quad \text{or} \quad l_{21} = \frac{1}{u_{11}} = \frac{1}{2}$$

$$l_{31}u_{11} = 3 \quad \text{or} \quad l_{31} = \frac{3}{u_{11}} = \frac{3}{2}$$

$$l_{21}u_{12} + u_{22} = 3$$

$$\Rightarrow \qquad u_{22} = 3 - l_{21}u_{12} = 3 - \frac{1}{2}(1) = 3 - \frac{1}{2} = \frac{5}{2}$$

$$l_{21}u_{13} + u_{23} = 2$$

$$\Rightarrow \qquad u_{23} = 2 - l_{21}u_{13} = 2 - \frac{1}{2}(1) = 2 - \frac{1}{2} = \frac{3}{2}$$

$$l_{31}u_{12} + l_{32}u_{22} = 1$$

$$\Rightarrow \qquad l_{32} = \frac{1 - l_{31}u_{12}}{u_{22}} = \frac{1 - \dfrac{3}{2}(1)}{\dfrac{5}{2}} = -\frac{1}{5}$$

$$l_{31}u_{13} + l_{32}u_{23} + u_{33} = 2$$

$$\Rightarrow \qquad u_{33} = 2 - l_{31}u_{13} - l_{32}u_{23}$$

$$= 2 - \frac{3}{2}(1) - \left(-\frac{1}{5}\right)\left(\frac{3}{2}\right) = 2 - \frac{3}{2} + \frac{3}{10} = \frac{8}{10} = \frac{4}{5}$$

Thus

$$L = \begin{bmatrix} 1 & 0 & 0 \\ l_{21} & 1 & 0 \\ l_{31} & l_{32} & 1 \end{bmatrix} = \begin{bmatrix} 1 & 0 & 0 \\ \dfrac{1}{2} & 1 & 0 \\ \dfrac{3}{2} & -\dfrac{1}{5} & 1 \end{bmatrix}$$

and

$$U = \begin{bmatrix} u_{11} & u_{12} & u_{13} \\ 0 & u_{22} & u_{23} \\ 0 & 0 & u_{33} \end{bmatrix} = \begin{bmatrix} 2 & 1 & 1 \\ 0 & \dfrac{5}{2} & \dfrac{3}{2} \\ 0 & 0 & \dfrac{4}{5} \end{bmatrix}$$

Since

$$LY = B \Rightarrow \begin{bmatrix} 1 & 0 & 0 \\ \dfrac{1}{2} & 1 & 0 \\ \dfrac{3}{2} & -\dfrac{1}{5} & 1 \end{bmatrix} \begin{bmatrix} y_1 \\ y_2 \\ y_3 \end{bmatrix} = \begin{bmatrix} 2 \\ 2 \\ 2 \end{bmatrix}$$

From these equation, we get

$$y_1 = 2$$

$$\frac{1}{2}y_1 + y_2 = 2$$

$$\frac{3}{2}y_1 - \frac{1}{5}y_2 + y_3 = 2$$

On solving, we get $y_1 = 2, y_2 = 1, y_3 = -\frac{4}{5}$

and since, we have $UX = Y$

$$\begin{bmatrix} 2 & 1 & 1 \\ 0 & \dfrac{5}{2} & \dfrac{3}{2} \\ 0 & 0 & \dfrac{4}{5} \end{bmatrix} \begin{bmatrix} x_1 \\ x_2 \\ x_3 \end{bmatrix} = \begin{bmatrix} y_1 \\ y_2 \\ y_3 \end{bmatrix} = \begin{bmatrix} 2 \\ 1 \\ -\dfrac{4}{5} \end{bmatrix}$$

$$\therefore \qquad 2x_1 + x_2 + x_3 = 2$$

$$\frac{5}{2}x_2 + \frac{3}{2}x_3 = 1$$

$$\frac{4}{5}x_3 = -\frac{4}{5}$$

By back substitution, we get

$$x_3 = -1, x_2 = 1, x_1 = 1$$

Hence, solution is $x_1 = 1, x_2 = 1, x_3 = -1$.

EXAMPLE 2. *Solve the following equations by LU decomposition method*

$$2x + 3y + z = 9$$
$$x + 2y + 3z = 6$$
$$3x + y + 2z = 8$$

SOLUTION. Above equation can be written as follows.

$$AX = B \qquad \qquad \qquad \qquad ...(1)$$

where $\qquad A = \begin{bmatrix} 2 & 3 & 1 \\ 1 & 2 & 3 \\ 3 & 1 & 2 \end{bmatrix}, X = \begin{bmatrix} x \\ y \\ z \end{bmatrix}, B = \begin{bmatrix} 9 \\ 6 \\ 8 \end{bmatrix}$

Let $\qquad A = LU$

$$\therefore \quad \begin{bmatrix} 2 & 3 & 1 \\ 1 & 2 & 3 \\ 3 & 1 & 2 \end{bmatrix} = \begin{bmatrix} 1 & 0 & 0 \\ l_{21} & 1 & 0 \\ l_{31} & l_{32} & 1 \end{bmatrix} \begin{bmatrix} u_{11} & u_{12} & u_{13} \\ 0 & u_{22} & u_{23} \\ 0 & 0 & u_{33} \end{bmatrix}$$

$$\begin{bmatrix} 2 & 3 & 1 \\ 1 & 2 & 3 \\ 3 & 1 & 2 \end{bmatrix} = \begin{bmatrix} u_{11} & u_{12} & u_{13} \\ l_{21}u_{11} & l_{21}u_{12} + u_{22} & l_{21}u_{13} + u_{23} \\ l_{31}u_{11} & l_{31}u_{12} + l_{32}u_{22} & l_{31}u_{13} + l_{32}u_{23} + u_{33} \end{bmatrix}$$

Comparing, we get

$$u_{11} = 2, u_{12} = 3, u_{13} = 1$$

and $\qquad l_{21}u_{11} = 1 \qquad$ or $\qquad l_{21} = \dfrac{1}{2}$

$$l_{21}u_{12} + u_{22} = 2$$

$$\Rightarrow \quad u_{22} = 2 - l_{21}u_{12} = 2 - \frac{1}{2}(3) = 2 - \frac{3}{2} = \frac{1}{2}$$

$$l_{21}u_{13} + u_{23} = 3$$

or $$u_{23} = 3 - l_{21}u_{13} = 3 - \frac{1}{2}(1) = \frac{5}{2}$$

$$l_{31}u_{11} = 3$$

$$\Rightarrow \quad l_{31} = \frac{3}{u_{11}} = \frac{3}{2}$$

and $$l_{31}u_{12} + l_{32}u_{22} = 1$$

$$\Rightarrow \quad l_{32} = \frac{1 - l_{31}u_{12}}{u_{22}} = \frac{1 - \frac{3}{2}(3)}{\frac{1}{2}} = \frac{2-9}{1} = -7$$

Now $l_{31}u_{13} + l_{32}u_{23} + u_{33} = 2$

$$\Rightarrow \quad u_{33} = 2 - l_{31}u_{13} - l_{32}u_{23}$$

$$= 2 - \frac{3}{2}(1) - (-7)\left(\frac{5}{2}\right) = 2 - \frac{3}{2} + \frac{35}{2} = \frac{36}{2} = 18$$

Thus L and U are given by

$$L = \begin{bmatrix} 1 & 0 & 0 \\ \frac{1}{2} & 1 & 0 \\ \frac{3}{2} & -7 & 1 \end{bmatrix}, U = \begin{bmatrix} 2 & 3 & 1 \\ 0 & \frac{1}{2} & \frac{5}{2} \\ 0 & 0 & 18 \end{bmatrix}$$

Since $\qquad LY = B$

$$\therefore \quad \begin{bmatrix} 1 & 0 & 0 \\ \frac{1}{2} & 1 & 0 \\ \frac{3}{2} & -7 & 1 \end{bmatrix} \begin{bmatrix} y_1 \\ y_2 \\ y_3 \end{bmatrix} = \begin{bmatrix} 9 \\ 6 \\ 8 \end{bmatrix}$$

From above equation , we get

$$y_1 = 9$$

$$\frac{1}{2}y_1 + y_2 = 6$$

$$\frac{3}{2}y_1 - 7y_2 + y_3 = 8$$

Solving these equation, we get

$$y_1 = 9, y_2 = \frac{3}{2}, y_3 = 5.$$

Further since $\qquad UX = Y$

$$\therefore \quad \begin{bmatrix} 2 & 3 & 1 \\ 0 & \frac{1}{2} & \frac{5}{2} \\ 0 & 0 & 18 \end{bmatrix} \begin{bmatrix} x \\ y \\ z \end{bmatrix} = \begin{bmatrix} 9 \\ \frac{3}{2} \\ 5 \end{bmatrix}$$

$$\therefore \qquad 2x + 3y + z = 9$$

$$\frac{1}{2}y + \frac{5}{2}z = \frac{3}{2}$$

$$\Rightarrow \qquad\qquad 18z = 5$$

By back substitution, we get

$$z = \frac{5}{18} \text{ and } \frac{1}{2}y + \frac{5}{2}\left(\frac{5}{18}\right) = \frac{3}{2}$$

$$\Rightarrow \qquad\qquad y = 3 - \frac{25}{18} = \frac{29}{18}$$

$$\text{and} \quad 2x + 3\left(\frac{29}{18}\right) + \left(\frac{5}{18}\right) = 9$$

$$\Rightarrow \qquad\qquad 2x = 9 - \frac{29}{6} - \frac{5}{18} = 9 - \frac{92}{18} \Rightarrow \quad x = \frac{35}{18}$$

Hence, the solution is

$$x = \frac{35}{18}, y = \frac{29}{18}, z = \frac{5}{18}.$$

4.6 JORDAN'S METHOD

This method is a modification of Gauss's elimination method. In Jordan's method, the elimination takes places not only below the equations but above also so in this way we get a diagonal matrix. From this values of unknown are found. For understanding this method. Let us consider a system of three equations:

$$a_{11}x_1 + a_{12}x_2 + a_{13}x_3 = b_1 \qquad\qquad …(1)$$

$$a_{21}x_1 + a_{22}x_2 + a_{23}x_3 = b_2 \qquad\qquad …(2)$$

$$a_{31}x_1 + a_{32}x_2 + a_{33}x_3 = b_3 \qquad\qquad …(3)$$

Eliminating x_1 from (2) and (3) with the help of (1), we get

$$a_{11}x_1 + a_{12}x_2 + a_{13}x_3 = b_1{}' \qquad\qquad …(4)$$

$$b_{22}x_2 + b_{23}x_3 = b_2{}' \qquad\qquad …(5)$$

$$a_{32}x_2 + a_{33}x_3 = b_3{}' \qquad\qquad …(6)$$

Now eliminating x_2 from (4) and (6) with the help of (5) we get

$$a'_{11}x_1 + a'_{13}x_3 = b_1{}' \qquad\qquad …(7)$$

$$b_{22}x_2 + b_{23}x_3 = b_2{}' \qquad\qquad …(8)$$

$$c'_{33}x_3 = b_3{}' \qquad\qquad …(9)$$

Finally eliminating x_3 from (7) and (8) using (9) we get

$$a''_{11}x_1 = b''_1, \quad b'_{22}x_2 = b''_2, \quad c'_{33}x_3 = b''_3$$

$$\Rightarrow \qquad \begin{bmatrix} a''_{11} & 0 & 0 \\ 0 & b'_{22} & 0 \\ 0 & 0 & c'_{33} \end{bmatrix} \begin{bmatrix} x_1 \\ x_2 \\ x_3 \end{bmatrix} = \begin{bmatrix} b''_1 \\ b''_2 \\ b''_3 \end{bmatrix} \qquad\qquad …(10)$$

Hence, the coefficient matrix reduced to diagonal matrix from (10) we can easily find x_1, x_2 and x_3.

SOLVED EXAMPLES

EXAMPLE 1. *Apply Jordan's method to solve*

$$x + 2y + z = 8$$
$$2x + 3y + 4z = 20$$
$$4x + 3y + 2z = 16$$

SOLUTION. First, eliminating x from last two equations using first equation, we get

$$x + 2y + z = 8 \qquad \text{...(1)}$$
$$-y + 2z = 4 \qquad \text{...(2)}$$
$$-5y - 2z = -16 \qquad \text{...(3)}$$

Now eliminating y from (1) and (3) with the help of (2), we get

$$x + 5z = 16 \qquad \text{...(4)}$$
$$-y + 2z = 4 \qquad \text{...(5)}$$
$$-12z = -36 \qquad \text{...(6)}$$

Now eliminating z from (4) and (5) with the help of (6), we get

$$12x = 12 \qquad \text{...(7)}$$
$$-6y = -12 \qquad \text{...(8)}$$
$$-12z = -36 \qquad \text{...(9)}$$

From (7), (8) and (9), we get

$$x = 1, y = 2, z = 3.$$

EXAMPLE 2. *Apply Jordan's method to solve the equations*

$$x + y + z = 9$$
$$2x - 3y + 4z = 13$$
$$3x + 4y + 5z = 40 \qquad \text{(VTU–2009, PTU–2005)}$$

SOLUTION. First of all eliminating x from last two equations with the help of first equation, we get

$$x + y + z = 9 \qquad \text{...(1)}$$
$$-5y + 2z = -5 \qquad \text{...(2)}$$
$$y + 2z = 13 \qquad \text{...(3)}$$

Now eliminating y from (1) and (3) with the help of (2), we get

$$5x + 7z = 40 \qquad \text{...(4)}$$
$$-5y + 2z = -5 \qquad \text{...(5)}$$
$$12z = 60 \qquad \text{...(6)}$$

Further, eliminating z from (4) and (5) using (6), we get

$$60x = 60 \qquad \text{...(7)}$$
$$-30y = -90 \qquad \text{...(8)}$$
$$12z = 60 \qquad \text{...(9)}$$

From (7), (8) and (9), we get

$$x = 1, y = 3, z = 5.$$

4.7 CROUT'S METHOD

We shall explains this method by considering three equations. Let us consider the equations as follows:

$$\left. \begin{array}{l} a_{11}x_1 + a_{12}x_2 + a_{13}x_3 = b_1 \\ a_{21}x_1 + a_{22}x_2 + a_{23}x_3 = b_2 \\ a_{31}x_1 + a_{32}x_2 + a_{33}x_3 = b_3 \end{array} \right\} \qquad \text{...(1)}$$

The agumented matrix of (1) is

$$(A/B) = \begin{bmatrix} a_{11} & a_{12} & a_{13} & b_1 \\ a_{21} & a_{22} & a_{23} & b_2 \\ a_{31} & a_{32} & a_{33} & b_3 \end{bmatrix} \qquad \dots(2)$$

Now we consider derived matrix as follows:

$$(A'/B') = \begin{bmatrix} a'_{11} & a'_{12} & a'_{13} & b'_1 \\ a'_{21} & a'_{22} & a'_{23} & b'_2 \\ a'_{31} & a'_{32} & a'_{33} & b'_3 \end{bmatrix} \qquad \dots(3)$$

ALGORITHM

Step 1.	First column of (3) is same as first column of (2)
	i.e., $\quad a'_{11} = a_{11}, a'_{21} = a_{21}, a'_{31} = a_{31} \quad$ or $\quad a'_{i_1} = a_{i_1} \ \forall i = 1, 2, 3.$
Step 2.	Elements of first row to the right of first column in (3) are given by
	$$a'_{12} = \frac{a_{12}}{a_{11}}, a'_{13} = \frac{a_{13}}{a_{11}}, b'_1 = \frac{b_1}{a_{11}}$$
	i.e., $\quad a'_{1j} = \dfrac{a_{1j}}{a_{11}}, j = 2, 3.$
Step 3.	Elements of second column except a'_{12} are given by
	$$a'_{22} = a_{22} - a'_{12}a'_{21}$$ $$a'_{32} = a_{32} - a'_{12}a'_{31}$$
	i.e., $\quad a'_{j2} = a_{j2} - a'_{12}a'_{j1}, j = 2, 3$
Step 4.	Elements of second row except a'_{21}, a'_{22} are given by
	$a'_{23} = \dfrac{a_{23} - a'_{13}a'_{21}}{a'_{22}}, b'_2 = \dfrac{b_2 - b'_1 a'_{21}}{a'_{22}}$ *i.e.,* $\quad a'_{2j} = \dfrac{a_{1j} - a'_{1j}a'_{21}}{a'_{22}}, j = 3.$
Step 5.	Elements of third column except a'_{13}, a'_{23} given by $a_{33} = a_{33} - a'_{23}a'_{32} - a'_{13}a'_{31}.$
Step 6.	Element of third row except $a'_{31}, a'_{32}, a'_{33}$ is $b'_3 = \dfrac{b_3 - b'_2 a'_{32} - b'_1 a'_{31}}{a'_{33}}$

Thus the solution of the given equation is given by

$$x_3 = b'_3, x_2 = b'_2 - a'_{23}x_3, x_1 = b'_1 - a'_{13}x_3 - a'_{12}x_2.$$

This Crout's method is explained properly like below in which the coefficient a' and constants b' are discussed how they are obtained.

Crout established a method in which Gauss elimination is often performed.

Let us consider a system of three equations.

$$\left. \begin{array}{l} a_{11}x_1 + a_{12}x_2 + a_{13}x_3 = b_1 \\ a_{21}x_1 + a_{22}x_2 + a_{23}x_3 = b_2 \\ a_{31}x_1 + a_{32}x_2 + a_{33}x_3 = b_3 \end{array} \right\} \qquad \dots(1)$$

Above equations can be written as $\qquad AX = B$

where

$$A = \begin{bmatrix} a_{11} & a_{12} & a_{13} \\ a_{21} & a_{22} & a_{23} \\ a_{31} & a_{32} & a_{33} \end{bmatrix}, X = \begin{bmatrix} x_1 \\ x_2 \\ x_3 \end{bmatrix}, B = \begin{bmatrix} b_1 \\ b_2 \\ b_3 \end{bmatrix} \qquad \dots(2)$$

$\therefore$ Augmented matrix

$$(A\,/\,B) = \begin{bmatrix} a_{11} & a_{12} & a_{13} & b_1 \\ a_{21} & a_{22} & a_{23} & b_2 \\ a_{31} & a_{32} & a_{33} & b_3 \end{bmatrix} \qquad \ldots(3)$$

Equation (1) becomes after Guass elimination process as follows:

$$\left.\begin{array}{r} x_1 + a'_{12}x_2 + a'_{13}x_3 = b'_1 \\ x_2 + a'_{23}x_3 = b'_2 \\ x_3 = b'_3 \end{array}\right\} \qquad \ldots(4)$$

Now the first equation of (1) is obtained by multiplication of first equation in (4) by a constant a'_{11} and second equation of (1) is obtained through multiplication of first and second equation in (4) by a'_{21} and a'_{22} respectively, and adding. Similarly, the third equation of (1) is obtained through multiplication of first, second and third equations in (4) by a'_{31}, a'_{32} and a'_{33} and then adding. Thus, we get the following equations

$$\left.\begin{array}{r} a'_{11}b'_1 = b_1 \\ a'_{21}b'_1 + a'_{22}b'_2 = b_2 \\ a'_{31}b'_1 + a'_{32}b'_2 + a'_{33}b'_3 = b_3 \end{array}\right\} \qquad \ldots(5)$$

Let us introduce the matrices P and Q as follows:

$$P = \begin{bmatrix} a'_{11} & 0 & 0 \\ a'_{21} & a'_{22} & 0 \\ a'_{31} & a'_{32} & a'_{33} \end{bmatrix}, Q = \begin{bmatrix} 0 & a'_{12} & a'_{13} \\ 0 & 0 & a'_{23} \\ 0 & 0 & 0 \end{bmatrix}$$

or $\qquad\qquad\qquad\qquad P + Q = A'\,\text{(say)}.$

Now equation (1), (4) and (5) take the form

$$\left.\begin{array}{r} AX = B \\ (Q+I)X = B' \\ PB' = B \end{array}\right\} \qquad \ldots(6)$$

From (6), we obtain

$$P(\,Q + I)X = AX$$

and hence $\qquad\qquad\qquad P(Q + I) = A.$

Augmenting the matrix $(Q + I)$ with new column B' and augmenting the matrix A by column B. Thus we get

$$\begin{bmatrix} a'_{11} & 0 & 0 \\ a'_{21} & a'_{22} & 0 \\ a'_{31} & a'_{32} & a'_{33} \end{bmatrix}\begin{bmatrix} 1 & a'_{12} & a'_{13} & b'_1 \\ 0 & 1 & a'_{23} & b'_2 \\ 0 & 0 & 1 & b'_3 \end{bmatrix} = \begin{bmatrix} a_{11} & a_{12} & a_{13} & b_1 \\ a_{21} & a_{22} & a_{23} & b_2 \\ a_{31} & a_{32} & a_{33} & b_3 \end{bmatrix}$$

From above equation, we get

$$a'_{11} = a_{11}, a'_{21} = a_{21}, a'_{31} = a_{31},$$

$$a'_{12} = \frac{a_{12}}{a'_{11}} = \frac{a_{12}}{a_{11}}, a'_{13} = \frac{a_{13}}{a'_{11}} = \frac{a_{13}}{a_{11}}, b'_1 = \frac{b_1}{a_{11}}$$

$$a'_{22} = a_{22} - a'_{21}a'_{12}, a'_{23} = \frac{1}{a'_{22}}(a_{23} - a'_{21}a'_{13})$$

$$a'_{21}b'_1 + a'_{22}b'_2 = b_2 \qquad \text{or} \qquad b'_2 = \frac{b_2 - a'_{21}b'_1}{a'_{22}}$$

$$a'_{31}a'_{12} + a'_{32} = a_{32} \qquad \text{or} \qquad a'_{32} = a_{32} - a'_{31}a'_{12}$$

$$a'_{31}a'_{13} + a'_{32}a'_{23} + a'_{33} = a_{33} \quad \text{or} \quad a'_{33} = a_{33} - a'_{31}a'_{13} - a'_{32}a'_{23}$$

$$a'_{31}b'_1 + a'_{32}b'_2 + a'_{33}b'_3 = b_3 \quad \text{or} \quad b'_3 = \frac{b_3 - a'_{31}b'_1 - a'_{32}b'_2}{a'_{33}}$$

Thus after getting all a' and b', with the help of (4), we get the solution given by

$$x_3 = b'_3, x_2 = b'_2 - a'_{23}x_3, x_1 = b'_1 - a'_{12}x_2 - a'_{13}x_3.$$

SOLVED EXAMPLES

EXAMPLE 1.   *Solve the following equations by Crout's method :*

$$x_1 + x_2 + x_3 = 1$$
$$3x_1 + x_2 - 3x_3 = 5$$
$$x_1 - 2x_2 - 5x_3 = 10$$

SOLUTION.   Above equations can be written as

$$AX = B \qquad \qquad \qquad ...(1)$$

where

$$A = \begin{bmatrix} 1 & 1 & 1 \\ 3 & 1 & -3 \\ 1 & -2 & -5 \end{bmatrix}, X = \begin{bmatrix} x_1 \\ x_2 \\ x_3 \end{bmatrix}, B = \begin{bmatrix} 1 \\ 5 \\ 10 \end{bmatrix}$$

$$\therefore \quad \begin{bmatrix} a_{11} & a_{12} & a_{13} \\ a_{21} & a_{22} & a_{23} \\ a_{31} & a_{32} & a_{33} \end{bmatrix} = \begin{bmatrix} 1 & 1 & 1 \\ 3 & 1 & -3 \\ 1 & -2 & -5 \end{bmatrix}$$

Now derived matrix is given by

$$\begin{bmatrix} a'_{11} & a'_{12} & a'_{13} & b'_1 \\ a'_{21} & a'_{22} & a'_{23} & b'_2 \\ a'_{31} & a'_{32} & a'_{33} & b'_3 \end{bmatrix}$$

where

$$a'_{11} = a_{11} = 1, a'_{21} = a_{21} = 3, a'_{31} = a_{31} = 1$$

and

$$a'_{12} = \frac{a_{12}}{a_{11}} = \frac{1}{1} = 1, a'_{13} = \frac{a_{13}}{a_{11}} = \frac{1}{1} = 1,$$

$$b'_1 = \frac{b_1}{a_{11}} = \frac{1}{1} = 1$$

and

$$a'_{22} = a_{22} - a'_{12}.a'_{21} = 1 - 1.3 = 1 - 3 = -2,$$
$$a'_{32} = a_{32} - a'_{12}.a'_{31} = -2 - 1.1 = -3$$

$$a'_{23} = \frac{a_{23} - a'_{13}.a'_{21}}{a'_{22}} = \frac{-3 - 3}{-2} = 3,$$

$$b'_2 = \frac{b_2 - a'_{21}.b'_1}{a'_{22}} = \frac{5 - 3}{-2} = -1$$

$$a'_{33} = a_{33} - a'_{31}a'_{13} - a'_{32}a'_{23} = -5 - 1 + 9 = 3$$

and

$$b'_3 = \frac{b_3 - a'_{31}b'_1 - a'_{32}b'_2}{a'_{33}} = \frac{10 - 1 - 3}{3} = \frac{6}{3} = 2$$

Thus the solution is

$$x_3 = b'_3 = 2,$$

$$x_2 = b'_2 - a'_{23}x_3 = -1 - 3(2) = -7$$

$$x_1 = b_1' - a_{12}'x_2 - a_{13}'x_3 = 1 - 1(-7) - 1(2)$$

$$= 1 + 7 - 2 = 6$$

Hence, $\qquad x_1 = 6, x_2 = -7, x_3 = 2$

EXAMPLE 2. *Solve the following equations by Crout's method :*

$$2x_1 + 3x_2 + x_3 = -1$$
$$5x_1 + x_2 + x_3 = 9$$
$$3x_1 + 2x_2 + 4x_3 = 11$$

SOLUTION. Above equations can be written as

$$AX = B$$

where $\qquad A = \begin{bmatrix} 2 & 3 & 1 \\ 5 & 1 & 1 \\ 3 & 2 & 4 \end{bmatrix}, X = \begin{bmatrix} x_1 \\ x_2 \\ x_3 \end{bmatrix}, B = \begin{bmatrix} -1 \\ 9 \\ 11 \end{bmatrix}$

Now derived matrix is given by

$$\begin{bmatrix} a_{11}' & a_{12}' & a_{13}' & b_1' \\ a_{21}' & a_{22}' & a_{23}' & b_2' \\ a_{31}' & a_{32}' & a_{33}' & b_3' \end{bmatrix}$$

And

$$(A/B) = \begin{bmatrix} 2 & 3 & 1 & -1 \\ 5 & 1 & 1 & 9 \\ 3 & 2 & 4 & 11 \end{bmatrix} = \begin{bmatrix} a_{11} & a_{12} & a_{13} & b_1 \\ a_{21} & a_{22} & a_{23} & b_2 \\ a_{31} & a_{32} & a_{33} & b_3 \end{bmatrix}$$

Now find a_{ij}' and b_j

$$a_{11}' = a_{11} = 2, a_{21}' = a_{21} = 5, a_{31}' = a_{31} = 3$$

$$a_{12}' = \frac{a_{12}}{a_{11}} = \frac{3}{2}, a_{13}' = \frac{a_{13}}{a_{11}} = \frac{1}{2}, b_1' = \frac{b_1}{a_{11}} = \frac{-1}{2}$$

$$a_{22}' = a_{22} - a_{12}'a_{21}' = 1 - \frac{3}{2}(5) = 1 - \frac{15}{2} = -\frac{13}{2}$$

$$a_{32}' = a_{32} - a_{12}'a_{31}' = 2 - \frac{3}{2}(3) = 2 - \frac{9}{2} = -\frac{5}{2}$$

$$a_{23}' = \frac{a_{23} - a_{13}'a_{21}'}{a_{22}'} = \frac{1 - \left(\dfrac{1}{2}\right)5}{-13/2} = \frac{-3/2}{-13/2} = \frac{3}{13}$$

$$b_2' = \frac{b_2 - a_{21}'b_1'}{a_{22}'} = \frac{9 - 5\left(-\dfrac{1}{2}\right)}{-13/2} = \frac{23/2}{-13/2} = -\frac{23}{13}$$

$$a_{33}' = a_{33} - a_{31}'a_{13}' - a_{32}'a_{23}'$$

$$= 4 - 3\left(\frac{1}{2}\right) - \left(-\frac{5}{2}\right)\left(\frac{3}{13}\right) = 4 - \frac{3}{2} + \frac{15}{26}$$

$$= \frac{104 - 39 + 15}{26} = \frac{80}{26} = \frac{40}{13}$$

$$b_3' = \frac{b_3 - a_{31}'b_1' - a_{32}'b_2'}{a_{33}'}$$

$$= \frac{11 - 3\left(-\dfrac{1}{2}\right) - \left(-\dfrac{5}{2}\right)\left(-\dfrac{23}{13}\right)}{\dfrac{40}{13}}$$

$$= \frac{11 + \dfrac{3}{2} - \dfrac{115}{26}}{\dfrac{40}{13}} = \frac{(286 + 39 - 115)/26}{40/13} = \frac{210/26}{40/13} = \frac{210}{80}$$

$$\Rightarrow \qquad b_3' = \frac{21}{8}$$

Thus solution is

$$x_3 = b_3' = \frac{21}{8}$$

$$x_2 = b_2' - a_{23}' x_3 = -\frac{23}{13} - \left(\frac{3}{13}\right)\frac{21}{8} = -\frac{19}{8}$$

and

$$x_1 = b_1' - a_{12}' x_2 - a_{13}' x_3$$

$$= -\frac{1}{2} - \frac{3}{2}\left(-\frac{19}{8}\right) - \frac{1}{2}\left(\frac{21}{8}\right)$$

$$= -\frac{1}{2} + \frac{57}{16} - \frac{21}{16} = \frac{-8 + 57 - 21}{16} = \frac{14}{8} = \frac{7}{4}.$$

Hence, solution is

$$x_1 = \frac{7}{4}, x_2 = -\frac{19}{8}, x_3 = \frac{21}{8}.$$

EXAMPLE 3.   *Solve by Crout's method :*
$$x + 2y + 3z = 6$$
$$2x + 3y + z = 9$$
$$3x + y + 2z = 8$$

SOLUTION.   Above equations can be expressed as
$$AX = B$$

where
$$A = \begin{bmatrix} 1 & 2 & 3 \\ 2 & 3 & 1 \\ 3 & 1 & 2 \end{bmatrix}, X = \begin{bmatrix} x \\ y \\ z \end{bmatrix}, B = \begin{bmatrix} 6 \\ 9 \\ 8 \end{bmatrix}$$

Thus augmented matrix (A/B) is given by
$$\begin{bmatrix} a_{11} & a_{12} & a_{13} & b_1 \\ a_{21} & a_{22} & a_{23} & b_2 \\ a_{31} & a_{32} & a_{33} & b_3 \end{bmatrix} = \begin{bmatrix} 1 & 2 & 3 & 6 \\ 2 & 3 & 1 & 9 \\ 3 & 1 & 2 & 8 \end{bmatrix}$$

It's derived matrix is given by
$$\begin{bmatrix} a_{11}' & a_{12}' & a_{13}' & b_1' \\ a_{21}' & a_{22}' & a_{23}' & b_2' \\ a_{31}' & a_{32}' & a_{33}' & b_3' \end{bmatrix}$$

Now finding a_{ij}' and b_i'

$$\therefore \qquad a_{11}' = a_{11} = 1, a_{21}' = a_{21} = 2, a_{31}' = a_{31} = 3$$

$$a_{12}' = \frac{a_{12}}{a_{11}} = 2, a_{13}' = \frac{a_{13}}{a_{11}} = 3, b_1' = \frac{b_1}{a_{11}} = 6$$

$$a'_{22} = a_{22} - a'_{21}a'_{12} = 3 - 2(2) = 3 - 4 = -1$$

$$a'_{32} = a_{32} - a'_{31}a'_{12} = 1 - 3(2) = -5$$

$$a'_{23} = \frac{a_{23} - a'_{21}a'_{13}}{a'_{22}} = \frac{1 - 2(3)}{-1} = 5$$

$$b'_2 = \frac{b_2 - a'_{21}b'_1}{a'_{22}} = \frac{9 - 2(6)}{-1} = 3$$

$$a'_{33} = a_{33} - a'_{31}a'_{13} - a'_{32}a'_{23}$$
$$= 2 - 3(3) - (-5)(5) = 2 - 9 + 25 = 18$$

$$b'_3 = \frac{b_3 - a'_{31}b'_1 - a'_{32}b'_2}{a'_{33}}$$

$$= \frac{8 - 3(6) - (-5)3}{18} = \frac{8 - 18 + 15}{18} = \frac{5}{18}$$

Now solution is

$$z = x_3 = b'_3 = \frac{5}{18}$$

$$y = x_2 = b'_2 - a'_{23}x_3 = 3 - 5\left(\frac{5}{18}\right) = 3 - \frac{25}{18} = \frac{29}{18}$$

$$x = x_1 = b'_1 - a'_{12}x_2 - a'_{13}x_3$$

$$= 6 - 2\left(\frac{29}{18}\right) - 3\left(\frac{5}{18}\right) = 6 - \frac{58}{18} - \frac{15}{18} = \frac{35}{18}$$

Hence, solution is given by

$$x = \frac{35}{18}, y = \frac{29}{18}, z = \frac{5}{18}.$$

4.8 ITERATIVE METHODS

The previous methods for finding the solutions of simultaneous linear equations are known as direct methods, because after certain amount of fixed calculation, we obtain the solutions. On the other hand an iterative method is that method in which we assume initial approximations to the solution and obtain better and better solution with the desired degree of accuracy through number of iterations. Thus the amount of calculation depends on the desired degree of accuracy.

In this section, we shall discuss some iterative methods as Jacobi method, Gauss-seidel method and relaxation method.

4.8.1 JACOBI METHOD

Let us consider the system of linear simultaneous equations as

$$\left.\begin{array}{l} a_{11}x_1 + a_{12}x_2 + a_{13}x_3 + ... + a_{1n}x_n = b_1 \\ a_{21}x_1 + a_{22}x_2 + a_{23}x_3 + ... + a_{2n}x_n = b_2 \\ a_{31}x_1 + a_{32}x_2 + a_{33}x_3 + ... + a_{3n}x_n = b_3 \\ \dotfill \\ \dotfill \\ a_{n1}x_1 + a_{n2}x_2 + a_{33}x_3 + ... + a_{nn}x_n = b_n \end{array}\right\} \qquad ...(1)$$

provided all the diagonal elements are not equal to zero. If so then rearrange the equations in such a way that all $a_{ii} \neq 0$.

Now the system of equation (1) can be written as

$$
\left.
\begin{aligned}
x_1 &= \frac{1}{a_{11}}(b_1 - a_{12}x_2 - a_{13}x_3 - \ldots - a_{1n}x_n) \\[2mm]
x_2 &= \frac{1}{a_{22}}(b_2 - a_{21}x_1 - a_{23}x_3 - \ldots - a_{2n}x_n) \\[2mm]
x_3 &= \frac{1}{a_{33}}(b_3 - a_{31}x_1 - a_{32}x_2 - \ldots - a_{3n}x_n) \\[2mm]
&\qquad\ldots\ldots\ldots\ldots\ldots\ldots\ldots\ldots \\[2mm]
x_n &= \frac{1}{a_{nn}}(b_n - a_{n1}x_1 - a_{n2}x_2 - \ldots - a_{n(n-1)}x_{n-1})
\end{aligned}
\right\} \qquad \ldots(2)
$$

Let us assume the first approximations $x_1^{(1)}, x_2^{(1)}, \ldots, x_n^{(1)}$ for the values of $x_1, x_2, \ldots, x_n$ respectively. Substituting these values on R.H.S. of (2), we get a system of second approximation

$$
\left.
\begin{aligned}
x_1^{(2)} &= \frac{1}{a_{11}}(b_1 - a_{12}x_2^{(1)} - a_{13}x_3^{(1)} - \ldots - a_{1n}x_n^{(1)}) \\[2mm]
x_2^{(2)} &= \frac{1}{a_{22}}(b_2 - a_{21}x_1^{(1)} - a_{23}x_3^{(1)} - \ldots - a_{2n}x_n^{(1)}) \\[2mm]
x_3^{(2)} &= \frac{1}{a_{33}}(b_3 - a_{31}x_1^{(1)} - a_{32}x_2^{(1)} - \ldots - a_{3n}x_n^{(1)}) \\[2mm]
&\qquad\ldots\ldots\ldots\ldots\ldots\ldots\ldots\ldots \\[2mm]
x_n^{(2)} &= \frac{1}{a_{nn}}(b_n - a_{n1}x_1^{(1)} - a_{n2}x_2^{(1)} - \ldots - a_{n(n-1)}x_{n-1}^{(1)})
\end{aligned}
\right\} \qquad \ldots(3)
$$

Proceeding in the same way, we obtain third, fourth, etc. approximation. Let $x_1^{(n)}, x_2^{(n)}, \ldots, x_n^{(n)}$ be the nth approximation, then we get $(n + 1)$th approximation as

$$
\left.
\begin{aligned}
x_1^{(n+1)} &= \frac{1}{a_{11}}(b_1 - a_{12}x_2^{(n)} - a_{13}x_3^{(n)} - \ldots - a_{1n}x_n^{(n)}) \\[2mm]
x_2^{(n+1)} &= \frac{1}{a_{22}}(b_2 - a_{21}x_1^{(n)} - a_{23}x_3^{(n)} - \ldots - a_{2n}x_n^{(n)}) \\[2mm]
x_3^{(n+1)} &= \frac{1}{a_{33}}(b_3 - a_{31}x_1^{(n)} - a_{32}x_2^{(n)} - \ldots - a_{3n}x_n^{(n)}) \\[2mm]
&\qquad\ldots\ldots\ldots\ldots\ldots\ldots\ldots\ldots \\[2mm]
x_n^{(n+1)} &= \frac{1}{a_{nn}}(b_n - a_{n1}x_1^{(n)} + a_{n2}x_2^{(n)} + \ldots + a_{n(n-1)}x_{n-1}^{(n)})
\end{aligned}
\right\} \qquad \ldots(4)
$$

Now equation (2) can be written in matrix form as

$$
X^{(2)} = BX + C \qquad \ldots(5)
$$

Similarly, equation (4) can be written as

$$
X^{(n+1)} = BX^{(n)} + C \qquad \ldots(6)
$$

▼ REMARKS

▸ This method is also known as method of simultaneous-displacement and it is sufficiently convergent provided $|B| < 1$.

▸ In the absence of any better approximation, we can take each equal to zero.

SOLVED EXAMPLES

EXAMPLE 1. *Solve the following equation by Jacobi method.*

$$27x + 6y - z = 85$$
$$6x + 15y + 2z = 72$$
$$x + y + 54z = 110$$

SOLUTION. Above equations can be written as

$$\left. \begin{array}{l} x = \dfrac{1}{27}(85 - 6y + z) \\[2mm] y = \dfrac{1}{15}(72 - 6x - 2z) \\[2mm] z = \dfrac{1}{54}(110 - x - y) \end{array} \right\} \qquad \qquad ...(1)$$

Let us take the first approximation to be $x^{(1)} = 0, y^{(1)} = 0$ and $z^{(1)} = 0$ we obtain second approximation from (1)

$$x^{(2)} = \frac{1}{27}(85 - 6y^{(1)} + z^{(1)}) = \frac{1}{27}(85 - 0 + 0) = \frac{85}{27} = 3.15$$

$$y^{(2)} = \frac{1}{15}(72 - 6x^{(1)} - 2z^{(1)}) = \frac{1}{15}(72 - 0 - 0) = \frac{72}{15} = 4.80$$

and

$$z^{(2)} = \frac{1}{54}(110 - x^{(1)} - y^{(1)}) = \frac{110}{54} = 2.04 \cdot$$

Now we obtain third approximation as follows:

$$x^{(3)} = \frac{1}{27}(85 - 6y^{(2)} + z^{(2)})$$

$$= \frac{1}{27}(85 - 6(4.8) + 2.04) = \frac{58.24}{27} = 2.16$$

$$y^{(3)} = \frac{1}{15}(72 - 6x^{(2)} - 2z^{(2)})$$

$$= \frac{1}{15}(72 - 6(3.15) - 2(2.04)) = \frac{49.02}{15} = 3.27$$

$$z^{(3)} = \frac{1}{54}(110 - x^{(2)} - y^{(2)})$$

$$= \frac{1}{54}(110 - 3.15 - 4.80) = \frac{102.05}{54} = 1.89$$

Now fourth approximation is

$$x^{(4)} = \frac{1}{27}(85 - 6y^{(3)} + z^{(3)}) = \frac{1}{27}(85 - 6(3.27) + 1.89) = 2.49$$

$$y^{(4)} = \frac{1}{15}(72 - 6x^{(3)} - 2z^{(3)}) = \frac{1}{15}(72 - 6(2.16) - 2(1.89)) = 3.68$$

$$z^{(4)} = \frac{1}{54}(110 - x^{(3)} - y^{(3)}) = \frac{1}{54}(110 - 2.16 - 3.27) = 1.95$$

The fifth approximation is

$$x^{(5)} = \frac{1}{27}(85 - 6y^{(4)} + z^{(4)}) = \frac{1}{27}(85 - 6(3.68) + 1.95) = 2.40$$

$$y^{(5)} = \frac{1}{15}(72 - 6x^{(4)} - 2z^{(4)}) = \frac{1}{15}(72 - 6(2.49) - 2(1.95)) = 3.54$$

$$z^{(5)} = \frac{1}{54}(110 - x^{(4)} - y^{(4)}) = \frac{1}{54}(110 - 2.49 - 3.68) = 1.92$$

Hence, approximate solution is

$$x = 2.4, y = 3.54, z = 1.92.$$

EXAMPLE 2.   *Solve the following equations by Jacobi method :*

$$20x + y - 2z = 17$$
$$3x + 20y - z = -18$$
$$2x - 3y + 20z = 25$$

(Bhopal–2009)

SOLUTION.   Above equations can be written as

$$\left. \begin{array}{l} x = \dfrac{1}{20}(17 - y + 2z) \\[2mm] y = \dfrac{1}{20}(-18 - 3x + z) \\[2mm] z = \dfrac{1}{20}(25 - 2x + 3y) \end{array} \right\} \qquad \dots(1)$$

Let us take first approximation $x^{(1)} = 0$, $y^{(1)} = 0$ and $z^{(1)} = 0$. Then the second approximation using (1), we get

$$x^{(2)} = \frac{1}{20}(17 - y^{(1)} + 2z^{(1)}) = \frac{17}{20} = 0.85$$

$$y^{(2)} = \frac{1}{20}(-18 - 3x^{(1)} + z^{(1)})$$

$$= \frac{1}{20}(-18 - 3x^{(1)} + z^{(1)}) = -0.9$$

$$z^{(2)} = \frac{1}{20}(25 - 2x^{(1)} + 3y^{(1)})$$

$$= \frac{1}{20}(25 - 0 + 0) = \frac{25}{20} = 1.25 \cdot$$

Substituting these values in R.H.S. of (1), we get third approximation

$$x^{(3)} = \frac{1}{20}(17 - y^{(2)} + 2z^{(2)}) = \frac{1}{20}(17 + 0.9 + 2(1.25)) = 1.02$$

$$y^{(3)} = \frac{1}{20}(-18 - 3x^{(2)} + z^{(2)})$$

$$= \frac{1}{20}(-18 - 3(0.85) + 1.25) = -\frac{19.3}{20} = -0.965$$

$$z^{(3)} = \frac{1}{20}(25 - 2x^{(2)} + 3y^{(2)})$$

$$= \frac{1}{20}(25 - 2(0.85) + 3(0.9)) = \frac{20.6}{20} = 1.03$$

Now substitute these values into (1), we get fourth approximation

$$x^{(4)} = \frac{1}{20}(17 - y^{(3)} + 2z^{(3)})$$

$$= \frac{1}{20}(17 + 0.965 + 2(1.03)) = 1.0013$$

$$y^{(4)} = \frac{1}{20}(-18 - 3x^{(3)} + z^{(3)})$$

$$= \frac{1}{20}(-18 - 3(1.02) + 1.03)$$

$$= -\frac{20.03}{20} = -1.0015$$

$$z^{(4)} = \frac{1}{20}(25 - 2x^{(3)} + 3y^{(3)}) = \frac{1}{20}(25 - 2(1.02) + 3(-0.965))$$

$$= \frac{20.065}{20} = 1.0032$$

Substitute these values into (1), we get fifth approximation

$$x^{(5)} = \frac{1}{20}(17 - y^{(4)} + 2z^{(4)}) = \frac{1}{20}(17 + 1.0015 + 2(1.0032))$$

$$= \frac{20.0079}{20} = 1.0004$$

$$y^{(5)} = \frac{1}{20}(-18 - 3x^{(4)} + z^{(4)}) = \frac{1}{20}(-18 - 3(1.0013) + 1.0032)$$

$$= -\frac{20.0007}{20} = -1.0003$$

$$z^{(5)} = \frac{1}{20}(25 - 2x^{(4)} + 3y^{(4)}) = \frac{1}{20}(25 - 2(1.0013) + 3(-1.005))$$

$$= \frac{19.9929}{20} = 0.9996$$

Substituting these values into (1), we get sixth approximation

$$x^{(6)} = \frac{1}{20}(17 - y^{(5)} + 2z^{(5)}) = \frac{1}{20}(17 + 1.00003 + 2(0.9996))$$

$$= \frac{19.99923}{20} = 0.99996$$

$$y^{(6)} = \frac{1}{20}(-18 - 3x^{(5)} + z^{(5)}) = \frac{1}{20}(-18 - 3(1.0004) + 0.9996)$$

$$= -\frac{20.0016}{20} = -1.00008$$

$$z^{(6)} = \frac{1}{20}(25 - 2x^{(5)} + 3y^{(5)}) = \frac{1}{20}(25 - 2(1.0004) + 3(-1.00003))$$

$$= \frac{19.99911}{20} = 0.99995$$

The values in the fifth and sixth approximation are nearly same. Hence, the solution is $x = 1, y = -1, \ z = 1$.

4.8.2 Gauss-Seidel Method

This method is a modification of Jacobi Method and sometimes it gives faster convergence. To explain this method, let us consider a system of n equations in which $a_{ij} \neq 0$.

$$\left.\begin{array}{l} a_{11}x_1 + a_{12}x_2 + a_{13}x_3 + ... + a_{1n}x_n = b_1 \\ a_{21}x_1 + a_{22}x_2 + a_{23}x_3 + ... + a_{2n}x_n = b_2 \\ a_{31}x_1 + a_{32}x_2 + a_{33}x_3 + ... + a_{3n}x_n = b_3 \\ \cdots\cdots\cdots\cdots\cdots\cdots\cdots\cdots\cdots\cdots\cdots\cdots \\ \cdots\cdots\cdots\cdots\cdots\cdots\cdots\cdots\cdots\cdots\cdots\cdots \\ a_{n1}x_1 + a_{n2}x_2 + a_{33}x_3 + ... + a_{nn}x_n = b_n \end{array}\right\} \qquad ...(1)$$

Above equations can be written as

$$\left.\begin{array}{l} x_1 = \dfrac{1}{a_{11}}(b_1 - a_{12}x_2 - a_{13}x_3 - ... - a_{1n}x_n) \\[2mm] x_2 = \dfrac{1}{a_{22}}(b_2 - a_{21}x_1 - a_{23}x_3 - ... - a_{2n}x_n) \\[2mm] x_3 = \dfrac{1}{a_{33}}(b_3 - a_{31}x_1 - a_{32}x_2 - ... - a_{3n}x_n) \\[2mm] \cdots\cdots\cdots\cdots\cdots\cdots\cdots\cdots\cdots\cdots\cdots\cdots \\[2mm] x_n = \dfrac{1}{a_{nn}}(b_n - a_{n1}x_1 - a_{n2}x_2 - ... - a_{n(n-1)}x_{n-1}) \end{array}\right\} \qquad ...(2)$$

Now let us assume first approximations $x_1^{(1)}, x_2^{(1)}, x_3^{(1)}, ..., x_n^{(1)}$. Substitute these values into the first equation of (2), we get

$$x_1^{(2)} = \frac{1}{a_{11}}(b_1 - a_{12}x_2^{(1)} - a_{13}x_3^{(1)} - ... - a_{1n}x_n^{(1)}).$$

Now substitute $x_1^{(2)}, x_2^{(1)}, x_3^{(1)}, ..., x_n^{(1)}$ into the R.H.S. of second equation of (2), we get

$$x_2^{(2)} = \frac{1}{a_{22}}(b_2 - a_{21}x_1^{(2)} - a_{23}x_3^{(1)} - ... - a_{2n}x_n^{(1)}).$$

In the third equation of (2) substitute $x_1^{(2)}, x_2^{(2)}, x_3^{(1)}, ..., x_n^{(1)}$ we get

$$x_3^{(2)} = \frac{1}{a_{33}}(b_3 - a_{31}x_1^{(2)} - a_{32}x_2^{(3)} - ... - a_{3n}x_n^{(1)}).$$

Proceeding in this way, we get $x_n^{(2)}$ and this completes the first stage of iteration. The whole process is repeated till we get the values of $x_1, x_2, x_3, ..., x_n$ to the desired degree of accuracy. Therefore, Gauss-Seidel method is also known as a method of successive displacement.

▶ REMARKS

▶ Both the methods Jacobi and Gauss-Seidel converge for any type of first approximation if every equation of the system (2) satisfies the condition that the sum of the absolute values of the coefficients $\dfrac{a_{ij}}{a_{ii}}$ is atmost equal to or in atleast one equation less than unity. That is, $\displaystyle\sum_{\substack{j=1\\(j\neq i)}}^{n}\left|\dfrac{a_{ij}}{a_{ii}}\right| \leq 1, i = 1, 2, 3, \dots n.$

▶ The sign < holds good in the case of atleast one equation.

▶ Gauss-Seidel Method converges twice as fast as the Jacobi Method.

▶ This method is also known as method of successive displacement.

SOLVED EXAMPLES

EXAMPLE 1. *Solve by following equation by Gauss-Seidel method* :

$$27x + 6y - z = 85$$
$$6x + 15y + 2z = 72$$
$$x + y + 54z = 110$$

(Anna–2006)

SOLUTION. Above equations can be written as

$$\left.\begin{aligned} x &= \frac{1}{27}(85 - 6y + z)\\[4pt] y &= \frac{1}{15}(72 - 6x - 2z)\\[4pt] z &= \frac{1}{54}(110 - x - y) \end{aligned}\right\} \qquad \dots(1)$$

Assuming first approximation $x^{(1)} = 0, y^{(1)} = 0, z^{(1)} = 0$.

Substituting these values into the first equation of (1), we get

$$x^{(2)} = \frac{1}{27}(85 - 6y^{(1)} + z^{(1)}) = \frac{1}{27}(85) = 3.14$$

Now substitute $x^{(2)} = 3.14, y^{(1)} = 0, z^{(1)} = 0$ into second equation of (1), we get

$$y^{(2)} = \frac{1}{15}(72 - 6x^{(2)} + 2z^{(1)}) = \frac{1}{15}(72 - 6(3.14) - 0) = 3.54$$

Substitute $x^{(2)} = 3.14, y^{(2)} = 3.54, z^{(1)} = 0$ into third equation of (1), we get

and $$z^{(2)} = \frac{1}{54}(110 - x^{(2)} - y^{(2)}) = \frac{1}{54}(110 - 3.14 - 3.54)$$

$$= \frac{103.32}{54} = 1.91$$

Thus second approximation are $x^{(2)} = 3.14, y^{(2)} = 3.54, z^{(2)} = 1.91$. Now, we proceed to obtain third approximations. For this substitute the values $x^{(2)} = 3.14, y^{(2)} = 3.54, z^{(2)} = 1.91$ into the first equation of (1), we get

$$x^{(3)} = \frac{1}{27}(85 - 6y^{(2)} + z^{(2)}) = \frac{1}{27}(85 - 6(3.54) + 1.91)$$

$$= \frac{65.67}{27} = 2.43$$

Now, substitute $x^{(3)} = 2.43, y^{(2)} = 3.54, z^{(2)} = 1.91$ into second equation of (1), we get

$$y^{(3)} = \frac{1}{15}(72 - 6x^{(3)} - 2z^{(2)}) = \frac{1}{15}(72 - 6(2.43) - 2(1.91)) = 3.57$$

Now substitute $x^{(3)} = 2.43, y^{(3)} = 3.57, z^{(2)} = 1.91$ into third equation of (1), we get

$$z^{(3)} = \frac{1}{54}(110 - x^{(3)} - y^{(3)}) = \frac{1}{54}(110 - 2.43 - 3.57) = 1.92$$

Thus these values are close to $x^{(2)}, y^{(2)}, z^{(2)}$ respectively. Hence the solution is $x = 2.43, y = 3.57, z = 1.92$.

EXAMPLE 2. *Solve the following system by Gauss-Seidel method*:

$$10x + 2y + z = 9$$
$$2x + 20y - 2z = -44$$
$$-2x + 3y + 10z = 22$$

(Manipal-2000)

SOLUTION. Above equations can be written as

$$\left. \begin{array}{l} x = \dfrac{1}{10}(9 - 2y - z) \\[2mm] y = \dfrac{1}{20}(-44 - 2x + 2z) \\[2mm] z = \dfrac{1}{10}(22 + 2x - 3y) \end{array} \right\} \qquad \ldots(1)$$

Let us start with first approximation $x^{(1)} = 0, \ y^{(1)} = 0, z^{(1)} = 0$.
Substitute these values into first equation of (1), we get

$$x^{(2)} = \frac{1}{10}(9 - 2y^{(1)} - z^{(1)}) = \frac{1}{10}(9 - 0 - 0) = 0.9$$

Now substitute $x^{(2)} = 0.9, y^{(1)} = 0, z^{(1)} = 0$ into second equation of (1), we get

$$y^{(2)} = \frac{1}{20}(-44 - 2x^{(2)} + 2z^{(1)}) = \frac{1}{20}(-44 - 2(0.9) + 0) = -2.29$$

and substitute the values $x^{(2)} = 0.9, y^{(2)} = -2.29, z^{(1)} = 0$ into third equation of (1), we get

$$z^{(2)} = \frac{1}{10}(22 + 2x^{(2)} + 3y^{(2)})$$

$$= \frac{1}{10}(22 + 2(0.9) - 3(-2.29)) = \frac{30.67}{10} = 3.067$$

Here first stage of iteration is completed. For next iteration, substitute the values $x^{(2)} = 0.9, y^{(2)} = -2.29, z^{(2)} = 3.067$ into first equation of (1), we get

$$x^{(3)} = \frac{1}{10}(9 - 2y^{(2)} - z^{(2)})$$

$$= \frac{1}{10}(9 - 2(-2.29) - 3.067) = \frac{10.513}{10} = 1.051$$

Substitute $x^{(3)} = 1.051, y^{(2)} = -2.29, z^{(2)} = 3.067$ into second equation of (1), we get

$$y^{(3)} = \frac{1}{20}(-44 - 2x^{(3)} + 2z^{(2)})$$

$$= \frac{1}{20}(-44 - 2(1.051) + 2(3.067)) = -\frac{39.968}{20} = -1.99$$

Now substitute $x^{(3)} = 1.051$, $y^{(3)} = -1.99$, $z^{(2)} = 3.067$ into third equation of (1), we get

$$z^{(3)} = \frac{1}{10}(22 + 2x^{(3)} - 3y^{(3)})$$

$$= \frac{1}{10}(22 + 2(1.051) - 3(-1.99)) = \frac{30.072}{10} = 3.007$$

Second stage of iteration is now completed. For next iteration, substitute $x^{(3)} = 1.051$, $y^{(3)} = -1.99$, $z(3) = 3.007$ into first equation of (1), we get

$$x^{(4)} = \frac{1}{10}(9 - 2y^{(3)} - z^{(3)})$$

$$= \frac{1}{10}(9 - 2(-1.99) - (3.007)) = \frac{9.973}{10} = 0.997$$

Substitute $x^{(4)} = 0.997$, $y^{(3)} = -1.99$, $z^{(3)} = 3.007$ into second equation of (1), we get

$$y^{(4)} = \frac{1}{20}(-44 - 2x^{(4)} + 2z^{(3)})$$

$$= \frac{1}{20}(-44 - 2(0.997) + 2(3.007)) = \frac{39.98}{20} = -1.99$$

Now substitute $x^{(4)} = 0.997$, $y^{(4)} = -1.99$, $z^{(3)} = 3.007$ into third equation of (1), we get

$$z^{(4)} = \frac{1}{10}(22 + 2x^{(4)} - 3y^{(4)})$$

$$= \frac{1}{10}(22 + 2(0.997) - 3(-1.99)) = \frac{29.964}{10} = 2.99$$

Thus $x^{(4)}, y^{(4)}, z^{(4)}$ are very close to $x^{(3)}, y^{(3)}, z^{(3)}$ respectively. Hence the solution is

$$x = 1.0, y = -2.0, z = 3.0$$

EXAMPLE 3. *Find the solution of the system of Gauss-Seidel method:*

$$83x + 11y - 4z = 95$$
$$7x + 52y + 13z = 104$$
$$3x + 8y + 29z = 71 \qquad \text{(Hazaribagh–2009)}$$

SOLUTION. Given equations can be written as

$$x = \frac{1}{83}(95 - 11y + 4z) \qquad \text{...(1)}$$

$$y = \frac{1}{52}(104 - 7x - 13z) \qquad \text{...(2)}$$

$$z = \frac{1}{29}(71 - 3x - 8y) \qquad \text{...(3)}$$

Taking first approximation $x^{(1)} = 0, y^{(1)} = 0, z^{(1)} = 0$ and substitute these values in (1), we get

$$x^{(2)} = \frac{1}{83}(95 - 11y^{(1)} + 4z^{(1)}) = \frac{1}{83}(95 - 0 + 0) = 1.14$$

Put $x^{(2)} = 1.14$, $y^{(1)} = 0$, $z^{(1)} = 0$ into (2), we get

$$y^{(2)} = \frac{1}{52}(104 - 7x^{(2)} - 13z^{(1)})$$

$$= \frac{1}{52}(104 - 7(1.14) - 0) = \frac{96.02}{52} = 1.85$$

Put $x^{(2)} = 1.14$, $y^{(2)} = 1.85$, $z^{(1)} = 0$ into (3), we get

$$z^{(2)} = \frac{1}{29}(71 - 3x^{(2)} - 8y^{(2)})$$

$$= \frac{1}{29}(71 - 3(1.14) - 8(1.85)) = \frac{52.78}{29} = 1.82$$

Next, put $x^{(2)} = 1.14$, $y^{(2)} = 1.85$, $z^{(2)} = 1.82$ into (1), we get

$$x^{(3)} = \frac{1}{83}(95 - 11y^{(2)} + 4z^{(2)})$$

$$= \frac{1}{83}(95 - 11(1.85) + 4(1.82)) = 0.99$$

Put $x^{(3)} = 0.99$, $y^{(2)} = 1.85$, $z^{(2)} = 1.82$ into (2), we get

$$y^{(3)} = \frac{1}{52}(104 - 7x^{(3)} - 13z^{(3)})$$

$$= \frac{1}{52}(104 - 7(0.99) - 13(1.82)) = 1.41$$

Put $x^{(3)} = 0.99$, $y^{(3)} = 1.41$, into (3), we get

$$z^{(3)} = \frac{1}{29}(71 - 3x^{(3)} - 8y^{(3)})$$

$$= \frac{1}{29}(71 - 3(0.99) - 8(1.41)) = 1.95$$

Again put $x^{(3)} = 0.99$, $y^{(3)} = 1.41$, $z^{(3)} = 1.95$ into (1), we get

$$x^{(4)} = \frac{1}{83}(95 - 11y^{(3)} + 4z^{(3)})$$

$$= \frac{1}{83}(95 - 11(1.41) + 4(1.95)) = \frac{87.29}{83} = 1.05$$

Now put $x^{(4)} = 1.05$, $z^{(3)} = 1.95$ into (2) we get

$$y^{(4)} = \frac{1}{52}(104 - 7x^{(4)} - 13z^{(3)})$$

$$= \frac{1}{52}(104 - 7(1.05) - 13(1.951)) = 1.37$$

Put $x^{(4)} = 1.05$, $y^{(4)} = 1.37$, into (3), we get

$$z^{(4)} = \frac{1}{29}(71 - 3x^{(4)} - 8y^{(4)}) = \frac{1}{29}(71 - 3(1.05) - 8(1.37)) = 1.96$$

Thus $x^{(4)}, y^{(4)}, z^{(4)}$ are sufficiently close to $x^{(3)}, y^{(3)}, z^{(3)}$ respectively. Hence, the solution is

$$x = 1.05, y = 1.37, z = 1.96$$

4.8.3 RELAXATION METHOD

This method was originally established by R.V. Southwll in 1935 for finding the solution related to engineering problems. To explain this method let us consider a system of three equations

$$\left.\begin{array}{l} a_{11}x_1 + a_{12}x_2 + a_{13}x_3 = b_1 \\ a_{21}x_1 + a_{22}x_2 + a_{23}x_3 = b_2 \\ a_{31}x_1 + a_{32}x_2 + a_{33}x_3 = b_3 \end{array}\right\} \qquad \text{...(1)}$$

First we define the residuals Rx_1, Rx_2, Rx_3 by the given relations

$$\left.\begin{array}{l} Rx_1 = b_1 - a_{11}x_1 - a_{12}x_2 - a_{13}x_3 \\ Rx_2 = b_2 - a_{21}x_1 - a_{22}x_2 - a_{23}x_3 \\ Rx_3 = b_3 - a_{31}x_1 - a_{32}x_2 - x_{33}x_3 \end{array}\right\} \qquad \text{...(2)}$$

Let us take the initial approximation $x_1 = 0, x_2 = 0, x_3 = 0$ and find the initial residuals. These residuals are further reduced step by step, by giving some increments to the value of x_1, x_2 and x_3. At each step the magnitude of the largest residual is reduced to almost zero. In particular to reduce Rx_1, the value of the corresponding variable x_1 is changed, *i.e.*, to reduce Rx_1 by a , we shall increase x_1 by a/a_{11}. This process is repeated till all the residuals have been reduced to almost zero, then to obtain the solution add all the increments in x_1, x_2, x_3 separately. To find the particular residual by giving increment to the corresponding variable, we use following operating table:

	δR_{x_1}	δR_{x_2}	δR_{x_3}
$\delta R_{x_1} = 1$	$-a_{11}$	$-a_{21}$	$-a_{31}$
$\delta R_{x_2} = 1$	$-a_{12}$	$-a_{22}$	$-a_{32}$
$\delta R_{x_3} = 1$	$-a_{13}$	$-a_{23}$	$-a_{33}$

We observe from the above table that if x_1 is increased by 1 keeping y and z constant, then R_{x_1}, R_{x_2} and R_{x_3} decreases by a_{11}, a_{21} and a_{31} respectively. Similar effect has been shown in above table for the variables y and z respectively.

▶ **REMARKS**

▸ If $|a_{11}| \ge |a_{12}| \ge |a_{13}|, |a_{21}| \ge |a_{22}| \ge |a_{23}|, |a_{33}| \ge |a_{31}| \ge |a_{32}|$ hold, then relaxation Method can be applied successfully.

▸ In relaxation method, during the process, if all residuals are not negligible, then there is some mistake and the whole process should be rechecked.

SOLVED EXAMPLES

EXAMPLE 1. *Solve by Relaxation method, the equations :*

$$9x - 2y + z = 50$$
$$x + 5y - 3z = 18$$
$$-2x + 2y + 7z = 19 \qquad \text{(Madras–2000S)}$$

SOLUTION. The residuals are given by

$$R_x = 50 - 9x + 2y - z$$

$$R_y = 18 - x - 5y + 3z$$
$$R_z = 19 + 2x - 2y - 7z$$

The operations table is

	δR_x	δR_y	δR_z
$\delta R_x = 1$	−9	−1	2
$\delta R_y = 1$	2	−5	−2
$\delta R_z = 1$	−1	3	−7

The relaxation table is

	R_x	R_y	R_z
$x = y = z = 0$	50	18	19
$\delta x = 5$	5	13	29
$\delta z = 4$	1	25	1
$\delta y = 5$	11	0	−9
$\delta x = 1$	2	−1	−7
$\delta z = -1$	3	−4	0
$\delta y = -0.8$	1.4	0	1.6
$\delta z = 0.23$	1.17	0.69	−0.09
$\delta x = 0.13$	0	0.56	0.17
$\delta y = 0.112$	0.224	0	−0.054

From the above table we observe that initial residuals are 50, 18, 19, out of which 50 is the largest so to reduce it, we give an increment $\delta x = 5$ (or $50/9 \approx 5.5$) and the resulting residuals are 5, 13, 29 out of which 29 is the largest so to reduce it give an increment $\delta z = 4$ and thus 1, 25, 1 are resulting residuals. Continue this process till all the residuals become almost zero. Now find the addition of increments separately. That is,

$$\Sigma \delta x = 5 + 1 + 0.13 = 6.13,$$
$$\Sigma \delta y = 5 - 0.8 + 0.112 = 4.312$$
and $$\Sigma \delta z = 4 + 0.23 = 4.23.$$

Hence, the solution is

$$x = 6.13, y = 4.31, \quad z = 4.23.$$

EXAMPLE 2.   *Solve the following equations by relaxation method :*

$$10x - 2y - 3z = 205$$
$$-2x + 10y - 2z = 154$$
$$-2x - y + 10z = 120$$

(VTU–2011S, Rohtak–2005)

SOLUTION.   The residuals are given by

$$R_x = 205 - 10x + 2y + 3z$$
$$R_y = 154 + 2x - 10y + 2z$$
$$R_z = 120 + 2x + y - 10z$$

The operations table is

	δR_x	δR_y	δR_z
$\delta R_x = 1$	–10	2	2
$\delta R_y = 1$	2	–10	1
$\delta R_z = 1$	3	2	–10

Now, the relaxation table is

	R_x	R_y	R_z
$x = y = z = 0$	205	154	120
$\delta x = 20$	5	194	160
$\delta y = 19$	43	4	179
$\delta z = 18$	97	40	–1
$\delta x = 10$	–3	60	19
$\delta y = 6$	9	0	25
$\delta z = 2$	15	4	5
$\delta x = 2$	–5	8	9
$\delta z = 1$	–2	10	–1
$\delta y = 1$	0	0	0

From above table, we observe that the initial residuals are 205, 154 and 120 out of these 205 is the largest so to reduce it give an increment $\delta x = 20$(which is obtained by $205/10 \approx 20$) and the resulting residual are 5, 194 and 160. Out of which 194 is the largest so to reduce it give and increment $\delta y = 19$ (again by $194/10 \approx 19$) and the resulting residual are 43, 4, 179. Proceeding in this way till all the residuals becomes almost zero. Thus

$$\Sigma \delta x = 20 + 10 + 2 = 32$$
$$\Sigma \delta y = 19 + 6 + 1 = 26$$
$$\Sigma \delta z = 18 + 2 + 1 = 21.$$

Hence, the solution is $x = 32, y = 26, z = 21$.

EXERCISE 4.1

1. Solve the equations by Gauss-elimination Method:

(i) $x_1 + x_2 + 2x_3 = 4$
$3x_1 + x_2 - 3x_3 = -4$
$2x_1 - 3x_2 - 5x_3 = -5$

(ii) $2x_1 + x_2 + 4x_3 = 12$
$8x_1 - 3x_2 + 2x_3 = 20$
$4x_1 + 11x_2 - x_3 = 33$

(iii) $x_1 + x_2 + x_3 = 10$
$2x_1 + x_2 + 2x_3 = 17$
$3x_1 + 2x_2 + x_3 = 17$

(iv) $2x + 3y - z = 5$
$4x + 4y - 3z = 3$
$2x - 3y + 2z = 2$

(v) $2x_1 + 4x_2 + x_3 = 2$
$3x_1 + 2x_2 - 2x_3 = -2$
$x_1 - x_2 + x_3 = 6$

(vi) $2x + 2y + z = 12$
$3x + 2y + 2z = 8$
$5x + 10y - 8z = 10$ (WBTU–2004)

(vii) $2x - y + 3z = 9$

$x + y + z = 6$

$x - y + z = 2$ (Bhopal–2009)

2. Solve the following equations by Jordan's Method:

(i) $2x - 3y + z = -1$ (Kerala–2003)

$x + 4y + 5z = 25$

$3x - 4y + z = 2$

(ii) $x_1 + 2x_2 + x_3 = 8$

$2x_1 + 3x_2 + 4x_3 = 20$

$4x_1 + 3x_2 + 2x_3 = 16$

(iii) $10x + y + z = 12$

$x + 10y + z = 12$

$x + y + 10z = 12$

(iv) $2x + y + z = 10$

$3x + 2y + 3z = 18$

$x + 4y + 9z = 16$ (VTU–2008)

(v) $x + y + z = 9$

$2x + y - z = 0$

$2x + 5y + 7z = 52$ (VTU–2009)

(vi) $x + 3y + 3z = 16$

$x + 4y + 3z = 18$

$x + 3y + 4z = 19$ (Anna–2005)

(vii) $2x_1 + x_2 + 5x_3 + x_4 = 5$

$x_1 + x_2 - 3x_3 + 4x_4 = -1$

$3x_1 + 6x_2 - 2x_3 + x_4 = 8$

$2x_1 + 2x_2 + 2x_3 - 3x_4 = 2$ (Madras–1997)

3. Solve by LU decomposition Method:

(i) $2x + y + 2z = 2$

$x + 5y + 3z = 4$

$x + y - z = 0$

(ii) $x + y + 3z = 10$

$3x + 2y + 4z = 20$

$3x + 5y - z = 30$

(iii) $2x - 3y + 10z = 3$

$-x + 4y + 2z = 20$

$5x + 2y + z = -12$

4. Solve by Crout's Method:

(i) $2x - 6y + 8z = 24$

$5x + 4y - 3z = 2$

$3x + y + 2z = 16$

(ii) $2x_1 + 4x_2 + x_3 = 5$

$4x_1 + 4x_2 + 3x_3 = 8$

$4x_1 + 8x_2 + x_3 = 9$

(iii) $10x + y + 2z = 13$

$3x + 10y + z = 14$

$2x + 3y + 10z = 15$

5. Solve the equations by Jacobi's method :

(i) $5x - y + z = 10$

$2x + 4y = 12$

$x + y + 5z = -1$

(ii) $5x + 2y + z = 12$

$x + 4y + 2z = 15$

$x + 2y + 5z = 20$

6. Solve the following equations by Gauss-Seidel Method:

(i) $1.2x + 2.1y + 4.2z = 9.9$

$5.3x + 6.1y + 4.7z = 21.6$

$9.2x + 8.3y + z = 15.2$

(ii) $10x_1 - 2x_2 - x_3 - x_4 = 3$

$-2x_1 + 10x_2 - x_3 - x_4 = 15$

$-x_1 - x_2 + 10x_3 - 2x_4 = 27$

$-x_1 - x_2 - 2x_3 + 10x_4 = -9$

    (Bhopal–2009, JNTU–2004)

7. Solve the following equations by relaxation method:

(i) $3x + 9y - 2z = 11$

$4x + 2y + 13z = 24$

$4x - 4y + 3z = -8$ (Bhopal–2002)

(ii) $-9x + 3y + 4z + 100 = 0$

$x - 7y + 3z + 80 = 0$

$2x + 3y - 5z + 60 = 0$

(iii) $10x - 2y - 2z = 6$

$-x + 10y - 2z = 7$

$-x - y + 10z = 8$ (PTU–2001)

8. Solve the following equation by Gauss-Seidal method

(i) $10x + y + z = 12$

$2x + 10y + z = 13$

$2x + 2y + 10z = 14$ (VTU–2007)

(ii) $7x_1 + 52x_2 + 13x_3 = 104$

$83x_1 + 11x_2 - 4x_3 = 95$

$3x_1 + 8x_2 + 29x_3 = 71$ (Hazaribagh–2009)

(iii) $3x_1 - 0.1x_2 - 0.2x_3 = 7.85$

$0.1x_1 + 7x_2 - 0.3x_3 = -19.3$

$0.3x_1 - 0.2x_2 + 10x_3 = 17.4$ (Mumbai–2004)

9. Solve by Jacobi's iteration method, the equations

$10x + y - z = 11.19,$

$x + 10y + z = 28.8,$

$-x + y + 10z = 35.61,$

correct to two decimal places.   (Anna–2007)

Answers

1. (i) $x_1 = 1, x_2 = -1, x_3 = 2$; (ii)$x_1 = 3, x_2 = 2, x_3 = 1$; (iii) $x_1 = 2, x_2 = 3, x_3 = 5$;
(iv) $x = 1, y = 2, z = 3$; (v) $x_1 = 1, x_2 = -1, x_3 = 3$;
(vi) $x = -12.75, y = 14.375, z = 8.75$ (vii) $x = 1, y = 2, z = 3$

2. (i) $x = 8.7, y = 5.7, z = -1.3$;(ii)$x_1 = 1, x_2 = 2, x_3 = 3$; (iii) $x = 1, y = 5, z = 1$;
(iv) $x = 7, y = -9, z = 5$ (v) $x = 1, y = 3, z = 5$
(vi) $x = 1, y = 2, z = 3$ (vii) $x_1 = 2, x_2 = 1/5, x_3 = 0, x_4 = 4/5$

3. (i) $x = 1/5, y = 2/5, z = 3/5$(ii) $x = 2, y = 5, z = 1$ (iii) $x = -4, y = 3, z = 2$

4. (i) $x = 1, y = 3, z = 5$ (ii) $x_1 = 1/2, x_2 = 3/4, x_3 = 1$ (iii) $x = y = z = 1$;

5. (i) $x = 2.56, y = 1.72, z = -1.06$ (ii) $x = 1.08, y = 1.95, z = 3.16$

6. (i) $x = -13.22, y = 16.76, z = -2.30$ (ii) $x_1 = 0.99, x_2 = 1.99, x_3 = 2.99, x_4 = -0.001$

7. (i) $x = -13.22, y = 16.76, z = -2.30$; (ii) $x = 52.5, y = 44.5, z = 59.7$ (iii) $x = y = z = 1$

8. (i) $x = y = z = 1$(ii) $x_1 = 1.058, x_2 = 1.367, x_3 = 1.962$ (iii) $x_1 = 3, x_2 = -2.5, x_3 = 7$

9. $x = 1.23, y = 2.34, z = 3.45$

MISCELLANEOUS EXERCISE

1. Solve
$10x - 7y + 3z + 5u = 6$,
$-6x + 8y - z - 4u = 5$,
$3x + y + 4z + 11u = 2$,
$5x - 9y - 2z + 4u = 7$
by Gauss elimination method. (SVTU–2007)

2. Solve the equation by Gauss elimination method.
$2x + 2y + z = 12$;
$3x + 2y + 2z = 8; 5x + 10y - 8z = 10$.
(WBTU–2004)

3. Solve the following equations by Gauss-Jordan method.
(i) $2x + 5y + 7z = 52$;
$2x + y - z = 0, x + y + z = 9$ (VTU–2010)
(ii) $2x + y + z = 10$;
$3x + 2y + 3z = 18, x + 4y + 9z = 16$
(VTU–2008)

4. Solve by Jacobi's iteration method, the equations,
$10x + y - z = 11.19, x + 10y + z = 28.08$,
$-x + y + 10z = 35.61$, correct to two decimal places. (Anna–2007)

5. Apply Gauss-Seidel iteration method to solve the equations.
$20x + y - 2z = 17$;
$3x + 20y - z = -18, 2x - 3y + 20z = 25$
(VTU–2011, Rohtak–2005, Madras–2003, Bhopal–2009)

6. Solve the following equations by Gauss-Seidel method.
$28x + 4y - z = 32$,
$x + 3y + 10z = 24, 2x + 17y + 4z = 35$
(Mumbai–2009)

Answers

1. $u = 1, z = -7, y = 4$ and $x = 5$ **2.** $x = -51/4, y = 115/8, z = 35/4$
3. (i) $x = 1, y = 3, z = 5$ (ii) $x = 7, y = -9, z = 5$ **4.** $x = 1.23, y = 2.34, z = 3.45$
5. $x = 1, y = -1, z = 1$ **6.** $x = 0.998, y = 1.723, z = 2.024$

Objective Evaluations

FILL IN THE BLANKS

1. The system of equations $AX = B$ are called homogeneous if $B =$ _____.

2. If number of variables are greater than number of linear equations then number of solutions are _____.

3. In the Gauss elimination method if any one of the pivots element is zero, then the method will _____.

4. In the Gauss elimination method, one of the pivots elements must be _____.

5. In LU decomposition method, if $A = LU$, then L is a _____.

6. If $A = LU$, then U is a _____.

7. LU decomposition method is _____ to Gauss elimination method.

8. In LU decomposition method, all the principal minors of A are _____.

9. In Jordan's method, the system of equations are reduced to _____.

10. In Jacobi's Iterative method, the diagonal coefficient a_{ii} of the system of linear equations must be _____.

11. Jacobi's method converges if $|B|$ is _____ where $x^{(n+1)} = BX^{(n)} + C$.

12. If $X = B(X) + C$ holds in Gauss-Seidal method, then this method converges if the sum of the absolute values of the coefficients $\dfrac{a_{ij}}{a_{ii}}$ is atmost equal to or in at least one equation less than _____.

13. In Relaxation method the process is completed if all the residuals reduced to _____.

TRUE/FALSE

Write 'T' for True and 'F' False statement.

1. The system of equations $AX = 0$ are called homogeneous. **(T/F)**

2. In Gauss-elimination method, one of the pivot element must be non-zero. **(T/F)**

3. If a square matrix A can not be expressed at the product of lower and upper triangular matrices, then LU method is also applicable. **(T/F)**

4. In LU decomposition method, L is an upper triangular matrix while U is lower triangular matrix. **(T/F)**

5. In LU decomposition method all the principal minors of A must be non-zero. **(T/F)**

6. In Jordan's method, in $AX = B$, A reduces to column matrix. **(T/F)**

7. The diagonal coefficient a_{ii} in $AX = B$ must be non-zero in Jacobi's method. **(T/F)**

8. In Jacobi's Iterative method, we obtain $X^{(n+1)} = BX^{(n)} + C$ in which $|B| > 1$, then method converges. **(T/F)**

9. In Relaxation method, the residuals R_x, R_y and R_z are given by:
$$R_x = b_1 - a_{11}x - a_{12}y - a_{13}z,$$
$$R_y = b_2 - a_{21}x - a_{22}y - a_{23}z$$
and $R_z = b_3 - a_{31}x - a_{32}y - a_{33}z.$ **(T/F)**

10. In question no-9 above if $\delta x = 1$ (increment in x), then $\delta R_x = -a_{11}$, $\delta R_y = -a_{21}$, $\delta R_z = -a_{31}$. **(T/F)**

MULTIPLE CHOICE QUESTIONS

Choose the most appropriate one.

1. The system of equations $AX = B$ are non-homogeneous if B equals:
   (a) 0
   (b) x
   (c) y
   (d) none of these

2. If $AX = B$, the number of variables are greater than number of equations then the solutions are :
   (a) 0
   (b) 1
   (c) 2
   (d) none of these

3. In Jacobi's method $X^{(n+1)} = BX^{(n)} + C$ converges if $|B|$ is less than :

(a) 1   (b) 0

(c) 2   (d) 3

4. In LU decomposition method, the diagonal element in L are all :

(a) 0   (b) 1

(b) –1   (d) 2

5. In Relaxation method, $R_x = b_1 - a_{11}x - a_{12}y - a_{13}z$ and $\delta x = 1$, then δR_x :

(a) a_{11} (b) $-a_{11}$

(b) a_{12} (d) a_{13}

6. In the end of the relaxation method, all the residuals reduce to :

(a) 1   (b) 2

(c) 0   (d) –1

7. If $L^{-1} = X$, then LX equals to :

(a) I (b) L

(c) X (d) X^{-1}

8. In the Relaxation method if $a_{11}x + a_{12}y + a_{13}z = b_1$, $a_{21}x + a_{22}y + a_{23}z = b_2$ and $a_{31}x + a_{32}y + a_{33}z = b_3$ and $|a_{11}| \geq |a_{12}| + |a_{13}|$, $|a_{22}| \geq |a_{21}| + |a_{23}|$, then $|a_{33}|$ is greater than or equal to :

(a) $|a_{31}| + |a_{32}|$ (b) $|a_{31}|$

(c) $|a_{32}|$ (d) $|a_{31}| - |a_{32}|$

9. In Gauss elimination method, $AX = B$, then A reduces to :

(a) column matrix

(b) row matrix

(c) diagonal matrix

(d) upper triangular matrix

10. In Crout's method the elements of the first column of derived matrix are given if $A = [a_{ij}]_{3 \times 3}$:

(a) a_{11}, a_{21}, a_{31} (b) a_{11}, a_{12}, a_{13}

(c) a_{11}, a_{22}, a_{33} (d) None of these

11. The matrix $(A) = \begin{bmatrix} 2 & 1 \\ 4 & -1 \end{bmatrix}$ is decomposed into a product of a lower triangular matrix (L) and an upper triangular matrix (U). The properly decompose (L) and (U) matrices respectively are :   (GATE(EE)–2011)

(a) $\begin{bmatrix} 2 & 0 \\ 4 & -3 \end{bmatrix}$ and $\begin{bmatrix} 1 & 0.5 \\ 0 & 1 \end{bmatrix}$

(b) $\begin{bmatrix} 1 & 0 \\ 4 & 1 \end{bmatrix}$ and $\begin{bmatrix} 2 & 1 \\ 0 & -1 \end{bmatrix}$

(c) $\begin{bmatrix} 2 & 0 \\ 4 & -1 \end{bmatrix}$ and $\begin{bmatrix} 1 & 1 \\ 0 & 1 \end{bmatrix}$

(d) none of these

12. In the solution of the following set of linear equations by Gauss elimination using partial pivoting $5x + y + 2z = 34$; $4y - 3z = 12$ and $10x - 2y + z = -4$; the pivots for elimination of x and y are :   (GATE(CE)–2009)

(a) 10 and 2   (b) 10 and 4

(c) 5 and 4   (d) none of these

Answers

FILL IN THE BLANKS

1. zero   2. infinite   3. fail   4. non-zero   5. lower triangular matrix

6. upper triangular matrix   7. superior   8. non-zero   9. diagonal matrix

10. non-zero   11. less than one   12. one   13. almost zero

TRUE/ FALSE

1. T   2. T   3. F   4. F   5. T   6. F   7. T   8. F   9. T

10. T

MULTIPLE CHOICE QUESTIONS

1. (d)   2. (d)   3. (a)   4. (b)   5. (b)   6. (c)   7. (a)   8. (a)   9. (d)

10. (a)   11. (d)   12. (b)

COMPUTATIONAL TECHNIQUE LAB

1. GAUSS ELIMINATION METHOD

FLOW CHART FOR GAUSS-ELIMINATION METHOD

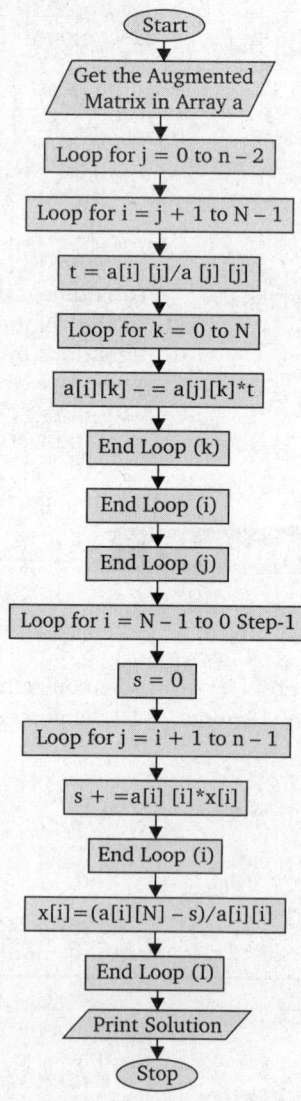

Start

Get the Augmented Matrix in Array a

Loop for j = 0 to n – 2

Loop for i = j + 1 to N – 1

t = a[i] [j]/a [j] [j]

Loop for k = 0 to N

a[i][k] – = a[j][k]*t

End Loop (k)

End Loop (i)

End Loop (j)

Loop for i = N – 1 to 0 Step-1

s = 0

Loop for j = i + 1 to n – 1

s + =a[i] [i]*x[i]

End Loop (i)

x[i] = (a[i] [N] – s)/a[i] [i]

End Loop (I)

Print Solution

Stop

PROGRAM

```
/* Gauss elimination method */
        # include <studio .h>
        # define N 4
        main ( )
        {
                float a [N] [N + 1], x  [N], t, s;
                int i, j, k;
                printf ("Enter the elements of the"
                        augmented matrix rowwise/n");
                for (i = 0; i < N; i++)
                    for (j = 0; j < N + 1; J++)
                    scanf (" %", &a [i] [j]);
                for (j = 0 ; j < N - 1; j++)
                    for i = j + 1; i < N; i++)
                    {
                            t = a [i] [j] / a [j] [j] ;
                    for (k = 0 ; k < n + 1 ; k++)
                            a [i] [k] - = a [j] [k] * t;
                        }
                printf ("The upper triangular matrix"
                        is :-\n") ;
                for ( i = 0; i < N; j++)
        {

                for (j = 0 ; j < N; i++)
                        printf (" %8 . 4f ", a [i] [j] ;
                printf (" \ n ") ;
        }
        for (i = N - 1 ; i > = 0; i - -)
        {

                s = 0;
                for ( j = i + 1; j < N; j++)
                    s + = a [i] [j] *x [j];
                x [i] = (a [i] (N) - S) / a [i] [i] ;
        }
        printf ( "The solution is : - \n") ;
                for (i = 0; i < N; i++)
                        printf ("x [% 3d] = % 7.4f\n",  i + 1,
        x [i]);
                }
```

2. GAUSS-JORDAN METHOD

FLOW CHART

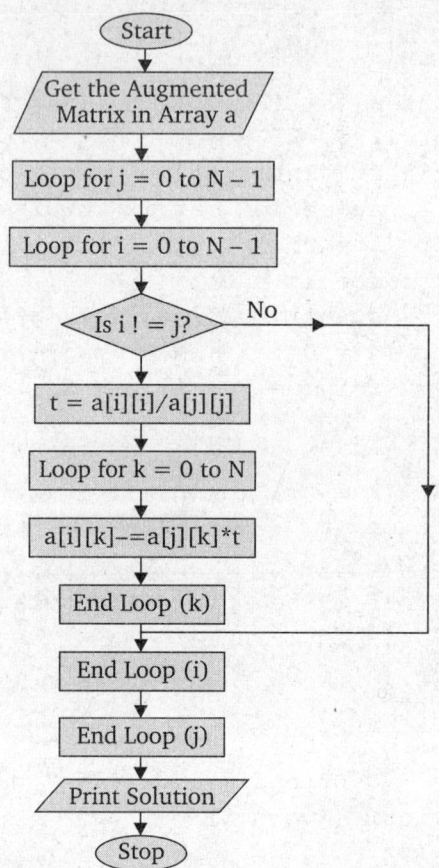

PROGRAM

```
/ * Gauss jordan method */

            # include <studio.h>
            #define N 3
            main ( )
            {
                float a [N]  [N + 1], t ;
                int i, j, k;
                printf (Enter the elements of the"
                        "augmented matrix rowwise\n") ;
                for (i = 0; i < N; i++)
                        for (j = 0 ; j < N + 1 ; j ++)
```

```
                    scanf ( " % f " , &a [i] [j]) ;
        for (j = 0; j < N ; j++)
            for (i = 0; i < N; i++)
                if (i ! = j)
                {
                    t = a [i] [j]/ a [j] [j];
                    for (k = 0; k < N + 1; k++)
                    a [i] [k] - = a [j] [k] * t ;
                }
        printf ("The diagonal matrix is :-\n") ;
        for (i = 0; i < N; i++)
        {
            for (j = 0; j < N + 1 ; j++)
                    print f (% 9.4f ", a [i] [j] ;
            printf ( "\n") ;
        }
        printf ( " The solution is : - \n") ;
        for (i = 0; i < N; i++)
            printf ("x [% 3d] = % 7.4 f \ n" , i + 1, a [i]
                                [N] a [i] [i] ;

}
```

3. CROUT'S TRI-ANGULARISATION METHOD

(1) FLOW CHART

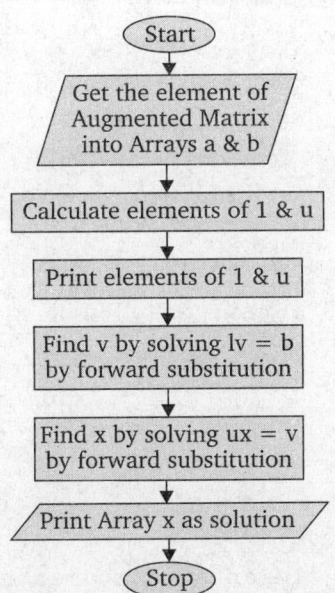

```
/* Crout triangularization method */
        # include <studio .h>
        # define N 4
        typedef float matrix [N] [N] ;
        matrix 1, u, a ;
        float b [N], x [N], v [N] ;
        void urow (int i)
        {
            float s ;
            int j, k ;
            for (j = 1; j < N; j++)
            {
            s = 0;
                for (k = 0 ; k < N - 1 ; k++)
            s + = u [k] [j] * 1 [i] [k] ;
                u [i] [j] = a [i] [j] - s ;
            }
}
void lcol (int j)
{
    float s ;
    int i, k ;
    for (i = j + 1 ; i < N; i++)
{
            s = 0 ;
    for (k = 0; k < = j - 1 ; k++)
            s + = u [k] [j] * 1 [i] [k];
    1 [i] [j] = (a [i] [j] - s) / u [j] [j] ;
      }
            }
void printmat (matrix x)
{
    int i, j ;
    for (i = 0 ; i < N ; i ++)
{
    for (j = 0; j < N; j++)
    printf ( " % 8 . 4f", x [i] [j] ;
    printf ("\n") ;
}
}
main ( )
{
int i, j, m;
float s;
printf ("Enter the elements of augmented"
            "matrix rowwise \n") ;
```

```
        for   (i = 0;  i < N;  i++)
        {
              for (j = 0;  j < N;  j++)
                  scanf (" % f " , & a [i] [j]) ;
                  scanf ( "% f" , & b [i]) ;
        }
1 and u */
for (i = 0;  i < N;            i++)
        1 [i] [i] = 1 . 0 ;
for (m = 0 ; m < N; m++)
{
        uprow (m);
        if (m < N - 1) lcol (m) ;
}
printf ("\t\tU\n") ; printmat (u);
printf ("\t\tL\n") ; printmat (l);
for (i = 0; i < N; i++)
{
        s = 0;
        for (j = 0; j < = i - 1; j++)
              s + = 1 [i] [j] * v [j];
        v [i] = b [i] - s ;
}
for (i = N - 1 ; i > = 0 ; i - )
{
        s = 0 ;
        for (j = i + 1 ; j < N; j++)
              s + = u [i] [j] * x [j] ;
        x [i] = (v [i] - s/u [i] [i] ;
}
printf , ( "The solution is : - \n") ;
for (i = 0 ; i < N ; i++)
        printf (" X [% 3d] = % 6 . 4f/n" , i + 1, x [i]) ;
```

4. /* LU DECOMPOSITION METHOD */

```
# include <conio .h>
# include <studio. h>
# include <math .h.>
main ( )
{
        float a [15] [15] al [15] [15], b [15], X [15], au [15] [15], z [15];
              float sum, t, big, ab, p ;
              int n, m, 1i, 1j, k, j, i, 1k, 13, m2, jj, kpl, kk, 1 ;
              clrscr (   ) ;
              printf ("enter the value of n/n") ;
              scanf (" % d", & n) ;
              printf ("enter the matrix row wise /n") ;
              for (i = 1 ; j < = n ; i++)
              for (j = 1 ; j < = n; j++)
```

```
            scanf ("% f " , & a [i] [j] ) ;
            printf ("enter the matrix B\n" ) ;
            for (j = 1 ; j < = n; j++)
            scanf ( " % f" , & b [j] ) ;
for (i = 1 ; i < = n; i++)
for (j = 1 ; j < = n ; j++)
{
        a1 [i] [j] = 0 ;
        au [i] [j] = 0 ;
}
for (i = 1 ; i < = n ; i++)
{
          au [i] [i] = 1 ;
          a1 [i] [1] = a [i] [1] ;
        au [1] [i] = a [1] [i] / a1 [1] [1];
}
for (j = 2 ; j < = n; j++)
{
for (i = j ; i < = n ; i++)
{
        sum = 0 ;
            for (k = 1 ; k ≤ j - 1 ; k++)
            sum=sum+a1 [i] [k] * au [k] [j] ;
            a1 [i] [j] = a [i] [j] - sum ;
}
if ( j ! =n)
(
        for (jj = j + 1; jj < = n; jj++)
        {
            sum=0;
            for (kk = 1; kk < j - 1 ; jj++)
            sum=sum + a1 [j] [kk] * au [kk] [jj];
            au [j] [jj] = (a [j][jj] - sum) / a1 [j][j];
            }
                }
}
z [1] = b [1] / a1 [1] [1] ;
for (i = 2; i < = n; i++)
{
        sum=0;
        for (k = 1; k < i - 1; k++)
        sum=sum+a1 [i] [k] * z [k];
        z [i] = (b [i] - sum)/a1 [i] [i] ;
}
```

```
        x [n] = z [n] ;
        for (i = 2 ; i < = n ; i++)
        {
            1 = n - i + 1;
                sum=0;
            for (k = 1 + 1; k < = n; k++)
                sum= sum+au [1] [k] * x [k];
            x [1] = z [1] - sum;
    }

    printf ("\n") ;
    printf ( "lower triangular matrix \n") ;
    for (i = 1; i < = n ; i++)
    {
        for (j = 1 ; j < = n; j++)
        printf ("% 5. 2f ", a1 [i] [j]) ;
        printf ("\n") ;
    }
    printf ( "upper triangular matrix \n") ;
    for ( i = 1 ; i < = n ; i++)
    {
        for (j = 1; j < = n; j++)
        printf ( "% 5. 3f " , au [i] [j]) ;
        printf ( ''\n'' ) ;
    }
    printf ( "solution vector\n") ;
    for (i = 1 ; i < = n ; i++)
    printf ( "% 5. 2f " , x [i] ) ;
    printf ( "\n") ;
    getch (  ) ;
    }
```

5. SOLUTION OF SYSTEM OF EQUATIONS USING GAUSS SEIDAL METHOD

```
    */
        # include <conio .h>
        # include <studio .h>
        # include <math .h>
        main (  )


        {
            float a [15] [15] , b [15] , x [15] , oldx [15], eps,
c , big, sum;
            int n, niter1, i, j, k, l, ii ;
            clrscr (  ),
            printf ( "enter the value of N, NITHER, ESP\n") ;
            scanf ( "% d% d% f" , &n, &niteral, &eps) ;
```

```
            printf ("enter the matrix A\n") ;
            for (i = 1 ; i < = n; i++)
            for (j = 1; j < = n; j++)
                scanf "%f", a[i][j];
                scanf ("enter the array B\n") ;
                for (i = 1; i < = n; i++)
            scanf ( " % f, & b [i] ;
                printf ("enter the array 01dx1n"
                for (i = 1; i < = n; i ++);
                    {
                    for (j = 1; j < = n, j + 1)
                    {
                printf ( " % f " , a [i] , [j] ) ;
                    }
                print ( " % / n") ;
    printf (" the array B/n") ;
    for (i = 1, i < = n; i ++)
        {
            x (i) = 0/dx [i] ;
            printf ( "% f " , b [i]) ;
        }
                printf ("\n") ;
                for (i = 1 ; i < = n; i ++) ;
                        {
            for (i = 1 ; i < = n iter ; i ++)
            {
                sum = 0
            for (j = 1 : j < = n ; j + j)
            if ([i - j] ; = 0)
            sum = sum + a [i] [j] *x [j] ;
            x [i] = b [i] - sum/ a [i] [i] ;
        }
    printf (n iter = % d, i) ;
    for (i = 1 ; i < = n; i ++)
    printf (" % 12.6 f" , x [i]) ;
    printf ("\n") ;
    big = abb (x [i] - old x [i]) ;
    for (k = 2; k < = n, k++)
    {
            c = abb (xck) - old x [k] ;
            if (big < = c)
            big = c;
    }
    For (i = 1; i < = n ; i++)
    old x [i] = x [i] ;
    }
}
```

❋❋❋❋❋❋

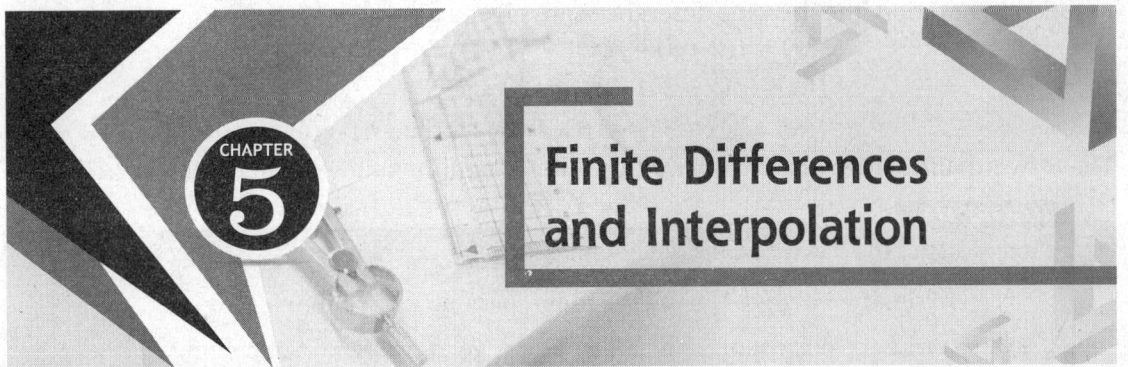

CHAPTER 5

Finite Differences and Interpolation

5.1 INTRODUCTION

Let $y = f(x)$ be a function. Then a rational integral function of the nth degree in x, i.e., $f(x) = a_0 + a_1 x + a_2 x^2 + ... + a_n x^n$ where n is a positive integer and $a_0, a_1, ..., a_n$ are constant with $a_n \neq 0$, is called polynomial of degree n. It is denoted by $P_n(x)$.

▼ REMARKS

‣ A polynomial $P_n(x)$ with real coefficients which cannot be written as the product of two polynomials of degree less than n is called a prime polynomial.
‣ A polynomial $P_n(x)$ is called interpolating polynomial, if the values of $P_n(x)$ and its certain order derivatives coincide with those of f_{xy} and its some certain order derivatives at one or more tabular points.

5.2 DIFFERENCE SCHEMES

Let $y = f(x)$ be a function (discrete), which takes the values $f(x_0), f(x_0 + h), f(x_0 + 2h) ... f(x_0 + nh)$, for the equidistant values $x_0, x_0 + h, x_0 + 2h ... x_0 + nh$ of x respectively. Here, the value of the independent variable x is usually called the argument and the corresponding functional value is called entry.

Generally, there are following three types of differences :
(1) Forward differences   (2) Backward differences   (3) Central differences.

5.2.1 FORWARD DIFFERENCES

The difference $(f(x_0 + h) - f(x_0)), (f(x + 2h) - f(x_0 + h)), ... (f(x_0 + nh) - f(x_0 + (n-1)h))$ are called first forward differences and are denoted by $\Delta f(x_0), \Delta f(x_0 + h) ... \Delta f(x_0 + (n-1)h)$. Here, Δ is known as forward difference operator.

Therefore, $$\Delta f(x_0) = f(x_0 + h) - f(x_0)$$
or $$\Delta y_0 = y_1 - y_0$$
The first forward differences are given by the following formula,
$$\Delta y_k = y_{k+1} - y_k$$
Similarly, the second forward differences are
$$\Delta^2 y_k = \Delta y_{k+1} - y_k$$
$$= y_{k+2} - 2y_{k+1} - y_k$$
In general $$\Delta^r y_k = \Delta^{r-1} y_{k+1} - \Delta^{r-1} y_k$$

In function notation, the forward differences are given below:

$$\Delta f(x) = f(x+h) - f(x)$$
$$\Delta^2 f(x) = f(x+2h) - 2f(x+h) - f(x)$$
$$\Delta^3 f(x) = f(x+3h) - 3f(x+2h) + 3f(x+h) - f(x) \dots \text{ and so on.}$$

The forward differences are usually arranged in form in the following manner.

ALGORITHM

(To construct the forward difference table)	
Step 1.	Write the value of the independent variable x in column (1).
Step 2.	Write the corresponding value of y for given value of x in column (2).
Step 3.	Subtracting each value of $f(x)$ in column (ii) from the succeeding value of $f(x)$ and write this value in column (iii).
Step 4.	The operation of step-3 is applied on the figures of column (iii) to get column (iv) and so on till all the figures in a column become constant.

Forward Difference Table

Argument x	Entry y	First Diff. Δ	Second Diff. Δ^2	Third Diff. Δ^3	Fourth Diff. Δ^4	Fifth Diff. Δ^5
x_0	y_0					
		Δy_0				
$x_1 = x_0 + h$	y_1		$\Delta^2 y_0$			
		Δy_1		$\Delta^3 y_0$		
$x_2 = x_0 + 2h$	y_2		$\Delta^2 y_1$		$\Delta^4 y_0$	
		Δy_2		$\Delta^3 y_1$		$\Delta^5 y_0$
$x_3 = x_0 + 3h$	y_3		$\Delta^2 y_2$		$\Delta^4 y_1$	
		Δy_3		$\Delta^3 y_2$		
$x_4 = x_0 + 4h$	y_4		$\Delta^2 y_3$			
		Δy_4				
$x_5 = x_0 + 5h$	y_5					

In the above table y_0 is called leading term and the difference $\Delta y_0, \Delta^2 y_0, \Delta^3 y_0, \dots,$ are called the leading differences.

▶ REMARK

▸ The above difference table is known as forward difference table or diagonal difference table.

5.2.2 PROPERTIES OF FORWARD DIFFERENCE OPERATOR Δ

The forward operator Δ satisfies the following properties:

(i) $\Delta[f(x) \pm g(x)] = \Delta f(x) \pm \Delta g(x)$

(ii) $\Delta[a.f(x)] = a.\Delta f(x), a \in \mathbf{R}$

(iii) $\Delta^m \Delta^n f(x) = \Delta^{m+n} f(x) = \Delta^{n+m} f(x) = \Delta^n \Delta^m f(x), m, n \in \mathbf{Z}^+$

(iv) $\Delta[f(x).g(x)] = [\Delta f(x)]g(x) + f(x)[\Delta g(x)]$

5.2.3 BACKWARD DIFFERENCES

The differences $y_1 - y_0, y_2 - y_1, ..., y_n - y_{n-1}$ denoted by $\nabla y_1, \nabla y_2, ..., \nabla y_n$ respectively are called the first backward differences, when ∇ is called backward difference operator.

Therefore, $\qquad \nabla y_1 = y_1 - y_0$

$$\nabla y_2 = y_2 - y_1$$
$$\cdots\cdots\cdots\cdots\cdots$$
$$\nabla y_n = y_n - y_{n-1}$$

Now, the second backward differences are defined as follows :

$$\begin{aligned}
\nabla^2 y_2 &= \nabla(\nabla y_2) = \nabla(y_2 - y_1) \\
&= \nabla y_2 - \nabla y_1 \\
&= (y_2 - y_1) - (y_1 - y_0) \\
&= y_2 - 2y_1 + y_0 \\
\nabla^3 y_3 &= \nabla y_3 - \nabla y_2 \\
&= y_3 - 2y_2 + y_1
\end{aligned}$$

and so on.

In general $\qquad \nabla^n y_k = \nabla^{n-1} y_k - \nabla^{n-1} y_{k-1}$

In functional notation, the backward differences are given below :

$$\nabla f(x) = f(x) - f(x - h)$$
$$\nabla f(x + h) = f(x + h) - f(x)$$
$$\nabla^2 f(x + 2h) = f(x + 2h) - 2f(x + h) + f(x) \text{; where } h \text{ is the interval of differences.}$$

and so on.

The backward differences are usually arranged in a tabular from in the following manner.

Backward Difference Table

Argument x	Entry y	First Diff. (∇)	Second Diff. (∇^2)	Third Diff. (∇^3)	Fourth Diff. (∇^4)	Fifth Diff. (∇^5)
x_0	y_0					
		∇y_1				
$x_1 = x_0 + h$	y_1		$\nabla^2 y_2$			
		∇y_2		$\nabla^3 y_3$		
$x_2 = x_0 + 2h$	y_2		$\nabla^2 y_3$		$\nabla^4 y_4$	
		∇y_3		$\nabla^3 y_4$		$\nabla^5 y_5$
$x_3 = x_0 + 3h$	y_3		$\nabla^2 y_4$		$\nabla^4 y_5$	
		∇y_4		$\nabla^3 y_5$		
$x_4 = x_0 + 4h$	y_4		$\nabla^2 y_5$			
		∇y_5				
$x_5 = x_0 + 5h$	y_5					

5.2.4 CENTRAL DIFFERENCES

If $\qquad y_1 - y_0 = \delta y_{1/2}, y_2 - y_1 = \delta y_{3/2}, y_n - y_{n-1} = \delta y_{n-2}$

Then these differences called central difference and δ is called central difference operator.

Similarly, we can define higher order central differences as

$$\delta y_{3/2} - \delta y_{1/2} = \delta^2 y_1, \delta y_{5/2} - \delta y_{3/2} = \delta^2 y_2$$

and $\delta^2 y_2 - \delta^2 y_1 = \delta^3 y_{3/2}$ and so on.

The central difference table is given below.

Central Difference Table

Argument x	Entry y	First Diff.	Second Diff.	Third Diff.	Fourth Diff.	Fifth Diff.
x_0	y_0					
		$\delta y_{1/2}$				
x_1	y_1		$\delta^2 y_1$			
		$\delta y_{3/2}$		$\delta^3 y_{3/2}$		
x_2	y_2		$\delta^2 y_2$		$\delta^4 y_2$	
		$\delta y_{5/2}$		$\delta^3 y_{5/2}$		$\delta^5 y_{5/2}$
x_3	y_3		$\delta^2 y_3$		$\delta^4 y_3$	
		$\delta y_{7/2}$		$\delta^3 y_{7/2}$		
x_4	y_4		$\delta^2 y_4$			
		$\delta y_{9/2}$				
x_5	y_5					

5.2.5 SHIFT OPERATOR

Define the operator which increases the argument x by h such that

$$Ef(x) = f(x+h)$$

and $$E^2 f(x) = f(x+2h)$$

$$E^3 f(x) = f(x+3h) \text{ ... etc.}$$

This operator E is called shift operator. The inverse operator E^{-1} is defined as

$$E^{-1} f(x) = f(x-h), E^{-2} f(x) = f(x-2h) \text{ ... etc.}$$

In general, $E^n f(x) = f(x+nh)$ where n any real number.

Then these differences are called central difference and δ is called central difference operator.

5.3 RELATION BETWEEN THE OPERATORS

We have the following identities :

  (i) $\Delta = E - 1$ **(ii)** $\nabla = 1 - E^{-1}$

 (iii) $\delta = E^{1/2} - E^{-1/2}$ **(iv)** $\Delta = E\nabla = \nabla E = \delta E^{-1/2}$

  (v) $E = e^{hD}$

Proof :

  (i) Since $\Delta y_x = y_{x+h} - y_x$ $= Ey_x - y_x$;   for all x

                                $= (E-1)y_x$;   for all x

    $\therefore$ $\Delta = E - 1$ or $E = 1 + \Delta$

(ii) Since $\qquad \nabla y_x = y_x - y_{x-h} \qquad\qquad\qquad = y_x - E^{-1}y_x \qquad\qquad$ for all x

$$\nabla y_x = (1 - E^{-1})y_x \qquad\qquad \nabla = 1 - E^{-1} \qquad\qquad \text{for all } x$$

(iii) Since $\qquad \delta y_x = y_{x+\frac{h}{2}} - y_{x-\frac{1}{2}h} \qquad\qquad = E^{1/2}y_x - E^{-1/2}y_x \qquad$ for all x

$$= (E^{1/2} - E^{-1/2})y_x \qquad\qquad\qquad\qquad\qquad \text{for all } x$$

$$\therefore \qquad\qquad \delta = E^{1/2} - E^{-1/2}$$

(iv) Since $\qquad E\nabla y_x = E(y_x - y_{x-h}) \quad = Ey_x - Ey_{x-h} \qquad\qquad$ for all x

$$= y_{x+h} - y_x \qquad\qquad\qquad\qquad\qquad\qquad \text{for all } x$$

$$= \Delta y_x \qquad\qquad\qquad\qquad\qquad\qquad\qquad\quad \text{...(1)}$$

$$\therefore \qquad\qquad E\nabla = \Delta$$

and $\qquad\qquad \nabla E y_x = \Delta y_{x+h} \qquad\qquad\qquad\qquad\qquad\qquad$ for all x

$$= y_{x+h} - y_x \qquad\qquad\qquad\qquad\qquad\qquad\quad \text{for all } x$$

$$\therefore \qquad\qquad \nabla E = \Delta \qquad\qquad\qquad\qquad\qquad\qquad\qquad\qquad \text{...(2)}$$

from (1) and (2), $\qquad \Delta = E\nabla = \nabla E$

(v) Since $\qquad Ef(x) = f(x + h)$

$$= f(x) + h'f(x) + \frac{h^2}{2!}f''(x) + ... \qquad\qquad \text{By Taylor's Theorem}$$

$$= f(x) + hD(x) + \frac{h^2}{2!}D^2 f(x) + ... = \left(1 + hD + \frac{h^2}{2!}D^2 + ...\right)f(x)$$

$$Ef(x) = e^{hD}f(x)$$

$$\therefore \qquad\qquad E = e^{hD} \qquad\qquad\qquad\qquad\qquad\qquad\qquad\qquad \text{for all } x.$$

5.4 FACTORIAL NOTATION

A product of the form $x(x-1)(x-2)...(x-n+1)$ is denoted by $x^{(n)}$ and is called a factorial. In particular,

$$x^{(1)} = x, x^{(2)} = x(x-1), x^{(3)} = x(x-1)(x-2) ... \text{ etc.}$$

If the interval of differencing is h, then,

$$x^{(n)} = x(x-h)(x-2h)...[x-(n-1)h]$$

5.4.1 DIFFERENCE OF X $^{(n)}$

Consider, $\quad \Delta x^{(n)} = (x+h)^n - x^n$

$$= (x+h)x(x-h)...[x-(n-2)h] - x(x-h)...[x-(n-1)h]$$

$$= x(x-h)...[x-(n-1)h]\{x+h-x-(n+1)h\} = x^{(r-1)}.n.h = nhx^{(n-1)}$$

Similarly, $\Delta^2 x^{(n)} = \Delta[\Delta x^{(n)}] = \Delta[nhx^{(n-1)}] = nh(n-1)nx^{(n-2)} = n(n-1)h^2 x^{(n-2)}$

$$\Delta^3 x^{(n)} = n(n-1)(n-2)h^3 x^{(n-3)}$$

In general,

$$\Delta^n x^{(n)} = n(n-1)(n-2)...3.2.1h^n x^{(n-n)} = n!h^n$$

▶ **Remarks**

▸ The result of difference $x^{(n)}$ is analogous to that of differentiating x^n.

▸ If $h = 1$, then $\Delta^n x^{(n)} = n!$.

5.4.2 Reciprocal Factorial

The reciprocal factorial function $x^{(-n)}$ is defined by

$$x^{(-n)} = \frac{1}{(x+h)(x+2h)...(x+nh)} \qquad \text{where } n \in \mathbf{Z}^+$$

▶ **Remark**

▸ Clearly $\Delta x^{(-n)} = (-n)h x^{[-(n+1)]}, \Delta^2 x^{(-n)} = (-1)^2 n(n+1)h^2 x^{[-(n+2)]}$

In general $\Delta^k x^{(-n)} = (-1)^k n(n+1)h^2 x^{[-(n+2)]}$

5.5 TO EXPRESS A GIVEN POLYNOMIAL INTO FACTORIAL NOTATION

(*Unity being the differencing interval*)

Method I.

This can be done by using the finite differences of the function. Let $f(x)$ be the polynomial of degree n which is to be expressed in factorial notation.

Let $\qquad f(x) = A + B x^{(1)} + C x^{(2)} + ... + K x^{(n)}$ ...(1)

where $A, B, C, ..., K$ are some unknown constants, $K \neq 0$.

Then $\Delta f(x) = \Delta[A + B x^{(1)} + C x^{(2)} + ... + K x^{(n)}] = B.1 + C.2 x^{(2)} + ... + K.n x^{(n-1)}$

$\Delta^2 f(x) = \Delta[B + C.2 x^{(1)} + ... + K n.x^{(n-1)}] = C.2.1 + D.3.2.x^{(1)} + ... + K n.(n-1)x^{(n-2)}$

$$... \quad ... \quad ... \quad ... \quad ... \quad ... \quad ... \quad ... \quad ...$$

$\Delta^n f(x) = K.n.(n-1)...2.1.x^{(0)} = K.n.!$

Substitution of $x = 0$ in all above equations yields

$$f(0) = A, \Delta f(0) = \frac{B}{1!}$$

$$\Delta^2 f(0) = C.2! \implies C = \frac{\Delta^2 f(0)}{2!} \ ; \ \Delta^3 f(0) = D.3! \implies D = \frac{\Delta^3 f(0)}{3!}$$

$$... \quad ... \quad ... \quad ... \quad ... \quad ... \quad ... \quad ... \quad ... \quad ... \quad ... \quad ...$$
$$... \quad ... \quad ... \quad ... \quad ... \quad ... \quad ... \quad ... \quad ... \quad ... \quad ... \quad ...$$

$$\Delta^n f(0) = K.n! \implies K = \frac{\Delta^n f(0)}{n!}$$

Putting the value of $A, B, C, ... , K$, in (1) we get

$$f(x) = f(0) + \frac{\Delta f(0)}{1!} x^{(1)} + \frac{\Delta^2 f(0)}{2!} x^{(2)} + ... + \frac{\Delta^n f(0)}{n!} x^{(n)}$$

SOLVED EXAMPLES

EXAMPLE 1. *Express* $f(x) = 2x^3 - 3x^2 + 3x - 10$ *and its differences in factorial notation, the interval of differencing being unity.*
(Bhopal–2007)

SOLUTION. **Method 1.** The given polynomial is of degree 3. Hence the maximum power of the function in factorial notation can be three only. Thus using upto $x = 3$ only, we get

$$f(x) = f(0) + \Delta f(0)x^{(1)} + \frac{\Delta^2 f(0)}{2!} x^{(2)} + \frac{\Delta^3 f(0)}{3!} x^{(3)}$$

Now to find the values of $f(0), \Delta f(0), \Delta^2 f(0), \Delta^3 f(0)$ we must have four successive values of the function $f(x)$, *i.e.*, at $x = 0$, $x = 1$, $x = 2$ and $x = 3$. From $f(x) = 2x^3 - 3x^2 + 3x - 10$ the values for $x = 0, 1, 2, 3$ respectively are $f(0) = -10, f(1) = -8, f(2) = 0$ and $f(3) = 26$.

Now to find $\Delta f(0)$, $\Delta^2 f(0)$, and $\Delta^3 f(0)$ we can construct the difference table in the usual manner.

x	$f(x)$	$\Delta f(x)$	$\Delta^2 f(x)$	$\Delta^3 f(x)$
0	– 10			
		2		
1	– 8		6	
		8		12
2	0		18	
		26		
3	26			

Here $f(0) = -10, \Delta f(0) = 2, \Delta^2 f(0) = 6, \Delta^3 f(0) = 12$.

Hence, we get

$$f(x) = -10 + \frac{2}{1!} x^{(1)} + \frac{6}{2!} x^{(2)} + \frac{12}{3!} x^{(3)}$$

$$= -10 + 2x^{(1)} + 3x^{(2)} + 2x^{(3)}$$

The problem can also be solved by alternative methods given below:

Method 2.
Let us write

$$2x^3 - 3x^2 + 3x - 10 = Ax^{(3)} + Bx^{(2)} + Cx^{(1)} + D$$
$$= Ax(x-1)(x-2) + Bx(x-1) + Cx + D \qquad ...(1)$$

Now putting $x = 0$ in (1), we get $D = -10$.

Again putting $x = 1$ in (1), we get

$$2 - 3 + 3 - 10 = C + D$$

$\therefore$ $C = 2$ as $D = -10$

Similarly putting $x = 2$ in (1), we obtain

$$16 - 12 + 6 - 10 = 2B + 2C + D$$

$$0 = 2B - 6 \quad \text{or } B = 3.$$

Equating the coefficients of x^3 on both the sides of (i) we get $A = 2$. Putting the values of A, B, C and D in R.H.S. of (i), the required polynomial in factorial notation will be

$$f(x) = 2x^{(3)} + 3x^{(2)} + 2x - 10$$

and by the rule of simple differentiation, we get

$$\Delta f(x) = 6x^{(2)} + 6x + 2$$
$$\Delta^2 f(x) = 12x + 6 \text{ and } \Delta^3 f(x) = 12.$$

ALGORITHM

Step 1.	Express the given function term by term in factorial functions with certain unknown coefficients as shown in (1).
Step 2.	The values of the unknown coefficients are calculated by putting $x = 0, 1, 2,$ ... successively in L.H.S. and R.H.S. of (1) and then these are solved to find the values of A, B, C, etc.
Step 3.	The values of A, B, C, etc. are substituted from step 2 in R.H.S. of (1) to get the given polynomial in factorial notation.

▼ REMARK

▶ The above method though simple is not a convenient method for expressing a polynomial in factorial form. The procedure is lengthily and there is every possibility of committing some error.

Method 3. Detached Coefficient Method

Let
$$2x^3 - 3x^2 + 3x - 10 = Ax^{(3)} + Bx^{(2)} + Cx + D$$
$$= Ax(x-1)(x-2) + Bx(x-1) + Cx + D$$

If we divide the given function by x then the remainder will be -10 and the quotient $2x^2 - 3x + 3$. The value of D in (i) is taken as -10.

Again divide the quotient $2x^2 - 3x + 3$ by $(x - 1)$, *i.e.*,

$$\begin{array}{r} 2x - 1 \\ x - 1 \overline{)2x^2 - 3x + 3} \\ \underline{2x^2 - 2x} \\ -x + 3 \\ \underline{-x + 1} \\ 2 \end{array}$$

∴ The quotient now is $2x - 1$ and the remainder is the value of C, *i.e.*, $C = 2$

Again divide $2x - 1$ by $x - 2$ so that

$$\begin{array}{r} 2 \\ x - 2 \overline{)2x - 1} \\ \underline{2x - 4} \\ 3 \end{array}$$

The quotient 2 is the value of A and remainder 3 is B. Thus the required polynomial is

$$2x^{(3)} + 3x^{(2)} + 2x - 10$$

The above method can be simplified by the procedure of detached coefficients, which is as follows.
Taking the coefficients of the various powers of x in the given polynomial, we have

1		2	−3	3	−10 = D	... (2)
		0	2	−1		
2		2	−1	2 = C		... (3)
		0	4			
3		2	3 = B			... (4)
		0				
	2 = A					

ALGORITHM

The following are the steps in the method of detached coefficients:

Step 1.	Write the coefficients of various power of x in the order starting with the coefficient of the highest power of x.
Step 2.	Write the coefficients of various power of x in the order starting with the coefficient of the highest power of x.
Step 3.	Multiply 2 by 1 and 0 by 1 and add to get the sum 2 and write it below –3 as shown in. Similarly multiply –3 by 1 and 2 by 1 and add to get –1 and write it below 3. The remainder –10 is the value of D.
Step 4.	Add the values of corresponding column of (2) to get 2, –1 and 2 of (3).
Step 5.	Write 2 in the left hand column of (3) and repeat the steps (2) and (3) to get $(2 \times 2) + (0 \times 2) = 4$ and write below –1. The remainder 2 of (3) is equal to C.
Step 6.	Apply operation of step 4 on (3) to get 2 and 3 of (4).
Step 7.	Write 3 in left hand column of (4) and repeat the steps 2, 3 and 4 to get 2 which is equal to A and remainder 3 of (4) is equal to B.

5.6 FUNDAMENTAL THEOREM OF DIFFERENCE CALCULUS

THEOREM-1. *If $y(x)$ is a polynomial of nth degree, then its nth differences are constant and the $(n+1)^{th}$ differences are zero.*   (Avadh-2005, 07, 09; Delhi-2005, 08; Rohtak-2007, 09, UPTU-2002, 04; UPTU MCA-2004)

Proof.   Let $y(x) = a_0 x^n + a_1 x^{n-1} + ... + a_{n-1} x + a_n$ ...(1)

be the polynomial of nth degree, where $a_0, a_1, ..., a_n$ are constants and $a_0 \neq 0$. Let h be the interval of differencing.

Then $y(x+h) = a_0[(x+h)^n - a^n + a_1(x+h)^{n-1} - x^{n-1}]e + ... + a_{n-1}[(x+h) - x] + a_n$

Now $\Delta y(x) = y(x+h) - y(x)$

$$= a_0[(x+h)^n - x^n] + a_1[(x+h)^{n-1} - x^{n-1}]e + ... + a_{n-1}[(x+h) - x]$$

$$= a_0 nh x^{n-1} + b'x^{n-2} + c'x^{n-3} + ... + k'x + l'$$ ...(2)

Where $b', c' ..., k', l'$ are constants.

Clearly (2) is a polynomial of $(n-1)^{th}$ degree.

Now $\Delta^2 y(x) = \Delta(\Delta y(x)) = \Delta y(x+h) - \Delta y(x)$

$$= a_0 nh[(x+h)^{n-1} - x^{n-1}] + b'[(x+h)^{n-2} - x^{n-2}] + ... + k'(x+h-x)$$

$$= a_0 n(n-1)h^2 x^{n-2} + b''x^{n-3} + c''x^{n-4} + ... + k''$$

$$= \text{A Polynomial of degree } (n-2) \text{ proceeding in the same way, we get}$$

$$\Delta^n y(x) = a_0(n-1)(n-2)...(2.1) \quad h^n x^{n-x}$$

$$= a_0 h^n . n! = \text{A constant, (independent of } x)$$

Therefore, the nth difference is constant.

Hence, $\Delta^{n+1} y(x) = 0$

$\Delta^{n+2} y(x) = 0$ and so on.

▶ REMARK

▶ The converse of the above theorem is also true, *i.e.*, if the nth differences of a function tabulated at equally spaced intervals are constant, the function is a polynomial of degree n.

SOLVED EXAMPLES

EXAMPLE 1. *Evaluate* $\Delta \log x$.

SOLUTION. Consider $\Delta \log x = \log(x+h) - \log x$ $\qquad\qquad$ $[\because \nabla f(x) = f(x+h) - f(x)]$

$$= \log\left(\frac{x+h}{x}\right)$$

$$= \log\left(1+\frac{h}{x}\right) \text{ , } h\text{, being the differences.}$$

EXAMPLE 2. *Evaluate*

    (i) $\Delta \tan^{-1} x$ $\qquad\qquad\qquad\qquad\qquad\qquad\qquad$ (Patna-2006)

    (ii) $\Delta^2 \cos 2x$ $\qquad\qquad\qquad$ (Meerut-2006; UPPCS-2006; PTU–2001)

SOLUTION. (i) $\Delta \tan^{-1} x = \tan^{-1}(x+h) - \tan^{-1} x$

$$= \tan^{-1}\left[\frac{x+h-x}{1+(x+h)x}\right] = \tan^{-1}\left[\frac{h}{1+hx+x^2}\right]$$

    (ii) $\Delta^2 \cos 2x = \Delta[\Delta \cos 2x] = \Delta[\cos(2x+2h) - \cos 2x]$

$$= \Delta \cos(2x+2h) - \Delta \cos 2x$$

$$= [\cos(2x+4h) - \cos(2x+2h)] - [\cos(2x+2h) - \cos 2x]$$

$$= 2\sin(2x+3h).\sin(-h) - 2\sin(2x+h).\sin(-h)$$

$$= -2\sinh[\sin(2x+3h) - \sin(2x+h)]$$

$$= -2\sinh[2\cos(2x+2h)\sin h]$$

$$= -4\sin^2 h \cos(2x+2h)$$

$\therefore \quad \Delta^2 \cos 2x = -4\sin^2 h \cos(2x+2h)$

EXAMPLE 3. *Prove that* $e^x = \left(\dfrac{\Delta^2}{E}\right)e^x . \dfrac{Ee^x}{\Delta^2 e^x}$, *the interval of differencing being unity.*

SOLUTION. We know that $\qquad\qquad\qquad\qquad$ (Meerut-2003; Rajasthan-2002, 09)

$$Ef(x) = f(x+1)$$

$\therefore \qquad Ee^x = e^{x+1}$

and $\qquad \Delta^2 e^x = \Delta(\Delta e^x) = \Delta(e^{x+1} - e^x)$

$$\Delta(e-1)e^x = (e-1)\Delta e^x = (e-1)(e^{x+1} - e^x) = (e-1)^2 e^x$$

and $\left(\dfrac{\Delta^2}{E}\right)e^x = (\Delta^2 E^{-1})e^x$

$$= (\Delta^2 E^{-1} e^x) = \Delta^2(e^{x-1}) = \Delta[\Delta(e^{x-1})] = \Delta[e^x - e^{x-1}]$$

$$= \Delta e^x - \Delta e^{x-1} = e^{x+1} - e^x - e^x + e^{x-1} = e^x\left(e - 2 + \frac{1}{e}\right) = \frac{(e-1)^2}{e}e^x$$

Consider $\left(\dfrac{\Delta^2}{E}\right)e^x . \dfrac{Ee^x}{\Delta^2 e^x} = \dfrac{(e-1)^2}{e}e^x . \dfrac{e^{x+1}}{(e-1)^2 e^x} = e^x$.

EXAMPLE 4. *Evaluate* $\Delta^n \sin(ax + b)$. [UPTU-2003, 04 (Co)]

SOLUTION. Consider $\Delta \sin(ax + b) = \sin(a(x + h) + b) - \sin(ax + b)$

$$= 2\sin\frac{ah}{2}\cos\left[a\left(x + \frac{h}{2}\right) + b\right] = 2\sin\frac{ah}{2}\sin\left[ax + b + \frac{ah + \pi}{2}\right]$$

$\therefore \quad \Delta^2 \sin(ax + b) = \Delta\left[2\sin\frac{ah}{2}\sin\left(ax + b + \frac{ah + \pi}{2}\right)\right] = \left(2\sin\frac{ah}{2}\right)\left(2\sin\frac{ah}{2}\right)$

$$\sin\left[ax + b + \frac{ah + \pi}{2} + \frac{ah + \pi}{2}\right] = \left(2\sin\frac{ah}{2}\right)^2 \sin\left[ax + b + 2\left(\frac{ah + \pi}{2}\right)\right]$$

Proceeding in the same way, we get

$$\Delta^3 \sin(ax + b) = \left(2\sin\frac{ah}{2}\right)^3 \sin\left[ax + b + 3\left(\frac{ah + \pi}{2}\right)\right]$$

$$\cdots \quad \cdots \quad \cdots \quad \cdots \quad \cdots \quad \cdots \quad \cdots \quad \cdots$$
$$\cdots \quad \cdots \quad \cdots \quad \cdots \quad \cdots \quad \cdots \quad \cdots \quad \cdots$$

$$\Delta^n \sin(ax + b) = \left(2\sin\frac{ah}{2}\right)^n \sin\left[ax + b + n\left(\frac{ah + \pi}{2}\right)\right]$$

EXAMPLE 5. *If* $f(x) = e^{ax}$, *evaluate* $\Delta^n f(x)$. [UPTU(MCA)–2004]

SOLUTION. If the interval of differencing is h, then

$$\Delta f(x) = f(x + h) - f(x)$$

$\Rightarrow \quad \Delta[e^{ax}] = e^{a(x+h)} - e^{ax} = e^{ax}.e^{ah} - e^{ax}$

$\therefore \quad \Delta e^{ax} = e^{ax}(e^{ah} - 1)$

Again $\quad \Delta^2 f(x) = \Delta[\Delta f(x)] = \Delta[(e^{ah} - 1)e^{ax}] = (e^{ah} - 1)\Delta e^{ax}$

$$\Delta^2 e^{ax} = (e^{ah} - 1)^2 e^{ax}$$

Similarly, $\quad \Delta^3 e^{ax} = (e^{ah} - 1)^3 e^{ax}$

$$\cdots \quad \cdots \quad \cdots \quad \cdots \quad \cdots \quad \cdots \quad \cdots$$
$$\cdots \quad \cdots \quad \cdots \quad \cdots \quad \cdots \quad \cdots \quad \cdots$$

$$\Delta^n e^{ax} = (e^{ah} - 1)^n e^{ax}$$

EXAMPLE 6. *Find* $\Delta^{10}[(1 - ax)(1 - bx^2)(1 - cx^3)(1 - dx^4)]$. [UPTU B.Tech (Ag.)–2004]

SOLUTION. Let $\quad f(x) = (1 - ax)(1 - bx^2)(1 - cx^3)(1 - dx^4)$

Then $f(x)$ is a polynomial of degree 10 and coefficient of x^{10} is $abcd$. Since, we know that if $f(x)$ is a polynomial of degree n with leading coefficient a_0, then

$$\Delta^n f(x) = a_0 h^n n!$$

Here $n = 10$, $a_0 = abcd$, $h=1$ Then $\Delta^{10} f(x) = abcd10!$

EXAMPLE 7. *Prove that* $\Delta \log f(x) = \log\left[1 + \frac{\Delta f(x)}{f(x)}\right]$. (JNTU–2009; Rajasthan–2001, 05; Bangluru–2007)

SOLUTION. We know that

$$\Delta f(x) = f(x + h) - f(x)$$

$$\therefore \qquad \Delta \log f(x) = \log f(x+h) - \log f(x) = \log \left[\frac{f(x+h)}{f(x)} \right]$$

$$= \log \left[\frac{Ef(x)}{f(x)} \right] \qquad\qquad \because Ef(x) = f(x+h)$$

$$= \log \left[\frac{(1+\Delta)f(x)}{f(x)} \right] \qquad\qquad (\because E = 1 + \Delta)$$

$$= \log \left[1 + \frac{\Delta f(x)}{f(x)} \right]$$

EXAMPLE 8. *Show that* $\Delta^p y_k = \nabla^p y_{k+p}$.

SOLUTION. Since we know that

$$\Delta = \nabla E$$

Then $\qquad \Delta^p y_k = (\nabla E)^p y_k = \nabla^p E^p y_k = \nabla^p y_{k+p}$.

EXAMPLE 9. *If* $y = (3x+1)(3x+4)...(3x+22)$

Show that $\Delta^4 y = 136080(3x+13)(3x+16)(3x+19)(3x+22)$.

SOLUTION. Clearly, the given equation

$$y = (3x+1)(3x+4)...(3x+22)$$

contains eight factors.

Therefore, $\qquad y = 3^8 \left(x + \frac{1}{3} \right) \left(x + \frac{4}{3} \right) ... \left(x + \frac{22}{3} \right) = 3^8 \left(x + \frac{22}{3} \right)^{(8)}$

$$\Rightarrow \qquad\qquad \Delta y = 3^8.8 \left(x + \frac{22}{3} \right)^{(7)}$$

$$\Delta^2 y = 3^8.8.7 \left(x + \frac{22}{3} \right)^{(6)}$$

$$\Delta^3 y = 3^8.8.7.6 \left(x + \frac{22}{3} \right)^{(5)}$$

and $\qquad\qquad \Delta^4 y = 3^8.8.7.6.5 \left(x + \frac{22}{3} \right)^{(4)}$

Hence, $\qquad \Delta^4 y = 11022480 \left(x + \frac{22}{3} \right) \left(x + \frac{22}{3} - 1 \right) \left(x + \frac{22}{3} - 2 \right) \left(x + \frac{22}{3} - 3 \right)$

$$= 136080(3x+22)(3x+19)(3x+16)(3x+13)$$

EXAMPLE 10. *Show that* $hD = -\log(1-\nabla) = \sinh^{-1}(\mu\delta)$.

(UPTU–2005, 06, MDU–2005)

SOLUTION. We have $\qquad hD = \log E = -\log(E^{-1}) = -\log(1-\nabla)$

Also, $\qquad\qquad \mu = \frac{1}{2}(E^{1/2} + E^{-1/2})$ and $\delta = E^{1/2} - E^{-1/2}$

$$\therefore \qquad\qquad \mu\delta = \frac{1}{2}(E - E^{-1}) = \frac{1}{2}(e^{hD} - e^{-hD}) = \sinh(hD)$$

Hence, $\qquad\qquad hD = \sinh^{-1}(\mu\delta) = -\log(1-\nabla)$

EXAMPLE 11. *Show that* $(E^{1/2} + E^{-1/2})(1+\Delta)^{1/2} = 2 + \Delta$. (Bhopal–2009, UPTU–2004, 09)

SOLUTION. We have $(E^{1/2} + E^{-1/2})E^{1/2} = E + 1 = 1 + \Delta + 1 = \Delta + 2$.

EXAMPLE 12. *Show that* $\Delta + \nabla = \dfrac{\Delta}{\nabla} - \dfrac{\nabla}{\Delta}$ (UPTU MCA–2003, 07, 09, B.Tech. (Ag.)–2004, B.Tech.–2008)

SOLUTION. We have $\left(\dfrac{\Delta}{\nabla} - \dfrac{\nabla}{\Delta}\right)y_x = \left(\dfrac{E-1}{1-E^{-1}} - \dfrac{1-E^{-1}}{E-1}\right)y_x = \left\{\dfrac{E-1}{\left(\dfrac{E-1}{E}\right)} - \dfrac{\left(\dfrac{E-1}{E}\right)}{E-1}\right\}y_x$

$$= \left(E - \dfrac{1}{E}\right)y_x = (E - E^{-1})y_x$$

$$= \left\{(1+\Delta) - (1-\nabla)\right\}y_x = (\Delta + \nabla)y_x$$

Hence, $\dfrac{\Delta}{\nabla} - \dfrac{\nabla}{\Delta} = \Delta + \nabla$.

EXAMPLE 13. *Show that* $\nabla - \Delta = \nabla\Delta$. [Mumbai–2005, UPTU(MCA)–2005; Bilaspur-2010, Agra-2004, 07; Bhopal-2005, 07, 09]

SOLUTION. We have $\Delta - \nabla = (1 - E^{-1}) - (E-1) = \left(\dfrac{E-1}{E}\right) - (E-1)$

$$= -(E-1)(E^{-1} - 1) = -(E-1)(1 - E^{-1}) = -\nabla\Delta$$

EXAMPEL 14. *Evaluate* $\Delta^2\left[\dfrac{5x+12}{x^2 + 5x + 6}\right]$. (Mumbai–2003)

SOLUTION. We have

$$\Delta^2\left[\dfrac{5x+12}{x^2+5x+6}\right] = \Delta^2\left[\dfrac{2(x+3) + 3(x+2)}{(x+2)(x+3)}\right]$$

$$= \Delta^2\left[\dfrac{2}{x+2} + \dfrac{3}{x+3}\right] = \Delta.\Delta\left[\dfrac{2}{x+2} + \dfrac{3}{x+3}\right] \qquad \ldots(1)$$

Now, we have

$$\Delta\left[\dfrac{2}{x+2} + \dfrac{3}{x+3}\right] = \left[\dfrac{2}{x+3} - \dfrac{2}{x+2}\right] + \left[\dfrac{3}{x+4} - \dfrac{3}{x+3}\right]$$

$$= -\dfrac{2}{(x+2)(x+3)} - \dfrac{3}{(x+4)(x+3)}$$

Then from (1)

$$\Delta^2\left[\dfrac{5x+12}{x^2+5x+6}\right] = \Delta\left[-\dfrac{2}{(x+2)(x+3)} - \dfrac{3}{(x+4)(x+3)}\right]$$

$$= -2\left[-\dfrac{1}{(x+3)(x+4)} - \dfrac{1}{(x+2)(x+3)}\right]$$

$$-3\left[\dfrac{1}{(x+5)(x+4)} - \dfrac{1}{(x+4)(x+3)}\right]$$

$$= -2\left[\dfrac{(x+2) - (x+4)}{(x+2)(x+3)(x+4)}\right] - 3\left[\dfrac{(x+3) - (x+5)}{(x+3)(x+4)(x+5)}\right]$$

$$= \dfrac{4}{(x+2)(x+3)(x+4)} + \dfrac{6}{(x+3)(x+4)(x+5)}$$

EXAMPLE 15. *Prove that* $\Delta^n \log ax = \log a (x+nh) - {}^nC_1 \Delta \log a[x+(n-1)h]$.   (Patna–2007, 09)

SOLUTION.   We have $\Delta(\log ax) = \log a(x+h) - \log ax$ ...(1)

$$\Delta^2[\log(ax)] = \Delta[\Delta \log ax] = \Delta[\log a(x+h) - \log ax]$$

$$= \Delta \log a(x+h) - \Delta \log ax$$

$$= [\log[a(x+2h) - \log a(x+h) \ -[\log a(x+h) - \log ax]$$

$$= \log a(x+2h) - 2\log a(x+h) + \log ax \qquad ...(2)$$

From (1) and (2), we conclude that

$$\Delta^n \log ax = \log a(x+nh) \ - {}^nC_1 \cdot \log a(x+(n-1))h \qquad ...(3)$$

is true for $n = 1, 2$

We shall prove that relation (3) is true for all positive integral value of n.

Consider, $\Delta^{n+1} \log ax = \Delta[\Delta^n \log ax]$

$$= \Delta \log a(x+nh) - {}^nC_1 \Delta \log a[x+(n-1)h]$$

$$+ ... + (-1)^n \cdot {}^nC_r \Delta \log a[x+(n-r)h] + ... + (-1)^n \Delta \log ax$$

$$= \log[a+(x+(n-1)h] - \log a(x+nh)$$

$$- {}^nC_1[\log a(x+nh) - \log a[x+(n-1)h + ...$$

$$+ (-1)^r \cdot {}^nC_r[\log a\{x+(n-r+1)h]$$

$$- \log a[x+(n-r)h] + ... + (-1)^n$$

$$\log a[(x+h) - \log ax]$$

$$= \log a[x+(n+1)h] - (1 + {}^nC_1)\log a(x+nh)$$

$$+ ... + (-1)^r[{}^nC_{r-1} + {}^nC_r[\log a\{x+(n-r+1)h\}]$$

$$+ ... + +(-1)^{n+1} \log ax \qquad ...(4)$$

(4) $\Rightarrow$ Relation (3) is true for $n+1$.

Hence by principle of mathematical induction, result is true for all positive integral values of n.

EXAMPLE 16. *Evaluate* $\Delta\left[\dfrac{2^x}{(x+1)!}\right] : h = 1$.

(Rajasthan-2002, 07)

SOLUTION.   Let $f(x) = 2^x$ and $g(x) = (x+1)!$

Then $\Delta f(x) = 2^{x+1} - 2^x = 2^x(2-1) = 2^x$

and $\Delta g(x) = (x+1+1)! - (x+1)! = (x+2)! - (x+1)! = (x+1)(x+1)!$

Therefore, $\Delta\left[\dfrac{f(x)}{g(x)}\right] = \dfrac{g(x)\Delta f(x) - f(x)\Delta g(x)}{g(x+h)g(x)} = \dfrac{(x+1)!2^x - (x+1)!2^x(x+1)}{(x+2)!(x+1)!}$

$$= \dfrac{(x+1)!2^x[1-(x+1)]}{(x+2)!(x+1)!} = -\dfrac{x}{(x+2)!}2^x .$$

EXAMPLE 17. *Show that* $\displaystyle\sum_{k=\infty}^{n-1} \Delta^2 f_k = \Delta f_n - \Delta f_0$.   (Meerut-2003, 2014; UKTU–2014)

SOLUTION.   Consider   L.H.S. $= \displaystyle\sum_{k=\infty}^{n-1} \Delta^2 f_k = \sum_{k=0}^{n-1} (E-1)^2 f_k$

$$= \sum_{k=0}^{n-1}(E^2 - 2E + 1)f_k = \sum_{k=0}^{n-1}(f_{k+2} - 2f_{k+1} + f_k)$$

$$= (f_2 - 2f_1 + f_0) + (f_3 - 2f_2 + f_1) + (f_4 - 2f_3 + f_2) + (f_5 - 2f_4 + f_3)$$

$$+ (f_{n-1} - 2f_{n-2} + f_{n-3}) + (f_n - 2f_{n-1} + f_{n-2}) + (f_{n+1} - 2f_n + f_{n-1})$$

$$= f_{n+1} - f_n + f_0 - f_1 = (f_{n+1} - f_n) - (f_1 - f_0) = \Delta f_n - \Delta f_0$$

EXAMPLE 18. *Prove that* $u_x = u_{x-1} + \Delta u_{x-2} + \Delta^2 u_{x-3} + \dots + \Delta^{n-1} u_{x-n} + \Delta^n u_{x-n}$

SOLUTION. Consider

$$u_x - \Delta^n u_{x-n} = u_x - \Delta^n . E^{-n} u_x$$

$$= \left(1 - \frac{\Delta^n}{E^n}\right)u_x = E^{-n}(E^n - \Delta^n)u_x$$

$$= E^{-n}\left[\frac{E^n - \Delta^n}{E - \Delta}\right]u_x \qquad\qquad (\because E = 1 + \Delta)$$

$$= E^{-n}[E^{n-1} + \Delta E^{n-2} + \Delta^2 E^{n-3} + \dots + \Delta^{n-1}]u_x$$

$$= [E^{-1} + \Delta E^{-2} + \Delta^2 E^{-3} + \dots + \Delta^{n-1} E^{-n}]u_x$$

$$\therefore \qquad u_x - \Delta^n u_{x-n} = u_{x-1} + \Delta u_{x-2} + \Delta^2 u_{x-3} + \dots + \Delta^{n-1} u_{x-n}$$

Hence, $\qquad u_x = u_{x-1} + \Delta u_{x-2} + \Delta^2 u_{x-3} + \Delta^{n-1} u_{x-n} + \Delta^n u_{x-n}.$

EXAMPLE 19. *Prove that*

(i) $u_0 + \dfrac{u_1 x}{1!} + \dfrac{u_2 x^2}{2!} \dots = e^x\left[u_0 + x\Delta u_0 + \dfrac{x^2}{2!}\Delta^2 u_0 + \dots\right]$

(Rajasthan-2008; Delhi-2005, 09)

(ii) $u_0 + u_1 + u_2 + \dots + u_n = {}^{n+1}C_1 u_0 + {}^{n+1}C_2 \Delta u_0 + {}^{n+1}C_3 \Delta^2 u_0 + \dots + \Delta^n u_0$

(Karnataka-2004, 06; Mysore-2007)

SOLUTION. **(i)** $u_0 + \dfrac{u_1 x}{1!} + \dfrac{u_2 x^2}{2!} + \dots = u_0 + \dfrac{xEu_0}{1!} + \dfrac{x^2 E^2 u_0}{2!} + \dfrac{x^3 E^3 u_0}{3!} + \dots$

$$= \left[1 + \frac{xE}{1!} + \frac{x^2 E^2}{2!} + \frac{x^3 E^3}{3!} + \dots\right]u_0$$

$$= [e^{xE}]u_0 = [e^{x(1+\Delta)}]u_0 \qquad\qquad (\because E = 1 + \Delta)$$

$$= (e^x . e^{x\Delta})u_0 = e^x\left[1 + \frac{x\Delta}{1!} + \frac{x^2\Delta^2}{2!} + \frac{x^3\Delta^3}{3!} + \dots\right]u_0$$

$$= e^x\left[u_0 + x\Delta u_0 + \frac{x^2}{2!}\Delta^2 u_0 + \frac{x^3}{3!}\Delta^3 u_0 + \dots\right]$$

(ii) $u_0 + u_1 + u_2 + \dots + u_n = u_0 + Eu_0 + E^2 u_0 + \dots + E^n u_0$

$$= (1 + E + E^2 + \dots + E^n)u_0 = \left[\frac{E^{n+1} - 1}{E - 1}\right]u_0$$

$$= \left[\frac{(1+\Delta)^{1+n} - 1}{\Delta} \right] u_0 \qquad (\because E = 1 + D)$$

$$= \frac{1}{\Delta} [^{n+1}C_0 + {}^{n+1}C_1 \Delta + {}^{n+1}C_2 \Delta^2$$

$$+ {}^{n+1}C_3 \Delta^3 + \dots + {}^{n+1}C_{n+1} \Delta^{n+1} - 1] u_0$$

$$= [^{n+1}C_1 + {}^{n+1}C_2 \Delta + {}^{n+1}C_3 \Delta + \dots + \Delta^n] u_0$$

$$= {}^{n+1}C_1 u_0 + {}^{n+1}C_2 \Delta u_0 + {}^{n+1}C_3 \Delta^2 u_0 + \dots + \Delta^n u_0.$$

EXAMPLE 20. *If* $\Delta^3 u_x = 0$, *prove that*

$$u_{x+\frac{1}{2}} = \frac{1}{2}(u_x + u_{x+1}) - \frac{1}{16}(\Delta^2 u_{x+1} + \Delta^2 u_x) \qquad \text{(Agra-2006, 09)}$$

SOLUTION. Consider

$$u_{x+\frac{1}{2}} = E^{1/2} u_x = (1+\Delta)^{1/2} u_x \qquad (\because E = 1 + \Delta)$$

$$= \left(1 + \frac{1}{2}\Delta - \frac{1}{8}\Delta^2 + \frac{1}{16}\Delta^3 - \dots \right) u_x$$

(Using Binomial Expansion)

$$= u_x + \frac{1}{2}\Delta u_x - \frac{1}{8}\Delta^2 u_x \qquad (\because \Delta^3 u_x = 0, \dots \text{ etc.})$$

$$\therefore \qquad u_{x+\frac{1}{2}} = u_x + \frac{1}{2}\Delta u_x - \frac{1}{8}\Delta^2 u_x \qquad \dots(1)$$

Now $\qquad \Delta^3 u_x = 0 \qquad$ (given)

$$\therefore \qquad \Delta^2(\Delta u_x) = 0$$

or $\qquad \Delta^2(u_{x+1} - u_x) = 0 \qquad$ or $\qquad \Delta^2 u_{x+1} - \Delta^2 u_x = 0$

Also $\qquad \Delta u_x = u_{x+1} - u_x$

Substituting these values in (1), we get

$$u_{x+\frac{1}{2}} = u_x + \frac{1}{2}[u_{x+1} - u_x] - \frac{1}{8}\left[\frac{2\Delta^2 u_x}{2} \right]$$

$$= \frac{1}{2}(u_x + u_{x+1}) - \frac{1}{8}\left[\frac{\Delta^2 u_x + \Delta^2 u_x}{2} \right]$$

$$\therefore \qquad u_{x+\frac{1}{2}} = \frac{1}{2}(u_x + u_{x+1}) - \frac{1}{16}[\Delta^2 u_x + \Delta^2 u_{x+1}].$$

EXAMPLE 21. *Prove the following identity :*

$$u_1 x + u_2 x^2 + u_3 x^3 + \dots = \frac{x}{1-x} u_1 + \frac{x^2}{(1-x)^2} \Delta u_1 + \frac{x^3}{(1-x)^3} \Delta^2 u_1 + \dots$$

where $y = u_x$ *is any function of x and* $0 < x < 1$. $\qquad$ (Agra-2005; Rajasthan-2002, 06)

SOLUTION.

$$\text{R.H.S.} = \frac{x}{1-x}u_1 + \frac{x^2}{(1-x)^2}(E-1)u_1 + \frac{x^3}{(1-x)^3}(E-1)^2 u_1 + \dots$$

$$= \left[\frac{x}{1-x} - \frac{x^2}{(1+x)^2} + \frac{x^3}{(1-x)^3} + \dots\right]u_1 + \left[\frac{x^2}{(1-x)^2} - \frac{2x^3}{(1-x)^3} + \dots\right]Eu_1$$

$$+ \left[\frac{x^3}{(1-x)^3}\right]E^2 u_1 + \dots$$

$$= \frac{x}{1-x}\left(1 + \frac{x}{1-x}\right)^{-1} u_1 + \frac{x^2}{(1-x)^2}\left(1 + \frac{x}{1-x}\right)^{-2} u_2$$

$$+ \frac{x^3}{(1-x)^3}\left(1 + \frac{x}{(1-x)}\right)^{-3} u_3 + \dots$$

$$= u_1 x + u_2 x^2 + u_3 x^3 + \dots = \text{L.H.S.}$$

EXAMPLE 22. *Prove the following identity:*

$$u_{x+n} = u_n + {}^xC_1 \Delta u_{n-1} + {}^{x+1}C_2 \Delta^2 u_{n-2} + {}^{x+2}C_3 \Delta^3 u_{n-3} + \dots$$

(Bangaluru–2002, 03, 07)

SOLUTION.

$$\text{R.H.S.} = u_n + {}^xC_1 \Delta u_{n-1} + {}^{x+1}C_2 \Delta^2 u_{n-2} + {}^{x+2}C_3 \Delta^3 u_{n-3} + \dots$$

$$= u_n + {}^xC_1 \Delta E^{-1} u_n + {}^{x+1}C_2 \Delta^2 E^{-2} u_n + {}^{x+2}C_3 \Delta^3 E^{-3} u_n + \dots$$

$$= [1 + {}^xC_1 \Delta E^{-1} + {}^{x+1}C_2 \Delta^2 E^{-2} + {}^{x+2}C_3 \Delta^3 E^{-3} + \dots]u_n$$

$$= [1 - \Delta E^{-1}]^{-x} u_n = \left(1 - \frac{\Delta}{E}\right)^{-x} u_n = \left(\frac{E - \Delta}{E}\right)^{-x} u_n$$

$$= \left(\frac{1}{E}\right)^{-x} u_n = E^x u_n = u_{n+1} = \text{L.H.S.}$$

EXAMPLE 23. *Prove the following identity :*

$$u_x - \frac{1}{8}\Delta^2 u_{x-1} + \frac{1.3}{8.16}\Delta^4 u_{x-2} - \frac{1.3.5}{8.16.24}\Delta^6 u_{x-3} + \dots$$

$$= u_{x+\frac{1}{2}} - \frac{1}{2}\Delta u_{x+\frac{1}{2}} + \frac{1}{4}\Delta^2 u_{x+\frac{1}{2}} + \dots \qquad \text{(Bhopal-2002, 06)}$$

SOLUTION. To prove $u_x - \dfrac{1}{8}\Delta^2 u_{x-1} + \dfrac{1.3}{8.16}\Delta^4 u_{x-2} - \dfrac{1.3.5}{8.16.24}\Delta^6 u_{x-3} + \dots$

$$= u_{x+\frac{1}{2}} - \frac{1}{2}\Delta u_{x+\frac{1}{2}} + \frac{1}{4}\Delta^2 u_{x+\frac{1}{2}} + \dots$$

$$= \left[1 - \frac{1}{8}\Delta^2 E^{-1} + \frac{1.3}{8.16}\Delta^4 E^{-2} - \frac{1.3.5}{8.16.24}\Delta^6 E^{-3} + \dots\right]u_x$$

$$
= \left[1 + \left(-\frac{1}{2} \right) \left(\frac{\Delta^2 E^{-1}}{4} \right) + \frac{\left(-\frac{1}{2} \right) \left(-\frac{3}{2} \right)}{2!} \left(\frac{\Delta^2 E^{-1}}{4} \right)^2 \right.
$$

$$
\left. + \frac{\left(-\frac{1}{2} \right) \left(-\frac{3}{2} \right) \left(-\frac{5}{2} \right)}{3!} \left(\frac{\Delta^2 E^{-1}}{4} \right)^3 + \dots \right] u_x
$$

$$
= \left[1 + \frac{\Delta^2 E^{-1}}{4} \right]^{-1/2} u_x = E^{1/2} \left[E + \frac{\Delta^2}{4} \right]^{-1/2} u_x
$$

$$
= E^{1/2} \left[1 + \Delta + \frac{\Delta^2}{4} \right]^{-1/2} u_x \qquad\qquad (\because E = 1 + \Delta)
$$

$$
= E^{1/2} \left[\left(1 + \frac{\Delta}{2} \right)^2 \right]^{-1/2} u_x = E^{1/2} \left(1 + \frac{\Delta}{2} \right)^{-1} u_x
$$

$$
= E^{1/2} \left[1 - \frac{\Delta}{2} + \frac{\Delta^2}{2^2} - \dots \right] u_x
$$

$$
= u_{x+\frac{1}{2}} - \frac{1}{2} \Delta u_{x+\frac{1}{2}} + \frac{1}{4} \Delta^2 u_{x+\frac{1}{2}} - \dots = \text{R.H.S.}
$$

EXAMPLE 24.   *Prove the following identity :*

$$
u_x - u_{x+1} + u_{x+2} - u_{x+3} + \dots
$$

(Lucknow-2003; Agra-2005, Bhopal-2001)

$$
= \frac{1}{2} \left[u_{x+\frac{1}{2}} - \frac{1}{8} \Delta^2_{x-\frac{3}{2}} + \frac{1.3}{2!} \left(\frac{1}{8} \right)^2 \Delta^4 u_{x-\frac{5}{2}} - \frac{1.3.5}{3!} \left(\frac{1}{8} \right)^3 \Delta^6 u_{x-\frac{7}{2}} + \dots \right].
$$

SOLUTION.   R.H.S. $= \dfrac{1}{2} \left[u_{x-\frac{1}{2}} - \dfrac{1}{8} \Delta^2 u_{x-\frac{3}{2}} + \dfrac{1.3}{2!} \left(\dfrac{1}{8} \right)^2 \Delta^4 u_{x-\frac{5}{2}} - \dfrac{1.3.5}{3!} \left(\dfrac{1}{8} \right)^3 \Delta^6 u_{x-\frac{7}{2}} + \dots \right]$

$$
= \frac{1}{2} \left[E^{-1/2} u_x - \left(\frac{1}{2} \right) \left(\frac{1}{4} \right) \Delta^2 E^{-3/2} u_x + \frac{\left(\frac{1}{2} \right) \left(\frac{3}{2} \right)}{2!} \left(\frac{1}{4} \right)^2 \Delta^4 E^{-5/2} u_x \right.
$$

$$
\left. - \frac{\left(\frac{1}{2} \right) \left(\frac{3}{2} \right) \left(\frac{5}{2} \right)}{3!} \left(\frac{1}{4} \right)^3 \Delta^6 E^{-7/2} u_x + \dots \right]
$$

$$= \frac{1}{2} E^{-1/2} \left[1 + \left(-\frac{1}{2} \right) \left(\frac{1}{4} \Delta^2 E^{-1} \right) + \frac{\left(-\frac{1}{2} \right) \left(-\frac{3}{2} \right)}{2!} \left(\frac{1}{4} \Delta^2 E^{-1} \right)^2 \right.$$

$$\left. + \frac{\left(-\frac{1}{2} \right) \left(-\frac{3}{2} \right) \left(-\frac{5}{2} \right)}{3!} \left(\frac{1}{4} \Delta^2 E^{-1} \right)^3 + ... \right] u_x$$

$$= \frac{1}{2} E^{-1/2} \left[1 + \frac{1}{4} \Delta^2 E^{-1} \right]^{-1/2} u_x$$

$$= \frac{1}{2} E^{-1/2} \left[\frac{4E + \Delta^2}{4E} \right]^{-1/2} u_x = \frac{1}{2} E^{-1/2} \left[\frac{4 + 4\Delta + \Delta^2}{4E} \right]^{-1/2} u_x$$

$$= \frac{1}{2} E^{-1/2} . 2 E^{1/2} [(2 + \Delta)^2]^{-1/2} u_x = (2 + \Delta)^{-1} u_x = (1 + E)^{-1} u_x$$

$$= [1 - E + E^2 - E^3 + E^4 - E^5 + ...] u_x$$

$$= u_x - u_{x+1} + u_{x+2} - u_{x+3} + u_{x+4} - u_{x+5} + ... = \text{L.H.S.}$$

EXAMPLE 25. *Show that*

$$\Delta x^m - \frac{1}{2} \Delta^2 x^m + \frac{1.3}{2.4} \Delta^3 x^m - \frac{1.3.5}{2.4.6} \Delta^4 x^m + ... + m \text{ terms} \qquad \text{(Gujrat-2008)}$$

$$= \left(x + \frac{1}{2} \right)^m - \left(x - \frac{1}{2} \right)^m$$

SOLUTION. We have $\Delta^r x^m = 0$ for $r > m$. Therefore, the sum of m terms of the series will be same as the sum of infinite terms of the series.

Now, LHS $= \Delta x^m - \frac{1}{2} \Delta^2 x^m + \frac{1.3}{2.4} \Delta^3 x^m - \frac{1.3.5}{2.4.6} \Delta^4 x^m + ... + m$ terms

$$= \Delta \left[1 - \frac{\Delta}{2} + \frac{1.3}{2.4} \Delta^2 - \frac{1.3.5}{2.4.6} \Delta^3 + ... \quad + m \text{ terms} \right] x^m$$

$$= \Delta \left[1 - \frac{1}{2} \Delta + \frac{1.3}{2.4} \Delta^2 - \frac{1.3.5}{2.4.6} \Delta^3 + ... \text{ upto } \infty \right] x^m$$

$$= \Delta (1 + \Delta)^{-1/2} x^m = \Delta E^{-1/2} x^m$$

$$= \Delta \left(x - \frac{1}{2} \right)^m = \left(x + \frac{1}{2} \right)^m - \left(x - \frac{1}{2} \right)^m = \text{R.H.S.}$$

EXAMPLE 26. *Find the function whose first difference is $9x^2 + 11x + 5$.*

SOLUTION. Let $f(x)$ be the required function. As per given

$$\Delta f(x) = 9x^2 + 11x + 5 = 9x(x - 1) + Ax + B \qquad \text{(Meerut-2002,04)}$$

Putting $x = 0$, we get $\qquad B = 5$

Putting $x = 1$, we get $\qquad A = 20$

Therefore, $\qquad \Delta f(x) = 9x(x - 1) + 20x + 5 = 9x^{(2)} + 20x^{(1)} + 5$

$$\Rightarrow \qquad f(x) = \frac{9x^{(3)}}{3} + \frac{20x^{(2)}}{2} + 5x^{(1)} + C$$

$$\Rightarrow \qquad f(x) = 3x^{(3)} + 10x^{(2)} + 5x^{(1)} + C.$$

EXAMPLE 27. *Find the relation between* α, β *and* γ *in order that* $\alpha + \beta x + \gamma x^2$ *may be expressible in one term of factorial notations.*   (IAS-2004)

SOLUTION.   Let $f(x) = \alpha + \beta x + \gamma x^2 = (a + bx)^{(2)}$

where a and b are unknown constants

Now $(a + bx)^2 = (a + bx)(a + b(x - 1))$

$$= (a^2 - ab) + (2ab - b^2)x + b^2 x^2$$

Comparing with $\alpha + \beta x + \gamma x^2$ we get

$$\alpha = a^2 - ab, \beta = 2ab - b^2, \gamma = b^2$$

On eliminating a, b from these relations we get

$$\gamma^2 + 4\alpha\gamma = \beta^2$$

which is required relation.

EXAMPLE 28. *Find the function whose first difference is* $x^3 + 3x^2 + 5x + 12$ *if 1 be the interval of differencing.*

SOLUTION.   Here, it is clear that required function is of degree four in x.

∴ Let $f(x) = Ax^{(4)} + Bx^{(3)} + Cx^{(2)} + Dx^{(1)} + E$ be the required function.

According to the given question, we have

$$\Delta f(x) = x^3 + 3x^2 + 5x + 12$$
$$= \Delta(Ax^{(4)} + Bx^{(3)} + Cx^{(2)} + Dx^{(1)} + E)$$

The given expression can be written in the factorial notation as follows:

$$x^3 + 3x^2 + 5x + 12 = x^{(3)} + 6x^{(2)} + 9x^{(1)} + 12$$

Now, $\Delta f(x) = x^{(3)} + 6x^{(2)} + 9x^{(1)} + 12$

Therefore $f(x) = \dfrac{x^{(4)}}{4} + \dfrac{6x^{(3)}}{3} + \dfrac{9x^{(2)}}{2} + 12x + f$

Which is the required function.

EXAMPLE 29. *Express* $f(x) = x^4 - 12x^3 + 24x^2 - 30x + 9$ *and its successive difference in factorial notation, interval of differencing being unity. Hence show that* $\Delta^5 f(x) = 0$.

[Mumbai–2004; Himachal–2006, 10; UPTU (Sum) 2004]

SOLUTION.   Let $f(x) = Ax^{(4)} + Bx^{(3)} + Cx^{(2)} + Dx^{(1)} + E$ ...(1)

Using the method of synthetic division, we divide the given $f(x)$ by x, $x - 1$, $x - 2$ and $x - 3$ successively.

1	1	−12	24	−30	9 = E
		1	−11	13	
2	1	−11	13	−17 = D	
		2	−18		
3	1	−9	−5 = C		
		3			
4	1	−6 = B			
	1 = A				

Hence $\qquad f(x) = x^{(4)} - 6x^{(3)} - 5x^{(2)} - 17x^{(1)} + 9$

$\therefore \qquad \Delta f(x) = 4x^{(3)} - 18x^{(2)} - 10x^{(1)} - 17$

[$\because \Delta$ is treated as differential operator on factorial polynomial.]

$\Rightarrow \qquad \Delta^2 f(x) = 12x^{(2)} - 36x^{(1)} - 10$

$\Rightarrow \qquad \Delta^3 f(x) = 24x^{(1)} - 36$

$\Rightarrow \qquad \Delta^4 f(x) = 24$

Hence, $\qquad \Delta^5 f(x) = 0$

EXAMPLE 30. *Prove that* $\Delta^2 x^{(m)} = m(m-1)x^{(m-2)}$ *where* m *is a positive integer and interval of differencing is* 1.

SOLUTION. We have $x^{(m)} = x(x-1)...[x-(m-1)]$

Therefore,

$$\Delta x^{(m)} = [(x+1)x...(x+1-(m-1)] - ... - [x(x-1)...(x-(m-1)]$$

$$= mx^{(m-1)}$$

and $\qquad \Delta^2 x^{(m)} = \Delta . \Delta x^{(m)} = m\Delta x^{(m-1)} = m(m-1)x^{(m-2)}$.

EXAMPLE 31. *Estimate the missing term in the following table :*

x	0	1	2	3	4
$f(x)$	1	3	9	?	81

(SVTU–2007, 04; Meerut-2004, 14; Avadh-2005)

SOLUTION. There are 4 values of $f(x)$, which are given.

Then, we have $\quad \Delta^4 f(x) = 0$ for all x.

Here, the interval of differencing is 1.

Now, $\qquad\qquad \Delta^4 f(x) = 0$

$\Rightarrow \qquad\qquad (E-1)^4 f(x) = 0 \qquad\qquad\qquad (\because E = 1 + \Delta)$

$(E^4 - 4E^3 + 6E^2 - 4E + 1)f(x) = 0$

$E^4 f(x) - 4E^3 f(x) + 6E^2 f(x) - 4Ef(x) + f(x) = 0$

$f(x+4) - 4f(x+3) + 6f(x+2) - 4f(x+1) + f(x) = 0$

Putting $x = 0$, we get

$f(4) - 4f(3) + 6f(2) - 4f(1) + f(0) = 0 \Rightarrow 81 - 4f(3) + 6(9) - 4(3) + 1 = 0$

$124 - f(3) = 0 \Rightarrow \qquad\qquad f(3) = 31.$

EXAMPLE 32. *Estimate the production for 1964 and 1966 from the following table:*

Year (x_i)	1961	1962	1963	1964	1965	1966	1967
Production (y_i)	200	220	260	–	350	–	430

(UPTU-2002, 03)

SOLUTION. Since five pairs are complete, therefore we assume all the differences of order fifth are zero, *i.e.*, $\Delta^5 y_0 = 0$ and $\qquad \Delta^5 y_1 = 0$

$$\Rightarrow \qquad (E-1)^5 y_0 = 0 \text{ and } (E-1)^5 y_1 = 0$$

$$\Rightarrow \quad (E^5 - 5E^4 + 10E^3 - 10E^2 + 5E - 1)y_0 = 0$$

$$y_5 - 5y_4 + 10y_3 - 10y_2 + 5y_1 - y_0 = 0 \qquad \qquad \dots(1)$$

and $\quad y_6 - 5y_5 + 10y_4 - 10y_3 + 5y_2 - y_1 = 0 \qquad \qquad \dots(2)$

Using the given table in (1) and (2), we get

$$y_5 - 1750 + 10y_3 - 2600 + 1100 - 260 = 0$$

$$\Rightarrow \qquad \qquad y_5 + 10y_3 = 3450 \qquad \qquad \dots(3)$$

$$430 - 5y_5 + 3500 - 10y_3 + 1300 - 220 = 0$$

$$\Rightarrow \qquad \qquad y_5 + 10y_3 = -5010 \qquad \qquad \dots(4)$$

On solving (3) and (4), we get

$$y_3 = 306 \text{ and } y_5 = 390.$$

EXAMPLE 33. *Find the first term of the series where second and subsequent terms are 8, 3, 0, –1, 0.*

SOLUTION.   We know that   (Agra-2001, 03; Avadh-2005)

$$f(1) = E^{-1} f(2) \text{ , the interval of differencing being unity}$$

$$= (1 + \Delta)^{-1} f(2) = (1 - \Delta + \Delta^2 - \Delta^3 + \dots)f(2)$$

Since five observations are given so that 4th difference will be constant and fifth difference will be zero.

Difference Table

x	$f(x)$	$\Delta f(x)$	$\Delta^2 f(x)$
2	**8**		
		–5	
3	3		**2**
		–3	
4	0		2
		–1	
5	–1		2
		1	
6	0		

$$f(1) = f(2) - \Delta f(2) + \Delta^2 f(2)$$

                        ($\because$ Other higher order differences are zero)

$$= 8 - (-5) + 2 = 15.$$

EXAMPLE 34. *If $u_0 + u_8 = 1.9243$, $u_1 + u_7 = 1.9590$ and $u_2 + u_6 = 1.9823$, $u_3 + u_5 = 1.9956$.*

Find u_4.   (Mumbai-2003, Delhi-2005, 09)

SOLUTION.   Since there are 8 values of $f(x)$, are given.

Therefore, $\qquad \Delta^8 u_0 = 0 \quad \Rightarrow (E-1)^8 u_0 = 0 \qquad \qquad (\because E = 1 + \Delta)$

$$\Rightarrow \quad (^8C_0 E^8 - {}^8C_1 E^7 + {}^8C_2 E^6 - {}^8C_3 E^5 + {}^8C_4 E^4 - {}^8C_5 E^3 + {}^8C_6 E^2 - {}^8C_7 E$$

$$+ {}^8C_8 E^0)u_0 = 0$$

$\Rightarrow \quad (E^8 - 8E^7 + 28E^6 - 56E^5 + 70E^4 - 56E^3 + 28E^2 - 8E + 1)u_0 = 0$

$\Rightarrow \quad (u_0 + E^8 u_0) - 8(Eu_0 + E^7 u_0) + 28$

$\Rightarrow \quad (E^2 u_0 + E^6 u_0) - 56(E^3 u_0 + E^5 u_0) + 70E^4 u_0 = 0$

$\Rightarrow \quad (u_0 + u_8) - 8(u_1 + u_7) + 28(u_2 + u_6) - 56(u_3 + u_5) + 70u_4 = 0$

$\Rightarrow \quad 1.9243 - 8(1.9590) + 28(1.9823) - 56(1.9956) + 70u_4 = 0$

$\Rightarrow \quad 70u_4 + 1.9243 - 15.672 + 55.5044 - 111.7536 = 0$

$\Rightarrow \quad 70u_4 - 69.9969 = 0$

$\Rightarrow \quad u_4 = \dfrac{69.9969}{70} = 0.999955$

5.7 DIFFERENCES OF ZERO

Let m and n be positive integers. Then, we know that

$$\Delta^n x^m = (E - 1)^n x^m$$

$$= E^n - {}^n C_1 E^{n-1} + {}^n C_2 E^{n-2} + \dots + (-1)^{n-1} {}^n C_{n-1} E + (-1)^n x^m$$

$$= (x + n)^m - {}^n C_1 (x + n - 1)^m + {}^n C_2 (x + n - 2)^m - \dots + {}^n C_{n-1}(-1)^{n-1}$$

$$(x + 1)^m + (-1)^n x^m$$

Putting $x = 0$, we get

$$[\Delta^n x^m]_{x=0} = n^m - {}^n C_1 (n-1)^m \, {}^n C_2 (n-2)^m - \dots + {}^n C_{n-1}(-1)^{n-1} \qquad \dots(1)$$

Here, we write $[\Delta^n x^m]_{x=0}$ as $\Delta^n 0^m$ and $\Delta^n 0^m$ are called the nth differences of zero.

5.8 DIVIDED DIFFERENCE

Let $f(x_0), f(x_1), \dots, f(x_n)$ be the values of the function corresponding to the values of $x_0, x_1, \dots, x_n$ which are not equally spaced. We know that the difference of the function values with respect to the difference of the arguments are called divided differences.

We define the first divided difference of $f(x)$ between x_0 and x_1 as follows

$$f(x_0, x_1) = \underset{x_i}{\Delta} \, f(x_0) = \frac{f(x_1) - f(x_0)}{x_1 - x_0}$$

Similarly, the second divided difference is given by

$$f(x_0, x_1, x_2) = \underset{x_1, x_2}{\Delta^2} \, f(x_0) = \frac{f(x_1, x_2) - f(x_0, x_1)}{x_2 - x_0}$$

and so on.

In general, the nth divided difference is given by

$$f(x_0, x_1, \dots, x_n) = \underset{x_1, x_2, \dots, x_n}{\Delta^n} \, f(x_0) = \frac{f(x_1, x_2, \dots x_n) - f(x_0, x_1, \dots x_{n-1})}{x_n - x_0}$$

Divided Difference Table

(1) x	(2) $f(x)$	(3) $\Delta f(x)$	(4) $\Delta^2 f(x)$	(5) $\Delta^3 f(x)$
a	$f(a)$	$\dfrac{f(b)-f(a)}{b-a} = \underset{b}{\Delta}\, \boldsymbol{f(a)}$		
b	$f(b)$		$\dfrac{\underset{c}{\Delta} f(b) - \underset{b}{\Delta} f(a)}{c-a} = \underset{bc}{\Delta^2}\, \boldsymbol{f(a)}$	
		$\dfrac{f(c)-f(b)}{c-b} = \underset{c}{\Delta}\, \boldsymbol{f(b)}$		$\dfrac{\underset{cd}{\Delta^2} f(b) - \underset{bc}{\Delta^2} f(a)}{d-a} = \underset{bcd}{\Delta^3}\, \boldsymbol{f(a)}$
c	$f(c)$		$\dfrac{\underset{d}{\Delta} f(c) - \underset{c}{\Delta} f(b)}{d-b} = \underset{cd}{\Delta^2}\, \boldsymbol{f(b)}$	
d	$f(d)$	$\dfrac{f(d)-f(c)}{d-c} = \underset{d}{\Delta}\, \boldsymbol{f(c)}$		$\dfrac{\underset{de}{\Delta^2} f(c) - \underset{cd}{\Delta^2} f(d)}{c-b} = \underset{cde}{\Delta^3}\, \boldsymbol{f(d)}$
e	$f(e)$	$\dfrac{f(e)-f(d)}{e-d} = \underset{e}{\Delta}\, \boldsymbol{f(d)}$	$\dfrac{\underset{e}{\Delta} f(d) - \underset{d}{\Delta} f(c)}{e-c} = \underset{de}{\Delta^2}\, \boldsymbol{f(c)}$	

ALGORITHM

(Construction of divided difference table)

Step 1. Write the values of x and corresponding values of $f(x)$ in columns (1) and (2) of the table.

Step 2. Calculate the values of column (3) by subtracting the successive values of column (2) and dividing the differences by the corresponding differences of the arguments.

Step 3. Calculate the values of column (4) by taking the difference between two successive values of column (3) and divide each difference by the difference between the lowest and highest values of the arguments corresponding to the two differences, of column (3), *e.g.*, corresponding to the differences $\underset{b}{\Delta} f(a)$ and $\underset{c}{\Delta} f(b)$, the lowest value of the argument is a and the highest value is c, hence in column (4) we get the first observation as $\dfrac{\underset{c}{\Delta} f(b) - \underset{b}{\Delta} f(a)}{c-a}$

Step 4. Repeat the operations of step 3 on the observation of column (4) to get column (5) and so on.

▶ **Remark**

▸ Some commonly used Nationals for Divided Difference are

These are $\quad f(a,b), f(a,b,c), f(a,b,c,d)$, etc.

Here $\quad f(a,b) = \dfrac{f(b) - f(a)}{b - a}$ = first order divided difference.

$$f(a,b,c) = \dfrac{f(b,c) - f(a,b)}{c - a}$$ = second order divided difference.

$$f(a,b,c,d) = \dfrac{f(b,c,d) - f(a,b,c)}{d - a}$$ = third order divided difference.

In general,

$$f(a,b,c,...,l,m,n) = \dfrac{f(b,c,...,m,n) - f(a,b,...,l,m)}{n - a}$$ = m^{th} order divided difference.

5.9 DIFFERENCES BETWEEN DIVIDED DIFFERENCE AND ORDINARY DIFFERENCE

The following are the basic differences between divided difference and ordinary differences :

(i) An ordinary difference is not affected by the changes in the values of the argument, whereas in divided difference, the changes in the values of the argument are to be taken into consideration.

(ii) An ordinary difference being the difference between two successive values of the entry has numerator only whereas, in the case of divided difference though the numerator will be same as that of ordinary difference but it will also contain a term in denominator which is equal to the difference between the values of the argument corresponding to the entries in the numerator of the difference e.g.,

Given

$x = a$	b	c	d ...
$y = f(a)$	$f(b)$	$f(c)$	$f(d)$ ...

Ordinary difference $\Delta y = f(b) - f(a)$ and Divided difference $\Delta\, y = \dfrac{f(b) - f(a)}{b - a} = \dfrac{\Delta y}{b - a}$

(iii) The operator and the operand of ordinary differences contain no suffixes whereas the suffixes of the operator and the operand in divided difference have special significance e.g., $\underset{b}{\Delta}\, f(a)$ implies that it is the divided difference of the function $f(x)$ at the point $x = a$ taking into consideration the entry at $x = b$. Similarly $\underset{bc}{\Delta^2}\, f(a)$ means that it is the divided difference between first order differences at $x = a$ and $x = b$.

Similarly, $\quad \underset{bc}{\Delta}\, f(a) = \underset{c}{\Delta}\left[\underset{b}{\Delta}\, f(a) \right] = \underset{c}{\Delta}\left[\dfrac{f(a)}{a - b} + \dfrac{f(b)}{b - a} \right] = \dfrac{\left[\dfrac{f(a)}{a - b} + \dfrac{f(b)}{b - a} \right]}{a - c} + \dfrac{\left[\dfrac{f(c)}{c - b} + \dfrac{f(b)}{b - c} \right]}{c - a}$

$$...(2)$$

In second order divided difference there are three suffixes a, b and c and we consider it as the operation of $\underset{c}{\Delta}$ on $\underset{b}{\Delta}\, f(a)$, i.e., in the first term of (2) $\underset{b}{\Delta}\, f(a)$ is written as it is and is divided by $(a - c)$ and in second term put c wherever there is 'a' in $\underset{b}{\Delta}\, f(a)$ and then divide by $c - a$.

Simplifying (2) we get

$$\Delta^2_{bc} f(a) = \frac{f(a)}{(a-b)(a-c)} + \frac{f(b)}{(b-a)(b-c)} + \frac{f(c)}{(c-a)(c-b)}$$

$$= \sum \frac{f(a)}{(a-b)(a-c)} \qquad \qquad ...(3)$$

Now $\qquad \Delta^3_{bcd} f(a) = \Delta_d \left(\Delta^2_{bc} f(a) \right)$

$$= \frac{1}{(a-d)} \left[\frac{f(a)}{(a-b)(a-c)} + \frac{f(b)}{(b-a)(b-c)} + \frac{f(c)}{(c-a)(c-b)} \right]$$

$$+ \frac{1}{(d-a)} \left[\frac{f(d)}{(d-b)(d-c)} + \frac{f(b)}{(b-d)(b-c)} + \frac{f(c)}{(c-d)(c-b)} \right]$$

$$= \frac{f(a)}{(a-b)(a-c)(a-d)} + \frac{f(b)}{(b-a)(b-c)(b-d)}$$

$$+ \frac{f(c)}{(c-a)(c-b)(c-d)} + \frac{f(d)}{(d-a)(d-b)(d-c)}$$

$$= \sum \frac{f(a)}{(a-b)(a-c)(a-d)} \qquad \qquad ...(4)$$

Similarly we can get the higher order divided differences in terms of the values of the entry as

$$\Delta^{n-1}_{bc...n} f(a) = \sum \frac{f(a)}{(a-b)(a-c)...(a-n)} \qquad \qquad ...(5)$$

THEOREM 1. *Divided differences are symmetric with respect to the arguments.*

PROOF. We known that, $f(x_0, x_1) = \dfrac{f(x_1) - f(x_0)}{x_1 - x_0}$

$$= \frac{-[f(x_0) - f(x_1)]}{-[x_0 - x_1]} = \frac{f(x_0) - f(x_1)}{x_0 - x_1}$$

$$f(x_0, x_1) = f(x_1, x_0) \qquad \qquad ...(1)$$

Also $f(x_0, x_1, x_2) = \dfrac{f(x_1, x_2) - f(x_1, x_0)}{x_2 - x_0} = \dfrac{f(x_1, x_0) - f(x_2, x_1)}{x_0 - x_2} = f(x_2, x_1, x_0)$

[From (1)]

Similarly, we can show $f(x_0, x_1, x_2 ... x_n) = f(x_n, x_{n-1}, ..., x_1, x_0)$

Hence, divided differences are symmetric.

THEOREM 2. *The nth divided differences of a polynomial of degree n are constant.*

PROOF. Let $f(x) = A_0 x^n + A_1 x^{n-1} + A_2 x^{n-2} + ... + A_{n-1} x + A_n$ be a polynomial of degree n provided $a_0 \neq 0$. Then the first divided difference between the arguments x and $x + p_1$, is given by

$$f(x, x + p_1) = \frac{f(x + p_1) - f(x)}{x + p_1 - x}$$

$$= \frac{1}{p_i}[A_0(x+p_1)^n + A_1(x+p_1)^{n-1} + ... + A_{n-1}(x+p_1)$$

$$+ A_n - A_0(x^n - a_1 x^{n-1})... - A_{n-1}x - A_n]$$

$$= \frac{1}{p_1}[\{A_0\{x+p_1)^n + x^n\} + A_1\{(x+p_1)^{n-1} - x^{n-1}\} + ...$$

$$+ A_{n-1}(x+p_1-x)]$$

$$= \frac{1}{p_1}\left[A_0\left\{nx^{n-1}p_1 + \frac{n(n-1)}{2!}x^{n-2}p_1^2 + ...\right\}\right.$$

$$+ A_1\left\{(n-1)x^{n-2}p_1 + \frac{(n-1)(n-2)}{2!}\right.$$

$$\left.\left. x^{n-3}p_1^2 + ...\right\}... + A_{n-1}p_1\right]$$

(Expand by binomial theorem)

$$= nA_0 x^{n-1} + \left\{p_1\frac{n(n-1)}{2!}A_0 + A_1(n-1)p_1\right\}x^{n-2}... + A_{n-1}$$

$$f(x,x+p_1) = b_0 x^{n-1} + b_1 x^{n-1} + ... + b_{n-1} \qquad ...(1)$$

where $B_0, B_1, ..., B_{n-1}$ are the constants in terms of $p_1, A_0, A_1, ..., A_n$. Clearly, equation (1) is a polynomial of degree $(n-1)$.

Now the second divided difference between x and $x+p_2$ is given by

$$f(x,x+p_1,x+p_2) = \frac{f(x+p_1,x+p_2) - f(x,x+p_1)}{x+p_2-x}$$

$$= \frac{1}{p_2}\left[\left(B_0 x^{n-1} + B_1' x^{n-2} + ... + B_{n-1}'\right) - \left(B_0 x^{n-1} + B_1 x^{n-2} + ... + B_{n-1}\right)\right]$$

$$= \frac{1}{p_2}\left[(B_1' - B_1)x^{n-2} + (B_2' - B_2)x^{n-3} + ... + (B_{n-1}' - B_{n-1})\right]$$

$$f(x,x+p_1,x+p_2) = c_0 x^{n-2} + c_1 x^{n-3} + ... + c_{n-2} \qquad ...(2)$$

where $c_0, c_1, ..., c_{n-2}$ are the constants in terms of $b_1', b_1, b_2', b_2, ...$, etc. Clearly, equation (2) is of degree $(n-2)$. Proceeding in the same way we can say that $(n-1)$th divided difference will linear in x. That is,

$$f(x,x+p_1,...,x+P_{n-1}) = Cx + D \qquad ...(3)$$

∴ n^{th} divided difference is given by

$$f(x,x+p_1,x+p_2,...x+p_n) = \frac{1}{p_n}[(Cx+D') - (Cx+D)]$$

$$= \frac{D'-D}{p_n} \text{ constant}$$

Hence, n^{th} divided difference of $f(x)$ is constant.

THEOREM 3.   *The nth divided difference can be expressed as the quotient of two determinants of order $(n+1)$.*

PROOF.   The first divided difference for the arguments x_0 and x_1 is given by

$$f(x_0,x_1) = \frac{f(x_1)-f(x_0)}{(x_1-x_0)} = \frac{f(x_1)}{(x_1-x_0)} - \frac{f(x)}{(x_1-x_0)}$$

$$\Rightarrow f(x_0,x_1) = \begin{vmatrix} f(x_1) & f(x_0) \\ 1 & 1 \end{vmatrix} \div \begin{vmatrix} x_1 & x_0 \\ 1 & 1 \end{vmatrix}$$

R.H.S. of (1) is the quotient of two determinants of order 2. Now the second divided difference for the arguments is given by

$$f(x_0,x_1,x_2) = \frac{f(x_1,x_2)-f(x_0,x_1)}{(x_2-x_0)} = \frac{1}{(x_2-x_0)}\left[\frac{f(x_2)-f(x_1)}{x_2-x_1} - \frac{f(x_1)-f(x_0)}{x_1-x_0}\right]$$

$$= \frac{1}{x_2-x_0}\left[\frac{f(x_0)}{x_1-x_0} - f(x_1)\left\{\frac{x_1-x_0+x_2-x_1}{(x_2-x_1)(x_1-x_0)}\right\} + \frac{f(x_2)}{(x_2-x_1)}\right]$$

$$= \frac{f(x_0)}{(x_2-x_0)(x_1-x_0)} + \frac{f(x_1)}{(x_1-x_2)(x_1-x_0)} + \frac{f(x_2)}{(x_2-x_0)(x_2-x_1)}$$

$$= \frac{(x_1-x_2)f(x_0)+(x_2-x_0)f(x_1)+(x_0-x_1)f(x_2)}{(x_1-x_0)(x_1-x_2)(x_2-x_0)}$$

$$= \begin{vmatrix} f(x_0) & f(x_1) & f(x_2) \\ x_0 & x_1 & x_2 \\ 1 & 1 & 1 \end{vmatrix} \div \begin{vmatrix} x_0^2 & x_1^2 & x_2^2 \\ x_0 & x_1 & x_2 \\ 1 & 1 & 1 \end{vmatrix}$$

Similarly third divided difference is given by

$$f(x_0,x_1,x_2,x_3) = \begin{vmatrix} f(x_0) & f(x_1) & f(x_2) & f(x_3) \\ x_0^2 & x_1^2 & x_2^2 & x_3^2 \\ x_0 & x_1 & x_2 & x_3 \\ 1 & 1 & 1 & 1 \end{vmatrix} \div \begin{vmatrix} x_0^3 & x_1^3 & x_2^3 & x_3^3 \\ x_0^2 & x_1^2 & x_2^2 & x_3^2 \\ x_0 & x_1 & x_2 & x_3 \\ 1 & 1 & 1 & 1 \end{vmatrix}$$

proceeding in this way, we conclude that the nth divided difference can be expressed as the quotient of two determinants of order $(n+1)$.

THEOREM 4.   *The nth divided difference can be expressed as the product of multiple integrals such that*

$$f(x_1,x_2,...,x_n) = \int_0^1 dt_1 \int_0^{t_1} dt_2 \int_0^{t_2} dt_3 ... \int_0^{t_{n-2}} f^{n-1}(u_n)dt_{n-1} \qquad ...(1)$$

where, $u_n = (1-t_1)x_1 + (t_1-t_2)x_2 + ... + (t_{n-2}-t_{n-1})x_{n-1} + t_{n-1}x_n$ *and*

$t_1,t_2,...,t_n$ *are independent variables, and* f^{n-1} *means the* $(n-1)^{th}$ *derivative of f.*

PROOF.   We will prove this result by the principle of Mathematical induction. First, we shall prove the result for $n = 2$ and $n = 3$.

For $n = 2$, Then

$$\text{R.H.S.} = \int_0^1 f'(u_2)dt_1 = \int_0^1 f'[(1-t_1)x_1 + t_1x_2]dt_1$$

$$= \left[\frac{f[(1-t_1)x_1 + t_1x_2]}{-x_1+x_2}\right]_0^1 = \frac{f(x_2)-f(x_1)}{x_2-x_1} = f(x_1,x_2)$$

For $\quad n = 3$.

R.H.S. $= \int_0^1 dt_1 \int_0^{t_1} f''(u_3) dt_2 = \int_0^1 dt_1 \int_0^{t_1} f''[(1-t_1)x_1 + (t_1 - t_2)x_2 + t_2 x_3] dt_2$

$= \int_0^1 \left[\dfrac{f'[(1-t_1)x_1 + (t_1 - t_2)x_2 + t_2 x_3]}{-x_2 + x_3} \right]_0^{t_1} dt_1$

$= \dfrac{1}{x_3 - x_2} \int_0^1 [f'[(1-t_1)x_1 + t_1 x_3] - f'[(1-t_1)x_1 + t_1 x_2]] dt_1$

$= \dfrac{1}{(x_3 - x_2)} \left[\dfrac{f(1-t_1)x_1 + t_1 x_3}{-x_1 + x_3} - \dfrac{f[(1-t_1)x_1 + t_1 x_2]}{-x_1 + x_2} \right]_0^1$

$= \dfrac{1}{(x_3 - x_2)} \left[\dfrac{f(x_3)}{x_3 - x_1} - \dfrac{f(x_2)}{x_2 - x_1} - \dfrac{f(x_1)}{x_3 - x_1} + \dfrac{f(x_1)}{x_2 - x_1} \right]$

$= \dfrac{f(x_1)}{(x_1 - x_2)(x_1 - x_3)} + \dfrac{f(x_2)}{(x_2 - x_1)(x_2 - x_3)} + \dfrac{f(x_3)}{(x_3 - x_1)(x_3 - x_2)}$

$= f(x_1, x_2, x_3)$

Let the above result (1) be true for n arguments. We shall now prove that the result is true for $(n + 1)$ arguments.

Consider, $\int_0^{t_{n-1}} f^n(u_{n+1}) dt_n$

$= \int_0^{t_{n-1}} f^n[(1-t_1)x_1 + (t_1 - t_2)x_2 + ... + (t_{n-1} - t_n)x_n + t_n x_{n+1}] dt_n$

$= \left[\dfrac{f^{n-1}[(1-t_1)x_1 + (t_1 - t_2)x_2 + ... + (t_{n-1} - t_n)x_n + t_n x_{n+1}]}{-x_n + x_{n+1}} \right]_0^{t_{n-1}}$

$= \dfrac{1}{(x_{n+1} - x_n)} [f^{n-1}[(1-t_1)x_1 + (t_1 - t_2)x_2 + ... + (t_{n-2} - t_{n-1})x_{n-1} + t_{n-1}x_{n+1}]$

$\qquad -f^{n-1}[(1-t_1)x_1 + (t_1 - t_2)x_2 + ... + (t_{n-2} - t_{n-1})x_{n-1} + t_{n-1}x_n]$

$\therefore \quad \int_0^1 dt_1 \int_0^{t_1} dt_2 \int_0^{t_2} dt_3 ... \int_0^{t_{n-1}} f^n(u_{n+1}) dt_n$

$= \dfrac{1}{(x_{n+1} - x_n)} \left[\int_0^1 dt_1 \int_0^{t_1} dt_2 ... \int_0^{t_{n-2}} \left[f^{n-1}[(1-t_1)x_1 + (t_1 - t_2)x_2 + ... \right. \right.$

$\left. + (t_{n-2} - t_{n-1})x_{n-1} + t_{n-1}x_{n+1} \right]$

$\left. -f^n[(1-t_1)x_1 + (t_1 - t_2)x_2 + ... + (t_{n-2} - t_{n-1})x_{n-1} + t_{n-1}x_n] \right]$

$= \dfrac{f(x_1, x_2, ..., x_{n-1}, x_{n+1}) - f(x_1, x_2, ..., x_{n-1}, x_n)}{x_{n+1} - x_n}$

$= f(x_1, x_2, x_3, ..., x_{n+1})$

i.e., result is true for $(n+1)$ arguments. Hence, by principle of mathematical induction we can show that the nth divided difference can be expressed as the product of multiple integrals.

Recapitulations ||||||||||||||||||||||||||||||||||||||||||||||||||||||||||||||||||

* The order of any divided difference is one less than the number of values of the argument in it.

SOLVED EXAMPLES

EXAMPLE 1.   *Evaluate* $\underset{y}{\Delta} x^2$ *and* $\underset{y,z}{\Delta}^2 x^3$.   (Indore-2007, 09; Meerut-2001, 04)

SOLUTION.   Since we know that $\underset{y}{\Delta} f(x) = \dfrac{f(y) - f(x)}{y - x}$

$$\therefore \quad \underset{y}{\Delta} x^2 = \frac{y^2 - x^2}{y - x} = \frac{(y-x)(y+x)}{(y-x)} = y + x$$

and $\underset{y,z}{\Delta}^2 x^3 = \dfrac{\underset{z}{\Delta} y^3 - \underset{y}{\Delta} x^3}{z - x} = \dfrac{\dfrac{z^3 - y^3}{z - y} - \dfrac{y^3 - x^3}{y - x}}{z - x}$

$$= \frac{(z^2 + y^2 + yz) - (y^2 + x^2 + xy)}{z - x}$$

$$= \frac{z^2 - x^2 + yz - xy}{z - y} = \frac{(z-x)(z+x) + y(z-x)}{z - x}$$

$$\therefore \quad \underset{y,z}{\Delta}^2 x^3 = x + y + z.$$

EXAMPLE 2.   *Show that* $\underset{bcd}{\Delta}^3 \left(\dfrac{1}{a} \right) = -\dfrac{1}{abcd}$.   (Meerut-2005, 08, 10, 13, 16)

SOLUTION.   The divided difference table is as follows :

x	$f(x)$	$\Delta f(x)$	$\Delta^2 f(x)$	$\Delta^3 f(x)$
a	$\dfrac{1}{a}$	$\dfrac{\dfrac{1}{b} - \dfrac{1}{a}}{b - a} = -\dfrac{1}{ab}$		
			$\dfrac{(-1)\dfrac{1}{bc} - \dfrac{1}{ab}}{c - a} = (-1)^2 \dfrac{1}{abc}$	
b	$\dfrac{1}{b}$	$\dfrac{\dfrac{1}{c} - \dfrac{1}{b}}{c - b} = -\dfrac{1}{bc}$		$\dfrac{(-1)^2 \dfrac{1}{bcd} - \dfrac{1}{abc}}{d - a} = (-1)^3 \dfrac{1}{abcd}$
			$\dfrac{(-1)\dfrac{1}{dc} - \dfrac{1}{bc}}{d - b} = (-1)^2 \dfrac{1}{bcd}$	
c	$\dfrac{1}{c}$	$\dfrac{\dfrac{1}{d} - \dfrac{1}{c}}{d - c} = -\dfrac{1}{dc}$		
d	$\dfrac{1}{d}$			

Hence,

$$\underset{bcd}{\Delta}^3 \left(\frac{1}{a} \right) = (-1)^3 \frac{1}{abcd} = -\frac{1}{abcd}$$

EXAMPLE 3. *Find the third difference with arguments* 2, 4, 9, 10 *of the function* $f(x) = x^3 - 2x$.

SOLUTION. The divided difference table is as under :

x	$f(x)$	$\Delta f(x)$	$\Delta^2 f(x)$	$\Delta^3 f(x)$
2	4	$\dfrac{56-4}{4-2} = 26$		
4	56		$\dfrac{13-26}{9-2} = 15$	
9	711	$\dfrac{711-56}{9-4} = 131$		$\dfrac{23-15}{10-2} = 1$
10	980	$\dfrac{980-711}{10-9} = 269$	$\dfrac{269-131}{10-4} = 23$	

Hence, the third divided difference is 1.

EXAMPLE 4. *Evaluate :*

    (i) $\Delta(e^x \log 2x)$ **(ii)** $\Delta(x^2/\cos 2x)$ **(iii)** $\Delta^2 \cos 2x$ (PTU–2001)

    (iv) $\Delta^n(e^x)$ (Rohtak–2003)

SOLUTION.   **(i)** We have

$$\Delta(e^x \log 2x) = e^{x+h} \log 2(x+h) - e^x \log 2x$$

$$= e^{x+h} \log 2(x+h) - e^{x+h} \log 2x + e^{x+h} \log 2x - e^x \log 2x$$

$$= e^{x+h} \log \frac{x+h}{x} + (e^{x+h} - e^x) \log 2x$$

$$= e^x \left[e^h \log\left(1 + \frac{h}{x}\right) + + (e^h - 1) \log 2x \right]$$

  (ii) We have

$$\Delta\left(\frac{x^2}{\cos 2x}\right) = \frac{(x+h)^2}{\cos 2(x+h)} - \frac{x^2}{\cos 2x} = \frac{(x+h)^2 \cos 2x - x^2 \cos 2(x+h)}{\cos 2(x+h) \cos 2x}$$

$$= \frac{[(x+h)^2 - x^2]\cos 2x + x^2[\cos 2x - \cos 2(x+h)]}{\cos 2(x+h) \cos 2x}$$

$$= \frac{(2hx + h^2)\cos 2x + 2x^2 \sin(h)\sin(2x+h)}{\cos 2(x+h)\cos 2x}$$

  (iii) We have

$$\Delta^2 \cos 2x = \Delta\{\cos 2(x+h) - \cos 2x\}$$

$$= \Delta \cos 2(x+h) - \Delta \cos 2x$$

$$= [\cos 2(x+2h) - \cos 2(x+h)] - [\cos 2(x+h) - \cos 2x]$$

$$= -2\sin(2x+3h)\sin h + 2\sin(2x+h)\sin h$$

$$= -2\sin h[\sin(2x+3h) - \sin(2x+h)]$$

$$= -2\sin h[2\cos(2x+2h)\sin h]$$

$$= -4\sin^2 h \cos(2x+2h)$$

(iv) 　　　　　$\Delta e^x = e^{x+1} - e^x = (e-1)e^x$

$$\Delta^2 e^x = \Delta(\Delta e^x) = \Delta[(e-1)e^x] = (e-1)\Delta e^x$$
$$= (e-1)(e-1)e^x = (e-1)^2 e^x$$

Similarly, $\Delta^3 e^x = (x-1)^3 e^x$

and 　　　$\Delta^n e^x = (x-1)^n e^x$

EXAMPLE 5. 　If $y = a(3)^x + b(-2)^x$ and $h = 1$, prove that $(\Delta^2 + \Delta - 6)y = 0$. 　　(Mumbai–2003)

SOLUTION. 　Here, we have

$$y = a(3)^x + b(-2)^x$$
$$\Delta y = [a(3)^{x+1} + b(-2)^{x+1}] - [a(3)^x + b(-2)^x]$$
$$= 2a(3)^x - 3b(-2)^x$$
$$\Delta^2 y = [2a(3)^{x+1} - 3b(-2)^{x+1}] - [2a(3)^x - 3b(-2)^x]$$
$$= 4a(3)^x + 9b(-2)^x$$

Therefore, $(\Delta^2 + \Delta - 6)y = [4a(3)^x + 9b(-2)^x] + [2a(3)^x - 3b(-2)^x]$
$$- 6[a(3)^x + b(-2)^x] = 0$$

EXERCISE 5.1

1. Prove the following:

(i) $(1+\Delta)(1-\nabla) = 1$

(ii) $\mu\delta = \dfrac{1}{2}(\Delta + \nabla)$

(iii) $1 + \dfrac{\delta^2}{2} = \sqrt{1 + \delta^2\mu^2}$

　　　　[UPTU(MCA)–2008, Madras–2001]

(iv) $\nabla = \Delta E^{-1} = E^{-1}\Delta = 1 - E^{-1}$
　　　　　　　　　　　(Rohilkand-2003)

(v) $E^{1/2} = \mu + \dfrac{\delta}{2}$

(vi) $\delta(E^{1/2} + E^{-1/2}) = \Delta E^{-1} + \Delta$

(vii) $\Delta\nabla = \nabla\Delta = \delta^2$

(viii) $\delta = \Delta(1+\Delta)^{-1/2} = \nabla(1-\nabla)^{-1/2}$

2. Evaluate the following :

(i) $\Delta \cot 2^x$

(ii) $\dfrac{\Delta^2 x^3}{Ex^3}$ 　　　　　(Meerut-2007)

(iii) $\Delta \tan^{-1} a.x$

3. Prove the following :

(i) For any positive integer m,

$$\frac{(x+1)^{(m)}}{m!} = \frac{x^{(m)}}{m!} + \frac{x^{(m-1)}}{(m-1)!}$$

(ii) $\Delta \sin^{-1} x$
$$= \sin^{-1}[(x+1)\sqrt{1-x^2} - x\sqrt{1-(x+1)^2}]$$

(iii) $\Delta x^{(n)} = nx^{(n-1)}$ for all integers n.

4. Prove the following :

(i) $y_4 = y_3 + \Delta y_2 + \Delta^2 y_1 + \Delta^3 y_1$

(ii) $y_4 = y_0 + 4\Delta y_0 + 6\Delta^2 y_{-1} + 10\Delta^3 y_{-1}$

5. Prove: $\dfrac{\Delta^2}{E}\sin(x+h) + \dfrac{\Delta^2 \sin(x+h)}{E\sin(x+h)}$
$$= 2\cos(h-1)[\sin(x+h)+1] .$$
　　　　　　　　　　　(Nagpur-2001, 05)

6. Find the missing term in the following :

x	100	101	102	103	104
$f(x)$	2.000	2.0043	–	2.0128	2.0170

7. Evaluate the missing term in the following :

x	1	2	3	4	5	6	7
$f(x)$	2	4	8	–	32	64	128

8. Find the missing term in the following table :

x	2.0	2.1	2.2	2.3	2.4	2.5	2.6
$f(x)$	0.135	–	0.111	0.100	–	0.082	0.024

　　　　　　　　(UPTU–2004, Agra-2006)

9. Prove the following :

(i) $(2\Delta^2 + \Delta - 1)(x^2 + 2x + 1)$
$$= 5h^2 + 2hx + 2h - x^2 - 2x - 1$$

(ii) $(\Delta + 1)(2\Delta - 1)(x^2 + 2x + 1)$
$$= 5h^2 + 2hx + 2h - x^2 - 2x - 1$$

(iii) $(E-2)(E-1)(2^{x/h} + x) = -h$

(iv) $(E^2 - 3E + 2)(2^{x/h} + x) = -h$

10. Find the lowest degree polynomial which takes the following values:

x	0	1	2	3	4	5
$f(x)$	0	3	8	15	24	35

11. Find the lowest degree polynomial which satisfying the following numbers:

0, 7, 26, 63, 124, 215, 342, 511

12. Prove the following :

(i) $u_0 + {}^xC_1\Delta u_1 + {}^xC_2\Delta^2 u_2 + {}^xC_3\Delta^3 u_3 + ...$

$= u_x + {}^xC_1\Delta^2 u_{x+1} + {}^xC_2\Delta^4 u_{x+2}$

(ii) $u_x = u_{x-1} + \Delta u_{x-2} + \Delta^2 u_{x-3} + ...$

$+ \Delta^{n-1} u_{x-n} + \Delta^n u_{x-n}$

13. Find the eighth term of the series 1, 1.095, 1.179, 1.251, 1.310.

14. Find the successive differences of $x^4 - 12x^3 + 42x^2 - 30x + 9$ when the interval of differencing is unity. *(Himachal-2006,10)*

15. Prove the following :

(i) $\Delta^n 0^m = n^m - {}^nC_1(n-1)^m$

$+ {}^nC_2(n-2)^m + ...$

(ii) $\dfrac{\Delta^n 0^m}{n!} = \dfrac{n\Delta^n 0^{m-1}}{n!} + \dfrac{\Delta^{n-1} 0^{m-1}}{(n-1)!}$

(iii) $(n+1)\Delta^n 0^n = 2[\Delta^{n-1} 0^n + \Delta^n 0^n]$

(iv) $\Delta^m 0^n + {}^nC_1\Delta^m 0^{n-1} + {}^nC_2\Delta^m 0^{n-2} + ...$

$+ \dfrac{n!}{(n-m)!} = \dfrac{1}{m+1} \cdot \Delta^{m+1} 0^{n+1}$

16. Express the following polynomial into factorial notations :

(i) $x^4 + x + 1$ (ii) $2x^4 + 5x^2 + 4x + 6$

17. Show that :

(i) $\Delta\left[\dfrac{1}{f(x)}\right] = \dfrac{-\Delta f(x)}{f(x)f(x+1)}$ *(JNTU-2009)*

(ii) $D = \dfrac{2}{h}\sin^{-1}\left(\dfrac{\delta}{2}\right)$ *(Madras-2001)*

(iii) $\Delta(f_k^2) = (f_k + f_{k+1})\Delta f_k$ *(JNTU-2006)*

18. Evaluate :

(i) $\Delta \tan^{-1}\left(\dfrac{n-1}{n}\right)$ *(JNTU-2008)*

(ii) $\Delta\left\{\dfrac{1}{x(x+4)(x+8)}\right\}$ *(Madras-2001)*

(iii) $\Delta^3 \cos 3x$ *(Mumbai-2003)*

(iv) $\Delta^2\left(\dfrac{1}{x^2+5x+6}\right)$ *(PTU-2001)*

(v) $\Delta^4[(1-x)(1-2x)(1-3x)(1-4x)]$, $(h=1)$ *(Madras-2001)*

(vi) $\Delta^{10}[(1-x)(1-2x^2)(1-3x^3)(1-4x^4)]$, if the interval of differencing is 2.   *(JNTU-2009)*

19. If $f(x) = e^{ax+b}$, show that its leading differences form a geometric progression. *(Mumbai-2003)*

20. Express $x^3 - 2x^2 + x - 1$ into factorial polynomial. Hence, show that $\Delta^4 f(x) = 0$. *(PTU-2001)*

21. Explain the difference between $\left(\dfrac{\Delta^2}{E}\right)u_x$ and $\left(\dfrac{\Delta^2 u_x}{Eu_x}\right)$.   *(Gujrat-2004, 08; Madras-2003)*

22. Obtain the estimate of the missing figures in the following table :

x	1	2	3	4	5	6	7	8
y	2	4	8	...	32	...	128	256

(Mumbai-2004)

23. If $u_{13} = 1, u_{14} = -3, u_{15} = -1, u_{16} = 13$, find u_8.   *(Mumbai-2004)*

24. Prove that $\Delta = \dfrac{1}{2}\delta^2 + \delta\sqrt{\left(1+\dfrac{\delta^2}{4}\right)}$.   *(Mumbai-2004)*

25. If $u_{10} = 3, u_{11} = 6, u_{12} = 11, u_{13} = 18$, $u_{14} = 27$ find u_4.   *(Mumbai-2005)*

26. If y_x is a polynomial for which fifth differences is constant and $y_1 + y_7 = 784, y_2 + y_6 = 686$, $y_3 + y_5 = 1088$, find y_4.   *(Mumbai-2004)*

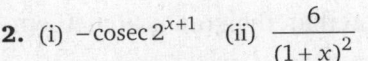

Answers

2. (i) $-\operatorname{cosec} 2^{x+1}$ (ii) $\dfrac{6}{(1+x)^2}$ (iii) $\tan^{-1}\left(\dfrac{1}{1+a^2 x + a^2 x^2}\right)$ **6.** 2.0086   **7.** 17

8. 0.132, 0.90   **10.** $x^2 + 2x$ **11.** $x^3 + 3x^2 + 3x$ **13.** 1.399

14. $\Delta f(x) = 4x^3 - 30x^2 + 52x + 1$, $\Delta^2 f(x) = 12x^2 - 48x + 26$, $\Delta^3 f(x) = 24x - 36$, $\Delta^4 f(x) = 24$

16. (i) $x^{(4)} + 6x^{(3)} + 7x^{(2)} + 2x^{(1)} + 1$ (ii) $2x^{(4)} + 12x^{(3)} + 19x^{(2)} + 11x^{(1)} + 6$

18. (i) $\tan^{-1}\left(\dfrac{1}{2n^2}\right)$ (ii) $192/[x(x+4)(x+8)(x+12)(x+16)]$ (iii) $(e^2 - 1)^n e^{2x+3}$

　　　(iv) $(-1)^n n!/x[x(x+1)(x+2) \dots (x+n)]$ **20.** $[x]^3 + [x]^2 - 1$

22. $y(4) = 16.26, y(6) = 63.47$ **23.** -99 **25.** 27 **26.** 571

5.10 INTERPOLATION

Any one who has had occasion to consult tables of mathematical functions is familiar with the method of linear interpolation and probably face the situation in which the method of "reading between the lines of the table" has appeared.

For example, if we are to find out the population of India in 1998 when we know that the census in India is done is 1941, 1951, ... 2001, *i.e.*, the figure of population are available. Then the process of finding the figure is known as interpolation.

Definition. *The method of obtaining the value of a function for any intermediate value of the argument from the given set of values of the function for certain values of arguments is known n as interpolation. On the other hand, the process of computing the value of the function outside the given range is called extrapolation.*

▶ **REMARK**

▸ The study of interpolation is based on the calculus of finite differences. The calculus of finite differences plays an important role in numerical analysis. It deals with the variations in a function when independent variable changes by finite jumps which may be equal or unequal.

5.10.1 ASSUMPTIONS OF INTERPOLATION

　(1) The values of the function should be either increasing or decreasing.
　(2) The rise or fall in the given values should be uniform.
　(3) The given set of observations should be capable of being expressed in a polynomial form.

5.11 METHODS OF INTERPOLATION

5.11.1 METHOD OF GRAPH

　　Let $y = f(x)$ be a given function. Plot a graph between the values of x and the corresponding values of $y = f(x)$. With the help of the graph we can obtained the unknown values of $f(x)$ for the given values of x.

5.11.2 THE METHOD OF CURVE FITTING

　　This method can be used only when the form of the function is known to us. Then by the method of least squares we can fit the curve of known form to the given set of observations and with the help of the fitted curve we can calculate the unknown value.

Merit

The only merit of this method lies in its closer approximation than the graphical method.

Demerits

　　(1) The form of the function for the given set of observations is assumed to be known.
　　(2) When some additional observations are included in the data then the calculation for finding the unknown constants are to be done afresh.

(3) The method is not exact.

(4) The method is complicated when the number of observations is sufficiently large.

5.11.3 FINITE DIFFERENCE METHOD

We can find the value of entry for a given value of argument without actually knowing the form of the function. The use of these methods though approximate have distinct advantages over the methods of graphs and curve fitting. The merits and demerits of this method are as follows.

Merits

(i) The method does not assume the form of the function to be known.

(ii) It is less approximate than the method of graphs.

(iii) The calculations remain simple even if some additional observations are included in the given data.

Demerit

There is no definite way to verify whether the assumptions for the application of finite difference calculus are valid for the given set of observations.

5.12 FINITE DIFFERENCE CALCULUS

We now proceed to study the use of finite difference calculus for the purpose of interpolation. This we shall do in three cases which are as follows :

(1) The value of the argument in given data varies by an equal interval. The technique is known as interpolation with equal intervals.

(2) The values of the argument are not at equal intervals. This is known as interpolation with unequal intervals.

(3) The technique of central differences.

5.13 NEWTON'S FORMULAE FOR INTERPOLATION

(1) Newton-Gregory Formula for Forward Interpolation with Equal Intervals.

Newton's-Gregory's formula for forward interpolation with equal interval is

$$f(x_0 + hu) = f(x_0) + u\Delta f(x_0) + \frac{u(u-1)}{2!}\Delta^2 f(x_0) + \frac{u(u-1)(u-2)}{3!}\Delta^3 f(x_0) + \dots$$

$$+ \frac{u(u-1)(u-2)\dots(u-(n-1)h)}{n!}\Delta^n f(x_0)$$

where $u = \dfrac{x - x_0}{h}$.

Proof. Let $y = f(x)$ be a function which takes the values $f(x_0), f(x_0 + h), f(x_0 + 2h)$, ... $f(x_0 + nh)$, for $x = x_0, (x_0 + h), (x_0 + 2h)\dots(x_0 + nh)$, *i.e,* for $(n + 1)$ equidistant values $(x_0), (x_0 + 2h)\dots(x_0 + nh)$ of the independent variable x.

Here, we assume that $f(x)$ is a polynomial of nth degree, therefore, $f(x)$ can be written as

$$f(x_0) = A_0 + A_1(x - x_0) + A_2(x - x_0)(x - x_0 - h) + A_3(x - x_0)(x - x_0 - h)(x - x_0 - 2h)\dots$$
$$+ A_n(x - x_0)(x - x_0 - h)\dots(x - x_0 - (n-1)h) \qquad \dots(1)$$

where $A_0, A_1, \dots, A_n$ are constants.

Putting $x = x_0, x_0 + h, x_0 + 2h, \dots$ in succession in (1), we get

$$f(x_0) = A_0 \qquad \dots(2)$$

and $\qquad f(x_0 + h) = A_0 + A_1.h$

$\Rightarrow \qquad\qquad A_1.h = f(x_0 + h) - A_0 = f(x_0 + h) - f(x_0)$

$\Rightarrow \qquad\qquad A_1 = \dfrac{f(x_0 + h) - f(x_0)}{h} = \dfrac{\Delta f(x_0)}{h}$

Also, $\quad f(x_0 + 2h) = A_0 + A_1(2h) + A_2(2h)(h)$

$\Rightarrow \qquad\qquad 2h^2 A_2 = f(x_0 + 2h) - A_0 - 2A_1.h = f(x_0 + 2h) - f(x_0) - 2\Delta f(x_0)$

$\qquad\qquad\qquad = f(x_2 + 2h) - f(x_0) - 2f[[(x_0 + h) + f(x_0)]$

$\qquad\qquad\qquad = f(x_2 + 2h) - 2f(x_0 + h) + f(x_0) = \Delta^2 f(x_0)$

$\Rightarrow \qquad\qquad A_2 = \dfrac{\Delta^2 f(x_0)}{2!h^2}$

Similarly $\qquad\qquad A_3 = \dfrac{\Delta^3 f(x_0)}{3!h^3} \,\ldots\ldots\, A_n = \dfrac{\Delta^n f(x_0)}{n!h^n}$ and so on.

Substituting the values of $A_0, A_1, A_2, \ldots$ from (2), (3) ... (5) in (1), we get

$$f(x) = f(x_0) + \frac{\Delta f(x_0)}{h} + \frac{\Delta^2 f(x_0)}{2!h^2}(x - x_0)(x - x_0 - h)$$

$$+ \frac{\Delta^3 f(x_0)}{3!.h^3}(x - x_0)(x - x_0 - h)(x - x_0 - 2h)\ldots$$

$$+ \frac{\Delta^n f(x_0)}{n!h^n}(x - x_0)(x - x_0 - h)\ldots\{x - x_0(n-1)h\} \qquad \ldots(6)$$

Now let $\qquad\qquad u = \dfrac{x - x_0}{h}$. Then $(u-1) = \dfrac{x - x_0 - h}{h}; \quad (u-2) = (x - x_0 - 2h)$

$$\vdots \qquad\qquad \vdots$$

$$[u - (n-1)] = [x - x_0 - (n-1)h]$$

Substituting these values in (6), we get

$$f(x + hu) = f(x_0) + u\Delta f(x_0) + \frac{u(u-1)}{2!}\Delta^2 f(x_0) + \frac{u(u-1)(u-2)}{3!}\frac{\Delta^3 f(x_0)}{2} + \ldots$$

$$+ \frac{u(u-1)(u-2)\ldots(u-(n-1)h)}{n!}\Delta^2 f(x_0)$$

which is the required Newton-Gregory's formula for forward interpolation with equal interval.

▶ **REMARKS**

▶ This formula is particularly useful for interpolating the value of $f(x)$ near the beginning of the set of given values.

▶ This formula can be expressed in terms of factorials, as follows

$$f(x_0 + hu) = f(x_0) + \frac{u^{(1)}}{1!}\Delta f(x_0) + \frac{u^{(2)}}{2!}\Delta^2 f(x_0) + \frac{u^{(3)}}{3!}\Delta^3 f(x_0) + \ldots + \frac{u^{(n)}}{n!}\Delta^n f(x_0)$$

Remainder Term

The remainder term in Newton-Gregory's forward interpolation formula is given by

$$R_n = \frac{u(u-1)(u-2)\ldots(u-n)}{(n+1)!}h^{n+1}f^{n+1}(\theta) \text{ where } x_0 \le \theta \le x$$

Since R_n contains derivative of $(n+1)^{\text{th}}$ order of function $f(x)$ at $x = \theta$ so we may not say something about the error when the form of the function is not known. In this case, expressed $f^{n+1}(\theta)$ in terms of the difference of the function

$$\therefore \qquad R_n = \frac{u^{(n+1)}}{(n+1)!} \Delta^{n+1} f(x_0).$$

(2) Newton-Gregory's Backward Interpolation formula with Equal Intervals.

Newton-Gregory's formula for backward interpolation formula with equal interval is

$$f(a + nh + uh) = f(a + nh) + u\nabla f(a + nh) + \frac{u(u+1)}{2!}\nabla^2 f(a + nh) + \dots$$

$$+ \frac{u(u+1)(u+2)\dots(u+n-1)}{n!}\nabla^n f(a + nh)$$

[UPTU MCA-2008 (BP)]

Proof. Let $y = f(x)$ be a function, which assumes the values $f(a), f(a+h), f(a+2h), \dots, f(a+nh)$ corresponding to the values of $x = a, (a+h), (a+2h), \dots, + (a+nh)$ respectively. The values of x are equidistant.

Assume the function $f(x)$ such that

$$f(x) = A_0 + A_1(x - a - nh) + A_2(x - a - nh)[x - a - (n-1)h]$$

$$+ A_3(a - x - nh)(x - a - (n-1)h)[(x - a(n-2)h]$$

$$+ \dots + A_n(x - a - nh)[x - a - (n-1)h]\dots(x - a - h) \quad \dots(1)$$

where $A_0, A_1, A_2, \dots, A_n$ are constants.

Putting $x = a + nh$ in (1), we get

$$f(a + nh) = A_0 \qquad \dots(2)$$

Now, putting $x = a + (n-1)h$ in (1), we get

$$f[a + (n-1)h] = A_0 + A_1[a + (n+1)h - a - nh] = A_0 - A_1h$$

$$\Rightarrow \qquad A_1h = A_0 - f[a + (n+1)h]$$

$$= f(a + nh) - f[a + (n-1)h] = \Delta f(a + nh)$$

i.e.,
$$A_1 = \frac{\nabla f(a + nh)}{h} \qquad \dots(3)$$

Similarly, putting $x = a + (n-2)h$ in (1), we get

$$f[a + (n-2)h] = A_0 + A_1[a + (n-2)h - a - nh]$$

$$+ A_2[a + (n-2)h - a - nh][a + (n-2)h - a - (n-1)]$$

$$= A_0 + A_1(-2h) + A_2(-2h)(-h)$$

$$A_2 \cdot 2h^2 = f[a + (n-2)h] - A_0 + 2A_1h$$

$$= f[a + (n-2)h] - f(a + nh) + 2\nabla f(a + nh)$$

$$= f[a + (n-2)h] - f(a + nh) + 2\{f(a + nh) - f[a + (n-1)h]\}$$

$$= \{f[a + (n-2)h - f(a + (n-1)h]\} - \{f[a + (n-1)h] - f(a + nh)\}$$

$$= \nabla f[a + (n-1)h] - \nabla f(a + nh) = \nabla^2 f(a + nh)$$

$$\Rightarrow \qquad A_2 = \frac{\nabla^2 f(a + nh)}{h^2 \cdot 2!} \qquad \dots(4)$$

Proceeding in the same way, by putting $x = a + (n-3)h, a + (n-4)h$ ..., etc. in (1), we get

$$A_3 = \frac{\nabla^3 f(a+nh)}{h^3.3!} ... A_n = \frac{\nabla^n f(a+nh)}{h^n.n!} \qquad ...(5)$$

Substituting all these values of $A_0, A_1, A_2,$ ... in (1), we get

$$f(x) = f(a+nh) + (x-a-nh)\frac{\nabla f(a+nh)}{h.1!} + (x-a-nh)[x-a(n-1)h]\frac{\nabla^2 f(a+nh)}{2!} + ...$$

$$+ (x-a-nh)[x-a(n-1)h]...(x-a-h)\frac{\nabla^n f(a+nh)}{h^n.n!} \qquad ...(6)$$

Let us define
$$u = \frac{x-(a-nh)}{h} \quad \text{so,} \quad x-a-(n-1)h = uh+h = (u+1)h$$

$$x-a-h = (a+nh+uh)-a-h = (u+n-1)h$$

Put all these values in equation (6), we get

$$f(a+nh+uh) = f(a+nh) + u\nabla f(a+nh) + \frac{u(u+1)}{2!}\nabla^2 f(a+nh)$$

$$+...+ \frac{u(u+1)...(u+n-1)}{n!}\nabla^2 f(a+nh)$$

which is the required Newton Gregory's backward interpolation formula for equal intervals.

▶ **REMARK**

▶ This formula is particularly useful for interpolating the values of $f(x)$ near the end of the set of given values.

ALGORITHM

Step 1.	Put the values of $x_i, f(x_i)$ and value of x for which $f(x)$ is to be calculated.
Step 2.	Construct the difference table by using. $\nabla^n f(x_i) = \nabla^{n-1} f(x_i + 1) - \nabla^{n-1} f(x_i - 1)$
Step 3.	Obtain $u = \dfrac{x-(a-nh)}{h}$
Step 4.	Put all these values in Newton's backward interpolation formula.

Remainder Term

The remainder term R_n is given by

$$R_n = u(u+1)...(u+n)\frac{h^{n+1}}{(n+1)!}f^{n+1}(\theta) \quad \text{where} \quad u = \frac{x-x_n}{h}$$

▶ **REMARK**

▶ If the analytical form of the function $f(x)$ is not known then expressed $f^{n+1}(\theta)$ in terms of difference, i.e., $\dfrac{\nabla^{n+1} f(x_n)}{h^{n+1}}$ such that $R_n = \dfrac{\nabla^{n+1} f(x_n)}{(n+1)!}u(u+1)(u+2)...(u+n)$.

Recapitulations

● In Newton's forward interpolation formula, the starting point may be any tabular value but then the formula will contain only those values of y which come after the value chosen as starting point.

SOLVED EXAMPLES

EXAMPLE 1. *For a certain town the population was given below*:

Year	1931	1941	1951	1961
Populations in thousands	172	171.7	157.2	183.9

Find the population for 1946.

SOLUTION. Here $'a' = 1931$, $'h' = 10$ and $'a + hu' = 1946$

$\therefore$ $\qquad a + hu = 1946 \qquad$ or $\qquad 1931 + 10u = 1946$

or $\qquad u = 1.5$ $\qquad\qquad\qquad\qquad\qquad\qquad$ ...(1)

For the given data we have the following difference table:

Year	x	Population (y_x)	Δy_x	$\Delta^2 y_x$	$\Delta^3 y_x$
1931	0	172			
1941	1	171.7	−0.3	−14.2	
1951	2	157.2	−14.5	41.2	55.4
1961	3	183.9	26.7		

Now from Newton-Gregory's formula, we have

$$y(a+hu) = y_a + u\Delta y_a + \frac{u(u-1)}{2!}\Delta^2 y_a + \frac{u(u-1)(u-2)}{3!}\Delta^3 y_a$$

$$\Rightarrow \quad y_{1946} = y_{1931} + u\Delta y_{1931} + \frac{u(u-1)}{2!}\Delta^2 y_{1931} + \frac{u(u-1)(u-2)}{3!}\Delta^3 y_{1931}$$

$$= 172 + \frac{1.5}{1!}(-0.3) + \frac{(1.5)(1.5-1)}{2!}(-14.2) + \frac{(1.5)(1.5-1)(1.5-2)}{3!}(55.4)$$

$$= 172 - \frac{(1.5)(0.3)}{1} - \frac{(1.5)(0.5)(14.2)}{(2\times1)} - \frac{(1.5)(0.5)(0.5)(55.4)}{3\times2\times1}$$

$$= 172 - 0.45 - 5.325 - 3.4625 = 162.7625.$$

Hence, the required population for 1946 is 162.7625 or 162.8 thousands.

EXAMPLE 2. *From the following table, find the number of students who obtained less than 45 marks*:

Marks	Number of Students
30-40	31
40-50	42
50-60	51
60-70	35
70-80	31

(VTU–2007; UKTU, 2011 S; SVTU 2007; Madras, 2006; UPPCS-2006; Avadh-2005, Rohilkhand-2004; Madras-2009)

SOLUTION. The difference table for the given data is as follows:

Marks	No. of student $f(x)$	$\Delta f(x)$	$\Delta^2 f(x)$	$\Delta^3 f(x)$	$\Delta^4 f(x)$
Less 40	**31**				
Less 50	73	**42**	**9**		
Less 60	124	51	-16	**-25**	
Less 70	159	35	-4	12	37
Less 80	190	31			

Here $h = 10$, $a = 40$ and $x = 45$

For Newton-Gregory forward interpolation formula

Let $\quad u = \dfrac{x-a}{h} = \dfrac{45-40}{10} = \dfrac{5}{10} = \dfrac{1}{2}$

Therefore,

$$f(45) = f(40) + \frac{1}{2}\Delta f(40) + \frac{\frac{1}{2}\left(\frac{1}{2}-1\right)}{2!}\Delta^2 f(40) + \frac{\frac{1}{2}\left(\frac{1}{2}-1\right)\left(\frac{1}{2}-2\right)}{3!}\Delta^3 f(40) +$$

$$\frac{\frac{1}{2}\left(\frac{1}{2}-1\right)\left(\frac{1}{2}-2\right)\left(\frac{1}{2}-3\right)}{4!}\Delta^4 f(40)$$

$$= 31 + \frac{1}{2}\times 42 - \frac{1}{8}\times 9 - \frac{1}{16}\times 25 - \frac{5}{128}\times 37$$

$$= 31 + 21 - 1.125 - 1.563 \times 1.445 = 47.867 = 48$$

Hence, the number of the students who obtain less than 45 marks are 48.

EXAMPLE 3. *The following table gives the marks secured by 100 students in the numerical subject:*

Range of Marks	30-40	40-50	50-60	60-70	70-80
No. of Students	25	35	22	11	7

Use Newton's forward difference interpolation formula to find :

(i) the number of students who got more than 55 marks.

(ii) the number of students who secured marks in the range from 36 to 45.

[UPTU(MCA)–2002]

SOLUTION. The difference table for the given data is as follows :

Marks obtained	Number of Students
Less than 40	25
Less than 50	60
Less than 60	82
Less than 70	93
Less than 80	100

(i) Here $\quad a = 40$, $h = 10$, $a + uh = 55$

$\Rightarrow \quad 40 + 10 u = 55$

$\Rightarrow \quad u = 1.5$

Now, we find the number of students who got less than 55 marks. The difference table is given as under :

Marks obtained less than	No. of students, y	Δy	$\Delta^2 y$	$\Delta^3 y$	$\Delta^4 y$
40	**25**				
50	60	**35**	−13		
60	82	22	−11	2	
70	93	11	−4	7	5
80	100	7			

Then, by Newton's forward interpolation formula, we have

$$y_{55} = y_{40} + u\Delta y_{40} + \frac{u(u-1)}{2!}\Delta^3 y_{40} + \frac{u(u-1)(u-2)(u-3)}{4!}\Delta^4 y_{40}$$

$$= 25 + (-0.4)(35) + \frac{(-4.0)(-1.4)}{2!}(-13) + \frac{(-0.4)(-1.4)(-2.4)}{3!}(2)$$

$$+ \frac{(-0.4)(-1.4)(-2.4)(-3.4)}{4!}(5)$$

$$= 7.864 = 8$$

Similarly

$$y_{45} = y_{40} + u\Delta y_{40} + \frac{u(u-1)}{2!}\Delta^2 y_{40} + \frac{u(u-1)(u-2)}{3!}\Delta^3 y_{40}$$

$$+ \frac{u(u-1)(u-2)(u-3)}{4!}\Delta^4 y_{40}$$

$$= 25 + (0.5)(35) + \frac{(0.5)(-0.5)}{2} + \frac{(0.5)(-0.5)(-1.5)}{6}(2)$$

$$+ \frac{(0.5)(-0.5)(-1.5)(-2.5)}{24}(5)$$

$$= 44.0546 = 44$$

Hence, the number of students who secured marks in the range from 36 to 45 is given by

$$y_{45} - y_{36} = 44 - 8 = 36.$$

EXAMPLE 4. *From the following table of half yearly premium for policies maturing at different ages, estimate the premium for policies maturing at the age of 46.* (UPTU-2004, 10)

Age	45	50	55	60	65
Premium (in rupees)	114.84	96.16	83.32	74.48	68.48

SOLUTION. We have the following difference table.

Age x	Premium in rupees y	Δy	$\Delta^2 y$	$\Delta^3 y$	$\Delta^4 y$
45	**114.84**				
50	96.16	**− 16.68**			
55	83.32	− 12.84	**5.84**	−1.84	
60	74.48	− 8.84	4	−1.16	0.68
65	68.48	− 6	2.84		

We have $h = 5, a = 45, a + hu = 46$

$\therefore$ $45 + 5u = 46 \Rightarrow u = 0.2$

Then, by Newton's forward difference formula, we have

$$y_{46} = y_{45} + u\Delta y_{45} + \frac{u(u-1)}{2!}\Delta^2 y_{45} + \frac{u(u-1)(u-2)}{3!}\Delta^3 y_{45} +$$

$$+ \frac{u(u-1)(u-2)(u-3)}{4!}\Delta^4 y_{45}$$

$$= 114.84 + (0.1)(-18.68) + \frac{(0.2)(0.2-1)}{2!}(5.84)$$

$$+ \frac{(0.2)(0.2-1)(0.2-2)}{3!}(-1.84)$$

$$+ \frac{(0.2)(0.2-1)(0.2-2)(0.2-3)}{4!}(0.68)$$

$$= 110.525632$$

Hence, the premium for policies maturing at the age of 46 is Rs. 110.52.

EXAMPLE 5.   *Find the value of* sin 52° *from the given table.*

θ	45°	50°	55°	60°
sin θ	0.7071	0.7660	0.8192	0.8660

(UPTU–2004, JNTU–2006; UPPCS-2004, 08, Avadh-2004)

SOLUTION.   The difference table is given below :

θ	$f(x) = \sin\theta$	$\Delta f(\theta)$	$\Delta^2 f(\theta)$	$\Delta^3 f(\theta)$
45°	**0.7071**			
		0.0589		
50°	0.7660		**– 0.0057**	
		0.0532		**–0.0007**
55°	0.8192		–0.0064	
		0.0468		
60°	0.8660			

By Newton's forward interpolation, we have,

$$f(a+uh) = f(a) + u\Delta f(a) + \frac{u(u-1)}{2!}\Delta^2 f(a) + \frac{u(u-1)(u-2)}{3!}\Delta^3 f(a) + \dots \quad \dots(1)$$

Here $a = 45°, a + uh = 52°, h = 5°,$

$$u = \frac{52° - 45°}{5} = \frac{7}{5} = 1.4$$

Put all these values in (1), we get

$$\therefore f(52°) = f(45°) + 1.4\Delta f(45°) + \frac{1.4(1.4-1)}{2!}\Delta^2 f(45°) + \frac{1.4(1.4-1)(1.4-2)}{3!}\Delta^3 f(45°)$$

$$= 0.7071 + 1.4 \times 0.0589 - \frac{1.4 \times 0.4}{2} \times 0.0057 + \frac{1.4 \times 0.4 \times (-0.6)}{6} \times (0.0007)$$

$$+ \frac{1.4 \times 0.4 \times (-0.6)}{6} \times (0.0007)$$

$$= 0.7071 + 0.08246 - 0.001596 - 0.0000392 = 0.78880032$$

Hence, sin 52° = 0.7880032

EXAMPLE 6. *Find the value of the area of the circle of diameter from the following given data.*

(Diameter)	80	85	90	95	100
A (Area)	5026	5674	6362	7088	7854

(UPPCS-2005, 10, Ajmer–2002, 05, 08; Meerut-2002, 06)

SOLUTION. The difference table of the above data is given as follows :

d (Dia-meter)	$A = f(d)$ (Area)	$\Delta f(x)$	$\Delta^2 f(x)$	$\Delta^3 f(x)$	$\Delta^4 f(x)$
80	5026				
85	5674	**648**	**40**		
90	6362	688	38	**−2**	**4**
95	7088	726	40	2	
100	7854	766			

Here $a = 80, h = 5$ and $x = 82$,

then $u = \dfrac{x-a}{h} = \dfrac{82-80}{5} = \dfrac{2}{5} = 0.4$

From Newton's forward interpolation formula, we get

$$f(82) = f(80) + 0.4\Delta f(80) + \frac{0.4(0.4-1)}{2!}\Delta^2 f(80)$$

$$+ \frac{0.4(0.4-1)(0.4-2)(0.4-3)}{4!}\Delta^4 f(80) + \frac{0.4(0.4-1)(0.4-2)}{3!}\Delta^3 f(80)$$

$$= 5026 + 0.4(648) + \frac{0.4(0.4-1)}{2}(40) + \frac{0.4(0.4-1)(0.4-2)}{6}(-2)$$

$$= 5026 + 259.2 - 4.8 - 0.128 - 0.1664 = 5280.10$$

Hence, the required area = 5280.10.

EXAMPLE 7. *Find the number of men getting wages between Rs. 10 and Rs. 15 from the following table :*

Weight in Rs.	0-10	10-20	20-30	30-40
Frequency	9	30	35	42

(Nagarjuna–2001; Delhi-2008; Avadh-2004)

SOLUTION. The difference table is given as follows :

Wages in Rs. x	Frequency $f(x)$	$\Delta f(x)$	$\Delta^2 f(x)$	$\Delta^3 f(x)$
Less than 10	**9**			
Less than 20	39	**30**	**5**	
Less than 30	74	35	7	**2**
Less than 40	116	42		

Here, $a = 10, h = 10, x = 15$

then $u = \dfrac{x-a}{h} = \dfrac{15-10}{10} = \dfrac{5}{10} = \dfrac{1}{2} = 0.5$

By Newton's forward interpolation formula, we get

$$f(15) = f(10) + 0.5\Delta f(10) + \frac{0.5(0.5-1)}{2!}\Delta^2 f(10) + \frac{0.5(0.5-1)(0.5-2)}{3!}\Delta^3 f(10)$$

$$= 9 + 0.5(30) - \frac{0.5(0.5)}{2}(5) + \frac{0.5(0.5)(1.5)}{6}(2)$$

$$= 9 + 15 - 0.625 + 0.125 = 23.5 = 24 \text{ approximately.}$$

Hence, number of men getting between Rs. 10 and Rs. 15 = 24 − 9 = 15.

EXAMPLE 8.   *The population of a country in decennial census were as under. Estimate the population for the year 1925.*

Year (x)	1891	1901	1911	1921	1931
Population (y) (in thousand)	46	66	81	93	101

(Meerut-2005)

SOLUTION.   The difference table is given as follows :

x	$y = f(x)$	$\Delta f(x)$	$\Delta^2 f(x)$	$\Delta^3 f(x)$	$\Delta^4 f(x)$
1891	46				
1901	66	20	−5		
1911	81	15	−3	2	−3
1921	93	12	−4	−1	
1931	101	8			

Here, $a = 1891, h = 10, x = 1925$

then $u = \dfrac{x-a}{h} = \dfrac{1925 - 1891}{10} = 3.4$

By Newton's forward interpolation formula, we get

$$\log_{10} 337.5 = \log 310 + 2.75\,\Delta\log 310 + \frac{2.75(2.75-1)}{2!}\Delta^2 \log 310$$

$$+ \frac{3.4(3.4-1)(3.4-2)}{3!}\Delta^3 f(1891)$$

$$+ \frac{3.4(3.4-1)(3.4-2)(3.4-3)}{4!}\Delta^4 f(1891)$$

$$= 46 + 3.4(20) - \frac{3.4(2.4)}{2}(-5) + \frac{3.4(2.4)(1.4)}{6}(2) + \frac{3.4(2.4)(1.4)(0.4)}{24}(-3)$$

$$= 46 + 68 - 20.4 + 3.808 - 0.5712 = 96.8368$$

Hence, the population for 1925 is 96.8368 thousands.

EXAMPLE 9.   *Find the value of $\log_{10} 337.5$ from the following data:*

x	310	320	330	340	350	360
$\log_{10} x$	2.4913	2.5051	2.5185	2.5315	2.5440	2.5563

(UPTU–2003)

SOLUTION. The difference table is given as follows :

x	$\log_{10}x$	$\Delta\log_{10}x$	$\Delta^2\log_{10}x$	$\Delta^3\log_{10}x$	$\Delta^4\log_{10}x$	$\Delta^5\log_{10}x$
310	**2.4913**					
320	2.5051	**0.0138**				
		0.0134	**– 0.0004**			
330	2.5185	0.0129	– 0.0005	**– 0.0001**		
340	2.5314	0.0126	– 0.0003	0.0002	**0.0003**	**– 0.0005**
350	2.5440	0.0123	– 0.0003	0.0000	– 0.0002	
360	2.5563					

Here, $a = 310$, $h = 10$. Let us assume $x = 337.5$, then

then $u = \dfrac{x-a}{h} = \dfrac{337.5-310}{10} = 2.75$

By Newton's forward interpolation formula, we get

$$\log_{10}337.5 = \log 310 + 2.75\,\Delta\log 310 + \frac{2.75(2.75-1)}{2!}\Delta^2\log 310$$

$$+\frac{2.75(2.75-1)(2.75-2)}{3!}\Delta^3\log 310$$

$$+\frac{2.75(2.75-1)(2.75-2)(2.75-3)}{4!}\Delta^4\log 310$$

$$+\frac{2.75(2.75-1)(2.75-2)(2.75-3)(2.75-4)}{5!}\Delta^5\log 310$$

$$= 2.4913 + 2.75(0.0138) + \frac{2.75(1.75)}{2}(-0.0004)$$

$$+\frac{2.75(1.75)(0.75)}{6}(-0.0001) + \frac{2.75(1.75)(0.75)(-0.25)}{24}(0.0003)$$

$$+\frac{2.75(1.75)(0.75)(-0.25)(-1.25)}{120}(-0.0005)$$

$$= 2.4913 + 0.03795 - 0.0009625 - 0.000062 - 0.000011$$
$$- 0.0000046$$

$$= 2.5282$$

(Since, $\log_{10}337.5 = 2.5282$ $\qquad\qquad \therefore\log_{10}3375 = 3.5282$)

EXAMPLE 10. *Find the form of the function from the following given data :*

x	0	1	2	3	4
$f(x)$	3	6	11	18	27

SOLUTION. The difference table is given as follows : (Bangaluru-2008; Mysore-2002,06)

x	Frequency $f(x)$	$\Delta f(x)$	$\Delta^2 f(x)$	$\Delta^3 f(x)$
0	**3**			
1	6	**3**	**+2**	
2	11	5	+2	0
3	18	7	+2	0
4	27	9		

Here $a = 10$, $h = 1$, $x = 15$ and $u = \dfrac{x-a}{h} = x$

By Newton's formula is given by

$$f(x) = f(0) + x\Delta f(0) + \frac{x(x-1)}{2!}\Delta^2 f(0) + \frac{x(x-1)(x-2)}{3!}\Delta^3 f(0)$$

$$= 3 + x.3 + \frac{x(x-1)}{2}(2) + 0 \ = \ 3 + 3x + x^2 - x$$

$$f(x) = x^2 + 2x + 3.$$

EXAMPLE 11. *Given*

x	1	2	3	4	5	6	7	8
f(x)	1	8	27	64	125	216	343	512

Find $f(7.5)$.   [UPTU–2003, MKU(Tamilnadu)-2005; Avadh-2004]

SOLUTION.   The backward difference table is given as follows :

x	$f(x)$	$\nabla f(x)$	$\Delta^2 f(x)$	$\Delta^3 f(x)$
1	1			
2	8	7	12	
3	27	19	18	6
4	64	37	24	6
5	125	61	30	6
6	216	91	36	6
7	343	129	42	**6**
8	**512**	**169**		

Here $a + nh = 8$, $x = 7.5$, $h = 1$

then $u = \dfrac{x-(a+nh)}{h} = \dfrac{7.5-8}{1} = -0.5$

Now using Newton's Backward interpolation, we get

$$f(7.5) = f(a+nh) + u\nabla f(a+nh) + \frac{u(u+1)}{2!}\nabla^2 f(a+nh)$$

$$+ \frac{u(u+1)(u+2)}{3!}\nabla^3 f(a+nh)$$

$$f(7.5) = f(8) + (-0.5)\nabla f(8) + \frac{(-0.5)(-0.5+1)}{2}\nabla^2 f(8)$$

$$= 512 - 0.5(169) - \frac{(0.5)(0.5)}{2}(42) - \frac{(0.5)(0.5)(1.5)}{6}(6)$$

$$= 512 - 84.5 - 5.25 - 0.375$$

Hence, $f(7.8) = 421.875$.

EXAMPLE 12. *If l_x represents the number of persons living at age x in a life table, find as accurately as the data will permit the value of l_{47}. Given that*

$$l_{20} = 512, l_{30} = 439, l_{40} = 346, l_{50} = 243$$

SOLUTION.   The backward difference table is given as follows:

x	l_x	∇l_x	$\nabla^2 l_x$	$\nabla^3 l_x$
20	512			
30	439	−73	−20	**10**
40	346	−93	**−10**	
50	**243**	**−103**		

Here $50 = a+nh$, $h = 10$ and $u = \dfrac{47-50}{10} = -0.3$

Now applying Newton's backward interpolation formula, we get

$$l_{47} = f(47) = l_{50} + u\nabla l_{50} + \frac{u(u+1)}{2!}\nabla^2 l_{50} + \frac{u(u+1)(u+2)}{3!}\nabla^3 l_{50}$$

$$= 243 + (-0.3)(-103) + \frac{(-0.3)(-0.3+1)}{2}(-10)$$

$$-\frac{10(-0.3)(-0.2+1)(-0.3+2)}{6}$$

$$= 243 + 30.9 + 1.05 - 0.595$$

$$l_{47} = 274.355$$

Hence, the number of persons living at age 47 are 276.

EXAMPLE 13. *Marks obtained by the candidates in an examination are given as under :*

Marks obtained	10-20	20-30	30-40	40-50	50-60
No. of Students	42	45	39	48	19

Find the number of candidates who obtained less than 44 marks.

SOLUTION.   The backward difference table is as under :

Marks less than x	Candidate (y)	Δy	$\Delta^2 y$	$\Delta^3 y$	$\Delta^4 y$
20	42				
30	87	45	−6		
40	126	39	9	−3	**−17**
50	174	48	**−29**	**−20**	
60	193	**19**			

Here $a+nh = 60$, $h=10$, $a+nh+uh = 44$, then

$$u = \frac{44-60}{10} = -1.6$$

Now from backward interpolation formula,

$$y_{44} = y_{60} + u\nabla y_{60} + \frac{u(u+1)}{2!}\nabla^2 y_{60} + \frac{u(u+1)(u+2)}{3!}\nabla^3 y_{60}$$

$$+ \frac{u(u+1)(u+2)(u+3)}{4!}\nabla^4 y_{60}$$

$$= 193 + (-1.6)19 + \frac{(-1.6)(-1.6+1)}{2!}(-29) + \frac{(-1.6)(-1.6+1)(-1.6+2)}{6}(-20)$$

$$+ \frac{(-1.6)(-1.6+1)(-1.6+2)(-1.6+3)}{24}(-17)$$

$$= 193 - 30.4 - 13.92 - 1.28 - 0.3808$$

Hence, $y_{44} = 147.01$

Hence, the number of candidates who obtained less than 44 marks are 147.

EXAMPLE 14. *In an examination the number of candidates who secured marks between certain limits were as follows :*

Marks	0-19	20-39	40-59	60-79	80-99
No. of students	41	62	65	50	17

Estimate the number of candidates getting marks less than 70.

SOLUTION. Here we construct the following table:

Marks less than (x)	Cumulative frequency f(x)
19	41
39	103
59	168
79	218
99	255

∴ The backward difference table is given by below:

x	$f(x)$	$\nabla f(x)$	$\nabla^2 f(x)$	$\nabla^3 f(x)$	$\nabla^4 f(x)$
19	41				
		62			
39	103		3		
		65		−18	
59	168		−15		0
		50		**−18**	
79	218		**−33**		
		17			
99	**235**				

Here, $a + nh = 99$, $h = 20$, $a + nh + uh$

$$= 70 = 99 + u(20)$$

$$\Rightarrow \quad 204 = 77 - 99 = -29$$

$$\Rightarrow \quad u = -\frac{29}{20} = -1.45$$

Now from backward interpolation formula, we have

$$f(x) = 235 + (-1.45)0.17 + \frac{(-1.45)(-1.45+1)}{2!}(-33)$$

$$+ \frac{(-1.45)(-1.45+1)(-1.45+2)}{3!}(-18)$$

$$= 235 - 24.65 - \frac{(1.45)(0.45)}{2}(33) + \frac{(1.45)(0.45)(0.55)}{6}(-18)$$

$$= 235 - 24.65 - 10.766 + 1.077 - 200.661 \approx 201.$$

Hence, the number of candidates getting marks less than 70 is 201.

EXAMPLE 15. *Find the cubic polynomial which takes the following values:*

x	0	1	2	3
$f(x)$	1	2	1	10

(Bhopal-2009, Rohtak-2005, WBTU-2005; UPTU-2002, 08)

SOLUTION. Here, the difference table is given as under

x	$f(x)$	$\Delta f(x)$	$\Delta^2 f(x)$	$\Delta^3 f(x)$
0	**1**			
1	2	**1**	**–2**	**12**
		–1	10	
2	1	9		
3	10			

Here $x_0 = 0, u = \dfrac{x-0}{h} = x$ $\qquad\qquad\qquad (\because h = 1)$

By using the Newton's forward interpolation formula, we get

$$f(x) = f(0) + u\Delta f(0) + \frac{u(u-1)}{2!}\Delta^2 f(0) + \frac{u(u-1)(u-2)}{3!}\Delta^3 f(0) + \dots$$

$$= 1 + x.1 + \frac{x(x-1)}{2!}(-2) + \frac{x(x-1)(x-2)}{6}(12) = 2x^3 - 7x^2 + 6x + 1$$

which is the required polynomial.

EXAMPLE 16. *The values of $f(x)$ are given below*

x	0	1	2	3	4	5	6
$f(t)$	1	3	11	31	69	131	223

Find the value of $f(34)$, using only four of the given values.

SOLUTION. We choose the last four values since $f(3.4)$ is required.

Here $a = 3$, $h = 1$ and $a + hu = 3.4$

$\Rightarrow\qquad\qquad\qquad 3 + u = 3.4$

$\Rightarrow\qquad\qquad\qquad u = 0.4$

The difference table is given as under:

x	$f(x)$	$\Delta f(x)$	$\Delta^2 f(x)$	$\Delta^3 f(x)$
3	**31**			
4	69	**38**	**24**	**6**
		62	30	
5	131	92		
6	223			

Apply Newton-Gregory formula, we have

$$f(3.4) = f(3) + \frac{(0.4)^1}{1!}\Delta.f(3) + \frac{(0.4)^2}{2!}\Delta^2 f(3)\; \frac{(0.4)^3}{3!}\Delta^3 f(3)$$

$$= 31 + 0.4 \times 38 + \frac{(0.4)(-0.6)}{2} \times 24 + \frac{(0.4)(-0.6)(-1.6)}{6} \times 6$$

$$= 31 + 15.2 - 2.88 + 0.384 = 43.704$$

Hence, $f(3.4) = 43.704$.

EXAMPLE 17. *The table gives the distance in nautical miles, of the visible horizon for the given heights m feet above the earth's surface.*

x (height)	100	150	200	250	300	350	400
f distance	10.63	13.03	15.04	16.81	18.42	19.96	21.27

Use Newton Gregory's forward and backward interpolation formula to find values of y where x = 218 ft and 410 ft.

(UPTU-2009)

SOLUTION.

x	$y = f(x)$	$\Delta f(x)$	$\Delta^2 f(x)$	$\Delta^3 f(x)$	$\Delta^4 f(x)$
100	10.63				
150	13.03	2.40			
200	**15.04**	2.01	−0.39	−0.15	−0.07
250	16.81	**1.77**	−0.24	−0.08	−0.05
300	18.42	1.61	**−0.16**	**−0.03**	**−0.01**
350	19.9	1.48	−0.13	−0.02	
400	21.27	1.37	−0.11		

If we take $x_0 = 200$, then

$$y_0 = 15.04, \Delta y_0 = 1.77, \Delta^2 y_0 = -0.16, \Delta^3 y_0 = 0.03, \Delta^4 y_0 = -0.01$$

Since $\quad x = 218, h = 50, u = \dfrac{x - x_0}{h} = \dfrac{18}{15} = 0.36$

Using Newton's forward interpolation formula, we get

$$y_{218} = y_0 + u\Delta y_0 + \frac{u(u-1)}{2!}\Delta^2 y_0 + \frac{u(u-1)(u-2)}{3!}\Delta^3 y_0$$

$$+ \frac{u(u-1)(u-2)(u-3)}{4!}\Delta^4 y_0$$

$$= 15.04 + 0.36(1.77) + \frac{0.36(-0.64)(-0.16)}{2!}$$

$$+ \frac{0.36(-0.64)(-1.64)}{6}(0.3) + \frac{0.36(-0.64)(-1.64)(-2.64)}{24}(-0.01)$$

$$= 15.04 + 0.637 + 0.018 + 0.002 + 0.0004 = 15.697$$

$\therefore \quad y_{218} = 15.7$ (app)

Again $x = 410$ is near to the end of table

$$x_n = 400$$

$$n = \frac{x - x_n}{h} = \frac{10}{50} = 0.2$$

Using the following values in Newton's backward interpolation formula,

$$y_n = 21.27, \nabla y_n = 1.37,$$

$$\nabla^2 y_n = -0.11, \nabla^3 y_n = 0.02, \nabla y_n = 0$$

We have, $y_{410} = y_{400} + u \nabla y_{400} + \dfrac{u(u+1)}{2!} \nabla^2 y_{100}$

$$+ \dfrac{u(u+1)(u+2)}{3!} \nabla^3 y_{400} + \dfrac{u(u+1)(u+2)(u+3)}{4!} \nabla^4 y_{400}$$

$$= 21.27 + 0.2(1.37) + \dfrac{0.2(1.2)}{2!}(-0.11) + \dfrac{0.2(1.2)(2.2)}{6}(0.02)$$

$$+ \dfrac{0.2(1.2)(2.2)(3.2)}{24}(-0.01)$$

$$= 21.27 + 0.274 - 0.0132 + 0.0018 - 0.0007$$

$\Rightarrow \qquad y_{410} = 21.53$ approximately.

EXAMPLE 18. *Consider the following table:*

x	3	4	5	6	7
$f(x)$	3	6.6	15	22	25

Obtain interpolating polynomial of degree 2 or less using Newton's backward difference interpolation method. Hence compute $f(5.5)$. [UPTU(MCA)-2007]

SOLUTION. The difference table is given as under

x	$f(x)$	$\Delta f(x)$	$\Delta^2 f(x)$	$\Delta^3 f(x)$	$\Delta^4 f(x)$
3	3				
4	6.6	3.6	4.8	−6.2	
5	15	8.4	−1.4	7.4	13.6
6	22	7	**6**		
7	25	13			

Newton's backward difference formula is given by

$$f(x) = f(x_n) + u \nabla f(x_n) + \dfrac{u(u-1)}{2!} \nabla^2 f(x_n) + \dots$$

To get the polynomial of degree 2 or less we use only upto second difference term.

Now $\quad u = \dfrac{x - x_n}{n} = \dfrac{x - 7}{1}$

$\therefore \quad f(x) = 35 + (x-7)1 + \dfrac{(x-7)(x-8)}{2!} \cdot 6$

$$= 35 + 13x - 91 + 3(x^2 + 56 - 15x)$$

$$= 3x^2 - 32x + 112$$

Hence, $\quad f(5.5) = 3(5.5)^2 - 32 \times 5.5 + 112 = 26.75$

EXAMPLE 19. *The following table gives the values of y which is a polynomial of degree five. It is known that $f(3)$ is in error. Correct the error.*

x	0	1	2	3	4	5	6
$f(x)$	1	2	33	254	1025	3126	7777

(Mumbai–2004)

SOLUTION. The difference table is given as under

x	$f(x)$	$\Delta f(x)$	$\Delta^2 f(x)$	$\Delta^3 f(x)$	$\Delta^4 f(x)$	$\Delta^5 f(x)$	$\Delta^6 f(x)$
0	1						
1	2	1					
2	33	31	30				
3	a	$a-33$	$a-64$	$a-94$			
4	1025	$1025-a$	$1058-2a$	$1122-3a$	$1216-4a$	$-2320-10a$	$4880-20a$
5	3126	2101	$1076+a$	$18+3a$	$-1104+6a$	$2560-10a$	
6	7777	4651	2550	$1474-a$	$1456-4a$		

Here $f(x)$ is a polynomial of 6th degree therefore we get $\nabla^6 y = 0$

$\Rightarrow$ $\qquad$ $4880 - 20a = 0$

Hence, $\qquad$ $a = 244$

and $\qquad$ the error $= 254 - 244 = 10$

EXERCISE 5.2

1. The value of $f(x)$ for $x = 0, 1, 2, \ldots 6$ are given below:

x	0	1	2	3	4	5	6
$f(x)$	2	4	10	16	20	24	38

Estimate the value of $f(3.2)$ using only four of given values. Choose the four values that you think will give the best approximations.

2. From the following table find the value of $e^{0.24}$:

x	0.1	0.2	0.3	0.4	0.5
e^x	1.10517	1.22140	13.4986	1.49182	1.64872

3. If p, q, r, s be the successive entries corresponding to equidistant arguments in a table. Show that when third difference are taken into account the entry corresponding to the arguments half way between the arguments of q and r is $A + \left(\dfrac{B}{24}\right)$, where

A is the arithmetic mean of q and r and B is the arithmetic mean of $3q - 2p - s$ and $3r - 2s - p$. $\qquad$ (Nagpur-2006, 09, Vanketeswara-2004, 07, 09, MKU-2005, 07)

4. From the following table :

x	10°	20°	30°	40°
$\cos x$	0.9848	0.9397	0.8660	0.7660
x	50°	60°	70°	80°
$\cos x$	0.6428	0.5000	0.3420	0.1737

Calculate cos 25° and cos 73° with Gregory's Newton formula. $\qquad$ (UPTU-2006)

5. The table below gives values of $\tan x$ for $0.10 \le x \le 0.30$:

x	0.10	0.15	0.20	0.25	0.30
$\tan x$	0.1003	0.1511	0.2027	0.2523	0.3093

Evaluate tan 0.12 using Newton's forward difference formula.

6. In the following table, values of y are consecutive terms of a series of which 23.6 is the 6th term. Find the first and tenth term of the series. $\qquad$ (MDU(B.E.)-2004)

x	3	4	5	6	7	8	9
y	4.8	8.4	14.5	23.6	36.2	52.8	73.9

(Anna-2007)

7. Find the following table of half yearly premium for policies maturing at different ages, estimate the premium for policy maturing at the age of 63. (UPTU–2004, 10)

Age	45	50	55	60	65
Premium	114.84	96.16	83.32	74.48	68.48

8. Find the value of $e^{-1.9}$ from the following table of values of e^{-x}.

x	1	1.25	1.50	1.75	2.00
e^{-x}	0.3679	0.2865	0.2231	0.1738	0.1353

9. Using Newton's backward difference formula, construct an interpolating polynomial of degree 3 for the data :

$f(-0.75) = -0.0718125, f(-0.5) = -0.02475,$ $f(-0.25) = 0.3349375, f(0) = 1.10100.$ Hence, find $f(-1/3)$. (Anna–2003)

10. Using Newton's forward formula, find the value of $f(1.6)$, if

x	1	1.4	1.8	2.2
$f(x)$	3.49	4.82	5.96	6.5

(JNTU–2006)

11. From the following table, find $f(x)$ when $x = 1.85$ and 2.4 by Newton's interpolation formula :

x	1.7	1.8	1.9	2.0	2.1	2.2	2.3
$f(x)$ $= e^x$	5.474	6.050	6.686	7.389	8.166	9.025	9.974

(Kottayam–2005)

12. From the following table :

x :	0.1	0.2	0.3	0.4	0.5	0.6
$f(x)$:	2.68	3.04	3.38	3.68	3.96	4.21

(VTU–2001)

13. A test performed on NPN transistor gives the following result :

Base Current $f(mA)$	0	0.01	0.02	0.03	0.04	0.05
Collector current $I_c(mA)$	0	1.2	2.5	3.6	4.3	5.34

Evaluate :

(i) the value of the collector current for the base current of 0.05 mA.

(ii) the value of base current required for a collector correct of 4.0 mA. (Pune–2004)

14. Construct Newton's forward interpolation polynomial for the following data :

x	4	6	8	10
y	1	3	8	16

Hence, evaluate y for $x = 5$. (Madras-2006)

15. Find the cubic polynomial which takes the following values :

$y(0) = 1, y(1) = 0, y(2) = 1$ and $y(3) = 10$. Hence or otherwise, obtain $y(4)$.

(UPTU–2005; Madras-2006)

16. In the following table, the value of y are consecutive terms of a series of which 12.5 is the 5th term. Find the first and tenth terms of the series.

x	3	4	5	6	7	8	9
y	2.7	6.4	12.5	21.6	34.3	51.2	72.9

(PTU–2001)

17. Using a polynomial of the third degree, complete the record given below of the export of a certain commodity during five years :

Year	1989	1990	1991	1992	1993
Export	443	384	–	397	467

18. Given $u_1 = 40, u_3 = 45, u_5 = 54$, find u_2 and u_4. (Nagarjuna–2003)

 **Answers**

1. 17.28 2. 1.271249088   5. 0.1205 6. 3.1, 100 7. 70.585152   8. 0.1496
9. $x^3 + 4.001x^2 + 4.002x + 1.101$, 0.1725   10. 5.54   11. 6.36, 11.02   12. 4.43
13. (i) $I_c = 0.5878$ (ii) $f_b = 0.363$ 14. 1.625 15. 33   16. 0.1; 100
17. 369 tons   18. $u_2 = 42, u_4 = 49$

5.14 HERMITE'S INTERPOLATION FORMULA

[Rohilkhand-2010; Meerut-2004, 12; UPTU(MCA)–2008]

The Hermitian interpolation is similar to that of Lagrange's Interpolation. The difference is that in the Lagrange's Interpolation the interpolating polynomial $p(x)$ considers with $f(x)$ at the inerpolation points $x_0, x_1, x_2, \dots x_n$, whereas in Hermite interpolation $p(x)$ and $f(x)$ as well as $p'(x)$ and $f'(x)$ coincide. That is,

$$p(x_k) = f(x_k)$$
and $$p'(x_k) = f'(x_k) \text{ for } k = 1, 2, 3 \dots n$$ ...(1)

We have $(2n+2)$ conditions given by (1) and we know that a polynomial of degree $2n+1$ has

$2n+2$ coefficients to be determined. Thus we have the form

$$p(x) = \sum_{k=0}^{n} u_k(x)f(x_k) + \sum_{k=0}^{n} v_k(x)f'(x_k) \qquad ...(2)$$

Here $u_k(x)$ and $v_k(x)$ are polynomial in x of degree $(2n+1)$. From (1), we obtain,

$$\left.\begin{array}{ll} u_k(x_i) = 1, \text{ if } k = i & v_k(x_i) = 0 \\ \qquad\quad = 0, \text{ if } k \neq i, & v_k'(x_i) = 1, \text{ if } k = i \\ u_k'(x_i) = 0 & \qquad\quad = 0, \text{ if } k \neq i \end{array}\right\} \qquad ...(3)$$

Now we may choose $u_k(x) = w_k(x)[L_k(x)]^2$ $\qquad ...(4)$

and $v_k(x) = z_k(x)[L_k(x)]^2$

where $L_k(x) = \dfrac{(x-x_0)(x-x_1)...(x-x_{k-1})(x-x_{k+1})...(x-x_n)}{(x_k-x_0)(x_k-x_1)...(x_k-x_{k-1})(x_k-x_{k+1})...(x_k-x_n)}$

Since $L_k(x)$ is a polynomial of degree n. Then $w_k(x)$ and $z_k(x)$ are both linear. Therefore using condition (3), we get

and $\left.\begin{array}{l} w_k(x_k) = 1, w_k'(x_k) = -2L_k'(x_k) \\ z_k(x_k) = 0, z_k'(x_k) = 1 \end{array}\right\} \qquad ... (5)$

From (5), we find $w_k(x) = 1 - 2L_k'(x_k)(x-x_k)$ and $z_k(x) = x - x_k$

From (2), (4) and (5), we get

$$p(x) = \sum_{k=0}^{n} \{1 - 2L_k'(x_k)(x-x_k)\}[L_k(x)]^2 f(x_k) + \sum_{k=0}^{n} (x-x_k)[L_k(x)]^2 f'(x_k)$$

This is Hermite interpolation formula.

▶ **REMARK**

▸ Hermite interpolation formula is sometimes called osculating interpolation formula.

ALGORITHM

Step 1.	Put the values of $x_i, f(x_i), f'(x_i)$ and value of x for which $f(x)$ is to be determined.
Step 2.	Obtained $L_k(x) = \prod_{\substack{r=0 \\ k \neq r}}^{n} \dfrac{(x-x_k)}{(x_l - x_r)}, l \neq k$,
Step 3.	Obtained $L_k'(x)$ at $x = x_i$.
Step 4.	Put all these values in Hermite's interpolation formula.

SOLVED EXAMPLES

EXAMPLE 1.   *Apply Hermite interpolation formula to find the value of* sin (1.05) *from the following data :*

x	1.00	1.10
$\sin x$	0.84147	0.89121
$\cos x$	0.54030	0.45360

SOLUTION.   Here $f(x) = \sin x, f'(x) = \cos x$

and $x_0 = 1.00, x_1 = 1.10$

$$L_0(x) = \frac{x - x_1}{x_0 - x_1} = \frac{x - 1.10}{1.00 - 1.10}$$

$$L_0(x) = -10x + 11$$

$$\therefore \quad L_0'(x) = -10$$

and $\quad L_1(x) = \dfrac{x - x_0}{x_1 - x_0} = \dfrac{x - 1.00}{1.10 - 1.00}$

$$L_0(x) = 10x - 10$$

$$\therefore \quad L_1'(x_1) = 10$$

Since $f(x_0) = 0.84147, f'(x_0) = 0.54030$

$$f(x_1) = 0.89121, f'(x_1) = 0.45360$$

Now by Hermite interpolation formula, we have

$$P(x) = \sum_{k=0}^{1} \{1 - 2L_k'(x_k)(x - x_k)\}[L_k(x)]^2$$

$$f(x_k) + \sum_{k=0}^{1} (x - x_k)[L_k(x)]^2 f'(x_k)$$

$$= \{1 - 2L_0'(x_0)(x - x_0)\}[L_0(x)]^2 f(x_0)$$

$$+ \{1 - 2L_1'(x_1)(x - x_1)\}[L_1(x)]^2 f(x_1)$$

$$+ (x - x_0)[L_0(x)]^2 f'(x_0) + (x - x_0)[L_1(x)]^2 f'(x_1)$$

$$P(x) = \{1 + 20(x - 1.00)\}(-10x + 11)^2 (0.84147)$$

$$+ \{1 - 20(x - 1.10)\}(10x - 10)^2 (0.89121)$$

$$+ (x - 1.00)(-10x + 11)^2 (0.54030)$$

$$+ (x - 1.10)(10x - 10)^2 (0.45360)$$

Putting $x = 1.05$ in this equation, we get

$$P(1.05) = \sin(1.05) = \{(1 + 20(1.05 - 1.00)\}(-10(1.05) + 11)^2 (0.84147)$$

$$+ \{1 - 20(1.05 - 1.10)\} (10 (1.05) - 10)^2 (0.89121)$$

$$+ (1.05 - 1.00) (-10(1.05) - 11)^2 (54030)$$

$$+ (1.05 - 1.10)\{(1.05) - 10\}^2 (0.45360)$$

$$= (2)(0.25) (0.84147) + (2)(0.25)(0.89121)$$

$$+ (0.05)(0.25) (0.54030) - (0.5) (0.25) (0.45360)$$

$$= 0.420735 + 0.445605 + 0.00675375 - 0.00567$$

Hence, $\sin(1.05) = 0.86742$.

EXAMPLE 2. *Apply Hermite's formula to find a polynomial from the following data :*

x	0	1	2
$f(x)$	0	1	0
$f'(x)$	0	0	0

SOLUTION. Here $\quad x_0 = 0, x_1 = 1, x_2 = 2$

and $\quad f(x_0) = 0, f(x_1) = 1, f(x_2) = 0$

and $\quad f'(x_0) = 0, f'(x_1) = 0, f'(x_2) = 0$

$$L_0(x) = \frac{(x-x_1)(x-x_2)}{(x_0-x_1)(x_0-x_2)}$$

$$L_0(x) = \frac{(x-1)(x-2)}{(0-1)(0-2)} = \frac{1}{2}(x^2 - 3x + 2)$$

$$\therefore \quad L_0'(x_0) = \frac{1}{2}(2x_0 - 3) = \frac{1}{2}(-3) = -\frac{3}{2}$$

$$L_1(x) = \frac{(x-x_0)(x-x_2)}{(x_1-x_0)(x-x_2)} \Rightarrow L_1(x) = \frac{(x)(x-2)}{(1-0)(1-2)} = -(x^2 - 2x)$$

$$\therefore \quad L_1'(x_1) = -(2x_1 - 2) = -(2-2) = 0$$

$$L_2(x) = \frac{(x-x_0)(x-x_1)}{(x_2-x_0)(x_2-x_1)} \Rightarrow L_2(x) = \frac{x(x-1)}{(2-0)(x_2-x_1)} = \frac{1}{2}(x^2 - x)$$

and $\quad L_2'(x) = \frac{1}{2}(2x_2 - 1) = \frac{1}{2}(4-1) = \frac{3}{2}$

Now by Hermite interpolation formula, we get

$$P(x) = \sum_{K=0}^{2} \{1 - 2L_k'(x_k)(x-x_k)\}[L_k(x)]^2 \, f(x_k) + \sum_{k=0}^{2} (x-x_k)[L_k(x)]^2 f'(x_k)]$$

$$= \{1 - 2L_0'(x_0)(x-x_0)\}[L_0(x)]^2 f(x_0) + \{1 - 2L_1'(x_1)(x-x_1)\}[L_1(x)]^2 f(x_1)$$

$$+ \{1 - 2L_2'(x_2)(x-x_2)\}[L_2(x)]^2 f(x_2)$$

$$+ (x-x_0)[L_0(x)]^2 f'(x_0) + (x-x_1)[L_1(x)]^2$$
$$f'(x_1) + (x-x_2)[L_2(x)]f'(x_2)$$

$$\Rightarrow P(x) = \{1 + 3(x-0)\}\frac{1}{4}(x^2 - 3x + 2)^2 \times 0 + \{1 - 0\}(x^2 - 2x)^2.1 + \{1 - 3(x-2)\}$$

$$\frac{1}{4}(x^2 - x)^2 \times 0 + (x-0)\frac{1}{4}(x^2 - 3x + 2)^2 \times 0 + (x-1)(x^2 - 2x)^2 \times 0 + (x-2)$$

$$\left[\frac{1}{4}(x^2 - x)^2\right] \times 0$$

$$P(x) = (x^2 - 2x)^2.$$

This is required polynomial.

EXAMPLE 3. *Apply Hermite's interpolation formula to obtain cubic polynomial, using the following data :*

x_i	0	1
$f(x_i)$	0	1
$f'(x_i)$	0	1

SOLUTION. Here, we have

$k = 0$ and 1 then Hermite formula becomes

$$P(x) = \{1 - 2L_0'(x_0)(x-x_0)\}[L_0(x)]^2 \cdot f(x_0) + (x-x_0)[L_0(x)]^2 f'(x_0)$$

$$+ [1 - 2L_1'(x_1)(x-x_1)][u(x)]^2 f(x_1) + (x-x_0)\{L_1(x)\}^2 f'(x_1) \quad ...(1)$$

Now, $\quad L_0(x) = \frac{x-x_1}{x_0-x_1} = 1 - x$

$\Rightarrow \quad L_0'(x_0) = -1$

Also, $L_1(x) = \dfrac{x - x_0}{x_1 - x_0} = 2$

$\Rightarrow \qquad L_1'(x) = 1 \Rightarrow L_1'(x_1) = 1$

Put all these values in (1), we get

$$P(x) = (3 - 2x)x^2 + x^3 - x^2 = 2x^2 - x^3.$$

EXAMPLE 4. *Apply Hermite's interpolation to find $f(x)$ at $x = 0.5$, using the following data :*

x_i	−1	0	1
$f(x_i)$	1	1	3
$f'(x_i)$	−5	1	7

SOLUTION. Here, we have $k = 0, 1, 2$

(Anna–2005)

Then $\qquad P(x) = \displaystyle\sum_{k=0}^{2} \{1 - 2L_k'(x_k)(x - x_k)\}[L_k(x)]^2$

$$f(x_k) + \sum_{k=0}^{2} (x - x_k)[L_k(x)]^2 f'(x_k)] \qquad \text{...(1)}$$

Now, $\qquad L_0(x) = \dfrac{x(x-1)}{(-1)(-2)} = \dfrac{1}{2}x(x-1) \Rightarrow L_0'(x_0) = -\dfrac{3}{2}$

$$L_1(x) = \dfrac{(x+1)(x-1)}{1.(-1)} = (1 - x^2) \Rightarrow L_1'(x_1) = 0$$

and $\qquad L_2(x) = \dfrac{(x+1)(x)}{2.1} = \dfrac{1}{2}x(x+1) \Rightarrow L_2'(x_2) = \dfrac{3}{2}$

Put all these values in equation (1), we get

$$P(x) = \dfrac{1}{4}(3x^5 - 2x^4 - 5x^3 + 4x^2).(1) + (x^4 - 2x^2 + 1).(1) + \dfrac{1}{4}(-3x^5 - 2x^4$$

$$+ 5x^3 + 4x^2).3 + \dfrac{1}{4}(x^5 - x^4 - x^3 + x^2)(-5) + (x^4 - 2x^3 + x)(1)$$

$$+ \dfrac{1}{4}(x^5 + x^4 - x^3 - x^2).7$$

$$= 2x^4 - x^2 + x + 1$$

Hence, $P(0.5) = \dfrac{11}{8}$.

5.15 NEWTON'S DIVIDED DIFFERENCE FORMULA

[UPTU(MCA)–2002, 09]

Let $f(x_0), f(x_1), ..., f(x_n)$ be the values of the function $f(x)$ for the values of the arguments $x_0, x_1, ..., x_n$ respectively which are not equally spaced.

The first divided difference of $f(x)$ is given by

$$f(x, x_0) = \dfrac{f(x_0) - f(x)}{(x_0 - x)}$$

$\Rightarrow \qquad f(x) = f(x_0) + (x - x_0)f(x, x_0) \qquad \text{...(1)}$

Also, the second divided difference is given by

$$f(x, x_0, x_1) = \frac{f(x_0, x_1) - f(x, x_0)}{(x_1 - x)}$$

$$f(x, x_0) = f(x_0, x_1) + (x - x_1)f(x, x_0, x_1) .$$

$\Rightarrow \qquad \dfrac{f(x) - f(x_0)}{(x - x_0)} = f(x_0, x_1) + (x - x_1)f(x, x_0, x_1)$ \hfill [From (1)]

$\Rightarrow \qquad f(x) = f(x_0) + (x - x_0)f(x_0, x_1) + (x - x_0)(x - x_1)f(x_0, x_1, x_2)$ \hfill ... (2)

Similarly, $\qquad f(x) = f(x_0) + (x - x_0)f(x_0, x_1) + (x - x_0)(x - x_1)f(x_0, x_1, x_2)$

$$+ (x - x_0)(x - x_1)(x - x_2)f(x_0, x_1, x_2, x_3) \qquad ... (3)$$

Proceeding in the same way, we get

$$f(x) = f(x_0) + (x - x_0)f(x_0, x_1) + (x - x_0)(x - x_1)f(x_0, x_1, x_2)$$
$$+ (x - x_0)(x - x_1)(x - x_2)f(x_0, x_1, x_2, x_3) + ...$$
$$+ (x - x_0)(x - x_1)(x - x_2)...(x - x_{n-1})f(x_0, x_1, x_2, x_3, ..., x_n)$$
$$+ (x - x_0)(x - x_1)(x - x_2)...(x - x_n)f(x_0, x_1, ..., x_{n+1}) \qquad ...(4)$$

Since the function $f(x)$ is a polynomial of degree n, therefore

$\therefore \qquad f(x, x_0, x_1, ..., x_{n+1}) = 0$

Hence (4) becomes,

$$f(x) = f(x_0) + (x - x_0)f(x_0, x_1) + (x - x_0)(x - x_1)f(x_0, x_1, x_2)$$
$$+ ... + (x - x_0)(x - x_1)(x - x_2)...(x - x_{n-1})f(x_0, x_1, x_2, ..., x_n)$$

This is known as Newton's divided difference formula.

▼ **REMARK**

▸ Newton's divided difference formula reduces to Newton's Gregory's forward difference formula if the values of the arguments are equally spaced.

ALGORITHM

Step 1.	Obtain the values of $x_i, f(x_i)$ and values of x for which $f(x)$ is to be calculated.
Step 2.	Construct the divided difference table.
Step 3.	Substitute the values of divided differences in the Newton's divided difference formula.

SOLVED EXAMPLES

EXAMPLE 1. *By means of Newton's divided difference formula, find the value of $f(8)$ and $f(15)$ from the following table:* (UPTU MCA-2009, VTU-2008)

SOLUTION. Here $x_0 = 4, x_1 = 5, x_2 = 7, x_3 = 10, x_4 = 11, \ x_5 = 13$

The divided difference table is given below :

x	$f(x)$	$\Delta f(x)$	$\Delta^2 f(x)$	$\Delta^3 f(x)$	$\Delta^4 f(x)$
4	48				
		$\dfrac{100-48}{5-4}$ $=52$	$\dfrac{97-57}{7-4}$ $=15$		
5	100			$\dfrac{21-15}{10-4}$ $=1$	
		$\dfrac{294-100}{7-5}$ $=97$	$\dfrac{202-97}{10-5}$ $=21$		
7	294				0
				$\dfrac{27-21}{11-5}$ $=1$	
10	900	$\dfrac{900-294}{10-7}$ $=202$	$\dfrac{310-202}{11-7}$ $=27$		
					0
11	1210	$\dfrac{1210-900}{11-10}$ $=310$		$\dfrac{33-27}{13-7}$ $=1$	
13	2028		$\dfrac{409-310}{13-10}$ $=33$		
		$\dfrac{2028-1210}{13-11}$ $=409$			

Now using Newton's divided difference formula, we get

$$f(x) = f(x_0) + (x - x_0)f(x_0, x_1) + (x - x_0)(x - x_1)f(x_0, x_1, x_2)$$
$$+ (x - x_0)(x - x_1)(x - x_2)\, f(x_0, x_1, x_2, x_3) \qquad \ldots(1)$$

$\Rightarrow \quad f(8) = f(4) + (8-4) \times 52 + (8-4)(8-5)15 + (8-4)(8-5)(8-7) \times 1$

$\qquad = 48 + 208 + 180 + 12 = 448$

and $f(5) = 48 + (15-4) \times 52 (15-4)(15-5)(15) + (15-4)(15-5)(15-7).1$

$\qquad = 48 + 572 + 1650 + 880$

$\therefore \quad f(5) = 3150.$

EXAMPLE 2. *Find the polynomial of the lowest possible degree which takes the values 3, 12, 15, –21, when x has the value 3, 2, 1, –1 respectively.* (Rohtak-2005, 08, Avath-2003; UPTU-2008)

SOLUTION. The divided difference table is as under :

x	$f(x)$	$\Delta f(x)$	$\Delta^2 f(x)$	$\Delta^3 f(x)$
–1	–21			
		$\dfrac{15+21}{1-1} = 18$	$-\dfrac{3-18}{2-1} = -7$	
1	15			$\dfrac{-3+7}{3-1} = 1$
		$\dfrac{12-15}{2-1} = -3$	$\dfrac{-9-3}{3-1} = -3$	
2	12			
		$\dfrac{3-12}{3-2} = -9$		
3	3			

Here $x_0 = -1, x_1 = 1, x_2 = 2, x_3 = 3$.

Now applying Newton's divided difference formula, we get

$$f(x) = f(x_0) + (x - x_0)f(x_0, x_1) + (x - x_0)$$

$$(x - x_1)f(x_0, x_1, x_2) + (x - x_0)(x - x_1)(x - x_2)f(x_0, x_1, x_2, x_3)$$
$$= f(-1) + (x+1)(18) + (x+1)(x-1)(-7) + (x+1)(x-1)(x-2)(1)$$
$$f(x) = x^3 - 9x^2 + 17x + 6$$

EXAMPLE 3. *Given*

$$\log_{10} 654 = 2.8156, \log_{10} 658 = 2.8182 \ \log_{10} 659 = 2.8189, \log_{10} 661 = 2.8202$$

find the value of $\log_{10} 656$ *by the divided difference formula.*

(Agra-2002, 07, Rohilkhand-2007, Lucknow-2005, 09)

SOLUTION. Here $x_0 = 654, x_1 = 658, x_2 = 659, x_3 = 661$ and $f(x) = \log_{10} x$. The divided difference table is as under :

x	$f(x)$	$\Delta f(x)$	$\Delta^2 f(x)$	$\Delta^3 f(x)$
654	2.8156	$\dfrac{2.8182 - 2.8156}{658 - 654}$ $= 0.00065$	$\dfrac{0.00070 - 0.00065}{659 - 654}$ $= 0.00001$	
658	2.8182	$\dfrac{2.8189 - 2.8182}{659 - 658}$ $= 0.00070$	$\dfrac{0.00065 - 0.00070}{661 - 658}$ $= -0.000017$	$\dfrac{-0.000017 - 0.00001}{661 - 654}$ $= -0.000004$
2	12			
661	2.8202	$\dfrac{2.8202 - 2.8189}{661 - 659}$ $= 0.00065$		

Now apply Newton's divided difference formula

$$f(x) = f(x_0) + (x - x_0)f(x_0, x_1) + (x - x_0)(x - x_1)f(x_0, x_1, x_2)$$
$$+ (x - x_0)(x - x_1)(x - x_2)f(x_0, x_1, x_2, x_3)$$

we get

$$f(656) = f(654) + (656 - 654) \times (0.00065) + (656 - 654) - (656 - 658) \times (0.00001)$$
$$+ (656 - 654)(656 - 658)(656 - 659) \times (-0.000004)$$
$$= 2.8156 + 0.0013 - 0.0004 - 0.000048 = 2.816812$$

$\therefore \log_{10} 656 = 2.8168$

EXAMPLE 4. *Using Newton's divided difference formula, calculate the value of* $f(6)$ *from the following data:* (UPTU-2002, Meerut-2009, Garhwal-2014)

x	1	2	7	8
$f(x)$	1	5	5	4

SOLUTION.

x	$f(x)$	$\Delta f(x)$	$\Delta^2 f(x)$	$\Delta^3 f(x)$
1	1			
		4		
2	5		–2 / 3	
		0		1 / 14
7	5		–1 / 6	
		–1		
8	4			

Put the values in the Newton's divided difference formula

$$f(x) = f(x_0) + (x - x_0)f(x_0, x_1) + (x - x_0)(x - x_1)f(x_0, x_1, x_2)$$
$$+ (x - x_0)(x - x_1)(x - x_2) f(x_0, x_1, x_2, x_3)$$

$$\Rightarrow f(6) = 1 + (6-1).4 + (6-1)(6-2)(-2/3) + (6-1)(6-2)(6-7)\frac{1}{14}$$

$$= 1 + 20 + \frac{5 \times 4 \times -2}{3} + 5 \times 4 \times -1 \times \frac{1}{14}$$

$$= 21 - \frac{40}{3} - \frac{10}{7} = \frac{441 - 280 - 30}{21} = \frac{131}{31} = 6.2 .$$

EXAMPLE 5. *Certain corresponding values of x and* $\log_{10} x$ *are* (300, 2.4771), (304. 2.4829), (305, 2.4843)*and* (307, 2.4871). *Find* $\log_{10} 301$ *using Newton's divided difference formula.* (UPTU MCA-2001)

SOLUTION. Using the given data, we construct the following divided difference table:

x	$f(x)$	$\Delta f(x)$	$\Delta^2 f(x)$
300	**2.4771**		
		0.00145	
			– 0.00001
304	2.4829		
		0.00140	
			0
305	2.4843		
		0.00140	
307	2.4871		

Put the values of $f(x)$, $\Delta f(x)$, $\Delta^2 f(x)$ from the table in Newton's divided difference formula, we get

$$f(x) = \log_{10} 301 = f(x_0) + (x - x_0)\Delta f(x_0) + (x - x_0)(x - x_1)\Delta^2 f(x_0)$$
$$= 2.4771 + 0.00145 + (-3)(-0.00001) = 2.4786.$$

EXAMPLE 6. *Using the Newton's divided difference formula, find a polynomial function satisfying the following data :* [UPTU-2004, MDU (B.E.)-2007, VTU-2007]

x	–4	–1	0	2	5
$f(x)$	1245	33	5	9	1355

SOLUTION. The divided difference table is given by

x	$f(x)$	$\Delta f(x)$	$\Delta^2 f(x)$	$\Delta^3 f(x)$	$\Delta^4 f(x)$
– 4	1245				
		– 404			
– 1	33		94		
		– 28		–14	
0	5		10		**3**
		2		13	
2	9		88		
		442			
5	1355				

Putting the values in Newton-divided difference formula,

$$f(x) = f(x_0) + (x-x_0)\Delta\ f(x_0) + (x-x_0)(x-x_1)\Delta^2 f(x_0) + (x-x_0)(x-x_1)$$
$$(x-x_2)\Delta^3 f(x_0) + (x-x_0)(x-x_1)(x-x_2)(x-x_3)\Delta^4 f(x_0)$$

we get $f(x) = 1245 + (x+4)(-404) + (x+4)(x+1)(94) + (x+4)(x+1)$
$$x(-14) + (x+4)(x+1)x(x-2)(3)$$
$$= 3x^4 - 5x^3 + 6x^2 - 14x + 5.$$

EXAMPLE 7.   *Using the following table, obtain f(x) as a polynomial in powers of (x − 5).*

x	0	2	3	4	7	9
$f(x)$	4	26	58	112	466	922

SOLUTION.   We have the following difference table:

x	$f(x)$	$\Delta\ f(x)$	$\Delta^2 f(x)$	$\Delta^3 f(x)$
0	4			
2	26	11	7	
3	58	32	11	1
4	112	54	16	1
7	466	118	22	1
9	922	228		

Putting these values in the Newton's divided difference formula, we get

$$f(x) = 4 + (x-0)(11) + (x-0)(x-2)(7) + (x-0)(x-2)(x-3)(1)$$
$$= x^3 + 2x^2 + 3x + 4.$$

To express $f(x)$ in power of $(x - 5)$ we use the following synthetic division scheme.

```
5 | 1    2     3      4
  |      5     35     190
5 | 1    7     38    194
  |      5     60
5 | 1   12    98
  |      5
    1   17
```

Hence, $x^3 + 2x^2 + 3x + 4 = (x-5)^3 + 17(x-5)^2 + 98\ (x-5) + 194.$

EXAMPLE 8.   *Find approximately the real root of the equation* $x^3 - 2x - 5 = 0$.

SOLUTION.   Let $f(x) = x^3 - 2x - 5$

The real root of $f(x) = 0$ lies between 2 and 2.1.

The values of $f(x)$ at $x = 1.9, 2, 2.1, 2.2$ are $-1.941\ -1.0000, 0.061$ and 1.248 respectively.

Let

x	−1.941	−1.000	0.061	1.248
y_x	1.9	2.0	2.1	2.2

We have to find y_x at $x = 0$.

Now, we have the following difference table:

x	$f(x)$	$\Delta\, f(x)$	$\Delta^2\, f(x)$	$\Delta^3\, f(x)$
-1.941	1.9			
-1.000	2.0	**0.1062699**	**–0.0060035**	**0.0004869**
0.061	2.1	0.0942507	–0.0044505	
1.248	2.2	0.0842459		

Putting these values in Newton's divided difference formula, we get

$$y_x = 1.9 + (x + 1.941) \times 0.1062699 + (x + 1.941)(x + 1)(-0.0060035)$$
$$+ (x + 1.941)(x + 1)(x - 0.061\,(0.0004869)$$

Putting $x = 0$, we get

$y_{x,\text{ at } x = 0}$, $1.9 + 0.2062698 - 0.0116527 - 0.0000576 = 2.0945595$

Hence, the required root is given by 2.0945595.

EXAMPLE 9. *Using the following table, find the value of $f(x)$ at $x = 4$.*

x	1.5	3	6
$f(x)$	-0.25	2	20

(UPTU MCA (Co)-2003)

SOLUTION. Using the above data, we have the following difference table:

x	$f(x)$	$\Delta\, f(x)$	$\Delta^2\, f(x)$
1.5	-0.25		
3	2	**1.5**	1
6	20	6	

Putting all these values in Newton's divided difference formula, we get

$$f(x) = -0.25 + (x - 1.5)(1.5) + (2 - 1.5)(x - 3)(1)$$

Putting $x = 4$, we get

$$f(4) = 6$$

EXAMPLE 10. *Using Newton's divided difference formula, show that*

$$f(x) = f(0) + x\Delta\, f(-1) + \frac{x(x+1)}{2!}\Delta^2\, f(-1) + \frac{x(x-1)(x+1)}{3!}\Delta^3 f(-2) + \dots$$

SOLUTION. Taking the arguments 0, –1, 1, –2, ... the Newton's divided difference formula is given by

$$f(x) = f(0) + x \underset{-1}{\Delta}\, f(0) + x(x+1) \underset{-1,1}{\Delta^2}\, f(0) + x(x+1)(x-1) \underset{-1,-2}{\Delta^3}\, f(0) + \dots \quad \dots(1)$$

$$= f(0) + x \underset{0}{\Delta}\, f(-1) + x(x+1) \underset{0,1}{\Delta^2}\, f(-1) + x(x+1)(x-1) \underset{-1,0,1}{\Delta^3}\, f(-2) + \dots$$

Now, $\underset{0}{\Delta}\, f(x) = \dfrac{f(0) - f(-1)}{0 - (-1)} = \Delta f(-1)$

$$\underset{0,1}{\Delta^2}\, f(-1) = \frac{1}{1 - (-1)}[\underset{1}{\Delta}\, f(0) - \underset{0}{\Delta}\, f(-1)]$$

$$= \frac{1}{2}[\Delta f(0) - \Delta f(0)] = \frac{1}{2}\Delta^2 f(-1)$$

$$\Delta^3_{-1,0,1} f(-2) = \frac{1}{1-(-2)}[\Delta^2_{0,1} f(-1) - \Delta^2_{-1,0} f(-2)]$$

$$= \frac{1}{3}\left[\frac{\Delta^2 f(-1)}{2} - \frac{\Delta^2 f(-2)}{2}\right] = \frac{\Delta^3 f(-2)}{3.2} = \frac{\Delta^3 f(-2)}{3!}$$

.... and so on.

Finally, putting all these values in (1), we get

$$f(x) = f(0) + x\Delta f(-1) + \frac{x(x+1)}{2!}\Delta^2 f(-1) + \frac{2(x+1)(x-1)}{3!}\Delta^3 f(-2) + ...$$

EXAMPLE 11. *Using Newton's divided difference method, compute $f(3)$ from the following table :*

x	0	1	2	4	5	6
y	1	14	15	5	6	19

[UPTU (MCA)-2005]

SOLUTION.   Constructing the divided difference table, we have

x	$f(x)$	$\Delta f(x)$	$\Delta^2 f(x)$	$\Delta^3 f(x)$	$\Delta^4 f(x)$
0	1				
		$\frac{14-1}{1}=13$			
1	14		$\frac{1-13}{2}=-6$		
		$\frac{15-14}{1}=1$	$\frac{-5-1}{3}=-2$	$\frac{-2+6}{4}=1$	**0**
2	15				
		$\frac{5-15}{2}=-5$	$\frac{1+5}{3}=2$	$\frac{2+2}{4}=1$	**0**
4	5				
		$\frac{6-5}{1}=1$	$\frac{13-1}{2}=6$	$\frac{6-2}{4}=1$	
5	6				
		$\frac{19-6}{1}=13$			
6	19				

From the Newton's divided formula, we get

$$f(x) = f(x_0) + (x-x_0)f(x_0,x_1) + (x-x_0)(x-x_1)f(x_0,x_1,x_2) + (x-x_0)$$
$$(x-x_1)(x-x_2)f(x_0,x_1,x_2,x_3)$$

$$f(3) = 1 + (3-0)\,13 + (3-0)(3-1)(-6) + (3-0)(3-1)(3-2)\times 1$$
$$= 1 + 39 + 3\times 2\,(-6) + 3\times 2\times 1$$
$$= 1 + 39 + (-36) + 6 = 40 - 36 + 6 = 10$$

EXAMPLE 12. *Using Newton's divided differences formula, evaluate $f(8)$ and $f(15)$ using the following table :*

x	4	5	7	10	11	13
$f(x)$	48	100	294	900	1210	2028

(VTU–2008, UPTU–2004; Rohilkhand-2007, Agra-2000, 04, Delhi-2008)

SOLUTION.

Difference Table

x	$f(x)$	$\Delta\, f(x)$	$\Delta^2\, f(x)$	$\Delta^3\, f(x)$	$\Delta^4\, f(x)$
4	48				
5	100	52	15		
7	294	97	21	1	
10	900	202	27	1	0
11	1210	310	33	1	0
13	2028	409			

Putting $x = 8$, in Newton's divided difference formula, we obtain

$$f(8) = 48 + (8-4)52 + (8-4)(8-5)15 + (8-4)(8-5)(8-7)1$$
$$= 448$$

and $f(15) = 3150$

5.16 LAGRANGE'S INTERPOLATION FORMULA

[Meerut-2016; Delhi-2011; Tamilnadu (MKU)-2005; UPTU(MCA)-2006]

Let $y = f(x)$ be a function, which takes the values $f(x_0)$, $f(x_1)$, ... $f(x_n)$, corresponding to the values of $x = x_0, x_1, ..., x_n$, not necessarily equally spaced. Then

$$f(x) = \frac{(x-x_1)(x-x_2)...(x-x_n)}{(x_0-x_1)(x_0-x_2)...(x_0-x_n)} f(x_0) + \frac{(x-x_0)(x-x_2)...(x-x_n)}{(x_1-x_0)(x_1-x_2)...(x_1-x_n)} f(x_1) + ...$$
$$+ \frac{(x-x_0)(x-x_1)...(x-x_{n-1})}{(x_n-x_0)(x_n-x_1)...(x_n-x_{n-1})} f(x_n)$$

Proof. Let $y = f(x)$ be a function which can assume the values $f(x_0), f(x_1) ... f(x_n)$ corresponding to the values of the arguments $x_0, x_1, ..., x_n$, respectively.

Now, since there are $(n+1)$ pairs of values of x and y, therefore, we can represent $f(x)$ by a polynomial in x of degree n.

Let this polynomial be of the form

$$f(x) = A_0(x-x_1)(x-x_2)...(x-x_n) + A_1(x-x_0)(x-x_2)...(x-x_n)$$
$$A_2(x-x_0)(x-x_1)(x-x_3)...(x-x_n) + ... + A_n(x-x_0)(x-x_1)(x-x_2)...(x-x_{n-1})$$
$$...(1)$$

where $A_0, A_1, A_2, ..., A_n$ are constants and can be determined by putting

$$f(x) = f(x_0) \text{ at } x = x_0, f(x) = f(x_1) \text{ at } x = x_1 ... \text{ etc.}$$

Now put $x = x_0$ and $f(x) = f(x_0)$ in (1), we get

$$f(x_0) = A_0(x_0-x_1)(x_0-x_2)...(x_0-x_n)$$

$$\Rightarrow \qquad A_0 = \frac{f(x_0)}{(x_0-x_1)(x_0-x_2)...(x_0-x_n)}$$

Similarly, putting $f(x) = f(x_1)$ and $x = x_1$ in (1), we get

$$f(x_1) = A_1(x_1-x_0)(x_1-x_2)...(x_1-x_n)$$

$$A_1 = \frac{f(x_1)}{(x_1-x_0)(x_1-x_2)...(x_1-x_n)}$$

Proceeding in the same way, we get $A_2, A_3, ..., A_n$,

i.e.,
$$A_i = \frac{f(x_i)}{(x_i - x_0)(x_i - x_1)...(x_i - x_{i-1})(x_i - x_{i+1})...(x_i - x_n)}$$

Put all these values of $A_0, A_1, ..., A_n$ in (1), we get

$$f(x) = \frac{(x - x_1)(x - x_2)...(x - x_n)}{(x_0 - x_1)(x_0 - x_2)...(x_0 - x_n)} f(x_0) + \frac{(x - x_0)(x - x_2)...(x - x_n)}{(x_1 - x_0)(x_1 - x_2)...(x_1 - x_n)} f(x_1) + ... +$$

$$+ \frac{(x - x_0)(x - x_1)...(x - x_{n-1})}{(x_n - x_0)(x_n - x_1)...(x_n - x_{n-1})} f(x_n)$$

which is the required Lagrange's interpolation formula.

▶ **REMARKS**

▸ The Lagrange's formula can be applied whether the values x_i are equally spaced or not.
▸ If role of x and y are interchanged, then this formula can also be applied.
▸ This formula can also be used to split the given function into partial fractions.
▸ Lagrange's interpolation formula is easy to remember but its application is not speedily.
▸ The main drawback of it is that if another interpolation value is inserted, then the interpolation coefficient are required to be calculated.

ALGORITHM

Step 1.	Put the values of x_i, $f(x_i)$ and values of x for which $f(x)$ is to be calculated.
Step 2.	Calculate the values of $\displaystyle\prod_{\substack{r=0 \\ k \neq 0}}^{n} \frac{(x - x_r)}{(x_i - x_r)}$.
Step 3.	Find $f(x)$ by using the Lagrange's interpolation formula.

5.16.1 REMAINDER TERM IN LAGRANGE'S INTERPOLATION FORMULA

Let $f(x)$ be a function which is approximated by means of some polynomial $P_n(x)$ in x of nth degree.

Let us suppose $f(x)$ satisfying all the conditions of Rolle's theorem.

Let
$$f(x) = P_n(x) + g(x) \qquad ...(1)$$

where $g(x)$ is a polynomial with root $x_0, x_1, ..., x_n$.

$\Rightarrow \qquad f(x) = P_n(x) + k(x)(x - x_0)(x - x_1)...(x - x_n) \qquad ...(2)$

We want to find the value of $k(x)$.

Let
$$\phi(t) = f(t) - P_n(t) - k(x)(t - x_0)(t - x_1)...(t - x_n) \qquad ...(3)$$

We observe that $\phi(t)$ vanishes for $(n+1)$ values of $t = (x_0, x_1, ... x_n)$ and $t = x$, which implies $\phi(t)$ vanishes for $(n+2)$ real roots $x, x_0, x_1, ..., x_n$. Then by Rolle's theorem $\phi'(t)$ has at least $(n+1)$ roots lying between the smallest and greatest of the above roots. Similarly, $\phi'(t)$ has at least n roots and $\phi^{n+1}(t)$ has at least one root, say $t = \theta$ in same interval (x_0, x_n).

$$\phi^{n+1}(t) = f^{n+1}(t) - k(x)[(n+1)!] \qquad (\because P_n^{n+1}(t) = 0)$$

$\Rightarrow \qquad \phi^{n+1}(\theta) = f^{n+1}(\theta) - k(x)(n+1)!$

but $\phi^{n+1}(\theta) = 0$, therefore
$$f^{n+1}(\theta) - k(x)(n+1)! = 0$$
$$\Rightarrow \qquad k(x) = \frac{1}{(n+1)!} f^{n+1}(\theta) \qquad\qquad 0 < \theta < x_n$$

Now putting the value of $k(x)$ in (2), we get
$$f(x) = P_n(x) + f^{n+1}(\theta) \frac{(x - x_0)(x - x_1)...(x - x_n)}{(n+1)!}$$

$\therefore$ The required truncation error in $f(x)$ is $R_n = f(x) - P_n(x)$.

$$\Rightarrow \qquad R_n = \frac{f^{n+1}(\theta)}{(n+1)!} \prod_{i=0}^{n}(x - x_i)$$

5.16.2 MERIT AND DEMERITS OF LAGRANGE'S INTERPOLATION FORMULA

(1) Lagrange's method is simple and easy to remember.

(2) While applying Lagrange's interpolation formula, there is no need to construct the difference table.

(3) The calculation is more complicated than the divided difference formula.

(4) There are more chances of errors in computation.

(5) The calculations provide no check whether the functional values used are taken correctly or not.

Recapitulations

- Lagrange's formula is simply a relation between two variables either of which may be taken as the independent variable, it is evident that by considering y as the independent variable we can write a formula giving x as a function of y.

- The main uses of Lagrange's formula are two: (1) to find any value of a function when the given values of the independent variable are not equidistant, and (2) to find the value of the independent variable corresponding to a given value of the function.

SOLVED EXAMPLES

EXAMPLE 1. *The value of x and y are given as below :*

x	5	6	9	11
$f(x)$	12	13	14	16

Find the value of y at x = 10. (Madras–2000, UPTU-2003, JNTU-2008, 09)

SOLUTION. Here the values of x are not equally spaced.

We have, $x_0 = 5, x_1 = 6, x_2 = 9, x_3 = 11$

and $y_0 = 12, y_1 = 13, y_2 = 14, y_4 = 16$

$\therefore \quad y_{10} = \frac{(10-6)(10-9)(10-11)}{(5-6)(5-9)(5-11)} \times 12 + \frac{(10-5)(10-9)(10-11)}{(6-5)(6-9)(6-11)} \times 13$

$\qquad + \frac{(10-5)(10-6)(10-11)}{(9-5)(9-6)(9-11)} \times 14 + \frac{(10-5)(10-6)(10-9)}{(11-5)(11-6)(11-9)} \times 16$

$= \frac{4 \times 1 \times -1}{-1 \times -4 \times -6} \times 12 + \frac{5 \times 1 \times -1}{1 \times -3 \times -5} \times 13$

$= \frac{5 \times 4 \times -1}{4 \times 3 \times -2} \times 14 + \frac{5 \times 4 \times 1}{6 \times 5 \times 2} \times 16 = 2 - 4.33 + 11.67 + 5.33$

Hence, $y_{10} = 14.67$

$$\therefore \quad \underset{y}{\Delta} x^2 = \frac{y^2 - x^2}{y - x} = \frac{(y-x)(y+x)}{(y-x)} = y + x$$

EXAMPLE 2. *Find the form of the function given below :*

x	3	2	1	-1
$f(x)$	3	12	15	-21

SOLUTION. Here, $x_0 = 3, x_1 = 2, x_2 = 1, x_3 = -1$ (Kurukeshetra-2008)

$\therefore$ By Lagrange's formula, we have

$$f(x) = \frac{(x-2)(x-1)(x+1)}{(3-2)(3-1)(3+1)} \times 3 + \frac{(x-2)(x-1)(x+1)}{(2-3)(2-1)(2+1)} \times 12$$

$$+ \frac{(x-3)(x-2)(x+1)}{(1-3)(1-2)(1+1)} \times 15 + \frac{(x-3)(x-2)(x-1)}{(-1-3)(-1-2)(-1-1)} \times -21$$

$$= \frac{3}{8}(x^3 - 2x^2 - x + 2) - 4(x^3 - 3x^2 - x + 3) + \frac{15}{4}(x^3 - 4x^2 + x + 6)$$

$$+ \frac{21}{24}(x^3 - 6x^2 + 11x - 6)$$

Hence, $f(x) = x^3 - 9x^2 + 17x + 6$.

EXAMPLE 3. *Find the form of the function given by*

x	0	1	2	3	4
$f(x)$	3	6	11	18	27

SOLUTION. Here given that

$x_0 = 0, x_1 = 1, x_2 = 2, x_3 = 3, x_4 = 4$; $y_0 = 3, y_1 = 6, y_2 = 11, y_3 = 18, y_4 = 27$

$\therefore$ By Lagrange's formula, we get

$$f(x) = \frac{(x-1)(x-2)(x-3)(x-4)}{(0-1)(0-2)(0-3)(0-4)} \times 3 + \frac{(x-0)(x-2)(x-3)(x-4)}{(1-0)(1-2)(1-3)(1-4)} \times 6$$

$$+ \frac{(x-0)(x-1)(x-3)(x-4)}{(2-0)(2-1)(2-3)(2-4)} \times 11 + \frac{(x-0)(x-1)(x-2)(x-4)}{(3-0)(3-1)(3-2)(3-4)} \times 18$$

$$+ \frac{(x-0)(x-1)(x-2)(x-3)}{(4-0)(4-1)(4-2)(4-3)} \times 27$$

$$= \frac{1}{8}[x^4 - 10x^3 + 35x^2 - 50x + 24] - [x^4 - 9x^3 + 26x^2 - 24x]$$

$$+ \frac{11}{4}[x^4 - 8x^3 + 19x^2 + 2x] - 3[x^4 - 7x^3 + 14x^2 - 8x]$$

$$+ \frac{27}{24}[x^4 - 6x^3 + 11x^2 - 6x]$$

Hence, $f(x) = x^2 + 2x + 3$.

EXAMPLE 4. *Find $f(4)$ from the following table :*

x	0	1	2	3
$f(x)$	2	5	7	8

        (Avadh- 2004, 09)

SOLUTION. Here, $x_0 = 0, x_1 = 1, x_2 = 2, x_3 = 5$; $y_0 = 2, y_1 = 5, y_2 = 7, y_3 = 8$

$\therefore$ By Lagrange's interpolation formula, we get

$$f(4) = \frac{(4-1)(4-2)(4-5)}{(0-1)(0-2)(0-2)} \times 2 + \frac{(4-0)(4-2)(4-5)}{(1-0)(1-2)(1-5)} \times 5$$

$$+ \frac{(4-0)(4-1)(4-5)}{(2-0)(2-1)(2-5)} \times 7 + \frac{(4-0)(4-1)(4-2)}{(5-0)(5-1)(5-2)} \times 8$$

$$= \frac{3.2.(-1)}{(-1)(-2)(-5)} \times 2 + \frac{4.2.(-1)}{1.(-1).(-4)} \times 5 + \frac{4.3.(-1)}{2.1.(-3)} \times 7 + \frac{4.3.2}{5.4.3} \times 8$$

$$= 1.2 - 10 + 14 + 3.2$$

$\therefore$ $f(4) = 8.4$.

EXAMPLE 5. *Find the value of $f(x)$ for $x = 2.55$ from the following data*

x	1	2	3	4
$f(x)$	1	8	27	64

(UPTU MCA-2001)

SOLUTION.   Here, $x_0 = 1, x_1 = 2, x_2 = 3, x_3 = 4$

$y_0 = 1, y_1 = 8, y_2 = 27, y_3 = 64$

Putting all these values in Lagrange's interpolation formula, we get

$$f(x) = \frac{(x-2)(x-3)(x-4)}{(1-2)(1-3)(1-4)}(1) + \frac{(x-1)(x-3)(x-4)}{(2-1)(2-3)(2-4)}(8)$$

$$+ \frac{(x-1)(x-2)(x-4)}{(3-1)(3-2)(3-4)}(27) + \frac{(x-1)(x-2)(x-3)}{(4-1)(4-2)(4-3)}(64)$$

$$= -\frac{1}{6}(x-2)(x-3)(x-4) + 4(x-1)(x-3)(x-4) - \frac{27}{2}(x-1)(x-2)(x-4)$$

$$+ \frac{32}{3}(x-1)(x-2)(x-3)$$

Putting $x = 2.5$, we get

$$f(2.5) = -\frac{1}{6}(2.5-2)(2.5-3)(2.5-4) + 4(2.5-1)(2.5-3)(2.5-4) - \frac{27}{2}(2.5-1)$$

$$(2.5-2)(2.5-4) + \frac{32}{3}(2.5-1)(2.5-2)(2.5-3)$$

$$= 15.625.$$

EXAMPLE 6. *Find the Lagrange's interpolating polynomial from the following data:*

x	0	1	2	5
$f(x)$	2	3	12	147

(Avadh-2004, 09; Anna–2005, UPTU–2004)

SOLUTION.   We have

$x_0 = 0$ $x_1 = 1$ $x_2 = 2$ $x_3 = 5$; $y_0 = 2$ $y_1 = 3$ $y_2 = 12$ $y_3 = 147$

Putting all these values in Lagrange's interpolation formula, we get

$$f(x) = \frac{(x-1)(x-2)(x-5)}{(0-1)(0-2)(0-5)}(2) + \frac{(x-0)(x-2)(x-5)}{(1-0)(1-2)(1-5)}(3)$$

$$+ \frac{(x-0)(x-1)(x-5)}{(2-0)(2-1)(2-5)}(12) + \frac{(x-0)(x-1)(x-2)}{(5-0)(5-1)(5-2)}(147)$$

$$= \frac{1}{5}(x-1)(x-2) + \frac{3}{4}x(x-2)(x-5) - 2x(x-1)(x-5) + \frac{49}{20}x(x-1)(x-2)$$

$$= -\frac{1}{5}(x^3 - 8x^2 + 17x - 10) + \frac{3}{4}(x^3 - 7x^2 + 10x) - 2(x^3 - 6x^2 + 5x)$$
$$+ \frac{49}{20}(x^3 - 3x^2 + 2x)$$

$\Rightarrow f(x) = x^5 + x^2 - x + 2$, which is the required polynomial.

EXAMPLE 7. *Find the unique polynomial P(x) of degree 2 such that $P(1) = 1, P(3) = 27, P(4) = 64$ using Lagrange's method.*

(UPTU MCA-2003)

SOLUTION. We have $x_0 = 1, x_1 = 3, x_2 = 4$

$\qquad f(x_0) = 1, f(x_1) = 27, f(x_2) = 64$

Putting all these values in Lagrange's interpolation formula, we get

$$P(x) = \frac{(x-3)(x-4)}{(1-3)(1-4)}(1) + \frac{(x-1)(x-4)}{(3-1)(3-4)}(27) + \frac{(x-1)(x-3)}{(4-1)(4-3)}(64)$$

$$= \frac{1}{6}(x^2 - 7x + 12) - \frac{27}{2}(x^2 - 5x + 4) + \frac{64}{3}(x^2 - 4x + 3)$$

$$= 8x^2 - 19x - 12, \text{ which is the required polynomial.}$$

EXAMPLE 8. *Find the value of y at x = 5 given that*

x	1	3	4	8	10
y	8	15	19	32	40

SOLUTION. Let, $x_0 = 1, x_1 = 3, x_2 = 4, x_3 = 8, x_4 = 10$

and $y_0 = 8, y_1 = 15, y_2 = 19, y_3 = 32, y_4 = 40$

We have to find the value of y at $x = 5$. Then by Lagrange's formula, we get

$$y_5 = \frac{(5-3)(5-4)(5-8)(5-10)}{(1-3)(1-4)(1-8)(1-10)} \times 8 + \frac{(5-1)(5-4)(5-8)(5-10)}{(3-1)(3-4)(3-8)(3-10)} \times 15$$

$$+ \frac{(5-1)(5-3)(5-8)(5-10)}{(4-1)(4-3)(4-8)(4-10)} \times 19 + \frac{(5-1)(5-3)(5-4)(5-10)}{(8-1)(8-3)(8-4)(8-10)} \times 32$$

$$+ \frac{(5-1)(5-3)(5-4)(5-8)}{(10-1)(10-3)(10-4)(10-8)} \times 40$$

$$= \frac{240}{378} - \frac{900}{70} + \frac{2280}{72} + \frac{1280}{280} - \frac{960}{756}$$

$$= 0.635 - 12.857 + 31.667 + 4.571 - 12.698$$

Hence, $y_5 = 11.318$.

EXAMPLE 9. *Find u_3, given $u_0 = 580, u_1 = 556, u_2 = 520$ and $u_4 = 385$*

SOLUTION. Here $x_0 = 0, x_1 = 1, x_2 = 2, x_4 = 4$ and $x = 3$ and $u_0 = 580, u_1 = 566, u_2 = 520, u_4 = 365$.

Now, the value of u_3 by Lagrange's formula is given by

$$u_3 = \frac{(3-1)(3-2)(3-4)}{(0-1)(0-2)(0-4)} \times 580 + \frac{(3-0)(3-2)(3-4)}{(1-0)(1-2)(1-4)} \times 385$$

$$+ \frac{(3-0)(3-1)(3-4)}{(2-0)(2-1)(2-4)} \times 520 + \frac{(3-0)(3-2)(3-4)}{(1-0)(1-2)(1-4)} \times 556$$

$$= \frac{580}{4} - 556 + 780 + \frac{385}{4} = 241.25 + 224$$

Hence, $u_3 = 465.25$.

EXAMPLE 10. *Using Lagrange's formula, prove that*

$$y_0 = \frac{1}{2}(y_1 + y_{-1}) - \frac{1}{8}\left[\frac{1}{2}(y_3 - y_1) - \frac{1}{2}(y_{-1} - y_{-3})\right].$$

(Agra-2004, 06; Kanpur-2004, 07)

SOLUTION. Here, $x_0 = -3, x_1 = -1, x_2 = 1, x_3 = 3$

Then by Lagrange's formula, we get

$$y_x = \frac{(x+1)(x-1)(x-3)}{(-3+1)(-3-1)(-3-3)} \times y_{-3} + \frac{(x+3)(x-1)(x-3)}{(-1+3)(-1-1)(-1-3)} \times y_{-1}$$
$$+ \frac{(x+3)(x+1)(x-3)}{(1+3)(1+1)(1-3)} \times y_1 + \frac{(x+3)(x+1)(x-1)}{(3+3)(3+1)(3-1)} \times y_3$$

Now putting $x = 0$ we have

$$y_0 = -\frac{3}{48}y_{-3} + \frac{9}{16}y_{-1} + \frac{9}{16}y_1 - \frac{3}{48}y_3$$

$$= -\frac{1}{16}y_{-3} + \left(\frac{1}{16} + \frac{8}{16}\right)y_{-1} + \left(\frac{1}{16} + \frac{8}{16}\right)y_1 - \frac{1}{16}y_3$$

$$= \frac{8}{16}(y_1 + y_{-1}) - \frac{1}{16}y_3 + \frac{1}{16}y_1 - \frac{1}{16}y_{-3} + \frac{1}{16}y_{-1}$$

Hence, $y_0 = \frac{1}{2}(y_1 + y_{-1}) - \frac{1}{8}\left[\frac{1}{2}(y_3 - y_1) - \frac{1}{2}(y_{-1} - y_{-3})\right]$

EXAMPLE 11. *By Lagrange's formula, prove that*

$$y_1 = y_3 - 0.3(y_5 - y_{-3}) + 0.2(y_{-3} - y_{-5}).$$

(UPTU–2003; Avadh-2007, UPPCS-2006)

SOLUTION. Here, the values of the arguments are given by $x_0 = -5, x_1 = -3, x_2 = 3, x_3 = 5$
and corresponding values of functions are y_{-5}, y_{-3}, y_3, y_5
Now using Lagrange's formula, we get

$$y_x = \frac{(x+3)(x-3)(x-5)}{(-5+3)(-5-3)(-5-5)} \times y_{-5} + \frac{(x+5)(x-3)(x-5)}{(-3+5)(-3-3)(-3-5)} \times y_{-3}$$
$$+ \frac{(x+5)(x+3)(x-5)}{(3+5)(3+3)(3-5)} \times y_3 + \frac{(x+5)(x+3)(x-3)}{(5+5)(5+3)(5-3)} \times y_5$$

Putting $x = 1$, we get

$$y_1 = -\frac{32}{160}y_{-5} + \frac{48}{96}y_{-3} + \frac{96}{96}y_3 - \frac{48}{160}y_5$$

$$= y_3 - 0.3y_5 + 0.5y_{-3} - 0.2y_{-5} = y_3 - 0.3y_5 + 0.3y_{-3} + 0.2y_{-3} - 0.2y_{-5}$$

Hence, $y_1 = y_3 - 0.3(y_5 - y_{-3}) + 0.2(y_{-3} - y_{-5})$.

EXAMPLE 12. *Using Lagrange's method, prove that*

$$y_3 = 0.05(y_0 + y_6) - 0.3(y_1 + y_5) + 0.75(y_2 + y_4).$$

(Meerut-2001, Avadh-2005)

SOLUTION. Here, the arguments are

$$x_0 = 0, x_1 = 1, x_2 = 2, x_3 = 4, x_4 = 5, x_5 = 6$$

and their corresponding values of functions are given by

$$y_0, y_1, y_2, y_4, y_5 \text{ and } y_6$$

$\therefore$ Lagrange's formula is given by

$$y_x = \frac{(x-1)(x-2)(x-4)(x-5)(x-6)}{(0-1)(0-2)(0-4)(0-5)(0-6)} y_0$$

$$+ \frac{(x-0)(x-2)(x-4)(x-5)(x-6)}{(1-0)(1-2)(1-4)(1-5)(1-6)} y_1$$

$$+ \frac{(x-0)(x-1)(x-4)(x-5)(x-6)}{(2-0)(2-1)(2-4)(2-5)(2-6)} \cdot y_2$$

$$+ \frac{(x-0)(x-1)(x-2)(x-5)(x-6)}{(4-0)(4-1)(4-1)(4-5)(4-6)} \cdot y_4$$

$$+ \frac{(x-0)(x-1)(x-2)(x-4)(x-6)}{(5-0)(5-1)(5-2)(5-4)(5-6)} \cdot y_5$$

$$+ \frac{(x-0)(x-1)(x-2)(x-4)(x-5)}{(6-0)(6-1)(6-2)(6-4)(6-5)} \cdot y_6$$

Putting $x = 3$, we get

$$y_3 = \frac{12}{240} y_0 - \frac{18}{60} y_1 + \frac{36}{48} y_2 + \frac{36}{48} y_4 + \frac{18}{60} y_5 + \frac{12}{240} y_6$$

$$= \frac{12}{240}(y_0 + y_6) - \frac{18}{60}(y_1 + y_5) + \frac{36}{48}(y_2 + y_4)$$

Hence, $y_3 = 0.05(y_0 + y_6) - 0.3(y_1 + y_5) + 0.75(y_2 + y_4)$.

EXAMPLE 13. *Find the form of the function $f(x)$ for the following table:*

x	0	1	4	5
$f(x)$	8	11	68	123

SOLUTION. Here given that

$$x_0 = 0, x_1 = 1, x_2 = 4, x_3 = 5$$

and $y_0 = 8, y_1 = 11, y_2 = 68, y_3 = 123$

Now, using Lagrange's formula, we get

$$f(x) = \frac{(x-1)(x-4)(x-5)}{(0-1)(0-4)(0-5)} \times 8 + \frac{(x-0)(x-4)(x-5)}{(1-0)(1-4)(1-5)} \times 11$$

$$+ \frac{(x-0)(x-1)(x-5)}{(4-0)(4-1)(4-5)} \times 68 + \frac{(x-0)(x-1)(x-4)}{(5-0)(5-1)(5-4)} \times 123$$

$$= -\frac{2}{5}(x^3 - 10x^2 + 29x - 20) + \frac{11}{12}(x^3 - 9x^2 + 20x)$$

$$- \frac{17}{3}(x^3 - 6x^2 + 5x) + \frac{123}{20}(x^3 - 5x^2 + 4x)$$

$$= x^3 \left(-\frac{2}{5} + \frac{11}{12} - \frac{17}{3} + \frac{123}{20}\right) + x^2 \left(4 - \frac{33}{4} + 34 - \frac{123}{4}\right) + x$$

$$\left(-\frac{58}{5} + \frac{110}{6} - \frac{85}{3} + \frac{123}{5}\right) + 8$$

Hence, $f(x) = x^3 - x^2 + 3x + 8$.

EXAMPLE 14. *The function $y = f(x)$ is given at the point $(7, 3)$, $(8, 1)$, $(9,1)$ and $(10, 9)$ Find $f(9, 5)$ using Lagrange's formula.* (UPTU-2004)

SOLUTION. It is given that

$$x_0 = 7, x_1 = 8, x_2 = 9, x_3 = 10 \text{ and } y_0 = 3, y_1 = 1, y_2 = 1, y_3 = 9$$

Lagrange's formula is given by

$$f(x) = \frac{(x-8)(x-9)(x-10)}{(7-8)(7-9)(7-10)} \times 3 + \frac{(x-7)(x-9)(x-10)}{(8-7)(8-9)(8-10)} \times 1$$

$$+ \frac{(x-7)(x-8)(x-10)}{(9-7)(9-8)(9-10)} \times 1 + \frac{(x-7)(x-8)(x-9)}{(10-7)(10-8)(10-9)} \times 9$$

Putting $x = 9.5$ we get

$$f(9.5) = \frac{(9.5-8)(9.5-9)(9.5-10)}{(-1)(-2)(-3)} \times 3 + \frac{(9.5-7)(9.5-9)(9.5-10)}{(1)(-1)(-2)} \times 1$$

$$+ \frac{(9.5-7)(9.5-8)(9.5-10)}{(2)(1)(-1)} \times 1 + \frac{(9.5-7)(9.5-8)(9.5-9)}{(3)(2)(1)} \times 9$$

$$+ \frac{(1.5)(0.5)(-0.5)}{-6} \times 3 + \frac{(2.5)(0.5)(-0.5)}{2} \times 1$$

$$+ \frac{(2.5)(1.5)(-0.5)}{-2} \times 1 + \frac{(2.5)(1.5)(0.5)}{6} \times 9$$

$$= 0.1875 - 0.3125 + 0.9375 + 2.8125$$

Hence, $f(9.5) = 3.625$.

EXAMPLE 15. *Using Lagrange's interpolation formula, find the form of the function $y(x)$ from the following table:* (MDU(B.E.)-2003)

x	0	1	3	4
y	−12	0	12	24

SOLUTION. Here, we have

$$x_0 = 0, \ y_0 = -12; \ x_1 = 1, y_1 = 0; \ x_2 = 3, \ y_2 = 12; \ x_3 = 4, y_3 = 24$$

Now applying Lagrange's formula

$$y = f(x) = \frac{(x-x_1)(x-x_2)(x-x_3)}{(x_0-x_1)(x_0-x_2)(x_0-x_3)}(y_0)$$

$$+ \frac{(x-x_0)(x-x_2)(x-x_3)}{(x_1-x_0)(x_1-x_2)(x_1-x_3)}y_1 + \frac{(x-x_0)(x-x_1)(x-x_3)}{(x_2-x_0)(x_2-x_1)(x_2-x_3)}y_2$$

$$+ \frac{(x-x_0)(x-x_1)(x-x_2)}{(x_3-x_0)(x_3-x_1)(x_3-x_2)}y_3$$

We get

$$y(x) = \frac{(x-1)(x-3)(x-4)}{(-1)(-3)(-4)} \times (-12) + \frac{(x-0)(x-3)(x-4)}{4 \times 3 \times 1} \times 0$$

$$+ \frac{(x-0)(x-1)(x-4)}{3 \times 2 \times -1} \times 12 + \frac{(x-0)(x-1)(x-3)}{4 \times 3 \times 1} \times 24$$

$$= (x-1)(x-3)(x-4) - 2x(x-1)(x-4) + 2x(x-1)(x-3)$$

$$= x^3 - 8x^2 + 19x - 12 - 2x^3 - 8x + 10x^2 + 2x^3 - 8x^2 + 6x$$

$$= x^3 - 6x^2 + 17x - 12$$

$$\therefore y_{(x)} = x^3 - 6x^2 + 17x - 12.$$

EXAMPLE 16. *If* $y(1) = -3, y(3) = 9, y(4) = 30, y(6) = 132$ *find the Lagrange's interpolation polynomial that takes the same values as 'y' at the given points.*

SOLUTION. Here, we have $x_0 = 1, x_1 = 3, x_2 = 4, x_3 = 6$ (MDU(B.E.)-2004; VTU-2006)

$$y_0 = -3, y_1 = 9, y_2 = 30, y_3 = 132$$

Let the polynomial be of the form

$$y = a_0(x - x_1)(x - x_2)(x - x_3) + a_1(x - x_0)$$
$$(x - x_2)(x - x_3) + a_2(x - x_0)(x - x_1)(x - x_3)$$
$$+ a_3(x - x_0)(x - x_1)(x - x_2)$$
$$= a_0(x - 3)(x - 4)(x - 6) + a_1(x - 1)(x - 4)(x - 6) + a_2(x - 1)(x - 3)(x - 6)$$
$$+ a_3(x - 1)(x - 3)(x - 4) \qquad \qquad \dots(1)$$

Now $a_0 = \dfrac{y_0}{(x_0 - x_1)(x_0 - x_2)(x_0 - x_3)} = \dfrac{-3}{(1 - 3)(1 - 4)(1 - 6)} = 0.1$

$a_1 = \dfrac{y_1}{(x_1 - x_0)(x_1 - x_2)(x_1 - x_3)} = \dfrac{9}{(3 - 1)(3 - 4)(3 - 6)} = 1.5$

$a_2 = \dfrac{y_2}{(x_2 - x_0)(x_2 - x_1)(x_2 - x_3)} = \dfrac{30}{(4 - 1)(4 - 3)(4 - 6)} = -5$

$a_3 = \dfrac{y_3}{(x_3 - x_0)(x_3 - x_1)(x_3 - x_2)} = \dfrac{132}{(6 - 1)(6 - 3)(6 - 4)} = 4.4$

Substituting these above values in (1), we get

$$y = 0.1(x^3 - 13x^2 + 54x - 72) + 1.5 \ (x^3 - 11x^2 + 34x - 24) - 5(x^3$$
$$-10x^2 + 27x - 18) \ + 4.4(x^3 - 8x^2 + 9x - 12)$$

$\therefore \qquad y = x^3 - 3x^2 + 5x - 6$

EXAMPLE 17. *From the given table.*

x	20	25	30	35
$y(x)$	0.342	0.423	0.500	0.650

find the value of x from $y(x) = 0.390$.   [UPTU (MCA)-2003]

SOLUTION. Here,

$$x_0 = 20 \qquad \Rightarrow \qquad y_0 = 0.342$$
$$x_1 = 25 \qquad \Rightarrow \qquad y_1 = 0.423$$
$$x_2 = 30 \qquad \Rightarrow \qquad y_2 = 0.500$$
and $\quad x_3 = 35 \qquad \Rightarrow \qquad y_3 = 0.650$

Using Lagrange's formula, we get

$$x = \frac{(y - y_1)(y - y_2)(y - y_3)}{(y_0 - y_1)(y_0 - y_2)(y_0 - y_3)} x_0 + \frac{(y - y_0)(y - y_2)(y - y_3)}{(y_1 - y_0)(y_1 - y_2)(y_1 - y_3)} x_1$$
$$+ \frac{(y - y_0)(y - y_1)(y - y_3)}{(y_2 - y_0)(y_2 - y_1)(y_2 - y_3)} x_2 + \frac{(y - y_0)(y - y_1)(y - y_2)}{(y_3 - y_0)(y_3 - y_1)(y_3 - y_2)} x_3$$

Now, at $y = 0.390$

$$x = \frac{(0.390 - 0.423)(0.390 - 0.500)(0.390 - 0.650)}{(0.342 - 0.423)(0.342 - 0.500)(0.342 - 0.650)}(20)$$

$$+ \frac{(0.390 - 0.342)(0.390 - 0.500)(0.390 - 0.650)}{(0.432 - 0.342)(0.432 - 0.500)(0.432 - 0.650)}(25)$$

$$+ \frac{(0.390 - 0.342)(0.390 - 0.500)(0.390 - 0.650)}{(0.500 - 0.342)(0.500 - 0.432)(0.500 - 0.650)}(30)$$

$$+ \frac{(0.390 - 0.342)(0.390 - 0.423)(0.390 - 0.500)}{(0.650 - 0.342)(0.650 - 0.423)(0.650 - 0.500)}(35)$$

$$x = \frac{(-0.033)(-0.110)(-0.260)}{(-0.081)(-0.158)(-0.308)}(20) + \frac{(0.048)(-0.110)(-0.260)}{(0.081)(-0.077)(-0.227)}(25)$$

$$+ \frac{(0.048)(-0.033)(-0.260)}{(0.158)(-0.077)(-0.150)}(30) + \frac{(0.048)(-0.033)(-0.110)}{(0.308)(0.227)(0.150)}(35)$$

$$\Rightarrow \quad x = 23.67.$$

EXAMPLE 18. *Given*

$$\log_{10} 654 = 2.8156 \, , \log_{10} 658 = 2.8182, \ \log_{10} 659 = 2.8189, \log_{10} 661 = 2.8202$$

Find $\log_{10} 656$. 　　　　(Agra-2007; Rohtak-2006, 10, Bangluru-2008, Hazaribagh-2009)

SOLUTION. Putting the given values in Lagrange's interpolation formula, we get

$$\log_{10} 654 = \frac{(656 - 658)(656 - 659)(656 - 661)}{(654 - 658)(654 - 659)(654 - 661)} \times (2.8156)$$

$$+ \frac{(656 - 654)(656 - 659)(656 - 661)}{(658 - 654)(658 - 659)(658 - 661)} \times (2.8182)$$

$$+ \frac{(656 - 654)(656 - 658)(656 - 661)}{(659 - 654)(659 - 658)(659 - 661)} \times (2.8189)$$

$$+ \frac{(656 - 654)(656 - 658)(656 - 659)}{(661 - 654)(661 - 658)(661 - 659)} \times (2.8202)$$

$$= \frac{3}{14}(2.8156) + \frac{5}{2}(2.8182) - 2(2.8189) + \frac{2}{7}(2.8202) = 2.8170.$$

EXAMPLE 19. *Four equidistant values* u_{-1}, u_0, u_1 *and* u_2 *being given, a value is interpolated by Lagrange's formula, show that it may be written in the form*

$$u_x = yu_0 + xu_1 + \frac{y(y^2 - 1)}{3!}\Delta^2 u_{-1} + \frac{x(x^2 - 1)}{3!}\Delta^2 u_0 \text{ where } x + y = 1.$$

　　　　(Baroda-2004, Nagpur-2006, 09, Bangluru-2005, 08, Lucknow-2005, Delhi-2008,
　　　　Kurukshetra-2006, 10, Rohilkhand-2006, 09)

SOLUTION. To prove

$$u_x = yu_0 + xu_1 + \frac{y(y^2 - 1)}{3!}\Delta^2 u_{-1} + \frac{x(x^2 + 1)}{3!}\Delta^2 u_0$$

$$= (1 - x)u_0 + xu_1 + \frac{(1 - x)\{(1 - x^2) - 1\}}{6} + (u_2 - 2u_0 + u_{-1}) + \frac{x(x^2 - 1)}{6}(u_2 - 2u_1 + u_0)$$

$$= (1 - x)u_0 + xu_1 + \frac{(1 - x)x(x - 2)}{6}(u_1 + 2u_0 + u_{-1}) + \frac{x(x^2 - 1)}{6}(u_2 - 2u_1 + u_0)$$

$$= u_{-1}\left[\frac{(1-x)x(x-2)}{6}\right] + u_0\left[(1-x) - \frac{(1-x)x(x-2)}{3}\right]$$

$$+ \frac{x(x-1)(x+1)}{6}\right] + u_1\left[x + \frac{(1-x)x(x-2)}{6} - \frac{x(x^2-1)}{3}\right] + u_2\left[\frac{x(x-1)(x+1)}{6}\right]$$

$$= -\frac{x(x-1)(x-2)}{6}u_{-1} + \frac{(x+1)}{6}[-6 + 2x(x-2) + x(x+1)]u_0$$

$$+ \frac{1}{6}x[6 + (1-x)(x-2) - 2(x^2-1)]u_1 + \frac{(x+1)x(x-1)}{6}u_2 \qquad \dots(1)$$

Further using u_{-1}, u_0, u_1, u_2 Lagrange's interpolation formula, we have

$$u_x = \frac{(x-0)(x-1)(x-2)}{(-1-0)(-1-1)(-1-2)}u_{-1} + \frac{(x+1)(x-1)(x-2)}{(0+1)(0-1)(0-2)}u_0$$

$$+ \frac{(x+1)(x-0)(x-2)}{(1+1)(1-0)(1-2)}u_1 + \frac{(x+1)(x-0)(x-1)}{(2+1)(2-0)(2-1)}u_2$$

$$= \frac{-x(x-1)(x-2)}{6}u_{-1} + \frac{(x+1)(x-1)(x-2)}{2}u_0$$

$$- \frac{(x+1)x(x-2)}{2}u_1 + \frac{(x+1)x(x-1)}{6}u_2 \qquad \dots(2)$$

Clearly (1) and (2) are same and hence the result.

EXAMPLE 20. *Values of $f(x)$ are given at a, b and c, show that the maximum is obtained by*

$$x = \frac{f(a)(b^2-c^2) + f(b)(c^2-a^2) + f(c)(a^2-b^2)}{[f(a)(b-c) + f(b)(c-a) + f(c)(a-b)]}$$

(Meerut-2006; IAS-2006, UPPCS-2005, 07)
(Delhi-2009, 10; Vanketshwara-2006; Nagpur-2005, 08, 10)

SOLUTION. For the arguments a, b and c the Lagrange's formula is given by

$$f(x) = \frac{(x-b)(x-c)}{(a-b)(a-c)}f(a) + \frac{(x-a)(x-c)}{(b-a)(b-c)}f(b) + \frac{(x-a)(x-b)}{(c-a)(c-b)}f(c)$$

$$= \frac{x^2-(b+c)x+bc}{(a-b)(a-c)}f(a) + \frac{x^2-(a+c)x+ac}{(b-a)(b-c)}f(b) + \frac{x^2-(a+b)x+ab}{(c-a)(c-b)}f(c)$$

For maxima and minima of $f(x)$ we must have $f'(x) = 0$

i.e., $\dfrac{2x-(b+c)}{(a-b)(a-c)}f(a) + \dfrac{2x-(a+c)}{(b-a)(b-c)}f(b) + \dfrac{2x-(a+b)}{(c-a)(c-b)}f(c) = 0$

On solving for x, we get

$$x = \frac{f(a)(b^2-c^2) + f(b)(c^2-a^2) + f(c)(a^2-b^2)}{2[f(a)(b-c) + f(b)(c-a) + f(c)(a-b)]}$$

EXAMPLE 21. *Values of $y(x)$ are given for all integral values of x from 0 to $n-1$. Show that y_x is capable of expression in the form*

$$\frac{x!}{(x-n)!(n-1)!}\left(\frac{y_{n-1}}{x-n+1} - {}^{n-1}C_1\frac{y_{n-2}}{x-n+2}\right.$$

$$\left. + {}^{n-1}C_2\frac{y_{n-1}}{x-n+3} + (-1)^{n-1}.{}^{n-1}C_{n-1}\frac{y_0}{x_0}\right)$$

(Banguluru-2004, 08; Patna-2006, 08; Meerut-2003; Delhi-2004, 07)

SOLUTION. We know that

$$y = \frac{(x-0)(x-1)(x-2)-(x-(n-2))}{[(n-1-0)(n-1-1)(n-1-2)-(n-1-(n-2))]}y_{n-1}$$

$$+\frac{(x-0)(x-1)...(x-(n-3)(x-(n+1))}{(n-2-0)(n-2-1)...(n-2-(n-3)).(n-2)-(n-1)}y_{n-2}$$

$$+...+\frac{(x-1)(x-2)...(x-(n-1))}{(0-1)(0-2)...(0-(n-1))}y_0$$

$$\Rightarrow\ y_x = \frac{x(x-1)(x-2)...(x-n+2)}{(n-1)!}y_{n-1}$$

$$+\frac{x(x-1)...(x-n+3)(x-n+1)}{(n-2)!(-1)}y_{n-2}$$

$$+\frac{x(x-1)...(x-n+4)(x-n+2)(x-n+1)}{(n-3)!(-1)(-2)}y_{n-3}$$

$$+\frac{(x-1)(x-2)...(x-n+1)}{(-1)(-2)...(-(n-1))}y_0$$

$$y_x = \frac{x(x-1)...(x-n+2)(x-n+1)}{(n-1)!}\cdot\frac{y_{n-1}}{(x-n+1)}$$

$$+\frac{x(x-1)...(x-n+3)(x-n+2)(x-n+1)}{(n-2)!(-1)}\cdot\frac{y_{n-2}}{x-n+2}$$

$$+\frac{x(x-1)...(x-n+1)}{(n-3)!(-1)(-2)}\cdot\frac{y_{n-3}}{x-n-3}+...$$

$$+x\frac{(x-1)(x-2)...(x-n+1)}{(-1)(-2)...[-(n-1)]}\cdot\frac{y_0}{x}$$

$$=\frac{x!}{(x-n)!}\cdot\frac{1}{(n-1)!}\frac{y_{n-1}}{x-n+1}+(-1)\cdot\frac{x!}{(x-n)!}\frac{1}{(x-2)}\cdot\frac{1}{1!}\cdot\frac{y_{n-2}}{x-n+2}$$

$$+(-1)^2\frac{x!}{(x-n)!}\frac{1}{0!}\frac{1}{(n-1)!}\frac{y_0}{x}$$

$$=\frac{x!}{(x-n)!}\frac{1}{(n-1)!}\cdot\frac{y_{n-1}}{x-n+1}+(-1)\frac{x!}{(x-n)!}\frac{1}{(n-1)!}\,{}^{n-1}C_1\frac{y_{n-2}}{x-n+2}+(-1)^2$$

$$\frac{x!}{(x-n)!}\cdot\frac{1}{(n-1)!}\,{}^{n-1}C_2\frac{y_{n-3}}{x-n+3}+...\ +(-1)^{n-1}\cdot\frac{x!}{(x-n)!}\frac{1}{(n-1)!}\cdot{}^{n-1}C_{n-1}\frac{y_0}{x}$$

$$=\frac{x!}{(x-n)!(n-1)!}\left(\frac{y_{n-1}}{x-n+1}+(-1)^{n-1}C_1\frac{y_{n-2}}{x-n+2}\right)+(-1)^2.{}^{n-1}C_2.\frac{y_{n-3}}{x-n+3}$$

$$+...+(-1)^{n-1}.{}^{n-1}C_{n-1}\frac{y_0}{x_0}.$$

EXAMPLE 22. *Prove that Lagrange's formula can be put in the form of* $P_n(x) = \sum\limits_{r=1}^{n}\dfrac{\phi(x)f(x_r)}{(x-x_r)\phi'(x_r)}$

where $\phi(x) = \prod\limits_{r=0}^{n}(x-x_r).$ (Rohilkhand-2005, Garhwal-2005, 14, 15, MDU-2008)

SOLUTION.   Clearly, the Lagrange's formula is given by

$$P_n(x) = \sum_{r=0}^{n} \frac{(x-x_0)(x-x_1)...(x-x_n)}{(x_r-x_0)(x_r-x_1)...(x_r-x_n)} f(x)$$

$$= \sum_{r=0}^{n} \left[\frac{\phi(x)f(x_r)}{(x-x_r)} \right] \left[\frac{1}{(x_r-x_0)(x_r-x_1)...(x_r-x_n)} \right]$$

where $\phi(x) = \prod_{r=0}^{n} (x-x_r)$

$$\Rightarrow \qquad \phi'(x_r) = \left[\frac{d}{dx} \phi(x) \right]_{x=x_r} = (x_r-x_0)(x_r-x_1)...(x_r-x_n)$$

Hence, $\quad P_n(x) = \sum_{r=0}^{n} \frac{\phi(x)f(x)}{(x-x_r)\phi'(x_r)}$.

EXAMPLE 23.   *If all terms except y_5 of the sequence $y_1, y_2, y_3, ... y_9$ be given, show that the values*

of y_5 is $\left[\dfrac{56(y_4+y_6) - 28(y_3+y_7) + 8(y_2+y_8) - (y_1+y_9)}{70} \right]$.   (Meerut-2007)

SOLUTION.   Lagrange's formula is given by

$$f(x) = \frac{(x-x_1)(x-x_2)...(x-x_n)}{(x_0-x_1)(x_0-x_2)...(x_0-x_n)} f(x) + \frac{(x-x_0)(x-x_2)...(x-x_n)}{(x_1-x_0)(x_1-x_2)...(x_1-x_n)}$$

$$+...+ \frac{(x-x_0)(x-x_1)...(x-x_{n-1})}{(x_n-x_0)(x_n-x_1)...(x_n-x_{n-1})} f(x_n)$$

$$\Rightarrow \quad \frac{f(x)}{(x-x_0)(x-x_1)...(x-x_n)} = \frac{f(x_0)}{(x-x_0)(x_0-x_1)(x_0-x_2)...(x_0-x_n)}$$

$$+ \frac{f(x_1)}{(x-x_1)(x_1-x_0)(x_1-x_2)...(x_1-x_n)} +$$

$$...+ \frac{f(x_n)}{(x-x_n)(x_n-x_0)(x_n-x_1)...(x_n-x_{n-1})} \qquad ... (1)$$

Putting

$x_0 = 1, x_1 = 2, x_3 = 4, x_4 = 6, x_5 = 7, x_6 = 8, x_7 = 9, x = 5$

and $\quad f(1) = y_1, f(2) = y_2, f(3) = y_3, f(4) = y_4,$

$\qquad f(5) = y_5, f(6) = y_6, f(7) = y_7, f(8) = y_8, f(9) = y_9$

in (1), we get

$$\frac{y_5}{(5-1)(5-2)(5-3)(5-4)(5-6)(5-7)(5-8)(5-9)}$$

$$= \frac{y_1}{(5-1)(1-2)(1-3)(1-4)(1-6)(1-7)(1-8)(1-9)}$$

$$+ \frac{y_2}{(5-2)(2-1)(2-3)(2-4)(2-6)(2-7)(2-8)(2-9)}$$

$$+ \frac{y_3}{(5-3)(3-1)(3-2)(3-4)(3-6)(3-7)(3-8)(3-9)}$$

$$+ \frac{y_4}{(5-4)(4-1)(4-2)(4-3)(4-6)(4-7)(4-8)(4-9)}$$

$$+ \frac{y_6}{(5-6)(6-1)(6-2)(6-3)(6-4)(6-7)(6-8)(6-9)}$$

$$+ \frac{y_7}{(5-7)(7-1)(7-2)(7-3)(7-4)(7-6)(7-8)(7-9)}$$

$$+ \frac{y_8}{(5-8)(8-1)(8-2)(8-3)(8-4)(8-6)(8-7)(8-9)}$$

$$+ \frac{y_9}{(5-9)(9-1)(9-2)(9-3)(9-4)(9-6)(9-7)(9-8)}$$

$$\Rightarrow \quad \frac{y_5}{24} = -\frac{y_1}{1680} + \frac{y_2}{210} - \frac{y_3}{60} + \frac{y_4}{30} + \frac{y_6}{30} + \frac{y_7}{60} + \frac{y_8}{210} - \frac{y_9}{1680}$$

$$= \frac{1}{30}(y_4 + y_6) - \frac{1}{60}(y_3 + y_7) + \frac{1}{210}(y_2 + y_8) - \frac{1}{1680}(y_1 + y_9)$$

Hence, $y_5 = \frac{4}{5}(y_4 + y_6) - \frac{2}{5}(y_3 + y_7) + \frac{8}{70}(y_2 + y_8) - \frac{1}{70}(y_1 + y_9)$

$$= \frac{56(y_4 + y_6) - 28(y_3 + y_7) + 8(y_2 + y_8) - (y_1 + y_9)}{70}$$

EXAMPLE 24. *Show that the sum of Lagrangian coefficient is unity.*

SOLUTION. Define $\phi(x)$ such that

$$\phi(x) = \prod_{r=0}^{n} (x - x_r) = (x - x_0)(x - x_1)...(x - x_n) \qquad ... (1)$$

$$\Rightarrow \quad \frac{1}{\phi(x)} = \frac{1}{(x - x_0)(x - x_1)...(x - x_n)} = \frac{A_0}{x - x_0} + \frac{A_1}{x - x_1} + ... + \frac{A_n}{x - x_n}$$

$$= \sum_{r=0}^{n} \frac{A_r}{(x - x_r)} \qquad ...(2)$$

where $A_r = \dfrac{1}{(x_r - x_0)(x_r - x_1)...(x_r - x_{r-1})(x_r - x_{r+1})...(x_r - x_n)} = \dfrac{1}{\phi'(x_r)}$ ...(3)

Using (3) in (2), we get

$$\frac{1}{\phi(x)} = \sum_{r=0}^{n} \frac{1}{(x - x_r)\phi'(x_r)}$$

$$\Rightarrow \quad 1 = \sum_{r=0}^{n} \frac{\phi(x)}{(x - x_r)\phi(x_r)}.$$

EXERCISE 5.3

1. Find the function u_x in powers of $(x - 1)$ where, $u_0 = 8, u_1 = 11, u_4 = 68, u_5 = 123$.

2. Find u_x in powers of $(x - 4)$ where, $u_0 = 8, u_1 = 11, u_4 = 68, u_5 = 125$.

3. Given the values

x	5	7	11	13	17
$f(x)$	150	392	1452	2366	5202

Find $f(9)$ by using Newton's divided difference formula. (VTU-2006, 10; PTU-2005; Anna-2006)

4. Given that

x	1	3	4	6	7
y_x	1	27	81	729	2187

Find f_5. Why it does differ from 3^5?

5. Given that

x	300	304	305	307
$\log_{10}x$	2.4771	2.4829	2.4823	2.4871

Find $\log_{10}310$ by Newton's divided difference formula.

6. Find the value of t when $A = 85$ from the following data, using Lagrange's interpolation formula

t	2	5	8	14
A	94.8	87.9	81.3	68.7

7. The observed value of a function are respectively 168, 120, 72 and 63 at the four positions 3, 7, 9 and 20 of the independent variable. What is the best estimate you can give for values of the function at the position 6 of independent variable.

8. Given that

x	1.2	2.1	2.8	4.1	4.9	6.2
y	4.2	6.8	9.8	13.4	15.5	19.6

Find the value of x corresponding to $y=12$ using Lagrange's interpolation formula.

(VTU–2009)

9. Given that

x	5	6	9	11
$f(x)$	12	13	14	18

Find $f(x)$ as a polynomial in x using Newton's divided difference formula.

10. Use Lagrange's formula to find $f(0.4)$ given that $f(-1) = 1, f(0) = 3, f(1) = 2, f(2) = 5$

11. Use Lagrange's formula to find u_2 given that
$$u_0 = 6, u_1 = 9, u_3 = 33, u_7 = -15$$

12. The following are the measurements T made on a curve recorded by oscilograph representing a change of current I due to a change in the conditions of an electric current.

T	1.2	2.0	2.5	3.0
I	1.36	0.58	0.34	0.20

Using Lagrange's formula, find I and $T = 1.6$

(JNTU–2009)

13. Using Lagrange's interpolation, calculate the profit in the year 2000 from the following data :

(Anna-2004)

Year	1997	1999	2001	2002
Profit in Lakhs of Rs	43	65	159	248

14. Use Lagrange's formula to find the form of $f(x)$, given :

x	0	2	3	6
$f(x)$	648	704	729	792

(Madras–2003)

15. Given $f(0) = -18, f(1) = 0, f(3) = 0, f(5) = -248, f(6) = 0, f(9) = 13104$, find $f(x)$.

(Nagarjuna–2003)

16. Using Newton's divided difference interpolation, find $u(3)$ given that $u(1) = -26, u(2) = 12, u(4) = 256, u(6) = 844$.

(Anna–2004)

17. Using Newton's divided difference interpolation, find the polynomial of the given data :

x	–1	0	1	3
$f(x)$	2	1	0	–1

(Anna–2007)

18. Apply Lagrange's formula inversely to obtain a root of the equation $f(x) = 0$, given that $f(30) = -30, f(34) = -13, f(38) = 3$ and $f'(42) = 18$.

(VTU–2009)

19. For the following data :

x	$f(x)$	$f'(x)$
0.5	4	–16
1	1	–2

Find the Hermite interpolating polynomial.

(UPTU–2008)

Answers

1. $(x-1)^3 + 2(x-1)^2 + 4(x-1) + 11$

2. $\frac{1}{10}[11(x-4)^3 + 117(x-4)^2 + 447(x-4) + 680]$

3. 810   **4.** 208.82222, 3^x is not a polynomial

5. 2.4786   **6.** 6.5928

7. 147   **8.** 3.55   **10.** 1.88   **11.** 9   **12.** 0.89   **13.** 100   **14.** $648 + 30x - x^2$

15. $x^5 - 9x^4 + 18x^3 - x^2 + 9x - 18$ **16.** 100   **17.** $f(x) = \frac{1}{24}(x^3 - 25x + 24)$

18. 37.23

5.17 CENTRAL DIFFERENCES INTERPOLATION FORMULAE

Let $y = f(x)$ be a function, which takes the values ... $y_{-2}, y_{-1}, y_0, y_1, y_2 \cdots$ corresponding to the values of $x = x_0 - 2h, x_0 - h, x_0, x_0 + h, x_0 + 2h \ldots$

(i) Newton's Forward Difference Formula.

We want to find the value of the function y_u for $x = x_0 + uh$, where, in general $-1 < u < 1$. Then

$$y_u = E^u y_0 = (1 + \Delta)^u y_0 \qquad (\because E = 1 + \Delta)$$

$$= ({}^u c_0 + {}^u c_1 \Delta + {}^u c_2 \Delta^2 + {}^u c_3 \Delta^3 + \ldots + {}^u c_4 \Delta^4) y_0$$

$$y_u = 1 + u \Delta y_0 + \frac{u(u-1)}{2!} \Delta^2 y_0 + \frac{u(u-1)(u-2)}{3!} \Delta^3 y_0 + \ldots$$

This is called Newton's forward difference formula.

(ii) Gauss-forward Difference formula.

This formula is obtained by Newton's forward-difference formula, we have

$$y_u = (E^{-1})^{-u} y_0$$

$$= (1 - \nabla)^{-u} y_0 \qquad (\because E^{-1} = 1 - \nabla)$$

$$= [1 + u\nabla + \frac{u(u+1)}{2!} \nabla^2 + \frac{u(u+1)(u+2)}{3!} \nabla^3 y_0 + \ldots] y_0$$

$$y_u = y_0 + u \nabla y_0 + \frac{u(u+1)}{2!} \nabla^2 y_0 + \frac{u(u+1)(u+2)}{3!} \nabla^3 y_0 + \ldots$$

This is called Newton's backward-difference formula.

(iii) Gauss-forward Difference Formula.

This Formula is obtained by Newton's forward-difference formula, we have

$$y_u = 1 + u \nabla y_0 + \frac{u(u-1)}{2!} \Delta^2 y_0 + \frac{u(u-1)(u-2)}{3!} \nabla^3 y_0 + \cdots \quad \ldots (1)$$

Since we have

$$\Delta^2 y_0 - \Delta^2 y_{-1} = \Delta^3 y_{-1}$$

or

$$\Delta^2 y_0 = \Delta^2 y_{-1} + \Delta^3 y_{-1} \qquad \ldots (2)$$

Similarly,

$$\Delta^3 y_0 = \Delta^3 y_{-1} + \Delta^4 y_{-1} \qquad \ldots (3)$$

$$\Delta^4 y_0 = \Delta^4 y_{-1} + \Delta^5 y_{-1} \quad \ldots \text{etc} \qquad \ldots (4)$$

Also

$$\Delta^3 y_{-1} = \Delta^3 y_{-2} + \Delta^4 y_{-2}$$

and

$$\Delta^4 y_{-1} = \Delta^4 y_{-2} + \Delta^5 y_{-2} \quad \ldots \text{etc.} \qquad \ldots (5)$$

Substituting the values of $\Delta^2 y_0, \Delta^3 y_0, \Delta^4 y_0$, from (2), (3), (4) ... in(1), we get

$$y_u = 1 + u \Delta y_0 + \frac{u(u-1)}{2!} (\Delta^2 y_{-1} + \Delta^3 y_{-1})$$

$$+ \frac{u(u-1)(u-2)}{3!} (\Delta^3 y_{-1} + \Delta^4 y_{-1}) + \frac{u(u-1)(u-2)(u-3)}{4!} (\Delta^4 y_{-1} + \Delta^5 y_{-1}) + \ldots$$

$$= 1 + u \Delta y_0 + \frac{u(u+1)}{2!} \Delta^2 y_{-1} + \frac{(u+1)u(u-1)}{3!} \Delta^3 y_{-1} + \frac{(u+1)u(u-1)(u-2)}{4!} \Delta^4 y_{-2}$$

This is called Gauss's-Forward difference formula.

(iv) Gauss's backward Difference Formula.

The Newton's-forward difference formula is given by

$$y_u = y_0 + u\Delta y_0 + \frac{u(u-1)}{2!}\Delta^2 y_0 + \frac{u(u-1)(u-2)}{3!}\Delta^3 y_0 + \dots \qquad \dots (1)$$

Since, we have

$$\Delta y_0 = \Delta y_{-1} + \Delta^2 y_{-1} \qquad \dots (2)$$

$$\Delta^2 y_0 = \Delta^2 y_{-1} + \Delta^3 y_{-1} \qquad \dots (3)$$

$$\Delta^3 y_0 = \Delta^3 y_{-1} + \Delta^4 y_{-1} \dots \text{ etc.} \qquad \dots (4)$$

Also $\quad \Delta^3 y_{-1} = \Delta^3 y_{-2} + \Delta^4 y_{-2} \qquad \dots (5)$

$$\Delta^4 y_{-1} = \Delta^4 y_{-2} + \Delta^5 y_{-2} \dots \text{ etc.} \qquad \dots (6)$$

With the help of (1) to (6), we get

$$y_u = y_0 + u(\Delta y_{-1} + \Delta^2 y_{-1}) + \frac{u(u-1)}{2!}(\Delta^2 y_{-1} + \Delta^3 y_{-1})$$

$$+ \frac{u(u-1)(u-2)}{3!}(\Delta^3 y_{-1} + \Delta^4 y_{-1}) + \frac{u(u-1)(u-2)(u-3)}{4!}$$

$$(\Delta^4 y_{-1} + \Delta^5 y_{-1}) + \dots$$

or $\quad y_u = y_0 + u\Delta y_{-1} + \frac{(u+1)u}{2!}\Delta^2 y_{-1} + \frac{(u+1)u(u-1)}{3!}\Delta^3 y_{-1}$

$$+ \frac{(u+1)u(u-1)(u-2)}{4!}\Delta^4 y_{-1} + \frac{u(u-1)(u-2)(u-3)}{4!}\Delta^4 y_{-1}$$

$$= y_0 + u\Delta y_{-1} + \frac{(u+1)u}{2!}y_{-1} + \frac{u(u+1)(u-1)}{3!}(\Delta^3 y_{-2} + \Delta^4 y_{-2})$$

$$+ \frac{(u+1)u(u-1)(u-2)}{4!}(\Delta^4 y_{-2} + \Delta^5 y_{-2}) + \dots$$

$$y_u = y_0 + u\Delta y_{-1} + \frac{(u+1)u}{2!}\Delta^2 y_{-1} + \frac{(u+1)u(u-1)}{3!}\Delta^3 y_{-2} + \frac{(u+2)(u+1)u(u-1)}{4!}\Delta^4 y_{-2} \dots$$

This is called Gauss's-backward difference formula.

5.18 STIRLING'S DIFFERENCE FORMULA

The Stirling's formula can be obtained by taking the average of Gauss's forward difference formula and Gauss's backward difference formula.

The Gauss's-forward difference formula is given by

$$y_u = y_0 + u\Delta y_0 + \frac{u(u-1)}{2!}\Delta^2 y_{-1} + \frac{(u+1)u(u-1)}{3!}\Delta^3 y_{-1}$$

$$+ \frac{(u+1)(u)(u-1)(u-2)}{4!}\Delta^4 y_{-2} + \dots \qquad \dots (1)$$

Also, Gauss's backward formula is given by

$$y_u = y_0 + u\Delta y_{-1} + \frac{(u+1)u}{2!}\Delta^2 y_{-1} + \frac{(u+1)u(u-1)}{3!}\Delta^3 y_{-2}$$

$$+ \frac{(u+2)(u+1)u(u-1)}{4!}\Delta^4 y_{-2} \dots \qquad \dots (2)$$

Taking average of (1) and (2), we get

$$y_u = y_0 + u\left(\frac{\Delta y_0 + \Delta y_{-1}}{2}\right) + \frac{u^2}{2!}\Delta^2 y_{-1} + \frac{u(u^2-1)}{3!}\left(\frac{\Delta^3 y_{-1} + \Delta^3 y_{-2}}{2}\right) + \frac{u^2(u^2-1)}{4!}\Delta^4 y_{-2} + \dots$$

This is called Stirling's difference formula.

5.18.1 REMAINDER TERM IN STIRLING'S FORMULA

Let the function $f(x)$ be approximated by a polynomial $P_{2n}(x)$.

Assume

$$f(x) = P_{2n}(x) + k(x)(x-x_0)(x-x_1)(x-x_{-1})\dots(x-x_n)(x-x_{-n}) \qquad \dots(1)$$

where $k(x)$ is obtained.

Consider the function,

$$g(t) = f(t) - P_{2n}(t) - k(x)(t-x_0)(t-x_1)(t-x_{-1})\dots(t-x_n)(t-x_{-n}) \qquad \dots(2)$$

Here, $g(t)$ vanishes for $(2n+2)$ values which are given by

$$t = x, x_0, x_1, x_{-1}, \dots, x_n, x_{-n}$$

Then, by Rolle's theorem, we have $\phi^{2n+1}(\theta)$ has at least one root $t = \theta$ in (x_{-n}, x_n)

But $\qquad \phi^{2n+1}(t) = f^{2n+1}(t) - k(x)\{(2n+1!)\}$

$\Rightarrow \qquad \phi^{2n+1}(\theta) = f^{2n+1}(\theta) - k(x)\{(2n+1)\}$

Now $\qquad \phi^{2n+1}(\theta) = 0 \quad \Rightarrow \quad k(x) = \dfrac{f^{2n+1}(\theta)}{(2n+1)!} \qquad\qquad x_{-n} < \theta < x_n$

Now putting the value of $k(x)$ in (1), we get

$$f(x) = P_{2x}(x) + f^{2n+1}(\theta)\left[\frac{(x-x_0)(x-x_1)(x-x_{-1})\dots(x-x_n)(x-x_{-n})}{(2n+1)!}\right]$$

New error $R_n = f(x) - P_{2n}(x) = f^{2n+1}(\theta)\dfrac{(x-x_0)(x-x_1)(x-x_{-1})\dots(x-x_n)(x-x_{-n})}{(2n+1)!}$

Now using, $u = \dfrac{x - x_0}{h}$ we get

$$R_n = \frac{h^{2n+1}f^{2n+1}(\theta)}{(2n+1)!}u(u^2-1^2)(u^2-2^2)\dots(u^2-n^2)$$

Now, the values of R_n, in terms of difference is given by

$$R_n = \frac{\Delta^{2n+1}f(x_{-n-1}) + \Delta^{2n+1}f(x_{-n})}{\{2.(2n+1)!\}}u(u^2-1^2)(u^2-2^2)\dots(u^2-n^2).$$

5.19 CHOICE TO SELECT THE SUITABLE INTERPOLATION FORMULA

- (i) **Newton's forward difference formula.** To find a tabulated value near. the beginning of the data.
- (ii) **Newton's backward difference formula.** To find a tabulated value near the end of the data.
- (iii) **Stirling formula.** If interpolation is required for u, lying between $-\dfrac{1}{4}$ and $\dfrac{1}{4}$.
- (iv) **Bessel's or Everett's Formula.** If interpolation is required for u lying between $-\dfrac{1}{4}$ and $\dfrac{3}{4}$.

(v) **Lagrange's and Newton's divided difference formula.** For unequal intervals.

(vi) **Hermite interpolation formula.** To interpolate the value of $f(x)$ and its derivative.

5.20 BESSEL'S DIFFERENCE FORMULA

This formula is obtained with the help of Gauss's forward difference formula as follows, Gauss's forward difference formula is given by

$$y_u = y_0 + u\Delta y_0 + \frac{u(u-1)}{2!}\Delta^2 y_{-1} + \frac{(u+1)u(u-1)}{3!}\Delta^3 y_{-1} + \frac{(u+1)u(u-1)(u-2)}{4!}\Delta^4 y_{-2} + \dots$$

using $\Delta^2 y_{-1} = \Delta^2 y_{-0} - \Delta^3 y_{-1}, \Delta^4 y_{-2} = \Delta^4 y_{-1} - \Delta^5 y_{-2}$ and so on we get,

$$y_u = y_0 + u\Delta y_0 + \frac{u(u-1)}{2!}\cdot\left(\frac{\Delta^2 y_{-1} + \Delta^2 y_0}{2}\right) + \frac{\left(u-\frac{1}{2}\right)u(u-1)}{3!}\Delta^3 y_{-1}$$

$$+ \frac{(u+1)u(u-1)(u-2)}{4!}\left(\frac{\Delta^4 y_{-2} + \Delta^4 y_{-1}}{2}\right) + \dots$$

This is called Bessel's difference formula.

▶ REMARKS

▶ Bessel's formula is used to find the entry against any arguments between 0 and 2. This formula gives better result if $\frac{1}{4} \le p \le \frac{3}{4}$.

▶ Bessel's formula is most convenient for bisection of the interval.

5.20.1 REMAINDER OR ERROR TERM IN BESSEL'S FORMULA

Since there are $(2n+2)$ terms in Bessel's formula. Therefore assume

$$f(x) = P_{2n+1}(x) + k(x)(x - x_0)(x - x_1)(x - x_{-1})\dots(x - x_n)(x - x_{-n})(x - x_{n+1}) \qquad \dots(1)$$

when $k(x)$ is to be determined.

Now, consider the function

$$g(t) = f(t) - P_{2n+1}(t) - k(x)(t - x_0)(t - x_1)(t - x_{-1})\dots(t - x_n)(t - x_{-n})(t - x_{n+1}) \quad \dots (2)$$

Here, $g(t)$ vanishes at $(2n+3)$ points namely

$$t = x, x_0, x_1, x_{-1}, \dots, x_{-n}, x_{n+1}$$

Then, by Rolle's theorem $\phi^{2n+1}(\theta) = 0$ for $\theta \in]x_{-n}, x_{n+1}[$

Now $\quad \phi^{2n+2}(t) = f^{2n+2}(t) - k(x)[(2n+2)!]$

$$\phi^{2n+2}(\theta) = 0 \Rightarrow k(x) = \frac{f^{2n+2}(\theta)}{(2n+2)!} \qquad\qquad x_{-n} < \theta < x_{n-1}$$

Putting this value of $k(x)$ in (1), we get

$$f(x) = P_{2n+1}(x) + \frac{f^{2n+2}(\theta)}{(2n+2)!}(x - x_0)(x - x_1)(x - x_{-1})\dots (x - x_n)(x - x_{-n})(x - x_{n-1})$$

Now error $\qquad\qquad R_n = f(x) - P_{2n+1}(x)$

$$\Rightarrow \quad R_n = \frac{f^{2n+2}(\theta)}{(2n+2)!}(x-x_0)(x-x_1)(x-x_{-1})...(x-x_n)(x-x_{-n})(x-x_{n-1})$$

The value of R_n in terms of $u = \left(\frac{x-x_0}{h}\right)$ is given by

$$R_n = \frac{h^{2n+2}}{(2n+2)!}f^{2n+2}(\theta).u(u-1)(u+1)...(u-1)(u+n)(u-n-1).$$

5.21 EVERETT'S DIFFERENCE FORMULA

After eliminating the odd differences from Gauss's difference formula we obtain Everett's difference formula. The Gauss's formula is given by

$$y_u = y_0 + u\Delta y_0 + \frac{u(u-1)}{2!}\Delta^2 y_{-1} + \frac{(u+1)u(u-1)}{4!}\Delta^3 y_{-1} + \frac{(u+1)u(u-1)(u-2)}{4!}\Delta^4 y_{-2} + ...$$

$$... (1)$$

Using the differences $\Delta y_0 = y_1 - y_0, \Delta^3 y_{-1} = \Delta^2 y_0 - \Delta^2 y_{-1}, \Delta^5 y_{-2} = \Delta^4 y_{-1} - \Delta^4 y_{-2}...$ etc.

Then, equation (1) becomes

$$y_u = 1 - uy_0 + uy_1 - \frac{u(u-1)(u-2)}{3!}\Delta^2 y_{-1} + \frac{(u+1)u(u-1)}{3!}\Delta^2 y_0$$

$$- \frac{(u+1)u(u-1)(u-2)(u-3)}{5!}\Delta^4 y_{-2} + ...$$

This is known as Everett's difference formula.

▶ **REMARK**

▸ All the differences formulae can be explained with help of Fraser or Lozenge diagrams, which is as follows:

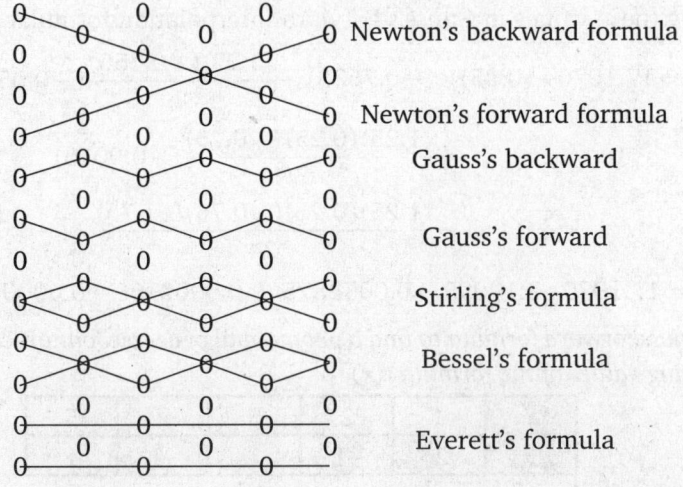

SOLVED EXAMPLES

EXAMPLE 1.   *Use Gauss's forward formula to find* y_{30} *for the following data*
$y_{21} = 18.4708, y_{25} = 17.8144, \ y_{29} = 17.1070, y_{33} = 16.3432, \ y_{37} = 15.5154$.

[UPTU(MCA)–2002, 06; Meerut-2004, 06; Avadh-2003; ISA-2006]

SOLUTION.   From the above data, we have the following table.

x	21	25	29	33	37
y	18.4708	17.8144	17.1070	16.3432	15.5154

Let us take the origin at $x = 29$ and $h = 4$.

We have to find the value of y for $u = \dfrac{30 - 29}{4} = 0.25$

Difference Table

x	u	y_u	Δy_u	$\Delta^2 y_u$	$\Delta^3 y_u$	$\Delta^4 y_u$
21	–2	18.4708				
			– 0.6564			
25	–1	17.8144		–0.0510		
			– 0.7074		–0.0054	
29	0	**17.1070**	**–0.7638**	**–0.0564**	**0.0076**	**–0.0022**
33	1	16.3432		–0.0640		
			– 0.8278			
37	2	15.5154				

Putting these values in Gauss's Forward interpolation formulae, we get

$$y_{0.25} = 17.1070 + (0.25) \times (-0.7638) + \frac{(0.25)(-0.750)}{2}(-0.0564)$$

$$+ \frac{(1.25)(0.25)(-0.75)}{6}(-0.0076)$$

$$+ \frac{(1.25)(0.25)(-0.75)(-1.75)}{24} \times (-0.0022)$$

$$= 17.1070 - 0.19095 + 0.0052875 + 0.0002968 - 0.00000375 \approx 16.9216$$

EXAMPLE 2.   *Use Gauss forward formula to find a polynomial of degree four or less which takes the following values of the formula* $f(x)$.

x	1	2	3	4	5
f(x)	1	–1	1	–1	1

(UPTU-2003)

SOLUTION. Let $u = x - 3$. Then we construct the difference table as follows:

x	u	$f(x)$	$\Delta f(x)$	$\Delta^2 f(x)$	$\Delta^3 f(x)$	$\Delta^4 f(x)$
1	-2	1				
			-2			
2	-1	-1		4		
			2		-8	
3	0	1		-4		16
			-2		8	
4	1	-1		4		
			2			
5	2	1				

Now Gauss's forward formula is given by

$$f(x) = f_0 + u\Delta f_0 + \frac{u(u-1)}{2!}\Delta^2 f_{-1} + \frac{(u+1)u(u-1)}{6}\Delta^3 f_{-1}$$
$$+ \frac{u(u-1)(u+1)(u+2)}{24}\Delta^4 f_{-2}$$

$$\Rightarrow \quad f(x) = 1 + u(-2) + \frac{u(u-1)}{2}(-4) + \frac{(u+1)u(u-1)}{6}(8)$$
$$+ \frac{u(u-1)(u+1)(u+2)}{24}(16)$$

$$= 1 - 2u + 2u(u-1) + \frac{4u(u^2-1)}{3} + \frac{2(u^2-1)(u^2-2u)}{3}$$

$$= 1 - 2u - 2u^2 + 2u + \frac{4u^3}{3} - \frac{4u}{3} + \frac{2u^4}{3} - \frac{4u^3}{3} - \frac{2u^2}{3} + \frac{4u}{3}$$

$$= \frac{2}{3}u^4 + \left(\frac{4}{3} - \frac{4}{3}\right)u^3 + \left[-2 + \frac{2}{3}\right]u^2 + \left[-2 + 2 - \frac{4}{5} + \frac{4}{3}\right]u + 1$$

$$= \frac{2}{3}u^4 + 0.u^3 + \left(-\frac{8}{3}u^2\right) + 0.u + 1$$

$$= \frac{2}{3}u^4 - \frac{8}{3}u^2 + 1 = \frac{1}{3}[2u^4 - 8u^2 + 3]$$

Hence, the required function is

$$f(x) = \frac{1}{3}(2(x-3)^4 - 8(x-3)^2 + 3) = \frac{2}{3}x^4 - 8x^3 + \frac{100}{3}x^2 - 56x + 31$$

EXAMPLE 3. *The value of e^{-x} at $x = 1.72$ to $x = 1.76$ are given in the following table:*

x	1.72	1.73	1.74	1.75	1.76
e^{-x}	0.17907	0.17728	0.17552	0.17377	0.17204

Find the value of $e^{-1.7425}$ using Gauss's forward interpolation formula.

SOLUTION. Let us take the origin at 1.74. Also $h = 0.01$

Therefore, $x = a + uh$

$$\Rightarrow \quad u = \frac{x - a}{h} = \frac{1.7425 - 1.7400}{0.01} = 0.25$$

Now, we have the following difference table:

u	x	$10^5 f(x)$	$10^5 \Delta f(x)$	$10^5 \Delta^2 f(x)$	$10^5 \Delta^3 f(x)$	$10^5 \Delta^4 f(x)$
-2	1.72	17907				
			-179			
-1	1.73	17728		3		
			-176		-2	
0	1.74	**17552**		1		3
			-175		1	
1	1.75	17377		2		
			-173			
2	1.76	17204				

Gauss formula is given by

$$f(u) = f(0) + u\Delta f(0) + \frac{u(u-1)}{2!}\Delta^2 f(-1) - \frac{(u+1)u(u-1)}{3!}\Delta^3 f(-1)$$
$$+ \frac{(u+1)u(u-1)(u-2)}{4!}\Delta^4 f(-2)$$

$\Rightarrow \quad 10^5 f(0.25) = 17552 + (0.25)(-175)$

$$+ \frac{(0.25)(-0.75)}{2}(1) + \frac{(1.25)(0.25)(-0.75)}{6}(1)$$
$$+ \frac{(1.25)(0.25)(-0.75)(-1.75)}{24}(3)$$

$$= 17508.16846$$

Hence, $f(0.25) = e^{-1.7425} = 0.1750816846$.

EXAMPLE 4. *Apply a central difference formula to obtain f(32), if*

$f(25) = 0.2707, f(35) = 0.3386 ; f(30) = 0.3027, f(40) = 0.3794$

SOLUTION. We have $a + hu = 32$ and $h = 5$.
Let us take origin at 30

$\therefore \qquad a = 30, u = 0.4$

Then we have the following difference table :

u	x	$f(x)$	$\Delta f(x)$	$\Delta^2 f(x)$	$\Delta^3 f(x)$
-1	25	0.2707			
			0.032		
0	30	**0.3027**		**0.0039**	
			0.0359		**0.0010**
1	35	0.3386		0.0049	
			0.0408		
2	40	0.3794			

The Gauss's forward difference formula is given by

$$f(u) = f(0) + u\Delta f(0) + \frac{u(u-1)}{2!}\Delta^3 f(-1) + \frac{(u+1)u(u-1)}{3!}\Delta^3 f(-1)$$

$$\Rightarrow \quad f(0.4) = 0.3027 + (0.4)(0.0359) + \frac{(0.4)(0.4-1)}{2!}(0.0039)$$

$$+ \frac{(1.4)(0.4)(0.4-1)}{3!}(0.0010)$$

$$\therefore \quad f(0.4) = 0.316536$$

EXAMPLE 5. *Using the following table, find by Gauss's backward formula the sales of a concern for the year 1936,*

Year	1901	1911	1921	1931	1941	1951
Sales (in thousand)	12	15	20	27	39	52

[UPTU–2004(Sum)–2005; Meerut-2003, 06, 08)]

SOLUTION. Let us take the origin = 1931 and h = 10 years then $u = \dfrac{1936-1931}{10} = 0.5$

Then we have the following difference table :

x	u	y_u	Δy_u	$\Delta^2 y_u$	$\Delta^3 y_u$	$\Delta^4 y_u$	$\Delta^5 y_u$
1901	– 3	12					
			3				
1911	– 2	15		2			
			5		0		
1921	– 1	20		2		3	
			7		3		10
1931	0	**27**		5		–7	
			12		– 4		
1941	1	39		1			
			1				
1951	2	52					

Now using Gauss's backward interpolation formula

$$y_u = y_0 + {}^4c_1\Delta y_{-1} + {}^{4+1}c_2\Delta^2 y_{-1} + {}^{4+1}c_3\Delta^3 y_{-2} + \dots$$

We get,

$$y_u = 27 + 0.5 \times 7 + \frac{(1.5)\times(.5)}{2}\times 5 + \frac{(1.5)(0.5)(-0.5)}{6}\times 3 + \frac{(2.5)(1.5)(0.5)(-0.5)}{24}$$

$$\times(-7) + \frac{(2.5)(1.5)(0.5)(-0.5)}{120}\times(-10)$$

$$= 27 + 3.5 + 1.875 - 0.1875 + 0.2734 - 0.11718$$

$$= 32.3437 \text{ thousands.}$$

EXAMPLE 6. *Given that* $\sqrt{12500} = 111.803399, \sqrt{12510} = 111.848111$

$\sqrt{12520} = 111.892806, \sqrt{12530} = 111.937483$

Show by Gauss's backward formula that $\sqrt{12516} = 87.49301$ (Rajasthan-2008)

SOLUTION.   Let us take the origin at 12520

$$\therefore \quad u = \frac{x-a}{h} = \frac{12516-12520}{10} = -\frac{4}{10} = -0.4$$

The difference table is given as below:

u	x	$10^6 f(x)$	$10^6 \Delta f(x)$	$10^6 \Delta^2 f(x)$	$10^6 \Delta^3 f(x)$
-2	12500	11803399			
			44712		
-1	12510	111848111		-17	
			44695		-1
0	12520	**11892806**		-18	
			44677		
1	12530	111937483			

Putting the values in Gauss's backward formula

$$f(u) = u\Delta f(-1) + \frac{u(u+1)}{2!}\Delta^2 f(-1) + \frac{u(u+1)(u-1)}{3!}\Delta^3 f(-2) + \dots$$

we get $10^6 f(-4) = 111892806 + (-0.4)(44695)$

$$+ \frac{(0.6)(-0.4)}{2!}(-18) + \frac{(0.6)(-0.4)(-1.4)}{3!}(-1) = 1118749301$$

$$\Rightarrow \qquad f(-4) = 111.8749301$$

Hence $\quad \sqrt{12516} = 111.8749301$.

EXAMPLE 7.   *$f(x)$ is a polynomial of degree 4 and given that $f(4) = 270, f(5) = 648, \Delta f(5) = 682$, $\Delta^3 f(-4) = 132$ Find the value of (5.8) using Gauss's backward formula.*

SOLUTION.   We know that

$$\Delta f(5) = f(6) - f(5)$$

$$\therefore \quad f(6) = f(5) + \Delta f(5) = 648 + 682 = 1330$$

$$\Delta^3 f(4) = (E-1)^3 f(4)$$

$$= f(7) - 3f(6) + 3f(5) - f(4) = 132$$

$$f(7) = 3f(6) - 3f(5) + f(4) + 132 = 3 \times 1330 - 3 \times 648 + 270 + 132 = 2448$$

Taking origin at 6, the difference table is given below

u	x	$f(x)$	$\Delta f(x)$	$\Delta^2 f(x)$	$\Delta^3 f(x)$
-2	4	270			
			378		
-1	5	678		304	
			682		**132**
0	6	**1330**		**436**	
			1118		
1	7	2448			

We have $\quad a = 6, h = 1, a + hu = 5.8$

$$6 + u = 5.8 \quad \Rightarrow \quad u = -0.2$$

By Gauss's backward formula, we have

$$f(-0.2) = f(0) + u\Delta f(-1) + \frac{u(u+1)}{2!}\Delta^2 f(-1) + \frac{u(u+1)(u-1)}{3!}\Delta^2 f(-2)$$

$$= 1330 + (-0.2)(682) + \frac{(0.8)(-0.2)}{2}(436) + \frac{(0.8)(-0.2)(-1.2)}{6}132$$

$$= 1162.944$$

Hence, $f(5.8) = 1162.94$.

EXAMPLE 8. Use Stirling's formula to find y_{35}, given $y_{20} = 512, y_{30} = 439, y_{40} = 346$, and $y_{50} = 243$.

(Agra-2003; UPTU-2004)

SOLUTION. Let us assume origin $= 30$, $h = 10$

Then $a + hu = 35 \Rightarrow 30 + 10u = 35 \Rightarrow u = 0.5$

The difference table is given as under:

u	x	y	Δy	$\Delta^2 y$	$\Delta^3 y$
-1	20	512			
			-73		
0	30	**439**		-20	
			-93		10
1	40	346		-10	
			-1.3		
2	50	243			

By Stirling's formula, we have

$$f(0.5) = 439 + (0.5)\left(\frac{-93-73}{2}\right) + \frac{(0.5)^2}{2!}(-20) + \frac{(1.5)(0.5)(-0.5)}{3!}\left(\frac{10}{2}\right)$$

$$= 394.6875$$

Hence, $y_{35} = 394.6875$

EXAMPLE 9. Use Stirling's formula to find y_{28}, given $y_{20} = 49225, y_{25} = 48316, y_{30} = 47236$ $y_{35} = 45926, y_{40} = 4306$.

(UPTU (Co)-2004, 2007; Rohilkhand-2004, 08; Agra-2006, Bhopal-2002, Avadh-2005, 08)

SOLUTION. Let $x = 30$, $h = 5$ then, $u = \dfrac{28-30}{5} = -0.4$

x	u	y_u	Δy_u	$\Delta^2 y_u$	$\Delta^3 y_u$	$\Delta^4 y_u$
20	-2	49225				
			-909			
25	-1	48316		-171		
			-1080		-59	
30	0	**47236**		-230		-21
			-1310		-80	
35	1	45926		-310		
			-1620			
40	2	44306				

The Stirling's formula is given by

$$y_u = y_0 + u\left[\frac{\Delta y_0 + \Delta y_{-1}}{2}\right] + \frac{u^2}{2}[\Delta^2 y_{-1}] + \frac{u(u^2-1)}{6}\left[\frac{\Delta^3 y_{-1} + \Delta^3 y_{-2}}{2}\right]$$

$$+ u^2 \frac{(u^2-1)}{24}\Delta^4 y_{-2} \qquad\qquad ...(1)$$

On putting the values in (1), we get

$$y_{28} = 47236 \times (-0.4)\left[\frac{-1310 - 1080}{2}\right] + \frac{(0.16)}{2}(-230) + \frac{(-0.4)(0.16-1)}{6}$$

$$\frac{(-80-59)}{2} + \frac{(0.16)(0.16-1)}{24} \times (-21)$$

$$= 47236 + 478 - 18.4 - 3.8920 + 0.1176$$

Hence, $y_{28} = 47692$.

EXAMPLE 10. *Given* $y_{20} = 24, y_{24} = 32, y_{28} = 35, y_{32} = 40$. *Find* y_{25} *by Bessel's formula.*

(JNTU–2006, UPTU-2002, 05, Delhi-2004, 2008; Meerut-2003, Agra-2006)

SOLUTION.   Let $x = 24, h = 4$

$$\Rightarrow \qquad 4 = \frac{25-24}{4} = \frac{1}{4} = 0.25$$

The difference table is given as under

x	u	y_u	Δy_u	$\Delta^2 y_u$	$\Delta^3 y_u$
20	– 1	24			
			8		
24	0	**32**		–5	
			3		7
28	1	**34**		– 2	
			5		
32	2	40			

Now Bessel's formula is given by

$$y_0 = \frac{1}{2}(y_0 + y_1) + (u - \frac{1}{2})\Delta y_0 + \frac{u(u-1)}{2}\cdot\frac{\Delta^2 y_{-1} + \Delta^2 y_0}{2!} + \frac{\left(u - \frac{1}{2}\right)u(u-1)}{3!}\Delta^3 y_{-1}$$

Putting the values in Bessel's formula, we get

$$y_{0.25} = \frac{32+34}{2} + \left(\frac{1}{4} - \frac{1}{2}\right)3 + \frac{\frac{1}{4}\left(\frac{1}{4}-1\right)}{2}\left(\frac{-5+2}{2}\right) + \frac{\left(\frac{1}{4} - \frac{1}{2}\right)\left(\frac{1}{4} - 1\right)}{6}7$$

$$= 33.5 - 0.75 + 0.1416 + 0.054687 = 32.945287$$

EXAMPLE 11. *Given that*

x	4	6	8	10	12	14
$f(x)$	3.5460	5.0753	6.4632	7.7217	8.8633	9.8986

Apply Bessel's formula to find the value of $f(9)$.

SOLUTION.   Let us take the origin at 8 and $h = 2$.

$$\therefore \qquad 9 = 8 + 2u$$

$$\Rightarrow \qquad u = \frac{1}{2}$$

Then, we have the following difference table:

u	$10^4 y_u$	$10^4 \Delta y_u$	$10^4 \Delta^2 y_u$	$10^4 \Delta^3 y_u$	$10^4 \Delta^4 y_u$	$10^4 \Delta^5 y_u$
-2	35460					
		15293				
-1	50753		-1414			
		13879		120		
0	**64632**		1294		5	
		1258		**125**		-24
1	**77217**		-1169		-19	
		11416		106		
2	88633		-1063			
		10353				
3	98986					

Putting the values from above table in Bessel's formula

$$y_u = \frac{1}{2}(y_1 + y_0) + \left(u - \frac{1}{2}\right)\Delta y_0 + \frac{u(u-1)}{2!} \cdot \frac{1}{2}(\Delta^2 y_0 + \Delta^2 y_{-1})$$

$$+ \frac{u(u-1)\left(u - \frac{1}{2}\right)}{3!}\Delta^3 y_{-1} - \frac{u(u+1)(u-1)(u-2)}{4!} \times \frac{1}{2}(\Delta^4 y_{-3} + \Delta^3 y_{-2})$$

$$+ \frac{u(u-1)(u-2)(u+1)\left(u - \frac{1}{2}\right)}{5!}\Delta^5 y_{-2}$$

we get

$$10^4 y_{1/2} = \frac{1}{2}(77217 + 64632) + 0 + \frac{\frac{1}{2}\left(-\frac{1}{2}\right)}{2} \cdot \frac{1}{2}(-1169 - 1294) + 0 +$$

$$\frac{\frac{3}{2} \cdot \frac{1}{2}\left(-\frac{1}{2}\right)\left(-\frac{3}{2}\right)}{24} \cdot \frac{1}{2}(-19 + 5) + 0$$

$$= 71078.27344$$

$$\Rightarrow \qquad y_{1/2} = 7.107827344$$

Hence $y(9) = 7.107827344$

EXAMPLE 12.   *If third differences are constant, show that*

$$y_{x+\frac{1}{2}} = \frac{1}{2}(y_x + y_{x+1}) - \frac{1}{16}(\Delta^2 y_{x-1} + \Delta^2 y_x) \qquad \text{(Avadh-2004)}$$

SOLUTION. Putting $u = \dfrac{1}{2}$ in Bessel's formula, we get

$$y_{1/2} = \frac{1}{2}(y_0 + y_1) - \frac{1}{16}(\Delta^2 y_0 + \Delta^2 y_{-1})$$

Shifting the origin to x we get

$$y_{x+1/2} = \frac{1}{2}(y_x + y_{x+1}) - \frac{1}{16}(\Delta^2 y_x + \Delta^2 y_{x-1})$$

EXAMPLE 13. *Find the value of $f(27.4)$ from the following table:* (UPTU-2009)

x	25	26	27	28	29	30
$f(x)$	4.000	3.846	3.704	3.571	3.448	3.333

SOLUTION. Let

$$u = \frac{27.4 - 27.0}{1} = 0.4 \qquad \text{(Because origin is at 27.0 and } h = 1)$$

Also $w = 1 - u = 0.6$

Difference table is given as under:

u	$10^3 f(u)$	$10^3 \Delta f(u)$	$10^3 \Delta^2 f(u)$	$10^3 \Delta^3 f(u)$	$10^3 \Delta^4 f(u)$
-2	4000				
		-154			
-1	3846		12		
		-142		-3	
0	**3704**		$\rightarrow$ **9**		$\rightarrow$ **4**
		-133		1	
1	**3571**		$\rightarrow$ **10**		$\rightarrow$ **-3**
		-123		2	
2	3448		8		
		-115			
3	3333				

Putting these values in Everett's formula, we get

$$f(0.4) = (0.4)(3571) + \frac{(1.4)(0.4)(-0.6)}{3!}(10) + \frac{(2.4)(1.4)(0.4)(-0.6)(-1.6)}{5!}(-3)$$

$$+ (0.6)(3704) + \frac{(1.6)(0.6)(-0.4)}{3!}(9)$$

$$+ \frac{(2.6)(1.6)(-0.6)(-0.4)(-1.4)}{5!}(4)$$

$$= 3649.678336$$

Hence, $f(27.4) = 3649.678336$.

EXAMPLE 14. *Obtain y_{25} by using Everett's formula, using the following data:*

$$y_{20} = 2854, y_{24} = 3162, y_{28} = 3544, \ y_{32} = 3992$$

(UPTU–2003, SVTU–2007, VTU–2000, 01; Kanpur–2004; Bangluru–2008; Meerut–2000, 01)

SOLUTION. Let us take $x = 24$, $h = 4$ then

$$u = \frac{25 - 24}{4} = \frac{1}{4} = 0.25$$

The difference table is given as under:

x	u	y_u	Δy_u	$\Delta^2 y_u$	$\Delta^3 y_u$
20	-1	2854			
			308		
24	0	3162		74	
			382		-8
28	1	3544		66	
			448		
32	2	3992			

Everett's formula is given by

$$y_u = \frac{u(u^2 - 1)}{3!} \Delta^2 y_0 + \frac{u(u^2 - 1)(u^2 - 4)}{5!} \Delta^4 y_{-1} + \ldots + v y_0 + \frac{v(v^2 - 1)}{3!} \Delta^2 y_{-1} \ldots$$

$$(v = 1 - u)$$

$$\therefore y_{25} = \frac{1}{4} \times 3544 + \frac{\frac{1}{4}\left(\frac{1}{16} - 1\right)}{6} \times 66 + \frac{3}{4} \times 3162 \times \frac{\frac{3}{4}\left(\frac{9}{16} - 1\right)}{6} 74$$

$$= 886 - 2.5781 + 2371.5 - 4.0469 = 3250.875.$$

EXAMPLE 15. *Given y_0, y_1, y_2, y_4, y_5 assuming fifth difference constant and using Bessel's formula, show that*

$$y_{25} = \frac{1}{2} c + \frac{25 c.(-b) + 3(a - c)}{256} \quad \text{where } a = y_0 + y_5, b = y_1 + y_4, c = y_2 + y_3.$$

(Agra-2001, 04, Rohilkhand-2010, Patna-2006, 08; UPTU-2003)

SOLUTION. Bessel's formula is given by

$$y_u = \frac{1}{2}(y_0 + y_1) + \left(u - \frac{1}{2}\right)\Delta y_0 + \frac{u(u-1)}{2!}\left(\frac{\Delta^2 y_0 + \Delta^2 y_{-1}}{2}\right)$$

$$+ \frac{\left(u - \frac{1}{2}\right)u(u-1)}{3!} \Delta^3 y_{-1} + \frac{(u+1)u(u-1)(u-2)}{4!}\left[\frac{\Delta^4 y_{-1} + \Delta^4 y_{-2}}{2}\right] + \ldots$$

Putting $u = \frac{1}{2}$ and taking upto fifth differences, we have

$$y_{1/2} = \frac{1}{2}(y_0 + y_1) - \frac{1}{16}(\Delta^2 y_0 + \Delta^2 y_{-1}) + \frac{3}{256}(\Delta^4 y_{-1} + \Delta^4 y_{-2})$$

Now shifting the origin to 2, we have

$$y_{2\frac{1}{2}} = \frac{1}{2}(y_2 + y_3) - \frac{1}{16}(\Delta^2 y_2 + \Delta^2 y_1) + \frac{3}{256}(\Delta^4 y_1 + \Delta^4 y_0)$$

$$= \frac{1}{2}(y_2 + y_3) - \frac{1}{16}(y_4 - 2y_3 + y_2 + y_3 - 2y_2 + y_1) + \frac{3}{256}(y_5 - 4y_4 + 6y_3$$

$$- 4y_2 + y_1 + y_4 - 4y_3 + 6y_2 - 4y_1 + y_0)$$

$$= \frac{1}{2}(y_2 + y_3) - \frac{1}{16}(y_4 - 2y_3 - y_2 + y_1)$$

$$+ \frac{3}{256}(y_5 - 3y_4 + 2y_3 + 2y_2 - 3y_1 + y_0)$$

$$= \frac{1}{2}(y_2 + y_3) - \frac{1}{16}[(y_2 + y_4) - (y_2 + y_3)]$$

$$+ \frac{3}{256}[(y_0 + y_5) - 3(y_1 + y_4) + 2(y_2 + y_3)]$$

$$= \frac{1}{2}c + \frac{1}{16}(b - c) + \frac{3}{256}(9 - 3b + 2c)$$

$$= \frac{1}{2}c + \frac{1}{256}[16(c - b) + 3(a - 3b + 2c)]$$

$$= \frac{1}{2}c + \frac{1}{256}(16c - 16b + 3a - 9b + 6c)$$

$$= \frac{1}{2}c + \frac{1}{256}(22c - 25b + 3a)$$

$$= \frac{1}{2}c + \frac{1}{256}(25c - 25b - 3c + 3a)$$

$$= \frac{1}{2}c + \frac{1}{256}[25c(-b) + 3(a - c)]$$

EXAMPLE 16. *Given*

$\theta°$	0	5	10	15	20	25	30
$\tan\theta$	0	0.875	0.1763	0.2679	0.3640	0.4663	'0.5774

Using Stirling formula, estimate the value of $\tan 16°$.

(Anna–2005; MKU (Tamil Nadu)-2008)

SOLUTION.   Let $\theta = 15°$, $h = 5$ and $u = \dfrac{\theta - 15}{5}$

u	$\tan\theta = y$	Δy	$\Delta^2 y$	$\Delta^3 y$	$\Delta^4 y$	$\Delta^5 y$
−3	0.0000					
		0.0875				
−2	0.0875		0.0013			
		0.0888		0.0015		
−1	0.1763		0.0028		0.0002	
		0.0916		**0.0017**		**−0.0002**
0	**0.2679**		**0.0045**		**0.0000**	
		0.0961		**0.0017**		**0.0009**
1	0.3640		0.0062		0.0009	
		0.1023		0.0026		
2	0.4663		0.0088			
		0.1111				
3	0.5774					

The Stirling formula is

$$y_u = y_0 + u\left[\frac{\Delta y_0 + \Delta y_{-1}}{2}\right] + \frac{u^2}{2}[\Delta^2 y_{-1}] + \frac{u(u^2 - 1)}{6}\left[\frac{\Delta^3 y_{-1} + \Delta^3 y_{-2}}{2}\right]$$
$$+ \frac{u^2(u^2 - 1)}{24} \Delta^4 y_{-2} \;\ldots(1)$$

On putting the values in (1), we get

$$y_{0.2} = 0.2679 + (0.2)\left[\frac{0.0961 + 0.0916}{2}\right] + \frac{(0.2)^2}{2!}(0.0045) + \ldots$$
$$= 0.2679 + 0.01877 + 0.00009 + \ldots = 0.28676$$

Hence, $\tan 16° = 0.28676$.

EXERCISE 5.4

1. By means of Lagrange's formula, Prove that

$$y_0 = \frac{1}{2}(y_1 + y_{-1}) - \frac{1}{8}\left[\frac{1}{2}(y_3 - y_1) - \frac{1}{2}(y_1 - y_3)\right]$$

2. Apply Lagrange's formula to find $f(1.5)$ using the following table:

x	1.0	1.2	1.4	1.6	1.8	2.0
$f(x)$	0.2420	0.1942	0.1497	0.1109	0.079	0.0540

3. Using Lagrange's formula find the cubic polynomial from given data:

x	0	1	4	6
$f(x)$	1	1	1	−1

4. Obtain the value of t when, $A = 85$ from the following table, using Lagrange's method:

t	2	5	8	14
A	94.8	87.9	81.3	68.7

5. Using Newton's forward formula find the value of $f(1.6)$, if:

x	1	1.4	1.8	2.2
$f(x)$	3.49	4.82	5.96	6.5

6. Determine the values of $f(22)$ and $f(42)$ from the following table: (JNTU-2007)

x	20	25	30	35	40	45
$f(x)$	354	332	291	260	231	204

7. Apply Newton's backward difference formula to the data given below to obtain a polynomial of degree 4 in x.

x	1	2	3	4	5
$f(x)$	1	−1	1	−1	1

8. Evaluate log 5875 from the following table:

x	40	45	50
$f(x)$	1.60206	1.65321	1.69897

x	55	60	65
$f(x)$	1.74036	1.77815	1.81291

9. Using Hermite formula for points oscillation, derive the following formula:

$$y\left(\frac{1}{2}\right) = \frac{1}{2}(y_0 + y_1) + \frac{1}{3}L(y_0' - y_1')$$

where $L = (x_1 - x_0)$.

10. Given that $f(0) = 9$, $f(1) = 68$, $f(5) = 123$. Find $f(2)$ with the help of Newton's divided difference formula.

11. Apply Newton's divided difference formula to find the value of $f(8)$ if $f(1) = 3, f(3) = 31, f(6) = 223, f(10) = 1011$, $f(11) = 1343$.

12. Find the form of $f(x)$ from the following table:

x	0	1	2	5
$f(x)$	2	3	12	147

13. If $f(x) = \frac{1}{x^2}$ find the divided difference $f(a, b)$ and $f(a, b, c)$.

14. (i) Use Gauss's forward interpolation formula to find y_{41} with the help of following data:

$$y_{30} = 3678.2, y_{35} = 2995.1, y_{40} = 2400.1$$
$$y_{45} = 1876.2, y_{50} = 1416.3$$

(ii) Use Gauss's forward formula to find the annuity value for 27 years from the following data:

Years	15	20	25
Annuity	10.3797	12.4622	14.0939

Years	30	35	40
Annuity	15.3725	16.3742	17.1591

15. From the following table find the value of $f(0.5437)$ by Gauss, Stirling, Bessel and Everett's Formula.

x	0.51	0.52	0.53
$f(x)$	0.529244	0.537895	0.546464

x	0.54	0.55	0.56	0.57
$f(x)$	0.554939	0.663323	0.571616	0.579816

16. Eliminate odd differences from the Gauss's forward formula to derive Everett's formula

$$y_\alpha = (-\alpha)f_0 + \alpha f_1 - \frac{\alpha(\alpha-1)(\alpha-2)}{3!}\delta^2 f_0 + \frac{(\alpha+1)(\alpha)(\alpha-1)}{3!}\delta^2 f_1 + \dots$$

where $y = f(x_0 + \alpha h), \alpha = \dfrac{x - x_0}{h}$.

17. Apply (i) Stirling's formula, (ii) Bessel's formula to find the value of $f(0.44)$ from the following table which gives the values of $f(t) = \dfrac{1}{\sqrt{2\pi}} \int_0^t \exp\left(-\dfrac{t^2}{2}\right) dt$ at t = 0.5 from t = 0 to t = 3.

x	0	0.5	1.0	1.5
$f(x)$	0	0.19146	0.34134	0.43319

x	2.0	2.5	3.0
$f(x)$	0.47725	0.49379	0.49865

18. Find the missing value of the following data:

x	1	2	3	4	5
$f(x)$	7	–	13	21	37

19. Using Hermite interpolation formula, find polynomial which meets the following requirements:

x	0	1	2
$f(x)$	1	0	9
$f'(x)$	0	0	0

20. Find the polynomial, for following data, using Hermite formula:

x	–1	0	1
$f(x)$	–1	0	1
$f'(x)$	0	0	0

21. Find the cubic polynomial which takes following values $y(0) = 1, y(1) = 1, y(2) = 1$, and $y(3) = 10$. Hence, or otherwise obtain $y(4)$.

22. Using Evertt's formula, evaluate $f(30)$ if $f(20) = 2854, f(28) = 3162, f(36) = 7088, f(44) = 7984$.   (UPTU–2006)

23. Using Gauss's forward formula, evaluate $f(3.75)$ from the table :

x	2.5	3.0	3.5	4.0	4.5	5.0
y	24.145	22.043	20.225	18.644	17.262	16.047

(MKU-2001, 05, 06; Bhopal–2002)

24. Using Gauss's backward formula, estimate the number of persons earning wages between Rs. 60 and Rs. 70 from the following data :

Wages (Rs)	Below 40	40-60	60-80
No. of persons (in thousands)	250	120	100

Wages (Rs)	80-100	100-120
No. of persons (in thousands)	70	50

(Tirchirapalli–2001)

25. Using Stirling formula find y_{35}, given $y_{20} = 512, y_{30} = 439, y_{40} = 346, y_{50} = 243$, where y_x represents the number of persons at age x years in a life table.   (Nagarjuna–2003)

26. Using Stirling's formula to evaluate $f(1.22)$, given   (Tirchirapalli-2001)

x	1.0	1.1	1.2	1.3	1.4
$f(x)$	0.841	0.891	0.932	0.963	0.985

27. Calculate the value of $f(1.5)$, using Bessel's interpolation formula, from the tables :

x	0	1	2	3
$f(x)$	3	6	12	15

(UPTU–2008)

28. Use Bessel's formula to obtain y_{25}, given $y_{20} = 24, y_{24} = 32, y_{28} = 35, y_{32} = 40$.   (Gurukul–2000)

29. From the following table :

x	20	25	30
f(x)	11.4699	12.7834	13.7648

x	35	40	
f(x)	14.4982	15.0463	

find f(34) using Everett's formula.

(Madras–2000)

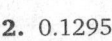

Answers

2. 0.1295

3. $\dfrac{1}{30}(-x^3 + 5x^2 - 4x + 30)$

4. 6.5928

5. 5.54

6. 352 ; 219

7. $\dfrac{2}{3}x^4 - 8x^3 + \dfrac{100}{3}x^2 - 56x + 31$

8. 3.76905

9. 109.5 **11.** 521

12. $x^3 + x^2 - x + 2$

13. $\dfrac{-(a+b)}{a^2b^2}, \dfrac{ab + bc + ca}{a^2b^2c^2}$

14. (i) $y_{41} = 2290.1$ (ii) 14.643

15. 0.558052

17. (i) 0.38891 (ii) 0.38873

18. 9.5

19. $x^4 - 4x^3 + 4x^2$

20. $\dfrac{1}{2}x^3(5 - 3x^2)$

22. 3884.4

23. 19.4 **24.** 54000

25. 395 **26.** 0.934

27. 9

28. 32.945

29. 14.368

MISCELLANEOUS EXERCISE

1. Express $y = 2x^3 - 3x^2 + 3x - 10$ in a factorial notation and hence show that $\Delta^3 y = 12$.

(Bhopal–2007, PTU–2005)

2. Find the missing values in the following table:

x	45	50	55	60	65
y	3.0	–	2.0	–	2.4

(Bhopal–2007, VTU–2001)

3. Find the missing term in the table:

x	2	3	4	5	6
y	45.0	49.2	54.1	–	67.4

4. Prove that (UPTU–2008)

(i) $\mu^2 = 1 + \dfrac{\delta^2}{4}$ (UPTU–2009)

(ii) $\Delta f_k^2 = (f_k + f_{k+1})\Delta f_k$ (JNTU MCA–2006)

5. Find the missing term in the following table:

x	0	1	2	3	4
f(x)	1	3	9	–	81

(SVTU–2007)

6. Find the missing term in the following data:

x	1	1.5	2	2.5	3	3.5	4
f(x)	6	?	10	20	?	1.5	5

(UPTU–2010)

7. Find the missing term in the following table:

x	0	1	2	3	4	5	6
f(x)	5	11	22	40	–	140	–

(VTU–2006)

8. Find the polynomial interpolating the data:

x	0	1	2
f(x)	0	5	2

(UPTU–2008)

9. Construct the difference table for the following data:

x	0.1	0.3	0.5	0.7	0.9	1.1	1.36
f(x)	0.003	0.067	0.148	0.248	0.370	0.518	0.697

Evaluate f(0.6). (JNTU–2007)

10. Estimate from following table f(3.8) to three significant figures using Gregory Newton backward interpolation formula:

x	0	1	2	3	4
f(x)	1	1.5	2.2	3.1	4.6

(UPTU–2009)

11. The following table gives the population of a town during the last six censuses. Estimate the increase in the population during the period from 1976 to 1978.

year	1941	1951	1961	1971	1981	1991
Population (in thousands)	12	15	20	27	39	52

(UPTU–2009)

12. Find by means of Gauss's backward formula, the population of a town for the year 1974, given that:

year	1939	1949	1959	1969	1979	1989
Population (in thousands)	12	15	20	27	39	52

(Kottayam–2005, Madras–2003)

13. Apply Stirling's formula to compute $y_{12.2}$ from the following table:

$x°$	10	11	12	13	14
$10^5 y_x$	23.967	28.060	31.788	35.209	38.368

(VTU-2004)

14. Using Gauss's backward difference formula, find $y(8)$ from the following table:

x	0	5	10	15	20	25
y	7	11	14	18	24	32

(JNTU-2007)

15. From the following table:

x	1.00	1.05	1.10	1.15
e^x	2.7183	2.8577	3.0042	3.1582

x	1.20	1.25	1.30
e^x	3.3201	3.4903	3.6693

Find $e^{1.17}$ using Gauss's forward formula.

(UPTU-2006)

16. Calculate the value of $f(1.5)$ using Bessel's interpolation formula, from the following table:

x	0	1	2	3
$f(x)$	3	6	12	15

(UPTU-2008)

17. Curve passes through the point $(0,18)$, $(1,10)$, $(3, -18)$ and $(6, 90)$. Find the slope of the curve at $x = 2$.　(JNTU-2009)

18. The following are the measurements T made on a curve recorded by oscilograph representing a change of current I due to change in the conditions of an electric current.

T	1.5	2.0	2.5	3.0
I	1.36	0.58	0.34	0.20

(JNTU-2009)

Using Lagrange's formula find I at $T = 16$

19. Find the third divided difference with arguments 2, 4, 9, 10 of the function $f(x) = x^3 - 2x$.　(UPTU-2005)

20. Use Newton's divided difference method to compute $f(5.5)$ from the following data:

x	0	1	4	5	6
$f(x)$	1	14	15	6	3

(UPTU-2010)

21. Obtain the Newton's divided difference interpolation polynomial and hence find $f(6)$:

x	3	7	9	10
$f(x)$	168	120	72	63

(UPTU-2007)

22. Using the following table, find $f(x)$ as a polynomial in x.

x	-1	0	3	6	7
$f(x)$	3	-6	39	822	1611

(UPTU-2009)

Answers

1. $y = 2[x]^3 + 3[x]^2 + 2[x] - 10$
2. 2.925, 0.225
3. 60.05
5. 31

6. $f(1.5) = 0.222, f(5) = 22.022$
7. $y(4) = 74, y(6) = 261$

8. $f(x) = 9x - 4x^2$
9. 0.1955
10. 4.219
11. 2530

12. 32.345 thousands approx
13. 0.32497
14. 12.826
15. 3.2219

16. 9
17. -16
18. 0.89
19. 1　**20.** 3.09

21. 133.19
22. $f(x) = x^4 - 3x^3 + 5x^2 - 6$

Objective Evaluations

FILL IN THE BLANKS

1. The technique for computing the value of the function inside the given arguments is called _____.

2. The technique for computing the value of the function outside the given arguments is called _____.

3. Lagrange's interpolation formula is also used for the arguments which are _____ spaced.

4. Lagrange's formula can also be used to split the given function into _____.

5. Lagrange's formula is also useful for _____.

6. Lagrange' interpolation polynomial is _____.

7. Newton's forward interpolation formula is mainly used to determine the value of function in the _____ of the arguments.

8. Newton's backward formula is mainly used to interpolate the value of the function nearly at the _____ of the arguments.

9. The nth divided difference of $f(x) = a_0 x^n$ is equal to _____.

10. The nth divided difference can be expressed as the quotient of two determinants each of order _____.

11. The ratio of nth divided difference to n^{th} ordinary differences is _____.

12. $E - 1 =$ _____

13. $\delta =$ _____ $E^{-1/2}$.

14. $\Delta \nabla = \Delta -$ _____ .

TRUE OR FALSE

Write 'T' for True and 'F' for False statement.

1. $1 + \Delta \neq E$.

2. The $(n + 1)^{th}$ divided difference of $f(x)$ of degree n is zero. **(T/F)**

3. There are more than one Lagrange interpolating polynomials. **(T/F)**

4. Newton's forward and backward interpolation formula can be used for unequally spaced arguments. **(T/F)**

5. The nth divided difference cannot be expressed as the ratio of two determinants each of order $(n + 1)$. **(T/F)**

6. The nth divided difference can be expressed as the product of multiple integrals. **(T/F)**

7. The error in a difference table propagates in a triangular pattern. **(T/F)**

8. The algebraic sum of the errors in any difference column is zero. **(T/F)**

9. $\Delta^2 y_0 = \Delta^2 y_{-1} + \Delta^3 y_{-1}$ **(T/F)**

10. $\Delta^2 f(x) = f(x + 2h) + 2f(x + h) - f(x)$ **(T/F)**

11. Bessel's formula is the arithmetic mean of Gauss's forward and Gauss's backward formula. **(T/F)**

12. Stirling's formula is used for interpolation near the middle of the arguments. **(T/F)**

13. Bessel's formula is most efficient near $u = \dfrac{1}{2}$. **(T/F)**

14. If n values of $f(x)$ are given. Then $\Delta^n f(x) = 0$ **(T/F)**

MULTIPLE CHOICE QUESTIONS

Problem Set – 1

Choose the most appropriate one

1. The nth divided difference of $f(x)$ of degree n is:
   (a) 0
   (b) constant
   (c) –1
   (d) none of these

2. The nth divided difference can be expressed as the ratio of two determinants each of order:
   (a) n
   (b) $n + 1$
   (c) $n - 1$
   (d) n^2

3. $1 + \Delta$ is equal to:
   (a) $E - 1$ (b) ∇
   (c) E (d) δ

4. If n values of $f(x)$ are given. Then $\Delta^n f(x)$ is:
   (a) 1   (b) 0
   (c) 2   (d) n

5. $L_2(x)$ is a Lagrange polynomial of degree:
   (a) 1   (b) 2
   (c) 3   (d) 0

6. $L_k(x_k)$ is equal to:
   (a) 0   (b) 1
   (c) 2   (d) 3

7. $L_k(x_l)$ for $k \neq l$ is equal to:
   (a) 0   (b) 1
   (c) -1 (d) 2

8. In a difference table $\Delta^k y_0$ lie on a:
   (a) parabola
   (b) straight line sloping down
   (c) straight line sloping upward
   (d) circle

9. In a difference table $\delta^{2k} y_n$ lie on a:
   (a) horizontal line   (b) parabola
   (c) circle   (d) triangle

10. Δ^2 is equal to:
    (a) $E^2 - I$ (b) $E^2 - 2E + I$
    (c) $E - 1$ (d) E

11. If $f(x) = -6x^3 + 11x^2 - 6x + 1$ then $\Delta^3 f(x)$ is equal to:
    (a) 36   (b) -36
    (c) 32   (d) 1

12. $(I + \Delta)(I - \nabla)$ is equal to:
    (a) I (b) E
    (c) Δ (d) ∇

13. If $x^{(3)} = x(x - 1)(x - 2)$ then $\Delta^3 x^{(3)}$ is equal to:
    (a) 2   (b) 6
    (c) 1   (d) 0

14. The value of $\Delta^2_{y,z} x^2$ is equal to:
    (a) $x + y + z$ (b) $x^2 + y^2 + z^2$
    (c) 0   (d) 1

Problem Set – 2

1. The method of obtaining the values of a function for any intermediate value of the argument for the given set of values of the function for certain values of arguments is called :
   (a) extrapolation   (b) interpolation
   (c) interrelation   (d) none of these

2. For the function $y = f(x)$, the value $f(x + h) - f(x)$ is called :
   (a) $\Delta f(x)$ (b) $\nabla f(x)$
   (c) $E f(x)$ (d) none of these

3. For the function $y = f(x)$, the value $f(x + h) - f(x)$ is called :
   (a) $\Delta f(x)$ (b) $\nabla f(x + h)$
   (c) $E f(x)$ (d) none of these

4. For the function $y = f(x)$, $\nabla f(x)$ is defined by
   (a) $f(x) - f(x - h)$ (b) $f(x) + f(x - h)$
   (c) $f(x + h) - f(x)$ (d) none of these

5. $\Delta^{r-1} y_{k+1} - \Delta^{r-1} y_k =$
   (a) $\Delta^r y_k$ (b) $\Delta^r y_{k+1}$
   (c) $\nabla^r y_k$ (d) none of these

6. For the function $y = f(x)$, $E f(x) =$
   (a) $f(x - h)$ (b) $f(x + h)$
   (c) $f(x)$ (d) none of these

7. If $a \in \mathbf{R}$, then $\Delta[a \cdot f(x)] =$
   (a) $a \Delta f(x)$ (b) $\Delta a \cdot \Delta f(x)$
   (c) $\Delta a \cdot f(x)$ (d) none of these

8. $\Delta[f(x) \pm g(x)] =$
   (a) $\Delta f(x) \pm \Delta g(x)$ (b) $\Delta f(x) \pm g(x)$
   (c) $\Delta g(x) \pm f(x)$ (d) none of these

9. $\Delta[f(x) \cdot g(x)]$
   (a) $\Delta f(x) \cdot \Delta g(x)$
   (b) $f(x) \Delta g(x) + g(x) \Delta f(x)$
   (c) $f(x) \Delta g(x)$
   (d) none of these

10. The value of $f(x + h) - f(x) =$
    (a) $\Delta f(x)$
    (b) $\nabla f(x + h)$
    (c) both (a) and (b) are true
    (d) none of these

11. The value of $f(x + 2h) - 2f(x + h) + f(x) =$
    (a) $\Delta^2 f(x)$
    (b) $\nabla^2 f(x + 2h)$
    (c) both (a) and (b) are true
    (d) none of these

12. The value of $\delta y_{1/2} =$
    (a) $y_0 - y_1$ (b) $y_1 - y_0$
    (c) $y_1 - y_{1/2}$ (d) none of these

13. For the shift operator E, which of the following is not true
   (a) $E^n f(x) = f(x + nh)$
   (b) $E^{-n} f(x) = f(x - nh)$
   (c) $\Delta = E - 1$
   (d) all are true

14. The value of $1 - E^{-1}$ is equal to :
   (a) Δ
   (b) ∇
   (c) $1/\Delta$
   (d) none of these

15. The value of $E^{1/2} - E^{-1/2} =$
   (a) Δ
   (b) ∇
   (c) δ
   (d) none of these

16. Which of the following value is not equal to Δ
   (a) $E\nabla$
   (b) ∇E
   (c) $\delta E^{1/2}$
   (d) none of these

17. The value of $x^{(3)} =$　　(If $h = 1$)
   (a) $x \cdot x \cdot x$
   (b) $x(x - 1)(x - 2)$
   (c) $x^2(x - 1)$
   (d) none of these

18. If $h = 1$, then value of $\Delta^n x^{(n)} =$
   (a) 0
   (b) n
   (c) $n!$
   (d) none of these

19. The value of $\Delta^{n + 1} x^{(n)} =$
   (a) 0
   (b) n
   (c) $n!$
   (d) none of these

20. $\dfrac{1}{(x-1)(x-2)(x-3)} =$
   (a) $x^{(3)}$
   (b) $x^{-(3)}$
   (c) both (a) and (b) are true
   (d) none of these

21. The value of $\Delta \log x =$
   (a) $\log(1 + hx)$
   (b) $\log(1 + h/x)$
   (c) $\log(1 - h/x)$
   (d) none of these

22. The value of $\Delta \tan^{-1} x =$
   (a) $\tan^{-1}\left[\dfrac{h}{1 + hx + x^2}\right]$
   (b) $\tan^{-1}(1 + hx + x^2)$
   (c) $\tan^{-1}(hx + x^2)$
   (d) none of these

23. The value of $\Delta^2 \cos 2x =$
   (a) $4 \sin^2 h \cos(2x + 2h)$
   (b) $-4 \sin^2 h \cos(2x + 2h)$
   (c) $\sin^2 h (\cos(2x + 2h))$
   (d) none of these

24. The value of $\left(\dfrac{\Delta^2}{E}\right) e^x \cdot \dfrac{Ee^x}{\Delta^2 e^x} =$

25. If $\Delta f(x) = x^3 + 3x^2 + 5x + 12, h = 1$, then $f(x) =$
   (a) $\dfrac{x^{(4)}}{4} + \dfrac{6x^{(3)}}{3} + \dfrac{9x^{(2)}}{2} + 12x + c$
   (b) $\dfrac{x^4}{4} + \dfrac{6x^3}{3} + \dfrac{9x^2}{2} + c$
   (c) both (a) and (b) are true
   (d) none of these

26. The value of $\Delta^n[\sin(ax + b)] =$
   (a) $\left(2\sin\dfrac{ah}{2}\right)^n \sin\left(ax + b + n\left(\dfrac{ah + \pi}{2}\right)\right)$
   (b) $2\sin\dfrac{ah}{2} \cdot \sin\left[ax + b + n\left(\dfrac{ah + \pi}{2}\right)\right]$
   (c) both (a) and (b) are true
   (d) none of these

27. If $f(x) = e^{ax}$, then value of $\Delta^n f(x) =$
   (a) $n e^{ax}$
   (b) $(e^{ah} - 1)^n e^{ax}$
   (c) $(e^{ah})^n \cdot e^{ax}$
   (d) none of these

28. The value of $\Delta^{10}[(1 - ax)(1 - bx^2)(1 - cx^3)(1 - dx^4)] =$
   (a) $10!$
   (b) $abcd$
   (c) $abcd \cdot 10!$
   (d) none of these

29. The value of $\Delta \log f(x) =$
   (a) $\log[1 + \Delta f(x)]$
   (b) $\log\left(1 + \dfrac{\Delta f(x)}{f(x)}\right)$
   (c) $\log[f(x) + \Delta f(x)]$
   (d) none of these

30. The value of $\Delta^p y_k =$
   (a) $\Delta^p y_k$
   (b) $\nabla^p y_{k + p}$
   (c) $\nabla^p y_{k - p}$
   (d) none of these

31. The value of $hD =$
   (a) $1 - \log(1 - \nabla)$
   (b) $\sinh^{-1}(\mu\delta)$
   (c) both (a) and (b) are true
   (d) none of these

32. The value of $(E^{1/2} + E^{-1/2})(1 + \Delta)^{1/2} =$
   (a) $1 + \Delta$
   (b) $2 + \Delta$
   (c) $3 + \Delta$
   (d) none of these

33. The value of $\dfrac{\Delta}{\nabla} - \dfrac{\nabla}{\Delta} =$
   (a) $\Delta + \nabla$
   (b) $\Delta\nabla$
   (c) $\Delta - \nabla$
   (d) none of these

34. The value of $\Delta - \nabla =$
   (a) $\nabla\Delta$
   (b) $-\nabla\Delta$
   (c) Δ/∇
   (d) none of these

(a) e^x
(b) e^{-x}
(c) e^{2x}
(d) none of these

35. The value of $\mu\delta =$
   (a) $\Delta + \nabla$
   (b) $\frac{1}{2}(\Delta + \nabla)$
   (c) $\frac{1}{2}(\Delta - \nabla)$
   (d) none of these

36. The value of $\nabla =$
   (a) ΔE^{-1}
   (b) $E^{-1}\Delta$
   (c) $1 - E^{-1}$
   (d) all are true

37. The value of $\delta(E^{1/2} + E^{-1/2}) =$
   (a) $\Delta + \Delta E^{-1}$
   (b) ΔE^{-1}
   (c) Δ/E^{-1}
   (d) none of these

38. The value of $\delta =$
   (a) $\Delta(1 + \Delta)^{-1/2}$
   (b) $\nabla(1 - \nabla)^{-1/2}$
   (c) both (a) and (b) are true
   (d) none of these

39. The value of $1 + \frac{\delta^2}{2} =$
   (a) $1 + \delta^2\mu^2$
   (b) $\sqrt{1 + \delta^2\mu^2}$
   (c) $\sqrt{1 - \delta^2\mu^2}$
   (d) none of these

40. The value of $E^{1/2} =$
   (a) $\mu + \delta/2$
   (b) $\mu - \delta/2$
   (c) $\mu(\delta/2)$
   (d) none of these

41. The value of $\Delta\nabla =$
   (a) $\nabla\Delta$
   (b) δ^2
   (c) both (a) and (b) are true
   (d) none of these

Answers

FILL IN THE BLANKS

1. interpolation	**2.** extrapolation	**3.** equality	**4.** partial fraction
5. inverse interpolation	**6.** unique	**7.** beginning	**8.** end **9.** a_0
10. $(n + 1)$	**11.** constant	**12.** Δ	**13.** Δ **14.** ∇

TRUE OR FALSE

1. F	**2.** T	**3.** F	**4.** F	**5.** F	**6.** T	**7.** T	**8.** T	**9.** T
10. F	**11.** T	**12.** T	**13.** T	**14.** T				

MULTIPLE CHOICE QUESTIONS

Problem Set – 1

1. (b)	**2.** (b)	**3.** (c)	**4.** (b)	**5.** (b)	**6.** (b)	**7.** (a)	**8.** (b)	**9.** (a)
10. (b)	**11.** (b)	**12.** (a)	**13.** (b)	**14.** (d)				

Problem Set – 2

1. (b)	**2.** (a)	**3.** (b)	**4.** (a)	**5.** (a)	**6.** (b)	**7.** (a)	**8.** (a)	**9.** (b)
10. (c)	**11.** (c)	**12.** (b)	**13.** (d)	**14.** (b)	**15.** (c)	**16.** (d)	**17.** (b)	**18.** (c)
19. (a)	**20.** (b)	**21.** (b)	**22.** (a)	**23.** (b)	**24.** (a)	**25.** (a)	**26.** (a)	**27.** (b)
28. (c)	**29.** (b)	**30.** (b)	**31.** (c)	**32.** (b)	**33.** (a)	**34.** (b)	**35.** (b)	**36.** (d)
37. (a)	**38.** (c)	**39.** (b)	**40.** (a)	**41.** (c)				

COMPUTATIONAL TECHNIQUE LAB

1. NEWTON'S FORWARD INTERPOLATION FORMULA

SYMBOLS USED

x_0 = initial value of x

h = length of interval

n = number of subintervals

x = value of x at which we have to find the value of y

y = value of y at x

ay = array to store the different values of y

t = temporary variable

ALGORITHM : NEWTON'S FORWARD INTERPOLATION FORMULA

Step 1.	Start
Step 2.	Input n, x_0, h, x
Step 3.	Input values of ay
Step 4.	y = ay [0], t = 1
Step 5.	u = (x–x_0)/h
Step 6.	For i = 1 to n
Step 7.	For j = 0 to (n–i)
Step 8.	y [j] = y [j+1] – y [j]
Step 9.	End of j loop.
Step 10.	t = t * (p – i + 1)/i
Step 11.	y = y + t * ay [0]
Step 12.	End of i loop
Step 13.	Print y
Step 14.	STOP

FLOW CHART : NEWTON'S FORWARD INTERPOLATION FORMULA

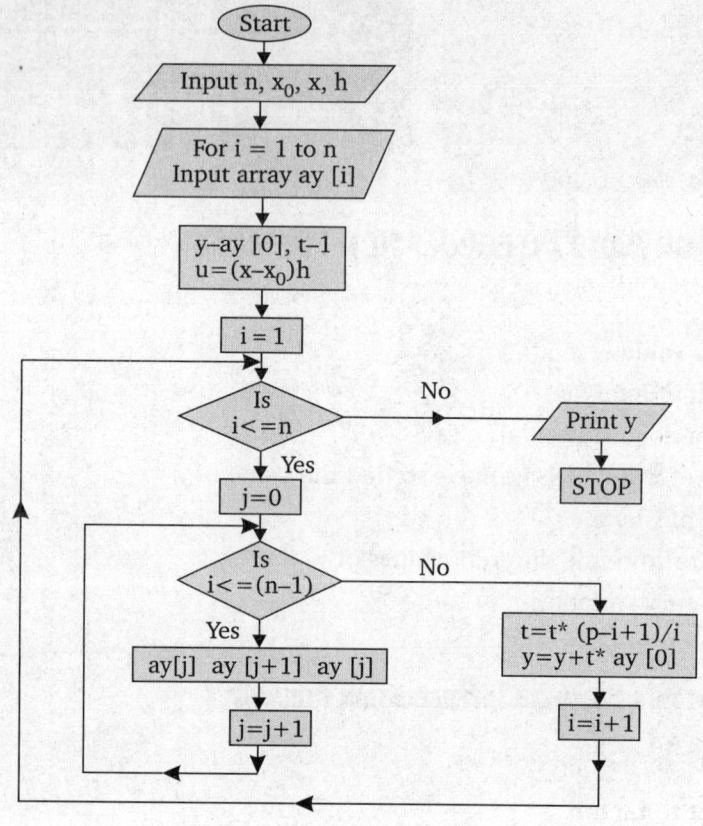

PROGRAM .   *Following is a program to show the Newton's forward interpolation formula.*

NEWTON'S FORWARD INTERPOLATION FORMULA   (UPTUMCA-2004)

```
# include<studio.h>
#include<conio.h>
 void main ( )
     {
     clrscr ( );
     float ay [30], x0, h, x, y, t = 1, u;
     int n, i, j;
     printf ("Enter the value of n\n");
     scanf ("% d", & n);
     printf ("Enter the initial value of x\n");
     scanf ("%d", &x0);
     printf ("\n enter length of each interval\n");
     scanf ("%f", &h);
     for (i = 0; i < n; i++)
```

```
{           print f ("Enter the value of y (%d) = " , i);
            scan f ( "%f", & ay [i]);

}
printf("Enter the value of x for which value of y is wanted \n");
scanf ("%f", &x)';
y=ay [0];
u=(x-x0)/h;
for (i=1; i<=n; i++)
    {
            for (j=0; j<=n-1; j++)
            ay [j] = ay [j+1]-ay [j];
            t=t* (u-i+1)/i;
            y=y+t*ay [0]

    }
printf ("\n Value of y at x=%. 2 fis is % .2f " , x, y); getch ( ) ;
        }
```

Output : NEWTON'S FORWARD INERPOLATION FORMULA

```
Enter the value of n
6
Enter the initial value of x
0
enter length of each interval
1
Enter the value of       y (0) =        1
Enter the value of       y (1) =        3
Enter the value of       y (2) =        11
Enter the value of       y (3) =        31
Enter the value of       y (4) =        69
Enter the value of       y (5) =        131
Enter the value of       y (6) =        223
Enter the value of x for which value of y is wanted
3.4
Value of y at x = 3.40 is 43.70
```

LAB ASSIGNMENT : NEWTON'S FORWARD INTERPOLATION FORMULA :

1. Write a C program for the Newton's forward interpolation formula to find the value of y at x = 2.7 from the following data

x	1	2	3	4	5	6	7	8
f(x)	1	8	27	64	125	216	343	512

Hint: Input $x_0 = 1, h = 1, n = 7$

        Output $f(2.7) = 50.65$

2. Write a C program to find f (3.4) using the following values by Newton's forward interpolation formula

x	0	1	2	3	4	5	6
f(x)	1	3	11	31	69	131	223

    Hint: Input $x_0 = 0, h = 1, n = 6$

        Output $f(3.4) = 43.704$

2. NEWTON'S BACKWARD INTERPOLATION FORMULA

SYMBOLS USED

    n = number of sub-intervals

    h = length of interval

    x_n = last value of x_i

    x = values of x at which we have to find the value of y

    a [y] = an array to store different values of y

    y = value of y at x

ALGORITHM : NEWTON'S BACKWARD INTERPOLATION FORMULA

Step 1 :   Start

Step 2 :   Input n, x_n, h, x

Step 3 :   For i = 0 to n

Step 4 :   Input ay [i]

Step 5 :   ay = ay [n], t = 1

Step 6 :   u = (x–x_n)/h

Step 7 :   For i = 1 to n

Step 8 :   For j = 0 to (n–i)

Step 9 :   ay [j] = ay [j+1] – ay [j]

Step 10 :   End of j loop

Step 11 :   t = t* (u+i–1)/i

Step 12 :   y = y + t* ay [n – i]

Step 13 :   End of j loop

Step 14 :   Print y

Step 15 :   STOP.

FLOW CHART : NEWTON'S BACKWARD INTERPOLATION FORMULA

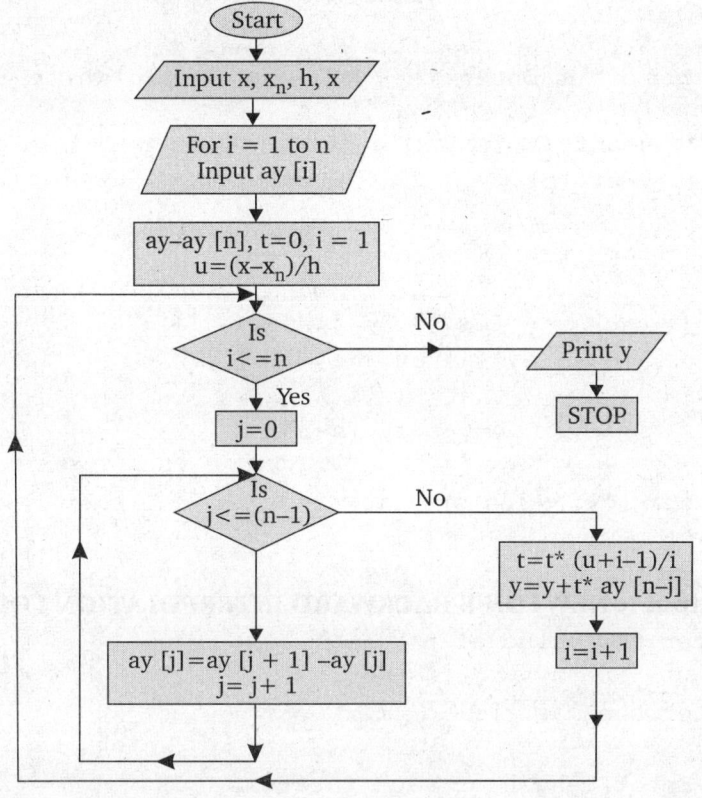

PROGRAM : *Following is a program to demonstrate the Newton's Backward Interpolation Formula.*

NEWTON'S BACKWARD INTERPOLATION FORMULA

```
#include<studio.h>
#include<conio .h>
void main ( )
    {
    clrscr ( );
    float ay [30], xn, h, x, y, t=1, u;
    int n, i, j;
    printf ("Enter the value of n\n");
    scanf ("%d", &n) ;
    printf ("Enter the last value of x\n");
    scanf ("%f " , & xn);
    printf ("\n enter length of each interval\n") ;
    scanf ("%f " , &h) ;
    for (i=0; i<=n; i++)
```

```
{          print f ("Enter the value of y(%d) = ", i) ;
           scanf ("% f ", & ay [i]) ;
}
printf ("\n Enter the value of x for which value of y is
watned\n");
     scanf ("%f", &x) ;
     y=ay [n] ;
     u=(x-xn)/h;
     for (i=1; i = n;i++)
       {
              for (j=0; j<=n-i; j++)
           ay [j]=ay[j+1]-ay [j];
           t =t* (u+i-1)/i;
           y=y+t*ay [n-i];
       }
printf ("\n value of y at x = % .2f   is%.2f", x, y) ; getch
( ) ;
       }
```

Output : NEWTON'S BACKWARD INTERPOLATION FORMULA

```
Enter the value of n
4
Enter the last value of x
60
   Enter length of each interval
10
Enter the value of    y (0)   =        42
Enter the value of    y (1)   =        87
Enter the value of    y (2)   =        126
Enter the value of    y (3)   =        174
Enter the value of    y (4)   =        193

   Enter the value of x for which value of y is wanted
44
value of y at x=44.00 is   145.06
```

LAB ASSIGNMENT : NEWTON'S BACKWARD INTERPOLATION FORMULA

1. Write a C program for Newton's Backward Interpolation formula to find the value of f (7.5) for the data given below :

x	1	2	3	4	5	6	7	8
f(x)	1	8	27	64	125	216	343	512

Hint: Input $x_n = 5, h = 1, n = 7. \ x = 7.5$

Output $f (7.5) = 421.875$

2. Write a C program to find f (4.4) by the Newton backward interpolation formula from the given data.

x	0	1	2	3	4	5
y	5	20	81	224	485	900

Hint: Input : $x_n = 5, h = 1, n = 5, x = 4.4$

Output: $y (4.4) = 630.5041$

3. GAUSS FORWARD INTERPOLATION FORMULA

SYMBOLS USED

n = number of subintervals

h = length of interval

x = value of x at which we have to find the value of y

k = location of x_0. x_0 is that value which is closed to x

y = value of y at x

a [x] = an array to store different values of x

a [y] = an array to store the different values of y

ALGORITHM : GAUSS FORWARD INTERPOLATION FORMULA

Step 1 :	Start
Step 2 :	Input n, h, x, k
Step 3 :	For i = 0 to n
Step 4 :	Input ax [i], ay [i]
Step 5 :	y = ay [k], t = 1
Step 6 :	u = (x–ax [k])/h
Step 7 :	For i = 1 to n
Step 8 :	For j = 0 to n – i
Step 9 :	ay [j] = ay [j + 1] – ay [j]
Step 10 :	End of j loop
Step 11 :	If (i% 2! = 0)
	t = (t * (u + i/2))/i
	else
	t = (t * (u – i/2))/i ;
Step 12 :	y = y + t * ay [k – i/2]
Step 13 :	End of i loop
Step 14 :	Print "Value of y at x is" . y
Step 15 :	STOP

FLOW CHART : GAUSS FORWARD INTERPOLATION METHOD

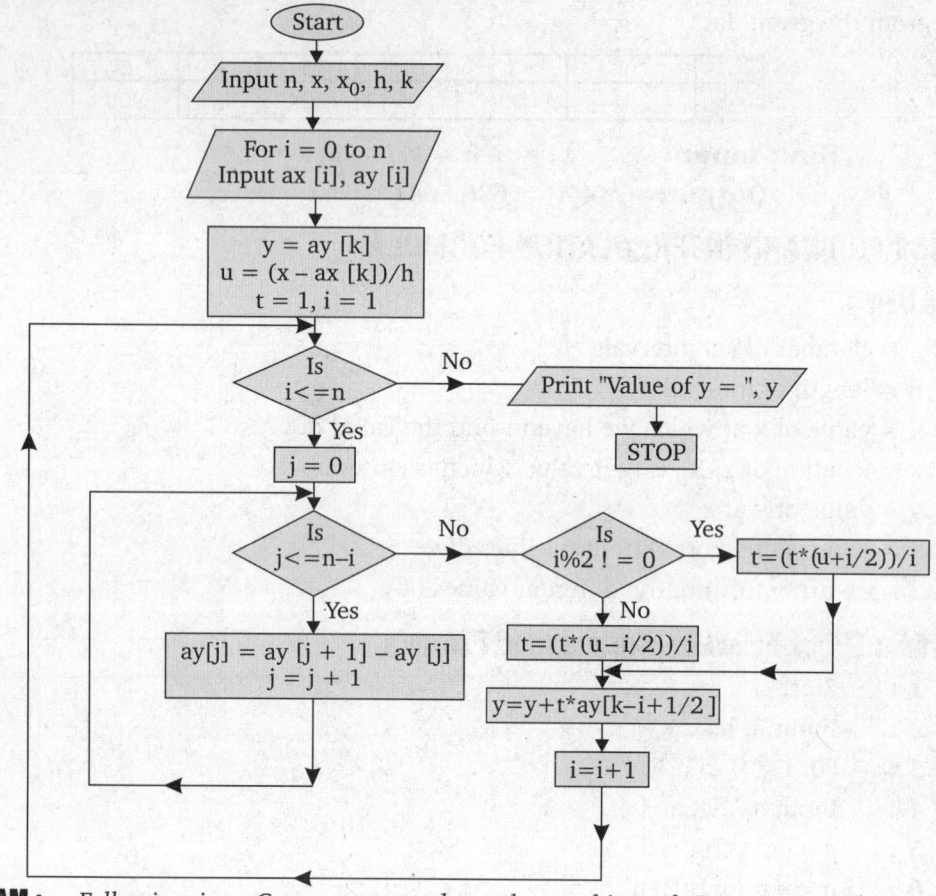

PROGRAM : *Following is a C program to show the working of Gauss Forward Interpolation formula.*

//GAUSS FORWARD INTERPOLATION FORMULA

```c
#include<studio .h>
#include<conio ,h>
void main ( )
    {
    clrscr ( ) ;
    float ax [30], ay [30] , h, x, y, t=1, u;
    int n, i, j, k;
    printf ("Enter the value n\n");
    scanf ("%d", &n) ;
    printf ("\n enter length of each interval\n");
    scanf ("%f", &h) ;
        printf ("Enter the value of x and y/n") ;
```

```
        for (i=0; i<=n; i++)
                scanf ("%f %f ", & ax [i], & ay [i] ;
        printf ("\n Enter the value of x for which value of y
is wanted\n");
        scanf ("%f", &x);
        printf ("\n enter the location of x0 i.e. k/n");
        scanf ("%d", &k);
        y=ay [k];
        u=(x-ax [k]) /h;
        for (i=1; i<=n; i++)
            {
                for (j=0; j<=n-i; j++)
                  ay [j]=ay[j+1]-ay [j];
                if (i%2 ! = 0)
                        t=(t* (u+i/2))/i;
                else
                        t=(t*(u-i/2))/i;
                y=y+t*ay[k-i/2];
                }
        printf ("\n value of y at x=%. 2f is % .2f ", x,
y) ; getch ( ) ;
                }
```

Output : GAUSS FORWARD INTERPOLATION FORMULA

```
Enter the value of n
5
    enter length of each interval
.5
Enter the value of x and y
2.5    24.145
3      22.043
3.5    20.225
4      18.644
4.5    17.262
5      .16.047
Enter the value of x for which value of y is wanted
3.75
    enter the location of x0 i.e. k
2
value of y at x=3.75 is 19.41
```

LAB ASSIGNMENT : GAUSS FORWARD INTERPOLATION FORMULA

1. Write a C program to find value of y (30) by Gauss Forward Interpolation formula from the data given below :

x	21	25	29	33	37
f(x)	18.4708	17.8144	17.1070	16.3432	15.5154

Hint :Input : n = 4, h = 4, x = 30, k = 2

Output : Value of y = 16.92

2. Write a C program for Gauss Forward interpolation formula to find value of f (2.3) from the given data

x	1	2	3	4	5
f(x)	1	−1	1	−1	1

Hint: Input : n = 4, h = 1, x = 2.3, k = 1

Output: Value of y = − 0.146600.

4. GAUSS BACKWARD INTERPOLATION FORMULA

SYMBOLS USED

(Note : same as used in Gauss forward Interpolation formula).

ALGORITHM: GAUSS BACKWARD INTERPOLATION FORMULA

Step 1 : Start

Step 2 : Input x, h, x_0, k

Step 3 : For i = 0 to n

Step 4 : Input ax [i], ay [i]

Step 5 : y = ay [k], t = 1

Step 6 : u = (x–ax [k])/h

Step 7 : For i = 1 to n

Step 8 : For j = 0 to n – i

Step 9 : ay [j] = ay [j + 1] – ay [j]

Step 10 : End of j loop

Step 11 : if (1%2 ! = 0)

$$t = (t * (u – i/2))/i$$

else

$$t = (t * (u + i/2))/i$$

Step 12 : y = y + t * ay [k – (i + 1)/2]

Step 13 : End of i loop.

Step 14 : Print "value of y=", y

Step 15 : Stop.

Flow Chart : Gauss Backward Interpolation Method

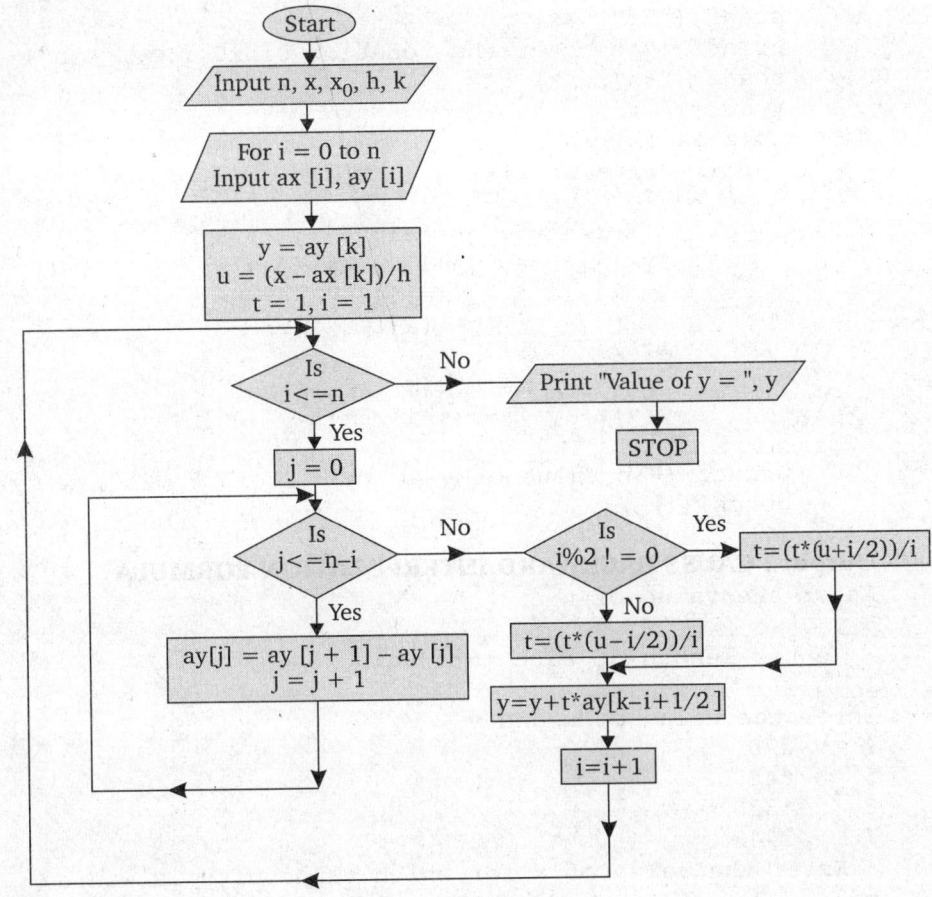

PROGRAM : *Following is a C program to show the working of Gauss Backward Interpolation Method.*

GAUSS BACKWARD INTERPOLATION FORMULA

```
#include<studio .h>
#include<conio .h>
void main ( )
    {
    clrscr ( ) ;
    float   ax [30], ay [30] , h, x, y, t = 1, u;
    int n, i, j, k;
    printf ("Enter the value of n\n") ;
    scanf ("%d", &n) ;
    printf ("\n enter length of each interval\n") ;
    scanf ("%f", &h) ;
        printf ("Enter the value of x and y\n ") ;
    for (i=0; i<=n;i++)
    {
        scanf ("%f %f", &ax [i], &ay [i]);
    }
```

```
        printf ("\n Enter the value of x for which value of y
is wanted\n") ;
        scanf ("%f", &x) ;
        printf ("\n enter the location of x0 i.e., k\n") ;
        scanf ("%d", &k) ;
        y=ay [k] ;
        u=(x-ax [k])/h ;
        for (i=1;i<= n; i++)
            {
                for (j=0; j<=n-i; j++)
                ay [j] = ay [j+1] - ay [j];
                    if (i % 2 ! = 0)
                        t=(t* (u-i/2))/i;
                else
                        t=(t* (u+i/2))/i;
                y=y+t*ay [k- (i+1)/2];
                }
        printf ("\n Value of y at x=% .2f is % .2f", x, y) ;
        getch ( ) ;
        }
```

Output : GAUSS BACKWARD INTERPOLATION FORMULA

```
Enter the value of n
3
    enter length of each interval
1
Enter the value of x and y
4       270
5       648
6       1330
7       2448
    Enter the value of x for which value of y is wanted
5.8
    enter the location of x0 i.e. k
1
Value of y at  x=5.80  is 1169.28
```

LAB ASSIGNMENT : GAUSS BACKWARD INTERPOLATION METHOD

1. Write a C program for Gauss Backward interpolation method to find y (1.15) form data given below :

x	1	1.10	1.20	1.30
y	1.0	1.04881	1.09544	1.14017

Hint : **Input :** n = 3, h = 0.10, k = 1, x = 1.15

Output : Value of y = 1.072397

2. Write a C program for Gauss Backward Interpolation Method to find the value for 1936 from the given data.

x	1901	1911	1921	1931	1941	1951
f(x)	12	15	20	27	39	52

Hint : **Input :** n = 5, h = 10, x = 1936, k = 3

output : Value of y = 32.3437.

5. STIRLING'S DIFFERENCE FORMULA

SYMBOLS USED

n = number of subintervals

h = length of interval

x = value of x at which we have to find the value of y

k = location of x_0. x_0 is that value which is closed to x

y = value of y at x

a [x] = an array to store the different values of x

a [y] = an array to store the different values of y

ALGORITHM : STERLING'S DIFFERENCE FORMULA

Step 1 :   Start

Step 2 :   Input n, h

Step 3 :   For i = 1 to n

Step 4 :   Input ax [i], ay [i]

Step 5 :   Input x, k

Step 6 :   u = $(x-x_k)/h$

Step 7 :   y = ay [k], m = n

Step 8 :   if (k<=n/2)

$$n = 2k$$

else

$$n = 2 (n - k)$$

Step 9 :   For i = 1 to n

Step 10 :   For j = 0 to (m – i)

Step 11 :   ay [j] = ay [j + 1] – ay [j]   .

Step 12 :   End of j loop

Step 13 :   if (i%2 ! = 0)

$$t_1 = (t_1 * (u - i/2))/i$$
$$t_2 = (t_2 * (u + i/2))/i$$

else

$$t_1 = (t_1 * (u + i/2))/i$$
$$t_2 = (t_2 * (u - i/2))/i$$

Step 14 :   y = y + $(t_1 * ay [k - (i + 1)/2] + t_2 * ay [k - i/2])/2$

Step 15 :   End of i loop

Step 16 :   Print y

Step 17:   Stop

FLOW CHART : STERLING DIFFERENCE FORMULA

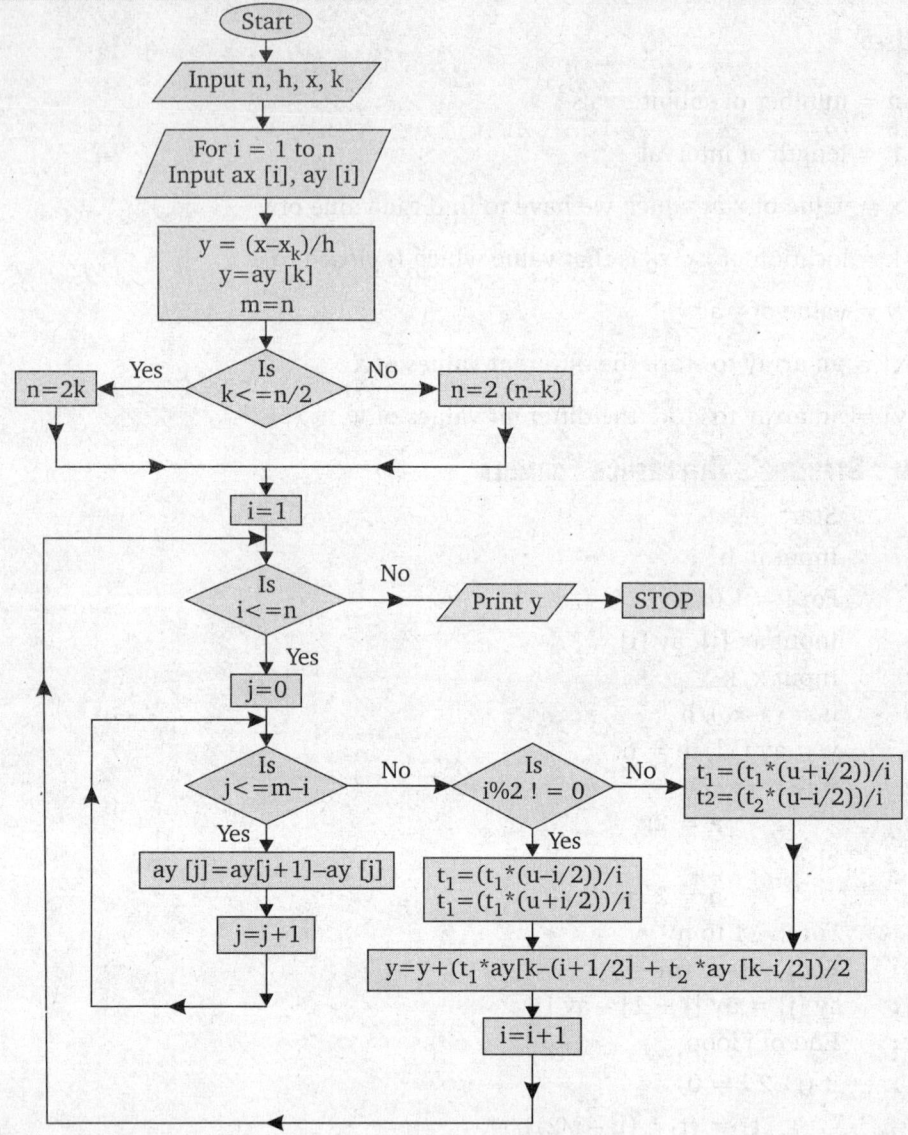

PROGRAM :   *Following is a C program showing the working of Stirling Difference Formula for interpolation.*

//STIRLING'S INTERPOLATION FORMULA

```
#include<studio.h>
#include<conio.h>
void main ( )
    {
    clrscr ( );
    float ax [30], ay [30], h, x, y, t1=1, t2=1, u;
    int n, i, j, m, k;
```

```
        printf ("Enter the value of n\n");
        Scanf ("%f", & h);
        printf ("\n enter length of each interval\n");
        scanf ("%f", &n);
                printf ("Enter the value of x and y\n ") ;
        for (i=0, i< = n;++)
        {
                scanf ("%f %f ", &ax [i], &ay [i]) ;
        }
        printf ("Enter the value of x for which value of y is
wanted\n") ;
        scan ("%f", &x) ;
        printf ("\n enter the location of x0 i.e. k\n") ;
        scanf ("%d", & k);
        printf ("\n enter the location of x0 i.e. k\n') ;
        scanf ("%d", &k) ;
        y=ay [k] ;
        u=(x-ax [k])/h;
        m=n;
        if     (k< = n/2)
        n=2*k;
        else
           n = 2* (n-k);
        for (i=1; i<=n; i++)
           {
                for (j=0; j<=m-i; j++)
                ay[j]=ay [j+1]-ay [j];
                   if (i%2 ! = 0)
           {   t1=(t1*(u-i/2)) /i;
                t2=(t2* (u+i/2)) /i;
           }
        else
             {
                t1=(t1*(u+i/2))/i;
                t2=(t2*(u-i/2))/i;
             }
        y=y+(t1*ay[k-(i+1) / 2] + t2*ay[k-i/2)) /2 ;
           }
        print f ("\n Value of y at x=% .2f is    % .2f ", x, y) ;
        getch ( ) ;
        }
```

Output : STIRLING'S INTERPOLATION FORMULA
```
Enter the value of n
4
enter length of each interval
5
Enter the value of x and y
   10  492
15    483
20    472
25    459
```

```
30     453
Enter the value of x for which value of y is wanted
19
    enter the location of x0   i.e.   k
2
    Value of y at x=19.00   is   474.49
```

LAB ASSIGNMENT : STIRLING DIFFERENCE FORMULA

1. Write a C program to find the value of y [28] from the given data using Stirling Difference Formula.

x	20	25	30	35	40
y	49225	48316	47236	45926	44306

Hint : Input : $x = 28$, $n = 4$, $h = 5$, $x_0 = 30$, $k = 2$
Output : $y(28) = 47692$.

2. Using Stirling difference formula, write a C program to find y (25) by given data :

x	20	24	28	32
y	24	32	34	40

Hint : Input : $h = 4$, $n = 3$, $x = 25$, $k = 1$
Output : $y(25) = 32.945287$

6. LAGRANGE'S INTERPOLATION FORMULA FOR UNEQUAL INTERVAL

SYMBOLS USED

n = number of subintervals
x = value of x at which we have to find the value of y
nr = numerator of each term of Lagrange's formula
dr = denominator of each term of Lagrange's formula
y = value of y at x
a [x] = an array to store different values of x
a [y] = an array to store the difference vaues of y'

ALGORITHM : LAGRANGE'S INTERPOLATION FORMULA

Step 1 :	Start
Step 2 :	Input n, x
Step 3 :	For i = 0 to n
Step 4 :	Input ax [i], ay [i]
Step 5 :	For i = 0 to n
Step 6 :	nr=dr=1
Step 7 :	For j = 0 to n
Step 8 :	if (i ! = j)
	nr = nr * (x–ax [j])
	dr = dr * (ax [i] – ax [j])
Step 9 :	End of j loop.
Step 10 :	y = y + (nr/dr) * ay [i]
Step 11 :	End of i loop
Step 12 :	Print y
Step 13 :	Stop.

FLOW CHART : LAGRANGE'S INTERPOLATION FORMULA

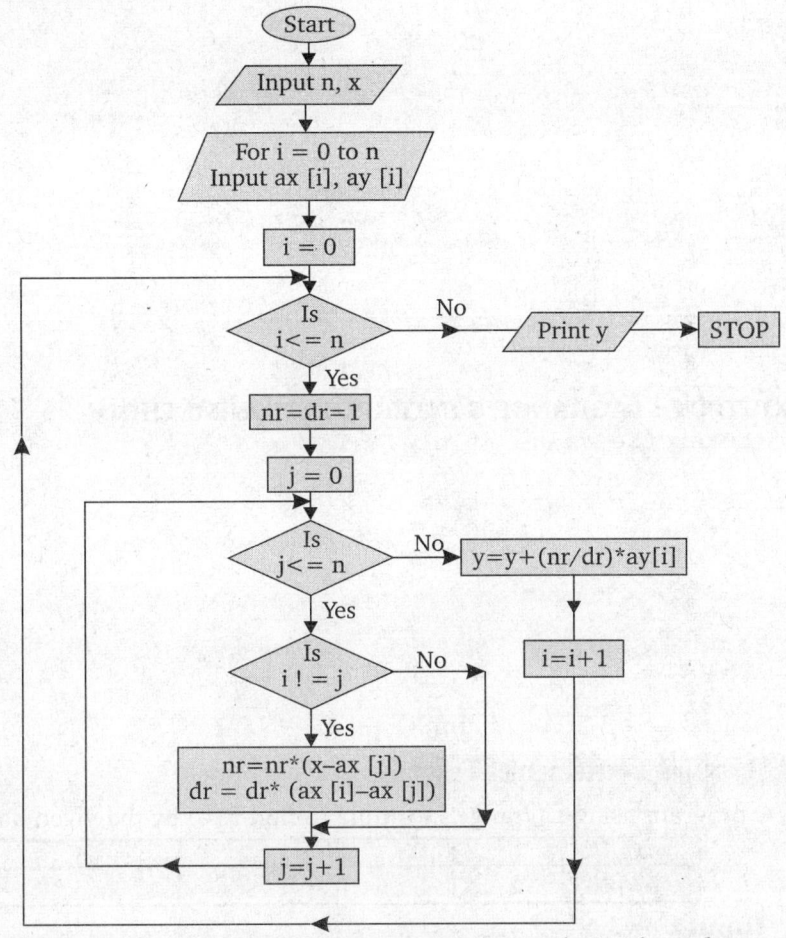

PROGRAM : *Following is a program for the Lagrange's interpolation formula.*

//Lagrange's Interpolation Method

```
#include<<conio .h>
#include<studio .h>
#define MAX 100

void main ( )
{
      float ax [MAX+1], ay [MAX+1], nr, dr, x, y=0;
      int i, j, n ;
      clrscr ( ) ;
      printf ("\n\Enter the value of n\n") ;
      scanf ("%d", &n) ;
      printf ("Enter the set of values of x and y\n") ;
      for (i=0; i< = n =; i++)
              scanf ("%f%f", &ax [i], &ay [i]);
      puts ("\nEnter the value of x for which value of y is
required");
```

```
scanf ("%f", &x);
for (i=0; i<=n;'i++)
{
        nr=dr=1;
        for (j=0; j<= n; j++)
        if (i ! = j)
        {
                nr*=x-ax [j];
                dr*=ax [i] -ax[j] ;
        }
        y=(nr/dr) *ay [i] ;
}
printf ("\n when x=%f,      y=%f", x, y) ;
getch ( ) ;
}
```

OUTPUT : LAGRANGE'S INTERPOLATION METHOD

```
  Enter the value of n
4
Enter the set of values of x and y
1     8
2     15
4     19
8     32
10    40
Enter the value of x for which value of y is required
5
when  x = 5,  y = 22.7460
```

1. Write a C program using Lagrange's formula to find f (4) by the given data

x	0	1	2	5
f(x)	2	5	7	8

Hint :   **Input :** n=3, x=4

         Output : f (4)=8.4

2. Write a C program to find y (5) using Langrange's interpolation formula for the data given below :

x	1	3	4	8	10
f(x)	8	15	19	32	40

Hint :   **Input :** n = 4, x=5

Output :   y (5)= 11.318

7. NEWTON'S DIVIDED DIFFERENCE FORMULA

SYMBOLS USED :

      n = number of subintervals

      x = value of x at which we have to find the value of y

      y = value of y at x

    a [x] = an array to store the different values of x

    a [y] = an array to store the different values of y.

ALGORITHM : NEWTON'S DIVIDED DIFFERENCE FORMULA

Step 1 :	Start
Step 2 :	Input n, x
Step 3 :	For i = 0 to n
Step 4 :	Input ax [i], ay [i]
Step 5 :	y = ay [0], t = 1
Step 6 :	For i = 1 to n
Step 7 :	For j = 0 to (n–i)
Step 8 :	ay [j] = (ay [j + 1] – ay [j])/ax [j + 1] – ax [j])
Step 9 :	End of j loop
Step 10 :	t = t* (x – ax [i – 1])
Step 11 :	y = y + t* ay [0]
Step 12 :	End of i loop
Step 13 :	Print "value of y=", y
Step 14 :	Stop.

FLOW CHART : NEWTON'S DIVIDED DIFFERENCE FORMULA

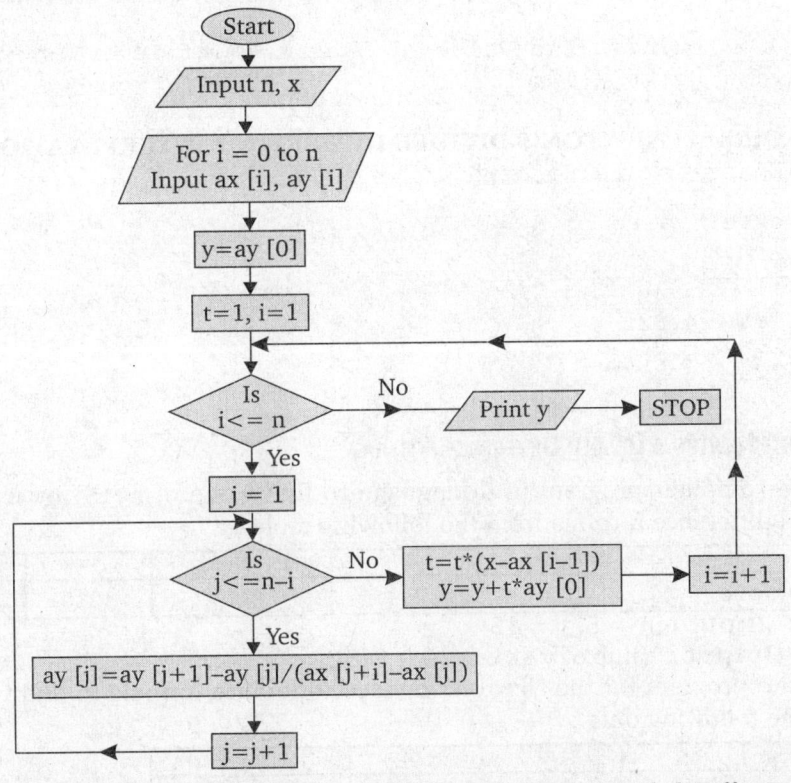

PROGRAM : *Flowing program shows the working of Newton's Divided Difference formula for interpolation.*

//NEWTON'S DIVIDED DIFFERENCE INTERPOLATION FORMULA

```
#include<studio .h>
#include<conio .h>
```

```
void main ( )
    {
    clrscr ( ) ;
    float  ax [30], ay [30], x, y, t=1;
    printf ("Enter the value of n\n") ;
    scanf ("%d", &n) ;
    printf ("\n enter value of x \n") ;
    scanf (" %f ", &x) ;
     printf ("Enter the value of x and y\n ") ;
    for (i=0;  i<=n;i++)
    {
    scanf ("%f  %f", &ax [i],  &ay [i]) ;
    }
    y=ay [0] ;
    for (i=1;  i<=n;  i++)
      {
    for (j=0;j<=n;j++)
            ay [j]=(ay [j+1] – ay [j]/ (ax [j+i] – ax [j]) ;
    t = t* (x–ax [i–1]) ;
    y=y+t*ay [0] ;
    }
    printf ("\n Value of y at x=% .2f is % .2f ", x, y) ;
    getch ( ) ;

    }
```

Output : NEWTON'S DIVIDED DIFFERENCE INTERPOLATION FORMULA

```
Enter the value of n
3
enter value of x
1.6
Enter the value of x and y
1    3.49
1.4  4.82
1.8  5.96
2.2  6.5
Value of y at x=1.60 is 5.44
```

LAB ASSIGNMENT: NEWTON'S DIVIDED DIFFERENCE FORMULA

1. Write a computer program in C language to find value of f (15) by using Newtons divided difference formula from the following table :

x	4	5	7	10	11	13
$f(x)$	48	100	294	900	1210	2028

    Hint : Input : $n = 5, x = 15$

        Output : Value of y at x = 15 is 3150.

2. Write a C program for the Newton's divided difference formula to find value of $f(10)$ from the following data :

x	5	6	9	11
$f(x)$	12	13	14	16

Hint : Input : $n = 3, x = 10$

Output : Value of y at $x = 10 = 14.667.$

⌘⌘⌘⌘⌘⌘

Numerical Differentiation and Integrations

CHAPTER
6

6.1 INTRODUCTION

Let $y = f(x)$ be the given function. The process of evaluating the derivatives of a function $f(x)$ with the help of the given set of values of that functions is known as numerical differentiation. This may be done by first approximating the function by a suitable approximation formula and then differentiating it as many times as desired.

The choice of the formula is the same as discussed for interpolation in chapter 5, *i.e.*, if the derivative at a point near the beginning of a set of values given by a table is required then we use Newton forward formula, and if the same is required at a point near the end of the set of given tabular values, then we use Newton's backward interpolation formula. The central difference formula (Bessel's and Stirling's) used to calculate the value for points near the middle of the set of given tabular values. If the values of x are not equally spaced, we use Newton's divided difference interpolation formula or Lagrange's interpolation formula to get the required value of the derivative.

▼ REMARK
▸ While analytic methods give exact answers, the numerical techniques provide only approximations to derivatives. Numerical differentiation methods are very sensitive to roundoff errors, in addition to the truncation error introduced by the methods themselves.

6.2 DERIVATIVE USING NEWTON'S FORWARD INTERPOLATION FORMULA

Newton's forward interpolation formula is given by

$$y = y_0 + u\Delta y_0 + \frac{u(u-1)}{2!}\Delta^2 y_0 + \frac{u(u-1)(u-2)}{3!}\Delta^3 y_0 + \dots \qquad \dots(1)$$

where
$$u = \frac{x - x_0}{h}$$

Differentiating equation (1) with respect to u, we get

$$\frac{dy}{du} = \Delta y_0 + \frac{2u-1}{2!}\Delta^2 y_0 + \frac{3u^2 - 6u + 2}{3!}\Delta^3 y_0 + \dots \qquad \dots(2)$$

Now
$$\frac{dy}{dx} = \frac{dy}{du} \cdot \frac{du}{dx} = \frac{1}{h} \cdot \frac{dy}{du}$$

Therefore

$$\frac{dy}{dx} = \frac{1}{h}\left[\Delta y_0 + \frac{2u-1}{2!}\Delta^2 y_0 + \frac{3u^2 - 6u + 2}{3!}\Delta^3 y_0 + \frac{4u^3 - 18u^2 + 22u - 6}{4!}\Delta^4 y_0 + \dots\right] \dots(3)$$

As $x = x_0$, $u = 0$, therefore, putting $u = 0$ in (3), we get

$$\left[\frac{dy}{dx}\right]_{x=x_0} = \frac{1}{h}\left[\Delta y_0 - \frac{1}{2}\Delta^2 y_0 + \frac{1}{3}\Delta^3 y_0 - \frac{1}{4}\Delta^4 y_0 + ...\right] \qquad ...(4)$$

Differentiating equation (3) again w.r.t. x, we get

$$\frac{d^2 y}{dx^2} = \frac{d}{du}\left(\frac{dy}{dx}\right)\frac{du}{dx} = \frac{1}{h}\times\frac{d}{du}\left(\frac{dy}{dx}\right)$$

$$= \frac{1}{h^2}\left[\Delta^2 y_0 + (u-1)\Delta^3 y_0 + \frac{6u^2 - 18u + 11}{12}\Delta^4 y_0 + ...\right] \qquad ...(5)$$

Putting $u = 0$ in (5), we get

$$\left[\frac{d^2 y}{dx^2}\right]_{x=x_0} = \frac{1}{h^2}\left[\Delta^2 y_0 - \Delta^3 y_0 + \frac{11}{12}\Delta^4 y_0 - ...\right] \qquad ...(6)$$

Similarly $\left[\dfrac{d^3 y}{dx^3}\right]_{x=x_0} = \dfrac{1}{h^3}\left[\Delta^3 y_0 - \dfrac{3}{2}\Delta^4 y_0 - ...\right]$ and so on

Aliter : We know that   [UPTU(MCA)–2005, 07]

$$1 + \Delta = E = e^{hD}, D = \frac{d}{dx}$$

$$\therefore \qquad hD = \log(1+\Delta) = \Delta - \frac{\Delta^2}{2} + \frac{\Delta^3}{3} - \frac{\Delta^4}{4} + \frac{\Delta^5}{5} - \frac{\Delta^6}{6} + ...$$

or

$$D = \frac{1}{h}\left[\Delta - \frac{\Delta^2}{2} + \frac{\Delta^3}{3} - \frac{\Delta^4}{4} + \frac{\Delta^5}{5} - \frac{\Delta^6}{6} + ...\right] \qquad ...(1)$$

$$\Rightarrow \qquad D^2 = \frac{1}{h^2}\left[\Delta - \frac{\Delta^2}{2} + \frac{\Delta^3}{3} - \frac{\Delta^4}{4} + \frac{\Delta^5}{5} - \frac{\Delta^6}{6} + ...\right]^2$$

$$= \frac{1}{h^2}\left[\Delta^2 - \Delta^3 + \frac{11}{12}\Delta^4 + ...\right] \qquad ...(2)$$

also $\qquad D^3 = \dfrac{1}{h^3}\left[\Delta - \dfrac{\Delta^2}{2} + \dfrac{\Delta^3}{3} - \dfrac{\Delta^4}{4} + ...\right]^3 = \dfrac{1}{h^3}\left[\Delta^3 - \dfrac{3}{2}\Delta^4 + ...\right] \qquad ...(3)$

Applying equations (1), (2) and (3) for y_0, we get

$$\left[\frac{dy}{dx}\right]_{x=x_0} = \frac{1}{h}\left[\Delta y_0 - \frac{1}{2}\Delta^2 y_0 + \frac{1}{3}\Delta^3 y_0 - \frac{1}{4}\Delta^4 y_0 + ...\right] \qquad ...(4)$$

$$\left[\frac{d^2 y}{dx^2}\right]_{x=x_0} = \frac{1}{h^2}\left[\Delta^2 y_0 - \Delta^3 y_0 - \frac{11}{12}\Delta^4 y_0 - ...\right] \qquad ...(5)$$

and $\qquad \left[\dfrac{d^3 y}{dx^3}\right]_{x=x_0} = \dfrac{1}{h^3}\left[\Delta^3 y_0 - \dfrac{3}{2}\Delta^4 y_0 + ...\right] \qquad ...(6)$

6.3 DERIVATIVES USING NEWTON'S BACKWARD DIFFERENCE FORMULA

Newton's backward interpolation formula is given by

$$y = y_n + u\nabla y_n + \frac{u(u+1)}{2!}\nabla^2 y_n + \frac{u(u+1)(u+2)}{3!}\nabla^3 y_n + \ldots; \quad \text{where} \quad u = \frac{x - x_n}{n} \quad \ldots(1)$$

Differentiating both sides of equation (1) w.r.t. x, we get

$$\frac{dy}{dx} = \frac{1}{h}\left[\nabla y_n + \frac{2u+1}{2!}\nabla^2 y_n + \frac{3u^2 + 6u + 2}{3!}\nabla^3 y_n + \ldots\right] \quad \ldots(2)$$

At $x = x_n$, $u = 0$. Therefore putting $u = 0$ in (2), we get

$$\left[\frac{dy}{dx}\right]_{x=x_n} = \frac{1}{h}\left[\nabla y_n + \frac{1}{2}\nabla^2 y_n + \frac{1}{3}\nabla^3 y_n + \ldots\right] \quad \ldots(3)$$

Again differentiating both sides of equation (2) w.r.t. x, we get

$$\frac{d^2 y}{dx^2} = \frac{1}{h^2}\left[\nabla^2 y_n + \frac{6u+6}{6}\nabla^3 y_n + \frac{12u^2 + 36u + 22}{24}\nabla^4 y_n + \ldots\right]$$

At $x = x_n$, $u = 0$. Therefore putting $u = 0$ in (3), we get

$$\left[\frac{d^2 y}{dx^2}\right]_{x=x_n} = \frac{1}{h^2}\left[\nabla^2 y_n + \nabla^3 y_n + \frac{11}{12}\nabla^4 y_n + \ldots\right] \quad \ldots(4)$$

Similarly, $$\left[\frac{d^3 y}{dx^3}\right]_{x=x_n} = \frac{1}{h^3}\left[\nabla^3 y_n + \frac{3}{2}\nabla^4 y_n + \ldots\right] \text{and so on.} \quad \ldots(5)$$

Aliter : We know that

$$1 - \nabla = E^{-1} = e^{-hD}$$

$$\therefore \quad -hD = \log(1 - \nabla) = -\left[\nabla + \frac{1}{2}\nabla^2 + \frac{1}{3}\nabla^3 + \frac{1}{4}\nabla^4 + \frac{1}{5}\nabla^5 + \ldots\right]$$

or $$D = \frac{1}{h}\left[\nabla + \frac{1}{2}\nabla^2 + \frac{1}{3}\nabla^3 + \frac{1}{4}\nabla^4 + \frac{1}{5}\nabla^5 + \ldots\right] \quad \ldots(6)$$

$$D^2 = \frac{1}{h^2}\left[\nabla^2 + \nabla^3 + \frac{11}{12}\nabla^4 + \ldots\right] \quad \ldots(7)$$

and $$D^3 = \frac{1}{h^3}\left[\nabla^3 + \frac{3}{2}\nabla^4 + \ldots\right] \quad \ldots(8)$$

Applying these identities to y_n, we get

$$\left[\frac{dy}{dx}\right]_{x=x_n} = \frac{1}{h}\left[\nabla y_n + \frac{1}{2}\nabla^2 y_n + \frac{1}{3}\nabla^3 y_n + \frac{1}{4}\nabla^4 y_n + \ldots\right] \quad \ldots(9)$$

$$\left[\frac{d^2 y}{dx^2}\right]_{x=x_n} = \frac{1}{h^2}\left[\nabla^2 y_n + \nabla^3 y_n + \frac{11}{12}\nabla^4 y_n + \ldots\right] \quad \ldots(10)$$

and $$\left[\frac{d^3 y}{dx^3}\right]_{x=x_n} = \frac{1}{h^3}\left[\nabla^3 y_n + \frac{3}{2}\nabla^4 y_n + \ldots\right] \quad \ldots(11)$$

6.4 DERIVATIVES USING STIRLING'S FORMULA

Here we have to determine the values of the derivatives of the function near the middle of the given set of arguments. We may apply any central difference formula. Therefore using Stirling's formula, we get

$$y_n = y_0 + u\left(\frac{\Delta y_0 + \Delta y_{-1}}{2}\right) + \frac{u^2}{2!}\Delta^2 y_{-1} + \frac{u(u^2-1)}{3!}\left(\frac{\Delta^3 y_{-1} + \Delta^3 y_{-2}}{2}\right)$$

$$+ \frac{u^2(u^2-1)}{4!}\Delta^4 y_{-2} + \dots \quad \dots(1)$$

where $u = \dfrac{x - x_0}{h}$

Now differentiating w.r.t. x, we get

$$\frac{dy}{dx} = \left[\left(\frac{\Delta y_0 + \Delta y_{-1}}{2}\right) + u\Delta^2 y_{-1} + \frac{(3u^2-1)}{3!}\left(\frac{\Delta^3 y_{-1} + \Delta^3 y_{-2}}{2}\right) + \frac{(4u^3-2u)}{4!}\Delta^4 y_{-2} + \dots\right]\frac{du}{dx}$$

since $u = \dfrac{x-x_0}{h}$ $\Rightarrow$ $\dfrac{du}{dx} = \dfrac{1}{h}$

$$\therefore \quad \frac{dy}{dx} = \frac{1}{h}\left[\left(\frac{\Delta y_0 + \Delta y_{-1}}{2}\right) + u\Delta^2 y_{-1} + \frac{3u^2-1}{3!}\left(\frac{\Delta^3 y_{-1} + \Delta^3 y_{-2}}{2}\right) + \frac{(4u^3-2u)}{4!}\Delta^4 y_{-2} + \dots\right]$$

$$\dots(2)$$

At $x = x_0$, $u = 0$, therefore, putting $u = 0$ in (2), we get

$$\left[\frac{dy}{dx}\right]_{x=x_0} = \frac{1}{h}\left[\frac{\Delta y_0 + \Delta y_{-1}}{2} - \frac{1}{6}\left(\frac{\Delta^3 y_{-1} + \Delta^3 y_{-2}}{2}\right) + \dots\right]$$

Again differentiating, we get

$$\frac{d^2 y}{dx^2} = \frac{1}{h^2}\left[\Delta^2 y_{-1} + \frac{6u}{6}\left(\frac{\Delta^3 y_{-1} + \Delta^3 y_{-2}}{2}\right) + \frac{12u^2-2}{4!}\Delta^4 y_{-2} + \dots\right] \dots(3)$$

At $x = x_0$, $u = 0$, therefore, putting $u = 0$ in (3), we get

$$\therefore \quad \left[\frac{d^2 y}{dx^2}\right]_{x=x_0} = \frac{1}{h^2}\left[\Delta^2 y_{-1} - \frac{1}{12}\Delta^4 y_{-2} + \dots\right] \text{ and so on.}$$

6.5 DERIVATIVE USING NEWTON'S DIVIDED DIFFERENCE FORMULA

In this case we apply Newton's divided difference formula for finding the successive differentiation at given value of x. Let us consider a function $f(x)$ of degree n, then

$$y = f(x) = f(x_0) + (x - x_0)\Delta f(x_0) + (x - x_0)(x - x_1)\Delta^2 f(x_0)$$

$$\Rightarrow \quad (x - x_0)(x - x_1)(x - x_2)\Delta^3 f(x_0) + (x - x_0)(x - x_1)\dots(x - x_{n-1})\Delta^n f(x_0)$$

Differentiate this equation w.r.t. 'x' as many times as we require and put $x = x_i$, we get the required derivatives.

6.6 MAXIMA AND MINIMA OF TABULATED FUNCTION

If we differentiate the Newton's forward interpolation formula with respect to x, then we get

$$\frac{dy}{dx} = \frac{1}{h}\left[\Delta y_0 + \frac{2u-1}{2}\Delta^2 y_0 + \frac{3u^2 - 6u + 2}{3!}\Delta^3 y_0 + \dots\right] \qquad \dots(1)$$

We know that the maximum and minimum values of a function $y = f(x)$ can be found by equating $\frac{dy}{dx}$ to zero and solving it for x.

Put $\frac{dy}{dx} = 0$, then equation (1) gives

$$\Delta y_0 + \frac{2u-1}{2}\Delta^2 y_0 + \frac{3u^2 - 6u + 2}{6}\Delta^2 y_0 + \dots = 0$$

Solving this for u, and substituting $\Delta y_0, \Delta^2 y_0, \Delta^3 y_0$ (using difference table), we get x as $x_0 + ah$ at which y is maximum or minimum.

Recapitulations |||||||||||||||||||||||||||||||||||||||||||||||||||||||||||||||

- $|A - \lambda I| = 0$ is known as characteristic equation and the roots of this equation are called characteristic roots or eigen values.

- To find the value of the derivative near the end of the data, use Newton's backward interpolation formula.

- To find the values of the derivatives of the function near the middle of the given set of arguments, use Stirling's formula.

- In case of unequal interval, use the derivatives of Newton's divided difference formula.

SOLVED EXAMPLES

EXAMPLE 1. *Using following table :*

x	1.0	1.1	1.2	1.3	1.4	1.5	1.6
y	7.989	8.403	8.781	9.129	9.451	9.750	10.031

Find $\frac{dy}{dx}$ and $\frac{d^2 y}{dx^2}$ at $x = 1.1$. (Delhi- 2005, 07, 10; Rohtak-2006; VTU–2006, Madras–2003S, MDU(B.E.)–2007)

SOLUTION. Since the values are at equidistant and we want to find the value of y at $x = 1.1$. Therefore, we apply Newton's forward difference formula.

Difference table

x	y	Δy	$\Delta^2 y$	$\Delta^3 y$	$\Delta^4 y$	$\Delta^5 y$	$\Delta^6 y$
1.0	7.989						
		0.414					
1.1	**8.403**		−0.036				
		0.378		0.006			
1.2	8.781		**−0.030**		−0.002		
		0.348		**0.004**		+0.001	
1.3	9.129		−0.026		**−0.001**		0.002
		0.322		0.003		**−0.003**	
1.4	9.451		−0.023		0.002		
		0.299		0.005			
1.5	9.750		−0.018				
		0.281					
1.6	10.031						

We have

$$\left(\frac{dy}{dx}\right)_{x=x_0} = \frac{1}{h}\left[\Delta y_0 - \frac{1}{2}\Delta^2 y_0 + \frac{1}{3}\Delta^3 y_0 - \frac{1}{4}\Delta^4 y_0 + \frac{1}{5}\Delta^5 y_0 - \frac{1}{6}\Delta^6 y_0 + ...\right]$$

Putting $x_0 = 1.1, \Delta y_0 = 0.378, \Delta^2 y_0 = 0.030$ and so on and $h = 0.1$, we get

$$\left(\frac{dy}{dx}\right)_{1.1} = \frac{1}{0.1}\left[0.378 - \frac{1}{2}(-0.030) + \frac{1}{3}(0.004) - \frac{1}{4}(-0.001) + \frac{1}{5}(-0.001)\right]$$

$$= 10[0.378 + 0.015 + 0.0013 + 0.00025 - 0.0002]$$

$$= 10[0.39435] = 3.9435$$

and

$$\left(\frac{d^2 y}{dx^2}\right)_{x_0} = \frac{1}{h^2}\left[\Delta^2 y_0 - \Delta^3 y_0 + \frac{11}{12}\Delta^4 y_0 - \frac{5}{6}\Delta^5 y_0 + \frac{137}{180}\Delta^6 y_0 + ...\right]$$

$$\Rightarrow \left(\frac{d^2 y}{dx^2}\right)_{1.1} = \frac{1}{(0.1)^2}\left[-0.030 - 0.004 - \frac{11}{12}\times 0.001 + \frac{5}{6}\times 0.001\right]$$

$$= 100[-0.030 - 0.004 - 0.0009 + 0.0008]$$

$$= 100(-0.0341) = -0.341 \ .$$

EXAMPLE 2.   *Using the given table, find* $\dfrac{dy}{dx}$ *at* $x = 1.2$.   (UPTU-2003)

x	1.0	1.2	1.4	1.6	1.8	2.0	2.2
y	2.7183	3.3201	4.0552	4.9530	6.0496	7.3891	9.0250

SOLUTION.

Difference table

x	y	Δy	$\Delta^2 y$	$\Delta^3 y$	$\Delta^4 y$	$\Delta^5 y$	$\Delta^6 y$
1.0	2.7183						
		0.6018					
1.2	**3.3201**		0.1333				
		0.7351		0.0294			
1.4	4.0552		**0.1627**		0.0067		
		0.8978		**0.0361**		0.0013	
1.6	4.9530		0.1988		**0.0080**		0.0001
		1.0966		0.0441		**0.0014**	
1.8	6.0496		0.2429		0.0094		
		1.3395		0.0535			
2.0	7.3891		0.2964				
		1.6359					
2.2	9.0250						

We have

$$\left(\frac{dy}{dx}\right)_{x=x_0} = \frac{1}{h}\left[\Delta y_0 - \frac{1}{2}\Delta^2 y_0 + \frac{1}{3}\Delta^3 y_0 - \frac{1}{4}\Delta^4 y_0 + \frac{1}{5}\Delta^5 y_0 - \frac{1}{6}\Delta^6 y_0 + ...\right]$$

Here $x_0 = 1.2, h = 0.2, \Delta y_0 = 0.7351, \Delta^2 y_0 = 0.1627$ etc.,
we get

$$\left(\frac{dy}{dx}\right)_{1.2} = \frac{1}{0.2}\left[0.7351 - \frac{1}{2}(0.1627) + \frac{1}{3}(0.0361) - \frac{1}{4}(0.0080) + \frac{1}{5}(0.0014)\right]$$

$$= 5[0.7351 - 0.08135 + 0.0120 - 0.002 + 0.00028]$$

$$= 3.32015.$$

EXAMPLE 3. *Find $f'(1.5)$ and $f''(1.5)$ from the following table :*

x	1.5	2.0	2.5	3.0	3.5	4.0
f(x)	3.375	7.000	13.625	24.000	38.875	59.000

(MDU(B.E.)–2005, 06, SVTU–2007; Raipur-2009)

SOLUTION. **Difference table**

x	f(x)	$\Delta f(x)$	$\Delta^2 f(x)$	$\Delta^3 f(x)$	$\Delta^4 f(x)$
1.5	**3.375**				
		3.625			
2.0	7.000		**3.000**		
		6.625		**0.75**	
2.5	13.625		3.75		**0**
		10.375		0.75	
3.0	24.000		4.5		0
		14.875		0.75	
3.5	38.875		5.25		
		20.125			
4.0	59.000				

Here $x_0 = 1.5$, $h = 0.5$

Now, $f'(1.5) = \dfrac{1}{0.5}\left[\Delta y_0 - \dfrac{1}{2}\Delta^2 y_0 + \dfrac{1}{3}\Delta^3 y_0 - \dfrac{1}{4}\Delta^4 y_0 + ...\right]$

$\qquad = \dfrac{1}{0.5}\left[3.625 - \dfrac{1}{2}(3.000) + \dfrac{1}{3}(0.75)\right]$

$\qquad = 2[3.625 - 1.5 + 0.25] = 2[2.375]$

$f'(1.5) = 4.75$

and $\qquad f''(1.5) = \dfrac{1}{(0.5)^2}\left[\Delta^2 y_0 - \Delta^3 y_0 + \dfrac{11}{12}\Delta^4 y_0\right]$

$\qquad = 4[3.000 - 0.75] = 4[2.25]$

$\therefore \qquad f''(1.5) = 9$.

EXAMPLE 4. *Find $\dfrac{dy}{dx}$ and $\dfrac{d^2 y}{dx^2}$ of $y = x^{1/3}$ at $x = 50$ from the following table :*

x	50	51	52	53	54	55	56
$y = x^{1/3}$	3.6840	3.7084	3.7325	3.7563	3.7798	3.8030	3.8259

SOLUTION. **Difference table**

x	$y = x^{1/3}$	Δy	$\Delta^2 y$	$\Delta^3 y$
50	**3.6840**			
		0.0244		
51	3.7084		**–0.0003**	
		0.0241		**0**
52	3.7325		–0.0003	
		0.0238		0
53	3.7563		–0.0003	
		0.0235		0
54	3.7798		–0.0003	
		0.0232		0
55	3.8030		–0.0003	
		0.0229		
56	3.8259			

Here $x_0 = 50$, $h = 1$. Then

$$\left(\frac{dy}{dx}\right)_{x=x_0} = \frac{1}{h}\left[\Delta y_0 - \frac{1}{2}\Delta^2 y_0 + \frac{1}{3}\Delta^3 y_0 + \ldots\right]$$

$$\left(\frac{dy}{dx}\right)_{50} = \frac{1}{1}\left[0.0244 - \frac{1}{2}(-0.0003) + \frac{1}{3}(0)\right] = [0.0244 + 0.00015] = 0.02455$$

$$\left(\frac{d^2 y}{dx^2}\right)_{x=x_0} = \frac{1}{h^2}[\Delta^2 y_0 - \Delta^3 y_0 + \ldots] = \frac{1}{(1)^2}[(-0.0003)]$$

$$\left(\frac{d^2 y}{dx^2}\right)_{50} = -0.0003 .$$

EXAMPLE 5.   *Find* $f'(1)$ *for* $f(x) = \dfrac{1}{1+x^2}$ *using the following table :*

x	1.0	1.1	1.2	1.3	1.4
f(x)	0.5000	0.4524	0.4098	0.3717	0.3378

SOLUTION.

Difference table

x	f(x)	$\Delta f(x)$	$\Delta^2 f(x)$	$\Delta^3 f(x)$	$\Delta^4 f(x)$
1.0	**0.5000**				
		−0.0476			
1.1	0.4524		0.0050		
		−0.0426		−0.0005	
1.2	0.4098		0.0045		0.0002
		−0.0381		−0.0003	
1.3	0.3717		0.0042		
		−0.0339			
1.4	0.3378				

Here $x_0 = 1.0$, $h = 0.1$. Therefore

$$\left(\frac{dy}{dx}\right)_{x=x_0} = \frac{1}{h}\left[\Delta y_0 - \frac{1}{2}\Delta^2 y_0 + \frac{1}{3}\Delta^3 y_0 - \frac{1}{4}\Delta^4 y_0 + \ldots\right]$$

$$\Rightarrow \quad f'(1) = \frac{1}{0.1}\left[-0.0476 - \frac{1}{2}(0.050) + \frac{1}{3}(-0.0005) - \frac{1}{4}(0.0002)\right]$$

$$= 10[-0.0476 - 0.0025 - 0.00016 - 0.00005] = -0.5031$$

EXAMPLE 6.   *Find* $f'(1.1)$ *and* $f''(1.1)$ *from the following table :*

x	1.0	1.2	1.4	1.6	1.8	2.0
f(x)	0	0.1280	0.5440	1.2960	2.4320	4.0000

(Indore-2010; UPTU–2004, 10, Bhopal–2007, 09)

SOLUTION.

Difference table

x	f(x)	$\Delta f(x)$	$\Delta^2 f(x)$	$\Delta^3 f(x)$	$\Delta^4 f(x)$
1.0	**0**				
		0.1280			
1.2	0.1280		**0.288**		
		0.4160		**0.048**	
1.4	0.5440		0.336		**0**
		0.7520		0.048	
1.6	1.2960		0.384		0
		1.1360		0.048	
1.8	2.4320		0.432		
		1.5680			
2.0	4.0000				

Here, we have to find the derivatives at $x = 1.1$ which lies between given arguments 1.0 and 1.2. So apply Newton's forward formula,

$$u = \frac{x - x_0}{h} = \frac{x - 1}{0.2} = 5(x - 1) \qquad \ldots(1)$$

$$f(x) = f(x_0) + u\Delta f(x_0) + \frac{u(u-1)}{2!}\Delta^2 f(x_0) + \frac{u(u-1)(u-2)}{3!}\Delta^3 f(x_0) + \ldots$$

$$\qquad \ldots(2)$$

Differentiating w.r.t. x, we get

$$f'(x) = \left[\Delta f(x_0) + \frac{(2u-1)}{2!}\Delta^2 f(x_0) + \frac{3u^2 - 6u + 2}{3!}\Delta^3 f(x_0)\right]\frac{du}{dx}$$

Also, from (1) $\dfrac{du}{dx} = 5$

$$\therefore \qquad f'(x) = 5\left[\Delta f(x_0) + \frac{(2u-1)}{2}\Delta^2 f(x_0) + \frac{3u^2 - 6u + 2}{3!}\Delta^3 f(x_0)\right]$$

At $x = 1.1$, $\quad u = 5(1.1 - 1) = 0.5$

$$f'(1.1) = 5\left[0.128 + \frac{2(0.5) - 1}{2}(0.288) + \frac{3(0.5)^2 - 6(5.0) + 2}{6}(0.048)\right]$$

$$= 5[0.128 + 0 - 0.002] \Rightarrow \quad f'(1.1) = 0.63$$

Differentiating equation (2) again w.r.t. 'x', we get

$$f''(x) = 5\left[\Delta^2 f(x_0) + \frac{6u - 6}{6}\Delta^3 f(x_0)\right]\frac{du}{dx}$$

$$\therefore \qquad f''(1.1) = 25[0.288 + (0.5 - 1) \times 0.048] = 25[0.288 - 0.024]$$

Hence, $\quad f''(1.1) = 6.6$

EXAMPLE 7. *Find $f'(1.5)$ from the following table :*

x	0.0	0.5	1.0	1.5	2.0
$f(x)$	0.3989	0.3521	0.2420	0.1295	0.0540

SOLUTION. Here, we want to find the derivatives of $f(x)$ at $x = 1.5$, which is near the end of the arguments. Therefore, apply Newton's backward formula.

Difference table

x	$f(x)$	$\Delta f(x)$	$\Delta^2 f(x)$	$\Delta^3 f(x)$	$\Delta^4 f(x)$
0.0	0.3989				
		−0.468			
0.5	03521		−0.0633		
		−0.1101		**0.0609**	
1.0	0.2420		**−0.0024**		−0.0215
		−0.1125		0.0394	
1.5	**0.1295**		0.037		
		−0.0755			
2.0	0.0540				

Here $x_n = 1.5$, $h = 0.5$

Therefore,

$$\left(\frac{dy}{dx}\right)_{x=x_n} = \frac{1}{h}\left[\nabla y_n + \frac{1}{2}\nabla^2 y_n + \frac{1}{3}\nabla^3 y_n + \frac{1}{4}\nabla^4 y_n + \ldots\right]$$

$$\Rightarrow \quad f'(1.5) = \frac{1}{0.5}\left[-0.1125 + \frac{1}{2}(-0.0024) + \frac{1}{3}(0.0609)\right]$$

$$= 2[-0.1125 - 0.0012 + 0.015225] = 2[-0.098475]$$
$$\Rightarrow \quad f'(1.5) = -0.19695$$

EXAMPLE 8. *From the following table, find the values of* $\dfrac{dy}{dx}$ *and* $\dfrac{d^2y}{dx^2}$ *at* $x = 2.03$.

x	1.96	1.98	2.00	2.02	2.04
y	0.7825	0.7739	0.7651	0.7563	0.7473

(Anna–2005)

SOLUTION.

Difference table

x	y	Δy	Δ²y	Δ³y	Δ⁴y
1.96	0.7825				
		−0.0086			
1.98	0.7739		−0.0002		
		−0.0088		0.0002	
2.00	0.7651		−0		**−0.0004**
		−0.0088		**−0.0002**	
2.02	0.7563		**−0.0002**		
		−0.0090			
2.04	0.7473				

Here $\quad x_n = 2.04, h = 0.02, u = \dfrac{x - x_n}{h}$

$$\Rightarrow \quad \frac{du}{dx} = \frac{1}{h} \text{ at } x = 2.03$$

$$\therefore \quad u = \frac{2.03 - 2.04}{0.02} = -\frac{0.01}{0.02} = -\frac{1}{2}$$

Then by Newton's backward formula, we have

$$y(x) = y_n + u\nabla y_n + \frac{u(u+1)}{2}\nabla^2 y_n + \frac{u(u+1)(u+2)}{6}\nabla^3 y_n$$

$$+ \frac{u(u+1)(u+2)(u+3)}{24}\nabla^3 y_n + \ldots \qquad \ldots(1)$$

Differentiating w.r.t. x, we get

$$y'(x) = \frac{1}{h}\left[\nabla y_n + \frac{2u+1}{2}\nabla^2 y_n + \frac{3u^2 + 6u + 2}{6}\nabla^3 y_n\right.$$

$$\left. + \frac{4u^3 + 18u^2 + 22u + 6}{24}\nabla^4 y_n + \ldots\right] \qquad \ldots(2)$$

$$y'(2.03) = \frac{1}{0.02}\left[-0.0090 + 0 + \frac{3(-1/2)^2 + 6(-1/2) + 2}{6}(-0.0002)\right.$$

$$\left. + \frac{4(-1/2)^3 + 18(-1/2)^2 + 22(-1/2) + 6}{24}(-0.0004)\right]$$

$$= 50[-0.0090 + 0.000008 + 0.000017] = -0.44875$$

Again differentiating equation (2) w.r.t. x,

$$y''(x) = \frac{1}{h^2}\left[\nabla^2 y_n + \frac{6u+6}{6}\nabla^3 y_n + \frac{12u^2+36u+22}{24}\nabla^4 y_n + \ldots\right]$$

$$y''(2.03) = \frac{1}{(0.02)^2}\left[-0.0002 + \left(-\frac{1}{2}+1\right)(-0.0002)\right.$$

$$\left. + \frac{12\left(-\frac{1}{2}\right)^2 + 36\left(-\frac{1}{2}\right)+22}{24}(-0.0004)\right]$$

$$= 2500[-0.0002 - 0.0001 - 0.00012]$$

Hence, $y''(2.03) = -1.05$.

EXAMPLE 9. *Find $f'(2.5)$ from the following table :* (Allahabad-2008)

x	1.5	1.9	2.5	3.2	4.3	5.9
$f(x)$	3.375	6.059	13.625	29.368	73.907	196.579

SOLUTION. Here, arguments are not equally spaced. Therefore, apply Newton's divided difference formula.

Difference table

x	$f(x)$	$\Delta f(x)$	$\Delta^2 f(x)$	$\Delta^3 f(x)$
1.5	3.375			
		6.71		
1.9	6.059		**5.90**	
		12.61		**1**
2.5	13.625		7.60	
		22.49		1
3.2	29.368		10.00	
		40.49		1
4.3	73.907		13.40	
		76.67		
5.9	196.579			

We know that, Newton's divided difference formula is given by

$$f(x) = f(x_0) + (x-x_0)\Delta f(x_0) + (x-x_0)(x-x_1)\Delta^2 f(x_0)$$

$$+ (x-x_0)(x-x_1)(x-x_2)\Delta^3 f(x_0) + \ldots$$

Differentiating w.r.t. x, we get

$$f'(x) = \Delta f(x_0) + [(x-x_1)+(x-x_0)]\Delta^2 f(x_0) + [(x-x_1)(x-x_2)$$

$$+ (x-x_0)(x-x_1) + (x-x_0)(x-x_2)]\Delta^3 f(x_0)$$

At $x = 2.5$,

$$f'(2.5) = 6.71 + (2 \times 2.5 - 1.5 - 1.9)5.90 + [(2.5-1.9)(2.5-2.5)$$

$$+ (2.5-1.5)(2.5-1.9) + (2.5-1.5)(2.5-2.5)] \cdot 1$$

$$= 6.71 + 9.44 + 0.6$$

Hence, $f'(2.5) = 16.75$.

EXAMPLE 10. *Find $f'(5)$ from the following table :*

x	1	2	4	8	10
f(x)	0	1	5	21	27

(Meerut-2005, Allahabad-2008; JNTU–2009)

SOLUTION. Here, the arguments are not equally spaced. So we use Newton's divided difference formula.

Difference table

x	f(x)	$\Delta f(x)$	$\Delta^2 f(x)$	$\Delta^3 f(x)$	$\Delta^4 f(x)$
1	0				
		1			
2	1		1/3		
		2		0	
4	5		1/3		–1/144
		4		–1/16	
8	21		–1/6		
		3			
10	27				

Newton's divided difference formula is given by

$$f(x) = f(x_0) + (x - x_0)\, \Delta f(x_0) + (x - x_0)(x - x_1)\, \Delta^2 f(x_0)$$

$$+ (x - x_0)(x - x_1)(x - x_2)\, \Delta^3 f(x_0)$$

$$+ (x - x_0)(x - x_1)(x - x_2)(x - x_3)\, \Delta^4 f(x_0) \quad \ldots(1)$$

Differentiating (1) w.r.t. x, we get

$$f'(x) = \Delta f(x_0) + (2x - x_0 - x_1)\, \Delta^2 f(x_0) + [(x - x_1)(x - x_2) + (x - x_0)(x - x_2)$$

$$+ (x - x_0)(x - x_1)]\, \Delta^3 f(x_0) + [(x - x_1)(x - x_2)(x - x_3)$$

$$+ (x - x_0)(x - x_2)(x - x_3) + (x - x_0)(x - x_1)(x - x_3)$$

$$+ (x - x_0)(x - x_1)(x - x_2)]\, \Delta^4 f(x_0)$$

At $x = 5$

$$f'(5) = 1 + (10 - 1 - 2)\frac{1}{3} + [(5 - 2)(5 - 4) + (5 - 1)(5 - 4) + (5 - 1)(5 - 2)] \times 0$$

$$+ [(5 - 2)(5 - 4)(5 - 8) + (5 - 1)(5 - 4)(5 - 8) + (5 - 1)(5 - 2)(5 - 8)$$

$$+ (5 - 1)(5 - 2)(5 - 4)]\left(\frac{-1}{144}\right)$$

$$= 1 + 7\left(\frac{1}{3}\right) - \frac{1}{144}[-9 - 12 - 36 + 12] = 1 + \frac{7}{3} + \frac{45}{144}$$

Hence, $f'(5) = 3.6458$.

EXAMPLE 11. *Find $f'(0.6)$ and $f''(0.6)$ from the following table :*

x	0.4	0.5	0.6	0.7	0.8
f(x)	1.5836	1.7974	2.0442	2.3275	2.6510

(Rajasthan-2004, 06)

SOLUTION. Here, the derivatives are required at the central point $x = 0.6$, so we use Stirling's formula.

Difference table

u	x	f(x)	$\Delta f(x)$	$\Delta^2 f(x)$	$\Delta^3 f(x)$	$\Delta^4 f(x)$
–2	0.4	1.5836				
			0.2138			
–1	0.5	1.7974		0.0330		
0	0.6	(2.0442)	$\begin{pmatrix}0.2468\\0.2833\end{pmatrix}$	(0.0365)	$\begin{pmatrix}0.0035\\0.0037\end{pmatrix}$	(0.0002)
1	0.7	2.3275		0.0402		
			0.3235			
2	0.8	2.6510				

Here, we have $x_0 = 0.6, h = 0.1, u = \dfrac{x - x_0}{h}$

at $x = 0.6, \quad u = 0$

∴ Stirling's formula is

$$f(x) = y = y_0 + u\left(\frac{\Delta y_0 + \Delta y_{-1}}{2}\right) + \frac{u^2}{2}\Delta^2 y_{-1}$$

$$+ \frac{u^3 - u}{6}\left(\frac{\Delta^3 y_{-1} + \Delta^3 y_{-2}}{2}\right) + \frac{u^4 - u^2}{24}\nabla^4 y_{-2} \qquad ...(1)$$

Differentiating (1) w.r.t. x, we get

$$f'(x) = \frac{1}{h}\left[\left(\frac{\Delta y_0 + \Delta y_{-1}}{2}\right) + u\Delta^2 y_{-1} + \frac{3u^2 - 1}{6}\left(\frac{\Delta^3 y_{-1} + \Delta^3 y_{-2}}{2}\right)\right.$$

$$\left. + \frac{4u^3 - 2u}{24}\Delta^4 y + ...\right]$$

Using Difference table, we have

$$\Delta y_0 = 0.2468, \Delta y_{-1} = 0.2833, \Delta^2 y_{-1} = 0.0365, \Delta^3 y_{-1} = 0.0035,$$

$$\Delta^3 y_{-2} = 0.0037, \Delta^4 y_{-2} = 0.0002$$

$$\therefore \quad f'(0.6) = \left[\left(\frac{0.2468 + 0.2833}{2}\right) + 0 - \frac{1}{6}\left(\frac{0.0035 + 0.0037}{2}\right) + 0\right]\frac{1}{0.1}$$

$$= 10[0.26505 - 0.0006]$$

$$f'(0.6) = 2.6445$$

Again differentiating (1), we get

$$f''(x) = \frac{1}{h^2}\left[\Delta^2 y_{-1} + u\left(\frac{\Delta^3 y_{-1} + \Delta^3 y_{-2}}{2}\right) + \frac{12u^2 - 2}{24}\Delta^4 y_{-2} + ...\right]$$

$$f''(0.6) = \frac{1}{(0.1)^2}\left[0.0365 + 0 - \frac{1}{12} \times 0.0002\right] = 100[0.0365 - 0.000016]$$

Hence, $f''(0.6) = 3.6484$.

EXAMPLE 12. *Find f′(93) from the following table :*

x	60	75	90	105	120
f(x)	28.2	38.2	43.2	40.9	37.7

SOLUTION.

Difference table

u	x	f(x)	$\Delta f(x)$	$\Delta^2 f(x)$	$\Delta^3 f(x)$	$\Delta^4 f(x)$
−2	60	28.2				
			10			
−1	75	38.2		−5		
			$\binom{5}{2.3}$		$\binom{-2.3}{6.4}$	
0	90	(43.2)		(−7.3)		(8.7)
1	105	40.9		−0.9		
			−3.2			
2	120	37.7				

Here, we have $x_0 = 90, x = 93, h = 15$

$$u = \frac{x - x_0}{h} = \frac{93 - 90}{15} = \frac{3}{15} = \frac{1}{5} = 0.2$$

Now stirling's formula is given by

$$f(x) = y = y_0 + 4\left(\frac{\Delta y_0 + \Delta y_{-1}}{2}\right) + \frac{u^2}{2}\Delta^2 y_{-1} + \frac{u^3 - u}{6}\left(\frac{\Delta^3 y_{-1} + \Delta^3 y_{-2}}{2}\right)$$

$$+ \frac{u^4 - u^2}{24}\Delta^4 y_{-2} + ... \qquad ...(1)$$

Differentiating w.r.t. to x, we get

$$f'(x) = \left[\left(\frac{\Delta y_0 + \Delta y_{-1}}{2}\right) + u\Delta^2_{y-1} + \frac{3u^2 - 1}{6}\left(\frac{\Delta^3 y_{-1} + \Delta^3 y_{-2}}{2}\right)\right.$$

$$\left. + \frac{4u^3 - 2u}{24}\Delta^4 y_{-2} + ...\right] \qquad \left(\because \frac{du}{dx} = \frac{1}{h}\right)$$

Putting the value of $x = 93, u = 0.2, h = 15$ and $\Delta y_0 = 5, \Delta y_{-1} = -2.3,$
$\Delta^2 y_{-1} = -7.3, \Delta^3 y_{-1} = -2.3, \Delta^3 y_{-2} = 6.4, \Delta^4 y_{-2} = 8.7$, we get

$$f'(93) = \frac{1}{15}\left[\frac{5 - 2.3}{2} + 0.2(-7.3) + \frac{3(0.2)^2 - 1}{6}\left(\frac{-2.3 + 6.4}{2}\right)\right.$$

$$\left. + \frac{4(0.2)^3 - 2(0.2)}{24}(8.7)\right.$$

$$= \frac{1}{15}\left[\frac{2.7}{2} - 1.46 - \frac{3.608}{6 \times 2} - \frac{3.2016}{24}\right] = \frac{1}{15}[1.35 - 1.46 - 0.30065 - 0.1334]$$

$$= \frac{1}{15}\left[1.35 - 1.46 - \frac{0.6013}{2} - 0.1334\right] = -0.03627.$$

EXAMPLE 13. *From the following table, for what value of x, y is minimum. Also find this value of y.*
(Mumbai-2008)

x	3	4	5	6	7	8
y	0.205	0.240	0.259	0.260	0.250	0.224

SOLUTION.

Difference table

x	y	Δ	Δ^2	Δ^3
3	**0.205**			
		0.035		
4	0.240		**–0.016**	
		0.019		**0.000**
5	0.259		–0.016	
		0.003		0.001
6	0.262		–0.015	
		–0.012		0.001
7	0.250		–0.014	
		–0.026		
8	0.224			

Now taking $x_0 = 3$, we have $y_0 = 0.205$, $\Delta y_0 = 0.035$, $\Delta^2 y_0 = -0.016$ and $\Delta^3 y_0 = 0$

Therefore, Newton's forward interpolation formula gives

$$y = 0.205 + u(0.035) - \frac{u(u-1)}{2}(-0.016)$$

Differentiating (1) w.r.t. to u, we get

$$\frac{dy}{du} = 0.035 + \frac{2u-1}{2}(-0.016)$$

For y to be minimum put $\dfrac{dy}{du} = 0$

$\Rightarrow$ $\qquad\qquad$ 0.035 – 0.008 (2u –1) = 0

$\Rightarrow$ $\qquad\qquad\qquad\qquad$ u = 2.6875

Therefore, $\qquad\qquad\qquad\qquad$ $x = x_0 + uh = 3 + 2.6875 \times 1 = 5.6875$

Hence, y is minimum when $x = 5.6875$.

Putting $u = 2.6875$ in (1), we get the minimum value of y given by

$$= 0.205 + 2.6875 \times 0.035 + \frac{1}{2}(2.6875 \times 1.6875)(-0.016)$$

$$= 0.2628.$$

EXAMPLE 14. *The table given below reveals the velocity v of a body during the time t. Find its acceleration at t = 1.1.* (UPTU–2002, JNTU–2009)

x	1.0	1.1	1.2	1.3	1.4
y	43.1	47.7	52.1	56.4	60.8

SOLUTION.

Difference table

t	v	Δ	Δ^2	Δ^3	Δ^4
1.0	43.1				
		4.6			
1.1	47.7		–0.2		
		4.4		0.1	
1.2	52.1		–0.1		0.1
		4.3		0.2	
1.3	56.4		0.1		
		4.4			
1.4	60.8				

We have, $t_0 = 1.1, v_0 = 4.77$ and $h = 0.1$

Then acceleration at $t = 1.1$ is given by

$$\left(\frac{du}{dt}\right) = \frac{1}{0.1}\left[4.4 - \frac{1}{2}(-0.1) + \frac{1}{3}(0.2)\right]$$

(By Newton's forward interpolation formula)

$$= 10 - \left(4.4 + 0.05 + \frac{0.2}{3}\right)$$

$$= 51.2 \text{(approx.)}$$

EXAMPLE 15. *The distance covered by an athlete for the 50 metre race is given in the following table :*

Time (Sec.)	0	1	2	3	4	5	6
Distance (metre)	0	2.5	8.5	15.5	24.5	36.5	50

Determine the speed of athlete at $t = 5$ sec. correct to two decimals. (UPTU–2003, 09)

SOLUTION. We have to find the derivative at $t = 5$, which is near to the end of the table. Therefore we shall use Newton's backward interpolation formula

Difference table

t	s	∇s	$\nabla^2 s$	$\nabla^3 s$	$\nabla^4 s$	$\nabla^5 s$	$\nabla^6 s$
0	0						
		2.5					
1	2.5		3.5				
		6		–2.5			
2	8.5		1		3.5		
		7		1		–3.5	
3	15.5		2		0		1
		9		1		–2.5	
4	24.5		3		–2.5		
		12		–1.5			
5	36.5		–1.5				
		13.5					
6	50						

The speed of athelete at $t = 5$ sec. is given by

$$\left(\frac{ds}{dt}\right)_{t=5} = \frac{1}{h}\left[\nabla s_5 + \frac{1}{2}\nabla^2 s_5 + \frac{1}{3}\nabla^3 s_5 + \frac{1}{4}\nabla^4 s_5 + \frac{1}{5}\nabla^5 s_5\right]$$

$$= \frac{1}{1}\left[12 + \frac{1}{2}(3) + \frac{1}{3}(1) + \frac{1}{4}(0) + \frac{1}{5}(-3.5)\right]$$

$= 13.1333 = 13.13$ meter/sec. (correct to two decimal places.)

EXAMPLE 16. *The table below gives the result of an observations. θ is the observed temperature in degrees centigrade of a vessel of cooling water, t is the time of minutes from the beginning of the observations.*

t	1	3	5	7	9
θ	85.3	74.5	67.0	60.5	54.3

Find the approximate rate of cooling at $t = 3$ and 3.5.

SOLUTION.

Difference table

t	θ	$\Delta\theta$	$\Delta^2\theta$	$\Delta^3\theta$	$\Delta^4\theta$
1	85.3				
		−10.8			
3	74.5		3.3		
		−7.5		−2.3	
5	67.0		1.0		1.6
		−6.5		−0.7	
7	60.5		0.3		
		−6.2			
9	54.3				

When $t = 3, \theta_0 = 7.45$. Also, $h = 2$

Rate of cooling $= \dfrac{d\theta}{dt}$

$$\therefore \quad \left(\frac{d\theta}{dt}\right)_{t=3} = \frac{1}{h}\left[\Delta\theta_0 - \frac{1}{2}\Delta^2\theta_0 + \frac{1}{3}\Delta^3\theta_0 - \frac{1}{4}\Delta^4\theta_0\right]$$

$$= \frac{1}{2}\left[-7.5 - \frac{1}{2}\times 1 + \frac{1}{3}\times(-0.7)\right]$$

$$= -4.11667°C / \min$$

when $t = 3.5$. Since, t is the non-tabular value of t, so by Newton's forward interpolation formula, we have

$$\frac{dy}{dx} = \frac{1}{h}\left[\Delta y_0 + \left(\frac{2u-1}{2}\right)\Delta^2 y_0 + \left(\frac{3u^2 - 6u + 2}{6}\right)\Delta^3 y_0\right.$$

$$\left. + \left(\frac{2u^3 - 9u^2 + 11u - 3}{12}\right)\Delta^4 y_0 + ...\right]$$

$$\Rightarrow \quad \frac{d\theta}{dt} = \frac{1}{h}\left[\Delta\theta_0 + \left(\frac{2u-1}{2}\right)\Delta^2\theta_0 + \left(\frac{3u^2 - 6u + 2}{6}\right)\Delta^3\theta_0\right.$$

$$\left. + \left(\frac{2u^3 - 9u^2 + 11u - 3}{12}\right)\Delta^4\theta_0 + ...\right] \quad ...(1)$$

Here, we have $a = 3.0$ and $h = 2$.

At t = 3.5, $u = \dfrac{3.5 - 3.0}{2} = \dfrac{0.5}{2} = 0.25$

Then, from (1)

$$\left(\frac{d\theta}{dt}\right)_{t=3.5} = \frac{1}{2}\left[-7.5 + \left\{\frac{2(0.25) - 1}{2}\right\}(1) + \left\{\frac{3(0.25)^2 - 6(0.25) + 2}{6}\right\}(-0.7)\right]$$

$$= -3.915 \,°C/min$$

EXAMPLE 17. *Find $f'''(5)$ using the following data :*

x	2	4	9	13	16	21	29
f(x)	57	1345	66340	402052	1118209	4287844	21242820

Find the approximate rate of cooling at t = 3 and 3.5.

SOLUTION. Since the values of the argument x are not equally spaced, so we apply the Newton's divided difference formula, The difference table is given as follows :

x	f(x)	$\Delta f(x)$	$\Delta^2 f(x)$	$\Delta^3 f(x)$	$\Delta^4 f(x)$	$\Delta^5 f(x)$	$\Delta^6 f(x)$
2	57						
		644					
4	1345		1765				
		12999		556			
9	66340		7881		45		
		83928		1186		1	
13	402052		22113		64		0
		238719		2274		1	
16	1118209		49401		89		
		633927		4054			
21	42877844		114265				
		2119372					
29	21242820						

Putting all these values in Newton's divided difference formula, we get

$f(x) = 57 + (x-2)\ (644) + (x-2)\ (x-4)\ (1765) + (x-2)\ (x-4)\ (x-9)\ (556)$
$\qquad + (x-2)(x-4)(x-9)\ (x-13)(45) + (x-2)(x-4)(x-9)(x-13)(x-16)1$

$\qquad = 57 + 644\ (x-2) + 1765\ (x^2-6x+8) + 556\ (x^3-15x^2+62x-72)$

$\qquad\qquad +45\ (x^4-28x^3 + 257x^2 -878\ x + 936) + x^5 - 44x^4 + 705x^3$

$$\qquad\qquad\qquad - 4990\ x^2 + 14984\ x - 14976.$$

$\Rightarrow f'(x) = 644 + 1765\ (2x-6) + 556(3x^2 - 30x + 62)$
$\qquad\qquad +45(4x^3 - 84x^2 + 514x - 878) + 5x^4 - 176\ x^3 + 2115\ x^2$

$$\qquad\qquad\qquad - 9980\ x + 14984$$

$\Rightarrow f''(x) = 3530 + 556(6x - 30) + 45(12x^2 - 168x + 514)$

$$\qquad\qquad\qquad + 20x^3 - 528x^2 + 4230x - 9980$$

$\qquad f(x) = 3336 + 45(24\ x - 168) + 60x^2 - 1056\ x + 4230$

$$\qquad = 60x^2 + 24\ x + 6$$

$\Rightarrow f'''(5) = 60(5)^2 + 24(5) + 6 = 1626$

EXAMPLE 18. *The following data gives the velocity of a particle for 20 seconds at an interval of 5 seconds. find the initial acceleration using the entire data :*

Time t (sec)	0	5	10	15	20
Velocity θ(m/sec)	0	3	14	69	228

(Anna-2004)

SOLUTION.

Difference table

t	θ	Δθ	$\Delta^2\theta$	$\Delta^3\theta$	$\Delta^4\theta$
0	0				
		3			
5	3		8		
		11		36	
10	14		44		24
		55		60	
15	69		104		
		159			
20	228				

When $t = 0$

$$\therefore \left(\frac{d\theta}{dt}\right) = \frac{1}{h}\left[\Delta\theta - \frac{1}{2}\Delta^2\theta + \frac{1}{3}\Delta^3\theta - \frac{1}{4}\Delta^4\theta + ...\right]$$

$$= \frac{1}{5}\left[3 - \frac{1}{2}(8) + \frac{1}{3}(36) - \frac{1}{4}(24)\right] = \frac{1}{5}(3 - 4 + 12 - 6) = 1 \text{ m/sec}^2.$$

EXAMPLE 19. *A slider in a machine moves along a fixed straight rod. Its distance x cm along the rod is given below for various values of the time t seconds. Find the velocity of the slider and its acceleration when t = 0.3 second.*

t	0	0.1	0.2	0.3	0.4	0.5	0.6
x	30.13	31.62	32.87	33.64	39.95	33.81	33.24

(Agra-2003; VTU-2003, 09)

SOLUTION.

Difference table

t	x	Δ	Δ^2	Δ^3	Δ^4	Δ^5	Δ^6
0	30.13						
		1.49					
0.1	31.62		– 0.24				
		1.25		– 0.24			
0.2	32.87		– 0.48		0.26		
		0.77		– 0.02		– 0.27	
0.3	33.64		– 0.46		– 0.01		0.29
		0.31		0.01		0.02	
0.4	33.95		– 0.45		0.01		
		–0.14		0.02			
0.5	33.81		– 0.43				
		–0.57					
0.6	33.24						

Now, using Striling's formula, we have

$$\left(\frac{dx}{dt}\right)_{t_0} = \frac{1}{h}\left(\frac{\Delta x_0 + \Delta x_{-1}}{2}\right) - \frac{1}{6}\left(\frac{\Delta^3 x_{-1} + \Delta^3 x_{-2}}{2}\right) + \frac{1}{30}\left(\frac{\Delta^5 x_{-2} + \Delta^5 x_{-3}}{2}\right) + \dots \quad ..(1)$$

and $\quad \left(\frac{d^2 x}{dt^2}\right)_{t_0} = \frac{1}{h^2}\left[\Delta^2 x_{-1} - \frac{1}{12} x_{-2} + \frac{1}{90}\Delta^6 x_{-3}\dots\right]$ ...(2)

Here, $\quad t_0 = 0.3,\ h = 0.1,\ \Delta_{x-1} = 0.77,\ \Delta x_0 = 0.31$ and $\Delta^2 x_{-1} = -0.46$

From eqn. (1) and (2), we get

$$\left(\frac{dx}{dt}\right)_{0.3} = \frac{1}{0.1}\left[\frac{0.31 + 0.77}{2} - \frac{1}{6}\left(\frac{0.01 + 0.02}{2}\right) + \frac{1}{30}\left(\frac{0.02 - 0.27}{2}\right) + \dots\right] = 5.33$$

$$\left(\frac{d^2 x}{dt^2}\right)_{0.3} = \frac{1}{(0.1)^2}\left[-0.46 - \frac{1}{12}(-0.01) + \frac{1}{90}(0.29) - \dots\right] = -45.6$$

Hence, the velocity is 5.33 cm/sec and acceleration is -45.6 cm/sec^2.

6.7 ERROR ANALYSIS IN NUMERICAL DIFFERENTIATION

Let $E_r(h)$ be the rounded off error in an approximated derivative, then, the total error is
$$E(h) = E_r(h) + E_t(h) \qquad ...(1)$$
where E_r and E_h denotes the rounding off and truncation error respectively.

By definition of the derivative of a function, we have

$$f'(x) = \frac{f(x+h) - f(x)}{h} = \frac{f_1 - f_0}{h}$$

Let us suppose e_1 and e_0 is the rounding off errors in f_1 and f_0 respectively, then we have

$$f'(x) = \frac{(f_1 + e_1) - (f_0 + e_0)}{h} = \frac{f_1 - f_0}{h} + \frac{e_1 - e_0}{h}$$

Let e be the magnitude of e_1 and e_0, then we have

$$[E_r(h)] \le \frac{2e}{h}$$

Since, the truncation error for two points formula is

$$\left|E_t(h)\right| = -\frac{h}{2} f''(\theta) \quad \Rightarrow \quad \left|E_t(h)\right| \le \frac{Mh}{2}$$

where $M = \max\left|f''(\theta)\right|, x \le \theta \le x + h$

Thus, the bound for the total error in the derivative is

$$\left|E(h)\right| \le \frac{Mh}{2} + \frac{2e}{h} \qquad ...(2)$$

6.8 ERROR IN HIGHER ORDER DERIVATIVES

Let f_n denotes the $f(x + nh)$ and f_{-n} denote the $f(x - nh)$. Then we have the following difference table :

6.8.1 FORWARD DIFFERENCE DERIVATIVES

S. No.	Derivative	Formula	Error
1.	$f'(x_0)$	$\left(-\dfrac{f_0 - f_1}{h}\right)$	$E(h) = -\dfrac{h}{2} f''(\theta)$ $E_r = (2e / h)$
		$\left(-\dfrac{3f_0 + 4f_1 - f_2}{h}\right)$	$E(h) = \dfrac{h^2}{3} f'''(\theta)$ $E_r = (4e / h)$
		$\left(-\dfrac{11f_0 + 18f_1 - 9f_2 + 2f_3}{6h}\right)$	$E(h) = -\dfrac{h^3}{4} f^{iv}(\theta)$ $E_r = (20e / 3h)$
		$\left(-\dfrac{25f_0 + 48f_1 - 36f_2 + 16f_3 - 3f_4}{12h}\right)$	$E(h) = -\dfrac{h^4}{5} f^{v}(\theta)$ $E_r = (32e / h)$
2.	$f''(x_0)$	$\left(\dfrac{f_0 - 2f_1 + f_2}{h^2}\right)$	$E(h) = -hf'''(\theta)$ $E_r = (4e / h^2)$
		$\left(\dfrac{2f_0 - 5f_1 + 4f_2 - f_3}{h^2}\right)$	$E(h) = \dfrac{11h^2}{12} f^{iv}(\theta)$ $E_r = (12e / h^2)$

6.8.2 BACKWARD DIFFERENCE DERIVATIVES

S. No.	Derivative	Formula	Error
1.	$f'(x_0)$	$\left(-\dfrac{f_{-1} - f_1}{h}\right)$	$E(h) = -\dfrac{h}{2} f''(\theta)$ $E_r = (2e / h)$
		$\left(-\dfrac{f_{-2} + 4f_{-1} - 3f_0}{2h}\right)$	$E(h) = \dfrac{h^4}{3} f'''(\theta)$ $E_r = (4e / h)$
		$\left(-\dfrac{2f_{-3} + 9f_{-2} - 18f_{-1} + 11f_0}{6h}\right)$	$E(h) = -\dfrac{h^3}{3} f^{iv}(\theta)$ $E_r = (20e / 3h)$
		$\left(\dfrac{3f_{-4} - 16f_{-3} + 36f_{-2} - 48f_{-1} + 25f_0}{12h}\right)$	$E(h) = -\dfrac{h^3}{5} f^{v}(\theta)$ $E_r = (32e / h)$
2.	$f''(x_0)$	$\left(\dfrac{f_{-2} - 2f_{-1} + f_0}{h^2}\right)$	$E(h) = hf''(\theta)$ $E_r = (4e / h)$
		$\left(-\dfrac{f_{-3} + 4f_{-2} - 5f_{-1} + 2f_0}{h^2}\right)$	$E(h) = \dfrac{11h^2}{12} f^{iv}(\theta)$ $E_r = (12e / h^2)$

6.8.3 CENTRAL DIFFERENCE DERIVATIVES

S. No.	Derivative	Formula	Error
1.	$f'(x_0)$	$\left(-\dfrac{f_{-1} - f_1}{2h} \right)$	$E(h) = -\dfrac{h^2}{2} f'''(\theta)$ $E_r = (e/h)$
		$\left(\dfrac{f_{-2} - 8f_{-1} + 8f_1 - f_2}{12h} \right)$	$E(h) = \dfrac{h^4}{30} f^v(\theta)$ $E_r = (3e/2h)$
		$\left(\dfrac{f_{-3} + 9f_{-2} - 45f_{-1} + 45f_1 - 9f_2 + f_3}{6h} \right)$	$E(h) = -\dfrac{h^6}{140} f^{vii}(\theta)$ $E_r = (11e/6h)$
2.	$f''(x_0)$	$\left(\dfrac{f_{-1} - 2f_0 + f_1}{h^2} \right)$	$E(h) = -\dfrac{h^2}{12} f^{iv}(\theta)$ $E_r = (4e/h^2)$
		$\left(\dfrac{-2f_{-2} + 16f_{-1} - 30f_0 + 16f_1 - f_2}{h^2} \right)$	$E(h) = \dfrac{h^4}{90} f^{vi}(\theta)$ $E_r = (16e/3h^2)$

SOLVED EXAMPLES

EXAMPLE 1. *The equation for deflection of a beam is given by*

$$y''(x) = -e^{x^2} \text{ with } y(0) = 0, \ y(1) = 0.$$

Find, using a second order derivatives, the approximate deflection at $x = 0.25$, 0.5 and 0.75. Note that $y(x)$ is the deflection at x.

SOLUTION. Since, we know that

$$y''(x) = \frac{y(x+h) - 2y(x) + y(x-h)}{h^2} = e^{x^2}$$

Take $h = 0.25$ ...(1)

Now, (1) $\Rightarrow \quad \dfrac{y(x+0.25) - 2y(x) + y(x-0.25)}{0.0625} = e^{x^2}$

$\Rightarrow \quad y(x+0.25) - 2y(x) + y(x-0.25) = 0.0625 e^{x^2}$ ...(2)

Putting $x = 0.25$, 0.5 and 0.75 in eq. (2), we get

$$y(0.5) - 2y(0.25) + y(0) = 0.0665$$
$$y(0.75) - 2y(0.5) + y(0.25) = 0.0803$$
$$y(1.0) - 2y(0.75) + y(0.5) = 0.1097$$

Using the given conditions $y(0) = y(1) = 0$

we get
$$\left. \begin{array}{l} 0 + y_2 - 2y_1 = 0.0665 \\ y_3 - 2y_2 + y_1 = 0.0803 \\ -2y_3 + y_2 = 0.1097 \end{array} \right\}$$ ...(3)

Solving eq. (3) for y_1, y_2 and y_3, we get

$$y_1 = y(0.25) = -0.1175,$$

$$y_2 = y(0.5) = -0.1684$$

and $\qquad y_3 = y(0.75) = -0.1391.$

EXAMPLE 2. *Find the approximate derivative of $f(x) = \sin x$ at $x = 0.45$ radians at increasing values of h from 0.01 to 0.04 with a step size of 0.05. Analyse the total error. What is the optimum step size?*

SOLUTION. Here, we have $\qquad f(x) = \sin x$

Then we know that $\quad f'(x) = \dfrac{f(x+h) - f(x)}{h}$

Given $\qquad\qquad\qquad x = 0.45$ radians

$\Rightarrow \qquad\qquad\qquad f(x) = \sin(0.45) = 0.4350$

The exact value of $\;f'(x) = \cos x$

$$f'(0.45) = \cos(0.45) = 0.9004$$

Now, we have the following table :

h	$f(x + h)$	$f'(x)$	Error
0.010	0.4439	0.8900	0.0104
0.015	0.4484	0.8933	0.0071
0.020	0.4529	0.8950	0.0054
0.025	0.4573	0.8935	0.0069
0.030	0.4618	0.8933	0.0071
0.035	0.4662	0.8914	0.0090
0.040	0.4706	0.8900	0.0104

From table, we observe that the total error decreases from 0.0104 (at $h = 0.01$) till $h = 0.002$ and again increased.

Now, the bound M is given by

$$M = \max|f''(\theta)| \qquad\qquad 0.41 \le \theta \le 0.49$$

$$= |\sin(0.49)| = 0.4706$$

Now, the optimum size is

$$h_{opt} = 2\sqrt{\frac{e}{M}} = 2\sqrt{\frac{0.5 \times 10^{-4}}{0.4706}} = 0.0206.$$

EXERCISE 6.1

1. Use the following data to find $f'(3)$:

x	3	5	11	27	34
$f(x)$	-13	28	899	17315	35606

2. Find $f'(6)$ from the following table :

x	0	1	3	4	5	7	9
$f(x)$	150	108	0	-54	-100	-144	-84

3. Using divided differences, find the value of $f'(8)$, given that $f(6) = 1.556$, $f(7) = 1.690$, $f(9) = 1.908$, $f(12) = 2.158$.

4. Find $f'(5)$ from the following table :

x	2	4	9	10
$f(x)$	4	56	711	980

5. The population of a certain town is shown in tabular form below :

Year	1921	1931	1941	1951	1961
Population (in thousands)	19.96	39.65	58.81	77.21	94.61

Find the rate of growth of population in 1951.

6. Find $y'(0)$ and $y''(0)$ from the following table :

x	0	1	2	3	4	5
y	4	8	15	7	6	2

7. Given that

$\theta°$	0	10	20	30	40
$\sin \theta°$	0.000	0.1736	0.3420	0.5000	0.6428

Find $\cos \theta$ when $\theta = 10°$.

8. A rod is rotating in a plane. The following table given the angle θ in radians through which the rod has turned for various values of the time t second.

t	0	0.2	0.4	0.6	0.8	1.0	1.2
θ	0	0.12	0.49	1.12	2.02	3.20	4.67

Calculate the angular velocity and the angular acceleration of the rod at $t = 0.6$ second.

(UPTU(MCA)–2002, UPTU(B. tech.)–2003, 04, Rohtak–2004, VTU–2004)

9. The following table of values of x and y is given :

x	0	1	2	3	4	5	6
y	6.9897	7.4036	7.7815	8.1291	8.4510	8.7306	9.0309

Find $\dfrac{d}{dx}$ at

  (i) $x = 1$ (ii) $x = 3$ (iii) $x = 6$

10. From the following values of x and y, find $y'(6)$.

x	4.5	5.0	5.5	6.0	6.5	7.0	7.5
y	9.69	12.90	16.71	21.18	26.37	32.34	39.15

11. Find $f'(10)$ from the following data :

x	3	5	11	27	34
$f(x)$	–13	23	899	17315	35606

12. Find the first derivative for the following table of data at $x = 0.75$, 1.00 and 1.25. Use $h = 0.05$ and 0.1

x	0.5	0.7	0.9	1.1	1.3	1.5
y	1.48	1.64	1.78	1.89	1.96	1.00

Compare the result with $h = 0.05$ and $h = 0.1$.

13. Find $\dfrac{dy}{dx}$ and $\dfrac{d^2y}{dx^2}$ at $x = 6$ using the following table :

x	4.5	5.0	5.5	6.0	6.5	7.0	7.5
y	9.69	12.90	16.71	21.18	26.37	32.34	39.15

14. Using the following table, find x for which y to minimum, also find this value of y :

x	0.60	0.65	0.70	0.75
y	0.6221	0.6155	0.6138	0.6170

15. Using the following table, find the value of x, for which y is maximum and find this value of y.

x	1.2	1.3	1.4	1.5	1.6
y	0.9320	0.9636	0.9855	0.9975	0.9996

16. A slider in a machine moves along a fixed straight rod. Its distance x (in cm) along the rod is given at various times t (in secs).

t	0	0.1	0.2	0.3	0.4	0.5	0.6
x	30.28	31.43	32.98	33.54	33.97	33.48	32.13

Evaluate $\dfrac{dx}{dt}$ at t = 0.1 and at t = 0.5. $\hspace{2cm}$ (UPTU–2005, VTU–2009)

17. The population of a certain town is given below. Find the rate of growth of the population in 1941 and 1961.

Year	1931	1941	1951	1961	1971
Population in lacs	40.62	60.80	17.95	103.56	132.65

18. Find $f'(7.50)$ from the following table :

x	7.47	7.48	7.49	7.50	7.51	7.52	7.53
f(x)	0.193	0.195	0.198	0.201	0.203	0.206	0.208

$\hspace{2cm}$ (Rajasthan-2005, 07; Rohilkhand-2004, 06, 10; JNTU–2006, Rohatk–2005, 10)

19. Find $f'''(15)$ from the following data :

x	2	4	9	13	16	21	29
f(x)	57	1345	66340	402052	1118209	4287844	21242820

$\hspace{2cm}$ (IFS-2007, Lucknow-2008)

20. Find the value of $f'(8)$, given that $f(6) = 1.556, f(7) = 1.690, f(9) = 1.908, f(12) = 2.158$.

$\hspace{2cm}$ (Rajasthan-2006, 10)

21. Find $f'(1)$ for $f(x) = \dfrac{1}{(1+x^2)}$ using the following data :

x	1.0	1.1	1.2	1.3	1.4
f(x)	0.2500	0.2268	0.2066	0.1890	0.1736

$\hspace{2cm}$ (Meerut-2002, 04)

22. Given the following table of values of x and y

x	1.00	1.05	1.10	1.15	1.20	1.25	1.30
f(x)	1.000	1.025	1.049	1.072	1.095	1.118	1.140

Find $\dfrac{dy}{dx}$ and $\dfrac{d^2y}{dx^2}$ at (VTU–2008)

   (a) $x = 1.05$ (b) $x = 1.25$ (c) $x = 1.15$

23. Find the values of cos 1.74 from the following table :

x	1.7	1.74	1.78	1.82	1.86
$\sin x$	0.9916	0.9857	0.9781	0.9691	0.9584

(JNTU–2009)

24. If $y = f(x)$ and y_n denotes $f(x_0 + nh)$ prove that, if powers of h above h^6 be neglected.

$$\left(\frac{dy}{dx}\right)_{x_0} = \frac{3}{4h}\left[(y_1 - y_{-1}) - \frac{1}{5}(y_2 - y_{-2}) + \frac{1}{45}(y_3 - y_{-3})\right]$$

25. Find the $f'(6)$ from the following data :   (UPTU–2006)

x	0	2	3	4	7	8
$f(x)$	4	26	58	112	466	922

(IFS-2009; JNTU–2009: UPTU–2008)

26. For the following values of x and y, find the first derivative at $x = 4$

x	1	2	4	8	10
y	0	1	5	21	27

(JNTU–2009)

27. Find the derivative of $f(x)$ at $x = 0.4$ from the following table :

x	0.1	0.2	0.3	0.4
y	1.10517	1.22140	1.34986	1.49182

(UPPCS-2008; Nagarjuna-2001)

28. The population of a certain town (as obtained from census data) is shown in the following table :

Year	1961	1971	1981	1991	2001
Population (in thousands)	19.96	39.65	58.81	77.21	94.61

Estimate the population in the years 1976 and 2003. Also find the rate of growth of population in 1991.   (Delhi-2002)

29. If $y = f(x)$ and y_n denotes $f(x_0+nh)$, prove that, if powers of h above h^6 be neglected.

$$\left(\frac{dy}{dx}\right)_{x_0} = \frac{3}{4h}\left[(y_1 - y_{-1}) - \frac{1}{5}(y_2 - y_{-2}) + \frac{1}{45}(y_3 - y_{-3})\right]$$

(UPTU–2006)

30. Find the value of $f(8)$ from the table given below :

x	6	7	9	12
$f(x)$	1.556	1.690	1.908	2.158

(Anna-2007)

31. Find $f'(6)$ from the following data :

x	0	2	3	4	7	8
$f(x)$	4	26	58	112	466	922

(JNTU-2009)

32. Find the maximum and minimum value of y from the following data :

x	-2	-1	0	1	2	3	4
y	2	-0.25	0	-0.25	2	15.75	56

(Anna-2004)

Answers

1. 1.8828 **2.** -23 **3.** 0.109 **4.** 2097.69 **5.** 1.80 thousands per year **6.** -27.9, 117.67
7. 0.9848 **8.** 3.28 rad/sec., 6.75 rad/sec^2 **9.** (i) 0.3950 (ii) 0.3341 (iii) 0.2719 **10.** 9.66
11. 2.33 **13.** 9.64, 2.88 **14.** 0.692, 0.6137 **15.** 1.58, 1.00 **18.** 0.22666 **19.** 1626
20. 0.10859 **21.** -0.5031 **22.** (a) 0.493, -1.165 (b) 0.4473, -0.1583 (c) 0.4662, -0.2043
23. 0.175 **25.** 135 **26.** 2.8326 **27.** 1.4913 **28.** 49.3, 97.68, 1.8 thousand 1 years
30. 0.1086 **31.** 135 **32.** maximum value $= 0$, minimum value $= -0.25$

6.9 NUMERICAL QUADRATURE

Given a set of tabulated values of the integrand $f(x)$, to find the value of $\int_{x_0}^{x_n} f(x)dx$ is known as numerical integration. We divide the given interval into a large number of subintervals of equal width h and replace the function tabulated at the points of sub-division by any one of the interpolating polynomial, over each of the subintervals and calculate the integral.

If the integrand has no singular points in the field of the domain then we can calculate the value of the definite integral of that integrand by any numerical method. The process for finding the values of the integral of a function of a single variable, is called quadrature.

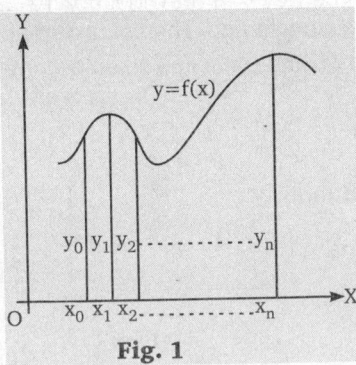

Fig. 1

▶ REMARKS
‣ The problem for finding the value of the integral of $f(x)$ is solved by first approximating the function $f(x)$ by an interpolating polynomial and then integrating it between the desired limits.
‣ Numerical integration method yield much better results compared to the numerical differentiation. This is due to the fact that the error introduced in separate subintervals tends to cancel each other.

6.10 QUADRATURE FORMULA FOR EQUALLY SPACED ARGUMENTS

Let $y = f(x)$ be a function which assume the values $y_0, y_1, \ldots y_n$ corresponding to the values of the arguments $x = x_0, x_0 + h, x_0 + 2h, \ldots x_0 + nh$.

Divide the interval (a, b) into n subintervals of width h, such that

$$x_0 = a, x_1 = x_0 + h, x_2 = x_0 + 2h, \ldots, x_n = x_0 + nh = b$$

Consider

$$I = \int_a^b f(x)dx = \int_{x_0}^{x_0+nh} f(x)dx \qquad \qquad \ldots(1)$$

Newton's forward interpolation formula is given by

$$y = f(x) = y_0 + u\Delta y_0 + \frac{u(u-1)}{2!}\Delta^2 y_0 + \frac{u(u-1)(u-2)}{3!}\Delta^3 y_0 + \ldots$$

where $u = \dfrac{x - x_0}{h}$

$$\therefore \quad \frac{du}{dx} = \frac{1}{h}dx \Rightarrow dx = hdu$$

Now, equation (1) becomes

$$I = h\int_0^n \left[y_0 + u\Delta u_0 + \frac{u(u-1)}{2!}\Delta^2 y_0 + \frac{u(u-1)(u-2)}{3!}\Delta^3 y_0 + ... \right] du$$

$$= nh\left[y_0 + \frac{n}{2}\Delta y_0 + \frac{n(2n-3)}{12}\Delta^2 y_0 + \frac{n(n-2)^2}{24}\Delta^3 y_0 + ... + \text{upto } (n+1) \text{ terms} \right]$$

$$\therefore \int_{x_0}^{x_n} f(x)dx = nh\left[y_0 + \frac{n}{2}\Delta y_0 + \frac{n(2n-3)}{12}\Delta^2 y_0 + \frac{n(n-2)^3}{24}\Delta^3 y_0 + ... \text{ upto } (n+1) \text{ terms} \right] \quad ...(2)$$

This is called general quadrature formula.

6.11 THE TRAPEZOIDAL RULE

Putting $n = 1$ into equation (2) and taking the curve $y = f(x)$ between the points (x_0, y_0) and (x_1, y_1) as a straight line. That is curve is approximated by a polynomial of degree one so that differences of order two and three and so on becomes zero, we get

$$\int_{x_0}^{x_1} f(x)dx = h\left[y_0 + \frac{1}{2}\Delta y_0 \right] = h\left[y_0 + \frac{1}{2}(y_1 - y_0) \right] = \frac{h}{2}[y_0 + y_1]$$

Similarly,
$$\int_{x_1}^{x_2} f(x)dx = \frac{h}{2}[y_1 + y_2]$$

$$\int_{x_2}^{x_3} f(x)dx = \frac{h}{2}[y_2 + y_3]$$

....

$$\int_{x_{n-2}}^{x_{n-1}} f(x)dx = \frac{h}{2}[y_{n-2} + y_{n-1}]$$

and
$$\int_{x_{n-1}}^{x_n} f(x)dx = \frac{h}{2}[y_{n-1} + y_n]$$

Now adding these n integrals, we get

$$\int_{x_0}^{x_n} f(x)dx = \frac{h}{2}[y_0 + 2(y_1 + y_2 + ... + y_{n-1}) + y_n] \quad ...(3)$$

This is known as the trapezoidal rule.

▶ **REMARKS**

▸ $\int_a^b f(x)dx$ gives the area of $y = f(x)$ bounded by $x = a$ and $x = b$ and axis of x.

▸ The shape of each strip between any consecutive points is taken to trapezium. The area of each strip is found separately. That the area bounded by $y = f(x)$ and $x = x_0$ and $x = x_n$ is approximately equal to the sum of the areas of n trapezium. That is how this formula is called trapezoidal rule.

▸ In trapezoidal rule, $f(x)$ is a linear function of x. It is the simplest rule but least accurate.

ALGORITHM

Step 1.	Obtain the values of $f(x)$, limits of the integral x_0 and x_n and the number of subintervals n.
Step 2.	Calculate $\dfrac{x_n - x_0}{n} = h$.
Step 3.	Evaluate the value of x_i as $x_0 + ih$ and corresponding value of $f(x_i)$.
Step 4.	Substitute the above values in trapezoidal rule

6.11.1 ERROR ESTIMATION IN TRAPEZOIDAL RULE

Let $y = f(x)$ be continuous and possess continuous derivatives in $[x_0, x_n]$. Expand y by Taylor's theorem about $x = x_0$, we get

$$y = y_0 + (x - x_0)y_0' + \frac{(x - x_0)^2}{2!} y_0'' + \ldots$$

Now we obtain

$$\int_{x_0}^{x_1} y\,dx = \int_{x_0}^{x_1} \left[y_0 + (x - x_0)y_0' + \frac{(x - x_0)^2}{2!} y_0'' + \ldots \right] dx$$

$$= \left[xy_0 + \left(\frac{x^2}{2} - x_0 x \right) y_0' + \left(\frac{x^3}{3} - x_0 x^2 + x_0^2 x \right) \frac{y_0''}{2!} + \ldots \right]_{x_0}^{x_1}$$

$$= (x_1 y_0 - x_0 y_0) + y_0' \left[\frac{x_1^2}{2} - x_1 x_0 - \frac{x_0^2}{2} + x_0^2 \right]$$

$$+ \frac{y_0''}{2!} \left[\frac{x_1^3}{2} - x_0 x_1^2 + x_0^2 x_1 - \frac{x_0^3}{3} + x_0^3 - x_0^3 \right] + \ldots$$

$$= (x_1 - x_0)y_0 + \frac{(x_1 - x_0)^2}{2} y_0' + \frac{(x_1 - x_0)^3}{3!} y_0'' + \ldots$$

Since $h = x_1 - x_0$, then

$$\int_{x_0}^{x_1} y\,dx = hy_0 + \frac{h^2}{2} y_0' + \frac{h^3}{6} y_0'' + \ldots \qquad \ldots(1)$$

Now find

$$\frac{h}{2}[y_0 + y_1] = \frac{h}{2} \left[y_0 + y_0 + hy_0' + \frac{h^2}{2} y_0'' + \frac{h^3}{6} y_0''' + \ldots \right] \qquad \ldots(2)$$
$$[\because y_1 = y(x_0 + h)]$$

From eq. (1) and (2), we get

$$\int_{x_0}^{x_1} y\,dx - \frac{h}{2}[y_0 + y_1] = -\frac{1}{12} h^3 y_0'' + \ldots \qquad \ldots(3)$$

This is an error in the interval $[x_0, x_1]$. Proceeding in the same way, we get the errors in the sub-interval $[x_1, x_2]$, $[x_2, x_3] \ldots [x_{n-1}, x_n]$. Thus the total error is given by

$$\varepsilon = -\frac{1}{12} h^3 (y_0'' + y_1'' + \ldots + y_{n-1}'')$$

Let $y''(\xi)$ be the largest value of the $y_0'', y_1'' \ldots y_{n-1}''$ where $a < \xi < b$.

$$\therefore \qquad \varepsilon \approx -\frac{1}{12} h^3 [y''(\xi) + y''(\xi) + \ldots + y''(\xi)]$$

$$\varepsilon \approx -\frac{1}{12} h^3 n y''(\xi) \qquad (a < \xi < b)$$

$$\varepsilon \approx -\frac{b-a}{12} h^2 y''(\xi) \qquad (\because nh = b - a)$$

This is an error in trapezoidal rule.

6.12 SIMPSON'S 1/3 RULE

Putting $n = 2$ in general quadrature formula and taking the curve through (x_0, y_0), (x_1, y_1) and (x_2, y_2) as a polynomial of degree 2 so that differences of order three and greater than three becomes zero, therefore, we get

$$\int_{x_0}^{x_2} y\,dx = 2h\left[y_0 + \Delta y_0 + \frac{1}{6}\Delta^2 y_0\right]$$

$$= 2h\left[y_0 + y_1 - y_0 + \frac{1}{6}(y_2 - 2y_1 + y_0)\right] = \frac{2h}{6}[y_2 - 2y_1 + y_0 + 6y_1]$$

$$= \frac{h}{3}[y_0 + 4y_1 + y_2]$$

Similarly,

$$\int_{x_2}^{x_4} y\,dx = \frac{h}{3}[y_2 + 4y_3 + y_4]$$

$$\int_{x_4}^{x_6} y\,dx = \frac{h}{3}[y_4 + 4y_5 + y_6]$$

....

$$\int_{x_{n-4}}^{x_{n-2}} y\,dx = \frac{h}{3}[y_{n-4} + 4y_{n-3} + y_{n-2}]$$

and

$$\int_{x_{n-2}}^{x_n} y\,dx = \frac{h}{3}[y_{n-2} + 4y_{n-1} + y_n]$$

Adding all these integrals, we get

$$\int_{x_0}^{x_n} f(x)\,dx = \int_{x_0}^{x_n} y\,dx = \frac{h}{3}[y_0 + 2(y_2 + y_4 + y_6 + \dots + y_{n-2}) + 4(y_1 + y_3 + y_5 + \dots + y_{n-1}) + y_n]$$

This is known as Simpson's 1/3 Rule.

▶ **REMARKS**

▶ Simpson's 1/3-Rule requires the division of the interval into an even number of subintervals of width h.

▶ In this rule, the interpolating polynomial is of degree 2. Therefore this rule is also known as parabolic rule.

ALGORITHM

Step 1.	Obtain the values of $f(x), x_0, x_n$ and n.
Step 2.	Calculate $\dfrac{x_n - x_0}{n} = h$
Step 3.	Calculate the value of x_i as $x_0 + ih$ and corresponding value of $f(x_i)$.
Step 4.	Substitute the above obtained values in Simpson's 1/3 Rule.

6.12.1 ERROR ESTIMATION IN SIMPSON'S 1/3 RULE

Let $y = f(x)$ be a continuous function and having continuous derivatives in $[x_0, y_0]$. Expand y by Taylor's theorem about $x = x_0$, we get

$$y = y_0 + (x - x_0)y_0' + \frac{(x - x_0)^2}{2!}y_0'' + \frac{(x - x_0)^3}{3!}y_0''' + \frac{(x - x_0)^4}{4!}y_0^{iv} + \dots$$

Now, find

$$\int_{x_0}^{x_2} y\,dx = \int_{x_0}^{x_2}\left[y_0 + (x - x_0)y_0' + \frac{(x - x_0)^2}{2!}y_0'' + \frac{(x - x_0)^3}{3!}y_0''' + \frac{(x - x_0)^4}{4!}y_0^{iv} + \dots\right]dx$$

$$= (x_2 - x_0)y_0 + \frac{(x_2 - x_0)^2}{2}y_0' + \frac{(x_2 - x_0)^3}{3!}y_0'' + \frac{(x_2 - x_0)^4}{4!}y_0''' + \frac{(x_2 - x_0)^5}{5!}y_0^{iv} + \dots$$

$$\int_{x_0}^{x_2} y\,dx = 2hy_0 + 4h^2 y_0' + \frac{8h^3}{3!}y_0'' + \frac{16h^4}{4!}y_0''' + \frac{32h^5}{5!}y_0^{iv} + \dots \qquad \dots(1)$$

where $x_2 - x_0 = 2h$.

Again using Taylor's Theorem, we get

$$y_1 = y(x_0 + h) = y_0 + hy_0' + \frac{h^2}{2!}y_0'' + \frac{h^3}{3!}y_0''' + \frac{h^4}{4!}y_0^{iv} + \dots$$

and

$$y_2 = y(x_0 + 2h) = y_0 + 2hy_0' + \frac{4h^2}{2!}y_0'' + \frac{8h^3}{3!}y_0''' + \frac{16h^4}{4!}y_0^{iv} + \dots$$

Now find

$$\frac{h}{3}[y_0 + 4y_1 + y_2] = 2hy_0 + 2h^2 y_0' + \frac{8h^3}{3!}y_0'' + \frac{16h^4}{4!}y_0''' + \frac{20h^5}{3.4!}y_0^{iv} + \dots \qquad \dots(2)$$

From eqs. (1) and (2), we get

$$\int_{x_0}^{x_2} y\,dx - \frac{h}{3}[y_0 + 4y_1 + y_2] = \left(\frac{32}{120} - \frac{20}{72}\right)h^5 y_0^{iv} + \dots = -\frac{1}{90}h^5 y_0^{iv} + \dots$$

which is the error in the interval $[x_0, x_2]$. Similarly, we obtain the errors in the intervals $[x_2, x_4], [x_4, x_6] \dots [x_{2n-2}, x_{2n}]$. Thus the total error is given by

$$\varepsilon = -\frac{1}{90}h^5[y_0^{iv} + y_2^{iv} + \dots + y_{2n-2}^{iv}]$$

Let $y^{iv}(\xi)$ be the largest value of each $y_0^{iv}, y_1^{iv}, \dots y_{n-2}^{iv}$ for $a < \xi < b$, we get

$$\varepsilon = -\frac{h^5}{90}[y^{iv}(\xi) + y^{iv}(\xi) + \dots + y^{iv}(\xi)] = -\frac{h^5}{90}ny^{iv}(\xi)$$

$$\therefore \qquad \varepsilon = -\frac{b-a}{180}h^4 y^{iv}(\xi) \qquad\qquad (\because 2nh = b - a)$$

This is an error in the Simpson's 1/3 rule.

Recapitulations ||||||||||||||||||||||||||||||||||||||||||||||||||||||||

- The geometric significance of Simpson's rule is that we replace the graph of the given function by $n/2$ arcs of second degree polynomials, or parabolas with vertical axes.

6.13 SIMPSON'S 3/8 RULE

Putting $n = 3$ into equation (2) and taking the curve through the points (x_0, y_0), (x_1, y_1), (x_2, y_2) and (x_3, y_3) as a polynomial of degree 3 so that the differences of order greater than three becomes zero. We get

$$\int_{x_0}^{x_3} f(x)dx = 3h\left[y_0 + \frac{3}{2}\Delta y_0 + \frac{3(3)}{12}\Delta^2 y_0 + \frac{3.1}{24}\Delta^3 y_0\right]$$

$$= 3h\left[y_0 + \frac{3}{2}(y_1 - y_0) + \frac{3}{4}(y_2 - 2y_1 + y_0) + \frac{1}{8}(y_3 - 3y_2 + 3y_1 - y_0)\right]$$

$$= \frac{3h}{8}\left[y_0 + 3y_1 + 3y_2 + y_3\right]$$

Similarly in the intervals $[x_3, x_6], [x_6, x_9]...[x_{n-3}, x_n]$, we get

$$\int_{x_3}^{x_6} f(x)dx = \frac{3h}{8}[y_3 + 3y_4 + 3y_5 + y_6]$$

$$\int_{x_6}^{x_9} f(x)dx = \frac{3h}{8}[y_6 + 3y_7 + 3y_8 + y_9]$$

$$\cdots\cdots\cdots\cdots\cdots\cdots$$

$$\int_{x_{n-6}}^{x_{n-3}} f(x)dx = \frac{3h}{8}[y_{n-6} + 3y_{n-5} + 3y_{n-4} + y_{n-3}]$$

and $$\int_{x_{n-3}}^{x_n} f(x)dx = \frac{3h}{8}[y_{n-3} + 3y_{n-2} + 3y_{n-1} + y_n]$$

Adding all these integrals, we get

$$\int_{x_0}^{x_n} f(x)dx = \frac{3h}{8}[y_0 + 3(y_1 + y_2 + y_4 + y_5 + ... + y_{n-1}) + 2(y_3 + y_6 + ... + y_{n-3}) + y_n]$$

This is known as Simpson 3/8 rule.

ALGORITHM

Step 1.	Obtain the values of $f(x), x_0, x_n$ and n.
Step 2.	Calculate $\dfrac{x_n - x_0}{n} = h$
Step 3.	Calculate the value of x_i as $x_0 + ih$ and corresponding value of $f(x_i)$.
Step 4.	Substitute the above obtained values in Simpson's 3/8 Rule.

▶ REMARKS

▸ In this rule, the number of subintervals should be taken as multiple of 3.

▸ In Simpson's 3/8 rule, the interpolating polynomial is taken to be of degree 3. That is $y = ax^3 + bx^2 + cx + d = 0$.

6.13.1 ERROR ESTIMATION IN SIMPSON'S 3/8 RULE

Expand y by Taylor's theorem about $x = x_0$, we get

$$y = y_0 + (x + x_0)y_0' + \frac{(x - x_0)^2}{2!}y_0'' + \frac{(x - x_0)^3}{3!}y_0''' + \frac{(x - x_0)^4}{4!}y_0^{iv} + ...$$

Now

$$\int_{x_0}^{x_3} ydx = \int_{x_0}^{x_3}\left[y_0 + (x - x_0)y_0' + \frac{(x - x_0)^2}{2!}y_0'' + \frac{(x - x_0)^3}{3!}y_0''' + \frac{(x - x_0)^4}{4!}y_0^{iv} + ...\right]dx$$

$$= (x_3 - x_0)y_0 + \frac{(x_3 - x_0)^2}{2}y_0' + \frac{(x_3 - x_0)^3}{3!}y_0'' + \frac{(x_3 - x_0)^4}{4!}y_0''' + \frac{(x_3 - x_0)^5}{5!}y_0^{iv} + ...$$

Since $x_3 - x_0 = 3h$

$$\therefore \int_{x_0}^{x_3} ydx = 3hy_0 + \frac{9h^2}{2}y_0' + \frac{27h^3}{3!}y_0'' + \frac{81h^4}{4!}y_0''' + \frac{243h^5}{5!}y_0^{iv} + ... \qquad ...(1)$$

Also using Taylor's theorem for y_1, y_2 and y_3, we get

$$y_1 = y(x_0 + h) = y_0 + hy_0' + \frac{h^2}{2!}y_0'' + \frac{h^3}{3!}y_0''' + \frac{h^4}{4!}y_0^{iv} + ...$$

$$y_2 = y(x_0 + 2h) = y_0 + 2hy_0' + \frac{4h^2}{2!}y_0'' + \frac{8h^3}{3!}y_0''' + \frac{16h^4}{4!}y_0^{iv} + \dots$$

and $$y_3 = y(x_0 + 3h) = y_0 + 3hy_0' + \frac{9h^2}{2!}y_0'' + \frac{27h^3}{3!}y_0''' + \frac{81h^4}{4!}y_0^{iv} + \dots$$

Thus, the error in the interval $[x_0, x_3]$ is given by

$$\varepsilon = \int_{x_0}^{x_3} ydx - \frac{3h}{8}[y_0 + 3y_1 + 3y_2 + y_3] = \left(\frac{81}{40} - \frac{33}{16}\right)h^5 y_0^{iv} + \dots = \left(\frac{162-165}{80}\right)h^5 y_0^{iv} + \dots$$

$$= -\frac{3}{80}h^5 y_0^{iv} + \dots$$

Similarly, we can find the errors in the intervals $[x_3, x_6], [x_6, x_9], \dots [x_{n-3}, x_n]$. Thus the total error is given by

$$\varepsilon = -\frac{3}{80}h^5[y_0^{iv} + y_3^{iv} + \dots + y_{3n-3}^{iv}]$$

Let $y^{iv}(\xi)$ be the largest value of $y_0^{iv}, y_3^{iv}, \dots y_{n-3}^{iv}$ for some $a < \xi < b$, we get

$$\varepsilon = -\frac{3}{80}h^5[y^{iv}(\xi) + y^{iv}(\xi) + \dots + y^{iv}(\xi)] = -\frac{3}{80}h^5[ny^{iv}(\xi)]$$

$$= -\frac{3}{80}nh^5 y^{iv}(\xi)]$$

$$= -\frac{(b-a)}{80}h^4 y^{iv}(\xi)] \qquad (\because 3nh = b-a)$$

This is the error in the Simpson's 3/8 rule.

6.14 WEDDLE'S RULE

Putting $n = 6$ in general quadrature formula and taking the curve through the points (x_k, y_k); $k = 0, 1, 2, 3, 4, 5, 6$ as a polynomial of degree six so that the differences of order greater than 6 becomes zero. We obtain

$$\int_{x_0}^{x_6} ydx = 6h\left[y_0 + 3\Delta y_0 + \frac{9}{2}\Delta^2 y_0 + 4\Delta^3 y_0 + \frac{123}{60}\Delta^4 y_0 + \frac{11}{20}\Delta^5 y_0 + \frac{41}{840}\Delta^6 y_0\right]$$

$$= \frac{3h}{10}[y_0 + 5y_1 + y_2 + 6y_3 + y_4 + 5y_5 + y_6]$$

Similarly, in the intervals $[x_6, x_{12}], [x_{12}, x_{18}], \dots [x_{n-6}, x_n]$

$$\int_{x_6}^{x_{12}} ydx = \frac{3h}{10}[y_6 + 5y_7 + y_8 + 6y_9 + y_{10} + 5y_{11} + y_{12}]$$

$$\dots\dots\dots\dots\dots\dots\dots\dots\dots\dots\dots\dots\dots\dots\dots\dots\dots\dots\dots$$

$$\int_{x_{n-6}}^{x_n} ydx = \frac{3h}{10}[y_{n-6} + 5y_{n-5} + y_{n-4} + 6y_{n-3} + y_{n-2} + 5y_{n-1} + y_n]$$

Adding all these integrals, we get

$$\int_{x_0}^{x_n} ydx = \frac{3h}{10}[y_0 + 5y_1 + y_2 + 6y_3 + y_4 + 5y_5 + 2y_6 + 5y_7 + y_8 + \dots]$$

This is known as Weddle's rule.

ALGORITHM

Step 1.	Obtain the values of $f(x), x_0, x_n$ and n.
Step 2.	Calculate $\dfrac{x_n - x_0}{n} = h$
Step 3.	Calculate the value of x_i as $x_0 + ih$ and corresponding value of $f(x_i)$.
Step 4.	Substitute the above obtained values in Weddle's Rule.

▶ REMARKS

▸ While applying this rule, the number of subintervals should be taken as a multiple of 6.
▸ In this rule the interpolating polynomial is taken to be of degree 6. That is
$y = ax^6 + bx^5 + cx^4 + dx^3 + ex + f = 0$

6.14.1 ERROR ESTIMATION IN WEDDLE'S RULE

Expand y by Taylor's theorem about $x = x_0$, we get

$$y = y_0 + (x - x_0)y_0' + \frac{(x - x_0)^2}{2!} y_0'' + \frac{(x - x_0)^3}{3!} y_0''' + \frac{(x - x_0)^4}{4!} y_0^{iv}$$

$$+ \frac{(x - x_0)^5}{5!} y_0^{v} + \frac{(x - x_0)^6}{6!} y_0^{vi} + \dots$$

$$\therefore \quad \int_{x_0}^{x_6} y\, dx = (x_6 - x_0)y_0 + \frac{(x_6 - x_0)^2}{2!} y_0' + \frac{(x_6 - x_0)^3}{3!} y_0'' + \frac{(x_6 - x_0)^4}{4!} y_0'''$$

$$+ \frac{(x_6 - x_0)^5}{5!} y_0^{iv} + \frac{(x_6 - x_0)^6}{6!} y_0^{v} + \frac{(x_6 - x_0)^7}{7!} y_0^{vi} + \dots$$

Since $x_6 - x_0 = 6h$

$$\therefore \quad \int_{x_0}^{x_6} y\, dx = 6h y_0 + \frac{36h^2}{2} y_0' + \frac{216h^3}{3!} y_0'' + \frac{1296h^4}{4!} y_0'''$$

$$+ \frac{7776h^5}{5!} y_0^{iv} + \frac{46656h^6}{6!} y_0^{v} + \frac{279936h^7}{7!} y_0^{vi} + \dots \qquad \dots(1)$$

Now applying Taylor's Theorem for $y_1, y_2, y_3, y_4, y_5, y_6$, we get

$$y_1 = y(x_0 + h) = y_0 + h y_0' + \frac{h^2}{2!} y_0'' + \frac{h^3}{3!} y_0''' + \frac{h^4}{4!} y_0^{iv} + \frac{h^5}{5!} y_0^{v} + \frac{h^6}{6!} y_0^{vi} + \dots$$

$$y_2 = y(x_0 + 2h) = y_0 + 2h y_0' + \frac{4h^2}{2!} y_0'' + \frac{8h^3}{3!} y_0''' + \frac{16h^4}{4!} y_0^{iv} + \frac{32h^5}{5!} y_0^{v} + \frac{64h^6}{6!} y_0^{vi} + \dots$$

$$y_3 = y(x_0 + 3h) = y_0 + 3h y_0' + \frac{9h^2}{2!} y_0'' + \frac{27h^3}{3!} y_0''' + \frac{81h^4}{4!} y_0^{iv} + \frac{243h^5}{5!} y_0^{v} + \frac{729h^6}{6!} y_0^{vi} + \dots$$

$$y_4 = y(x_0 + 4h) = y_0 + 4h y_0' + \frac{16h^2}{2!} y_0'' + \frac{64h^3}{3!} y_0''' + \frac{256h^4}{4!} y_0^{iv} + \frac{1024h^5}{5!} y_0^{v} + \frac{4096h^6}{6!} y_0^{vi} + \dots$$

$$y_5 = y(x_0 + 5h) = y_0 + 5h y_0' + \frac{25h^2}{2!} y_0'' + \frac{125h^3}{3!} y_0''' + \frac{625h^4}{4!} y_0^{iv} + \frac{3125h^5}{5!} y_0^{v} + \frac{15625h^6}{6!} y_0^{vi} + \dots$$

$$y_6 = y(x_0 + 6h) = y_0 + 6h y_0' + \frac{36h^2}{2!} y_0'' + \frac{216h^3}{3!} y_0''' + \frac{1296h^4}{4!} y_0^{iv} + \frac{7776h^5}{5!} y_0^{v} + \frac{46656h^6}{6!} y_0^{vi} + \dots$$

The error in the interval $[x_0, x_6]$ is given by

$$\varepsilon = \int_{x_0}^{x_6} y\, dx - \frac{3h}{10}[y_0 + 5y_1 + y_2 + 6y_3 + y_4 + 5y_5 + y_6]$$

$$= \frac{h^7 y_0^{vi}}{7!}[279936 - 279972] = -\frac{36h^7 y_0^{vi}}{7 \times 720} = -\frac{h^7 y_0^{vi}}{140}$$

Similarly, we can find the error in each sub-intervals $[x_6, x_{12}] \dots [x_{n-6}, x_n]$. Hence, the total

error is given by

$$\varepsilon = -\frac{h^7}{140}[y_0^{vi} + y_6^{vi} + \dots + y_{6n-6}^{vi}]$$

Let $y^{vi}(\xi)$ be the largest value of $y_0^{vi}, y_6^{vi}, \dots y_{6n-6}^{vi}$ for some $a < \xi < b$. Then

$$\varepsilon = -\frac{h^7 n}{140}y^{vi}(\xi) \qquad \Rightarrow \qquad \varepsilon = -\frac{(b-a)h^6}{840}y^{vi}(\xi) \qquad (\because 6nh = b - a)$$

This is the error in the Weddle's rule.

Recapitulations ‖‖

- In general, the Weddle's rule is more accurate than Simpson's rule but it requires at least seven consecutive values of the function.
- The geometric meaning of Weddle's rule is that we replace the graph of the given function by $n/6$ acrs of fifth-degree polynomials.

6.15 ROMBERG'S METHOD

We know that, for an interval of width h, the error in trapezoidal rule is

$$= -\frac{(b-a)h^2}{12}f''(\xi), a < \xi < b = Ch^2 \text{ where } C = \frac{-(b-a)}{12}f''(\xi)$$

Suppose, we calculate $I = \int_a^b f(x)dx$ by trapezoidal rule with two different subintervals h_1 and

h_2. Let I_1, I_2 be the approximations with errors E_1 and E_2 respectively. Then, clearly

$$I = I_1 + E_1 = I_1 + Ch_1^2 \qquad \text{and} \qquad I = I_2 + E_2 = I_2 + Ch_2^2$$

Therefore, $\qquad\qquad I_1 = Ch_1^2 + I_2 + Ch_2^2 \qquad\qquad \Rightarrow \qquad\qquad C = \frac{I_1 - I_2}{h_2^2 - h_1^2}$

$$\therefore \qquad\qquad I = I_1 + \left(\frac{I_1 - I_2}{h_2^2 - h_1^2}\right)h_1^2 = \frac{I_1 h_2^2 - I_2 h_1^2}{h_2^2 - h_1^2}$$

which will be a better approximation to I than I_1 or I_2. To calculate systematically we take

$h_1 = h$ and $h_2 = \dfrac{h}{2}$, then, we have

$$I = \frac{I_1\left(\frac{h}{2}\right)^2 - I_2 h^2}{\left(\frac{h}{2}\right)^2 - h^2} = \frac{4I_2 - I_1}{3} = I_2 + \frac{1}{3}(I_2 - I_1)$$

We may find this result by applying Trapezoidal rule two times. By applying the rule several

times, we get a sequence of results $A_1, A_2 \dots$ in which the error is reduced by $\dfrac{1}{4}$ every time.

We apply the formula (1) again to each pair of A's to get improved results $B_1, B_2,...$ etc. Again applying equation (1) to the pairs $B_1, B_2; B_2, B_3$ etc. we get still better result. Continue this process until two successive values are closed to each other.

▶ **REMARK**

▶ The above method is called Richardson's deferred approach to the limit and its systematic improvement is called Romberg integration or method.

6.16 NEWTON-COTE'S FORMULA

The usual strategy is developing formulae of numerical integration is similarly to that for numerical differentiation. We pass a polynomial through points defined by the function, and then integrate this polynomial approximation to the function. This allow us to integrate a function known only as a table of values.

If $f(x)$ has numerical values at certain points, then we can compute the value of the integral of $f(x)$ over (a, b) numerically. As in the problem related to the interpolation, we approximate $f(x)$ by a suitable function $P(x)$, which is usually a polynomial. Suppose the value of $f(x)$ are given at equidistant points of x. Let $f(x)$ be approximated by $P(x)$ given by Lagrange's formula.

$$\therefore \qquad f(x) = P(x) = \sum_{k=0}^{n} L_k(x) y_k$$

where $L_k(x) = \dfrac{(x - x_0)(x - x_1)...(x - x_{k-1})(x - x_{k+1})...(x - x_n)}{(x_k - x_0)(x_k - x_1)...(x_k - x_{k-1})(x_k - x_{k+1})...(x_k - x_n)}$

Here $\qquad x_k = x_0 + xh$ and $x = x_0 + uh$

$\therefore \qquad L_k = L_k(x_0 + uh)$

$$= \dfrac{(x_0 + uh - x_0)(x_0 + uh - x_1)...(x_0 + uh - x_{k-1})(x_0 + uh - x_{k+1})...(x_0 + uh - x_n)}{(x_k - x_0)(x_k - x_1)...(x_k - x_{k-1})(x_k - x_{k+1})...(x_k - x_n)}$$

$$\therefore \qquad L_k = \dfrac{u(u-1)...(u-k+1)(u-k-1)...(u-n)}{k(k-1)...(+1)(-1)...(k-n)} \qquad ...(1)$$

Now find

$$\int_{x_0}^{x_n} f(x)dx = \int_{x_0}^{x_n} \sum_{k=0}^{n} L_k(x) y_k dx = \int_{x_0}^{x_n} [L_0(x)y_0 + L_1(x)y_1 + ... + L_n(x)y_n]dx$$

$$= y_0 \int_{x_0}^{x_n} L_0(x)dx + y_1 \int_{x_0}^{x_n} L_1(x)dx + ... + y_n \int_{x_0}^{x_n} L_n(x)dx$$

$$= \sum_{k=0}^{n} y_k \int_{x_0}^{x_n} L_k(x)dx$$

Since $x = x_0 + uh$ and $\qquad x_n = x_0 + nh$

$$dx = hdu$$

$$\therefore \qquad \int_{x_0}^{x_n} f(x)dx = \sum_{k=0}^{n} y_k h.\int_0^n L_k du = nh \sum_{k=0}^{n} y_k . \frac{1}{n} \int_0^n L_k du,$$

where L_k is given by eq. (1)

Let $\qquad \dfrac{1}{n} \int_0^n L_k du = C_k^n$, we get

$$\int_{x_0}^{x_n} f(x)dx = nh \sum_{k=0}^{n} y_k C_k^n \qquad ...(2)$$

This is known as Newton-Cote's formula, where C_k^n are called Cote's numbers.

6.17 PROPERTIES OF COTE'S NUMBERS

(i) $C_k^n = C_{n-k}^n$

(ii) $\sum\limits_{k=0}^{n} C_k^n = 1$

PROOF.

(i) Consider LHS

$$C_k^n = \frac{1}{n}\int_0^n L_k \, du = \frac{1}{n}\int_0^n \left[\frac{u(u-1)\ldots(u-k+1)(u-k-1)\ldots(u-n)}{k(k-1)\ldots(+1)(-1)\ldots(k-n)} \right] du$$

$$= \frac{1}{n}\int_0^n \frac{u(u-1)\ldots(u-k+1)(u-k-1)\ldots(u-n)}{k!(-1)^{n-k}(n-k)!}$$

Let $u - n = -t \Rightarrow u = n - t$ $\therefore du = -dt$, we get

$$C_k^n = \frac{1}{n}\int_n^0 \frac{(n-t)(n-t-1)\ldots(n-t-k+1)(n-t-k-1)\ldots(-t)}{k!(-1)^{n-k}(n-k)!}(-dt)$$

$$= \frac{1}{n}\int_0^n \frac{(-1)^n t(t-1)\ldots\{t-(n-k)+1\}\{t-(n-k)-1\}\ldots(t-n)}{(n-k)!k!(-1)^{n-(n-k)}(-1)^n} dt$$

$$= \frac{1}{n}\int_0^n \frac{t(t-1)\ldots\{t-(n-k)+1\}\{t-(n-k)-1\}\ldots(t-n)}{(n-k)!k!(-1)^{n-(n-k)}} dt$$

$$= \frac{1}{n}\int_0^n \frac{u(u-1)\ldots\{u-(n-k)-1\}\{u-(n-k)+1\}\ldots(u-n)}{(n-k)!k!(-1)^{n-(n-k)}} du$$

$$= C_{n-k}^n = \text{RHS} \qquad \left(\because \int_0^a f(x)dx = \int_0^a f(t)dt \right)$$

$\therefore \qquad C_k^n = C_{n-k}^n$.

(ii) If $n = 1$, then

$$\sum_{k=0}^{1} C_k^1 = C_0^1 + C_1^1 = \frac{1}{1}\int_0^1 L_0 \, du + \frac{1}{1}\int_0^1 L_1 \, du \qquad \left(\because C_k^n = \frac{1}{n}\int_0^n L_k \, du \right)$$

Now $L_0 = \dfrac{u-1}{0-1} = (1-u)$

$L_1 = \dfrac{u-0}{1-0} = u = \int_0^1 [L_0 + L_1] \, du = \int_0^1 [1 - u + u] \, du = \int_0^1 du = 1$

$\therefore \quad \sum\limits_{k=0}^{1} C_k^1 = 1$

If $n = 2$, then

$$\sum_{k=0}^{2} C_k^2 = C_0^2 + C_1^2 + C_2^2$$

$$= \frac{1}{2}\int_0^2 L_0 \, du + \frac{1}{2}\int_0^2 L_1 \, du + \frac{1}{2}\int_0^2 L_2 \, du = \frac{1}{2}\int_0^2 [L_0 + L_1 + L_2] \, du$$

Now $L_0 = \dfrac{(u-1)(u-2)}{(0-1)(0-2)} = \dfrac{1}{2}(u^2 - 3u + 2)$

$L_1 = \dfrac{u(u-2)}{1(1-2)} = -(u^2 - 2u)$; $L_2 = \dfrac{u(u-1)}{2(2-1)} = \dfrac{1}{2}(u^2 - u)$

$\therefore$ $L_0 + L_1 + L_2 = \dfrac{1}{2}(u^2 - 3u + 2) - (u^2 - 2u) + \dfrac{1}{2}(u^2 - u)$

$$= \dfrac{1}{2}[u^2 - 3u + 2 - 2u^2 + 4u + u^2 - u]$$

$$= \dfrac{1}{2}[2u^2 - 2u^2 - 4u + 4u + 2] = 1$$

$\therefore$ $\displaystyle\sum_{k=0}^{2} C_k^2 = \dfrac{1}{2}\int_0^2 [L_0 + L_1 + L_2]du = \dfrac{1}{2}\int_1^2 1.du = \dfrac{1}{2}(u)_0^2 = \dfrac{1}{2}(2) = 1$

$\therefore$ $\displaystyle\sum_{k=0}^{2} C_k^2 = 1$

Now taking $n = 3$

$$\sum_{k=0}^{3} C_k^3 = C_0^3 + C_1^3 + C_2^3 + C_3^3 = 2C_0^3 + 2C_1^3 \qquad\qquad \left(\because C_{n-k}^n = C_k^n\right)$$

$$= 2[C_0^3 + C_1^3] = 2\left[\dfrac{1}{3}\int_0^3 L_0 du + \dfrac{1}{3}\int_0^3 L_1 du\right] = \dfrac{2}{3}\int_0^3 [L_0 + L_1]du$$

Also,

$$L_0 = \dfrac{(u-1)(u-2)(u-3)}{(0-1)(0-2)(0-3)} = -\dfrac{1}{6}[u^3 - 6u^2 + 11u - 6]$$

$$L_1 = \dfrac{u(u-2)(u-3)}{1(1-2)(1-3)} = \dfrac{1}{2}[u^3 - 5u^2 + 6u]$$

$$L_0 + L_1 = -\dfrac{1}{6}(u^3 - 6u^2 + 11u - 6) + \dfrac{1}{2}(u^3 - 5u^2 + 6u)$$

$$= \dfrac{1}{6}[-u^3 + 6u^2 - 11u + 6 + 3u^3 - 15u^2 + 18u]$$

$$= \dfrac{1}{6}[2u^3 - 9u^2 + 7u + 6]$$

$\therefore$ $\displaystyle\sum_{k=0}^{3} C_k^3 = \dfrac{2}{3}\int_0^3 \dfrac{1}{6}(2u^3 - 9u^2 + 7u + 6)du$

$$= \dfrac{1}{9}\left[\dfrac{u^4}{2} - 3u^3 + \dfrac{7}{2}u^2 + 6u\right]_0^3 = \dfrac{1}{9}\left[\dfrac{81}{2} - 81 + \dfrac{63}{2} + 18\right]$$

$$= \dfrac{1}{9}[72 + 18 - 81] = \dfrac{1}{9}(9) = 1$$

$\therefore$ $\displaystyle\sum_{k=0}^{3} C_k^3 = 1$

Hence, in general we can say that $\displaystyle\sum_{k=0}^{n} C_k^n = 1$.

6.18 DEDUCTIONS FROM NEWTON-COTE'S FORMULA

We known that

$$\int_{x_0}^{x_n} f(x)dx = nh \sum_{k=0}^{n} y_k C_k^n$$

Putting $n = 1$ into this equation, we get

$$\int_0^{x_1} f(x)dx = n \sum_{k=0}^{1} y_k C_k^1 = h[y_0 C_0^1 + y_1 C_1^1] \qquad ...(1)$$

Now find
$$C_0^1 = \frac{1}{1}\int_0^1 L_0 du = \int_0^1 \frac{u-1}{0-1} du \qquad \left(\because L_0 = \frac{u-1}{0-1}\right)$$

$$= \int_0^1 (1-u)du = \left(u - \frac{u^2}{2}\right)_0^1 = 1 - \frac{1}{2} = \frac{1}{2}$$

and
$$C_1^1 = \int_0^1 L_1 du = \int_0^1 \frac{u}{1-0} du = \left(\frac{u^2}{2}\right)_0^2 = \frac{1}{2}$$

Substituting the values of C_0^1 and C_1^1 into eq. (1), we get

$$\int_{x_0}^{x_1} f(x)dx = \frac{h}{2}[y_0 + y_1]$$

This is the Trapezoidal rule.
Now putting $n = 2$ into Newton-Cote's formula, we get

$$\int_{x_0}^{x_2} f(x)dx = 2h \sum_{k=0}^{2} y_k C_k^2 = 2h[y_0 C_0^2 + y_1 C_1^2 + y_2 C_2^2] \qquad ...(2)$$

Find
$$C_0^2 = \frac{1}{2}\int_0^2 L_0 du = \frac{1}{2}\int_0^2 \frac{(u-1)(u-2)}{(0-1)(0-2)} du = \frac{1}{2} \cdot \frac{1}{2}\int_0^2 (u^2 - 3u + 2)du$$

$$= \frac{1}{4}\left[\frac{u^3}{3} - \frac{3u^2}{2} + 2u\right]_0^2 = \frac{1}{4}\left[\frac{8}{3} - \frac{12}{2} + 4\right] = \frac{1}{4}\left[\frac{8}{3} - 2\right] = \frac{1}{6}$$

and
$$C_1^2 = \frac{1}{2}\int_0^2 L_1 du = \frac{1}{2}\int_0^2 \frac{u(u-2)}{1(1-2)} du = -\frac{1}{2}\int_0^2 (u^2 - 2u)du = -\frac{1}{2}\left[\frac{u^3}{3} - u^2\right]_0^2$$

$$= -\frac{1}{2}\left[\frac{8}{3} - 4\right] = -\frac{1}{2}\left[-\frac{4}{3}\right] = \frac{2}{3}$$

and
$$C_2^2 = C_0^2 = \frac{1}{6} \qquad (\because C_k^n = C_{n-k}^n)$$

Substituting the values of C_0^2, C_1^2, C_2^2 into equation (1), we get

$$\int_{x_0}^{x_2} f(x)dx = 2h\left[\frac{1}{6}y_0 + \frac{2}{3}y_1 + \frac{1}{6}y_2\right] = \frac{h}{3}[y_0 + 4y_1 + y_2]$$

This gives Simpson's 1/3 Rule.
Similarly, putting $n = 3$ into Newton-Cote's formula, we get

$$\int_{x_0}^{x_3} f(x)dx = 3h \sum_{k=0}^{3} y_k C_k^3 = 3h[y_0 C_0^3 + y_1 C_1^3 + y_2 C_2^3 + y_3 C_3^3] \qquad ...(3)$$

Now find
$$C_0^3 = \frac{1}{3}\int_0^3 L_0 du = \frac{1}{3}\int_0^3 \left[\frac{(u-1)(u-2)(u-3)}{(0-1)(0-2)(0-3)}\right] du$$

$$= \frac{1}{3}\left(-\frac{1}{6}\right)\int_0^3 (u^3 - 6u^2 + 11u - 6)du$$

$$= -\frac{1}{18}\left[\frac{u^4}{4} - 2u^3 + \frac{11u^2}{2} - 6u\right]_0^3 = -\frac{1}{18}\left[\frac{81}{4} - 54 + \frac{99}{2} - 18\right]$$

$$= -\frac{1}{18}\left[\left(\frac{81+198}{4}\right) - 72\right]$$

$$= -\frac{1}{18}\left[\frac{279}{4} - 72\right] = -\frac{1}{18}\left[\frac{279-288}{4}\right] = -\frac{1}{18}\left[\frac{-9}{4}\right] = \frac{1}{2}\left(\frac{1}{4}\right) = \frac{1}{8}$$

$$\therefore \qquad C_0^3 = C_3^3 = \frac{1}{8} \qquad\qquad (\because C_k^n = C_{n-k}^n)$$

Now

$$C_1^3 = \frac{1}{3}\int_0^3 L_1 du = \frac{1}{3}\int_0^3 \frac{u(u-2)(u-3)}{1(1-2)(1-3)} du = \frac{1}{6}\int_0^3 (u^3 - 5u^2 + 6u)du$$

$$= \frac{1}{6}\left[\frac{u^4}{4} - \frac{5u^3}{3} + 3u^2\right]_0^3 = \frac{1}{6}\left[\frac{81}{4} - \frac{135}{3} + 27\right] = \frac{1}{6}\left[\frac{81}{4} - 45 + 27\right]$$

$$= \frac{1}{6}\left[\frac{81}{4} - 18\right] = \frac{1}{6}\left[\frac{81-72}{4}\right] = \frac{9}{24} = \frac{3}{8}$$

$$\therefore \qquad C_2^3 = C_1^3 = \frac{3}{8}$$

Substituting the values of C_0^3, C_1^3, C_2^3 and C_3^3 in equation (2), we get

$$\int_{x_0}^{x_3} f(x)dx = 3h\left[\frac{y_0}{8} + \frac{3y_1}{8} + \frac{3y_2}{8} + \frac{1}{8}y_3\right] = 3h[y_0 + 3y_1 + 3y_2 + y_3]$$

This gives the Simpson's 3/8 Rule.

The Cote's numbers for some values of $f(x)$ are given in the following table :

n	C_0^n	C_1^n	C_2^n	C_3^n	C_4^n	C_5^n	C_6^n
1	$\dfrac{1}{2}$	$\dfrac{1}{2}$					
2	$\dfrac{1}{6}$	$\dfrac{4}{6}$	$\dfrac{1}{6}$				
3	$\dfrac{1}{8}$	$\dfrac{3}{8}$	$\dfrac{3}{8}$	$\dfrac{1}{8}$			
4	$\dfrac{7}{90}$	$\dfrac{32}{90}$	$\dfrac{12}{90}$	$\dfrac{32}{90}$	$\dfrac{7}{90}$		
5	$\dfrac{19}{288}$	$\dfrac{75}{288}$	$\dfrac{50}{288}$	$\dfrac{50}{288}$	$\dfrac{75}{288}$	$\dfrac{19}{288}$	
6	$\dfrac{41}{840}$	$\dfrac{216}{840}$	$\dfrac{27}{840}$	$\dfrac{272}{840}$	$\dfrac{27}{840}$	$\dfrac{216}{840}$	$\dfrac{41}{840}$

6.19 GAUSS'S QUADRATURE FORMULA

Gauss derived a formula for the integration of a function whose values are given for the values of x which are not equally spaced but are symmetrically placed with respect to the middle point of the interval of integration.

Let $I = \int_a^b y\,dx$ where $y = f(x)$ be computed. Now change the variable x to u by the substitution given by

$$x = (b-a)u + \frac{1}{2}(a+b) \qquad \qquad ...(1)$$

Thus the limit of integration becomes $u = -\frac{1}{2}$ where $x = a$ and $u = \frac{1}{2}$ when $x = b$ and also

$$dx = (b-a)du$$

and

$$y = f(x) = f\left[(b-a)u + \frac{1}{2}(a+b)\right]$$

$$y = \phi(u), \text{ (say)}$$

Then the integral becomes

$$I = (b-a)\int_{-1/2}^{1/2} \phi(u)\,du \qquad \qquad ...(2)$$

Gauss quadrature formula is

$$I = \int_{-1/2}^{1/2} \phi(u)\,du = R_1\phi(u_1) + R_2\phi(u_2) + ... + R_n\phi(u_n)... \qquad ...(3)$$

where $u_1, u_2..., u_n$ are the points of subdivision of the interval $\left(-\frac{1}{2}, \frac{1}{2}\right)$. Thus the corresponding values of x are given by

$$x_k = (b-a)u_k + \frac{1}{2}(a+b), \qquad k = 1, 2, 3, ... n.$$

Therefore, the value of the integral from equation (2) is obtained as

$$I = \int_a^b f(x)dx = (b-a)[R_1\phi(u_1) + R_2\phi(u_2) + ... + R_n\phi(u_n)] \quad ...(4)$$

The derivation of Gauss' formula is out of scope, therefore, we shall not give a detailed derivation but the method for finding the values of $u_1, u_2..., u_n$ and $R_1, R_2, ..., R_n$ be given.

Let us consider a convergent power series $\phi(u)$ in the interval $\left[-\frac{1}{2}, \frac{1}{2}\right]$. We therefore write $\phi(u)$ as follows :

$$\phi(u) = a_0 + a_1u + a_2u^2 + ... + a_mu^m + ... \qquad \qquad ...(5)$$

and also consider that the integral can be expressed as a linear function of the ordinate of the form equation (3).

Now integrating equation (5) from $u = -\frac{1}{2}$ to $\frac{1}{2}$, we get

$$I = \int_{-1/2}^{1/2} \phi(u)du = \int_{-1/2}^{1/2}[a_0 + a_1u + a_2u^2 + ... + a_mu^m + ...]du$$

$$= \left[a_0u + a_1\frac{u^2}{2} + a_2\frac{u^3}{2} + ... + a_m\frac{u^{m+1}}{m+1} + ...\right]_{-1/2}^{1/2}$$

$$= a_0\left[\frac{1}{2} + \frac{1}{2}\right] + \frac{1}{2}a_1\left[\frac{1}{4} - \frac{1}{4}\right] + \frac{1}{3}a_2\left[\frac{1}{8} + \frac{1}{8}\right] + ... + \frac{a_m}{m+1}\left[\left(\frac{1}{2}\right)^{m+1} - \left(-\frac{1}{2}\right)^{m+1}\right] + ...$$

or

$$I = a_0 + \frac{1}{12}a_2 + \frac{1}{80}a_4 + \frac{1}{448}a_6 + \frac{1}{2304}a_8 + ... \qquad \qquad ...(6)$$

Now from equation (5), we also have

$$\phi(u_1) = a_0 + a_1 u_1 + a_2 u_1^2 + ... + a_m u_1^m + ...$$

$$\phi(u_2) = a_0 + a_1 u_2 + a_2 u_2^2 + ... + a_m u_2^m + ...$$

$$\phi(u_3) = a_0 + a_1 u_3 + a_2 u_3^2 + ... + a_m u_3^m + ...$$

....

$$\phi(u_n) = a_0 + a_1 u_n + a_2 u_n^2 + ... + a_m u_n^m + ...$$

Substituting these values into equation (3), we get

$$I = R_1(a_0 + a_1 u_1 + a_2 u_1^2 + ... + a_m u_1^m + ...) + R_2(a_0 + a_1 u_2 + a_2 u_2^2 + ... + a_m u_2^m + ...) + ...$$

$$... + R_n(a_0 + a_1 u_n + a_2 u_n^2 + ... + a_m u_n^m + ...)$$

$$I = a_0(R_1 + R_2 + ... + R_n) + a_1(R_1 u_1 + R_2 u_2 + ... + R_n u_n) + a_2(R_1 u_1^2 + R_2 u_2^2 + ... + R_n u_n^2) + ...$$

$$... + a_m(R_1 u_1^m + R_2 u_2^m + ... + R_n u_n^m) \quad ...(7)$$

From eqs. (6) and (7), we get identically the same for all values of $a_0, a_1, ..., a_m$ etc.

$$\left.\begin{array}{r} R_1 + R_2 + R_3 + ... + R_n = 1 \\ R_1 u_1 + R_2 u_2 + R_3 u_3 + ... + R_n u_n = 0 \\ R_1 u_1^2 + R_2 u_2^2 + R_3 u_3^2 + ... + R_n u_n^2 = \dfrac{1}{12} \\ R_1 u_1^3 + R_2 u_2^3 + R_3 u_3^3 + ... + R_n u_n^3 = 0 \\ R_1 u_1^4 + R_2 u_2^4 + R_3 u_3^4 + ... + R_n u_n^4 = \dfrac{1}{80} \\ .... \\ .... \end{array}\right\} \quad ...(8)$$

If we take $2n$ of these equations and solve them simultaneously, we can find them theoretically because of much labour of solving these equations even for small values of n. Therefore we adopt following method. It can be shown that if $\phi(u)$ is a polynomial of degree not higher than $2n - 1$, then $u_1, u_2, ..., u_n$ are the roots of Legendre polynomial $P_n(u) = 0$. These roots can be calculated from the equation given by

$$\frac{d^n}{du^n}\left[u^2 - \left(\frac{1}{2}\right)^2\right]^n = 0 \quad ...(9)$$

All n roots $u_1, u_2, ..., u_n$ of this nth degree equation are all real. Substituting these values into (8), we get $R_1, R_2, ..., R_n$. Let us take $n = 3$, then from eq. (9)

$$\frac{d^3}{du^3}\left[u^2 - \left(\frac{1}{2}\right)^2\right]^3 = 0$$

$$\Rightarrow \quad \frac{d^3}{du^3}\left[u^6 - \left(\frac{1}{2}\right)^6 - 3u^4\left(\frac{1}{2}\right)^2 + 3u^2\left(\frac{1}{2}\right)^4\right] = 0$$

or $\quad \dfrac{d^3}{du^3}\left[u^6 - \dfrac{3}{4}u^4 + \dfrac{3}{16}u^2 - \dfrac{1}{64}\right] = 0$ or $\quad \dfrac{d^2}{du^2}\left[6u^5 - 3u^3 + \dfrac{3}{8}u\right] = 0$

or $\quad \dfrac{d}{du}\left[30u^4 - 9u^2 + \dfrac{3}{8}\right] = 0$ or $\quad 120u^3 - 18u = 0$

or
$$6u(20u^2 - 3) = 0$$

or
$$u = 0, u = \pm\sqrt{\frac{3}{20}} \pm \left(\frac{1}{2}\right)\sqrt{\frac{3}{5}}$$

i.e.,
$$u_1 = -\frac{1}{2}\sqrt{\frac{3}{5}}, u_2 = 0, u_3 = \frac{1}{2}\sqrt{\frac{3}{5}}$$

Now equation (8) takes the form for $n = 3$,
$$R_1 + R_2 + R_3 = 1; \quad R_1 u_1 + R_2 u_2 + R_3 u_3 = 0; \quad R_1 u_1^2 + R_2 u_2^2 + R_3 u_3^2 = \frac{1}{12}$$

Substitute the value of u_1, u_2 and u_3 into above equations, we get
$$R_1 + R_2 + R_3 = 1; \qquad R_3 - R_1 = 0$$

$$(R_1 + R_2)\frac{3}{20} = \frac{1}{12}; \qquad R_1 + R_2 = \frac{20}{36}$$

On solving these equations, we get
$$R_1 = \frac{5}{18}, R_2 = \frac{4}{9}, R_3 = \frac{5}{18}$$

From the values of u_1, u_2 and u_3 it is observed that u's are symmetrically placed with respect to the middle point of the interval [–1/2, 1/2]. Also we observe that R's are symmetrically placed with respect to each pair of u's. In the above case u_0 is taken to be u_2 and u_1 and u_3 are taken to be u_{-1} and u_{+1} respectively. Thus u_1, u_2, u_3 can be taken as u_{-1}, u_0, u_{+1}. Similarly $u_{\pm 2}$ are taken next pair of symmetric points.

The numerical values of the u's and corresponding R's for $n = 2$ to $n = 10$ are given in the following table, where $u_{\pm k} = N$ means $u_k = N, u_{-k} = -N$.

n	u	R
$n = 2$	$u_1 = \pm 0.2886751346$	$R = 1/2$
$n = 3$	$u_0 = 0$	$R = 4/9$
	$u_1 = \pm 0.3872983346$	$R = 5/18$
$n = 4$	$u_1 = \pm 0.1699905218$	$R = 0.3260725744$
	$u_2 = \pm 0.4305681558$	$R = 0.1739274226$
$n = 5$	$u_0 = 0$	$R = 64/225$
	$u_1 = \pm 0.2692346551$	$R = 0.2393143352$
	$u_2 = \pm 0.4530899230$	$R = 0.1184634425$
$n = 6$	$u_1 = \pm 0.1193095930$	$R = 0.2339569673$
	$u_2 = \pm 0.3306046932$	$R = 0.1803807865$
	$u_3 = \pm 0.4662347571$	$R = 0.0856622419$
$n = 7$	$u_0 = 0$	$R = 0.2089795918$
	$u_1 = \pm 0.2029225757$	$R = 0.1909150253$
	$u_2 = \pm 0.3707655928$	$R = 0.1398526957$
	$u_3 = \pm 0.4745539562$	$R = 0.06474248308$

$n = 8$	$u_1 = \pm 0.0917173212$	$R = 0.1813418917$
	$u_2 = \pm 0.2627662050$	$R = 0.1568533229$
	$u_3 = \pm 0.39883332387$	$R = 0.1111905172$
	$u_4 = \pm 0.4801449282$	$R = 0.05061426815$
$n = 9$	$u_0 = 0$	$R = 0.1651196775$
	$u_1 = \pm 0.1621267117$	$R = 0.1561735385$
	$u_2 = \pm 0.3066857164$	$R = 0.1303053482$
	$u_3 = \pm 0.4180155537$	$R = 0.09032408035$
	$u_4 = \pm 0.4840801198$	$R = 0.4063719418$
$n = 10$	$u_1 = \pm 0.0744371695$	$R = 01477621124$
	$u_2 = \pm 0.2166976971$	$R = 0.1346333597$
	$u_3 = \pm 0.3397047841$	$R = 0.1095431813$
	$u_4 = \pm 0.4325316833$	$R = 0.07472567458$
	$u_5 = \pm 0.4869532643$	$R = 0.3333567215$

REMARKS

▶ In Simpson's and Weddle's formulae the ordinates are equally spaced while in Gauss formula ordinates are not equally spaced.

▶ In this formula we shall subdivide the interval (a, b) by means of points which shall not equidistant but shall be symmetrically placed with respect to the middle points of the interval.

6.20 CHEBYCHEV'S FORMULA

Chebychev's derived a quadrature formula in which the coefficients of the y's are all equal while the points of subdivision of the interval of integration are symmetrically placed with respect to the middle point of the interval.

Let
$$I = \int_a^b y\,dx\,, \qquad \text{where } y = f(x) \qquad \qquad ...(1)$$

be calculated. Now change the variable x to u by

$$x = (b - a)u + \frac{1}{2}(a + b)$$

and let
$$y = f(x) = f\left[(b - a)u + \frac{1}{2}(a + b)\right] = \phi(u)\,, \text{ say}$$

Then the integral eq. (1) becomes

$$I = (b - a)\int_{-1/2}^{1/2} \phi(u)\,du$$

Thus Chebychev's formula is

$$I = \int_a^b y\,dx = \frac{b - a}{n}[\phi(u_1) + \phi(u_2) + ... + \phi(u_n)] \qquad ...(2)$$

where $u_1, u_2, ..., u_n$ are the points of subdivision of the interval $\left(-\frac{1}{2}, \frac{1}{2}\right)$ and related with x by

$$x_k = (b - a)u_k + \frac{1}{2}(a + b) \quad \text{for } k = 1, 2, 3, ..., n.$$

The u's are the zeros of certain polynomials, some of which are equated to zero, we get

$$(2u)^2 - \frac{1}{3} = 0 \quad \Rightarrow \quad (2u)^3 - \frac{1}{2}(2u) = 0$$

$$(2u)^4 - \frac{2}{3}(2u)^2 + \frac{1}{45} = 0 \quad \Rightarrow \quad (2u)^5 - \frac{5}{6}(2u)^3 + \frac{7}{22}(2u) = 0$$

$$(2u)^6 - (2u)^4 + \frac{1}{5}(2u)^2 - \frac{1}{105} = 0$$

The values of u's are given below for $n = 2$ to $n = 7$ and $n = 9$.

n	u's
$n = 2$	$u_1 = \pm 0.288675$
$n = 3$	$u_0 = 0; u_1 = \pm 0353553$
$n = 4$	$u_1 = \pm 0.0937962; u_2 = \pm 0.397327$
$n = 5$	$u_0 = 0; u_1 = \pm 0.187271; u_2 = \pm 0.416249$
$n = 6$	$u_1 = \pm 0.133318; u_2 = \pm 0.211259; u_3 = \pm 0.433123$
$n = 7$	$u_0 = 0; u_1 = \pm 0.161956; u_2 = \pm 0.264828; u_3 = \pm 0.441931$
$n = 9$	$u_0 = 0; u_1 = \pm 0.0839531; u_2 = \pm 0.264381; u_3 = \pm 0.300509; u_4 = \pm 0.455795$

Chebychev's formula can also be obtained directly in terms of y's at the points of subdivision. For this let us consider a polynomial

$$y = a_0 + a_1 x + a_2 x^2 + a_3 x^3 + a_4 x^4 + a_5 x^5 + a_6 x^6 + a_7 x^7 + a_8 x^8 \qquad ...(3)$$

with the help of two given conditions

(i) the coefficients of all y's are to be equal and

(ii) the arguments are taken to be symmetrically placed with respect to the mid point of interval of integration so that take the middle point of the interval to be origin and let $x_1, -x_1, x_2, -x_2, x_3, -x_3$ etc. are the distances from the origin to the points of subdivision and let $y_1, -y_1, y_2, -y_2, y_3, -y_3$ and so on be the values of the function $y = f(x)$ corresponding the given arguments. Let A be the area of region bounded by $y = f(x)$ given in (3) and x-axis between the limits of the integration.

Thus, we get $\qquad A = k(y_0 + y_1 + y_{-1} + y_2 + y_{-2} + y_3 + y_{-3} + y_4 + y_{-4}) \qquad ...(4)$

where k is the common coefficient of y's. Let l be the length of the interval of integration i.e., $l = b - a$. Then

$$A = \int_{-l/2}^{l/2} y \, dx$$

$$= \int_{-l/2}^{l/2} (a_0 + a_1 x + a_2 x^2 + a_3 x^3 + a_4 x^4 + a_5 x^5 + a_6 x^6 + a_7 x^7 + a_8 x^8) \, dx$$

$$= \left[a_0 x + \frac{a_1}{2} x^2 + \frac{a_2}{3} x^3 + \frac{a_3}{4} x^4 + \frac{a_4}{5} x^5 + \frac{a_5}{6} x^6 + \frac{a_6}{7} x^7 + \frac{a_7}{8} x^8 + \frac{a_8}{9} x^9 \right]_{-l/2}^{l/2}$$

$$A = a_0 l + \frac{a_2}{12} l^3 + \frac{a_4}{80} l^5 + \frac{a_6}{448} l^7 + \frac{98}{2304} l^9 \qquad ...(5)$$

Now find the values of $y_0, y_1, y_{-1}, \ldots$ by putting $x = 0, x_1, -x_1 \ldots$ into eq. (3), we get

$$y_0 = a_0$$

$$y_1 + y_{-1} = 2(a_0 + a_2x_1^2 + a_4x_1^4 + a_6x_1^6 + a_8x_1^8)$$

$$y_2 + y_{-2} = 2(a_0 + a_2x_2^2 + a_4x_2^4 + a_6x_2^6 + a_8x_2^8)$$

$$y_3 + y_{-3} = 2(a_0 + a_2x_3^2 + a_4x_3^4 + a_6x_3^6 + a_8x_3^8)$$

and
$$y_4 + y_{-4} = 2(a_0 + a_2x_4^2 + a_4x_4^4 + a_6x_4^6 + a_8x_4^8)$$

Substituting these values of y's into eq. (4), we get

$$A = k[9a_0 + 2a_2(x_1^2 + x_2^2 + x_3^2 + x_4^2) + 2a_4(x_1^4 + x_2^4 + x_3^4 + x_4^4)$$
$$+2a_6(x_1^6 + x_2^6 + x_3^6 + x_4^6) + 2a_8(x_1^8 + x_2^8 + x_3^8 + x_4^8)] \quad ...(6)$$

Now equating the coefficient of a's in equations (5) and (6), we get $k = \dfrac{l}{9}$

and
$$\left.\begin{aligned} x_1^2 + x_2^2 + x_3^2 + x_4^2 &= \frac{9l^2}{24} \\ x_1^4 + x_2^4 + x_3^4 + x_4^4 &= \frac{9l^2}{160} \\ x_1^6 + x_2^6 + x_3^6 + x_4^6 &= \frac{9l^2}{896} \\ x_1^8 + x_2^8 + x_3^8 + x_4^8 &= \frac{9l^2}{4608} \end{aligned}\right\} \quad ...(7)$$

The equations given in (7) cannot be solved easily but for the values

$$x_1 = 0.083951l, x_2 = 0.264381l, x_3 = 0.300509l, x_4 = 0.455795l$$

above equation can easily verify.

Hence, for $n = 9$, we have $u_1 = \dfrac{x_1}{l}, u_2 = \dfrac{x_2}{l}$ etc.

Consequently, Chebychev's formula in terms of y's is thus for $n = 9$

$$A = \frac{l}{9}(y_0 + y_1 + y_{-1} + y_2 + y_{-2} + y_3 + y_{-3} + y_4 + y_{-4})$$

where all the y's are measured at the subdivision of the interval of integration.

REMARKS

- In this formula, the coefficients of y's are taken to be same while in Gauss quadrature formula these are different.
- This formula is particularly appropriate for use when the functional values are found by measurement.
- In this case the positive and negative errors of the measurement will largely cancel one another.

Recapitulations

- The Trepezodial rule gives the exact result when the integral is a linear polynomial.
- In Simpson's 1/3 rule, curve is a parabola through (x_0, y_0), (x_1, y_1) and (x_2, y_2).
- Geometrically in 1/3 rule, we find the area of two strips at a time, hence the given interval is divided into even number of equal subintervals.
- In Simpson's 1/3 rule, curve is a parabola through (x_0, y_0), (x_1, y_1) and (x_2, y_2).

SOLVED EXAMPLES

(A) BASED ON TRAPEZOIDAL RULE

EXAMPLE 1. *Evaluate $\int_0^6 \frac{dx}{1+x^2}$.* (Kurukshetra-2004, 06; Rohilkhand-2008, 10; Meerut-2010, 14; Bhopal-2005, 09; (Rohtak–2004, 06)

SOLUTION. Divide the interval [0, 6] into six parts each of width $h = 1$. The values of $f(x) = \frac{1}{1+x^2}$ are given below :

x	0	1	2	3	4	5	6
$y = f(x)$	1	0.5	0.2	0.1	0.0588	0.0385	0.027

Here $y_0 = 1, y_1 = 0.5, y_2 = 0.2, y_3 = 0.1, \ y_4 = 0.0588,$
$y_5 = 0.0385$ and $y_6 = 0.027$.

By Trapezoidal Rule

$$\int_0^6 \frac{dx}{1+x^2} = \frac{h}{2}[y_0 + 2(y_1 + y_2 + y_3 + y_4 + y_5) + y_6]$$

We have

$$\int_0^6 \frac{dx}{1+x^2} = \frac{1}{2}[1 + 2(0.5 + 0.2 + 0.1 + 0.0588 + 0.0385) + 0.027] = 1.7573.$$

EXAMPLE 2. *Calculate the value of the integral $\int_4^{5.2} \log x dx$.* (JNTU–2006, Kerala–2003, VTU–2008)

SOLUTION. Taking $h = 0.2$ and divide [4, 5.2] into six equal parts. Then the values of $\log x$ for each points of subdivision are given below :

x	4	4.2	4.4	4.6	4.8	5.0	5.2
$y = \log x$	1.38629	1.43508	1.48160	1.52605	1.56861	1.60943	1.64865

Here $y_0 = 1.38629, y_1 = 1.43508, y_2 = 1.48160, y_3 = 1.52605,$

$y_4 = 1.56861, y_5 = 1.60943$ and $y_6 = 1.64865$.

Then by Trapezoidal rule

$$\int_{4.0}^{5.2} \log x dx = \frac{h}{2}[y_0 + 2(y_1 + y_2 + y_3 + y_4 + y_5) + y_6]$$

$$= \frac{0.2}{2}[1.38629 + 2(1.43508 + 1.48160 + 1.52605 + 1.56861 + 1.60943) + 1.64865]$$

$$= 0.1[3.03494 + 2(7.62077)] = 0.1[18.27648] = 1.827648$$

EXAMPLE 3. *Evaluate the integral $\int_{0.2}^{1.4} (\sin x - \log_e x + e^x)dx$.* (IAS-2009; Kurukshetra-2005, 07) (Kurukshatra–2005, 07, Mumbai–2005)

SOLUTION. Divide the internal [0.2, 1.41] into 12 equal parts each of width $h = 0.1$. Then $x_0 = 0.2, x_1 = 0.3, x_2 = 0.4, x_3 = 0.5, x_4 = 0.6, x_5 = 0.7, x_6 = 0.8, x_7 = 0.9, x_8 = 1.0, x_9 = 1.1, x_{10} = 1.2, x_{11} = 1.3, x_{12} = 1.4$. Then the values of

$y = f(x) = \sin x - \log_e x + e^x$ are given below :

x	$\sin x$	$\log_e x$	e^x	$y = f(x) = \sin x - \log_e x + e^x$
0.2	0.19867	−1.60943	1.22140	$3.02950 = y_0$
0.3	0.29552	−1.20347	1.34986	$2.84935 = y_1$
0.4	0.38942	−0.91629	1.49182	$2.79753 = y_2$
0.5	0.47943	−0.69315	1.64872	$2.82130 = y_3$
0.6	0.56464	−0.51083	1.82212	$2.89759 = y_4$
0.7	0.64422	−0.35667	2.01375	$3.01464 = y_5$
0.8	0.71736	−0.22314	2.52255	$3.16604 = y_6$
0.9	0.78333	−0.10536	2.45960	$3.34829 = y_7$
1.0	0.84147	0.0000	2.71828	$3.55975 = y_8$
1.1	0.89121	0.09531	3.00417	$3.80007 = y_9$
1.2	0.93204	0.18252	3.32012	$4.06984 = y_{10}$
1.3	0.96356	0.26236	3.66930	$4.37050 = y_{11}$
1.4	0.98545	0.33647	4.05520	$4.70418 = y_{12}$

Then by Trapezoidal Rule

$$\int_{0.2}^{1.4} (\sin x - \log_e x + e^x)dx = \frac{h}{2}[y_0 + 2\{y_1 + y_2 + y_3 + y_4 + y_5 + y_6 + y_7 + y_8$$
$$+ y_9 + y_{10} + y_{11}) + y_{12}\}]$$

$$= \frac{0.1}{2}[3.02950 + 2(2.84935 + 2.79753 + 2.82130 + 2.89759$$
$$+ 3.01464 + 3.16604 + 3.34829 + 3.55975 + 3.80007 + 4.06984$$
$$+ 4.37050) + 4.70418]$$

$$= \frac{0.1}{2}[7.73368 + 2(36.6949)] = 0.05(81.12348) = 4.056174$$

EXAMPLE 4. *From the following table, find the area bounded by the curve and the x-axis from*
x = 7.47 to x = 7.52. (UPTU–2004)

x	7.47	7.48	7.49	7.50	7.51	7.52
$f(x)$	1.93	1.95	1.98	2.01	2.03	2.06

SOLUTION. The required area is given by the integral
Here $h = 0.01$, $y_0 = 1.93$, $y_1 = 1.95$, $y_2 = 1.98$, $y_3 = 2.01$, $y_4 = 2.03$ and
$y_5 = 2.06$. Then by trapezoidal rule

$$A = \int_{7.47}^{7.52} f(x)dx$$

$$= \frac{h}{2}[y_0 + 2(y_1 + y_2 + y_3 + y_4) + y_5]$$

$$= \frac{0.01}{2}[1.93 + 2(1.95 + 1.98 + 2.01 + 2.03) + 2.06]$$

$$= 0.005[3.99 + 2(7.97)] = 0.005[19.93] = 0.09965.$$

EXAMPLE 5. *Calculate an approximate value of integral $\int_0^{\pi/2} \sin x\, dx$.*

SOLUTION. Divide $[0, \pi/2]$ into ten equal parts each of width $h = \dfrac{\pi}{20}$. Then the values of $y = \sin x$ are given in table below :

x	$y = \sin x$	x	$y = \sin x$
0	0.00000	$6\pi/20$	0.80902
$\pi/20$	0.15643	$7\pi/20$	0.89101
$2\pi/20$	0.30902	$8\pi/20$	0.95106
$3\pi/20$	0.45399	$9\pi/20$	0.98769
$4\pi/20$	0.58779	$10\pi/20$	1.00000
$5\pi/20$	0.70711		

Hence, $y_0 = 0.00000, y_1 = 0.15643, y_2 = 0.30902, y_3 = 0.45399, y_4 = 0.58779,$
$\quad\quad y_5 = 0.70711,\ y_6 = 0.80902, y_7 = 0.89101, y_8 = 0.95106, y_9 = 0.98769$
and $\ y_{10} = 1.00000$.
Then by trapezoidal rule

$$\int_0^{\pi/2} \sin x\, dx = \frac{h}{2}[y_0 + 2(y_1 + y_2 + y_3 + y_4 + y_5 + y_6 + y_7 + y_8 + y_9) + y_{10}]$$

$$= \frac{\pi/20}{2}[0 + 2(0.15643 + 0.30902 + 0.45399 + 0.58779$$

$$+ 0.70711 + 0.80902 + 0.89101 + 0.95106 + 0.98764) + 1.00000]$$

$$= \frac{\pi}{40}[1.00000 + 2(5.85312)] \ = \frac{\pi}{40}(12.70624) = 0.99795\,.$$

EXAMPLE 6. *Evaluate the integral $\int_0^1 \dfrac{1}{1+x}\, dx$.* $\hspace{1cm}$ (UPTU(B. Tech.)–2005, MCA–2008)

SOLUTION. Let $h = 0.125$ and $y = f(x) = \dfrac{1}{1+x}$, then the values of y are given for the arguments which are obtained by dividing the interval $[0, 1]$ into eight equal parts as given below :

x	0	0.125	0.250	0.375	0.5	0.625	0.750	0.875	1.0
$y = \dfrac{1}{1+x}$	1.0	0.8889	0.8000	0.7273	0.6667	0.6154	0.5714	0.5333	0.5

Hence $y_0 = 1.0, y_1 = 0.8889, y_2 = 0.8000, y_3 = 0.7273, y_4 = 0.6667, y_5 = 0.6154,$
$y_6 = 0.5714, y_7 = 0.5333, y_8 = 0.5$.
Now, by Trapezoidal Rule

$$\int_0^1 \frac{1}{1+x}\, dx = \frac{h}{2}[y_0 + 2(y_1 + y_2 + y_3 + y_4 + y_5 + y_6 + y_7) + y_8]$$

$$= \frac{0.125}{2}[1.0 + 2(0.8889 + 0.8000 + 0.7273 + 0.6667$$

$$+ 0.6154 + 0.5714 + 0.5333) + 0.5]$$

$$= \frac{0.125}{2}[1.5 + 2(4.803)] \ = \frac{0.125}{2}[11.106] = 0.69413\,.$$

(B) BASED ON SIMPSON'S RULE

EXAMPLE 7. *Find the value of the integral $\int_0^1 \dfrac{dx}{1+x^2}$ by using Simpson's $\dfrac{1}{3}$ and $\dfrac{3}{8}$ rule. Hence obtain the approximate value of π in each case.* (Delhi-2010; Meerut-2001, 03, Avadh-2004,10; Rohtak–2006, UPTU(MCA)–2006, 07, JNTU–2008, UPTU–2010, VTU–2007, Bhopal–2009; IAS-2000,06)

SOLUTION. Divide the interval [0, 1] into six equal parts each of width $h = \dfrac{1}{6}$. Then the values of $y = f(x) = \dfrac{1}{1+x^2}$ at each points of subdivisions are given below :

x	$y = \dfrac{1}{1+x^2}$
0	1.0000
1/6	0.9729
2/6	0.9000
3/6	0.8000
4/6	0.6923
5/6	0.5901
1	0.5000

Here, $y_0 = 1.000, y_1 = 0.9729, y_2 = 0.9000, y_3 = 0.8000, y_4 = 0.6923, y_5 = 0.5901$, and $y_6 = 0.5000$.

By Simpson's $\dfrac{1}{3}$ Rule

$$\int_0^1 \frac{dx}{1+x^2} = \frac{h}{3}[y_0 + 4(y_1 + y_3 + y_5) + 2(y_2 + y_4) + y_6)]$$

$$= \frac{1}{18}[1.000 + 4(0.9729 + 0.8000 + 0.5901)$$
$$+ 2(0.9000 + 0.6923) + 0.5000] = 0.785367$$

By Simpson's $\dfrac{3}{8}$ Rule

$$\int_0^1 \frac{dx}{1+x^2} = \frac{3h}{8}[y_0 + 4(y_1 + y_2 + y_4 + y_5) + 2y_3 + y_6]$$

$$= \frac{3 \times \dfrac{1}{6}}{8}[1.0000 + 3(0.9729 + 0.9000 + 0.6923 + 0.5901) + 2(0.8000) + 0.5000]$$

$$= \frac{1}{16}[1.5 + 3(3.1553) + 1.6] = \frac{1}{16}(12.5659) = 0.785369$$

But $\int_0^1 \dfrac{dx}{1+x^2} = [\tan^{-1} x]_0^1 = \tan^{-1} 1 - \tan^{-1} 0 = \dfrac{\pi}{4}$

In case of Simpson's $\dfrac{1}{3}$ Rule

$$\frac{\pi}{4} = 0.785367 \quad \text{or} \quad \pi = 4(0.785367) = 3.141468$$

In case of Simpson's $\frac{3}{8}$ Rule

$$\frac{\pi}{4} = 0.785395 \quad \text{or} \quad \pi = 4(0.785395) = 3.141476$$

EXAMPLE 8. *Calculate the value of the integral $\int_2^{10} \frac{dx}{1+x}$ using Simpson's $\frac{1}{3}$ rule by dividing the interval [2, 10] into eight equal parts upto 4 decimal places.*

(Meerut-2001, 07; Rajasthan-2008, 10)

SOLUTION. Divide the interval [2, 10] into 8 equal parts so that width $h = \frac{10-2}{8} = \frac{8}{8} = 1$ and the values of $y = f(x) = \frac{1}{1+x}$ for each point of subdivision are given below :

x	$y = \dfrac{1}{1+x^2}$	x	$y = \dfrac{1}{1+x^2}$
2	0.33333	7	0.12500
3	0.25000	8	0.11111
4	0.20000	9	0.10000
5	0.16667	10	0.09091
6	0.14286		

Here, $y_0 = 0.33333$, $y_1 = 0.250000$, $y_2 = 0.20000$, $y_3 = 0.16667$, $y_4 = 0.14286$, $y_5 = 0.125000$, $y_6 = 0.11111$, $y_7 = 0.10000$, $y_8 = 0.09091$.

By Simpson's $\frac{1}{3}$ Rule, we get

$$\int_2^{10} \frac{dx}{1+x} = \frac{h}{3}[y_0 + 4(y_1 + y_3 + y_5 + y_7) + 2(y_2 + y_4) + y_6]$$

$$= \frac{1}{3}[0.33333 + 4(0.25000 + 0.16667 + 0.12500$$

$$+ 0.10000) + 2(0.20000 + 0.14286 + 0.11111) + 0.09091]$$

$$= \frac{1}{3}[0.42424 + 4(0.64167) + 2(0.45397)]$$

$$= \frac{1}{3}[0.42424 + 2.56668 + 0.90794] = \frac{1}{3}[3.89886] = 1.29962 \cdot$$

EXAMPLE 9. *Evaluate $\int_0^6 \frac{dx}{1+x^2}$ by Simpson's $\frac{1}{3}$ Rule.*

(UPTU–2002, MDU(B.E.)–2004, 06, Kurukshetra–2004, 07, Mumbai–2005;
Bhopal-2005, 09; Rohilkhand-2008, 10)

SOLUTION. Divide the interval [0, 6] into 6 equal parts so that width of each subinterval $h = \frac{6-0}{6} = 1$. The values of $y = \frac{1}{1+x^2}$ at each points of subdivision are given in the table :

x	0	1	2	3	4	5	6
y	1.00000	0.50000	0.20000	0.10000	0.58824	0.03846	0.02702

Here, $y_0 = 1.00000, y_1 = 0.50000, y_2 = 0.20000, y_3 = 0.10000, y_4 = 0.58824,$ $y_5 = 0.03846$ and $y_6 = 0.02702$.

Then by Simpson's Rule

$$\int_0^6 \frac{dx}{1+x^2} = \frac{h}{3}[y_0 + 4(y_1 + y_3 + y_5) + 2(y_2 + y_4) + y_6]$$

$$= \frac{1}{3}[1.0000 + 4(0.50000 + 0.10000 + 0.03846)$$

$$+ 2(0.20000 + 0.58824) + 0.02702]$$

$$= \frac{1}{3}[1.02702 + 4(0.63846) + 2(0.78824)]$$

$$= \frac{1}{3}[1.02702 + 2.55384 + 1.57648] = \frac{1}{3}[5.15734] = 1.71911 \,.$$

EXAMPLE 10. *Evaluate the integral $\int_0^4 e^x dx$ by Simpson's 1/3 rule.*

SOLUTION. Divide the interval [0 ,4] into 4 equal parts so that with $h = \dfrac{4-0}{4} = 1$. The value of $y = f(x) = e^x$ are given at each points of subdivision as follows :

x	0	1	2	3	4
y	1	27.2	7.39	20.09	54.60

Here, $y_0 = 1, y_1 = 27.2, y_2 = 7.39, y_3 = 20.09$ and $y_4 = 54.60$

By Simpson's 1/3 rule, we get

$$\int_0^4 e^x dx = \frac{h}{3}[y_0 + 4(y_1 + y_3) + 2y_2 + y_4]$$

$$= \frac{1}{3}[1 + 4(2.72 + 20.09) + 2(7.39) + 54.60]$$

$$= \frac{1}{3}[55.60 + 4(22.81) + 14.78] = 53.87$$

EXAMPLE 11. *Use Simpson's 1/3 rule to prove that $\log_e 7$ is approximately 1.9587 using $\int_1^7 \frac{dx}{x}$.*

(UPPCS-2006, IAS-2003, Lucknow-2005, 09; Allahabad-2003, 07; Bangluru-2008)

SOLUTION. Divide the interval [1, 7] into six equal parts of width $h = \dfrac{7-1}{6} = \dfrac{6}{6}$. The value of $y = f(x) = \dfrac{1}{x}$ for each points of subdivision are given as under:

x	1	2	3	4	5	6	7
y = 1/x	1	0.5	0.3333	0.25	0.2	0.1667	0.1429

Here , $y_0 = 1, y_1 = 0.5, y_2 = 0.3333, y_3 = 0.25, y_4 = 0.2, y_5 = 0.1667$ and $y_6 = 0.1429$. Then by Simpson's 1/3 rule, we get

$$\int_1^7 \frac{dx}{x} = \frac{h}{3}[y_0 + 4(y_1 + y_2 + y_3) + 2(y_2 + y_4) + y_6]$$

$$= \frac{1}{3}[1 + 4(0.5 + 0.25 + 0.1667) + 2(0.3333 + 0.2) + 0.1429]$$

$$= \frac{1}{3}[1.1429 + 4(0.9167) + 2(0.5333)]$$

$$= \frac{1}{3}[1.1429 + 3.6668 + 1.0666] = \frac{1}{3}(5.8763) = 1.9587$$

The exact value of $\int_1^7 \dfrac{dx}{x} = [\log x]_1^7 = \log_e 7$

$$= \log e = 1.9587$$

EXAMPLE 12. *Evaluate the integral $\int_{0.5}^{0.7} x^{1/2} e^{-x} dx$ by Simpson's 1/3 rule.*

(MDU-2003, 07, Rohilkhand-2001, 04, 09)

SOLUTION. Divide the interval [0.5, 0.7] into four equal parts of width $h = \dfrac{0.7 - 0.5}{4} = 0.05$.

The value of $y = x^{1/2} e^{-x}$ for each points of sub-division are given as under

x	0.5	0.55	0.60	0.65	0.70
$y = e^{-x} x^{1/2}$	0.42888	0.42787	0.42511	0.42088	0.41547

Here, $y_0 = 0.42888, y_1 = 0.42787, y_2 = 0.42511, y_3 = 0.42088 , y_4 = 0.41547$

The Simpson's 1/3 rule, we get

$$\int_{0.5}^{0.7} x^{1/2} e^{-x} dx = \frac{h}{3}[y_0 + 4(y_1 + y_3) + 2y_2 + y_4]$$

$$= \frac{0.05}{3}[0.42888 + 4(0.42787 + 0.42088 + 2(0.42511) + 0.41547]$$

$$\left[\frac{d^2 y}{dx^2}\right] = \frac{1}{h^2}\left[\Delta^2 y_{-1} - \frac{1}{12}\Delta^4 y_{-2} + ...\right]$$

$$= \frac{0.05}{3}(5.08957) = 0.08483$$

EXAMPLE 13. *Using Simpson's $\dfrac{3}{8}$ rule, evaluate $\int_0^1 \dfrac{1}{1+x} dx$ with $h = \dfrac{1}{6}$.*   (UPTU–2010, VTU–2007)

SOLUTION. Divide the interval [0, 1] into 6 equal parts of width $h = \dfrac{1}{6}$. The values of

$y = \dfrac{1}{1+x}$ for each points of sub division are given below :

x	0	1/6	2/6	3/6	4/6	5/6	1
$y = \dfrac{1}{1+x}$	1	0.85714	0.75	0.66667	0.6	0.54545	0.5

Here, $y_0 = 1, y_1 = 0.85714, y_2 = 0.75, y_3 = 0.66667, y_4 = 0.6, y_5 = 0.54545$

and $y_6 = 0.5$.

Now by Simpson's $\dfrac{3}{8}$ rule, we get

$$\int_0^1 \frac{1}{1+x} dx = \frac{3h}{8}[y_0 + 3(y_1 + y_2 + y_4 + y_5) + 2y_3 + y_6]$$

$$= \frac{3 \times \dfrac{1}{6}}{8}[1 + 3(0.85714 + 0.75 + 0.6 + 0.54545) + 2(0.66667) + 0.5]$$

$$= \frac{1}{16}[1.5 + 3(2.75259) + 1.33334] = \frac{1}{16}[11.091111] = 0.69319$$

EXAMPLE 14. *Evaluate the integral $\int_1^2 \frac{1}{x}dx$ by Simpson's $\frac{1}{3}$ rule with 4 strips and 8 strips respectively. Determine the error by direct integration.*   (UPTU-2003)

SOLUTION.   **Case I :** Divide the interval [1, 2] into 4 equal parts of each of width $h = \frac{2-1}{4} = \frac{1}{4} = 0.25$. The values of $y = \frac{1}{x}$ at each points of subintervals are given below :

x	1	5/4	6/4	7/4	2
$y = 1/x$	1	0.8	0.66667	0.57143	0.5

Here, $y_0 = 1, y_1 = 0.8, y_2 = 0.66667, y_3 = 0.57143$ and $y_4 = 0.5$.

By Simpson's $\frac{1}{3}$ rule, we get

$$\int_1^2 \frac{1}{x}dx = \frac{h}{3}[y_0 + 4(y_1 + y_3) + 2y_2 + y_4]$$

$$= \frac{0.25}{3}[1 + 4(0.8 + 0.57143) + 2(0.66667) + 0.5]$$

$$= \frac{0.25}{3}[1.5 + 4(1.37143) + 1.33334] = 0.69325.$$

Case II : Now divide the interval [1, 2] into eight equal parts. In this case $h = \frac{2-1}{8} = \frac{1}{8} = 0.125$. The values of y for each points of subdivision are given by

x	$y = 1/x$	x	$y = 1/x$
1	1	13/8	0.61538
9/8	0.88889	14/8	0.57143
10/8	0.8	15/8	0.53333
11/8	0.72727	2	0.5
12/8	0.66667		

Here $y_0 = 1, y_1 = 0.88889, y_2 = 0.8, y_3 = 0.72727, y_4 = 0.66667, y_5 = 0.61538, y_6 = 0.57143, y_7 = 0.53333$ and $y_8 = 0.5$.

Now applying Simpson's $\frac{1}{3}$ rule, we get

$$\int_1^2 \frac{1}{x}dx = \frac{h}{3}[y_0 + 4(y_1 + y_3 + y_5 + y_7) + 2(y_2 + y_4 + y_6) + y_8]$$

$$= \frac{0.125}{3}[1 + 4(0.88889 + 0.72727 + 0.61538 + 0.53333)$$

$$+ 2(0.8 + 0.66667 + 0.57143) + 0.5]$$

$$= \frac{0.125}{3}[1.5 + 4(2.76487) + 2(2.0381)] = \frac{0.125}{3}[1.5 + 11.05948 + 4.0762]$$

$$= \frac{0.125}{3}[16.63568] = 0.69315 .$$

The exact Integration is

$$\int_1^2 \frac{1}{x}dx = [\log x]_1^2 = \log_e 2 = 0.69314$$

Therefore, error in case I
$$= 0.69314 - 0.69325 = -0.00011$$
and error in case II
$$= 0.69314 - 0.69315 = -0.00001.$$

EXAMPLE 15. *Using Simpson's one-third rule, find* $\int_0^6 \dfrac{dx}{(1+x)^2}$.

SOLUTION. Divide the interval [0, 6] in 6 equal parts with $h = \dfrac{6-0}{6} = 1$. The values of $y = \dfrac{1}{(1+x)^2}$ at each points of sub-divisions are given below :

x	0	1	2	3	4	5	6
$y = \dfrac{1}{(1+x)^2}$	1	0.25	0.11111	0.0625	0.04	0.02778	0.02041

Now, we have $y_0 = 1, y_1 = 0.25, y_2 = 0.11111, y_3 = 0.0625, y_4 = 0.04,$
$$y_5 = 0.02778 \text{ and } y_6 = 0.02041.$$

By Simpson's $\dfrac{1}{3}$ rule, we get

$$\int_0^6 \frac{dx}{(1+x)^2} = \frac{h}{3}[y_0 + 4(y_1 + y_3 + y_5) + 2(y_2 + y_4) + y_6]$$

$$= \frac{1}{3}[1 + 4(0.25 + 0.0625 + 0.02778) + 2(0.11111 + 0.04) + 0.02041]$$

$$= \frac{1}{3}[1.02041 + 4(0.34028) + 2(0.15111)] = \frac{1}{3}(2.68375) = 0.89458$$

EXAMPLE 16. *Calculate an approximate value of the integral* $\int_0^{\pi/2} \sin x\, dx$ *by Simpson's 1/3 rule, using 11 ordinates.*

SOLUTION. We have to use 11 ordinates. So divide the interval $[0, \pi/2]$ into 10 equal parts with $h = \pi/20$. The value $y = \sin x$ at each point of sub-divisons are given below:

x	$y = \sin x$	x	$y = \sin x$
0	0.00000	$6\pi/20$	0.80902
$\pi/20$	0.15643	$7\pi/20$	0.89101
$2\pi/20$	0.30902	$8\pi/20$	0.95106
$3\pi/20$	0.45399	$9\pi/20$	0.98769
$4\pi/20$	0.58779	$10\pi/20$	1.00000
$5\pi/20$	0.70711		

Here $y_0 = 0.00000, y_1 = 0.15643, y_2 = 0.30902, y_3 = 0.45399, y_4 = 0.58779,$
$$y_5 = 0.70711, y_6 = 0.80902, y_7 = 0.89101, y_8 = 0.95106, y_9 = 0.98769$$
and $y_{10} = 1.00000.$

By Simpson's 3/8 rule, we get

$$\int_0^{\pi/2} \sin x\, dx = \frac{3h}{8}[y_0 + 3(y_1 + y_2 + y_4 + y_5 + y_7 + y_8) + 2(y_3 + y_6 + y_9 + y_{10})]$$

$$= \frac{3\left(\dfrac{\pi}{20}\right)}{8}[0.00000 + 3(0.15643 + 0.30902 + 0.58779 + 70711 + 0.89101 + 0.95106)$$

$$+ 2(0.45399 + 0.80902 + 0.98769) + 1.00000]$$

$$= \frac{3\pi}{10}[1 + 3(3.59942) + 2(2.2507)] \quad = \frac{3\pi}{160}[1 + 10.7826 + 4.5014]$$

$$= \frac{3\pi}{160}[16.29966] = \pi(0.305618625) = 0.96013.$$

EXAMPLE 17. *The speed v meter per second of a car, t seconds after it starts, is shown in the following table :*

t	0	12	24	36	48	60	72	84	96	108	120
v	0	3.60	10.08	18.90	21.60	18.54	10.26	5.40	4.50	5.40	9.00

Using Simpson's Rule, find the distance travelled by the car in 2 seconds.

SOLUTION . Let s be the distance covered in t seconds, then

$$\frac{ds}{dt} = v$$

$$\Rightarrow \quad [s]_{t=0}^{t=120} = \int_0^{120} v\, dt$$

Here, the number of subintervals is 10, which is even, so applying Simpson's 1/3 rule, we get

$$\int_0^{120} v\, dt = \frac{h}{3}[(v_0 + v_{10}) + 4(v_1 + v_3 + v_5 + v_7 + v_9) + 2(v_2 + v_4 + v_6 + v_8)]$$

$$= \frac{12}{3}[(0 + 9) + 4(3.6 + 18.9 + 18.54 + 5.4 + 5.4)$$

$$+ 2(10.08 + 21.6 + 10.26 + 4.5)]$$

$$= 1236.96 \text{ meters.}$$

which is the required distance travelled by car in 2 minutes.

EXAMPLE 18. *Evaluate $\int_1^2 e^{-\frac{1}{2}x}\, dx$ using four intervals.*   (UPTU–2003)

SOLUTION. Here, we have the following table :

x	1	1.25	1.5	1.75	2
$y = e^{-\frac{1}{2}x}$	0.60653	0.53526	0.47237	0.41686	0.36788

Here, we have 4 (even) subintervals, so applying Simpson's $\frac{1}{3}$ rule, we get

$$\int_1^2 e^{-\frac{1}{2}x}\, dx = \frac{h}{3}[(y_0 + y_4) + 4(y_1 + y_3) + 2y_2]$$

$$= \frac{0.25}{3}[(0.60653 + 0.36788) + 4(0.53526 + 0.41868) + 2(0.47237)]$$

$$= 0.4773025$$

EXAMPLE 19. *Find $\int_0^6 \frac{e^x}{1+x}$ approximately using Simpson's $\frac{3}{8}$ rule.*   (UPTU–2006)

SOLUTION. Divide the given interval of integration into 6 equal subintervals, the arguments and 0, 1, 2, 3, 4, 5, 6 and $h = 1$.

Now, $f(x) = \dfrac{e^x}{1+x}, \, y_0 = f(0) = 1,$

$$y_1 = f(1) = \frac{e}{2}, \, y_2 = f(2) = \frac{e^2}{3}, \quad y_3 = f(3) = \frac{e^3}{4},$$

$$y_4 = f(4) = \frac{e^4}{5}, \quad y_5 = f(5) = \frac{e^5}{6}, y_6 = f(6) = \frac{e^6}{7}$$

Putting all these values in Simpson's $\frac{3}{8}$ rule, we get

$$\int_0^6 \frac{e^x}{1+x} dx = \frac{3h}{8}[(y_0 + y_6) + 3(y_1 + y_2 + y_4 + y_5) + 2y_3]$$

$$= \frac{3}{8}\left[\left(1 + \frac{e^6}{7}\right) + 3\left(\frac{e}{2} + \frac{e^2}{3} + \frac{e^4}{5} + \frac{e^5}{6}\right) + \frac{2e^3}{4}\right]$$

$$= \frac{3}{8}[(1 + 57.6327) + 3(1.3591 + 2.463 + 10.9196$$
$$+ 24.7355 + 2(5.0214)]$$

$$= 70.1652 \cdot$$

EXAMPLE 20. *Using Simpson's rule for evaluating $\int_{-0.6}^{0.3} f(x)dx$ from the following table :*

x	−0.6	−0.5	−0.4	−0.3	−0.2	−0.1	0	0.1	0.2	0.3
f(x)	4	2	5	3	−2	1	6	4	2	8

SOLUTION. Since, the number of subintervals is 9 (multiple of 3) so we will use Simpson's $\frac{3}{8}$ rule. Therefore,

$$\int_{-0.6}^{0.3} f(x)dx = \frac{(3)(0.1)}{8}[(4 + 8) + 3\{2 + 5 + (-2) + 1 + 4 + 2\} + 2(3 + 6)]$$

$$= 2.475$$

EXAMPLE 21. *A train in moving at the speed of 30 m/sec. Suddenly breaks are applied. The speed of the train per second after t second is given by :* (UPTU-2003)

Time (t)	0	5	10	15	20	25	30	35	40	45
Speed (v)	30	24	19	16	13	11	10	8	7	5

Applying Simpson's $\frac{3}{8}$ rule to determine the distance moved by the train in 45 seconds.

SOLUTION. Let *s* be the distance in meter covered in *t* seconds, then

$$\frac{ds}{dt} = v \qquad \Rightarrow \qquad [s]_{t=0}^{t=45} = \int_0^{45} v dt$$

Then, using Simpson's $\frac{3}{8}$ rule, we get

$$\int_0^{45} v dt = \frac{3h}{8}[(v_0 + v_9) + 3(v_1 + v_2 + v_4 + v_5 + v_7 + v_8) + 2(v_3 + v_6)]$$

$$= \frac{15}{8}[(30 + 5) + 3(24 + 19 + 13 + 11 + 8 + 7)$$
$$+ 2(16 + 10)] = 624.375 \text{ meters.}$$

(C) Based on Weddle's Rule

EXAMPLE 22. *Evaluate the integral $\int_0^6 \frac{dx}{1+x^2}$.*

(Kurukshetra-2004, 07; Rohilkhand-2008, 10; Bhopal-2005, 09)

SOLUTION. Divide [0, 6] into six equal sub intervals with each of width $h = \frac{6-0}{6} = 1$ and the values of $y = \frac{1}{1+x^2}$ at each points of the sub division are given below :

x	0	1	2	3	4	5	6
y	1.00000	0.50000	0.20000	0.10000	0.05882	0.03846	0.02702

Here $y_0 = 1.00000$, $y_1 = 0.50000$, $y_2 = 0.20000$, $y_3 = 0.10000$, $y_4 = 0.05882$, $y_5 = 0.03846$ and $y_6 = 0.02702$.

By Weddle's rule, we get

$$\int_0^6 \frac{dx}{1+x^2} = \frac{3h}{10}[y_0 + 5(y_1 + y_5) + y_2 + y_4 + 6y_3 + y_6]$$

$$= \frac{3}{10}[1.0000 + 5(0.50000 + 0.03846) + 0.20000$$
$$+ 0.05882 + 6(0.10000) + 0.27002]$$

$$= \frac{3}{10}[1.28584 + 5(0.53846) + 6(0.10000)]$$

$$= \frac{3}{10}[1.28584 + 2.6923 + 0.6] \ = \frac{3}{10}(4.57814) = 1.37344$$

EXAMPLE 23. *Evaluate the integral* $\int_4^{5.2} \log x \, dx$ *, using Weddle's rule.*

(Lucknow-2008, Bhopal-2002; VTU–2008)

SOLUTION. Divide the interval [4, 5.2] into six equal sub-intervals each of width $h = \frac{5.2 - 4}{6} = 0.2$ and the values of $y = \log x$ are given below at the arguments.

x	4.0	4.2	4.4	4.6	4.8	5.0	5.2
y	1.3862	1.4350	1.4816	1.5261	1.5686	1.6094	1.6486

Here $y_0 = 1.3862$, $y_1 = 1.4350$, $y_2 = 1.4816$, $y_3 = 1.5261$, $y_4 = 1.5686$,
$y_5 = 1.6094$ and $y_6 = 1.6486$.

By Weddle's rule, we get

$$\int_4^{5.2} \log x \, dx = \frac{3h}{10}[y_0 + 5(y_1 + y_5) + y_2 + y_4 + 6y_3 + y_6]$$

$$= \frac{3(0.2)}{10}[1.3862 + 5(1.4350 + 1.6094)$$
$$+ 1.4816 \quad + 1.5686 + 6(1.5261) + 1.6486]$$

$$= \frac{0.6}{10}[1.3862 + 5(3.0444) + 1.4816 \ + 1.5686 + 6(1.5261) + 1.6486]$$

$$= \frac{0.6}{10}[6.085 + 5(3.0444) + 6(1.5261)]$$

$$= \frac{0.6}{10}[6.085 + 15.222 + 9.1566]$$

$$= \frac{0.6}{10}(30.4636) = 1.8278$$

EXAMPLE 24. *Evaluate the integral* $\int_0^5 \frac{dx}{4x+5}$ *.*   (Anna-2005)

SOLUTION. Divide the interval [0, 5] into 12 equal parts of width each $h = \frac{5-0}{12} = \frac{5}{12}$ and the values of $y = \dfrac{1}{4x+5}$ are given below :

x	$y = \dfrac{1}{4x+5}$	x	$y = \dfrac{1}{4x+5}$
0	0.2	35/12	0.06
5/12	0.15	40/12	0.0545
10/12	0.12	45/12	0.05
15/12	0.10	50/12	0.0461
20/12	0.0857	55/12	0.0428
25/12	0.0750	60/12	0.04
30/12	0.0667		

Here $y_0 = 0.2$, $y_1 = 0.15$, $y_2 = 0.12$, $y_3 = 0.10$, $y_4 = 0.0857$, $y_5 = 0.0750$, $y_6 = 0.0667$, $y_7 = 0.06$, $y_8 = 0.0545$, $y_9 = 0.05$, $y_{10} = 0.0461$, $y_{11} = 0.0428$, $y_{12} = 0.04$.

By Weddle's Rule, we get

$$\int_0^5 \frac{dx}{4x+5} = \frac{3h}{10}[y_0 + 5(y_1 + y_5 + y_7 + y_{11}) + y_2 + y_4 + y_8$$

$$+ y_{10} + 6(y_3 + y_9) + 2y_6 + y_{12}]$$

$$= \frac{3 \times \dfrac{5}{12}}{10}[0.2 + 5(0.15 + 0.0750 + 0.06 + 0.0428)$$

$$+ 0.12 + 0.0857 + 0.0545 + 0.0461 + 6(0.10 + 0.05)$$

$$+ 2(0.0667) + 0.04]$$

$$= \frac{1}{8}[0.5463 + 5(0.3278) + 6(0.15) + 2(0.0667)]$$

$$= \frac{1}{8}[0.5463 + 1.639 + 0.9 + 0.1334]$$

$$= \frac{1}{8}(3.2187) = 0.4023$$

EXAMPLE 25. *Evaluate the integral* $\int_0^6 \dfrac{dx}{1+x^3}$.

SOLUTION. Divide the interval [0, 6] into 6 equal parts each of width $h = \dfrac{6-0}{6} = 1$. The values of $y = \dfrac{1}{1+x^3}$ at each points of sub-divisions are given below :

x	0	1	2	3	4	5	6
y	1.0000	0.5000	0.1111	0.0357	0.0153	0.0079	0.0046

Here $y_0 = 1.0000$, $y_1 = 0.5000$, $y_2 = 0.1111$, $y_3 = 0.0357$, $y_4 = 0.0153$, $y_5 = 0.0079$ and $y_6 = 0.0046$.

By Weddle's rule, we get

$$\int_0^6 \frac{dx}{1+x^3} = \frac{3h}{10}[y_0 + 5(y_1 + y_5) + y_2 + y_4 + 6y_3 + y_6]$$

$$= \frac{3h}{10}[1.0000 + 5(0.5000 + 0.0079) + 0.1111 + 0.0153 + 6(0.0357) + 0.0046]$$

$$= \frac{3}{10}[1.131 + 5(0.5079) + 6(0.0357)]$$

$$= \frac{3}{10}[1.131 + 2.5395 + 0.2142]$$

$$= \frac{3}{10}(3.8847) = 1.1654$$

EXAMPLE 26. *Evaluate the integral* $\int_0^{1.5} \frac{x^3}{e^x - 1} dx$.

SOLUTION. Divide the interval [0, 1.5] into 6 equal parts each of width $h = \frac{1.5 - 0}{6} = 0.25$

and the values of $y = \frac{x^3}{e^x - 1}$ at each points of sub divisions are given by the

following table :

x	0	0.25	0.50	0.75	1.0	1.25	1.50
y	0	0.0549	0.1927	0.3777	0.5820	0.7843	0.9694

Here $y_0 = 0, y_1 = 0.0549, y_2 = 0.1927, y_3 = 0.3777, y_4 = 0.5820, y_5 = 0.7843$
and $y_6 = 0.9694$.

By Weddle's rule, we get

$$\int_0^{1.5} \frac{x^3}{e^x - 1} dx = \frac{3h}{10}[y_0 + 5(y_1 + y_5) + y_2 + y_4 + 6y_3 + y_6]$$

$$= \frac{3(0.25)}{10}[0 + 5(0.0549 + 0.7843) + 0.1927$$

$$+ 0.5820 + 0.9694 + 6(0.3777)]$$

$$= 0.075[1.7441 + 5(0.8392) + 6(0.3777)]$$

$$= 0.075[1.7441 + 4.196 + 2.2662]$$

$$= 0.075(8.2063) = 0.6155 \cdot$$

EXAMPLE 27. *A curve passes through the points given by the following table :*

x	1	1.5	2	2.5	3	3.5	4
y	2	2.4	2.7	2.8	3	2.6	2.1

Find the area bounded by the curve, the x-axis and the lines $x = 1, x = 4$.

SOLUTION. Now divide the interval [1,4] into 6 equal parts of width $h = \frac{4-1}{6} = \frac{3}{6} = 0.5$. Therefore,
form the above table, we have

$y_0 = 2, y_1 = 2.4, y_2 = 2.7, y_3 = 2.8, y_4 = 3, y_5 = 2.6$ and $y_6 = 2.1$.
Now by Weddle's rule, we get

$$\int_1^4 y dx = \frac{3h}{10}[y_0 + 5(y_1 + y_5) + y_2 + y_4 + 6y_3 + y_6]$$

$$= \frac{3(0.5)}{10}[2 + 5(2.4 + 2.6) + 2.7 + 3 + 6(2.8) + 2.1]$$

$$= \frac{1.5}{10}[9.8 + 5(5.0) + 6(2.8)] = 4.365$$

(D) BASED ON NEWTON'S-COTE'S FORMULA

EXAMPLE 28. *Find the value of* C_3^4, *where* $C_k^n = \dfrac{1}{n}\int_0^n L_k du$ *and*

$$L_k = \frac{u(u-1)...(u-k+1)(u-k-1)...(u-n)}{k(k-1)...(+1)(-1)...(k-n)}$$

SOLUTION. Since $\qquad C_k^n = \dfrac{1}{n}\int_0^n L_k du \qquad \therefore \qquad C_3^4 = \dfrac{1}{4}\int_0^4 L_3 du$

and $\qquad L_3 = \dfrac{u(u-1)(u-2)(u-4)}{3(3-1)(3-2)(3-4)} = \dfrac{u(u^3-7u^2+14u-8)}{3.2.1.(-1)}$

$$= -\frac{1}{6}(u^4 - 7u^3 + 14u^2 - 8u)$$

$\therefore \qquad C_3^4 = \dfrac{1}{4}\int_0^4 -\dfrac{1}{6}(u^4 - 7u^3 + 14u^2 - 8u)du$

$$= -\frac{1}{24}\left[\frac{u^5}{5} - \frac{7u^4}{4} + \frac{14}{3}u^3 - 4u^2\right]_0^4 = -\frac{1}{24}\left[\frac{1024}{5} - \frac{1792}{4} + \frac{896}{3} - 64\right]$$

$$= -\frac{1}{24}\left[\frac{12288 - 26880 + 17920 - 3840}{60}\right]$$

$$= -\frac{1}{24}\left[\frac{30208 - 30720}{60}\right] = -\frac{1}{24}\left[-\frac{512}{60}\right] = \frac{512}{24 \times 60} = \frac{64}{180} = \frac{32}{90}$$

$\therefore \qquad C_3^4 = \dfrac{32}{90}$

EXAMPLE 29. *Prove that for* $n = 3$; $L_0 + L_1 + L_2 + L_3 = 1$ *where*

$$L_k = \frac{u(u-1)...(u-k+1)(u-k-1)...(u-n)}{k(k-1)...(+1)(-1)...(k-n)}$$

SOLUTION. We have

$\therefore \qquad L_0 = \dfrac{(u-1)(u-2)(u-3)}{(0-1)(0-2)(0-3)} = -\dfrac{1}{6}(u^3 - 6u^2 + 11u - 6)$

$\qquad L_1 = \dfrac{u(u-2)(u-3)}{1(1-2)(1-3)} = \dfrac{1}{2}(u^3 - 5u^2 + 6u)$

$\qquad L_2 = \dfrac{u(u-1)(u-3)}{2(2-1)(2-3)} = -\dfrac{1}{2}(u^3 - 4u^2 + 3u)$

and $\qquad L_3 = \dfrac{u(u-1)(u-2)}{3(3-1)(3-2)} = \dfrac{1}{6}(u^3 - 3u^2 + 2u)$

Now adding all L's we get

$$L_0 + L_1 + L_2 + L_3 = -\frac{1}{6}(u^3 - 6u^2 + 11u - 6)$$

$$+ \frac{1}{2}(u^3 - 5u^2 + 6u) - \frac{1}{2}(u^3 - 4u^2 + 3u) + \frac{1}{6}(ku^3 - 3u^2 + 2u)$$

$$= \frac{1}{6}[-u^3 + 6u^2 - 11u + 6 + 3u^3 - 15u^2 + 18u$$

$$-3u^3 + 12u^2 - 9u + u^3 - 3u^2 + 2u]$$

$$= \frac{1}{6}[-4u^3 + 4u^3 + 18u^2 - 18u^2 - 20u + 20u + 6] = \frac{1}{6}(6) = 1$$

Hence, $L_0 + L_1 + L_2 + L_3 = 1$.

EXAMPLE 30. *Prove for* $n = 2$, $L_0 = -\dfrac{L_1 L_2}{(L_1 + 2L_2)^2}$ *where*

$$L_k = \frac{u(u-1)\ldots(u-k+1)(u-k-1)\ldots(u-n)}{k(k-1)\ldots(+1)(-1)\ldots(k-n)}$$

SOLUTION. Since

$$L_k = \frac{u(u-1)\ldots(u-k+1)(u-k-1)\ldots(u-n)}{k(k-1)\ldots(+1)(-1)\ldots(k-n)}$$

If $n = 2$

Then

$$L_0 = \frac{(u-1)(u-2)}{(0-1)(0-2)} = \frac{1}{2}(u-1)(u-2) \qquad \ldots(1)$$

$$L_1 = \frac{u(u-2)}{(1-0)(1-2)} = -u(u-2) \qquad \ldots(2)$$

and

$$L_2 = \frac{u(u-1)}{2(2-1)} = \frac{1}{2}u(u-1) \qquad \ldots(3)$$

From equation (2) and (3), we get

$$\frac{2L_2}{L_1} = -\frac{(u-1)}{(u-2)}$$

Adding 1 of both sides

$$\frac{L_1 + 2L_2}{L_1} = -\frac{u+1+u-2}{u-2} = -\frac{1}{u-2}$$

$$\therefore \qquad u - 2 = -\frac{L_1}{L_1 + 2L_2}$$

$$\therefore \qquad u = -\frac{L_1}{L_1 + 2L_2} + 2 \implies u = \frac{L_1 + 4L_2}{L_1 + 2L_2}$$

Substituting the value of u into eq. (1), we get

$$L_0 = \frac{1}{2}\left[\frac{L_1 + 4L_2}{L_1 + 2L_2} - 1\right]\left[\frac{L_1 + 4L_2}{L_1 + 2L_2} - 2\right]$$

$$L_0 = \frac{1}{2}\left[\frac{2L_2}{L_1 + 2L_2}\right]\left[-\frac{L_1}{L_1 + 2L_2}\right] = \frac{1}{2} \cdot \frac{-2L_1 L_2}{(L_1 + 2L_2)^2} = -\frac{L_1 L_2}{(L_1 + 2L_2)^2}.$$

EXAMPLE 31. *Evaluate the integral* $\int_0^1 \dfrac{dx}{1+x}$.

SOLUTION. Divide the interval [0, 1] into 6 equal parts with width $h = \dfrac{1-0}{6} = \dfrac{1}{6}$. The value

of $y = \dfrac{1}{1+x}$ at each points of the subintervals are given below :

x	0	1/6	2/6	3/6	4/6	5/6	6/6
y	1	6/7	6/8	6/9	6/10	6/11	1/2

Here $y_0 = 1, y_1 = 6/7, y_2 = 6/8, y_3 = 6/9, y_4 = 6/10, y_5 = 6/11$ and $y_6 = 1/2$.
Now Cote's number for some values of n are given below :

n	C_0^n	C_1^n	C_2^n	C_3^n	C_4^n	C_5^n	C_6^n
1	1/2	1/2					
2	1/6	4/6	1/6				
3	1/8	3/2	3/8	1/8			
4	7/90	32/90	12/90	32/90	7/90		
5	19/288	75/288	50/288	50/288	75/288	19/288	
6	41/840	216/840	27/840	272/840	27/840	216/840	41/840

By Newton-Cote's formula, we get

$$\int_0^1 \frac{dx}{1+x} = nh \sum_{k=0}^{n} y_k C_k^n$$

Here $\qquad n = 6, h = \dfrac{1}{6}$.

$$\therefore \quad \int_0^1 \frac{dx}{1+x} = 6\left(\frac{1}{6}\right) \sum_{k=0}^{6} y_k C_k^6 = 1.[y_0 C_0^6 + y_1 C_1^6 + y_2 C_2^6 + y_3 C_3^6$$
$$+ y_4 C_4^6 + y_5 C_5^6 + y_6 C_6^6]$$

$$= \left[1\left(\frac{41}{840}\right) + \frac{6}{7}\left(\frac{216}{840}\right) + \frac{6}{8}\left(\frac{27}{840}\right) + \frac{6}{9}\left(\frac{272}{840}\right)\right.$$

$$\left. + \frac{6}{10}\left(\frac{27}{840}\right) + \frac{6}{11}\left(\frac{216}{840}\right) + \frac{1}{2}\left(\frac{41}{840}\right)\right]$$

$$= \left[\left(1 + \frac{1}{2}\right)\frac{41}{840} + \left(\frac{6}{7} + \frac{6}{11}\right)\frac{216}{840} + \left(\frac{6}{8} + \frac{6}{10}\right)\frac{27}{840} + \frac{6}{9}\left(\frac{272}{840}\right)\right]$$

$$= \frac{3}{2}\left(\frac{41}{840}\right) + \frac{108}{77}\left(\frac{216}{840}\right) + \frac{54}{40}\left(\frac{27}{840}\right) + \frac{6}{9}\left(\frac{272}{840}\right)$$

$$= \frac{1}{840}[61.5 + 302.96 + 36.45 + 181.35]$$

$$= \frac{1}{840}(582.24) = 0.693142 \cdot$$

EXAMPLE 32. *Evaluate the integral $\int_0^{1.2} e^x dx$ by Newton Cote's formula.*

SOLUTION. Divide the interval [0, 1.2] into 6 equal parts with width $\dfrac{1.2 - 0}{6} = 0.2$. The value of $y = e^x$ at each points of sub-division are given below

x	0	0.2	0.4	0.6	0.8	1.0	1.2
y	1	1.2214	1.4918	1.8221	2.2255	2.7182	3.3201

Here $y_0 = 1, y_1 = 1.2214, y_2 = 1.4918, y_3 = 1.8221, y_4 = 2.2255, y_5 = 2.7182$ and
$y_6 = 3.3201$.
The Cote's number for $n = 6$ are given by

$$C_0^6 = \frac{41}{840}, C_1^6 = \frac{216}{840}, C_2^6 = \frac{27}{840}, C_3^6 = \frac{272}{840}, C_4^6 = \frac{27}{840}, C_5^6 = \frac{216}{840}, C_6^6 = \frac{41}{840}$$

Now by Newton-Cote's formula, we get

$$\int_0^{1.2} e^x dx = nh \sum_{k=0}^{n} y_k c_k^n$$

Now $n = 6, h = 0.2$

$$\int_0^{1.2} e^x dx = 6(0.2) \sum_{k=0}^{6} y_k c_k^6$$

$$= 1.2[y_0 c_0^6 + y_1 c_1^6 + y_2 c_2^6 + y_3 c_3^6 + y_4 c_4^6 + y_5 c_5^6 + y_6 c_6^6]$$

$$= 1.2\left[1\left(\frac{41}{840}\right) + 1.2214\left(\frac{217}{840}\right) + 1.4918\left(\frac{27}{840}\right) + .18221\left(\frac{272}{840}\right)\right.$$

$$\left. + 2.225\left(\frac{27}{840}\right) + 2.7182\left(\frac{216}{840}\right) + 3.3201\left(\frac{41}{840}\right)\right]$$

$$= \frac{1.2}{840}[41 + 263.8224 + 40.2786 + 495.6112 + 60.0885$$

$$+ 587.1312 + 136.1241]$$

$$= \frac{1.2}{840}[1624.056] = \frac{1948.8672}{840} = 2.32008.$$

EXAMPLE 33. *Find the value of the integral* $\int_1^7 \frac{dx}{x}$.

SOLUTION. Divide the interval $[1, 7]$ into 6 equal parts with width $h = \frac{7-1}{6} = 1$. The value

of $y = \frac{1}{x}$ at each points of sub-division are given in the following table :

x	1	2	3	4	5	6	7
y	1	1/2	1/3	1/4	1/5	1/6	1/7

Here $y_0 = 1$, $y_1 = 1/2$, $y_2 = 1/3, y_3 = 1/4, y_4 = 1/5, y_5 = 1/6$, and $y_6 = 1/7$.
The Cote's number for $n = 6$ are given by

$$c_0^6 = \frac{41}{840}, c_1^6 = \frac{216}{840}, c_2^6 = \frac{27}{840}, c_3^6 = \frac{272}{840}, c_4^6 = \frac{27}{840}, c_5^6 = \frac{216}{840}, c_6^6 = \frac{41}{840}$$

By Newton-Cote's formula, we get

$$\int_1^7 \frac{dx}{x} = nh \sum_{k=0}^{n} y_k c_k^n$$

Here $n = 6, h = 1$

$$\therefore \quad \int_1^7 \frac{dx}{x} = 6(1) \sum_{k=0}^{6} y_k c_k^6$$

$$= 6[y_0 c_0^6 + y_1 c_1^6 + y_2 c_2^6 + y_3 c_3^6 + y_4 c_4^6 + y_5 c_5^6 + y_6 c_6^6]$$

$$= 6\left[1\left(\frac{41}{840}\right) + \frac{1}{2}\left(\frac{217}{840}\right) + \frac{1}{3}\left(\frac{27}{840}\right) + \frac{1}{4}\left[\frac{272}{840}\right] + \frac{1}{5}\left(\frac{27}{840}\right) + \frac{1}{6}\left(\frac{216}{840}\right)\right.$$

$$\left. + \frac{1}{7}\left(\frac{41}{840}\right)\right]$$

$$= \frac{6}{840}[41 + 108 + 9 + 68 + 5.4 + 36 + 5.8571]$$

$$= \frac{6}{840}[273.2571] = \frac{1639.5426}{840} = 1.95183.$$

(E) Based on Gauss Quadrature Formula

EXAMPLE 34. *Evaluate the integral* $I = \int_5^{12} \frac{dx}{x}$.

SOLUTION. Here taking

$$x = (b-a)u + \frac{1}{2}(a+b) = (12-5)u + \frac{1}{2}(12+5)$$

$\Rightarrow \qquad x = 7u + 8.5$

Since $\qquad y = \frac{1}{x} = \frac{1}{7u+8.5} = \phi(u)$

Now taking $n = 5$, we have

$\therefore \qquad y_0 = \phi(u_0) = \frac{1}{8.5} = 0.117647058$

$\qquad\qquad\qquad\qquad\qquad\qquad\qquad\qquad (\because u_0 = 0)$

$$y_1 = \phi(u_1) = \frac{1}{7u_1 + 8.5}$$

$$= \frac{1}{7(0.2692346551) + 8.5} \qquad (\because u_1 = 0.2692346551)$$

$$= \frac{1}{10.3846426} = 0.0962960439$$

$$y_{-1} = \phi(u_{-1}) = \frac{1}{7u_{-1} + 8.5}$$

$$= \frac{1}{6.61535741} \qquad (\because u_{-1} = -0.2692346551)$$

$$= 0.1511634112$$

$$y_2 = \phi(u_2) = \frac{1}{7u_2 + 8.5} = \frac{1}{11.67162946}$$

$$\qquad\qquad\qquad\qquad\qquad\qquad (\because y_2 = 0.4530899230)$$

$$= 0.0856778399$$

$$y_{-2} = \phi(u_{-2}) = \frac{1}{7u_{-2} + 8.5}$$

$$= \frac{1}{5.32837054} \qquad (\because y_{-2} = -0.4530899230)$$

$$= 0.187674636$$

For $n = 5$, we have taken the values of $u_0, u_1, u_{-1}, u_2, u_{-2}$ and $R_0, R_1 = R_{-1}$ and $R_2 = R_{-2}$ from table given in previous section.

$\therefore \qquad R_0 = \frac{64}{225}, R_1 = 0.2393143352,$

$\qquad R_2 = 0.1184634425$

By Gauss quadrature formula, we get

$$I = \int_5^{12} \frac{dx}{x} = (b-a)[R_0\phi(u_0) + R_1\phi(u_1) + R_{-1}\phi(u_{-1}) + R_2\phi(u_2) + R_{-2}\phi_2(u_{-2})]$$

$$= (12-5)\left[\frac{64}{225}(0.117647058) + 0.2393143352(0.0962960439\right.$$

$$\left. + 0.151163412) + 0.1184634425(0.0856778399 + 0.187674636)\right]$$

$$= 7[0.033464052 + 0.059220595 + 0.032382275]$$

$$= 7(0.125066922) = 0.875468454$$

$$\therefore \int_5^{12} \frac{dx}{x} = 0.875468454 .$$

EXAMPLE 35.   *Find the value of the integral $\int_0^1 x\,dx$.*

SOLUTION.   Here taking

$$x = (b-a)u + \frac{1}{2}(a+b) \quad \text{or} \quad x = (1-0)u + \frac{1}{2}(0+1)$$

$$\Rightarrow \qquad x = u + \frac{1}{2}$$

Since $y = x = u + \dfrac{1}{2} = \phi(u)$ ...(1)

Now taking $n = 4$ so that the values of u_1, u_{-1}, u_2, u_{-2} and $R_1 = R_{-1}$ and $R_2 = R_{-2}$ are taken from table given in previous section.

From (1),

$$\therefore \quad \phi(u_{-1}) = u_{-1} + 0.5 = -0.1699905218 + 0.5$$

$$= 0.330009479 \qquad\qquad (\because u_{-1} = -0.1699905218)$$

$$\phi(u_1) = u_1 + 0.5 = 0.1699905218 + 0.5 \qquad (\because u_1 = 0.1699905218)$$

$$= 0.669990521$$

$$\phi(u_{-2}) = u_{-2} + 0.5$$

$$= -0.4305681558 + 0.5$$

$$= 0.069431845 \qquad\qquad (\because u_{-2} = -0.4305681558)$$

$$\phi(u_2) = u_2 + 0.5 = 0.4305681558 + 0.5$$

$$= 0.930568155 \qquad\qquad (\because u_2 = 0.4305681558)$$

and values of R's are

$$R_1 = R_{-1} = 0.3260725774$$

and $R_2 = R_{-2} = 0.1739274226$

Now by Gauss quadrature formula, we get

$$\int_0^1 x\,dx = (b-a)[R_{-1}\phi(u_{-1}) + R_1\phi(u_1) + R_{-2}\phi(u_{-2}) + R_2\phi(u_2)]$$

$$= (1-0)[0.3260725774(0.330009479 + 0.669990521)$$

$$+ 0.1739274226(0.069431845 + 0.930568155)]$$

$$= 1[0.3260725774(1) + 0.1739274226(1)]$$

$$= 0.499999999 .$$

(F) Based on Chebychev's Formula

EXAMPLE 36. *Evaluate the integral $\int_5^{12} \frac{dx}{x}$ for $n = 5$.*

SOLUTION. Putting

$$x = (b-a)u + \frac{1}{2}(a+b) \quad i.e., \quad x = (12-5)u + \frac{1}{2}(12+5)$$

Since $\quad y = \frac{1}{x} \quad \therefore \quad y = \frac{1}{7u+8.5} = \phi(u)$

$$\Rightarrow \quad \phi(u) = \frac{1}{7u+8.5}$$

For $n = 5$, taking the value of u's from the table given with article of Chebychev's formula, we have

$u_0 = 0, u_{-1} = -0.187271, u_1 = 0.187271, u_{-2} = -0.416249$ and $u_2 = 0.416249$

Then $\quad \phi(u_0) = \frac{1}{8.5} = 0.117647$

$$\phi(u_{-1}) = \frac{1}{7(-0.187271)+8.5} = \frac{1}{7.189103} = 0.139099$$

$$\phi(u_1) = \frac{1}{7u_1+8.5} = \frac{1}{7(0.187271)+8.5} = \frac{1}{9.810897} = 0.101927$$

$$\phi(u_{-2}) = \frac{1}{7u_{-2}+8.5} = \frac{1}{7(-0.416249)+8.5} = \frac{1}{5.586257} = 0.179011$$

and $\quad \phi(u_2) = \frac{1}{7u_2+8.5} = \frac{1}{7(0.416249)+8.5} = \frac{1}{11.413743} = 0.087613$

By Chebychev's formula, we get

$$\int_5^{12} \frac{dx}{x} = \frac{b-a}{n}[\phi(u_0)+\phi(u_{-1})+\phi(u_1)+\phi(u_{-2})+\phi(u_2)]$$

$$= \frac{12-5}{5}[0.117647+0.139099+0.101927+0.179011+0.087613]$$

$$= \frac{7}{5}[0.625297] = 0.875416.$$

EXAMPLE 37. *Evaluate the integral $\int_0^1 x \, dx$ for $n = 5$.*

SOLUTION. $x = (b-a)u + \frac{1}{2}(a+b) = u + \frac{1}{2} = u + 0.5$

Since $\quad y = x = u + 0.5 = \phi(u)$

$\therefore \quad \phi(u) = u + 0.5$

Now for $n = 5$, taking the values of u's from the previous example, we have

$u_0 = 0, u_{-1} = -0.187271, u_1 = 0.187271, u_{-2} = -0.416249$ and $u_2 = 0.416249$.

Then $\quad \phi(u_0) = 0 + 0.5 = 0.5$

$$\phi(u_{-1}) = u_{-1} + 0.5 = -0.187271 + 0.5 = 0.312726$$

$$\phi(u_1) = u_1 + 0.5 = 0.187271 + 0.5 = 0.687271$$

$$\phi(u_{-2}) = u_{-2} + 0.5 = -0.416249 + 0.5 = 0.083751$$

$$\phi(u_2) = u_2 + 0.5 = 0.416249 + 0.5 = 0.916249$$

By Chebychev's formula, we get

$$\int_0^1 x\,dx = \frac{b-a}{n}[\phi(u_0) + \phi(u_{-1}) + \phi(u_1) + \phi(u_{-2}) + \phi(u_2)]$$

$$= \frac{1-0}{5}[0.5 + 0.312726 + 0.687271 + 0.083751 + 0.916249]$$

$$= \frac{1}{5}[2.499997] = 0.499999 \ .$$

6.21 HIGHER ORDER RULES

6.21.1 BOOLE'S RULE

Putting $n = 4$ in the general quadrature formula and take the differences upto fourth order. Then, we get

$$\int_{x_0}^{x_0+4h} f(x)\,dx = 4h\left(y_0 + 2\Delta y_0 + \frac{5}{3}\Delta^2 y_0 + \frac{2}{3}\Delta^3 y_0 + \frac{7}{90}\Delta^4 y_0\right)$$

$$= \frac{2h}{45}\left(7y_0 + 32y_1 + 12y_2 + 32y_3 + 7y_4\right)$$

Similarly,

$$\int_{x_0+4h}^{x_0+8h} f(x)\,dx = \frac{2h}{45}[7y_4 + 32y_5 + 12y_6 + 32y_7 + 7y_8]$$

Adding all such integrals from x_0 to $(x_0 + nh)$, (where n must be a multiple of 4), we get

$$\int_{x_0}^{x_0+nh} f(x)\,dx = \frac{2h}{45}[7y_0 + 32y_1 + 12y_2 + 32y_3 + 14y_4 + 32y_5 + 12y_6 + 32y_7 + 14y_8 + \dots$$

$$+ 14y_{n-4} + 32y_{n-3} + 12y_{n-2} + 32y_{n-1} + 7y_n]$$

which is known as Boole's Rule.

6.21.2 ERROR IN BOOLE'S RULE

The error in Boole's Rule is given by $E(h) = -\dfrac{8h^7}{945}y^{vi}$.

6.21.3 THE EULAR-MACLAURIN'S SUMMATION FORMULA

Let $\Delta F(x) = f(x)$, then we have an operator Δ^{-1} defined by

$$F(x) = \Delta^{-1}f(x)$$

Again we have $\qquad \Delta F(x) = f(x_0)$

$\Rightarrow \qquad\qquad F(x_1) - F(x_0) = f(x_0)$

$$F(x_2) - F(x_1) = f(x_1)$$

$$\dots\dots\dots\dots\dots\dots\dots\dots$$

$$F(x_n) - F(x_{n-1}) = f(x_{n-1})$$

Adding all these, we get

$$F(x_n) - F(x_0) = \sum_{i=0}^{n-1} f(x_i)$$

$$\dots(1)$$

where $x_0, x_1, ..., x_n$ are the $(n + 1)$ equidistant values of x with interval h.

Now $\quad F(x) = \Delta^{-1} f(x) = (E - 1)^{-1} f(x) = (e^{hD} - 1)^{-1} f(x)$

$$= \left\{ \left(1 + hD + \frac{h^2 D^2}{2!} + \frac{h^3 D^3}{3!} + ... \right) - 1 \right\}^{-1} f(x)$$

$$= \left(hD + \frac{h^2 D^2}{2!} + \frac{h^3 D^3}{3!} + ... \right)^{-1} f(x) = (hD)^{-1} \left\{ 1 + \left(\frac{hD}{2!} + \frac{h^2 D^2}{3!} + ... \right) \right\}^{-1} f(x)$$

$$= \frac{1}{h} D^{-1} \left\{ 1 + \left(\frac{hD}{2!} + \frac{h^2 D^2}{3!} + ... \right) + \frac{(-1)(-2)}{2!} \left(\frac{hD}{2!} + \frac{h^2 D^2}{3!} + ... \right)^2 + ... \right\} f(x)$$

$$= \frac{1}{h} D^{-1} \left(1 - \frac{hD}{2!} + \frac{h^2 D^2}{12} - \frac{h^4 D^4}{720} ... \right) f(x) = \frac{1}{h} \int f(x) dx - \frac{1}{2} f(x) + \frac{h}{12} f'(x) - \frac{h^3}{720} f'''(x) ...(2)$$

Between limits $x = x_0$ and $x = x_n$ from eq. (2), we have

$$F(x_n) - F(x_0) = \frac{1}{h} \int_{x_0}^{x_n} f(x) dx - \frac{1}{2} [f(x_n) - f(x_0)] + \frac{h}{12} [f'(x_n) + f'(x_0)]$$

$$- \frac{h^3}{720} [f'''(x_n) - f'''(x_0)] + ... \qquad ...(3)$$

From eqs. (1) and (3), we have

$$\sum_{i=1}^{n-1} f(x_i) = \frac{1}{h} \int_0^{x_n} f(x) dx - \frac{1}{2} [f(x_n) - f(x_0)] + \frac{h}{12} [f'(x_n) - f'(x_0)]$$

$$- \frac{h^3}{720} [f'''(x_n) - f'''(x_0)] + ...$$

But $\sum\limits_{i=0}^{n-1} f(x_i) = \sum\limits_{i=0}^{n} [f(x_i) - f(x_n)]$ and $x_n = x_0 + nh$. Then the above relation reduces to

$$\frac{1}{h} \int_{x_0}^{x_0 + nh} f(x) dx = \sum_{i=0}^{n} f(x_i) - \frac{1}{2} [f(x_0) + f(x_n)] - \frac{h}{12} [f'(x_0 + nh) - f'(x)]$$

$$- \frac{h^3}{720} [f'''(x_0 + nh) - f^{iv}(x_0)] - ... \qquad ...(4)$$

$\Rightarrow \qquad \int_{x_0}^{x_n} f(x) dx = \frac{h}{2} [f(x_0) + 2f(x_1) + 2f(x_2) + ... + 2f(x_{n-1}) + f(x_n)]$

$$- \frac{h^2}{12} [f'(x_n) - f'(x_0)] + \frac{h^4}{720} [f'''(x_n) - f''(x_0)] + ...$$

$\Rightarrow \qquad \int_{x_0}^{x_n} y dx = \frac{h}{2} [y_0 + 2y_1 + y_2 + ... + 2y_{n-1} + y_n] - \frac{h^2}{12} (y_n' - y_0') + \frac{h^2}{720} (y_n''' - y_0''') + ...$

which is known as Euler's Maclaurin's summation formula.

6.22 NUMERICAL EVALUATION OF THE SINGULAR INTEGRAL

We know that the numerical integration formulae are valid only if the given function $f(x)$ can be represented by a polynomial or it can be expanded by Taylor's series. If the function $f(x)$ have the singularity at any point then these formulae cannot be appeared. Here, we describe some methods which can be applied in such type of situations.

6.22.1 EVALUATION OF PRINCIPAL VALUE INTEGRALS

Consider
$$I(f) = \int_a^b \frac{f(x)}{x-t}dx \qquad \qquad \text{...(1)}$$

Clearly it is singular at $x = t$. The principal value, $P(I)$ of the integral is defined by

$$P(I) = \lim_{\varepsilon \to 0}\left[\int_a^{t-\varepsilon} \frac{f(x)}{x-t}dx + \int_{t+\varepsilon}^b \frac{f(x)}{x-t}dx\right], a < t < b \qquad \text{...(2)}$$

$$= I(f) \qquad \qquad \text{for } t < a \text{ or } t > b$$

Let $x = a + uh$ and $t = a + kh$ in equation (1), we get

$$P(I) = P\int_0^P \frac{f(a+uh)}{u-k}du \qquad \qquad \text{...(3)}$$

By Newton's forward interpolation formula, we have

$$f(a+uh) = f(a) + u\Delta f(x) + \frac{u(u-1)}{2!}\Delta^2 f(a) + \frac{u(u-1)(u-2)}{3!}\Delta^3 f(a) + ...$$

Put this value in equation (3), and simplify, we get

$$I_n(f) = \sum_{j=0}^n \frac{\Delta^j f(a)}{j!}c_j$$

When C_j are the constant given by

$$C_j = P\int_0^p \frac{(u)_j}{u-k}du$$

Now put n = 1, 2, 3, ... in equation (4), we may obtain the following formulae :

(a) Two-point Rule ($n = 1$)

$$I_1(f) = \sum_{j=0}^1 \frac{\Delta^j f(a)}{j!}c_j = c_0 f(a) + c_1 \Delta f(a) = (c_0 - c_1)f(a) + c_1 f(a+h)$$

(b) Three-point Rule ($n = 2$)

$$I_2(f) = \sum_{j=0}^2 \frac{\Delta^j f(a)}{j!}c_j = c_0 f(a) + c_1 \Delta f(a) + c_2 \Delta^2 f(a)$$

$$= \left(c_0 - c_1 + \frac{1}{2}c_2\right)f(a) + (c_1 - c_2)f(a+h) + \frac{1}{2}c_2 f(a+2h)$$

▶ **REMARKS**

▶ In the above relations, the values of c_j are given by

$$c_0 = \log_e\left(\frac{p-k}{k}\right) ; \quad c_1 = p + c_0 k \text{ and } c_2 = \frac{p^2}{2} + p(k-1) + c_0 k(k-1)$$

▶ A function $f(x)$ is said to have a singularity at a point if $f(x)$ or any of its derivatives is infinite at the that point.

6.22.2 GENERALIZED QUADRATURE

Consider the numerical quadrature of integrals of the formula

$$I(s) = \int_a^b f(t)\phi(t-s)dt \qquad \qquad \text{...(1)}$$

where $f(t)$ is continuous, but $\phi(a)$ may have an integrable singularity.
Divide the range (a, b) such that

$$t_j = a + jh (j = 0,1,2,...,n) \text{ with } b = a + nh$$

Then (1) can be written as

$$I(s) = \sum_{j=0}^{n-1} \int_{t_j}^{t_{j+1}} f(t)\phi(t-s)dt \qquad \ldots(2)$$

Now approximate the function $f(t)$ by the interpolating function $f_n(t)$, where

$$f_n(t) = \frac{1}{h}[(t_{j+1}-t)f(t_j) + (t-t_j)(t_{j+1})] \qquad \ldots(3)$$

Putting this value in (2), we get

$$I(s) = \frac{1}{h}\sum_{j=0}^{n-1} \int_{t_j}^{t_{j+1}} [(t_{j+1}-t)f(t_j) + (t-t_j)f(t_{j+1})]\phi(t-s)dt$$

getting $\quad t = t_j + ph$

$$\Rightarrow \quad I(s) = h\sum_{j=0}^{n-1}\int_0^1 [(1-p)f(t_j)+pf(t_{j+1})]\phi(t_j+ph-s)dp = \frac{1}{h}\sum_{j=0}^{n-1}[\alpha_j f(t_j)+\beta_j f(t_{j+1})] \ldots(4)$$

when $\quad \begin{aligned}\alpha_j &= h\int_0^1 (1-p)\phi(t_j+ph-s)dp \\ \beta_j &= h\int_0^1 p\phi(t_j+ph-s)dp\end{aligned}\Bigg] \qquad \ldots(5)$

If $\phi(u) = 1$. Then from equation 5, we have $\alpha_j = \beta_j = \dfrac{1}{2}h$.

Hence equation (4) becomes

$$I(s) = \frac{h}{2}[f(t_0) + 2f(t_1) + 2f(t_2) + \ldots + 2f(t_{n-1}) + f(t_n)]$$

which is the trapezoidal rule.

▼ **REMARK**
▶ In a simple manner we can deduce the Simpson's rule.

SOLVED EXAMPLES

EXAMPLE 1. *Use Boole's formula to compute $\int_0^{\pi/2}\sqrt{\sin x}\,dx$.* (UPTU–2008)

SOLUTION. Here $\qquad f(x) = \sqrt{\sin x}$

$\Rightarrow \qquad f_0 = 0$

$$f_1 = f\left(\frac{\pi}{8}\right) = 0.61861 \; ; \qquad f_2 = f\left(\frac{\pi}{4}\right) = 0.84090$$

$$f_3 = f\left(\frac{3\pi}{8}\right) = 0.96119 \; ; \qquad f_4 = f\left(\frac{\pi}{2}\right) = 1.0$$

$$\therefore \int_{x_0}^{x_0+4h} f(x)dx = \frac{\pi}{180}[0 + 32(0.61861 + 0.96119)$$
$$+ 12(0.84090) + 7(1.0)]$$

$$= 1.18062.$$

EXAMPLE 2. *Find the sum of the fourth powers of the first n natural numbers by means of the Euler Maclaurin's formula.* (Meerut-2000, 01, 03, 11, 14)

SOLUTION. The Euler Maclaurin's formula is :

$$\frac{1}{h}\int_{x_0}^{x_0+nh} f(x)dx = \sum_{0}^{n} f(x+nh) - \frac{1}{2}[f(x_0+nh) + f(x_0)]$$

$$-\frac{h}{12}[f'(x_0+nh) - f'(x_0)] + \frac{h^3}{720}[f'''(x_0+nh) - f'''(x_0)] + ... \quad ...(1)$$

Putting $x_0 = 0, h = 1, f(x) = x^4$, in eq. (1), we get

$$\int_0^n x^4 dx = \sum_0^n x^4 - \frac{1}{2}[n^4] - \frac{1}{12}[5n^3] + \frac{1}{720}[24n]$$

$$\Rightarrow \qquad \frac{n^5}{5} = \sum_0^n x^4 - \frac{1}{2}n^4 - \frac{1}{3}n^3 + \frac{n}{30}$$

$$\Rightarrow \qquad \sum_0^n x^4 = \frac{n^5}{5} + \frac{1}{2}n^4 + \frac{1}{3}n^3 - \frac{n}{30}.$$

EXAMPLE 3. *A river is 80 m wide. The depth y of the river at a distance x from bank is given by the following table :*

x	0	10	20	30	40	50	60	70	80
y	0	4	7	9	12	15	14	8	3

Using Boole's rule find the approximate area of cross section of the river.

(Rohtak–2005, UPTU(MCA)–2002)

SOLUTION. It is known that, the required area of the cross-section of the river

$$= \int_0^{80} y dx$$

Then by Boole's rule, we have

$$\int_0^{80} y dx = \frac{2h}{45}[7y_0 + 32y_1 + 12y_2 + 32y_3 + 7y_4$$

$$+ 7y_4 + 32y_5 + 12y_6 + 32y_7 + 7y_8]$$

$$= \frac{2 \times 10}{45}[7 \times 0 + 32 \times 4 + 12 \times 7 + 32 \times 9 +$$

$$7 \times 12 + 7 \times 12 + 32 \times 15 + 12 \times 14 + 32 \times 8 + 7 \times 3]$$

$$= 708.$$

EXAMPLE 4. *Using Euler-Maclaurin's formula, obtain the value of $\log_e 2$ from $\int_0^1 \frac{dx}{1+x}$.*

SOLUTION. Let $\qquad y = f(x) = \frac{1}{1+x}$

Here $\quad x_0 = 0, n = 10, h = 0.1, x_n = 1$

$$\Rightarrow \qquad y' = -\frac{1}{(1+x)^2}, y'' = \frac{2}{(1+x)^3}, \quad y''' = -\frac{6}{(1+x)^4}$$

Then, by Euler-Maclaurin's formula, we have

$$\int_0^1 y dx = \frac{1}{2}\left[\frac{1}{1+0} + \frac{2}{1+0.1} + \frac{2}{1+0.2} + \frac{2}{1+0.3}\right.$$

$$\left. + \frac{2}{1+0.4} + \frac{2}{1+0.5} + \frac{2}{1+0.6} + ... + \frac{2}{1+1.0}\right]$$

$$-\frac{(0.1)^2}{12}\left[-\frac{1}{(1+1)^2}-\frac{-1}{(1+0)^2}\right]+\frac{(0.1)^4}{720}\left[-\frac{6}{(1+1)^4}-\frac{-6}{(1+0)^4}\right]$$

$$= 0.693773 - 0.000625 + 0.0000010 = 0.631149.$$

EXAMPLE 5. Use Euler-Maclaurin's formula to prove that

$$\sum_{1}^{n} x^2 = \frac{n(n+1)(2n+1)}{6}.$$

SOLUTION. Putting $f(x) = x^2, x_0 = 0, h = 1, x_0 + nh = n$, $x_i = x_0 + ih = i$ in the Euler-Maclaurin's summation formula, we get

$$\int_0^n x^2 dx = \sum_{i=0}^{n} i^2 - \frac{1}{2}(n^2 + 0) - \frac{1}{12}(2n - 0)$$

$$\Rightarrow \quad \frac{n^3}{3} = \sum_{i=0}^{n} i^2 - \frac{1}{2}n^2 - \frac{n}{6}$$

$$\Rightarrow \quad \sum_{i=0}^{n} i^2 = \frac{n^3}{3} + \frac{1}{2}n^2 + \frac{n}{6} = \frac{2n^3 + 3n^2 + n}{6} = \frac{n(2n^2 + 3n + 1)}{6} = \frac{n(n+1)(2n+1)}{6}$$

$$\Rightarrow \quad \sum_{i=0}^{n} x^2 = \frac{n(n+1)(2n+1)}{6}.$$

EXAMPLE 6. Evaluate $\int_0^1 \frac{dx}{1+x}$ to five places of decimal, using Euler-Maclaurin's formula.

SOLUTION. Let $\quad y = \frac{1}{1+x}$

Here, we have $x_0 = 0, n = 10$ and $h = 0.1$

Then we want to evaluate $\int_{x_0}^{x_1} y dx$

where, $y' = -\frac{1}{(1+x)^2}$ and $y''' = -\frac{6}{(1+x)^4}$

Using Euler-Maclaurin's formula, we get

$$\int_0^1 \frac{dx}{1+x} = \frac{h}{2}(y_0 + 2y_1 + 2y_2 + ... + y_n)$$

$$-\frac{h^2}{12}[f'(x_{10}) - f'(x_0)] + \frac{h^2}{720}[f''(x_{10}) - f'''(x_0)]$$

$$= \frac{0.1}{2}\left[\frac{1}{1} + \frac{2}{1.01} + \frac{2}{1.02} + \frac{2}{1.03} + \frac{2}{1.04}\right.$$

$$\left. + \frac{2}{1.05} + \frac{2}{1.06} + \frac{2}{1.07} + \frac{2}{1.08} + \frac{2}{1.09} + \frac{1}{2}\right]$$

$$= -\frac{(0.1)^2}{1^2}\left[-\frac{1}{2^2} + \frac{1}{1^2}\right] + \frac{(0.1)^4}{720}\left[-\frac{6}{2^4} + \frac{6}{1^4}\right]$$

$$= 0.693773 - 0.000625 + 0.000001 = 0.693149.$$

EXAMPLE 7. Find the sum of the series using Euler-Maclaurin's formula

$$\frac{1}{51^2} + \frac{1}{53^2} + \frac{1}{55^2} + ... + \frac{1}{99^2}.$$

(UPTU-2003, 09)

SOLUTION. Here, we have $y = \dfrac{1}{x^2}, x_0 = 51, n = 24, h = 2$

Then $\qquad y' = -\dfrac{2}{x^3}, y''' = -\dfrac{24}{x^5}$ .... and so on.

Using Euler-Maclaurin's formula, we get

$$\int_{51}^{99} \frac{1}{x^2}dx = \frac{h}{2}[y_0 + 2y_1 + 2y_2 + ... + 2y_{23} + y_{24}]$$

$$-\frac{h^2}{12}[y'_{24} - y'_0] + \frac{h^4}{720}[y'''_{24} - y'''_0] + ...$$

$$= \left[\frac{1}{51^2} + \frac{2}{53^2} + \frac{2}{55^2} + ... + \frac{2}{97^2} + \frac{1}{99^2}\right]$$

$$-\frac{4}{12}\left[-\frac{2}{99^3} + \frac{2}{51^3}\right] + \frac{16}{720}\left[-\frac{24}{99^5} + \frac{24}{51^5}\right] + ...$$

which gives

$$\frac{1}{51^2} + \frac{2}{53^2} + \frac{2}{55^2} + ... + \frac{2}{97^2} + \frac{1}{99^2}$$

$$= \int_{51}^{99}\frac{1}{x^2}dx + \frac{2}{3}\left[\frac{1}{51^3} - \frac{1}{99^3}\right] - \frac{8}{15}\left[\frac{1}{51^5} - \frac{1}{99^5}\right] + ...$$

$$\Rightarrow 2\left[\frac{1}{51^2} + \frac{1}{53^2} + \frac{1}{55^2} + ... + \frac{1}{99^2}\right] = \int_{51}^{99}\frac{1}{x^2}dx + \left(\frac{1}{51^2} + \frac{1}{99^2}\right) + \frac{2}{3}$$

$$\left(\frac{1}{51^3} - \frac{1}{99^3}\right) - \frac{8}{15}\left(\frac{1}{51^5} - \frac{1}{99^5}\right) + ...$$

$$\Rightarrow \quad \frac{1}{51^2} + \frac{1}{53^2} + \frac{1}{55^2} + ... + \frac{1}{99^2} = \frac{1}{2}\left[-\frac{1}{x}\right]_{51}^{99} + \frac{1}{2}\left(\frac{1}{51^2} + \frac{1}{99^2}\right)$$

$$+ \frac{1}{3}\left(\frac{1}{51^3} - \frac{1}{99^3}\right) - \frac{4}{15}\left(\frac{1}{51^5} - \frac{1}{99^5}\right) + ...$$

$$= 0.00475 + 0.00024 + 0.000002 + ... = 0.00499.$$

EXAMPLE 8. Show that $\displaystyle\sum_{k=1}^{n} k^7 + \sum_{k=1}^{n} k^5 = 2\left(\sum_{k=1}^{n} k^3\right)^2$.

SOLUTION. The Euler-Maclaurin's formula is given by

$$\frac{1}{h}\int_{x_0}^{x_n} f(x)dx = \sum_{x=x_0}^{x_{n-1}} f(x) + \frac{1}{2}(f_n - f_0) - \frac{h}{12}[f'(x_n) - f'(x_0)]$$

$$+ \frac{h^3}{720}[f'''(x_n) - f'''(x_n)] - \frac{h^5}{30240}[f^v(x_n) - f^v(x_0)] + ... \qquad ...(1)$$

Put $f(x) = x^7, x_0 = 0$ and $h = 1$ in (1), we get

$$\int_0^n x^7 dx = \sum_{x=0}^{n-1} x^7 + \frac{1}{2}(n^7) - \frac{1}{12}(7n^6) + \frac{1}{720}(210n^4) - \frac{1}{30240}(2520n^2)$$

$$\Rightarrow \sum_{x=0}^{n-1} x^7 = \frac{1}{8}n^8 - \frac{1}{2}n^7 + \frac{7}{12}n^6 - \frac{7}{24}n^4 + \frac{1}{12}n^2$$

$$\Rightarrow \sum_{x=0}^{n} x^7 = \frac{1}{7}n^8 + \frac{1}{2}n^7 + \frac{7}{12}n^6 - \frac{7}{24}n^4 + \frac{1}{12}n^2 \qquad \text{...(2)}$$

[On adding x^7 of both sides]

In a similar way, we may find by substituting $f(x) = x^5, x_0 = 0$ and $h = 1$, we get

$$\sum_{x=1}^{n} x^7 + \sum_{x=1}^{n} x^5 = \sum_{x=0}^{n} x^7 + \sum_{x=0}^{n} x^5$$

$$= \frac{1}{8}n^8 + \frac{1}{2}n^7 + \left(\frac{7}{12} + \frac{1}{6}n^6\right) + \frac{1}{2}n^5 - \left(\frac{7}{24} - \frac{5}{12}\right)n^4$$

$$= \frac{1}{8}n^8 + \frac{1}{2}n^7 + \frac{3}{4}n^6 + \frac{1}{2}n^5 + \frac{1}{8}n^4 = \frac{1}{8}(n^8 + 4n^7 + 6n^6 + 4n^5 + n^4)$$

$$= \frac{n^4}{8}(n^4 + 4n^3 + 6n^2 + 4n + 1) = \frac{n^4}{8} \times (n+1)^4$$

$$= \frac{2[n^2(n+1)^2]^2}{4} = 2\left(\sum_{x=1}^{n} x^3\right)^2$$

MISCELLANEOUS SOLVED EXAMPLES

EXAMPLE 1. *By applying method of undetermined coefficients, derive the formula*

$$\int_0^h y(x)dx = h(a_0 y_0 + a_1 y_1) + h^2(b_0 y_0' + b_1 y_1')$$

SOLUTION. Since, there are four coefficients, so we make the formula exact for $y(x) = 1, x, x^2$ and x^3.

For y = 1 $y_0 = 1 = y_1, y_0' = 0 = y_1'$

So $\qquad \int_0^h dx = h(a_0 + a_1) \Rightarrow [x]_0^h = h(a_0 + a_1)$

$\Rightarrow \qquad h = h(a_0 + a_1)$

So, $\qquad a_0 + a_1 = 1$ $\qquad\qquad\qquad\qquad\qquad\qquad$ ...(1)

For y = x $\int_0^h x dx = h(a_1) + h^2(b_0 + b_1)$

$$\Rightarrow \qquad \left(\frac{x^2}{2}\right)_0^h = ha_1 + h^2(b_0 + b_1) \Rightarrow \frac{h^2}{2} = ha_1 + h^2(b_0 + b_1)$$

$$\Rightarrow \qquad a_0 + h(b_0 + b_1) = \frac{h}{2} \qquad\qquad\qquad\qquad\qquad \text{...(2)}$$

For y = x²

$$y_0 = (0)^2 = 0, y_1 = (1)^2 = 1, y_0' = 0 = y_1' = 2$$

So, $\qquad \int_0^h x^2 dx = h(a_0 y_0 + a_1 y_1) + h^2(b_0 y_0' + b_1 y_1')$

$$\Rightarrow \qquad \left(\frac{x^3}{3}\right)_0^h = h(a_1) + h^2(2b_1) \Rightarrow \frac{h^3}{3} = ha_1 + 2h^2 b_1$$

So, $a_1 + 2hb_1 = \dfrac{h^2}{3}$ ...(3)

For $y = x^3$ $\int_0^h x^3 dx = h(a_1) + h^2(3b_1)$

$\Rightarrow \qquad \left(\dfrac{x^4}{4}\right)_0^h = ha_1 + 3b_1 h^2$

$\Rightarrow \qquad a_1 + 3b_1 h = \dfrac{h^3}{4}$ ...(4)

Putting $h = 1$ in (1) – (4) and then solve, we get

$$a_0 = \frac{1}{2}, a_1 = \frac{1}{2}, b_0 = \frac{1}{12}, b_1 = -\frac{1}{12}$$

Hence, $\int_0^h y(x)dx = \dfrac{h}{2}(y_0 + y_1) + \dfrac{h^2}{12}(y_0' - y_1')$

EXAMPLE 2. *What is the effect of change of origin and the change of scale on Simpson's $\dfrac{1}{3}$ rule?*

SOLUTION. It is known that the step size h and the functional values y_0, y_1 and y_2 do not change when the origin is changed, therefore, there is no effect of change of origin on Simpson's $\dfrac{1}{3}$ rule.

Now, let $\dfrac{h}{3}[y_0 + 4y_1 + y_2]$ be the value of the integral by Simpson's $\dfrac{1}{3}$ rule.

Now, suppose on changing the scale the new value of h becomes kh and the new values y_0, y_1, y_2 are y_0', y_1', y_2'. The new value of I is given by

$$I_1 = \frac{kh}{3}(y_0' + 4y_1' + y_2')$$

But $\quad y_0' = y_0, y_1' = y_1, y_2' = y_2,$

then $\quad I_1 = \dfrac{kh}{3}(y_0 + 4y_1 + y_2) = KI$

Hence, on changing the scale, Simpson's $\dfrac{1}{3}$ rule changes.

EXAMPLE 3. *If $V_x = a + bx + cx^2$, show that $\int_1^3 V_x dx = 2V_2 + \dfrac{1}{12}(V_0 - 2V_2 + V_4)$*

and hence, find an approximate value for   (Bhopal-2002, 07)

$$\int_{-1/2}^{1/2} e^{-x^2/10} dx.$$

SOLUTION. Firstly, shifting the origin to –2. Then, we have to prove that

$$\int_{-1}^1 V_x dx = 2V_0 + \frac{1}{12}(V_{-2} - 2V_0 + V_2) \qquad \text{...(1)}$$

$$\text{L.H.S.} = \int_{-1}^1 V_x dx = \int_{-1}^1 (a + bx + cx^2)dx \left[ax + \frac{bx^2}{2} + \frac{cx^3}{3}\right]_{-1}^1$$

$$= \left[2a + \frac{2c}{3}\right]$$

Also, $\qquad V_0 = a, V_{-2} = a - 2b + 4c, V_2 = a + 2b + 4c$

Now,

$$\text{R.H.S.} = 2V_0 + \frac{1}{12}(V_{-2} - 2V_0 + V_2)$$

$$= 2a + \frac{1}{12}(a - 2b + 4c - 2a + a + 2b + 4c)$$

$$= 2a + \frac{1}{12}(8c) = 2a + \frac{2c}{3}$$

Hence,   L.H.S. = R.H.S.

Further, changing the scale to 1/2 in (1), we get

$$\int_{-1/2}^{1/2} V_x dx = \frac{1}{2}\left[2V_0 + \frac{1}{12}(V_{-1} - 2V_0 + V_1)\right]$$

Taking $V_x = e^{-x^2/10}$,

then $V_0 = e^0 = 1, V_{-1} = e^{-1/10}$ and $V_1 = e^{-1/10}$

so that

$$\int_{-1/2}^{1/2} e^{-x^2/10} dx = \frac{1}{2}\left[2 + \frac{1}{12}(e^{-1/10} - 2 + e^{-1/10})\right] = 1 + \frac{1}{12}(e^{-1/10} - 1).$$

EXAMPLE 4.   If $f(x) = a + bx + cx^2$, show that

$$\int_1^3 f(x)dx = \frac{1}{12}[f(0) + 22f(2) + f(4)]$$

SOLUTION.   Here, $f(x)$ is a polynomial of degree 2 in x, therefore its third and higher order differences are zero. In this case we shall take $h = 2$.

Now, $f(x) = f\left(\frac{x}{2}.2\right) = f\left(\frac{x}{2}.h\right)$

$$= f\left(0 + \frac{x}{2}h\right) = E^{x/2}f(0)$$

$$= (1 + \Delta)^{x/2} f(0)$$

$$= \left[1 + \frac{x}{2}\Delta + \frac{x}{2}\left(\frac{x}{2} - 1\right)\frac{\Delta^2}{2!} + \ldots\right]f(0) = f(0) + \frac{x}{2}\Delta f(0) + \frac{x}{2}\left(\frac{x}{2} - 1\right)\frac{\Delta^2 f(0)}{2!}$$

(Neglecting all the other terms)

So $\int_1^3 f(x)dx = \int_1^3 \left[f(0) + \frac{x}{2}\Delta f(0) + \left(\frac{x^2}{4} - \frac{x}{2}\right)\frac{1}{2}\Delta^2 f(0)\right]dx$

$$= \left[x.f(0) + \frac{x^2}{4}\Delta f(0) + \left(\frac{x^3}{12} - \frac{x^2}{4}\right)\frac{1}{2}\Delta^2 f(0)\right]_1^3$$

$$= 2f(0) + \left(\frac{9}{4} - \frac{1}{4}\right)\Delta f(0) + \left(\frac{27}{12} - \frac{9}{4} - \frac{1}{12} + \frac{1}{4}\right)\frac{1}{2}\Delta^2 f(0)$$

$$= 2f(0) + 2\Delta f(0) + \frac{1}{12}\Delta^2 f(0)$$

$$= 2f(0) + 2[f(0 + h) - f(0)] + \frac{1}{12}[f(0 + 2h) - 2f(0 + h) + f(0)]$$

$$= 2f(0) + 2[f(2) - f(0)] + \frac{1}{12}[f(4) - 2f(2) + f(0)]$$

$$[\because h = 2]$$

$$= 2f(2) + 2[f(2) - f(0)] + \frac{1}{12}[f(4) - 2f(2) + f(0)]$$

$$= 2f(2) + \frac{1}{12}[f(4) - 2f(2) + f(0)] \quad = \frac{1}{12}[f(0) + 22f(2) + f(4)]$$

EXAMPLE 5. *If $f(x)$ is a polynomial in x of the third degree, find an expression for $\int_0^t f(x)dx$ in terms of $f(0)$, $f(1)$, $f(2)$ and $f(3)$. Use this result, show that*

$$\int_0^2 f(x)dx = \frac{1}{24}[-f(0) + 13f(1) + 13f(2) - f(3)]$$

SOLUTION. Using the given values in Lagrange's interpolation formula, we have

$$f(x) = \frac{(x-1)(x-2)(x-3)}{(0-1)(0-2)(0-3)}f(0) + \frac{(x-0)(x-2)(x-3)}{(1-0)(1-2)(1-3)}f(1)$$

$$+ \frac{(x-0)(x-1)(x-3)}{(2-0)(2-1)(2-3)}f(2) + \frac{(x-0)(x-1)(x-2)}{(3-0)(3-1)(3-2)}f(3)$$

$$= \frac{1}{6}[x^3 - 6x^2 + 11x - 6]f(0) + \frac{1}{2}[x^3 - 5x^2 + (6x)]f(1)$$

$$- \frac{1}{2}[x^3 - 4x^2 + 3x)]f(2) + \frac{1}{6}[x^3 - 3x^2 + 2x]f(3)$$

$$\Rightarrow \quad \int_0^t f(x)dx = \left[-\frac{1}{6}\left(\frac{x^4}{4} - \frac{6x^3}{3} + 11\frac{x^2}{2} - 6x \right)f(0) \right.$$

$$+ \frac{1}{2}\left(\frac{x^4}{4} - \frac{5x^3}{3} + \frac{6x^2}{2} \right)f(1) - \frac{1}{2}\left(\frac{x^4}{4} - \frac{4x^3}{3} + \frac{3x^2}{2} \right)f(2)$$

$$\left. + \frac{1}{6}\left(\frac{x^4}{4} - \frac{3x^3}{3} + \frac{2x^2}{2} \right)f(3) \right]_0^t$$

$$= t^4\left[-\frac{1}{24}f(0) + \frac{1}{8}f(1) - \frac{1}{8}f(2) + \frac{1}{24}f(3) \right]$$

$$+ t^3\left[\frac{1}{3}f(0) - \frac{5}{6}f(1) + \frac{2}{3}f(2) - \frac{1}{6}f(3) \right]$$

$$+ t^2\left[-\frac{11}{12}f(0) + \frac{2}{3}f(1) - \frac{3}{4}f(2) + \frac{1}{6}f(3) \right] + tf(0) \qquad \ldots(1)$$

Putting $t = 2$, in (1) and after some simplification, we get

$$\int_0^2 f(x)dx = f(0)\left[-\frac{16}{24} + \frac{8}{3} - \frac{44}{12} + 2 \right]$$

$$+ f(1)\left[\frac{16}{8} - \frac{40}{6} + \frac{12}{2} \right] + f(2)\left[-\frac{16}{8} + \frac{16}{3} - \frac{12}{4} \right] + f(3)\left[\frac{16}{24} - \frac{3}{6} + \frac{4}{6} \right]$$

$$\Rightarrow \quad \int_0^2 f(x)dx = \frac{1}{3}f(0) + \frac{4}{3}f(1) + \frac{1}{3}f(2) \qquad \ldots(2)$$

Similarly putting $t = 1$ in (1), we get

$$\int_0^1 f(x)dx = \frac{3}{8}f(0) + \frac{19}{24}f(1) - \frac{5}{24}f(2) + \frac{1}{24}f(3) \qquad \text{...(3)}$$

Subtracting (3) from (2), we get

$$\int_0^2 f(x)dx - \int_0^1 f(x)dx = -\frac{1}{24}f(0) + \frac{13}{24}f(1) + \frac{13}{24}f(2) - \frac{1}{24}f(3)$$

$$\Rightarrow \qquad \int_1^2 f(x)dx = \frac{1}{24}[-f(0) + 13f(1) + 13f(2) - f(3)]$$

EXERCISE 6.2

1. Use trapezoidal rule to evaluate $\int_0^1 x^3 dx$ considering five sub intervals.

2. Calculate the approximate value of $\int_{-3}^3 x^4 dx$ by trapezoidal rule. (UPPCS-2004, 08; (IAS-2008; Indore-2006, 10, Meerut-2002,05)

3. Use Simpson's rule dividing the range into ten equal parts, to show that
$\int_0^1 \frac{\log(1+x^3)}{(1+x^2)}dx = 0.1730$.

4. Evaluate $\int_3^5 \frac{4}{2+x^2}dx$ by dividing the range into eight equal parts.

5. Use Simpson's rule to find the value of $\int_1^5 f(x)dx$ given

x	1	2	3	4	5
$f(x)$	10	50	70	80	100

6. State and prove Simpson's $\frac{1}{3}$ rule. What is the effect of
(i) change of origin and (ii) change of scale on this rule?

7. Calculate $\int_0^{\pi/2} e^{\sin x}dx$ correct to four decimal places by Simpson's $\frac{3}{8}$ rule, dividing the range of integration $(0, \pi/2)$ into 3 equal parts. (UPTU–2007)

8. Find by Weddle's rule the value of the integral $\int_{0.4}^{1.5}\frac{x}{\sinh x}dx$ by taking 12 subintervals.

9. (i) The velocities of a car running on a straight road at interval of 2 minutes are given below :

Time (in min.)	0	2	4	6	8	10	12
Velocities (in km/ hr)	0	22	33	27	18	7	0

Apply Simpson's rule to find the distance covered by car.

(ii) A rocket is launched from the ground. Its acceleration is registered during the first 80 seconds and is given in the table below. Find the velocity of the rocket at $t = 80$ seconds by Simpson's $\frac{1}{3}$ rule. (Mumbai-2004)

t (sec)	0	10	20	30	40	50	60	70	80
f (cm/ sec^2)	30	31.63	33.34	35.47	37.75	40.33	43.25	46.69	50.67

10. Integrate numerically $\int_0^{\pi/2}\sqrt{\cos\theta}\,d\theta$. (Rohtak–2004; VTU-2009)

11. Prove that $\sum\limits_{k=0}^{n} L_k = 1$ where
$$L_k = \frac{u(u-1)...(u-k+1)(u-k-1)...(u-n)}{k(k-1)...(+1)(-1)...(k-n)}$$

12. For $n = 1$, prove that $\frac{1}{L_0} + \frac{1}{L_1} = \frac{1}{L_0 L_1}$.

13. Prove $C_k^n = C_{n-k}^n$ where $C_k^n = \frac{1}{n}\int_0^n L_k\,du$.

14. Prove that $C_1^1.C_1^2 = \frac{C_0^2}{C_0^1}$ where $C_k^n = \frac{1}{n}\int_0^n L_k\,du$

15. Compute the integral $\int_0^1\frac{dx}{1+x^2}$ by Gauss quadrature formula, taking $n = 5$.

16. Using Gauss quadrature formula evaluate the integral $I = \int_1^2 \frac{dx}{x}$. Taking $n = 5$.

17. Find the value of the integral $\int_1^0 \frac{dx}{1+x^2}$ by Chebychev's formula, taking $n = 4$.

18. Evaluate the integral $\int_1^7 \frac{dx}{x}$ by Chebychev's formula taking $n = 5$.

19. Evaluate $\int_0^1 \frac{dx}{1+x}$ correct to five decimal places by using Euler-Maclaurin's formula.

20. Find the sum of the fifth powers of the first n natural numbers by means of Euler-Maclaurin's formula.

21. Use Boole's Rule, find the value of the integral
$$I = \int_{0.4}^{1.6} \frac{x}{\sinh x} dx$$
By taking 12 sub intervals.

22. Using Euler-Maclaurin summation formula, sum the following series :

(i) $\dfrac{1}{400} + \dfrac{1}{402} + ... + \dfrac{1}{498} + \dfrac{1}{505}$

(ii) $\dfrac{1}{100} + \dfrac{1}{101} + \dfrac{1}{102} + \dfrac{1}{103} + \dfrac{1}{104}$

23. Show that $\sum\limits_1^n x^3 = \left(\dfrac{n(n+1)}{2}\right)^2$, applying Euler-Maclaurin's formula.

24. Show that $\sum\limits_0^n i^4 = \dfrac{n^5}{5} + \dfrac{n^4}{2} + \dfrac{n^3}{3} + \dfrac{n}{30}$, by applying Euler-Maclaurin's summation formula.

25. Evaluate the integral $\int_0^1 \frac{x^2}{1+x^3} dx$, using Simpson's rule and hence find the value of $\log_e 3$. (Agra-2003; Avadh-2003; Mumbai-2004)

26. The velocity of v of a particle at distance s from a point on its path is given by the following table :

s	0	10	20	30	40	50	60
v	47	58	64	65	61	52	38

Estimate the time taken to travel 60 meters by using Simpson's rule.

(UPTU–2007, Madras–2003)

27. Evaluate $\int_0^2 \frac{dx}{1+x+x^2}$ to three decimal places dividing the range of integration into 8 equal parts.

28. If third differences are constant, prove that
$$\int_{-1}^1 f(x)dx = \frac{2}{3}\left[f(0) + f\left(\frac{1}{\sqrt{2}}\right) + f\left(\frac{-1}{\sqrt{2}}\right)\right]$$
(Bhopal-2010, Avadh-2004, 10, Allahabad-2010)

29. If $f(x) = a + bx + cx^2$, prove that
$$\int_1^2 f(x)dx = \frac{1}{12}[f(0) + 22f(2) + f(4)]$$

30. Obtain the approximate quadrature formula
$$\int_0^1 \frac{dx}{1+x^2} = \frac{1}{4a}\left(1 + \frac{1}{6a}\right) + \sum_{x=1}^a \frac{a}{a^2+x^2}$$

31. Use the trapezoidal rule to estimate the integral $\int_0^2 e^{x^2} dx$ taking 10 intervals.

(UPTU–2008)

32. Use Simpson's 1/3rd rule to find $\int_0^{0.6} e^{-x^2} dx$ by taking seven ordinates.

(VTU–2011, Bhopal–2009)

33. The velocity v of a particle at distance s from a point on its linear path is given by the following table :

s (m)	0	2.5	5.0	7.5	10.0	12.5	15.0	17.5	20.0
v (m/sec)	16	19	21	22	20	17	13	11	9

Estimate the time taken by the particle to traverse the distance of 20 meters using Boole's rule. (UPTU–2007)

34. Evaluate $\int_0^6 x \sec x\, dx$ using eight intervals by trapezoidal rule. (UPTU–2009)

35. Evaluate using Simpson's 1/3rd rule $\int_0^2 e^{-x^2} dx$ (take $h = 0.25$). (JNTU–2007)

36. Evaluate using Simpson's 1/3rd rule $\int_0^1 \frac{dx}{x^3+x+1}$, choose step length 0.25 (UPTU–2009)

37. Evaluate using Simpson's 1/3rd rule, $\int_0^{\pi/2} \sqrt{\cos\theta}\, d\theta$ taking 9 ordinates. (VTU–2009)

38. Evaluate correct to 4 decimal places, by Simpson's 3/8th rule.
$$\int_0^9 \frac{dx}{1+x^3}$$ (UPTU(M.Tech)–2010)

39. A curve is drawn to pass through the points given by following table :

x	1	1.5	2	2.5	3	3.5	4
y	2	2.4	2.7	2.8	3	2.6	2.1

Using Weddle's rule, estimate the area bounded by the curve, the x-axis and the lines $x = 1, x = 4$.   (VTU–2011S; Bhopal-2007)

40. The following table gives the velocity v of a particle at time t :

s (sec)	0	2	4	6	8	10	12
v (m/sec)	4	6	16	34	60	94	136

Find the distance moved by the particle in 12 seconds and also the acceleration at $t = 2$ sec.   (SVTU–2007)

41. Evaluate $\int_0^4 e^x \, dx$ by Simpson's rule, given that e = 2.72, $e^2 = 7.39$, $e^3 = 20.09$, $e^4 = 54.6$

and compare it with the actual value.

(Nagarjuna-2003; Lucknow-2009; Avadh-2005)

42. Calculate the value of $\int_0^\pi \sin x \, dx$ by Simpson's 1/3 rule, using 11-ordinates.   (JNTU-2009)

43. Given that

x	4.0	4.2	4.4	4.6	4.8	5.0	5.2
log x	1.3863	1.4351	1.4816	1.5261	1.5686	1.6094	1.6487

Evaluate $\int_4^{5.2} \log x \, dx$ by

(i) Trapezoidal rule
(ii) Simpson's 3/8 rule   (JNTU-2006)
(iii) Weddle's rule   (VTU-2008)
(iv) Simpson's 1/3 rule   (Kerala-2003)

44. The table below shows the temperature $f(t)$ as a function of time :

t	1	2	3	4	5	6	7
f(t)	81	75	80	83	78	70	60

Using Simpson's 1/3 rd rule to estimate $\int_1^7 f(t) \, dt$.   (JNTU-2007)

45. A river is 80 ft. wide. The depth d in feet at a distance x ft from one bank is given by the following table :

x	0	10	20	30	40	50	60	70	80
y	0	4	7	9	12	15	14	8	3

Find approximately the area of the cross-section.

46. A curve is drawn to pass through the points given by the following table :

x	1	1.5	2	2.5	3	3.5	4
y	2	2.4	2.7	2.8	3	2.6	2.1

Using Weddle's rule, estimate the area bounded by the curve, the x axis and the lines $x=1, x=4$.   (VTU–2005)

47. The following table gives the velocity v of a particle at time t :

t	0	2	4	6	8	10	12
v	4	6	16	34	60	94	136

Find the distance moved by the particle in 12 seconds and also the acceleration at $t = 2$ sec.   (SVTU–2007)

Answers

1. 0.26   **2.** 115   **4.** 26.716   **5.** 256.66   **7.** 3.1017 **8.** 1.01019

9. (i) $3\frac{5}{9}$ km   (ii) 30.87 m/sec   **10.** 1.1873   **15.** 0.785398   **16.** 0.6931

17. 0.785396   **18.** 1.945910   **19.** 0.69315   **20.** $\left(\dfrac{n^6}{6}+\dfrac{n^5}{2}+\dfrac{5n^4}{12}-\dfrac{n^2}{12}\right)$

21. 1.010784   **22.** (i) 0.11382114   (ii) 0.0490291   **25.** 0.23108   **31.** 17.0621

32. 0.5351   **33.** 1.35 sec   **34.** – 6.436   **35.** 0.635   **36.** 0.6305

37. 1.1873   **38.** 1.1249   **39.** 3.032   **40.** 552m; 3m/sec^2

41. 53.87, 53.6   **42.** 2.0009

43. (i) 1.8276551, 0.0001924 (ii) 1.8278470, 0.000005 (iii) 1.8278474, 0.0000001

(iv) 1.8278472, 0.0000003.

44. 403.67   **45.** 710 sq. ft.   **46.** 3.032   **47.** 552 m; 3m/sec^2

Objective Evaluations

Problem Set–1

FILL IN THE BLANKS

1. The differential operator $D = $ _____.
2. If $y = ax + b$, then $Dy = $ _____.
3. $\dfrac{1}{h}\left[\nabla + \dfrac{1}{2}\nabla^2 + \dfrac{1}{3}\nabla^3 = \underline{\hspace{3cm}}\right]$

   $= \underline{\hspace{2cm}}$.

4. If $y = f(x)$ is a polynomial of degree 3, then

   $D^2 y = \dfrac{1}{h^2}[\Delta^2 y - \underline{\hspace{2cm}}]$.

5. If $1 - \nabla = e^{-hD}$, then $D^3 = \dfrac{1}{h^3}[\underline{\hspace{2cm}}]$.

6. In general $D^r = \underline{\hspace{2cm}}[\log(1 + D)]^r$.
7. If $y = f(x)$ and y' is required at the middle of the given arguments, then we will use _____ for differentiation.
8. If the arguments are not equally spaced, then we will use _____ for finding the derivative.
9. $f(x) = ax^2 + b$ and differencing of interval is h. Then the value of $D^2 y = $ _____.
10. $D^r = \dfrac{}{h^r}[\log(1 - \nabla)]^r$.

TRUE / FALSE

Write 'T' for True and 'F' for False statement.

1. If $y = x^k$, then $D^k y = k!$ **(T/F)**
2. $D^r = (-1)^r \cdot h^{-r}[\log(1 - \nabla)]^r$ **(T/F)**
3. If y is a polynomial of degree 3 in x. Then

   $D^3 y = \dfrac{1}{h^2}\left[\Delta^3 y - \dfrac{3}{2}\Delta^4 y\right]$ **(T/F)**

4. $\left(\dfrac{dy}{dx}\right)_{x_0} = \dfrac{1}{h}\left[\Delta y_0 - \dfrac{1}{2}\Delta^2 y_0 + \dfrac{1}{3}\Delta^3 y_0 - ...\right]$ **(T/F)**

5. If second order divided difference of a function is zero. Then $f'(x) = -$ (First divided difference). **(T/F)**

6. If $f(0) = 0$, $f(1) = 1$, $f(2) = 8$, $f(3) = 27$. Then $f'(2) = 12$. **(T/F)**

7. $\left(\dfrac{dy}{dx}\right)_{x_n} = \dfrac{1}{h}\left[\nabla y_n + \dfrac{1}{2}\nabla^2 y_n + \dfrac{1}{3}\nabla^3 y_n + ...\right]$ **(T/F)**

8. $D = \dfrac{1}{h}\left[\Delta - \dfrac{1}{2}\Delta^2 + \dfrac{1}{3}\Delta^3 - ...\right]$ **(T/F)**

9. $D \neq \dfrac{1}{h}\left[\nabla + \dfrac{1}{2}\nabla^2 + \dfrac{1}{3}\nabla^3 + ...\right]$ **(T/F)**

10. If $y = y_0 + u\Delta y_0 + \dfrac{u(u-1)}{2!}\Delta^2 y_0$, $x = x_0 + uh$

    Then $\left(\dfrac{d^2 y}{dx^2}\right)_{x_0} = \dfrac{\Delta^2 y_0}{h^2}$ **(T/F)**

MULTIPLE CHOICE QUESTIONS

Choose the most appropriate one.

1. If $y = y_0 + u\Delta y_0 + \dfrac{u(u-1)}{2!}\Delta^2 y_0$, $x_0 + uh$.

   Then $h^2\left(\dfrac{d^2 y}{dx^2}\right)_{x_0}$ is:

   (a) Δy_0 (b) $\Delta^2 y_0$

   (c) $\Delta^3 y_0$ (d) y_0

2. If $y = x^k$, then $(D^{k-1} y)x = 0$ is equal to :

   (a) 0   (b) $(k - 1)!$

   (c) $k!$ (d) $(k - 1)!.x$

3. If $y = y_0 + u\Delta y_0 + \dfrac{u(u-1)}{2!}\Delta^2 y_0$ and

   $x = x_0 + uh$ when $u = 1/2$ then $\left(\dfrac{dy}{dx}\right)$ is equal to:

   (a) $\Delta^2 y_0$ (b) 0

   (c) 1   (d) $\dfrac{1}{h}\Delta y_0$

4. If $y = ax + b$, then Dy is equal to:

   (a) a (b) b

   (c) $a + b$ (d) $a - b$

5. If $f(x) = 5$ for all values of x, then $f(5)$ is:

(a) 1   (b) 0   (c) –1   (d) 5

Problem Set– 2

FILL IN THE BLANKS

1. $\int_{x_0}^{x_1} y\,dx = \dfrac{h}{2}[y_0 + \underline{\hspace{2cm}}]$.

2. To obtain trapezoidal rule $y = f(x)$ must be \underline{\hspace{2cm}}.

3. In the trapezoidal rule, in each sub interval $[x_0, x_1], [x_1, x_2]...(x_{n-1}, x_n)$ the area of \underline{\hspace{2cm}} is calculated.

4. In the Simpson's 1/3 rule, the number of subinterval must be \underline{\hspace{2cm}}.

5. To obtain Simpson's 1/3 rule, the function $y = f(x)$ should be a polynomial of degree \underline{\hspace{2cm}}.

6. In Simpson's 3/8 rule, the number of division of interval of integration should be taken as a multiple of \underline{\hspace{2cm}}.

7. To obtain Simpson's 3/8 rule, the function is to be taken a polynomial of degree \underline{\hspace{2cm}}.

8. The number of sub intervals in Weddle's rule should be taken as a multiple of \underline{\hspace{2cm}}.

9. In Weddle's rule, the polynomial is taken of degree \underline{\hspace{2cm}}.

10. $\sum\limits_{k=0}^{n} L_k = \underline{\hspace{2cm}}$.

11. If $C_x^n = C_y^n$, then $x + y = \underline{\hspace{2cm}}$ where C_k^n are Cote's numbers.

12. $\sum\limits_{k=0}^{n} C_k^n = \underline{\hspace{2cm}}$.

13. $\int_0^n L_k\,du = \underline{\hspace{2cm}} C_k^n$.

14. In the Chebychev's formula, the coefficients of y's are \underline{\hspace{2cm}}.

TRUE / FALSE

Write 'T' for True and 'F' for False statement.

1. In trapezoidal rule $y = f(x)$ is taken to be linear. **(T/F)**

2. In Simpson's 1/3 rule $\int_{x_0}^{x_2} y\,dy = \dfrac{3h}{4}[y_0 + 4y_1 + y_2]$. **(T/F)**

3. To obtain Simpson's 1/3 rule, the function $y = f(x)$ is taken to be of degree 3. **(T/F)**

4. In Simpson's 1/3 rule the number of division of $[x_0, x_n]$ should be even. **(T/F).**

5. In Weddle's rule, the function $y = f(x)$ should be of degree 6. **(T/F)**

6. $L_0 + L_1 + L_2 \neq 1$ for $n = 2$. **(T/F)**

7. $\sum\limits_{k=0}^{n} C_k^n = 1$ **(T/F)**

8. $C_k^n = \dfrac{1}{n} \int_0^n L_k\,du$ then $C_0^6 = \dfrac{41}{840}$ **(T/F)**

9. $C_k^n \neq C_{n-k}^n$ **(T/F)**

10. In Gauss quadrature formula coefficients of y's are same. **(T/F)**

11. $C_0^1 . C_1^1 . C_1^2 = C_0^2$ **(T/F)**

12. The number of subintervals are taken to be multiple of 3 in Simpson's 1/3 rule. **(T/F)**

13. In trapezoidal rule, in each subinterval the area of trapezium is calculated. **(T/F)**

14. Simpson's 1/3 is more accurate than Simpson's 3/8 rule. **(T/F)**

MULTIPLE CHOICE QUESTIONS

Choose the most appropriate one

1. In trapezoidal rule, the area of which shape is calculated :

(a) square   (b) rectangle
(c) trapezium   (d) circle

2. The number of sub-intervals should be taken in Simpson's 1/3 rule :

(a) odd   (b) even
(c) prime   (d) none of these

3. In Trapezoidal rule, the function must be :

(a) linear   (b) quadratic
(c) cubic   (d) biquadratic

4. In Simpson's 1/3 rule, function $y = f(x)$ is taken to be :

(a) circle   (b) ellipse
(c) parabola   (d) line

5. If $y = ax^6 + bx^5 + cx^4 + dx^3 + ex^2 + fx + g$

is taken, then we obtain the formula :
   (a) Simpson's 1/3 rule (b) Simpson 3/8 rule
   (c) Weddle's rule   (d) Trapezoidal rule

6. C_2^6 is equal to :

   (a) C_0^6 (b) C_4^6

   (c) C_1^6 (d) C_3^6

7. The value of $\dfrac{\int_0^n L_k du}{C_k^n}$ is:

   (a) n^2 (b) $n - 1$
   (c) n (d) $n + 1$

8. In trapezoidal rule, the value of $\int_{x_3}^{x_4} y\,dx$ is :

   (a) $\dfrac{h}{2}[y_0 + y_1]$ (b) $\dfrac{h}{2}[y_3 + y_4]$

   (c) $\dfrac{h}{2}[y_1 + y_2]$ (d) $\dfrac{h}{2}[y_0 + y_2]$

9. The value of $\sum\limits_{k=0}^{n} C_k^n - \sum\limits_{k=0}^{n} L_k$ is :
   (a) 1   (b) – 1
   (c) 2   (d) 0

10. The value of $C_0^1 + C_1^1 + C_0^2$ is :
    (a) 6/7   (b) 7/6
    (c) 0   (d) 1

11. The value of $16 C_0^2 C_2^2$ is :
    (a) $(C_0^1)^2$ (b) $(C_0^2)^2$
    (c) $(C_0^2)^3$ (d) C_1^1

12. The number of subintervals in Weddle's rule should be taken as a multiple of :
    (a) 3   (b) 4
    (c) 6   (d) 5

Problem Set – 3

1. The process of evaluating the derivatives of a function with the help of the given set of values of that function is called :
   (a) numerical integration
   (b) numerical differentiation
   (c) quadrature
   (d) none of these

2. Which one of the following is not true :

   (a) $D = \dfrac{1}{h}\left[\Delta - \dfrac{\Delta^2}{2} + \dfrac{\Delta^3}{3} - \dfrac{\Delta^4}{4} + \dfrac{\Delta^5}{5} - \dfrac{\Delta^6}{6} + \ldots\right]$

   (b) $D = \dfrac{1}{h^2}\left[\Delta^2 - \Delta^3 - \dfrac{11}{12}\Delta^4 + \ldots\right]$

   (c) $D^3 = \dfrac{1}{h^3}\left[\Delta^3 - \dfrac{3}{2}\Delta^4 + \ldots\right]$

   (d) are all true

3. Which one of the following is not true

   (a) $\left(\dfrac{dy}{dx}\right)_{at\ x = x_0}$

   $= \dfrac{1}{h}\left[\nabla y_n + \dfrac{1}{2}\nabla^2 y_n + \dfrac{1}{3}\nabla^3 y_n + \ldots\right]$

   (b) $\left(\dfrac{d^2 y}{dx^2}\right)_{at\ x = x_0} = \dfrac{1}{h^2}\left[\begin{array}{c}\nabla^2 y_n + \nabla^3 y_n \\ + \dfrac{11}{12}\nabla^4 y_n + \ldots\end{array}\right]$

   (c) $\left(\dfrac{d^3 y}{dx^3}\right)_{at\ x = x_0} = \dfrac{1}{h^3}\left[\nabla^3 y_n + \dfrac{3}{2}\nabla^4 y_n + \ldots\right]$

   (d) all are true

4. Given a set of tabulated values of the integrand $f(x)$, to find the value of $\int_{x_0}^{x_n} f(x)\,dx$ is called :
   (a) numerical differentiation
   (b) numerical integration
   (c) both (a) and (b) are true
   (b) none of these

5. The process of finding the values of the integral of a function of a single variable is called :
   (a) differentiation   (b) integration
   (c) quadrature   (d) none of these

6. The Trapezoidal formula can be obtained from quadrature formula by putting :
   (a) $n = 1$ (b) $n = 2$
   (c) $n = 3$ (b) none of these

7. The Simpson' 1/3 rule can be obtained from quadrature formula by putting
   (a) $n = 1$ (b) $n = 2$
   (c) $n = 3$ (b) none of these

8. The Simpson's 1/3 rule can be obtained from quadrature formula by putting :
   (a) $n = 1$ (b) $n = 2$
   (c) $n = 3$ (b) none of these

9. In Simpson's 1/3 rule, the interpolating polynomial is of degree :
   (a) 2
   (b) 3
   (b) 4
   (d) none of these

10. Simpson's 1/3 rule is also known as:
    (a) Cubic rule
    (b) parabolic rule
    (c) quadrature rule
    (d) none of these

11. In Simpson's 1/3 rule, the division of interval (x_0, x_n) into sub intervals should be:
    (a) odd
    (b) even
    (c) may or may note be even
    (d) none the these

12. In Simpson's 3/8 rule the interpolating polynomial is of degree :
    (a) 2
    (b) 3
    (b) 4
    (d) none of these

13. In Simpson's 3/8 rule, the divisions of interval into subintervals should be :
    (a) odd
    (b) even
    (c) multiple of 3
    (d) none of these

14. In trapezoidal rule, the area of which shape is calculated :
    (a) Square
    (b) rectangle
    (c) trapezium
    (d) none of these

15. In trapezoidal rule the function must be:
    (a) linear
    (b) quadratic
    (c) cubic
    (d) none of these

16. In Simpson's 1/3 rule, function $y = f(x)$ is taken to be :
    (a) circle
    (b) parabola
    (c) ellipse
    (d) none of these

17. In Trapezodial rule, the value of $\int_{x_3}^{x_4} y\, dx =$
    (a) $\frac{h}{2}(y_3 + y_4)$
    (b) $\frac{h}{3}(y_3 + y_4)$
    (c) $\frac{h}{2}(y_1 + y_3)$
    (d) none of these

18. The number of subinterval in Weddel's rule should be taken as a multiple of:
    (a) 2
    (b) 3
    (c) 6
    (d) 4

19. If $y = y_0 + \dfrac{4(4-1)}{2!}\Delta^2 y_0$, $x = x_0 + 4h$, then value of $h^2\left(\dfrac{d^2 y}{dx^2}\right)_{x_0} =$
    (a) Δy_0
    (b) $\Delta^2 y_0$
    (c) $\Delta^3 y_0$
    (d) none of these

20. If $y = x^k$, then $(D^{k-1}y)_{x=0} =$
    (a) h
    (b) k
    (c) 0
    (d) $k-1$

21. If $f(x) = 6$ for all values of x then $f'(6) =$
    (a) 6
    (b) 6!
    (c) 0
    (d) none of these

22. In the Weddle's rule the interpolating polynomial is taken of be of degree:
    (a) 6
    (b) 4
    (c) 3
    (d) none of these

23. Formula $\int_{x_0}^{x_n} f(x)\,dx = nh \sum_{k=0}^{n} y_k c_k^n$ is called:
    (a) Quadrature formula
    (b) Newton's Cote's formula
    (c) Newton's formula
    (d) none of these

24. The value of $c_k^n =$
    (a) c_n^k
    (b) c_{n-k}^n
    (c) c_{-k}^n
    (d) none of these

25. The value of $\sum_{k=0}^{n} c_k^n =$
    (a) 1
    (b) 0
    (c) n
    (d) none of these

26. Trapezoidal rule for evaluation of $\int_a^b f(x)\,dx$ requires the interval (a, b) to be divided into:
    (a) $2n$ subinterval of equal width
    (b) $(2n+1)$ subinterval of equal width
    (c) any number of subinterval of equal width
    (d) none of these

27. By Trapezoidal rule, the value of the interval $\int_1^5 x^2\,dx$ on dividing the interval into four equal parts is :
    (a) 40
    (b) 41
    (c) 42
    (d) none of these

28. Simpson's rule is applied to a function for its :
    (a) expansion
    (b) numerical integration
    (c) differentiation
    (d) none of these

29. By Simpson's rule the value of the integral $\int_1^6 x\,dx$ on dividing the interval into five equal parts is :
    (a) 17
    (b) 17.5
    (c) 18
    (d) 18.5

30. By Simpson's rule, the value of $\int_{-3}^{3} x^4 \, dx$ by taking 6 subinterval is :

(a) 64   (b) 48

(c) 98   (d) none of these

31. The value of $\int_{x_0}^{x_0+nh} y \, dx$, n is even number, by Simpson's one-third rule is :

(a) $\frac{h}{3}[(y_0 + y_n) + 4(y_1 + y_3 + ... + y_{n-1})$
$+ 2(y_2 + y_4 + ... + y_{n-2})]$

(b) $\frac{h}{2}[(y_0 + y_n) + 3(y_1 + y_3 + ... + y_{n-1}) + ...]$

(c) both (a) and (b) are true

(d) none of these

32. The value of $f(x)$ given only at $x = 0, \frac{1}{3}, \frac{2}{3}$ and 1. Which of the following can be used to evaluate $\int_0^1 f(x) \, dx$ approximately :

(a) Trapezoidal rule   (b) Simpson's rule

(c) both (a) and (b)   (d) none of these

33. $\int_{-1}^{1} f(x) \, dx = \lambda_0 f(x_0) + \lambda_1 f(x_1)$. In this formula, the values of $\lambda_0, \lambda_1, x_0, x_1$ are respectively given by :

(a) $1, 1, -\frac{1}{\sqrt{3}}, \frac{1}{\sqrt{3}}$ (b) $1, -1, \frac{1}{\sqrt{3}}, -\frac{1}{\sqrt{3}}$

(c) $1, \frac{1}{\sqrt{3}}, -1, -\frac{1}{\sqrt{3}}$ (d) none of these

34. The remainder of the following rule

$\int_{x_0}^{x_1} f(x) \, dx = \frac{3h}{8}(f(x_0) + 3f(x_1) + 3f(x_2) + f(x_3)]$

is of order :

(a) 4   (b) 5

(c) 6   (d) none of these

35. The interpolating polynomial of highest degree which corresponds to function values $f(-1) = 9$, $f(0) = 5, f(2) = 3, f(5) = 15$ is :

(a) $x^2 - 3x + 5$ (b) $x^2 + 3x + 5$

(c) $4x^2 + 6x + 5$ (d) none of these

36. From the following table :

x	1.0	1.2	1.4	1.6	1.8	2.0	2.2
y	2.7813	3.3201	4.0552	4.9530	6.0496	7.3891	9.0250

The value of dy/dx at $x = 1.2$ is :

(a) 3.320   (b) 3.420

(c) 3   (d) none of these

37. From the following table :

x	1.5	2.0	2.5	3.0	3.5	4.0
f(x)	3.375	7.000	13.625	24.000	38.875	59.000

Which of the following is/are true :

(a) $f'(1.5) = 4.75$

(b) $f''(1.5) = 9$

(c) both (a) and (b) are true

(d) none of these

38. For the table of $y = x^{1/3}$ given by :

x	50	51	52	53	54	55	56
y	3.6840	3.7084	3.7325	3.7563	3.7798	3.8030	3.8259

at $x = 50$, which of the following is/are true :

(a) $\frac{dy}{dx} = 0.02455$

(b) $\frac{d^2y}{dx^2} = -0.0003$

(c) both (a) and (b) are true

(d) none of these

39. The value of $f'(1)$ for $f(x) = \frac{1}{1 + x^2}$, using the following table is :

x	1.0	1.1	1.2	1.3	1.4
y	0.5000	0.4524	0.4098	0.3717	0.3378

is given by :

(a) 0.5031   (b) −0.5031

(c) 5.0   (d) none of these

40. For the following table :

x	1.0	1.2	1.4	1.6	1.8	2.0
y	0	0.1280	0.5440	1.2960	2.4320	4.000

which of the following is/are true :

(a) $f'(1.1) = 0.63$

(b) $f''(1.1) = 6.6$

(c) both (a) and (b) are true

(d) none of these

41. For the following table :

x	0.0	0.5	1.0	1.5	2.0
f(x)	0.3989	0.3521	0.2420	0.1245	0.0540

the value of $f'(1.5) =$

(a) 0.19695   (b) − 0.19695

(c) both (a) and (b) are true

(d) none of these

42. The value of $f'(2.5)$ from the following table :

x	1.5	1.9	2.5	3.2	4.3	5.9
y	3.375	6.059	13.625	29.368	73.907	196.579

is given by :

(a) 16.75   (b) 167.5

(c) 1.675   (d) none of these

43. The value of $f'(5)$ from the following table :

x	1	2	4	8	10
f(x)	0	1	5	21	27

is given by :

(a) 3.6458　　　　　　(b) 3.78

(c) 3.87　　　　　　　(d) none of these

44. The value of $f'(93)$ from the table :

x	60	75	90	105	120
f(x)	28.2	38.2	43.2	40.9	37.7

is given by :

(a) 0.03627

(b) −0.03627

(c) both (a) and (b) are true

(d) none of these

45. A table given below reveals the velocity v of a body during the time t is :

x	1.0	1.1	1.2	1.3	1.4
y	43.1	47.7	52.1	56.4	60.8

Then the acceleration at $t = 1.1$ is given by :

(a) 45　　　　　　　　(b) 45.167

(c) 46　　　　　　　　(d) 48

46. The distance covered by an athlete for the 50 metre race is given in the following table :

Time (sec)	0	1	2	3	4	5	6
Distance (meter)	0	2.5	8.5	15.5	24.5	36.5	50

Then speed of athlete at $t = 5$ sec correct two decimals is given by :

(a) 13.18 m/s　　　(b) 13.13 m/s

(c) 14.13 m/s　　　(d) none of these

47. The value of $f'(3)$ from the following table :

x	3	5	11	27	34
f(x)	−13	28	899	17315	35606

is given by :

(a) 1.8828　　　　　(b) 1.98

(c) 2.08　　　　　　(d) none of these

48. The value of $f'(6)$ from the following table :

x	0	1	3	4	5	7	9
f(x)	150	108	0	−54	−100	−144	−84

is given by :

(a) 23　　　　　　　(b) −23

(c) 20　　　　　　　(d) none of these

49. If $f(6) = 1.556$, $f(7) = 1.690$, $f(9) = 1.908$, $f(12) = 2.158$. Then value $f'(8) =$ which of the following is/are true :

(a) 1.09　　　　　　(b) 0.109

(c) 1.109　　　　　(d) none of these

50. The value of $f'(5)$ from the following table :

x	2	4	9	10
f(x)	4	56	711	980

is given by :

(a) 2097.68　　　　(b) 2197.68

(c) 2098.70　　　　(d) none of these

51. From the following table,

x	4.5	5.0	5.5	6.0	6.5	7.0	7.5
y	9.69	12.90	16.71	21.18	26.37	32.34	39.15

The value of $f'(6)$ is given by :

(a) 8.66　　　　　　(b) 9.66

(c) 10.66　　　　　(d) none of these

52. The value of $f'(10)$ from the following data :

x	3	5	11	27	34
f(x)	−13	23	899	17315	35606

(a) 3.33　　　　　　(b) 2.33

(c) 4.33　　　　　　(d) none of these

53. By trapezoidal rule, the value of $\int_0^6 \dfrac{dx}{1+x^2}$ is given by :

(a) 1.4108　　　　　(b) 1.5108

(c) 1.6108　　　　　(d) none of these

54. The value of integral $\int_4^{5.2} \log x \, dx$ is given by :

(a) 1.827648　　　　(b) 1.98

(c) 2.01　　　　　　(d) none of these

55. The value of the integral $\int_0^{\pi/2} \sin x \, dx$ is given by :

(a) 0.99795　　　　(b) 1.997

(c) 0.986　　　　　(d) none of these

56. The value of the integral $\int_0^1 \dfrac{1}{1+x} dx$ is given by :

(a) 0.69413　　　　(b) 0.79413

(c) 0.894　　　　　(d) none of these

57. The value of the integral $\int_2^{10} \dfrac{dx}{1+x}$ using Simpson's rule is given by :

(a) 1.396　　　　　(b) 1.29962

(c) 1.49　　　　　　(d) none of these

58. The value of the integral $\int_0^6 \dfrac{dx}{1+x^2}$ by Simpson's rule is given by :

(a) 1.7911　　　　　(b) 1.8119

(c) 1.911　　　　　(d) none of these

59. By Simpson's rule, the value of the integral $\int_0^4 e^x\,dx$ is given by :
   (a) 54.87
   (b) 53.87
   (c) 63.87
   (d) none of these

60. The value of the integral $\int_0^1 \frac{1}{1+x}\,dx$, by Simpson's $\frac{3}{8}$ rule with $h = 1/6$ is given by :
   (a) 0.69319
   (b) 0.79319
   (c) 0.89
   (d) none of these

61. By Simpson's rule, the value of the integral $\int_0^6 \frac{dx}{(1+x)^2}$ is given by :
   (a) 0.89458
   (b) 0.884
   (c) 0.874
   (d) none of these

62. The speed v meter per second of a car, t seconds after it starts is shown in the following table :

x	0	12	24	36	48	60
y	0	3.60	10.08	18.90	21.60	18.54

x	72	84	96	108	120	
y	10.26	5.40	4.50	5.40	9.00	

Then, distance travelled by the car in 2 seconds is given by :
   (a) 1236 meters
   (b) 1236.96 meters
   (c) 1236.78
   (d) none of these

63. By Simpson's 3/8 rule, the value of the integral $\int_0^6 \frac{e^x}{1+x}\,dx$ is given by :
   (a) 70
   (b) 70.1652
   (c) 71.1652
   (d) none of these

64. A train is moving at the speed of 30 m/sec. Suddenly breaks are applied. The speed of the train per second after t second is given by

t	0	5	10	15	20	25	30	35	40	45
v	30	24	19	16	13	11	10	8	7	5

Then distance moved by the train in 45 seconds is given by :
   (a) 624
   (b) 624.36
   (c) 624.375
   (d) none of these

65. By Weddle's rule, the value of the integral $\int_0^6 \frac{dx}{1+x^2}$ is given by :
   (a) 1.37
   (b) 1.37344
   (c) 1.47
   (d) none of these

66. By Weddle's rule, the value of the integral $\int_0^5 \frac{dx}{4x+5}$ is given by :
   (a) 0.4023
   (b) 0.5023
   (c) 0.6023
   (d) none of these

67. By Weddle's rule, the value of the integral $\int_0^6 \frac{dx}{1+x^3}$ is given by :
   (a) 1.1654
   (b) 1.2654
   (c) 1.0654
   (d) none of these

68. A curve passing through the points given by the following table :

x	1	1.5	2	2.5	3	3.5	4
y	2	2.4	2.7	2.8	3	2.6	2.1

Then area bounded by the curve, the x-axis and the lines $x = 1, x = 4$ is :
   (a) 17.06
   (b) 7.74
   (c) 7.04
   (d) none of these

69. The value of $c_3^4 =$
   (a) 32
   (b) 90
   (c) 32/90
   (d) none of these

70. For $n = 3$, the value of $L_0 + L_1 + L_2 + L_3 =$
   (a) 1
   (b) 2
   (c) 3
   (d) 10

71. The sum of the fourth powers of the first n natural numbers by means of the Euler's Maclaurin's formula is :
   (a) $\frac{n^5}{5} + \frac{n^2}{2} + \frac{n^3}{3} - \frac{n}{3}$
   (b) $n^5 + n^4 + \frac{n^3}{3}$
   (c) $\frac{n^5}{5} + \frac{n^4}{2}$
   (d) none of these

72. The approximate value of $\int_{-3}^3 x^4\,dx$ by Trapezoidal rule is given by :
   (a) 110
   (b) 114
   (c) 115
   (d) none of these

73. The value of $\sum_{k=0}^n L_k =$
   (a) 1
   (b) 0
   (c) k
   (d) n

74. The value of the integral $\int_3^5 \frac{4}{2+x^2}\,dx$ by dividing the range into eight equal parts is given by :
   (a) 26
   (b) 26.716
   (c) 27.716
   (d) none of these

75. If

x	1	2	3	4	5
$f(x)$	10	50	70	80	100

then, by Simpson's 1/3 rule, the value of $\int_1^5 f(x)\,dx =$

(a) 256.66

(b) 257.66

(c) 258.66

(d) none of these

76. The estimate of $\int_{0.5}^{1.5} \dfrac{dx}{x}$ obtained using Simpson's rule with three point function evaluation exceeds the exact value by: (GATE (CE)–2012)

(a) 0.235

(b) 0.012

(c) 0.068

(d) none of these

77. The table below gives values of a function $F(x)$ obtained for values of x at intervals of 0.25.

x	0	0.25	0.5	0.75	1.0
$f(x)$	1	0.9412	0.8	0.64	0.50

The value of the integral of the function between the limits 0 to 1 using Simpson's rule is: (GATE (CE)–2010)

(a) 2.3562

(b) 0.7854

(c) 7.5000

(d) none of these

78. The accuracy of Simpson's quadrature for a step size h is: (GATE (ME)–2003)

(a) $0(h^2)$

(b) $0(h^4)$

(c) $0(h^3)$

(d) none of these

79. A 2nd degree polynomial, $f(x)$ has values of 1, 4 and 15 at $x = 0$, 1 and 2 respectively. The integral $\int_0^2 f(x)\,dx$ is to be estimated by applying the trapezoidal rule to this data. What is the error identified as "true value-approximate value" in the estimate? (GATE (CE)–2006)

(a) 0

(b) $-\dfrac{2}{3}$

(c) $\dfrac{2}{3}$

(d) $-\dfrac{4}{3}$

80. A calculator has accuracy upto 8 digits after decimal place. The value of $\int_0^{2\pi} \sin x\,dx$ when evaluated using this calculator by trapezoidal method with 8 equal intervals, to 5 significant

digits is: (GATE (ME)–2007)

(a) 1.0000

(b) 0.00000

(c) 0.00500

(d) 0.00025

81. Torque exerted on a flywheel over a cycle is listed in the table. Flywheel energy (in J per unit cycle) using Simpson's rule is: (GATE (ME)–2010)

Angle (degree)	0	60	120	180	240	300	360
Torque (Nm)	0	1066	–323	0	323	–355	0

(a) 993

(b) 542

(c) 1444

(d) none of these

82. The magnitude as the error (correct to two decimal places) in the estimation of following integral using Simpson 1/3 rule. Take the step length as 1. (GATE (CE)–2013)

$$\int_0^4 (x^4 + 10)\,dx$$

(a) 0.53

(b) 0.10

(c) 0.42

(d) none of these

83. The integral $\int_1^3 \dfrac{1}{x}\,dx$ when evaluated by using Simpson's 1/3 rule on two equal sub-intervals each of length 1, equals. (GATE (ME)–2011)

(a) 1.111

(b) 1.098

(c) 1.120

(d) none of these

84. The minimum number of equal length subintervals needed to approximate $\int_1^2 xe^x\,dx$ to an accuracy of at least $1/3 \times 10^{-6}$ using the trapezoidal rule is: (GATE (CS)–2008)

(a) 1000

(b) 1000e

(c) 100

(d) none of these

85. Match the correct pairs. (GATE (ME)–2013)

Numerical integration Order of Fitting scheme   **Polynomial**

P. Simpson's 3/8 rule   1. First

Q. Trapezoidal Rule   2. Second

R. Simpson's 1/3 Rule   3. Third

(a) $P-3, Q-1, R-2$ (b) $P-1, Q-2, R-3$

(c) $P-2, Q-1, R-3$ (d) none of these

86. The error in $\dfrac{d}{dx} f(x)\bigg|_{x=x_0}$ for a continuous function estimated with $h = 0.03$ using the central difference formula,

$$\dfrac{d}{dx} f(x)\bigg|_{x=x_0} = \dfrac{f(x_0+h) - f(x_0-h)}{2n} \text{ is}$$

2×10^{-3}. The values of x_0 and $f(x_0)$ are 19.78 and 500.01 respectively. The corresponding error in the central difference estimate for $h = 0.02$ approximately.   (GATE(CE)–2012)

(a) 9.0×10^{-4} (b) 1.3×10^{-4}

(c) 3.0×10^{-4} (d) none of these

Answers

Problem Set – 1

FILL IN THE BLANKS

1. $\frac{1}{h}\log(1+\Delta)$ **2.** a **3.** D **4.** $\Delta^3 y$ **5.** $\nabla^3 + \frac{3}{2}\nabla^4 + \ldots$

6. h^{-r} **7.** Stirling's formula   **8.** Divided difference formula   **9.** $2a$ **10.** $(-1)^r$

TRUE/FALSE

1. T   **2.** T   **3.** F   **4.** T   **5.** T   **6.** T   **7.** T   **8.** T   **9.** T
10. T

MULTIPLE CHOICE QUESTIONS

1. (b)   **2.** (a)   **3.** (d)   **4.** (a)   **5.** (b)

Problem Set – 2

FILL IN THE BLANKS

1. y_1 **2.** linear   **3.** trapezium   **4.** even   **5.** two   **6.** three
7. 3   **8.** six   **9.** 6   **10.** 1   **11.** n **12.** 1   **13.** n **14.** same

TRUE/FALSE

1. T   **2.** F   **3.** F   **4.** T   **5.** T   **6.** F   **7.** T   **8.** T   **9.** F
10. F   **11.** T   **12.** F   **13.** T   **14.** T

MULTIPLE CHOICE QUESTIONS

1. (c)   **2.** (b)   **3.** (a)   **4.** (c)   **5.** (c)   **6.** (b)   **7.** (c)   **8.** (b)   **9.** (d)
10. (b)   **11.** (b)   **12.** (c)

Problem Set – 3

MULTIPLE CHOICE QUESTIONS

1. (b)   **2.** (d)   **3.** (d)   **4.** (b)   **5.** (c)   **6.** (a)   **7.** (b)   **8.** (c)   **9.** (a)
10. (b)   **11.** (b)   **12.** (b)   **13.** (c)   **14.** (c)   **15.** (a)   **16.** (b)   **17.** (a)   **18.** (c)
19. (b)   **20.** (c)   **21.** (c)   **22.** (a)   **23.** (b)   **24.** (b)   **25.** (a)   **26.** (c)   **27.** (c)
28. (b)   **29.** (b)   **30.** (c)   **31.** (a)   **32.** (c)   **33.** (a)   **34.** (a)   **35.** (a)   **36.** (a)
37. (c)   **38.** (c)   **39.** (b)   **40.** (c)   **41.** (b)   **42.** (a)   **43.** (a)   **44.** (b)   **45.** (b)
46. (b)   **47.** (a)   **48.** (b)   **49.** (b)   **50.** (a)   **51.** (b)   **52.** (b)   **53.** (a)   **54.** (a)
55. (a)   **56.** (a)   **57.** (b)   **58.** (a)   **59.** (b)   **60.** (a)   **61.** (a)   **62.** (b)   **63.** (b)
64. (c)   **65.** (b)   **66.** (a)   **67.** (a)   **68.** (b)   **69.** (c)   **70.** (a)   **71.** (a)   **72.** (c)
73. (a)   **74.** (b)   **75.** (a)   **76.** (b)   **77.** (b)   **78.** (b)   **79.** (d)   **80.** (b)   **81.** (a)
82. (a)   **83.** (a)   **84.** (b)   **85.** (a)   **86.** (a)

COMPUTATIONAL TECHNIQUE LAB

1. TRAPEZOIDAL RULE

SYMBOLS USED

n = number of subdivision

x_0 = lower limit of integral

x_n = upper limit of integral

h = length of subinterval

$y(x)$ = function to be integrated

ALGORITHM : TRAPEZOIDAL RULE

Step 1 :	Start
Step 2 :	Input x_0, x_n, n
Step 3 :	h = $(x_n - x_0)/n$
Step 4 :	s = y (x_0) + y(xn)
Step 5 :	For i = 1 to n – 1
Step 6 :	s = s + 2 * y $(x_0 + i * h)$
Step 7 :	End of i loop
Step 8 :	Sum = s * (h/2)
Step 9 :	Print "Value of integral = ", sum
Step 10 :	Stop.

FLOW CHART : TRAPEZOIDAL RULE

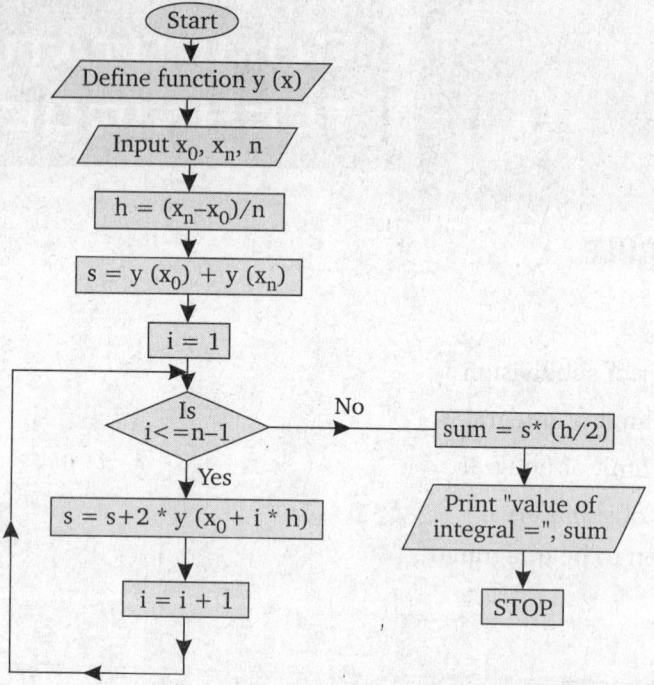

PROGRAM : *Following is a C program to find the value of the integral* $\int_0^6 \frac{dx}{1+x^2}$ *by trapezoidal rule.*

Define a function y as $1/(1+x*x)$

```c
// Trapezoidal Rule
# include<stdio.h>
#include<conio.h>
float y (float x)
{
        return 1/(1+x*x);
}
void main()
{
     float x0,xn,s,h,sum;
     int i,n;
     clrscr();

     puts("\n Enter number of subdivisions i.e n  ");
     scanf(" %d",&n);
     puts("\n Enter lower limit of integral i.e. x0 ");
     scanf(" %f",&x0);
     puts("\n Enter upper limit of integral i.e. xn ");
```

```
        scanf(" %f",&xn);


    h=(xn-x0)/n;
            s=y(x0)+y(xn);
            for(i=1;i<=n-1;i++)
                s+=2*y(x0+i*h);
            sum=s*(h/2);
            printf("\n Value of Integral is %.3lf\n",sum);
            getch();


    }
```

Output: TRAPEZOIDAL RULE

```
enter number of subdivisions i.e. n
6
enter lower limit of integral i.e. x0
0
enter upper limit of integral i.e. xn
6
Value of Integral is  1.411
```

LAB ASSIGNMENT : TRAPEZOIDAL RULE

1. Write a C program to solve the following integral by trapezoidal rule

$$\int_{0.2}^{1.4} (\sin x - \log_e x + e^x)\,dx$$

Hint : Define a function y
$$\sin (x - \log_e (x) + \exp (x)$$
Input : $x_0 = 0.2$ $x_n = 1.4$ n = 12
Output : Value of integral = 4.056174

2. Write a C program to calculate the value of integral $\int_4^{5.2} \log x\,dx$ by trapezoidal rule.
Hint : Define function y = log (x)
Input : $x_0 = 4$, $x_n = 5.2$ n = 6
Output : Value of integral = 1.827648

2. SIMPSON'S 1/3 RULE

SYMBOLS USED

x_0 = lower limit of integral
x_n = upper limit of integral
n = no. of subdivisions
h = length of subinterval
y (x) = function to be integrated

ALGORITHM : SIMPSON'S 1/3 RULE

Step 1 :	Start
Step 2 :	Input x_0, x_n, n
Step 3 :	h = $(x_n - x_0)/n$
Step 4 :	s = y (x_0) + y (x_n) + 4 * y $(x_0 + h)$
Step 5 :	For i = 3 to (n – 1)
Step 6 :	s = s – 4 * y $(x_0 + i * h)$ + 2 * y $(x_0 + (i – 1) * h$
Step 7 :	i = i + 2
Step 8 :	End of i loop.
Step 9 :	Sum = s * (h/3)
Step 10 :	Print "Value of integral = ", sum
Step 11 :	Stop.

FLOW CHART : SIMPSON'S 1/3 RULE

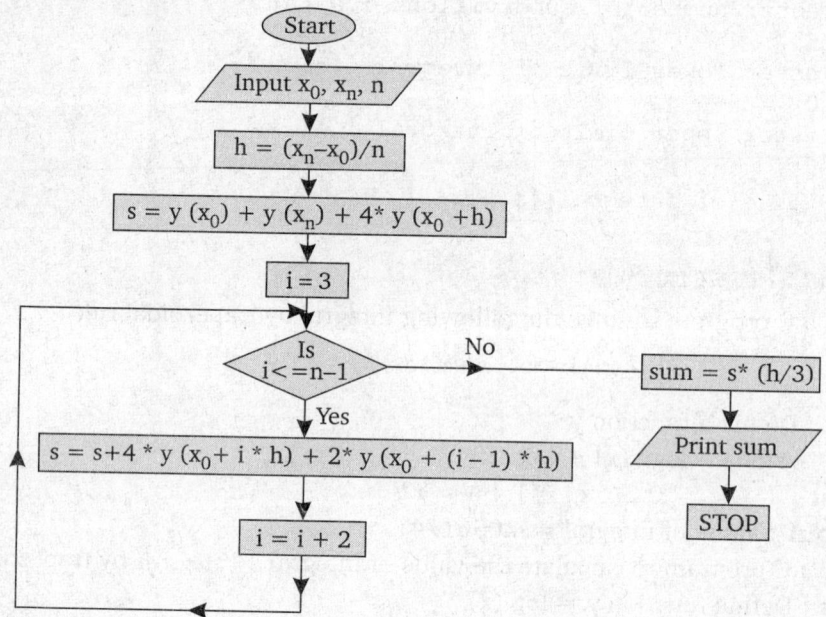

PROGRAM : *Following is a C program to the integral* $\int_0^6 \frac{dx}{1+x^2}$ *by Simpson's 1/3 rule.*

```
//SIMPSON 1/3rd RULE
#include<stdio.h>
#include<conio.h>
float y(float x)
{
    return 1/(1+x*x);
}
void main()
```

```
{
        float x0,xn,h,s,sum;
        int i,n;
        clrscr();
        puts("\n enter number of subdivisions i.e. n ");
        scanf("%d",&n);
        puts("\n enter lower limit of integral i.e. x0 ");
        scanf("%f",&x0);
        puts("\n enter upper limit of integral i.e. xn ");
        scanf("%f",&xn);

        h=(xn-x0)/n;
        s=y(x0)+y(xn)+4*y(x0+h);

        for(i=3;i<=n-1;i+=2)
              s+=4*y(x0+i*h)+2*y(x0+(i-1)*h);
        sum=s*(h/3);
        printf("\n Value of Integral is %.3lf\n",sum);
        getch();
}
```

Output: SIMPSON 1/3 rd RULE

```
enter number of subdivisions i.e. n
6
enter lower limit of integral i.e. x0
0
enter upper limit of integral i.e. xn
6
Value of Integral is   1.366
```

LAB ASSIGNMENT : SIMPSON'S 1/3 RULE

1. Write a C program to solve the integral $\int_0^4 e^x \, dx$ using Simpson's 1/3 rule.
   Hint : Define function y = exp (x)
   Input : $x_0 = 0$, $x_n = 4$, n = 4
   Output : Value of integral = 53.87

2. Write a C program to solve the integral $\int_0^7 \frac{1}{x} . dx$ using Simpson's 1/3 rule.
   Hint : Define function y = (1/x)
   Input : $x_0 = 1$, $x_n = 7$, n = 6
   Output : Value of integral = 1.9587

3. Write a C program to solve the integral $\int_{0.5}^{0.7} x^{1/2} e^{-x} \, dx$ by Simpson's 1/3 rule.
   Hint : Define function y = sqrt (x) * exp (–x)
   Input : $x_0 = 5$, $x_n = 0.7$, n = 4
   Output : Value of integral = 0.08483

3. SIMPSON'S 3/8 RULE

SYMBOLS USED

n = number of subdivisions
x_0 = lower limit of integral
x_n = upper limit of integral
h = length of each subinterval
y (x) = function to be integrated

ALGORITHM : SIMPSON'S 3/8 RULE

Step 1 :   Start
Step 2 :   Input x_0, x_n, n
Step 3 :   h = $(x_n - x_0)/n$, $s_1 = 0$, $s_2 = 0$
Step 4 :   s = y (x_0) + y (x_n)
Step 5 :   For i = 1 to (n – 1)
Step 6 :   If (i%3 ! = 0)
                    $s_1 = s_1 + 3 * y (x0 + i * h)$
            else
                    $s_2 = s_2 + 2 * y (x_0 + i * h)$
Step 7 :   End of i loop
Step 8 : $s = s + s_1 + s_2$
Step 9 :   Sum = s * (3 * h)/8
Step 10 :   Print "Value of integral = ", sum
Step 11 :   STOP.

FLOW CHART : SIMPSON'S 3/8 RULE.

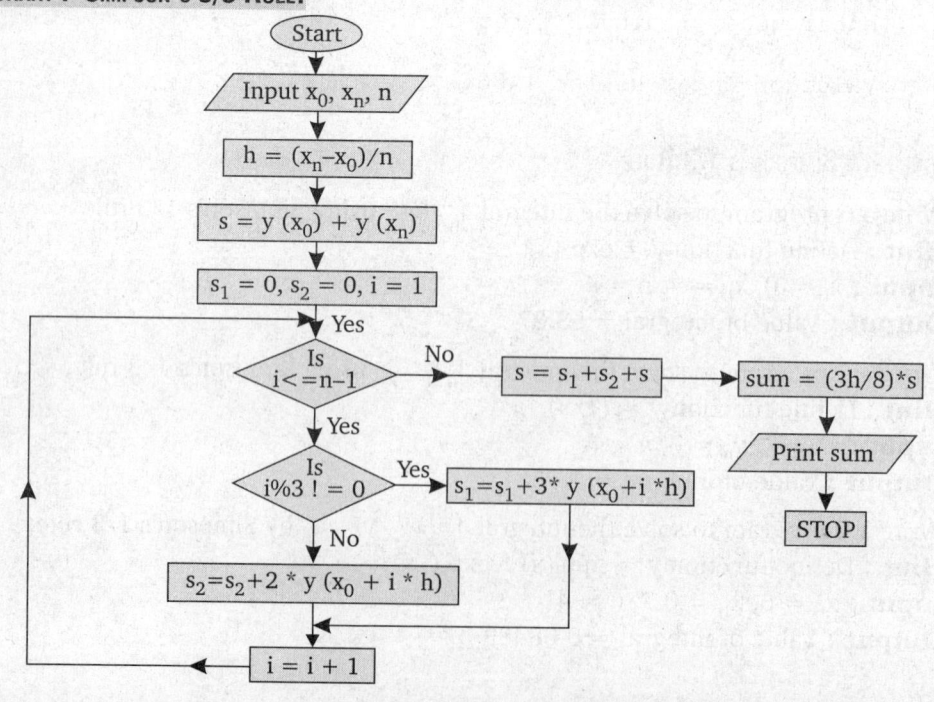

PROGRAM : *Following is a C program to the integral $\int_0^1 \frac{dx}{1+x}$ by Simpson's 3/8 rule.*

```
//SIMPSON 3/8th RULE

#include<stdio.h>
#include<conio.h>

float y(float x)
{
    return 1/(1+x);
}
void main()
{
    float x0,xn,h,s,s1=0,s2=0,sum;
    int i,n;
    clrscr();
    puts("\n enter number of subdivisions i.e. n ");
    scanf("%d",&n);
    puts("\n enter lower limit of integral i.e. x0 ");
    scanf("%f",&x0);
    puts("\n enter upper limit of integral i.e. xn ");
    scanf("%f",&xn);
    h=(xn-x0)/n;
    s=y(x0)+y(xn);
    for(i=1;i<=n-1;i++)
        if(i%3!=0)
          s1+=3*y(x0+i*h);
        else
           s2+= 2*y(x0+i*h);
    s=s+s1+s2;
    sum=s*(3*h/8);
    printf("\n Value of Integral is %.3lf\n",sum);
    getch();
}
```

Output: SIMPSON 3/8th RULE
```
enter number of subdivisions i.e. n
6
enter lower limit of integral i.e. x0
0
enter upper limit of integral i.e. xn
1
Value of Integral is  0.693
```

LAB ASSIGNMENT : SIMPSON'S 3/8TH RULE

1. Write a C program to solve the integral $\int_0^1 \frac{dx}{1+x^2}$ using Simpson 3/8 rule.

Hint : Define function y = 1/(1 + x * x)

Input : $x_0 = 0$, $x_1 = 1$, n = 6

Output : Value of integral = 0.785396

2. Write a C program to solve the integral $\int_0^6 (1+x^2)dx$ using Simpson 3/8th rule.

Hint : Define function y (x) = (1 + x * x)

Input : $x_0 = 0$, $x_n = 6$, n = 6

Output : Value of integral = 78

4. BOOLE'S RULE

SYMBOLS USED

n = number of subdivision

x_0 = lower limit of integral

x_n = upper limit of integral

h = length of subinterval

y (x) = function to be integrated

ALGORITHM : BOOLE'S RULE

Step 1 :	Start
Step 2 :	Define function y (x)
Step 3 :	Input n , x_0, x_n
Step 4 :	$s_2 = s_3 = s_4 = 0$
Step 5 :	h = $(x_n - x_0)$/n
Step 6 :	$s_1 = 7 * y (x_0) + y (x_n)$
Step 7 :	For i = 1 to (n – 1)
Step 8 :	$s_2 = s_2 + 32 * y (x_0 + i * h)$
Step 9 :	i = i + 2
Step 10 :	End of i loop
Step 11 :	For i = 2 to (n – 1)
Step 12 :	$s_3 = s_3 + 12 * y (x_0 + i */ h)$
Step 13 :	i = i + 4
Step 14 :	End of i loop
Step 15 :	For o = 4 to (n – 1)
Step 16 :	$s_4 = s_4 + 14 * y (x_0 + i * h)$
Step 17 :	i = i + 4

Step 18 :   End of i loop

Step 19 :   Sum = $(s_1 + s_2 + s_3 + s_4) * (2 *h) / 45$

Step 20 :   Print "Value of integral = ", sum

Step 21 :   STOP.

FLOW CHART : BOOGLE'S RULE

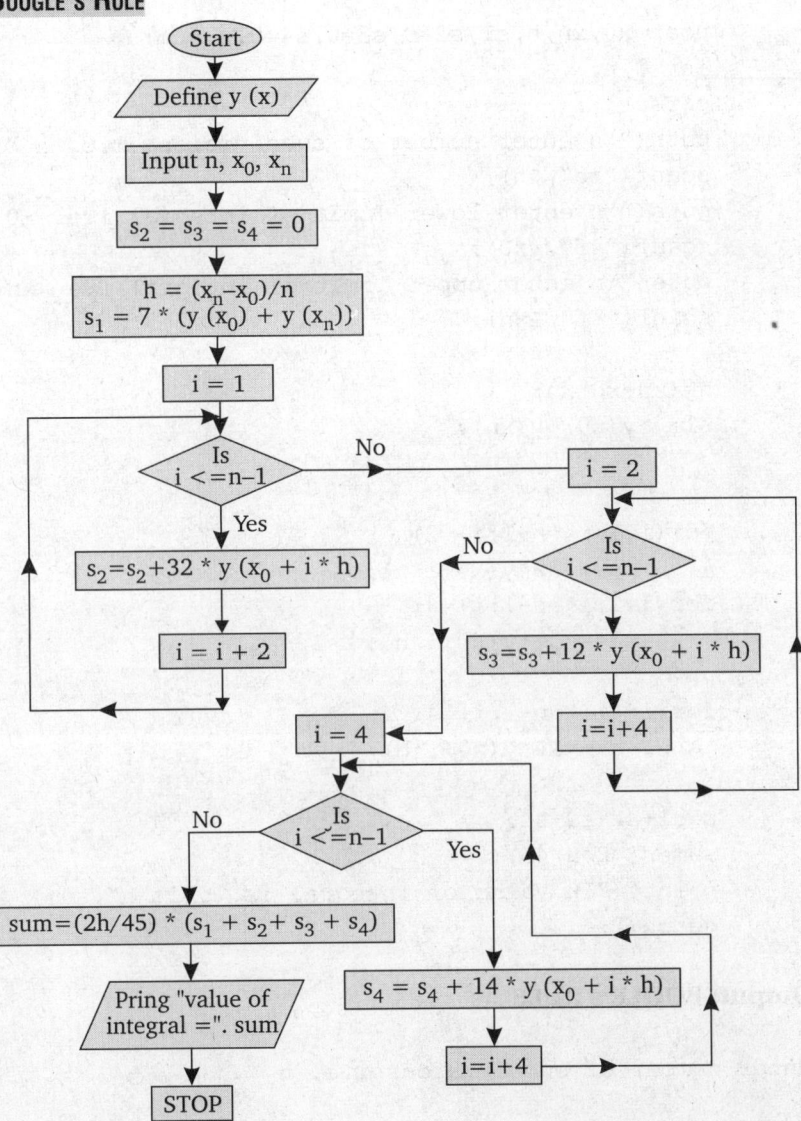

PROGRAM :   *Following is a C program to solve the integral $\int_0^6 (1+x^2)\,dx$ by the Boole's rule.*

```
// BOOLE's RULE
#include<stdio.h>
#include<conio.h>
```

```
float y(float x)
{
    return (1+x*x);
}
void main()
{
        float x0,xn,h,s1,s2=0,s3=0,s4=0,s,sum;
        int i,n;
        clrscr();
        puts("\n enter number of subdivisions i.e. n ");
        scanf("%d",&n);
        puts("\n enter lower limit of integral i.e. x0 ");
        scanf("%f",&x0);
        puts("\n enter upper limit of integral i.e. xn ");
        scanf("%f",&xn);

        h=(xn-x0)/n;
        s1=7*y(x0)+y(xn);


        for(i=1;i<=n-1;i+=2)
            s2+=32*y(x0+i*h);
        for(i=1;i<=n-1;i+=4)
            s3+= 12*y(x0+i*h);

        for(i=4;i<=n-1;i+=4)
            s4+=14*y(x0+i*h);

        s=s1+s2+s3+s4;
        sum=s*(2*h/45);
        printf("\n Value of Integral is %.3lf\n",sum);
        getch();
}
```

Output: BOOLE's RULE

```
enter number of subdivisions i.e. n
8
enter lower limit of integral i.e. x0
0
enter upper limit of integral i.e. xn
6
Value of Integral is  67.450
```

Lab Assignment : Boole's Rule

1. Write a C program to solve the integral $\int_0^4 \frac{dx}{1+x^2}$ by Boole's rule. Divide the interval into 4 equal parts.

   Hint : Define a function y (x) = 1/(1 + x * x)

       Input   : $x_0 = 0$, $x_n = 4$, n = 4

       Output : Value of integral = 1.28941

2. Write a C program to solve the integral $\int_1^7 \frac{1}{x}$ dx by Boole's rule.

   Hint : Define a function y (x) = (1/x)

       Input : $x_0 = 1$, $x_n = 7$, n = 8

       Output : Value of integral = 1.949018

5. WEDDLE'S RULE

Symbols Used

    n = number of subdivision

    x_0 = lower limit of integral

    x_n = upper limit of integral

    h = length of subinterval

    y (x) = function to be integrated

Algorithm : Wedding's Rule

Step 1 :	Start
Step 2 :	Define function y (x)
Step 3 :	Input x_0, x_n, n
Step 4 :	$s_1 = s_2 = s_3 = s_4 = 0$
Step 5 :	$h = (x_n - x_0)/n$
Step 6 :	If (n %6 = 0)
Step 7 :	$s_5 = y(x0) + y(xn)$
Step 8 :	For i = 1 to n – 1
Step 9 :	If (1% 2! = 0 and i% 3! = 0)
	$s_1 = s_1 + 5 * y(x_0 + i * h)$
	else
	if (i%2 = 0 and i%3 ! = 0)
	$s_2 = s_2 + y(x_0 + i * h)$
	else
	if (i%2 = 0 and i%3 = 0)
	$s_3 = s_3 + 3 * y(x_0 + i * h)$
	else
	$s_4 = s_4 + 6 * y(x_0 + i * h)$
Step 10 :	End of i loop
Step 11 :	Sum = $3 * h * (s_1 + s_2 + s_3 + s_4 + s_5)/10$
Step 12 :	print "Value of integral = ", sum

Step 13 :	else
Step 14 :	Print "Weedle's rule is not applicable"
Step 15 :	Stop.

FLOW CHART : WEDDLE'S RULE

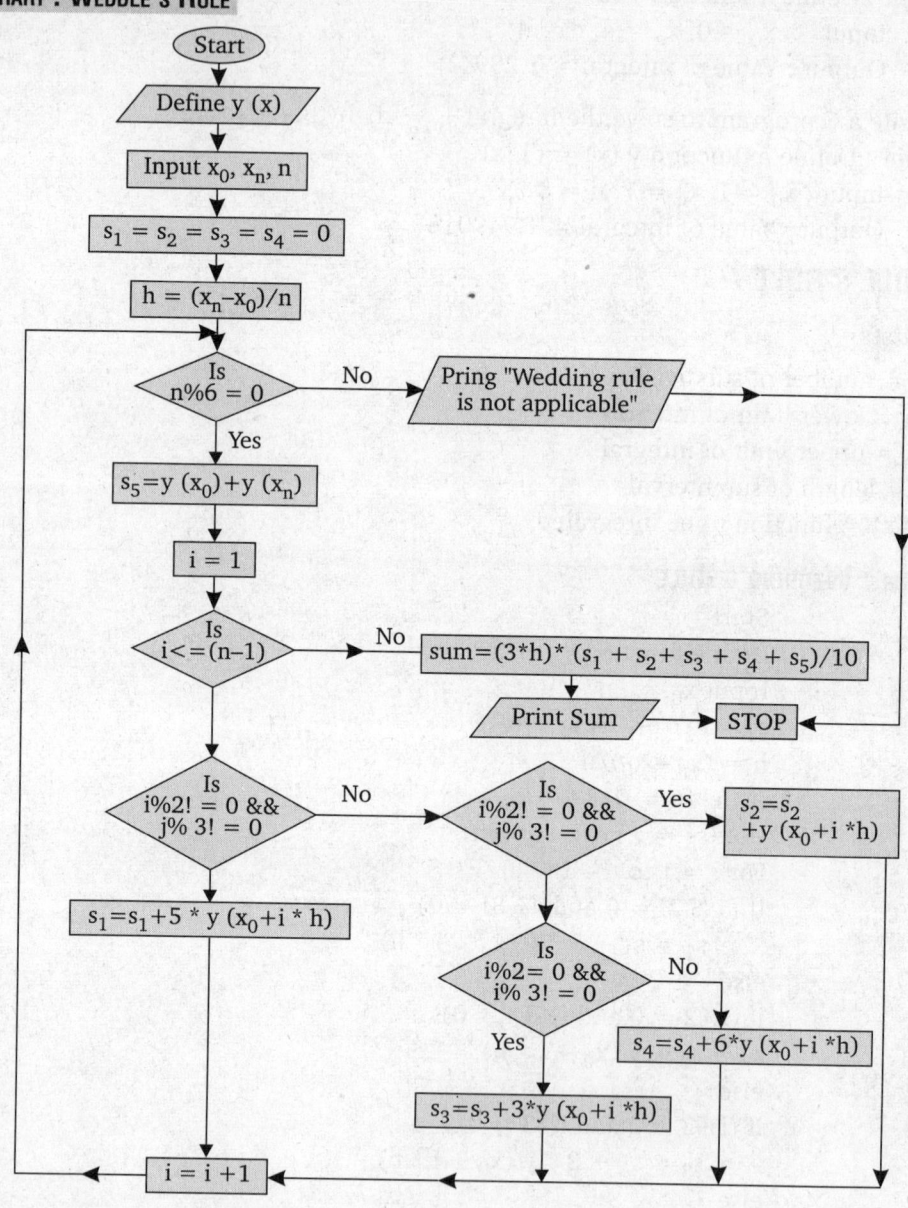

PROGRAM : *Following is a C program to solve the integral $\int_0^6 (1+x^2)\,dx$ by the Weddle's rule.*

```c
// WEDDLE's RULE
#include<stdio.h>
#include<conio.h>
float y(float x)
{
    return (1+x*x);
}
void main()
{
    float x0,xn,h,s1=0,s2=0,s3=0,s4=0,s5,s,sum;
    int i,n;
    clrscr();
    puts("\n enter number of subdivisions i.e. n ");
    scanf("%d",&n);
    puts("\n enter lower limit of integral i.e. x0 ");
    scanf("%f",&x0);
    puts("\n enter upper limit of integral i.e. xn ");
    scanf("%f",&xn);
    h=(xn-x0)/n;
    if(n%6==0)
        s5=y(x0)+y(xn);
    else
        {printf("\n Weddle's rule is not applicable\n");
         goto end;
        }
    for(i=1;i<=n-1;i++)
    {
        if(i%2!=0 && i%3!=0)
            s1+=5*y(x0+i*h);
        else
            if(i%2==0 && i%3=0)
                s2+=y(x0+i*h);
            else
                if(i%2==0 && i%3==0)
                    s3+= 3*y(x0+i*h);
                else
                    s4+=6*y(x0+i*h);
    }
    s=s1+s2+s3+s4+s5;
    sum=s*(3*h/10);
    printf("\n Value of Integral is %.3lf\n",sum);
```

```
end:   getch();
}
```

Output: WEDDLE's RULE

```
 enter number of subdivisions i.e. n
 6


 enter lower limit of integral i.e. x0
 0


 enter upper limit of integral i.e. xn
 6


 Value of Integral is 78.000
```

LAB ASSIGNMENT : WEDDLE'S RULE

1. Write a C program to solve the integral $\int_0^1 \frac{dx}{1+x^2}$ by Weddle's rule. Divide the range into six equal parts.

   Hint : Define function y (x) = 1/(1 + x * x)

   Input : $x_0 = 0$, $x_n = 1$, n = 6

   Output : Value of integral = 0.7854

2. Write a C program to solve the integral $\int_0^{1.5} \frac{x^3}{e^x - 1}$ dx by dividing the interval into six equal parts using Weddle's rule.

   Hint : Define function y (x) = (x * x * x)/exp (x) – 1)

   Input : $x_0 = 0$, $x_n = 1.5$, n = 6

   Output : Value of integral = 0.6155

3. Write a C program to solve the integral $\int_0^5 \frac{dx}{4x+5}$ by Weddle's rule.

   Hint : Define function y (x) = 1/(4 * x + 5)

   Input : $x_0 = 0$, $x_0 = 5$, n = 12

   Output : Value of integral = 0.4023.

⌘⌘⌘⌘⌘⌘

Matrix Inversion and Eigen Values Problems

CHAPTER 7

7.1 INTRODUCTION

If A is a square matrix of order $n \times n$ provided $|A| \neq 0$, then a square matrix B of order $n \times n$ such that $AB = BA = I$, is called the inverse of A, where I is a unit matrix of order $n \times n$. The inverse of a matrix, if exist, will be unique. There are several methods for finding the inverse of a matrix. We shall now, describe some methods for finding the inverse of a given matrix.

7.2 GAUSS ELIMINATION METHOD

To apply the Gauss-elimination method, let us take a unit matrix I of the same order as the given matrix A has and put this unit matrix with A as AI.

Now performing simultaneous row transformation on AI and convert the matrix A into an upper triangular matrix and then to a unit matrix. In the whole process the matrix A reduced to unit matrix and the adjacent matrix gives the inverse of A. To increase the accuracy, the largest element at the first position of the row is taken as the pivot element by using row transformations only.

7.3 GAUSS-JORDAN METHOD

It is similar to the Gauss elimination method. In Gauss elimination method we convert the matrix A into an upper triangular matrix and then to unit matrix. But matrix A is directly converted into identity matrix by elementary row transformations.

SOLVED EXAMPLES

EXAMPLE 1. *Find the inverse of* $A = \begin{bmatrix} 0 & 2 & 4 \\ 2 & 4 & 6 \\ 6 & 2 & 2 \end{bmatrix}$ *by Gauss-elimination method.*

(Agra-2009, Garhwal-2010)

SOLUTION. Let us put $A = AI$

$$\therefore \qquad A = \begin{bmatrix} 0 & 2 & 4 \\ 2 & 4 & 6 \\ 0 & 2 & 4 \end{bmatrix} \begin{bmatrix} 1 & 0 & 0 \\ 0 & 1 & 0 \\ 0 & 0 & 1 \end{bmatrix}$$

Taking the pivot element $a_{11} = 6$ (the largest element) so interchanging R_1 and R_3.

$$\Rightarrow \qquad A = \begin{bmatrix} 6 & 2 & 2 \\ 2 & 4 & 6 \\ 0 & 2 & 4 \end{bmatrix} \begin{bmatrix} 0 & 0 & 1 \\ 0 & 1 & 0 \\ 1 & 0 & 0 \end{bmatrix}$$

Now multiply R_1 by $\dfrac{1}{3}$.

$$\sim \begin{bmatrix} 2 & \dfrac{2}{3} & \dfrac{2}{3} \\ 2 & 4 & 6 \\ 0 & 2 & 4 \end{bmatrix} \begin{bmatrix} 0 & 0 & \dfrac{1}{3} \\ 0 & 1 & 0 \\ 1 & 0 & 0 \end{bmatrix}$$

Performing $R_2 \rightarrow R_2 - R_1$, we get

$$\sim \begin{bmatrix} 2 & \dfrac{2}{3} & \dfrac{2}{3} \\ 0 & \dfrac{10}{3} & \dfrac{16}{3} \\ 0 & 2 & 4 \end{bmatrix} \begin{bmatrix} 0 & 0 & \dfrac{1}{3} \\ 0 & 1 & -\dfrac{1}{3} \\ 1 & 0 & 0 \end{bmatrix}$$

Performing $\dfrac{3}{5} R_2$, we get $\quad \sim \begin{bmatrix} 2 & \dfrac{2}{3} & \dfrac{2}{3} \\ 0 & 2 & \dfrac{16}{5} \\ 0 & 2 & 4 \end{bmatrix} \begin{bmatrix} 0 & 0 & \dfrac{1}{3} \\ 0 & \dfrac{3}{5} & -\dfrac{1}{5} \\ 1 & 0 & 0 \end{bmatrix}$

Performing $R_3 \rightarrow R_3 - R_2$, we get

$$\sim \begin{bmatrix} 2 & \dfrac{2}{3} & \dfrac{2}{3} \\ 0 & 2 & \dfrac{16}{5} \\ 0 & 2 & \dfrac{4}{5} \end{bmatrix} \begin{bmatrix} 0 & 0 & \dfrac{1}{3} \\ 0 & \dfrac{3}{5} & -\dfrac{1}{5} \\ 1 & -\dfrac{3}{5} & \dfrac{1}{5} \end{bmatrix}$$

Thus A is now reduced to an upper triangular matrix. Again applying row transformation to convert upper triangular matrix into unit matrix.

Multiplying R_3 by $5/2$, we get

$$\sim \begin{bmatrix} 2 & \dfrac{2}{3} & \dfrac{2}{3} \\ 0 & 2 & \dfrac{16}{5} \\ 0 & 0 & 2 \end{bmatrix} \begin{bmatrix} 0 & 0 & \dfrac{1}{3} \\ 0 & \dfrac{3}{5} & -\dfrac{1}{5} \\ \dfrac{5}{2} & -\dfrac{3}{2} & \dfrac{1}{2} \end{bmatrix}$$

Now performing $R_1 \rightarrow R_1 - \dfrac{1}{3} R_3$ and $R_2 \rightarrow R_2 - \dfrac{8}{5} R_3$, we get

$$\sim \begin{bmatrix} 2 & \dfrac{2}{3} & 0 \\ 0 & 2 & 0 \\ 0 & 0 & 2 \end{bmatrix} \begin{bmatrix} -\dfrac{5}{6} & \dfrac{1}{2} & \dfrac{1}{6} \\ -4 & 3 & -1 \\ \dfrac{5}{2} & -\dfrac{3}{2} & \dfrac{1}{2} \end{bmatrix}$$

Performing $R_1 \rightarrow R_1 - \dfrac{1}{3} R_2$, we get

$$\sim \begin{bmatrix} 2 & 0 & 0 \\ 0 & 2 & 0 \\ 0 & 0 & 2 \end{bmatrix} \begin{bmatrix} 1/2 & -1/2 & 1/2 \\ -4 & 3 & -1 \\ 5/2 & -3/2 & 1/2 \end{bmatrix}$$

Now divide R_1, R_2 and R_3 by 2, we get

$$\sim \begin{bmatrix} 1 & 0 & 0 \\ 0 & 1 & 0 \\ 0 & 0 & 1 \end{bmatrix} \begin{bmatrix} 1/4 & -1/4 & 1/4 \\ -2 & 3/2 & -1/2 \\ 5/2 & -3/4 & 1/4 \end{bmatrix}$$

Hence, $\qquad A^{-1} = \begin{bmatrix} 1/4 & -1/4 & 1/4 \\ -2 & 3/2 & -1/2 \\ 5/2 & -3/4 & 1/4 \end{bmatrix}$

EXAMPLE 2. *Find the inverse of* $A = \begin{bmatrix} 1 & 1 & 2 \\ 1 & 2 & 3 \\ 2 & 3 & 1 \end{bmatrix}$ *by Gauss-elimination method.*

SOLUTION. Let us put

$$A = AI$$

$$\therefore \qquad A = \begin{bmatrix} 1 & 1 & 2 \\ 1 & 2 & 3 \\ 2 & 3 & 1 \end{bmatrix} \begin{bmatrix} 1 & 0 & 0 \\ 0 & 1 & 0 \\ 0 & 0 & 1 \end{bmatrix}$$

Interchanging R_1 and R_3

$$A = \sim \begin{bmatrix} 2 & 3 & 1 \\ 1 & 2 & 3 \\ 1 & 1 & 2 \end{bmatrix} \begin{bmatrix} 0 & 0 & 1 \\ 0 & 1 & 0 \\ 1 & 0 & 0 \end{bmatrix}$$

Performing $R_2 \rightarrow R_2 - \dfrac{1}{2} R_1$, $R_3 \rightarrow R_3 - \dfrac{1}{2} R_1$

$$A = \sim \begin{bmatrix} 2 & 3 & 1 \\ 0 & 1/2 & 5/2 \\ 0 & -1/2 & 3/2 \end{bmatrix} \begin{bmatrix} 0 & 0 & 1 \\ 0 & 1 & -1/2 \\ 1 & 0 & -1/2 \end{bmatrix}$$

Performing $R_3 \rightarrow R_3 + R_2$

$$A = \sim \begin{bmatrix} 2 & 3 & 1 \\ 0 & 1/2 & 5/2 \\ 0 & 0 & 4 \end{bmatrix} \begin{bmatrix} 0 & 0 & 1 \\ 0 & 1 & -1/2 \\ 1 & 0 & -1 \end{bmatrix}$$

Now A has been converted into an upper triangular matrix. By row transformation again, convert this upper triangular matrix into Identity matrix.
Multiplied R_2 by 2

$$A = \begin{bmatrix} 2 & 3 & 1 \\ 0 & 1 & 5 \\ 0 & 0 & 4 \end{bmatrix} \begin{bmatrix} 0 & 0 & 1 \\ 0 & 2 & -1 \\ 1 & 1 & -1 \end{bmatrix}$$

Performing $R_1 \rightarrow R_1 - 3R_2$, we get

$$A = \begin{bmatrix} 2 & 0 & -14 \\ 0 & 1 & 5 \\ 0 & 0 & 4 \end{bmatrix} \begin{bmatrix} 0 & -6 & 4 \\ 0 & 2 & -1 \\ 1 & 1 & -1 \end{bmatrix}$$

Divide R_1 by 2 and R_3 by 4, we get

$$A = \begin{bmatrix} 1 & 0 & -7 \\ 0 & 1 & 5 \\ 0 & 0 & 1 \end{bmatrix} \begin{bmatrix} 0 & -3 & 2 \\ 0 & 2 & -1 \\ 1/4 & 1/4 & -1/4 \end{bmatrix}$$

Performing $R_1 \to R_1 + 7R_3$, $R_2 \to R_2 - 5R_3$, we get

$$A = \begin{bmatrix} 1 & 0 & 0 \\ 0 & 1 & 0 \\ 0 & 0 & 1 \end{bmatrix} \begin{bmatrix} 7/2 & -5/4 & 1/4 \\ -5/4 & 3/4 & 1/4 \\ 1/4 & 1/4 & -1/4 \end{bmatrix}$$

Hence,

$$A^{-1} = \begin{bmatrix} 7/4 & -5/4 & 1/4 \\ -5/4 & 3/4 & 1/4 \\ 1/4 & 1/4 & -1/4 \end{bmatrix}$$

EXAMPLE 3. *Apply Gauss-elimination method to find the inverse of*

$$A = \begin{bmatrix} 2 & 6 & 6 \\ 2 & 8 & 6 \\ 2 & 6 & 8 \end{bmatrix}$$

SOLUTION. Let up put

$$A = AI$$

$\therefore$

$$A = \begin{bmatrix} 2 & 6 & 6 \\ 2 & 8 & 6 \\ 2 & 6 & 8 \end{bmatrix} \begin{bmatrix} 1 & 0 & 0 \\ 0 & 1 & 0 \\ 0 & 0 & 1 \end{bmatrix}$$

Performing $R_2 \to R_2 - R_1$, $R_3 \to R_3 - R_1$, we get

$$A = \begin{bmatrix} 2 & 6 & 6 \\ 0 & 2 & 0 \\ 0 & 0 & 2 \end{bmatrix} \begin{bmatrix} 1 & 0 & 0 \\ -1 & 1 & 0 \\ -1 & 0 & 1 \end{bmatrix}$$

Performing $R_1 \to R_1 - 3R_2$, we get

$$A = \begin{bmatrix} 2 & 0 & 6 \\ 0 & 2 & 0 \\ 0 & 0 & 2 \end{bmatrix} \begin{bmatrix} 4 & -3 & 0 \\ -1 & 1 & 0 \\ -1 & 0 & 1 \end{bmatrix}$$

Performing $R_1 \to R_1 - 3R_3$, we get

$$A = \begin{bmatrix} 2 & 0 & 0 \\ 0 & 2 & 0 \\ 0 & 0 & 2 \end{bmatrix} \begin{bmatrix} 7 & -3 & -3 \\ -1 & 1 & 0 \\ -1 & 0 & 1 \end{bmatrix}$$

Divide R_1, R_2 and R_3 by 2 respectively, we get

$$A = \begin{bmatrix} 1 & 0 & 0 \\ 0 & 1 & 0 \\ 0 & 0 & 1 \end{bmatrix} \begin{bmatrix} 7/2 & -3/2 & -3/2 \\ -1/2 & 1/2 & 0 \\ -1/2 & 0 & 1/2 \end{bmatrix}$$

Hence,

$$A^{-1} = \begin{bmatrix} 7/2 & -3/2 & -3/2 \\ -1/2 & 1/2 & 0 \\ -1/2 & 0 & 1/2 \end{bmatrix}$$

EXAMPLE 4. *Find the inverse of* $A = \begin{bmatrix} 0 & 1 & 2 \\ 1 & 2 & 3 \\ 3 & 1 & 1 \end{bmatrix}$ *by Gauss-elimination method.*

SOLUTION. Let us put

$$A = AI$$

$$\therefore \qquad A = \begin{bmatrix} 0 & 1 & 2 \\ 1 & 2 & 3 \\ 3 & 1 & 1 \end{bmatrix} \begin{bmatrix} 1 & 0 & 0 \\ 0 & 1 & 0 \\ 0 & 0 & 1 \end{bmatrix}$$

Interchanging R_1 and R_3, we get

$$A \sim \begin{bmatrix} 3 & 1 & 1 \\ 1 & 2 & 3 \\ 0 & 1 & 2 \end{bmatrix} \begin{bmatrix} 0 & 0 & 1 \\ 0 & 1 & 0 \\ 1 & 0 & 0 \end{bmatrix}$$

Performing $R_2 \to R_2 - \dfrac{1}{3} R_1$, we get

$$A \sim \begin{bmatrix} 3 & 1 & 1 \\ 0 & 5/3 & 8/3 \\ 0 & 1 & 2 \end{bmatrix} \begin{bmatrix} 0 & 0 & 1 \\ 0 & 1 & -1/3 \\ 1 & 0 & 0 \end{bmatrix}$$

Multiply R_2 by $\dfrac{3}{5}$, we get

$$A \sim \begin{bmatrix} 3 & 1 & 1 \\ 0 & 1 & 8/5 \\ 0 & 1 & 2 \end{bmatrix} \begin{bmatrix} 0 & 0 & 1 \\ 0 & 3/5 & -1/5 \\ 1 & 0 & 0 \end{bmatrix}$$

Performing $R_3 \to R_3 - R_2$, we get

$$A \sim \begin{bmatrix} 3 & 1 & 1 \\ 0 & 1 & 8/5 \\ 0 & 0 & 2/5 \end{bmatrix} \begin{bmatrix} 0 & 0 & 1 \\ 0 & 3/5 & -1/5 \\ 1 & -3/5 & 1/5 \end{bmatrix}$$

Performing $R_1 \to R_1 - R_2$, we get

$$A \sim \begin{bmatrix} 3 & 0 & -3/5 \\ 0 & 1 & 8/5 \\ 0 & 0 & 2/5 \end{bmatrix} \begin{bmatrix} 0 & -3/5 & 6/5 \\ 0 & 3/5 & -1/5 \\ 1 & -3/5 & 1/5 \end{bmatrix}$$

Performing $R_1 \to R_1 + \dfrac{3}{2} R_3$, we get

$$A \sim \begin{bmatrix} 3 & 0 & 0 \\ 0 & 1 & 8/5 \\ 0 & 0 & 2/5 \end{bmatrix} \begin{bmatrix} 3/2 & -3/2 & 3/2 \\ 0 & 3/5 & -1/5 \\ 1 & -3/5 & 1/5 \end{bmatrix}$$

Performing $R_2 \to R_2 - 4R_3$, we get

$$A \sim \begin{bmatrix} 3 & 0 & 0 \\ 0 & 1 & 0 \\ 0 & 0 & 2/5 \end{bmatrix} \begin{bmatrix} 3/2 & -3/2 & 3/2 \\ -4 & 3 & -1 \\ 1 & -3/5 & 1/5 \end{bmatrix}$$

Divide R_1 by 3 and R_3 by 2/5

$$A \sim \begin{bmatrix} 1 & 0 & 0 \\ 0 & 1 & 0 \\ 0 & 0 & 1 \end{bmatrix} \begin{bmatrix} 1/2 & -1/2 & 1/2 \\ -4 & 3 & -1 \\ 5/2 & -3/2 & 1/2 \end{bmatrix}$$

Hence $\qquad A^{-1} = \begin{bmatrix} 1/2 & -1/2 & 1/2 \\ -4 & 3 & -1 \\ 5/2 & -3/2 & 1/2 \end{bmatrix}$

EXAMPLE 5. *Find the inverse of $A = \begin{bmatrix} 1 & 3 & 3 \\ 1 & 4 & 3 \\ 1 & 3 & 4 \end{bmatrix}$ by Gauss-elimination method.*

SOLUTION. Let

$$A = AI$$

$$\therefore \quad A = \begin{bmatrix} 1 & 3 & 3 \\ 1 & 4 & 3 \\ 1 & 3 & 4 \end{bmatrix} \begin{bmatrix} 1 & 0 & 0 \\ 0 & 1 & 0 \\ 0 & 0 & 1 \end{bmatrix}$$

Performing $R_2 \to R_2 - R_1, R_3 \to R_3 - R_1$, we get

$$A \sim \begin{bmatrix} 1 & 3 & 3 \\ 0 & 1 & 0 \\ 0 & 0 & 1 \end{bmatrix} \begin{bmatrix} 1 & 0 & 0 \\ -1 & 1 & 0 \\ -1 & 0 & 1 \end{bmatrix}$$

Performing $R_1 \to R_1 - 3R_2$

$$A \sim \begin{bmatrix} 1 & 0 & 3 \\ 0 & 1 & 0 \\ 0 & 0 & 1 \end{bmatrix} \begin{bmatrix} 4 & -3 & 0 \\ -1 & 1 & 0 \\ -1 & 0 & 1 \end{bmatrix}$$

Performing $R_1 \to R_1 - 3R_3$

$$A \sim \begin{bmatrix} 1 & 0 & 0 \\ 0 & 1 & 0 \\ 0 & 0 & 1 \end{bmatrix} \begin{bmatrix} 7 & -3 & -3 \\ -1 & 1 & 0 \\ -1 & 0 & 1 \end{bmatrix}$$

Hence,

$$A^{-1} = \begin{bmatrix} 7 & -3 & -3 \\ -1 & 1 & 0 \\ -1 & 0 & 1 \end{bmatrix}$$

EXAMPLE 6. *Using Gauss-Jordon method, find the inverse of the matrix*

$$\begin{bmatrix} 1 & 1 & 3 \\ 1 & 3 & -3 \\ -2 & -4 & -4 \end{bmatrix}$$

(Kurukshetra–2006)

SOLUTION. Let

$$A = AI$$

$$\Rightarrow \quad A = \begin{bmatrix} 1 & 1 & 3 \\ 1 & 3 & -3 \\ -2 & -4 & -4 \end{bmatrix} \begin{bmatrix} 1 & 0 & 0 \\ 0 & 1 & 0 \\ 0 & 0 & 1 \end{bmatrix}$$

Performing $R_2 - R_1$ and $R_3 + 2R_1$

$$A \sim \begin{bmatrix} 1 & 1 & 3 \\ 0 & 2 & -6 \\ 0 & -2 & 2 \end{bmatrix} \begin{bmatrix} 1 & 0 & 0 \\ -1 & 1 & 0 \\ 2 & 0 & 1 \end{bmatrix}$$

Performing $\frac{1}{2}R_2$ and $\frac{1}{2}R_3$

$$A \sim \begin{bmatrix} 1 & 1 & 3 \\ 0 & 1 & -3 \\ 0 & -1 & 1 \end{bmatrix} \begin{bmatrix} 1 & 0 & 0 \\ -1/2 & 1/2 & 0 \\ 0 & 0 & 1/2 \end{bmatrix}$$

Performing $R_1 - R_2$ and $R_3 + R_2$, we get

$$A \sim \begin{bmatrix} 1 & 0 & 6 \\ 0 & 1 & -3 \\ 0 & 0 & -2 \end{bmatrix} \begin{bmatrix} 3/2 & -1/2 & 0 \\ -1/2 & 1/2 & 0 \\ 1/2 & 1/2 & 1/2 \end{bmatrix}$$

Performing $R_1 + 3R_3, R_2 - \frac{3}{2}R_3$ and $\frac{1}{2}R_2$

$$A \sim \begin{bmatrix} 1 & 0 & 0 \\ 0 & 1 & 0 \\ 0 & 0 & 1 \end{bmatrix} \begin{bmatrix} 3 & 1 & 3/2 \\ -5/4 & -1/4 & -3/4 \\ -1/4 & -1/4 & -1/4 \end{bmatrix}$$

Hence,
$$A^{-1} = \begin{bmatrix} 3 & 1 & 3/2 \\ -5/4 & -1/4 & -3/4 \\ -1/4 & -1/4 & -1/4 \end{bmatrix}$$

EXAMPLE 7. *Using Gauss-Jordan method, find the inverse of the matrix*

$$\begin{bmatrix} 2 & 2 & 3 \\ 2 & 1 & 1 \\ 1 & 3 & 5 \end{bmatrix}$$

(Anna–2004)

SOLUTION. Let $A = AI$

$$A = \begin{bmatrix} 2 & 2 & 3 \\ 2 & 1 & 1 \\ 1 & 3 & 5 \end{bmatrix} \begin{bmatrix} 1 & 0 & 0 \\ 0 & 1 & 0 \\ 0 & 0 & 1 \end{bmatrix}$$

Performing $\dfrac{1}{2}R_1$, we get $\quad A \sim \begin{bmatrix} 1 & 1 & 3/2 \\ 2 & 1 & 1 \\ 1 & 3 & 5 \end{bmatrix} \begin{bmatrix} 1/2 & 0 & 0 \\ 0 & 1 & 0 \\ 0 & 0 & 1 \end{bmatrix}$

Performing $R_2 - 2R_1, R_3 - R_1, A \sim \begin{bmatrix} 1 & 1 & 3/2 \\ 0 & -1 & -2 \\ 0 & 2 & 7/2 \end{bmatrix} \begin{bmatrix} 1/2 & 0 & 0 \\ -1 & 1 & 0 \\ -1/2 & 0 & 1 \end{bmatrix}$

Performing $R_1 + R_2, R_3 + 2R_2,$

$$A = \begin{bmatrix} 1 & 0 & -1/2 \\ 0 & -1 & -2 \\ 0 & 0 & -1/2 \end{bmatrix} \begin{bmatrix} -1/2 & 1 & 0 \\ -1 & 1 & 0 \\ -5/2 & 2 & 1 \end{bmatrix}$$

$$= \begin{bmatrix} 1 & 0 & -1/2 \\ 0 & 1 & 2 \\ 0 & 0 & -1/2 \end{bmatrix} \begin{bmatrix} -1/2 & 1 & 0 \\ 1 & -1 & 0 \\ -5/2 & 2 & 1 \end{bmatrix}$$

Performing $(-2)R_3, \quad A = \begin{bmatrix} 1 & 0 & -1/2 \\ 0 & 1 & 2 \\ 0 & 0 & 1 \end{bmatrix} \begin{bmatrix} -1/2 & 1 & 0 \\ 1 & -1 & 0 \\ 5 & -4 & -2 \end{bmatrix}$

Performing $R_1 + \dfrac{1}{2}R_3, R_2 - 2R_3, A = \begin{bmatrix} 1 & 0 & 0 \\ 0 & 1 & 0 \\ 0 & 0 & 1 \end{bmatrix} \begin{bmatrix} 2 & -1 & -1 \\ -9 & 7 & 4 \\ 5 & -4 & -2 \end{bmatrix}$

Hence, $\quad A^{-1} = \begin{bmatrix} 2 & -1 & -1 \\ -9 & 7 & 4 \\ 5 & -4 & -2 \end{bmatrix}$

7.4 TRIANGULARISATION METHOD

This method is also known as 'Factorization' method. Thus in this method, we factorize the given matrix A as LU.
i.e.,
$$A = LU \qquad \qquad ...(1)$$
where L is a lower triangular matrix having diagonal elements unity and U is an upper triangular matrix. That is,

$$L = \begin{bmatrix} 1 & 0 & 0 \\ l_{21} & 1 & 0 \\ l_{31} & l_{32} & 1 \end{bmatrix}, U = \begin{bmatrix} u_{11} & u_{12} & u_{13} \\ 0 & u_{22} & u_{23} \\ 0 & 0 & u_{33} \end{bmatrix}$$

From (1), we get

$$A^{-1} = (LU)^{-1} = U^{-1}L^{-1}$$

Determination of L^{-1}:

Let us take $L^{-1} = X$, where X is also a lower triangular matrix

$$\therefore \qquad LX = I$$

$$\therefore \quad \begin{bmatrix} 1 & 0 & 0 \\ l_{21} & 1 & 0 \\ l_{31} & l_{32} & 1 \end{bmatrix} \begin{bmatrix} x_{11} & 0 & 0 \\ x_{21} & x_{22} & 0 \\ x_{31} & x_{32} & x_{33} \end{bmatrix} = \begin{bmatrix} 1 & 0 & 0 \\ 0 & 1 & 0 \\ 0 & 0 & 1 \end{bmatrix}$$

Multiplying the matrices on the R.H.S. and equating of both sides of matrices, we get

$$x_{11} = 1, x_{22} = 1, x_{33} = 1$$
$$l_{21}x_{11} + x_{21} = 0, l_{31}x_{11} + l_{32}x_{21} + x_{31} = 0$$
$$l_{32}x_{22} + x_{32} = 0.$$

From above equations, we get all the elements of X and hence we obtain $L^{-1} = X$.

Determination of U^{-1}:

Let us take $U^{-1} = Y$, where Y is also an upper triangular matrix. That is

$$UY = I$$

$$\therefore \quad \begin{bmatrix} u_{11} & u_{12} & u_{13} \\ 0 & u_{22} & u_{23} \\ 0 & 0 & u_{33} \end{bmatrix} \begin{bmatrix} y_{11} & y_{12} & y_{13} \\ 0 & y_{22} & y_{23} \\ 0 & 0 & y_{33} \end{bmatrix} = \begin{bmatrix} 1 & 0 & 0 \\ 0 & 1 & 0 \\ 0 & 0 & 1 \end{bmatrix}$$

Multiplying the matrices on R.H.S. and equating of both sides the corresponding elements of the matrices, we get

$$u_{11}y_{11} = 1, u_{11}y_{12} + u_{12}y_{22} = 0,$$
$$u_{11}y_{13} + u_{12}y_{23} + u_{13}y_{33} = 0$$
$$u_{22}y_{22} = 1, u_{22}y_{23} + u_{23}y_{33} = 0, u_{33}y_{33} = 1$$

From these equations, we get all the elements of Y and hence we obtain

$$U^{-1} = Y.$$

Consequently from (2), we obtain A^{-1}.

SOLVED EXAMPLES

EXAMPLE 1. *Using the triangularisation method, find the inverse of the matrix*

$$A = \begin{bmatrix} 50 & 107 & 36 \\ 25 & 54 & 20 \\ 31 & 66 & 21 \end{bmatrix}.$$

SOLUTION. Let $\qquad A = LU$, where

$$\therefore \qquad A = \begin{bmatrix} 1 & 0 & 0 \\ l_{21} & 1 & 0 \\ l_{31} & l_{32} & 1 \end{bmatrix}, U = \begin{bmatrix} u_{11} & u_{12} & u_{13} \\ 0 & u_{22} & u_{23} \\ 0 & 0 & u_{33} \end{bmatrix}$$

$$\therefore \quad \begin{bmatrix} 50 & 107 & 36 \\ 25 & 54 & 20 \\ 31 & 66 & 21 \end{bmatrix} = \begin{bmatrix} 1 & 0 & 0 \\ l_{21} & 1 & 0 \\ l_{31} & l_{32} & 1 \end{bmatrix} \begin{bmatrix} u_{11} & u_{12} & u_{13} \\ 0 & u_{22} & u_{23} \\ 0 & 0 & u_{33} \end{bmatrix}$$

$$u_{11} = 50, u_{12} = 107, u_{13} = 36$$

$$l_{21}u_{11} = 25 \Rightarrow l_{21} = \frac{25}{50} = \frac{1}{2}$$

$$\therefore \qquad l_{21}u_{12} + u_{22} = 54 \Rightarrow u_{22} = 54 - l_{21}u_{12}$$

$$= 54 - \frac{107}{2} = \frac{1}{2}$$

$$\therefore \qquad l_{21}u_{13} + u_{23} = 20 \Rightarrow u_{23} = 20 - l_{21}u_{13}$$

$$= 20 - 18 = 2$$

$$\therefore \qquad l_{31}u_{11} = 31 \Rightarrow l_{31} = \frac{31}{50}$$

$$\therefore \qquad l_{31}u_{12} + l_{32}u_{22} = 66$$

$$\Rightarrow \qquad l_{32} = \frac{1}{u_{22}}(66 - l_{31}u_{12})$$

$$= 2\left(66 - \frac{31}{50} \cdot 107\right) = 132 - \frac{3317}{25} = -\frac{17}{25}$$

$$\therefore \ \ l_{31}u_{13} + l_{32}u_{23} + u_{33} = 21$$
$$\Rightarrow \qquad u_{33} = 21 - l_{31}u_{13} - l_{32}u_{23}$$

$$\Rightarrow \qquad u_{33} = 21 - \frac{31}{50} \cdot 36 + \frac{17}{25} \cdot 2 = 21 - \frac{558}{25} + \frac{34}{25} = \frac{1}{25}$$

Thus
$$L = \begin{bmatrix} 1 & 0 & 0 \\ 1/2 & 1 & 0 \\ 31/50 & -17/25 & 1 \end{bmatrix}$$

and
$$U = \begin{bmatrix} 50 & 107 & 36 \\ 0 & 1/2 & 2 \\ 0 & 0 & 1/25 \end{bmatrix}$$

Determination of L^{-1} and U^{-1} :

Let
$$L^{-1} = X$$
$$\therefore \qquad LX = I$$

$$\begin{bmatrix} 1 & 0 & 0 \\ 1/2 & 1 & 0 \\ 31/50 & -17/25 & 1 \end{bmatrix}\begin{bmatrix} x_{11} & 0 & 0 \\ x_{21} & x_{22} & 0 \\ x_{31} & x_{32} & x_{33} \end{bmatrix} = \begin{bmatrix} 1 & 0 & 0 \\ 0 & 1 & 0 \\ 0 & 0 & 1 \end{bmatrix}$$

$$\therefore \qquad x_{11} = 1$$

and
$$\frac{1}{2}x_{11} + x_{21} = 0 \qquad \Rightarrow \qquad -\frac{1}{2}x_{11} = -\frac{1}{2}, \ x_{22} = 1$$

and
$$\frac{31}{50}x_{11} - \frac{17}{25}x_{21} + x_{31} = 0$$

$$\Rightarrow \qquad x_{31} = \frac{17}{25}x_{21} - \frac{31}{50}x_{11}$$

$$= \frac{17}{25}\left(-\frac{1}{2}\right) - \frac{31}{30}(1) = -\frac{48}{50} = -\frac{24}{25}$$

and
$$-\frac{17}{25}x_{22} + x_{32} = 0$$

$$\Rightarrow \qquad x_{32} = \frac{17}{25}x_{22} = \frac{17}{25}(1) = \frac{17}{25}$$

and $\qquad x_{33} = 1.$

$\therefore \qquad L^{-1} = \begin{bmatrix} 1 & 0 & 0 \\ -1/2 & 1 & 0 \\ -24/25 & 17/25 & 1 \end{bmatrix}$

Now let $\qquad U^{-1} = Y$

$\therefore \qquad UY = I$

$\begin{bmatrix} 50 & 107 & 36 \\ 0 & 1/2 & 2 \\ 0 & 0 & 1/25 \end{bmatrix} \begin{bmatrix} y_{11} & y_{12} & y_{13} \\ 0 & y_{22} & y_{23} \\ 0 & 0 & y_{33} \end{bmatrix} = \begin{bmatrix} 1 & 0 & 0 \\ 0 & 1 & 0 \\ 0 & 0 & 1 \end{bmatrix}$

$\therefore \qquad 50y_{11} = 1 \Rightarrow y_{12} = -\frac{107}{50} y_{22} = -\frac{107}{50} \times 2 = -\frac{107}{25}$

$\frac{1}{2} y_{22} = 1 \Rightarrow y_{22} = 2, \frac{1}{25} y_{33} = 1 \Rightarrow y_{33} = 25$

and $\qquad 50y_{12} + 107y_{22} = 0 \Rightarrow y_{12} = -\frac{107}{50} y_{22} = -\frac{107}{50} \times 2 = -\frac{107}{25}$

and $\qquad \frac{1}{2} y_{23} + 2y_{33} = 0$

$\Rightarrow \qquad y_{23} = -4y_{33} = -100$

and $\qquad 50y_{13} + 107y_{23} + 36y_{33} = 0$

$\Rightarrow \qquad y_{13} = +\frac{1}{50}(-107y_{23} - 36y_{33})$

$= +\frac{1}{50}[-107(-100) - 35(25)]$

$= +\frac{1}{50}(10700 - 900) = +196$

Thus $\qquad U^{-1} = Y = \begin{bmatrix} 1/50 & -107/25 & 196 \\ 0 & 2 & -100 \\ 0 & 0 & 25 \end{bmatrix}$

Hence $\qquad A^{-1} = U^{-1}L^{-1}$

and $\begin{bmatrix} 1/50 & -107/25 & 196 \\ 0 & 2 & -100 \\ 0 & 0 & 25 \end{bmatrix} \begin{bmatrix} 1 & 0 & 0 \\ -1/2 & 1 & 0 \\ -24/25 & 17/25 & 1 \end{bmatrix} = \begin{bmatrix} -186 & 129 & 196 \\ 95 & -66 & -100 \\ -24 & 17 & 25 \end{bmatrix}$

EXAMPLE 2. *Apply Triangularisation method to find the inverse of*

$$A = \begin{bmatrix} 2 & -2 & 4 \\ 2 & 3 & 2 \\ -1 & 1 & -1 \end{bmatrix} \qquad \text{(Kumaun-2000, 01, 10)}$$

SOLUTION. Let $\qquad A = LU$

$\therefore \qquad \begin{bmatrix} 2 & -2 & 4 \\ 2 & 3 & 2 \\ -1 & 1 & -1 \end{bmatrix} = \begin{bmatrix} 1 & 0 & 0 \\ l_{21} & 1 & 0 \\ l_{31} & l_{32} & 1 \end{bmatrix} \begin{bmatrix} u_{11} & u_{12} & u_{13} \\ 0 & u_{22} & u_{23} \\ 0 & 0 & u_{33} \end{bmatrix}$

$\therefore \qquad u_{11} = 2, u_{12} = -2, u_{13} = 4$

$$l_{21}u_{11} = 2 \Rightarrow l_{21} = \frac{2}{u_{11}} = \frac{2}{2} = 1$$

$$l_{21}u_{12} + u_{22} = 3 \Rightarrow u_{22} = 3 - l_{21}u_{12} = 3 - 1(-2) = 5$$
$$l_{21}u_{13} + u_{23} = 2 \Rightarrow u_{23} = 2 - l_{21}u_{13} = 2 - 1(4) = -2$$

$$l_{31}u_{11} = -1 \Rightarrow l_{31} = -\frac{1}{u_{11}} = -\frac{1}{2}$$

$$l_{31}u_{12} + l_{32}u_{22} = 1 \Rightarrow l_{32} = \frac{1}{u_{12}}(1 - l_{31}u_{12}) = \frac{1}{5}\left(1 + \frac{1}{2}(-2)\right) = 0$$

$$l_{31}u_{13} + l_{32}u_{23} + u_{33} = -1$$
$$\Rightarrow \qquad u_{33} = -1 - l_{31}u_{13} - l_{32}u_{23}$$

$$= -1 + \frac{1}{2}(4) - 0 = 1$$

Thus
$$L = \begin{bmatrix} 1 & 0 & 0 \\ 1 & 1 & 0 \\ -1/2 & 0 & 1 \end{bmatrix}, U = \begin{bmatrix} 2 & -2 & 4 \\ 0 & 5 & -2 \\ 0 & 0 & 1 \end{bmatrix}$$

Determination L^{-1} and U^{-1}

Let
$$L^{-1} = X$$
$$\therefore \qquad LX = I$$

$$\begin{bmatrix} 1 & 0 & 0 \\ 1 & 1 & 0 \\ -1/2 & 0 & 1 \end{bmatrix}\begin{bmatrix} x_{11} & 0 & 0 \\ x_{21} & x_{22} & 0 \\ x_{31} & x_{32} & x_{33} \end{bmatrix} = \begin{bmatrix} 1 & 0 & 0 \\ 0 & 1 & 0 \\ 0 & 0 & 1 \end{bmatrix}$$

$$x_{11} = 1, x_{11} + x_{21} = 0 \Rightarrow x_{21} = -x_{11} = -1$$

$$x_{22} = 1, \frac{1}{2}x_{11} + x_{31} = 0 \Rightarrow x_{31} = \frac{1}{2}x_{11} = \frac{1}{2}$$

$$x_{32} = 0, x_{33} = 1$$

Thus
$$L^{-1} = \begin{bmatrix} 1 & 0 & 0 \\ -1 & 1 & 0 \\ -1/2 & 0 & 1 \end{bmatrix}$$

Now let
$$U^{-1} = Y$$
$$\therefore \qquad UY = I$$

$$\begin{bmatrix} 2 & -2 & 4 \\ 0 & 5 & -2 \\ 0 & 0 & 1 \end{bmatrix}\begin{bmatrix} y_{11} & y_{12} & y_{13} \\ 0 & y_{22} & y_{23} \\ 0 & 0 & y_{33} \end{bmatrix} = \begin{bmatrix} 1 & 0 & 0 \\ 0 & 1 & 0 \\ 0 & 0 & 1 \end{bmatrix}$$

$$\therefore \qquad 2y_{11} = 1 \Rightarrow y_{11} = \frac{1}{2},$$

$$2y_{22} = 1 \Rightarrow y_{22} = \frac{1}{5}$$

$$2y_{12} - 2y_{22} = 0 \Rightarrow y_{12} = y_{22} = \frac{1}{5}$$

$$y_{33} = 1$$
$$5y_{23} - 2y_{33} = 0 \Rightarrow y_{23} = 2/5$$
and $2y_{13} - 2y_{23} + 4y_{33} = 0$

$\Rightarrow$ $\qquad y_{13} = \dfrac{1}{2}(2y_{23} - 4y_{33}) = \dfrac{1}{2}\left(\dfrac{4}{5} - 4\right) = -\dfrac{8}{5}$

Thus $\qquad U^{-1} = Y = \begin{bmatrix} 1/2 & 1/5 & -8/5 \\ 0 & 1/5 & 2/5 \\ 0 & 0 & 1 \end{bmatrix}$

Hence, $\qquad A^{-1} = U^{-1}L^{-1}$

$$= \begin{bmatrix} 1/2 & 1/5 & -8/5 \\ 0 & 1/5 & 2/5 \\ 0 & 0 & 1 \end{bmatrix}\begin{bmatrix} 1 & 0 & 0 \\ -1 & 1 & 0 \\ 1/2 & 0 & 1 \end{bmatrix}$$

$$= \begin{bmatrix} -1/2 & 1/5 & -8/5 \\ 0 & 1/5 & 2/5 \\ 1/2 & 0 & 1 \end{bmatrix}$$

EXAMPLE 3. *Use factorization method to find the inverse of*

$$A = \begin{bmatrix} 1 & 2 & 3 \\ 3 & 2 & 1 \\ 2 & 1 & 3 \end{bmatrix}$$

SOLUTION. Let $\qquad A = LU$

$\therefore \qquad \begin{bmatrix} 1 & 2 & 3 \\ 3 & 2 & 1 \\ 2 & 1 & 3 \end{bmatrix} = \begin{bmatrix} 1 & 0 & 0 \\ l_{21} & 1 & 0 \\ l_{31} & l_{32} & 1 \end{bmatrix}\begin{bmatrix} u_{11} & u_{12} & u_{13} \\ 0 & u_{22} & u_{23} \\ 0 & 0 & u_{33} \end{bmatrix}$

$\therefore \qquad u_{11} = 1,\ u_{12} = 2,\ u_{13} = 3$

$\qquad l_{21}u_{11} = 3 \Rightarrow l_{21} = 3,\ l_{21}u_{12} + u_{22} = 2$

$\Rightarrow \qquad u_{22} = -4$

$\qquad l_{21}u_{13} + u_{23} = 1 \Rightarrow u_{23} = -8$

$\qquad l_{31}u_{11} = 2 \Rightarrow l_{31} = 2,\ l_{31}u_{12} + l_{32}u_{22} = 1$

$\Rightarrow \qquad l_{32} = 3/4$

$l_{31}u_{13} + l_{32}u_{23} + u_{23} = 3 \Rightarrow u_{23} = 3.$
Thus

$$L = \begin{bmatrix} 1 & 0 & 0 \\ 3 & 1 & 0 \\ 2 & 3/4 & 1 \end{bmatrix}, U = \begin{bmatrix} 1 & 2 & 3 \\ 0 & -4 & -8 \\ 0 & 0 & 3 \end{bmatrix}$$

Determination of L^{-1} and U^{-1}.

Let $\qquad L^{-1} = X$ *i.e.,* $\qquad LX = I$

$\therefore \qquad \begin{bmatrix} 1 & 0 & 0 \\ 3 & 1 & 0 \\ 2 & 3/4 & 1 \end{bmatrix}\begin{bmatrix} x_{11} & 0 & 0 \\ x_{21} & x_{22} & 0 \\ x_{31} & x_{32} & x_{33} \end{bmatrix} = \begin{bmatrix} 1 & 0 & 0 \\ 0 & 1 & 0 \\ 0 & 0 & 1 \end{bmatrix}$

$\therefore \qquad x_{11} = 1,$

$\qquad 3x_{11} + x_{21} = 0 \Rightarrow x_{21} = -3,\ x_{22} = 1$

$\qquad 2x_{11} + \dfrac{3}{4}x_{21} + x_{31} = 0$

$$\Rightarrow \qquad x_{31} = \frac{1}{4}, \frac{3}{4}x_{22} + x_{32} = 0 \Rightarrow x_{32} = -\frac{3}{4}, x_{33} = 1$$

$$\therefore \qquad L^{-1} = \begin{bmatrix} 1 & 0 & 0 \\ -3 & 1 & 0 \\ 1/4 & -3/4 & 1 \end{bmatrix}.$$

Now let $\qquad U^{-1} = Y$

$$\therefore \qquad UY = I$$

$$\therefore \qquad \begin{bmatrix} 1 & 2 & 3 \\ 0 & -4 & -8 \\ 0 & 0 & 3 \end{bmatrix}\begin{bmatrix} y_{11} & y_{12} & y_{13} \\ 0 & y_{22} & y_{23} \\ 0 & 0 & y_{33} \end{bmatrix} = \begin{bmatrix} 1 & 0 & 0 \\ 0 & 1 & 0 \\ 0 & 0 & 1 \end{bmatrix}$$

$$y_{11} = 1, -4y_{22} = 1 \Rightarrow y_{22} = -\frac{1}{4}, 3y_{33} = 1 \Rightarrow y_{33} = \frac{1}{3}$$

$$y_{12} + 2y_{22} = 0 \Rightarrow y_{12} = \frac{1}{2}, -4y_{23} - 8y_{33} = 0 \Rightarrow y_{23} = -\frac{2}{3}$$

$$y_{13} + 2y_{23} + 3y_{33} = 0 \Rightarrow y_{13} = \frac{1}{3}.$$

Thus $\qquad U^{-1} = Y = \begin{bmatrix} 1 & 1/2 & 1/3 \\ 0 & -1/4 & -2/3 \\ 0 & 0 & 1/3 \end{bmatrix}$

Consequently, $A^{-1} = U^{-1}L^{-1}$

$$= \begin{bmatrix} 1 & 1/2 & 1/3 \\ 0 & -1/4 & -2/3 \\ 0 & 0 & 1/3 \end{bmatrix}\begin{bmatrix} 1 & 0 & 0 \\ -3 & 1 & 0 \\ 1/4 & -3/4 & 1 \end{bmatrix} = \begin{bmatrix} 5/12 & 1/4 & 1/3 \\ 7/12 & 1/4 & -2/3 \\ 1/12 & -1/4 & 1/3 \end{bmatrix}$$

EXAMPLE 4. *By Triangularisation method, find the inverse of*

$$A = \begin{bmatrix} 5 & -2 & 1 \\ 7 & 1 & -5 \\ 3 & 7 & 4 \end{bmatrix}$$

SOLUTION. Let $\qquad A = LU$

$$\begin{bmatrix} 5 & -2 & 1 \\ 7 & 1 & -5 \\ 3 & 7 & 4 \end{bmatrix} = \begin{bmatrix} 1 & 0 & 0 \\ l_{21} & 1 & 0 \\ l_{31} & l_{32} & 1 \end{bmatrix}\begin{bmatrix} u_{11} & u_{12} & u_{13} \\ 0 & u_{22} & u_{23} \\ 0 & 0 & u_{33} \end{bmatrix}$$

$$\therefore \qquad u_{11} = 5, u_{12} = -2, u_{13} = 1$$

$$l_{21}u_{11} = 7 \Rightarrow l_{21} = \frac{7}{5}, l_{21}u_{12} + u_{22} = 1$$

$$\Rightarrow \qquad u_{22} = 19/5$$

$$l_{21}u_{23} = -5 \Rightarrow u_{23} = -\frac{32}{5}$$

$$l_{31}u_{11} = 3 \Rightarrow l_{31} = \frac{3}{5}, l_{31}u_{12} + l_{32}u_{22} = 7$$

$$\Rightarrow \qquad l_{32} = 41/19$$

$$l_{31}u_{13} + l_{32}u_{23} + u_{33} = 4 \Rightarrow u_{33} = \frac{327}{19}$$

Thus we get

$$L = \begin{bmatrix} 1 & 0 & 0 \\ 7/5 & 1 & 0 \\ 3/5 & 41/19 & 1 \end{bmatrix}, U = \begin{bmatrix} 5 & -2 & 1 \\ 0 & 19/5 & -32/5 \\ 0 & 0 & 327/19 \end{bmatrix}$$

Determination of L^{-1} and U^{-1}.

Let $\qquad L^{-1} = X \qquad i.e., \quad LX = I$

$$\therefore \begin{bmatrix} 1 & 0 & 0 \\ 7/5 & 1 & 0 \\ 3/5 & 41/19 & 1 \end{bmatrix}\begin{bmatrix} x_{11} & 0 & 0 \\ x_{21} & x_{22} & 0 \\ x_{31} & x_{32} & x_{33} \end{bmatrix} = \begin{bmatrix} 1 & 0 & 0 \\ 0 & 1 & 0 \\ 0 & 0 & 1 \end{bmatrix}$$

$$\therefore \quad x_{11} = 1, \frac{7}{5}x_{11} + x_{21} = 0 \Rightarrow x_{21} = -\frac{7}{5}, x_{22} = 1$$

$$\frac{3}{5}x_{11} + \frac{41}{19}x_{21} + x_{31} = 0 \Rightarrow x_{31} = \frac{46}{19}$$

$$\frac{3}{5} \times 0 + \frac{41}{19}x_{22} + x_{32} = 0 \Rightarrow x_{32} = -\frac{41}{19}, x_{33} = 1$$

Thus we get

$$L^{-1} = \begin{bmatrix} 1 & 0 & 0 \\ -7/5 & 1 & 0 \\ 46/19 & -41/19 & 1 \end{bmatrix}$$

Let $\qquad U^{-1} = Y \quad i.e., UY = 1$

$$\therefore \begin{bmatrix} 5 & -2 & 1 \\ 0 & 19/5 & -32/5 \\ 0 & 0 & 327/19 \end{bmatrix}\begin{bmatrix} y_{11} & y_{12} & y_{13} \\ 0 & y_{22} & y_{23} \\ 0 & 0 & y_{33} \end{bmatrix} = \begin{bmatrix} 1 & 0 & 0 \\ 0 & 1 & 0 \\ 0 & 0 & 1 \end{bmatrix}$$

$$5y_{11} = 1 \Rightarrow y_{11} = \frac{1}{5}, \frac{19}{5}y_{22} = 1 \Rightarrow y_{22} = \frac{5}{19},$$

$$\frac{327}{19}y_{33} = 1 \Rightarrow y_{33} = \frac{19}{327}$$

$$5y_{12} - 2y_{22} = 0 \Rightarrow y_{12} = \frac{2}{19},$$

$$\frac{19}{5}y_{23} - \frac{32}{5}y_{33} = 0 \Rightarrow y_{23} = \frac{32}{327}$$

$$5y_{13} - 2y_{23} + y_{33} = 0 \Rightarrow y_{13} = \frac{9}{327} = \frac{3}{109}$$

Thus $\qquad U^{-1} = \begin{bmatrix} 1/5 & 2/19 & 3/109 \\ 0 & 5/19 & 32/327 \\ 0 & 0 & 19/327 \end{bmatrix}$

Consequently, we obtain

$$A^{-1} = U^{-1}L^{-1}$$

$$= \begin{bmatrix} 1/5 & 2/19 & 3/109 \\ 0 & 5/19 & 32/327 \\ 0 & 0 & 19/327 \end{bmatrix} \begin{bmatrix} 1 & 0 & 0 \\ -7/5 & 1 & 0 \\ 46/19 & -41/19 & 1 \end{bmatrix}$$

$$= -\begin{bmatrix} \dfrac{13}{109} & \dfrac{5}{109} & \dfrac{3}{109} \\ \dfrac{43}{327} & \dfrac{17}{327} & \dfrac{32}{327} \\ \dfrac{46}{327} & \dfrac{41}{327} & \dfrac{19}{327} \end{bmatrix}$$

7.5 CHOLESKI'S METHOD

(Agra-2006, Patna-2008, Delhi-2000, 04 , 09)

In this method, the matrix A has been assumed a symmetric matrix. That is

$$A' = A.$$

Let us suppose that

$$A = LL'. \qquad \qquad ...(1)$$

where L is a lower triangular matrix and L' is its transpose. Then the inverse is obtained from

$$A^{-1} = (LL')^{-1} = (L')^{-1}L^{-1}$$

Further let

$$A = \begin{bmatrix} a_{11} & a_{12} & a_{13} \\ a_{21} & a_{22} & a_{23} \\ a_{31} & a_{32} & a_{33} \end{bmatrix} \text{ and } L = \begin{bmatrix} l_{11} & 0 & 0 \\ l_{21} & l_{22} & 0 \\ l_{31} & l_{32} & l_{33} \end{bmatrix}$$

Then from (1), we get

$$\begin{bmatrix} a_{11} & a_{12} & a_{13} \\ a_{21} & a_{22} & a_{23} \\ a_{31} & a_{32} & a_{33} \end{bmatrix} = \begin{bmatrix} l_{11} & 0 & 0 \\ l_{21} & l_{22} & 0 \\ l_{31} & l_{32} & l_{33} \end{bmatrix} \begin{bmatrix} l_{11} & l_{21} & l_{31} \\ 0 & l_{22} & l_{32} \\ 0 & 0 & l_{33} \end{bmatrix}$$

Multiplying the two matrices on R.H.S. and equating both sides the corresponding elements, we obtain

$$l_{11}^2 = a_{11} \Rightarrow l_{11} = \sqrt{a_{11}}$$

$$l_{11}l_{21} = a_{12} \Rightarrow l_{21} = \frac{a_{12}}{l_{11}}$$

$$l_{11}l_{31} = a_{13} \Rightarrow l_{31} = \frac{a_{13}}{l_{11}}$$

$$l_{21}^2 + l_{22}^2 = a_{22} \Rightarrow l_{22} = \sqrt{a_{22} - l_{21}^2}$$

$$l_{21}l_{31} + l_{22}l_{32} = a_{23} \Rightarrow l_{32} = (a_{23} - l_{21}l_{31}) / l_{22}$$

$$l_{31}^2 + l_{32}^2 + l_{33}^2 = a_{33} \Rightarrow l_{33} = \sqrt{a_{33} + l_{31}^2 + l_{32}^2} \ .$$

Thus L is obtained.

As pointed out above, some elements of L may be imaginary. This will cause no extra trouble.

SOLVED EXAMPLES

EXAMPLE 1. *Find the inverse of*

$$A = \begin{bmatrix} 1 & 2 & 6 \\ 2 & 5 & 15 \\ 6 & 15 & 46 \end{bmatrix}$$

by Choleski's method. (Meerut-2001, 03, Rohilkhand-2008, Garhwal-2009, Kumaun-2003, 05, 10)

SOLUTION. Above matrix A is real symmetric matrix, hence this method is applicable. Let us consider

$$A = LL'$$

$$\therefore \quad \begin{bmatrix} 1 & 2 & 6 \\ 2 & 5 & 15 \\ 6 & 15 & 46 \end{bmatrix} = \begin{bmatrix} l_{11} & 0 & 0 \\ l_{21} & l_{22} & 0 \\ l_{31} & l_{32} & l_{33} \end{bmatrix} \begin{bmatrix} l_{11} & l_{21} & l_{31} \\ 0 & l_{22} & l_{32} \\ 0 & 0 & l_{33} \end{bmatrix}$$

$$\therefore \quad l_{11}^2 = 1 \Rightarrow l_{11} = 1$$

$$l_{11} l_{21} = 2 \Rightarrow l_{21} = \frac{2}{l_{11}} = 2$$

$$l_{11} l_{31} = 6 \Rightarrow l_{31} = 6$$

$$l_{21}^2 + l_{22}^2 = 5 \Rightarrow l_{22} = \sqrt{5 - 4} = 1$$

$$l_{21} l_{31} + l_{22} l_{32} = 15 \Rightarrow l_{32} = (15 - 12) = 3$$

$$l_{31}^2 + l_{32}^2 + l_{33}^2 = 46 \Rightarrow l_{33} = \sqrt{46 - 6^2 - 3^2}$$
$$= \sqrt{46 - 45} = 1.$$

Thus we obtain

$$L = \begin{bmatrix} 1 & 0 & 0 \\ 2 & 1 & 0 \\ 6 & 3 & 1 \end{bmatrix}$$

Determination of L^{-1} : Let $L^{-1} = X$, then $LX = I$

$$\therefore \quad \begin{bmatrix} 1 & 0 & 0 \\ 2 & 1 & 0 \\ 6 & 3 & 1 \end{bmatrix} \begin{bmatrix} x_{11} & 0 & 0 \\ x_{21} & x_{22} & 0 \\ x_{31} & x_{32} & x_{33} \end{bmatrix} = \begin{bmatrix} 1 & 0 & 0 \\ 0 & 1 & 0 \\ 0 & 0 & 1 \end{bmatrix}$$

$$x_{11} = 1$$
$$2x_{11} + x_{21} = 0 \Rightarrow x_{21} = -2, \ x_{22} = 1$$
$$5x_{11} + 3x_{21} + x_{31} = 0 \Rightarrow x_{31} = 0$$
$$3x_{22} + x_{32} = 0 \Rightarrow x_{32} = -3$$
$$x_{33} = 1$$

$$\therefore \quad L^{-1} = \begin{bmatrix} 1 & 0 & 0 \\ -2 & 1 & 0 \\ 0 & -3 & 1 \end{bmatrix}$$

$$\therefore \quad (L^{-1})' = \begin{bmatrix} 1 & -2 & 0 \\ 0 & 1 & -3 \\ 0 & 0 & 1 \end{bmatrix}$$

Thus we obtain

$$A^{-1} = (L')^{-1}(L^{-1}) = (L^{-1})'(L^{-1})$$

EXAMPLE 2. *Find the inverse of the matrix A given below by Choleski's Method*

$$A = \begin{bmatrix} 1 & 1 & 1 \\ 1 & 2 & 3 \\ 1 & 3 & 6 \end{bmatrix}$$

SOLUTION. Let us put $\quad A = LL'$

$$\therefore \quad \begin{bmatrix} 1 & 1 & 1 \\ 1 & 2 & 3 \\ 1 & 3 & 6 \end{bmatrix} = \begin{bmatrix} l_{11} & 0 & 0 \\ l_{21} & l_{22} & 0 \\ l_{31} & l_{32} & l_{33} \end{bmatrix} \begin{bmatrix} l_{11} & l_{21} & l_{31} \\ 0 & l_{22} & l_{32} \\ 0 & 0 & l_{33} \end{bmatrix}$$

$\therefore \qquad l_{11}^2 = 1 \Rightarrow l_{11} = 1$

$\qquad l_{11}l_{21} = 1 \Rightarrow l_{21} = 1, \; l_{11}l_{31} = 1 \Rightarrow l_{31} = 1,$

$\qquad l_{21}^2 + l_{22}^2 = 2 \Rightarrow l_{22} = 1$

$\qquad l_{21}l_{31} + l_{22}l_{32} = 3 \Rightarrow l_{32} = 2$

$\qquad l_{31}^2 + l_{32}^2 + l_{33}^2 = 6 \Rightarrow l_{33}^2 = 6 - 5 = 1 \quad \text{or} \quad l_{33} = 1.$

Thus we obtain

$$L = \begin{bmatrix} 1 & 0 & 0 \\ 1 & 1 & 0 \\ 1 & 2 & 1 \end{bmatrix}$$

Determination of L^{-1} : Let $L^{-1} = X$, then $LX = 1$

$$\therefore \quad \begin{bmatrix} 1 & 0 & 0 \\ 1 & 1 & 0 \\ 1 & 2 & 1 \end{bmatrix} \begin{bmatrix} x_{11} & 0 & 0 \\ x_{21} & x_{22} & 0 \\ x_{31} & x_{32} & x_{33} \end{bmatrix} = \begin{bmatrix} 1 & 0 & 0 \\ 0 & 1 & 0 \\ 0 & 0 & 1 \end{bmatrix}$$

$\qquad x_{11} = 1$

$\qquad x_{11} + x_{21} = 0 \Rightarrow x_{21} = -1, \; x_{22} = 1$

$x_{11} + 2x_{21} + x_{31} = 0 \Rightarrow x_{31} = 1$

$\qquad 2x_{22} + x_{32} = 0 \Rightarrow x_{32} = -2$

and $\qquad x_{33} = 1$

Thus, $\qquad L^{-1} = \begin{bmatrix} 1 & 0 & 0 \\ -1 & 1 & 0 \\ 1 & -2 & 1 \end{bmatrix}$

Consequently,

$$A^{-1} = (LL')^{-1} = (L')^{-1}(L^{-1}) = (L^{-1})'(L^{-1})$$

$$= \begin{bmatrix} 1 & -1 & 1 \\ 0 & 1 & -2 \\ 1 & 0 & 1 \end{bmatrix} \begin{bmatrix} 1 & 0 & 0 \\ -1 & 1 & 0 \\ 1 & -2 & 1 \end{bmatrix} = \begin{bmatrix} 3 & -3 & 1 \\ -3 & 5 & -2 \\ 1 & -2 & 1 \end{bmatrix}.$$

EXAMPLE 3. *By Choleski's method, find the inverse of*

$$A = \begin{bmatrix} 1 & 2 & 2 \\ 2 & 1 & 2 \\ 2 & 2 & 1 \end{bmatrix}$$

SOLUTION. Let us write $\quad A = LL'$, where L is a lower triangular matrix.

$$\therefore \quad \begin{bmatrix} 1 & 2 & 2 \\ 2 & 1 & 2 \\ 2 & 2 & 1 \end{bmatrix} = \begin{bmatrix} l_{11} & 0 & 0 \\ l_{21} & l_{22} & 0 \\ l_{31} & l_{32} & l_{33} \end{bmatrix} \begin{bmatrix} l_{11} & l_{21} & l_{31} \\ 0 & l_{22} & l_{32} \\ 0 & 0 & l_{33} \end{bmatrix}$$

From above, we get

$$\therefore \quad l_{11}^2 = 1 \Rightarrow l_{11} = 1$$
$$l_{11}l_{21} = 2 \Rightarrow l_{21} = 2,\ l_{11}l_{31} = 2 \Rightarrow l_{31} = 2,$$
$$l_{21}^2 + l_{22}^2 = 1 \Rightarrow l_{22} = \sqrt{1-4}$$
$$= \sqrt{-3} = i\sqrt{3} \qquad (\because i^2 = -1)$$
$$l_{21}l_{31} + l_{22}l_{32} = 2 \Rightarrow l_{32} = -\frac{2}{i\sqrt{3}} = \frac{2i}{\sqrt{3}}$$
$$l_{31}^2 + l_{32}^2 + l_{33}^2 = 1 \Rightarrow l_{33} = \sqrt{1-4+\frac{4}{3}} = \sqrt{-\frac{5}{3}} = i\sqrt{\frac{5}{3}}.$$

Thus we obtain

$$L = \begin{bmatrix} 1 & 0 & 0 \\ 2 & i\sqrt{3} & 0 \\ 2 & 2i/\sqrt{3} & i\sqrt{5/3} \end{bmatrix}$$

Determination of L^{-1} :

Let
$$L^{-1} = X = \begin{bmatrix} x_{11} & 0 & 0 \\ x_{21} & x_{22} & 0 \\ x_{31} & x_{32} & x_{33} \end{bmatrix}$$

Then $\quad LX = I$

$$\therefore \begin{bmatrix} 1 & 0 & 0 \\ 2 & i\sqrt{3} & 0 \\ 2 & 2i/\sqrt{3} & i\sqrt{5/3} \end{bmatrix} \begin{bmatrix} x_{11} & 0 & 0 \\ x_{21} & x_{22} & 0 \\ x_{31} & x_{32} & x_{33} \end{bmatrix} = \begin{bmatrix} 1 & 0 & 0 \\ 0 & 1 & 0 \\ 0 & 0 & 1 \end{bmatrix}$$

$$x_{11} = 1$$
$$2x_{11} + (i\sqrt{3})x_{21} = 0 \Rightarrow x_{21} = -\frac{2}{i\sqrt{3}} = \frac{2i}{\sqrt{3}}$$
$$(i\sqrt{3})x_{22} = 1 \Rightarrow x_{22} = -\frac{i}{\sqrt{3}}$$
$$2x_{11} + (2i/\sqrt{3})x_{21} + (i\sqrt{5/3})x_{31} = 0 \Rightarrow x_{31} = \frac{2i}{\sqrt{15}}$$
$$(2i/\sqrt{3})x_{22} + (i\sqrt{5/3})x_{32} = 0 \Rightarrow x_{32} = 2i/\sqrt{15}$$
$$(i\sqrt{5/3})x_{33} = 1 \Rightarrow x_{33} = -i\sqrt{3/5}$$

Thus we obtain

$$L^{-1} = \begin{bmatrix} 1 & 0 & 0 \\ 2i/\sqrt{3} & -1/\sqrt{3} & 0 \\ 2i/\sqrt{15} & 2i/\sqrt{15} & -i\sqrt{3/5} \end{bmatrix}$$

Hence $\quad A^{-1} = (LL')^{-1} = (L')^{-1}(L^{-1}) = (L^{-1})'(L^{-1})$

$$= \begin{bmatrix} 1 & 2i/\sqrt{3} & 2i/\sqrt{15} \\ 0 & -1/\sqrt{3} & 2i/\sqrt{15} \\ 0 & 0 & -i/\sqrt{3/5} \end{bmatrix} \begin{bmatrix} 1 & 0 & 0 \\ 2i/\sqrt{3} & -1/\sqrt{3} & 0 \\ 2i/\sqrt{15} & 2i/\sqrt{15} & -i\sqrt{3/5} \end{bmatrix}$$

$$= \begin{bmatrix} -3/5 & 2/5 & 2/5 \\ 2/5 & -3/5 & 2/5 \\ 2/5 & 2/5 & -3/5 \end{bmatrix}$$

EXAMPLE 4. *Find the inverse of the given matrix by Choleski's method*

$$A = \begin{bmatrix} 1 & 2 & 3 \\ 2 & 8 & 5 \\ 3 & 5 & 6 \end{bmatrix}.$$

SOLUTION. Let us write $A = LL'$, where L is a lower triangular matrix given by

$$L = \begin{bmatrix} l_{11} & 0 & 0 \\ l_{21} & l_{22} & 0 \\ l_{31} & l_{32} & l_{33} \end{bmatrix}$$

$$\therefore \quad \begin{bmatrix} 1 & 2 & 3 \\ 2 & 8 & 5 \\ 3 & 5 & 6 \end{bmatrix} = \begin{bmatrix} l_{11} & 0 & 0 \\ l_{21} & l_{22} & 0 \\ l_{31} & l_{32} & l_{33} \end{bmatrix}\begin{bmatrix} l_{11} & l_{21} & l_{31} \\ 0 & l_{22} & l_{32} \\ 0 & 0 & l_{33} \end{bmatrix}$$

From above matrix equation, we get

$$\therefore \quad l_{11}^2 = 1 \Rightarrow l_{11} = 1, \; l_{11}l_{21} = 2 \Rightarrow l_{21} = 2$$

$$l_{11}l_{31} = 3 \Rightarrow l_{31} = 3,$$

$$l_{21}^2 + l_{22}^2 = 8 \Rightarrow l_{22} = \sqrt{8-4} = \sqrt{4} = 2$$

$$l_{21}l_{31} + l_{22}l_{32} = 5 \Rightarrow l_{32} = \frac{(5-6)}{2} = -\frac{1}{2}$$

$$l_{31}^2 + l_{32}^2 + l_{33}^2 = 6 \Rightarrow l_{33} = \sqrt{6-9-\frac{1}{4}} = \sqrt{-\frac{13}{4}} = \frac{i\sqrt{13}}{2}.$$

Hence, $\quad L = \begin{bmatrix} 1 & 0 & 0 \\ 2 & 2 & 0 \\ 3 & -1/2 & i\sqrt{13}/2 \end{bmatrix}$

Determination of L^{-1} :

Let $\quad L^{-1} = X$

Then $\quad LX = I$

$$\therefore \quad \begin{bmatrix} 1 & 0 & 0 \\ 2 & 2 & 0 \\ 3 & -1/2 & i\sqrt{13}/2 \end{bmatrix}\begin{bmatrix} x_{11} & 0 & 0 \\ x_{21} & x_{22} & 0 \\ x_{31} & x_{32} & x_{33} \end{bmatrix} = \begin{bmatrix} 1 & 0 & 0 \\ 0 & 1 & 0 \\ 0 & 0 & 1 \end{bmatrix}$$

This gives $\quad x_{11} = 1$

$$2x_{11} + 2x_{21} = 0 \Rightarrow x_{21} = -1, \; 2x_{22} = 1 \Rightarrow x_{22} = \frac{1}{2}$$

$$3x_{11} - \frac{1}{2}x_{21} + \frac{i\sqrt{13}}{2}x_{31} = 0 \Rightarrow x_{31} = \frac{7i}{\sqrt{13}}$$

$$-\frac{1}{2}x_{22} + \frac{i\sqrt{13}}{2}x_{32} = 0 \Rightarrow x_{32} = -\frac{i}{2\sqrt{13}}$$

$$\frac{i\sqrt{13}}{2}x_{33} = 1 \Rightarrow x_{33} = -\frac{2i}{\sqrt{13}}.$$

Therefore, we get

$$L^{-1} = X = \begin{bmatrix} 1 & 0 & 0 \\ -1 & 1/2 & 0 \\ 7i/\sqrt{13} & -i/2.\sqrt{13} & -2i\sqrt{13} \end{bmatrix}$$

Hence we obtain

$$A^{-1} = (LL')^{-1} = (L')^{-1}(L^{-1}) = (L^{-1})'(L^{-1})$$

$$= \begin{bmatrix} 1 & -1 & 7i/\sqrt{13} \\ 0 & 1/2 & -i/2\sqrt{13} \\ 0 & 0 & -2i/\sqrt{13} \end{bmatrix} \begin{bmatrix} 1 & 0 & 0 \\ -1 & 1/2 & 0 \\ 7i/\sqrt{13} & -i/2.\sqrt{13} & -2i\sqrt{13} \end{bmatrix}$$

$$\therefore \qquad A^{-1} = \begin{bmatrix} -23/13 & -3/13 & 14/13 \\ -3/13 & 3/13 & -1/13 \\ 14/13 & -1/13 & -4/13 \end{bmatrix}$$

EXERCISE 7.1

1. Find the inverse of the following matrices using Gauss-elimination method :

(a) $\begin{bmatrix} 2 & 1 & 2 \\ 2 & 2 & 1 \\ 1 & 2 & 2 \end{bmatrix}$ (b) $\begin{bmatrix} 10 & 2 & 1 \\ 2 & 20 & -2 \\ -2 & 3 & 10 \end{bmatrix}$

(c) $\begin{bmatrix} 1 & -1 & 1 \\ 1 & -2 & 4 \\ 1 & 2 & 2 \end{bmatrix}$ (d) $\begin{bmatrix} 1 & 4 & 0 \\ -1 & 2 & 2 \\ 0 & 0 & 2 \end{bmatrix}$

2. Find the inverse of the following matrices by Triangularisation method :

(a) $\begin{bmatrix} 10 & 2 & 1 \\ 2 & 20 & -2 \\ -2 & 3 & 10 \end{bmatrix}$ (b) $\begin{bmatrix} 2 & 2 & 3 \\ 1 & -2 & 3 \\ 9 & 1 & -1 \end{bmatrix}$

(c) $\begin{bmatrix} 2 & 5 & 2 \\ 3 & 1 & 2 \\ 1 & 2 & 1 \end{bmatrix}$

3. Find the inverse of the following matrices by Choleski's method :

(a) $\begin{bmatrix} 1 & 3 & 2 \\ 3 & 4 & 7 \\ 2 & 7 & 6 \end{bmatrix}$ (b) $\begin{bmatrix} 1 & 3 & -5 \\ 3 & 25 & 9 \\ -5 & 9 & 63.25 \end{bmatrix}$

(c) $\begin{bmatrix} 1 & 1/3 & 1/5 \\ 1/3 & 1/5 & 1/7 \\ 1/3 & 1/7 & 1/9 \end{bmatrix}$ (Agra-2004)

(d) $\begin{bmatrix} 0 & 1 & 1 \\ 1 & 0 & 1 \\ 1 & 1 & 0 \end{bmatrix}$ (e) $\begin{bmatrix} 2 & 2 & 2 \\ 2 & 5 & 5 \\ 2 & 5 & 11 \end{bmatrix}$

(f) $\begin{bmatrix} 3 & -2 & 1 & 0 \\ -2 & 5 & -1 & 1 \\ 1 & -1 & -11 & -2 \\ 0 & 1 & -2 & 12 \end{bmatrix}$

4. Using Cayley-Hamilton theorem, find the inverse of the matrix $A = \begin{bmatrix} 0 & 1 & 2 \\ 0 & -2 & 0 \\ 0 & 0 & 3 \end{bmatrix}$

(UPTU–2006)

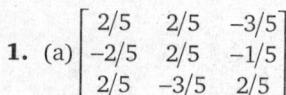

Answers

1. (a) $\begin{bmatrix} 2/5 & 2/5 & -3/5 \\ -2/5 & 2/5 & -1/5 \\ 2/5 & -3/5 & 2/5 \end{bmatrix}$ (b) $\dfrac{1}{2078}\begin{bmatrix} 206 & -17 & -24 \\ -16 & 102 & 22 \\ 46 & 44 & 196 \end{bmatrix}$ (c) $\begin{bmatrix} 1.2 & -0.4 & 0.2 \\ -0.2 & -0.1 & 0.3 \\ -0.4 & 0.3 & 0.1 \end{bmatrix}$

(d) $\begin{bmatrix} 2/6 & -4/6 & 4/6 \\ 1/6 & 1/6 & -1/6 \\ 0 & 0 & 3/6 \end{bmatrix}$ **2.** (a) $\dfrac{1}{2078}\begin{bmatrix} 206 & -17 & -24 \\ -16 & 102 & 22 \\ 46 & 34 & 196 \end{bmatrix}$ (b) $\dfrac{1}{111}\begin{bmatrix} -1 & 5 & 12 \\ 28 & -29 & -3 \\ 19 & 16 & -6 \end{bmatrix}$

(c) $\dfrac{1}{13}\begin{bmatrix} -41 & 13 & -104 \\ -7 & 1 & -26 \\ 65 & -13 & 169 \end{bmatrix}$ **3.** (a) $\dfrac{1}{11}\begin{bmatrix} 25 & 4 & -13 \\ 4 & -2 & 1 \\ -13 & 1 & 5 \end{bmatrix}$ (b) $\dfrac{1}{144}\begin{bmatrix} 6001 & -939 & 6089 \\ -939 & 153 & -96 \\ 608 & -96 & 64 \end{bmatrix}$

(c) $\dfrac{15}{64}\begin{bmatrix} 15 & -70 & 63 \\ -70 & 588 & -630 \\ 63 & -630 & 735 \end{bmatrix}$ (d) $\begin{bmatrix} -1/2 & 1/2 & 1/2 \\ 1/2 & -1/2 & 1/3 \\ 1/2 & 1/2 & -1/2 \end{bmatrix}$ (e) $\dfrac{1}{6}\begin{bmatrix} 5 & -2 & 0 \\ -2 & 3 & -1 \\ 0 & -1 & 1 \end{bmatrix}$

(f) $\dfrac{1}{1506}\begin{bmatrix} 677 & 282 & 39 & -17 \\ 282 & 396 & -6 & -36 \\ 39 & -6 & -129 & -25 \\ -17 & -36 & -25 & 125 \end{bmatrix}$

4. $\begin{bmatrix} 1 & 1/2 & -2/3 \\ 0 & -1/2 & 0 \\ 0 & 0 & 1/3 \end{bmatrix}$

7.6 ESCALATER METHOD

(Agra-2003, 07, 09)

We know that if the inverse of matrix of order n is given, then we can find the inverse of the matrix of order $(n + 1) \times (n + 1)$.

Let the inverse of a $n \times n$ matrix say A_1, be given. We want to find the inverse of $(n + 1) \times (n + 1)$ matrix say A, which is obtained from A_1 by adding $(n + 1)^{th}$ row and $(n + 1)^{th}$ column to A_1.

Now, let $\qquad A = \begin{bmatrix} A_1 & A_2 \\ A_3^T & k \end{bmatrix}$ and $\qquad A^{-1} = \begin{bmatrix} B_1 & B_2 \\ B_3^T & x \end{bmatrix}$

where A_2, B_2 are column vectors A^T_3, B^T_3 are row vectors, k and x are scalars and B_1 is $n \times n$ matrix.

Then, we have
$$AA^{-1} = I_{n+1} = \begin{bmatrix} I_n & \cdots & Q_{n\times 1} \\ \vdots & & \vdots \\ 0_{1\times n} & \cdots & 1 \end{bmatrix}$$

$$= \begin{bmatrix} A_1 & A_2 \\ A_3^T & k \end{bmatrix}\begin{bmatrix} B_1 & B_2 \\ B_3^T & x \end{bmatrix}$$

On solving, we get
$$A_1 B_1 + A_2 B_3^T = I \qquad \qquad \text{...(1)}$$
$$A_2 B_2 + A_2 x = 0 \qquad \qquad \text{...(2)}$$
$$A_3^T B_1 + k B_3^T = 0 \qquad \qquad \text{...(3)}$$
$$A_3^T B_2 + kx = 1 \qquad \qquad \text{...(4)}$$

Now (2) implies
$$B_2 = -A_1^{-1} A_2 x$$
$\Rightarrow \qquad [k - A_3^T A_1^{-1} A_2]x = 1 \qquad \qquad$ [by (4)] $\qquad$ ...(5)

Therefore, we can find x and B_2. Also, from (1), we get
$$B_1 = A_1^{-1}(I - A_2 B_3^T)$$
$$(k - A_3^T A_1^{-1} A_2)B_3^T = -A_3^T A_1^{-1}$$

From (6), we can obtain B_3^T.

Hence, we can find B_1 from (1) and hence A^{-1} can be obtained.

SOLVED EXAMPLES

EXAMPLE 1. *Find the inverse of* $A = \begin{bmatrix} 1 & 3 & 3 & 2 \\ 1 & 4 & 3 & 4 \\ 1 & 3 & 4 & 5 \\ 2 & 5 & 3 & 2 \end{bmatrix}$, *given that if* (Agra-2004, 05, 08, 09)

$$A_1 = \begin{bmatrix} 1 & 3 & 3 \\ 1 & 4 & 3 \\ 1 & 3 & 4 \end{bmatrix}, \textit{ then } A^{-1} = \begin{bmatrix} 7 & -3 & -3 \\ -1 & 1 & 0 \\ -1 & 0 & 1 \end{bmatrix},$$

SOLUTION.   The given matrix A, can be written as

$$A = \begin{bmatrix} 1 & 3 & 3 & : & 2 \\ 1 & 4 & 3 & : & 4 \\ 1 & 3 & 4 & : & 5 \\ \hline 2 & 5 & 3 & \cdots & 2 \end{bmatrix} = \begin{bmatrix} A_1 & : & A_2 \\ \cdots & & \cdots \\ A_3^T & : & k \end{bmatrix}$$

where $A_2 = (2\ 4\ 5)^T$

$A_3^T = (2\ 5\ 3)$

$k = 2$

Now let its inverse be $A^{-1} = \begin{bmatrix} B_1 & : & B_2 \\ \hline B_3^T & : & x \end{bmatrix}$

which implies

$$A_3^T A_1^{-1} A_2 = [2\ \ 5\ \ 3] = \begin{bmatrix} 7 & -3 & -3 \\ -1 & 1 & 0 \\ -1 & 0 & 1 \end{bmatrix}\begin{bmatrix} 2 \\ 4 \\ 5 \end{bmatrix} = -7$$

Also, $(5) \Rightarrow [2 - (-7)]\, x = 1 \Rightarrow x = \dfrac{1}{9}$

$(6) \Rightarrow 9B_3^T = -A_3^T A_1^{-1} = -[6\ \ -1\ \ -3]$

$\Rightarrow B_3^T = -\dfrac{1}{9}[6\ \ -1\ \ -1]$

and $B_2 = -A_1^{-1} A_2 x = -(1/9)(-13\ \ 2\ \ 3)^T$ [using(2)]

Now using (1), we have

$$B_1 = A_1^{-1}(I - A_2 B_3^T) = A_1^{-1} - A_1^{-1} A_2 A_3^T$$

$$= \begin{bmatrix} 7 & -3 & -3 \\ -1 & 1 & 0 \\ -1 & 0 & 1 \end{bmatrix} + \begin{bmatrix} -13 \\ 2 \\ 3 \end{bmatrix}\begin{bmatrix} \dfrac{2}{3} & -\dfrac{1}{9} & -\dfrac{1}{3} \end{bmatrix}$$

$$= \begin{bmatrix} -7 & 3 & 3 \\ 1 & -1 & 0 \\ 1 & 0 & -1 \end{bmatrix} + \begin{bmatrix} -\dfrac{26}{9} & \dfrac{13}{9} & \dfrac{13}{3} \\ \dfrac{4}{3} & -\dfrac{2}{9} & -\dfrac{2}{3} \\ 2 & -\dfrac{1}{3} & -1 \end{bmatrix} = \begin{bmatrix} -\dfrac{5}{3} & -\dfrac{14}{9} & \dfrac{4}{3} \\ \dfrac{1}{3} & \dfrac{7}{9} & -\dfrac{2}{3} \\ 1 & -\dfrac{1}{3} & 3 \end{bmatrix}$$

Hence, we get

$$A^{-1} = \begin{bmatrix} -5/3 & -14/9 & 4/3 & : & 13/9 \\ 1/3 & 7/9 & -2/2 & : & -2/9 \\ 1 & -1/3 & 0 & : & -1/3 \\ \hline -2/3 & -1/3 & 1/3 & : & 1/9 \end{bmatrix}$$

EXAMPLE 2.   *Using escalator method, find the inverse of the given matrix*

$$A = \begin{bmatrix} 13 & 14 & 6 & 4 \\ 8 & -1 & 13 & 9 \\ 6 & 7 & 3 & 2 \\ 9 & 5 & 16 & 11 \end{bmatrix}.$$ (Agra-2003, Meerut-2003)

SOLUTION. The given matrix can be written as

$$A = \begin{bmatrix} 13 & 14 & 6 & : & 4 \\ 8 & -1 & 13 & : & 9 \\ 6 & 7 & 3 & : & 2 \\ \cdots & \cdots & \cdots & \cdots & \cdots \\ 9 & 5 & 16 & : & 11 \end{bmatrix} A = \begin{bmatrix} A_1 & | & A_2 \\ A_3^T & | & k \end{bmatrix}$$

where $A_1 = \begin{bmatrix} 13 & 14 & 6 \\ 8 & -1 & 13 \\ 6 & 7 & 3 \end{bmatrix}, A_2 = \begin{bmatrix} 4 \\ 9 \\ 2 \end{bmatrix}$

$$A_3^T = [9 \quad 5 \quad 16]$$

and $k = 11$

Proceeding as usual, we can find

$$A_1^{-1} = \frac{1}{94} \begin{bmatrix} 94 & 0 & -188 \\ -54 & -3 & 121 \\ -62 & 7 & 125 \end{bmatrix}$$

Then, we can write

$$A_1^{-1} = \begin{bmatrix} B_1 & : & B_2 \\ B_3^T & : & x \end{bmatrix} \text{ such that } AA^{-1} = I$$

Pull all these values in the following equations,

$$A_1 B_1 + A_2 B_3^T = I \qquad \qquad \ldots(1)$$

$$A_1 B_2 + A_2 x = 0 \qquad \qquad \ldots(2)$$

$$A_3^T B_1 + k B_3^T = 0 \qquad \qquad \ldots(3)$$

$$A_3^T B_2 + kx = 1 \qquad \qquad \ldots(4)$$

We get $A_1 B_2 + A_2 x = 0$

$$\Rightarrow \qquad B_2 = -A_1^{-1} A_2 x,$$

$$= \frac{1}{94} \begin{bmatrix} 94 & 0 & -188 \\ -54 & -3 & 121 \\ -62 & 7 & 125 \end{bmatrix} \begin{bmatrix} 4 \\ 9 \\ 4 \end{bmatrix} x = -\frac{1}{94} \begin{bmatrix} 0 \\ -1 \\ 65 \end{bmatrix} x$$

Now $A_3^T B_2 + kx = 1$ gives

$$(9 \quad 5 \quad 16) \left(-\frac{1}{94}\right) \begin{pmatrix} 0 \\ -1 \\ 65 \end{pmatrix} x + 11x = 1$$

$$\Rightarrow \quad -\frac{1}{94}(9.0 - 5 + 1040)x + 11x = 1$$

$$\Rightarrow \quad -\frac{1}{94}(1035)x + 11x = 1, \text{ i.e., } -1035x + 11 \times 94x = 94$$

$$\Rightarrow \quad -1035x + 1034x = 94 \Rightarrow x = -94$$

Therefore, $B_2 = \frac{-94}{-94} \begin{bmatrix} 0 \\ -1 \\ 65 \end{bmatrix} = \begin{bmatrix} 0 \\ -1 \\ 65 \end{bmatrix}$

Also, $B_1 = A_1^{-1}[I - A_2 B_3^T]$

Put this value in (3), we get

$$A_3^T A_1^{-1}(I - A_2 B_3^T) + k B_3^T = 0$$

$$\Rightarrow \quad A_3^T A_1^{-1} - A_3^T A_1^{-1} A_2 B_3^T + k B_3^T = 0$$

$$\Rightarrow A_3^T A_1^{-1} = [A_3^T A_1^{-1} A_2 - k] B_3^T$$

$$= \frac{1}{94}(-1664 + 873 + 1826) - 11 = \frac{1}{94}$$

Hence, from (5), we get

$$B_3^T = \frac{A_3^T A_1^{-1}}{(A_3^T A_1^{-1} A_2 - k)} = \frac{(1/94)(-416 \quad 97 \quad 913)}{1/94}$$

$$= (-416 \quad 97 \quad 913)$$

Also $B_1 = A_1^{-1}(I - A_2 B_3^T)$

$$= A_1^{-1} - A_1^{-1} A_2 B_3^T$$

$$= \frac{1}{94}\begin{bmatrix} 94 & 0 & -188 \\ -54 & -3 & 121 \\ -62 & 7 & 125 \end{bmatrix} - \frac{1}{94}\begin{bmatrix} 94 & 0 & -188 \\ -54 & -3 & 121 \\ -62 & 7 & 125 \end{bmatrix} \times \begin{bmatrix} 4 \\ 9 \\ 2 \end{bmatrix}[-416 \quad 97 \quad 913]$$

$$= \frac{1}{94}\begin{bmatrix} 94 & 0 & -188 \\ -54 & -3 & 121 \\ -62 & 7 & 125 \end{bmatrix} - \frac{1}{94}\begin{bmatrix} 94 & 0 & -188 \\ -54 & -3 & 121 \\ -62 & 7 & 125 \end{bmatrix} \times \begin{bmatrix} -1664 & 388 & 3652 \\ -3744 & 873 & 821 \\ -832 & 194 & 1826 \end{bmatrix}$$

$$= \frac{94}{94}\begin{bmatrix} 1 & 0 & -2 \\ -5 & 1 & 11 \\ 287 & -67 & -630 \end{bmatrix} = \begin{bmatrix} 1 & 0 & -2 \\ -5 & 1 & 11 \\ 287 & -67 & -630 \end{bmatrix}$$

Hence, the inverse of the given matrix is given by

$$A^{-1} = \begin{bmatrix} 1 & 0 & -2 & : & 0 \\ -5 & 1 & 11 & : & -1 \\ 287 & -67 & -630 & : & 65 \\ \cdots & \cdots & \cdots & \cdots & \cdots \\ -416 & 97 & 213 & : & -94 \end{bmatrix}$$

7.7 INVERSION OF COMPLEX MATRICES

(Agra-2004)

Let $A + iB$ be the given complex matrix, in which at least one of the matrices A and B is non-singular.

Let us suppose, inverse of matrix $(A + iB)$ be $(X + iY)$.

Therefore, by definition of inverse we have

$$(A + iB)(X + iY) = I$$

Which implies

$$(AX - BY) + i(BX + AY) = I + iO$$

Therefore, $$AX - BY = I \qquad \qquad \dots(1)$$

and $$AY + BX = O \qquad \qquad \dots(2)$$

From (1) we may get

$$B^{-1}AX - B^{-1}BY = B^{-1}I$$

$$B^{-1}AX - Y = B^{-1}I = B^{-1} \qquad \qquad \dots(3)$$

Also, $AY + BX = 0 \Rightarrow A^{-1}AY + A^{-1}BX = 0 \Rightarrow Y = -A^{-1}BX$ ...(4)

Adding (3) and (4), we get

$$B^{-1}AX = B^{-1} - A^{-1}BX$$

$\Rightarrow \qquad B^{-1}AX + A^{-1}BX = B^{-1}$

$\Rightarrow \qquad (B^{-1}A + A^{-1}B)X = B^{-1}$

From (5), we obtained

$$(BB^{-1}A + BA^{-1}B)X = I \Rightarrow (A + BA^{-1}B)X = I$$

$\Rightarrow \qquad\qquad X = (A + BA^{-1}B)^{-1}I$

$$= A^{-1} + (BA^{-1}B)^{-1}$$

$$= A^{-1} + B^{-1}AB^{-1} \qquad\qquad ...(6)$$

Similarly, when B^{-1} exists, we can find

$$X = B^{-1}A(AB^{-1}A + B)^{-1}$$

and $\qquad\qquad Y = -(AB^{-1}A + B)^{-1}$

and

If A and B both are regular, the two expressions for X and Y given by (6) and (7) are identical.

$\therefore \qquad B^{-1}A(AB^{-1}A + B)^{-1} = (A^{-1}B^{-1}) \cdot (AB^{-1}A + B)^{-1}$

$$= [(AB^{-1}A + B)(A^{-1}B)]^{-1} = [A + BA^{-1}B]^{-1}$$

If both A and B are singular matrices and $A + iB$ is regular, then we can find the inverse of $A + iB$ as follows :

Let $\quad F = A + rB \quad$ and $\quad G = B - rA$, where r is real.

Then there must exist a number r such that F becomes regular to prove this, let us write

$$\det.F = \det. (A + rB)$$

$$= f(r), \text{ where } f(r) \text{ is a polynomial of degree } n$$

Let $f(r) = 0$ for all r, then we have $f(r) = 0$ against the assumption that $F = A + rB$ is regular.

Now, $\qquad\qquad F + iG = (A + rB) + i(B - rA)$

$$= (A + iB) - ir(A + iB) = (1 - ir)(A + iB)$$

$\Rightarrow \qquad\qquad (F + iG)^{-1} = (1 - ir)^{-1}(A + iB)^{-1}$

$\Rightarrow \qquad\qquad (A + iB)^{-1} = (1 - ir)(F + iG)^{-1}$

SOLVED EXAMPLES

EXAMPLE 1. *Evaluate the inverse of the following matrix*

$$M = \begin{pmatrix} 5+i & 4+2i \\ 10+3i & 8+6i \end{pmatrix} \qquad \text{(Agra-2004, 06, 09, Meerut-2002)}$$

SOLUTION. The given matrix can be written as

$$M = A + iB \begin{pmatrix} 5+i & 4+2i \\ 10+3i & 8+6i \end{pmatrix} = \begin{pmatrix} 5 & 4 \\ 10 & 8 \end{pmatrix} + i\begin{pmatrix} 1 & 2 \\ 3 & 6 \end{pmatrix} \qquad ...(1)$$

Clearly det $A = 0$ and det $B = 0$

$\Rightarrow$ Both the matrices are singular.

Consider $F = A + rB$ and $G = B - rA$, where r is real

Now
$$F + iG = (1 - ir)(A + iB)$$
$$(A + iB)^{-1} = (1 - ir)(F + iG)^{-1} \qquad \qquad ...(2)$$

Also,
$$F = \begin{pmatrix} 5 & 4 \\ 10 & 8 \end{pmatrix} + r\begin{pmatrix} 1 & 2 \\ 3 & 6 \end{pmatrix}$$

$$= \begin{pmatrix} 5 & 4 \\ 10 & 8 \end{pmatrix} + \begin{pmatrix} r & 2r \\ 3r & 6r \end{pmatrix} = \begin{pmatrix} 5+r & 4+2r \\ 10+3r & 8+6r \end{pmatrix}$$

and
$$G = B - rA = \begin{pmatrix} 1 & 2 \\ 3 & 6 \end{pmatrix} - r\begin{pmatrix} 5 & 4 \\ 10 & 8 \end{pmatrix} = \begin{pmatrix} 1 & 2 \\ 3 & 6 \end{pmatrix} - \begin{pmatrix} 5r & 4r \\ 10r & 8r \end{pmatrix}$$

$$= \begin{pmatrix} 1-5r & 2-4r \\ 3-10r & 6-8r \end{pmatrix}$$

Let us choose $r = 1$, therefore, we get

$$F = \begin{pmatrix} 6 & 6 \\ 13 & 14 \end{pmatrix} \text{ and } G = \begin{pmatrix} -4 & -2 \\ -7 & -2 \end{pmatrix}$$

Clearly, det. $F \neq 0$ and det $G \neq 0$

$\Rightarrow$ F and G both are non-singular matrices.

Now, we want to find the inverse of $F + iG$.

Suppose inverse of $G + iG = X + iY$

i.e., $(F + iG)^{-1} = X + iY$

Proceed same as article, we can find

$$X = (F + GF^{-1}G)^{-1}$$
and
$$Y = -F^{-1}G(F + GF^{-1}G)^{-1}$$

$$F = \begin{pmatrix} 6 & 6 \\ 13 & 14 \end{pmatrix} \Rightarrow F^{-1} = \begin{pmatrix} 7/3 & -1 \\ -13/6 & 1 \end{pmatrix}$$

Therefore
$$F^{-1}G = \begin{pmatrix} 7/3 & -1 \\ -13/6 & 1 \end{pmatrix}\begin{pmatrix} -4 & -2 \\ -7 & -2 \end{pmatrix} = \begin{pmatrix} -7/3 & -8/3 \\ 5/3 & 7/3 \end{pmatrix}$$

and
$$GF^{-1}G = \begin{pmatrix} -4 & -2 \\ -7 & -2 \end{pmatrix} = \begin{pmatrix} -7/3 & -8/3 \\ 5/3 & 7/3 \end{pmatrix} = \begin{pmatrix} 6 & 6 \\ 13 & 14 \end{pmatrix}$$

$\Rightarrow$
$$F + GF^{-1}G = \begin{pmatrix} 6 & 6 \\ 13 & 14 \end{pmatrix} + \begin{pmatrix} 6 & 6 \\ 13 & 14 \end{pmatrix} = \begin{pmatrix} 12 & 12 \\ 26 & 28 \end{pmatrix}$$

$\Rightarrow$
$$[F + GF^{-1}G]^{-1} = \text{inverse of } \begin{pmatrix} 12 & 12 \\ 26 & 28 \end{pmatrix} = \begin{pmatrix} 7/6 & -1/2 \\ -13/12 & 1/2 \end{pmatrix} = X$$

Also, $(F^{-1}G)(F + GF^{-1}G)^{-1} = \begin{pmatrix} -7/3 & -8/3 \\ 5/3 & 7/3 \end{pmatrix}\begin{pmatrix} 7/6 & -1/2 \\ -13/12 & 1/2 \end{pmatrix}$

$$= \begin{pmatrix} 1/6 & -1/6 \\ -7/12 & 1/3 \end{pmatrix}$$

$\Rightarrow$
$$Y = -(F^{-1}G)(F + GF^{-1}G)^{-1} = \begin{pmatrix} 1/6 & -1/6 \\ -7/12 & 1/3 \end{pmatrix}$$

$\Rightarrow$
$$X + iY = \begin{pmatrix} 7/6 & -1/2 \\ -13/12 & 1/2 \end{pmatrix} + i(-1)\begin{pmatrix} 1/6 & -1/6 \\ -7/12 & 1/3 \end{pmatrix}$$

$$= \frac{1}{6}\begin{pmatrix} 7-i & -3+i \\ -6.5+3.5i & 3-2i \end{pmatrix}$$

Now, $(A + iB)^{-1} = (1 - ir)(F + iG)^{-1} = (1 - i)(F + iG)^{-1}$

$$= (1 - i)(X + iY) = \frac{1}{6}(1 \cdot i)\begin{pmatrix} 7 - i & -3 + i \\ -6.5 + 3.5i & 3 - 2i \end{pmatrix}$$

$$= \frac{1}{6}\begin{pmatrix} 6 - 8i & -2 + 4i \\ -3 + 10i & 1 - 5i \end{pmatrix}$$

which is the required inverse.

EXERCISE 7.2

1. Find the inverse of the given matrix

$$A = \begin{bmatrix} 3 & -1 & 10 & 2 \\ 5 & 1 & 20 & 3 \\ 9 & 7 & 39 & 4 \\ 1 & -2 & 2 & 1 \end{bmatrix}$$

using Escalator method.

2. Find the inverse of the given matrix

$$A = \begin{bmatrix} 13 & 14 & 6 & 4 \\ 8 & -1 & 13 & 9 \\ 6 & 7 & 3 & 2 \\ 9 & 5 & 16 & 11 \end{bmatrix}$$

using the Escalator method.

7.8 EIGEN VALUES AND EIGEN VECTORS

Eigen value problems are arises in several engineering situation and applied Mathematics. Let a problem be denoted by a system of equations in matrix form as

$$Y = AX = \lambda X$$

where A = square matrix of order n

X = a real vector never equal to '0'

λ = eigen values or latent root of matrix A, which is real and positive,

the vector Y is real, has similar sense and direction when the system has an eigen value λ.

7.8.1 CHARACTERISTICS EQUATION

Let $AX = \lambda X$ be the given eigen value problem then

$$(A - \lambda I)X = 0$$

where

$$A = \begin{bmatrix} a_{11} & a_{12} & \cdots & a_{1n} \\ a_{21} & a_{22} & \cdots & a_{2n} \\ \vdots & \vdots & \vdots & \vdots \\ a_{n1} & a_{n2} & \cdots & a_{nn} \end{bmatrix}$$

$$\lambda I = \begin{bmatrix} \lambda & 0 & \cdots & 0 \\ 0 & \lambda & \cdots & 0 \\ \vdots & \vdots & \vdots & \vdots \\ 0 & 0 & \cdots & \lambda \end{bmatrix}$$

Now, $(A - \lambda I)X = 0 \Rightarrow \begin{bmatrix} a_{11} - \lambda & a_{12} & \cdots & a_{1n} \\ a_{21} & a_{22} - \lambda & \cdots & a_{2n} \\ \vdots & \vdots & \vdots & \vdots \\ a_{n1} & a_{n2} & \cdots & a_{nn} - \lambda \end{bmatrix} X = 0$

It will have a non-trivial solution when the coefficient of determinant does not vanishes.

i.e., $\det(A - \lambda I) = \begin{bmatrix} a_{11} - \lambda & a_{12} & \cdots & a_{1n} \\ a_{21} & a_{22} - \lambda & \cdots & a_{2n} \\ \vdots & \vdots & \vdots & \vdots \\ a_{n1} & a_{n2} & \cdots & a_{nn} - \lambda \end{bmatrix} = 0$ $[\because X \neq 0]$

which is known as '**characteristic equation**'.

On expanding, it gives the polynomial equation or characteristic equation in λ such that

$$P_n(\lambda) = (-1)^n \lambda^n + a_1 \lambda^{n-1} + \ldots + a_n = 0$$

where
$$a_1 = -(a_{11} + a_{22} + a_{nn}) \ldots \text{ and } a_n = \det A.$$

It is of degree n, has n roots say $\lambda_1, \lambda_2, \ldots, \lambda_n$ and also $AX_i = \lambda_i X_i$ is known as eigen vector or characteristic vector.

ALGORITHM

Step 1.	First evaluate the characteristic equation by putting $$\det (A - \lambda I) = 0 \qquad \ldots(1)$$
Step 2.	Find the coefficient or co-factors by expansion of equation (1)
Step 3.	Put each value of λ into matrix $(A - \lambda_i)X_i = 0$.
Step 4.	Find the different values of x_i for each λ_i.

7.9 RELATION BETWEEN EIGENVALUES AND EIGENVECTORS

THEOREM 1. *λ is an eigenvalue of a matrix A if and only if there exists a non-zero vector X such that $AX = \lambda X$.*

PROOF. Suppose that λ is an eigenvalue of A, then $|A - \lambda I| = 0$

$\Rightarrow \quad A - \lambda I$ is a singular matrix.

$\therefore$ The matrix equation $(A - \lambda I) X = O$ has a non-zero solution, thus there exists a non-zero vector X such that

$$(A - \lambda I) X = O \text{ or } AX = \lambda X$$

Conversely, Suppose that there is a non-zero vector X such that

$$AX = \lambda X$$

$\Rightarrow \qquad (A - \lambda I)X = O$

Since the matrix equation $(A - \lambda I)X = O$ has a non-zero solution, then the coefficient matrix $A - \lambda I$ is singular, therefore

$$|A - \lambda I| = 0$$

Hence A is an eigenvalue of A.

THEOREM 2. *If X is an eigenvector of a matrix A corresponding to an eigenvalue of A, then kX is also an eigenvector of A corresponding to the same eigenvalue A, where k is any non-zero number.*

PROOF. Since X is an eigenvector of a matrix A corresponding to an eigenvalue of A, then we have

$$AX = \lambda X \qquad \ldots(1)$$

Since $k \neq 0$, then multiplying both sides of (1) by k, we get

$$k(AX) = k(\lambda X)$$

$\Rightarrow \qquad A(kX) = \lambda(kX) \qquad \ldots(2)$

From equation (2), it follows that kX is also an eigenvector of A corresponding to the same eigenvalue λ.

THEOREM 3. *If X is a non-zero eigenvector of a matrix A, then X cannot correspond to more than one eigenvalue of A.*

PROOF. If possible, let X be an eigenvector corresponding to eigenvalues λ_1 and λ_2, then

$$AX = \lambda_1 X \qquad \ldots(1)$$

and
$$AX = \lambda_2 X \qquad \ldots(2)$$

From (1) and (2) we have
$$\lambda_1 X = \lambda_2 X$$
$$\Rightarrow \quad (\lambda_1 - \lambda_2) X = O$$
$$\Rightarrow \quad \lambda_1 - \lambda_2 = 0 \qquad\qquad \because X \neq O$$
$$\Rightarrow \quad \lambda_1 = \lambda_2$$

THEOREM 4. *If X_1 and X_2 be non-zero eigenvectors of a matrix A corresponding to an eigenvalue λ of A, then $k_1 X_1 + k_2 X_2$ is also an eigenvector of A corresponding to eigenvalue λ, where k_1 and k_2 are non-zero number.*

PROOF. Since $X_1 \neq 0, X_2 \neq 0$ and $k_1 \neq 0, k_2 \neq 0$, then $k_1 X_1 + k_2 X_2 \neq O$. Also X_1 and X_2 are eigenvectors of A corresponding to an eigenvalue λ of A, then we have

$$AX_1 = \lambda X_1 \qquad\qquad ...(1)$$
and
$$AX_2 = \lambda X_2 \qquad\qquad ...(2)$$

Multiplying (1) by k_1 and (2) by k_2 and then adding, we get

$$Ak_1 X_1 + Ak_2 X_2 = \lambda(k_1 X_1 + k_2 X_2)$$
$$\Rightarrow \quad Ak_1 X_1 + Ak_2 X_2 = \lambda(k_1 X_1 + k_2 X_2) \qquad\qquad ...(3)$$

As $k_1 X_1 + k_2 X_2 \neq O$ then from (3) it follows that $k_1 X_1 + k_2 X_2$ is also an eigenvector of A corresponding to an eigenvalue of A.

THEOREM 5. *Let A be an $n \times n$ matrix. Then the distinct eigenvectors corresponding to distinct eigenvalues of A are linearly independent.*

PROOF. Since A is an $n \times n$ matrix so it will have atmost n eigenvalues. Let $\lambda_1, \lambda_2, \lambda_3, ..., \lambda_m$ be m distinct eigenvalues of A out of n eigenvalues and let $X_1, X_2, X_3, ..., X_m$ be m distinct eigenvectors corresponding to eigenvalues $\lambda_1, \lambda_2, ..., \lambda_m$ respectively. Then we have

$$AX_1 = \lambda_1 X_1, AX_2 = \lambda_2 X_2, ..., AX_m = \lambda_m X_m$$

Let $S = \{X_1, X_2, ..., X_m\}$. Then we have to show that S is linearly independent. We shall prove it by induction hypothesis on m.

If $m = 1$, then there is only one non-zero vector, which is obviously linearly independent.

Suppose the result is true for $m = k$, i.e., $\{X_1, X_2, ..., X_m\}$ is linearly independent. Let this set be denoted by S_1, then

$$S_1 = \{X_1, X_2, ..., X_k\}$$

Finally we shall prove that the set $S_1 \cup \{X_{k+1}\}$ is linearly independent.

For scalars $a_1, a_2, a_3, ..., a_k, a_{k+1}$ such that

$$a_1 X_1 + a_2 X_2 + ... + a_k X_k + a_{k+1} X_{k+1} = O \qquad\qquad ...(1)$$
$$\Rightarrow \quad A(a_1 X_1 + a_2 X_2 + ... + a_k X_k + a_{k+1} X_{k+1}) = AO = O$$
$$\Rightarrow \quad a_1 AX_1 + a_2 AX_2 + ... + a_k AX_k + a_{k+1} AX_{k+1} = O$$
$$\Rightarrow \quad a_1 \lambda_1 X_1 + a_2 \lambda_2 X_2 + ... + a_k \lambda_k X_k + a_{k+1} \lambda_{k+1} X_{k+1} = O \qquad\qquad ...(2)$$

Multiplying (1) by λ_{k+1} and then substracting it from (2), we get

$$a_1(\lambda_1 - \lambda_{k+1})X_1 + a_2(\lambda_2 - \lambda_{k+1})X_2 + ... + a_k(\lambda_k - \lambda_{k+1})X_k = O$$

Since S_1 is linearly independent, hence

$$\Rightarrow \quad a_1(\lambda_1 - \lambda_{k+1}) = a_2(\lambda_2 - \lambda_{k+1}) = ... = a_k(\lambda_k - \lambda_{k+1}) = 0$$
$$\Rightarrow \quad a_1 = a_2 = ... = a_k = 0 \qquad\qquad [\because \lambda_1, \lambda_2, ..., \lambda_m \text{ are all distinct.}]$$

Putting $a_1 = a_2 = \ldots = a_k = 0$ in (1), we get

$$a_{k+1}X_{k+1} = O$$

$\Rightarrow \qquad a_{k+1} = 0 \qquad\qquad\qquad\qquad [\because X_{k+1} \neq O]$

$\therefore$ The set $S_1 \cup \{X_{k+1}\}$ is linearly independent.

Hence, the result is proved by induction.

7.10 EIGENVALUES OF SPECIAL TYPE OF MATRICES

THEOREM 1. *The eigenvalues of a Hermitian matrix are real.*

PROOF. Let A be a Hermitian matrix. Let λ be an eigenvalue of A and let X be its corresponding eigenvector.

Then we have $\qquad AX = \lambda X \qquad\qquad\qquad\qquad\qquad$ ...(1)

Pre-multiplying both sides of (1) by X^θ, we have

$$X^\theta A X = X^\theta \lambda X = \lambda X^\theta X \qquad\qquad\qquad\qquad ...(2)$$

Taking conjugate transpose of both sides of (2), we have

$$(X^\theta A X)^\theta = (\lambda X^\theta X)^\theta$$

$\Rightarrow \qquad X^\theta A^\theta (X^\theta)^\theta = \bar{\lambda} X^\theta (X^\theta)^\theta$

$\Rightarrow \qquad X^\theta A^\theta X = \bar{\lambda} X^\theta X \qquad\qquad\qquad\qquad \because (X^\theta)^\theta = X$

$\Rightarrow \qquad X^\theta A X = \bar{\lambda} X^\theta X \qquad\qquad\qquad [\because A \text{ is Hermitian} \Rightarrow A^\theta = A]$

$\Rightarrow \qquad X^\theta \lambda X = \bar{\lambda} X^\theta X \qquad\qquad\qquad\qquad\qquad [\text{Using (1)}]$

or $\qquad \lambda X^\theta X = \bar{\lambda} X^\theta X$

or $\qquad (\lambda - \bar{\lambda}) X^\theta X = O$

or $\qquad (\lambda - \bar{\lambda}) = 0 \qquad\qquad\qquad [\because X \text{ is non-zero} \Rightarrow X^\theta X \neq O]$

or $\qquad \lambda = \bar{\lambda}$

Hence λ is real.

THEOREM 2. *The eigenvalues of a real symmetric matrix are all real.*

PROOF. Let A be a real symmetric matrix, then

$$A' = A \qquad\qquad\qquad\qquad\qquad ...(1)$$

Let λ be any eigenvalue of A and let X be its corresponding eigenvector, then we have

$$AX = \lambda X \qquad\qquad\qquad\qquad\qquad ...(2)$$

Pre-multiplying both sides of (2) by X' we get

$$X'AX = \lambda X'X \qquad\qquad\qquad\qquad\qquad ...(3)$$

Taking transpose of both sides of (3), we get

$$(X'AX)' = (\lambda X'X)'$$

$\Rightarrow \qquad X'A'(X')' = \bar{\lambda} X'(X')'$

$\Rightarrow \qquad X'A'X = \bar{\lambda} X'X \qquad\qquad\qquad\qquad [\because (X')' = X]$

$\Rightarrow \qquad X'AX = \bar{\lambda} X'X \qquad\qquad\qquad\qquad [\because A' = A]$

$\Rightarrow \qquad X'\lambda X = \bar{\lambda} X'X \qquad\qquad\qquad\qquad [\because AX = \lambda X]$

$\Rightarrow \qquad \lambda X'X = \bar{\lambda} X'X$

$\Rightarrow \qquad (\lambda - \bar{\lambda})X'X = O$

$\Rightarrow \qquad (\lambda - \bar{\lambda}) = 0 \qquad\qquad\qquad\qquad [\because X \neq O \Rightarrow X'X \neq O]$

$\Rightarrow \qquad \lambda = \bar{\lambda}$

$\Rightarrow \qquad \lambda$ is real.

THEOREM 3. *The eigenvalues of a skew-Hermitian matrix are either purely imaginary or zero.*

PROOF. Let A be a skew-Hermitian matrix, then

$$A^\theta = -A$$

Now $$(iA)^\theta = -iA^\theta$$

$\Rightarrow$ $$(iA)^\theta = iA$$

$\Rightarrow$ iA is a Hermitian matrix

Let λ be an eigenvalue of A and X be its corresponding eigenvector, then

$$AX = \lambda X$$

$\Rightarrow$ $$iAX = i\lambda X$$

$\therefore$ $i\lambda$ is an eigenvalue of a Hermitian matrix. By theorem 1, we can say that $i\lambda$ is real. It follows that either λ is purely real or zero.

COROLLARY. *The eigenvalues of a real skew-symmetric matrix are either purely imaginary or zero.*

PROOF. If the elements of a skew-Hermitian matrix are all real, then it is a real skew-symmetric. Therefore, a real skew-symmetric matrix is skew-Hermitian matrix, hence the result follows from theorem 3.

THEOREM 4. *The eigenvalues of a unitary matrix are of unit modulus.*

PROOF. Let A be a unitary matrix, then

$$A^\theta A = I \qquad \qquad ...(1)$$

Let λ be an eigenvalue of A and let X be its corresponding eigenvector, then

$$AX = \lambda X \qquad \qquad ...(2)$$

Taking conjugate transpose of both sides of (2), we have

$$(AX)^\theta = (\lambda X)^\theta$$

or $$X^\theta A^\theta = \bar{\lambda} X^\theta \qquad \qquad ...(3)$$

Now $$(X^\theta A^\theta)(AX) = (\bar{\lambda} X^\theta)(\lambda X) \qquad \text{[Using (2) and (3)]}$$

$\Rightarrow$ $$X^\theta (A^\theta A)X = \bar{\lambda}\lambda X^\theta X$$

$\Rightarrow$ $$X^\theta IX = \bar{\lambda}\lambda X^\theta X \qquad \qquad \text{[Using (1)]}$$

$\Rightarrow$ $$X^\theta X = \bar{\lambda}\lambda X^\theta X$$

$\Rightarrow$ $$(1 - \bar{\lambda}\lambda)X^\theta X = O$$

$\Rightarrow$ $$\bar{\lambda}\lambda = 1 \qquad \qquad [\because X \neq O \Rightarrow X^\theta X \neq O]$$

$\Rightarrow$ $$|\lambda|^2 = 1$$

$\Rightarrow$ $$|\lambda| = 1$$

THEOREM 5. *The eigenvalues of an orthogonal matrix are of unit modulus.*

PROOF. We know that if the elements of a unitary matrix are all real, then it is an orthogonal matrix, therefore an orthogonal matrix is a unitary matrix hence the result follows from theorem 4.

Recapitulations ||||||||||||||||||||||||||||||||||||||||||||||||||||||||||||

- $|A - \lambda I| = 0$ is known as characteristic equation and the roots of this equation are called characteristic roots or eigen values.

- The value of λ for which a non-zero vector, *i.e.*, $X \neq O$ satisfies $AX = \lambda X$ is called an eigenvalues of the matrix A and the non-zero vector X is called an eigenvector of A corresponding to that eigenvalue λ.

- λ is an eigenvalue of a matrix A if and only if there exists a non-zero vector X such that $AX = \lambda X$.

SOLVED EXAMPLES

EXAMPLE 1.   *If λ is a non-zero eigenvalue of a matrix A, then show that $\dfrac{1}{\lambda}$ is an eigenvalue of A^{-1}.*

SOLUTION.   Let $X \neq O$ be an eigenvector corresponding to the eigenvalue λ of A, then

$$AX = \lambda X$$
$$\Rightarrow \qquad\qquad A^{-1}(AX) = A^{-1}(\lambda X) \qquad\qquad [\because A^{-1} \text{ exists.}]$$
$$\Rightarrow \qquad\qquad (A^{-1}A)X = \lambda(A^{-1}X)$$
$$\Rightarrow \qquad\qquad IX = \lambda(A^{-1}X) \qquad\qquad [\because A^{-1}A = I]$$
$$\Rightarrow \qquad\qquad X = \lambda(A^{-1}X) \qquad\qquad [\because IX = X]$$

$$\Rightarrow \qquad\qquad A^{-1}X = \left(\frac{1}{\lambda}\right)X$$

Hence, $\dfrac{1}{\lambda}$ is an eigenvalue of A^{-1}.

EXAMPLE 2.   *Let A be an $n \times n$ matrix. Then show that zero is an eigenvalue of A iff A is singular.*

SOLUTION.   Let $X \neq O$ be an eigenvector corresponding to the eigenvalue 0 of A, then

$$AX = 0X = O \qquad\qquad\qquad ...(1)$$

Since (1) represents a system of homogeneous equations, it will have non-zero solution if and only if $\rho(A) < n$

i.e., $\qquad\qquad\qquad$ iff $|A| = 0$

i.e., iff A is singular.

EXAMPLE 3.   *If $\lambda_1, \lambda_2, ..., \lambda_n$ are the eigenvalues of A, then show that $k\lambda_1, k\lambda_2, ..., k\lambda_n$ are eigenvalues of kA, where k is any number.*

SOLUTION.   If $k = 0$, then $kA = 0A = O$. Since each eigenvalue of a zero matrix is zero, therefore $0\lambda_1, 0\lambda_2, ..., 0\lambda_n$ are the eigenvalues of kA if $\lambda_1, \lambda_2, ..., \lambda_n$ are eigenvalues of A.

Next, suppose that $k \neq 0$, then we have

$$|kA - k\lambda I| = k^n |A - \lambda I|$$

Now $\qquad\qquad\qquad |kA - k\lambda I| = 0$ iff $|A - \lambda I| = 0$

It follows that $k\lambda$ is an eigenvalue of kA.

Hence, if $\lambda_1, \lambda_2, ..., \lambda_n$ are the eigenvalues of A then $k\lambda_1, k\lambda_2, ..., k\lambda_n$ are the eigenvalues of kA.

EXAMPLE 4.   *If X be a non-zero eigenvector of an $n \times n$ matrix A, then prove that for each positive integer n, X is an eigenvector of A^n corresponding to the eigenvalue λ^n.*

SOLUTION.   Since $X \neq O$ is an eigenvector corresponding eigenvalue λ of A, then we have

$$AX = \lambda X \qquad\qquad\qquad ...(1)$$

Now we have to show that $A^n X = \lambda^n X$.

We shall prove this by induction on n.

If $n = 1$, then the result is true by virtue of (1).

Suppose that the result is true for $n = k$, then we have

$$A^k X = \lambda^k X \qquad\qquad\qquad ...(2)$$

Now $\qquad\qquad A^{k+1}X = (A^k A)X$
$$= A^k(AX)$$
$$= A^k(\lambda X) \qquad\qquad\qquad [\text{Using (1)}\}$$
$$= \lambda(A^k X)$$

$$= \lambda(\lambda^k X)$$ [Using (2)]
$$= \lambda^{k+1} X$$
$$A^{k+1} X = \lambda^{k+1} X$$

Thus, the result is true for $n = k+1$.

Hence by induction the result is true for all positive integers n.

EXAMPLE 5. *Show that similar matrices have the same eigenvalues.*

SOLUTION. Two matrices A and B of the same order are said to be similar if there exists a non-singular matrix P such that

$$B = P^{-1}AB$$

Let λ be an eigenvalue of A, then X is a root of $|A - \lambda I| = 0$.

Now
$$B - \lambda I = P^{-1}AP - \lambda I$$
$$= P^{-1}AP - P^{-1}(\lambda I)P \qquad [\because P^{-1}(\lambda I)P = \lambda P^{-1}P = \lambda I]$$
$$= P^{-1}(A - \lambda I)P$$
$$\Rightarrow \qquad |B - \lambda I| = |P^{-1}||A - \lambda I||P|$$
$$\Rightarrow \qquad |B - \lambda I| = |A - \lambda I||P^{-1}||P|$$
$$\Rightarrow \qquad |B - \lambda I| = |A - \lambda I||P^{-1}P|$$
$$\Rightarrow \qquad |B - \lambda I| = |A - \lambda I| \qquad [\because |P^{-1}P| = |I| = 1]$$

Since λ is a root of $|A - \lambda I| = 0$, therefore λ is also a root of $|B - \lambda I| = 0$, it follows that λ is an eigenvalue of B.

Hence similar matrices have the same eigenvalues.

EXAMPLE 6. *Let A and B be two matrices of order $n \times n$. Let $X \neq O$ be an eigenvector of A and B corresponding to the eigenvalues λ_1 and λ_2 respectively, then show that X is an eigenvector of AB corresponding to the eigenvalue $\lambda_1 \lambda_2$ of AB.*

SOLUTION. Since $X \neq O$ is an eigenvector of A and B corresponding to the eigenvalues λ_1 and λ_2 respectively, then we have
$$AX = \lambda_1 X \qquad \qquad ...(1)$$
and
$$BX = \lambda_2 X \qquad \qquad ...(2)$$

Now
$$(AB)X = A(BX)$$
$$= A(\lambda_2 X) \qquad \qquad [\text{Using (2)}]$$
$$= \lambda_2(AX)$$
$$= \lambda_2(\lambda_1 X) \qquad \qquad [\text{Using (1)}]$$
$$= (\lambda_2 \lambda_1)X$$
$$(AB)X = (\lambda_1 \lambda_2)X \text{ with } X \neq O$$

It follows that X is an eigenvector of AB corresponding to the eigenvalue $\lambda_1 \lambda_2$.

EXAMPLE 7. *Determine the eigenvalues of the matrix :*

$$A = \begin{bmatrix} 1 & 2 & 3 \\ 0 & -4 & 2 \\ 0 & 0 & 7 \end{bmatrix}$$ (Bhopal-2002, Agra-2002,05,08)

SOLUTION. The characteristic equation of A is given by
$$|A - \lambda I| = 0$$

i.e.,
$$\begin{vmatrix} 1-\lambda & 2 & 3 \\ 0 & -4-\lambda & 2 \\ 0 & 0 & 7-\lambda \end{vmatrix} = 0$$

i.e.,
$$(1-\lambda)(-2-\lambda)(7-\lambda) = 0$$

The roots of this characteristic equation are given by $\lambda = 1, -4, 7$.

These are the required eigenvalues of A.

▼ REMARK

▶ It is clear that the given matrix A is an upper triangular matrix so that the principal diagonal elements $1, -4, 7$ will be the eigenvalues of A.

EXAMPLE 8.   *Determine the eigenvalues of the matrix :*

$$A = \begin{bmatrix} 0 & 1 & 2 \\ 1 & 0 & -1 \\ 2 & -1 & 0 \end{bmatrix}.$$

SOLUTION.   The characteristic equation of A is given

$$|A - \lambda I| = 0$$

or

$$\begin{vmatrix} 0-\lambda & 1 & 2 \\ 1 & 0-\lambda & -1 \\ 2 & -1 & 0-\lambda \end{vmatrix} = 0$$

or $-\lambda(\lambda^2 - 1) - 1(-\lambda + 2) + 2(-1 + 2\lambda) = 0$ or $-\lambda^3 + 6\lambda - 4 = 0$

$\Rightarrow$ $(\lambda - 2)(\lambda^2 + 2\lambda - 2) = 0$

$\Rightarrow$ $\lambda = 2$ and $\lambda = -1 \pm \sqrt{3}$

Hence the eigenvalues of A are $2, -1 \pm \sqrt{3}$.

EXAMPLE 9.   *Determine the eigenvalues and eigenvectors of the matrix*

$$A = \begin{bmatrix} 5 & 4 \\ 1 & 2 \end{bmatrix}.$$

SOLUTION.   The characteristic equation of A is given by

$$|A - \lambda I| = 0$$

or

$$\begin{vmatrix} 5-\lambda & 4 \\ 1 & 2-\lambda \end{vmatrix} = 0$$

or $(5-\lambda)(2-\lambda) - 4 = 0$

or $\lambda^2 - 7\lambda + 10 - 4 = 0$

or $\lambda^2 - 7\lambda + 6 = 0$

The roots of this equation are $\lambda = 6, 1$. Thus, the eigenvalues of A are $6, 1$.

Eigenvector corresponding to $\lambda_1 = 6$:

Let $X_1 = \begin{bmatrix} x_1 \\ x_2 \end{bmatrix} \neq O$ be an eigenvector of A corresponding to $\lambda_1 = 6$, then we have

$$AX_1 = 6X_1$$

or $(A - 6I)X_1 = O$

or

$$\begin{bmatrix} 5-6 & 4 \\ 1 & 2-6 \end{bmatrix} \begin{bmatrix} x_1 \\ x_2 \end{bmatrix} = \begin{bmatrix} 0 \\ 0 \end{bmatrix}$$

or

$$\begin{bmatrix} -1 & 4 \\ 1 & -4 \end{bmatrix} \begin{bmatrix} x_1 \\ x_2 \end{bmatrix} = \begin{bmatrix} 0 \\ 0 \end{bmatrix} \qquad \qquad ...(1)$$

The non-zero solution of (1) will give X_1.
Applying $R_2 \rightarrow R_2 + R_1$, we have

$$\begin{bmatrix} -1 & 4 \\ 1 & -4 \end{bmatrix} \begin{bmatrix} x_1 \\ x_2 \end{bmatrix} = \begin{bmatrix} 0 \\ 0 \end{bmatrix} \qquad \qquad ...(2)$$

The coefficient matrix of equation (1) is of rank 1, *i.e.*, $\rho(A - 6I) = 1$, therefore the system of equations (1) will have $2 - 1 = 1$ linearly independent solution.
From (2), we have

$$-x_1 + 4x_2 = 0$$

Clearly, $x_1 = 4$ and $x_2 = 1$ satisfy the above equation.
Hence the eigenvector corresponding to eigenvalue $\lambda_1 = 6$ is

$$X_1 = \begin{bmatrix} 4 \\ 1 \end{bmatrix}$$

Eigenvector corresponding to $\lambda_2 = 1$:

Let $X_2 = \begin{bmatrix} x_1 \\ x_2 \end{bmatrix} \neq O$ be an eigenvector of A corresponding to eigen value $\lambda_2 = 1$, then we have

$$AX_2 = \lambda_2 X_2$$
or $$AX_2 = IX_2$$
or $$(A - I)X_2 = O$$
or $$\begin{bmatrix} 5-1 & 4 \\ 1 & 2-1 \end{bmatrix} \begin{bmatrix} x_1 \\ x_2 \end{bmatrix} = \begin{bmatrix} 0 \\ 0 \end{bmatrix}$$
or $$\begin{bmatrix} 4 & 4 \\ 1 & 1 \end{bmatrix} \begin{bmatrix} x_1 \\ x_2 \end{bmatrix} = \begin{bmatrix} 0 \\ 0 \end{bmatrix} \qquad \qquad ...(3)$$

The non-zero solution of (3) will give X_2.

Applying $R_2 \rightarrow R_2 - \dfrac{1}{4} R_1$, we get

$$\begin{bmatrix} 4 & 4 \\ 0 & 0 \end{bmatrix} \begin{bmatrix} x_1 \\ x_2 \end{bmatrix} = \begin{bmatrix} 0 \\ 0 \end{bmatrix} \qquad \qquad ...(4)$$

Clearly, $\rho(A - I) = 1$, therefore the system of equations (3) will have $2 - 1 = 1$ linearly independent solution.
From (4), we get

$$4x_1 + 4x_2 = 0$$

Clearly, $x_1 = 1$ and $x_2 = -1$, satisfy above equation.
Hence the eigenvector corresponding to eigenvalue $\lambda_2 = 1$ is

$$X_2 = \begin{bmatrix} 1 \\ -1 \end{bmatrix}.$$

EXAMPLE 10. *Determine the eigenvalues and eigenvectors of the matrix*

$$A = \begin{bmatrix} 8 & -6 & 2 \\ -6 & 7 & -4 \\ 2 & -4 & 3 \end{bmatrix}.$$

SOLUTION. The characteristic equation of A is given by

$$|A - \lambda I| = 0$$

or

$$\begin{vmatrix} 8-\lambda & -6 & 2 \\ -6 & 7-\lambda & -4 \\ 2 & -4 & 3-\lambda \end{vmatrix} = 0$$

or $(8-\lambda)\{(7-\lambda)(3-\lambda)-16\} + 6\{-18+6\lambda+8\} + 2\{24-14+2\lambda\} = 0$

or $\lambda^3 - 18\lambda^2 + 45\lambda = 0$

or $\lambda(\lambda-3)(\lambda-15) = 0$

The roots of this equation are $\lambda = 0, 3, 15$.

Thus, the eigenvalues of A are $\lambda_1 = 0$, $\lambda_2 = 3$, $\lambda_3 = 15$.

Eigenvector corresponding to $\lambda_1 = 0$:

Let $X_1 = \begin{bmatrix} x_1 \\ x_2 \\ x_3 \end{bmatrix} \neq O$ be an eigenvector corresponding to the eigen value $\lambda_1 = 0$,

then we have $AX_1 = \lambda_1 X_1$

or $AX_1 = 0X_1$

or $(A - 0I)X_1 = O$

or $\begin{bmatrix} 8 & -6 & 2 \\ -6 & 7 & -4 \\ 2 & -4 & 3 \end{bmatrix} \begin{bmatrix} x_1 \\ x_2 \\ x_3 \end{bmatrix} = \begin{bmatrix} 0 \\ 0 \\ 0 \end{bmatrix}$ ...(1)

The non-zero solution of (1) will give X_1.

Reducing the coefficient matrix of (1) in Echeleon form by applying elementary row transformations.

Applying $R_1 \leftrightarrow R_3$, we get

$$\begin{bmatrix} 2 & -4 & 3 \\ -6 & 7 & -4 \\ 8 & -6 & 2 \end{bmatrix} \begin{bmatrix} x_1 \\ x_2 \\ x_3 \end{bmatrix} = \begin{bmatrix} 0 \\ 0 \\ 0 \end{bmatrix}$$

Applying $R_2 \to R_2 + 3R_1$, $R_3 \to R_3 - 4R_1$, we get

$$\begin{bmatrix} 2 & -4 & 3 \\ 0 & -5 & 5 \\ 0 & 10 & -10 \end{bmatrix} \begin{bmatrix} x_1 \\ x_2 \\ x_3 \end{bmatrix} = \begin{bmatrix} 0 \\ 0 \\ 0 \end{bmatrix}$$

Applying $R_3 \to R_3 + 2R_2$, we get

$$\begin{bmatrix} 2 & -4 & 3 \\ 0 & -5 & 5 \\ 0 & 0 & 0 \end{bmatrix} \begin{bmatrix} x_1 \\ x_2 \\ x_3 \end{bmatrix} = \begin{bmatrix} 0 \\ 0 \\ 0 \end{bmatrix}$$ ...(2)

Clearly $\rho(A - 0.I) = 2$, therefore the system of equations (2) will have $3 - 2 = 1$ (unknowns – rank) linearly independent solution.

From (2), we have

$$2x_1 - 4x_2 + 3x_3 = 0$$
$$- 5x_2 + 5x_3 = 0$$

Clearly, $x_1 = \dfrac{1}{2}$, $x_2 = 1$ and $x_3 = 1$ satisfy the above equations.

Hence the eigenvector corresponding to eigenvalue $\lambda_1 = 0$ is

$$X_1 = \begin{bmatrix} 1/2 \\ 1 \\ 1 \end{bmatrix}$$

Eigenvector corresponding to $\lambda_2 = 3$:

Let $X_2 = \begin{bmatrix} x_1 \\ x_2 \\ x_3 \end{bmatrix} \neq O$ be an eigenvector of A corresponding to $\lambda_2 = 3$, then we have

$$AX_2 = \lambda_2 X_2$$

or $\qquad\qquad (A - \lambda_2 I)X_2 = O$

or $\qquad\qquad (A - 3I)X_2 = O$

or $\qquad \begin{bmatrix} 8-3 & -6 & 2 \\ -6 & 7-3 & -4 \\ 2 & -4 & 3-3 \end{bmatrix} \begin{bmatrix} x_1 \\ x_2 \\ x_3 \end{bmatrix} = O$

or $\qquad \begin{bmatrix} 5 & -6 & 2 \\ -6 & 4 & -4 \\ 2 & -4 & 0 \end{bmatrix} \begin{bmatrix} x_1 \\ x_2 \\ x_3 \end{bmatrix} = O \qquad\qquad\dots(3)$

The non-zero solution of (3) will give X_2.

Applying $R_1 \to R_1 + R_2$, we get

$$\begin{bmatrix} -1 & -2 & -2 \\ -6 & 4 & -4 \\ 2 & -4 & 0 \end{bmatrix} \begin{bmatrix} x_1 \\ x_2 \\ x_3 \end{bmatrix} = \begin{bmatrix} 0 \\ 0 \\ 0 \end{bmatrix}$$

Applying $R_2 \to R_2 - 6R_1$, $R_3 \to R_3 + 2R_1$, we get

$$\begin{bmatrix} -1 & -2 & -2 \\ 0 & 16 & 8 \\ 0 & -8 & -4 \end{bmatrix} \begin{bmatrix} x_1 \\ x_2 \\ x_3 \end{bmatrix} = \begin{bmatrix} 0 \\ 0 \\ 0 \end{bmatrix}$$

Applying $R_2 \to \dfrac{1}{8}R_2$, we get

$$\begin{bmatrix} -1 & -2 & -2 \\ 0 & 2 & 1 \\ 0 & -8 & -4 \end{bmatrix} \begin{bmatrix} x_1 \\ x_2 \\ x_3 \end{bmatrix} = \begin{bmatrix} 0 \\ 0 \\ 0 \end{bmatrix}$$

Again applying $R_3 \to R_3 + 4R_2$, we get

$$\begin{bmatrix} -1 & -2 & -2 \\ 0 & 2 & 1 \\ 0 & 0 & 0 \end{bmatrix} \begin{bmatrix} x_1 \\ x_2 \\ x_3 \end{bmatrix} = \begin{bmatrix} 0 \\ 0 \\ 0 \end{bmatrix} \qquad\qquad\dots(4)$$

Clearly $\rho(A - 3I) = 2$, therefore the system of equations (3) will have $3 - 2 = 1$ linearly independent solution.

From (4), we have

$$-x_1 - 2x_2 - 2x_3 = 0$$
$$2x_2 + x_3 = 0$$

Clearly, $x_1 = -2$, $x_2 = -1$ and $x_3 = 2$ satisfy the above equations.
Hence the eigenvector corresponding to eigenvalue $\lambda_2 = 3$ is

$$X_2 = \begin{bmatrix} -2 \\ -1 \\ 2 \end{bmatrix}$$

Eigenvector corresponding to $\lambda_3 = 15$:

Let $X_3 = \begin{bmatrix} x_1 \\ x_2 \\ x_3 \end{bmatrix} \neq O$ be an eigenvector of A corresponding to $\lambda_3 = 15$, then we

have

$$AX_3 = \lambda_3 X_3$$

or $$(A - \lambda_3 I)X_3 = O$$

or $$(A - 15I)X_3 = O$$

or $$\begin{bmatrix} 8-15 & -6 & 2 \\ -6 & 7-15 & -4 \\ 2 & -4 & 3-15 \end{bmatrix} \begin{bmatrix} x_1 \\ x_2 \\ x_3 \end{bmatrix} = O$$

or $$\begin{bmatrix} -7 & -6 & 2 \\ -6 & -8 & -4 \\ 2 & -4 & -12 \end{bmatrix} \begin{bmatrix} x_1 \\ x_2 \\ x_3 \end{bmatrix} = O \qquad \qquad ...(5)$$

The non-zero solution of (5) will give X_3.
Applying $R_1 \leftrightarrow R_3$, we get

$$\begin{bmatrix} 2 & -4 & -12 \\ -6 & -8 & -4 \\ -7 & -6 & 2 \end{bmatrix} \begin{bmatrix} x_1 \\ x_2 \\ x_3 \end{bmatrix} = \begin{bmatrix} 0 \\ 0 \\ 0 \end{bmatrix}$$

Applying $R_1 \rightarrow \dfrac{1}{2} R_1$, we get

$$\begin{bmatrix} 1 & -2 & -6 \\ -6 & -8 & -4 \\ -7 & -6 & 2 \end{bmatrix} \begin{bmatrix} x_1 \\ x_2 \\ x_3 \end{bmatrix} = \begin{bmatrix} 0 \\ 0 \\ 0 \end{bmatrix}$$

Applying $R_2 \rightarrow R_2 + 6R_1$, $R_3 \rightarrow R_3 + 7R_1$, we get

$$\begin{bmatrix} 1 & -2 & -6 \\ 0 & -20 & -40 \\ 0 & -20 & -40 \end{bmatrix} \begin{bmatrix} x_1 \\ x_2 \\ x_3 \end{bmatrix} = \begin{bmatrix} 0 \\ 0 \\ 0 \end{bmatrix}$$

Applying $R_3 \rightarrow R_3 - R_2$, we get

$$\begin{bmatrix} 1 & -2 & -6 \\ 0 & -20 & -40 \\ 0 & 0 & 0 \end{bmatrix} \begin{bmatrix} x_1 \\ x_2 \\ x_3 \end{bmatrix} = \begin{bmatrix} 0 \\ 0 \\ 0 \end{bmatrix} \qquad \qquad ...(6)$$

Clearly $\rho(A - 15I) = 2$, therefore the system of equations (5) will have $3 - 2 = 1$ linearly independent solution.

From (6), we have $\quad x_1 - 2x_2 - 6x_3 = 0$

$$- 20x_2 - 40x_3 = 0$$

Clearly, $x_1 = 2$, $x_2 = -2$ and $x_3 = 1$ satisfy the above equations.

Hence the eigenvector corresponding to eigenvalue $\lambda_3 = 15$ is

$$X_3 = \begin{bmatrix} 2 \\ -2 \\ 1 \end{bmatrix}$$

EXAMPLE 11. *Determine the eigenvalues and eigenvectors of the matrix*

$$A = \begin{bmatrix} 2 & 1 & 0 \\ 0 & 1 & -1 \\ 0 & 2 & 4 \end{bmatrix}.$$

SOLUTION. The characteristic equation of A is given by

$$|A - \lambda I| = 0$$

or

$$\begin{vmatrix} 2 - \lambda & 1 & 0 \\ 0 & 1 - \lambda & -1 \\ 0 & 2 & 4 - \lambda \end{vmatrix} = 0$$

or $\quad (2 - \lambda)\{(1 - \lambda)(4 - \lambda) + 2\} = 0$

or $\quad (2 - \lambda)(\lambda^3 - 5\lambda + 6) = 0$

or $\quad (2 - \lambda)(2 - \lambda)(3 - \lambda) = 0$

The roots of this equation are $\lambda = 2, 2, 3$.

Thus the eigenvalues of A are $\lambda_1 = 2$, $\lambda_2 = 2$, $\lambda_3 = 3$.

Eigenvector corresponding to the eigenvalue $\lambda_1 = \lambda_2 = 2$:

Let $X_1 = \begin{bmatrix} x_1 \\ x_2 \\ x_3 \end{bmatrix} \neq O$ be an eigenvector corresponding to the eigenvalue 2, then

we have $\qquad\qquad AX_1 = 2X_1.$

or $\qquad\qquad (A - 2I) X_1 = O$

or

$$\begin{bmatrix} 2 - 2 & 1 & 0 \\ 0 & 1 - 2 & -1 \\ 0 & 2 & 4 - 2 \end{bmatrix} \begin{bmatrix} x_1 \\ x_2 \\ x_3 \end{bmatrix} = \begin{bmatrix} 0 \\ 0 \\ 0 \end{bmatrix}$$

or

$$\begin{bmatrix} 0 & 1 & 0 \\ 0 & -1 & -1 \\ 0 & 2 & 2 \end{bmatrix} \begin{bmatrix} x_1 \\ x_2 \\ x_3 \end{bmatrix} = \begin{bmatrix} 0 \\ 0 \\ 0 \end{bmatrix} \qquad\qquad \ldots(1)$$

The non-zero solution of (1) will give X_1.

Applying $R_3 \to R_3 + 2R_2$, we get

$$\begin{bmatrix} 0 & 1 & 0 \\ 0 & -1 & -1 \\ 0 & 0 & 0 \end{bmatrix} \begin{bmatrix} x_1 \\ x_2 \\ x_3 \end{bmatrix} = \begin{bmatrix} 0 \\ 0 \\ 0 \end{bmatrix}$$

Applying $R_2 \to R_2 + R_1$, we get

$$\begin{bmatrix} 0 & 1 & 0 \\ 0 & 0 & -1 \\ 0 & 0 & 0 \end{bmatrix}\begin{bmatrix} x_1 \\ x_2 \\ x_3 \end{bmatrix} = \begin{bmatrix} 0 \\ 0 \\ 0 \end{bmatrix} \qquad \ldots(2)$$

From (2), it is clear that $\rho(A - 2I) = 2$, therefore the system of equations (1) will have $3 - 2 = 1$ linearly independent solution.

From (2), we have $x_2 = 0$ and $-x_3 = 0$

Clearly, $x_1 = 1, x_2 = 0, x_3 = 1$ satisfy the above equations.

Hence the eigenvector corresponding to the eigenvalue $\lambda_1 = \lambda_2 = 2$ is

$$X_1 = \begin{bmatrix} 1 \\ 0 \\ 0 \end{bmatrix}$$

Eigenvector corresponding to the eigenvalue $\lambda_3 = 3$:

Let $X_2 = \begin{bmatrix} x_1 \\ x_2 \\ x_3 \end{bmatrix} \neq O$ be an eigenvector corresponding to $\lambda_3 = 3$, then we have

$$AX_2 = 3X_2$$

or

$$(A - 3I)X_3 = O$$

or

$$\begin{bmatrix} 2-3 & 1 & 0 \\ 0 & 1-3 & -1 \\ 0 & 2 & 4-3 \end{bmatrix}\begin{bmatrix} x_1 \\ x_2 \\ x_3 \end{bmatrix} = \begin{bmatrix} 0 \\ 0 \\ 0 \end{bmatrix}$$

or

$$\begin{bmatrix} -1 & 1 & 0 \\ 0 & -2 & -1 \\ 0 & 2 & 1 \end{bmatrix}\begin{bmatrix} x_1 \\ x_2 \\ x_3 \end{bmatrix} = \begin{bmatrix} 0 \\ 0 \\ 0 \end{bmatrix} \qquad \ldots(3)$$

The non-zero solution of (3) will give X_2.

Applying $R_3 \to R_3 + R_2$, we get

$$\begin{bmatrix} -1 & 1 & 0 \\ 0 & -2 & -1 \\ 0 & 0 & 0 \end{bmatrix}\begin{bmatrix} x_1 \\ x_2 \\ x_3 \end{bmatrix} = \begin{bmatrix} 0 \\ 0 \\ 0 \end{bmatrix} \qquad \ldots(4)$$

Clearly $\rho(A - 3I) = 2$, therefore the system of equations (3) will have $3 - 2 = 1$ linearly independent solution.

From (4), we have

$$-x_1 + x_2 = 0$$
$$-2x_2 - x_3 = 0$$

Clearly, $x_1 = 1, x_2 = 1$ and $x_3 = -2$ satisfy the above equations. Hence the eigenvector corresponding to eigenvalue $\lambda_3 = 3$ is

$$X_2 = \begin{bmatrix} 1 \\ 1 \\ -2 \end{bmatrix}$$

EXAMPLE 12. *Find the eigenvalues and eigenvectors of the matrix*

$$A = \begin{bmatrix} 5 & 4 & 2 \\ 4 & 5 & 2 \\ 2 & 2 & 2 \end{bmatrix}.$$

SOLUTION. The characteristic equation of A is given by

$$|A - \lambda I| = 0$$

$$\Rightarrow \qquad \begin{vmatrix} 5-\lambda & 4 & 2 \\ 4 & 5-\lambda & 2 \\ 2 & 2 & 2-\lambda \end{vmatrix} = 0$$

or $(5-\lambda)\{(5-\lambda)(2-\lambda)-4\}-4\{4(2-\lambda)-4\}+2\{8-2(8-\lambda)\}=0$

or $\qquad\qquad\qquad\qquad\qquad\qquad -\lambda^3 + 12\lambda^2 - 21\lambda + 10 = 0$

or $\qquad\qquad\qquad\qquad\qquad\qquad -(\lambda-1)^2(\lambda-10) = 0$

The roots of this equation are 1, 1, 10.

Thus the eigenvalues of A are $\lambda_1 = 1, \lambda_2 = 1, \lambda_3 = 10$.

Eigenvector corresponding to the eigenvalue $\lambda_1 = \lambda_2 = 1$:

Let $X = \begin{bmatrix} x_1 \\ x_2 \\ x_3 \end{bmatrix} \neq O$ be an eigenvector of A corresponding to the eigenvalue $\lambda_1 =$

$\lambda_2 = 1$, then we have

$$AX_1 = IX$$

or $\qquad\qquad (A - I)X_1 = O$

or $\qquad \begin{bmatrix} 5-1 & 4 & 2 \\ 4 & 5-1 & 2 \\ 2 & 2 & 2-1 \end{bmatrix} \begin{bmatrix} x_1 \\ x_2 \\ x_3 \end{bmatrix} = \begin{bmatrix} 0 \\ 0 \\ 0 \end{bmatrix}$

or $\qquad \begin{bmatrix} 4 & 4 & 2 \\ 4 & 4 & 2 \\ 2 & 2 & 2 \end{bmatrix} \begin{bmatrix} x_1 \\ x_2 \\ x_3 \end{bmatrix} = \begin{bmatrix} 0 \\ 0 \\ 0 \end{bmatrix}$ ...(1)

Applying $R_1 \leftrightarrow R_3$, we get

$$\begin{bmatrix} 2 & 2 & 1 \\ 4 & 4 & 2 \\ 4 & 4 & 2 \end{bmatrix} \begin{bmatrix} x_1 \\ x_2 \\ x_3 \end{bmatrix} = \begin{bmatrix} 0 \\ 0 \\ 0 \end{bmatrix}$$

Applying $R_2 \to R_2 - 2R_1$, we get

$$\begin{bmatrix} 2 & 2 & 1 \\ 0 & 0 & 0 \\ 0 & 0 & 0 \end{bmatrix} \begin{bmatrix} x_1 \\ x_2 \\ x_3 \end{bmatrix} = \begin{bmatrix} 0 \\ 0 \\ 0 \end{bmatrix} \qquad\qquad ...(2)$$

From (2), it is clear that $\rho(A - I) = 1$, therefore the system of equation (1) will have $3 - 2 = 1$ linearly independent solution.

Now from (2), we have
$$x_2 = 0 \quad \text{and} \quad -x_3 = 0$$
Clearly $x_1 = 0$, $x_2 = 0$, $x_3 = 0$ satisfy the above equations. .

Hence the eigenvector corresponding to the eigenvalue $\lambda_1 = \lambda_2 = 2$ is

$$X_1 = \begin{bmatrix} 1 \\ 0 \\ 0 \end{bmatrix}$$

Eigenvector corresponding to the eigenvalue $\lambda_1 = \lambda_2 = 1$

Let $X = \begin{bmatrix} x_1 \\ x_2 \\ x_3 \end{bmatrix} \neq 0$ be an eigenvector corresponding to the eigenvalue $\lambda_1 = \lambda_2 = 1$,

then we have

$$AX = IX$$

or $$(A - I)X = O$$

or $$\begin{bmatrix} 5-1 & 4 & 2 \\ 4 & 5-1 & 2 \\ 2 & 2 & 2-1 \end{bmatrix} \begin{bmatrix} x_1 \\ x_2 \\ x_3 \end{bmatrix} = \begin{bmatrix} 0 \\ 0 \\ 0 \end{bmatrix}$$

or $$\begin{bmatrix} 4 & 4 & 2 \\ 4 & 4 & 2 \\ 2 & 2 & 1 \end{bmatrix} \begin{bmatrix} x_1 \\ x_2 \\ x_3 \end{bmatrix} = \begin{bmatrix} 0 \\ 0 \\ 0 \end{bmatrix} \qquad \qquad ...(1)$$

Applying $R_1 \rightarrow R_3$, we get

$$\begin{bmatrix} 2 & 2 & 1 \\ 4 & 4 & 2 \\ 4 & 4 & 2 \end{bmatrix} \begin{bmatrix} x_1 \\ x_2 \\ x_3 \end{bmatrix} = \begin{bmatrix} 0 \\ 0 \\ 0 \end{bmatrix}$$

Applying $R_2 \rightarrow R_2 - 2R_1$, $R_3 \rightarrow R_3 - 2R_1$, we get

$$\begin{bmatrix} 2 & 2 & 1 \\ 0 & 0 & 0 \\ 0 & 0 & 0 \end{bmatrix} \begin{bmatrix} x_1 \\ x_2 \\ x_3 \end{bmatrix} = \begin{bmatrix} 0 \\ 0 \\ 0 \end{bmatrix} \qquad \qquad ...(2)$$

From (2), it is clear that $\rho(A-I) = 1$, therefore the equation (1) will have $3-1 = 2$ linearly independent solutions.

From (2), we have

$$2x_1 + 2x_2 + x_3 = 0$$

Since this equation has two linearly independent solutions so we take $x_2 = c_1$ and $x_3 = c_2$, where c_1 and c_2 are non-zero scalars, then $x_1 = -c - \dfrac{c_2}{2}$.

Therefore, $$\begin{bmatrix} x_1 \\ x_2 \\ x_3 \end{bmatrix} = \begin{bmatrix} -c_1 - \dfrac{c_2}{2} \\ c_1 \\ c_2 \end{bmatrix} = c_1 \begin{bmatrix} -1 \\ 1 \\ 0 \end{bmatrix} + c_2 \begin{bmatrix} -1/2 \\ 0 \\ 1 \end{bmatrix}$$

Hence the eigenvectors corresponding to the eigenvalue $\lambda_1 = \lambda_2 = 1$ are

$$X_1 = \begin{bmatrix} -1 \\ 0 \\ 1 \end{bmatrix} \text{ and } X_2 = \begin{bmatrix} -1/2 \\ 0 \\ 1 \end{bmatrix}$$

Eigenvector corresponding to the eigenvalue $\lambda_3 = 10$:

Let $X_3 = \begin{bmatrix} x_1 \\ x_2 \\ x_3 \end{bmatrix} \neq O$ be an eigenvector corresponding to $\lambda_3 = 10$, then we have

$$AX_3 = 10X_3$$

or $$(A - 10I)X_3 = O$$

or $$\begin{bmatrix} 5-10 & 4 & 2 \\ 4 & 5-10 & 2 \\ 2 & 2 & 2-10 \end{bmatrix}\begin{bmatrix} x_1 \\ x_2 \\ x_3 \end{bmatrix} = \begin{bmatrix} 0 \\ 0 \\ 0 \end{bmatrix}$$

or $$\begin{bmatrix} -5 & 4 & 2 \\ 4 & -5 & 2 \\ 2 & 2 & -8 \end{bmatrix}\begin{bmatrix} x_1 \\ x_2 \\ x_3 \end{bmatrix} = \begin{bmatrix} 0 \\ 0 \\ 0 \end{bmatrix} \qquad \ldots(3)$$

Applying $R_1 \rightarrow R_1 + R_2$, we get

$$\begin{bmatrix} -1 & -1 & 4 \\ 4 & -5 & 2 \\ 2 & 2 & -8 \end{bmatrix}\begin{bmatrix} x_1 \\ x_2 \\ x_3 \end{bmatrix} = \begin{bmatrix} 0 \\ 0 \\ 0 \end{bmatrix}$$

Applying $R_2 \rightarrow R_2 + 4R_1, R_3 \rightarrow R_3 + 2R_1$, we get

$$\begin{bmatrix} -1 & -1 & 4 \\ 0 & -9 & 18 \\ 0 & 0 & 0 \end{bmatrix}\begin{bmatrix} x_1 \\ x_2 \\ x_3 \end{bmatrix} = \begin{bmatrix} 0 \\ 0 \\ 0 \end{bmatrix} \qquad \ldots(4)$$

From (4), it is clear that $\rho(A - 10I) = 2$, therefore the system of equations (3) will have $3 - 1 = 2$ linearly independent solutions.

From (4), we have

$$-x_1 - x_2 + 4x_3 = 0$$
$$-9x_2 + 18x_3 = 0$$

Let us take $x_3 = c$, then $x_2 = 2c$ and $x_1 = 2c$.

Therefore,

$$\begin{bmatrix} x_1 \\ x_2 \\ x_3 \end{bmatrix} = \begin{bmatrix} 2c \\ 2c \\ c \end{bmatrix} = c\begin{bmatrix} 2 \\ 2 \\ 1 \end{bmatrix}$$

Hence the eigenvector corresponding to eigenvalue $\lambda_3 = 10$ is

$$X_3 = \begin{bmatrix} 2 \\ 2 \\ 1 \end{bmatrix}.$$

EXAMPLE 13. *Find the eigenvalues and eigenvectors of the matrix*

$$A = \begin{bmatrix} 2 & 0 & 1 & -3 \\ 0 & 2 & 10 & 4 \\ 0 & 0 & 2 & 0 \\ 0 & 0 & 0 & 3 \end{bmatrix}.$$

SOLUTION.   The characteristic equation of A is given by

$$|A - \lambda I| = 0$$

or

$$A = \begin{vmatrix} 2-\lambda & 0 & 1 & -3 \\ 0 & 2-\lambda & 10 & 4 \\ 0 & 0 & 2-\lambda & 0 \\ 0 & 0 & 0 & 3-\lambda \end{vmatrix} = 0$$

or

$$(\lambda - 2)^2 (\lambda - 3) = 0$$

The roots of characteristic equations are 2, 2, 2, 3.

Thus, the eigenvalues of A are $\lambda_1 = 2$, $\lambda_2 = 2$, $\lambda_3 = 2$, $\lambda_4 = 3$.

Eigenvector corresponding to $\lambda_1 = \lambda_2 = \lambda_3 = 2$:

Let $X = \begin{bmatrix} x_1 \\ x_2 \\ x_3 \\ x_4 \end{bmatrix} \neq O$ be an eigenvector of A corresponding to the eigenvalue 2, then we have

$$AX = 2X.$$

or

$$(A - 2I) X = O$$

or

$$\begin{bmatrix} 2-2 & 0 & 1 & -3 \\ 0 & 2-2 & 10 & 4 \\ 0 & 0 & 2-2 & 0 \\ 0 & 0 & 0 & 3-2 \end{bmatrix} \begin{bmatrix} x_1 \\ x_2 \\ x_3 \\ x_4 \end{bmatrix} = \begin{bmatrix} 0 \\ 0 \\ 0 \\ 0 \end{bmatrix}$$

or

$$\begin{bmatrix} 0 & 0 & 1 & -3 \\ 0 & 0 & 10 & 4 \\ 0 & 0 & 0 & 0 \\ 0 & 0 & 0 & 1 \end{bmatrix} \begin{bmatrix} x_1 \\ x_2 \\ x_3 \\ x_4 \end{bmatrix} = \begin{bmatrix} 0 \\ 0 \\ 0 \\ 0 \end{bmatrix} \qquad \ldots(1)$$

Applying $R_3 \leftrightarrow R_4$, we get

$$\begin{bmatrix} 0 & 0 & 1 & -3 \\ 0 & 0 & 10 & 4 \\ 0 & 0 & 0 & 1 \\ 0 & 0 & 0 & 0 \end{bmatrix} \begin{bmatrix} x_1 \\ x_2 \\ x_3 \\ x_4 \end{bmatrix} = \begin{bmatrix} 0 \\ 0 \\ 0 \\ 0 \end{bmatrix}$$

Applying $R_2 \rightarrow R_2 - 10R_1$, we get

$$\begin{bmatrix} 0 & 0 & 1 & -3 \\ 0 & 0 & 0 & 34 \\ 0 & 0 & 0 & 1 \\ 0 & 0 & 0 & 0 \end{bmatrix} \begin{bmatrix} x_1 \\ x_2 \\ x_3 \\ x_4 \end{bmatrix} = \begin{bmatrix} 0 \\ 0 \\ 0 \\ 0 \end{bmatrix}$$

Applying $R_2 \to \dfrac{1}{34} R_2$, we get

$$\begin{bmatrix} 0 & 0 & 1 & -3 \\ 0 & 0 & 0 & 1 \\ 0 & 0 & 0 & 1 \\ 0 & 0 & 0 & 0 \end{bmatrix}\begin{bmatrix} x_1 \\ x_2 \\ x_3 \\ x_4 \end{bmatrix} = \begin{bmatrix} 0 \\ 0 \\ 0 \\ 0 \end{bmatrix}$$

Applying $R_2 \to R_3 - R_2$, we get

$$\begin{bmatrix} 0 & 0 & 1 & -3 \\ 0 & 0 & 0 & 1 \\ 0 & 0 & 0 & 0 \\ 0 & 0 & 0 & 0 \end{bmatrix}\begin{bmatrix} x_1 \\ x_2 \\ x_3 \\ x_4 \end{bmatrix} = \begin{bmatrix} 0 \\ 0 \\ 0 \\ 0 \end{bmatrix} \qquad \text{...(2)}$$

From (2), it is clear that $\rho(A - 2I) = 2$, therefore the equation (1) will have $4 - 2$ = 2 linearly independent solution.

Equation (2) reduces to

$$\left.\begin{array}{r} x_3 - 3x_4 = 0 \\ x_4 = 0 \end{array}\right\} \Rightarrow x_4 = 0, x_3 = 0$$

Let us take $x_1 = c_1, x_2 = c_2$, then we have

$$X = \begin{bmatrix} x_1 \\ x_2 \\ x_3 \\ x_4 \end{bmatrix} = \begin{bmatrix} c_1 \\ c_2 \\ 0 \\ 0 \end{bmatrix} = c_1 \begin{bmatrix} 1 \\ 0 \\ 0 \\ 0 \end{bmatrix} + c_2 \begin{bmatrix} 0 \\ 1 \\ 0 \\ 0 \end{bmatrix}$$

Hence the eigenvector corresponding to eigenvalue 2 are

$$X_1 = \begin{bmatrix} 1 \\ 0 \\ 0 \\ 0 \end{bmatrix}, X_2 = \begin{bmatrix} 0 \\ 1 \\ 0 \\ 0 \end{bmatrix}$$

Eigenvector corresponding to the eigenvalue $\lambda_4 = 3$:

Let $X = \begin{bmatrix} x_1 \\ x_2 \\ x_3 \\ x_4 \end{bmatrix} \neq O$ be an eigenvector corresponding to the eigenvalue 3, then we have

$$AX = 3X$$

or

$$(A - 3I)X = O$$

or

$$\begin{bmatrix} 2-3 & 0 & 1 & -3 \\ 0 & 2-3 & 10 & 4 \\ 0 & 0 & 2-3 & 0 \\ 0 & 0 & 0 & 3-3 \end{bmatrix}\begin{bmatrix} x_1 \\ x_2 \\ x_3 \\ x_4 \end{bmatrix} = \begin{bmatrix} 0 \\ 0 \\ 0 \\ 0 \end{bmatrix}$$

or
$$\begin{bmatrix} -1 & 0 & 1 & -3 \\ 0 & -1 & 10 & 4 \\ 0 & 0 & -1 & 0 \\ 0 & 0 & 0 & 0 \end{bmatrix} \begin{bmatrix} x_1 \\ x_2 \\ x_3 \\ x_4 \end{bmatrix} = \begin{bmatrix} 0 \\ 0 \\ 0 \\ 0 \end{bmatrix}$$
...(3)

From (3), it is clear that $\rho(A - 3I) = 3$, therefore the equation (3) will have $4 - 3 = 1$ linearly independent solution.

Equation (3) reduces to
$$-x_1 + x_3 - 3x_4 = 0$$
$$-x_2 + 10x_3 + 4x_4 = 0$$
$$-x_3 = 0$$

Let us take $x_4 = c$, then from above equaions, we get
$$x_1 = -3c, \ x_2 = 4c, \ x_3 = 0, \ x_4 = c$$
Therefore,
$$\begin{bmatrix} x_1 \\ x_2 \\ x_3 \\ x_4 \end{bmatrix} = \begin{bmatrix} -3c \\ 4c \\ 0 \\ c \end{bmatrix} = c \begin{bmatrix} -3 \\ 4 \\ 0 \\ 1 \end{bmatrix}$$

Hence the eigenvector corresponding to eigenvalue 3 is
$$X = \begin{bmatrix} -3 \\ 4 \\ 0 \\ 1 \end{bmatrix}.$$

EXAMPLE 14. *Find the eigenvalues and eigenvectors of the matrix*
$$A = \begin{bmatrix} 2 & -1 \\ 5 & -2 \end{bmatrix}.$$

SOLUTION. The characteristic equation of A is given by
$$|A - \lambda I| = 0$$
or
$$\begin{vmatrix} 2 - \lambda & -1 \\ 5 & -2 - \lambda \end{vmatrix} = 0$$
or
$$(2 - \lambda)(-2 - \lambda) + 5 = 0$$
or
$$\lambda^2 + 1 = 0$$
The roots of this equation are $\lambda = \pm i$.
Thus, the eigenvalues of A are $\lambda_1 = -i, \ \lambda_2 = i$.

Eigenvector corresponding to the eigenvalaue $\lambda_1 = -i$:

Let $X = \begin{bmatrix} x_1 \\ x_2 \end{bmatrix} \neq O$ be an eigenvector of A corresponding to the eigenvalue

$\lambda_1 = -i$, then we have

$$AX = \lambda_1 X$$

or $$AX = (-i)X$$

or $$(A - (-i)I)X_1 = O$$

or $$\begin{bmatrix} 2-(-i) & -1 \\ 5 & -2-(-i) \end{bmatrix}\begin{bmatrix} x_1 \\ x_2 \end{bmatrix} = \begin{bmatrix} 0 \\ 0 \end{bmatrix}$$

or $$\begin{bmatrix} 2+i & -1 \\ 5 & -2+i \end{bmatrix}\begin{bmatrix} x_1 \\ x_2 \end{bmatrix} = \begin{bmatrix} 0 \\ 0 \end{bmatrix} \qquad \text{... (1)}$$

Applying $R_1 \rightarrow \left(\dfrac{1}{2+i}\right)R_1$, we get

$$\begin{bmatrix} 1 & -\dfrac{1}{2+i} \\ 5 & -2+i \end{bmatrix}\begin{bmatrix} x_1 \\ x_2 \end{bmatrix} = \begin{bmatrix} 0 \\ 0 \end{bmatrix}$$

Now $$\frac{-1}{2+i} = \frac{-1(2-i)}{(2+i)(2-i)} = \frac{-2+i}{5} = \frac{-2}{5} + \frac{i}{5}$$

$\therefore$ $$\begin{bmatrix} 1 & -\dfrac{2}{5}+\dfrac{i}{5} \\ 5 & -2+i \end{bmatrix}\begin{bmatrix} x_1 \\ x_2 \end{bmatrix} = \begin{bmatrix} 0 \\ 0 \end{bmatrix}$$

Applying $R_2 \rightarrow R_2 - 5R_1$, we get

$$\begin{bmatrix} 1 & -\dfrac{2}{5}+\dfrac{i}{5} \\ 0 & 0 \end{bmatrix}\begin{bmatrix} x_1 \\ x_2 \end{bmatrix} = \begin{bmatrix} 0 \\ 0 \end{bmatrix} \qquad \text{...(2)}$$

From (2), it is clear that $\rho(A - (-i)I) = 1$, therefore the equation (1) will have $2 - 1 = 1$ linearly independent solution.

Equation (2) reduces to

$$x_1 + \left(-\frac{2}{5} + \frac{i}{5}\right)x_2 = 0$$

Let us take $x_2 = c$, then we get

$$x_1 = -c\left(\frac{-2}{5} + \frac{i}{5}\right) = c\left(\frac{2}{5} - \frac{i}{5}\right)$$

$\therefore$ $$\begin{bmatrix} x_1 \\ x_2 \end{bmatrix} = \begin{bmatrix} c\left(\dfrac{2}{5} - \dfrac{i}{5}\right) \\ c \end{bmatrix} = \frac{c}{5}\begin{bmatrix} 2-i \\ 5 \end{bmatrix}$$

Hence the eigenvector corresponding to eigenvalue $\lambda_1 = -i$ is

$$X_1 = \begin{bmatrix} 2-i \\ 5 \end{bmatrix}$$

Eigenvector corresponding to the eigenvalue $\lambda_2 = i$:

Let $X = \begin{bmatrix} x_1 \\ x_2 \end{bmatrix} \neq O$ be an eigenvector corresponding to the eigenvalue $\lambda_2 = i$, then we have

$$AX = \lambda_2 X$$
$$AX = i(X)$$

or

$$(A - (i)I)X = O$$

or

$$\begin{bmatrix} 2-i & -1 \\ 5 & -2-i \end{bmatrix}\begin{bmatrix} x_1 \\ x_2 \end{bmatrix} = \begin{bmatrix} 0 \\ 0 \end{bmatrix} \qquad \qquad ...(3)$$

Applying $R_1 \to \left(\dfrac{1}{2-i}\right)R_1$, we get

$$\begin{bmatrix} 1 & -\dfrac{1}{2-i} \\ 5 & -2-i \end{bmatrix}\begin{bmatrix} x_1 \\ x_2 \end{bmatrix} = \begin{bmatrix} 0 \\ 0 \end{bmatrix}$$

Now

$$\frac{-1}{2-i} = \frac{-1(2+i)}{(2-i)(2+i)} = \frac{-2-i}{5} = \frac{-2}{5} - \frac{i}{5}$$

$\therefore$

$$\begin{bmatrix} 1 & -\dfrac{2}{5} - \dfrac{i}{5} \\ 5 & -2-i \end{bmatrix}\begin{bmatrix} x_1 \\ x_2 \end{bmatrix} = \begin{bmatrix} 0 \\ 0 \end{bmatrix}$$

Applying $R_2 \to R_2 - 5R_1$, we get

$$\begin{bmatrix} 1 & -\dfrac{2}{5} - \dfrac{i}{5} \\ 0 & 0 \end{bmatrix}\begin{bmatrix} x_1 \\ x_2 \end{bmatrix} = \begin{bmatrix} 0 \\ 0 \end{bmatrix} \qquad \qquad ...(4)$$

From (4) it is clear that $\rho(A - (i)I) = 1$, therefore the system of equation (3) will have $2 - 1 = 1$ linearly independent solution.

Equation (4) reduces to

$$x_1 + \left(-\frac{2}{5} - \frac{i}{5}\right)x_2 = 0$$

Let us take $x_2 = c$, then $x_1 = \left(\dfrac{2+i}{5}\right)c$

$\therefore$

$$\begin{bmatrix} x_1 \\ x_2 \end{bmatrix} = \begin{bmatrix} \left(\dfrac{2+i}{5}\right)c \\ c \end{bmatrix} = \frac{c}{5}\begin{bmatrix} 2+i \\ 5 \end{bmatrix}$$

Hence the eigenvector corresponding to eigenvalue $\lambda_2 = i$ is

$$X_1 = \begin{bmatrix} 2+i \\ 5 \end{bmatrix}$$

▶ REMARK

▶ Let A be an $n \times n$ matrix with real entries. If λ is a complex eigenvalue of A with associated eigenvector X, then $\bar{\lambda}$ is also an eigenvalue of A with associated eigenvector $\bar{X}$.

EXERCISE 7.3

1. Prove that a square matrix A and its transpose A' have the same set of eigenvalues.

2. Let A be an $n \times n$ matrix and let $g(x)$ be any polynomial. If λ is an eigenvalue of A, then prove that $g(\lambda)$ is an eigenvalue of $g(A)$.

3. Show that the eigenvalues of a triangular matrix are just the diagonal elements of the matrix.

4. Let $A = $ dig. $(\lambda_1, \lambda_2,..., \lambda_n)$ be a diagonal matrix. Prove that each λ_i $(i = 1, 2, 3,..., n)$ is an eigenvalue of A.

5. Let A be an 3×3 matrix. If $\lambda_1, \lambda_2, \lambda_3$ are the eigenvalues of A, then find the eigenvalues of the matrix $(I + aA)^{-1} (1 + bA)$, where a, b are scalars such that $a\lambda_i \neq -1$ for $i = 1, 2, 3$.

6. Let A and B be two $n \times n$ matrices. Let X be an eigenvector of A and B both. Show that X is also an eigenvector of $aA + bB$, where a, b are scalars.

7. Prove that the eigenvectors of a real symmetric matrix corresponding to two distinct eigenvalues are orthogonal.

8. Prove that the eigenvectors of a Hermitian matrix corresponding to two distinct eigenvalues are orthogonal.

9. (i) If λ is an eigenvalue of a matrix A, then show that $k + \lambda$ is an eigenvalue of $A + kI$.

   (ii) If the matrix A has characteristic roots $\lambda_1, \lambda_2,..., \lambda_n$ show that the matrix A^2 has such roots as $\lambda_1^2, \lambda_2^2,..., \lambda_n^2$.

10. (i) Find the eigenvalues of a matrix $\begin{bmatrix} 1 & 4 \\ 2 & 3 \end{bmatrix}$.

    (ii) Find the eigenvalues of the matrix $A = \begin{bmatrix} a & h & g \\ 0 & b & f \\ 0 & 0 & c \end{bmatrix}$.

11. Find the eigenvalues and eigenvectors of the following matrices :

    (i) $\begin{bmatrix} 2 & -4 \\ -1 & -1 \end{bmatrix}$ (ii) $\begin{bmatrix} -1 & 0 \\ 0 & 1 \end{bmatrix}$

    (iii) $\begin{bmatrix} 1 & 1 \\ -2 & 4 \end{bmatrix}$ (iv) $\begin{bmatrix} 10 & -18 \\ 6 & -11 \end{bmatrix}$

12. Find the eigenvalues and eigenvectors of the following matrices :

    (i) $\begin{bmatrix} 0 & 1 & 0 \\ 0 & 0 & 1 \\ 1 & -3 & 3 \end{bmatrix}$ (ii) $\begin{bmatrix} 5 & 8 & 16 \\ 4 & 1 & 8 \\ -4 & -4 & -11 \end{bmatrix}$

    (iii) $\begin{bmatrix} 1 & -1 & -1 \\ -1 & 1 & -1 \\ -1 & -1 & 1 \end{bmatrix}$ (iv) $\begin{bmatrix} 1 & 2 & 2 \\ 1 & 2 & -1 \\ -1 & 1 & 4 \end{bmatrix}$

    (v) $\begin{bmatrix} 6 & -2 & 2 \\ -2 & 3 & -1 \\ 2 & -1 & 3 \end{bmatrix}$ (vi) $\begin{bmatrix} -2 & 2 & -3 \\ 2 & 1 & -6 \\ -1 & -2 & 0 \end{bmatrix}$

    (vii) $\begin{bmatrix} 1 & 2 & 3 \\ 0 & 2 & 3 \\ 0 & 0 & 2 \end{bmatrix}$ (viii) $\begin{bmatrix} 1 & 1 & 0 \\ 0 & 2 & 2 \\ 0 & 0 & 3 \end{bmatrix}$

    (ix) $\begin{bmatrix} 3 & 1 & 1 \\ 2 & 4 & 2 \\ 1 & 1 & 3 \end{bmatrix}$ (x) $\begin{bmatrix} 2 & 1 & 0 \\ 0 & 2 & 1 \\ 0 & 0 & 2 \end{bmatrix}$

13. Find the eigenvalues and eigenvectors of the matrix $A = \begin{bmatrix} 1 & 1 & 0 & 0 \\ 0 & 2 & 0 & 0 \\ 0 & 0 & 1 & 1 \\ 0 & 0 & -2 & 4 \end{bmatrix}$

14. Find all the characteristic roots and the corresponding characteristic vectors of the matrix $A = \begin{bmatrix} 2 & 1 & -1 \\ 0 & 3 & -2 \\ 2 & 4 & -3 \end{bmatrix}$

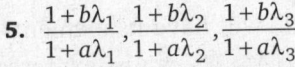

Answers

5. $\dfrac{1+b\lambda_1}{1+a\lambda_1}, \dfrac{1+b\lambda_2}{1+a\lambda_2}, \dfrac{1+b\lambda_3}{1+a\lambda_3}$

10. (i) $-1, 5$ (ii) a, b, c

11. (i) $\lambda_1 = -2, X_1 = \begin{bmatrix} 1 \\ 1 \end{bmatrix}; \lambda_2 = 3, X_2 = \begin{bmatrix} -4 \\ 1 \end{bmatrix}$ (ii) $\lambda_1 = 1, X_1 = \begin{bmatrix} 0 \\ 1 \end{bmatrix}; \lambda_2 = -1, X_2 = \begin{bmatrix} 1 \\ 0 \end{bmatrix}$

    (iii) $\lambda_1 = 2, X_1 = \begin{bmatrix} 1 \\ 1 \end{bmatrix}; \lambda_2 = 3, X_2 = \begin{bmatrix} 1 \\ 2 \end{bmatrix}$ (iv) $\lambda_1 = -2, X_1 = \begin{bmatrix} 3 \\ 2 \end{bmatrix}; \lambda_2 = 1, X_2 = \begin{bmatrix} 2 \\ 1 \end{bmatrix}$

12. (i) $\lambda_1 = \lambda_2 = \lambda_3 = 1, X = \begin{bmatrix} 1 \\ 1 \\ 1 \end{bmatrix}$ (ii) $\lambda_1 = 1, X_1 = \begin{bmatrix} -2 \\ -1 \\ 1 \end{bmatrix}; \lambda_2 = -3, X_2 = \begin{bmatrix} -1 \\ 1 \\ 0 \end{bmatrix}; \lambda_3 = -3, X_3 = \begin{bmatrix} -2 \\ 0 \\ 1 \end{bmatrix}$

(iii) $\lambda_1 = -1, X_1 = \begin{bmatrix} 1 \\ 1 \\ 1 \end{bmatrix}; \lambda_2 = 2, X_2 = \begin{bmatrix} -1 \\ 1 \\ 0 \end{bmatrix}; \lambda_3 = 2, X_3 = \begin{bmatrix} -1 \\ 0 \\ 1 \end{bmatrix}$

(iv) $\lambda_1 = 1, X_1 = \begin{bmatrix} 2 \\ -1 \\ 1 \end{bmatrix}; \lambda_2 = 3, X_2 = \begin{bmatrix} 1 \\ 1 \\ 0 \end{bmatrix}; \lambda_3 = 3, X_3 = \begin{bmatrix} 1 \\ 0 \\ 1 \end{bmatrix}$

(v) $\lambda_1 = 2, X_1 = \begin{bmatrix} -1 \\ 0 \\ 2 \end{bmatrix}; \lambda_2 = 2, X_2 = \begin{bmatrix} 1 \\ 2 \\ 0 \end{bmatrix}; \lambda_3 = 8, X_3 = \begin{bmatrix} 2 \\ -1 \\ 1 \end{bmatrix}$

(vi) $\lambda_1 = -3, X_1 = \begin{bmatrix} -2 \\ 1 \\ 0 \end{bmatrix}; \lambda_2 = -3, X_2 = \begin{bmatrix} 3 \\ 0 \\ 1 \end{bmatrix}; \lambda_3 = 5, X_3 = \begin{bmatrix} 1 \\ 2 \\ 1 \end{bmatrix}$

(vii) $\lambda_1 = \lambda_2 = 1, X = \begin{bmatrix} 1 \\ 0 \\ 0 \end{bmatrix}; \lambda_3 = 2, X_1 = \begin{bmatrix} 2 \\ 1 \\ 0 \end{bmatrix}$ (viii) $\lambda_1 = 1, X_1 = \begin{bmatrix} 1 \\ 0 \\ 0 \end{bmatrix}; \lambda_2 = 2, X_2 = \begin{bmatrix} 2 \\ 1 \\ 0 \end{bmatrix}; \lambda_3 = 3, X_3 = \begin{bmatrix} 1 \\ 2 \\ 1 \end{bmatrix}$

(ix) $\lambda_1 = 2, X_1 = \begin{bmatrix} -1 \\ 1 \\ 0 \end{bmatrix}; \lambda_2 = 2, X_2 = \begin{bmatrix} -1 \\ 0 \\ 1 \end{bmatrix}; \lambda_3 = 6, X_3 = \begin{bmatrix} 1 \\ 2 \\ 1 \end{bmatrix}$ (x) $\lambda_1 = \lambda_2 = \lambda_3 = 2, X = \begin{bmatrix} 1 \\ 0 \\ 0 \end{bmatrix}$

13. $\lambda_1 = 1, X_1 = \begin{bmatrix} 1 \\ 0 \\ 0 \\ 0 \end{bmatrix}; \lambda_2 = 2, X_2 = \begin{bmatrix} 1 \\ 1 \\ 0 \\ 0 \end{bmatrix}; \lambda_3 = 2, X_3 = \begin{bmatrix} 0 \\ 0 \\ 1 \\ 1 \end{bmatrix}; \lambda_4 = 3, X_4 = \begin{bmatrix} 0 \\ 0 \\ 0 \\ 1 \end{bmatrix}$

7.11 THE CAYLEY-HAMILTON THEOREM

THEOREM 1.   *Every square matrix satisfies its characteristic equation.*

PROOF.   Let A be a square matrix of order n and the characteristic equation of A is

$$|A - \lambda I| = (-1^n) \,[\lambda^n + a_1\lambda^{n-1} + a_2\lambda^{n-2} + ... + a_{n-1}\lambda + a_n] = 0$$

then we have to show that its matrix equation

$$X^n + a_1 X^{n-1} + a_2 X^{n-2} + ... + a_{n-1}X + a_n I = O$$

is satisfied by the matrix X = A

i.e., $A^n + a_1 A^{n-1} + a_2 A^{n-2} + ... + a_{n-1}A + a_n I = O$

where I is a unit matrix of order n and O is null matrix of order n.

Since A and I are two square matrices of order n and λ is any characteristic root of A, then the matrix $(A - \lambda I)$ is also a square matrix of order n whose elements are at most of degree one in λ. Therefore Adj. $(A - \lambda I)$ will have its elements a polynomials in λ of degree $n - 1$ or less and thus Adj. $(A - \lambda I)$ can be expressed as a matrix polynomial in λ as follows :

$$\text{Adj. } (A - \lambda I) = B_0\lambda^{n-1} + B_1\lambda^{n-2} + ... + B_{n-2}\lambda + B_{n-1} \qquad ...(1)$$

where $B_0, B_1, ..., B_{n-1}$ are the square matrices of order n.

Since we know that $A(\text{Adj } A) = |A| I_n$

$\therefore \qquad (A - \lambda I) \text{ Adj. } (A - \lambda I) = |A - \lambda I| I$

or $\qquad (A - \lambda I) \text{ Adj. } (A - \lambda I) = (-1^n) [\lambda^n + a_1 \lambda^{n-1} + a_2 \lambda^{n-2} + \dots + a_{n-1} \lambda + a_n] I$

$$\dots (2)$$

Multiplying both sides of (1) by $(A - \lambda I)$, we get

$$(A - \lambda I) \text{ Adj.} (A - \lambda I) = (A - \lambda I) [B_0 \lambda^{n-1} + B_1 \lambda^{n-2} + \dots + B_{n-2} \lambda + B_{n-1}]$$

$$\dots (3)$$

From (2) and (3), we get

$$(A - \lambda I) [B_0 \lambda^{n-1} + B_1 \lambda^{n-2} + \dots + B_{n-2} \lambda + B_{n-1}]$$

$$= (-1^n) [\lambda^n + a_1 \lambda^{n-1} + a_2 \lambda^{n-2} + \dots + a_{n-1} \lambda + a_n] I$$

Now comparing the coefficients of like powers of λ, we get

$$\left.\begin{array}{l} -I B_0 = (-1)^n I \\[2mm] A B_0 - I B_1 = (-1)^n a_1 I \\[2mm] A B_1 - I B_2 = (-1)^n a_2 I \\[2mm] \dots\dots\dots\dots\dots\dots \\[2mm] A B_{n-2} - I B_{n-3} = (-1)^n a_{n-1} I \\[2mm] A B_{n-1} = (-1)^n a_n I \end{array}\right\} \qquad \dots (4)$$

Premultiplying first, second, third etc. equations of (4) by A^n, A^{n-1}, A^{n-2}, etc. respectively and then adding, we get

$$-A^n B_0 + A^n B_0 - A^{n-1} B_1 + A^{n-1} B_1 + \dots$$

$$= (-1)^n [A^n + a_1 A^{n-1} + \dots + a_n I]$$

or $\qquad 0 = (-1)^n [A^n + a_1 A^{n-1} + \dots + a_n I]$

Hence $\qquad A^n + a_1 A^{n-1} + \dots + a_n I = O$

COROLLARY 1. *If A be a non-singular matrix of order $n \times n$ and its characteristic polynomial is*

$$|A - \lambda I| = (-1)^n [\lambda^n + a_1 \lambda^{n-1} + \dots + a_{n-1} \lambda + a_n]$$

then $\qquad det (A) = (-1)^n a_n.$

PROOF. We have

$$|A - \lambda I| = (-1)^n [\lambda^n + a_1 \lambda^{n-1} + \dots + a_{n-1} A + a_n]$$

Putting $\lambda = 0$, we get

$$|A - 0I| = (-1)^n a_n$$

$\Rightarrow \qquad |A| = (-1)^n a_n$

COROLLARY 2. *If $\lambda_1, \lambda_2,\dots, \lambda_n$ are eigenvalues of a square matrix of order $n \times n$, then*

$$det (A) = \lambda_1 \lambda_2 \lambda_3 \dots \lambda_n.$$

PROOF. If λ is an eigenvalue of A, then it is a root of characteristic equation of A. The characteristic equation of A is given by

$$|A - \lambda I| = (-1)^n [\lambda^n + a_1 \lambda^{n-1} + \dots + a_{n-1} \lambda + a_n] = 0$$

Since $\lambda_1, \lambda_2,\dots, \lambda_n$ are the eigenvalues of A, hence

$$(-1)^n [\lambda^n + a_1 \lambda^{n-2} + \dots + a_{n-1} \lambda + a_n] = (\lambda_1 - \lambda)(\lambda_2 - \lambda)\dots(\lambda_n - \lambda) = 0$$

Comparing the constant terms of both sides, we get

$$(-1)^n a_n = \lambda_1 \lambda_2 \lambda_3 \dots \lambda_n$$

From Corollary 1,

$$\det(A) = (-1)^n a_n$$

Hence $\qquad \det(A) = \lambda_1 \lambda_2 \lambda_3 \ldots \lambda_n$

COROLLARY 3. *Let A be an $n \times n$ matrix with characteristic polynomial*

$$f(t) = (-1)^n [t^n + a_1 t^{n-1} + \ldots + a_{n-1} t + a_n]$$

Then A is invertible iff $a_n \neq 0$ and its inverse is

$$A^{-1} = \left(\frac{-1}{a_n}\right) [A^{n-1} + a_1 A^{n-2} + \ldots + a_{n-2} A + a_{n-1} I]$$

PROOF.   By Corollary 1,

$$|A| = (-1)^n a_n$$

$\therefore \qquad |A| \neq 0 \Leftrightarrow a_n \neq 0$

i.e., A is invertible iff $a_n \neq 0$

By Cayley-Hamilton theorem, we have

$$f(A) = O$$

$\Rightarrow \qquad (-1)^n [A^n + a_1 A^{n-1} + \ldots + a_{n-1} A + a_n I] = O$

$\Rightarrow \qquad A^n + a_1 A^{n-1} + \ldots + a_{n-1} A + a_n I = O$

$\Rightarrow \qquad A^{-1} [A^n + a_1 A^{n-1} + \ldots + a_{n-1} A + a_n I] = A^{-1} O = O$

$\Rightarrow \qquad A^{n-1} + a_1 A^{n-2} + \ldots + a_{n-2} A + a_{n-1} I + a_n A^{-1} = O$

Hence $\qquad A^{-1} = \left(\dfrac{-1}{a_n}\right) [A^{n-1} + a_1 A^{n-2} + \ldots + a_{n-2} A + a_{n-1} I]$

COROLLARY 4. *If $\lambda_1, \lambda_2, \ldots, \lambda_n$ are the eigenvalue of a matrix A of order $n \times n$, then*

$$Tr(A) = \text{Trace of } A = \sum_{i=1}^{n} \lambda_i$$

PROOF.   Let $A = [a_{ij}]_{n \times n}$. Then $\text{Tr}(A) = \sum_{i=1}^{n} \lambda_i$

Now $\qquad |A - \lambda I| = \begin{vmatrix} a_{11-\lambda} & a_{12} & \cdots & a_{1n} \\ a_{21} & a_{22-\lambda} & \cdots & a_{2n} \\ \vdots & & & \\ a_{n1} & a_{n2} & \cdots & a_{nn-\lambda} \end{vmatrix}$

$$= (a_{11} - \lambda) A_{11} - a_{12} A_{12} + a_{13} A_{13} + \ldots + (-1)^n a_{1n} A_n$$

where A_{11} = Minor of $a_{11-\lambda}$, A_{12} = Minor of a_{12} and so on.

Clearly A_{11} is a polynomial of degree $n - 1$ in λ.

Therefore

$$A_{11} = (a_{22} - \lambda)(a_{33} - \lambda) \ldots (a_{nn} - \lambda) + O(\lambda^{n-3})$$

$\Rightarrow \qquad (a_{11} - \lambda) A_{11} = (a_{11} - \lambda)(a_{22} - \lambda)(a_{33} - \lambda) \ldots (a_{nn} - \lambda) + O(\lambda^{n-2})$

Similarly,

$$A_{12} = A \text{ polynomial of degree } n - 2 \text{ in } \lambda$$

$$A_{13} = A \text{ polynomial of degree } n - 2 \text{ in } \lambda$$

$$\ldots\ldots\ldots\ldots\ldots\ldots\ldots\ldots\ldots\ldots\ldots\ldots\ldots\ldots\ldots\ldots\ldots\ldots$$

$$A_{1n} = A \text{ polynomial of degree } n - 2 \text{ in } \lambda$$

$\therefore$ $$|A - \lambda I| = (-1)^n \left[\lambda^n - \left(\sum_{i=1}^{n} a_{ii} \right) \lambda^{n-1} + O\left(\lambda^{n-2} \right) \right] \qquad \text{...(1)}$$

Since $\lambda_1, \lambda_2, ..., \lambda_n$ are eigenvalues of A, hence

$$|A - \lambda I| = (-1)^n \left[\lambda^n - \left(\sum_{i=1}^{n} \lambda_i \right) \lambda^{n-1} + O\left(\lambda^{n-2} \right) \right] \qquad \text{...(2)}$$

From (1) and (2), we get

$$(-1)^n \left[\lambda^n - \left(\sum_{i=1}^{n} a_{ii} \right) \lambda^{n-1} + O\left(\lambda^{n-2} \right) \right] = (-1)^n \left[\lambda^n - \left(\sum_{i=1}^{n} \lambda_i \right) \lambda^{n-1} + O\left(\lambda^{n-2} \right) \right]$$

Equating the coefficient of λ^{n-1} of both sides, we get

$$\sum_{i=1}^{n} a_{ii} = \sum_{i=1}^{n} \lambda_i$$

Hence $$\text{Tr}(A) = \sum_{i=1}^{n} a_{ii} = \sum_{i=1}^{n} \lambda_i$$

COROLLARY 5. *If the characteristic equation of a matrix A of order $n \times n$ is*

$$|A - \lambda I| = (-1)^n [\lambda^n + a_1 \lambda^{n-1} + a_2 \lambda^{n-2} + ... + a_{n-1} \lambda + a_n] = 0$$

then $Tr (A) = -a_1$

PROOF. If $\lambda_1, \lambda_2, ..., \lambda_n$ are the eigenvalues of a matrix A, then

$$|A - \lambda I| = (-1)^n (\lambda - \lambda_1)(\lambda - \lambda_2)...(\lambda - \lambda_n) = 0$$

$$= (-1)^n \left[\lambda^n - \left(\sum_{i=1}^{n} \lambda_i \right) \lambda^{n-1} + ... \right] = 0 \qquad \text{...(1)}$$

But $$|A - \lambda I| = (-1)^n [\lambda^n + a_1 \lambda^{n-1} + a_2 \lambda^{n-2} + ... + a_{n-1} \lambda + a_n] = 0 \qquad \text{...(2)}$$

From (1) and (2) we get

$$(-1)^n \left[\lambda^n - \left(\sum_{i=1}^{n} \lambda_i \right) \lambda^{n-1} + ... \right] = (-1)^n [\lambda^n + a_1 \lambda^{n-1} + a_2 \lambda^{n-2} + ... + a_{n-1} \lambda + a_n]$$

Taking the coefficient of λ^{n-1} on both sides, we get

$$\sum_{i=1}^{n} \lambda_i = -a_1$$

By Corollary 4, we have

$$\text{Tr}(A) = \sum_{i=1}^{n} \lambda_i$$

Hence $$\text{Tr} (A) = -a_1$$

COROLLARY 6. *Let A be a matrix of order $n \times n$. If m be a positive integer such that $m \geq n$, then Am is linearly expressible in terms of those of lower order of A.*

PROOF.   By Cayley-Hamilton theorem,

$$A^n + a_1 A^{n-1} + a_2 A^{n-2} + \ldots + a_{n-1} A + a_n I = O \qquad \ldots(1)$$

Multiplying (1) by A^{m-n}, we get

$$A^m + a_1 A^{m-1} + a_2 A^{m-2} + \ldots + a_n A^{m-n} = O$$

or $A^m = (-a_1)A^{m-1} + (-a_2)A^{m-2} + \ldots + (-a_n)A^{m-n}$

Hence the result.

SOLVED EXAMPLES

EXAMPLE 1.   *Find the characteristic equation of the matrix*

$$A = \begin{bmatrix} 1 & 0 & 2 \\ 0 & 2 & 1 \\ 2 & 0 & 3 \end{bmatrix}$$

and verify that it is satisfied by A and hence find its inverse.

SOLUTION.   The characteristic equation of A is given by

$$|A - \lambda I| = 0$$

or

$$\begin{vmatrix} 1-\lambda & 0 & 2 \\ 0 & 2-\lambda & 1 \\ 2 & 0 & 3-\lambda \end{vmatrix} = 0$$

or $(1-\lambda)\{(2-\lambda)(3-\lambda)-0\} + 2\{0 - 2(2-\lambda)\} = 0$

or $-\lambda^3 + 6\lambda^2 - 7\lambda - 2 = 0$

or $\lambda^3 - 6\lambda^2 + 7\lambda + 2 = 0$

Next we have to show that

$$A^3 - 6A^2 + 7A + 2I = O$$

Now $A^2 = A.A$

$$= \begin{bmatrix} 1 & 0 & 2 \\ 0 & 2 & 1 \\ 2 & 0 & 3 \end{bmatrix}\begin{bmatrix} 1 & 0 & 2 \\ 0 & 2 & 1 \\ 2 & 0 & 3 \end{bmatrix}$$

$$= \begin{bmatrix} 5 & 0 & 8 \\ 2 & 4 & 5 \\ 8 & 0 & 13 \end{bmatrix}$$

and $A^3 = A^2.A = \begin{bmatrix} 5 & 0 & 8 \\ 2 & 4 & 5 \\ 8 & 0 & 13 \end{bmatrix}\begin{bmatrix} 1 & 0 & 2 \\ 0 & 2 & 1 \\ 2 & 0 & 3 \end{bmatrix} = \begin{bmatrix} 21 & 0 & 34 \\ 12 & 8 & 23 \\ 34 & 0 & 55 \end{bmatrix}$

$\therefore$ $A^3 - 6A^2 + 7A + 2I$

$$= \begin{bmatrix} 21 & 0 & 34 \\ 12 & 8 & 23 \\ 34 & 0 & 55 \end{bmatrix} - 6\begin{bmatrix} 5 & 0 & 8 \\ 2 & 4 & 5 \\ 8 & 0 & 13 \end{bmatrix} + 7\begin{bmatrix} 1 & 0 & 2 \\ 0 & 2 & 1 \\ 2 & 0 & 3 \end{bmatrix} + 2\begin{bmatrix} 1 & 0 & 0 \\ 0 & 1 & 0 \\ 0 & 0 & 1 \end{bmatrix}$$

$$= \begin{bmatrix} 21-30+7+2 & 0-0+0+0 & 34-48+14+0 \\ 12-12+0+0 & 8-24+14+2 & 23-30+7+0 \\ 34-48+14+0 & 0-0+0+0 & 55-78+21+2 \end{bmatrix}$$

$$= \begin{bmatrix} 0 & 0 & 0 \\ 0 & 0 & 0 \\ 0 & 0 & 0 \end{bmatrix} = O$$

Hence $\quad A^3 - 6A^2 + 7A + 2I = O$ ...(1)

To find A^{-1} :

Since the characteristic equation of A is

$$\lambda^3 - 6\lambda^2 + 7\lambda + 2 = 0$$

$\therefore \qquad |A| = (-1)^3 2 = -2 \neq 0$ $[\because |A| = (-1)^n a_n]$

$\Rightarrow A^{-1}$ exist.

Premultiplying (1) by A^{-1}, we get

$$A^2 - 6A + 7I + 2A^{-1} = O$$

$\Rightarrow \qquad A^{-1} = -\dfrac{1}{2}[A^2 - 6A + 7I]$

$\Rightarrow \qquad A^{-1} = -\dfrac{1}{2}\left\{ \begin{bmatrix} 5 & 0 & 8 \\ 2 & 4 & 5 \\ 8 & 0 & 13 \end{bmatrix} - 6\begin{bmatrix} 1 & 0 & 2 \\ 0 & 2 & 1 \\ 2 & 0 & 3 \end{bmatrix} + 7\begin{bmatrix} 1 & 0 & 0 \\ 0 & 1 & 0 \\ 0 & 0 & 1 \end{bmatrix} \right\}$

$$= -\dfrac{1}{2}\begin{bmatrix} 6 & 0 & -4 \\ 2 & -1 & -1 \\ -4 & 0 & 2 \end{bmatrix}$$

Hence $\quad A^{-1} = -\dfrac{1}{2}\begin{bmatrix} 6 & 0 & -4 \\ 2 & -1 & -1 \\ -4 & 0 & 2 \end{bmatrix} = \dfrac{1}{2}\begin{bmatrix} -6 & 0 & 4 \\ -2 & 1 & 1 \\ 4 & 0 & -2 \end{bmatrix}$

EXAMPLE 2.   *Find the characteristic equation of the matrix*

$$A = \begin{bmatrix} 2 & -1 & 1 \\ -1 & 2 & -1 \\ 1 & -1 & 2 \end{bmatrix}$$

and verify that it is satisfied by A and hence find A^{-1}.

SOLUTION.   The characteristic equation of A is given by

$$|A - \lambda I| = 0$$

or

$$\begin{vmatrix} 2-\lambda & -1 & 2 \\ -1 & 2-\lambda & -1 \\ 1 & -1 & 2-\lambda \end{vmatrix} = 0$$

or $(2-\lambda)\{(2-\lambda)(2-\lambda)-1\} + 1\{-2+\lambda+2\} + 1\{1-2+\lambda\} = 0$

or $\qquad\qquad\qquad\qquad -\lambda^3 + 6\lambda^2 - 9\lambda + 4 = 0$

or $\qquad\qquad\qquad\qquad \lambda^3 - 6\lambda^2 + 9\lambda - 4 = 0$

Next we have to show that
$$A^3 - 6A^2 + 9A - 4I = O$$

Now
$$A^2 = A.A$$

$$= \begin{bmatrix} 2 & -1 & 1 \\ -1 & 2 & -1 \\ 1 & -1 & 2 \end{bmatrix} \begin{bmatrix} 2 & -1 & 1 \\ -1 & 2 & -1 \\ 1 & -1 & 2 \end{bmatrix} = \begin{bmatrix} 6 & -5 & 5 \\ -5 & 6 & -5 \\ 5 & -5 & 6 \end{bmatrix}$$

and $\quad A^3 = A^2.A = \begin{bmatrix} 6 & -5 & 5 \\ -5 & 6 & -5 \\ 5 & -5 & 6 \end{bmatrix} \begin{bmatrix} 2 & -1 & 1 \\ -1 & 2 & -1 \\ 1 & -1 & 2 \end{bmatrix} = \begin{bmatrix} 22 & -21 & 21 \\ -21 & 22 & -21 \\ 21 & -21 & 22 \end{bmatrix}$

Now $A^3 - 6A^2 + 9A + 2I$

$$= \begin{bmatrix} 22 & -21 & 21 \\ -21 & 22 & -21 \\ 21 & -21 & 22 \end{bmatrix} - 6 \begin{bmatrix} 6 & -5 & 5 \\ -5 & 6 & -5 \\ 5 & -5 & 6 \end{bmatrix} + 9 \begin{bmatrix} 2 & -1 & 1 \\ -1 & 2 & -1 \\ 1 & -1 & 2 \end{bmatrix} - 4 \begin{bmatrix} 1 & 0 & 0 \\ 0 & 1 & 0 \\ 0 & 0 & 1 \end{bmatrix}$$

$$= \begin{bmatrix} 22-36+18-4 & -21+30-9-0 & 21-30+9-0 \\ -21+30-9-0 & 22-36+18-4 & -21+30-9-0 \\ 21-30+9-0 & -21+30-9-0 & 22-36+18-4 \end{bmatrix} = \begin{bmatrix} 0 & 0 & 0 \\ 0 & 0 & 0 \\ 0 & 0 & 0 \end{bmatrix} = O$$

Hence $\quad A^3 - 6A^2 + 9A + 4I = O$ $\qquad$ ...(1)

Since $|A| = 2(4-1) + 1(-2+1) + 1(1-2) = 6 - 1 - 1 = 4 \neq 0$

$\Rightarrow A^{-1}$ exist.

Premultiplying (1) by A^{-1}, we get
$$A^2 - 6A + 9I - 4A^{-1} = O$$

$\Rightarrow \qquad A^{-1} = +\dfrac{1}{4}[A^2 - 6A + 9I]$

$\Rightarrow \qquad A^{-1} = \dfrac{1}{4} \left\{ \begin{bmatrix} 6 & -5 & 5 \\ -5 & 6 & -5 \\ 5 & -5 & 6 \end{bmatrix} - 6 \begin{bmatrix} 2 & -1 & 1 \\ -1 & 2 & -1 \\ 1 & -1 & 2 \end{bmatrix} + 9 \begin{bmatrix} 1 & 0 & 0 \\ 0 & 1 & 0 \\ 0 & 0 & 1 \end{bmatrix} \right\}$

$$= \dfrac{1}{4} \begin{bmatrix} 6-12+9 & -5+6+0 & 5-6+0 \\ -5+6+0 & 6-12+9 & -5+6+0 \\ 5-6+0 & -5+6+0 & 6-12+9 \end{bmatrix}$$

$\therefore \qquad A^{-1} = \dfrac{1}{4} \begin{bmatrix} 3 & 1 & -1 \\ 1 & 3 & 1 \\ -1 & 1 & 3 \end{bmatrix}$

EXAMPLE 3. *Find the characteristic equation of the matrix*

$$A = \begin{bmatrix} 1 & 2 & 0 \\ 2 & -1 & 0 \\ 0 & 0 & -1 \end{bmatrix}$$

and hence find A^{-1}.

SOLUTION. The characteristic equation of A is given by
$$|A - \lambda I| = 0$$

or
$$\begin{vmatrix} 1-\lambda & 2 & 0 \\ 2 & -1-\lambda & 0 \\ 0 & 0 & -1-\lambda \end{vmatrix} = 0$$

or $(1-\lambda)\{(-1-\lambda)(-1-\lambda) - 0\} - 2\{2(-1-\lambda) - 0\} = 0$

or $\qquad (1-\lambda)(1+\lambda)^2 + 4(1+\lambda) = 0$

or $\qquad 1 + \lambda^2 + 2\lambda - \lambda - \lambda^3 - 2\lambda^2 + 4 + 4\lambda = 0$

or $\qquad -\lambda^3 - \lambda^2 + 5\lambda + 5 = 0$

or $\qquad \lambda^3 + \lambda^2 - 5\lambda - 5 = 0$

By Cayley-Hamilton theorem, we have
$$A^3 + A^2 - 5A - 5I = O \qquad \qquad \dots(1)$$

Since $\qquad |A - \lambda I| = -\lambda^3 - \lambda^2 + 5\lambda + 5$

$\Rightarrow \qquad |A| = 5 \neq 0 \qquad\qquad \text{(Putting } \lambda = 0\text{)}$

$\Rightarrow A^{-1}$ exists.

Premultiplying (1) by A^{-1}, we get
$$A^2 + A - 5I - 5A^{-1} = 0$$

$\Rightarrow \qquad A^{-1} = \dfrac{1}{5}[A^2 + A - 5I] \qquad\qquad \dots(2)$

Now $\qquad A^2 = A.A = \begin{bmatrix} 1 & 2 & 0 \\ 2 & -1 & 0 \\ 0 & 0 & -1 \end{bmatrix}\begin{bmatrix} 1 & 2 & 0 \\ 2 & -1 & 0 \\ 0 & 0 & -1 \end{bmatrix} = \begin{bmatrix} 5 & 0 & 0 \\ 0 & 5 & 0 \\ 0 & 0 & 5 \end{bmatrix}$

So $\qquad A^{-1} = \dfrac{1}{5}\left\{ \begin{bmatrix} 5 & 0 & 0 \\ 0 & 5 & 0 \\ 0 & 0 & 1 \end{bmatrix} + \begin{bmatrix} 1 & 2 & 0 \\ 2 & -1 & 0 \\ 0 & 0 & -1 \end{bmatrix} - 5\begin{bmatrix} 1 & 0 & 0 \\ 0 & 1 & 0 \\ 0 & 0 & 1 \end{bmatrix} \right\}$

$\therefore \qquad A^{-1} = \dfrac{1}{5}\begin{bmatrix} 1 & 2 & 0 \\ 2 & -1 & 0 \\ 0 & 0 & -5 \end{bmatrix}$

EXAMPLE 4. *Show that the matrix*
$$A = \begin{bmatrix} 0 & c & -b \\ -c & 0 & a \\ b & -a & 0 \end{bmatrix}$$

satisfies Cayley-Hamilton Theorem.

SOLUTION. The characteristic equation of A is given by
$$|A - \lambda I| = 0$$

or
$$\begin{vmatrix} -\lambda & c & -b \\ -c & -\lambda & a \\ b & -a & -\lambda \end{vmatrix} = 0$$

or $\qquad -\lambda(\lambda^2 + a^2) - c(c\lambda - ab) - b(ca + b\lambda) = 0$

or $\qquad\qquad \lambda^3 + \lambda(a^2 + b^2 + c^2) = 0$

We have to show that $A^3 + A(a^2 + b^2 + c^2) = O$

Now

$$A^2 = A.A = \begin{bmatrix} 0 & c & -b \\ -c & 0 & a \\ b & -a & 0 \end{bmatrix}\begin{bmatrix} 0 & c & -b \\ -c & 0 & a \\ b & -a & 0 \end{bmatrix} = \begin{bmatrix} -(c^2+b^2) & ab & ac \\ ab & -(c^2+a^2) & bc \\ ac & bc & -(a^2+b^2) \end{bmatrix}$$

and $\qquad A^3 = A^2.A = \begin{bmatrix} -(c^2+b^2) & ab & ac \\ ab & -(c^2+a^2) & bc \\ ac & bc & -(a^2+b^2) \end{bmatrix}\begin{bmatrix} 0 & c & -b \\ -c & 0 & a \\ b & -a & 0 \end{bmatrix}$

$$= \begin{bmatrix} 0 & -c^3 - b^2c - a^2c & bc^2 + b^2 + a^2b \\ c^3 + a^2c + b^2c & 0 & -ab^2 - ac^2 - a^3 \\ -bc^2 - b^3 - a^2b & ac^2 + ab^2 + a^3 & 0 \end{bmatrix}$$

$$= \begin{bmatrix} 0 & -c(a^2 + b^2 + c^2) & b(a^2 + b^2 + c^2) \\ c(a^2 + b^2 + c^2) & 0 & -a(a^2 + b^2 + c^2) \\ -b(a^2 + b^2 + c^2) & a(a^2 + b^2 + c^2) & 0 \end{bmatrix}$$

$$A^3 = -(a^2 + b^2 + c^2)\begin{bmatrix} 0 & c & -b \\ -c & 0 & a \\ b & -a & 0 \end{bmatrix}$$

$$= -(a^2 + b^2 + c^2)A$$

Hence, $\qquad A^3 + (a^2 + b^2 + c^2)A = O$

EXAMPLE 5.   *Verify Cayley-Hamilton theorem for the matrix*

$$A = \begin{bmatrix} 1 & 1 & 0 & 0 \\ 0 & 2 & 0 & 0 \\ 0 & 0 & -1 & 1 \\ 0 & 0 & -2 & 4 \end{bmatrix}$$

SOLUTION.   The characteristic equation of the matrix A is given by

$$|A - \lambda I| = 0$$

or

$$\begin{vmatrix} 1-\lambda & 1 & 0 & 0 \\ 0 & 2-\lambda & 0 & 0 \\ 0 & 0 & 1-\lambda & 1 \\ 0 & 0 & -2 & 4-\lambda \end{vmatrix} = 0$$

or
$$(1-\lambda)\begin{vmatrix} 2-\lambda & 0 & 0 \\ 0 & 1-\lambda & 1 \\ 0 & -2 & 4-\lambda \end{vmatrix} = 0$$

[Expanding along first column]

or $(1-\lambda)\left[(2-\lambda)\{(1-\lambda)(4-\lambda)+2\}\right] = 0$

or $(1-\lambda)\left[(2-\lambda)(1-\lambda)(4-\lambda)+2(2-\lambda)\right] = 0$

or $(1-\lambda)(2-\lambda)(\lambda^2 - 5\lambda + 6) = 0$

or $(1-\lambda)(2-\lambda)(2-\lambda)(3-\lambda) = 0$

or $(\lambda-1)(\lambda-2)^2(\lambda-3) = 0$

We have to show that
$$(A-I)(A-2I)^2(A-3I) = O$$

Now $A - I = \begin{bmatrix} 1 & 1 & 0 & 0 \\ 0 & 2 & 0 & 0 \\ 0 & 0 & 1 & 1 \\ 0 & 0 & -2 & 4 \end{bmatrix} - \begin{bmatrix} 1 & 0 & 0 & 0 \\ 0 & 1 & 0 & 0 \\ 0 & 0 & 1 & 0 \\ 0 & 0 & 0 & 1 \end{bmatrix} = \begin{bmatrix} 0 & 1 & 0 & 0 \\ 0 & 1 & 0 & 0 \\ 0 & 0 & 0 & 1 \\ 0 & 0 & -2 & 3 \end{bmatrix}$

and $A - 2I = \begin{bmatrix} 1 & 1 & 0 & 0 \\ 0 & 2 & 0 & 0 \\ 0 & 0 & 1 & 1 \\ 0 & 0 & -2 & 4 \end{bmatrix} - 2\begin{bmatrix} 1 & 0 & 0 & 0 \\ 0 & 1 & 0 & 0 \\ 0 & 0 & 1 & 0 \\ 0 & 0 & 0 & 1 \end{bmatrix} = \begin{bmatrix} -1 & 1 & 0 & 0 \\ 0 & 0 & 0 & 0 \\ 0 & 0 & -1 & 1 \\ 0 & 0 & -2 & 2 \end{bmatrix}$

$\therefore$ $(A - 2I)^2 = \begin{bmatrix} -1 & 1 & 0 & 0 \\ 0 & 0 & 0 & 0 \\ 0 & 0 & -1 & 1 \\ 0 & 0 & -2 & 2 \end{bmatrix}\begin{bmatrix} -1 & 1 & 0 & 0 \\ 0 & 0 & 0 & 0 \\ 0 & 0 & -1 & 1 \\ 0 & 0 & -2 & 2 \end{bmatrix} = \begin{bmatrix} 1 & -1 & 0 & 0 \\ 0 & 0 & 0 & 0 \\ 0 & 0 & -1 & 1 \\ 0 & 0 & -2 & 2 \end{bmatrix}$

$A - 3I = \begin{bmatrix} 1 & 1 & 0 & 0 \\ 0 & 2 & 0 & 0 \\ 0 & 0 & 1 & 1 \\ 0 & 0 & -2 & 4 \end{bmatrix} - 3\begin{bmatrix} 1 & 0 & 0 & 0 \\ 0 & 1 & 0 & 0 \\ 0 & 0 & 1 & 0 \\ 0 & 0 & 0 & 1 \end{bmatrix} = \begin{bmatrix} -2 & 1 & 0 & 0 \\ 0 & -1 & 0 & 0 \\ 0 & 0 & -2 & 1 \\ 0 & 0 & -2 & 1 \end{bmatrix}$

Now $(A-I)(A-2I)^2(A-3I)$

$= \begin{bmatrix} 0 & 1 & 0 & 0 \\ 0 & 1 & 0 & 0 \\ 0 & 0 & 0 & 1 \\ 0 & 0 & -2 & 3 \end{bmatrix}\begin{bmatrix} 1 & -1 & 0 & 0 \\ 0 & 0 & 0 & 0 \\ 0 & 0 & -1 & 1 \\ 0 & 0 & -2 & 2 \end{bmatrix}\begin{bmatrix} -2 & 1 & 0 & 0 \\ 0 & -1 & 0 & 0 \\ 0 & 0 & -2 & 1 \\ 0 & 0 & -2 & 1 \end{bmatrix}$

$= \begin{bmatrix} 0 & 1 & 0 & 0 \\ 0 & 1 & 0 & 0 \\ 0 & 0 & 0 & 1 \\ 0 & 0 & -2 & 3 \end{bmatrix}\begin{bmatrix} -2 & 2 & 0 & 0 \\ 0 & 0 & 0 & 0 \\ 0 & 0 & 0 & 0 \\ 0 & 0 & 0 & 0 \end{bmatrix} = \begin{bmatrix} 0 & 0 & 0 & 0 \\ 0 & 0 & 0 & 0 \\ 0 & 0 & 0 & 0 \\ 0 & 0 & 0 & 0 \end{bmatrix} = O$

$\therefore$ $(A-I)(A-2I)^2(A-3I) = O$

Hence the Cayley-Hamilton theorem is verified.

EXAMPLE 6.   _Use Cayley-Hamilton theorem to express $2A^5 - 3A^4 + A^2 - 4I$ as a linear polynomial in A, where :_

$$A = \begin{bmatrix} 3 & 1 \\ -1 & 2 \end{bmatrix}$$

SOLUTION.   The characteristic equation of A is given by

$$|A - \lambda I| = 0$$

or

$$\begin{vmatrix} 3-\lambda & 1 \\ -1 & 2-\lambda \end{vmatrix} = 0$$

or

$$(3-\lambda)(2-\lambda) + 1 = 0$$

or

$$\lambda^2 - 5\lambda + 7 = 0$$

By Cayley-Hamilton theorem, we have

$$A^2 - 5A + 7I = O \hspace{6cm} ...(1)$$

$$\Rightarrow \hspace{2cm} A^2 = 5A - 7I \hspace{5cm} ...(2)$$

Now $\hspace{2cm} A^3 = A^2.A = (5A - 7I)A = 5A^2 - 7A$

$$\therefore \hspace{2.5cm} A^3 = 5A^2 - 7A \hspace{4.5cm} ...(3)$$

Again, $\hspace{1.5cm} A^4 = A^3.A = (5A^2 - 7A)A = 5A^3 - 7A^2$

$$\Rightarrow \hspace{1.5cm} A^4 = 5(5A^2 - 7A) - 7(5A - 7I) \hspace{2cm} \text{[Using (2) and (3)]}$$

$$\Rightarrow \hspace{1.5cm} A^4 = 25A^2 - 35A - 35A + 49I$$

$$\Rightarrow \hspace{1.5cm} A^4 = 25(5A - 7I) - 70A + 49I \hspace{2.5cm} \text{[Using (2)]}$$

$$\Rightarrow \hspace{1.5cm} A^4 = 125A - 175I - 70A + 49I$$

$$\Rightarrow \hspace{1.5cm} A^4 = 55A - 126I \hspace{5cm} ...(4)$$

Also $\hspace{1.7cm} A^5 = A^4.A = (55A - 126I)A$

$$A^5 = 55A^2 - 126A$$

$$\Rightarrow \hspace{1.5cm} A^5 = 55(5A - 7I) - 126A \hspace{3cm} \text{[Using (2)]}$$

$$\therefore \hspace{1.8cm} A^5 = 149A - 385I \hspace{4.5cm} ...(5)$$

Now $\hspace{1cm} 2A^5 - 3A^4 + A^2 - 4I$

$$= 2(149A - 385I) - 3(55A - 126I) + 5A - 7I - 4I$$

$$\hspace{3cm} \text{[Using (2), (4) and (5)]}$$

$$= 298A - 770I - 165A + 378I + 5A - 11I$$

$$= 138A - 403I$$

$\therefore \hspace{1cm} 2A^5 - 3A^4 + A^2 - 4I = 138A - 403I$ which is a linear polynomial in A.

EXAMPLE 7.   _Using Cayley-Hamilton theorem, find the inverse of the matrix_

$$A = \begin{bmatrix} 2 & 1 & 1 \\ 0 & 1 & 0 \\ 1 & 1 & 2 \end{bmatrix} \hspace{2cm} \text{(Rajasthan–2005)}$$

SOLUTION.   Here, $\hspace{2cm} A = \begin{bmatrix} 2 & 1 & 1 \\ 0 & 1 & 0 \\ 1 & 1 & 2 \end{bmatrix}$

Now, the characteristic equation of the matrix is

$$\begin{bmatrix} 2-\lambda & 1 & 1 \\ 0 & 1-\lambda & 0 \\ 1 & 1 & 2-\lambda \end{bmatrix} = 0$$

$\Rightarrow \qquad \lambda^3 - 5\lambda^2 + 7\lambda - 3 = 0$

From Cayley-Hamilton theorem, we have

$$A^3 - 5A^2 + 7A - 3I = 0 \qquad \qquad ...(1)$$

$$A^2 - 5A + 7I - 3A^{-1} = 0 \qquad \qquad \text{[Multiplying by } A^{-1}\text{]}$$

Since,

$$A^2 = \begin{bmatrix} 2 & 1 & 1 \\ 0 & 1 & 0 \\ 1 & 1 & 2 \end{bmatrix} \begin{bmatrix} 2 & 1 & 1 \\ 0 & 1 & 0 \\ 1 & 1 & 2 \end{bmatrix} = \begin{bmatrix} 5 & 4 & 4 \\ 0 & 1 & 0 \\ 4 & 4 & 5 \end{bmatrix}$$

$$A^2 - 5A + 7I = \begin{bmatrix} 5 & 4 & 4 \\ 0 & 1 & 0 \\ 4 & 4 & 5 \end{bmatrix} - 5\begin{bmatrix} 2 & 1 & 1 \\ 0 & 1 & 0 \\ 1 & 1 & 2 \end{bmatrix} + 7\begin{bmatrix} 1 & 0 & 0 \\ 0 & 1 & 0 \\ 0 & 0 & 1 \end{bmatrix}$$

$$= \begin{bmatrix} 2 & -1 & -1 \\ 0 & 3 & 0 \\ -1 & -1 & 2 \end{bmatrix}$$

Hence,

$$A^{-1} = \frac{1}{3} \begin{bmatrix} 2 & -1 & -1 \\ 0 & 3 & 0 \\ -1 & -1 & 2 \end{bmatrix}$$

EXERCISE 7.4

1. Verify Cayley-Hamilton theorem for the matrix

$$A = \begin{bmatrix} 1 & 1 \\ 8 & 1 \end{bmatrix}$$

and use it to find A^{-1}.

2. Use Cayley-Hamilton theorem to find the inverse of the matrix

$$A = \begin{bmatrix} 2 & 1 \\ 5 & 3 \end{bmatrix}$$

3. Verify Cayley-Hamilton theorem for the matrix

$$A = \begin{bmatrix} 0 & 0 & 0 \\ 3 & 1 & 0 \\ -2 & 1 & 4 \end{bmatrix}$$

and hence find A^{-1}.

4. Verify Cayley-Hamilton theorem for the following matrix:

$$A = \begin{bmatrix} 2 & 0 \\ 0 & 1 \end{bmatrix}$$

5. Show that the matrix

$$A = \begin{bmatrix} 1 & 2 \\ 1 & 1 \end{bmatrix}$$

satisfies Cayley-Hamilton theorem.

6. State the Cayley-Hamilton theorem and verify it for the matrix

$$A = \begin{bmatrix} 1 & 0 & -2 \\ 0 & 0 & 0 \\ -2 & 0 & 4 \end{bmatrix}$$

7. Verify Cayley-Hamilton theorem for the matrix

$$A = \begin{bmatrix} 1 & 4 \\ 2 & 3 \end{bmatrix}$$

and hence obtain A^{-1}.

8. Verify Cayley-Hamilton theorem for the matrix

$$A = \begin{bmatrix} 1 & 2 & 1 \\ 0 & 1 & -1 \\ 3 & -1 & 1 \end{bmatrix}$$

and hence find A^{-1}.

9. Verify that the matrix

$$A = \begin{bmatrix} 1 & 2 & 1 \\ -1 & 0 & 3 \\ 2 & -1 & 1 \end{bmatrix}$$

satisfies its characteristic equation.

10. Show that the matrix

$$A = \begin{bmatrix} 2 & 2 & 1 \\ 1 & 3 & 1 \\ 1 & 2 & 2 \end{bmatrix}$$

satisfies Cayley-Hamilton theorem.

11. Verify Cayley-Hamilton theorem for the matrix

$$A = \begin{bmatrix} 1 & \sqrt{2} & 0 \\ \sqrt{2} & -1 & 0 \\ 0 & 0 & 1 \end{bmatrix}$$

and hence find A^{-1}.

12. Verify Cayley-Hamilton theorem for the matrix

$$A = \begin{bmatrix} 1 & 0 & 2 \\ 0 & -1 & 1 \\ 0 & 1 & 0 \end{bmatrix}$$

and hence find A^{-1}.

13. Verify Cayley-Hamilton theorem for the matrix

$$A = \begin{bmatrix} 1 & 3 & 7 \\ 4 & 2 & 3 \\ 0 & 2 & 1 \end{bmatrix}$$

and hence find A^{-1}.

14. Verify Cayley-Hamilton theorem for the matrix

$$A = \begin{bmatrix} 1 & 2 & 3 \\ 3 & -2 & 1 \\ 4 & 2 & 1 \end{bmatrix}$$

and hence find A^{-1}.

15. Verify Cayley-Hamilton theorem for the matrix

$$A = \begin{bmatrix} 1 & 1 & 3 \\ 5 & 2 & 6 \\ -2 & -1 & -3 \end{bmatrix}$$

16. Verify Cayley-Hamilton theorem for the matrix

$$A = \begin{bmatrix} 3 & 2 & 4 \\ 4 & 3 & 2 \\ 2 & 4 & 3 \end{bmatrix}$$

17. Verify Cayley-Hamilton theorem for the matrix

$$A = \begin{bmatrix} 2 & 3 & -2 \\ 0 & 5 & 4 \\ 1 & 0 & 1 \end{bmatrix}$$

18. If $A = \begin{bmatrix} 1 & 2 \\ -1 & 3 \end{bmatrix}$ express $A^6 - 4A^5 + 8A^4 - 12A^3 + 14A^2$ as a linear polynomial in A.

Answers

1. $A^{-1} = -\dfrac{1}{7}\begin{bmatrix} 1 & -1 \\ -8 & 1 \end{bmatrix}$

2. $A^{-1} = \begin{bmatrix} 3 & -1 \\ -5 & 2 \end{bmatrix}$

3. $A^{-1} = \dfrac{1}{5}\begin{bmatrix} 4 & 1 & -1 \\ -12 & 2 & 3 \\ 5 & 0 & 0 \end{bmatrix}$

7. $A^{-1} = -\dfrac{1}{3}\begin{bmatrix} 3 & -4 \\ -2 & 1 \end{bmatrix}$

8. $A^{-1} = \dfrac{1}{9}\begin{bmatrix} 0 & 3 & 3 \\ 3 & 2 & -1 \\ 3 & -7 & -1 \end{bmatrix}$

11. $A^{-1} = -\dfrac{1}{3}\begin{bmatrix} -1 & -\sqrt{2} & 0 \\ -\sqrt{2} & 1 & 0 \\ 0 & 0 & -3 \end{bmatrix}$

13. $A^{-1} = \dfrac{1}{10}\begin{bmatrix} -4 & 11 & -5 \\ -4 & 1 & 25 \\ 8 & -2 & -10 \end{bmatrix}$

14. $A^{-1} = \dfrac{1}{36}\begin{bmatrix} -4 & 4 & 8 \\ 1 & -11 & 0 \\ 14 & 6 & -8 \end{bmatrix}$

18. $-4A + 5I$

7.12 POWER METHOD

It is an iterative technique which gives largest eigen value and its eigen vector conveniently. Let us suppose A be a square matrix of order n having eigen values $\lambda_1, \lambda_2, \lambda_3, ..., \lambda_n$ such that

$$|\lambda_1| > |\lambda_2| > ... > |\lambda_n|.$$

Let
$$X = c_1 x_1 + c_2 x_2 + ... + c_n x_n$$
$$AX = c_1 A x_1 + c_2 A x_2 + ... + c_n A x_n.$$

Since $\lambda_1, \lambda_2, ..., \lambda_n$ be the eigen values corresponding to vectors $x_1, x_2, ..., x_n$, then

$$AX = c_1 \lambda_1 x_1 + c_2 \lambda_2 x_2 + ... + c_n \lambda_n x_n$$
$$A^2 X = c_1 \lambda_1^2 x_1 + c_2 \lambda_2^2 x_2 + ... + c_n \lambda_n^2 x_n$$
$$\cdots$$
$$A^m X = c_1 \lambda_1^m x_1 + c_2 \lambda_2^m x_2 + ... + c_n \lambda_n^m x_n$$
$$= c_1 \lambda_1^m x_1 + \lambda_1^m \sum_{i=2}^{n} c_i \left(\frac{\lambda_i}{\lambda_1}\right)^m x_i$$

$$= c_1 \lambda_1^m x_1 \text{ as } m \to \infty \left(\frac{\lambda_i}{\lambda_1} \right) \to 0 .$$

The power method gives the largest eigen value and eigen vector λ_1 and x_1 respectively. Now avoid this procedure, let us set an iterative procedure as

$$Y^{(i)} = AX^{(i-1)} = \lambda^{(i)} X^{(i)}$$

Now, we repeat this process till $[X^{(i)} - X^{(i-1)}]$ be equal to zero.

On this condition, $\lambda_1^{(i)}$ will be largest eigen value and corresponding eigen vector is $x^{(i)}$.

Also $A^{-1}X = \dfrac{1}{\lambda}X$ gives the smallest eigen value.

SOLVED EXAMPLES

EXAMPLE 1. *Find the largest eigen value of the matrix* $A = \begin{bmatrix} 4 & 1 \\ 1 & 3 \end{bmatrix}$, *by power method.*

SOLUTION. Since $\qquad A = \begin{bmatrix} 4 & 1 \\ 1 & 3 \end{bmatrix}$

Let eigen vector $X = [1,0]^T = \begin{bmatrix} 1 \\ 0 \end{bmatrix}$.

First Iteration. $Y^{(1)} = AX = \begin{bmatrix} 4 & 1 \\ 1 & 3 \end{bmatrix} \begin{bmatrix} 1 \\ 0 \end{bmatrix} = \begin{bmatrix} 4 \\ 1 \end{bmatrix} = 4 \begin{bmatrix} 1 \\ 0.25 \end{bmatrix}$

Here eigen value is 4 and eigen vector is $\begin{bmatrix} 1 \\ 0.25 \end{bmatrix}$.

Second Iteration. $Y^{(2)} = AX^{(1)} = \begin{bmatrix} 4 & 1 \\ 1 & 3 \end{bmatrix} \begin{bmatrix} 1 \\ 0.25 \end{bmatrix} = \begin{bmatrix} 4.25 \\ 1.75 \end{bmatrix} = 4.25 \begin{bmatrix} 1 \\ 0.41 \end{bmatrix}$

Here $\qquad \lambda_2 = 4.25 X^{(2)} = \begin{bmatrix} 1 \\ 0.41 \end{bmatrix}$.

Third Interation. $Y^{(3)} = AX^{(2)} = \begin{bmatrix} 4 & 1 \\ 1 & 3 \end{bmatrix} \begin{bmatrix} 1 \\ 0.41 \end{bmatrix} = \begin{bmatrix} 4.41 \\ 2.23 \end{bmatrix} = 4.41 \begin{bmatrix} 1 \\ 0.50 \end{bmatrix}$

Here $\qquad \lambda_3 = 4.41, x^{(3)} = \begin{bmatrix} 1 \\ 0.50 \end{bmatrix}$

Fourth Iteration. $Y^{(4)} = AX^{(3)} = \begin{bmatrix} 4 & 1 \\ 1 & 3 \end{bmatrix} \begin{bmatrix} 1 \\ 0.50 \end{bmatrix} = \begin{bmatrix} 4.5 \\ 2.5 \end{bmatrix} = 4.5 \begin{bmatrix} 1 \\ 0.55 \end{bmatrix}$

Here $\qquad \lambda_4 = 4.5, X^{(4)} = \begin{bmatrix} 1 \\ 0.55 \end{bmatrix}$

Fifth Iteration. $Y^{(5)} = AX^{(4)} = \begin{bmatrix} 4 & 1 \\ 1 & 3 \end{bmatrix} \begin{bmatrix} 1 \\ 0.55 \end{bmatrix} = \begin{bmatrix} 4.55 \\ 2.66 \end{bmatrix} = 4.55 \begin{bmatrix} 1 \\ 0.58 \end{bmatrix}$

Here $\qquad \lambda_5 = 4.55, X^{(5)} = \begin{bmatrix} 1 \\ 0.58 \end{bmatrix}$

Sixth Iteration. $Y^{(6)} = AX^{(5)} = \begin{bmatrix} 4 & 1 \\ 1 & 3 \end{bmatrix} \begin{bmatrix} 1 \\ 0.58 \end{bmatrix} = \begin{bmatrix} 4.58 \\ 2.74 \end{bmatrix} = 4.58 \begin{bmatrix} 1 \\ 0.59 \end{bmatrix}$

Here $\qquad \lambda_6 = 4.58, X^{(6)} = \begin{bmatrix} 1 \\ 0.59 \end{bmatrix}$

Seventh Iteration. $Y^{(7)} = AX^{(6)} = \begin{bmatrix} 4 & 1 \\ 1 & 3 \end{bmatrix} \begin{bmatrix} 1 \\ 0.59 \end{bmatrix} = \begin{bmatrix} 4.59 \\ 2.77 \end{bmatrix} = 4.59 \begin{bmatrix} 1 \\ 0.60 \end{bmatrix}$

Here $\qquad \lambda_7 = 4.59, X^{(7)} = \begin{bmatrix} 1 \\ 0.60 \end{bmatrix}$

Eighth Iteration. $\quad Y^{(8)} = AX^{(7)} = \begin{bmatrix} 4 & 1 \\ 1 & 3 \end{bmatrix} \begin{bmatrix} 1 \\ 0.60 \end{bmatrix} = \begin{bmatrix} 4.6 \\ 2.8 \end{bmatrix} = 4.6 \begin{bmatrix} 1 \\ 0.60 \end{bmatrix}$

Here $\qquad \lambda_8 = 4.6, X^{(8)} = \begin{bmatrix} 1 \\ 0.60 \end{bmatrix}$

Since here $\qquad Y^{(8)} = AX^7 = \lambda_8 X^{(8)}$ and $\quad X^{(8)} - X^{(7)} = 0$.

Hence, largest eigen value = 4.6, and corresponding eigen vector will be

$$X = \begin{bmatrix} 0 \\ 0.60 \end{bmatrix}.$$

EXAMPLE 2. *Find the largest eigen value of the matrix*

$$A = \begin{bmatrix} 1 & 2 & 3 \\ 0 & -4 & 2 \\ 0 & 0 & 7 \end{bmatrix}, \text{ using power method.}$$

SOLUTION. Let the eigen vector $X = \begin{bmatrix} 1 & 1 & 1 \end{bmatrix}^T = \begin{bmatrix} 1 \\ 1 \\ 1 \end{bmatrix}$.

First Iteration. $Y^{(1)} = AX = \begin{bmatrix} 1 & 2 & 3 \\ 0 & -4 & 2 \\ 0 & 0 & 7 \end{bmatrix} \begin{bmatrix} 1 \\ 1 \\ 1 \end{bmatrix} = \begin{bmatrix} 6 \\ -2 \\ 7 \end{bmatrix} = 7 \begin{bmatrix} 0.86 \\ -0.28 \\ 1 \end{bmatrix}$

Here $\qquad \lambda_1 = 7, X^{(1)} = \begin{bmatrix} 0.86 \\ -0.28 \\ 1 \end{bmatrix}$.

Second Iteration. $Y^{(2)} = AX^{(1)} = \begin{bmatrix} 1 & 2 & 3 \\ 0 & -4 & 2 \\ 0 & 0 & 7 \end{bmatrix} \begin{bmatrix} 0.86 \\ -0.28 \\ 1 \end{bmatrix} = \begin{bmatrix} 3.3 \\ 3.12 \\ 9 \end{bmatrix} = 7 \begin{bmatrix} 0.47 \\ 0.44 \\ 1 \end{bmatrix}$

Here $\qquad \lambda_2 = 7, X^{(2)} = \begin{bmatrix} 0.47 \\ 0.44 \\ 1 \end{bmatrix}$.

Third Iteration. $Y^{(3)} = AX^{(2)} = \begin{bmatrix} 1 & 2 & 3 \\ 0 & -4 & 2 \\ 0 & 0 & 7 \end{bmatrix} \begin{bmatrix} 0.47 \\ 0.44 \\ 1 \end{bmatrix} = 7 \begin{bmatrix} 0.62 \\ 0.034 \\ 1 \end{bmatrix} = 7$

Here $\qquad \lambda_3 = 7, X^{(3)} = \begin{bmatrix} 0.62 \\ 0.034 \\ 1 \end{bmatrix}$.

Fourth Iteration. $Y^{(4)} = AX^{(3)} = \begin{bmatrix} 1 & 2 & 3 \\ 0 & -4 & 2 \\ 0 & 0 & 7 \end{bmatrix} \begin{bmatrix} 0.62 \\ 0.034 \\ 1 \end{bmatrix} = \begin{bmatrix} 3.68 \\ 1.88 \\ 7 \end{bmatrix} = 7 \begin{bmatrix} 0.52 \\ 0.27 \\ 1 \end{bmatrix}$

Here $\qquad \lambda_4 = 7, X^{(4)} = \begin{bmatrix} 0.52 \\ 0.27 \\ 1 \end{bmatrix}$.

Fifth Iteration. $Y^{(5)} = AX^{(4)} = \begin{bmatrix} 1 & 2 & 3 \\ 0 & -4 & 2 \\ 0 & 0 & 7 \end{bmatrix} \begin{bmatrix} 0.52 \\ 0.27 \\ 1 \end{bmatrix} = \begin{bmatrix} 4.06 \\ 0.92 \\ 7 \end{bmatrix} = 7 \begin{bmatrix} 0.58 \\ 0.13 \\ 1 \end{bmatrix}$

Here $\qquad \lambda_5 = 7, X^{(5)} = \begin{bmatrix} 0.58 \\ 0.13 \\ 1 \end{bmatrix}$.

Sixth Iteration. $Y^{(6)} = AX^{(5)} = \begin{bmatrix} 1 & 2 & 3 \\ 0 & -4 & 2 \\ 0 & 0 & 7 \end{bmatrix} \begin{bmatrix} 0.58 \\ 0.13 \\ 1 \end{bmatrix} = \begin{bmatrix} 3.84 \\ 1.48 \\ 7 \end{bmatrix} = 7 \begin{bmatrix} 0.55 \\ 0.21 \\ 1 \end{bmatrix}$.

Here $\qquad \lambda_6 = 7, X^{(6)} = \begin{bmatrix} 0.55 \\ 0.21 \\ 1 \end{bmatrix}$

Seventh Iteration. $Y^{(7)} = AX^{(6)} = \begin{bmatrix} 1 & 2 & 3 \\ 0 & -4 & 2 \\ 0 & 0 & 7 \end{bmatrix} \begin{bmatrix} 0.55 \\ 0.21 \\ 1 \end{bmatrix} = \begin{bmatrix} 3.97 \\ 1.16 \\ 7 \end{bmatrix} = 7 \begin{bmatrix} 0.56 \\ 0.16 \\ 1 \end{bmatrix}$

Here $\qquad \lambda_7 = 7, X^{(7)} = \begin{bmatrix} 0.56 \\ 0.16 \\ 1 \end{bmatrix}$

Eight Iteration. $Y^{(8)} = AX^{(7)} = \begin{bmatrix} 1 & 2 & 3 \\ 0 & -4 & 2 \\ 0 & 0 & 7 \end{bmatrix} \begin{bmatrix} 0.56 \\ 0.16 \\ 1 \end{bmatrix} = \begin{bmatrix} 3.88 \\ 1.36 \\ 7 \end{bmatrix} = 7 \begin{bmatrix} 0.55 \\ 0.19 \\ 1 \end{bmatrix}$

Here $\qquad \lambda_8 = 7, X^{(8)} = \begin{bmatrix} 0.55 \\ 0.19 \\ 1 \end{bmatrix}$

Ninth Iteration. $Y^{(9)} = AX^{(8)} = \begin{bmatrix} 1 & 2 & 3 \\ 0 & -4 & 2 \\ 0 & 0 & 7 \end{bmatrix} \begin{bmatrix} 0.55 \\ 0.19 \\ 1 \end{bmatrix} = \begin{bmatrix} 3.93 \\ 1.24 \\ 7 \end{bmatrix} = 7 \begin{bmatrix} 0.56 \\ 0.177 \\ 1 \end{bmatrix}$

Here $\qquad \lambda_9 = 7, X^{(9)} = \begin{bmatrix} 0.56 \\ 0.177 \\ 1 \end{bmatrix}$

Tenth Iteration. $Y^{(10)} = AX^{(9)} = \begin{bmatrix} 1 & 2 & 3 \\ 0 & -4 & 2 \\ 0 & 0 & 7 \end{bmatrix} \begin{bmatrix} 0.56 \\ 0.177 \\ 1 \end{bmatrix} = \begin{bmatrix} 3.92 \\ 1.28 \\ 7 \end{bmatrix} = 7 \begin{bmatrix} 0.56 \\ 0.18 \\ 1 \end{bmatrix}$

Here $\qquad \lambda_{10} = 7, X^{(10)} = \begin{bmatrix} 0.56 \\ 0.18 \\ 1 \end{bmatrix}$

Eleventh Iteration. $Y^{(11)} = AX^{(10)} = \begin{bmatrix} 1 & 2 & 3 \\ 0 & -4 & 2 \\ 0 & 0 & 7 \end{bmatrix} \begin{bmatrix} 0.56 \\ 0.18 \\ 1 \end{bmatrix} = 7 \begin{bmatrix} 0.56 \\ 0.18 \\ 1 \end{bmatrix}$

$$\lambda_{11} = 7, X^{(11)} = \begin{bmatrix} 0.56 \\ 0.18 \\ 1 \end{bmatrix}$$

Here, $[X^{(11)} - X^{(10)}] = 0$.

Hence, largest eigen value is 7 and corresponding eigen vector is $\begin{bmatrix} 0.56 \\ 0.18 \\ 1 \end{bmatrix}$.

EXAMPLE 3. *Find the largest eigen value of the matrix* $A = \begin{bmatrix} 1 & 2 \\ 3 & 4 \end{bmatrix}$ *using power method.*

SOLUTION. Let eigen vector $X = [1 \quad 0]^T = \begin{bmatrix} 1 \\ 0 \end{bmatrix}$.

First Iteration. $Y^{(1)} = AX = \begin{bmatrix} 1 & 2 \\ 3 & 4 \end{bmatrix}\begin{bmatrix} 1 \\ 0 \end{bmatrix} = \begin{bmatrix} 1 \\ 3 \end{bmatrix} = 0\begin{bmatrix} 0.3 \\ 1 \end{bmatrix}$

Here $\lambda_1 = 3,\ X^{(1)} = \begin{bmatrix} 0.3 \\ 1 \end{bmatrix}$.

Second Iteration. $Y^{(2)} = AX^{(1)} = \begin{bmatrix} 1 & 2 \\ 3 & 4 \end{bmatrix}\begin{bmatrix} 0.3 \\ 1 \end{bmatrix} = \begin{bmatrix} 2.3 \\ 4.9 \end{bmatrix} = 4.9\begin{bmatrix} 0.46 \\ 1 \end{bmatrix}$

Here $\lambda_2 = 4.9,\ X^{(2)} = \begin{bmatrix} 0.46 \\ 1 \end{bmatrix}$.

Third Iteration. $Y^{(3)} = AX^{(2)} = \begin{bmatrix} 1 & 2 \\ 3 & 4 \end{bmatrix}\begin{bmatrix} 0.46 \\ 1 \end{bmatrix} = \begin{bmatrix} 2.46 \\ 5.38 \end{bmatrix} = 5.38\begin{bmatrix} 0.46 \\ 1 \end{bmatrix}$

Here $\lambda_3 = 5.38,\ X^{(3)} = \begin{bmatrix} 0.46 \\ 1 \end{bmatrix}$.

Since $[X^{(3)} - X^{(2)}] = 0$.

Hence, largest eigen value is 5.38 and corresponding eigen vector $[0.46 \quad 1]^T$.

EXAMPLE 4. *Find the largest eigen value of the following matrix, by using power method,*

$$A = \begin{bmatrix} 1 & 6 & 1 \\ 1 & 2 & 0 \\ 0 & 0 & 3 \end{bmatrix} as\ X = \begin{bmatrix} 1 \\ 0 \\ 0 \end{bmatrix}.$$ (Meerut-2002, Agra-2003)

SOLUTION. Let $X = \begin{bmatrix} 1 \\ 0 \\ 0 \end{bmatrix}$.

First Iteration. $Y^{(1)} = AX = \begin{bmatrix} 1 & 6 & 1 \\ 1 & 2 & 0 \\ 0 & 0 & 3 \end{bmatrix}\begin{bmatrix} 1 \\ 0 \\ 0 \end{bmatrix} = \begin{bmatrix} 1 \\ 1 \\ 0 \end{bmatrix} = 1\begin{bmatrix} 1 \\ 1 \\ 0 \end{bmatrix}$

Here $\lambda_1 = 1,\ X^{(1)} = \begin{bmatrix} 1 \\ 1 \\ 0 \end{bmatrix}$.

Second Iteration. $Y^{(2)} = AX^{(1)} = \begin{bmatrix} 1 & 6 & 1 \\ 1 & 2 & 0 \\ 0 & 0 & 3 \end{bmatrix}\begin{bmatrix} 1 \\ 1 \\ 0 \end{bmatrix} = \begin{bmatrix} 7 \\ 3 \\ 0 \end{bmatrix} = 7\begin{bmatrix} 1 \\ 0.42 \\ 0 \end{bmatrix}$

Here $\lambda_2 = 7,\ X^{(2)} = \begin{bmatrix} 1 \\ 0.42 \\ 0 \end{bmatrix}$.

Third Iteration. $Y^{(3)} = AX^{(2)} = \begin{bmatrix} 1 & 6 & 1 \\ 1 & 2 & 0 \\ 0 & 0 & 3 \end{bmatrix} \begin{bmatrix} 1 \\ 0.42 \\ 0 \end{bmatrix} = \begin{bmatrix} 3.52 \\ 1.84 \\ 0 \end{bmatrix} = 3.52 \begin{bmatrix} 1 \\ 0.52 \\ 0 \end{bmatrix}$

Here $\lambda_3 = 3.52, X^{(3)} = \begin{bmatrix} 1 \\ 0.52 \\ 0 \end{bmatrix}.$

Fourth Iteration. $Y^{(4)} = AX^{(3)} = \begin{bmatrix} 1 & 6 & 1 \\ 1 & 2 & 0 \\ 0 & 0 & 3 \end{bmatrix} \begin{bmatrix} 1 \\ 0.52 \\ 0 \end{bmatrix} = \begin{bmatrix} 4.12 \\ 2.04 \\ 0 \end{bmatrix} = 4.12 \begin{bmatrix} 1 \\ 0.49 \\ 0 \end{bmatrix}$

Here $\lambda_4 = 4.12, X^{(4)} = \begin{bmatrix} 1 \\ 0.49 \\ 0 \end{bmatrix}.$

Fifth Iteration. $Y^{(5)} = AX^{(4)} = \begin{bmatrix} 1 & 6 & 1 \\ 1 & 2 & 0 \\ 0 & 0 & 3 \end{bmatrix} \begin{bmatrix} 1 \\ 0.49 \\ 0 \end{bmatrix} = \begin{bmatrix} 3.94 \\ 1.98 \\ 0 \end{bmatrix} = 3.94 \begin{bmatrix} 1 \\ 0.50 \\ 0 \end{bmatrix}$

Here $\lambda^5 = 3.94, X^{(5)} = \begin{bmatrix} 1 \\ 0.50 \\ 0 \end{bmatrix}.$

Sixth Iteration. $Y^{(6)} = AX^{(5)} = \begin{bmatrix} 1 & 6 & 1 \\ 1 & 2 & 0 \\ 0 & 0 & 3 \end{bmatrix} \begin{bmatrix} 1 \\ 0.50 \\ 0 \end{bmatrix} = \begin{bmatrix} 4 \\ 2 \\ 0 \end{bmatrix} = 4 \begin{bmatrix} 1 \\ 0.50 \\ 0 \end{bmatrix}$

$$\lambda_6 = 4, X^{(6)} = \begin{bmatrix} 1 \\ 0.50 \\ 0 \end{bmatrix}.$$

Here $[X^{(6)} - X^{(5)}] = 0.$

Hence, the largest eigen value is 4 and corresponding eigen vector will be $\begin{bmatrix} 1 \\ 0.50 \\ 0 \end{bmatrix}.$

EXAMPLE 5. *Find the largest eigen value and corresponding eigen vector, using power method,*

$$A = \begin{bmatrix} 1 & -3 & 2 \\ 4 & 4 & -1 \\ 6 & 3 & 5 \end{bmatrix} where\ X = \begin{bmatrix} 1 \\ 1 \\ 1 \end{bmatrix}.$$

SOLUTION.

First Iteration. $Y^{(1)} = AX = \begin{bmatrix} 1 & -3 & 2 \\ 4 & 4 & -1 \\ 6 & 3 & 5 \end{bmatrix} \begin{bmatrix} 1 \\ 1 \\ 1 \end{bmatrix} = \begin{bmatrix} 0 \\ 7 \\ 14 \end{bmatrix} = 14 \begin{bmatrix} 0 \\ 0.5 \\ 1 \end{bmatrix},$

$$\lambda_1 = 14, X^{(1)} = \begin{bmatrix} 0 \\ 0.5 \\ 1 \end{bmatrix}.$$

Second Iteration. $Y^{(2)} = \begin{bmatrix} 1 & -3 & 2 \\ 4 & 4 & -1 \\ 6 & 3 & 5 \end{bmatrix} \begin{bmatrix} 0 \\ 0.50 \\ 1 \end{bmatrix} = \begin{bmatrix} 0.5 \\ 1 \\ 6.5 \end{bmatrix} = 6.5 \begin{bmatrix} 0.076 \\ 0.153 \\ 1 \end{bmatrix},$

$$\lambda_2 = 6.5, X^{(2)} = \begin{bmatrix} 0.076 \\ 0.153 \\ 1 \end{bmatrix}.$$

Third Iteration. $Y^{(3)} = \begin{bmatrix} 1 & -3 & 2 \\ 4 & 4 & -1 \\ 6 & 3 & 5 \end{bmatrix} \begin{bmatrix} 0.076 \\ 0.153 \\ 1 \end{bmatrix} = \begin{bmatrix} 1.617 \\ -0.084 \\ 5.915 \end{bmatrix} = 5.915 \begin{bmatrix} 0.273 \\ 0.014 \\ 1 \end{bmatrix},$

$$\lambda_3 = 5.915, X^{(3)} = \begin{bmatrix} 0.273 \\ 0.014 \\ 1 \end{bmatrix}.$$

Fourth Iteration. $Y^{(4)} = \begin{bmatrix} 1 & -3 & 2 \\ 4 & 4 & -1 \\ 6 & 3 & 5 \end{bmatrix} \begin{bmatrix} 0.273 \\ 0.014 \\ 1 \end{bmatrix} = \begin{bmatrix} 2.231 \\ 0.148 \\ 6.68 \end{bmatrix} = 6.68 \begin{bmatrix} 0.333 \\ 0.022 \\ 1 \end{bmatrix},$

$$\lambda_4 = 6.68, X^{(4)} = \begin{bmatrix} 0.333 \\ 0.022 \\ 1 \end{bmatrix}.$$

Fifth Iteration. $Y^{(5)} = \begin{bmatrix} 1 & -3 & 2 \\ 4 & 4 & -1 \\ 6 & 3 & 5 \end{bmatrix} \begin{bmatrix} 0.333 \\ 0.022 \\ 1 \end{bmatrix} = \begin{bmatrix} 2.267 \\ 0.42 \\ 7.064 \end{bmatrix} = 7.064 \begin{bmatrix} 0.320 \\ 0.059 \\ 1 \end{bmatrix},$

$$\lambda^5 = 7.064, X^{(5)} = \begin{bmatrix} 0.320 \\ 0.059 \\ 1 \end{bmatrix}.$$

Sixth Iteration. $Y_{(6)} = \begin{bmatrix} 1 & -3 & 2 \\ 4 & 4 & -1 \\ 6 & 3 & 5 \end{bmatrix} \begin{bmatrix} 0.320 \\ 0.059 \\ 1 \end{bmatrix} = \begin{bmatrix} 2.143 \\ 0.516 \\ 7.097 \end{bmatrix} = 7.097 \begin{bmatrix} 0.268 \\ 0.064 \\ 1 \end{bmatrix},$

$$\lambda_6 = 7.07, X^{(6)} = \begin{bmatrix} 0.268 \\ 0.064 \\ 1 \end{bmatrix}.$$

Seventh Iteration. $Y^{(7)} = \begin{bmatrix} 1 & -3 & 2 \\ 4 & 4 & -1 \\ 6 & 3 & 5 \end{bmatrix} \begin{bmatrix} 0.268 \\ 0.064 \\ 1 \end{bmatrix} = \begin{bmatrix} 2.076 \\ 0.0328 \\ 6.8 \end{bmatrix} = 6.8 \begin{bmatrix} 0.305 \\ 0.048 \\ 1 \end{bmatrix},$

$$\lambda_7 = 6.8, X^{(7)} = \begin{bmatrix} 0.305 \\ 0.048 \\ 1 \end{bmatrix}.$$

Eight Iteration. $Y^{(8)} = \begin{bmatrix} 1 & -3 & 2 \\ 4 & 4 & -1 \\ 6 & 3 & 5 \end{bmatrix} \begin{bmatrix} 0.305 \\ 0.048 \\ 1 \end{bmatrix} = \begin{bmatrix} 2.161 \\ 0.412 \\ 6.974 \end{bmatrix},$

$$\lambda_8 = 6.974, X^{(8)} = \begin{bmatrix} 0.309 \\ 0.059 \\ 1 \end{bmatrix}.$$

Proceeding in the same way, we get

$$Y^9 = \begin{bmatrix} 2.132 \\ 0.472 \\ 7.031 \end{bmatrix} \text{ where } \lambda_9 = 7.031, X^{(9)} = \begin{bmatrix} 0.303 \\ 0.067 \\ 1 \end{bmatrix}$$

$$Y^{(10)} = \begin{bmatrix} 2.102 \\ 0.48 \\ 7.019 \end{bmatrix} \text{ where } \lambda_{10} = 7.019, X^{(10)} = \begin{bmatrix} 0.299 \\ 0.068 \\ 1 \end{bmatrix}$$

$$Y^{(11)} = \begin{bmatrix} 2.0295 \\ 0.468 \\ 6.998 \end{bmatrix} \text{ where } \lambda_{11} = 6.998, X^{(11)} = \begin{bmatrix} 0.299 \\ 0.066 \\ 1 \end{bmatrix}$$

$$Y^{(12)} = \begin{bmatrix} 2.101 \\ 0.46 \\ 6.992 \end{bmatrix} \text{ where } \lambda_{12} = 6.992, X^{(12)} = \begin{bmatrix} 0.3004 \\ 0.065 \\ 1 \end{bmatrix}$$

$$Y^{(13)} = \begin{bmatrix} 2.1054 \\ 0.4614 \\ 7.9974 \end{bmatrix} \text{ where } \lambda_{13} = 6.9974, X^{(13)} = \begin{bmatrix} 0.3008 \\ 0.0659 \\ 1 \end{bmatrix}$$

$$Y^{(14)} = \begin{bmatrix} 2.1031 \\ 0.4668 \\ 7.0025 \end{bmatrix} \text{ where } \lambda_{14} = 7.0025, X^{(14)} = \begin{bmatrix} 0.3003 \\ 0.0666 \\ 1 \end{bmatrix}$$

$$Y^{(15)} = \begin{bmatrix} 2.1005 \\ 0.4676 \\ 7.0016 \end{bmatrix} \text{ where } \lambda_{15} = 7.0016, X^{(15)} = \begin{bmatrix} 0.3000 \\ 0.06678 \\ 1 \end{bmatrix}$$

$$Y^{(16)} = \begin{bmatrix} 2.0999 \\ 0.4668 \\ 7.001 \end{bmatrix} \text{ where } \lambda_{16} = 7.001, X^{(16)} = \begin{bmatrix} 0.2998 \\ 0.066 \\ 1 \end{bmatrix}$$

In the above iteration we saw that iteration is rounding off $Y^{(16)}$ thus let

$$\lambda = 0.7, X = \begin{bmatrix} 0.2998 \\ 0.066 \\ 1 \end{bmatrix}.$$

EXAMPLE 6. *Find the largest eigen value and the corresponding eigen vector of matrix*

$$A = \begin{bmatrix} 2 & -1 & 0 \\ -1 & 2 & -1 \\ 0 & -1 & 2 \end{bmatrix}.$$ (Agra-2008)

SOLUTION. Let the initial eigen vector be $X = \begin{bmatrix} 1 \\ 0 \\ 0 \end{bmatrix}$

First Iteration. $Y^{(1)} = AX = \begin{bmatrix} 2 & -1 & 0 \\ -1 & 2 & -1 \\ 0 & -1 & 2 \end{bmatrix} \begin{bmatrix} 1 \\ 0 \\ 0 \end{bmatrix} = \begin{bmatrix} 2 \\ -1 \\ 0 \end{bmatrix} = 2 \begin{bmatrix} 1 \\ -0.5 \\ 0 \end{bmatrix}$,

$$\lambda_1 = 2, X^{(1)} = \begin{bmatrix} 1 \\ -0.5 \\ 0 \end{bmatrix}.$$

Second Iteration. $Y^{(2)} = AX^{(1)} = \begin{bmatrix} 2 & -1 & 0 \\ +1 & 2 & -1 \\ 0 & -1 & 2 \end{bmatrix} \begin{bmatrix} 1 \\ -0.5 \\ 0 \end{bmatrix} = 2.5 \begin{bmatrix} 1 \\ -0.8 \\ 0.2 \end{bmatrix}$

$$\lambda_2 = 2.5, X^{(2)} = \begin{bmatrix} 1 \\ -0.8 \\ 0.2 \end{bmatrix}$$

Third Iteration. $Y^{(3)} = AX^{(2)} = \begin{bmatrix} 2 & -1 & 0 \\ -1 & 2 & -1 \\ 0 & -1 & 2 \end{bmatrix} \begin{bmatrix} 1 \\ -0.8 \\ 0.2 \end{bmatrix} = 2.8 \begin{bmatrix} 1 \\ -1 \\ 0.43 \end{bmatrix}$

$$\lambda_3 = 2.8, X^{(3)} = \begin{bmatrix} 1 \\ -1 \\ 0.43 \end{bmatrix}.$$

Fourth Iteration. $Y^{(4)} = AX^{(3)} = \begin{bmatrix} 2 & -1 & 0 \\ -1 & 2 & -1 \\ 0 & -1 & 2 \end{bmatrix} \begin{bmatrix} 1 \\ -1 \\ 0.43 \end{bmatrix} = \begin{bmatrix} 3 \\ -3.43 \\ 1.86 \end{bmatrix} = 3.43 \begin{bmatrix} 0.87 \\ -1 \\ 0.54 \end{bmatrix}$

$$\lambda_4 = 3.43, X^{(4)} = \begin{bmatrix} 0.87 \\ -1 \\ 0.54 \end{bmatrix}.$$

Fifth Iteration. $Y^{(5)} = AX^{(4)} = \begin{bmatrix} 2 & -1 & 0 \\ -1 & 2 & -1 \\ 0 & -1 & 2 \end{bmatrix} \begin{bmatrix} 0.87 \\ -1 \\ 0.54 \end{bmatrix} = \begin{bmatrix} 2.728 \\ -3.41 \\ 2.080 \end{bmatrix} = 3.41 \begin{bmatrix} 0.80 \\ -1 \\ 0.61 \end{bmatrix}$

$$\lambda_5 = 3.41, X^{(5)} = \begin{bmatrix} 0.81 \\ -1 \\ 0.61 \end{bmatrix}.$$

Sixth Iteration. $Y^{(6)} = AX^{(5)} = \begin{bmatrix} 2 & -1 & 0 \\ -1 & 2 & -1 \\ 0 & -1 & 2 \end{bmatrix} \begin{bmatrix} 0.81 \\ -1 \\ 0.61 \end{bmatrix} = \begin{bmatrix} 2.591 \\ -3.41 \\ 2.216 \end{bmatrix} = 3.41 \begin{bmatrix} 0.76 \\ -1 \\ 0.65 \end{bmatrix}$

$$\lambda_6 = 3.41, X^{(6)} = \begin{bmatrix} 0.76 \\ -1 \\ 0.65 \end{bmatrix}.$$

Seventh Iteration.

$$Y^{(7)} = AX^{(6)} = \begin{bmatrix} 2 & -1 & 0 \\ -1 & 2 & -1 \\ 0 & -1 & 2 \end{bmatrix} \begin{bmatrix} 0.76 \\ -1 \\ 0.65 \end{bmatrix} = \begin{bmatrix} 2.52 \\ -3.41 \\ 2.216 \end{bmatrix} = 3.41 \begin{bmatrix} 0.74 \\ -1 \\ 0.67 \end{bmatrix}$$

$$\lambda_2 = 3.41, X^{(7)} = \begin{bmatrix} 0.76 \\ -1 \\ 0.65 \end{bmatrix}.$$

Eight Iteration. $Y^{(8)} = AX^{(7)} = 3.41 \begin{bmatrix} 0.72 \\ -1 \\ 0.69 \end{bmatrix}, \lambda_8 = 3.41, X^8 = \begin{bmatrix} 0.72 \\ -1 \\ 0.69 \end{bmatrix}$

Ninth Iteration. $Y^{(9)} = AX^{(8)} = 3.41 \begin{bmatrix} 0.72 \\ -1 \\ 0.69 \end{bmatrix}, \lambda_9 = 3.41, X^9 = \begin{bmatrix} 0.72 \\ -1 \\ 0.69 \end{bmatrix}$

Since $\qquad \lambda_9 = 3.41, [X^{(9)} - X^{(8)}] = 0.$

Hence, the largest eigen value $\lambda_9 = 3.41$ and corresponding eigen vector is given by

$$X^{(9)} = \begin{bmatrix} 0.72 \\ -1 \\ 0.69 \end{bmatrix}.$$

7.13 INVERSE POWER METHOD

To find the smallest eigen value and the corresponding eigen vector of matrix using Inverse power method.

ALGORITHM

Step 1.	Find A^{-1} of the given matrix A.
Step 2.	Then find $Y^{(1)} = A^{-1}X$ where X is initial eigen vector.
Step 3.	Find largest eigen value and corresponding eigen vector and again find $Y^{(i)} = A^{-i}X^{(i-1)} = \lambda_i X^{(i)}$ till $[X^{(i)} - X^{(i-1)}] = 0$.
Step 4.	The smallest eigen value for A will be $\dfrac{1}{\lambda_i}$ and corresponding eigen vector will be $X^{(i)}$.

SOLVED EXAMPLES

EXAMPLE 1. *Find the smallest eigen value of given matrix*

$$A = \begin{bmatrix} 3 & 0 & 1 \\ 0 & -3 & 0 \\ 1 & 0 & 3 \end{bmatrix} as\, X = \begin{bmatrix} 0 \\ 1 \\ 2 \end{bmatrix}$$

using power method.

SOLUTION. The smallest eigen value of matrix will be equal to the reciprocal of largest eigen value for matrix A^{-1}

$$A = \begin{bmatrix} 3 & 0 & 1 \\ 0 & -3 & 0 \\ 1 & 0 & 3 \end{bmatrix} = -24 \neq 0.$$

Thus A^{-1} exist and we may easily find

$$A^{-1} = \begin{bmatrix} -3/8 & 0 & -1/8 \\ 0 & -1/3 & 0 \\ -1/8 & 0 & 3/8 \end{bmatrix}$$

Given that initial value $X = \begin{bmatrix} 0 \\ 1 \\ 2 \end{bmatrix}$.

First Iteration. $Y^{(1)} = A^{-1}X = \begin{bmatrix} -3/8 & 0 & -1/8 \\ 0 & -1/3 & 0 \\ -1/8 & 0 & 3/8 \end{bmatrix}\begin{bmatrix} 0 \\ 1 \\ 2 \end{bmatrix} = \begin{bmatrix} -1/4 \\ -1/3 \\ 3/4 \end{bmatrix}$

$$= \frac{3}{4}\begin{bmatrix} -1/3 \\ -4/9 \\ 1 \end{bmatrix}$$

Here $\lambda_1 = \dfrac{3}{4}, X^{(1)} = \begin{bmatrix} -1/3 \\ -4/9 \\ 1 \end{bmatrix}$

Second Iteration. $Y^{(2)} = A^{-1}X^{(1)} = \begin{bmatrix} -3/8 & 0 & -1/8 \\ 0 & -1/3 & 0 \\ -1/8 & 0 & 3/8 \end{bmatrix}\begin{bmatrix} -1/4 \\ -1/3 \\ 3/4 \end{bmatrix}$

$$= \begin{bmatrix} -1/4 \\ 4/27 \\ 5/12 \end{bmatrix} = \frac{5}{12} \begin{bmatrix} -3/5 \\ -16/45 \\ 1 \end{bmatrix}$$

Here $\qquad \lambda_2 = \dfrac{5}{12}, X^{(2)} = \begin{bmatrix} -3/5 \\ -16/45 \\ 1 \end{bmatrix}$

Third Iteration. $\quad Y^{(3)} = A^{-1}X^{(2)} = \begin{bmatrix} -3/8 & 0 & -1/8 \\ 0 & -1/3 & 0 \\ -1/8 & 0 & 3/8 \end{bmatrix} \begin{bmatrix} -3/5 \\ -16/45 \\ 1 \end{bmatrix}$

$$= \begin{bmatrix} -7/20 \\ -16/135 \\ 9/20 \end{bmatrix} = \frac{9}{20} \begin{bmatrix} -7/9 \\ -64/243 \\ 1 \end{bmatrix}$$

Here $\qquad \lambda_3 = \dfrac{9}{20}, X^{(3)} = \begin{bmatrix} -7/8 \\ -64/243 \\ 1 \end{bmatrix}$

Fourth Iteration. $\quad Y(4) = A^{-1}X^{(3)} = \begin{bmatrix} -3/8 & 0 & -1/8 \\ 0 & -1/3 & 0 \\ -1/8 & 0 & 3/8 \end{bmatrix} \begin{bmatrix} -7/9 \\ -64/243 \\ 1 \end{bmatrix}$

$$= \begin{bmatrix} -5/12 \\ -64/729 \\ 17/36 \end{bmatrix} = \frac{17}{36} \begin{bmatrix} -15/17 \\ -272/6561 \\ 1 \end{bmatrix}.$$

Here $\qquad \lambda_4 = \dfrac{17}{36}, X^{(4)} = \begin{bmatrix} -15/17 \\ -272/6561 \\ 1 \end{bmatrix}.$

Now by the successive iteration, the largest eigen value approaches to 1/2 and corresponding eigen vector is given by $\begin{bmatrix} 1 \\ 0 \\ -1 \end{bmatrix}$.

Hence, the largest eigen vector of matrix $A = 2$ and corresponding eigen vector will be $\begin{bmatrix} 1 \\ 0 \\ -1 \end{bmatrix}$.

EXAMPLE 2. *Find the smallest eigen value in magnitude of the matrix*

$$A = \begin{bmatrix} 2 & -1 & 0 \\ -1 & 2 & -1 \\ 0 & -1 & 2 \end{bmatrix}$$

using power method of inversion involving four iterations only.

SOLUTION.   Given $\qquad A = \begin{bmatrix} 2 & -1 & 0 \\ -1 & 2 & -1 \\ 0 & -1 & 2 \end{bmatrix}$

$\Rightarrow \qquad\qquad |A| = \begin{vmatrix} 2 & -1 & 0 \\ -1 & 2 & -1 \\ 0 & -1 & 2 \end{vmatrix} = 5 \neq 0.$

So A^{-1} exist.

Hence, $A^{-1} = \begin{bmatrix} 3/4 & 1/2 & 1/4 \\ 1/2 & 1 & 1/2 \\ 1/4 & 1/2 & 3/4 \end{bmatrix}$ let $X = \begin{bmatrix} 1 \\ 1 \\ 1 \end{bmatrix}$.

First Iteration. $Y^{(1)} = A^{-1}X = \begin{bmatrix} 3/4 & 1/2 & 1/4 \\ 1/2 & 1 & 1/2 \\ 1/4 & 1/2 & 3/4 \end{bmatrix} \begin{bmatrix} 1 \\ 1 \\ 1 \end{bmatrix} = \begin{bmatrix} 1.5 \\ 2 \\ 1.5 \end{bmatrix} = 2 \begin{bmatrix} 0.75 \\ 1 \\ 0.75 \end{bmatrix}$

Here $\lambda_1 = 2$ and $X^{(1)} = \begin{bmatrix} 0.75 \\ 1 \\ 0.75 \end{bmatrix}$.

Second Iteration. $Y^{(2)} = A^{-1}X^{(1)} = \begin{bmatrix} 3/4 & 1/2 & 1/4 \\ 1/2 & 1 & 1/2 \\ 1/4 & 1/2 & 3/4 \end{bmatrix} \begin{bmatrix} 0.75 \\ 1 \\ 0.75 \end{bmatrix} = \begin{bmatrix} 1.25 \\ 1.75 \\ 1.25 \end{bmatrix}$

$= 1.75 = \begin{bmatrix} 0.7143 \\ 1 \\ 0.7143 \end{bmatrix}$

Here $\lambda_2 = 1.75$ and $X^{(2)} = \begin{bmatrix} 0.7143 \\ 1 \\ 0.7143 \end{bmatrix}$.

Third Iteration. $Y^{(3)} = A^{-1}X^{(2)} = \begin{bmatrix} 3/4 & 1/2 & 1/4 \\ 1/2 & 1 & 1/2 \\ 1/4 & 1/2 & 3/4 \end{bmatrix} \begin{bmatrix} 0.743 \\ 1 \\ 0.7143 \end{bmatrix}$

$= \begin{bmatrix} 1.2143 \\ 1.7143 \\ 1.2143 \end{bmatrix} = 1.7143 \begin{bmatrix} 0.7083 \\ 1 \\ 0.7083 \end{bmatrix}$

Here $\lambda_3 = 1.7143$ and $X^{(3)} = \begin{bmatrix} 0.7073 \\ 1 \\ 0.7073 \end{bmatrix}$.

Fourth Iteration. $Y^{(4)} = A^{-1}X^{(3)} = \begin{bmatrix} 3/4 & 1/2 & 1/4 \\ 1/2 & 1 & 1/2 \\ 1/4 & 1/2 & 3/4 \end{bmatrix} \begin{bmatrix} 0.7083 \\ 1 \\ 0.7083 \end{bmatrix}$

$= \begin{bmatrix} 1.2083 \\ 1.7083 \\ 1.2083 \end{bmatrix} = 1.7083 \begin{bmatrix} 0.7083 \\ 1 \\ 0.7083 \end{bmatrix}$

Here $\lambda_4 = 1.7083$ and $X^{(4)} = \begin{bmatrix} 0.7073 \\ 1 \\ 0.7073 \end{bmatrix}$.

Since $1.7083 \approx 1.71$ is largest eigen value for A^{-1}.

Thus $\dfrac{1}{1.7083} \approx 0.5848$ is smallest eigen value for A and corresponding eigen

vector will be $[0.7073, 1, 0.7073]^T$.

The exact smallest eigen value of A is $2 - \sqrt{2}$.

7.14 RUTISHAUSER METHOD

This method is suggested by Rutishauser. It is based on LU-transformation, where L is a lower triangular matrix and U is an upper triangular matrix.

In this mehtod, first the given matrix split in the following form

$$A = A_1 = L_1 U_1, \text{ where } A \text{ is any square matrix of order } n$$

and

$$L_1 = \begin{bmatrix} l_{11} & 0 & \cdots & 0 \\ l_{21} & l_{22} & \cdots & 0 \\ \vdots & \vdots & \vdots & \vdots \\ l_{n1} & l_{n2} & \cdots & l_{nn} \end{bmatrix}, U_1 = \begin{bmatrix} u_{11} & u_{12} & \cdots & u_{1n} \\ 0 & u_{22} & \cdots & u_{2n} \\ \vdots & \vdots & \vdots & \vdots \\ 0 & 0 & \cdots & u_{nn} \end{bmatrix}$$

here

$$l_{ii} = 1 \text{ where } i = 1, 2, 3, \ldots, n.$$

Now from other matrix $A_2 = U_1 L_1$, since $L_1 = A_1 U_1^{-1}$.

Thus $A_2 = U_1 A_1 U_1^{-1}$ has the same eigen value as A_1 has. Therefore let $A_2 = L_2 U_2$.

Now proceeding in same way we get a sequence of reduced upper triangular matrix, $A_1, A_2, A_3, A_4, \ldots$ which have the same eigen values. This method is convenient only when it is applied to matrix in Hessenberg form (*i.e.*, when the matrix is in almost triangular form then, this method is used). The sequence converges to an upper triangular matrix after a number of transformation and elements converges.

SOLVED EXAMPLES

EXAMPLE. *Find the eigen values of matrix, applying Rutishauser Method.*

$$A = \begin{bmatrix} 1 & 1 & 1 \\ 2 & 1 & 2 \\ 1 & 3 & 2 \end{bmatrix}.$$

SOLUTION.

First Iteration. $A = A_1 = L_1 U_1 \Rightarrow \begin{bmatrix} 1 & 1 & 1 \\ 2 & 1 & 2 \\ 1 & 3 & 2 \end{bmatrix} = \begin{bmatrix} 1 & 0 & 0 \\ 2 & 1 & 0 \\ 1 & -2 & 1 \end{bmatrix} \begin{bmatrix} 1 & 1 & 1 \\ 0 & -1 & 0 \\ 0 & 0 & 1 \end{bmatrix}.$

Second Iteration. $A_2 = U_1 L_1 \Rightarrow \begin{bmatrix} 1 & 1 & 1 \\ 0 & -1 & 0 \\ 0 & 0 & 1 \end{bmatrix} \begin{bmatrix} 1 & 0 & 0 \\ 2 & 1 & 0 \\ 1 & -2 & 1 \end{bmatrix} = \begin{bmatrix} 4 & -1 & 1 \\ -2 & -1 & 0 \\ 1 & -2 & 1 \end{bmatrix}$

$$= \begin{bmatrix} 1 & 0 & 0 \\ -1/2 & 1 & 0 \\ 1/4 & 7/6 & 1 \end{bmatrix} \begin{bmatrix} 4 & -1 & 1 \\ 0 & -3/2 & 1/2 \\ 0 & 0 & 1/6 \end{bmatrix}$$

where

$$L_2 = \begin{bmatrix} 1 & 0 & 0 \\ -1/2 & 1 & 0 \\ 1/4 & 7/6 & 1 \end{bmatrix} \text{ and } U_2 = \begin{bmatrix} 4 & -1 & 1 \\ 0 & -3/2 & 1/2 \\ 0 & 0 & 1/6 \end{bmatrix}.$$

Third Iteration. $A_3 = U_2 L_2 = \begin{bmatrix} 4 & -1 & 1 \\ 0 & -3/2 & 1/2 \\ 0 & 0 & 1/6 \end{bmatrix} \begin{bmatrix} 1 & 0 & 0 \\ -1/2 & 1 & 0 \\ 1/4 & 7/6 & 1 \end{bmatrix}$

$$= \begin{bmatrix} 19/4 & 1/6 & 1 \\ 7/38 & -11/12 & 1/2 \\ 1/24 & 7/36 & 1/6 \end{bmatrix}$$

$$= \begin{bmatrix} 1 & 0 & 0 \\ 7/38 & 1 & 0 \\ 1/114 & -11/54 & 1 \end{bmatrix} \begin{bmatrix} 19/4 & 1/6 & 1 \\ 0 & -18/19 & 6/19 \\ 0 & 0 & 38/171 \end{bmatrix}$$

where $\quad L_3 = \begin{bmatrix} 1 & 0 & 0 \\ 7/38 & 1 & 0 \\ 1/114 & -11/54 & 1 \end{bmatrix}$ and $U_3 = \begin{bmatrix} 19/4 & 1/6 & 1 \\ 0 & -18/19 & 6/19 \\ 0 & 0 & 38/171 \end{bmatrix}$.

Fourth Iteration. Proceeding in same way, we have

$$A_4 = U_3 L_3 \begin{bmatrix} 4.7894 & -0.0370 & 1 \\ -0.1717 & -1.0116 & 0.3157 \\ -0.0019 & -0.0452 & 0.2222 \end{bmatrix}$$

$$= \begin{bmatrix} 1 & 0 & 0 \\ -0.0368 & 1 & 0 \\ 0.0004 & 0.0446 & 1 \end{bmatrix}\begin{bmatrix} 4.7894 & -0.0370 & 1 \\ 0 & -1.0130 & 0.3516 \\ 0 & 0 & 0.2061 \end{bmatrix}$$

where $\quad L_4 = \begin{bmatrix} 1 & 0 & 0 \\ -0.0368 & 1 & 0 \\ 0.0004 & 0.0446 & 1 \end{bmatrix}$ and $U_4 = \begin{bmatrix} 4.7894 & -0.0370 & 1 \\ 0 & -1.0130 & 0.3516 \\ 0 & 0 & 0.2061 \end{bmatrix}$

Fifth Iteration. Proceeding in same way, we have

$$A_5 = U_4 L_4 = \begin{bmatrix} 4.79120 & 0.00763 & 1 \\ 0.03646 & 0.99731 & 0.35164 \\ 0.0008 & 0.00920 & 0.20610 \end{bmatrix}$$

$$= \begin{bmatrix} 1 & 0 & 0 \\ 0.007612 & 1 & 0 \\ 0.000018 & -0.009231 & 1 \end{bmatrix}\begin{bmatrix} 4.79120 & 0.00763 & 1 \\ 0 & -0.99737 & 0.344036 \\ 0 & 0 & 0.209265 \end{bmatrix}$$

where $\quad L_5 = \begin{bmatrix} 1 & 0 & 0 \\ 0.007612 & 1 & 0 \\ 0.000018 & -0.009231 & 1 \end{bmatrix}$

and $\quad U_5 = \begin{bmatrix} 4.79120 & 0.00763 & 1 \\ 0 & -0.99737 & 0.344036 \\ 0 & 0 & 0.209265 \end{bmatrix}$

Sixth Iteration. Proceeding in same way, we have

$$A_6 = U_5 L_5 = \begin{bmatrix} 4.791285 & -0.001598 & 1 \\ -0.007586 & -1.000550 & 0.344036 \\ 0.000004 & -0.001932 & 0.209265 \end{bmatrix}$$

$$= \begin{bmatrix} 1 & 0 & 0 \\ -0.001583 & 1 & 0 \\ 0.000001 & 0.001931 & 1 \end{bmatrix}\begin{bmatrix} 4.791285 & -0.001598 & 1 \\ 0 & -1.000553 & 0.345619 \\ 0 & 0 & 0.208597 \end{bmatrix}$$

where $\quad L_6 = \begin{bmatrix} 1 & 0 & 0 \\ -0.001583 & 1 & 0 \\ 0.000001 & 0.001931 & 1 \end{bmatrix}$

and $\quad U_6 = \begin{bmatrix} 4.791285 & -0.001598 & 1 \\ 0 & -1.000553 & 0.345619 \\ 0 & 0 & 0.208597 \end{bmatrix}$.

Hence, the approximate values are 4.791285, –1.0005553, 0.208597, which approaches towards 4.791288, –1, 0.208717 respectively where

$$\lambda_1 = 4.791288 = (5 + \sqrt{21})/2, \lambda_2 = -1, \lambda_3 = 2.208717 = 5 - \sqrt{21}/2.$$

EXERCISE 7.5

1. Find the eigen values and eigen vectors of the matrix $A = \begin{bmatrix} 5 & 4 \\ 1 & 2 \end{bmatrix}$.

2. Find the eigen values and eigen vectors of the matrix $A = \begin{bmatrix} 1 & 2 & 3 \\ 0 & -4 & 2 \\ 0 & 0 & 7 \end{bmatrix}$.

3. Find the latent roots and corresponding vector of the given matrices

(i) $\begin{bmatrix} 6 & -2 & 2 \\ -2 & 3 & -1 \\ 2 & -1 & 3 \end{bmatrix}$ (ii) $\begin{bmatrix} 1 & 2 & 3 \\ 0 & 2 & 3 \\ 0 & 0 & 2 \end{bmatrix}$

(iii) $\begin{bmatrix} 3 & 2 & 2 \\ 2 & 5 & 2 \\ 2 & 2 & 3 \end{bmatrix}$ (iv) $\begin{bmatrix} 1 & 6 & 12 \\ 1 & 2 & 0 \\ 0 & 0 & 3 \end{bmatrix}$

4. (i) Locate the region of the eigen values of the matrix $A = \begin{bmatrix} 1 & 24 & 0 \\ -3 & 8 & 0 \\ -1 & 0 & 2 \end{bmatrix}$.

(ii) Estimate the eigen values and interval in which the eigen values exist of the matrix

$$A = \begin{bmatrix} 3 & 2 & 2 \\ 2 & 5 & 2 \\ 2 & 2 & 3 \end{bmatrix}.$$

5. Find the largest eigen value and corresponding eigen vector of given matrix using Rayleigh Power Method or Iterative Method :

(i) $\begin{bmatrix} 10 & -4 & 0 \\ -6 & 9 & -3 \\ 0 & -6 & 6 \end{bmatrix}$ (ii) $\begin{bmatrix} -15 & 4 & 3 \\ 10 & -12 & 6 \\ 20 & -4 & 2 \end{bmatrix}$

(iii) $\begin{bmatrix} 1 & 3 & -1 \\ 3 & 2 & 4 \\ -1 & 4 & 10 \end{bmatrix}$

7.15 JACOBI'S METHOD

(Agra-2004, 06, 09)

The method is used for a real square symmetric matrix. In this method an iterative procedure is used and the symmetric matrix is reduced into similar matrix and continuing this procedure till it not reduced into a diagonal matrix. The main diagonal matrix gives the eigen value, which is obtained by transformation of orthogonal matrix. Also the multiplication of orthogonal matrix which may be used gives the eigen vector.

Let us suppose A be a square symmetric matrix, that is

$$A = \begin{bmatrix} a_{ii} & a_{ij} \\ a_{ji} & a_{jj} \end{bmatrix}$$

and P_1 is an orthogonal invertiable matrix, then an orthogonal transformation is given by

$$B_1 = P_1^{-1} A P_1 \qquad \qquad \dots(1)$$

Since P_1 is orthogonal so inverse of P will be equal to its transpose that is $P_1^{-1} = P_1^T$, hence equation (1) becomes

$$B_1 = P_1^T A P_1 \qquad \qquad \dots(2)$$

Let us choose the element of P_1 such that

$$P_1 = \begin{bmatrix} \cos\theta & -\sin\theta \\ \sin\theta & \cos\theta \end{bmatrix} \quad \text{and} \quad P_1^T = \begin{bmatrix} \cos\theta & \sin\theta \\ -\sin\theta & \cos\theta \end{bmatrix}$$

Substituting these values in equation(2), we have

$$B_1 = \begin{bmatrix} \cos\theta & \sin\theta \\ -\sin\theta & \cos\theta \end{bmatrix} \begin{bmatrix} a_{ii} & a_{ij} \\ a_{ji} & a_{jj} \end{bmatrix} \begin{bmatrix} \cos\theta & -\sin\theta \\ \sin\theta & \cos\theta \end{bmatrix}$$

$$= \begin{bmatrix} a_{ii}\cos^2\theta + a_{ij}\sin 2\theta + a_{ji}\sin^2\theta & (a_{jj} - a_{ii})(\sin\theta)\cos\theta + a_{ij}\cos 2\theta \\ (a_{jj} - a_{ii})\sin\theta\cos\theta + a_{ij}\cos 2\theta & a_{ii}\sin^2\theta - a_{ij}\sin 2\theta + a_{ji}\cos^2\theta \end{bmatrix}$$

It will reduced as a diagonal matrix, when

$$(a_{ji} - a_{ii})\sin\theta\cos\theta + a_{ij}\cos 2\theta = 0$$

or

$$\frac{1}{2}(a_{jj} - a_{ii})\sin 2\theta + a_{ij}\cos 2\theta = 0$$

$\Rightarrow$

$$\tan 2\theta = \frac{2a_{ij}}{a_{ii} - a_{jj}} \quad \text{where } a_{ii} \neq a_{kk}$$

$$\tan 2\theta = \infty \implies -\frac{\pi}{4} \leq \theta \leq \frac{\pi}{4}$$

Let

$$\tan 2\theta = \frac{\alpha}{\beta} \quad \text{where} \quad \alpha = 2a_{ij}$$

$$\beta = a_{ii} - a_{jj}$$

So

$$\sin 2\theta = \frac{\alpha}{\sqrt{\alpha^2 + \beta^2}} \quad \text{where } \sqrt{\alpha^2 + \beta^2} = \sqrt{(a_{ii} - a_{ji})^2 + 4a_{ij}^2}$$

and

$$\cos 2\theta = \frac{\alpha}{\sqrt{\alpha^2 + \beta^2}}$$

We continue this process till we get diagonal elements upto r times,

$$B_r = P_r^{-1} P_{r-1}^{-1} ... P_1^{-1} A P_1 P_2 ... P_{r-1} P_r$$

$$= P^{-1} A P$$

where

$$P = P_1 ..... P_{r-1} P_r.$$

When r tends to infinite, it approaches to a diagonal matrix which gives eigen values and equation (3) gives the corresponding eigen vector of A.

SOLVED EXAMPLES

EXAMPLE 1. *Find the eigen values of the matrix A, using the Jacobi method. Iterate till the off diagonal elements are less than 0.00048, where*

$$A = \begin{bmatrix} 1 & 2 & -1 \\ 2 & 1 & 2 \\ -1 & 2 & 1 \end{bmatrix}.$$

SOLUTION. Since in given matrix $A, a_{12} = a_{21} = 2$ is the largest off diagonal element, therefore

$$\tan 2\theta = \frac{2a_{12}}{a_{11} - a_{22}} = \frac{4}{0} = \infty \quad [\text{where } a_{11} = a_{22} = 1]$$

$\Rightarrow$

$$-\frac{\pi}{4} \leq \theta \leq \frac{\pi}{4}$$

Let

$$\theta = \frac{\pi}{4}.$$

First Iteration. Let the orthogonal matrix in (1, 2) plane is given by

$$P_1 = \begin{bmatrix} \cos\dfrac{\pi}{4} & -\sin\dfrac{\pi}{4} & 0 \\ \sin\dfrac{\pi}{4} & \cos\dfrac{\pi}{4} & 0 \\ 0 & 0 & 1 \end{bmatrix} = \begin{bmatrix} 1/\sqrt{2} & -1/\sqrt{2} & 0 \\ 1/\sqrt{2} & 1/\sqrt{2} & 0 \\ 0 & 0 & 1 \end{bmatrix}$$

Therefore, $\qquad P_1^{-1} = P_1^T = \begin{bmatrix} \cos \pi/4 & \sin \pi/4 & 0 \\ -\sin \pi/4 & \cos \pi/4 & 0 \\ 0 & 0 & 1 \end{bmatrix} = \begin{bmatrix} 1/\sqrt{2} & -1/\sqrt{2} & 0 \\ 1/\sqrt{2} & 1/\sqrt{2} & 0 \\ 0 & 0 & 1 \end{bmatrix}$

$$B_1 = P_1^T A P_1$$

$$= \begin{bmatrix} 1/\sqrt{2} & 1/\sqrt{2} & 0 \\ -1/\sqrt{2} & 1/\sqrt{2} & 0 \\ 0 & 0 & 1 \end{bmatrix} \begin{bmatrix} 1 & 2 & -1 \\ 2 & 1 & 2 \\ -1 & 2 & 1 \end{bmatrix} \begin{bmatrix} 1/\sqrt{2} & -1/\sqrt{2} & 0 \\ 1/\sqrt{2} & 1/\sqrt{2} & 0 \\ 0 & 0 & 1 \end{bmatrix}$$

$$B_1 = \begin{bmatrix} 3 & 0 & 1/\sqrt{2} \\ 0 & -1 & 3/\sqrt{2} \\ 1/\sqrt{2} & 3/\sqrt{2} & 1 \end{bmatrix}$$

Now again, the largest off diagonal element in B_1 is $a_{23} = a_{32} = \dfrac{3}{\sqrt{2}} > 0$, therefore, we take orthogonal matrix in (2, 3), plane P_2 such that

$$P_2 = \begin{bmatrix} 1 & 0 & 0 \\ 0 & \cos \theta & -\sin \theta \\ 0 & \sin \theta & \cos \theta \end{bmatrix}$$

Here $\qquad \tan 2\theta = \dfrac{2a_{23}}{a_{22} - a_{33}} = -2.12132.$

$\Rightarrow \qquad\qquad \theta = -0.56514.$

Second Iteration. Therefore

$$P_2 = \begin{bmatrix} 1 & 0 & 0 \\ 0 & 0.844512 & 0.535537 \\ 0 & -0.535537 & 0.844512 \end{bmatrix}$$

$$P_2^{-1} = P_2^T = \begin{bmatrix} 1 & 0 & 0 \\ 0 & 0.844512 & -0.535537 \\ 0 & 0.535537 & 0.844512 \end{bmatrix}$$

$$B_2 = P_2^T B_1 P_2 = \begin{bmatrix} 1 & 0 & 0 \\ 0 & 0.844512 & -0.535537 \\ 0 & 0.535537 & 0.844512 \end{bmatrix} \begin{bmatrix} 3 & 0 & 1/\sqrt{2} \\ 0 & -1 & 3/\sqrt{2} \\ 1/\sqrt{2} & 3/\sqrt{2} & 1 \end{bmatrix}$$

$$\cdot \begin{bmatrix} 1 & 0 & 0 \\ 0 & 0.844512 & 0.535537 \\ 0 & 0.535537 & 0.844512 \end{bmatrix}$$

$$= \begin{bmatrix} 3 & -0.378682 & 0.59716 \\ -0.378682 & -2.345209 & -0.000003 \\ 0.59716 & -0.000003 & 2.345209 \end{bmatrix}$$

Here largest off diagonal element in B_2 is $a_{13} = 0.59716$, so

$$\tan 2\theta = \dfrac{2a_{13}}{a_{11} - a_{33}} = 1.82396$$

$\Rightarrow \qquad\qquad \theta = 0.534647$

$\qquad \cos \theta = 0.860449$ and $\sin \theta = 0.509537$

Thus we take a new orthogonal matrix in (1, 3) plane.

Third Iteration.

$$P_3 = \begin{bmatrix} \cos\theta & 0 & -\sin\theta \\ 0 & 1 & 0 \\ \sin\theta & 0 & \cos\theta \end{bmatrix} = \begin{bmatrix} 0.860449 & 0 & -0.509537 \\ 0 & 1 & 0 \\ 0.509537 & 0 & 0.860449 \end{bmatrix}$$

$$\Rightarrow \qquad P_3^{-1} = P_3^T = \begin{bmatrix} 0.860449 & 0 & 0.509537 \\ 0 & 1 & 0 \\ -0.509537 & 0 & 0.860449 \end{bmatrix}$$

Also, $\qquad B_3 = P_3^{-1} B_2 P_3 = \begin{bmatrix} 3.353625 & -0.325837 & 0.000001 \\ -0.325837 & -2.345208 & 0.192952 \\ 0.000001 & 0.192952 & 1.991585 \end{bmatrix}$

Here, the largest off diagonal element in magnitude in B_3 is $a_{12} = a_{21}$, which is equal to $|-0.35837| = 0.325837$.

So $\qquad \tan 2\theta = \dfrac{2a_{12}}{a_{11} - a_{22}} = -0.114352$

$\Rightarrow \qquad \theta = -0.056929.$

Fourth Iteration. The new orthogonal matrix P_4 in (1, 2) plane is

$$P_4 = \begin{bmatrix} \cos\theta & -\sin\theta & 0 \\ \sin & \cos\theta & 0 \\ 0 & 0 & 1 \end{bmatrix} = \begin{bmatrix} 0.998380 & 0.056898 & 0 \\ -0.056898 & 0.998380 & 0 \\ 0 & 0 & 1 \end{bmatrix}$$

$$\Rightarrow \qquad P_4^{-1} = P_4^T = \begin{bmatrix} 0.998380 & 0.056898 & 0 \\ -0.056898 & 0.998380 & 0 \\ 0 & 0 & 1 \end{bmatrix}$$

Also, $\qquad B_4 = P_4^{-1} B_3 P_4 = \begin{bmatrix} 3.372195 & 0 & -0.010970 \\ 0 & 1 & 0.192639 \\ -0.010979 & 0.192639 & 1.991585 \end{bmatrix}$

The largest off diagonal element in B_4 is $a_{23} = a_{32} = 0.192639$.

So $\qquad \tan 2\theta = -0.088462, \theta = -0.044116.$

Fifth Iteration.

$$P_5 = \begin{bmatrix} 1 & 0 & 0 \\ 0 & \cos\theta & -\sin\theta \\ 0 & \sin\theta & \cos\theta \end{bmatrix} = \begin{bmatrix} 1 & 0 & 0 \\ 0 & 0.999027 & 0.044102 \\ 0 & -0.044102 & 0.999027 \end{bmatrix}$$

$$P_5^{-1} = P_5^T = \begin{bmatrix} 1 & 0 & 0 \\ 0 & 0.999027 & -0.044102 \\ 0 & -0.044102 & 0.999027 \end{bmatrix}$$

$$B_5 = P_5^T B_4 P_5 = \begin{bmatrix} 3.372195 & 0.000484 & -0.010968 \\ 0.000484 & -2.372281 & 0.000003 \\ -0.010968 & -0.000003 & 2.000089 \end{bmatrix}$$

The largest off diagonal element in magnitude in B_5 is a_{13}, so

$$\tan 2\theta = -0.015987, \theta = -0.007992.$$

Sixth Iteration. $P_6 = \begin{bmatrix} 0.99996 & 0 & 0.00799 \\ 0 & 1 & 0 \\ -0.00799 & 0 & 0.99996 \end{bmatrix}$

and $\qquad P_6^{-1} = P_6^T \begin{bmatrix} 0.99996 & 0 & -0.00799 \\ 0 & 1 & 0 \\ 0.00799 & 0 & 0.99996 \end{bmatrix}$

$$B_6 = P_6^{-1} B_5 P_6 = P_6^T B_5 P_6 = \begin{bmatrix} 3.37228 & 0.00048 & 0 \\ 0.00048 & -2.37228 & 0 \\ 0 & 0 & 1.99999 \end{bmatrix}$$

which is diagonalised matrix and approximate eigen values for matrix A is 3.37228, –2.37228, 1.99999 and corresponding eigen vector will be find out as $P = P_1 P_2 P_3 P_4 P_5 P_6$.

EXAMPLE 2. *Find the eigen values and corresponding eigen vectors of the matrix using Jacobi's method*

$$A = \begin{bmatrix} 1 & \sqrt{2} & 2 \\ \sqrt{2} & 3 & \sqrt{2} \\ 2 & \sqrt{2} & 1 \end{bmatrix}.$$ (Meerut-2003)

SOLUTION. Since in given matrix, A, $a_{13} = a_{31} = 2$ is the largest off diagonal element, therefore

$$\tan 2\theta = \frac{2a_{13}}{a_{11} - a_{33}} = \frac{4}{0} = \infty \qquad [\text{where } a_{11} = a_{13} = 1]$$

$\Rightarrow \qquad -\dfrac{\pi}{4} \le \theta \le \dfrac{\pi}{4}.$

First Iteration. The orthogonal matrix is

$$P_1 = \begin{bmatrix} \cos\theta & 0 & -\sin\theta \\ 0 & 1 & 0 \\ \sin\theta & 0 & \cos\theta \end{bmatrix} = \begin{bmatrix} 1/\sqrt{2} & 0 & -1/\sqrt{2} \\ 0 & 1 & 0 \\ 1/\sqrt{2} & 0 & 1/\sqrt{2} \end{bmatrix}$$

$\Rightarrow \qquad P_1^{-1} = P_1^T = \begin{bmatrix} 1/\sqrt{2} & 0 & -1/\sqrt{2} \\ 0 & 1 & 0 \\ 1/\sqrt{2} & 0 & 1/\sqrt{2} \end{bmatrix}$

Also, $B_1 = P_1^{-1} A P_1 = \begin{bmatrix} 1/\sqrt{2} & 0 & 1/\sqrt{2} \\ 0 & 1 & 0 \\ -1/\sqrt{2} & 0 & 1/\sqrt{2} \end{bmatrix} \begin{bmatrix} 1 & \sqrt{2} & 2 \\ \sqrt{2} & 3 & \sqrt{2} \\ 2 & \sqrt{2} & 1 \end{bmatrix} \begin{bmatrix} 1/\sqrt{2} & 0 & -1/\sqrt{2} \\ 0 & 1 & 0 \\ 1/\sqrt{2} & 0 & 1/\sqrt{2} \end{bmatrix}$

$$= \begin{bmatrix} 3 & 2 & 0 \\ 2 & 3 & 0 \\ 0 & 0 & -1 \end{bmatrix}.$$

Now, the largest off diagonal element in B_1 is $a_{12} = a_{21} = 2$ and $a_{11} = a_{22} = 3$.

So, $\qquad \tan 2\theta = \dfrac{2a_{12}}{a_{11} - a_{22}} = \dfrac{4}{0} = \infty$

$\Rightarrow \qquad \theta = \dfrac{\pi}{4}.$

Second Iteration. The orthogonal matrix is $P_2 = \begin{bmatrix} \cos\theta & -\sin\theta & 0 \\ \sin\theta & \cos\theta & 0 \\ 0 & 0 & 1 \end{bmatrix}$

$$P_2 = \begin{bmatrix} 1/\sqrt{2} & -1/\sqrt{2} & 0 \\ 1/\sqrt{2} & 1/\sqrt{2} & 0 \\ 0 & 0 & 1 \end{bmatrix}$$

$$\Rightarrow \quad P_2^{-1} = P_2^T = \begin{bmatrix} 1/\sqrt{2} & 1/\sqrt{2} & 0 \\ -1/\sqrt{2} & 1/\sqrt{2} & 0 \\ 0 & 0 & 1 \end{bmatrix}$$

Also, $\quad B_2 = P_2^{-1}B_1P_2 = P_2^T B_1 P_2$

$$= \begin{bmatrix} 1/\sqrt{2} & 1/\sqrt{2} & 0 \\ -1/\sqrt{2} & 1/\sqrt{2} & 0 \\ 0 & 0 & 1 \end{bmatrix}\begin{bmatrix} 3 & 2 & 0 \\ 2 & 3 & 0 \\ 0 & 0 & -1 \end{bmatrix}\begin{bmatrix} 1/\sqrt{2} & -1/\sqrt{2} & 0 \\ 1/\sqrt{2} & 1/\sqrt{2} & 0 \\ 0 & 0 & 1 \end{bmatrix}$$

$$= \begin{bmatrix} 5 & 0 & 0 \\ 0 & 1 & 0 \\ 0 & 0 & -1 \end{bmatrix}$$

which is diagonalised matrix, thus eigen values are 5, 1, –1.
And the eigen vectors of matrix A is

$$P = P_1P_2 = \begin{bmatrix} 1/\sqrt{2} & -1/2 & -1/\sqrt{2} \\ 1/\sqrt{2} & 1/\sqrt{2} & 0 \\ 1/2 & -/\sqrt{2} & 1/\sqrt{2} \end{bmatrix}.$$

EXAMPLE 3. *Find the eigen values of the following system matrix*

$$A = \begin{bmatrix} 4 & 2 & 0 \\ 2 & 3 & 1 \\ 0 & 1 & 2 \end{bmatrix}.$$

SOLUTION. Since the largest off diagonal matrix of A is $a_{12} = a_{21} = 2$ and $a_{11} = 4$, $a_{22} = 3$.

Using $\quad \tan 2\theta = \dfrac{2a_{12}}{a_{11} - a_{22}} \Rightarrow \theta = 0.6629.$

First Iteration. The orthogonal matrix is

$$P_1 = \begin{bmatrix} \cos\theta & -\sin\theta & 0 \\ \sin\theta & \cos\theta & 0 \\ 0 & 0 & 1 \end{bmatrix} = \begin{bmatrix} 0.788 & -0.615 & 0 \\ +0.615 & 0.788 & 1 \\ 0 & 0 & 1 \end{bmatrix}$$

$$P_1^{-1} = P_1^T = \begin{bmatrix} 0.788 & 0.615 & 0 \\ -0.615 & 0.788 & 0 \\ 0 & 0 & 1 \end{bmatrix}$$

Therefóre,

$$B_1 = P_1^{-1}AP_1 = \begin{bmatrix} 0.788 & 0.615 & 0 \\ -0.615 & 0.788 & 0 \\ 0 & 0 & 1 \end{bmatrix}\begin{bmatrix} 4 & 2 & 0 \\ 2 & 3 & 1 \\ 0 & 1 & 2 \end{bmatrix}\begin{bmatrix} 0.788 & -0.615 & 0 \\ +0.615 & 0.788 & 0 \\ 0 & 0 & 1 \end{bmatrix}$$

Here, largest off diagonal element is $a_{23} = 0.788$ and $a_{22} = 1.437, a_{33} = 2$

$$\cos 2\theta = \frac{(a_{22} - a_{33})}{\sqrt{4a_{23}^2 + (a_{22} - a_{33})^2}} = -0.336$$

$$\cos \theta = \left(\frac{1 + \cos 2\theta}{2}\right)^{1/2} = 0.576 \ \ also \sin \theta = \left(\frac{1 - \cos 2\theta}{2}\right)^{1/2} = 0.817.$$

Second Iteration.

$$P_2 = \begin{bmatrix} 1 & 0 & 0 \\ 0 & \cos\theta & \sin\theta \\ 0 & -\sin\theta & \cos\theta \end{bmatrix} = \begin{bmatrix} 1 & 0 & 0 \\ 0 & 0.576 & 0.817 \\ 0 & -0.817 & 0.576 \end{bmatrix}$$

$$\Rightarrow \qquad P_2^{-1} = P_2^T = \begin{bmatrix} 1 & 0 & 0 \\ 0 & 0.576 & -0.817 \\ 0 & 0.817 & 0.576 \end{bmatrix}$$

Also, $\qquad B_2 = P_2^{-1} B_1 P_2 = \begin{bmatrix} 1 & 0 & 0 \\ 0 & 0.576 & -0.817 \\ 0 & 0.817 & 0576 \end{bmatrix}\begin{bmatrix} 4 & 2 & 0 \\ 2 & 3 & 1 \\ 0 & 1 & 2 \end{bmatrix}\begin{bmatrix} 1 & 0 & 0 \\ 0 & 0.576 & -0.817 \\ 0 & 0.817 & 0.576 \end{bmatrix}$

$$= \begin{bmatrix} 5.56 & 0.502 & 0.354 \\ 0.502 & 2.55 & 0 \\ 0.349 & 0 & 0.881 \end{bmatrix}$$

Proceeding in the same way, we have

Third Iteration. $\cos\theta = 0.987, \sin\theta = 0.160$ when $\cos 2\theta = 0.949$

$$B_3 = \begin{bmatrix} 5.64 & 0 & 0.349 \\ 0 & 2.47 & -0.057 \\ 0.349 & -0.057 & 0.881 \end{bmatrix}$$

where largest off diagonal element is $0.349 = a_{13}$.

So, $\cos\theta = 0.997, \sin\theta = 0.074$ when $\cos 2\theta = 0.997$

Fourth Iteration.

$$B_4 = \begin{bmatrix} 5.66 & -0.004 & -0.006 \\ -0.004 & 2.47 & -0.057 \\ -0.006 & -0.057 & 0.855 \end{bmatrix}$$

Here, the off diagonal element is approaches to zero. Hence,

$$\lambda_1 = 5.66, \lambda_2 = 2.47, \lambda_3 = 0.855 .$$

7.16   GIVEN'S METHOD

(Meerut-2003, delhi-2007)

In Jacobi's method, we face out a disadvantage that the element annihilate by a plane rotation may not necessarily remain zero during transformation. Given gives an algorithm involving plane rotations and use tridiagonal matrix instead of diagonal matrix used in Jacobi's Method. In this method three steps are used to find out eigen values and corresponding eigen vectors of matrix A :

    (i) By using plane rotations, reduce A into a tridiagonal matrix.

Let $\qquad A = \begin{bmatrix} a_{11} & a_{12} & a_{13} \\ a_{21} & a_{22} & a_{23} \\ a_{31} & a_{32} & a_{33} \end{bmatrix}$

where $a_{21} = a_{12}$ and $a_{23} = a_{32}$ also $a_{13} = a_{31}$

and $$P_1 = \begin{bmatrix} 1 & 0 & 1 \\ 0 & \cos\theta & -\sin\theta \\ 0 & \sin\theta & \cos\theta \end{bmatrix}$$

then we find $\quad B_1 = P_1^{-1}AP_1$.

Now, $-a_{12}\sin\theta + a_{13}\cos\theta = 0$ is reduce in tridiagonal matrix

$$\tan\theta = \frac{a_{13}}{a_{12}}$$

by this value of θ, we have zeroes in $(1, 3)$ and $(3, 1)$ positions by applying plane rotation. Again we perform rotation in $(2, 4)$ plane and put $(1, 4) = 0$. Proceeding same way, plane rotation is used to annihilate the elements $(1, 5)$, $(1, 6)$, ..., $(1, n)$; in $(2, 5)$, $(2, 6)$, ..., $(2, n)$ plane so that the zeros not affected, which have been obtained earlier. Then, we have a tridiagonal matrix

$$B = \begin{bmatrix} b_1 & c_1 & & & & 0 \\ c_1 & b_2 & c_2 & & & \\ & c_2 & b_3 & & c_3 & \\ & & & \cdot & \cdot & \\ 0 & & c_{n-2} & & b_{n-1} & c_{n-1} \\ & & & & c_{n-1} & b_n \end{bmatrix}$$

(ii) The eigen values of tridiagonal matrix B.

Let the tridiagonal matrix of A be given by

$$B = \begin{bmatrix} a_{11} & a_{12} & 0 \\ a_{12} & a_{22} & a_{23} \\ 0 & a_{23} & a_{33} \end{bmatrix}$$

Now, we have a characteristic equation is $|\lambda I - B| = 0$. Therefore,

$$f_3(\lambda) = \begin{vmatrix} \lambda - a_{11} & a_{12} & 0 \\ a_{12} & \lambda - a_{22} & a_{23} \\ 0 & a_{23} & \lambda - a_{33} \end{vmatrix} = 0.$$

Now, $\quad f_0(\lambda) = 1$

$$f_1(\lambda) = \lambda = a_{11} = \lambda f_0(\lambda) - a_{11}$$

$$f_2(\lambda) = \begin{vmatrix} \lambda - a_{22} & a_{23} \\ a_{23} & \lambda - a_{33} \end{vmatrix}$$

$$= (\lambda - a_{22})f_1(\lambda) - a_{12}^2 f_0(\lambda)$$

$$f_3(\lambda) = (\lambda - a_{33})\begin{vmatrix} \lambda - a_{11} & a_{12} \\ a_{12} & \lambda - a_{22} \end{vmatrix} - a_{23}\begin{vmatrix} \lambda - a_{11} & 0 \\ a_{12} & a_{23} \end{vmatrix}$$

$$f_3(\lambda) = (\lambda - a_{33})f_2(\lambda) - (a_{23})^2 f_1(\lambda).$$

Proceeding in same way, we have $f_r = (\lambda - b_r)f_{r-1} - c_{r-1}^2 f_{r-2}, 2 \le r \le n$, for general matrix B. Solving this equation we get the required eigen values.

(iii) The eigen vectors of the tridiagonal matrix. Let Y be the eigen vector of the tridiagonal matrix B and $P_1, P_2, ..., P_r$ are the orthogonal matrix performed the plane rotation and reduce A in tridiagonal matrix so the eigen vector is

$$X = P_1 P_2 ..... P_r Y.$$

▶ REMARKS

▸ For the matrix A of order 3, only one rotation is required to reduce matrix A into tridiagonal matrix and for matrix of order 4, only three rotation is required to reduce matrix A into tridiagonal matrix.

▸ We form a strum sequence $f_0(\lambda), f_1(\lambda), ..., f_r(\lambda)$. For this a table for various values of λ is prepared having the no. of change in sign and its difference for consecutive values of λ gives a approximation for eigen values where as, exact value is calculated by using any iterative method.

SOLVED EXAMPLES

EXAMPLE 1. *Use Given method, to find the eigen values of tridiagonal matrix*

$$B = \begin{bmatrix} 2 & -1 & 0 \\ -1 & 2 & -1 \\ 0 & -1 & 2 \end{bmatrix}$$

(Agra-2003, 07)

SOLUTION. Since, the Given matrix B is tridiagonal matrix and we have to find out the eigen values so characteristic equation is given by

$$f_3(\lambda) = |\lambda I - B| = \begin{vmatrix} \lambda - 2 & -1 & 0 \\ -1 & \lambda - 2 & -1 \\ 0 & -1 & \lambda - 2 \end{vmatrix} = 0$$

and $f_0(\lambda) = 1$

$f_1(\lambda) = \lambda - a_{11} = \lambda - 2$

$f_2(\lambda) = (\lambda - a_{11}) f_1 - f_0 = (\lambda - 2)^2 - 1$

$f_3(\lambda) = (\lambda - 2) \begin{vmatrix} \lambda - 2 & -1 \\ -1 & \lambda - 2 \end{vmatrix} + 1 \begin{vmatrix} \lambda - 2 & 0 \\ -1 & -1 \end{vmatrix}$

$= (\lambda - 2)^3 - 2(\lambda - 2).$

Now, we find out the value of λ, by putting successive values of λ and a table is formed as follows :

λ	f_0	f_1	f_2	f_3	$V(\lambda)$ (The no. of changes in sign)
−1	+	−	+	−	3
0	+	−	+	−	3
1	+	−	0	+	2
2	+	0	−	0	1
3	+	+	0	−	1
4	+	+	+	+	0

for $\lambda = 2, f_3(2) = 0$, so $\lambda = 2$ is an eigen value.

Hence, there exist value in interval $(0, 1)$ and $(3, 4)$.

First let in interval (0, 1), we have

λ	f_0	f_1	f_2	f_3	$V(\lambda)$
1/2	+	–	+	–	3

Thus in interval (0.5, 1), we have

λ	f_0	f_1	f_2	f_3	$V(\lambda)$
0.625	+	–	+	+	2

Now in interval (0.5, 0.625), we have

λ	f_0	f_1	f_2	f_3	$V(\lambda)$
0.5625	+	–	+	–	3

In the interval (0.5625, 0.625), we have

λ	f_0	f_1	f_2	f_3	$V(\lambda)$
0.5937	+	–	+	+	2

Proceeding in same way, we have the eigen value $0.5857 \simeq 2 - \sqrt{2}$.

EXAMPLE 2. *Reduce the matrix A into a tridiagonal matrix*

$$A = \begin{bmatrix} 1 & 2 & 3 \\ 3 & 1 & -1 \\ 3 & -1 & 1 \end{bmatrix}$$

SOLUTION. Since, the given matrix A is of order 3 so we use only rotational to change in tridiagonal matrix.

So, $$\tan\theta = \frac{a_{13}}{a_{12}} = \frac{3}{2}$$

$\Rightarrow$ $$\cos\theta = \frac{2}{\sqrt{3^2 + 2^2}} = \frac{2}{\sqrt{13}}, \sin\theta = \frac{3}{\sqrt{13}}$$

Now, $$P = \begin{bmatrix} 1 & 0 & 0 \\ 0 & \cos\theta & -\sin\theta \\ 0 & \sin\theta & \cos\theta \end{bmatrix} \begin{bmatrix} 1 & 0 & 0 \\ 0 & 2/\sqrt{13} & -3/\sqrt{13} \\ 0 & 3/\sqrt{13} & 2/\sqrt{13} \end{bmatrix},$$ an orthogonal matrix

Also, $B_1 = P_1^{-1}AP_1$

$$= \begin{bmatrix} 1 & 0 & 0 \\ 0 & 2/\sqrt{13} & 3/\sqrt{13} \\ 0 & -3/\sqrt{13} & 2/\sqrt{13} \end{bmatrix} \begin{bmatrix} 1 & 2 & 3 \\ 2 & 1 & -1 \\ 3 & -1 & 1 \end{bmatrix} \begin{bmatrix} 1 & 0 & 0 \\ 0 & 2/\sqrt{13} & -3/\sqrt{13} \\ 0 & 3/\sqrt{13} & 2/\sqrt{13} \end{bmatrix}$$

$$= \begin{bmatrix} 1 & \sqrt{13} & 0 \\ \sqrt{13} & 1/13 & -5/13 \\ 0 & 5/13 & 25/13 \end{bmatrix}$$

which is required tridiagonal matrix.

▶ **REMARK**
▶ If we want to find the eigen value and eigen vector we apply the step II and III.

EXAMPLE 3. *By using Given's Method reduce in tridiagonal matrix to the given matrix*

$$A = \begin{bmatrix} 2 & 1 & 3 \\ 1 & 4 & 2 \\ 3 & 2 & 3 \end{bmatrix}$$

[MKU(Tamilnade)-2005]

SOLUTION. Since, the given matrix is in the non-tridiagonal form and A has order 3. Now, the largest off diagonal element is $a_{13} = 3$.

We have, $\quad P_1 = \begin{bmatrix} 1 & 0 & 0 \\ 0 & \cos\theta & -\sin\theta \\ 0 & \sin\theta & \cos\theta \end{bmatrix}$

$$\tan\theta = \frac{a_{13}}{a_{12}} = 3, \cos\theta = \frac{3}{\sqrt{3^2+1^2}} = \frac{3}{\sqrt{10}}, \sin\theta = \frac{1}{\sqrt{3^2+1^2}} = \frac{1}{\sqrt{10}}$$

so $\quad P_1 = \begin{bmatrix} 1 & 0 & 0 \\ 0 & 3/\sqrt{10} & -1/\sqrt{10} \\ 0 & 1/\sqrt{10} & 3/\sqrt{10} \end{bmatrix}$ and $P_1^{-1} = \begin{bmatrix} 1 & 0 & 0 \\ 0 & 3/\sqrt{10} & 1/\sqrt{10} \\ 0 & -1/\sqrt{10} & 3/\sqrt{10} \end{bmatrix}$

$$B_1 = \begin{bmatrix} 1 & 0 & 0 \\ 0 & 3/\sqrt{10} & 1/\sqrt{10} \\ 0 & -1/\sqrt{10} & 3/\sqrt{10} \end{bmatrix} \begin{bmatrix} 2 & 1 & 3 \\ 1 & 4 & 2 \\ 3 & 2 & 3 \end{bmatrix} \begin{bmatrix} 1 & 0 & 0 \\ 0 & 3/\sqrt{10} & -1/\sqrt{10} \\ 0 & 1/\sqrt{10} & 3/\sqrt{10} \end{bmatrix}$$

for $\quad r = 2, 3, ... (n-1)$.

And also $\quad A = A_1$.

In first transformation, we find x_r's so that in the $(1, 3)$, $(1, 4)$, ..., $(1, n)$ plane, we have zeros in it (similar in Given's Method).

In second transformation we find x_r's in plane $(2, 4)$, $(2, 5)$, $(2, 6)$, ..., $(2, n)$ position so that we get zeroes. Finally we get a tridiagonal matrix.

By the above process, we see that it is an improvement procedure of Given's Method.

For example. Let us suppose an square matrix of order 3.

Where $\quad A = \begin{bmatrix} a_{11} & a_{12} & a_{13} \\ a_{21} & a_{22} & a_{23} \\ a_{31} & a_{32} & a_{33} \end{bmatrix}$ ...(1)

and $\quad W_2^T = [0, x_2, x_3]$, where $x_2^2 + x_3^2 = 1$ ...(2)

Then using (1) and (2) in $P = 1 - 2W_2 W_2^T$, we have

$$P = \begin{bmatrix} 1 & 0 & 0 \\ 0 & 1-2x_2^2 & -2x_2 x_3 \\ 0 & -2x_2 x_3 & 1-2x_3^2 \end{bmatrix}$$

If $(1, 3)$ element is zero in A.P. then corresponding element of PAP can be come zero.
First row elements of AP are

$$a'_{11} = a_{11}, a'_{12} = a_{12} - 2p_1 x_2, a'_{13} = a_{13} - 2p_1 x_3 \qquad ...(3)$$

and $\quad p_1 = a_{12} x_2 + a_{13} x_3$

and (1, 3) element of A.P. such that $a'_{13} = a_{13} - 2p_1x_3 = 0$. ...(4)

The sum of the squares of the elements in (1, 3) plane remain unaltered (also in any row) under the orthogonal transformation, so we have

$$a_{11}^2 + a_{12}^2 + a_{13}^2 = a'^2_{11} + a'^2_{12} + a'^2_{13} = a_{11}^2 + (a_{12} - 2p_1x_2)^2 + 0 \quad \text{[from equation (3)]}$$

$$\Rightarrow \qquad a_{12} - 2p_1x_2 = \sqrt{a_{12}^2 + a_{13}^2} = \pm s_1$$

Multiply equation (5) by x_2 and (4) by x_3 and after addition, we get

$$a_{12}x_2 + a_{13}x_3 - 2p_1(x_2^2 + x_3^2) = \pm s_1 x_2$$

$$\Rightarrow \qquad a_{12}x_2 + a_{13}x_2 - 2p_1 = \pm s_1 x_2 \qquad\qquad [\because x_2^2 + x_3^2 = 1]$$

$$\Rightarrow \qquad -p_1 = \pm s_1 x_2$$

$$\Rightarrow \qquad p_1 = \pm s_1 x_2$$

$$= \begin{bmatrix} 2 & 3.16 & 0 \\ 3.16 & 4.3 & -1.9 \\ 0 & -1.9 & 3.9 \end{bmatrix}$$

Aliter. We can find the matrix B_1 by using following values of its element $b_{11}, b_{12}, b_{22}, b_{23}, b_{33}$ in the form

$$b_{11} = a_{11},$$

$$b_{12} = a_{12}\cos\theta + a_{13}\sin\theta,$$

$$b_{22} = a_{22}\cos^2\theta + 2a_{23}\sin\theta\cos\theta + a_{33}\sin^2\theta$$

$$= a_{22}\cos^2\theta + a_{23}\sin 2\theta + a_{33}\sin^2\theta$$

$$b_{23} = (a_{33} - a_{22})\frac{1}{2}\sin 2\theta + a_{33}(\cos^2\theta - \sin^2\theta)$$

$$b_{33} = a_{22}\sin^2\theta + a_{33}\cos^2\theta.$$

7.17 HOUSE HOLDER'S METHOD

[Meerut-2002, MKU (Tamilnade-2007]

House Holder suggested a method to reduce a non-tridiagonal symmetric matrix into tridiagonal matrix. In this, we use an elementary orthogonal transformation in the form

$$P = I - 2ww^T \qquad\qquad ...(1)$$

where w is a column vector having real values such that

$$ww^T = x_1^2 + x_2^2 + ... + x_n^2 = 1 \qquad\qquad ...(2)$$

and

$$P^T = (I - 2ww^T)^T$$

$$= I - 2ww^T \qquad\qquad \text{[From equation (1)]}$$

$$= P.$$

Since P is an orthogonal matrix so its inverse will be equal to its transpose, so from equation (1) and (2), we have

$$P^{-1} = P^T = P(I - 2ww^T).$$

Now, let constructed column vector be as follows :

$$w_r^T = [0, 0, ..., 0, x_r, x_{r+1}, ..., x_n] \qquad ...(3)$$

$$w_r^T w_r = x_r^2 + x_{r+1}^2 + ... + x_n^2 - 1.$$

Then we have

$$P_r = I - 2w_r \cdot w_r^T.$$

And the transformation will be

$$B_r = P_r^{-1} A_{r-1} P_r = P_r^T A_{r-1} P_r = P_r A_{r-1} P_r \qquad ...(4)$$

[From equation (1), (2) and (3)]

Substituting this value in equation (4), we have

$$x_2^2 = \frac{1}{2}\left(1 \mp \frac{a_{12}}{s_1}\right)$$

and

$$x_3 = \mp \frac{a_{13}}{2 s_1 x_2}$$

Also let choose

$$x_2^2 = \frac{1}{2}\left(1 + \frac{a_{12} \operatorname{sign}(a_{12})}{s_1}\right)$$

also

$$x_3 = \frac{a_{13}(\operatorname{sign}(a_{12}))}{2 s_1 x_2}$$

The resulting matrix is tridiagonal.

For an square matrix of order 4.

Let us suppose A is an square matrix of order 4 such that

$$A = \begin{bmatrix} a_{11} & a_{12} & a_{13} & a_{14} \\ a_{21} & a_{22} & a_{23} & a_{24} \\ a_{31} & a_{32} & a_{33} & a_{34} \\ a_{41} & a_{42} & a_{43} & a_{44} \end{bmatrix}$$

And

$$w_2^T = [0, x_2, x_3, x_4] \text{ such that } x_2^2 + x_3^2 + x_4^2 = 1$$

and

$$P_2 = I - 2w_2 w_2^T = \begin{bmatrix} 1 & 0 & 0 & 0 \\ 0 & 1 - 2x_2^2 & -2x_2 x_3 & -2x_2 x_4 \\ 0 & -2x_2 x_3 & 1 - 2x_3^2 & -2x_3 x_4 \\ 0 & -2x_3 x_4 & -2x_3 x_4 & 1 - 2x_4^2 \end{bmatrix}$$

Now, the (1, 3) (1, 4) elements of $P_2 A P_2$ can become zero, when $A P_2$ has (1, 3), (1, 4) element corresponding zero. So the first row of $A P_2$ is

$$a_{11}, a_{12} - 2p_1 x_2, \ a_{13} - 2p_1 x_3, \ a_{14} - 2p_1 x_4$$

where

$$p_1 = a_{12} x_2 + a_{13} x_3 + a_{14} x_4.$$

We now have $\quad a_{13} - 2p_1 x_3 = 0, a_{14} - 2p1 x_4 = 0 \qquad$ (Same as for matrix of order 3)

Solving the above equations, we have

$$a_{12} - 2p_1 x_2 = \pm\sqrt{a_{12}^2 + a_{13}^2 + a_{14}^2} = \pm s_1$$

and
$$x_2^2 = \frac{1}{2}\left(1 \pm \frac{a_{12}}{s_1}\right) = \frac{1}{2}\left(1 + \frac{a_{12}\, sing(a_{12})}{s_1}\right)$$

$$x_3 = \frac{a_{13}\, sign(a_{12})}{2s_1 x_2}$$

$$x_4 = \frac{a_{14}\, sign(a_{12})}{2s_1 x_2}$$

This transformation gives two zero in the first column and first row.
And other transformation gives zero in (2, 4) and (4, 2) position.
Finally we get a tridiagonal matrix of order 4.

EXAMPLE 4. *Reduce the given matrix of order 3 into a tridiagonal matrix by using House Holder's Method, where*

$$A = \begin{bmatrix} 2 & 1 & 1 \\ 1 & 1 & 0 \\ 1 & 0 & 1 \end{bmatrix}$$

SOLUTION. Given that

$$A = \begin{bmatrix} 2 & 1 & 1 \\ 1 & 1 & 0 \\ 1 & 0 & 1 \end{bmatrix}$$

and a column vector w is such that

$$w_2^T = \begin{bmatrix} 0, x_2, x_3 \end{bmatrix}^T$$

and
$$P = I - 2w_2 w_2^T = \begin{bmatrix} 1 & 0 & 0 \\ 0 & 1-2x_2^2 & -2x_2 x_3 \\ 0 & -2x_2 x_3 & 1-2x_3^2 \end{bmatrix}. \ \dots \ \dots$$

Then (1, 3) element of PAP become zero only when AP has (1, 3) element equal to zero.
And we have, the first row element of AP

$$a_{11}, a_{12} - 2p_1 x_2, (a_{13} - 2p_1 x_3 = 0)$$

where $p_1 = a_{12}x_2 + a_{13}x_3$ ...(1)

such that $a_{12} - 2p_1 x_2 = \pm\sqrt{a_{12}^2 + a_{13}^2} = \pm s_1$ ...(2)

so we have the values of equation (1) and (2), from matrix A

$$p_1 = x_2 + x_3$$ ...(3)

$$s_1 = \sqrt{a_{12}^2 + a_{13}^2} = \sqrt{2}$$ ...(4)

Also we have $p_1 = \pm s_1 x_2$

$$p_1 = \pm\sqrt{2}x_2$$

and from equation (2), we have

$$x_2^2 = \frac{1}{2}\left[1 \pm \frac{a_{12}}{s_1}\right]$$

$$= \frac{1}{2}\left[1 + \frac{a_{12}\,\text{sign}(a_{12})}{s_1}\right]$$

$$= \frac{1}{2}\left[1 + \frac{1}{\sqrt{2}}\right] \qquad\qquad (\text{where } s_1 = \sqrt{2})$$

and $$x_3 = \frac{a_{13}[\text{sign}(a_{12})]}{2s_1 x_2}$$

$$= \frac{1}{2\sqrt{2}\left(\dfrac{\sqrt{2}+1}{2\sqrt{2}}\right)^{1/2}} = \frac{1}{[2\sqrt{2}(\sqrt{2}+1]^{1/2}}$$

So, $$1 - 2x_2^2 = -\frac{1}{\sqrt{2}}, 1 - 2x_3^2 = \frac{1}{\sqrt{2}} \text{ and } 2x_2 x_3 = \frac{1}{\sqrt{2}}$$

Hence, P becomes as

$$P = \begin{bmatrix} 1 & 0 & 0 \\ 0 & 1-2x_2^2 & -2x_2x_3 \\ 0 & -2x_2x_3 & 1-2x_3^2 \end{bmatrix} = \begin{bmatrix} 1 & 0 & 0 \\ 0 & -1/\sqrt{2} & -1/\sqrt{2} \\ 0 & -1/\sqrt{2} & 1/\sqrt{2} \end{bmatrix}$$

and $$B_1 = PAP = \begin{bmatrix} 1 & 0 & 0 \\ 0 & -1/\sqrt{2} & -1/\sqrt{2} \\ 0 & -1/\sqrt{2} & 1/\sqrt{2} \end{bmatrix}\begin{bmatrix} 2 & 1 & 1 \\ 1 & 1 & 0 \\ 1 & 0 & 1 \end{bmatrix}\begin{bmatrix} 1 & 0 & 0 \\ 0 & -1/\sqrt{2} & -1/\sqrt{2} \\ 0 & -1/\sqrt{2} & 1/\sqrt{2} \end{bmatrix}$$

$$= \begin{bmatrix} 2 & -\sqrt{2} & 0 \\ -\sqrt{2} & 1 & 0 \\ 0 & 0 & 1 \end{bmatrix}$$

which is the required tridiagonal matrix.

EXAMPLE 5. *Reduce the given matrix A of order 3 into the tridiagonal form by using Householder's Method, where*

$$A = \begin{bmatrix} 1 & 4 & 3 \\ 4 & 1 & 2 \\ 3 & 2 & 1 \end{bmatrix} \qquad\qquad \text{[Pune-2002, 04, MKU (Tamilnadu-2008)]}$$

SOLUTION. Here, in given matrix, $a_{11} = 1, a_{12} = 4, a_{13} = 3$ and $w = [0, x_2, x_3]^T$.

Now, $$s_1 = \sqrt{a_{12}^2 + a_{13}^2} = \sqrt{4^2 + 3^2} = \pm 5$$

Also, $$x_2^2 = \frac{1}{2}\left[1 + \frac{a_{12}\,\text{sign}(a_{12})}{s_1}\right]$$

$\Rightarrow$ $$x_2^2 = \frac{1}{2}\left[1 + \frac{4}{5}\right]$$

$\Rightarrow$ $$x_2 = \frac{3}{\sqrt{10}}$$

and
$$x_3 = \frac{a_{12}\,\text{sign}(a_{12})}{2s_1x_2} = \frac{3\times(1)}{2\times5\times\dfrac{3}{\sqrt{10}}} = \frac{1}{\sqrt{10}}$$

and $1 - 2x_2^2 = -\dfrac{4}{5}$ and $-2x_2x_3 = -\dfrac{3}{5}$.

So the orthogonal matrix becomes

$$P = \begin{bmatrix} 1 & 0 & 0 \\ 0 & 1-2x_2^2 & -2x_2x_3 \\ 0 & -2x_2x_3 & 1-2x_3^2 \end{bmatrix} = \begin{bmatrix} 1 & 0 & 0 \\ 0 & -4/5 & -3/5 \\ 0 & -3/5 & -4/5 \end{bmatrix}$$

So the first transformation is given by

$$PAP = \begin{bmatrix} 1 & 0 & 0 \\ 0 & -4/5 & -3/5 \\ 0 & -3/5 & -4/5 \end{bmatrix}\begin{bmatrix} 1 & 4 & 3 \\ 4 & 1 & 2 \\ 3 & 2 & 1 \end{bmatrix}\begin{bmatrix} 1 & 0 & 0 \\ 0 & -4/5 & -3/5 \\ 0 & -3/5 & -4/5 \end{bmatrix}$$

$$= \begin{bmatrix} 1 & -5 & 0 \\ -5 & 73/25 & -14/25 \\ 0 & -14/24 & -11/25 \end{bmatrix}$$

which is the required tridiagonal matrix.

EXERCISE 7.6

1. Using Jacobi's Method. Find all the eigen values and corresponding eigen vectors of following matrices

(i) $\begin{bmatrix} 5 & 0 & 1 \\ 0 & -2 & 0 \\ 1 & 0 & 5 \end{bmatrix}$ (ii) $\begin{bmatrix} 1 & -2 & 4 \\ -2 & 5 & -2 \\ 4 & -2 & 1 \end{bmatrix}$

2. Find the eigen values of following matrix

$A = \begin{bmatrix} 2 & 1 & 0 \\ 1 & 4 & 1 \\ 0 & 1 & 4 \end{bmatrix}$ use Jacobi's Method.

(Agra-2008, 09; Meerut-2003, 04, 06)

3. Find the eigen values and corresponding eigen vectors of followig matrix

$A = \begin{bmatrix} 3 & 2 & 2 \\ 2 & 5 & 2 \\ 2 & 2 & 3 \end{bmatrix}$ by using Jacobi's Method.

4. Reduce the matrix A in tridiagonal matrix and find the largest eigen value, by using Given's Method

$A = \begin{bmatrix} 1 & 2 & 4 \\ 2 & 1 & 2 \\ 4 & 2 & 1 \end{bmatrix}$.

For largest eigen value, use Newton Raphson Mathod.

5. Applying Given's Method, transform the matrix into tridiagonal form. Find the largest eigen value and corresponding eigen vector

$A = \begin{bmatrix} 1 & 2 & 2 \\ 2 & 1 & 2 \\ 2 & 2 & 1 \end{bmatrix}$.

6. Use the House holder's method to reduce the matrix A into tridiagonal form, if

$A = \begin{bmatrix} 4 & -1 & -2 & 2 \\ -1 & 4 & -1 & -2 \\ -2 & -1 & 4 & -1 \\ 2 & -2 & -1 & 4 \end{bmatrix}$

7. Use the House Holder's method to reduce the matrix A in to tridiagonal form also find the eigen value of given matrix, where

$A = \begin{bmatrix} -2 & -1 & -1 \\ -1 & 2 & -1 \\ -2 & -1 & 2 \end{bmatrix}$.   (Meerut-2002)

Answers

1. (i) Eigen values are 6, –2, 4 and corresponding eigen vectors are

$$[1/\sqrt{2}, 0, 1/\sqrt{2}]^T [0, 1, 0]^T, [-1\sqrt{2}, 0, 1\sqrt{2}]^T$$

(ii) Eigen values are $5 - 2\sqrt{2}$, $5 + 2\sqrt{2}$, –3 and corresponding eigen vectors are

$$[1/2, 1/\sqrt{2}, 1/2]^T [-1/2, 1/\sqrt{2}, -1/2]^T [-1/2, 0, 1/\sqrt{2}]^T$$

2. Eigen values are 1.52, 5.17, 3.3

3. Eigen values are $5 + 2\sqrt{2}, 5 - 2\sqrt{2}, 1$ and corresponding eigen vector is

$$[1/2 \quad 1/\sqrt{2} \quad 1/2]^T [-1/2 \quad 1/\sqrt{2} \quad -1/2]^T [-1/\sqrt{2} \quad 0 \quad 1/\sqrt{2}]^T$$

4. The tridiagonal matrix $B = \begin{bmatrix} 1 & 2/\sqrt{5} & 0 \\ -2/\sqrt{5} & 26 & -1.2 \\ 0 & -1.2 & -0.6 \end{bmatrix}$ and eigen value is $\lambda = 6.464$.

5. The tridiagonal matrix $B = \begin{bmatrix} 1 & 2\sqrt{2} & 0 \\ 2\sqrt{2} & 3 & 0 \\ 0 & 0 & 1 \end{bmatrix}$ and eigenvalue $h = 5$

and corresponding eigen vector $[1, 1, 1]^T$

6. The tridiagonal matrix is $B_2 = \begin{bmatrix} 4 & 3 & 0 & 0 \\ 3 & 16/3 & -5/3\sqrt{5} & 0 \\ 0 & -5/3\sqrt{5} & 16/3 & 9/5 \\ 0 & 0 & 9/5 & 12/5 \end{bmatrix}$

7. The tridiagonal matrix is $B = \begin{bmatrix} 2 & \sqrt{2} & 0 \\ \sqrt{2} & 1 & 0 \\ 0 & 0 & 3 \end{bmatrix}$ and eigen values are 0, 3, 3.

MISCELLANEOUS EXERCISE

1. Apply Crout's method to obtain the inverse of

$$\begin{bmatrix} 2 & -2 & 4 \\ 2 & 3 & 2 \\ -1 & 1 & -1 \end{bmatrix}$$ (Rohilkhand-2004,08)

2. Let $A = \begin{bmatrix} 3 & 12 & 9 \\ 2 & 10 & 12 \\ 1 & 12 & 2 \end{bmatrix}$, then find two triangular

matrices : L (lower triangular) and U (Upper triangular) as 3, 1, 5. Hence obtain A^{-1}.

(Lucknow-2007,09; Allahabad-2008, MDU(Rohtak-2010, Agra-2004, 09)

3. Find Eigen values and Eigen vectors of A, when

$$\begin{bmatrix} 10 & 7 & 8 & 7 \\ 7 & 5 & 6 & 5 \\ 8 & 6 & 10 & 9 \\ 7 & 5 & 9 & 10 \end{bmatrix} \text{ taking } X = \begin{bmatrix} 1 \\ 1 \\ 1 \\ 1 \end{bmatrix}$$

(Agra-2008, Bihar-2006)

4. Obtain λ_2, λ_3 and the corresponding Eigen vector

of A, when $\begin{bmatrix} -306 & -198 & 426 \\ 104 & 67 & -147 \\ -176 & -114 & 244 \end{bmatrix}$ given that

$\lambda_1 = 6$ and corresponding Eigen vector $= \begin{bmatrix} 2 \\ -1 \\ 1 \end{bmatrix}$.

(Agra-2004)

5. Find the Eigen values of the matrix

$$A = \begin{bmatrix} 2 & -i & 0 \\ i & 2 & 0 \\ 0 & 0 & 3 \end{bmatrix}.$$

Answers

1. $\begin{bmatrix} -1/5 & 1/5 & -8/5 \\ 0 & 1/5 & 2/5 \\ 1/2 & 0 & 1 \end{bmatrix}$ **2.** $\dfrac{1}{25} \begin{bmatrix} \dfrac{62}{3} & -14 & -9 \\ -\dfrac{4}{3} & \dfrac{1}{2} & 3 \\ -\dfrac{7}{3} & 4 & -1 \end{bmatrix}$ **3.** $\lambda_1 = 30.288662, \ x = \begin{bmatrix} x_1 \\ x_2 \\ x_3 \\ x_4 \end{bmatrix} = \begin{bmatrix} 0.95761 \\ 0.68892 \\ 1 \\ 0.94379 \end{bmatrix}$

4. $\lambda_2 = -2, \ \lambda_3 = 1, \ x_2 = \begin{pmatrix} 3 \\ 4 \\ 4 \end{pmatrix}, \ x_3 = \begin{pmatrix} 6 \\ -5 \\ 2 \end{pmatrix}$

5. Eigen values are 3, 1, 3.

Objective Evaluations

FILL IN THE BLANKS

1. Inverse of the matrix $\begin{bmatrix} 2 & 2 & 3 \\ 2 & 1 & 5 \\ 1 & 3 & 5 \end{bmatrix}$ is _____.

2. The eigenvalues of a real symmetric matrix are _____.

3. Inverse of the matrix $A = \begin{bmatrix} 1 & 1 \\ 8 & 1 \end{bmatrix}$ is _____.

4. The eigenvalues of a skew-Hermition matrix are either purely _____.

5. If $\lambda_1, \lambda_2, ..., \lambda_n$ are eigenvalues of a square matrix of order $n \times n$, then
$$\det (A) = \underline{\hspace{2cm}}$$

TRUE/FALSE

Write 'T' for True and 'F' for False statement.

1. l is an eigenvalue of a matrix A if and only if there exist a non-zero vector X such that
$$AX = \lambda X. \qquad \textbf{(T/F)}$$

2. Let A be an $n \times n$ matrix. Then the distinct eigenvector corresponding to distinct eigenvalues of A are linearly independent.
$$\textbf{(T/F)}$$

3. The eigenvalues of Hermitian matrix are not real. **(T/F)**

4. The eigenvalues of a unitary matrix are of unit modulus. **(T/F)**

5. The eigenvalues of the matrix
$$A = \begin{bmatrix} 0 & 1 & 2 \\ 1 & 0 & -1 \\ 2 & -1 & 0 \end{bmatrix} \text{ is } -2, 1, \pm\sqrt{3}. \quad \textbf{(T/F)}$$

6. If A be a non-singular matrix of order $n \times n$ and its characteristic polynomial is
$$|A-\lambda I| = (-1)^n [\lambda^n + a_1 \lambda^{n-1} + ... + a_{n-1}\lambda + a_n]$$
then $\det (A) = (-1)^n a_n$.

MULTIPLE CHOICE QUESTIONS

Choose the most appropriate one

1. The characteristic equation of the equation $AX = 0$ is given by :
   (a) $(A - \lambda I) = 0$ (b) $(A - \lambda I)X = 0$
   (c) $(A + \lambda I)X = 0$ (d) none of these

2. If λ be the eigen value of the matrix A, then eigen value of the matrix A^n is given by :
   (a) λ^n (b) $n\lambda$
   (c) $\lambda^{1/n}$ (d) none of these

3. The sum of the eigen values of matrix is equal to the sum of the elements of :
   (a) non-diagonal   (b) diagonal
   (c) principal diagonal (d) none of these

4. If λ be the eigen values of a matrix A and C be any similar matrix then λ is also an eigen value of :
   (a) AC (b) ACA
   (c) $C^{-1}AC$ (d) none of these

5. If C is any constant and λ_i is an eigen value of A then which of the following is/are true :
   (a) $C\lambda_i$ is an eigen value of CA

   (b) $C + \lambda i$ is an eigen value of $IC + A$
   (c) both (a) and (b) are true
   (d) none of these

6. Which of the following is/are true ?
   (a) Hermetian and real symmetric matrix has real eigen values
   (b) Hermetian matrix has zero or imaginary eigen value
   (c) both (a) and (b) are true
   (d) none of these

7. If λ be the eigen value of the matrix A then eigen values of A^{-1} is :
   (a) $\dfrac{1}{\lambda}$ (b) $1 - \lambda$
   (c) $\dfrac{1}{\lambda^{-1}}$ (d) none of these

8. Power method gives the :
   (a) eigen values
   (b) largest eigen values
   (c) both (a) and (b) are true
   (d) none of these

9. Jacobi's method is used to find the eigen values of :
(a) real symmetric matrix
(b) red non-symmetric matrix
(c) complex matrix
(d) none of these

10. The method which suggest to reduce a non-tridiagonal symmetric matrix into tridiagonal matrix is :
(a) Given method
(b) Householder method
(c) Power method
(d) None of these

11. Given's method transforms the given symmetric matrix to a :
(a) diagonal matrix
(b) tridiagonal matrix
(c) identity matrix
(d) none of these

12. The eigen values of the matrix $\begin{bmatrix} 5 & 4 \\ 1 & 2 \end{bmatrix}$ are given by :
(a) 1, 2   (b) 1, 6
(c) 2, 6   (d) 3, 4

13. The eigen values of a matrix A of order n are contained in region R which is made by the union of all the circles centred at a_{kk} and having the radius $\sum\limits_{\substack{j=1 \\ j \neq k}} |a_{kj}|$ is called :
(a) Cayley-Hamilton's theoprem
(b) Gerschogorin circle theorem
(c) Milne's circle theorem
(d) none of these

14. The eigen values of the matrix $\begin{bmatrix} 1 & -2 \\ -5 & 4 \end{bmatrix}$ are given by :
(a) 6, 1   (b) 6, -1
(c) 6, 2   (d) none of these

15. The eigen values of the matrix $\begin{bmatrix} 1 & -2 \\ -5 & 4 \end{bmatrix}$ are given by :
(a) (2, 5), (1, 1)   (b) (2, -5), (1, 1)
(c) (2, 5), (1, -1)   (d) none of these

16. The eigen values of the matrix $\begin{bmatrix} 2 & 0 & 1 \\ 0 & 2 & 0 \\ 1 & 0 & 2 \end{bmatrix}$ are given by :
(a) 1, -2, 3   (b) 1, -2, -3
(c) 1, 2, 3   (d) none of these

17. The eigen vectors, corresponding to the eigen values of the matrix $\begin{bmatrix} 2 & 0 & 1 \\ 0 & 2 & 0 \\ 1 & 0 & 2 \end{bmatrix}$ are :
(a) (1, 0, -1), (0, 1, 0), (1, 0, 1)
(b) (1, 2, -1), (1, 1, 0), (1, 0, 1)
(c) (1, 1, 1), (1, 2, 1), (1, 0, 0)
(d) none of these

18. The eigen values of the matrix $\begin{bmatrix} -2 & 2 & -3 \\ 2 & 1 & -6 \\ -1 & -2 & 0 \end{bmatrix}$ are :
(a) 3, 3, 5   (b) -3, -3, 5
(c) -3, 3, -5   (d) none of these

19. The eigen vectors corresponding to the eigen values of the matrix $\begin{bmatrix} -2 & 2 & -3 \\ 2 & 1 & -6 \\ -1 & -2 & 0 \end{bmatrix}$ are :
(a) (2, -1, 0), (1, 2, -1)
(b) (2, 1, 0), (1, 2, 1)
(c) (2, 1, 0), (1, 2, -1)
(d) none of these

20. The eigen values of the matrix $\begin{bmatrix} 8 & -6 & 2 \\ -6 & 7 & -4 \\ 2 & -4 & 3 \end{bmatrix}$ are given by :
(a) 0, 3, 15   (b) 1, 3, 5
(c) 1, 2, 15   (d) none of these

21. The eigen vectors corresponding to the eigen values of the matrix $\begin{bmatrix} 8 & -6 & 2 \\ -6 & 7 & -4 \\ 2 & -4 & 3 \end{bmatrix}$ are given by :
(a) (1, 2, 2), (2, 1, -2), (2, -2, 1)
(b) (1, 2, -2), (2, 1, -2), (2, 2, 1)
(c) (1, 2, 2), (2, 1, 2), (2, 2, 1)
(d) none of these

22. The eigen values of the matrix $\begin{bmatrix} -4 & 14 & 0 \\ -5 & 13 & 0 \\ -1 & 0 & 2 \end{bmatrix}$ are given by :
(a) 1, 2, 3   (b) 2, 3, 6
(c) 2, 3, 5   (d) none of these

23. The largest eigen value of the matrix $A = \begin{bmatrix} 4 & 1 \\ 1 & 3 \end{bmatrix}$ by power method upto third iteration is given by :
(a) 4.82   (b) 4.41
(c) 4.51   (d) none of these

24. The largest eigen value of the matrix
$$\begin{bmatrix} 1 & 2 & 3 \\ 0 & -4 & 2 \\ 0 & 0 & 7 \end{bmatrix}$$ upto third iteration is given by :

(a) 6   (b) 5

(c) 7   (d) 8

25. The largest eigen value of the matrix $\begin{bmatrix} 1 & 2 \\ 3 & 4 \end{bmatrix}$ is given by :

(a) 5.30   (b) 5.38

(c) 6.38   (d) none of these

26. The largest eigen value of the matrix
$$\begin{bmatrix} 1 & 6 & 1 \\ 1 & 2 & 0 \\ 0 & 0 & 3 \end{bmatrix}$$ upto third iteration is given by :

(a) 4.32   (b) 3.52

(c) 5.32   (d) none of these

27. The largest eigen values of the matrix
$$\begin{bmatrix} 1 & -3 & 2 \\ 4 & 4 & -1 \\ 6 & 3 & 5 \end{bmatrix}$$ upto third iteration is given by :

(a) 6.915   (b) 5.915

(c) 4.915   (d) none of these

28. The largest eigen value of the matrix
$$\begin{bmatrix} 2 & -1 & 0 \\ 1 & 2 & -1 \\ 0 & -1 & 2 \end{bmatrix}$$ upto third iteration is given by :

(a) 3.8   (b) 2.8

(c) 4.8   (d) none of these

29. The smallest eigen value of the matrix
$$\begin{bmatrix} 3 & 0 & 1 \\ 0 & -3 & 0 \\ 1 & 0 & 3 \end{bmatrix}$$ as $X = \begin{bmatrix} 0 \\ 1 \\ 2 \end{bmatrix}$ upto third iteration is given by :

(a) $\dfrac{9}{20}$ (b) $\dfrac{7}{20}$

(c) $\dfrac{8}{20}$ (d) none of these

30. The smallest eigen value in magnitude of the matrix $\begin{bmatrix} 2 & -1 & 0 \\ -1 & 2 & -1 \\ 0 & -1 & 2 \end{bmatrix}$ is given by :

(a) 0.5848   (b) 0.6848

(c) 0.7848   (d) none of these

Answers

FILL IN THE BLANKS

1. $\begin{bmatrix} 2 & -1 & -1 \\ -9 & 7 & 4 \\ 5 & -4 & -2 \end{bmatrix}$ 2. real   3. $-\dfrac{1}{7}\begin{bmatrix} 1 & -1 \\ -8 & 1 \end{bmatrix}$ 4. imaginary or zero.   5. $\lambda_1 \lambda_2 \lambda_3 \dots \lambda_n$.

TRUE/FALSE

1. T   2. T   3. F   4. T   5. F   6. T

MULTIPLE CHOICE QUESTIONS

1. (b)	2. (a)	3. (c)	4. (c)	5. (c)	6. (c)	7. (a)	8. (c)	9. (a)
10. (b)	11. (b)	12. (b)	13. (b)	14. (b)	15. (b)	16. (c)	17. (a)	18. (b)
19. (a)	20. (a)	21. (a)	22. (b)	23. (b)	24. (c)	25. (b)	26. (b)	27. (b)
28. (b)	29. (a)	30. (a)						

COMPUTATIONAL TECHNIQUE LAB

1. POWER METHOD FOR FINDING LARGEST EIGEN VALUE

SYMBOLS USED

a [ ] – Square matrix

x [ ] – Eigen vector at n^{th} iteration

r [ ] – Eigen vector at $(n + 1)^{th}$ iteration

n – No. of rows or columns in square matrix

e – Eigen value at the nth iteration

t – Eigen value at $(n + 1)^{th}$ iteration

aerr – allowed error in eigen value and eigen vector

maxitr – maximum number of iterations to be performed

err – error in an element of the eigen vector

maxe – maximum error in any element of the eigen vector

err – error in eigen value

i, j, k, itr – loop control reariables

ALGORITHM : TO FIND LARGEST EIGEN VALUE BY POWER METHOD

Step 1. Start.

Step 2. Input the values of square matrix a[n][n].

Step 3. Input initial approximation to eigen vector x[n].

Step 4. Input allowed error and maximum iteration in aerr and maxitr.

Step 5. e = x[0].

Step 6. for i = 1 to (n – 1)

if x[i] > e

e = x[i]

End of Step 6 for loop

Step 7. for itr = 1 to maxitr.

Step 8. for i = 0 to (n – 1)

s = 0

for k = 0 to (n – 1)

s = s + a[i][k] * x[k];

r[i] = s;

End of step 8 for loop.

Step 9.   t = r[0]

Step 10. for i = 1 to (n – 1)

   if (r[i] > t)

   t = r[i]

   End of step 10 for loop

Step 11. for i = 0 to n – 1

   r[i] = r[i]/t

   End of step 11 for loop

Step 12. max. = 0

Step 13. for i = 0 to (n – 1)

   err = abs (x[i] – r[i])

   if (err > max.)

   maxe = err;

   x[i] = r[i]

   End of step 13 for loop

Step 14. errv = abs (t – e)

Step 15. e = t

Step 16. Print itr, e

Step 17. for i = 0 to n – 1

   Print x[i]

   End of Step 17 for loop.

Step 18. if (errv < aerr) & (max. < = aerr)

   Print "solution converges in" itr "iterations"

   Pirnt "Largest eigen value =" e

   Print "Eigen vector"

   for i = 0 to n – 1

   Print i + 1, x[i]

   End of step 7 for loop.

Step 19. if (waxitr = itr)

   Print "solution does not converges"

Step 20. Stop.

FLOW CHART : LARGEST EIGEN VALUE BY POWER METHOD

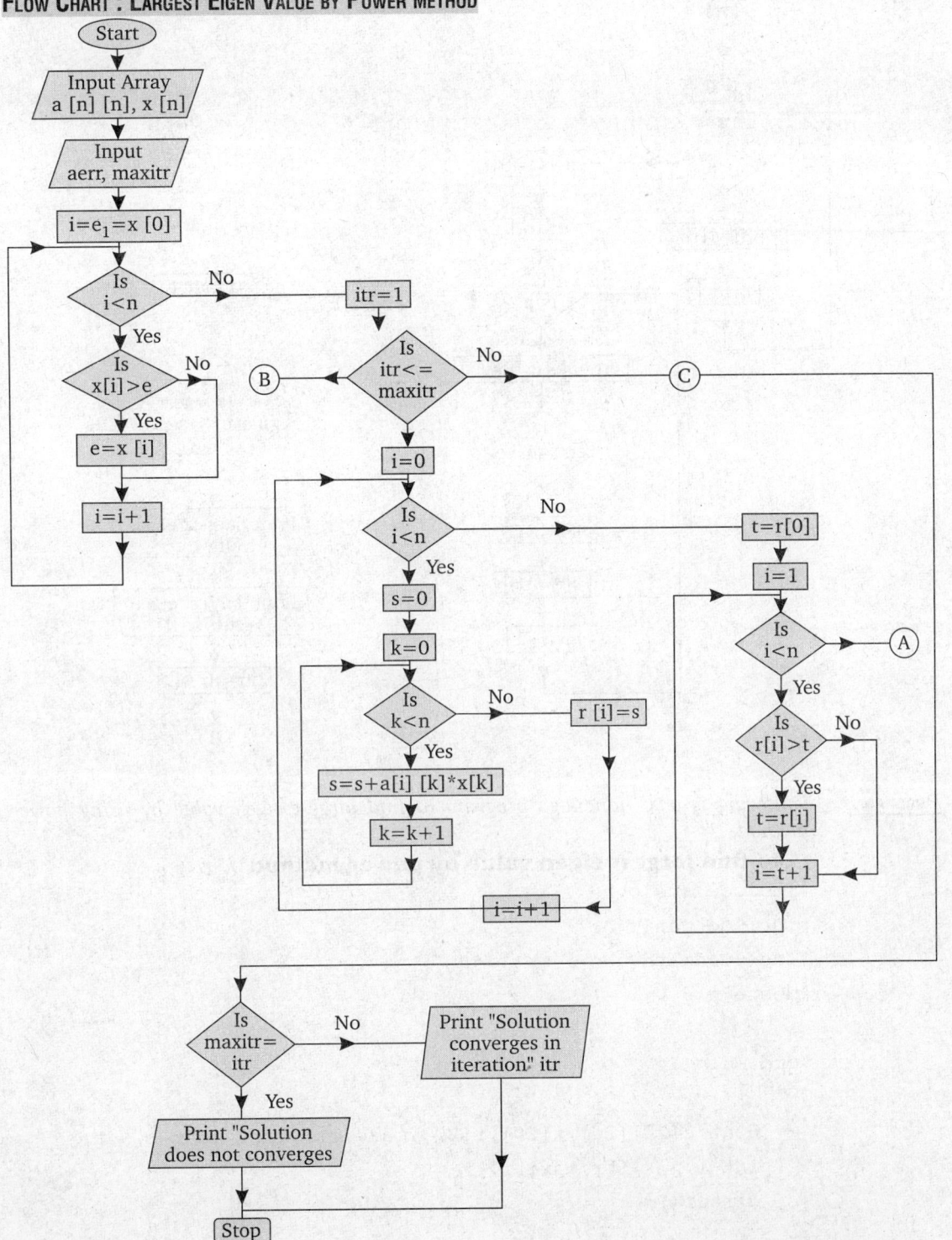

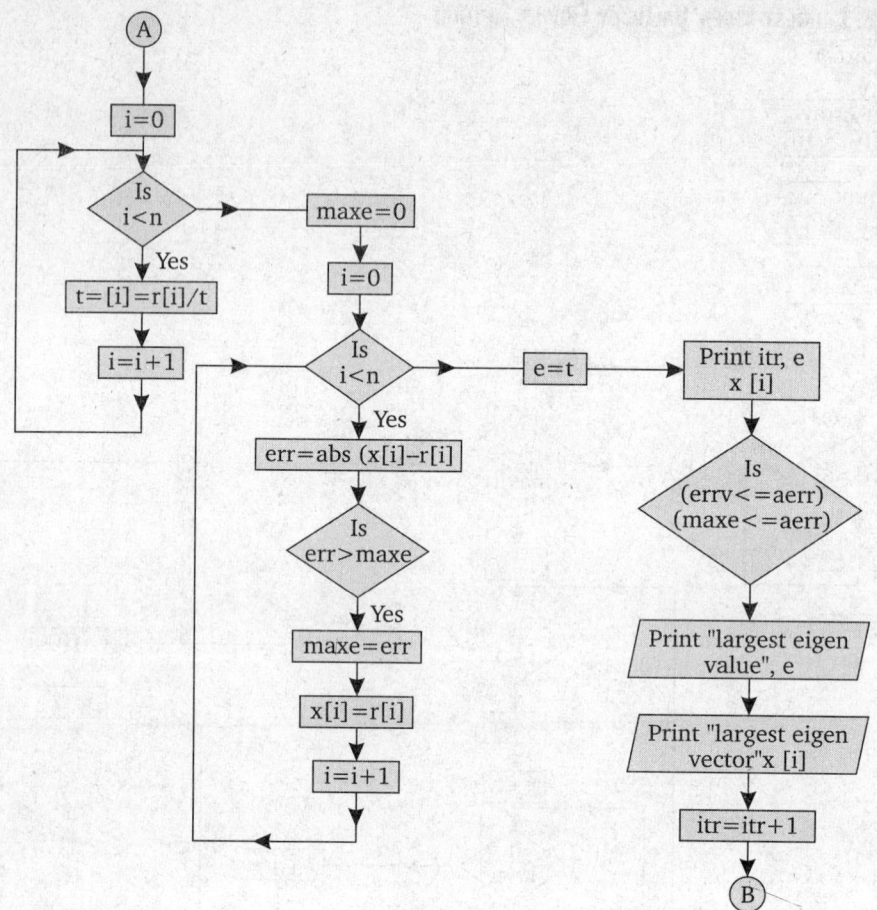

PROGRAM.   *Following is a 'C' language program to find largest eigen value by using power method.*

/* to find largest eigen value by power method*/

```c
#include<stdio.h>
#include<conio.h>
#include<math.h>
#define n 3

void main()
{
    float a[20][20],x[20],r[20],maxe,err,errv,aerr,e,s,t;
    int i,j,k,itr,maxitr;
    clrscr();
```

```c
    printf("\n enter elements of matrix \n");
    for(i=0;i<n;i++)
    for(j=0;j<n;j++)
  scanf("%f",&a[i][j]);
printf("\n enter initial approximation to eigen vector\n");
 for(i=0;i<n;i++)
  scanf("%f",&x[i]);
printf("\n enter allowed error, maximum iteration\n");
scanf("%f %d",&aerr,&maxitr);
printf(" itr no     eigen value    eigen vector\n");

e=x[0];
for(i=1;i<n;i++)
{
   if(x[i]>e)
   e=x[i];
}
for(itr=1;itr<=maxitr;itr++)
 {
  for(i=0;i<n;i++)
    {
     s=0;
     for(k=0;k<n;k++)
       s=s+a[i][k]*x[k];
     r[i]=s;
    }
     t=r[0];
for(i=1;i<n;i++)
{
   if(r[i]>t)
   t=r[i];
  }

    for(i=0;i<n;i++)
    r[i]=r[i]/t;
   maxe=0;
```

```
for(i=0;i<n;i++)
 {
  err=fabs(x[i]-r[i]);
  if(err > maxe)
     maxe=err;
  x[i]=r[i];
 }
 errv=fabs(t-e);
 e=t;
 printf("%4d %12.4f",itr,e);
 for(i=0;i<n;i++)
 printf("%9.2f",x[i]);
printf("\n");
if((errv<=aerr)&&(maxe<=aerr))
{
   printf("converges in %d"
   "iterations\n",itr);
   printf("\n largest eigen value="
   "%6.2f\n",e);
   printf("Eigen  Vector:\n");
   for(i=0;i<n;i++)
   printf("x[%d]=%6.2f\n",i+1,x[i]);
   printf("\n");
   break;

   }
 }
 if(maxitr==itr)
   printf("Solution does not converges"
   "iterations  not sufficient \n");
getch();
 }
```

Output: TO FIND THE LARGEST EIGEN VALUE BY POWER METHOD

```
enter elements of matrix
2 0 1
0 2 0
1 0 2
 enter initial approximation to eigen vector
1 0 0
 enter allowed error, maximum iteration
.001 10
 itr no. eigen value      eigen vector
    1         2.0000    1.00     0.00     0.50
    2         2.5000    1.00     0.00     0.80
    3         2.8000    1.00     0.00     0.93
    4         2.9286    1.00     0.00     0.98
    5         2.9756    1.00     0.00     0.99
    6         2.9918    1.00     0.00     1.00
    7         2.9973    1.00     0.00     1.00
    8         2.9991    1.00     0.00     1.00
    9         2.9997    1.00     0.00     1.00
Converges in 9iterations

largest eigen value=  3.00
Eigen   Vector:
x[1]=  1.00
x[2]=  0.00
x[3]=  1.00
```

LAB ASSIGNMENT

1. Write a computer program in 'C' language to find eigen value by using power method

$$\begin{bmatrix} 2 & -1 & 0 \\ -1 & 2 & -1 \\ 0 & -1 & 2 \end{bmatrix}$$

Hint: Input enter array elements

```
2  -1   0
-1   2  -1
 0  -1   2
```

Enter Initial approx to eigen vector.

   1   0   0

Enter allowed error, maximum iterations

   0.01   10

Output :

itr no.	eigen value		eigen vector	
1	2.00	1.00	−0.50	0.00
2	2.50	1.00	−0.80	0.20
3	2.80	1.00	−1.00	0.42
4	3.42	0.87	−1.00	0.54
5	3.41	0.80	−1.00	0.61
6	3.41	0.76	−1.00	0.65
7	3.41	0.74	−1.00	0.67
8	3.41	0.72	−1.00	0.68
9	3.41	0.71	−1.00	0.69

Converges in 9 iterations.

Largest eigen value = 3.41

⌘⌘⌘⌘⌘

CHAPTER 8

Difference Equations

8.1 INTRODUCTION

An equation that consists of an independent variable x, a dependent variable y_x and one or more of its difference $\Delta y_x, \Delta^2 y_x, ..., \Delta^n y_x$ is called a difference equation. It is of the form

$$F[x, y_x, \Delta y_x, \Delta^2 y_x, ..., \Delta^n y_x] = 0$$

A difference equation is therefore a relation involving differences.

For example :

(1) $\Delta y_x + 2y_x = 0$ (2) $\Delta^2 y_x + 3\Delta y_x - 7y_x = 0$

(3) $\Delta^3 y_x + 3\Delta^2 y_x - 6\Delta y_x + y_x = 3x + 2$ (4) $y_x \Delta^3 y_x = 6$

> ▶ **REMARKS**
> ▸ It is not necessary that the difference equations are defined over the set of all real numbers. They can be considered as difference equations over some other set.
> ▸ When difference equation defined over some set A, then the relation among the values of y_x, Δy_x, $\Delta^2 y_x$, ... given by the equation and the set of values, denoted by A for which this relation is said to hold.
> ▸ The set A consists of either a finite or infinite set of successive integers.
> ▸ It is often convenient to have this set start with zero but it is not necessary.

8.2 DIFFERENCE EQUATION AS A RELATION AMONG THE VALUE OF y_x

Using $\Delta^k = (E - 1)^k$ and noting that $E^h y_x = y_{x+h}$ under the assumption that the interval of differencing is one. We can express

$$\Delta y_x = (E - 1)y_x = Ey_x - y_x = y_{x+1} - y_x$$
$$\Delta^2 y_x = (E - 1)^2 y_x = (E^2 - 2E + 1)y_x = y_{x+2} - 2y_{x+1} + y_x$$
$$\Delta^3 y_x = (E - 1)^3 y_x = (E^3 - 3E^2 + 3E - 1)y_x = y_{x+3} - 3y_{x+2} + 3y_{x+1} - y_x$$

The difference equation (3) can be written as

$$y_{x+3} - 9y_{x+1} + 9y_x = 3x + 2$$

This equation can be written as

$$y(x + 3) - 9y(x + 1) + 9y(x) = 3x + 2$$
$$\Rightarrow \qquad (E^3 - 9E + 9)y_x = 3x + 2$$

Similarly we can solve other equation (1), (2) and (4), we get

$$(E + 1)y_x = 0 \qquad\qquad \Rightarrow \qquad\qquad (E^2 + E - 9)y_x = 0$$

or $\qquad (E^3 - 3E^2 + 3E - 1)y_x^2 = 6$

8.3 ORDER OF DIFFERENCE EQUATION

The order of a difference equation is the difference between the highest and the lowest subscripts of the y, it is free from Δ. Thus the order of equation (1) is $(x+1)-x=1$ equation (2) is $(x+2)-x=2$ equation (3) is $(x+3)-x=3$ and the equation (4) is $(x+4)-x=4$.

8.4 DEGREE OF DIFFERENCE EQUATION

The degree of difference equation is highest power of y and free from Δ. The degree of equation

$$y_{x+1}^2 y_{x+2}^3 - y_{x+1}y_x - y_x^2 = x$$

is 3 and the order of this equation is 2.

8.5 SOLUTION OF DIFFERENCE EQUATION

A function y is called a solution of a difference equation over a set A if the value of y make the difference equation a true statement for every point of A.

▼ REMARKS

▸ A general solution of a difference equation of order n involves n arbitrary constants.
▸ A particular solution of a difference equation is obtained from the general solution by giving particular values to the constants.

For example: $y_x = A2^x + B3^x$ is the general solution to $y_{x+2} - 5y_{x+1} + 6y_x = 0$ while $y_x = 2^x$ or $y_x = 3^x$ or $y_x = 52^x + 8(3^x)$ are particular solutions.

SOLVED EXAMPLES

EXAMPLE 1. *Form the difference equation corresponding to the family of curves $y = ax^2 + bx - 3$.*

SOLUTION. The given equation is

$$y_x = ax^2 + bx - 3 \qquad \ldots(1)$$

where, a and b are arbitrary constants to be determined

Now $y_{x+1} = a(x+1)^2 = b(x+1) - 3$

$\Rightarrow \quad y_{x+2} + a(x+2)^2 = b(x+2) - 3$

We know that

$$\Delta y_x = y_{x+1} - y_x = 2(x+1)a + b \qquad \ldots(2)$$
$$\Delta^2 y_x = y_{x+2} + 2y_{x-1} + y_x = 2a \qquad \ldots(3)$$

$$\Rightarrow \quad a = \frac{1}{2}\Delta^2 y_x$$

∴ From equation (2)

$$b = \Delta y_x - \frac{1}{2}(2x+1)\Delta^2 y_x \qquad \ldots(4)$$

Eliminating a, b from equation (1), (3) and (4), we get

$$(x+1)x\Delta^2 y_x - 2x\Delta y_x + 2y_x + 6 = 0$$

$$(x^2+x)y_{x+2} - 2(x^2+2x)y_{x+1} + (x^2+3x+2)y_x + 6 = 0.$$

EXAMPLE 2. *Form the difference equation given that $y_n = A3^n + B5^n$, where A and B are arbitrary constants.*

(Kurukshetra (NIT)–2013)

SOLUTION. Given equation is $y_n = A3^n + B5^n$ $\qquad \ldots(1)$

$y_{n+1} = A3^{n+1} + B5^{n+1} = 3A3^n + 5B5^n \qquad \ldots(2)$

$y_{n+2} = A3^{n+2} + B5^{n+2} = 9A3^n + 25B5^n \qquad \ldots(3)$

Eliminating A and B from equations (1) to (3), we get

$$\begin{vmatrix} y_n & 1 & 1 \\ y_{n+1} & 3 & 5 \\ y_{n+2} & 9 & 25 \end{vmatrix} = 0$$

or $y_{n+2} - 8y_{n+1} + 15y_n = 0$.

which is the required difference equation.

EXAMPLE 3. *Find the order of the difference equation $y_{x+2} - 7y_x = 5$*

SOLUTION. The given equation is $y_{x+2} - 7y_x = 5$

The difference between the highest and lowest subscripts of y is $x + 2 - x = 2$.

Hence the order of the equation is 2.

EXAMPLE 4. *Find the order of the following :*

    (i) $y_{x+4} - 5y_{x+2} + 6y_x = 0$ (ii) $\Delta^3 y_x + 2\Delta y_x + y_x = x + 3$

SOLUTION.   (i) The order of equation $y_{x+4} - 5y_{x+2} + 6y_x = 0$ is $x + 4 - x = 4$.

  (ii) The given equation

$$\Delta^3 y_x + 2\Delta y_x + y_x = x + 3$$

$$(y_{x+3} - 3y_{x+2} + 3y_{x+1} - y_x) + 2(y_{x+1} - y_x) + y_x = x + 3$$

$$y_{x+3} - 3y_{x+2} + 5y_{x+1} - 2y_x = x + 3$$

Order of this equation is the difference between the highest and lowest subscript of y is given by $(x + 3) - x = 3$.

EXAMPLE 5. *Show that $y_x = \dfrac{x(x-1)}{2}$ is a solution of the difference equation $y_{x+1} - y_x = x$.*

SOLUTION. We have $y_x = \dfrac{x(x-1)}{2}$.   Therefore $y_{x+1} = \dfrac{(x+1)x}{2}$

Substituting these values in $y_{x+1} - y_x = \dfrac{(x+1)x}{2} - \dfrac{x(x-1)}{2} = x$, we get right

hand side, *i.e.*, $y_x = x(x-1)$ satisfy the given difference equation. Hence it is a solution of given difference equation.

EXAMPLE 6. *Show that $y_x = 1 - \dfrac{2}{x}, x = 1, 2, 3, \ldots$ is a solution of the first order difference equation*

$$(x+1)y_{x+1} + xy_x = 2x - 3, x = 1, 2, 3, \ldots$$

SOLUTION. We have $y_x = 1 - \dfrac{2}{x} \Rightarrow y_{x+1} = 1 - \dfrac{2}{x+1}$

Substituting these values in LHS of given equation, we get

$$(x+1)y_{x+1} + xy_x$$

$$= (x+1)\left[1 - \frac{2}{x+1}\right] + x\left[1 - \frac{2}{x}\right] = x + 1\left[\frac{x+1-2}{x+1}\right] + x\left[\frac{x-2}{x}\right]$$

$$= x - 1 + x - 2 = 2x - 3 = \text{RHS}$$

$\therefore$ $y_x = 1 - \dfrac{2}{x}$ is the solution of the given first order difference equation.

EXAMPLE 7. *Show that $y_x = C_1 + C_2 2^x - x$ is a solution of the difference equation* $y_{x+2} - 3y_{x+1} + 2y_x = 1$.

SOLUTION. We have $y_x = C_1 + C_2 2^x - x$.

$\Rightarrow$ $y_{x+1} = C_1 + C_2 2^{x+1} - (x+1)$

and $y_{x+2} = C_1 + C_2 2^{x+2} - (x+2)$

Substituting these values in LHS of given equation, we get

$$y_{x+2} - 3y_{x+1} + 2y_x = (C_1 + C_2 2^{x+2} - x - 2)$$

$$-3(C_1 + C_2 2^{x+1} - x - 1) + 2(C_1 + C_2 2^x - x)$$
$$= (2^2 - 3.2 + 2)C_2 2^x + 1$$
$$= 1 = \text{RHS}$$

i.e., $y_x = C_1 + C_2 2^x - x$ satisfy the given difference equation. Hence, this is the solution of given difference equation.

EXERCISE 8.1

1. Form the difference equations by eliminating arbitrary constant

    (i) $y = C_1 x^2 + C_2 x + C_3$

    (ii) $y = C_1 3^x + C_2 8^x$

    (iii) $y = (C_1 + C_2 x) 2^x$

2. Find the order of the difference equation.

    (i) $\Delta^3 y_x + \Delta^2 y_x + \Delta y_x + y_x = 0$

    (ii) $y_{x+2} + 3y_x = 2$

    (iii) $\Delta^3 y_x + 2\Delta y_x + y_x = x$

3. Show that $C_1 + C_2 e^x$ is the solution of the difference equation $y_{x+2} - 3y_{x+1} + 2y_x = 0$, $x = 0, 1, 2, \ldots$

4. Show that the order of the difference equation $\Delta^2 y_x + 3\Delta y_x - 3y_x = x$ is 2.

5. Prove that $y_x = 3^x(A + Bx)$ satisfy $y_{x+2} - 6y_{x+1} + 9y_x = 0$.

6. Show that the difference equation $y_{x+2} - 4y_{x+1} + 4y_x = 0$, $x = 0, 1, 2, \ldots$ has the solution $y_x = 2^x(C_1 + xC_2)$, $x = 0, 1, 2, \ldots$ for any constants. Find the solution satisfying the initial conditions $y_0 = 1$ and $y_1 = 6$.

7. Show that $y_x = \dfrac{C}{1 + Cx}$ is the solution of the difference equation $y_{x+1} = \dfrac{y_x}{1 + y_x}$, $x = 0, 1, 2, \ldots$

8. Form the difference equation by eliminating the arbitrary constants a and b from the relation

    (i) $y_n = a \cos n\theta + b \sin n\theta$

    (ii) $y_n = an^2 + bn$ (Kurukshetra–2007, 08, Kurukshetra (NIT)–2009)

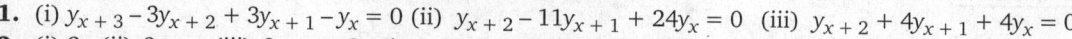

Answers

1. (i) $y_{x+3} - 3y_{x+2} + 3y_{x+1} - y_x = 0$ (ii) $y_{x+2} - 11y_{x+1} + 24y_x = 0$ (iii) $y_{x+2} + 4y_{x+1} + 4y_x = 0$

2. (i) 2 (ii) 2 (iii) 3   **9.** (i) $y_{n+2} - 2\cos\theta\, y_{n+1} + y_n = 0$

9. (ii) $n(n+1)\Delta^2 y_n - 2n\Delta y_n + 2y_n = 0$

8.6 LINEAR DIFFERENCE EQUATION

This is a most important type of difference equation and it has the general form

$$a_0 y_{x+n} + a_1 y_{x+n-1} + a_2 y_{x+n-2} + \ldots + a_n y_x = f(x) \qquad \ldots(1)$$

where $a_0, a_1, a_2, \ldots a_n$ and $f(x)$ are each functions of x (but not of y_x)

or $\qquad\qquad\qquad\qquad L(E)y_x = f(x) \qquad\qquad\qquad\qquad \ldots(2)$

where $L(E) = a_0 E^n + a_1 E^{n-1} + a_2 E^{n-2} + \ldots a_n$ is a polynomial expression in E and is known an non homogeneous linear equation. If $f(x) = 0$ in equation (2) then

$$L(E)y_x = 0 \qquad\qquad \ldots(3)$$

and is known as homogeneous linear equation.

The reader will notice an obvious analogy now with linear differential equations. In fact the general solution of (1) comprises a particular solution of it combined with the general solution of (3). This follows as an immediate consequence of the following theorems which arise from equation (1) and (3). Their proofs are so similar to those already given for corresponding linear difference equations.

    (i) If $y_x = f_1(x)$ is a solution of equation (3), then $y_x = A_1 f_1(x)$ is also a solution of equation (3), where A_1 is any constant.

(ii) If the homogeneous equation (3) is satisfied by equations $y_x = f_1(x)$, $y_x = f_2(x)$, ..., $y_x = f_n(x)$

then it is also satisfied by

$$y_x = A_1 f_1(x) + A_2 f_2(x) + \dots + A_r F_r(x)$$

where $A_1, A_2, \dots, A_r$ are constants.

(iii) If $y_x = f_1(x), y_x = f_2(x), y_x = f_3(x), \dots y_x = f_n(x)$

are n independent solutions of equation (3), then its general solution is

$$y_x = A_1 f_1(x) + A_2 f_2(x) + \dots + A_r f_r(x)$$

where $A_1, A_2, \dots, A_r$ are constants.

(iv) If $y_x = f_1(x), y_x = f_2(x)$ be solutions of the equation

$$L(E)y_x = g(x), L(E)y_x = h(x)$$

where $f(x) = g(x) + h(x)$, then $y_x = f_1(x) + f_2(x)$ is a solution of equation (2). This is the superposition principle and is valid only for linear equations.

(v) If $y_x = u_x$ is a particular solution of (2), then with the conditions of (iv), the general solution of (2) is

$$y_x = u_x + A_1 f_1(x) + A_2 f_2(x) + \dots A_r f_r(x)$$

We first study some homogeneous linear difference equations with constant coefficients, then some difference equations which are reducible to this type and finally some non-homogeneous types.

8.7 EXISTENCE AND UNIQUENESS THEOREM

(Rohilkhand-2005, 07, 10; Agra-2004)

Some difference equations have infinitely many solutions whereas others have no solution at all. In the case of linear difference equations we can always find at least one solution. Theorems establishing such results are referred to as existence and uniqueness theorems. Before starting and proving the theorem for the linear difference equations of order n, first look a case of order two. The analysis about the second order difference equation will be helpful to understand the general theorem. The equation of second order difference equation is

$$a_0 y_{x+2} + a_1 y_{x+1} + a_2 y_x = f(x) \quad x = 0, 1, 2 \dots \qquad \dots(1)$$

with $a_0 \neq 0$, $a_2 \neq 0$, $x = 0, 1, 2, \dots$

Now, let y_0 and y_1 be given. Then, with $x = 0$ (1) gives

$$a_0 y_2 + a_1 y_1 + a_2 y_0 = f(0)$$

$$\Rightarrow \qquad a_0 y_2 = f(0) - a_1 y_1 - a_2 y_0$$

Since $a_0 \neq 0$ for any x, $a_0(0)$ also $\neq 0$ and hence

$$y_2 = \frac{f(0)}{a_0(0)} - \frac{a_1(0)}{a_0(0)} y_1 - \frac{a_2(0)}{a_0(0)} y_0$$

Thus with the help of y_1 and y_0, we can find y_2. Now, we can find y_3. For that put $x = 1$ in (1), we get

$$a_0 y_3 = f(1) - a_1 y_2 - a_2 y_1$$

and since $a_0(1) \neq 0$.

$$y_3 = \frac{f(1)}{a_0(1)} - \frac{a_1(1)}{a_0(1)} y_2 - \frac{a_2(1)}{a_0(1)} y_1$$

Continue this way, generating the unique solution of the second order difference equation.

▶ **REMARK**

▸ The linear difference equation of order n

$$a_0 y_{x+n} + a_1 y_{x+n-1} + \ldots + a_{n-1} y_{x+1} + a_n y_x = f(x) \qquad \ldots(1)$$

over a set A of consecutive integral values of x has one and only one solution y for which values at n consecutive x-values are arbitrary prescribed.

8.8 SOLUTION OF THE EQUATION $y_{x+1} = Ay_x + B$

The linear first order difference equation is of the form

$$a_0(x) y_{x+1} + a_1(x) y_x = f(x), \, x = 0, 1, 2, \ldots \qquad \ldots (1)$$

Over the indicated set of x-values. The functions $a_0(x)$ and $a_1(x)$ according to the definition, are never zero, so if they are constant, they are non-zero.

Dividing (1) by $a_0(x)$, we get

$$y_{x+1} = \frac{-a_1(x)}{a_0(x)} y_x + \frac{f(x)}{a_0(x)}$$

If we now suppose a_0 and a_1 as well as f are constant function, we can write

$$y_{x+1} = Ay_x + B, \, x = 0, 1, 2, \ldots \qquad \ldots(2)$$

where A and B are constant and $A \neq 0$.

To find the solution of equation (2), put $x = 0$ in (2), then

$$y_1 = Ay_0 + B$$

At $x = 1$,

$$y_2 = Ay_1 = B$$
$$= A(Ay_0 + B) + B = A^2 y_0 + AB + B = A^2 y_0 + (A+1)B$$

Again at $x = 2$, we get

$$y_3 = Ay_2 + B$$
$$= A[A^2 y_0 + (A+1)B] + B = A^3 y_0 + (1 + A + A^2)B$$
$$\vdots \qquad \vdots \qquad \vdots \qquad \vdots$$
$$y_x = A^x y_0 + (1 + A + A^2 + \ldots + A^{x-1})B$$

We know that $1 + A + A^2 + \ldots A^{x-1}$ is geometric progression, hence

$$1 + A + A^2 + \ldots + A^{x-1} = \frac{1 - A^x}{1 - A}, \text{ if } A \neq 1$$
$$= x \qquad \text{if } A = 1$$

We can write

$$y_x = \begin{cases} A^x y_0 + \dfrac{B(1 - A^x)}{1 - A} & \text{if } A \neq 1 \\ y_0 + B_x & \text{if } A = 1 \end{cases} \qquad \ldots(3)$$

▶ **REMARKS**

▸ The function y given by equation (3) is a solution, and the only solution of the difference equation (2) with y_0 prescribed.

▸ The linear difference equations $y_{x+1} = Ay_x + B, \, x = i, i+1, i+2$ $\qquad \ldots(1)$
taken over the set of x-values has infinitely many solutions. If y is a solution and C is a constant such that

$$y_x = \begin{cases} CA^{x-1} + B\dfrac{1 - A^{x-1}}{1 - A} & \text{if } A \neq 1 \\ C + B(x - i) & \text{if } A = 1 \end{cases}, \, x = i, i+1, i+2 \qquad \ldots(2)$$

if a single value of y is prescribed for one of the h values $i, i+1, i+2, \ldots$, then a unique solution of (1) is determined. In particular, if y_i is prescribed then solution of (1) is given by (2) with $B = y_i$.

SOLVED EXAMPLES

EXAMPLE 1. *Solve the difference equation :* $y_{x+1} = 2y_x + 3$, $x = 1, 2, 3, \ldots$ *and* $y_0 = 0$.

SOLUTION. Comparing the given equation with $y_{x+1} = Ay_x + B$, where $A = 2$ and $B = 3$.

∴   The solution is

$$y_x = A^x y_0 + B\frac{1-A^x}{1-A} = 2^x.0 + 3\frac{1-2^x}{1-2} \qquad (\because y_0 = 0)$$

$$= -3(1 - 2^x) = 3(2^x - 1), x = 0, 1, 2, \ldots$$

Hence, we can write the sequence as 3, 9, 21, ...

EXAMPLE 2. *Solve the difference equation*

$y_{x+1} = -y_x + 1$, $x = 0, 1, 2, \ldots$ *and* $y_0 = 1$.

SOLUTION. Comparing the given equation with $y_{x+1} = Ay_x + B$, where $A = -1$ and $B = 1$, therefore solution is

$$y_x = A^x y_0 + B\frac{1-A^x}{1-A}$$

$$= (-1)^x y_0 + 1.\frac{1-(-1)^x}{1-(-1)} = (-1)^x y_0 + \frac{1-(-1)^x}{1+1}$$

$$= (-1)^x.1 + \frac{1-(-1)^x}{2} \qquad (\because y_0 = 1)$$

$$= (-1)^x + \frac{1}{2}[1-(-1)^x] = \frac{1}{2}[1+(-1)^x], x = 0, 1, 2, \ldots$$

Hence, we can write the sequence is 1, 0, 1, 0, 1, 0, ...

EXAMPLE 3. *Find the solution of* $y_{x+1} = 2y_x - 1$, $x = 0, 1, 2, \ldots$ *with initial condition* $y_0 = 5$.

SOLUTION. Comparing the given equation with $y_{x+1} = Ay_x + B$, we have, $A = 2$ and $B = -1$.

$$y_x = A^x y_0 + B\frac{1-A^x}{1-A} = 2^x.5 + (-1)\frac{1-2^x}{1-2} \qquad (\because y_0 = 5)$$

$$= 5.2^x + 1 - 2^x = 4.2^x + 1, x = 0, 1, 2, \ldots$$

It gives the sequences 5, 9, 17, 33, 65, 129,...

EXAMPLE 4. *Solve the difference equation over the set of x values* 0, 1, 2, ...

$3y_{x+1} = 2y_x + 3$, $y_0 = 2$.

SOLUTION. Comparing this equation with $y_{x+1} = Ay_x + B$.

We have $A = \frac{2}{3}, B = 1$ therefore we have

$$y_x = A^x y_0 + B\frac{1-A^x}{1-A} = \left(\frac{2}{3}\right)^x.2 + 1.\frac{1-\left(\frac{2}{3}\right)^x}{1-\frac{2}{3}}$$

$$= 2\left(\frac{2}{3}\right)^x.2 + 1.\frac{1-\left(\frac{2}{3}\right)^x}{1-\frac{2}{3}} = 2\left(\frac{2}{3}\right)^x + 3\left[1-\left(\frac{2}{3}\right)^x\right]$$

$$y_x = 3 - \left(\frac{2}{3}\right)^x, x = 0, 1, 2, \ldots$$

It gives the sequence $3, \dfrac{7}{3}, \dfrac{23}{3}, \ldots$

EXAMPLE 5. *Solve* $y_{x+1} = -y_x + 2, x = 0, 1, 2, \ldots$

SOLUTION. Comparing the given equation with $y_{x+1} = Ay_x + B$, where, $A = -1, B = 2$ solution is

$$y_x = A^x y_0 + B\frac{1 - A^x}{1 - A} = (-1)^x y_0 + 2.\frac{1 - (-1)^x}{1 - (-1)}$$

$$= (-1)^x y_0 + 1 - (-1)^x, x = 0, 1, 2, \ldots$$

$$= (-1)^x y_0 + [1 - (-1)^x], x = 0, 1, 2, \ldots$$

If $x = 0$ or even integer, then $(-1)^x = 1$ and if $x = $ an odd integer then $(-1)^x = -1$, we obtain the sequence $y_0, -y_0 + 2, y_0, -y_0 + 2, \ldots$

EXAMPLE 6. *Solve* $(E - a)(E - b)y_x = 0, a \neq b$.

SOLUTION. If we take $(E - b)y_x = y_x$, then equation reduces to

$$y_x = (E - b)y_x = Aa^x$$

$$\Delta^{x+1}\left[\frac{y_{x+1}}{b^{x+1}} - \frac{y_x}{b^x}\right] = Aa^x$$

$$b^{x+1}\Delta\left[\frac{y_x}{b^x}\right] = Aa^x$$

$$\Delta\left(\frac{y_x}{b^x}\right) = \frac{A}{b}\left(\frac{a}{b}\right)^x$$

$$\therefore \qquad \frac{y_x}{b^x} = B + \frac{A}{b}\Delta^{-1}\left(\frac{a}{b}\right)^x = B + \frac{A}{b}.\frac{\left(\frac{a}{b}\right)^x}{\frac{a}{b} - 1} = B + \frac{A}{a - b}\left(\frac{a}{b}\right)^x$$

$$y_x = \frac{A}{a - b}a^x + Bb^x.$$

EXAMPLE 7. *Solve the difference equation* $(E - a)^2 y_x = 0$.

SOLUTION. We may write

$$(E - a)^2 y_x = a^{x+2}\Delta^2\left(\frac{y_x}{a^x}\right) = 0$$

$$\Delta^2\left(\frac{y_x}{a^x}\right) = 0 \quad \Rightarrow \quad \Delta\left(\frac{y_x}{a^x}\right) = A$$

$$\Rightarrow \qquad \frac{y_x}{a^x} = Ax + B$$

Hence, $\quad y_x = a^x(Ax + B)$ is the solution of given equation.

EXAMPLE 8. *Solve* $y_{x+1} = f(x)y_x$.

SOLUTION. Dividing both sides by $\sum\limits_{r=1}^{n} f(r)$, we get

$$y_{x+1} \Big/ \sum_{r=1}^{n} f(r) = y_x \Big/ \sum_{r=1}^{n-1} f(r)$$

or $\quad \Delta\left[y_x \Big/ \sum_{r=1}^{n-1} f(r) \right] = 0 \Rightarrow y_x \Big/ \sum_{r=1}^{n-1} f(r) = A$

Hence, $\qquad y_x = A \sum\limits_{r=1}^{n-1} f(r)$

EXAMPLE 9. *Solve* $y_{x+1} = \sqrt{y_x}$

SOLUTION. We have $\log y_{x+1} = \sqrt{y_x}$

$$\log y_{x+1} = \frac{1}{2} \log y_x \quad \Rightarrow \quad \log y_{x+1} - \frac{1}{2} \log y_x = 0$$

$$\Rightarrow \quad \left(E - \frac{1}{2} \right) \log y_x = 0 \Rightarrow \quad \log y_x = A\left(\frac{1}{2} \right)^x = A2^{-x}$$

$$y_x = \exp(A2^{-x})$$

$$y_x = k^{2^{-x}} \text{ where } k = e^A$$

Alternatively we have

$$y_x = y_{x-1}^{1/2} = y_{x-2}^{(1/2)^2} = \dots = y_0^{(1/2)^x}$$

EXAMPLE 10. *Solve* $y_x y_{x+2} = y_{x+1}^2$

SOLUTION. We have $y_x y_{x+2} = y_{x+1}^2$

$$\Rightarrow \qquad \frac{y_{x+2}}{y_{x+1}} = \frac{y_{x+1}}{y_x}$$

$$\Rightarrow \qquad \Delta\left(\frac{y_{x+1}}{y_x} \right) = 0 \quad \Rightarrow \quad \frac{y_{x+1}}{y_x} = A$$

or $\quad (E - A)y_x = 0 \quad \Rightarrow \quad y_{x+1} = Ay_x$

Hence, $\qquad y_x = BA^x$

EXAMPLE 11. *Show that the solution* y_x *of* $2y_{x+2} + 3y_{x+1} - 2y_x = 0, y_0 = 1, y_1 = 1/2$ *is bounded and monotonically decreasing with limit zero.*

SOLUTION. From the given difference equation, we have

$$y_{x+2} = -\frac{3}{2} y_{x+1} + y_x$$

Put $x = 0, y_2 = -\frac{3}{2} y_1 + y_0 = -\frac{3}{2} \cdot \frac{1}{2} + 1 = \frac{1}{4}$

$$x = 1, y_3 = -\frac{3}{2}y_2 + y_1 = -\frac{3}{2}\cdot\frac{1}{4} + \frac{1}{2} = \frac{1}{8}$$

$$x = 2, y_4 = -\frac{3}{2}y_3 + y_2 = -\frac{3}{2}\cdot\frac{1}{8} + \frac{1}{4} = \frac{1}{16}$$

$$x = 3, y_5 = -\frac{3}{2}y_4 + y_3 = -\frac{3}{2}\cdot\frac{1}{16} + \frac{1}{8} = \frac{1}{32}$$

Hence, solution of the given difference equation is bounded and monotonically decreasing.

EXERCISE 8.2

1. Consider the linear difference equation of order 2. $y_{x+2} - xy_{x+1} - y_x = 0, x = 0, 1, 2,$ ... over the indicated set of x-values. Show that there is no solution of this difference equation for which $y_0 = 0$ and $y_2 = 1$. Show also that if the prescribed values y_0 and y_2 are equal, there are infinitely many different solutions of the difference equation. Why the existence and uniqueness theorem is not valid by these facts.

2. Prove that $a_0(x)\, y_{x+2} + a_1(x)\, y_{x+1} + a_2(x)$

   $= f(x), x = 0, 1, 2, ..., y_0 = \alpha, y_1 = \beta$ has one and only one solution where $a_0, a_1, a_2 \neq 0$. Hence, write down all possible solutions of

   $y_{x+2} + \sinh y_{x+1} + (2x+1)y_x = 0, x = 2,$ $3, 4, ..., y_2 = 0, y_3 = 0.$

3. Show that $y_x = \dfrac{x(x-1)}{2} + B$ is the solution of the difference equation $y_{x+1} - y_x = x$. Find the particular solution satisfying the initial conditions $y_0 = 1$.

4. Show that $y_x = A_1 + B_1(-1)^x$ is the solution of the difference equation $y_{x+2} - y_x = 0$. Find the particular solution satisfying the initial condition $y_0 = 1, y_1 = 2$.

5. Prove that the function y_x given by

   $$y_x = \begin{cases} A^x y_0 + B\left(\dfrac{1-A^x}{1-A}\right) & \text{if } A \neq 1 \\ y_0 + B & \text{if } A = 1 \end{cases}, k = 0,1,2...$$

   is a solution, and only solution of the following difference equation with y_0 prescribed.

6. Solve the following difference equation over the set of x-values 0, 1, 2, ...

   (i) $y_{x+1} = 3y_x - 1, y_0 = 6$

   (ii) $y_{x+1} = y_x + 2, y_0 = 2$

   (iii) $y_{x+1} = 2y_x + 1, y_0 = 5$

7. The difference equation $xy_{x+1} - y_x = 0,$ $x = 0, 1, 2, ...$ is linear but not of first order over the indicated set of x-values. Why? If the initial condition $y_0 = 0$ is given, show that y_1 is not uniquely determined and there are infinitely many solutions of the difference equation with $y_0 = 0$. Show also that if the value of y at any x-values different from zero is given, there is a unique solution of the difference equation.

8. Solve $y_{x+1} - 2\cos\alpha y_x + y_{x-1} = 0$
   (Kurukshetra–2013)

9. A series of values of y_n satisfy the relation $y_{n+2} + ay_{n+1} + by_n = 0,$ given $y_0 = 0, y_1 = 1,$
   $y_2 = y_3 = 2,$ show that $y_n = 2^{n/2}\sin\dfrac{n\pi}{4}.$
   (Kurukshetra–2012, Kurukshetra (NIT)–2006, 08)

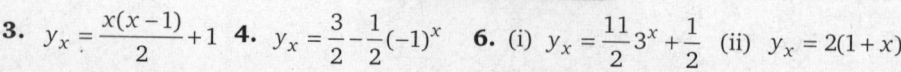

Answers

3. $y_x = \dfrac{x(x-1)}{2} + 1$ 4. $y_x = \dfrac{3}{2} - \dfrac{1}{2}(-1)^x$ 6. (i) $y_x = \dfrac{11}{2}3^x + \dfrac{1}{2}$ (ii) $y_x = 2(1+x)$

6. (iii) $y_x = 6.2^x - 1$ 8. $y_n = (1)^x[C_1\cos(x-1)\alpha + C_2\sin(x-1)\alpha]$

8.9 SOLUTION AS SEQUENCES

The linear first order difference equation

$$y_{x+1} = Ay_x + B, x = 0, 1, 2, \ldots \qquad \ldots(1)$$

The solution of above equation is

$$y_x = \begin{cases} A^x y_0 + B\dfrac{1-A^x}{1-A} & \text{if} \quad A \neq 1 \\ y_0 + Bx & \text{if} \quad A = 1 \end{cases}, x = 0, 1, 2, \ldots \qquad \ldots(2)$$

Thus, if the value of y_0, A and B are given, then equation (2) will give a sequence $<y_x>$. Two cases are here : Case I, $A = 1$ and Case II, $A \neq 1$.

Case I if $A = 1$. Then

If $<y_x>$ is the solution sequence of the difference equation $y_{x+1} = y_x + B$, $x = 0$, 1, 2, ... with y_0 prescribed then $<y_x>$ is a constant sequence, if $B = 0$ it diverges to $+\infty$ if $B > 0$ and diverges to $-\infty$ if $B < 0$.

PROOF.   If $A = 1$, we have

$$y_x = y_0 + Bx \qquad \qquad \text{[From equation (2)]}$$

Now, if $B = 0$, then

$$y_x = y_0 \text{ for } x = 0, 1, 2, \ldots$$

and $<y_x>$ is a constant sequence, as given.

If $B > 0$, we will prove that $<y_x>$ diverges to $+\infty$. By definition given, any positive number l, we must find a corresponding integer m such that

$$y_x > l \ \forall \ x > m$$

Suppose integer m is given then

$$l \leq y_0$$
$$y_x = y_0 + Bx > l$$

for all $x > 0$ we will take $m = 1$, therefore $y_x > l$ for all $x \geq m = 1$

Again if $l > y_0$, then we want $y_x = y_0 + Bx > l$

Or $\qquad \qquad B_x > l - y_0 \text{ for } B > 0$

Therefore, we choose $m > \dfrac{l - y_0}{C}$, we have

$$y_x > l \ \forall \ x \geq m$$

Similarly, we can show when $B < 0$, $<y_x>$ diverges to $-\infty$.

Case II if $A \neq 1$

The solution of linear difference equation is

$$y_x = A^x y_0 + \frac{B}{1-A} - \frac{B}{1-A} A^x$$

$$= A^x \left(y_0 - \frac{B}{1-A} \right) + \frac{B}{1-A} = A^x (y_0 - y^*) + y^*$$

$$y_x - y^* = A^x (y_0 - y^*), x = 0, 1, 2, \ldots$$

where $y^* = \dfrac{B}{1-A}$

Behaviour of the solution sequence
$$y_x = A^x(y_0 - y^*) + y^*$$
$$y_{x+1} = Ay_x + B, \ x = 0, 1, 2, \ldots$$

	Hypothesis					Conclusion
	A	B	y_0	for $x = 0, 1, 2, \ldots$		the sequence $\langle y_x \rangle$
(1)	$A \neq 1$		$y_0 = y^*$	$y_x = y^*$		constant $(= y^*)$
(2)	$A > 1$		$y_0 > y^*$	$y_x > y^*$		monotonic increasing diverges to $+\infty$
(3)	$A > 1$		$y_0 < y^*$	$y_x < y^*$		monotonic decreasing diverges to $-\infty$
(4)	$0 < A < 1$		$y_0 > y^*$	$y_x > y^*$		monotonic decreasing converges to y^*
(5)	$0 < A < 1$		$y_0 < y^*$	$y_x < y^*$		monotonic increasing converges to y^*
(6)	$-1 < A < 1$		$y_0 \neq y^*$			damped oscillatory converges to y^*
(7)	$A = -1$		$y_0 \neq y^*$			divergent, oscillates finitely
(8)	$A < -1$		$y_0 \neq y^*$	$y_x = y_0$		divergent, oscillates infinitely constant $(= y_0)$
(9)	$A = 1$	$B = 0$				
(10)	$A = 1$	$B > 0$		$y_x > 0$		monotonically increasing divergent
(11)	$A = 1$	$B < 0$		$y_x < 0$		monotonically decreasing divergent

SOLVED EXAMPLES

EXAMPLE 1.   *Solve* $2y_{x+1} - y_x = 2$, *when* $y_0 = 4$.

SOLUTION.   Comparing it with the equation $y_{x+1} = Ay_x + B$, we get $A = 1/2$ and $B = 1$. Then, solution will be

$$y_x = A^x y_0 + B\frac{1 - A^x}{1 - A} \ = \left(\frac{1}{2}\right)^x . 4 + 1 . \frac{1 - \left(\frac{1}{2}\right)^x}{1 - \frac{1}{2}}$$

$$y_x = 4\left(\frac{1}{2}\right)^x + 2\left[1 - \left(\frac{1}{2}\right)^x\right] = 2 + 2\left(\frac{1}{2}\right)^x$$

Putting $x = 0, 1, 2, \ldots$, we have

$$y_0 = 4, \ y_1 = 2 + 2\left(\frac{1}{2}\right)^1 = 3$$

$$y_2 = 2 + 2\left(\frac{1}{2}\right)^2 = \frac{5}{2} = 2\frac{1}{2}$$

$$y_3 = 2 + 2\left(\frac{1}{2}\right)^3 = 2\frac{1}{4}$$

$$y_4 = 2 + 2\left(\frac{1}{2}\right)^4 = 2\frac{1}{8}$$
$$\ldots \qquad \ldots \qquad \ldots$$
$$\ldots \qquad \ldots \qquad \ldots$$

The graphical representation is given below

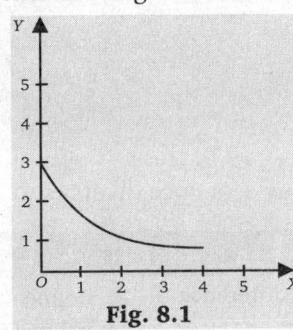

Fig. 8.1

EXAMPLE 2. *Solve $y_{x+1} = 3y_x - 1$, $y_0 = 1$.*

SOLUTION. On comparing with equation $y_{x+1} = Ay_x + B$, we get $A = 3$ and $B = -1$. The solution is

$$y_x = A^x y_0 + B\frac{1-A^x}{1-A} = 3^x.1 + (-1).\frac{1-3^x}{1-3}$$

$$y_x = 3^x + \frac{1}{2}\left[1 - 3^x\right] = 3^x + \frac{1}{2}[1 - 3^x]$$

$$y_x = \frac{1}{2}\left[3^x + 1\right], x = 0, 1, 2, \ldots$$

We have
$$y_0 = \frac{1}{2}\left[3^0 + 1\right] = 1 \; ; \quad y_1 = \frac{1}{2}\left[3^1 + 1\right] = 2$$

$$y_2 = \frac{1}{2}\left[3^2 + 1\right] = 5 \; ; \quad y_3 = \frac{1}{2}\left[3^3 + 1\right] = 14$$
$$\ldots \quad \ldots \quad \ldots$$
$$\ldots \quad \ldots \quad \ldots$$

The graphical representation is given below

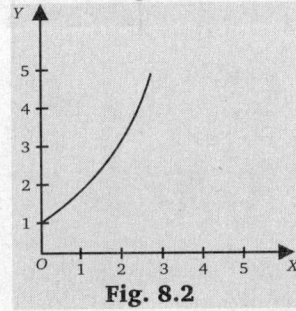

Fig. 8.2

EXAMPLE 3. *Solve $y_{x+1} = y_x - 1$, $y_0 = 5$.*

SOLUTION. On comparing this equation with $y_{x+1} = Ay_x + B$, we get $A = 1$ and $B = -1$. The solution is

$$y_x = y_0 + Bx$$
$$y_x = 5 + (-1)x = 5 - x, x = 0, 1, 2, \ldots$$
$$y_0 = 5$$
$$y_1 = 4$$
$$y_2 = 3$$

$$y_3 = 2$$
$$y_4 = 1$$
$$y_5 = 0$$
$$y_6 = -1$$
$$\cdots \quad \cdots$$
$$\cdots \quad \cdots$$

monotonically decreasing sequence diverges to $-\infty$.

EXERCISE 8.3

1. Discuss the sequence solution of the difference equation

$$y_{x+1} = \frac{1}{2}y_x + \frac{1}{2}, x = 0, 1, 2, \ldots$$

2. Define the term sequence and type of sequence.

3. Obtain the difference equation for a probability model of learning.

Answers

1. When $y_0 > 1, <y_x> = \left\{\left(\frac{1}{2}\right)^x (y_0 - 1) - 1\right\}$, When $y_0 < 1$, $<y_x> = 1 - \left(\frac{1}{2}\right)^x$, $y_0 = 1$, $<y_x> = 1$

8.10 LINEAR HOMOGENEOUS EQUATION WITH CONSTANT COEFFICIENTS

In order to obtain standard solutions of such equations, we first obtain alternative term for $(E - a)y_x$ and $(E - a)^r y_x$. First consider

$$(E - a)y_x = y_{x+1} - ay_x = a^{x+1}\left(\frac{y_{x+1}}{a^{x+1}} - \frac{y_x}{a^x}\right)$$

Thus, we have, $\qquad (E - a)y_x = a^{x+1}\Delta\frac{y_x}{a^x}$ ...(1)

From (1), we get $\qquad (E - a)^2 y_x = a^{x+1}\Delta\left\{\left(\frac{1}{a^x}\right)a^{x+1}\Delta\frac{y_x}{a^x}\right\} = a^{x+2}\Delta^2\frac{y_x}{a^x}$

and so, in general for $r = 1, 2, \ldots$

$$(E - a)^r y_x = a^{x+r}\Delta^r\frac{y_x}{a^x}$$ ...(2)

Case I : The equation $\Delta y_x = 0$

Here, $y_{x+1} - y_x = 0$ so that $y_{x+1} = y_x = y_{x-1} \cdots y_1 = y_0$

$$y_x = \text{constant}$$ ...(3)

Case II : The equation $\Delta^r y_x \ 0 \ (r = 1, 2, \ldots)$

Since, $\qquad \Delta^r y_x = 0; \ \Delta^{r-1}y_x = A_1$

$$\Delta^{r-2}y_x = A_1 x + A_2$$

$$\vdots \qquad \vdots \qquad \vdots$$

$$\Delta_x^{-1}(k) = \text{constant} + (x)^{k+1}/k + 1 \ (k = 0, 1, 2, \ldots)$$

So that integration of the last result gives

$$\Delta^{r-3}y_x = A_1\frac{x^{(2)}}{2!} + A_2 x^{(1)} + A_3$$

Proceeding, we get

$$\Delta^{r-4} y_x = \frac{A_1 x^{(3)}}{3!} + \frac{A_2 x^{(2)}}{2!} + A_3 x^{(1)} + A_y$$

$$\vdots \qquad \vdots \qquad \vdots \qquad \vdots$$

$$y_x = \frac{A_1 x^{(r-1)}}{(r-1)!} + \frac{A_2 x^{(r-2)}}{(r-2)!} + \dots + A_{r-1} x + A_r$$

This last equation is a polynomial in x of degree r and contains r constants. It may be rearranged in the more convenient form

$$y_x = B_0 + B_1 x + B_2 x^2 + \dots + B_{r-1} x^{r-1} \qquad \dots(4)$$

$B_0, B_1, \dots, B_{r-1}$ being constant.

Case III : The equation $y_{x+1} = a y_x$ can be written as

$$\frac{y_{x+1}}{a^{x+1}} - \frac{y_x}{a^x} = 0 \qquad \text{or} \qquad \Delta\left(\frac{y_x}{a^x}\right) = 0$$

$$\Rightarrow \qquad \frac{y_x}{a^x} = A \qquad \Rightarrow \qquad y_x = A a^x$$

$$\dots(5)$$

Case IV : The equation $(E - a)(E - b) y_x = 0 \;\; (a \neq b)$
Solving for $(E - b) y_x = A a^x$
Then above equation can be written as

$$b^{x+1} \Delta\left(\frac{y_x}{b^x}\right) = A a^x \qquad \Rightarrow \qquad \Delta \frac{y_x}{b^x} = \frac{A/b}{(a/b)^x}$$

$$\Rightarrow \qquad \frac{y_x}{b^x} = B + \frac{A}{b} \sum_{r=0}^{x-1} \left(\frac{a}{b}\right)^r$$

$$\sum_{r=0}^{x-1} \left(\frac{a}{b}\right)^r = \frac{(a/b)^x - 1}{(a/b) - 1} \text{ and so on}$$

$$y_x = B' b^x + \frac{A(a^x - b^x)}{a - b} \qquad \Rightarrow \qquad y_x = B' b^x + A' a^x$$

$$y_x = A' a^x + B' b^x$$

$$\dots(6)$$

where

$$A' = \frac{A}{a-b} \text{ and } B' = B - \frac{A}{a-b}$$

We can obtain general solution of the difference equations as the sum of the general solution of the homogeneous equation and particular solution of the complete equation. We will first discuss first order difference equation.

> **REMARK**
> ▸ Let us consider linear first order difference equation with constant coefficients
> $$y_{x+1} + a_1 y_x = f(x) \qquad \dots(1)$$
> (i) The function Y be given $y_x = \lambda (-a_1) x$ $\qquad \dots(2)$
> with λ an arbitrary constant is the general solution of the equation $y_{x+1} + a_1 y_x = 0$ $\qquad \dots(3)$
> (ii) If y' is any particular solution of the complete equation, then $Y + y'$ is the general solution of the complete equation. That is if y is any solution then there is a value of λ for which
> $$y_x = \lambda (a_1)^x + y_x' \qquad \dots(4)$$

8.11 LINEARLY INDEPENDENT SOLUTION OR FUNDAMENTAL SET OF SOLUTIONS

The rth order homogeneous difference equation

$$y_{x+r} + a_1 y_{x+r-1} + a_2 y_{x+r-2} + \dots + a_r y_x = 0 \qquad \dots(1)$$

$y^{(1)}, y^{(2)}, \dots, y^{(r)}$ are the solutions of equation (1) are said to form a fundamental set of solutions of (1) if the rth order determinant

$$\begin{vmatrix} y_0^{(1)} & y_0^{(2)} & \dots & y_0^{(r)} \\ y_1^{(1)} & y_1^{(2)} & \dots & y_1^{(r)} \\ \vdots & \vdots & \vdots & \vdots \\ y_{r-1}^{(1)} & y_{r-1}^{(2)} & \dots & y_{r-1}^{(r)} \end{vmatrix} \text{ is different from zero.}$$

SOLVED EXAMPLES

EXAMPLE 1. *Solve* $y_{x+1} + 3y_x = 8$.

SOLUTION. The corresponding homogeneous equation is

$$y_{x+1} + 3y_x = 0$$
$$y_{x+1} = -3y_x$$

On comparing it with $y_{x+1} = Ay_x + B$, we have $A = -3$ and $B = 0$, the solution is

$$y_x = \lambda A^x + B \frac{1 - A^x}{1 - A} = \lambda(-3)^x = (-3)^x \lambda$$

Particular solution : We will use hit and trial method. We try to see whether the constant function is a solution or not.

Let $y_x = A$ be a solution of the given equation.

$$y_{x+1} + 3y_x = 8$$
$$\Rightarrow \qquad A + 3A = 8$$
$$4A = 8$$
$$A = 2$$

Hence, $A = 2$ is a particular solution. Hence the general solution will be $y_x = \lambda(-3)^x + 2$.

EXAMPLE 2. *Solve* $y_{x+1} - y_x = 1$.

SOLUTION. The corresponding homogeneous equation is

$$y_{x+1} - y_x = 0$$

On comparing it with $y_{x+1} = A^x + B$, we get $A = 1$ and $B = 0$.

The solution will be

$$y_x = \lambda . A^x + B \frac{1 - A^x}{1 - A}$$
$$\Rightarrow \qquad y_x = \lambda(1)^x + 0$$
$$\Rightarrow \qquad y_x = \lambda$$

For particular solution, we shall try hit and trial method. We will see first whether a constant function is solution or not.

Let $y_x^* = A$ be a solution of the given equation, we get

$$A - A = 1$$
$$0 = -1, \text{ not true}$$

We now try
$$y_x^* = Ax + B \text{ so that}$$
$$y_{x+1}^* = A(x + 1) + B$$
from $y_{x+1} - y_x = 1$, we get
$$A(x + 1) + B - Ax - B = 1$$
$$A = 1$$
Hence, $y^* = x$ is solution of the given equation.

Hence, the general solution is
$$y_x = \lambda + x$$

EXAMPLE 3. Solve $y_{x+1} - 2y_x = 5$ ...(1)

SOLUTION. The corresponding equation is
$$y_{x+1} - 2y_x = 0 \qquad ...(2)$$
$$\Rightarrow \qquad y_{x+1} = 2y_x$$
On comparing it with $y_{x+1} = Ay_x + B$, we get $A = 2$ and $B = 0$.

The solution is
$$y_x = \lambda.A^x + B\frac{1 - A^x}{1 - A}$$
$$= \lambda 2^x + 0 = \lambda 2^x$$
To find the particular solution, we will use hit and trial method. We first try to see whether the constant function is a solution or not.

Let $y_x = A$ be a function of the given difference equation, we get
$$A - 2A = 5$$
$$\Rightarrow \qquad A = -5$$
Hence, $y_x^* = -5$ is particular solution.

Hence, general solution will be
$$y_x = \lambda 2^x - 5.$$

▼ **REMARK**

▶ From the last equation, it is always possible to find any solution satisfying given condition. Suppose we are interested in a solution for which
$$y_0 = 2 \qquad ...(3)$$
Then from
$$y_x = \lambda 2^x - 5 \qquad \Rightarrow \qquad y_0 = \lambda 2^0 - 5 = \lambda - 5$$
$$2 = \lambda - 5 \qquad \Rightarrow \qquad \lambda = 7$$
$$y_x = 72^x - 5$$
This satisfy both the difference equation (1) and the initial condition (3).

EXAMPLE 4. Solve $2y_{x+1} - y_x = 4$, $x = 0, 1, 2, ...$ ...(1)

SOLUTION. The corresponding homogeneous equation of (1) is
$$2y_{x+1} - y_x = 0$$

On comparing it with $y_{x+1} = Ay_x + B$, we get $A = \frac{1}{2}$ and $B = 0$. The solution is given by
$$y_x = \lambda.A^x + \frac{B(1 - A^x)}{1 - A} = \lambda.\left(\frac{1}{2}\right)^x + 0 = \lambda\left(\frac{1}{2}\right)^x$$

For particular solution, we will use hit and trial method. Let us first see whether a constant function will be solution or not.

Let $y_x^* = A$ be a solution of (1), we get

$$2A - A = 4$$

$$\Rightarrow \qquad A = 4$$

The general solution is given by

$$y_x = \lambda\left(\frac{1}{2}\right)^x + 4\,, x = 0, 1, 2$$

EXAMPLE 5. *Solve* $y_{x+1} - 2y_x = 6.$ \qquad\qquad ...(1)

SOLUTION. The corresponding homogeneous equation is

$$y_{x+1} - 2y_x = 6.$$

On comparing it with $y_{x+1} = Ay_x + B$, we get $A = 2$ and $B = 0$.

Then solution is

$$y_x = \lambda.A^x + B\frac{1 - A^x}{1 - A} = \lambda.2^x + C = \lambda.2^x$$

For particular solution, we will use hit and trial method. Let us first try for a constant function. Let $y_x^* = A$ is solution of the given difference equation, let $y_x^* = A$ be the solution of equation (1)

Then, $A - 2A = 6$

or $\qquad A = -6$

Hence, the general solution will be

$$y_x = \lambda.2^x - 6$$

EXAMPLE 6. *Show that* $y_x^* = x$ *is the particular solution of the difference equation*

$$y_{x+2} - 3y_{x+1} + 2y_x = -1 \qquad\qquad ...(1)$$

Also, show that the functions $y^{(1)}$ *and* $y^{(2)}$ *given by* $y_x^{(1)} = 1$ *and* $y_x^{(2)} = 2^x$ *are solution of homogeneous difference equation corresponding to this difference equation and they form a fundamental set. Find the general solution of the difference equation and two particulars solution for which*

(i) $y_0 = 0, y_1 = 3,$ *and* *(ii)* $y_0 = -3, y_1 = 5$

SOLUTION. Given $y_x^* = x$. Therefore

$$y_{x+1}^* = x + 1, y_{x+2}^* = x + 2$$

$$y_{x+2}^* - 3y_{x+1}^* + 2y_x^* = x + 2 - 3(x + 1) + 2x$$

$$= 3x - 3x + 2 - 3 = -1$$

Hence, $y_x^* = x$ is a solution of difference equation (1). The corresponding homogeneous equation is

$$y_{x+2} - 3y_x + 2y_x = 0 \qquad\qquad ...(2)$$

$y_x^{(1)} = 1, y_x^{(2)} = 2^x$ are the solution of equation (2) because they satisfy equation (2).

Now, $\begin{vmatrix} y_0^{(1)} & y_0^{(2)} \\ y_1^{(1)} & y_1^{(2)} \end{vmatrix} = \begin{vmatrix} 1 & 1 \\ 1 & 2 \end{vmatrix} = 2 - 1 = 1 \neq 0$

The function $y^{(1)}$ and $y^{(2)}$ form a fundamental set. The general solution is

$$y_x = C_1 y_x^{(1)} + C_2 y_x^{(2)} + y_x^*$$

$$y_x = C_1(1) + C_2 2^x + x$$

Now, $y_0 = 0$ and $y_1 = 3$ [from (i)], we get

$$y_0 = C_1 + C_2 = 0 \implies C_1 = -C_2$$
$$y_1 = C_1 + 2C_2 + 1$$
$$y_1 = -C_2 + 2C_2 + 1 = 3 \implies C_2 = 2$$

and $C_1 = -2$

The particular solution is

$$y_x = -2(1) + 2 \cdot 2^x + x$$
$$y_x = x + 2^{x+1} - 2$$

From (2), if $y_0 = -3$ and $y_1 = 5$, we get.

$$y_0 = C_1 + C_2 = -3$$
$$\implies C_1 = -3 - C_2$$
$$y_1 = C_1 + 2C_2 + 1$$
$$= -3 - C_2 + 2C_2 + 1 = 5$$
$$\implies C_2 = 7, C_1 = -10$$

Hence, the particular solution is given by

$$y_x = -10 + 72^x + x.$$

EXAMPLE 7. *Show that $y_x^* = x - 2$ is the particular solution of the difference equation*

$$2y_{x+2} + 3y_{x+1} - 2y_x = 3x + 1 \qquad \text{...(1)}$$

Also, show that the functions $y^{(1)}$ and $y^{(2)}$ given by $y^{(1)} = \left(\dfrac{1}{2}\right)^x$ and $y_2 = 2^x$ are

solutions of the homogeneous difference equation corresponding to the difference equation and they form a fundamental set. Find the general solution of the difference equation and the particular solution satisfying the initial conditions $y_0 = 0$ and

$y_1 = 1$.

SOLUTION. Given $y_x^* = x - 2$, therefore

$$\implies \quad y_{x+1}^* = x - 1 \text{ and } y_{x+2}^* = x$$

Now, from equation (1), we get

$$2y_{x+2}^* + 3y_{x+1}^* - 2y_x^* = 2x + 3(x-1) - 2(x-2) = 3x + 1$$

$y_x^* = x - 2$ is a particular solution of the difference equation (1). Now, the corresponding homogeneous equation is

$$2y_{x+2} + 3y_{x+1} - 2y_x = 0 \qquad \text{...(2)}$$

Given $y_x^{(1)} = \left(\dfrac{1}{2}\right)^x$, therefore

$$2\,2y_{x+2}^{(1)} + 3y_{x+1}^{(1)} - 2y_x^{(1)}$$

$$= 2\left(\frac{1}{2}\right)^{x+2} + 3\left(\frac{1}{2}\right)^{x+1} - 2\left(\frac{1}{2}\right)^{x}$$

$$= \left(\frac{1}{2}\right)^x \left[2 \cdot \frac{1}{4} + 3 \cdot \frac{1}{2} - 2\right] = \left(\frac{1}{2}\right)^x \left[\frac{1}{2} + \frac{3}{2} - 2\right] = 0$$

Hence, $y_x^1 = \left(\dfrac{1}{2}\right)^x$ is a solution of equation (2). Similarly we can show

$y_x^{(2)} = (-2)^x$ is also a solution of equation (2).

Now,

$$\begin{vmatrix} y_0^{(1)} & y_0^{(2)} \\ y_1^{(1)} & y_1^{(2)} \end{vmatrix} = \begin{vmatrix} 1 & 1 \\ 1/2 & -2 \end{vmatrix} = -2 - \frac{1}{2} = -\frac{5}{2} \neq 0$$

The functions $y^{(1)}$ and $y^{(2)}$ form a fundamental set. The general solution is

$$y_x = C_1 y_x^{(1)} + C_2 y_x^{(2)} + y_x^*$$

$$y_x = C_1 \left(\frac{1}{2}\right)^x + C_2(-2)^x + x - 2$$

Now, $y_0 = 0$ and $y_1 = 1$, we get

$$y_0 = C_1 + C_2 = 0 \Rightarrow C_1 = -C_2$$

$$y_1 = C_1\left(\frac{1}{2}\right) + C_2(-2) + 1 - 2 = 1$$

$\Rightarrow \qquad C_1 = \dfrac{12}{5}$ and $C_2 = -\dfrac{2}{5}$

The particular solution is given by

$$y_x = \frac{12}{5}\left(\frac{1}{2}\right)^x - \frac{2}{5}(-1)^x + x - 2 \,.$$

EXERCISE 8.4

Solve completely the following difference equations :

1. $2y_{x+1} - y_x = 2$
2. $y_{x+1} + y_x = 1$
3. Show that the function $y^{(1)}$ and $y^{(2)}$ given by $y_x^{(1)} = 1$ and $y_x^{(2)} = (-1)^x$ are solutions of the difference equation

$$y_{x+2} - y_x = 0$$

and they form a fundamental set. Find the general solution of the difference equation and a particular solution for which $y_0 = 0$ and $y_1 = 2$.

4. Solve $y_{x+1} - 2y_x = x + 1$.
5. Solve $y_{x+1} + 5y_x = 2x$.
6. Solve $y_{x+1} + 7y_x = 2, x = 0, 1, 2$.
7. Solve $2y_{x+1} - 5y_x = 3x + 1$.

Answers

1. $y_x = \lambda\left(\dfrac{1}{2}\right)^x + 2$

2. $y_x = \lambda(-3)^x + 2$

3. $C_1 + C_2(-1)^x, y_x = 1 - (-1)^x$

4. $y_x = -x - 2$

5. $y_x = \lambda(-5)^x + \dfrac{1}{3}x - \dfrac{1}{18}$

6. $y_x = \lambda(-1)^x + 1, x = 0, 1, 2, \ldots$

7. $y_x = A^x\lambda + B\dfrac{1-A^x}{1-A}$

8.12 GENERAL SOLUTION OF SECOND ORDER HOMOGENEOUS DIFFERENCE EQUATION

The second order homogeneous difference equation is given by

$$y_{x+2} + a_1 y_{x+1} + a_2 y_x = 0 \qquad \ldots(1)$$

where $a_2 \neq 0$.

Let us suppose $y_x = m^x$ satisfies equation (1), where $m\ (\neq 0)$ is any constant different from zero. Now, from (1), we get

$$m^{x+2} + a_1 m^{x+1} + a_2 m^x = 0$$

$$\Rightarrow \qquad m^2 + a_1 m + a_2 = 0 \qquad\qquad \dots(2)$$

It is auxiliary equation or characteristic equation of (1). If m is a root of (2) then $y_x = m^x$ is a solution of difference equation (1).

The auxiliary equation is quadratic algebraic equation. Therefore, it has two non-zero roots m_1 and m_2. Let $y_x^{(1)} = m_1^x$ and $y_x^{(2)} = m_2^{(x)}$ be the solutions. There are three cases that arise :

  (i) The roots m_1 and m_2 are real and unequal ($m_1 \neq m_2$).

  (ii) The roots m_1 and m_2 are real and equal ($m_1 = m_2$).

  (iii) The roots are complex.

Case I : The roots are complex.

$y_x^{(1)} = m_1^x$ and $y_x^{(2)} = m_2^x$ form a fundamental set, to prove this we will calculate

$$\begin{vmatrix} y_0^{(1)} & y_0^{(2)} \\ y_1^{(1)} & y_1^{(2)} \end{vmatrix} = \begin{vmatrix} 1 & 1 \\ m_1 & m_2 \end{vmatrix} = m_2 - m_1 \neq 0 \qquad\qquad [\because m_1 \neq m_2]$$

Hence, m_1 and m_2 are real and unequal, the general solution of difference equation (1) is given by

$$y_x = C_1 m_1^x + C_2 m_2^x$$

Case II : The roots are equal ($m_1 = m_2$)

Now $y^{(1)}$ and $y^{(2)}$ cannot form a fundamental set because $m_1 = m_2$, the value of determinant will be zero.

We have a function $y^{(1)}$ but find a function $y^{(2)}$ which form a fundamental set with function $y^{(1)}$.

Now, we claim that a function $y_x^{(2)}$ is given as

$$y_x^{(2)} = x\, m_1^x$$

Proof. Let m_1 and m_2 be the roots of second order difference equation, then we have

$$m_1^2 + a_1 m_1 + a_2 = 0$$

and $\qquad\qquad\qquad m_1 + m_2 = -a_1$

$\Rightarrow \qquad\qquad\qquad 2m_1 + a_1 = 0 \qquad\qquad\qquad (\because m_1 = m_2)$

Now, we will show that $y_x^2 = x\, m_1^x$ is a solution of the equation is given below

$$y_{x+2}^2 + a_1 y_{x+1}^{(2)} + a_2 y_x^{(2)} = (x+2)m_1^{x+2} + a_1(x+1)m_1^{x+1} + a_2 x m_1^x$$

$$= x m_1^x [m_1 + a_1 m_1 + a_2] = m_1^{x+1}[2m + a_1] = 0$$

Hence, $y_x^2 = x m_1^x$ is solution of the given equation.

Now, we will show $y_x^{(1)} = m_1^x\, y_x^{(2)} = x m_1^x$ form a fundamental set, we have

$$\begin{vmatrix} y_0^{(1)} & y_0^{(2)} \\ y_1^{(1)} & y_1^{(2)} \end{vmatrix} = \begin{vmatrix} 1 & 0 \\ m_1 & m_2 \end{vmatrix} = m_1 \neq 0$$

No root of auxiliary equation is zero. Thus, when $m_1 = m_2$, the general solution will be

$$y_x = C_1 m_1^x + C_2 x m_1^x = m_1^x (C_1 + x C_2)$$

Case III : Complex roots

We know that complex roots in quadratic equation always occur in conjugate pairs then $m_1 \neq m_2$. Let $y_x^{(1)} = m_1^x$ and $y_x^{(2)} = m_2^x$ form a fundamental set. The general solution is

$$y_x = C_1 m_1^x + C_2 m_2^x \qquad\qquad \dots(1)$$

Equation (1) will be complex if m_1 and m_2 are themselves complex. But if C_1 and C_2 are complex then y_x will be real. Let

$$m_1 = \alpha.(\cos\theta + i\sin\theta); \quad m_2 = \alpha.(\cos\theta - i\sin\theta)$$

and

$$C_1 = a(\cos B + i\sin B) ; \quad C_2 = a(\cos B - i\sin B)$$

Then

$$m_1^x = \alpha^x(\cos x\theta + i\sin x\theta)$$

$\Rightarrow$

$$C_1 m_1^x = a\alpha^x[(\cos(x\theta + B) + i\sin(x\theta + B)]$$

Similarly

$$C_2 m_1^x = a\alpha^x[(\cos(x\theta + B) - i\sin(x\theta + B)]$$

$$y_x = 2a\alpha^2[(\cos(x\theta + B)]$$

$$y_x = A\alpha^x[(\cos(x\theta + B)] \qquad\qquad [2a = A]$$

$$= \alpha^x[A\cos x\theta + B\sin x\theta]$$

We summarize the above result for second order difference equation

$$y_{x+2} + a_1 y_{x+1} + a_2 y_x = 0$$

Then, A. E is $m^2 + a_1 m + a_2 = 0$. Then solution is

$$y_x = C_1 m_1^x + C_2 m_2^x \text{ if } (m_1 \neq m_2)$$

$$y_x = (C_1 + C_2 x)m_1^2 \text{ if } m_1 = m_2$$

$$y_x = C_1\alpha^2 \cos(x\theta + C_2) , m_1 \text{ and } m_2 \text{ are complex.}$$

$$= \alpha^x(C_1\cos x\theta + C_2\sin x\theta)$$

8.13 GENERAL SOLUTION OF THE HOMOGENEOUS DIFFERENCE EQUATION OF ORDER n

The homogenous difference equation of nth order is

$$y_{x+n} + a_1 y_{x+n-1} + a_2 y_{x+n-2} + \dots + a_n y_x = 0 , a_n \neq 0 \qquad\qquad \dots(1)$$

The auxiliary equation is given by when $y_x = m^x$

$$m^x + a_1 m^{n-1} + a_2 m^{n-2} + \dots + a_n = 0 \qquad\qquad \dots(2)$$

Solve equation (1). Let $m_1, m_2, \dots, m_n$ are the n roots. The general solution of the homogeneous difference equation depends upon these roots. There are three cases that arise:

Case I : Roots are real and distinct.

In this case general solution will be

$$y_x = C_1 m_1^x + C_2 m_2^x + \dots + C_n m_n^x$$

Case II : Roots are Equal.

Some roots are repeated. Let root m_1 is repeated k times then the general solution is

$$(C_1 + C_2 x + C_3 x^2 + \dots + C_k x^{k-1})m_1^x$$

Case III : Roots are Complex.

Let m_1 and m_2 are the complex roots, then solution is $C_1\rho^x \cos(x\theta + C_2)$

or

$$\rho^x(C_1\cos x\theta + C_2\sin x\theta).$$

SOLVED EXAMPLES

EXAMPLE 1. *Solve* $y_{x+2} - 2y_{x+1} + 2y_x = 0$

SOLUTION. The corresponding auxiliary equation is
$$m^2 - 2m + 2 = 0, \text{ when } y_x = m^x$$

$$\therefore \qquad m = \frac{2 \pm \sqrt{4-8}}{2} = 1 \pm i$$

In polar form, $m = 1 + i = \rho(\cos\theta \pm i \sin\theta)$

On equating real and imaginary parts of both sides, we get
$$1 = \rho\cos\theta \text{ and } 1 = \rho\sin\theta$$

$$\Rightarrow \qquad \rho = \frac{1}{\sqrt{2}} \text{ and } \theta = \frac{\pi}{4}$$

Then, solution is $y_x = C_1(\sqrt{2})^x \cos\left(\dfrac{x\pi}{4} + C_2\right)$

EXAMPLE 2. *Solve the difference equation:* $y_{x+3} - 3y_{x+2} - 10y_{x+1} + 24y_x = 0.$

SOLUTION. The auxiliary equation, when $y_x = m^x$ is
$$m^3 - 3m^2 - 10m + 24 = 0 \Rightarrow (m-2)(m+3)(m-4) = 0$$

$$\therefore \qquad m = 2, -3, 4$$

Thus, the general solution is
$$y_x = C_1 2^x + C_2(-3)^x + C_3 4^x$$

EXAMPLE 3. *Solve* $y_{x+2} - 7y_{x+1} + 12y_x = 0.$

SOLUTION. The auxiliary equation, when $y_x = m^x$ is
$$m^2 - 7m + 12 = 0$$

$$\Rightarrow \qquad (m-3)(m-4) = 0$$

$$\therefore \qquad m = 3, 4$$

Thus, solution is $y_x = C_1 3^x + C_2 4^x$

EXAMPLE 4. *Solve the difference equation :* $y_{x+4} - 4y_{x+3} + 6y_{x+2} - 4y_{x+1} + y_x = 0.$

SOLUTION. When $y_x = m^x$, then auxiliary equation is
$$m^4 - 4m^3 + 6m^2 - 4m + 1 = 0$$

$$\Rightarrow \qquad (m-1)(m^3 - 3m^2 + 3m - 1) = 0$$

$$\Rightarrow \qquad (m-1)(m-1)(m^2 - 2m + 1) = 0$$

$$\Rightarrow \qquad (m-1)(m-1)(m-1)^2 = 0$$

$$m = 1, 1, 1, 1$$

Thus, solution is
$$y_x = (C_1 + C_2 x + C_3 x^2 + C_4 x^3).1^x.$$

EXAMPLE 5. *Solve the difference equation :* $y_{x+4} - 8y_{x+3} + 18y_{x+2} - 27y_x = 0.$

SOLUTION. The auxiliary equation is
$$m^4 - 8m^3 + 18m^2 - 27 = 0 \qquad\qquad (\because y_x = m^x)$$

$$\Rightarrow \qquad (m+1)(m-3)^3 = 0$$

$$\Rightarrow \qquad m = -1, 3, 3, 3$$

Thus, the general solution is
$$y_x = C_1(-1)^x + (C_2 + C_3 x + C_4 x^2)3^x.$$

EXAMPLE 6.   *Solve the difference equation:* $y_{x+3} + y_{x+2} - 8y_{x+1} - 12y_x = 0.$

SOLUTION.   The auxiliary equation is

$$m^3 + m^2 - 8m - 12 = 0 \qquad \qquad \text{(When } y_x = m^x)$$

$$\Rightarrow (m-3)(m^2 + 4m + 4) = 0 \Rightarrow (m-3)(m+2)^2 = 0$$

$$\Rightarrow \qquad \qquad m = 3, -2, -2$$

Thus, the general solution is

$$y_x = C_1 3^x + (C_2 + C_3 x)(-2)^x.$$

EXAMPLE 7.   *Solve* $2y_{x+2} - 5y_{x+1} + 2y_x = 0.$

(Agra-2006)

Also, find the particular solution satisfying the initial conditions $y_0 = 0$ *and* $y_1 = 1.$

SOLUTION.   The auxiliary equation is

$$2m^2 - 5m + 2 = 0 \Rightarrow \quad (2m-1)(m-2) = 0 \qquad (\because y_x = m^x)$$

$$\Rightarrow \qquad 2m - 1 = 0, m = 2 \quad i.e., \quad m = \frac{1}{2}, 2$$

Thus, solution is

$$y_x = C_1 \left(\frac{1}{2}\right)^x + C_2 2^x$$

When $x = 0$ and $x = 1$, we get

$$y_0 = C_1 + C_2 = 0 \quad \text{and} \quad y_1 = \frac{1}{2}C_1 + 2C_2 = 1$$

Solving these, we get $C_1 = -\dfrac{2}{3}$ and $C_2 = \dfrac{2}{3}$. Thus, particular solution with these conditions is

$$y_x = -\frac{2}{3}\left(\frac{1}{2}\right)^x + \frac{2}{3}2^x.$$

EXAMPLE 8.   *Solve the difference equation* $y_{x+2} + y_x = 0$ *with* $y_0 = 0$ *and* $y_1 = 1.$

SOLUTION.   When $y_x = m^x$, the auxiliary equation is

$$m^2 + 1 = 0$$

$$\Rightarrow \qquad \qquad m = \pm i = \cos\frac{\pi}{2} + i\sin\frac{\pi}{2} \; ; \quad \theta = \frac{\pi}{2} \quad \text{and} \quad \rho = 1$$

Thus, the general solution is

$$y_x = C_1 \cos\left[\frac{x\pi}{2} + C_2\right]$$

When $x = 0$ and $x = 1$, we get

$$y_0 = C_1 \cos C_2 = 0 \quad \text{and} \quad y_1 = C_1 \cos\left(\frac{\pi}{2} + C_2\right) = 1$$

Solving these, we get $C_1 = -1$ and $C_2 = \dfrac{\pi}{2}$

Hence, the solution is

$$y_x = -\cos\frac{\pi}{2}(x+1) = \sin\frac{x\pi}{2}.$$

EXERCISE 8.5

1. Solve the difference equation
$$9y_{x+2} - 6y_{x+1} + y_x = 0.$$
Also find the particular solution satisfying the initial conditions $y_0 = 0$ and $y_1 = 1$.

2. Solve $y_{x+1} - 2y_x \cos \alpha + y_{x-1} = 0.$

3. Solve the difference equation
$$3y_{x+2} - 6y_{x+1} + 7y_x = 0.$$

4. Solve $y_{x+4} - 4y_{x+3} + 8y_{x+2} - 8y_{x+1} + 4y_x = 0.$

5. Solve the difference equation
$$y_{x+2} + 6y_{x+1} + 25y_x = 0.$$

6. Form the Fibonacci difference equation and solve it.

(Kurukshetra–2009, Kurukshetra(NIT)–2013)

Answers

1. General solution is $y_x = (C_1 + C_2 x)\left(\dfrac{1}{3}\right)^x$ and particular solution is $y_x = 3x\left(\dfrac{1}{3}\right)^x$.

2. $y_x = C_1 \cos ax + C_2 \sin ax$ **3.** $y_x = C_1\left(\dfrac{2}{\sqrt{3}}\right)^x \cos\left(\dfrac{x\pi}{6} + C_2\right)$

4. $y_x = \left\{(C_1 + C_2 x)\cos\dfrac{\pi x}{4} + (C_3 + C_4 x)\sin\dfrac{\pi x}{4}\right\}(\sqrt{2})^x$

5. $y_x = C_1 (5)^x \cos(\theta x + C_2)$ when $\theta = \tan^{-1}\left(-\dfrac{4}{3}\right)$

6. $y_x = \left[\dfrac{5 - \sqrt{5}}{10}\right]\left[\dfrac{1 + \sqrt{5}}{2}\right]^x + \left[\dfrac{5 + \sqrt{5}}{10}\right]\left[\dfrac{1 - \sqrt{5}}{2}\right]^x$

8.14 PARTICULAR SOLUTION OF THE COMPLETE DIFFERENCE EQUATION

Let $y_{x+2} + a_1 y_{x+1} + a_2 y_x = f(x)$ is given equation, we can find the general solution of the corresponding homogeneous equation. Now, we will find the particular solution. There are many cases for finding particular solution.

Case I : When $f(x) = a^x$, where a is a constant

$$\phi(E)a^x = (a_0 E_n + a_1 E^{n-1} + \dots + a_{n-1} E + a_n)a^x$$
$$= a_0 a^{x+n} + a_1 a^{x+n-1} + \dots + a_{n-1} a^{x+1} + a_n a^x$$
$$= (a_0 a^n + a_1 a^{n-1} + \dots + a_{n-1} a + a_n)\, a^x = \phi(a)a^x$$

$$\frac{1}{\phi(E)}\phi(E)a^x = \frac{1}{\phi(E)}\phi(a)a^x$$

or $$\frac{1}{\phi(E)}a^x = \frac{1}{\phi(a)}a^x \text{ provided } \phi(a) \neq 0.$$

Case of failure : If $\phi(a) = 0$, then $\phi(E)$ will have any one of the following factor
$$(E - a) \text{ or } (E - a)^2, \text{ etc. or } (E - a)^n$$

Now, let $$\frac{1}{E - a}a^x = y_x$$

$\Rightarrow$ $$y_{x+1} - ay_x = a^x \quad \text{or} \quad a^{(x+1)}y_{x+1} - a^{-x} = a^{-1}$$

$\Rightarrow$ $$\Delta(a^{-x}y_x) = a^{-1} \quad \Rightarrow \quad a^{-x}y_x = a^{-1}x$$

$\Rightarrow$ $$y_x = xa^{x-1}$$

$\Rightarrow$ $\dfrac{1}{E-a}a^x = xa^{x-1}$

Similarly, $\dfrac{1}{(E-a)^2}a^x = \dfrac{x(x-1)}{2!}a^{x-2}$

In general $\dfrac{1}{(E-a)^n}a^x = \dfrac{x(x-1)...(x+n-1)}{n!}a^{x-n}$

Case II : When f(x) = a^xF(x), where F(x) is some function of x.

Noting that $E^n a^x F(x) = a^{x+n}F(x+n) = a^x a^n E^n F(x)$

We have $\phi(E)\{a^x F(x)\} = a^x (a_0 a^n E^n + a_1 a^{n-1}E^{n-1} + ... + a_{n-1}aE + a_n F(x)) = a^x \phi(aE)F(x)$

The inverse result is $\dfrac{1}{\phi(E)}\{a^x F(x)\} = a^x \dfrac{1}{\phi(aE)}F(x)$

Case III: When f(x) is a polynomial in x of degree r(say)

In this case $\dfrac{1}{\phi(E)}f(x) = \dfrac{1}{\phi(1+\Delta)}f(x) = \phi(1+\Delta)^{-1}f(x)$

Here, we expand $[\phi(1+\Delta)^{-1}]$ in ascending powers of Δ and operate on $f(x)$.

SOLVED EXAMPLES

EXAMPLE1.   *Solve the following equation :* $y_{x+2} - 3y_{x+1} + 2y_x = 1$.

SOLUTION.   On substituting $y_x = m^x$, the auxiliary equation is

$m^2 - 3m + 2 = 0$

$\Rightarrow$ $(m-1)(m-2) = 0, m = 1, 2$

The general solution of the reducible homogeneous equation is

$y_x = C_1 + C_2 2^x$

Now,   P.I. $= \dfrac{1}{(E-1)(E-2)} = -\dfrac{1}{(E-1)^1}1^x$ $[\because \phi(a) = 0]$

$= -\dfrac{x}{1!}1^{x-1} = -x$

Hence, the general solution of the complete equation is

$y_x = C_1 + C_2 2^x - x$

EXAMPLE 2.   *Solve* $y_{x+2} - 4y_{x+1} + 4y_x = 3^x + 2^x + 4$.

SOLUTION.   The auxiliary equation of the given equation is

$m^2 - 4m + 4 = 0 \Rightarrow (m-2)^2 = 0, m = 2, 2$

Then general solution is

$y_x = (C_1 + C_2 x)2^x$

Now, particular solution of 3^x

$= \dfrac{1}{E^2 - 4E + 4}3^x = \dfrac{1}{(E-2)^2}3^x = \dfrac{1}{(3.2)^2}3^x = 3^x$

Particular solution of $2^x = \dfrac{1}{(E-2)^2}2^x = \dfrac{x(x-1)}{2!}2^{x-2}$

Particular solution of 4

$$= \dfrac{1}{(E-2)^2}4 = 4\dfrac{1}{(E-2)^2}1^x = 4$$

Hence, the complete solution is

$$y_x = (C_1 + C_2 x)2^x + 3^x + x(x-1)2^{x-3} + 4.$$

EXAMPLE 3. *Solve* $y_{x+2} - 3y_{x+1} + 2y_x = a^x$, *where a is some constant.*

SOLUTION. The auxiliary equation of the given equation is

$$m^2 - 3m + 2 = 0 \Rightarrow (m-1)(m-2) = 0$$

$$\Rightarrow \qquad m = 1, 2$$

The general solution is

$$y_x = C_1 + C_2 2^x$$

Particular solution of $a^x = \dfrac{1}{(E-1)(E-2)}a^x$

$$= \dfrac{1}{(a-1)(a-2)}a^x \qquad \text{(provided } a \neq 1 \text{ and } a \neq 2)$$

When $a = 1$ or $a = 2$, we will use Case of Failure:

When $a = 1$, particular solution of

$$a^x = \dfrac{1}{(E-1)(E-2)}a^x = \dfrac{1}{(a-2)(E-1)}a^x = \dfrac{1}{(1-2)(E-1)}1^x = -\dfrac{1}{1}x = -x$$

When $a = 2$, particular solution of

$$a^x = \dfrac{1}{(E-1)(E-2)}a^x = \dfrac{1}{(2-1)(E-2)}2^x = \dfrac{1}{(E-2)}2^x$$

$$= \dfrac{x.2^x}{2!} = \dfrac{1}{2}x2^x$$

Hence, the complete solution is

$$y_x = C_1 + C_2 2^x + \dfrac{1}{(a-1)(a-2)}a^x \qquad \text{[When } a \neq 1 \text{ and } a \neq 2]$$

$$\Rightarrow \quad y_x = C_1 + C_2 2^x - x \qquad \text{[When } a = 1]$$

and $\quad y_x = C_1 + C_2 2^x + \dfrac{1}{2}x2^x \qquad$ [When $a = 2$]

EXAMPLE 4. *Solve* $y_{x+2} - 4y_{x+1} + 3y_x = 3^x + 1.$

SOLUTION. The auxiliary equation of the given equation is

$$\Rightarrow \qquad m^2 - 4m + 3 = 0$$

$$(m-1)(m-3) = 0, m = 1, 3$$

The general solution is

$$y_x = C_1 + C_2 3^x$$

The particular solution of $3^x + 1$

$$= \dfrac{1}{(E-1)(E-3)}(3^x + 1) = \dfrac{1}{(E-1)(E-3)}3^x + \dfrac{1}{(E-1)E-3)}.1$$

$$= \frac{1}{3-1}\frac{1}{E-3}3^x + \frac{1}{(1-3)(E-1)}1^x = \frac{1}{2}x3^{x-1} - \frac{1}{2}x(1)^{x-1} = \frac{1}{2}x(3^{x-1}-1)$$

Hence, the general solution of the complete equation is

$$y_x = C_1 + C_2 3^x + \frac{1}{2}x(3^{x-1}-1)$$

EXAMPLE 5. *Solve* $y_{x+2} - 5y_{x+1} + 6y_x = 2.$

Also, find the solution satisfying the initial conditions $y_0 = 1$ *and* $y_1 = -1.$

(Agra-2000, 09)

SOLUTION.   The auxiliary equation of the given homogeneous equation is

$$m^2 - 5m + 6 = 0$$
$$\Rightarrow \qquad (m-3)(m-2) = 0 \Rightarrow m = 3, 2$$

The general solution of the homogeneous equation is

$$y_x = C_1 3^x + C_2 2^x$$

Particular solution of $2 = \dfrac{1}{(E-3)(E-2)}.2 = 2.\dfrac{1}{(E-3)(E-2)}.1^x$

$$= 2.\frac{1}{(1-3)(1-2)} = \frac{2}{-2\times-1} = 1$$

Hence, the general solution of the complete equation is

$$y_x = C_1 3^x + C_2 2^x + 1$$

When $y_0 = 1$ and $y_1 = -1$, we get

$$y_0 = C_1 + C_2 + 1 = 1$$
$$y_1 = 3C_1 + 2C_2 + 1 = -1$$

We get $C_1 = -2$ and $C_2 = 2$

Hence, the solution is

$$y_x = -2.3^x + 2^{x+1} + 1.$$

EXAMPLE 6. *Solve* $\Delta y_x + \Delta^2 y_x = \sin x.$

SOLUTION.   The given equation in symbolic form is

$$\{(E-1) + (E-1)^2\}y_x = \sin x \Rightarrow (E^2 - E)y_x = \sin x$$
$$(E-1)y_{x+1} = \sin x$$

The auxiliary equation is

$$m - 1 = 0, m = 1$$

Then, general solution is $y_x = C_1 1^x = C_1$

Particular solution of $\sin x = \dfrac{1}{E-1}\sin x$

$$= \text{imaginary part of } \frac{1}{E-1}e^{ix}$$

$$= \text{imaginary part of} \frac{1}{E-1}(e^i)^x = \text{imaginary part of} \frac{e^{ix}}{e^i - 1}$$

$$= \text{imaginary part of} \frac{e^{ix}(e^{-i}-1)}{(e^i-1)(e^{-i}-1)}$$

$$= \text{imaginary part of} \frac{e^{i(x-1)} - e^{ix}}{1 - (e^i + e^{-i}) + 1} = \frac{\sin(x-1) - \sin x}{2(1 - \cos 1)}$$

Hence, the general solution is

$$y_{x+1} = C_1 + \frac{\sin(x-1) - \sin x}{2(1 - \cos 1)}$$

$$\Rightarrow \qquad y_x = C_1 + \frac{\sin(x-2) - \sin(x-1)}{2(1 \sin 1)}.$$

EXAMPLE 7. *Solve* : $y_{x+2} + y_x = \sin x \dfrac{\pi}{2}$. 　　　　　　　　　...(1)

SOLUTION. The auxiliary equation of the given equation is

$$m^2 + 1 = 0$$

$$\Rightarrow \qquad m = \pm i$$

$$\Rightarrow \qquad m = \cos\frac{\pi}{2} \pm i\sin\frac{\pi}{2}$$

The general solution of the reduced equation is

$$y_x = C_1 \cos\left(\frac{x\pi}{2} + C_2\right)$$

For particular solution, we will use hit and trial method. Let

$$y_x^* = A\sin\frac{x\pi}{2} + B\sin\frac{x\pi}{2}$$

Substituting this value in equation (1), we get

$$A\sin(x+2)\frac{\pi}{2} + B\sin(x+2)\frac{\pi}{2} + A\sin\frac{\pi x}{2} + B\sin\frac{\pi x}{2} = \sin\frac{\pi x}{2}$$

$$-A\sin\frac{\pi x}{2} - B\sin\frac{\pi x}{2} + A\sin\frac{\pi x}{2} + B\sin\frac{\pi x}{2} = \sin\frac{\pi x}{2}$$

$$0 = \sin\frac{\pi x}{2}$$

Hence, 　　　　$y_x^* = A\sin\dfrac{x\pi}{2} + B\cos\dfrac{x\pi}{2}$ fails.

Now, substituting 　$y_x^* = Ax\sin\dfrac{x\pi}{2} + Bx\cos\dfrac{x\pi}{2}$ in equation (1), we get

$$A(x+2)\sin(x+2)\frac{\pi}{2} + B(x+2)\cos(x+2)\frac{\pi}{2} - Ax\sin\frac{x\pi}{2} + Bx\cos\frac{x\pi}{2} = \sin\frac{x\pi}{2}$$

$$\Rightarrow \quad -A(x+2)\sin\frac{x\pi}{2} - B(x+2)\cos\frac{x\pi}{2} + Ax\sin\frac{x\pi}{2} + Bx\sin\frac{x\pi}{2} = \sin\frac{x\pi}{2}$$

$$\Rightarrow \quad \sin\frac{x\pi}{2}(-2A) + \cos\frac{x\pi}{2}(-2B) = \sin\frac{x\pi}{2}$$

On comparing the coefficient of $\cos\dfrac{x\pi}{2}$ and $\sin\dfrac{x\pi}{2}$, we get

$$-2A = 1, -2B = 0$$

$$\Rightarrow \qquad A = -\frac{1}{2}, B = 0$$

Hence, the general solution of the complete equation is

$$y_x = 4\cos\left(\frac{x\pi}{2} + C_2\right) - \frac{1}{2}x\sin\frac{x\pi}{2}.$$

EXAMPLE 8. *Solve* $y_{x+2} - 7y_{x+1} - 8y_x = (x^2 - x)2^x$.

SOLUTION. The auxiliary equation is

$$m^2 - 7m - 8 = 0 \Rightarrow m = 8, -1$$

The general solution is

$$y_x = C_1 8^x + C_2(-1)^x$$

Particular solution of $(x^2 - 1)2^x$

$$= \frac{1}{E^2 - 7E - 8}(x^2 - x)2^x = 2^x\frac{1}{(2E)^2 - 7(2E) - 8)}(x^2 - x)$$

$$= 2^x\frac{1}{4E^2 - 14E - 8}(x^2 - x) = 2^{x-1}\frac{1}{2(1+\Delta)^2 - 7(1+\Delta) - 8}(x^2 - x)$$

$$= 2^{x-1}\frac{1}{2\Delta^2 - 3\Delta - 9}(x^2 - x) = -\frac{2^{x-1}}{C_1}\left[1 - \frac{2\Delta^2 - 3\Delta}{9}\right]^{-1}(x^2 - x)$$

$$= -\frac{2^{x-1}}{9}\left[1 + \frac{2\Delta^2 - 3\Delta}{9} + \left(\frac{2\Delta^2 - 3\Delta}{9}\right)^2 + ...\right](x^2 - x)$$

$$= -\frac{2^{x-1}}{9}\left[1 + \frac{2}{9}\Delta^2 - \frac{\Delta}{9} + \frac{\Delta^2}{9}\right](x^2 - x)$$

$$= -\frac{2^{x-1}}{9}\left[1 - \frac{\Delta}{9} + \frac{\Delta^2}{9}\right](x^2 - x) \qquad \text{[Neglecting the higher terms]}$$

$$\Delta(x^2 - x) = \{(x+1)^2 - (x+1)\} - (x^2 - x) = 2x$$
$$\Delta^2(x^2 - x) = \Delta[\Delta(x^2 - x)] = \Delta(2x) = 2(x+1) - 2x = 2$$

$$\Rightarrow \quad \frac{-2^{x-1}}{9}\left[1 - \frac{\Delta}{9} + \frac{\Delta^2}{9}\right] = \frac{-2^{x-1}}{9}\left[x^2 - x - \frac{1}{3}(2x) - \frac{1}{3}(2)\right]$$

$$= \frac{-2^{x-1}}{27}(3x^2 - 5x + 2)$$

Hence, the general solution of the complete equation is

$$y_x = C_1 8^x + C_2(-1)^x - \frac{2^{x-1}}{27}(3x^2 - 5x + 2).$$

EXAMPLE 9. *Solve* : $y_{x+2} - 2\cos\alpha\, y_{x+1} + y_x = \cos\alpha x$.

(Nagpur–2008)

SOLUTION. The auxiliary equation of the given equation is

$$m^2 - 2\cos\alpha\, m + 1 = 0$$

$$m = \frac{2\cos\alpha \pm \sqrt{(4\cos^2\alpha - 4)^2}}{2} = \cos\alpha \pm i\sin\alpha$$

C.F. $= (1)^n[C_1\cos\alpha x + C_2\sin\alpha x]$
$\phantom{\text{C.F. }} = C_1\cos\alpha x + C_2\sin\alpha x$

Now, $P.I. = \dfrac{1}{E^2 - 2E\cos\alpha + 1} \cdot \cos\alpha x$

$= \dfrac{1}{E^2 - E(e^{i\alpha} + e^{-i\alpha}) + 1}\left(\dfrac{e^{i\alpha x} + e^{-i\alpha x}}{2}\right)$

$= \dfrac{1}{2}\left[\underbrace{\dfrac{1}{(E - e^{i\alpha})(E - e^{-i\alpha})} \cdot e^{i\alpha x}}_{(I)} + \underbrace{\dfrac{1}{(E - e^{i\alpha})(E - e^{-i\alpha})} \cdot e^{-i\alpha x}}_{(II)}\right]$

Now, puting $e^{i\alpha}$ and $e^{-i\alpha}$ in I and II part respectively.

$= \dfrac{1}{2}\left[\dfrac{1}{E - e^{i\alpha}} \cdot \dfrac{1}{e^{i\alpha} - e^{-i\alpha}} \cdot e^{i\alpha x} + \dfrac{1}{E - e^{-i\alpha}} \cdot \dfrac{1}{e^{-i\alpha} - e^{i\alpha}} \cdot e^{-i\alpha x}\right]$

$= \dfrac{1}{4i\sin\alpha}\left[\dfrac{1}{E - e^{i\alpha}} \cdot e^{i\alpha x} + \dfrac{1}{E - e^{-i\alpha}} \cdot e^{-i\alpha x}\right]$

$= \dfrac{1}{4i\sin\alpha}[n \cdot e^{i\alpha(x-1)} - n \cdot e^{-i\alpha(x-1)}]$

$= \dfrac{n}{2\sin\alpha}\left[\dfrac{e^{i\alpha(x-1)} - e^{-i\alpha(x-1)}}{2i}\right] = \dfrac{x\sin(x-1)\alpha}{2\sin\alpha}$

Hence, the complete solution is given by

$\qquad y = C.F. + P.I.$

$\qquad y = C_1 \cos\alpha x + C_2 \sin\alpha x + \dfrac{x\sin(x-1)\alpha}{2\sin\alpha}$

EXAMPLE 10. *Solve:* $y_{x+2} - 2y_{x+1} + y_x = x^2 \cdot 2^x$.   (Nagpur–2008)

SOLUTION.   Here, the auxiliary eq^n is

$\qquad m^2 - 2m + 1 = 0 \Rightarrow (m-1)(m-1) = 0$

$\qquad\qquad m = 1, 1$

$\therefore\qquad C.F. = (C_1 + C_2 x)e^x$

Now, $P.I. = \dfrac{1}{(E-1)^2} \cdot 2^x \cdot x^2 = 2^x \dfrac{1}{(2E-1)^2} x^2 = 2^x \dfrac{1}{(1+2\Delta)^2} x^2$

$= 2^x(1 + 2\Delta)^{-2}\{x(x-1) + x\} = 2^x(1 - 4\Delta + 12\Delta^2 \ldots)([x]^2 + (x))$

$= 2^x\{(x)^2 + (x) - 4[2(x) + 1] + 12 \times 2\}$

$= 2^x[(x)^2 - 7(x) + 20] = 2^x(x^2 - 8x + 20)$

Hence, the required solution is

$\qquad y = C_1 + C_2 x + 2^x(x^2 - 8x + 20)$

EXERCISE 8.6

1. Solve $y_{x+2} - 4y_{x+1} + 4y_x = 3x + 2^x$

(Meerut-2005, 07; Agra-2009)

2. Solve $y_{x+2} - 4y_x = 9x^2$.

3. Solve $y_{x+3} - 5y_{x+2} + 8y_{x+1} - 4y_x = x2^x$

4. Solve $y_{x+2} - 6y_{x+1} + 9y_x = 3x$

(Kurukshetra(NIT)–2008, VTU–2010)

5. Solve $y_{x+2} + y_x = \cos(x/2)$

(Madras–2001, Kurukshetra(NIT)–2008)

6. Solve $y_{x+2} - 5y_{x+1} - 6y_x = 4^x$, $y_0 = 0, y_1 = 1$.
(Madras–2003)

7. Solve $y_{x+2} + 6y_{x+1} + 9y_x = 2^x$, $y_0 = y_1 = 0$.
(VTU–2008)

8. Solve $y_{x+3} - 3y_{x+2} + 3y_{x+1} - y_x = 1$.
(Kottayam–2005)

9. Solve $y_{x+3} + y_x = x^2 + 1$, $y_0 = y_1 = y_2 = 0$.
(Trichirapalli–2001)

10. Solve $y_{x+3} - 5y_{x+2} + 3y_{x+1} + 9y_x = 2^x + 3x$.
(Nagpur–2009)

11. Solve $(4E^2 - 4E + 1)y = 2^x + 2^{-x}$
(Madras–2001)

12. Solve $y_{x+2} + 5y_{x+1} + 6y_x = x + 2^x$
(Nagpur–2006)

13. Solve $y_{x+3} + 8y_x = (2x+3)2^x$.
(Madras–2005)

14. Solve $(E^2 - 5E + 6)y_x = 4^x(x^2 - x + 5)$.
(Madras–2003)

Answers

1. $y_x = (C_1 + C_2 x)2^x + 6 + 3x + x(x-1)2^{x-3}$

2. $y_x = C_1(-2)^x + C_2 2^x - 3x^2 - 4x - \dfrac{20}{3}$

3. $y_x = C_1 + (C_2 + C_3 x)2^x + \left[\dfrac{x^{(3)}}{6} - x^{(2)}\right]2^{x-2}$

4. $y_x = (C_1 + C_2 x)3^x + \dfrac{1}{2}x(x-1)3^{x-2}$

5. $y_x = A\cos\dfrac{x\pi}{2} + \dfrac{1}{2}\cos\dfrac{x-1}{2}\sec\dfrac{1}{2}$

6. $y = C_1(-1)^x + C_2(6)^x - 2^x/12$

7. $y = \left(\dfrac{x}{15} - \dfrac{1}{25}\right)(-3)^x + \dfrac{2^x}{25}$

8. $y = C_1 + C_2 x + C_3 x^2 + \dfrac{1}{6}x(x-1)(x-2)$

9. $y = C_1(-1)^x + C_2\cos\dfrac{x\pi}{3} + C_3\sin\dfrac{x\pi}{3} + \dfrac{1}{2}x(x-3)$

10. $y = (C_1 + C_2 x)(3)^x + C_3(-1)^x + \dfrac{1}{3}(2)^x - \dfrac{3x}{4}$

11. $y = (C_1 + C_2 x)2^{-x} + \dfrac{2^x}{9} + x(x-1)\left(\dfrac{1}{2}\right)^{x-1}$

12. $y = C_1(-2)^x + C_2(-3)^x + \dfrac{x}{12} - \dfrac{7}{144}$

13. $y = C_1(-2)^x + 2^x(C_2\cos x\pi/3 + C_3\sin x\pi/3) + \dfrac{3}{16}(2)^x + 2^{x-4}(2x+3)$

14. $y = C_1 \cdot 2^x + C_2 \cdot 3^x + \dfrac{4^x}{2}(x^2 - 13x + 61)$

8.15   SOLUTION OF SIMULTANEOUS DIFFERENCE EQUATIONS

If two or more difference equations are given with the same number of unknown function, we can solve such equations simultaneously.

SOLVED EXAMPLES

EXAMPLE 1.   *Solve the simultaneous difference equations*
$$2y_{x+1} - 3y_x + 5z_x = 2$$
$$2y_x + z_{x+1} - 2z_x = 7$$

SOLUTION.   We can write these equations in the form
$$(2E - 3)y_x + 5z_x = 2 \qquad \qquad \ldots(1)$$
and $\quad 2y_x + (E - 2)z_x = 7 \qquad \qquad \ldots(2)$

Operate on (1) with $(E - 2)$, on (2) with 5 and subtract to give
$$(2E^2 - 7E - 4)y_x = (E - 2)z - 35 = -37$$
$$(2E + 1)(E - 4)y_x = -37$$

The auxiliary equation is $(2m + 1)(m - 4) = 0 \Rightarrow m = -\dfrac{1}{2}, 4$

General solution of A.E. is $y_x = C_1\left(-\dfrac{1}{2}\right)^x + C_2 4^x$

Particular solution of -37

$$= \frac{1}{(2E+1)(E-4)}(-37) = (-37)\frac{1}{(2E+1)(E-4)}1^x$$

$$= -37\frac{1}{(2+1)(1-4)} = \frac{-37}{3.(-3)} = \frac{37}{9}$$

Hence, the general solution of complete equation is

$$y_x = C_1\left(-\frac{1}{2}\right)^x + C_2 y^x + \frac{37}{9}$$

Now, we have from (1)

$$z_x = \frac{2}{5} - \frac{1}{5}(2E-3)y_x$$

$$= \frac{2}{5} - \frac{1}{5}(2E-3)\left[C_1\left(-\frac{1}{2}\right)^x + C_2 4^x + \frac{37}{9}\right]$$

$$= \frac{2}{5} - \frac{2}{5}\left[C_1\left(-\frac{1}{2}\right)^{x+1} + C_2 4^{x+1} + \frac{37}{9}\right]$$

$$+ \frac{3}{5}\left[C_1\left(-\frac{1}{2}\right)^x + C_2 4^x + \frac{37}{9}\right]$$

Hence, $\qquad z_x = \dfrac{11}{9} + \dfrac{4}{5}\left(-\dfrac{1}{2}\right)^x C_1 - C_2 4^x.$

EXAMPLE 2. *Solve the system :*

$$y_{x+1} - y_x + 2z_{x+1} = 0$$
$$z_{x+1} - z_x - 2y_x = z^x$$

SOLUTION. The given system of equation can be written as

$$(E-1)y_x + 2Ez_x = 0 \qquad\qquad\qquad ...(1)$$
$$-2y_x + (E-1)z_x = 2^x \qquad\qquad\qquad ...(2)$$

Operating on (3) with $E-1$ and (4) with $2E$, subtracting, we get

$$((E-1)^2 + 4E)y_x = -2E2^x = -22^{x+1}$$
$$(E+1)^2 y_x = -4.2^x$$

The auxiliary difference equation is

$$(m+1)^2 = 0$$
$$m = -1, -1$$

General solution of the homogeneous equation is

$$y_x = (C_1 + C_2 x)(-1)^x$$

Particular solution $= \dfrac{1}{(E+1)^2}(-4z^x)$

$$= -4\frac{1}{(E+1)^2}2^x = -4\frac{2^x}{(2+1)^2} = -\frac{4}{9}2^x$$

Hence, the general solution of the complete equation is

$$y_x = (C_1 + C_2 x)(-1)^x - \frac{4}{9} 2^x$$

given $\qquad 2y_{x+1} = y_x - y_{x+1}$

$$= (C_1 + C_2 x)(-1)^x - \frac{4}{9} 2^x - [C_1 + C_2(x+1)(-1)^{x+1} - \frac{4}{9} 2^{x+1}]$$

$$= (2C_1 + C_2 + 2C_2 x)(-1)^x + \frac{4}{9} z^x$$

$$z_{x+1} = (C_1 + \frac{1}{2} C_2 + C_2 x)(-1)^x + \frac{2}{9} 2^x$$

$$z_x = \left[C_1 + \frac{C_2}{2} + C_2(x-1) \right](-1)^{x-1} + \frac{2}{9} 2^{x-1}$$

$$z_x = \left[C_1 - \frac{C_2}{2} + C_2 x \right](-1)^{x-1} + \frac{1}{9} 2^x .$$

MISCELLANEOUS EXERCISE

1. Solve the following homogeneous equations :

(i) $y_{x+3} - 2y_{x+2} - y_{x+1} + 2y_x = 0$

(ii) $y_{x+2} - y_x = 0$

(iii) $y_{x+2} + 2y_{x+1} + y_x = 0$

(iv) $y_{x+4} + y_{x+3} - 13y_{x+2} - y_{x+1}$
$\qquad\qquad\qquad + 12y_x = 0$

(v) $y_{x+2} + 8y_{x+1} + 16y_x = 0$

(vi) $y_{x+2} - 5y_{x+1} + 6y_x = 0$

(vii) $y_{x+2} + 2y_{x+1} + 7y_x = 0$

(viii) $y_{x+2} - 3y_{x+1} + 2y_x = 0$

(ix) $y_{x+2} + 3y_{x+1} + y_x = 0$

(x) $y_{x+4} - 9y_{x+3} + 30y_{x+2} - 44y_{x+1}$
$\qquad\qquad\qquad + 24y_x = 0$

(xi) $2y_{x+3} - 7y_{x+2} + 5y_{x+1} + 2y_x = 0$

(xii) $y_{x+4} + y_x = 0$

(xiii) $y_{x+2} - y_{x+1} + y_x = 0$ given $y_0 = 1$ and

$$y_1 = \frac{1+\sqrt{3}}{2}$$

(xiv) $3y_{x+2} - 7y_{x+1} - 6y_x = 0, y_0 = 0, y_1 = 1$

(xv) $9y_{x+2} + 72y_{x+1} + 79y_x = 0, y_0 = 1,$
$\qquad y_1 = -7$

(xvi) $y_{x+2} - 5y_{x+1} + 6y_x = 0, y_0 = 1, y_1 = 2$

2. Solve the following difference equations

(i) $y_{x+2} - 6y_{x+1} + 8y_x = 4^x$

(ii) $y_{x+2} - 2y_{x+1} + y_x = 5 + 3x$

(iii) $y_{x+2} - 3y_{x+1} + 2y_x = 1, y_0 = 1, y_1 = -1$

(iv) $y_{x+2} - 3y_{x+1} + 2y_x = 5^x + 2^x$

(v) $8y_{x+2} - 6y_{x+1} + y_x = 2^x$

(vi) $8y_{x+2} - 6y_{x+1} + y_x = 5 \sin \frac{x\pi}{2}$

(vii) $y_{x+2} - 4y_{x+1} + 7y_x = 3.2^x + 5y^x$

(viii) $y_{x+3} - 3y_{x+1} + 2y_x = 3^x$

(ix) $y_{x+2} - y_{x+1} + 1/2 y_x = 1, y_0 = 2, y_1 = 3$

(x) $y_{x+2} - 4y_{x+1} + 3y_x = 2^x + 3^x + 7$

(xi) $y_{x+2} - 5y_{x+1} + 6y_x = x^2 + x + 1$

$\qquad\qquad\qquad\qquad$ (Kurukshetra–2012)

(xii) $(E^2 - 5E + 6)y_x = x^2, y_0 = 1, y(1) = -1$

(xiii) $y_{x+2} - 5y_{x+1} + 6y_x = x^2 - 1$

(xiv) $\Delta^2 y_x + 2\Delta y_x + y_x = 3x + 2$

(xv) $\Delta y_x + \Delta^2 y_x = \cos x$

(xvi) $y_{x+2} + y_{x+1} + 2y_x = 3^x - 10$

(xvii) $y_{x+2} - 4y_x = 9x^2$

(xviii) $y_{x+1} - y_x^2 = y_{x+1}^3$, if $y_1 = 1, y_2 = 2$

3. What do you understand by linear difference equation of the nth order?

4. If $y^{(1)}$ and $y^{(2)}$ constitute a fundamental set for $y_{x+2} + a_1 y_{x+1} + a^2 y_x = 0, x = 0, 1, 2,$ ..., prove that a necessary and sufficient condition for $\alpha_1 y^{(1)} + \beta_1 y^{(2)}$ and $\alpha_2 y^{(1)} + \beta_2 y^{(2)}$ to be a fundamental set is $\alpha_1 \beta_2 - \alpha_2 \beta_1 \neq 0$.

5. If u_x and v_x are the solutions of the homogeneous difference equation $y(x + 2) + p(x)y(x + 1) + q(x)y(x) = 0$, then $C_1 u_x + C_2 v_x$ is also a solution of the above equation.

6. Solve the systems of simultaneous equations

(i) $2x_{t+1} - 3x_t + 5y_t = 2$
$2x_t + y_{t+1} - 2y_t = 7$

(ii) $x_{t+1} - y_t = 2(t + 1)$
$y_{t+1} - y_t = -2(t + 1)$

(iii) $2x_{t+1} + y_{t+1} = x_t + 3y_t$
$x_{t+1} + y_{t+1} = x_t + y_t$

7. Solve $y_{x+1} - ay_x = 0$, $a \neq 1$.

(Rohilkhand-2006; Agra-2003, 06)

8. Solve $y_{x+1} - by_x = ca^x b$ where c is a periodic function of period 1, when (a) $b \neq a$ (b) $b = a$.

(Rohilkhand-2008, 10; Agra-2005, 08)

9. Show that $y_{x+2} - 4y_{x+1} + 4y_x \doteq 0$ ...(i) $(x=0, 1, ...)$ has the solution. (Agra-2005, 07, 10)

10. Solve $y_{x+2} - 4y_{x+1} + 13y_x = 0$. (Agra-2008, 09)

11. Find the particular solution of $y_{x+2} - 6y_{x+1} + 8y_x = 0$, such that $y_0 = 3$, $y_1 = 2$. Also obtain y_5. (Agra-2007)

12. Solve $y_{x+3} + y_{x+2} - 8y_{x+1} - 12y_x = 2x^2 + 5$.

(Bhopal-2002; Agra-2007)

13. Solve $y_{x+2} + y_{x+1} + y_x = x. 2^x$.

(Agra-2001, 07)

14. Solve $y_{x+2} - 5y_{x+1} + 6y_x = x^2$, by the method of variation of parameters. (Agra-2006)

15. Solve $y_{x+1} - e^{2x} y_x = 3x^2 e^{x2} + x$ (1/6)

(Agra-2002, 07)

16. Solve $y_{x+1} - ay_x = \cos nx$. (Bhopal-2002)

17. Solve the equation

$y_{x+4} - 5y_{x+3} + 9y_{x+2} + 7y_{x+1} + 2y_x = 0$.

(Agra-2003)

18. Solve the following equations :

(a) $y_{x+1} - 3y_x = 2^x$ (Garhwal-2006)

(b) $y_{x+2} - 2my_{x+1} + (m^2 + n^2) y_x = m^x$.

(Agra-2009)

(c) $y_{x+2} - 5y_{x+2} + 8y_{x+1} - 4y_x = x. 2^x$

(Gwalior - 2008)

19. Solve $y_{x+3} - 6y_{x+2} + 11y_{x+1} - 6y_x = 0$ subject to the given condition $y_0 = 1, y_1 = 1, y_2 = 1$.

(Gwalior-2001, 05)

20. Solve the difference equation $y_{x+1} + 3y_x = 0$, $y_0 = 2$.

21. Solve the difference equation $2y_{x+1} - y_x = 4$, $x = 0, 1, 2, ...$ with the $y_0 = 3$.

22. Solve the difference equation $y_{x+1} = -y_x + 3$, $y_0 = 6$ over the set of $x = 0, 1, 2, ...$

23. Solve $(E - a) y_x = 0$, $x = 0, 1, 2, ..., a$ is a constant.

Answers

1. (i) $y_x = C_1 + C_2(-1)^x + C_3 2^x$ (ii) $y_x = C_1 + C_2(-1)^x$ (iii) $y_x = (C_1 + C_2 x)(-1)^x$

(iv) $y_x = C_1 + C_2 (-1)^x + C_3 3^x + C_4(-4)^x$ (v) $y_x = (C_1 + C_2 x)(-4)^x$ (vi) $y_x = C_1 3^x + C_2 2^x$

(vii) $y_x = \left\{ C_1 \cos \dfrac{2n\pi}{3} + C_2 \sin \dfrac{2n\pi}{3} \right\} 2^x$ (viii) $y_x = C_1 + C_2 2^x$

(ix) $y_x = C_1 \left(\dfrac{-3 + \sqrt{5}}{2} \right)^x + C_2 \left(\dfrac{-3 - \sqrt{5}}{2} \right)^x$ (x) $y_x = (C_1 + C_2 x + C_3 x^2) 2^x + C_4 (-3)^x$

(xi) $y_x = C_1 2^x + C_2 \left(\dfrac{3 + \sqrt{17}}{2} \right)^x + C_3 \left(\dfrac{3 - \sqrt{17}}{2} \right)^x$

(xii) $y_x = C_1 \cos \left(\dfrac{3\pi}{4} x + C_2 \right) + C_3 \cos \left(\dfrac{\pi x}{4} + C_4 \right)$ (xiii) $y_x = \cos \dfrac{n\pi}{3} + \sin \dfrac{n\pi}{3}$

(xiv) $y_x = C_1 \left(\dfrac{3}{11} \right)^x + C_2 \left(\dfrac{3}{11} \right)^x$ (xv) $y_x = 3x \left(-\dfrac{7}{3} \right)^x$ (xvi) $y_x = 3^x - 2^x + 1$

2. (i) $y_x = C_1 2^x + C_2 4^x + \dfrac{x}{8} 4^x$ (ii) $y_x = (C_1 + C_2 x) + x^2 + \dfrac{x^3}{2}$ (iii) $y_x = 1 - 2^x$

(iv) $y_x = C_1 + C_2 2^x + \dfrac{5^x}{12} - x 2^{x-1}$ (v) $y_x = C_1 \left(\dfrac{1}{4} \right)^x + C_2 \left(\dfrac{1}{2} \right)^x + \dfrac{1}{21} 2^x$

(vi) $y_x = C_1\left(\dfrac{1}{4}\right)^x + C_2\left(\dfrac{1}{2}\right)^x + \dfrac{5}{3}\sin\dfrac{\pi x}{2}$ (vii) $y_x = (C_1 + C_2 x)2^x + 3x(x-1)2^{x-3} + 5.4^{x-1}$

(viii) $y_x = C_1 + C_2 2^x + \dfrac{1}{2}3^x$ (ix) $y_x = -(2)^{x/2}\cos\left(\dfrac{\pi x}{4} + \dfrac{\pi}{2}\right) + 2$

(x) $y_x = C_1 + C_2 3^x - 2^x + \dfrac{x}{2}3^{x-1} - \dfrac{7x}{2}$ (xi) $y_x = C_1 2^x + C_2 3^x + \dfrac{1}{4}(2x^2 + 8x + 15)$

(xii) $y_x = 2^{x+2} - 3^{x+1}$ (xiii) $y_x = C_1 2^x + C_2 3^x + 2 + \dfrac{3}{2}x + \dfrac{x^2}{2}$

(xiv) $y_x = C_1 3^x + C_2(-4)^x - 1 + \dfrac{1}{21}x3^x$ (xv) $y_x = C_1 2^x + C_2(-2)^x - (3x^2 + 3x + \dfrac{20}{3})$ (xvi) $y_x = 2^{2x-1} - 1$

6. (i) $x_t = -C_1 4^t + C_2\left(-\dfrac{1}{2}\right)^t + \dfrac{37}{9}$, $y_t = -C_1 4^t + \dfrac{4}{5}\left(-\dfrac{1}{2}\right)^t C_2 + \dfrac{11}{9}$

(ii) $x_t = C_1 + C_2(-1)^t + t, y_t = C_1 - C_2(-1)^t + (t+1)$

(iii) $x_t = C_1(-2)^t + C_2, y_t = -C_1(-2)^t + \dfrac{C_2}{2}$

7. $y_x = Ca^x$

8. (a) $y_x = \dfrac{bca^x}{a-b}$ (b) $y_x = xca^x$ **10.** $y_x = (13)^{x/2}[C_1\cos(x\theta) + C_2\sin(x\theta)]$ where $\theta = \tan^{-1}\left(\dfrac{3}{2}\right)$

11. $-1888.$ **12.** $y_x = [C_1 3^x + (C_2 + C_3 x)(-2)^x] - \dfrac{x^2}{9} + \dfrac{x}{27} - \dfrac{17}{54}$

13. $y_x = C_1\cos x\theta + C_2\sin x\theta + \dfrac{2^x}{7}\left(x - \dfrac{10}{7}\right)$ **14.** $y_x = \left[C_1 2^x + C_2 3^x + \dfrac{1}{2}x^2 + \dfrac{3}{2}k + \dfrac{3}{2}\right]$

15. $y_x = Ce^{x^2-x+(1/6)} + e^{x^2-x+(1/6)} + \{(x)^3 + \dfrac{3}{2}x^{(2)}\}$ **16.** $y_x = \dfrac{\cos(n-1)x - a\cos nx}{1 - 2a\cos n + a^2} + c(x)a^x$

17. $y_x = C_1 + C_2 x + C_3 x^2 + C_4(2)^x$

18. (a) $y_0 = (a.\,3^x - 2^x)$ (b) $y_x = a_1(-4)^x + [(5x-6)/25]$

(c) $y_x = C_1(1)^x + (C_2 + C_3 x)2^x + [2^x x^{(3)}/24]$.

19. $y_x = 5.\,2^x - 2.3^x - 3.$ **20.** $y_x = 2(-3)^x$ **21.** $3, \dfrac{7}{2}, \dfrac{15}{4}, \dfrac{31}{8}, \ldots$

22. $y_x = \dfrac{3}{2} + \dfrac{9}{2}(-1)^x$, $x = 0, 1, 2, \ldots)$ **23.** $a^x y_0$

Objective Evaluations

FILL IN THE BLANKS

1. The generating function of the sequence $\{B^h\}$ is _____.
2. The order of the difference equation $\Delta^2 yh + 3\Delta y_h - 3y_h = h$ is _____.
3. The linear difference equation of order n over a set A of consecutive integral values has _____.
4. The solution of the difference equation $y_{h+2} - 3y_{h+1} + 2y_h = b^h$ where b is any constant is _____.
5. The solution of the difference equation $y_{h+1} - y_h = 1$ is _____.
6. The equation involving the values of a function and their difference is called _____.

TRUE/FALSE

Write 'T' for True and 'F' for False statement.

1. The linear combination of two solutions of a given difference equation also the solution of the given equation. **(T/F)**
2. The order of the difference equation $y_{h+2} + 3y_h = 2$ is 2. **(T/F)**
3. The difference between the largest and smallest arguments for the function y involved in the equation is called degree. **(T/F)**
4. If $R(h) = b^h$, then particular solution is $\dfrac{1}{\phi(E)} b^h$, $\phi(b) \neq 0$. **(T/F)**
5. The order of the difference equation $\Delta^3 y_h + 2\Delta y_h + y_h = h + 3$ is 3. **(T/F)**

MULTIPLE CHOICE QUESTIONS

Choose the most appropriate one

1. The equation involving the values of a function and their differences is called :
   (a) differential equation
   (b) difference equation
   (c) integral equation
   (d) none of these

2. The difference between the largest and smallest arguments for the function y involved in the equation is called :
   (a) order (b) degree
   (c) power (d) none of these

3. A solution of a difference equation of order n is called general solution if it involved :
   (a) exactly n arbitrary constants
   (b) at least n arbitrary constants
   (c) at most n arbitrary constants
   (d) none of these

4. The solution obtained by assuming particular values to the arbitrary constant is called :
   (a) general solution
   (b) singular solution
   (c) particular solution
   (d) none of these

5. The linear difference equation of order n over a set A of consecutive integral values has :
   (a) unique solution
   (b) at least one solution
   (c) at most one solution
   (d) none of these

6. The solution of the equation $y_{h+1} = y_h + C$ is :
   (a) y_0
   (b) $y_0 + C$
   (c) both (a) and (b) are true
   (d) none of these

7. The linear combination of two solutions of a given difference equation :
   (a) also the solution of the given equation
   (b) not the solution of the given equation
   (c) may or may not be the solution
   (d) none of these

8. The solution $y^{(1)}, y^{(2)}, ..., y^{(n)}$ are said to form a fundamental set of solutions if n^{th} order determinant is :
   (a) $= 0$ (b) $\neq 0$
   (c) $\neq 1$ (d) none of these

9. If the roots m_1, m_2 of auxilliary equation are real and unequal then general solution is given by :

   (a) $y_h = C_1 m_1^h + C_2 m_2^h$

   (b) $y_h = (C_1 + C_2) m_1^h$

   (c) $y_h = (C_1 + C_2) m_2^h$

   (d) none of these

10. If $R(h) = b^h$, then particular solution is given by :

   (a) $\dfrac{1}{\phi(E)} b^h, \phi(b) \neq 0$

   (b) $\dfrac{b^h}{\phi(b)}, \phi(b) \neq 0$

   (c) both (a) and (b) are true

   (d) none of these

11. If $R(h) = P(h)$ a polynomial of degree m then particular solution is given by :

   (a) $\dfrac{1}{\phi(1 + \Delta)} P(h)$

   (b) $\dfrac{1}{\phi(\Delta)} P(h)$

   (c) both (a) and (b) are true

   (d) none of these

12. The generating function of the constant sequence 1, 1, 1, ... is given by :

   (a) $(1 - t)$ (b) $\dfrac{1}{1-t}$

   (c) $\dfrac{1}{1+t}$ (d) none of these

13. The generating function of the sequence $\{h\}$ is given by :

   (a) $(1 - t)$ (b) $\dfrac{t}{1-t}$

   (c) $\dfrac{t}{(1-t)^2}$ (d) none of these

14. The generating function of the sequence $\{1 + h\}$ is given by :

   (a) $(1 - t)$ (b) $\dfrac{1}{(1-t)^2}$

   (c) $\dfrac{t}{(1-t)^2}$ (d) none of these

15. The generating function of the sequence $\{B^h\}$ is given by :

   (a) $(1 - \beta t)$ (b) $\dfrac{1}{1 - \beta t}$

   (c) $\dfrac{1}{1 + \beta t}$ (d) none of these

16. The generating function for the general term y_{h+1} is :

   (a) $y(t)$ (b) $\dfrac{y(t) - y_0}{t}$

   (c) $\dfrac{y(t) + y_0}{t}$ (d) none of these

17. If the general term is y_{h+2}, then generating function is given by :

   (a) $\dfrac{y(t) - y_0}{t^2}$ (b) $\dfrac{y(t) - y_0 - y_1(t)}{t^2}$

   (c) $\dfrac{y(t) - y_1(t)}{t^2}$ (d) none of these

18. The order of the difference equation $\Delta^2 y_h + 3\Delta y_h - 3y_h = h$ is :

   (a) 2   (b) 3

   (c) 1   (d) 4

19. The order of the difference equation $\Delta^3 y_h + \Delta^2 y_h + \Delta y_h + y_h = 0$ is :

   (a) 1   (b) 2

   (c) 3   (d) 4

20. For which equation $y_h = \dfrac{1}{2} h(h-1)$ is the solution :

   (a) $y_{h+1} + y_h = h$ (b) $y_{h+1} - y_h = h$

   (c) $y_{h-1} - y_h = h$ (d) none of these

Answers

FILL IN THE BLANKS

1. $1/1 - \beta t$ **2.** 2   **3.** unique solution   **4.** $C_1 + C_2 \cdot 2^h + h \cdot 2^{h-1}$ **5.** $y_h = \lambda + h$

6. difference equation.

TRUE/FALSE

1. T   **2.** T   **3.** F   **4.** T   **5.** T

MULTIPLE CHOICE QUESTIONS

1. (b)   **2.** (a)   **3.** (a)   **4.** (c)   **5.** (a)   **6.** (b)   **7.** (a)   **8.** (b)   **9.** (a)

10. (c)   **11.** (a)   **12.** (b)   **13.** (c)   **14.** (b)   **15.** (b)   **16.** (b)   **17.** (b)   **18.** (a)

19. (b)   **20.** (b)

⌘⌘⌘⌘⌘⌘

CHAPTER 9

Numerical Solution of Ordinary Differential Equations

9.1 INTRODUCTION

Many problems of science and engineering can be formulated in terms of differential equations. A large number of motivation for building early computers came from the need to compute trajectories accurately and quickly. Today computers are used extensively to solve equations of physical situations.

A general equation of first order and first degree is given by

$$y' = f(x, y) \quad \text{with} \quad y(x_0) = y_0 \qquad \qquad ...(1)$$

Many analytical techniques exist for finding the solution of such equations. In most of these methods, we replace the differential equation by a difference equation and then solve it. These methods gives solutions either as a power series in x from which the values of y can be found by direct substitution or a set of values of x and y. The methods of Picard and Taylor series belong to the former class of solution. In these methods y in (1) is approximated by a truncated series, each term of which is a function of x. The information about the curve at one point is utilized and the solution is not interacted. As such, these are referred to as single-step methods. The methods of Euler, Runga-Kutta, Milne, Adams-Bashforth, etc. belong to the latter class of solutions. In these methods, the next point on the curve is evaluated in short steps ahead, by performing interactions till sufficient accuracy is achieved. As such these methods are called step-by-step methods.

▶ REMARK

▶ Euler and Runga-Kutta methods are used for computing y over a limited range of x-values whereas Milne and Adams methods may be applied for finding y over a wider range for x-values. Therefore Milne and Adams methods required starting values which are found by Picard's, Taylor series or Runga-Kutta methods.

9.2 EXISTENCE AND UNIQUENESS OF SOLUTION OF DIFFERENTIAL EQUATION

Consider the differential equation

$$\frac{dy}{dx} = f(x, y) \qquad \qquad ...(1)$$

Let us suppose, the function $f(x, y)$ be analytic and satisfies Lipschitz condition.

Now, by differentiating the given differential equation as many times as needed and using initial conditions $y = y_0$, for $x = x_0$, we get

$$y = y_0 + (x - x_0)y_0' + \frac{1}{2}(x - x_0)^2 y_0'' + ...$$

It is assumed that $|x - x_0|$ is very small, the series converges to a unique solution.

If $f(x, y)$ is bounded and continuous, the Lipschitz condition

$$|f(x, y) - f(x, z)| < L|y - z| \qquad \qquad ...(2)$$

where L is a constant is used to find a unique solution. Since, $f(x, y)$ is bounded, therefore $|f(x, y)| < M$.

Integrating (1) between x_0 and x, we get

$$\int dy = \int_{x_0}^{x} f(x, y) dx \quad \Rightarrow \quad \int_{x_0}^{x} \frac{dy}{dx} dx = y - y_0 = \int_{x_0}^{x} f(x, y) dx$$

$$\Rightarrow \qquad \qquad y = y_0 + \int_{x_0}^{x} f(x, y) dx \qquad \qquad ...(3)$$

Thus, the given differential equation has been transformed to an integral equation. Now, if $y_1(x)$ is the first approximation to the solution of (1), then a better solution is obtained by substituting $y_1(x)$ for y in the R.H.S. of (3). Therefore, we obtained an iterative formula.

$$y_{i+1}(x) = y_0 + \int_{x_0}^{x} f[t, y_i(t)] dt \qquad \qquad ...(4)$$

Now, we get

$$|y_{i+1}(x) - y_i(x)| \leq \int_{x_0}^{x} |f(t, y_i(t))| - |f(t, y_{i-1}(t)| dt \leq L \int |y_i(t) - y_{i-1}(t)| dt$$

[Using (2)]

Further, suppose that $x_0 \leq x \leq X$ and putting $X - x_0 = h$, we get

$$|y_2(x) - y_1(x)| = \left| y_0 + \int_{x_0}^{x} f[t, y_i(t) dt - y_i(x) \right| = \left| y_0 + y_1(x) + \int_{x_0}^{x} f[t, y_i(t)] dt \right|$$

$$= |y_1(x) - y_0| + \int_{x_0}^{x} M dt \qquad \qquad [\because f \text{ is bounded by } M.]$$

$$\leq 2Mh = N \quad \text{(say)}$$

Similarly, we can obtain

$$|y_3(x) - y_2(x)| \leq L \int_{x_0}^{x} N dt = NL \int_{x_0}^{x} dt = NL(x - x_0) \leq NLh$$

$$|y_4(x) - y_3(x)| \leq L \int_{x_0}^{x} NL(t - x_0) dt = NL^2 \frac{(x - x_0)^2}{2!} = \frac{NL^2 h^2}{2!}$$

and so on. Finally, we get

$$|y_{n+1}(x) - y_n(x)| \leq L \int_{x_0}^{x} \frac{NL^{n-2}(t - x_0)^{n-2}}{(n-2)!} dt = \frac{NL^{n-1}}{(n-1)!}(x - x_0)^{n-1} \leq \frac{NL^{n-1} h^{n-1}}{(n-1)!}$$

But $\qquad \qquad y_{n+1}(x) = y_1(x) + [y_2(x) - y_1(x)] + ... + [y_{n-1}(x) - y_n(x)] \qquad ...(5)$

Except the first term $y_1(x)$, every term of the series (5) is less than the terms of the series

$$N \left[1 + Lh + \frac{L^2 h^2}{2!} + ... + \frac{L^{n-1} h^{n-1}}{(n-1)!} \right]$$

which converges to Ne^{Lh}.

Therefore, the series (5) is absolutely and uniformly convergent to $y(x)$. Also,

$$\left| y(x) - \left(y_0 + \int_{x_0}^{x} [f(t, y(t) dt] \right) \right| = \left| y(x) - y_{n+1}(x) - \int_{x_0}^{x} [f\{t, y(t)\} - f\{t, y_n(t)\} dt] \right|$$

$$\leq |y(x) - y_{n+1}(x)| + L \int_{x_0}^{x} |y(t) - y_n(t)| dt$$

which tends to zero because $y_n(x)$ and $y_{n+1}(x)$ converges to $y(x)$ when $n \to \infty$.

Thus, we have

$$y(x) = y_0 + \int_{x_0}^{x} f[t, y(t)]dt \qquad \ldots(6)$$

Now, we shall prove the uniqueness of this solution. Let if possible $z = z(x)$ be another solution. Then

$$z = y_0 + \int_{x_0}^{x} f[t, z(t)]dt$$

$$y_{n+1} = y_0 + \int_{x_0}^{x} f[t, y_n(t)]dt$$

Thus, $\qquad \left| z - y_{n+1} \right| \le \int_{x_0}^{x} \left| f(t, z) - f(t, y_n) \right| dt \le L\int_{x_0}^{x} \left| z - y_n \right| dt \qquad$ [Using (2)]

But $\qquad \left| z - y_0 \right| \le Mh = \dfrac{N}{2}$ and therefore $\left| z - y_1 \right| \le \dfrac{N}{2}L(x - x_0) \le \dfrac{N}{2}Lh$

and $\qquad \left| z - y_2 \right| \le L^2 \dfrac{(x - x_0)^2}{2!} \le \dfrac{N}{2} \cdot \dfrac{L^2 h^2}{2!}$ and so on.

Finally, we get

$$\left| z - y_{n+1} \right| \le \dfrac{N}{2} \dfrac{L^{n+1} h^{n+1}}{(n+1)!} \text{ which tends to zero as } n \to \infty.$$

$\Rightarrow \qquad z(x)$ and $y(x)$ coincide.

Hence, the solution given by (6) is unique.

9.3 EULER'S METHOD

(Agra-2003, 06)

Consider the first order differential equation

$$y' = f(x, y) \text{ with } y = y_0 \text{ where } x = x_0. \qquad \ldots(1)$$

We wish to solve (1), for values of y at $x = x_i$ where $x_i = x_0 + ih$, $i = 1, 2, \ldots$

Integrating (1), we get

$$y_1 = y_0 + \int_{x_0}^{x} f(x, y)dx$$

Let $\qquad\qquad f(x, y) = f(x_0, y_0) \qquad \ldots(2)$

$$\text{in the range } x_0 \le x \le x_1$$

Now $\quad y_1 = y_0 + \int_{x_0}^{x_1} f(x_0, y_0)dx = y_0 + (x_1 - x_0)f(x_0, y_0) = y_0 + hf(x_0, y_0) \quad$ (Use $x_1 - x_0 = h$)

$$\ldots(3)$$

Similarly for the range $x_1 \le x \le x_2$, , we have

$$y_2 = y_1 + hf(x_1, y_1)$$

Proceeding in this manner, finally we obtain

$$y_{n+1} = y_n + hf(x_n, y_n).$$

Thus starting from x_0 when $y = y_0$, we can construct a table of y for given steps.

ALGORITHM

A new value of y is estimated using the previous value of y as the initial condition. The term $h.f(x_i, y_i)$, represents the increamental value of y and $f(x_i, y_i)$ is the slope of $y(x)$ at (x_i, y_i), i.e., the new value is obtained by extrapolating linearly over the step size h using the slope at its previous value. Hence,

New value = old value $\times$ slope $\times$ step size

▶ REMARK

▶ This process is very slow and to obtain desired accuracy with this method, h should be taken small. If h is not small then the method is too inaccurate.

SOLVED EXAMPLES

EXAMPLE 1. *Given* $y' = \dfrac{y - x}{y + x}$ *with* $y_0 = 1$ *find* y *for* $x = 0.1$ *in 4 steps. (By Euler's method).*

(PTU-2001; Avadh-2003)

SOLUTION.   Let $h = \dfrac{0.1}{4} = 0.025$, given $y_0 = 1$

We know that $y_{n+1} = y_n + hf(x_n, y_n)$.

By putting $n = 0, 1, 2, 3$, we obtain

$$y_1 = y_0 + hf(x_0, y_0)$$

$$= 1 + 0.025 \frac{(1-0)}{(1+0)} = 1.025$$

Again $y_2 = y_1 + hf(x_1, y_1)$

$$= 1.025 + 0.025 \frac{(1.025 - 0.025)}{(1.025 + 0.025)} = 1.0488$$

$$(\text{where } x_1 = x_0 + h = 0 + 0.025 \Rightarrow x_1 = 0.025)$$

Now again

$$y_3 = y_2 + hf(x_2, y_2) \, (\text{where } x_2 = x_0 + 2h = 0 + 2 \times 0.025 \Rightarrow x_2 = 0.05)$$

$$= 1.0488 + 0.025 \frac{(1.0488 - 0.05)}{(1.0488 + 0.05)} = 1.07152$$

and $y_4 = y_3 + hf(x_3, y_3)$ $(\text{where } x_3 = x_0 + 3h = 0 + 3 \times 0.025$

$$\Rightarrow x_3 = 0.075)$$

$$= 1.07152 + 0.025 \left(\frac{1.07152 - 0.075)}{1.07152 + 0.075} \right)$$

$\Rightarrow$ $y_4 = 1.09324$

at $x_4 = x_0 + 4h = 0 + 4 \times 0.025 = 0.01$

Hence, $y(0, 1) = 1.0932$

EXAMPLE 2. *Using Euler's method, compute* $y(0.5)$ *for differential equation* $\dfrac{dy}{dx} = y^2 - x^2$, *with* $y = 1$ *when* $x = 0$.

SOLUTION.   Let $h = \dfrac{0.5}{5} = 0.1$

$$x_0 = 0, y_0 = 1, f(x, y) = y^2 - x^2$$

Using Euler's method we have

$$y_{n+1} = y_n + hf(x_n, y_n)$$

By considering $n = 0, 1, 2, ...$ in succession, we get

$$y_1 = y_0 + hf(x_0, y_0) = 1 + 0.1[1^2 - 0] = 1.10000$$

$$y_2 = y_1 + hf(x_1, y_1) = 1.10000 + 0.1[(1.10000)^2 - (0.1)^2] = 1.22000$$

$$y_3 = y_2 + hf(x_2, y_2) = 1.22000 + 0.1[(1.22)^2 - (0.2)^2]$$

$$= 1.22000 - 0.14484 = 1.36484$$

$$y_4 = y_3 + hf(x_3, y_3) = 1.36484 + 0.1[(1.36484)^2 - (0.3)^2]$$

$$= 1.54212$$

and $y_5 = y_4 + hf(x_4, y_4) = 1.54212 + 0.1[(1.54212)^2 - (0.4)^2] = 1.76393$

Hence, the value of y at $x = 0.5$ is 1.76393.

EXAMPLE 3. *Find the solution of differential equation* $\dfrac{dy}{dx} = xy$ *with* $y(1) = 5$ *in the interval* [1, 1.5] *using* $h = 0.1$.

SOLUTION. As per given, we have $x_1 = 1, y_0 = 5, f(x, y) = xy$

Using Euler's method

$$y_{n+1} = y_n + hf(x_n, y_n)$$

By considering $n = 0, 1, 2, \ldots$ in succession, we get

For $n = 0$; $y_1 = y_0 + hf(x_0, y_0) = 5 + 0.1[1 \times 5] = 5.5$

For $n = 1$; $y_2 = y_1 + hf(x_1, y_1) = 5.5 + 0.1[1.1 \times 5.5] = 6.105$

For $n = 2$; $y_3 = y_2 + hf(x_2, y_2)$

$$= 6.105 + 0.1[1.2 \times 6.105] = 6.838$$

For $n = 3$; $y_4 = y_3 + hf(x_3, y_3)$

$$= 6.838 + 0.1[1.3 \times 6.838] = 7.727$$

For $n = 4$; $y_5 = y_4 + hf(x_4, y_4)$

$$= 7.727 + 0.1[1.4 \times 7.727] = 8.809$$

Hence, the value of y is 8.809.

EXAMPLE 4. *Find the solution of* $\dfrac{dy}{dx} = x^2 + y^2, y(0) = 0$ *in the range* $0 \le x \le 0.5$, *using Euler's method.*

SOLUTION. We have $x_0 = 0, y_0 = 0, h = 0.1$ and $0 \le x \le 0.5$

$\therefore$ $x_1 = 0.1, x_2 = 0.2, x_3 = 0.3, x_4 = 0.4, x_5 = 0.5$

Also $f(x, y) = x^2 + y^2$

$\Rightarrow$ $f(x_0, y_0) = f(0, 0) = 0$

$\therefore$ $y_1 = y_0 + hf(x_0, y_0) = 0 + 0.1 \times 0 = 0$

Now $f(x_1, y_1) = f(0.1, 0) = (0.1)^2 + 0^2 = 0.01$

$$y_2 = y_1 + hf(x_1, y_1) = 0 + 0.1(0.01) = 0.001$$

$$f(x_2, y_2) = f(0.2, 0.001) = (0.2)^2 + (0.001)^2 = 0.040001$$

$\therefore$ $y_3 = y_2 + hf(x_2, y_2) = 0.001 + 0.1(0.040001) = 0.005$

Now $f(x_3, y_3) = f(0.3, 0.005) = (0.3)^2 + (0.005)^2 = 0.090025$

$\Rightarrow$ $y_4 = y_3 + hf(x_3, y_3)$ $0.005 + 0.1(0.090025) = 0.014$

Now $f(x_4, y_4) = f(0.4, 0.014) = (0.4)^2 + (0.014)^2 = 0.160196$

$\Rightarrow$ $y_5 = y_4 + hf(x_4, y_4) = 0.014 + 0.1(0.160196) = 0.03$

Further

$$f(x_5, y_5) = f(0.5, 0.03) = (0.5)^2 + (0.03)^2 = 0.2509$$

$\Rightarrow$ $y_6 = y_5 + hf(x_5, y_5) = 0.03 + 0.1(0.2509) = 0.055$

Hence, $y(0.1) = 0.001, y(0.2) = 0.005,$

$y(0.03) = 0.055, y(0.4) = 0.03,$

$y(0.5) = 0.055.$

EXAMPLE 5. *Apply Euler's method to initial value problem* $\dfrac{dy}{dx} = x + y, y(0) = 0$, *when* $x = 0$ *to* $x = 1.0$ *taking* $h = 0.2$.

(Mumbai-2005, Rohtak-2003; Kattayam-2005)

SOLUTION. We have $h = 0.2, x_0 = 0, x_1 = 0.2, x_2 = 0.4,$

$x_3 = 0.6, x_4 = 0.8, x_5 = 1.0.$

Also, $f(x, y) = x + y$

By Euler's method,

$y_{n+1} = y_n + hf(x_n, y_n)$

and, $f(x_0, y_0) = f(0, 0) = 0 + 0 = 0$

Then $y_1 = y_0 + hf(x_0, y_0) = 0 + 0.2(0) = 0$

Now, $f(x_1, y_1) = f(0.2, 0) = 0.2 + 0 = 0.2$

So, $y_2 = y_1 + hf(x_1, y_1) = 0 + 0.2(0.2) = 0.04$

Now, $f(x_2, y_2) = f(0.4, 0.04) = 0.4 + 0.04 = 0.44$

So, $y_3 = y_2 + hf(x_2, y_2) = 0.04 + 0.2(0.44) = 0.128$

$f(x_3, y_3) = f(0.6, 0.128) = 0.6 + 0.128 = 0.728$

$y_4 = y_3 + hf(x_3, y_3) = 0.128 + 0.2(0.728) = 0.2736$

$f(x_4, y_4) = f(0.8, 0.2736) = 0.8 + 0.2736 = 1.0736$

$y_5 = y_4 + hf(x_4, y_4) = 0.2736 + 0.2(1.0736) = 0.48832$

and $f(x_5, y_5) = f(1.0, 0.48832) = 1.48832$

$y_6 = y_5 + hf(x_5, y_5)$

$= 0.48832 + 0.1(1.48832) = 0.63715$

Hence, $y_{(1)} = y_6 = 0.63715.$

9.4 EULER'S MODIFIED METHOD

(Bangluru-2004, 07)

Instead of approximating $f(x, y)$ by $f(x_0, y_0)$ in (2) in 9.2, the integral (3) in 9.2, is approximated by trapezoidal rule to obtain,

$$y_1 = y_0 + \frac{h}{2}[f(x_0, y_0) + f(x_1, y_1)] \qquad \text{...(4)}$$

We thus obtain the formula,

$$y_1^{n+1} = y_0 + \frac{h}{2}[f(x_0, y_0) + f(x_1, y_1^{(n)})] \qquad \text{...(5)}$$

(where $n = 0, 1, 2, ...$)

where $y_1^{(n)}$ is the nth approximation to y_1. The iteration formula (5) uses the initial value $y_1^{(0)}$ from Euler's method given by

$$y_1^{(0)} = y_0 + hf(x_0, y_0)$$

▼ REMARK

▸ The modified Euler's method gives a great improvement in accuracy over the original method. In this method we take the average of slopes at (x_0, y_0) and $(x_1, y_1^{(1)})$ instead of slope at (x_0, y_0) as used in Euler's method.

SOLVED EXAMPLES

EXAMPLE 1. *Using the Euler's modified method, solve numerically the equation $y' = x + \left|\sqrt{y}\right|$ with $y(0) = 1$ for $0 \leq x \leq 0.6$ in step of 0.2.*

(MKU-2004, 06, VTU-2007; Agra-2006, 08, Bhopal-2002)

SOLUTION. We have $y(0) = 1$ where $x_0 = 0$, and interval $h = 0.2$

By Euler's method $y_1 = y_0 + hf(x_0, y_0)$

$$= 1 + 0.2\left[0 + \left|\sqrt{1}\right|\right] = 1.2 \quad \Rightarrow \quad y_1 = 1.2$$

The value of y_1, thus obtained is improved by modified method,

$$y_1^{n+1} = y_0 + \frac{h}{2}[f(x_0, y_0) + f(x_1, y_1^{(n)})]$$

$$= 1 + \frac{0.2}{2}[(0 + \left|\sqrt{1}\right|) + (0.2 + \left|\sqrt{1.2}\right|)]$$

$$= 1.2295 \quad \Rightarrow \quad y_1^{(n)} = 1.2295$$

Now $n = 1$ gives

$$y_1^{(2)} = 1 + \frac{0.2}{2}[0 + \left|\sqrt{1}\right| + 0.2 + \left|\sqrt{1.2295}\right|]$$

$$= 1.2309 \quad \Rightarrow \quad y_1^{(2)} = 1.2309$$

Again $n = 2$, gives

$$y_1^{(3)} = 1 + \frac{0.2}{2}[0 + \left|\sqrt{1}\right| + 0.2 + \left|\sqrt{1.2309}\right|]$$

$$= 1.2309 \quad \Rightarrow \quad y_1^{(3)} = 1.2309$$

Hence, we take $y_1 = 1.2309$ at $x = 0.2$ and proceed to compute y at $x = 0.4$

Applying Euler's method, we get

$$y_2 = 1.2309 + 0.2[0.2 + \left|\sqrt{1.2309}\right|] = 1.4979$$

$$\Rightarrow \quad y_2 = 1.4979$$

Now, we apply modified method for more accurate approximations as follows.

$$y_2^{(1)} = 1.2309 + \frac{0.2}{2}[(0.2 + \left|\sqrt{1.2309}\right|) + (0.4 + \left|\sqrt{1.49279}\right|)] = 1.52402$$

$$y_2^{(2)} = 1.2309 + \frac{0.2}{2}[(0.2 + \left|\sqrt{1.2309}\right|) + (0.4 + \left|\sqrt{1.52402}\right|)] = 1.525297$$

$$y_2^{(3)} = 1.2309 + \frac{0.2}{2}[(0.2 + \left|\sqrt{1.2309}\right|) + (0.4 + \left|\sqrt{1.525297}\right|)] = 1.52535$$

$$y_2^{(4)} = 1.2309 + \frac{0.2}{2}[(0.2 + \left|\sqrt{1.2309}\right|) + (0.4 + \left|\sqrt{1.52535}\right|)] = 1.52535$$

Since $y_2^{(3)} = y_2^{(4)} = 1.52535$

To find the value of $y = y_3$ for $x = 0.6$ we apply Euler's method

$$y_3 = 1.52535 + 0.2[0.4 + |\sqrt{1.52535}|]$$

$$= 1.85236 \quad \Rightarrow \quad y_3 = 1.85236$$

For better approximations, we use Euler's modified method, as follows

$$y_3^{(1)} = 1.52535 + \frac{0.2}{2}[0.4 + |\sqrt{1.52535}|) + (0.6 + |\sqrt{1.855236}|)] = 1.88496$$

$$y_3^{(2)} = 1.52535 + \frac{0.2}{2}[0.4 + |\sqrt{1.52535}|) + (0.6 + |\sqrt{1.88496}|)] = 1.88615$$

$$y_3^{(3)} = 1.52535 + \frac{0.2}{2}[0.4 + |\sqrt{1.52535}|) + (0.6 + |\sqrt{1.88615}|)] = 1.88619$$

$$y_3^{(4)} = 1.52535 + \frac{0.2}{2}[0.4 + |\sqrt{1.52535}|) + (0.6 + |\sqrt{1.88619}|)] = 1.886193$$

$\Rightarrow \quad y_3^{(4)} = 1.88619$ correct to 5 decimal places

Hence, $y = 1.88619$ at $x = 0.6$

EXAMPLE 2. *Given* $\dfrac{dy}{dx} = x + y$ *with initial conditions* $y(0) = 1$. *Find* $y(0.05)$ *and* $y(0.1)$, *correct to 6 decimal places.* (MDU(B.E.)-2004, Rohtak-2005, Bhopal–2002S, Delhi–2002)

SOLUTION. Using Euler's method, we get

$$y_1^{(0)} = y_1 = y_0 + hf(x_0, y_0) = 1 + 0.05(0 + 1) = 1.05$$

Now, we improve this value of y by using Euler's modified method

$$y_1^{(1)} = y_0 + \frac{h}{2}[f(x_0, y_0) + f(x_1, y_1^{(0)})]$$

$$= 1 + \frac{0.05}{2}[(0 + 1) + (0.05 + 1.05)] = 1.0525$$

$$y_1^{(2)} = 1 + \frac{0.05}{2}[(0 + 1) + (0.05 + 1.0525)] = 1.0525625$$

$$y_1^{(3)} = 1 + \frac{0.05}{2}[(0 + 1) + (0.05 + 1.0525625)] = 1.052564$$

$$y_1^{(4)} = 1 + \frac{0.05}{2}[(0 + 1) + (0.05 + 1.052564)] = 1.052564$$

We observe that $y_1^{(3)} = y_1^{(4)} = 1.052564$, correct to 6 decimal places. Thus, we take $y_1 = 1.052564$, *i.e.*, we have $y(0.05) = 1.052564$. Again, using Euler's method, we get

$$y_2^{(0)} = y_2 = y_1 + hf(x_1, y_1)$$

$$= 1.052564 + 0.05(1.052564 + 0.05)$$

$$= 1.1076922$$

Now, we have to improve y_2 by using Euler's modified method

$$y_2^{(1)} = 1.52564 + \frac{0.05}{2}[(1.052564 + 0.05) + (1.1076922 + 0.1)] = 1.1120511$$

$$y_2^{(2)} = 1.052564 + \frac{0.05}{2}[(1.052564 + 0.05) + (1.1120511 + 0.1)] = 1.1104294$$

$$y_2^{(3)} = 1.052564 + \frac{0.05}{2}[(1.052564 + 0.05) + (1.1104294 + 0.1)] = 1.1103888$$

$$y_2^{(4)} = 1.052564 + \frac{0.05}{2}[(1.052564 + 0.05) + (1.1103888 + 0.1)] = 1.1103878$$

and $y_2^{(5)} = 1.052564 + \frac{0.05}{2}[(1.052564 + 0.05) + (1.1103878 + 0.1)] = 1.1103878$

Since, $y_2^{(4)} = y_2^{(5)} = 1.1103878$, correct to 7 decimal places. Hence, $y(0.1) = 1.116388$, correct to 6 decimal places.

EXAMPLE 3. Find $y(0.2)$. Given $\frac{dy}{dx} = f(x, y) = \log_{10}(x + y)$ with initial condition $y = 1$ for $x = 0$.   (UPTU–2007; Agra–2000)

SOLUTION. As per given, we have

$$\frac{dy}{dx} = f(x, y) = \log_{10}(x + y) \text{ and } h = 0.2$$

By Euler's formula, we have

$$y_1^{(0)} = y_1 = y_0 + hf(x_0, y_0) = 1 + 0.2 \log(0 + 1) = 1$$

Now, we improve this value by using Euler's modified formula as follows:

$$y_1^{(1)} = y_0 + \frac{h}{2}[f(x_0, y_0) + f(x_1, y_1^{(0)})]$$

$$= 1 + \frac{0.2}{2}[\log(0 + 1) + \log(0.2 + 1)] = 1.0079$$

and

$$y_1^{(2)} = 1 + \frac{0.2}{2}[\log(0 + 1) + \log(0.2 + 1.0079)] = 1.0082$$

Also,

$$y_1^{(3)} = 1 + \frac{0.2}{2}[\log(0 + 1) + \log(0.2 + 1.0082)] = 1.0082$$

Since, $y_1^{(2)} = y_1^{(3)} = 1.0082$. Hence, $y(0.2) = 1.0082$.

EXAMPLE 4. Find $y(2.2)$ by using Euler's method for $\frac{dy}{dx} = -xy^2$, where $y(2) = 1$.   (Take $h = 0.1$)

SOLUTION. By Euler's method, we get

$$y_1^{(0)} = y_1 = y_0 + hf(x_0, y_0) = 1 + 0.1(-2)(-1)^2 = 0.8$$

Now, this value y_1 is improved by Euler's method formula

$$y_1^{(1)} = y_0 + \frac{h}{2}[f(x_0, y_0) + f(x_1, y_1^{(0)})]$$

$$= 1 + \frac{0.1}{2}[-2(1)^2 + (-2.1)(-8)^2] = 0.8328.$$

$$y_1^{(2)} = 1 + \frac{0.1}{2}[-2(1)^2 + (-2.1)(0.8328)^2] = 0.8272$$

$$y_1^{(3)} = 1 + \frac{0.1}{2}[-2(1)^2 + (-2.1)(0.8272)^2] = 0.8281$$

$$y_1^{(4)} = 1 + \frac{0.1}{2}[-2(1)^2 + (-2.1)(0.8281)^2] = 0.8280$$

and $y_1^{(5)} = 1 + \frac{0.1}{2}[-2(1)^2 + (-2.1)(0.8280)^2] = 0.8280$

Since $y_1^{(4)} = y_1^{(5)} = 0.8280$. Hence $y_1 = 0.828$ at $x_1 = 2.1$.

If y_2 is the value of y at $x = 2.2$, then by Euler's method .
we get

$$y_2^{(0)} = y_2 = y_1 + hf(x_1, y_1)$$

$$= 0.828 + \frac{0.1}{2}[(-2.1)(0.828)^2 + (-2.2)(0.68402)^2]$$

$$= 0.70454$$

$$y_2^{(2)} = 0.828 + \frac{0.1}{2}[(-2.1)(0.828)^2 + (-2.2)(0.70454)^2]$$

$$= 0.70141$$

$$y_2^{(3)} = 0.828 + \frac{0.1}{2}[(-2.1)(0.828)^2 + (-2.2)(0.70141)^2]$$

$$= 0.70189$$

$$y_2^{(4)} = 0.828 + \frac{0.1}{2}[(-2.1)(0.828)^2 + (-2.2)(0.70189)^2]$$

$$= 0.70182$$

$$y_2^{(5)} = 0.828 + \frac{0.1}{2}[(-2.1)(0.828)^2 + (-2.2)(0.70182)^2]$$

$$= 0.70183$$

Since, $y_2^{(4)} = y_2^{(5)} = 0.7018$, correct to four decimal places.

Hence, $y(2.2) = 0.7018$.

EXAMPLE 5.   *Given that* $\dfrac{dy}{dx} = \log_{10}(x + y)$ *with the initial condition that* $y = 1$ *when* $x = 0$, *find y for* $x = 0.2$ *and* 0.5.

SOLUTION.   We have $x_0 = 0, x_1 = 0.2, x_2 = 0.5$

Then $y_0 = 1$

Also, $f(x, y) = \log(x + y)$

Let us take $h = 0.2$

$\therefore$ $f(x_0, y_0) = \log_{10}(x_0 + y_0) = \log(0 + 1) = \log 1 = 0$

Now $y_1^{(1)} = y_0 + hf(x_0, y_0) = 1 + 0.2(0) = 1$

$f(x_1, y_1^{(1)}) = \log_{10}(x_1 + y_1^{(1)}) = \log_{10}(0.2 + 1) = \log_{10}(1.2) = 0.07918.$

Now, $y_1^{(2)} = y_0 + \dfrac{h}{2}[f(x_0, y_0) + f(x_1, y_1^{(1)})]$

$$= 1 + \frac{1}{2}(0.2)[0.07918] = 1 + 0.007918 = 1.007918.$$

So, $f(x_1, y_1^{(2)}) = \log_{10}(x_1 + y_1^{(2)}) = \log_{10}[0.2 + 1.007918]$

$$= \log_{10}(1.207918) = 0.08204$$

Then $y_1^{(3)} = y_0 + \dfrac{h}{2}[f(x_0, y_0) + f(x_1, y_1^{(2)})]$

$$= 1 + \frac{1}{2}(0.2) + (0 + 0.08204) = 1 + 0.008204 = 1.008204$$

$\Rightarrow \quad f(x_1, y_1^{(3)}) = \log_{10}(x_1 + y_1^{(3)})$

$$= \log_{10}(0.2 + 1.008204) = \log_{10}(1.208204) = 0.08214$$

$$y_1^{(4)} = y_0 + \frac{h}{2}[f(x_0, y_0) + f(x_1, y_1^{(3)})]$$

$$= 1 + \frac{1}{2}(0.2)(0 + 0.08214) = 1.008214$$

We observe that $y_1^{(3)} \approx y_1^{(4)}$, therefore we stop here.

Now taking $\quad y_1 = 1.0082$, i.e., $y(0.2) = 1.0082$

To find y_2 i.e., $y(0.5)$, start with y_1 where first approximation is

$$y_2^{(1)} = y_1 + hf(x_1, y_1)$$

Here, $h = 0.5 - 0.2 = 0.3$

$$f(x_1, y_1) = \log_{10}(x_1 + y_1) = \log_{10}(0.2 + 1.0082)$$

$$= \log_{10}(1.2082) = 0.08214$$

Then $\quad y_2^{(1)} = y_1 + h_f(x_1, y_1) = 1.0082 + 0.3(0.8214) = 1.03284$

$\therefore \quad f(x_2, y_2^{(1)}) = \log_{10}(x_2 + y_2^{(1)}]$

$$= \log_{10}(0.5 + 1.03284) = \log_{10}(1.53284) - 0.18550$$

Now $\quad y_2^{(2)} = y_1 + \frac{h}{2}[f(x_1, y_1) + f(x_2, y_2^{(1)})]$

$$= 1.0082 + \frac{1}{2}(0.3)[0.08214 + 0.18550]$$

$$= 1.0082 + 0.04015 = 1.04835$$

Then $f(x_2, y_2^{(2)}) = \log_{10}(x_2, y_2^{(2)})$

$$= \log_{10}(0.5 + 1.04835)$$

$$= \log_{10}(1.54835) = 0.18987$$

Now, $\quad y_2^{(3)} = y_1 + \frac{h}{2}[f(x_1, y_1) + f(x_2, y_2^{(2)})]$

$$= 1.0082 + \frac{1}{2}(0.3)[0.08214 + 0.18987]$$

$$= 1.0082 + 0.04080 = 1.049.$$

$\therefore \quad f(x_2, y_2^{(3)}) = \log_{10}(x_2, y_2^{(3)})$

$$= \log_{10}(0.5 + 1.049) = \log_{10}(1.549) = 0.19005$$

so, $\quad y_2^{(4)} = y_1 + \frac{h}{2}[f(x_1, y_1) + f(x_2, y_2^{(3)}]$

$$= 1.0082 + \frac{1}{2}(0.3)[0.08214 + 0.19005]$$

$$= 1.0082 + 0.04083 = 1.04903$$

We observe that $y_2^{(3)} \approx y_2^{(4)}$ so we stop the process.

Hence, $\quad y_2^{(3)} \approx 1.0490$, i.e., $y(0.5) = 1.049$

and required values are $y(0.2) = 1.0082$ and $y(0.5) = 1.049$.

EXAMPLE 6.   *Let* $\dfrac{dy}{dx} = \dfrac{y - x}{x + y}$, *with boundary conditions* $y = 1$ *when* $x = 0$. *Find approximately* y

for $x = 0.1$ *by Euler's modified method (5 steps).*   (VTU-2007; Agra-2004, Avadh-2005)

SOLUTION.   We have $x_0 = 0, y_0 = 1, f(x, y) = \dfrac{y - x}{y + x}$

Taking $h = 0.1$, then $x_1 = 0.1$

$\therefore \qquad f(x_0, y_0) = f(0.1) = \dfrac{1 - 0}{1 + 0} = 1$

$\Rightarrow \qquad y_1^{(1)} = y_0 + h f(x_0, y_0) = 1 + 0.1(1) = 1.1$

Thus, $f(x_1, y_1^{(1)}) = f(0.1, 1.1)$

$$= \dfrac{1.1 - 0.1}{1.1 + 0.1} = \dfrac{1}{1.2} = 0.8333$$

Now, $\qquad y_1^{(2)} = y_0 + \dfrac{h}{2}[f(x_0, y_0) + f(x_1, y_1^{(1)})]$

$$= 1 + \dfrac{1}{2}(0.1)[1 + 0.8333] = 1.09170$$

$\Rightarrow \qquad f(x_1, y_1^{(2)}) = f(0.1, 1.09170)$

$$= \dfrac{1.09170 - 0.1}{10.9170 + 0.1} = \dfrac{0.9917}{1.1917} = 0.83217$$

So, $\qquad y_1^{(3)} = y_0 + \dfrac{h}{2}[f(x_0, y_0) + f(x_1, y_1^{(2)})]$

$$= 1 + \dfrac{1}{2}(0.1)(1 + 0.83217) = 1.09161$$

$\Rightarrow \qquad f(x_1, y_1^{(3)}) = f(0.1, 1.09161) = \dfrac{1.09161 - 0.1}{1.09161 + 0.1} = 0.83215$

Now, $\qquad y_1^{(4)} = y_0 + \dfrac{h}{2}[f(x_0, y_0) + f(x_1, y_1^{(3)})]$

$$= 1 + \dfrac{1}{2}(0.1)[1 + 0.83215]$$

So, $\qquad f(x_1, y_1^{(4)}) = f(0.1, 1.091607) = \dfrac{1.091607 - 0.1}{1.091607 + 0.1} = 0.83215$

$\Rightarrow \qquad y_1^{(5)} = y_0 + \dfrac{h}{2}[f(x_0, y_0) + f(x_1, y_1^{(4)})]$

$$= 1 + \dfrac{1}{2}(0.1)[1 + 0.83215] = 1.061607$$

We observe that $y_1^{(4)} = y_1^{(5)}$, Hence, the value of y at 0.1 is given by

$y(0.1) = 1.091607$.

9.5   SOLUTION BY TAYLOR SERIES

We consider the differential equation with initial condition

$$\left. \begin{array}{l} y' = f(x, y) \\ y_{x_0} = y_0 \end{array} \right\} \quad \text{where} \qquad x_0 = 0. \qquad \qquad \dots(1)$$

If $y(x)$ is the exact solution of (1), then the Taylor's series for $y(x)$ about the point $x = x_0$, we get

$$y_x = y_0 + (x - x_0)y_0' + \frac{(x - x_0)^2}{2!} y_0'' + \dots \qquad \dots(2)$$

Putting $x = x_0 + h$ in (2), we obtain

$$y_1 = y_0 + hy_0' + \frac{h^2}{2!} y_0'' + \dots \qquad \dots(3)$$

If the values of $y'_0, y''_0, \dots$ are known, then (3) gives a power series for y_1. The coefficient y'_0, $y''_0, \dots$ can be found from (1).

We can write (1) as

$$y' = f(x, y)$$
$$y'' = f' = f_x + f_y y' = f_x + f_y f$$

Similarly, we get,

$$y''' = f'' = f_{xx} + f_{xy}f + f[f_{yx} + f_{yy}f] + f_y[f_x + f_y f]$$

$$= f_{xx} + 2ff_{xy} + f^2 f_{yy} + f_x f_y + f^2 f_y$$

and other higher derivatives of y. The method is best understood by the following examples.

SOLVED EXAMPLES

EXAMPLE 1. *Using the Taylor series for $y_{(x)}$, find $y_{(0.1)}$ correct to four decimal places if $y_{(x)}$ satisfies $y' = x + (-y^2)$, $y_0 = 1$ where $x_0 = 0$.* (VTU-2010, Madras-2006)

SOLUTION. We know by Taylor series, for $y(x)$ that

$$y_x = 1 + xy_0' + \frac{x^2}{2!} y_0'' + \frac{x^3}{3!} y_0''' + \frac{x^4}{4!} y_0^{iv} + \frac{x^5}{5!} y_0^v + \dots$$

The derivatives $y'_0, y''_0, \dots$ etc. are obtained such that

$$y'_{(x)} = x - y^2 \qquad \text{(since } x = 0, y_0 = 1)$$
$$\Rightarrow \quad y_0' = -1$$
$$y''_x = 1 - 2xy' \qquad\qquad y'''_0 = 3$$
$$y'''_x = -2yy'' - 2y'^2 \qquad \Rightarrow y'''_0 = -8$$
$$y^{iv}_{(x)} = -2yy''' - 6y'y'' \qquad \Rightarrow \quad y_0^{iv} = 34$$
$$y^v_{(x)} = -2yy^{iv} + 8y'y''' - 6y''^2 \Rightarrow \quad y_0^v = -186$$

Using these values, the Taylor series becomes

$$y_x = 1 - x + \frac{3}{2}x^2 - \frac{4}{3}x^3 + \frac{17}{12}x^4 - \frac{31}{20}x^5 + \dots$$

At $x = 0.1$, we get

$$y_{(0.1)} = 1 - 0.1 + \frac{3}{2}(0.1) + 0.001\left(\frac{-4}{3}\right) + (0.0001)\left(\frac{17}{12}\right) = 0.9138$$

We can find the range for the value of x for which the above series truncated after term containing x^4 is convert to 4 decimal places.

$$\frac{31}{20}x^2 \le 0.00005 \quad i.e., \quad x \le 0.126.$$

EXAMPLE 2.   *Solve the differential equation $\dfrac{dx}{dy} = x + y$, with $y(0) = 1$, $x \in [0, 1]$ by Taylor series expansion to obtain y for $x = 0.1$.*   (UPTU-2006; Agra-2004)

SOLUTION.   As per given, we have

$$y' = x + y \quad\quad\quad \Rightarrow \quad\quad\quad y'(0) = 1$$
$$y'' = 1 + y' \quad\quad\quad \Rightarrow \quad\quad\quad y''(0) = 2$$
$$y''' = 0 + y'' \quad\quad\quad \Rightarrow \quad\quad\quad y'''(0) = 2$$

Therefore, we get

$$y(x) = 1 + x + \frac{2x^2}{2!} + \frac{2x^3}{3!} + \frac{2x^4}{4!} + \dots$$

Putting $x = 0.1$, we get

$$y(0.1) = 1 + 0.1 + (0.1)^2 \frac{2(0.1)^3}{6} + \frac{(0.1)^4}{12}$$
$$= 1 + 0.1 + 0.1 + 0.000333 + 0.0000083$$
$$= 1.11033$$

EXAMPLE 3.   *Using Taylor's series expansion, tabulate the solution $x = 4$ to $x = 4.4$ in steps of 0.1 of differential equation $5xy' + y^2 - 2 = 0$ with $y(4) = 1$.*

SOLUTION.   Differentiating the given equation, successively, we get

$$5xy'' + 5y' - 2yy' = 0$$
$$5xy''' + 10y'' + 2y''y - 2y'^2 = 0$$
$$5xy'''' + 15y''' + 2yy'' + 6y'y'' = 0$$
$$5xy''''' + 20y'''' + 2yy''' + 8y'y''' + 6y''^2 = 0$$

Therefore, the values of various derivatives at $x_0 = 4$, $y = 1$ are $y'_0 = 0.5$, $y''^2_0 = -0.175$, $y'''_0 = -0.00843$, $y''''_0 = 0.008998125$

Then, by Taylor series, we get

$$y(x) = 1 + 0.5(x - 4) - 0.00875(x - 4)^2 +$$
$$0.0017083(x - 4)^3 + (-0.0003521)(x - 4)^4 + \dots$$

Tabulating from $x = 4$ to $x = 4.4$, we get

$$y(4) = 1, y(4.1) = 1.004914, y(4.2) = 1.009663,$$
$$y(4.3) = 1.014256, y(4.4) = 1.018701.$$

EXAMPLE 4.   *Using Taylor's series expansion to find a solution of the differential equation $y' = (0.1)(x^3 + y^2)$ with $y(0) = 1$, correct to 4 decimal places.*

SOLUTION.   We have, the first few derivatives and their values at $x_0 = 0$, $y_0 = 1$ are

$$y' = (0.1)(x^3 + y^2) \quad\quad\quad \Rightarrow \quad\quad y'_0 = 0.1$$

$$y'' = (0.1)(3x^2 + 2yy') \quad\quad\quad \Rightarrow \quad\quad y''_0 = 0.02$$

$$y''' = 0.1[6x + 2yy'' + 6(y')^2] \quad \Rightarrow \quad y'''_0 = 0.006$$

$$y'''' = 0.1[6 + 2yy''' + 6y'y''] \quad \Rightarrow \quad y''''_0 \cong 0.6024$$

$$y''''' = 0.1[2yy'''' + 8y'y''' + 6(y'')^2] \quad \Rightarrow \quad y_0''''' = 0.1212$$

$$y''''' = 0.1[2yy''''' + 108y'y'''' + 20y''y'''] \quad \Rightarrow \quad y_0''''' = 0.8472$$

Now, using Taylor series expansion, we get

$$y = y_0 + xy_0' + \frac{x^2}{2!}y_0'' + \frac{x^3}{3!}y_0''' + \frac{x^4}{4!}y_0'''' + \frac{x^5}{5!}y_0''''' + \frac{x^6}{6!}y_0'''''' + ...$$

$$\Rightarrow y = 1 + 0.1x + 0.01x^2 + 0.001x^3 + 0.0251x^4 + 0.00101x^5 + 0.000117x^6.$$

EXAMPLE 5. *From the Taylor series for y(x), find y(0.1) correct to four decimal places, where y′*

$$= dy/dx = x - y^2, y(0) = 1.$$

SOLUTION. Here, we have $x_0 = 0, y_0 = 1$.

Taylor's series about $x = 0$ is given by

$$y(x) = y_0 + xy'(0) + \frac{x^2}{2!}y''(0) + \frac{x^3}{3!}y'''(0) \quad + \frac{x^4}{4!}y'''(0) + \frac{x^5}{5!}y''''(0) + ...$$

...(1)

Now, since, we have

$$y' = x - y^2 \qquad \Rightarrow \qquad y'(0) = x_0 - y_0^2 = -1$$

$$y'' = 1 - 2yy' \qquad \Rightarrow \qquad y''(0) = 1 - 2y_0 y'(0) = 3$$

$$y''' = -2yy'' - 2(y')^2 \quad \Rightarrow \qquad y'''(0) = -2y_0[y''(0)] - 2[y'(0)]^2 = -8$$

$$y'''' = -2yy''' - 6y'y'' \quad \Rightarrow \qquad y''''(0) = -2y_0[y'''(0)] - 6y'(0)y''(0) = 34$$

$$y''''' = -2yy'''' - 8y''y''' - 6y''^2$$

$$\Rightarrow y'''''(0) = -2y_0[y''''(0)] - 8y'(0)y'''(0) - 6[y''(0)]^2 = -180$$

Putting all these values in (1), we get

$$y(x) = 1 - x + \frac{3}{2}x^2 - \frac{4}{3}x^3 - \frac{17}{12}x^4 - \frac{31}{20}x^5 ...$$

...(2)

We have to find the value of $y(0.1)$ correct to four places of decimal. So we should consider the series upto the term x^4.

Thus,

$$y(0.1) = 1 - (0.1) + \frac{3}{2}(0.1)^2 - \frac{4}{3}(0.1)^3 - \frac{17}{12}(0.1)^4 = 0.9138$$

EXAMPLE 6. *Use Taylor's method, find approximate value of y at x = 0.2 for the differential*

equation $\frac{dy}{dx} = 2y + 3e^x$, y(0) = 0. Compare the numerical solution obtained with

the exact solution. (VTU–2009, PTU–2003)

SOLUTION. We have $x_0 = 0, y_0 = 0$ and $y' = \frac{dy}{dx} = 2y + 3e^x$ ...(1)

$$\therefore \qquad y'(0) = 2y_0 + 3e^{x_0} = 2 \times 0 + 3e^0 = 3$$

$$y'' = 2y' + 3e^x$$

$$\Rightarrow \qquad y''(0) = 2y'(0) = 3e^{x_0} = 9$$

$$y''' = 2y'' + 3e^x$$

$$\Rightarrow \qquad y'''(0) = 2y''(0) = 3e^{x_0} = 21$$

$$y'''' = 2y''' + 3e^x$$

$$\Rightarrow \qquad y''''(0) = 2y'''(0) = 3e^{x_0} = 45 \text{ .... etc.}$$

Taylor's series of $y(x)$ about $x = 0$ is given by

$$y(x) = y_0 + xy'(0) + \frac{x^2}{2!}y(0) + \frac{x^3}{3!}y'''(0) + \frac{x^4}{4!}y''''(0) + \dots \qquad \dots(2)$$

$$\Rightarrow \qquad y(x) = 3x + \frac{9}{2}x^2 + \frac{7}{2}x^3 + \frac{15}{8}x^4 + \dots$$

$$\Rightarrow \qquad y(0.2) = 3(0.2) + \frac{9}{2}(0.2)^2 + \frac{7}{2}(0.2)^3 + \frac{15}{8}(0.2)^4 + \dots = 0.8110 \qquad \dots(3)$$

Now exact solution of (1) can be written as $\frac{dy}{dx} = 2y + 3e^x$ which is a linear differential equation.

$$\text{I.F.} = e^{-2x}$$

Solution is

$$ye^{-2x} = \int e^{-2x}(3e^x)dx + c = 3(-e^{-x}) + c$$

$$\Rightarrow \qquad y(x) = -3e^x + ce^{2x}$$

when $x = 0, y = 0$ then $c = 3$

$$\therefore \qquad y(x) = 3e^{2x} - 3e^x$$

$$\Rightarrow \qquad y(0.2) = 3e^{0.4} = 0.8112$$

From (3) and (4), it is obvious that the approximate value is the same to the exact value upto 3 decimal places.

EXAMPLE 7. *Using Taylor's series method, find the value of y up to five places of decimal when*

$x = 1.02$, *where* $\dfrac{dy}{dx} = xy - 1, y(1) = 2.$ (Rohtak–2005)

SOLUTION. We have $x_0 = 1, y_0 = 2$

Also $\dfrac{dy}{dx} = xy - 1 = y'$

Now

$$y'(0) = x_0 y(0) - 1 = 2 - 1 = 1$$

$$y''(0) = x_0 y'(0) + y(0) = 1 \times 1 + 2 = 3$$

$$y'''(0) = xy'' + 2y' \quad \Rightarrow \quad y'''(0) = x_0 y''(0) - 2y'(0) = 1 \times 3 + 2 \times 1 = 5$$

$$y''''(0) = xy''' + 3y'' \quad \Rightarrow \quad y''''(0) = x_0 y'''(0) + 3y''(0) = 1 \times 5 + 3 \times 3 = 14$$

$$y'''''(0) = xy'''' + 4y''' \quad \Rightarrow \quad y'''''(0) = x_0 y''''(0) + 4y'''(0) = 1 \times 14 + 4 \times 5 = 34$$

Taylor's series about $x = 1$ is given by

$$y(x) = y_0 + (x-1)y'(0) + \frac{(x-1)^2}{2!}y''(0) + \frac{(x-1)^3}{3!}y'''(0)$$

$$+ \frac{(x-1)^4}{4!}y''''(0) + \frac{(x-1)^5}{5!}y'''''(0) + \dots \qquad \dots(2)$$

Putting the above values in (2), we get

$$\Rightarrow \qquad y(x) = 2 + (x-1) + \frac{3}{2}(x-1)^2 + \frac{5}{6}(x-1)^3 + \frac{7}{12}(x-1)^4 + \frac{17}{60}(x-1)^5 + \dots$$

$$\dots(3)$$

$$\Rightarrow \qquad y(1.02) = 2 + (0.02) + \frac{3}{2}(0.02)^2 + \frac{5}{6}(0.02)^3 + \frac{7}{12}(0.02)^4 + \frac{17}{60}(0.02)^5 + \dots$$

$$= 2.02 + 0.0006 + 0.0000067 + 0.0000000093 + \dots$$

$$= 2.02061 \text{ approximately.}$$

EXAMPLE 8. *Solve by Taylor's series method, the equation* $\dfrac{dy}{dx} = \log(xy)$ *for* $y(1.1)$ *and* $y(1.2)$,

given $y(1) = 2$. (Hazaribagh-2009)

SOLUTION. The given equation $\dfrac{dy}{dx} = \log(xy)$ can be written as $y' = \log(xy)$

We start with $y'(1) = \log 2$ ...(1)

Now, differentiating eq^n (1) and putting $x = 1$ and $y = 2$, we get

$$y'' = \frac{1}{x} + \frac{1}{y}y' \Rightarrow y''(1) = 1 + \frac{1}{2}\log 2$$

$$y''' = -\frac{1}{x^2} + \frac{1}{y}y'' + y'\left(-\frac{1}{y^2}\right)y$$

$\Rightarrow \qquad y'''(1) = -1 + \frac{1}{2}\left(1 + \frac{1}{2}\log 2\right) - \frac{1}{4}(\log 2)^2$

Taylor's series about $x = 1$ is given by

$$y(x) = y(1) + (x-1)y'(1) + \frac{(x-1)^2}{2!}y''(1) + \frac{(x-1)^3}{3!}y'''(1) + \dots$$

$$= 2 + (x-1)\log 2 + \frac{1}{2}(x-1)^2\left(1 + \frac{1}{2}\log 2\right) + \frac{1}{6}(x-1)^3\left[-\frac{1}{2} + \frac{1}{4}\log 2 - \frac{1}{4}(\log 2)^2\right]$$

$\Rightarrow \ y(1.1) = 2 + (0.1)\log 2 + \dfrac{(0.1)^2}{2}\left(1 + \dfrac{1}{2}\log 2\right) + \dfrac{(0.1)^3}{6}\left[-\dfrac{1}{2} + \dfrac{1}{4}\log 2 - \dfrac{1}{4}(\log 2)^2\right]$

and $y(1.2) = 2 + (0.2)\log 2 + \dfrac{(0.2)^2}{2}\left(1 + \dfrac{1}{2}\log 2\right) + \dfrac{(0.2)^3}{6}\left[-\dfrac{1}{2} + \dfrac{1}{4}\log 2 - \dfrac{1}{4}(\log 2)^2\right]$

9.6 PICARD'S METHOD OF SUCCESSIVE APPROXIMATIONS

(Avadh-2004, Agra-2003, 05; Bhopal-2000, 01; Rohilkhand-2008)

Consider the differential equation

$$y' = \frac{dy}{dx} = f(x, y) \qquad \text{...(1)}$$

Integrating (1), we get

$$y = y_0 + \int_{x_0}^{x} f(x, y)dx \qquad \text{...(2)}$$

Equation (2) in which the unknown function appears under the integral sign is called as integral equation. Such an equation can be solved by the method of successive approximation in which the first approximation to y is obtained by putting y_0 for y on the right hand side of (2), and we write,

$$y^{(1)} = y_0 + \int_{x_0}^{x} f(x, y_0)dx \qquad \text{...(3)}$$

The integral on the right side can now be solved and the resulting $y^{(1)}$ is substituted for y in the integration of (2) to obtain the second approximation $y^{(2)}$.

$$y^{(2)} = y_0 + \int_{x_0}^{x} f(x, y^{(1)})dx$$

Continuing in this manner, we obtain $y^{(3)}, y^{(4)} \ldots y^{(n-1)}$ and $y^{(n)}$ where

$$y^{(n)} = y_0 + \int_{x_0}^{x} f(x, y^{(n-1)}) dx \text{ with } y^{(0)} = y_0.$$

Hence, this method gives a sequence of approximation $y^{(1)}, y^{(2)} \ldots y^{(n)}$ and it can be proved $f(x, y)$ is bounded in some regions containing the point (x_0, y_0) and if $f(x, y)$ satisfies the Lipschitz condition, namely

$$[f(x, y) - f(x, \overline{y})] \le k[y - \overline{y}], \text{ where } k \text{ being a constant,} \qquad \ldots (4)$$

then the sequence $y^{(1)}, y^{(2)} \ldots$ converges to the solution (2).

SOLVED EXAMPLES

EXAMPLE 1.   *Solve the equation $y' = x + y^2$ with y when $x = 0$.*

SOLUTION.   The first approximation is given by

$$y^{(1)} = 1 + \int_0^x f(x+1) dx = 1 + x + \frac{x^2}{2}$$

Then the second approximation is

$$\Rightarrow \qquad y^{(2)} = 1 + \int_0^x \left[x + \left(1 + x + \frac{x^2}{2} \right)^2 \right] dx$$

$$= 1 + x + \frac{3}{2} x^2 + \frac{2}{3} x^3 + \frac{1}{4} x^4 + \frac{1}{20} x^5.$$

It is obvious that the integrations might become more and more difficult as, we proceed to higher approximation.

EXAMPLE 2.   *Using Picard's method to obtain y for $x = 0.1$ to 0.5 for the differential equation*

$$\frac{dy}{dx} = 1 + xy \text{ with } y(0) = 1.$$

SOLUTION.   As per given, we have $f(x, y) = 1 + xy$, $x_0 = 0$ and $y_0 = 1$.

Therefore, the first approximation y_1 of y is

$$y_1 = y_0 + \int_{x_0}^{x} f(x, y_0) dx = 1 + \int_0^x (1 + x) dx = 1 + x + \frac{x^2}{2}$$

The second approximation y_2 is given by

$$y_2 = y_0 + \int_{x_0}^{x} f(x, y_1) dx = 1 + \int_0^x \left[1 + x \left(1 + x + \frac{x^2}{2} \right) \right] dx$$

$$= 1 + \int_0^x \left[1 + x + x^2 + \frac{x^3}{2} \right] dx = 1 + x + \frac{x^2}{2} + \frac{x^3}{3} + \frac{x^4}{8}$$

Similarly, the third approximation y_3 is given by

$$y_3 = y_0 + \int_{x_0}^{x} f(x, y_2) dx$$

$$= 1 + \int_0^x \left[1 + x \left(1 + x + \frac{x^2}{2} + \frac{x^3}{3} + \frac{x^4}{8} \right) \right] dx$$

$$= 1 + x + \frac{x^2}{2} + \frac{x^3}{3} + \frac{x^4}{8} + \frac{x^5}{15} + \frac{x^6}{18}.$$

EXAMPLE 3. *Integrate the differential equation* $\dfrac{dy}{dx} = x \sin \pi y$ *with* $y = \dfrac{1}{2}$ *at* $x = 0$ *by Picard's method of successive approximation.*

SOLUTION. We have $y = y_0 + \int_{x_0}^{x} f(x, y) dx$ ...(1)

Putting $y = \dfrac{1}{2}$ in R.H.S. of (1), we get the first approximation y_1 as

$$y_1 = \frac{1}{2} + \int_0^x x \sin\left(\pi - \frac{1}{2}\right) dx = \frac{1}{2} + \frac{x^2}{2}$$

Similarly,

$$y_2 = \frac{1}{2} + \int_0^x x \sin\left[\pi \frac{(1+x^2)}{2}\right] dx$$

$$= \frac{1}{2} + \int_0^x x \cos\frac{\pi x^2}{2} dx = \frac{1}{2} + \int_0^x x\left(1 - \frac{\pi^2 x^4}{8} + ...\right) dx$$

$$y_3 = \frac{1}{2} + \int_0^x x \sin\pi\left[\frac{1}{2} + \frac{x^2}{2} - \frac{\pi^2 x^6}{48} + ...\right] dx = \frac{1}{2} + \int_0^x x \cos\pi\left(\frac{x^2}{2} - \frac{\pi^2 x^6}{48}\right) dx$$

$$= \frac{1}{2} + \int_0^x x\left\{1 - \frac{1}{2}\left(\frac{\pi x^2}{2} + \frac{\pi^2 x^6}{48}\right)^2 + ...\right\} dx$$

$$= \frac{1}{2} + \frac{x^2}{2} - \frac{\pi^2 x^6}{48}.$$

Here, we observe that y_2 agree with y_3 upto and including term in x^6.

EXAMPLE 4. *Given* $\dfrac{dy}{dx} = \dfrac{y - x}{y + x}$ *with* $y = 1$, *when* $x = 0$. *Find approximately the value of* y *for*

$x = 0.1$ *by Picard's method.* (PTU–2002, Bhopal-2002, Avadh-2005)

SOLUTION. The first approximation is given by y_1 as

$$y_1 = 1 + \int_0^x \frac{1-x}{1+x} dx = 1 + 2\log(1+x) - x$$

$\Rightarrow$ $y(0.1) = 1 + 2\log(1 + 0.1) - 0.1 = 0.9828$

Now, the second approximation y_2 is given by

$$y_2 = y_0 + \int_{x_0}^x f(x, y_1) dx$$

$$= 1 + \int_0^x \frac{1 + 2\log(1+x) - x - x}{1 + 2\log(1-x) - x + x} dx$$

$$= 1 + \int_0^x \left[1 - \frac{2x}{1 + 2\log(1+x)}\right] dx$$

$$= 1 + x - 2\int_0^x \frac{x}{1 + 2\log(1+x)} dx$$

$$= 1 + x - 2\int_0^t \frac{e^{2t}}{1 + 2t} dt + 2\int_0^t \frac{e^t}{1 + 2t} dt$$

where $t = \log(1+x)$.

EXAMPLE 5. *Find the series expansion that gives y as a function of x in the neighbourhood of*
$x = 0$, *when* $\dfrac{dy}{dx} = x^2 + y^2$ *with* $y(0) = 0$.

                                                                       (JNTU–2009)

SOLUTION. As per given, we have
$$f(x, y) = x^2 + y^2, \ x_0 = 0 \text{ and } y_0 = 0.$$
Also, we have the nth approximation y_n of y as
$$y_n = y_0 + \int_{x_0}^{x} f(x, y_{n-1})dx$$

Therefore, the first approximation y_1 is given by $y_1 = 0 + \int_{0}^{x}(x^2 + 0)dx = \dfrac{x^3}{3}$

The second approximation y_2 is given by $y_2 = 0 + \int_{0}^{x}\left[x^2 + \left(\dfrac{x^3}{3}\right)^2\right]dx = \dfrac{x^3}{3} + \dfrac{x^7}{63}$

Similarly, the higher order approximations are given by

$$y_3 = 0 + \int_{0}^{x}\left[x^2 + \left(\dfrac{x^3}{3} + \dfrac{x^7}{63}\right)^2\right]dx$$

$$= \dfrac{x^3}{3} + \dfrac{x^7}{63} + \dfrac{2x^{11}}{2079} + \dfrac{x^{15}}{59535} + \dots$$

and $$y_4 = 0 + \int_{0}^{x}\left[x^2 + \left(\dfrac{x^3}{3} + \dfrac{x^7}{63} + \dfrac{2x^{11}}{2079} + \dfrac{x^{15}}{59535}\right)^2\right]dx$$

$$= \dfrac{1}{3}x^3 + \dfrac{1}{63}x^7 + \dfrac{2}{2079}x^{11} + \dfrac{13}{218295}x^{15} + \dots$$

EXAMPLE 6. *Using Picard's method find a solution of* $\dfrac{dy}{dx} = x^4 y + x$, $y(0) = 3$.

SOLUTION. Let $\dfrac{dy}{dx} = x^4 y + x = f(x, y)$ with $y_0 = 3, x_0 = 0$
The first approximation is
$$y^{(1)} = y_0 + \int_{x_0}^{x} f(x, y_0)dx = 3 + \int_{0}^{x}(x + x^4 y_0)dx$$

$$= 3 + \int_{0}^{3}[x + 3x^4]dx = 3 + \dfrac{x^2}{2} + \dfrac{3x^5}{3}$$
The second approximation is
$$y^{(2)} = 3 + \int_{0}^{x} f(x, y^{(1)})dx = 3 + \int_{0}^{x}[x + x^4 y^{(1)}]dx$$

$$= 3 + \int_{0}^{x}\left[x + x^4\left(3 + \dfrac{x^2}{2} + \dfrac{3x^5}{3}\right)\right]dx$$

$$= 3 + \int_{0}^{x}\left[x + 3x^4 + \dfrac{x^6}{2} + \dfrac{3x^9}{3}\right]dx = 3 + \dfrac{x^2}{2} + \dfrac{3x^5}{5} + \dfrac{x^7}{14} + \dfrac{3x^{10}}{50}$$
The third approximation is
$$y^{(3)} = 3 + \int_{0}^{x} f(x, y^{(2)})dx = 3 + \int_{0}^{x}[x + x^4 y^{(2)}]dx$$

$$= 3 + \int_0^x \left[x + x^4 \left(3 + \frac{x^2}{2} + \frac{3x^5}{5} + \frac{x^7}{14} + \frac{3x^{10}}{52} \right) \right] dx$$

$$= 3 + \int_0^x \left[x + 3x^4 + \frac{x^6}{2} + \frac{3x^9}{3} + \frac{x^{11}}{14} + \frac{3x^{14}}{50} \right] dx$$

$$= 3 + \frac{x^2}{2} + \frac{3x^5}{5} + \frac{x^7}{14} + \frac{3x^{10}}{50} + \frac{x^{12}}{168} + \frac{x^{15}}{250}$$

which is the required solution of the given equation upto third approximation.

EXAMPLE 7. *Use Picard's method to obtain y for y = 0.25, 0.5 and 1.0 correct to three decimal places, where $\dfrac{dy}{dx} = \dfrac{x^2}{y^2+1}$ with $y(0) = 0$.*

SOLUTION. Let $\dfrac{dy}{dx} = f(x, y) = \dfrac{x^2}{y^2+1}$, $y(0) = 0$, $x(0) = 0$

The first approximation is

$$y^{(1)} = y(0) + \int_0^x f(x, y^{(0)}) dx = 0 + \int_0^x f(x, 0) dx = 0 + \int_0^x \frac{x^2}{0+1} = \int_0^x x^2 dx = \frac{x^3}{3}$$

The second approximation is

$$y^{(2)} = \int_0^x (x, y^{(1)}) dx = \int_0^x \frac{x^2}{(y^{(1)})^2 + 1} dx = \int_0^x \frac{x^2}{\left(\dfrac{x^3}{3}\right)^2 + 1} dx$$

$$= \tan^{-1}\left(\frac{x^3}{3}\right) - \tan^{-1} 0 = \tan^{-1}\left(\frac{x^3}{3}\right)$$

$$y^{(2)} = \frac{x^3}{3} - \frac{x^9}{81} + \dots$$

Observe that $y^{(1)}$ and $y^{(2)}$ both have first terms as $\dfrac{x^3}{3}$, so we shall find the range of values of x such that the term $\dfrac{x^3}{3}$ alone, will give the result to three decimal places.

EXAMPLE 8. *Using Picard's method of successive approximation, find a solution upto the fifth approximation of the equation $\dfrac{dy}{dx} = y + x, y(0) = 1$ and check your answer by finding the exact particular solution.*

SOLUTION. We have $\dfrac{dy}{dx} = f(x, y) = y + x$ with $y(0) = 1$, $x_0 = 0$

The first approximation is

$$y^{(1)} = y(0) + \int_{x_0}^x f(x, y_0) dx$$

$$= 1 + \int_0^x (y_0 + x) dx = \int_0^x (1 + x) dx = 1 + x + \frac{x^2}{2}$$

The second approximation is

$$y^{(2)} = y_0 + \int_{x_0}^x f(x, y^{(1)}) dx = 1 + \int_0^x (y^{(1)} + x) dx$$

$$= 1 + \int_0^x \left(1 + x + \frac{x^2}{2} + x\right) dx = 1 + \int_0^x \left[1 + 2x + \frac{x^2}{2}\right] dx$$

$$= 1 + x + x^2 + \frac{x^3}{6}$$

The third approximation is

$$y^{(3)} = 1 + \int_0^x f(x, y^{(2)}] dx = 1 + \int_0^x [x + y^{(2)}] dx$$

$$= 1 + \int_0^x \left[x + 1 + x + x^2 + \frac{x^3}{6}\right] dx = 1 + \int_0^x \left[1 + 2x + x^2 + \frac{x^3}{6}\right] dx$$

$$= 1 + x + x^2 + \frac{x^3}{3} + \frac{x^4}{24}$$

The fourth approximation is given by

$$y^4 = 1 + \int_0^x f(x, y^{(3)})] dx = 1 + \int_0^x [y^{(3)} + x] dx$$

$$= 1 + \int_0^x \left(1 + 2x + x^2 + \frac{x^3}{3} + \frac{x^4}{24}\right) dx = 1 + x + x^2 + \frac{x^3}{3} + \frac{x^4}{12} + \frac{x^5}{120}$$

The fifth approximation is given by

$$y^{(5)} = 1 + \int_0^x f\left[x, y^{(4)}\right] dx = 1 + \int_0^x [y^{(4)} + x] dx$$

$$= 1 + \int_0^x \left(1 + 2x + x^2 + \frac{x^3}{3} + \frac{x^4}{12} + \frac{x^5}{120}\right) dx$$

$$= 1 + x + x^2 + \frac{x^3}{3} + \frac{x^4}{12} + \frac{x^5}{60} + \frac{x^6}{720} \qquad \qquad ...(2)$$

Exact solution: Equation (1) can be written as $\frac{dy}{dx} - y = x$ which is a linear differential equation in y.

Therefore,   I.F. $= e^{-\int dx} = e^{-x}$

∴   The solution is

$$ye^{-x} = \int xe^{-x} dx + c = xe^{-x} - \int (-e^{-x}) dx + c$$

$$\Rightarrow \qquad\qquad y = (e^x - x - 1)$$

At $x = 0$, $y = 1 \Rightarrow c = 2$

Therefore required particular solution is

$$y = 2e^x - x - 1$$

$$= 2\left(1 + x + x^2 + \frac{x^3}{3} + \frac{x^4}{12} + \frac{x^5}{120} + \frac{x^6}{720} + ...\right) - x - 1$$

$$= 1 + x + x^2 + \frac{x^3}{3} + \frac{x^4}{12} + \frac{x^5}{60} + \frac{x^6}{360} + ...$$

On comparing (2) and (3) we conclude that, (2) approximates to the exact particular solution (3) upto the term x^5.

EXAMPLE 9. *Use Picard's method to approximate y when $x = 0.2$ given that $y = 1$ when $x = 0$ and $\dfrac{dy}{dx} = x - y$.*

(Agra-2003; Avadh-2003)

SOLUTION. Given $f(x, y) = x - y, x_0 = 0, y_0 = 1$,

Now,

$$y^{(1)} = y_0 + \int_{x_0}^{x} f(x, y_0)dx = 0 + \int_0^x (x-1)dx = \frac{x^2}{2} - x + 1$$

$$y^{(2)} = y_0 + \int_{x_0}^{x} f(x, y^{(1)})dx = 1 + \int_0^x [x - y^{(1)}]dx$$

$$= 1 + \int_0^x \left[x - \frac{x^2}{2} + x - 1 \right]dx = -\frac{x^3}{6} + x^2 - x + 1$$

$$y^{(3)} = 1 + \int_0^x f(x, y^{(2)})dx = 1 + \int_0^x (x - y^{(2)})dx$$

$$= 1 + \int_0^x \left[x + \frac{x^3}{6} - x^2 + x - 1 \right]dx = \frac{x^4}{24} - \frac{x^3}{3} + x^2 - x + 1$$

$$y^{(4)} = 1 + \int_0^x (x, y^{(3)})dx = 1 + \int_0^x (x - y^{(3)})dx$$

$$= 1 + \int_0^x \left[x - \frac{x^4}{24} + \frac{x^3}{3} - x^2 + x - 1 \right]dx = -\frac{x^5}{120} + \frac{x^4}{12} - \frac{x^3}{3} + x^2 - x + 1$$

and $y^{(5)} = 1 + \int_0^x (x, y^{(4)})dx$

$$= 1 + \int_0^x \left[\frac{x^5}{120} - \frac{x^4}{12} + \frac{x^3}{3} - x^2 + 2x - 1 \right]dx$$

$$= \frac{x^6}{720} - \frac{x^5}{60} + \frac{x^4}{12} - \frac{x^3}{3} + x^2 - x + 1$$

Now, when $x = 0.2$ we have

$$y_0 = 1, y^{(1)} = 0.82, y^{(2)} = 0.83867,$$

$$y^{(3)} = 0.83740, y^{(4)} = 0.83746, y^{(5)} = 0.83746.$$

EXAMPLE 10. *Use Picard's method to approximate y when $x = 0.1, x = 0.2$ given that $y = 0$ when $x = 0$, $\dfrac{dy}{dx} = x + y$.*

SOLUTION. Given that $\dfrac{dx}{dy} = f(x, y) = x + y; x_0 = 0, y_0 = 1$

Now,

$$y^{(1)} = y_0 + \int_{x_0}^{x} f(x, y_0)dx = 1 + \int_0^x (1 + x)dx = 1 + x + \frac{x^2}{2}$$

$$y^{(2)} = y_0 + \int_{x_0}^{x} f(x, y^{(1)})dx = 1 + \int_0^x \left[x + \left(\frac{x^2}{2} + x + 1 \right) \right]dx$$

$$= \frac{x^3}{6} + x^2 + x + 1$$

$$y^{(3)} = y_0 + \int_0^x \left[x + \left(\frac{x^3}{6} + x^2 + x + 1 \right) \right]dx = \frac{x^4}{24} + \frac{x^3}{3} + x^2 = x + 1$$

When $x = 0.1$

$$y^{(1)} = 1 + 0.1 + \frac{(0.1)^2}{2} = 1.105,$$

$$y^{(2)} = 1 + 0.1 + (0.1)^2 + \frac{1}{6}(0.1)^3 = 1.11016$$

$$y^{(3)} = 1 + 0.1 + (0.1)^2 + \frac{1}{3}(0.1)^3 + \frac{1}{24}(0.1)^4$$

$$= 1.1103$$

When $x = 0.2$

$$y^{(3)} = \frac{1}{24}(0.2)^4 + \frac{1}{3}(0.2)^3 + (0.2)^2 + 0.2 + 1$$

$$= 1.2427.$$

EXERCISE 9.1

1. Use Euler's modified method for :
   (i) Given $y' = x^2 + y$, with $y_0 = 1$ determine $y(0.02)$, $y(0.04)$ and $y(0.06)$.
   (Rohtak–2005)
   (ii) Given that $y' = 2 + \sqrt{xy}$ with $y_{(1)} = 1$. Compute $y_{(2)}$ in steps of (0.2).
   (Anna–2004)

2. Using Taylor Series method for :
   (i) The differential equation $y' = x^2 + y^2$ with $y(1) = 0$ obtain $y(1.3)$.
   (ii) The equation $y' = 2xy + 1$ with $y_0 = 0$ and taking $h = 0.2$ at $x = 0.4$.

3. Use Picard's method to solve $y' = 1 + xy$ with $x_0 = 2, y_0 = 0$.

4. Solve $y' = 1 - y$, $y_0 = 0$ by Euler's modified method and obtain y at $x = 0.1, 0.2, 0.3$.
   (Anna-2005)

5. Given that $dy/dx = x + y^2$ and $y = 1$ at $x = 0$. Find an approximate value of y at $x = 0.5$ by Euler's modified method. (Bhopal-2002)

6. Solve $y' = y^2 + x, y(0)) = 1$ using Taylor's series method and compute $y(0.1)$ and $y(0.2)$.
   (JNTU-2006)

7. Evaluate $y(0.1)$ correct to four decimal places using Taylor's series method if
   $dy/dx = x^2 + y^2$, $y(0) = 1$ (VTU-2006)

8. Find an approximate value of y when $x = 0.1$ if $dy/dx = x - y^2$ and $y = 1$ at $x = 0$, using
   (i) Picard's method   (ii) Taylor's series
   (Madras-2006)

9. Using Picard's method; solve $dy/dx = -xy$ with $x_0 = 0, y_0 = 1$ upto third approximation.
   (Mumbai-2005)

10. Using Euler's method, find an approximate value of y corresponding to $x = 1$, given that $dy/dx = x + y$ and $y = 1$ when $x = 1$.
    (Anna-2005)

11. Using Euler's modified method, find an approximate value of y when $x = 0.3$, given that $dy/dx = x + y$ and $y = 1$ when $x = 0$.
    (Rohtak-2005)

12. Solve the following by Euler's modified method
    $$\frac{dy}{dx} = \log(x + y), y(0) = 2$$
    at $x = 1.2$ and 1.4 with $h = 0.2$ (Bhopal-2009

Answers

1. (i) $y(0.2) = 1.0202, y(0.04) = 1.0408, y(0.06) = 1.0619$
   (ii) $y_2 = 5.0516, y_{(0.02)} = 1.6402, y_{(0.4)} = 2.3623, y_{(0.6)} = 3.1678, y_{(0.8)} = 4.0633$

2. (i) $y(1.3) = 0.4158$, (ii) $y(0.2) = 0.21$ app., $y(0.4) = 0.451$ 3. $y^{(3)} = \frac{x^5}{15} - \frac{x^4}{4} + \frac{x^3}{3} - \frac{x^2}{2} + x - \frac{22}{15!}$

4. $y_n = c + 2^n$ 5. $y_n = c(2)^n - cn + 1$ 6. $1.1164, 1.2725$ 7. 1.00035 8. (i) and (ii) $= 0.9138$

9. $y = 1 - \frac{x^2}{2} + \frac{x^4}{8} - \frac{x^6}{48}$ 10. 3.18 11. 1.4004 12. $y(1.2) = 2.5351, y(1.4) = 2.6531$

9.7 RUNGE-KUTTA METHOD

(Agra-2003, 05)

The Taylor's series method of solving differential equation's numerically is restricted by the labour involved in finding the higher order derivatives. However there is a class of methods known as Runge-Kutta methods which do not require the calculations of higher order derivatives and give greater accuracy. The Runge-Kutta formulae possess the advantage of requiring only the functional values at some selected points. These methods agree with Taylor's series solution upto the term h^r where r differs from method to method and is called the order of that method.

9.7.1 FIRST ORDER RUNGE-KUTTA METHOD

By Euler's method, we have

$$y_1 = y_0 + hf(x_0, y_0) \qquad\qquad [\because y' = f(x, y)]...(1)$$

Expanding by Taylor's series, we get

$$y_1 = y(x_0 + h) = y_0 + hy_0' + \frac{h^2}{2}.y_0'' + ...$$

It follows that the Euler's method agrees with the Taylor's series solution upto the term in h. Hence, the Euler's method is the Runge-Kutta method is of the first order.

9.7.2 SECOND ORDER RUNGE-KUTTA METHOD

We know that modified Euler's method is given by

$$y_1 = y_0 + \frac{h}{2}[f(x_0, y_0) + f(x_0 + h, y_0)]. \qquad\qquad ...(2)$$

Putting $y_1 = y_0 + hf(x_0, y_0)$ on the right hand side of (2), we obtain

$$y_1 = y_0 + \frac{h}{2}[f_0 + f(x_0 + h, y_0 + hf_0)]. \qquad\qquad ...(3)$$

Expanding Left hand side by Taylor's series, we get $\qquad$ [where $f_0 = f(x_0, y_0)$]

$$y_1 = y(x_0 + h) = y_0 + hy_0' + \frac{h^2}{2!}y_0'' + \frac{h^3}{3!}y_0''' + ... \qquad\qquad ...(4)$$

Expanding $f(x_0 + h, x_0 + hf_0)$ by Taylor's series for a function of two variables, then (3) gives

$$y_1 = y_0 + \frac{h}{2}\left[f_0 + \left\{f(x_0, y_0) + h\left(\frac{\partial f}{\partial x}\right)_0 + hf_0\left(\frac{\partial f}{\partial y}\right)_0 + 0(h^2)\right\}\right]$$

$$= y_0 + \frac{1}{2}\left[hf_0 + hf_0 + h^2\left\{\left(\frac{\partial f}{\partial x}\right)_0 + \left(\frac{\partial f}{\partial y}\right)_0\right\} + 0(h^3)\right]$$

$$= y_0 + hf_0 + \frac{h^2}{2}f_0' + 0(h^3) \qquad\qquad \left[\because \frac{\partial f(x,y)}{\partial x} = \frac{\partial f}{\partial x} = f\frac{\partial f}{\partial y}\right]$$

$$= y_0 + hy_0' + \frac{h^2}{2!}y_0'' + 0(h^3) \qquad\qquad ...(5)$$

From equation (4) and (5), it follows that the modified Euler's method agrees with the Taylor's series solution upto the term in h^2.

Hence, the modifies Euler's method is the Runge-Kutta method of the second order.
Therefore the second order Runge-Kutta formula is given by

$$y_1 = y_0 + \frac{1}{2}(k_1 + k_2) \text{ [where } k_1 = hf(x_0, y_0), k_2 = (x_0 + h, y_0 + k_1).$$

9.7.3 THIRD ORDER RUNGE-KUTTA METHOD

In a similar manner it can be seen that Runge method be agrees with the Taylor's series solution upto the term in h^3.

$$y_1 = y_0 + \frac{h}{6}\left[\begin{array}{c} f(x_0, y_0) + 4f\left(x_0 + \frac{h}{2}, y_0 + \frac{h}{2}f(x_0, y_0)\right) \\ + f(x_0 + h, y_0) + hf(x_0 + h, y_0) + hf(x_0, y_0)\end{array}\right] \quad \ldots(6)$$

As such, Runge-Kutta method is the Runge-Kutta of the third order.
The third order Runge-Kutta formula is

$$y_1 = y_0 + \frac{1}{6}(k_1 + 4k_2 + k_3)$$

where

$$k_1 = hf(x_0, y_0)$$

$$k_2 = hf\left(x_0 + \frac{h}{2}, y_0 + \frac{k_1}{2}\right)$$

and

$$k_3 = hf(x_0 + h, y_0 + k')$$

9.7.4 RUNGE-KUTTA METHOD OF ORDER 4

This method coincides with the Taylor's series solution upto terms of h^4. We know that the Taylor's series solution can be expressed in terms of $f(x, y)$ and its partial derivatives of various orders. But in this method we use a technique which avoids the calculation of derivatives.
We take

$$k_1 = hf(x, y)$$

$$\left.\begin{array}{l} k_2 = hf(x + mh, y + mk_1) \\ k_3 = hf(x + nh, y + nk_2) \\ k_4 = hf(x + hp, y + pk_3) \end{array}\right\} \quad \ldots(7)$$

Now

$$y(x + h) = y(x) + ak_1 + bk_2 + ck_3 + dk_4 \quad \ldots(8)$$

We know that by Taylor's series

$$y(x + h) = y_x + hy' + \frac{h^2}{2!}y'' + \frac{h^3}{3!}y''' + \frac{h^4}{4!}y^{iv} + 0(h^5) \quad \ldots(9)$$

First of all we shall put it in some convenient form

$$f_1 = f_x + ff_y, f_2 = f_{xx} + 2ff_{xy} + f^2 f_{yy}$$

$$f_3 = f_{xxx} + 3ff_{xxy} + 3f^3 f_{xyy} + f^3 f_{yyy}$$

Now differential equation $y' = \frac{dy}{dx} = f(x, y)$, we obtain

$$y'' = f(x) + f_y y' = f_x + ff_y = F_1$$

$$y''' = f_{xx} + 2ff_{xy} + f^2 f_{yy} + f_x f_y + ff_y^2$$

$$= (f_{xx} + 2ff_{xy} + f^2 f_{yy}) + f_y(f_x + ff_y) = F_2 + f_y F_1$$

Similarly
$$y^{iv} = F_3 + f_y F_2 + 3F_1(f_{xy} + ff_{yy}) + f_y^2 F_1$$

Putting these values in (7), we get

$$y(x+h) = y_x + hf + \frac{h^2}{2}F_1 + \frac{1}{6}h^3(F_2 + f_y F_1) + \frac{1}{24}h^4[F_3 + f_y F_2 + 3(f_{xy} + ff_{yy})F_1 + f_y^2 F_1] + \dots$$

...(10)

Using the above notation and the Taylor's theorems, we get, [where $f = f(x, y)$]

$$k_1 = hf$$

$$k_2 = h\left[f + mhF_1 + \frac{1}{2}m^2 h^2 F_2 + \frac{1}{6}m^3 h^3 F_3 + \dots \right]$$

$$k_3 = h\left[\begin{array}{l} f + mhF_1 + \frac{1}{2}h^2(n^2 F_2 + 2mn + f_y F_1) \\ + \frac{1}{6}h^3\{n^3 F_3 + 3m^2 n f_y F_2 + 6mn^2(f_{xy} + ff_{yy})F_1\} = \dots \end{array} \right]$$

and
$$k_4 = h\left[\begin{array}{l} f + phF_1 + \frac{1}{2}h^2(p^2 F_2 + 2npf_y F_1) + \frac{1}{6}h^3\{p^3 F_3 + 3n^2 pf_y F_2 \\ + 6np^2(f_{xy} + ff_{yy})F_1 + 6mnpf_y^2 F_1\} + \dots \end{array} \right]$$

Substituting these values in (8), we get

$$y(x+h) = y_{(x)} + (a+b+c+d)hf + (bm+cn+dp)h^2 F_1 + \frac{1}{2}(bm^2 + cn^2 + dp^2).$$

$$h^2 f_y F_1 + \frac{1}{2}(cm^2 n + dn^2 p)h^4 f_y F_2 + (cmn^2 + dnp^2)h^4(f_{xy} + ff_{yy})F_1 + dmnph^4 f_y^2 F_1 + 0(h^5)$$

...(11)

Equating this with (10), we get

$$a+b+c+d = 1, \qquad\qquad cmn + dnp = \frac{1}{6}$$

$$bm + cn + dp = \frac{1}{2}, \qquad\qquad cmn^2 + dnp^2 = \frac{1}{8}$$

$$bm^2 + cn^2 + dp^2 = \frac{1}{3}, \qquad\qquad cm^2 n + dn^2 p = \frac{1}{12}$$

$$bm^3 + cn^3 + dp^3 = \frac{1}{4}, \qquad\qquad dmnp = \frac{1}{24}.$$

These are eight equations in seven unknowns; A classical solution to these eight equations is

$$m = n = \frac{1}{2}, p = 1, a = d = \frac{1}{6}, b = c = \frac{1}{3}$$

Putting these values in (7) and (8), the Runge-Kutta formula reduces to

$$\left. \begin{array}{l} k_1 = hf(x, y) \\ k_2 = hf\left(x + \frac{h}{2}, y + \frac{1}{2}k_1\right) \\ k_3 = hf\left(x + \frac{h}{2}, y + \frac{1}{2}k_2\right) \\ k_4 = hf(x + h, y + k_3) \end{array} \right\}$$

...(12)

and
$$y(x+h) = y_{(x)} + \frac{1}{6}(k_1 + 2k_2 + 2k_3 + k_4)$$

From this formula, we have

where
$$y_1 = y(x_0 + h) = y_0 + \frac{1}{6}[k_1 + 2(k_2 + k_3) + k_4]$$

$$
\left.
\begin{aligned}
k_1 &= hf(x_0, y_0) \\
k_2 &= hf\left(x_0 + \frac{h}{2}, y_0 + \frac{k_1}{2}\right) \\
k_3 &= hf\left(x_0 + \frac{h}{2}, y_0 + \frac{k_2}{2}\right) \\
k_4 &= hf(x_0 + h, y_0 + k_3)
\end{aligned}
\right\} \qquad ...(13)
$$

▼ REMARK

▸ $0(h^2)$ means 'terms containing second and higher power of h and is read as order of h^2.'

SOLVED EXAMPLES

EXAMPLE 1.   *Apply Runge-Kutta method to find the solution of the differential equation*
$$y' = 3x + \frac{1}{2}y \text{ with } y_0 = 1, \text{ at } x = 0.1.$$
(VTU–2004)

SOLUTION.   We have $h = 0.1$

$$k_1 = hf(x_0, y_0) \qquad \text{(Given } y_0 = 1, \text{ when } x_0 = 0\text{)}$$

$$= 0.1\left(3x_0 + \frac{1}{2}y_0\right) = 0.1\left(3 \times 0 + \frac{1}{2} \times 1\right) = 0.05$$

Similarly $k_2 = hf\left(x_0 + \frac{h}{2}, y_0 + \frac{k_1}{2}\right) = 0.1f(0.05; 1.025) = 0.06625$

Now $k_3 = hf\left(x_0 + \frac{h}{2}, y_0 + \frac{k_2}{2}\right) = 0.1f(0.05, 1.033125) = 0.06665625$

and $k_4 = hf(x_0 + h, y_0 + k_3) = 0.1f(0.1, 1.0665625) = 0.0833328125$
$$k_4 = 0.0833328125$$

Obviously
$$k = \frac{1}{6}[k_1 + 2(k_2 + k_3) + k_4] = 0.066652421875$$

Now, $y_{(0.1)} = y_0 + k \Rightarrow y_{(0.1)} = 1 + 0.66652421875$

Hence, $y(0.1) = 1.066652421875$.

EXAMPLE 2.   *The unique solution of the problem $y' = -xy$ with $y_0 = 1$ is $y = e^{-x^2/2}$. Find approximately the value of $y_{(0.2)}$ using one application of Runge-Kutta method of order four.*

SOLUTION.   Let $h = 0.2$, we have $y_0 = 1$ when $x_0 = 0$
$$k_1 = hf(x_0, y_0) = 0.2[0(1)] = 0$$

$$k_2 = hf\left(x_0 + \frac{h}{2}, y_0 + \frac{k_1}{2}\right) = 0.2\left[-\left(0 + \frac{0.2}{2}\right).(1+0)\right] = -0.02$$

$$k_3 = hf\left(x_0 + \frac{h}{2}, y_0 + \frac{k_2}{2}\right) = 0.2\left[-\left(0 + \frac{0.2}{2}\right)\left(1 - \frac{0.02}{2}\right)\right] = -0.0198$$

$$k = \frac{1}{6}[k_1 + 2(k_2 + k_3) + k_4] = -0.198013$$

Hence, $y_{(0.02)} = 1.0000 - 0.198013 = 0.9801986 = 0.9802$

The exact value of $y_{(0.02)}$ is 0.9802.

EXAMPLE 3. *Solve the equation $y' = x + y$ with $y_0 = 1$ by Runge-Kutta method from $x = 0$ to $x = 0.4$ with $h = 0.1$.*   (VTU–2009, PTU–2007, SVTU–2007)

SOLUTION. Here $f(x, y) = x + y$, $h = 0.1$, given $y_0 = 1$, $x_0 = 0$.

We have

$$k_1 = hf(x_0, y_0) = 0.1[0 + 1] = 0.1$$

$$k_2 = hf\left(x_0 + \frac{h}{2}, y_0 + \frac{k_1}{2}\right) = 0.1[0.05 + 1.05] = 0.11$$

$$k_3 = hf\left(x_0 + \frac{h}{2}, y_0 + \frac{k_2}{2}\right) = 0.1[0.05 + 1.055] = 0.1105$$

and $k_4 = hf(x_0 + h, y_0 + k_3) = 0.1[0.1 + 1.1105] = 0.12105$

Hence, $y_1 = y_{(x=0.1)} = 1 + \dfrac{1}{6}[0.1 + 0.22 + 0.2210 + 0.12105] = 1.11034$

Similarly for finding $y_2 = y(x = 0.2)$, we get

$$k_1 = hf(x_1, y_1) = 0.[(0.1) + 1.11034] = 0.121034$$

$$k_2 = hf\left(x_1 + \frac{h}{2}, y_1 + \frac{k_1}{2}\right) = 0.1[0.15 + 1.11034 + 0.06604] = 0.13208$$

$$k_3 = hf\left(x_1 + \frac{h}{2}, y_1 + \frac{k_2}{2}\right) = 0.1[0.15 + 1.11034 + 0.06604] = 0.13208$$

$$k_4 = hf(x_1 + h, y_1 + k_3)$$
$$= 0.1[0.20 + 1.11034 + 0.13263] = 0.14263$$

Hence

$$y_2 = y(x = 0.2) = y_1 + \frac{1}{6}(k_1 + 2k_2 + 2k_3 + k_4)$$

$$= 1.11034 + \frac{1}{6}[0.121034 + 2(0.13208 + 0.13263) + 0.14429]$$

$$= 1.2428$$

Similarly, for finding $y_3 = y(x = 0.3)$, we have

$$k_1 = hf(x_2, y_2) = 0.1[0.2 + 1.2428] = 0.14428$$

$$k_2 = hf\left(x_2 + \frac{h}{2}, y_2 + \frac{h}{2}\right)$$
$$= (0.1)(0.25 + 1.2428 + 0.07214) = 0.15649$$

$$k_3 = hf\left(x_2 + \frac{h}{2}, y_2 + \frac{h}{2}\right)$$
$$= (0.1)(0.25 + 1.2428 + 0.07824) = 0.15710$$

$$k_4 = hf(x_2 + h, y_2 + k_3) = (0.1)(0.3 + 1.2428 + 0.15710) = 0.16999$$

Hence, similarly for $y_4 = y(x = 0.4)$

$$k_1 = (0.1)(0.3 + 1.3997) = 0.16997 \Rightarrow k_1 = 0.16997$$
$$k_2 = (0.1)[0.35 + 1.3997 + 0.8949] = 0.18347$$
$$k_3 = (0.1)[0.3 + 1.3997 + 0.9170] = 0.18414$$
$$k_4 = (0.1)[0.4 + 1.3997 + 0.18414) = 0.19838$$

Hence, $\quad y_4 = 1.3997 + \dfrac{1}{6}[0.16997 + 2(0.18347 + 0.18414) + 0.19838]$

$\Rightarrow \qquad y_4 = 1.5836$

EXAMPLE 4. *Apply Runge-Kutta method to solve* $\dfrac{dy}{dx} = xy^{1/3}, y(1) = 1$ *to obtain* $y(1.1)$.

SOLUTION. As per given, we have $x_0 = 1, y_0 = 1$ and $h = 0.1$. Then, we can find

$$k_1 = hf(x_0, y_0) = 0.1\ (1)\ (1)^{1/3} = 0.1$$

$$k_2 = hf\left(x_0 + \frac{h}{2},\ y_0 + \frac{k_1}{2} \right)$$

$$= (0.1) + \left(1 + \frac{0.1}{2} \right)\left(1 + \frac{0.1}{2} \right)^{1/3} = 0.10672$$

$$k_3 = hf\left(x_0 + \frac{h}{2},\ y_0 + \frac{k_2}{2} \right)$$

$$= 0.1\left(1 + \frac{0.1}{2} \right)\left(1 + \frac{0.10672}{2} \right)^{1/3} = 0.10684$$

$$k_4 = hf(x_0 + h,\ y_0 + k_3)$$

$$= 0.1\ (1 + 0.1)\ (1 + 0.10684)^{1/3} = 0.11378$$

$$k = \frac{1}{6}hf(x_0 + h, y_0 + k_3) \text{ which gives}$$

$$k = \frac{1}{6}(0.1 + 2 \times 0.10672 + 2 \times 1.0684 + 0.11378)$$

Hence $\qquad y(1.1) = 1 + 0.10682 = 1.10682.$

EXAMPLE 5. *Solve* $\dfrac{dy}{dx} = -2xy^2$ *with* $y(0) = 1$ *and* $h = 0.2$ *on the interval* $[0, 1]$, *using Runge-Kutta fourth order method.*

SOLUTION. As per given, we have

$$x_0 = 0, y_0 = 1, h = 0.2$$

Now, $\qquad k_1 = hf(x_0, y_0) = -2(0.2)\ (0)\ (1)^2 = 0$

$$k_2 = hf\left(x_0 + \frac{h}{2},\ y_0 + \frac{k_1}{2} \right) = -2(0.2)\left(\frac{0.2}{2} \right)(1)^2 = -0.4$$

$$k_3 = hf\left(x_0 + \frac{h}{2},\ y_0 + \frac{k_2}{2} \right) = -2(0.2)\left(\frac{0.2}{2} \right)(0.98)^2 = -0.38416$$

$$k_4 = hf\ (x_0 + h, y_0 + k_3)$$

$$= -2\ (0.2)\ (0.2)\ (0.961584)^2 = -0.0739715$$

Hence $\qquad y(0.2) = y_0 + \dfrac{1}{6}(k_1 + 2k_2 + 2k_3 + k_4)$

$$= 1 + \dfrac{1}{6}[0 - 0.08, 0.076832 - 0.0739715] = 0.9615328.$$

Further, we have $x_1 = 0.2$, $y_1 = 0.9615328$, $h = 0.2$, we get

$$k_1 = hf(x_1, y_1) = -2(0.2)(0.2)(0.9615328)^2 = -0.0739636$$

$$k_2 = hf\left(x_1 + \dfrac{h}{2}, y_1 + \dfrac{k_1}{2}\right) = -2(0.2)(0.3)(0.9102451)^2 = 0.1025754$$

$$k_3 = hf\left(x_1 + \dfrac{h}{2}, y_1 + \dfrac{k_2}{2}\right) = -2(0.2)(0.3)(0.9102451)^2 = -0.0994255$$

and $\qquad k_4 = hf(x_1 + h, y_1 + k_3) = -2(0.2)(0.4)(0.8621073)^2 = -0.1189166$

Thus $y(0.4) = y_1 + \dfrac{1}{6}(k_1 + 2k_2 + 2k_3 + k_4)$

$$= 0.9615328 + \dfrac{1}{6}(-0.0739636 - 0.2051508 - 0.1988510$$
$$- 0.1189166)$$

$$= 0.8620525$$

Similarly, we can obtained

$$y(0.6) = 0.7352784$$
$$y(0.8) = 0.6097519$$

and $\quad y(1.0) = 0.500073$.

EXAMPLE 6. *Given* $\dfrac{dy}{dx} = y - x$ *with* $y(0) = 2$, *find* $y(0.1)$ *and* $y(0.2)$ *correct to 4 decimal places.*

SOLUTION. We have $\qquad x_0 = 0$, $y_0 = 2$, $h = 0.1$

Then, we get $k_1 = hf(x_0, y_0) = 0.1(2 - 0) = 0.2$

$$k_2 = hf\left(x_0 + \dfrac{h}{2}, y_0 + \dfrac{k_1}{2}\right) = 0.1\left[2 + \dfrac{0.205}{2} - \left(0 + \dfrac{0.1}{2}\right)\right] = 0.20525$$

$$k_4 = hf(x_0 + h, y_0 + k_3) = 0.1[2 + 0.20525 - (0 + 0.1)] = 0.210525$$

Therefore, $\quad y = y_0 + \dfrac{1}{6}[k_1 + 2k_2 + 2k_3 + k_4] = 2 + 0.2051708 = 2.2051708$

$\Rightarrow \qquad y(0.1) = 2.2052$ (correct to four decimal places)

For $y(0.2)$, we have $x_0 = 0.1$, $y_0 = 2.2052$, we get

$$k_1 = hf(x_0, y_0) = 0.1[2.2052 - 0.1] = 0.21052$$

$$k_2 = hf\left(x_0 + \dfrac{h}{2}, y_0 + \dfrac{k_1}{2}\right)$$

$$= 0.1\left[2.2052 + \dfrac{0.21052}{2} - \left(0.1 + \dfrac{0.1}{2}\right)\right] = 0.216046$$

$$k_3 = hf\left(x_0 + \dfrac{h}{2}, y_0 + \dfrac{k_2}{2}\right)$$

$$= 0.1\left[2.2052 + \dfrac{0.216046}{2} - \left(0.1 + \dfrac{0.1}{2}\right)\right] = 0.2163223$$

$$k_4 = hf(x_0 + h, y_0 + k_3)$$

$$= 0.1 [2.2052 + 0.2163223 - (0.1 + 0.1)] = 0.22215223$$

$$y(0.2) = 2.2052 + \frac{1}{6}(k_1 + 2k_2 + 2k_3 + k_4)$$

$$= 2.2052 + 0.2162348 = 2.4214.$$

EXAMPLE 7. *Applying Runga-Kutta method to find an approximate value of y for x = 0.2 in step of 0.1 if* $\frac{dy}{dx} = x + y^2$. *Given that y = 1 when x = 0.*

(VTU–2009, Osmania–2007, Madras–2000; Agra-2003)

SOLUTION. We have $f(x, y) = x + y^2, x_0 = 0, y(0) = 1, h = 0.1$

We have to find $y(0.1)$ and $y(0.2)$.

By Runge-Kutta method, we have

$$k_1 = hf(x_0, y_0) = (0.1)f(0, 1) = (0.1)[0 + (1)^2] = 0.1$$

$$k_2 = hf\left(x_0 + \frac{h}{2}, y_0 + \frac{k_1}{2}\right) = (0.1)f\left(0 + \frac{0.1}{2}, 1 + \frac{0.1}{2}\right)$$

$$= (0.1)(0.05, 1.05) = (0.1)[0.05 + (1.05)^2] = 0.11525$$

$$k_3 = hf\left(x_0 + \frac{h}{2}, y_0 + \frac{k_2}{2}\right) = (0.1)f(0.05, 1.057625)$$

$$= (0.1)[0.05 + (1.05 + 625)^2] = 0.11686$$

and

$$k_4 = hf(x_0 + h, y_0 + k_3) = (0.1)f(0.1, 1.11686)$$

$$= (0.1)[0.1 + (1.11686)^2] = 0.13474$$

Therefore,

$$y(0.1) = y_0 + \frac{1}{6}(k_1 + 2k_2 + 2k_3 + k_4)$$

$$= 1 + \frac{1}{6}(0.1 + 2 \times 0.11525 + 2 \times 0.11686 + 0.13474)$$

$$= 1 + \frac{1}{6}(0.69896) = 1.11649$$

Further, to find $y(0.2)$, take $x_1 = 0.1, y_1 = 1.11649, h = 0.1$

Then, we have

$$k_1 = hf(x_1, y_1) = (0.1)f(0.1, 1.11649)$$

$$= (0.1)[0.1 + (1.11649)^2] = 0.13465$$

$$k_2 = hf\left(x_1 + \frac{h}{2}, y_1 + \frac{k_1}{2}\right) = (0.1)f(0.15, 1.118382)$$

$$= (0.1)[0.15 + (1.18382)^2] = 0.15514$$

$$k_3 = hf\left(x_1 + \frac{h}{2}, y_1 + \frac{k_2}{2}\right) = (0.1)f(0.15, 1.19406)$$

$$= (0.1)[0.15 + (1.19406)^2] = 0.15758$$

$$k_4 = hf(x_1 + h, y_1 + k_3) = (0.1)f(0.2, 1.27407)$$

$$= (0.1)[0.2 + (1.27407)^2] = 0.18233$$

Hence, $y(0.2) = y_1 + \dfrac{1}{6}[k_1 + 2(k_2 + k_3) + k_4]$

$$= 1.11649 + \dfrac{1}{6}[0.13465 + 2(0.15514 + 0.15758) + 0.18233]$$

$$= 1.11649 + 0.156707 = 1.27356.$$

EXAMPLE 8. *Using Runge-Kutta method of fourth order, solve* $\dfrac{dy}{dx} = \dfrac{y^2 - x^2}{y^2 + x^2}$ *with* $y(0) = 1$ *at* $x =$

0.2, 0.4.   [UPTU(MCA)–2004, UPTU(B.Tech.)–2010, JNTU–2009, VTU–2008]

SOLUTION.   We have

$$f(x) = \dfrac{y^2 - x^2}{y^2 + x^2}, \ x_0 = 0, y(0) = 1, h = 0.2$$

Firstly, we shall find $y(0.2)$.

Now, $k_1 = hf(x_0, y_0) = 0.2f(0, 1) = 0.2 \times 1 = 0.2$

$$k_2 = hf\left(x_0 + \dfrac{h}{2}, y_0 + \dfrac{k_1}{2}\right) = 0.2f(0.1, 1.1)$$

$$= 0.2 \times \left[\dfrac{(1.1)^2 - (0.1)^2}{(1.1)^2 + (0.1)^2}\right] = 0.2\left(\dfrac{1.2}{1.22}\right) = 0.19672$$

$$k_3 = hf\left(x_0 + \dfrac{h}{2}, y_0 + \dfrac{k_2}{2}\right) = 0.2f(0.1, 1.09836)$$

$$= 0.2\left[\dfrac{(1.09836)^2 - (0.1)^2}{(1.09836)^2 + (0.1)^2}\right] = 0.2\left(\dfrac{1.19639}{1.21639}\right) = 0.19671$$

$$k_4 = hf(x_0 + h, y_0 + k_3) = 0.2f(0.2, 1.19671)$$

$$= 0.2\left[\dfrac{(1.19671)^2 - (0.2)^2}{(1.19671)^2 + (0.2)^2}\right] = 0.2\left(\dfrac{1.39211}{1.47211}\right) = 0.18913$$

Therefore,
$$y(0.2) = y_0 + \dfrac{1}{6}(k_1 + 2k_2 + 2k_3 + k_4)$$

$$= 1 + \dfrac{1}{6}[0.2 + 2(0.19672 + 0.19671) + 0.18913]$$

$$= 1 + \dfrac{1}{6}(1.17599) = 1.1960$$

Further, to find $y(0.4)$, take $y_1 = 1.1960$, $x_1 = 0.2$ and $h = 0.2$.

Then $k_1 = hf(x_0, y_0) = 0.2f(0.2, 1.1960)$

$$= (0.2)\left[\dfrac{(1.960)^2 - (0.2)^2}{(1.960)^2 + (0.2)^2}\right] = (0.2)(0.94560) = 0.18912$$

$$k_2 = hf\left(x_1 + \frac{h}{2}, y_1 + \frac{k_1}{2}\right) = (0.2)f(0.3, 1.29056)$$

$$= (0.2)\left[\frac{(1.29056)^2 - (0.3)^2}{(1.29056)^2 + (0.3)^2}\right] = (0.2)\left(\frac{1.57555}{1.75555}\right) = 0.17949$$

$$k_3 = hf\left(x_1 + \frac{h}{2}, y_1 + \frac{k_2}{2}\right) = (0.2)f(0.3, 1.28575)$$

$$= (0.2)\left[\frac{(1.28575)^2 - (0.3)^2}{(1.28575)^2 + (0.3)^2}\right] = (0.2)\left(\frac{1.56315}{1.74315}\right) = 0.17935$$

and $\qquad k_4 = hf(x_1 + h, y_1 + k_3) = 0.2f(0.4, 1.37535)$

$$= (0.2)\left[\frac{(1.37535)^2 - (0.4)^2}{(1.37535)^2 + (0.4)^2}\right] = (0.2)\left(\frac{1.73159}{2.05159}\right) = 0.16880$$

Hence, $\qquad y_2 = y(0.4) = y_1 + \frac{1}{6}[k_1 + 2k_2 + 2k_3 + k_4]$

$$= 1.1960 + \frac{1}{6}[0.1892 + 2(0.17949 + 0.17935) + 0.16880]$$

$$= 1.1960 + \frac{1}{6}(1.0756) = 1.37527 \cdot$$

EXAMPLE 9. *Using Runge-Kutta method of fourth order, find $y(0.1)$ form $\dfrac{dy}{dx} = \dfrac{x^2 + y^2}{10}, y(0) = 1$,*

take $h = 0.1$. (Rohtak–2004, 05; Agra-2006; Meerut-2003)

SOLUTION. We have

$$f(x, y) = \frac{x^2 + y^2}{10} \text{ with } x_0 = 0, y(0) = 1, h = 0.1$$

Then,

$$k_1 = hf(x_0, y_0) = (0.1)f(0,1) = (0.1)\left[\frac{0^2 + 1^2}{10}\right] = 0.01$$

$$k_2 = hf\left(x_0 + \frac{h}{2}, y_0 + \frac{k_1}{2}\right) = (0.1)f(0.05, 1.005)$$

$$= (0.1)\left[\frac{(0.05)^2 + (1.005)^2}{10}\right] = 0.01013$$

$$k_3 = hf\left(x_0 + \frac{h}{2}, y_0 + \frac{k_2}{2}\right) = (0.1)f(0.05, 1.0051)$$

$$= (0.1)\left[\frac{(0.05)^2 + (1.0051)^2}{10}\right] = 0.01013$$

and $\qquad k_4 = hf(x_0 + h, y_0 + k_3) = (0.1)f(0.1, 1.01013)$

$$= (0.1)\left[\frac{(0.1)^2 + (1.01013)^2}{10}\right] = 0.01030$$

Hence, $y_1 = y(0.1) = y_0 + \dfrac{1}{6}[k_1 + 2k_2 + 2k_3 + k_4]$

$$= 1 + \frac{1}{6}[0.01 + 2(0.01013 + 0.01013) + 0.01030] = 1.01014.$$

9.8 SIMULTANEOUS DIFFERENTIAL EQUATIONS

9.8.1 SOLUTION OF PICARD'S METHOD

In this method, the first approximations are

$$y_1 = y_0 + \int f(x, y_0, z_0)dx, z_1 = z_0 + \int g(x, y_0, z_0)dx$$

and the second approximations are

$$y_2 = y_0 + \int f(x, y_1, z_1)dx, z_2 = z_0 + \int g(x, y_1, z_1)dx$$

In general, the $(n + 1)^{\text{th}}$ approximations are

$$y_{n+1} = y_0 + \int f(x, y_n, z_n)dx, z_{n+1} = z_0 + \int g(x, y_n, z_n)dx \text{ and so on.}$$

9.8.2 TAYLOR'S SERIES METHOD

If h be the step size, then

$$y_1 = y(x_0 + h), z_1 = z(x_0 + h)$$

Now this method gives

$$y_1 = y_0 + hy_0' + \frac{h^2}{2!}y_0'' + \frac{h^3}{3!}y_0''' + \dots \qquad \dots(4)$$

and

$$z_1 = z_0 + hz_0' + \frac{h^2}{2!}z_0'' + \frac{h^3}{3!}z_0''' + \dots \qquad \dots(5)$$

Here y_0', y_0'' and y_0''' etc. and z_0', z_0'' and z_0''' etc. are obtained by differentiating (1) and (2) successively.

Similarly, the next approximations are

$$y_2 = y_1 + hy_1' + \frac{h^2}{2!}y_1'' + \frac{h^3}{3!}y_1''' + \dots \qquad \dots(6)$$

and

$$z_2 = z_1 + hz_1' + \frac{h^2}{2!}z_1''' + \dots \qquad \dots(7)$$

Here y_1', y_1'' and y_1''' etc. z_1', z_1'' and z_1''', etc. are obtained by the differentiating (1) and (2) successively and putting x_1 for x and y_1 for y and z_1 for z.

Preceeding in this way, we can find other approximate values of x, y and z step by step.

9.8.3 RUNGE-KUTTA METHOD

Let h, k and l be the step sizes for x, y and z respectively. Now starting at (x_0, y_0, z_0) and taking the step sizes h, k and l for x, y and z respectively, Runge-Kutta method gives.

$$k_1 = hf(x_0, y_0, z_0), l_1 = hg(x_0, y_0, z_0)$$

$$k_2 = hf\left(x_0 + \frac{h}{2}, y_0 + \frac{k_1}{2}, z_0 + \frac{l_1}{2}\right), l_2 = hg\left(x_0 + \frac{h}{2}, y_0 + \frac{k_1}{2}, z_0 + \frac{l_1}{2}\right)$$

$$k_3 = hf\left(x_0 + \frac{h}{2}, y_0 + \frac{k_2}{2}, z_0 + \frac{l_2}{2}\right), l_3 = hg\left(x_0 + \frac{h}{2}, y_0 + \frac{k_2}{2}, z_0 + \frac{l_2}{2}\right)$$

$$k_4 = hf(x_0 + h, y_0 + k_3, z_0 + l_3), l_4 = hg(x_0 + h, y_0 + k_3, z_0 + l_3)$$

Hence
$$y_1 = y_0 + \frac{1}{6}[k_1 + 2k_2 + 2k_3 + k_4]$$

and
$$z_1 = z_0 + \frac{1}{6}[l_1 + 2l_2 + 2l_3 + l_4]$$

Again to find y_2 and z_2 we replace x_0, y_0, z_0 by x_1, y_1, z_1 in above formulae.
Using Runge-Kutta method, we can easily solve the simultaneous differential equation. The method will be clear by the following example.

SOLVED EXAMPLE

EXAMPLE. *Using Runge-Kutta method solve simultaneous differential equation*
$$\frac{dx}{dt} = f(x, y, t) = xy + t \text{ and} \frac{dy}{dt} = ty + x = g(x, y, t) \text{ where } t_0 = 0, x_0 = 1, y_0 = -1,$$
$h = 0.2.$

SOLUTION. The various k-values for x and y are
$$k_1 = hf(x_0, y_0, t_0) = (0.2)[0.1(-1) + 0] = -0.2$$
$$l_1 = hg(x_0, y_0, t_0) = (0.2)[0.1(-1) + 1] = 0.2$$
$$k_2 = hf\left(x_0 + \frac{k_1}{2}, y_0 + \frac{l_1}{2}, t_0 + \frac{h}{2}\right) = (0.2)\left[\left(1 - \frac{0.2}{2}\right)\left(-1 + \frac{0.2}{2}\right) + \left(0 + \frac{0.2}{2}\right)\right]$$
$$= -0.142$$
$$l_2 = hg\left(x_0 + \frac{k_1}{2}, y_0 + \frac{l_1}{2}, t_0 + \frac{h}{2}\right) = (0.2)\left[\left(\frac{0.2}{2}\right)\left(-1 + \frac{0.2}{2}\right) + \left(1 - \frac{0.2}{2}\right)\right]$$
$$= 0.162$$
$$k_3 = hf\left(x_0 + \frac{k_2}{2}, y_0 + \frac{l_2}{2}, t_0 + \frac{h}{2}\right) = (0.2)\left[\left(1 - \frac{0.142}{2}\right)\left(-1 + \frac{0.162}{2}\right) + \frac{0.2}{2}\right]$$
$$= -0.1508$$
$$l_3 = hf\left(x_0 + \frac{k_2}{2}, y_0 + \frac{l_3}{2}, t_0 + \frac{h}{2}\right) = (0.2)\left[\left(\frac{0.2}{2}\right)\left(-1 + \frac{0.162}{2}\right) + \left(1 - \frac{0.142}{2}\right)\right]$$
$$= 0.1674$$
$$k_4 = hf(x_0 + k_3, y_0 + l_3, t_0 + h) = (0.2)[(1 - 0.1508)(-1 + 0.1674) + 1.2]$$
$$= -0.1014$$
$$l_4 = hg(x_0 + k_3, y_0 + l_3, t_0 + h)$$
$$= (0.2)[(0.2)(-1 + 0.1674) + (1 - 0.1508)] = 0.1365$$

Hence, the new values of x and y at $t = 0.2$ are
$$x(0.2) = x_0 + \frac{1}{6}[k_1 + 2(k_2 + k_3) + k_4]$$
$$= 1 + \frac{1}{6}[-0.2 + 2\{(-0.142) + (-0.1508)\} - 0.1014] = 0.8522$$

and
$$y_{(0.2)} = y_0 + \frac{1}{6}[l_1 + 2(l_2 + l_3) + l_4] = -0.8341$$

9.9 SOLUTION OF SECOND ORDER DIFFERENTIAL EQUATIONS

Let $$\frac{d^2y}{dx^2} = f\left(x, y, \frac{dy}{dx}\right) \qquad \ldots(1)$$

be the second order differential equation with initial condition
$$y(x_0) = y_0 \text{ and } \qquad y'(x_0) = y_0' \qquad \ldots(2)$$

Let $$y' = \frac{dy}{dx} = z$$

Then (1) reduces to

$$\frac{dy}{dx} = z \qquad \ldots(3)$$

and $$\frac{dz}{dx} = f(x, y, z) \text{ with initial conditions } y(x_0) = y_0 \text{ and } z(x_0) = z_0. \qquad \ldots(4)$$

These equations can be solved easily by any method discussed earlier.

SOLVED EXAMPLES

EXAMPLE 1. *Using Runge-Kutta method, solve* $\dfrac{d^2y}{dx^2} = y'' = xy'^2 - y^2$ *for $x = 0.2$ correct to four*

decimal places, subject to the initial conditions $y(0) = 1$ and $y'(0) = 0$. (Delhi-2002)

SOLUTION. Let $$y' = \frac{dy}{dx} = z$$

Then given equation reduces to

$$\frac{dy}{dx} = z = f(x, y, z) \qquad \ldots(1)$$

and $$\frac{dz}{dx} = xz^2 - y^2 = g(x, y, z) \qquad \ldots(2)$$

with initial conditions $y(0) = 1$, $z(0) = 0$.

Here, we have $x_0 = 0$, $y_0 = 1$, $z_0 = 0$, $h = 0.2$

Using Runge-Kutta method, we have

$$k_1 = hf(x_0, y_0, z_0) = (0.2)f(0, 1, 0) = (2.0) \times 0 = 0$$

$$l_1 = hg(x_0, y_0, z_0) = (0.2)g(0, 1, 0) = (0.2) \times (-1) = -0.2$$

$$k_2 = hf\left(x_0 + \frac{h}{2}, y_0 + \frac{k_1}{2}, z_0 + \frac{l_1}{2}\right)$$
$$= (0.2)f(0.1, 1 - 0.1) = (0.2)(-0.1) = -0.02$$

$$l_2 = hg\left(x_0 + \frac{h}{2}, y_0 + \frac{k_1}{2}, z_0 + \frac{l_1}{2}\right) = (0.2)g(0.1, 1 - 0.1)$$

$$= (0.2)[(0.1)(-0.1)^2 - (1)^2] = -0.1998$$

$$k_3 = hf\left(x_0 + \frac{h}{2}, y_0 + \frac{k_2}{2}, z_0 + \frac{l_2}{2}\right)$$

$$= (0.2)f(0.1, 0.99 - 0.0999) = (0.2)(-0.0999) = -0.01998$$

$$l_3 = hg\left(x_0 + \frac{h}{2}, y_0 + \frac{k_2}{2}, z_0 + \frac{l_2}{2}\right)$$

$$= (0.2)g(0.1, 0.99 - 0.0999)$$

$$= (0.2)[(0.1)(-0.0999)^2 - (0.99)^2] = -0.1958$$

$$k_4 = hf\left(x_0 + h, y_0 + k_3, z_0 + l_3\right)$$

$$= (0.2)f(0.2, 0.98002, -0.1958) = (0.2)(-0.1958) = -0.03915$$

$$l_4 = hg\left(x_0 + h, y_0 + k_3, z_0 + l_3\right) = (0.2)g(0.2, 0.98002, -0.1958)$$

$$= (0.2)[(0.02)(-0.1958)^2 - (0.98002)^2] = -0.1906$$

At $x = 0.2$, we have

$$y(0.2) = y$$

$$= y(0) + \frac{1}{6}[k_1 + 2k_2 + 2k_3 + k_4]$$

$$= 1 + \frac{1}{6}[0 + 2(-0.02 - 0.01998) + (-0.3916)] = 0.9801$$

and

$$z = z_0 + \frac{1}{6}[l_1 + 2l_2 + 2l_3 + l_4]$$

$$= 0 + \frac{1}{6}[-0.2 + 2(-0.1998 - 0.1958) - 0.1906)] = -0.1970$$

EXAMPLE 2. *Using Picard's method find approximate values of y and z at $x = 0.1$, given that $y(0) = 2$, $z(0) = 1$ and $\dfrac{dy}{dx} = x + z, \dfrac{dz}{dx} = x - y^2$.*

SOLUTION. We have

$$x_0 = 0, y_0 = 2, z_0 = 1$$

and

$$\frac{dy}{dx} = f(x, y, z) = x + z$$

$$\frac{dz}{dx} = g(x, y, z) = x - y^2$$

So, $f(x, y_0, z_0) = f(x, 2, 1) = x + 1$

$g(x, y_0, z_0) = g(x, 2, 1) = x - 4$

Using Picard's method the first approximation is given by

$$y_1 = y_0 + \int_{x_0}^{x} f(x, y_0, z_0)dx = 2 + \int_0^x (x+1)dx = 2 + x + \frac{x^2}{2}$$

$$z_1 = z_0 + \int_{x_0}^{x} g(x, y_0, z_0)dx = 1 + \int_0^x (x-4)dx = 1 + \frac{x^2}{2} - 4x$$

The second approximations are

$$y_2 = y_0 + \int_{x_0}^{x} f(x, y_1, z_1)dx = 2 + \int_0^x \left(x + 1 + \frac{x^2}{2} - 4x\right)dx$$

$$= 2 + x - \frac{3x^2}{2} + \frac{x^3}{6}$$

and

$$z_2 = z_0 + \int_{x_0}^x g(x, y_1, z_1)dx = 1 + \int_0^x \left[x - \left(2 + x + \frac{x^2}{2} \right)^2 \right]dx$$

$$= 1 + \int_0^x \left[-4 - 3x - 3x^2 - x^3 - \frac{x^4}{4} \right]dx = 1 - 4x - \frac{3}{2}x^2 - x^3 - \frac{x^4}{4} - \frac{x^5}{20}$$

Similarly, the third approximations are

$$y_3 = y_0 + \int_0^x f(x, y_2, z_2)dx = 2 + \int_0^x (x + z_2)dx$$

$$= 2 + \int_0^x \left(x + 1 - 4x - \frac{3}{2}x^2 - x^3 - \frac{x^4}{4} - \frac{x^5}{20} \right)dx$$

$$= 2 + x - \frac{3}{2}x^2 - \frac{1}{2}x^3 - \frac{x^4}{4} - \frac{x^5}{20} - \frac{x^6}{120}$$

and

$$z_3 = z_0 + \int_0^x g(x, y_2, z_2)dx = 1 + \int_0^x (x - y_2^2)dx$$

$$= 1 + \int_0^x \left[x - \left(2 + x + \frac{3x^2}{2} + \frac{x^3}{6} \right)^2 \right]dx$$

$$= 1 + \int_0^x \left[x - \left(4 + x^2 + \frac{9x^4}{4} + \frac{x^6}{36} + 4x \right. \right.$$

$$\left. \left. - 6x^2 + \frac{2}{3}x^3 - 3x^2 + \frac{1}{3}x^4 - \frac{1}{2}x^5 \right) \right]dx$$

$$= 1 + \int_0^x \left[-4 - 3x + 5x^2 + \frac{7}{3}x^3 - \frac{31}{12}x^4 + \frac{1}{2}x^5 - \frac{1}{36}x^6 \right]dx$$

$$= 1 - 4x - \frac{3}{2}x^2 + \frac{5}{2}x^3 + \frac{7}{12}x^4 - \frac{31}{60}x^5 + \frac{1}{12}x^6 - \frac{1}{252}x^7 \quad \text{and so on.}$$

Putting $x = 0.1$, we get

$$y_1 = 2.105, \qquad z_1 = 0.605,$$

$$y_2 = 2.08517, \quad z_2 = 0.58397$$

$$y_3 = 2.08447, \quad z_3 = 0.58672$$

Hence, $y(0.1) = 2.0845, z(0.1) = 0.5867$ correct to four decimal places.

EXERCISE 9.2

1. Given $\frac{dy}{dx} = 1 + y^2$ where $y = 0$ when $x = 0$, find $y_{(0.2)}, y_{(0.4)}$ and $y_{(0.6)}$, using Runge-Kutta formula of order four.

2. Use Runge-Kutta formula of fourth order to find the numerical solution at $x = 0.8$ for $\frac{dy}{dx} = \sqrt{x + y}$, $y_{(0.4)} = 0.41$. Assume the step length $h = 0.2$. (SVTU–2007S)

3. Using Runge-Kutta method to solve $10y' = x^2 + y^2, y(0) = 1$ for the interval $0 < x \le 0.4$ with $h = 0.1$.

4. Use Runge-Kutta fourth order to approximate the solution of the initial value problem, given that $y' = y^2 - x^2$ with $y(0) = 1$, interval $0 \leq x \leq 0.2$, step length $h = 0.1$.

5. Solve the differential equation $\dfrac{dy}{dx} = \dfrac{2x-1}{x^2} y + 1$ when $x_0 = 1$, $y_0 = 2$, $h = 0.2$. Obtain $y_{(1.2)}$ and $y_{(1.4)}$ using Runge-Kutta method.

6. Using Runge-Kutta method compute the value of y for $x = 0.5$, 1, 1.5 given that $y' = 0.31 + 0.25y + 0.3x^2$ with $y(0) = 0.72$.

7. Use Runge-Kutta quadratic method to calculate three additional points on the solution curve of the problem $y' = 1 - 2xy$, $y_{(0)} = 0$ with $h = 0.1$.

8. Solve the simultaneous differential equation $y'' = (x^2 + y^2)(1 + y'^2)$ for $x = 0.5$, using Runge-Kutta method, initial value $x = 0$, $y = 1$, $y' = 0$. Take $h = 0.5$.
   where $y' = \dfrac{dy}{dx} = z$ so that $y'' = \dfrac{d^2y}{dx^2} = \dfrac{dz}{dx}$.

9. Solve the initial value problem $u' = 2tu^2$, $u(0) = 1$ with $h = 0.2$ on the interval $(0, 0.4)$. Use Runge-Kutta method.

10. Using fourth order Runge-Kutta method find the value of y when $x = 1$ given that $y = 1$ when $x = 0$ (taking $n = 2$) and $\dfrac{dy}{dx} = \dfrac{y-x}{y+x}$.
    (VTU–2011S)

11. Use Runge-Kutta method to approximate y when $x = 1.1$ given that $y = 1.2$ when $x = 1$ and $\dfrac{dy}{dx} = 3x + y^2$.   (Agra-2003, Avadh-2003)

12. Using fourth order Runge-Kutta method, compute $y(0.2)$ and $y(0.4)$ when $10\dfrac{dy}{dx} = x^2 + y^2$, $y(0) = 1$; taking $h = 0.1$.
    (Rohtak–2003, Bhopal–2006)

13. Using fourth order Runge-Kutta method, find the approximate solution at $x = 1.2$, $x = 1.4$ of the initial value problem $y' = xy$, $y(1) = 2$.
    (Bombay-2004)

14. Given $dy/dx = x^3 + y$, $y(0) = 2$. Compute $y_{(0.2)}$, $y_{(0.4)}$ and $y_{(0.6)}$ by Runge-Kutta method of fourth order.   (Anna-2004)

15. Using fourth order Runge-Kutta method, solve the following equation, taking each step of $h = 0.1$, given $y(0) = 3$, $dy/dx = (4x/y - xy)$. Calculate y for $x = 0.1$ and 0.2. (Anna-2007)

16. Find by the Runge-Kutta method an approximate value of y for $x = 0.8$, given that $y = 0.41$ when $x = 0.4$ and $dy/dx = \sqrt{(x+y)}$.   (SVTU-2007)

17. Using Runge-kutta method of order 4, find (0.2) for the equation $\dfrac{dy}{dx} = \dfrac{y-x}{y+x}$, $y(0) = 1$. Take $h = 0.2$ (SVTU-2007)

■■□   **⟨Answers⟩**   ■□□

1. $y_{(0.2)} = 0.2027$, $y_{(0.4)} = 0.4228$, $y_{(0.6)} = 0.6841$ **2.** $y_{(0.6)} = 0.61035$, $y_{(0.8)} = 0.84899$
3. 1.0101, 1.0207, 1.0318, 1.0438 **4.** $y_{(0.1)} = 1.1108$, $y_{(0.2)} = 1.24705$. $y_{(1.2)} = 2.658913$,
   $y_{(1.4)} = 3.432851$ **6.** $y_{(0.5)} = 0.8891$, $y_{(1)} = 0.6674$, $y_{(1.5)} = 0.5794$ **7.** $y_{(0.1)} = 0.99336$,
   $y_{(0.2)} = 1.194751$, $y_{(0.3)} = 0.282632$ **8.** $y_{(0.5)} = 1.1423$ **9.** 0.87343769 **10.** 1.5488
11. 1.7278 **12.** $1.0207 : 1.038$ **13.** $y_{(1.2)} = 2.4921$, $y_{(1.4)} = 2311$ **14.** $y_{(0.2)} = 2.44$, $y_{(0.4)} = 2.99$,
   $y_{(0.6)} = 3.68$ **15.** $y_{(0.1)} = 2.9917$, $y_{(0.2)} = 2.9627$ **16.** 1.1678 **17.** 1.1749.

9.10   MILNE'S METHODS

(Agra-2007)

Let $y' = f(x, y)$ with $y = y_0$, $x = x_0$. To find an approximate value of y for $x = x_0 + nh$ by Milne method, we proceed as follows :

The value $y_0 = y(x_0)$, is given, we may compute

$y_1 = y(x_0 + h)$, $y_2 = y(x_0 + 2h)$, $y_3 = y(x_0 + 3h)$ by Taylor's series and Picard's method.

Now we solve $f_0 = f(x_0, y_0)$, $f_1 = f(x_0 + h, y_1)$, $f_2 = f(x_0 + 2h, y_2)$, $f_3 = f(x_0 + 3h, y_3)$

Then to find $y_4 = y(x_0 + 4h)$, we putting Newton's forward interpolation formula

$$f_{(x,y)} = f_0 + n\Delta f_0 + \frac{n(n-1)}{2}\Delta^2 f_0 = \frac{n(n-1)(n-2)}{6}\Delta^3 f_{0+\ldots}$$

in the relation $y_4 = y_0 + \int_{x_0}^{x_0+4h} f(x,y)dx = y_0 + \int_{x_0}^{x_0+4h} \left(f_0 + n\Delta f_0 + \frac{n(n-1)}{2}\Delta^2 f_0 + ... \right) dx$

[Putting $x = x_0 + nh$, $dx = hdn$]

We get
$$y_4 = y_0 + h\int_0^4 \left(f_0 + n\Delta f_0 + \frac{n(n-1)}{2}\Delta^2 f_0 + ... \right) dn$$
$$= y_0 + \left(4f_0 + 8\Delta f_0 + \frac{20}{3}\Delta^2 f_0 + \frac{8}{3}\Delta^3 f_0 + ... \right)$$

Neglecting fourth and higher order differences and expressing Δf_0, $\Delta^2 f_0$ and $\Delta^3 f_0$ in terms of the function values, we obtain
$$y_4 = y_0 + \frac{4h}{3}(2f_1 - f_2 + 2f_3)$$
which is called a predictor formula.
To find y_4 we obtain a first approximation to $f_4 = f(x_0 + 4h, y_4)$
Then a best value of y_4 is found by Simpson's rule given by
$$y_4 = y_2 + \frac{h}{3}(f_2 + 4f_3 + f_4)$$
which is called a corrector formula.

REMARK

- Firstly, an improved value of f_4 is computed and again the corrector is applied to find a still better value of y_4. We repeat step until y_4 remains unchanged.

Once y_4 and f_4 are obtained to desired degree of accuracy, $y_5 = y(x_0 + 5h)$ is found from the predictor formula as
$$y_5 = y_1 + \frac{4h}{3}(2f_2 - 4f_3 + 2f_4)$$
Now $f_5 = f(x_0 + 5h_1, y_5)$ is solved. Then a better approximation to the value of y_5 is obtained from the corrector as
$$y_5 = y_3 + \frac{h}{3}(f_3 + 4f_4 + f_5)$$
We repeat this step till y_5 becomes same and we then proceed to find y_6 as before.
Thus in Milne's predictor-corrector method, to ensure greater accuracy, we must first improve the accuracy of the starting values and then sub-divide the intervals.

REMARK

- The general Milne's predictor and corrector formula are
$$y_{n+1} = y_{n-3} + \frac{4h}{3}[2y'_{n-2} - y'_{n-1} + 2y'_n]$$
and
$$y_{n+1} = y_{n-1} + \frac{h}{3}[y'_{n-1} + 4y'_n + y'_{n+1}]$$

SOLVED EXAMPLES

EXAMPLE 1. *Using Milne method obtain $y_{(4)}$ from the given table of tabulated values and $y' = y^2 - x^2$.*

x	0	0.1	0.2	0.3
y	1	1.11	1.25	1.42
f	1	1.22	1.52	1.92

SOLUTION.   By the predictor formula, we have

$$y_4 = y_0 + \frac{4h}{3}[2f_3 - f_2 + 2f_1]$$

$$= 1.0 + \frac{4 \times 0.1}{3}[2(1.92) - 1.52 + 2(1.22)] = 1.63$$

Now using this value we compute

$$f_4 = y_4^2 - x_4^2 = (1.63)^2 - (0.4)^2 = 2.50$$

Now by the corrector formula, we have

$$y_4 = y_2 + \frac{h}{3}[f_2 + 4f_3 + f_4]$$

$$= 1.25 + \frac{0.1}{3}[1.52 + 4(1.92) + 2.50] = 1.64$$

Hence, $y_{(0.4)} = 1.64$

EXAMPLE 2.   *Solve* $y' = 2e^x - y$ *at* $x = 0.4$ *and* $x = 0.5$ *by Milne's method, given their values at the four points.*

x	0	0.1	0.2	0.3
y	2	2.010	2.040	2.090

SOLUTION.   We want to find the value of f_1, f_2, f_3.

$$f_1 = 2e^{0.1} - 2.010 = 0.2003$$
$$f_2 = 2e^{0.2} - 2.040 = 0.4028$$
$$f_3 = 2e^{0.3} - 2.090 = 0.6097$$

By Milne's predictor formula, we have

$$y_4 = y_0 + \frac{4h}{3}[2f_1 - f_2 + 2f_3]$$

$$= 2 \times \frac{4 \times 0.1}{3}[2(0.2003) - (0.4028) + 2(0.6097)] \approx 2.1623$$

Now $f_4 = 2e^{0.4} - 2.1623 = 0.8213494 \approx 0.8213$

Now again by corrector formula, we get

$$y_4 = y_2 + \frac{h}{3}[f_2 + 4f_3 + f_4]$$

$$= 2.04 + \frac{0.1}{3}[0.4028 + 4(0.6097 + 0.8213]$$

$$= 2.162096 \approx 2.1621$$

Again by using predictor formula

$$y_5 = y_1 + \frac{4h}{3}[2f_2 - f_3 + 2f_4]$$

$$= 2.10 + \frac{4 \times 0.1}{3}[2(0.4028) - 0.6097 + 2(0.8215)]$$

$$= 2.2551867 \approx 2.2552$$

then $f_5 = 2e^{0.5} - 2.2552 = 1.0422425 \approx 1.0422$

By corrector formula, we get

$$y_5 = y_3 + \frac{h}{3}[f_3 + 4f_4 + f_5]$$

$$= 2.090 + \frac{0.1}{3}[0.6097 + 4(0.8215) + 1.0422]$$

$$= 2.2545967 \approx 2.255$$

EXAMPLE 3. *Apply Milne's method to find a solution of the differential equation* $dy/dx = x - y^2$ *in the range* $0 \leq x \leq 1$ *with* $y(0) = 0$.   (VTU–2009, Anna–2005, Rohtak–2005)

SOLUTION. Here, we use Picard's method to compute y_1, y_2 and y_3.

Picard's successive approximations are given by

$$y_n = y_0 + \int_{x_0}^{x} f(x, y_{n-1}) dx$$

for $n = 1$, we have

$$y_1 = 0 + \int_0^x x \, dx = \frac{x^2}{2}$$

for $n = 2$,

$$y_2 = 0 + \int_0^x \left[x - \left(\frac{x^2}{2} \right)^2 \right] dx = \frac{x^2}{2} - \frac{x^5}{20}$$

Similarly,

$$y_3 = 0 + \int_0^x \left[x - \left(\frac{x^2}{2} - \frac{x^5}{20} \right)^2 \right] dx = \frac{x^2}{2} - \frac{x^5}{20} + \frac{x^8}{160} - \frac{x^{11}}{4400}$$

Let us take $y = \frac{x^2}{2} - \frac{x^5}{20}$ for finding the various values of y_i's and f_i's

$$y_1 = y(0.2) = 0.019984 = 0.02,$$
$$f_1 = 0.1996$$
$$y_2 = y(0.4) = 0.079488 = 0.0795,$$
$$f_2 = 0.3937$$
$$y_3 = y(0.6) = 0.176112 = 0.176,$$
$$f_3 = 0.5690$$

Now, using predictor formula, we get

$$y_4 = y_0 + \frac{4h}{3}(2f_1 - f_2 + 2f_3)$$

$$= 0 + \frac{4 \times 0.2}{3}(2 \times 0.01996 - 0.3937 + 2 \times 0.5690)$$

$$= 0.3049333 = 0.3049$$

Further, using corrector formula, we get

$$y_4 = y_2 + \frac{h}{3}(f_2 + 4f_3 + f_4)$$

$$= 0.0795 + \frac{0.2}{3}(0.3937 + 4 \times 0.5690 + 0.7070)$$

$$[\because f_4 = x_4 + y_4 = 0.8 - (0.3049)^2 = 0.7070359 \approx 0.7070]$$

Hence, $y_4 = 0.3046$ at $x = 0.8$ and corrected $f_4 = 0.8 - (0.3046)^2 = 0.7072$

Again, using predictor formula, we get

$$y_5 = y_1 + \frac{4h}{3}(2f_2 - f_3 + 2f_4)$$

$$= 0.2 + \frac{4 \times 0.2}{3}(2 \times 0.3937 - 0.5690 + 2 \times 0.7072)$$

$$= 0.4554133 \approx 0.4554$$

Now, $\qquad f_5 = x_5 + y_5 = 1 - (0.4554)^2$

$$= 0.792610 \approx 0.7926$$

Using corrector formula, we get

$$y_5 = y_3 + \frac{h}{3}(f_3 + 4f_4 + f_5)$$

$$= 0.176 + \frac{0.2}{3}(0.5690 + 4 \times 0.7072 + 0.7926)$$

$$= 0.45536 \approx 0.4554$$

Hence, $\qquad y(1) = 0.4554.$

EXAMPLE 4.   *Compute $y(2)$ if $y(x)$ is the solution of $\dfrac{dx}{dy} = \dfrac{1}{2}(x + y)$ using Milne predictor-corrector*

method. Given $y(0) = 2$, $y(0.5) = 2.6336$, $y(1.0) = 3.595$, $y(1.5) = 4.968$.

SOLUTION.   Here, we have

$$x_0 = 0, y_0 = 2, \quad f_0 = \frac{1}{2}(0 + 2) = 1$$

$$x_1 = 0.5, y_1 = 2.636,$$

$$f_1 = \frac{1}{2}(0.5 + 2.636) = 1.568$$

$$x_2 = 1, y_2 = 3.595,$$

$$f_2 = \frac{1}{2}(1 + 3.595) = 2.2975$$

$$x_3 = 1.5, y_3 = 4.968,$$

$$f_3 = \frac{1}{2}(1.5 + 4.968) = 3.234$$

Using predictor formula, we get

$$y_4 = y_0 + \frac{4h}{3}(2f_1 - f_2 + 2f_3)$$

$$= 2.0 + \frac{4 \times 0.5}{3}(2 \times 1.568 - 2.2975 + 2 \times 3.234) = 6.871$$

and $\qquad f_4 = x_4 + y_4 = \dfrac{1}{2}(2 + 6.871) = 4.4355$

Using corrector formula, we get

$$y_4 = y_2 + \frac{h}{3}[f_2 + 4f_3 + f_4]$$

$$= 3.595 + \frac{0.5}{3}(2.2975 + 4 \times 3.234 + 4.4355)$$

$$= 6.873166 \approx 6.8732$$

Thus, corrected $f_4 = \dfrac{1}{2}(x_4 + y_4) = \dfrac{1}{2}(2.0 + 6.8732) = 4.4366$

Again, using corrector formula, we get

$$y_4 = y_2 + \frac{h}{3}(f_2 + 4f_3 + f_4)$$

$$= 3.595 + \frac{0.5}{3}(2.2975 + 4 \times 3.234 + 4.4366)$$

$$= 6.87335 \approx 6.8734$$

EXAMPLE 5. *The differential equation $\dfrac{dy}{dx} = 1 + y^2$ satisfies the following sets of values of x and y :*

x	0	0.2	0.4	0.6
y	0	0.2027	0.4228	0.6871

Compute the value of $f(1)$.

SOLUTION. Firstly, we calculate the following :

$$f_0 = 1 + y_0^2 = 1$$

$$f_1 = 1 + y_1^2 = 1.0411$$

$$f_2 = 1 + y_2^2 = 1.1787$$

$$f_3 = 1 + y_3^2 = 1.4681$$

Using predictor formula, we get

$$y_4 = y_0 + \frac{4h}{3}(2f_1 - f_2 + 2f_3)$$

$$= 0 + \frac{4 \times 0.2}{3}[2(1.0411) - 1.1787 + 2(1.4681)] \ = 1.0239$$

$$\therefore \quad f_4 = 1 + y_4^2 = 1 + (1.0239)^2 = 2.0480$$

Using corrector formula, we have

$$y_4 = y_2 + \frac{h}{3}(f_2 + 4f_3 + f_4)$$

$$= 0.4228 + \frac{0.2}{3}(1.1787 + 4 \times 1.4681 + 2.0480) \ = 1.0294$$

$$\Rightarrow \qquad y(0.8) = 1.0294$$

The corrected value of $f_4 = 1 + y_4^2 = 2.0597$.

Now, to find $f(1)$ we use predictor formula, such that

$$y_5 = y_1 + \frac{4h}{3}(2f_1 - f_3 + 2f_4)$$

$$= 0.2027 + \frac{4 \times 0.2}{3}[2(1.1787) - 1.4681 + 2(2.0597)]$$

$$= 1.5384$$

and $$f_5 = 1 + y_5^2 = 1 + (1.5384)^2 = 3.3667$$

Finally, using corrector formula, we get

$$y_5 = y_3 + \frac{h}{3}(f_3 + 4f_4 + f_5)$$

$$= 0.6841 + \frac{0.2}{3}[1.4681 + 4(2.0597) + 3.3667]$$

$$= 1.5556733 \approx 1.5557.$$

EXAMPLE 6. *Solve numerically the differential equation $\dfrac{dy}{dx} = x + y$ with $y(0) = 1$ by Milne's method from (a) $x = 0$ to $x = 0.4$ and (b) $x = 0.20$ to $x = 0.30$.* (Agra-2004, 08, 09)

SOLUTION.   We have, $\dfrac{dy}{dx} = y' = x + y \Rightarrow y'' = 1 + y'$

and $\quad y''' = y'', \ y'''' = y'''$

Hence $\quad y_0' = x_0 + y_0 = 1$

$$y_0'' = 1 + 1 = 2$$

$$y_0''' = y_0'' = 2$$

$$y_0'''' = y_0'''= 2$$

$$y_0''''' = y_0'''' = 2$$

Taking $h = 0.1$ and substituting these value in Taylor series

$$y_i = y_0 + (h)y_0' + \frac{1}{2!}(h)^2 y_0'' + \frac{1}{3!}(h)^3 y_0''' + \ldots.$$

we obtain $\quad y_1 = 1.1103, \ y_3 = 1.2428, \ y_3 = 1.3997.$

Now, we have the functional value

$$f_0 = y_0' = x_0 + y_0 = 1$$

$$f_1 = y_0' = x_1 + y_1 = 0.1 + 1.1103 = 1.2103$$

$$f_2 = y_2' = x_2 + y_2 = 0.2 + 1.2428 = 1.4428$$

and $\quad f_3 = y_3' = x_3 + y_3 = 0.3 + 1.3997 = 1.6997$

which when substituted in predictor formula

$$y_4 = y_0 + \frac{4h}{3}[2f_1 - f_2 + 2f_3]$$

$$= 1 + \frac{4 \times 0.1}{3}[2 \times 1.2103 - 1.4428 + 2 \times 1.6997) = 1.583627$$

Using corrector formula

$$y_4 = y_2 + \frac{h}{3}(f_2 + 4f_3 + f_4)$$

$$= 1.2428 + \frac{0.1}{3}[1.4428 + 4 \times 1.6997 + 1.983627] = 1.5836$$

(b) We calculate the value of y and f corresponding to $x = 0.05, 0.1, 0.15$ and $h = 0.5$.

x	y	f
0	1.0000	1.0000
0.05	1.0526	1.1026
1.0	1.1104	1.2104
0.15	1.1737	1.3237

Using predictor formula

$$y_4 = y_0 + \frac{4h}{3}[2f_1 - f_2 + 2f_3]$$

$$= 1 + \frac{4}{3}(0.5)[2 \times 1.1026 - 1.2104 + 2 \times 1.3237]$$

$$= 1.2428$$

and $f_4 = x_4 + y_4 = 0.2 + 1.2428 = 1.4428$.

y_4 is now corrected by corrector formula given by

$\therefore \qquad y_4 = y_0 + \dfrac{h}{3}[f_2 + 4f_3 + f_4]$

$\qquad = 1.1104 + \dfrac{1}{3}(0.5)(1.2104 + 4 \times 1.3237 + 1.4428)$

$\qquad = 1.2428$ (same as predicted value)

Now, we find y_5, *i.e.*, $y(0.25)$

Predictor formula is

$\qquad y_5 = y_1 + \dfrac{4h}{3}[2f_2 - f_3 + 2f_4]$

$\qquad = 1.0526 + \dfrac{4(0.05)}{3}[2 \times 1.2104 - 1.3237 + 2 \times 1.14428]$

$\qquad = 1.3181$

The value of $f_5 = x_5 + y_5 = 0.25 + 1.3181 = 1.5681$.

Using corrector formula

$\qquad y_5 = y_3 + \dfrac{h}{3}[f_3 + 4f_4 + f_5]$

$\qquad = 1.1737 + \dfrac{0.05}{3}[1.3237 + 4 \times 1.4428 + 1.5681]$

$\qquad = 1.3181$

Hence $y(0.25) = 1.3181$

Again $\qquad y_6 = y_2 + \dfrac{4h}{3}[2f_3 - f_4 + 2f_5]$

$\qquad = 1.1104 + \dfrac{4(0.05)}{3} - [2 \times 1.3237 - 1.4428 + 2 \times 1.5681]$

$\qquad = 1.3997$.

Using corrector formula, we obtain

$\qquad y_6 = y_4 + \dfrac{h}{3}[f_4 + 4f_5 + y_6]$

$\qquad = 1.2428 + \dfrac{0.05}{3}[1.4428 + 4 \times 1.5681 + 1.6997]$

$\qquad [\because f_6 = x_6 + y_6 = 0.3 + 1.3997 = 1.6997]$

$\qquad y_6 = 1.3997$

Hence, $y(0.3) = 1.3997$.

EXAMPLE 7. *Solve* $\dfrac{10dy}{dx} = x^2 + y^2$, *if* $y(0) = 1, y(0.1) = 1.0101, y(0.2) = 1.0206, y(0.3) = 1.0317$

and compute $y(0.4)$ *and* $y(0.5)$ *by Milne's method.*

SOLUTION. We have $10\dfrac{dy}{dx} = x^2 + y^2$

$\Rightarrow \qquad \dfrac{dy}{dx} = \dfrac{1}{10}(x^2 + y^2)$

$$f_n = \frac{1}{10}(x_n^2 + y_n^2)$$

$$f_0 = \frac{1}{10}(0+1) = \frac{1}{10} = 0.1$$

$$f_1 = \frac{1}{10}[(0.1)^2 + (1.0101)^2]$$

$$= \frac{1}{10}(0.01 + 1.0203) = \frac{1.0303}{10} = 0.10303$$

$$f_2 = \frac{1}{10}(0.2^2 + 1.0416)$$

$$= \frac{1}{10}(0.04 + 1.0416) = \frac{1.0816}{10} = 0.10816$$

$$f_3 = \frac{1}{10}(0.3^2 + 1.0717^2)$$

$$= \frac{1}{10}(0.09 + 1.0644) = \frac{1}{10}(1.1544) = 0.11544$$

Predicted value of $y(0.4)$

$$y_4^p = y_0 + \frac{4h}{3}[2f_1 - f_2 + 2f_3]$$

$$= 1 + \frac{4 \times 0.1}{3}[2 \times 0.10303 - 0.10816 + 2 \times 0.11544]$$

$$(\because h = 0.1)$$

$$= 1 + \frac{0.4}{3}[0.20606 - 0.10816 + 0.23088]$$

$$= 1 + \frac{0.4 \times 0.32878}{3} = 1 + 0.04384 = 1.04384$$

To improve upon this estimate of $y(0.4)$, we apply Milne's corrector formula

$$y_4^c = y_2 + \frac{1}{3}h[f_2 + 4f_3 + f_4^p]$$

where $x_0 = 0.2$, $y_4^c = y(0.4)$, $y_2 = 1.0206$, $f_2 = 0.10816$, $f_3 = 0.11544$
and $y_4{}^p$ is the value of f at predicted value 1.04384 of $y(0.4)$ at $x = 0.4$.

$$y_4^p = \frac{1}{10}[0.4^2 + 1.04384^2]$$

$$= \frac{1}{10}[0.16 + 1.0896] = \frac{1.2496}{10} = 0.12496$$

Hence, $y_4^c = y(0.4)$

$$= 1.0206 + \frac{1 \times 0.1}{3}[0.10876 + 4 \times 0.11544 + 0.12496]$$

$$= 1.0206 + \frac{0.1}{3}[0.10816 + 0.46176 + 0.12496]$$

$$= 1.0206 + \frac{0.1 \times 0.69488}{3} = 1.0206 + 0.1 \times 0.23162$$

$$= 1.0206 + 0.03162 = 1.043762 \approx 1.0438$$

$$f_4 = \frac{1}{10}[0.4^2 + 1.0438^2]$$

Computation of $y(0.5)$

$$y_5^p = y_1 + \frac{4h}{3}[2f_2 - f_3 + 3f_4]$$

$$= 1.0101 + \frac{4 \times 0.1}{3}[2 \times 0.10816 - 0.11544 + 2 \times 0.12495]$$

$$= 1.0101 + \frac{0.4}{3}[0.21632 - 0.11544 + 0.24990]$$

$$= 1.0101 + \frac{0.4 \times 0.35078}{3}$$

$$= 1.0101 + 0.04677 = 1.05687 \approx 1.0569$$

To improve upon this estimate of $y(0.5)$, we apply Milne's corrector formula

$$y_5^c = y_2 + \frac{h}{2}[f_3 + 4f_4 + f_5^p]$$

where $x_0 = 0.3$, $y_2 = 1.0317$, $f_3 = 0.11544$, $f_4 = 0.12495$

$$y_5^p = \frac{1}{10}(x_2^2 + (y_2^p)^2] = \frac{1}{10}[0.5^2 + 1.0569^2]$$

$$= \frac{1}{10}[0.25 + 1.1170] = \frac{1.3670}{10} = 0.13670$$

$\therefore \qquad y_5^c = y(0.5)$

$$= 1.0317 + \frac{0.1}{3}[0.11544 + 4 \times 0.12495 + 0.13670]$$

$$= 1.0317 + \frac{0.1 \times 0.75194}{3}$$

$$= 1.0317 + 0.02506 = 1.05676 \approx 1.0568.$$

EXAMPLE 8. *Find by Milne's method the numerical solution of $\frac{dy}{dx} = x + y$ for $x = 2.0$ and 3.0 when it is given that*

x	0	0.05	0.10	0.15
y	1	1.0526	1.1104	1.1737

SOLUTION. We have

$$\frac{dy}{dx} = x + y$$

$\Rightarrow \qquad f_n = x_n + y_n$

$\therefore \qquad f_0 = x_0 + y_0 = 0 + 1 = 1$

$$f_1 = x_1 + y_1 = 0.05 + 1.0526 = 1.1026$$

$$f_2 = x_2 + y_2 = 0.10 + 1.1104 = 1.2104$$

$$f_3 = x_3 + y_3 = 0.15 + 1.1737 = 1.3237$$

For y(0.2) : Predictor formula is

$$y_4^p = y_0 + \frac{4h}{3}[2f_1 - f_2 + 2f_3]$$

$$= 1 + \frac{4 \times 0.5}{3}[2 \times 1.1026 - 1.2104 + 2 \times 1.3237]$$

$$= 1 + \frac{0.2}{3}[2.205 - 1.2104 + 2.6474]$$

$$= 1 + \frac{0.2 \times 3.642}{3} = 1 + 0.2428 = 1.2428$$

The corrector formula is

$$y_4^c = y_3 + \frac{h}{3}[f_2 + 4f_3 + f_4^p]$$

$$= 1.1104 + \frac{0.05}{3}[1.2104 + 4 \times 1.3237 + (0.20 + 1.2428)]$$

$$= 1.1104 + \frac{0.05}{3}[1.2104 + 5.2948 + 1.4428] = 1.1104 + \frac{0.5 \times 7.9480}{3}$$

$$= 1.1104 + 0.1325 = 1.2429 = y(0.2) \qquad \qquad ...(1)$$

For y(0.25) :

$$f_4 = x_4 + y_4^c = 0.20 + 1.2429 = 1.4429$$

Predictor formula is

$$y_5^p = y_1 + \frac{4h}{3}[2f_2 - f_3 + 2f_4]$$

$$= 1.0526 + \frac{4 \times 0.05}{3}[2 \times 1.2104 - 1.3237 + 2 \times 1.4429]$$

$$= 1.0526 + \frac{0.2}{3}[2.4208 - 1.3237 + 2.8858] = 1.0526 + \frac{0.2 \times 3.9829}{3}$$

$$= 1.0526 + 0.2655 = 1.3181$$

The corrector formula is

$$y_5^c = y_3 + \frac{h}{3}[f_3 + 4f_4 + f_5^c]$$

$$= 1.1737 + \frac{0.05}{3}[1.3237 + 4 \times 1.4429 + (0.25) + 1.3181]$$

$$= 1.1737 + \frac{0.05}{3}[1.3237 + 5.7716 + 1.5681] = 1.1737 + \frac{0.05 \times 9.0639}{3}$$

$$= 1.1737 + 0.1510 = 1.3247 = y(2.5) \qquad \qquad ...(2)$$

For y(0.3) :

$$f_5 = x_5 + y_5^c = 0.25 + 1.3247 = 1.5747$$

Predictor formula is

$$y_6^p = y_2 + \frac{4h}{3}[2f_3 - f_4 + 2f_5]$$

$$= 1.1104 + \frac{4 \times 0.05}{3}[2 \times 1.3237 - 1.4424 + 2 \times 1.5747]$$

$$= 1.1104 + \frac{0.2}{3}[2.6474 - 1.4429 + 3.6494]$$

$$= 1.1104 + \frac{0.2 \times 4.8539}{3} = 1.1104 + 0.3236 = 1.4340$$

The corrector formula is

$$y_6^c = y_4 + \frac{h}{4}[f_4 + 4f_5 + f_6^p]$$

$$= 1.2429 + \frac{0.05}{3}[1.4429 + 4 \times 1.5747 + (0.30 + 1.4340]$$

$$= 1.2429 + \frac{0.05}{3}[1.4429 + 6.2988 + 1.4640]$$

$$= 1.2429 + \frac{0.05 \times 9.2057}{3}$$

$$= 1.2429 + 0.1534 = 1.3963 = y(0.3) \qquad \qquad \qquad \dots(3)$$

EXAMPLE 9. *Given $y'' + xy' + y = 0$, $y(0) = 1$, $y'(0) = 0$. Find y for x = 0.1, 0.2, 0.3, by any method.*

Further continue the solution of Milne's method to compute y(0.4). (Anna-2004)

SOLUTION. Let $\qquad \qquad \frac{dy}{dx} = y' = z$

Then given equation can be written as

$$y' = z \qquad \qquad \qquad \dots(1)$$
$$z' + xz + y = 0 \qquad \qquad \qquad \dots(2)$$

with initial conditions $y(0) = 1$, $z(0) = 0$.

Differentiating the given equation n times by Leibnitz's theorem, we get

$$y_{n+2} + xy_{n+1} + (n+1)y_n = 0$$

Putting $x = 0$, we get

$$y_{n+2}(0) + (n+1)y_n(0) = 0 \Rightarrow y_{n+2}(0) = -(n+1)y_n(0)$$

Since, $y(0) = 1$, then on putting $n = 0, 2, 4, \dots$ we get

$$y_2(0) = -1, y_4(0) = 3, y_6(0) = -15\dots$$

Also $y_1(0) = 0$, then on putting $n = 1, 3, 5, \dots$, we get

$$y_3(0) = 0, y_5(0) = 0, \dots$$

Then, by Taylor's series method, we have

$$y(x) = y(0) + xy_1(0) + \frac{x^2}{2!}y_2(0) + \frac{x^3}{3!}y_3(0) + \frac{x^4}{4!}y_4(0) + \frac{x^5}{5!}y_5(0) + \dots$$

$$\Rightarrow \qquad y(x) = 1 - \frac{x^2}{2!} + \frac{3}{4!}x^4 + \frac{15}{6!}x^6 + \dots$$

$$= 1 - \frac{x^2}{2} + \frac{1}{8}x^4 - \frac{1}{48}x^6 + \dots \qquad \qquad \dots(3)$$

Using (1) and (3), we get

$$z(x) = y'(x) = -x + \frac{1}{2}x^3 - \frac{1}{8}x^5 + \dots$$

$$= -x\left[1 - \frac{x^2}{2} + \frac{1}{8}x^4 - \dots\right] = -xy \qquad \qquad \dots(4)$$

Putting $x = 0.1$ in (3), we get

$$y(0.1) = 1 - \frac{(0.1)^2}{2} + \frac{1}{8}(0.1)^4 - \frac{1}{48}(0.1)^6$$
$$= 0.995$$

Further putting $x = 0.2$, 0.3 in (3), we get

$$y(0.2) = 1 - \frac{(0.2)^2}{2} + \frac{1}{8}(0.2)^4 - \frac{1}{48}(0.2)^6 + \ldots$$
$$= 0.9802$$

and

$$y(0.3) = 1 - \frac{(0.3)^2}{2} + \frac{1}{8}(0.3)^4 - \frac{1}{48}(0.3)^6 + \ldots$$
$$= 0.9560$$

From (4), we have

$$z(0.1) = (-0.1)y(0.1) = -0.1 \times 0.995 = -0.0995$$
$$z(0.2) = (-0.2)y(0.2) = -0.2 \times 0.9802 = -0.1960$$
$$z(0.3) = (-0.3)y(0.3) = -0.3 \times 0.9560 = -0.2863$$

Again from (1), we have

$$y'(0.1) = z(0.1) = -0.0995$$
$$y'(0.2) = z(0.2) = -0.1960$$
$$y'(0.3) = z(0.3) = -0.2863$$

Then from (2), $z'(z) = -(xz + y)$. Thus, we get

$$z'(0.1) = -[0.1z(0.1) + y(0.1)]$$
$$= -[0.1(-0.0995) + 0.995] = -0.985$$
$$z'(0.2) = -[0.2z(0.2) + y(0.2)]$$
$$= -[0.2(-0.1960) + 0.9802] = -0.941$$
$$z'(0.3) = -[0.3z(0.3) + y(0.3)]$$
$$= -[0.3(-0.2863) + 0.9560] = -0.87$$

Further, applying Milne's predictor formula to find $z(0.4)$ and then $y(0.4)$. Taking $h = 0.1$, we have

$$z(0.4) = z(0) + \frac{4}{3}h[2z'(0.1) - z'(0.2) + 2z'(0.3)]$$

$$= 0 + \frac{4}{3}(0.1)[2(-0.985) - (-0.941) + 2(-0.87)]$$

$$= \frac{4}{3}(0.1)(-2.769) = -0.3692$$

$$y(0.4) = y(0) + \frac{4h}{3}[2y'(0.1) - y'(0.2) + 2y'(0.3)]$$

$$= 1 + \frac{4}{3}(0.1)[2(-0.995) - (-0.1960) + 2(-0.2863)]$$

$$= 1 + \frac{4}{3}(0.1)(-0.5756) = 0.9233$$

Also

$$z'(0.4) = -[0.4z(0.4) + y(0.4)] = -[0.4(0.3692) + (0.9253)]$$
$$= -0.7754$$

Applying Milne's corrector formula, we get

$$z(0.4) = z(0.2) + \frac{h}{3}[z'(0.2) + 4z'(0.3) + z'(0.4)]$$

$$= -0.196 + \frac{0.1}{3}[-0.941 + 4(-0.87) + (-0.7754)] = -0.3692$$

and

$$y(0.4) = y(0.2) + \frac{h}{3}[y'(0.2) + 4y'(0.3) + y'(0.4)]$$

$$= y(0.2) + \frac{h}{3}[z(0.2) + 4z(0.3) + z(0.4)]$$

$$= 0.9802 = \frac{0.1}{3}[-0.196 + 4(-0.2863) + (-0.3692)] = 0.9232$$

Hence, $y(0.4) = 0.9232$ and $z(0.4) = -0.3692$.

EXERCISE 9.3

1. Solve the differential equation $y' = x^3 - y^2 - 2$, by Milne's method for $x = 0.3$ to $x = 0.6$ with initial value are $x(0) = 1$. The values of y for $x = -0.1, 0.1$ and 0.2 are to be computed by series expansion.

2. Using Milne's method, obtain $y_{(0.4)}$ and $y_{(0.5)}$ that satisfy the solution of $y' = 2 + y - 2x$ with $y(0) = 1$.

3. Find the value of y corresponding to $x = 0.08$ and $x = 0.10$ that satisfy the solution of $y' = x + y^2$ with $y_{(0)} = 1 (h = 0.02)$.

4. By Milne's method solve $y' = 2 - xy^2$ with $y(0) = 1$ for $x = 1$ taking $h = 0.2$.

5. By Milne's method solve $y_{(0.3)}$ from $y' = x^2 + y^2, y_{(0)} = 1$. Find the initial values $y_{(-0.1)}, y_{(0.1)}$ and $y_{(0.2)}$ from the Taylor's series method. (Rohtak–2004, 07)

6. Find the solution of $y' = x + y, y(0) = 0$ for $0.4 \leq x \leq 1.0$ with $h = 0.1$ by the predictor-corrector formulae.

7. The differential equation $\frac{dy}{dx} = x + y, y(0) = 1$, satisfies the following values :

x	0.1	0.2	0.3
y	1.1103	1.2428	1.3897

8. Solve, by Milne's method, the differential equation, $\frac{dy}{dx} = y - x^2$, when $y = 1$ at $x = 0$, $y = 1.1219$ at $x = 0.2$, $y = 1.4682$ at $x = 0.4$ and $y = 0.1737$ at $x = 0.6$ and compute at $x = 0.8$.

9. Solve the initial value problem
$$\frac{dy}{dx} = 1 + xy^2, y(0) = 1$$
for $x = 0.4, x = 0.5$, using Milne's method when it is given that $y(0.1) = 1.105$, $y(0.2) = 1.223, y(0.3) = 1.355$.

10. Given $2\frac{dy}{dx} = (1 + x^2)y^2$ and $y(0) = 1$, $y(0.1) = 1.06, y(0.2) = 1.12, y(0.3) = 1.21$.
Evaluate $y(0.4)$ by Milne's method.
(VTU–2011S, Madras–2003)

11. Using Milne's method, find $y(4.5)$ given $5xy' + y^2 - 2 = 0$ given $y(4) = 1$, $y(4.1) = 1.0049, y(4.2) = 1.0097, y(4.3) = 1.0143$; $y(4.4) = 1.0187$. (Anna-2007)

12. Given $y = x(x^2 + y^2)e^{-x}, y(0) = 1$, find y at $x = 0.1; 0.2$ and 0.3 by Taylor's series method and compute $y(0.4)$ by Milne's method.
(Anna-2007)

13. Using Runge-Kutta method of order 4, find y for $x = 0.1, 0.2, 0.3$ given that $dy/dx = xy + y^2$, $y(0) = 1$. Continue the solution at $x = 0.4$ using Milne's method. (SVTU-2007)

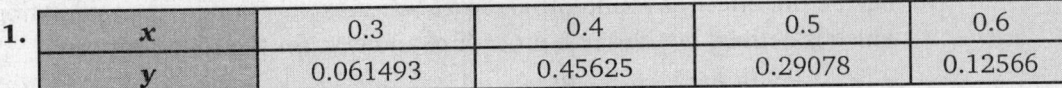

Answers

1.

x	0.3	0.4	0.5	0.6
y	0.061493	0.45625	0.29078	0.12566

2. 2.2918, 2.6488   **3.** 1.09035, 1.11649   **4.** 1.6505   **5.** 1.4392
6. 0.0918, 0.1487, 0.2221, 0.3138, 0.4255, 0.5596, 0.7183 **7.** 1.5836 **8.** 2.011 **9.**   1.538
10. 1.2797 **11.** $y(4.5) = 1.023$ **12.** $y(0.1) = 1.005, y(0.2) = 1.018, y(0.3) = 1.04, y(0.4) = 1.071$
13. $y(0.1) = 1.1169, y(0.2) = 1.2773, y(0.3) = 1.504, y(0.4) = 1.8392.$

9.11 ERROR ANALYSIS

Sometimes the numerical solutions of differential equations are different from their exact solution. The difference between the computed value y_i and the true value $y(x_i)$ at any stage is known as the total error. The total error at any stage is comprised of truncation error and round-off error.

We want to minimize the errors and obtain the solutions with the least errors. It is usually not possible to follow error development quite closely. We can make only rough estimates.

In some method, the truncation error can be reduced by taking smaller sub intervals. The round-off error cannot be minimized easily unless the computer used has the double precision arithmetic facility.

We know that the truncation error in Euler's method is $\frac{1}{2}h^2 y_n''$, i.e., of $O(h^2)$ while that of modified Euler's method is $\frac{1}{2}h^3 y_n'''$, i.e., of $O(h^3)$. Similarly in the fourth order Runge-Kutta method, the truncation error is of $O(h^5)$.

In the Milne's method, the truncation error due to predictor formula is $\frac{14}{45}y_n^v h^5$ and due to corrector formula $= -\frac{1}{90}y_n^v h^5$.

i.e., the truncation error in Milne's method is also of $O(h^5)$.

The relative error of an approximate solution is the ratio of the total error to the exact value.

SOLVED EXAMPLE

EXAMPLE. *Applying Euler's method to the differential equation,* $\dfrac{dy}{dx} = f(x, y), y(x_0) = y_0$ *estimate the total error. When* $f(x, y) = -y, y(0) = 1$, *compute this error neglecting the round-off error.*

SOLUTION.   The Euler's solution of the given differential equation is

$$y_{n+1} = y_n + hf(x_n, y_n) \text{ where } x_n = x_0 + nh$$

i.e., $$y_{n+1} = y_n + hy_n' \qquad \qquad ...(1)$$

Denoting the exact solution of the given equation at $x = x_n$ by $y(x_n)$ and expanding $y(x_{n+1})$ by Taylor's series, we obtain

$$y(x_{n+1}) = y(x_n) + hy'(x_n) + \frac{h^2}{2!}y^n(\xi_n), x_n \le \xi_n \le x_{n+1} \qquad \qquad ...(2)$$

The truncation error

$$T_{n+1} = y(x_{n+1}) - y_{n+1} = \frac{1}{2}h^2 y^n(\xi_n)$$

Thus the truncation error is of $O(h^2)$ as $h \to 0$.

To include the effect of round-off error R_n, we introduce a new approximation $\bar{y}_n$ which is defined by the same procedure allowing for the round-off error

$$\bar{y}_{n+1} = \bar{y}_n + hf(x_n, \bar{y}_n) - R_{n+1} \qquad \qquad ...(3)$$

The total error is defined by

$$E_{n+1} = y(x_{n+1}) - \bar{y}_{n+1} \hspace{3cm} [(2)-(3)]$$

$$= y(x_n) + hy'(x_n) + \frac{h^2}{2!} y''(\xi_n)$$
$$\{\bar{y}_n + hf(x_n, \bar{y}_n) - R_{n+1}\}$$
$$= [y(x_n) - \bar{y}_n] + h[y'(x_n)$$
$$- f(x_n, y_n)] + T_{n+1} - R_{n+1} \hspace{2cm} ...(4)$$

Assuming continuity of $\dfrac{\partial f}{\partial y}$ and using Mean-Value theorem, we have

$$f[x_n, y(x_n)] - f(x_n, y_n) = [y(x_n) - y]f_y(x_n, \xi_n),$$

where ξ_n lies between $y(x_n)$ and y_n.

$\therefore$ (4) takes the form

$$E_{n+1} = [y(x_n) - \bar{y}_n][1 + hf_y(x_n, \xi_n)] + T_{n+1} + R_{n+1}$$

or $$E_{n+1} = E_n[1 + hf_y(x_n, \xi_n)] + T_{n+1} + R_{n+1} \hspace{2cm} ...(5)$$

This is the recurrence formula for finding the total error. The first terms on the right hand side is the inherited error the propagation of the error from the previous step y_n to y_{n+1}.

Now, we have $\dfrac{dy}{dx} = -y, y(0) = 1$.

Taking $h = 0.01$ and applying (1) successively, we obtain

$y(0.01) = 1 + 0.01(-1) = 0.99$

$y(0.02) = 0.99 + 0.01(-0.99) = 0.9801$

$y(0.03) = 0.9703, y(0.04) = 0.9606$

The truncation error

$$T_{n+1} = \frac{1}{2} h^2 y''(\xi_n)$$

$$= 0.0005 \, y(\xi) \le 5 \times 10^{-5} \, y(x_n) \hspace{2cm} \left[\because \frac{dy}{dx} \text{ is } -ve\right]$$

i.e., $$T_1 \le 5 \times 10^{-5} y(0) = 5 \times 10^{-5}$$

$$T_2 \le 5 \times 10^{-5} y(0.01) = 5 \times 10^{-5}(0.99)$$
$$< 5 \times 10^{-5}$$

$$T_3 \le 5 \times 10^{-5} y(0.02) = 5 \times 10^{-5}(0.9801)$$
$$< 5 \times 10^{-5}$$

$$T_4 \le 5 \times 10^{-5} y(0.03) = 5 \times 10^{-5}(0.9703)$$
$$< 5 \times 10^{-5}, \text{ etc.}$$

Also $1 + hf_0(x_n, y_n) = 1 + 0.01(-1) = 0.99$.

Neglecting the round-off error and using the above results, (5) gives

$$E_0 = 0,$$

$$E_1 = E_0(0.99) + T_1 \le 5 \times 10^{-5} = 0.00005$$

$$E_2 = E_1(0.99) + T_2 < 5 \times 10^{-5} + 5 \times 10^{-5}$$
$$= 0.0001$$

$$E_3 = E_2(0.99) + T_3 < 10^{-4} + 5 \times 10^{-5}$$
$$= 0.00015$$

$$E_4 = E_3(0.99) + T_4 < 1.5 \times 10^{-4} + 5 \times 10^{-5}$$
$$= 0.0002 \qquad \text{etc.}$$

9.12 CONVERGENCE OF A METHOD

A numerical method for solving a differential equation is said to be convergent if the approximate solution y_n tends the exact solution $y(x_n)$ as h tends to zero provided the rounding errors arising from the initial conditions approach zero. This means that as a method is continually refined by taking smaller and smaller step-sizes, the sequence of approximate solutions must converge to the exact solution.

The Runge-Kutta methods are convergent under similar condition. Predictor-corrector methods are convergent if $f(x, y)$ satisfies Lipschitz condition given by

$$[f(x,y) - f(x,\overline{y})] \le k[y - \overline{y}],$$

k being a constant, then the sequence of approximations to the numerical solution converges to the exact solution.

9.13 STABILITY ANALYSIS

There is a limit to which the step size h can be reduced for minimizing the truncation error, beyond which a further reduction in h will result in the increase of round-off error and hence increase in the total error.

A method is said to be stable if it produces a bounded solution which initiates the exact solution. Otherwise it is said to be unstable. If a method is stable for all values of the parameter, it is said to be absolutely or unconditionally stable. If it is stable for some values of the parameter, it is said to be conditionally stable.

Euler's method and Runge-Kutta method are conditionally stable. The Milne's method is however, unstable since when the parameter is negative, each of the errors is magnified while the exact solution decays.

SOLVED EXAMPLE

EXAMPLE. *Applying Euler's method to the equation* $\dfrac{dy}{dx} = \lambda y$ *given* $y(x_0) = y_0,$ *determine its stability zone. What would be the range of stability when* $\lambda = -1$?
(UPTU-2006)

SOLUTION. We have $y' = \lambda y, y(x_0) = y_0$ ...(1)

By Euler's method,

$$y_n = y_{n-1} + hy'_{n-1}$$
$$= y_{n-1} + \lambda h y_{n-1} = (1 + \lambda h)y_{n-1}$$
$$y_{n-1} = (1 + \lambda h)y_{n-2}$$
$$\cdots \quad \cdots \quad \cdots \quad \cdots \quad \cdots$$
$$y_2 = (1 + \lambda h)y_1$$
$$y_1 = (1 + \lambda h)y_0$$

Multiplying all these equations, we obtain

$$y_n = (1 + \lambda h)^n y_0 \qquad \qquad ...(2)$$

Integrating (1), we get

$$y = ce^{\lambda x}$$

Using $y(x_0) = y_0, y_0 = ce^{\lambda x_0}$

$$y = y_0 e^{\lambda(x - x_0)}$$

In particular, the exact solution through (x_n, y_n) is

$$y_n = y_0 e^{\lambda(x_n - x_0)} = y_0 e^{\lambda nh}$$

$$y_n = y_0 (e^{\lambda h})^n = y_0 \left[1 + \lambda h + \frac{(\lambda h)^2}{2} + ... \right]^n \qquad [\because x_n \stackrel{.}{=} x_0 = nh] \qquad ...(3)$$

Clearly the numerical solution (2) agrees with exact solution (3) for small values of h. The solution (2) increase if $|1 + \lambda h| > 1$.

Hence $\qquad |1 + \lambda h| < 1$ defines a stable zone.

When λ is real, then the method is stable if $|1 + \lambda h| < 1$.

i.e., $\qquad -2 < \lambda, h < 0$

When λ is complex $(= a + ib)$, then it is stable if $|1 + (a + ib)h| < 1$

i.e., $(1 + ah)^2 + (bh)^2 < 1$

i.e., $\qquad (x + 1)^2 + y^2 < 1$, $\qquad \qquad$ [where $x = xh; y = bh$]

i.e., λh lies within the unit circle.

When λ is imaginary $(= ib)$, $|1 + \lambda h| = 1$, then we have a periodic stability.

Hence, Euler's method is absolutely stable if and only if

(i) $\quad$ real $\lambda : -2 < \lambda h < 0$.

(ii) $\quad$ complex $\lambda : \lambda h$ lies within the unit circle, i.e., Euler's method is conditionally convergent.

When $\lambda = -1$, the solution is stable in the range
$$-2 < -h < 0, i.e., 0 < h < 2.$$

EXERCISE 9.4

1. Show that the modified Euler's method is convergent.

2. Applying fourth order Runge-Kutta method to the equation $\dfrac{dy}{dx} = \mu y, y(x_0) = y_0$ and show that the range of absolute stability is $-2.78 < \mu h < 0$.

3. Starting with the equation $y' = \lambda y$, show that the modified Euler's method is relatively stable.

4. Show that the local truncation errors in the Milne's predictor and corrector formula are $\dfrac{14}{45} y^v h^5$ and $-\dfrac{1}{90} y^v h^5$ respectively.

MISCELLANEOUS EXERCISE

1. Find by Taylor's series method the value of y at $x = 0.1$ and $x = 0.2$ to five places of decimals from $\frac{dy}{dx} = x^2 y - 1$, $y(0) = 1$.

   (VTU–2009, Rohtak–2005)

2. Solve $y' = x + y$ given $y(1) = 0$. Find $y(1.1)$ and $y(1.2)$ by Taylor's method. Compare the result with its exact value.

   (JNTU–2008–Anna–2005)

3. Using modified Euler's method, Find $y(0.2)$ and $y(0.4)$ given $y' = y + e^x$, $y(0) = 0$.

   (JNTU–2009)

4. Given $y' = x + \sin y$, $y(0) = 1$. Compute $y(0.2)$ and $y(0.4)$ with $h = 0.2$ using Euler's modified method. (JNTU–2007)

5. Given that $\frac{dy}{dx} = x^2 + y$ and $y(0) = 1$. Find an approximate value of $y(0.1)$ taking $h = 0.05$ by modified Euler's method. (VTU–2010)

6. Find $y(0.1)$ and $y(0.2)$ using Runge–Kutta 4th order formula, given that $y' = x^2 - y$ and $y(0) = 1$. (JNTU–2006)

7. Use fourth order Runge–Kutta method to find y at $x = 0.1$, given that $\frac{dy}{dx} = 3e^x + 2y$, $y(0) = 0$ and $h = 0.1$. (VTU–2006)

8. Using Runge–Kutta method of order 4, find y for $x = 0.1, 0.2, 0.3$ given that $\frac{dy}{dx} = xy + y^2$, $y(0) = 1$. Continue the solution at $x = 0.4$ using Milne's method.

   (VTU–2008, SVTU–2007, Madras–2006)

9. From the data given below, find y at $x = 1.4$, using Milne's predictor–corrector formula :

   $$\frac{dy}{dx} = x^2 + \frac{y}{2}$$

x	1	1.1	1.2	1.3
y	2	2.2156	2.4549	2.7514

10. Find the solution of $\frac{dy}{dx} = 1 + xy$ which passes through $(0, 1)$ in the interval $(0, 0.5)$ such that the value of y is correct to three decimal places (use the whole interval) as one interval only. Take $h = 0.1$. (Agra-2002, 10)

11. Use Runge's method to approximate y at $x = 1.6$, when $y = 0.4$ at $x=1$ and $\frac{dy}{dx} = x - y$. (IAS-2002)

12. Apply Runge-Kutta fourth order method to find an approximate value of y when $x = 0.2$, given that $\frac{dy}{dx} = x + y$ and $y = 1$ when $x = 0$.

    (Agra-2004, Meerut-2002)

Answers

1. $y(0.1) = 0.90033$, $y(0.2) = 0.80227$
2. $y(1.1) = 0.1103$, $y = (1.2) = 0.2428$. Exact $y(1.1) = 0.1103$ and $y(1.2) = 0.2428$
3. $y(0.2) = 0.2468$ and $y(0.4) = 0.6031$ **4.** $y(0.2) = 1.2046$, $y(0.4) = 1.4644$ **5.** 2.2352,
6. $y(0.1) = 0.9052$, $y(0.2) = 0.8213$ **7.** 0.3487   **8.** $y(0.4) = 1.8392$
9. $y(1.4) = 3.0794$ **10.** $y(0) = 1.000$, $y(0.1) = 1.105$, $y(0.2) = 1.203$, $y(0.3) = 1.355$, $y(0.4) = 1.505$, $y(0.5) = 1.677$
11. $(y)_{x=1.6} = 0.8176$.   **12.** $y = 0.2428$

Objective Evaluations

FILL IN THE BLANKS

1. By Runga-Kutta method, the solution of the differential equation $y' = 3x + \dfrac{y}{2}$ with $y_0 = 1$ at $x = 0.1$, is _____ .

2. If $\dfrac{dy}{dx} = -xy^2$, $y(2) = 1$. Then $y(2.2) =$ _____ .

3. The value of $y(0.3)$ from $y' = x^2 + y^2$, $y(0) = 1$ is _____ .

4. If $\dfrac{dy}{dx} = 1 + y^2$, $y(0) = 0$ then $y(.2) =$ _____ .

5. Using Picard's method, the value of y when $y(0) = 1$ and $\dfrac{dy}{dx} = 1 + xy$ is _____ .

TRUE/FASLE

Write 'T' for True and 'F' for False statement.

1. The milne's predictor method is
$$y_4 = y_0 + \frac{4h}{3}(2f_1 - f_2 + 2f_3)$$ **(T/F)**

2. The milne's corrector method is
$$y_4 = y_0 + \frac{4h}{3}(2f_1 - f_2 + 2f_3) .$$ **(T/F)**

3. Runga-Kutta method of fourth order is a single step method. **(T/F)**

4. To solve the differential equation numerically the oldest method is Euler's method. **(T/F)**

5. If $\dfrac{dy}{dx} = y - x$ with $y(0) = 2$. Then $y(0.1) = 1.2052$. **(T/F)**

MULTIPLE CHOICE QUESTIONS

Choose the most appropriate one.

1. Euler's method is given by
   (a) $y_n = hf(x_n, y_n)$
   (b) $y_{n+1} = y_n + hf(x_n, y_n)$
   (c) both (a) and (b) are true
   (d) none of these

2. For modified Euler's method which of the following is/are true
   (a) $y_1^{(0)} = y_0 + hf(x_0, y_0)$
   (b) $y_1^{n+1} = y_0 + \dfrac{h}{2}f(x_0, y_0) = f(x_1, y_1^{(x)})$
   (c) both (a) and (b) are true
   (d) none of these

3. For Picard's method which of the following is/are true :
   (a) $y^{(1)} = y_0 + \int_{x_0}^{x} f(x, y_0)dx$
   (b) $y^{(n)} = y_0 + \int_{x_0}^{x} f(x, y^{(n-1)})dx$
   (c) both (a) and (b) are true
   (d) none of these

4. The Milne's predictor method is given by :
   (a) $y_4 = y_0 + \dfrac{4h}{3}(2f_1 - f_2 + 2f_3)$
   (b) $y_4 = y_2 + \dfrac{h}{3}(f_2 + 4f_3 + f_4)$
   (c) both (a) and (b) are true
   (d) none of these

5. Milne's corrector method is given by :
   (a) $y_4 = y_0 + \dfrac{4h}{3}(2f_1 - f_2 + 2f_3)$
   (b) $y_4 = y_2 + \dfrac{h}{3}(f_2 + 4f_3 + f_4)$
   (c) both (a) and (b) are true
   (d) none of these

6. Which method is not applicable for solving differential equation
   (a) Picard's method
   (b) Runge-Kutta method
   (c) Euler's method
   (d) Gause-Seidal method

7. The formula $y_1 = y_0 + \dfrac{h}{3}(f_0' + 4f_1' + f_2')$

   is called : [Agra–2012, Garhwal–2010]
   (a) Milne's Corrector method
   (b) Runge-Kutta method
   (c) Picard's formula
   (d) none of these

8. Runge-Kutta fourth order method is
   - (a) a single step method
   - (b) multiple step method
   - (c) both (a) and (b) are true
   - (d) none of these

9. To solve the differential equation numerically the oldest method is
   - (a) Euler's method
   - (b) Taylor's method
   - (c) Runge-Kutta method
   - (d) Picard's method

10. If $y' = \dfrac{y-x}{y+x}$ with $y_0 = 1$. Then value of y for $x = 0.1$ in 4 steps (By Euler's method) is given by :
    - (a) 1
    - (b) 0
    - (c) 1.0932
    - (d) none of these

11. If $y' = y^2 - x^2$, with $y = 1$ when $x=0$ then $y(0.5) =$
    - (a) 1.78
    - (b) 1.76393
    - (c) 1.89
    - (d) none of these

12. The solution of the differential equation $\dfrac{dy}{dx} = xy$ with $y(1) = 5$ in the interval $(1, 1.5)$, $h = 0.1$ is given by :
    - (a) 8.819
    - (b) 8.809
    - (c) 9.8
    - (d) none of these

13. By Euler's method the solution of $\dfrac{dy}{dx} = x + y$, $y = 0$ when $x = 0$ to $x = 1.0$ taking $h = 0.2$ is given by:
    - (a) 1.6
    - (b) 0.63715
    - (c) 1.67
    - (d) none of these

14. By Euler's method, the solution of $\dfrac{dy}{dx} = x^2 + y^2$, $y(0) = 0$ in the range $0 \le x \le 0.5$ is :
    - (a) 0.55
    - (b) 0.055
    - (c) 5.5
    - (d) none of these

15. If $\dfrac{dy}{dx} = \log_{10}(x + y)$ and $y(0) = 1$. Then $y(0.2) =$
    - (a) 1.6
    - (b) 1.0082
    - (c) 1.82
    - (d) none of these

16. If $\dfrac{dy}{dx} = -xy^2$, $y(2) = 1$. Then $y(2.2) =$
    - (a) 1.708
    - (b) 0.7018
    - (c) 7.18
    - (d) none of these

17. The value of y for $x = 0.1$, if $\dfrac{dy}{dx} = x + y$, $y(0) = 1$ (by Taylor series) is given by
    - (a) 1.10
    - (b) 1.11033
    - (c) 1.13
    - (d) none of these

18. If $5xy^1 + y^2 - 2 = 0$ with $y(4) = 1$. Then by Taylor series value of $y(4.4) =$
    - (a) 1.81
    - (b) 1.0187
    - (c) 1.83
    - (d) none of these

19. If $y' = x - y^2$, $y(0) = 1$. Then by Taylor's series, the value of $y(0.1) =$
    - (a) 0.9138
    - (b) 9.138
    - (c) 1.938
    - (d) none of these

20. If $y' = x + y^2$ with $y(0) = 1$. Then by Picard's method, solution is given by:
    - (a) $1 + x + \dfrac{3x^2}{2}$
    - (b) $1 + x + \dfrac{3}{2}x^2 + \dfrac{2}{3}x^3 + \dfrac{x^4}{4} + \dfrac{x^5}{20}$
    - (c) both (a) and (b) are true
    - (d) none of these

21. While numerically solving the differential equation $\dfrac{dy}{dx} + 2xy^2 = 0$, $y(0) = 1$ using Euler's predictor-corrector (improved Euler-Cauchy) with a step size of 0.2, the value of y after the first step is: (GATE (IN)–2013)
    - (a) 0.96
    - (b) 1.00
    - (c) 0.93
    - (d) none of these

22. The differential equation $\left(\dfrac{dy}{dx}\right) = 0.25y^2$ is to be solved using the backward (implicit) Euler's method with the boundary condition $y = 1$ at $x = 0$ and with a step size of 1. What would be the value of y at $x = 1$?   (GATE (CE)–2006)
    - (a) 1.33
    - (b) 2.00
    - (c) 2.33
    - (d) none of these

23. The differential equation $(dx/dt) = [(1-x)/\tau]$ is discretised using Euler's numerical integration method with a time step $\Delta T > 0$. What is the maximum permissible value of ΔT to ensure stability of the solution of the corresponding discrete one equation?   (GATE (EE)–2007)
    - (a) 1
    - (b) τ
    - (c) 2τ
    - (d) none of these

24. Consider a differential equation

$\dfrac{dy(x)}{dx} - y(x) = x$ with the initial condition

$y(0) = 0$. Using Euler's first order method

with a step size of 0.1, the value of $y(0.3)$ is:

(GATE (EC)–2010)

(a) 0.1   (b) 0.01

(c) 0.031   (d) none of these

Answers

FILL IN THE BLANKS

  1. 1.066652421875   **2.** 0.7018   **3.** 1.4392   **4.** 02027   **5.** $1 + x + \dfrac{x^2}{2} + \dfrac{x^3}{3} + \dfrac{x^4}{8} + \dfrac{x^5}{15} + \dfrac{x^6}{18}$

TRUE/FALSE

  1. T   **2.** F   **3.** T   **4.** T   **5.** F

MULTIPLE CHOICE QUESTIONS

  1. (b)   **2.** (c)   **3.** (c)   **4.** (a)   **5.** (b)   **6.** (d)   **7.** (a)   **8.** (a)   **9.** (a)

 10. (c)   **11.** (b)   **12.** (b)   **13.** (b)   **14.** (b)   **15.** (b)   **16.** (b)   **17.** (b)   **18.** (b)

 19. (a)   **20.** (b)   **21.** (a)   **22.** (b)   **23.** (c)   **24.** (c)

COMPUTATIONAL TECHNIQUE LAB

1. EULER'S METHOD

SYMBOLS USED

x_0, y_0 = Initial values of x and y

h = length of subintervals of step

x = the value of x at which we have to find y

n = number of subintervals

ALGORITHM : EULER'S METHOD

Step 1 :	Start
Step 2 :	Define df (x, y)
Step 3 :	Input n, x_0, y_0, x
Step 4 :	h = $(x-x_0)/n$
Step 5 :	$x_1 = x_0$, $y_1 = y_0$, i = 0
Step 6 :	Perform steps 7 to 9 while (i < n)
Step 7 :	$y_1 = y_1 + h * df (x_1, y_1)$
Step 8 :	$x_1 = x_1 + h$
Step 9 :	i = i + 1
Step 10 :	Print x_1, y_1
Step 11 :	Stop.

FLOW CHART : EULER'S METHOD

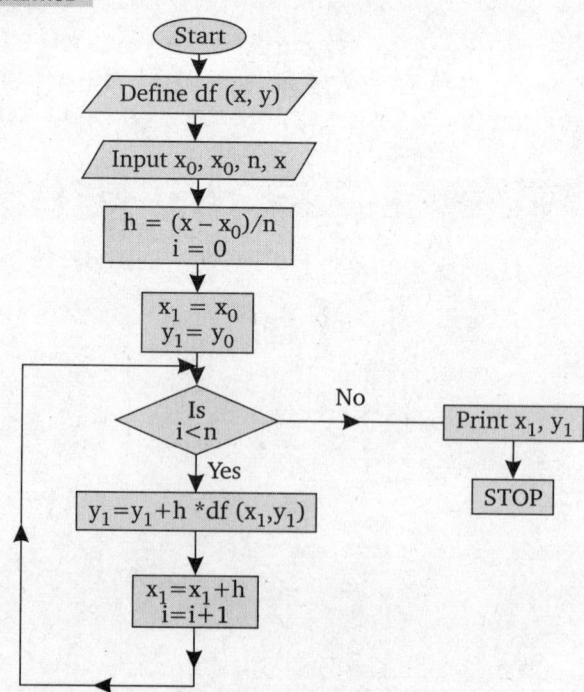

PROGRAM : *Following is a C program to find the value of y at x for the equation* $\frac{dy}{dx} = df(x, y) = x + y$ *by the Euler's method.*

```c
// EULER'S METHOD  TO FIND VALUE OF Y AT X=1 FOR EQUATION dy/dx = X+Y
#include<stdio.h>
#include<conio.h>
 float df(float x, float y)
 {
 return x+y;
 }

 void main()
   {
   clrscr();
   float x0,y0,h,x,x1,y1,n,i;

   puts("\n enter value of x0 ");
   scanf("%f", &x0);
   puts("\n enter the value of y0 ");
   scanf("%f", &y0);
   puts("\n enter the value of n ");
   scanf("%f", &n);
   puts("\n enter the value of x ");
   scanf("%f", &x);
   h=(x-x0)/n;
   x1=x0;
   y1=y0;

   for(i=0;i<n;i++)
    {
      y1+=h*df(x1,y1);
      x1+=h;
    }
      printf(" When x =%3.1f  y = %4.2f\n",x1,y1);
      getch();

   }
```

Output: EULER'S METHOD TO FIND THE VALUE OF Y(X)=1 FOR EQUATION X+Y = (dy/dx)

```
enter value of x0
0


enter value of y0
1
```

```
enter value of n
10
enter value of x
1

When x = 1   y = 3.18748
```

LAB ASSIGNMENT : EULER'S METHOD

1. Write a C program to find y for x = 0.1 for the equation $\dfrac{dy}{dx} = \dfrac{y-x}{y+x}$ with initial condition $y = 1$ at $x = 0$, dividing the interval into 5 equal parts.

   Hint : Define df = (y–x)/(y+x)

           Input : n = 5, $x_0 = 0$, $y_0 = 1$, x = 0.10

       Output : y (0.10) = 1.09283

2. Write a C program for the Euler's method to solve the differential equation $\dfrac{dy}{dx} = y^2 - x^2$ for y (0.5) with y = 1 when x = 0.

   Hint : Define df = y* y – x * x

           Input : $x_0 = 0$, $y_0 = 1$, x = 0.5, n = 5

       Output : y (0.5) = 1.76393

2. MODIFIED EULER'S METHOD

SYMBOLS USED

x_0, y_0 = Initial values of x and y
x = value of x at which we have to find y
n = number of subintervals
h = step size

ALGORITHM: MODIFIED EULER'S METHOD

Step 1 :	Start
Step 2 :	define df (x, y)
Step 3 :	Input x_0, y_0, x, n
Step 4 :	h = (x – x_0)/n
Step 5 :	for i = 0 to n – 1
Step 6 :	$y_1 = y_0 + h * f (x_0 + i * h, y_0)$
Step 7 :	$y = y_0 + h * (f (x_0 + i * h, y_0) + (f (x_0 + (i + 1) * h, y_1))/2$
Step 8 :	t = y_1
Step 9 :	y_1 = y
Step 10 :	Repeat step 7 to 9 until ($|y - t| <= 0.0005$)
Step 11 :	$y_0 = y_1$
Step 12 :	End of for loop
Step 13 :	Print y
Step 14 :	Stop.

FLOW CHART

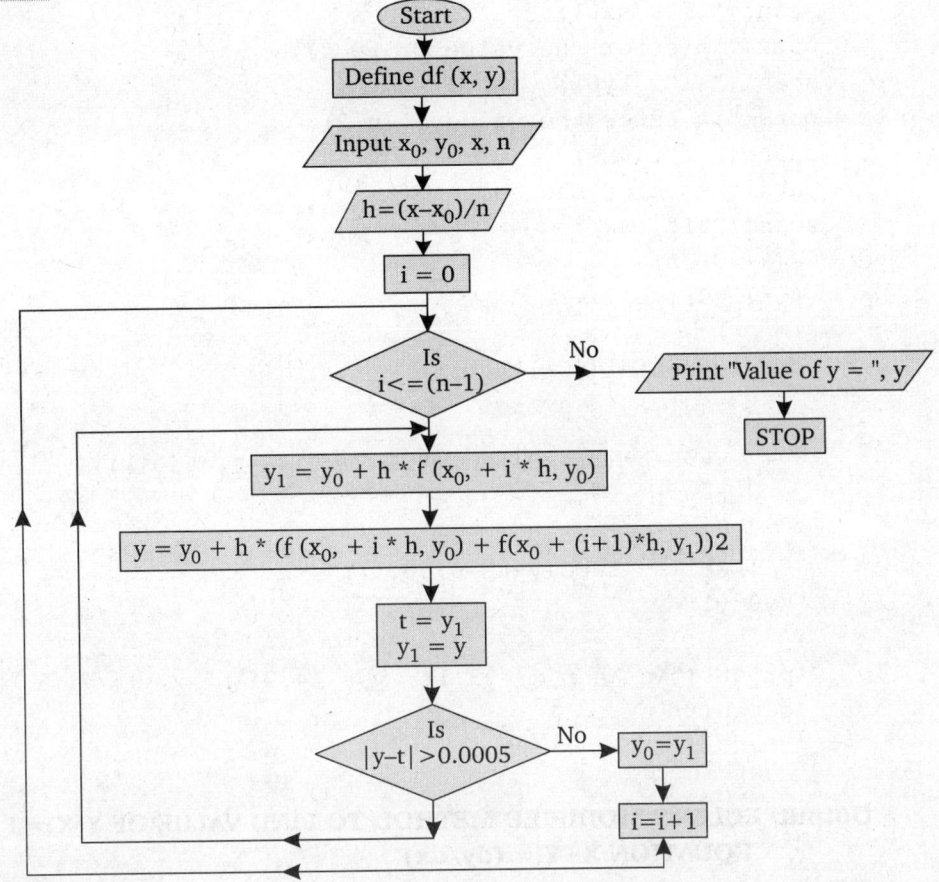

PROGRAM. *Following is a C program to find the value of y at x = 0.05 for the equation*
$\frac{dy}{dx} = x + y$ *by modified Euler's method with initial condition y(0) = 1.*

```
// Modified Euler's method for df = x + y.
// EULER'S MODIFIED METHOD  TO FIND VALUE OF Y AT X =1 FOR
// EQUATION dy/dx = X+Y
#include<stdio.h>
#include<conio.h>
#include<math.h>
 float df(float x, float y)
 {
 return  x+y;
 }
 void main()
   {
      clrscr();
   float x0,y0,h,x,x1,y1,y,t;
   int n,i;
```

```
puts("\n enter value of x0 ");
scanf("%f", &x0);
puts("\n enter the value of y0 ");
scanf("%f", &y0);
puts("\n enter the value of n ");
scanf("%f", &n);
puts("\n enter the value of x ");
scanf("%f", &x);
h=(x-x0)/n;
 for(i=0;i<n;i++)
 {
     y1=y0+h*df(x0+i*h,y0);

     do
     {   y=y0+h*(df(x0+i*h,y0)+df(x0+(i+1)*h,y1))/2;
         t=y1;
         y1=y;
     } while (fabs(y-t)>0.0005);
   y0=y1;
 }
  printf("\n When x =%3.1f  y = %4.2f\n",x,y);
  getch();


 }
```

Output: EULER'S MODIFIED METHOD TO FIND VALUE OF Y(X)=1 FOR EQUATION X+Y = (dy/dx)

```
enter value of x0
0
enter value of y0
1
enter value of n
5
enter value of x
.05


When x = .05  y = 1.116388
```

LAB ASSIGNMENT : MODIFIED EULER'S METHOD

1. Write a C program to find y (2.2) by Euler's modified method for $\frac{dy}{dx} = -xy^2$ where y (2) = 1.

   Hint : Define df = – x * y * y

Input : $x_0 = 2$, $y_0 = 1$, $x = 2.2$, $n = 2$

Output : $y(2.2) = 0.7018$

2. Write a C program to find $y(0.2)$ for $\dfrac{dy}{dx} = \log_{10}(x + y)$ with initial condition $y = 1$ for $x = 0$ by modified Euler's method.

Hint : Define df = log 10 (x + y)

Input : $x_0 = 0$, $y_0 = 1$, $x = 0.2$, $n = 1$

Output : $y(0.2) = 1.0082$

3. RUNGE-KUTTA METHOD

SYMBOLS USED

x_0, y_0 = initial values of x and y

h = length of subinterval

x = value of x at which we have to find y

n = number of subdivision

ALGORITHM : RUNGE-KUTTA 4TH ORDER

Step 1 :	Start
Step 2 :	Define f (x, y)
Step 3 :	Input n, x_0, y_0, x
Step 4 :	$h = (x - x_0)/n$
Step 5 :	For i = 0 to (n – 1)
Step 6 :	$x = x_0 + i * h$
Step 7 :	$k_1 = h * f(x, y_0)$
Step 8 :	$k_2 = h * f(x + h/2), y_0 + k_1/2)$
Step 9 :	$k_3 = h* f(x + h/2, y_0 + k_2/2)$
Step 10 :	$k_4 = h * f(x + h, y_0 + k_3)$
Step 11 :	$k = (k_1 + 2 * k_2 + 2 * k_3 + k_4)/6$
Step 12 :	$y + y_0 + k$
Step 13 :	$y_0 = y$
Step 14 :	Print y
Step 15 :	End of For Loop.
Step 16 :	Stop.

FLOW CHART : RUNGE-KUTTA 4TH ORDER

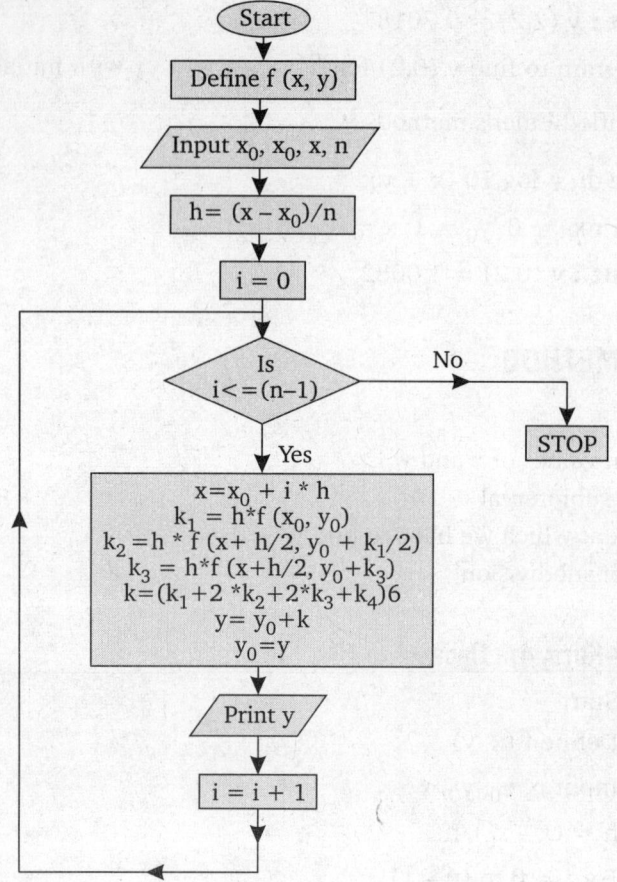

Start

Define f (x, y)

Input x_0, x_0, x, n

$h = (x - x_0)/n$

i = 0

Is
$i <= (n-1)$ → No → **STOP**

Yes

$$x = x_0 + i * h$$
$$k_1 = h*f(x_0, y_0)$$
$$k_2 = h * f(x + h/2, y_0 + k_1/2)$$
$$k_3 = h*f(x+h/2, y_0 + k_3)$$
$$k = (k_1 + 2*k_2 + 2*k_3 + k_4)6$$
$$y = y_0 + k$$
$$y_0 = y$$

Print y

i = i + 1

PROGRAM . *Following is a C program to find the value of the equation x+y for x = 0.2 for initial condition y = 1 at x = 0 by Runga Kutta 4th order.*

```
/* RUNGE KUTTA 4th ORDER METHOD FOR dy/dx = X+y²*/
#include<stdio.h>
#include<conio.h>

float f(float x, float y)
{
return x+y*y;
}
void main()
   {
   clrscr();
   float x0,y0,h,x,y,k1,k2,k3,k4,k;
   int i,n;
   puts("\n enter the value of x0 ");
   scanf("%f", &x0);
   puts("\n enter the value of y0 ");
   scanf("%f", &y0);
```

```
    puts("\n enter the value of n ");
    scanf("%f", &n);
    puts("\n enter the value of x ");
    scanf("%f", &x);
    h=(x-x0)/n;
        for(i=0;i<=n-1;i++)
        {
        x=x0+i*h;
        k1=h*f(x,y0);
        k2=h*f(x+h/2,y0+k1/2);
        k3=h*f(x+h/2,y0+k2/2);
        k4=h*f(x+h,y0+k3);
        k=(k1+2*(k2+k3)+k4)/6;
        y=y0+k;
        y0 =y;
        printf("\n  When x = %8.4f"
         "  y = %8.4f\n",x,y);
        getch();
        }

    }
```

Output: RUNGE KUTTA 4th ORDER METHOD FOR X+Y2
```
enter the value of x0
0
enter the value of y0
1
enter the value of n
2
enter the value of x
.2
When x = .2   y = 1.27356
```

LAB ASSIGNMENT : RUNGE-KUTTA 4TH ORDER

1. Write a C program to find y (0.4) dy/dx = $- 2xy^2$ with y (0) = 1 by Runga-Kutta 4th order.

   Hint : Define f (x, y) = $- 2 * x * y * y$

         Input : $x_0 = 0$, y_0, y = 1, x = 0.4, n = 2

         Output : y (0.4) = 0.86205

2. Write a C program to find the solution of the differential equation $\dfrac{dy}{dx} = 3x + \dfrac{1}{2}y$ with $y_0 = 1$ at x = 0.1 by Runge-Kutta 4th Order.

   Hint : Define f (x, y) = $3 * x + y/2$

         Input : $x_0 = 0$, $y_0 = 1$, x = 0.1, n =1

         Output : y (0.1) = 1.06665

3. Write a C program using Runga-Kutta 4th order method to solve $\dfrac{dy}{dx} = \dfrac{y^2 - x^2}{y^2 + x^2}$ with y (0) = 1 at x = 0.2, 0.4.

   Hint : Define f (x, y) = $(y * y - x * x)/(y * y + x * x)$

   (a) **Input :** $x_0 = 0$, $y_0 = 1$, x = 0.2, n = 1

         Output : y (0.2) = 1.1960

   (b) **Input :** $x_0 = 0$, $y_0 = 1$, x = 0.4, n = 2

         Output : y (0.4) = 1.37527

4. MILNE'S PREDICTOR CORRECTOR METHOD

SYMBOLS USED

x_0, y_0 = initial values of x and y

h = length of subinterval

x = the value of x at which we have to find y

n = number of subdivision

err = allowed error.

ALGORITHM : MILNE'S PREDICTOR-CORRECTOR METHOD

Step 1 :   Start

Step 2 :   define f (x, y)

Step 3 :   Input x_0, y_0, x, n

Step 4 :   h $(x - x_0)/n$

Step 5 :   ay [0] = y_0, err = 0.0005

Step 6 :   For i to 2

Step 7 :   x = x_0 + i * h

Step 8 : k_1 = h * f (x, y_0)

Step 9 : k_2 = h * f (x + h/2, y_0 + k_1/2)

Step 10 : k_3 = h * f (x + h/2, y_0 + k_2/2)

Step 11 : k_4 = h * f (x + h, y_0 + k_3)

Step 12 :   k = (k_1 + 2 * k_2 + 2 * k_3 + k_4)/6

Step 13 :   y = y_0 + k

Step 14 :   ay [i + 1] = y

Step 15 : y_0 = y

Step 16 :   End of For loop

Step 17 :   Print "Starting fair values for the Milne's method by Runga-Kutta method of order 4 are"

Step 18 :   For i = 0 to 3

Step 19 :   Print ay [i]

Step 20 : f_1 = f (x_0 + h, ay [i])

Step 21 : f_2 = f (x_0 + 2 * h, ay [2])

Step 22 : f_3 = f (x_0 + 3 * h, ay [3])

Step 23 :   Repeat steps 24 to 31 until (i > n)

Step 24 : y_4 = ay [i – 4] + (4 * h * (2 * f_1 – f_2 + 2 * f_3))/3

Step 25 :   Repeat steps 26 to 28 until ([y_4 – t| < err)

Step 26 :   t = y_4

Step 27 : f_4 = f (x_0 + i * h, y_4)

Step 28 : y_4 = ay [i – 2] + h*(f_2 + 4 * f_3 + f_4)/3

Step 29 :   print "Value of y at x =", x_0 + i * h, ay [i]

Step 30 :   i = i + 1

Step 31 : f_1 = f_2, f_2 = f_3, f_3 = f_4

Step 32 :   Stop.

FLOW CHART : MILNE'S PREDICTOR CORRECTOR METHOD

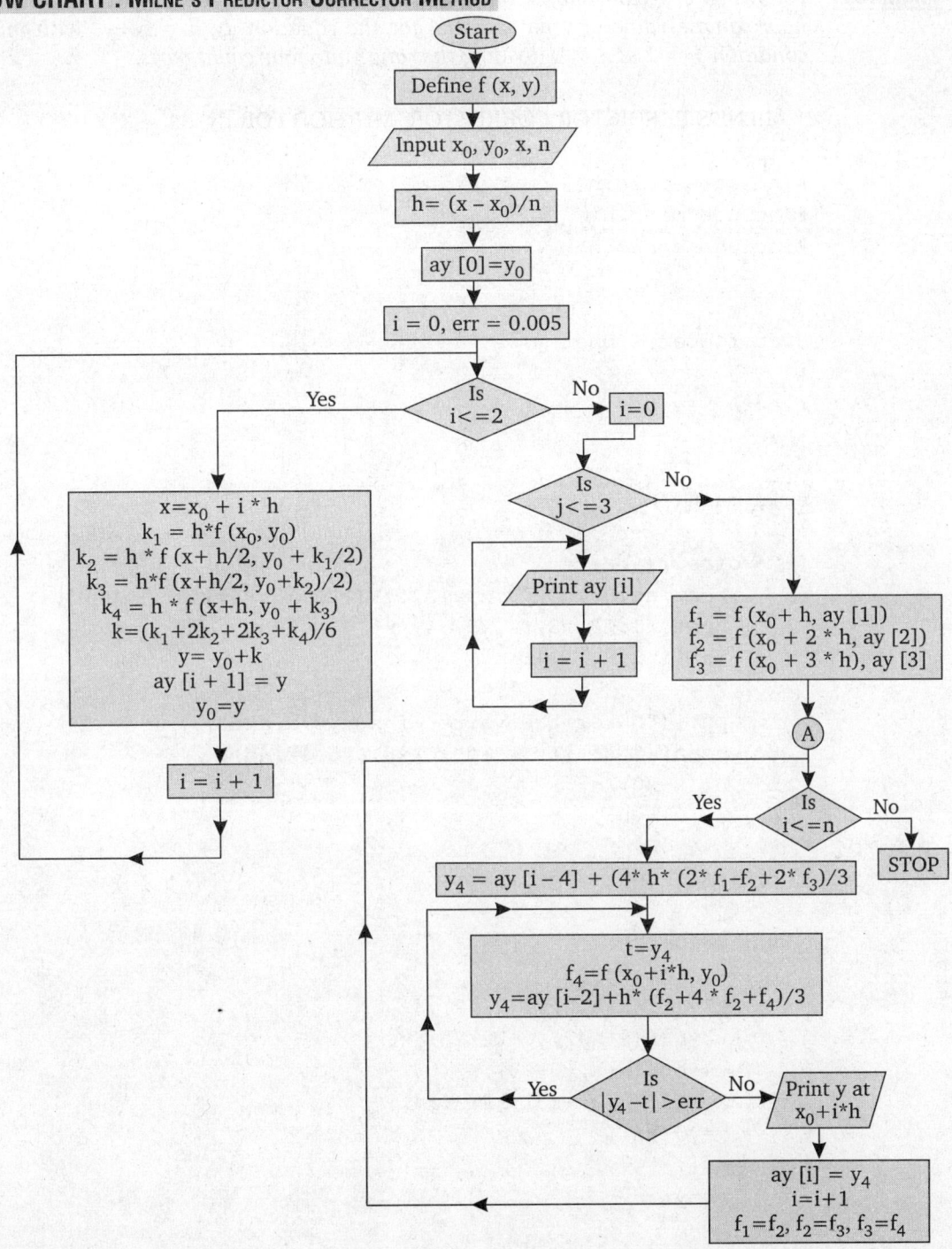

PROGRAM . *Following program shows the Milne's Predictor-Corrector Method to find the approximate value of y for x = 0.4 for the equation dy/dx=xy+y² with initial condition y = 1 at x = 0 dividing the range into four equal parts.*

```c
/* MILNE'S PREDICTOR CORRECTOR  METHOD FOR dy/dx = X*Y+Y*Y   */

#include<stdio.h>
#include<math.h>
#include<conio.h>


float f(float x,float y)
{
return x*y+y*y;
}

void main()
{
    clrscr();
    float ay[5],x0,y0,x,y,h,t,k1,k2,k3,k4,k,err=.0005;
    float f1,f2,f3,f4,y2,y4;
    int i,n;

    puts("Enter the value of x0,y0,x,n\n\n");
    scanf(" %f %f %f %d", &x0,&y0,&x,&n);
    h=(x-x0)/n;
    ay[0]=y0;

   for(i=0;i<n;i++)
    {
        x=x0+i*h;
        k1=h*f(x,y0);
        k2=h*f(x+h/2,y0+k1/2);
        k3=h*f(x+h/2,y0+k2/2);
        k4=h*f(x+h,y0+k3);
        k=(k1+2*(k2+k3)+k4)/6;
        y=y0+k;
        ay[i+1]=y;
        y0=y;
    }

    printf("\n Starting 4 values  by runge -kutta mrthod are\n");
```

```
for(i=0;i<=3;i++)
  printf("\n y= %d= %.5f",i,ay[i]);


  f1=f(x0+h,ay[1]);
  f2=f(x0+2*h,ay[2]);
  f3=f(x0+3*h,ay[3]);



while(i<=1)
   {
      y4=ay[i-4]+(4*h*(2*f1-f2+2*f3))/3;
      do
       {
         t=y4;
         f4=f(x0+i*h,y4);
         y4=ay[i-2]+h*(f2+4*f3+f4)/3;
         } while (fabs(y4-t)>err);
      printf("\n\n value of y at x=%.2f= %.5f",x0+i*h,y4);
      ay[i]=y4;
      i++;
      f1=f2; f2=f3, f3=f4;
   }
      getch();
      }
```

Output: MILNE'S PREDICTOR CORRECTOR METHOD FOR X*Y+Y*Y

```
Enter the value of x0,y0,x,n


0      1      .4       4


Starting 4 values  by runge -kutta mrthod are


y= 0= 1.00000
y= 1= 1.11689
y= 2= 1.27739
y= 3= 1.50412


Value of y at x=0.40    is  1.83941
```

LAB ASSIGNMENT : MILNE'S PREDICTOR CORRECTOR METHOD

1. Write a C program for Milne's Predictor Corrector method to find value of y for $x = 0.5$ for $dy/dx = 2e^x - y$ with initial condition $x = 0$, $y = 2$ by dividing range into 5 equal parts.

    Hint : Define $f(x, y) = 2 * \exp(x) - y$

    Input : $x_0 = 0$, $y_0 = 2$, $x = 0.5$, $n = 5$

    Output : $y(0.50) = 2.25525$

2. Write a C program to find value of y at $x = 0.5$ of the differential equation $dy/dx = x + y$ with initial condition $y(0) = 1$ by predictor-corrector method.

    Hint : Define $f(x, y) = x + y$

    Input : $x_0 = 0$, $y_0 = 1$, $x = 0.5$, $n = 5$

    Output : $y(0.5) = 1.7968$

3. Write a C program for Milne's method to find $y(1)$ for the equation $dy/dx = x - y^2$ with initial condition $y(0) = 0$.

    Hint : Define $f(x, y) = x - y * y$

    Input : $x_0 = 0$, $y_0 = 0$, $x = 1$, $n = 5$

    Output : $y(1) = 0.45552$

⌘⌘⌘⌘⌘⌘

Numerical Solution of Partial Differential Equations

10.1 INTRODUCTION

A partial differential equation is a differential equation involving more than one independent variable. These variables determine the behaviour of the dependent variable as described by their partial derivatives contained in the equation. Partial differential equations arise in the study of many branches of applied mathematics, *e.g.*, in the study of displacement of a vibrating string, in heat flow problems, in fluid flow analysis, in analysis of torsion in a bar subjected to twisting and in the study of diffusion of matter and so on. Only a few of these equations can be solved by analytical methods. In most cases, we go in for numerical methods to approximate solution. Of all the numerical methods available, for the solution of partial differential equations, the method of finite differences is most commonly used. In this method, the derivative appearing in the equation and the boundary conditions are replaced by their finite difference approximations. After that, the given equation is changed to a system of linear equations, which are solved by iterative procedures.

10.2 DIFFERENCE QUOTIENTS

Definition. *A difference quotient is the quotient obtained by dividing the difference between two values of a function by the difference between the two corresponding values of the independent variable.*

If $u(x)$ is a function, then difference coefficient is given by $\dfrac{u(x+h)-u(x)}{h}$.

The limiting value of above coefficient is the derivative of u w.r.t. x, *i.e.*, du/dx. .

If $u = u(x, y)$ is a function of two independent variables x and y, then we proceed as follows :
First, let us consider the difference in x-direction.
Consider the Taylor's series for $u(x, y_0)$ about the point (x_0, y_0) is

$$u(x,y_0) = u(x_0,y_0)+(x-x_0)u_x(x_0,y_0)+\frac{(x-x_0)^2}{2!}u_{xy}(x_0,y_0) \qquad ...(1)$$

where $x_0 \le t \le x$.
Put $x = x_0 + h$ in (1), we get

$$u(x_0+h,y_0) = u(x_0,y_0)+hu_x(x_0,y_0)+\frac{h^2}{2!}u_{xx}(t,y_0)$$

$$\Rightarrow \qquad \frac{u(x_0+h,y_0)-u(x_0,y_0)}{h} = u_x(x_0,y_0)+\frac{h}{2!}u_{xx}(t,y_0)$$

which gives that

$$u_x(x_0, y_0) = \frac{u(x_0 + h, y_0) - u(x_0, y_0)}{h} - \frac{h}{2!} u_{xx}(t, y_0) \qquad ...(2)$$

where $x_0 \le t \le x_0 + h$.

If we replace $u_x(x_0, y_0)$ by $\dfrac{u(x_0 + h, y_0) - u(x_0, y_0)}{h}$, then we get the truncation error

$$= -\frac{h}{2!} u_{xx}(t, y_0)$$

Therefore, we have the finite difference formula for the first partial derivative at (x_0, y_0) as

$$u_x(x_0, y_0) = \frac{u(x_0 + h, y_0) - u(x_0, y_0)}{h} \qquad ...(3)$$

which is called a forward difference approximation to $u_x(x_0, y_0)$.

Similarly by putting $x = x_0 - h$ in (1), we get

$$u_x(x_0, y_0) = \frac{u(x_0, y_0) - u_x(x_0 - h, y_0)}{h} \qquad ...(4)$$

which is known as backward approximation to $u_x(x_0, y_0)$.

Now, to find an approximation for u_{xx}, we use both forward and backward difference. Using forward difference for u_x, we get

$$u_{xx}(x_0, y_0) = \frac{u_x(x_0 + h, y_0) - u_x(x_0, y_0)}{h} \qquad ...(5)$$

Since there is a bias in the forward direction, we use backward difference for u_x to avoid this effect, *i.e.*,

$$u_{xx}(x_0, y_0) = \frac{u(x_0, y_0) - u_x(x_0 - h, y_0)}{h} \qquad ...(6)$$

Changing x_0 to $x_0 + h$ in (6), we get

$$u_x(x_0 + h, y_0) = \frac{u(x_0 + h, y_0) - u(x_0, y_0)}{h} \qquad ...(7)$$

Now, using (6) and (7) in (5), we get

$$u_{xx}(x_0, y_0) = \frac{u(x_0 + h, y_0) - 2u(x_0, y_0) + u(x_0 - h, y_0)}{h^2}$$

Proceeding in the same way, we have the following formula for the derivatives in y-direction taking the step size as k.

(i) $u_y(x_0, y_0) = \dfrac{u(x_0, y_0 + k) - 4(x_0, y_0)}{k}$ (Forward difference formula)

(ii) $u_y(x_0, y_0) = \dfrac{u(x_0, y_0) - 4(x_0, y_0 - k)}{k}$ (Backward difference formula)

(iii) $u_{yy}(x_0, y_0) = \dfrac{u(x_0, y_0 + k) - 24(x_0, y_0) + 4(x_0, y_0 - k)}{k^2}$

▼ REMARK

▸ The truncation error in the above formula will be given by $-\dfrac{k^2}{12} u_{yy}(x_0, \xi)$, where $y_0 - k \le \xi \le y_0 + k$

10.3 CLASSIFICATION OF PARTIAL DIFFERENTIAL EQUATIONS

Consider the general linear partial differential equation of the second order in two independent variables as given by

$$A\frac{\partial^2 u}{\partial x^2} + B\frac{\partial^2 u}{\partial x \partial y} + C\frac{\partial^2 u}{\partial y^2} + D\frac{\partial u}{\partial x} + E\frac{\partial u}{\partial y} + Fu = 0 \qquad \text{...(1)}$$

where A, B, C, D, E and F are in general functions of x and y.

Then, (1) is said to be
- (i) Elliptic if $B^2 - 4AC < 0$
- (ii) Parabolic if $B^2 - 4AC = 0$
- (iii) Hyperbolic if $B^2 - 4AC > 0$

10.4 ELLIPTIC EQUATIONS

Elliptic equations are governed by conditions on the boundary of closed domain. Here, we consider the two most commonly used elliptic equations, namely Laplace equations and Poisson's equation.

The Laplace equation is given by $\nabla^2 u = \dfrac{\partial^2 u}{\partial x^2} + \dfrac{\partial^2 u}{\partial y^2} = 0$

and the Poisson's equation is given by $\dfrac{\partial^2 u}{\partial x^2} + \dfrac{\partial^2 u}{\partial y^2} = f(x, y)$

10.4.1 SOLUTION OF LAPLACE EQUATION

Consider the Laplace equation $\dfrac{\partial^2 u}{\partial x^2} + \dfrac{\partial^2 u}{\partial y^2} = 0$ which can be written as

$$u_{xx} + u_{yy} = 0 \qquad \text{...(1)}$$

Replacing the derivative in (1) by their difference approximation, we get

$$\frac{1}{h^2}(u_{i-1,j} - 2u_{i,j} + u_{i+1,j}) + \frac{1}{k^2}(u_{i,j-1} - 2u_{i,j} + u_{i,j+1}) = 0 \qquad \text{...(2)}$$

Taking a square mesh and putting $h = k$ we get from (2)

$$u_{i,j} = \frac{1}{4}(u_{i+1,j} + u_{i-1,j} + u_{i,j+1} + u_{i,j-1}) \qquad \text{...(3)}$$

In solving a given boundary value problem we can use the above difference equation (3)

to complete the values at interior mesh points in terms of the values on the boundary. The formulae given by (3) is known as standard five point formula (SFPF).

▼ REMARKS

▸ We may also use the formula given below, in place of (3)

$$u_{i,j} = \frac{1}{4}\left[u_{i-1,j-1} + u_{i+1,j-1} + u_{i-1,j+1} + u_{i+1,j+1}\right]$$

which shows that the value of $u_{i,j}$ is the arithmetic mean of its values at the four neighbouring

diagonal mesh-points. This is called the diagonal five point formula (DFPF).

▸ The DFPF is valid, because we know that the Laplace eqaution remains invariant when the co-ordinate axes are rotated through $45°$. But the error in DFPF is four times the error in SFPF. Therefore, we prefer SFPF.

10.4.2 PICTORIAL REPRESENTATION

The SFPF and DFPF are represented in the figures given below :

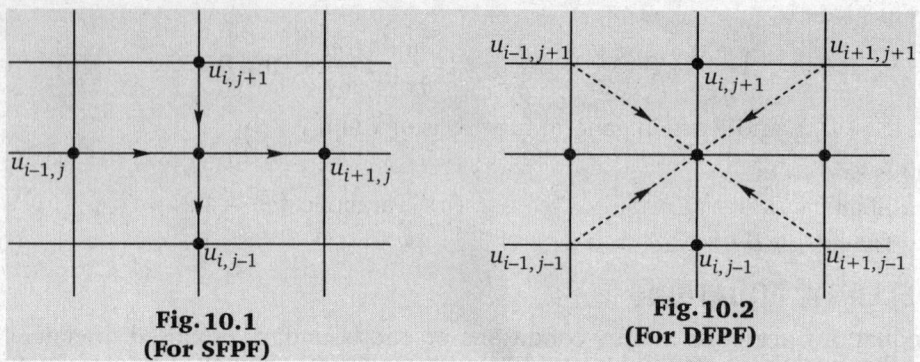

Fig. 10.1
(For SFPF)

          Fig. 10.2
          (For DFPF)

10.5 SOLUTION OF LAPLACE EQUATIONS BY LIBERMANN'S ITERATION PROCESS

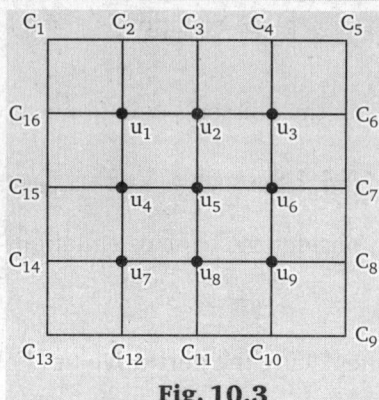

Fig. 10.3

Consider the Laplace equation $\dfrac{\partial^2 u}{\partial x^2} + \dfrac{\partial^2 u}{\partial y^2} = 0$ with the given boundary conditions. Let us assume the function $u(x, y)$ is required over a rectangular region with boundary C. Divide the rectangle R into a network of small squares of side h. Let the values of $u(x, y)$ on the boundary C be given by C_i and the interior mesh points and boundary points. Now to start the iteration process, initially we find rough values at interior points and then improve it by iterative process.

First of all, we find u_5, at the centre of the square by taking the average of four boundary values (SFPF)

$$\therefore \qquad u_5 = \frac{1}{4}\left[C_{15} + C_7 + C_3 + C_{11} \right]$$

Next, find the initial values at the centre of the four large inner squares, using DFPF. Therefore,

$$u_1 = \frac{1}{4}\left[u_5 + C_1 + C_3 + C_{15} \right], u_3 = \frac{1}{4}\left[u_5 + C_5 + C_3 + C_7 \right],$$

$$u_7 = \frac{1}{4}\left[u_5 + C_{13} + C_{11} + C_{15} \right], \ u_9 = \frac{1}{4}\left[u_5 + C_9 + C_7 + C_{11} \right]$$

The values of the remaining interior points are obtained by SFPF such that

$$u_2 = \frac{1}{4}\left[u_1 + u_3 + C_3 + u_5 \right], u_4 = \frac{1}{4}\left[C_{15} + u_5 + u_1 + u_7 \right],$$

$$u_6 = \frac{1}{4}\left[u_5 + C_7 + u_3 + u_9 \right], \ u_8 = \frac{1}{4}\left[u_7 + u_9 + u_5 + C_{11} \right]$$

REMARK

▸ To improve the accuracy, we start with u_1 and iterate it using the latest available values of the four adjacent points. Therefore, we iterate all the mesh-points systematically from left to right along successive rows.

(A) ITERATION FORMULAE

(1) Liebermann iteration formula

$$u_{i,j}^{(n+1)} = \frac{1}{4}\left[u_{i-1,j}^{(n+1)} + u_{i+1,j}^{(n)} + u_{i,j-1}^{(n)} + u_{i,j+1}^{(n+1)}\right]$$

(2) Gauss-Seidel method

$$u_{i,j}^{(n+1)} = \frac{1}{4}\left[u_{i-1,j}^{(n+1)} + u_{i+1,j}^{(n)} + u_{i,j+1}^{(n+1)} + u_{i,j-1}^{(n)}\right] \qquad \text{(Bhopal-2007)}$$

(3) Jacobi's method

$$u_{i,j}^{(n+1)} = \frac{1}{4}\left[u_{i-1,j}^{(n)} + u_{i+1,j}^{(n)} + u_{i,j+1}^{(n+1)} + u_{i,j-1}^{(n)}\right] \qquad \text{(Bhopal-2007)}$$

▼ REMARKS
- Here the subscript denotes the iteration number.
- Initial values may be obtained by either taking diagonal average or cross-average of the adjoining four points.
- The accuracy of calculation depends on the mesh size

SOLVED EXAMPLES

EXAMPLE 1. *Find the solution of Laplace equation $\dfrac{\partial^2 u}{\partial x^2} + \dfrac{\partial^2 u}{\partial y^2} = 0$ within a square with vertices at*

$A(0, 0)$, $B(1, 0)$, $C(1, 1)$ and $D(0, 1)$. The boundary conditions on the sides AB, BC and AD being zero and that along the side CD being 100 except that at the corners C and D, the values are 50. Take $h = 1/4$ and compute results, correct to three significant digits.

SOLUTION. We first construct the net with $h = k = 1/4$ and designate the interior mesh-points as shown below :

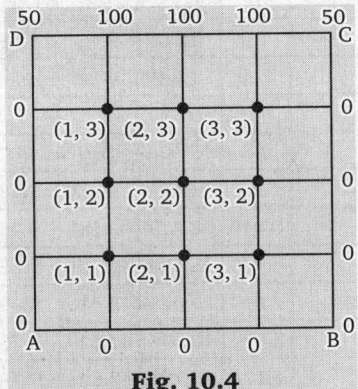

Fig. 10.4

By diagonal five points formula, we get

$$u_{i,j} = \frac{1}{4}\left[u_{i-1,j-1} + u_{i+1,j-1} + u_{i-1,j+1} + u_{i+1,j+1}\right] \qquad \ldots(1)$$

In order to solve the given boundary problem, we have to write the finite difference equation at all the nine interior points, But by symmetry of the problem, we have

$$u_{11} = u_{31}, u_{12} = u_{32}, u_{13} = u_{33} \qquad \ldots(2)$$

$\therefore$ We have the following six equations :

$$u_{11} = \frac{1}{4}\Big[u_{01} + u_{21} + u_{10} + u_{12}\Big], \quad u_{12} = \frac{1}{4}\Big[u_{02} + u_{22} + u_{11} + u_{13}\Big],$$

$$u_{13} = \frac{1}{4}\Big[u_{03} + u_{23} + u_{12} + u_{14}\Big], \quad u_{21} = \frac{1}{4}\Big[u_{11} + u_{31} + u_{20} + u_{22}\Big],$$

$$u_{22} = \frac{1}{4}\Big[u_{12} + u_{32} + u_{21} + u_{23}\Big], \quad u_{23} = \frac{1}{4}\Big[u_{13} + u_{33} + u_{22} + u_{24}\Big],$$

The above six equations involve values at the boundary points given by

$$u_{i,0} = 0, i = 1, 2, 3 ; \quad u_{0,j} = 0, j = 1, 2, 3 ; \quad u_{14} = u_{24} = u_{34} = 100$$

Taking into account these boundary conditions and the symmetry condition (2), we get the following six equations

$$u_{11} = \frac{1}{4}\Big[u_{12} + u_{21}\Big], \qquad u_{12} = \frac{1}{4}\Big[u_{11} + u_{13} + u_{22}\Big],$$

$$u_{13} = \frac{1}{4}\Big[u_{12} + u_{23} + 100\Big], \qquad u_{21} = \frac{1}{4}\Big[2u_{11} + u_{22}\Big],$$

$$u_{22} = \frac{1}{4}\Big[2u_{12} + u_{21} + u_{23}\Big], \qquad u_{23} = \frac{1}{4}\Big[2u_{13} + u_{22} + 100\Big].$$

These equations can be easily solved by Gauss-Seidel iteration method. The result of successive iteration are given below :

Iteration No.	u_{11}	u_{12}	u_{13}	u_{21}	u_{22}	u_{23}
0	0	0	25	0	0	25
1	0	6.25	33	0	8	43.5
2	1.6	10.6	38.5	2.8	11.6	47.15
3	3.4	13.4	40.1	4.6	19.6	49.95
4	4.5	16.05	41.5	7.15	22.3	51.32
5	5.8	17.405	42.18	8.475	23.469	52.00
6	6.54	18.09	42.52	9.182	24.340	52.34
7	6.818	18.419	42.680	9.494	24.668	52.572
8	6.975	18.584	42.774	9.656	24.834	52.596
9	7.060	18.670	42.816	9.698	24.865	52.624
10	7.092	18.693	42.829	9.762	24.943	52.650
11	7.114	18.721	42.843	9.793	24.971	52.664
12	7.128	18.736	42.850	9.807	24.986	52.671
13	7.136	18.743	42.803	9.814	24.993	52.675

Hence, the solution correct to three significant figures may be taken as

$$u_{11} = u_{31} = 7.14, u_{12} = u_{32} = 18.7,$$

$$u_{13} = u_{33} = 42.9, u_{21} = 9.81, u_{22} = 24.99,$$

$$u_{23} = 52.68.$$

EXAMPLE 2. *Solve the elliptic equation* $\dfrac{\partial^2 u}{\partial x^2} + \dfrac{\partial^2 u}{\partial y^2} = 0$ *for the following square mesh with boundary values given below :*

(Rohtak–2005, VTU–2005, 06)

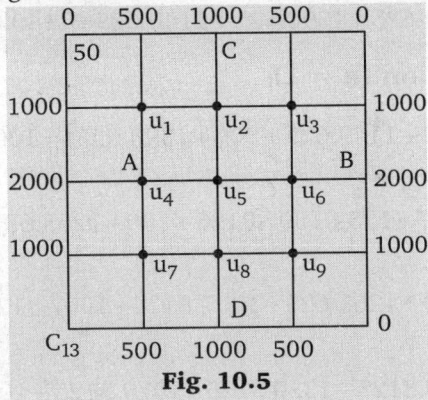

Fig. 10.5

SOLUTION. Let $u_1, u_2, ..., u_9$ be the values of x at the interior mesh-points. From the above figure, we have the following symmetry :

$u_7 = u_1, u_8 = u_2, u_9 = u_3, u_3 = u_1, u_6 = u_4, u_9 = u_7.$

Therefore, we find the values of u_1, u_2, u_3, u_4 and u_5.

Initial values. The initial values of u_1, u_2, u_3, u_4 and u_5 are given by

$$u_5^{(0)} = \frac{1}{4}\left[2000 + 2000 + 1000 + 1000\right] = 15000 \qquad \text{(SFPF)}$$

$$u_1^{(0)} = \frac{1}{4}\left[0 + 1500 + 1000 + 2000\right] = 1125 \qquad \text{(DFPF)}$$

$$u_2^{(0)} = \frac{1}{4}\left[1125 + 1125 + 1000 + 1500\right] = 1187.5 \qquad \text{(SFPF)}$$

$$u_4^{(0)} = \frac{1}{4}\left[2000 + 1500 + 1125 + 1125\right] = 1437.5 \qquad \text{(SFPF)}$$

Here, we have the following iteration formulae by SFPF :

$$u_1^{(n+1)} = \frac{1}{4}\left[1000 + u_2^{(n)} + 500 + u_4^{(n)}\right], \quad u_2^{(n+1)} = \frac{1}{4}\left[u_1^{(n+1)} + u_1^{(n+1)} + 1000 + u_5^{(n)}\right]$$

$$u_4^{(n+1)} = \frac{1}{4}\left[2000 + u_5^{(n)} + u_1^{(n+1)} + u_1^{(n+1)}\right],$$

$$u_5^{(n+1)} = \frac{1}{4}\left[u_4^{(n+1)} + u_4^{(n+1)} + u_2^{(n+1)} + u_2^{(n+1)}\right]$$

Now, we have the following iterations :

First iteration ($n = 0$)

$$u_1^{(1)} = \frac{1}{4}\left[1000 + 1187.5 + 500 + 1437.5\right] = 1031.25$$

$$u_2^{(1)} = \frac{1}{4}\left[1031.5 + 1031.25 + 1000 + 1500\right] = 1140.625$$

$$u_4^{(1)} = \frac{1}{4}[2000 + 1500 + 1031.25 + 1031.25] = 1390.625$$

$$u_5^{(1)} = \frac{1}{4}[1390.625 + 1390.625 + 1140.625 + 1140.625] = 1265.625$$

Second iteration ($n = 1$)

$$u_1^{(2)} = \frac{1}{4}[1000 + 1149.625 + 500 + 1390.625] = 1007.8125$$

$$u_2^{(2)} = \frac{1}{4}[1007.8125 + 1007.8125 + 100 + 1265.65] = 1070.3125$$

$$u_4^{(2)} = \frac{1}{4}[2000 + 1265.625 + 1007.8125 + 1007.8125] = 1320.3125$$

$$u_5^{(2)} = \frac{1}{4}[1320.3125 + 1320.3125 + 1070.3125 + 1070.3125] = 1195.3125$$

Third iteration ($n = 2$)

$$u_1^{(3)} = \frac{1}{4}[1000 + 1070.3125 + 500 + 1320.3125] = 972.6525$$

$$u_2^{(3)} = \frac{1}{4}[972.65625 + 972.65625 + 1000 + 1195.3125] = 1035.1563$$

$$u_4^{(3)} = \frac{1}{4}[2000 + 1195.3125 + 972.65625 + 972.65625] = 1285.1563$$

$$u_5^{(3)} = \frac{1}{4}[1285.1563 + 1285.1563 + 1035.1563 + 1035.1563] = 1160.1563$$

Fourth iteration ($n = 3$)

$$u_1^{(4)} = \frac{1}{4}[1000 + 1035.1563 + 500 + 1285.1563] = 955.07815$$

$$u_2^{(4)} = \frac{1}{4}[955.07815 + 955.07815 + 1000 + 1160.1563] = 1017.5782$$

$$u_4^{(4)} = \frac{1}{4}[2000 + 1160.1563 + 955.07815 + 955.07815] = 1267.5782$$

$$u_5^{(4)} = \frac{1}{4}[1267.5782 + 1267.5782 + 1017.5782 + 1017.5782] = 1142.5782$$

Fifth iteration ($n = 4$)

$$u_1^{(5)} = \frac{1}{4}[1000 + 1017.5782 + 500 + 1267.5782] = 946.2891$$

$$u_2^{(5)} = \frac{1}{4}[946.2891 + 946.2891 + 1000 + 1142.5782] = 1008.7891$$

$$u_4^{(5)} = \frac{1}{4}[2000 + 1142.5782 + 946.2891 + 946.2891] = 1258.7891$$

$$u_5^{(5)} = \frac{1}{4}[1258.7891 + 1258.7891 + 1008.7891 + 1008.7891] = 1133.7891$$

Sixth iteration ($n = 5$)

$$u_1^{(6)} = \frac{1}{4}[1000 + 1008.7891 + 500 + 1258.7891] = 941.89455$$

$$u_2^{(6)} = \frac{1}{4}[941.89455 + 941.89455 + 1000 + 1133.7891] = 1004.3946$$

$$u_4^{(6)} = \frac{1}{4}[2000 + 1133.7891 + 941.89455 + 941.89455] = 1254.3946$$

$$u_5^{(6)} = \frac{1}{4}[1254.3946 + 1254.3946 + 1004.3946 + 1004.3946] = 1129.3946$$

Seventh iteration ($n = 6$)

$$u_1^{(7)} = \frac{1}{4}[1000 + 1004.3946 + 500 + 1254.3946] = 939.6973$$

$$u_2^{(7)} = \frac{1}{4}[939.6973 + 939.6973 + 1000 + 1129.3946] = 1002.1973$$

$$u_4^{(7)} = \frac{1}{4}[2000 + 1129.3946 + 939.6973 + 939.6973] = 1252.1973$$

$$u_5^{(7)} = \frac{1}{4}[1252.1973 + 1252.1973 + 1002.1973 + 1002.1973] = 1127.1973$$

Eight iteration ($n = 7$)

$$u_1^{(8)} = \frac{1}{4}[1000 + 1002.1973 + 500 + 1252.1973] = 938.59865$$

$$u_2^{(8)} = \frac{1}{4}[938.59865 + 938.59865 + 1000 + 1127.1973] = 1001.0987$$

$$u_4^{(8)} = \frac{1}{4}[2000 + 1127.1973 + 938.59865 + 938.59865] = 1251.0987$$

$$u_5^{(8)} = \frac{1}{4}[1251.0987 + 1251.0987 + 1001.0987 + 1001.0987] = 1126.0987$$

Ninth iteration ($n = 8$)

$$u_1^{(9)} = \frac{1}{4}[1000 + 1001.0987 + 500 + 1251.0987] = 938.04935$$

$$u_2^{(9)} = \frac{1}{4}[938.04935 + 938.04935 + 1000 + 1126.0987] = 1000.5494$$

$$u_4^{(9)} = \frac{1}{4}[2000 + 1126.0987 + 938.04935 + 938.04935] = 1250.5494$$

$$u_5^{(9)} = \frac{1}{4}[1259.5494 + 1259.5494 + 1001.5494 + 1001.5494] = 1125.5494$$

Here, we observe that the difference between eighth and ninth iteration is negligible.

Hence, $u_1 = 939$, $u_2 = 1001$, $u_4 = 1251$ and $u_5 = 1126$ which implies $u_3 = 939$, $u_6 = 1251$, $u_7 = 939$, $u_8 = 1001$ and $u_9 = 939$.

EXAMPLE 3.   *Given the values of u(x, y) on the boundary of the square in the figure given below, calculate the function u(x, y) satisfying the Laplace equation $\nabla^2 u = 0$ at the pivotal points of this figure by*

     (a) *Jacobi's method*   (Mumbai-2005, 09; UPTU-2007; MKUCT-2008; Roorkee-2009)

     (b) *Gauss-Seidel method*   (Bhopal–2009, Madras–2003)

SOLUTION.   (a)   To find the initial values of u_1, u_2, u_3, u_4, we have assume that $u_4 = 0$. Then

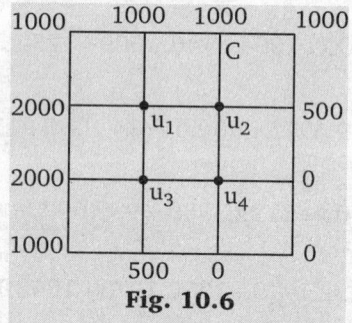

Fig. 10.6

$$u_1 = \frac{1}{4}[1000 + 0 + 1000 + 2000] = 1000 \; ; \; u_2 = \frac{1}{4}[1000 + 500 + 1000 + 0] = 625$$

$$u_3 = \frac{1}{4}[2000 + 0 + 1000 + 500] = 875 \; ; \; u_4 = \frac{1}{4}[875 + 0 + 625 + 0] = 375$$

By using Jacobi's iteration process, we have the following iteration :

$$u_1^{(n+1)} = \frac{1}{4}\left[2000 + u_2^{(n)} + 1000 + u_3^{(n)}\right]; \; u_2^{(n+1)} = \frac{1}{4}\left[u_1^{(n)} + 500 + 1000 + u_4^{(n)}\right]$$

$$u_3^{(n+1)} = \frac{1}{4}\left[2000 + u_4^{(n)} + u_1^{(n)} + 500\right] \; ; \; u_4^{(n)} = \frac{1}{4}\left[u_3^{(n)} + 0 + u_2^{(n)} + 0\right]$$

First iteration (n = 0)

$$u_1^{(1)} = \frac{1}{4}[2000 + 625 + 1000 + 875] = 1125$$

$$u_2^{(1)} = \frac{1}{4}[1000 + 500 + 1000 + 375] = 719$$

$$u_3^{(1)} = \frac{1}{4}[2000 + 375 + 1000 + 500] = 969 \; ; \; u_4^{(1)} = \frac{1}{4}[875 + 0 + 625 + 0] = 375$$

Second iteration (n = 1)

$$u_1^{(2)} = \frac{1}{4}[2000 + 719 + 1000 + 969] = 1172 \; \cdot$$

$$u_2^{(2)} = \frac{1}{4}[1125 + 500 + 1000 + 375] = 750$$

$$u_3^{(2)} = \frac{1}{4}[2000 + 375 + 1125 + 500] = 1000$$

$$u_4^{(2)} = \frac{1}{4}[969 + 0 + 719 + 0] = 422$$

Similarly, we may obtain

$$u_1^{(3)} = 1188, u_2^{(3)} = 774, u_3^{(3)} = 1024, u_4^{(3)} = 438,$$
$$u_1^{(4)} = 1200, u_2^{(4)} = 788, u_3^{(4)} = 1032, u_4^{(4)} = 450,$$

$$u_1^{(5)} = 1204, u_2^{(5)} = 788, u_3^{(5)} = 1038, u_4^{(5)} = 454,$$
$$u_1^{(6)} = 1206.5, u_2^{(6)} = 790, u_3^{(6)} = 1040, u_4^{(6)} = 456.5,$$

$$u_1^{(7)} = 1208, u_2^{(7)} = 791, u_3^{(7)} = 1041, u_4^{(7)} = 458,$$
$$u_1^{(8)} = 1208, u_2^{(8)} = 791.5, u_3^{(8)} = 1041.5, u_4^{(8)} = 458$$

Here, we observe that there is no significant difference between the seventh and eighth iteration.

Hence, $u_1 = 1208, u_2 = 792, u_3 = 1042$ and $u_4 = 458$.

(b) By using Gauss-Seidel method, we have the following iterations :

$$u_1^{(n+1)} = \frac{1}{4}\left[2000 + u_2^{(n)} + 1000 + u_3^{(n)}\right]$$

$$u_2^{(n+1)} = \frac{1}{4}\left[u_1^{(n+1)} + 500 + 1000 + u_4^{(n)}\right]$$

$$u_3^{(n+1)} = \frac{1}{4}\left[2000 + u_4^{(n)} + u_1^{(n+1)} + 500\right]$$

$$u_4^{(n+1)} = \frac{1}{4}\left[u_3^{(n+1)} + 0 + u_2^{(n+1)} + 0\right]$$

First iteration ($n = 0$)

$$u_1^{(1)} = \frac{1}{4}[2000 + 625 + 1000 + 875] = 1125$$

$$u_2^{(1)} = \frac{1}{4}[1125 + 500 + 1000 + 375] = 750$$

$$u_3^{(1)} = \frac{1}{4}[2000 + 375 + 1125 + 500] = 1000$$

$$u_4^{(1)} = \frac{1}{4}[1000 + 0 + 750 + 0] = 438$$

Second iteration ($n = 1$)

$$u_1^{(2)} = \frac{1}{4}[2000 + 750 + 1000 + 1000] = 1188$$

$$u_2^{(2)} = \frac{1}{4}[1188 + 500 + 1000 + 438] = 782$$

$$u_3^{(2)} = \frac{1}{4}[2000 + 438 + 1188 + 500] = 1032$$

$$u_4^{(2)} = \frac{1}{4}[1032 + 0 + 782 + 0] = 454$$

Similarly, we may obtain

$$u_1^{(3)} = 1204, u_2^{(3)} = 789, u_3^{(3)} = 1040,$$
$$u_4^{(3)} = 458, u_1^{(4)} = 1207, u_2^{(4)} = 791,$$

$$u_3^{(4)} = 1041, u_4^{(4)} = 458, u_1^{(5)} = 1208,$$
$$u_2^{(5)} = 791.5, u_3^{(5)} = 1041.5, u_4^{(4)} = 458.25$$

Here, we observe that there is no significant difference between the fourth and fifth iteration values.

Hence, $u_1 = 1208, u_2 = 792, u_3 = 1042$ and $u_4 = 458$.

EXAMPLE 4.   *Solve the partial differential equation* $\dfrac{\partial^2 u}{\partial x^2} + \dfrac{\partial^2 u}{\partial y^2} = 0$ *for figure given below by*

(a) *Jacobi's method*   **(b)** *Gauss-Seidel method*   (Chennai–2005, Roorkee–2006; Bangluru-2005; Andhra-2002, 04, 08; Bhopal-2003)

SOLUTION.

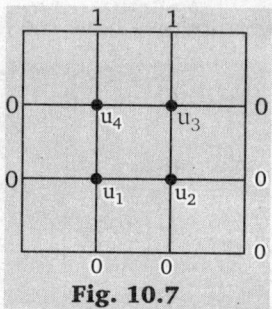

Fig. 10.7

Here, we have

$$u_1^{(1)} = \frac{1}{4}(0 + 0 + 0 + 1) = 0.25 \ ; \ u_2^{(1)} = \frac{1}{4}(0 + 0 + 0 + 1) = 0.25$$

$$u_3^{(1)} = \frac{1}{4}(1 + 1 + 0 + 0) = 0.50 \, ; \ u_4^{(1)} = \frac{1}{4}(1 + 1 + 0 + 0) = 0.50$$

(a)   Here, Jacobi's formula is

$$u_{i,j}^{(n+1)} = \frac{1}{4}[u_{i-1}^{(n)} + u_{i+1,j}^{(n)} + u_{i,j-1}^{(n)} + u_{i,j+1}^{(n)}] \text{ , then}$$

$$u_1^{(2)} = \frac{1}{4}[0 + u_2^{(1)} + 0 + u_3^{(1)}] = \frac{1}{4}(0.25 + 0.50) = 0.1875 \text{ , then}$$

$$u_2^{(2)} = \frac{1}{4}[u_1^{(1)} + 0 + 0 + u_4^{(1)}] = \frac{1}{4}(0.25 + 0.50) = 0.1875.$$

$$u_3^{(2)} = \frac{1}{4}[0 + u_4^{(1)} + u_1^{(1)} + 1] = \frac{1}{4}(0.50 + 0.25 + 1) = 0.4375.$$

$$u_4^{(2)} = \frac{1}{4}[u_3^{(1)} + 0 + u_2^{(1)} + 1] = \frac{1}{4}(0.50 + 0.25 + 1) = 0.4375.$$

$$u_1^{(3)} = \frac{1}{4}[u_3^{(1)} + 0 + u_2^{(1)} + 1] = \frac{1}{4}(0.1875 + 4375) = 1.5625.$$

$$u_2^{(3)} = \frac{1}{4}[u_1^{(2)} + 0 + 0 + u_4^{(2)}] = \frac{1}{4}(0.1875 + 4375) = 1.5625.$$

$$u_3^{(3)} = \frac{1}{4}[0 + u_4^{(2)} + u_1^{(2)} + 1] = \frac{1}{4}(0 + 0.4375 + 0.1875 + 1) = 0.40625$$

$$u_4^{(3)} = \frac{1}{4}[u_3^{(2)} + 0 + u_2^{(2)} + 1] = \frac{1}{4}(0.4375 + 0 + 0.1875 + 1) = 0.40625$$

$$u_1^{(4)} = \frac{1}{4}[0 + u_2^{(3)} + 0 + u_3^{(3)}] = \frac{1}{4}(0.15625 + 0.40625) = 0.14062$$

$$u_2^{(4)} = \frac{1}{4}[u_1^{(3)} + 0 + 0 + u_4^{(3)}] = \frac{1}{4}(0.15625 + 0.40625) = 0.14062$$

$$u_3^{(4)} = \frac{1}{4}[0 + u_4^{(3)} + u_1^{(3)} + 1] = \frac{1}{4}(0.40625 + 0.15625 + 1) = 0.39062$$

$$u_4^{(4)} = \frac{1}{4}[u_3^{(3)} + 0 + u_2^{(3)} + 1] = \frac{1}{4}(0.40625 + 0.15625 + 1) = 0.39062$$

$$u_1^{(5)} = \frac{1}{4}[0 + u_2^{(4)} + 0 + u_3^{(4)}] = \frac{1}{4}(0.14062 + 0.39062) = 0.13281$$

$$u_2^{(5)} = \frac{1}{4}[u_1^{(4)} + 0 + 0 + u_4^{(4)}] = \frac{1}{4}(0.14062 + 0.39062) = 0.13281$$

$$u_3^{(5)} = \frac{1}{4}[0 + u_4^{(4)} + u_1^{(4)} + 1] = \frac{1}{4}(0.39062 + 1.4062 + 1) = 0.38281$$

$$u_4^{(5)} = \frac{1}{4}[u_3^{(4)} + 0 + u_2^{(4)} + 1] = \frac{1}{4}(0.39062 + 0 + 1.4062 + 1) = 0.38281$$

$$u_1^{(6)} = \frac{1}{4}[0 + u_2^{(5)} + 0 + u_3^{(5)}] = \frac{1}{4}(0.13281 + 0.38281) = 0.12891$$

$$u_2^{(6)} = \frac{1}{4}[u_1^{(5)} + 0 + 0 + u_4^{(5)}] = \frac{1}{4}(0.13281 + 0.38281) = 0.12891$$

$$u_3^{(6)} = \frac{1}{4}[0 + u_4^{(5)} + u_1^{(5)} + 1] = \frac{1}{4}(0.38281 + 0.13281 + 1) = 0.37891$$

$$u_4^{(6)} = \frac{1}{4}[u_3^{(5)} + 0 + u_2^{(5)} + 1] = \frac{1}{4}(0.38281 + 0 + 0.13281 + 1) = 0.37891$$

(b) Here, Gauss-Seidal method is

$$\left[u_{i,j}^{(n+1)} = \frac{1}{4}\left(u_{i-1,j}^{(n+1)} + u_{i+1,j}^{(n)} + u_{i,j-1}^{(n+1)} + u_{i,j+1}^{(n)} \right) \right]$$

Now, $u_1^{(1)} = 0.25, u_2^{(1)} = 0.3125, u_3^{(1)} = 0.5625, u_4^{(1)} = 0.46875.$

$$u_1^{(1)} = \frac{1}{4}[0 + u_2^{(1)} + 0 + u_3^{(1)}] = \frac{0.3125 + 0.5625}{4} = 0.21875$$

$$u_2^{(2)} = \frac{1}{4}[u_1^{(2)} + 0 + 0 + u_4^{(1)}] = \frac{1}{4}(0.21875 + 0 + 0.46875) = 0.17187$$

$$u_3^{(2)} = \frac{1}{4}[0 + u_4^{(1)} + u_1^{(2)} + 1] = \frac{1}{4}(0.468750 + 0.21875 + 1) = 0.42187;$$

$$u_4^{(2)} = \frac{1}{4}[u_3^{(2)} + 0 + u_2^{(2)} + 1] = \frac{1}{4}(0.42187 + 0.17187 + 1) = 0.39844$$

$$u_1^{(3)} = \frac{1}{4}[u_2^{(2)} + u_3^{(2)}] = \frac{1}{4}(0.17187 + 0.42187) = 0.14844$$

$$u_2^{(3)} = \frac{1}{4}[u_1^{(3)} + u_4^{(2)}] = \frac{1}{4}(0.14844 + 0.39844) = 0.13672$$

$$u_3^{(3)} = \frac{1}{4}[u_4^{(2)} + u_1^{(3)} + 1] = \frac{1}{4}(0.39844 + 0.14844 + 1) = 0.38672,$$

$$u_4^{(3)} = \frac{1}{4}[u_3^{(3)} + u_2^{(3)} + 1] = \frac{1}{4}(0.38672 + 0.13672 + 1) = 0.38086,$$

$$u_1^{(4)} = \frac{1}{4}[u_2^{(3)} + u_3^{(3)}] = \frac{1}{4}(0.13672 + 0.38672) = 0.13086,$$

$$u_2^{(4)} = \frac{1}{4}[u_1^{(4)} + u_4^{(3)}] = \frac{1}{4}(0.13086 + 0.38086) = 0.12793,$$

$$u_3^{(4)} = \frac{1}{4}[u_4^{(3)} + u_1^{(4)} + 1] = \frac{1}{4}(0.38086 + 0.13086 + 1) = 0.37793,$$

$$u_4^{(4)} = \frac{1}{4}[u_3^{(4)} + u_2^{(4)} + 1] = \frac{1}{4}(0.37793 + 0.12793 + 1) = 0.37646,$$

$$u_1^{(5)} = \frac{1}{4}[u_2^{(4)} + u_3^{(4)}] = \frac{1}{4}(0.12793 + 0.37793) = 0.12646,$$

$$u_2^{(5)} = \frac{1}{4}[u_1^{(5)} + u_4^{(4)}] = \frac{1}{4}(0.12646 + 0.37646) = 0.12573,$$

$$u_3^{(5)} = \frac{1}{4}[u_4^{(4)} + u_1^{(5)} + 1] = \frac{1}{4}(0.37646 + 0.12646 + 1) = 0.37573,$$

$$u_4^{(5)} = \frac{1}{4}[u_3^{(5)} + u_2^{(5)} + 1] = \frac{1}{4}(0.37573 + 0.12573 + 1) = 0.37537,$$

$$u_1^{(6)} = \frac{1}{4}[u_2^{(5)} + u_3^{(5)}] = \frac{1}{4}(0.12573 + 0.37573) = 0.12536,$$

$$u_2^{(6)} = \frac{1}{4}[u_1^{(6)} + u_4^{(5)}] = \frac{1}{4}(0.12536 + 0.37537) = 0.12518,$$

$$u_3^{(6)} = \frac{1}{4}[u_4^{(5)} + u_1^{(6)} + 1] = \frac{1}{4}(0.37537 + 0.12536 + 1) = 0.37518,$$

$$u_4^{(6)} = \frac{1}{4}[u_3^{(6)} + u_2^{(6)} + 1] = \frac{1}{4}(0.37518 + 0.12518 + 1) = 0.37509.$$

(B) Solution of Poisson's Equation

Consider the Poisson's partial differential equation

$$\nabla^2 u = f(x, y) \;\Rightarrow\; \frac{\partial^2 u}{\partial x^2} + \frac{\partial^2 u}{\partial y^2} = f(x, y) \qquad\qquad ...(1)$$

where $f(x, y)$ is a given function of x and y.

To solve equation (1) numerically, the derivative in (1) are replaced by difference expressions at the points $x = ih, y = jk$ $(h = k)$,

Then we get

$$\frac{1}{h^2}[u_{i-1,j} - 2u_{i,j} + u_{i+1,j}] + \frac{1}{h^2}[u_{i,j-1} - 2u_{i,j} + u_{i,j+1}] = f(ih, jh)$$

$\Rightarrow$
$$[u_{i-1,j} + u_{i+1,j} + u_{i,j-1} + u_{i,j+1} - 4u_{i,j}] = h^2 f(ih, jh) \qquad ...(2)$$

Using the formula (2) at each mesh point, we get similar equations in the pivotal values i, j. These equations can be solved by iteration techniques such as Gauss-Seidel method.

▼ **REMARK**
▸ The error involved in (1) by difference method is of the order $O(h^2)$.

SOLVED EXAMPLES

EXAMPLE 1. *Solve the equation $\nabla^2 u = -10(x^2 + y^2 + 10)$ over the squares with sides $x = 0$, $y = 0$, $x = 3 = y$ with $u = 0$ on the boundary and mesh length $= 1$.*

(Anna–2007, PTU–2007, Delhi–2002;
MKU(Tamilnadu)-2005, 2008; Roorkee-2002, 05, 07)

SOLUTION. The standard five-point formula for the given equation is

$$u_{i-1,j} + u_{i+1,j} + u_{i,j+1} + u_{i,j-1} - 4u_{i,j}$$
$$= -10(i^2 + j^2 + 10) \qquad ...(1)$$

For $u_1(i = i, j = 2)$, equation (1) gives
$$0 + u_2 + 0 + u_3 - 4u_1 = -10(1 + 4 + 10)$$

$\Rightarrow$
$$u_1 = \frac{1}{4}(u_2 + u_3 + 150) \qquad ...(2)$$

For $u_2(i = 2, j = 2)$, equation (1) gives
$$u_2 = \frac{1}{4}(u_1 + u_4 + 180) \qquad ...(3)$$

For $u_3(i = 1, j = 1)$, equation (1) gives
$$u_3 = \frac{1}{4}(u_1 + u_4 + 120) \qquad ...(4)$$

For $u_4(i = 2, j = 1)$, equation (1) gives
$$u_4 = \frac{1}{4}(u_2 + u_3 + 150) \qquad ...(5)$$

Equation (2) and (5) show that $u_4 = u_1$. Therefore, the above equation reduces to

$$u_1 = \frac{1}{4}[u_2 + u_3 + 150], \ u_2 = \frac{1}{2}[u_1 + 90], \ u_3 = \frac{1}{2}[u_1 + 60]$$

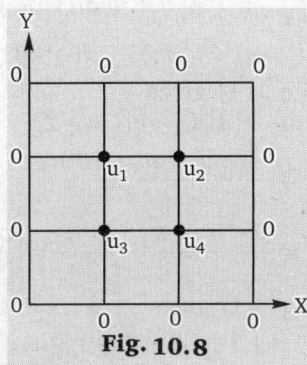

Fig. 10.8

Now, we solve these equation by Gauss-Seidel method, the iteration formulae are given below :

$$u_1^{(n+1)} = \frac{1}{4}(u_2^{(n)} + u_3^{(n)} + 150) \; ; \; u_2^{(n+1)} = \frac{1}{4}(u_1^{(n+1)} + 90) \; ; u_3^{(n+1)} = \frac{1}{4}(u_1^{(n+1)} + 60)$$

First iteration. Take initial approximation $u_2 = 0, u_3 = 0, u_1 = 0,$ then we get

$$u_1^{(1)} = 37.5 , u_2^{(1)} = \frac{1}{2}[37.5 + 90] = 64 ,$$

$$u_3^{(1)} = \frac{1}{2}[37.5 + 60] = 49$$

Second iteration.

$$u_1^{(2)} = \frac{1}{4}[64 + 49 + 150] = 66 , \; u_2^{(2)} = \frac{1}{2}[66 + 90] = 78 ,$$

$$u_3^{(2)} = \frac{1}{2}[66 + 60] = 63$$

Third iteration.

$$u_1^{(3)} = \frac{1}{4}[78 + 63 + 150] = 73 , \; u_2^{(3)} = \frac{1}{2}[73 + 90] = 82 ,$$

$$u_3^{(3)} = \frac{1}{2}[73 + 60] = 67$$

Fourth iteration.

$$u_1^{(4)} = \frac{1}{4}[82 + 67 + 150] = 75 , \; u_2^{(4)} = \frac{1}{2}[75 + 90] = 82.5 ,$$

$$u_3^{(4)} = \frac{1}{2}[75 + 60] = 67.5$$

Fifth iteration.

$$u_1^{(5)} = \frac{1}{4}[82.5 + 67.5 + 150] = 75 , \; u_2^{(5)} = \frac{1}{2}[75 + 90] = 82.5 ,$$

$$u_3^{(5)} = \frac{1}{2}[75 + 60] = 67.5$$

Hence, we observe that the values of fifth and fourth iteration are same. Therefore, $u_1 = 75, u_2 = 82.5, u_3 = 67.5$ and $u_4 = 75$.

EXAMPLE 2. *Solve the poisson equation $u_{xx} + u_{yy} = -81 xy$, $0 < x < 1$, $0 < y < 1$ given that $u(0, y) = 0, 4(x, 0) = 0, u(1, y) = 100, u(x, 1) = 100$ and $h = 1/3$.*   (Anna–2005)

SOLUTION. The standard five-point formula for the given equation is

$$u_{i-1,j} + u_{i+1,j} + u_{i,j+1} + u_{i,j-1} - 4u_{i,j} = h^2 f (ih, jh)$$

$$= h^2 [-81 (ih, jh)] = h^4 (-81) ij = -ij \qquad ...(1)$$

For u_1 $(i = 1, j = 2)$ eq$^{\text{n.}}$ (1) gives

$$0 + u_2 + u_3 + 100 - 4u_1 = -2$$

$$-u_1 = \frac{-2 - u_2 - u_3 - 100}{4}$$

$$\Rightarrow \qquad u_1 = \frac{2 + u_2 + u_3 + 100}{4} = \frac{u_2 + u_3 + 102}{4}$$

For u_2 $(i = 2, j = 2)$, eq$^{\text{n.}}$ (1) gives

$$u_1 + 100 + u_4 + 100 - 4u_2 = -4$$

$$\Rightarrow \qquad -u_2 = \frac{-204 - u_1 - u_4}{4} = \frac{u_1 + u_4 + 204}{4}$$

For u_3 $(i = 1, j = 1)$, eq$^{n.}$ (1) gives

$0 + u_4 + 0 + u_1 - 4u_3 = -1$

$$\Rightarrow \qquad u_3 = \frac{u_1 + u_4 + 1}{4}$$

For u_4 $(i = 2, j = 1)$, eqn. (1) gives

$u_3 + 100 + u_2 - 4u_4 = -2$

$$\Rightarrow \qquad u_4 = \frac{u_2 + u_3 + 102}{4}$$

Now, we solve these equations by Gauss Seidel method, then iteration formulae are given below :

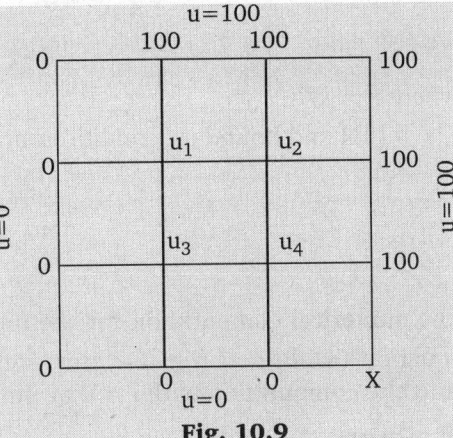

Fig. 10.9

$$u_1 = \frac{613}{12} = 51.0833 = u_4$$

$$u_2 = \frac{1}{2}(u_1 + 102) = 76.5477$$

$$u_3 = \frac{1}{2}\left(u_1 + \frac{1}{2}\right) = 25.7916.$$

10.6 PARABOLIC EQUATIONS

A popular case for parabolic type of equation is the study of heat flow in one- dimensional direction in an insulated rod. Such type of problems are governed by both boundary and initial conditions.

Elliptic equations studied previously describe problems that are time independent. Such problems are known as steady-state problems. But we come across problems that are not steady-state. This means that the function is dependent on both space and time. Parabolic equations for which $b^2 - 4ac = 0$ represents the problems that depend on space and time variables.

Definition. *An equation of the type $\dfrac{\partial u}{\partial t} = \alpha^2 \dfrac{\partial^2 u}{\partial x^2}$ is known as parabolic or heat equation.*

10.7 SOLUTION BY FORWARD DIFFERENCE METHOD

The simplest form of finite difference representation is obtained by approximating the time derivative by a forward difference and the space derivative by a central difference.

Now, consider the heat equation $\qquad \dfrac{\partial u}{\partial t} = \alpha^2 \dfrac{\partial^2 u}{\partial x^2}$ $\qquad$ ...(1)

Substitute the following values in (1)

$$\frac{\partial u}{\partial t}\bigg|_{(i,j)} = \frac{u_{i,j+1} - u_{i,j}}{k} + O(k) \qquad \text{...(2)}$$

$$\left.\frac{\partial^2 u}{\partial x^2}\right|_{(i,j)} = \frac{u_{i+1,j} - 2u_{i,j} + u_{i-1,j}}{h^2} + O(h^2) \qquad \qquad ...(3)$$

we get

$$u_{i,j+1} = r[u_{i+1,j} + u_{i-1,j}] + (1-2r)u_{i,j} \qquad \qquad ...(4)$$

where $r = k/h^2$.

The initial and boundary conditions are given by

$$\left.\begin{array}{l} u_{i,0} = g(ih) = g_1 \\ u_{0,j} = f_1(jk) = (f_1)_j \\ u_{n,j} = (f_2)_j \end{array}\right\} \qquad \qquad ...(5)$$

The numerical computation for the unknown values of u can be directly made by using the equation (4) for u at $(j + 1)^{\text{th}}$ time step using the known values of u at j^{th} time step. We will start the computation with $j = 0$ at time $t = 0$.

▶ **REMARK**

▸ The forward difference scheme (4) is convergent as well as stable for $r \le 1/2$, i.e., $\dfrac{k}{h^2} \le \dfrac{1}{2}$

10.8 SOLUTION BY BENDER-SCHMIDT'S METHOD

Consider one-dimensional heat equation

$$\frac{\partial u}{\partial t} = \alpha^2 \frac{\partial^2 u}{\partial x^2}, \qquad \text{where} \qquad \alpha^2 = \frac{k}{C_\rho} \qquad \qquad ...(1)$$

Equation (1) can be written as

$$u_{xx} = au_t, \qquad \text{where} \qquad a = \frac{1}{\alpha^2} \qquad \qquad ...(2)$$

Now, we want to solve equation (2) subject to the boundary conditions

$$u_{(0,\, t)} = T_0 \qquad \qquad ...(3)$$

$$u_{(l,\, t)} = T_l \qquad \qquad ...(4)$$

and the initial conditions

$$u(x, 0) = f(x) \qquad \qquad ...(5)$$

by finite difference method.

Consider the rectangular mesh in x-t plane with spacing h along x-direction and k along t-direction.

Since we know that

$$u_{xx} = \frac{1}{h^2}[u_{i+1,j} - 2u_{i,j} + u_{i-1,j}] \quad \text{and} \quad u_t = \frac{1}{k}[u_{i,j+1} - u_{i,j}]$$

Put all these values in (2), we get

$$\frac{1}{h^2}[u_{i+1,j} - 2u_{i,j} + u_{i-1,j}] = a.\frac{1}{k}[u_{i,j+1} - u_{i,j}]$$

$$\Rightarrow \qquad\qquad [u_{i,j+1} - u_{i,j}] = \lambda[u_{i+1,j} - 2u_{i,j} + u_{i-1,j}] \quad \text{where,}\ \lambda = \frac{k}{h^2}a$$

$$\Rightarrow \qquad\qquad u_{i,j+1} = \lambda u_{i+1,j} + (1-2\lambda)u_{i,j} + \lambda u_{i-1,j} \qquad \qquad ...(6)$$

The boundary conditions (3) and (4) can be put in the difference form as follows

$$u_{0,j} = T_0 \qquad \qquad ...(7)$$
$$u_{n,j} = T_l \qquad \qquad ...(8)$$

where, $j = 1, 2, ..., (nh = l)$
and the initial condition (5) can be written as

$$u_{i,\,0} = f(i, h), \, i = 1, 2, ... \qquad \qquad ...(9)$$

Equation (6) gives the value of u at $x = ih$ at time t_{j+k} in terms of values of u at $x = (i-1)h$, ih and $(i+1)h$ at a time t_1.

Since $u(x, 0) = f(x)$, u is known at $t = 0$.

Therefore, the recurrence relation (6) allows the evaluation of u at each pivotal point x_i at any t_j.

Hence, (6) is called the Schmidt formula which is valid for $0 \le \alpha \le 1/2$. Pictorial representation of this formula is given below :

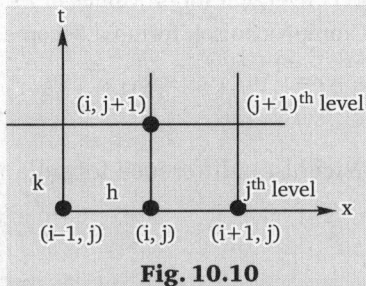

Fig. 10.10

▶ **REMARK**
▸ If h and k are chosen such that the coefficient of $u_{i,j}$ vanishes, i.e., $1 - 2\lambda = 0 \Rightarrow \lambda = 1/2$, then (6) gives

$$u_{i,j+1} = \frac{1}{2}[u_{i-1,j} + u_{i+1,j}] \qquad \qquad ...(10)$$

and

$$k = \frac{a}{2}h^2 \qquad \qquad ...(11)$$

This formula is known as Bender-Schmidt recurrence equation.

10.9 CRANK-NICHOLSON METHOD

Consider the parabolic equation $\qquad u_{xx} = au_t \qquad \qquad ...(1)$

subject to the conditions $\qquad u(0, t) = T_0 \qquad \qquad ...(2)$

$$u(l, t) = T_l \qquad \qquad ...(3)$$

and $\qquad u(x, 0) = f(x) \qquad \qquad ...(4)$

The finite difference approximation for u_{xx}, at point $u_{i,j}$ is given by

$$u_{xx} = \frac{1}{h^2}[u_{i+1,j} - 2u_{i,j} + u_{i-1,j}] \qquad \qquad ...(5)$$

Also the finite difference approximation for u_{xx} at point $u_{i,j+1}$ is

$$u_{xx} = \frac{1}{h^2}[u_{i+1,j+1} - 2u_{i,j+1} + u_{i-1,j+1}] \qquad \qquad ...(6)$$

Taking average of (5) and (6), we get

$$u_{xx} = \frac{1}{2h^2} \{u_{i+1,j+1} - 2u_{i,j+1} + u_{i-1,j+1} + u_{i+1,j} - 2u_{i,j} + u_{i-1,j}\} \qquad ...(7)$$

For u_t, the forward difference approximation is

$$u_t = \frac{1}{k}[u_{i,j+1} - u_{i,j}] \qquad ...(8)$$

Put the value of (7) and (8) in (1), we get

$$\frac{\lambda}{2}u_{i+1,j+1} - (\lambda+1)u_{i,j+1} + \frac{\lambda}{2}u_{i-1,j+1} = -\frac{\lambda}{2}u_{i+1,j} + (\lambda-1)u_{i,j} - \frac{\lambda}{2}u_{i-1,j}$$

(or) $$\lambda\{u_{i+1,j+1} + u_{i-1,j+1}\} - 2(\lambda+1)u_{i,j+1} = 2(\lambda-1)u_{i,j} - \lambda\{u_{i+1,j} + u_{i-1,j}\} \qquad ...(9)$$

Equation (9) is called Crank-Nicholson difference method.

▶ REMARKS

- ▶ The Crank-Nicholson formula is convergent for all values of λ.
- ▶ If we choose $\lambda - 1$ for λ, then Crank-Nicholson formula becomes

$$u_{i,j+1} = \frac{1}{4}[u_{i-1,j+1} + u_{i+1,j+1} + u_{i-1,j} + u_{i+1,j}]$$

COMPUTATIONAL MODEL

Computational model of Crank-Nicholson difference formula is given below :

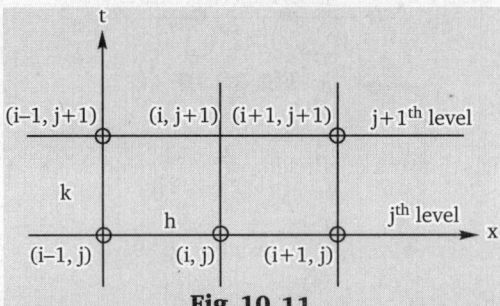

Fig. 10.11

Recapitulation ||||||||||||||||||||||||||||||||||||||||||||||||||||||||||||||||||

- Crank-Nicholson implicit scheme requires more computations per time step than the explicit scheme. However, a larger value of the time increment k can be used in the scheme, since its time derivative analogue is second order correct and the method is stable for every value of λ. Therefore, much fewer time steps and thus less computations are necessary to compute the values of the dependent variables for a given elapsed time.

10.10 DUFORT AND FRANKEL'S METHOD

Consider the one-dimensional heat equation

$$\frac{\partial u}{\partial t} = C^2 \frac{\partial^2 u}{\partial x^2} \qquad ...(1)$$

If we replace the derivatives in (1) by the central difference approximation

$$\frac{\partial u}{\partial t} = \frac{u_{i,j+1} - u_{i,j-1}}{2k} \quad \text{and} \quad \frac{\partial^2 u}{\partial x^2} = \frac{u_{i-1,j} - 2u_{i,j} + u_{i+1,j}}{h^2}$$

we get

$$u_{i,j+1} - u_{i,j-1} = \frac{2kC^2}{h^2}[u_{i-1,j} - 2u_{i,j} + u_{i+1,j}]$$

$$\Rightarrow \qquad u_{i,j+1} = u_{i,j+1} = 2\alpha[u_{i-1,j} - 2u_{i,j} + u_{i+1,j}] \qquad \ldots(2)$$

where,

$$\alpha = \frac{kC^2}{h^2}$$

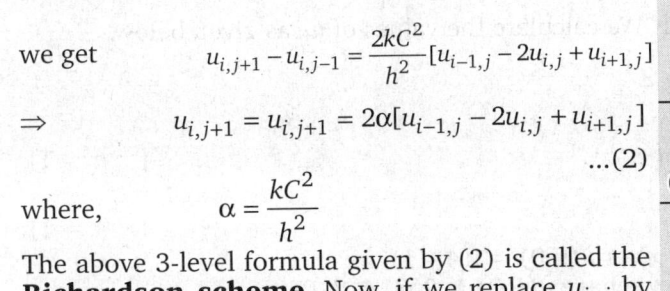

The above 3-level formula given by (2) is called the **Richardson scheme**. Now, if we replace $u_{i,j}$ by the mean of the values $u_{i,j-1}$ and $u_{i,j+1}$, *i.e.*,

$$u_{i,j} = \frac{1}{2}[u_{i,j-1} + u_{i,j+1}] \text{ in (2) then we get}$$

Fig. 10.12

$$u_{i,j+1} = u_{i,j-1} + 2\alpha[u_{i-1,j} - (u_{i,j-1} + u_{i,j+1}) + u_{i+1,j}]$$

This difference scheme is known as DuFort-Frankel formula.

10.11 ITERATIVE METHOD

By Crank-Nicholson method, we have

$$(1+\alpha)u_{i,j+1} = \frac{1}{2}\alpha(u_{i-1,j+1} + u_{i+1,j+1}) + u_{i,j} + \frac{1}{2}\alpha(u_{i-1,j} - 2u_{i,j} + u_{i+1,j}) \qquad \ldots(1)$$

Let

$$b_i = u_{i,j} + \frac{\alpha}{2}(u_{i-1,j} - 2u_{i,j} + u_{i+1,j}) \text{ and dropping } j\text{'s in (1)},$$

we get

$$u_1 = \frac{\alpha}{2(1+\alpha)}(u_{i-1} + u_{i+1}) + \frac{b_i}{1+\alpha}$$

which gives

$$u_i^{(n+1)} = \frac{\alpha}{2(1+\alpha)}\left\{u_{i-1}^{(n)} + u_{i+1}^{(n)})\right\} + \frac{b_i}{1+\alpha} \qquad \ldots(2)$$

Formula (2) expresses the $(n+1)^{th}$ iterations in terms of nth iterations only. This is known as Jacobi's iteration formula.

Since the latest value of $u_{i-1}^{(n+1)}$ of u_{i-1} is already available the convergence of the iteration formula (2) can be improved by replacing $u_{i-1}^{(n)}$ by $u_{i-1}^{(n+1)}$. Hence, (2) can be written as

$$u_i^{(n+1)} = \frac{\alpha}{2(1+\alpha)}\left\{u_{i-1}^{(n+1)} + u_{i+1}^{(n)})\right\} + \frac{b_i}{1+\alpha} \qquad \ldots(3)$$

which is known as Gauss-Seidel iteration formula.

SOLVED EXAMPLE

EXAMPLE 1. *Solve* $\dfrac{\partial u}{\partial t} = \dfrac{1}{2}\dfrac{\partial^2 u}{\partial x^2}$ *subject to the conditions* $u(0, t) = 0$, $u(4, t) = 0$, $u(x,0) = x(4-x)$

taking $h = 1$ *and using Bender-Schmidt recurrence equation.*

SOLUTION. Compare the given equation to general equation, we get

$$a = 2, h = 1 \Rightarrow \lambda = \frac{k}{h^2}, a = \frac{k}{2}$$

Therefore, $\lambda = \frac{1}{2}$ and $k = 1$. We calculate the values of u_{ij} as given below.

Since range for x is $0 \leq x \leq 4$

$$x = ih$$

$\Rightarrow \qquad x_i = ih = i$ $\qquad\qquad$ [$\because h = 1$]

$$t = jk$$

$\Rightarrow \qquad t_i = jk = j$ $\qquad\qquad$ [$\because k = 1$]

Given that $u(x, 0) = x(4 - x)$ or $u(i, 0) = i(4 - 1)$

For $0 \leq x \leq 4$, *i.e.*, $0 \leq i \leq 4$ we have $u(i, 0) = 0, 3, 4, 3, 0$.

Also given $u(0, t) = 0, \forall\, t$, *i.e.*, $u(0, j) = 0 \, \forall\, j$.

Hence, all entries in the first column are zero.

Now, $\qquad u(4, t) = 0 \,\forall\, t$, *i.e.*, $u(4, j) = 0 \,\forall\, j$.

Hence, all the entries in the last column are zero.

The Bender-Schmidt's recurrence relation is

$$u_{i,j+1} = \frac{1}{2}[u_{i+1,j} + u_{i-1,j}] \qquad ...(1)$$

Putting $\quad j = 0$ in (1), we get

$$u_{i,1} = \frac{1}{2}[u_{i+1,0} + u_{i-1,0}] \qquad ...(2)$$

Putting $\quad i = 1$ in (2), we get

$$u_{1,1} = \frac{1}{2}[u_{2,0} + u_{0,0}] = \frac{1}{2}[4 + 0] = 2$$

Putting $\quad i = 2$ in (2), we get

$$u_{2,1} = \frac{1}{2}[u_{3,0} + u_{1,0}] = \frac{1}{2}[3 + 3] = 3$$

Putting $\quad i = 3$ in (2), we get

$$u_{3,1} = \frac{1}{2}[u_{4,0} + u_{2,0}] = \frac{1}{2}[0 + 4] = 2$$

i / j	0	1	2	3	4
0	0	3	4	3	0
1	0	2	3	2	0
2	0	1.5	2	1.5	0
3	0	1	1.5	1	0
4	0	0.75	1	0.75	0
5	0	0.5	0.75	0.5	0
6	0	0.375	0.5	0.375	0
7	0	0.25	0.375	0.25	0
8	0	0.1875	0.25	0.1875	0
9	0	0.125	0.1875	0.125	0
10	0	0.094	0.125	0.094	0

Thus, the second row is filled. Similarly, we may fill all the rows. Hence, the values of $u_{i, j}$ can be tabulated as follows.

EXAMPLE 2. *Solve the equations $\dfrac{\partial u}{\partial t} = \dfrac{\partial^2 u}{\partial x^2}$ subject to the conditions*

$$u(x, 0) = \sin \frac{\pi x}{3}, 0 \leq x \leq 1, u(0, t) = u(1, t) = 0$$

Using (a) Schmidt method, (b) Crank-Nicholson method, (c) DuFort-Frankel method.

Carry out computations for two levels taking $h = \dfrac{1}{3}, k = \dfrac{1}{36}$.

[VTU–2005 (Similar); Rohtak–2003]

SOLUTION. Comparing the given equation to standard equation, we get

$$C^2 = 1, h = \frac{1}{3}, k = \frac{1}{36}$$

$$\Rightarrow \quad \alpha = \frac{kC^2}{h^2} = \frac{1}{4}$$

Also, $u_{1,0} = \sin\frac{\pi}{3} = \frac{\sqrt{3}}{2},$

$$u_{2,0} = \sin\frac{2\pi}{3} = \frac{\sqrt{3}}{2}$$

Then, we have the following figure :

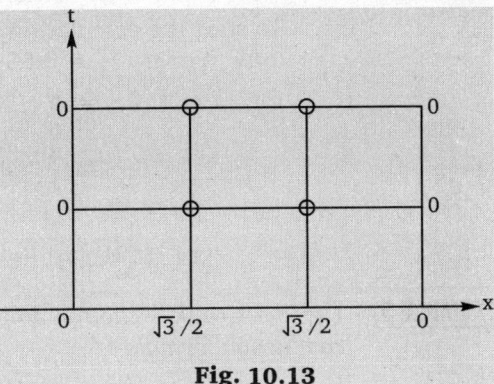

Fig. 10.13

(a) Schmidt formula

We have

$$u_{i,j+1} = \alpha u_{i-1,j} + (1 - 2\alpha)u_{i,j} + \alpha u_{i+1,j}$$

which implies

$$u_{i,j+1} = \frac{1}{4}[u_{i-1,j} + 2u_{i,j} + u_{i+1,j}]$$

Now, for $i = 1, 2; j = 0$

$$u_{1,1} = \frac{1}{4}[u_{0,0} + 2u_{1,0} + u_{2,0}] = \frac{1}{4}\left[0 + 2\frac{\sqrt{3}}{2} + \frac{3}{2}\right] = 0.65$$

$$u_{2,1} = \frac{1}{2}[u_{1,0} + 2u_{2,0} + u_{3,0}] = \frac{1}{4}\left[\frac{\sqrt{3}}{2} + 2\frac{\sqrt{3}}{2}\right] = 0.65$$

Similarly for $i = 1, 2, j = 1$, we get
$u_{1,2} = 0.49$ and $u_{2,2} = 0.49$

(b) Crank-Nicholson Method

Here, we have

$$-\frac{1}{4}u_{i-1,j+1} + \frac{5}{2}u_{i,j+1} - \frac{1}{4}u_{i+1,j+1} = \frac{1}{4}u_{i-1,j} + \frac{3}{2}u_{i,j} + \frac{1}{4}u_{i+1,j}$$

for $i = 1, 2, j = 0$, we get

$$-u_{0,1} + 10u_{1,1} - u_{2,1} = u_{0,0} + 6u_{1,0} + u_{2,0}$$

$$\Rightarrow \qquad 10u_{1,1} - u_{2,1} = \frac{7\sqrt{3}}{2}$$

and $\qquad -u_{1,1} + 10u_{2,1} - u_{3,1} = u_{1,0} + 6u_{2,0} + u_{3,0}$

$$\Rightarrow \qquad -u_{1,1} + 10u_{2,1} = \frac{7\sqrt{3}}{2}$$

On solving these two equations, we get $u_{1,1} = u_{2,1} = 0.67$

Similarly, for $i = 2, j = 1$, we get $u_{1,2} = u_{2,2} = 0.52$.

(c) DuFort Frankel method

Here, we have

$$u_{i,j+1} = \frac{1}{3}[u_{i,j-1} + u_{i-1,j} + u_{i+1,j}]$$

To start the calculation, we need $u_{1,\,1}$ and $u_{2,\,1}$, we may take

$$u_{1,-1} = u_{2,-1} = 0.65$$

for $i = 1, 2, j = 1$

(By Schmidt formula)

$$u_{1,2} = \frac{1}{3}[u_{1,0} + u_{0,1} + u_{2,1}] = \frac{1}{3}\left[\frac{\sqrt{3}}{2} + 0 + 0.65\right] = 0.5$$

$$u_{2,2} = \frac{1}{3}[u_{2,0} + u_{1,1} + u_{3,1}] = \frac{1}{3}\left[\frac{\sqrt{3}}{2} + 0.65 + 0\right] = 0.5$$

EXAMPLE 3.   *Use the Crank-Nicholson implicit method for solving the one- dimensional heat conduction formula*

$$\frac{\partial u}{\partial t} = \frac{\partial^2 u}{\partial x^2}$$

where $u(x,0) = \begin{cases} 2x & , \quad 0 \le x \le 1/2 \\ 2(1-x) & , \quad 1/2 \le x \le 1 \end{cases}$; $\dfrac{\partial u}{\partial x} = u$ *at* $x = 0$ *and* $\dfrac{\partial u}{\partial x} = -u$ *at*

$x = 1 \; \forall \; t$.   *Take $h = 0.2$ and $r = 1$ and compute the solution for the first time step.*

SOLUTION.   By Crank-Nicholson formula, we have

$$-u_{i-1,j+1} + 4u_{i,j+1} - u_{i+1,j+1} = u_{i-1,j} + u_{i+1,j} \qquad \qquad \ldots(1)$$

The finite difference representation of the boundary conditions

$$\frac{\partial u}{\partial x} = u \text{ at } x = 0 \text{ gives } \frac{u_{i+1,j} - u_{i-1,j}}{2h} = u_{ij}$$

$$\Rightarrow \qquad u_{i-1,j} = u_{i+1,j} - 2hu_{ij}, i = 0$$

$$\Rightarrow \qquad u_{-1,j} = u_{1,j} - 2hu_{0,j} \qquad \qquad \ldots(2)$$

Similarly, $\dfrac{\partial u}{\partial x} = -u$ at $x = 1$ gives

$$u_{i+1,j} - u_{i-1,j} + 2hu_{i,j}, i = 5$$

$$\Rightarrow \qquad u_{6,j} = u_{4,j} + 2hu_{5,j} \qquad \qquad \ldots(3)$$

Put $j = 0$ and $i = 0, 1, 2, 3, 4, 5$ in (1), we get

$$-u_{-1,1} + 4u_{0,1} - u_{1,1} = u_{-1,0} + u_{1,0} \qquad \qquad \ldots(4)$$

$$-u_{0,1} + 4u_{1,1} - u_{2,1} = u_{0,0} + u_{2,0} \qquad \qquad \ldots(5)$$

$$-u_{1,1} + 4u_{2,1} - u_{3,1} = u_{1,0} + u_{3,0} \qquad \qquad \ldots(6)$$

$$-u_{2,1} + 4u_{3,1} - u_{4,1} = u_{2,0} + u_{4,0} \qquad \qquad \ldots(7)$$

$$-u_{3,1} + 4u_{4,1} - u_{5,1} = u_{3,0} + u_{5,0} \qquad \qquad \ldots(8)$$

$$-u_{4,1} + 4u_{5,1} - u_{6,1} = u_{4,0} + u_{6,0} \qquad \qquad \ldots(9)$$

Now, substituting $j = 0$ in (2), we get

$$u_{-1,0} = u_{1,0} \qquad \qquad \ldots(10)$$

Similarly, putting $j = 0$ in (3), we get

$$u_{6,0} = u_{4,0} \qquad \qquad \ldots(11)$$

Also, for $j = 1$, equation (1) and (2) gives

$$u_{-1,1} = u_{1,1} - 4u_{0,1}$$

$$\Rightarrow \qquad u_{-1,1} = u_{1,1} - 0.4u_{0,1} \qquad \qquad \qquad ...(12)$$

$$u_{6,1} = u_{4,1} - 0.4u_{5,1} \qquad \qquad \qquad ...(13)$$

Further, the initial condition gives

$$u_{0,0} = 0, u_{1,0} = 0.4, u_{2,0} = 0.8,$$

$$u_{3,0} = 0.8, u_{4,0} = 0.4 \qquad \qquad \qquad ...(14)$$

Substituting the relation (10)-(14) in (3)-(9), we get, on simplification

$$4.4u_{0,1} - 2u_{1,1} = 0.8 \qquad \qquad \qquad ...(15)$$

$$-u_{0,1} + 4u_{1,1} - u_{2,1} = 0.8 \qquad \qquad \qquad ...(16)$$

$$-u_{1,1} + 4u_{2,1} - u_{3,1} = 1.2 \qquad \qquad \qquad ...(17)$$

$$-u_{2,1} + 4u_{3,1} - u_{4,1} = 1.2 \qquad \qquad \qquad ...(18)$$

$$-u_{3,1} + 4u_{4,1} - u_{5,1} = 0.8 \qquad \qquad \qquad ...(19)$$

$$-2u_{4,1} + 3.6u_{5,1} = 0.8 \qquad \qquad \qquad ...(20)$$

The system of equation (15)-(20) can be solved easily.

After solving, we get

$$u_{5,1} = 1.5990431, u_{4,1} = 1.6526411,$$

$$u_{3,1} = 2.0259389, u_{2,1} = 1.9738732,$$

$$u_{1,1} = 1.4080155, u_{0,1} = 0.821851$$

EXAMPLE 4. *Using Crank-Nicholson's method, solve $u_{xx} = 16u_t$, $0 < x < 1$, $t > 0$ given, $u(x, 0) = 0$, $u(0, t) = 0$, $u(1, t) = 50t$. Compute u for two steps in t-direction taking $h = 1/4$.*

SOLUTION. Here, we have $a = 16$, $h = 1/4$

$$k = ah^2 = 16\left(\frac{1}{16}\right) = 1$$

The crank-Nicholson scheme is given by

$$u_{i,j+1} = \frac{1}{4}[u_{i+1,j+1} + u_{i-1,j+1} + u_{i+1,j-1}, u_{i-1,j}] \qquad \qquad ...(1)$$

i $\diagdown$ j	0	0.25	0.5	0.75	1
0	0	0	0	0	0
1	0	u_1	u_2	u_3	50
2	0	u_4	u_5	u_6	100

Applying equation (1) at the mesh-points u_1, u_2, u_3, we get

$$u_1 = \frac{1}{4}[0 + 0 + 0 + u_2] = \frac{u_2}{4} \qquad \qquad \qquad ...(2)$$

$$u_2 = \frac{1}{4}[0 + 0 + u_1 + u_3] = \frac{1}{4}(u_1 + u_3) \qquad \qquad ...(3)$$

$$u_3 = \frac{1}{4}[0 + 0 + u_2 + 50] = \frac{1}{4}(u_2 + 50) \qquad \qquad ...(4)$$

Substituting (4) and (2) in (3), we get

$$u_2 = \frac{1}{4}\left[\frac{1}{4}u_2 + \frac{1}{4}[u_2 + 50]\right]$$

$$\Rightarrow \qquad 16u_2 = 2u_2 + 50$$

$$\Rightarrow \qquad u_2 = 3.5714$$

$$u_1 = 0.89285, u_3 = 13.39285$$

Applying equation (1) again at the mesh-points u_4, u_5 and u_6, we get

$$u_4 = \frac{1}{4}u_5 \qquad \qquad ...(5)$$

$$u_5 = \frac{1}{4}(u_6) \qquad \qquad ...(6)$$

$$u_6 = \frac{1}{4}(u_5 + 100) \qquad \qquad ...(7)$$

On solving, we get

$$u_4 = 1.7857, u_5 = 7.1429 \text{ and } u_6 = 26.7857 .$$

EXAMPLE 5. *Solve $\frac{\partial u^2}{\partial t^2} = \frac{\partial^2 u}{\partial x^2}$ in $0 < x < 5$, $t \geqslant 0$ given that $u(x,0) = 20$, $u(0, t) = 0$, $u(5, t) = 100$.*

Compute u for the time-step with $h = 1$ by Crank-Nicholson method.   (Anna-2006)

SOLUTION. Here, we have $a = 1$ and $h = 1$

$$k = ah^2 = 1 \Rightarrow k = 1$$

The crank-Nicholson scheme is given by

$$u_{i,j+1} = \frac{1}{4}[u_{i-1,j} + (1 - 2a)u_{i,j} + au_{i+1,j}] \qquad \qquad ...(1)$$

i / j	0	1	2	3	4	5
0	0	20	20	20	20	100
1	0	u_1	u_2	u_3	u_4	100

Applying equation (1) at the mesh-points u_1, u_2, u_3 and u_4, we get

$$u_1 = \frac{1}{4}[u_2 + 20] \qquad \qquad ...(2)$$

$$u_2 = \frac{1}{4}[u_1 + u_3 + 40] \qquad \qquad ...(3)$$

$$u_3 = \frac{1}{4}[u_2 + u_4 + 40] \qquad \qquad ...(4)$$

and $$u_4 = \frac{1}{4}[u_3 + 220] \qquad \qquad ...(5)$$

From eqn. (2), (3), (4) and (5), we get

$$u_1 = 10.05, u_2 = 20.2, u_3 = 30.75 \text{ and } u_4 = 62.69$$

which is the required values of u_1, u_2, u_3 and u_4.

EXAMPLE 6. *Solve the boundary value problem $u_t = u_{xx}$ under the conditions $u(0,t) = u(1,t) = 0$ and $u(x, 0) = \sin \pi x$, $0 \le x \le 1$ using Schmidt method (Take $h = 0.2$ and $a = 1/2$).*

(Chennai-2003)

SOLUTION. Here, we have $h = 0.2$ and $a = 1/2$

then $k = a \cdot h^2 = \dfrac{1}{2} \times 0.4 = 0.02$

We have Bendre-Schmidt relation

$$u_{i,j+1} = \frac{1}{4}[u_{i-1,j} + (1-2a)u_{i,j} + au_{i+1,j}] \qquad \ldots(1)$$

Therefore, $u(0, 0) = 0$, $u(0.2, 0) = \sin \pi/5 = 0.5875$

Similarly $u(0.4, 0) = 0.9511$, $u(0.6, 0) = 0.9511$

$$u(0.8, 0) = 0.5875, \ u(1, 0) = 0$$

Hence, the recurrence relations table as shown below :

$t \downarrow$		$x \to$	0	0.2	0.4	0.6	0.8	1.0
	j	i	0	1	2	3	4	5
0	0		0	0.5878	0.9511	0.9511	0.5878	0
0.02	1		0	0.4756	0.7695	0.7695	0.4756	0
0.04	2		0	0.3848	0.6225	0.6225	0.3848	0
0.06	3		0	0.3113	0.5036	0.5036	0.3113	0
0.08	4		0	0.2518	0.4074	0.4074	0.2518	0
0.1	5		0	0.2037	0.3296	0.3296	0.2037	0

10.12 SOLUTION OF TWO-DIMENSIONAL HEAT EQUATION: ADE METHOD

Consider a two-dimensional heat equation

$$\frac{\partial u}{\partial t} = C^2 \left(\frac{\partial^2 u}{\partial x^2} + \frac{\partial^2 u}{\partial y^2} \right) \qquad \ldots(1)$$

Consider a square region $0 \le x \le y \le a$ and assume that u is known at all points within and on the boundary of this square.

Let h be the step size, then a mesh-point $(x, y, t) = (ih, jh, nl)$ may be denoted as simply (i, j, n).

Now, replacing the derivatives in (1) by their finite difference approximation, we get

$$\frac{u_{i,j,n+1} - u_{i,j,n}}{l} = \frac{C^2}{h^2}[(u_{i-1,j,n} - 2u_{i,j,n} + u_{i+1,j,n}) + (u_{i,j-1,n} - 2u_{i,j,n} + u_{i,j+1,n})]$$

$$\Rightarrow \qquad u_{i,j,n+1} = u_{i,j,n} + \alpha(u_{i-1,j,n} + u_{i+1,j,n} + u_{i,j+1,n} + u_{i,j-1,n} - 4u_{i,j,n}) \qquad \ldots(2)$$

where, $\alpha = \dfrac{lC^2}{h^2}$

The above method is known as ADE (Alternating Direction Explicit) method.

SOLVED EXAMPLE

EXAMPLE. *Solve the equation $\dfrac{\partial u}{\partial t} = \dfrac{\partial^2 u}{\partial x^2} + \dfrac{\partial^2 u}{\partial y^2}$ subject to the initial conditions*

$u(x, y, 0) = \sin 2\pi x \sin 2\pi y$, $0 \le x, y \le 1$ and the conditions $u(x, y, t) = 0$, $t > 0$, on

the boundaries, using ADE method with $h = \dfrac{1}{3}$ and $\alpha = \dfrac{1}{8}$.

SOLUTION. By ADE method, we have

$$u_{i,j,n+1} = u_{i,j,n} + \frac{1}{8}(u_{i-1,j,n} + u_{i+1,j,n} + u_{i,j+1,n} + u_{i,j-1,n} - 4u_{i,j,n})$$

$$\Rightarrow \quad u_{i,j,n+1} = \frac{1}{2}u_{i,j,n} + \frac{1}{8}(u_{i-1,j,n} + u_{i+1,j,n} + u_{i,j+1,n} + u_{i,j-1,n}) \qquad \ldots(1)$$

Now, we shall find the mesh-points.

At the Zeroth level $(n = 0)$, the initial and boundary conditions are

$$u_{i,j,0} = \frac{1}{2}u_{i,j,0} + \frac{1}{8}(u_{i-1,j,0} + u_{i+1,j,0} + u_{i,j-1,0})$$

Step-1. Put $i = j = 1$ in (2), we get $\qquad \ldots(2)$

$$u_{1,1,1} = \frac{1}{2}u_{1,1,0} + \frac{1}{8}[u_{0,1,0} + u_{2,1,0} + u_{1,0,0}]$$

$$= \frac{1}{2}\left(\sin\frac{2\pi}{3}\right)^2 + \frac{1}{8}\left[0 + \sin\frac{4\pi}{3}\sin\frac{2\pi}{3} + \sin\frac{2\pi}{3}\sin\frac{4\pi}{3} + 0\right]$$

$$= \frac{3}{8} + \frac{1}{8}\left(-\frac{\sqrt{3}}{2} \times \frac{\sqrt{3}}{2} - \frac{\sqrt{3}}{2} \times \frac{\sqrt{3}}{2}\right) = \frac{3}{16}$$

Step-2. Put $i = 2, j = 1$ in (2), we get

$$u_{2,1,1} = \frac{1}{2}u_{2,1,0} + \frac{1}{8}[u_{1,1,0} + u_{3,1,0} + u_{2,2,0} + u_{2,2,0}]$$

$$= \frac{1}{2}\sin\frac{4\pi}{3}\sin\frac{2\pi}{3} + \frac{1}{8}\left\{\left(\sin\frac{2\pi}{3}\right)^2 + 0 + 0 + \left(\sin\frac{4\pi}{3}\right)^2 + 0\right\}$$

$$= -\frac{1}{2}\left(\frac{\sqrt{3}}{2}\right)^2 + \frac{1}{8}\left[\left(\frac{\sqrt{3}}{2}\right)^2 + \left(\frac{-\sqrt{3}}{2}\right)^2\right] = -\frac{3}{16}$$

Step-3. Put $i = 1, j = 2$ in (2), we get

$$u_{1,2,1} = \frac{1}{2}u_{1,2,0} + \frac{1}{8}(u_{0,2,0} + u_{2,2,0} + u_{1,1,0})$$

$$= \frac{1}{2}\sin\frac{2\pi}{3}\sin\frac{4\pi}{3} + \frac{1}{8}\left\{0 + \left(\sin\frac{4\pi}{3}\right)^2 + 0 + \left(\sin\frac{2\pi}{3}\right)^2\right\}$$

$$= -\frac{3}{8} + \frac{1}{8}\left(\frac{3}{4} + \frac{3}{4}\right) = -\frac{3}{16}$$

Step-4. Put $i = 2, j = 2$ in (2), we get

$$u_{2,2,1} = \frac{1}{2}u_{2,2,0} + \frac{1}{8}(u_{1,2,0} + u_{3,2,0} + u_{2,3,0} + u_{2,1,0})$$

$$= \frac{1}{2}\left(\sin\frac{4\pi}{3}\right)^2 + \frac{1}{8}\left(\sin\frac{2\pi}{3}\sin\frac{4\pi}{3} + 0 + 0 + \sin\frac{4\pi}{3}\sin\frac{2\pi}{3}\right)$$

$$= \frac{3}{8} + \frac{1}{8}\left(-\frac{3}{4} - \frac{3}{4}\right) = \frac{3}{16}$$

Similarly we may obtain the mesh values of the second and higher levels.

10.13 HYPERBOLIC EQUATION

An equation of the type $a^2 \dfrac{\partial^2 u}{\partial x^2} = \dfrac{\partial^2 u}{\partial t^2}$ *or* $a^2 u_{xx} - u_{it} = 0$ *is said to the hyperbolic equation.*

(RGPV Bhopal-2008)

10.13.1 SOLUTION BY METHOD OF FINITE DIFFERENCE

Consider the equation $\qquad a^2 \dfrac{\partial^2 u}{\partial x^2} = \dfrac{\partial^2 u}{\partial t^2}$ ...(1)

subject to the conditions

$$u(0, t) = 0 \qquad \text{...(2)}$$
$$u(l, t) = 0 \qquad \text{...(3)}$$

and
$$u_t(x, 0) = f(x) \qquad \text{...(4)}$$
$$u_t(x, 0) = 0 \qquad \text{...(5)}$$

Put $\qquad u_{xx} = \dfrac{1}{h^2}[u_{i+1,j} - 2u_{i,j} + u_{i-1,j}]$ and $u_{tt} = \dfrac{1}{k^2}[u_{i,j+1} - 2u_{i,j} + u_{i,j-1}]$

in (1), we get

$$\frac{a}{h^2}[u_{i+1,j} - 2u_{i,j} + u_{i-1,j}] - \frac{1}{k^2}[u_{i,j+1} - 2u_{i,j} + u_{i,j-1}] = 0$$

$$\Rightarrow \qquad \lambda^2 a^2[u_{i+1,j} - 2u_{i,j} + u_{i-1,j}] - [u_{i,j+1} - 2u_{i,j} + u_{i,j-1}] = 0$$

where, $\lambda = \dfrac{k}{h}$.

$$\Rightarrow \qquad u_{i,j+1} = 2(1 - \lambda^2 a^2)u_{i,j} + \lambda^2 a^2(u_{i+1,j} + u_{i-1,j}) - u_{i,j-1} \qquad \text{...(6)}$$

The boundary condition (2) and (3) can be put in the difference form as

$$u_{0,j} = 0 = u_{n,j}; j = 1, 2, 3$$

The initial condition (4) as

$$u_{i,0} = f(ih), i = 1, 2, ... \qquad \text{...(7)}$$

and (5) as

$$\frac{1}{k}(u_{i,j+1} - u_{i,0}) = 0 \quad \text{when } t = 0, i.e., j = 0$$

$$\therefore \qquad u_{i,1} - u_{i,0} = 0 \qquad \text{...(8)}$$

$$\Rightarrow \qquad u_{i,1} = u_{i,0} = f(ih) \qquad \text{...(9)}$$

Now, equation (7) and (8) give the values of u on the first two rows $j = 0$ and $j = 1$. Putting $j = 1$ in (6), we get

$$u_{i,2} = 2(1-\lambda^2a^2)u_{i,1} + \lambda^2a^2(u_{i+1,1} + u_{i-1,1}) - u_{i,0} \qquad ...(10)$$

Now, RHS of (10) involves the values of u on the first two rows $j = 0$ and $j = 1$. These are known from the initial condition (7) and (8). Hence, we can find $u_{i,\,2}$.

REMARK

- If $k < h$ solution(6) is convergent.

SOLVED EXAMPLES

EXAMPLE 1. *Use the finite difference method to find the solution of the problem*

$$\frac{\partial^2 u}{\partial t^2} - \frac{\partial^2 u}{\partial x^2} = 0$$

with $u(x,0) = x(\pi - x), \frac{\partial u}{\partial t}(x,0) = 0;\ \ u(0, t) = u(\pi, t) = 0$

SOLUTION. Here, we consider a square grid with spacing $h = k = \frac{\pi}{18}$. The given problem is symmetric over the interval $[0, \pi]$. Therefore, we find the solution in the interval $[0, \pi/2]$. The value of u_{10} are given by the initial condition $u_{10} = x_i(\pi - x)$.

$u_{00} = 0,\ u_{10} = 0.518,\ u_{20} = 0.975,\ u_{30} = 1.371,\ u_{40} = 1.706,\ u_{50} = 1.980,$
$u_{60} = 2.193,\ u_{70} = 2.346,\ u_{80} = 2.437,\ u_{90} = 2.467$

To find the solution for the next time step we use the Taylor series. We write

$$u_{i1} = u_{i0} + k\frac{\partial u_{i0}}{\partial t} + \frac{k^2}{2}\frac{\partial^2 u_{i0}}{\partial t^2}$$

$$\frac{\partial u_{i0}}{\partial t} = 0, \frac{\partial^2 u_{i0}}{\partial t^2} = \frac{\partial^2 u_{i0}}{\partial x^2} = -2$$

$$\therefore \quad u_{i1} = u_{i0} - k^2 = u_{i0} - 0.03048$$

Therefore, the values of u_{i1} are

$$u_{01} = 0, u_{11} = 0.487, u_{21} = 0.944,$$
$$u_{31} = 1.340, u_{41} = 1.675, u_{51} = 1.950,$$
$$u_{61} = 2.163, u_{71} = 2.315, u_{81} = 2.406,$$
$$u_{91} = 2.437$$

We now compute the values of $u_{i,\,j+1}$ for $j = 1, 2, ...$ using the finite difference representation

$$u_{i,\,j+1} - 2u_{ij} + u_{i,\,j-1} = u_{i+1,\,j} - 2u_{i,\,j} + u_{i-1,\,j}$$

$$\Rightarrow \quad u_{i,\,j+1} = u_{i+1,\,j} + u_{i-1,\,j} - u_{i,j-1}$$

Therefore,

$$u_{22} = u_{31} + u_{11} - u_{20} = 1.340 + 0.487 - 0.975 = 0.853$$

$$u_{32} = u_{41} + u_{21} - u_{30} = 1.675 + 0.944 - 1.371 = 1.249$$
$$u_{42} = u_{51} + u_{31} - u_{40} = 1.950 + 1.340 - 1.706 = 1.584$$

$$u_{52} = u_{61} + u_{41} - u_{50} = 2.163 + 1.675 - 1.980 = 1.858$$
$$u_{62} = u_{71} + u_{51} - u_{60} = 2.315 + 1.950 - 2.346 = 2.224$$
$$u_{82} = u_{91} + u_{71} - u_{80} = 2.437 + 2.315 - 2.437 = 2.315$$
$$u_{92} = u_{101} + u_{81} - u_{90} = 2.406 + 2.406 - 2.467 = 2.346$$

EXAMPLE 2. *Solve $u_{tt} = 4u_{xx}$ with the boundary conditions $u(0, t) = 0 = u(4, t)$, $u(x, 0)$ and $u(x, 0) = u(4 - x)$.*

SOLUTION. Here, the given equation is

$$\frac{\partial^2 u}{\partial t^2} = 4 \frac{\partial^2 u}{\partial x^2} \qquad \qquad ...(1)$$

Comparing with the standard equation, we get
$$a^2 = 4 \qquad \Rightarrow \qquad a = 2.$$

Taking $\qquad h = 1 \qquad \Rightarrow \qquad k = \dfrac{h}{a} = \dfrac{1}{2} = 0.5$.

From the initial conditions
$$u(0, t) = 0 \qquad \Rightarrow \qquad u = 0 \text{ along line } x = 0.$$
$$u(4, t) = 0 \qquad \Rightarrow \qquad u = 0 \text{ along line } x = 4.$$

which can be written in difference form as follows
$$u_{0, j} = 0 \qquad \text{and} \qquad u_{4, j} = 0, \forall j$$

Now, $\qquad u(x, 0) = x(4 - x)$

$\Rightarrow \qquad u(0, 0) = 0, u(1, 0) = 3, u(2, 0) = 4,$

$\qquad \qquad u(3, 0) = 3, u(4, 0) = 0$

In difference notation
$$u_{i, 0} = u(i, 0) = i(4 - 1), \text{ for different } i.$$

Putting $i = 0, 1, 2, 3, 4$, we get
$$u_{0, 0} = 0, u_{1, 0} = 3, u_{2, 0} = 4, \ u_{3, 0} = 3, u_{4, 0} = 0$$

The condition $u_t(x, 0)$ gives

$$\frac{1}{k}(u_{i, j+1} - u_{i,j}) = 0 \text{ when } j = 0.$$

$$u_{i, 1} = u_{i, 0} \ \forall_i$$

$\Rightarrow \quad u$ on the first two rows are equal.

Consider the recurrence relation
$$u_{i, j+1} = u_{i+1, j} + u_{i-1, j} - u_{i, j-1}$$

If we put $j = 1$, we get
$$u_{i, 2} = u_{i+1, 1} + u_{i-1, 1} - u_{i, 0}$$

Putting $i = 1, 2, 3, ...$ successively, we get
$$u_{1, 2} = u_{2, 1} + u_{0, 1} - u_{1, 0} = 4 + 0 - 3 = 1$$
$$u_{2, 2} = u_{3, 1} + u_{1, 1} - u_{2, 0} = 3 + 3 - 4 = 2$$
$$u_{3, 2} = u_{4, 1} + u_{2, 1} - u_{3, 0} = 0 + 4 - 3 = 1.$$

In the similar way, we can fill in the remaining rows as shown in following table :

j \\ i	0	1	2	3	4
0	0	3	4	3	0
1	0	3	4	3	0
2	0	1	2	1	0
3	0	–1	–2	–1	0
4	0	–3	–4	–3	0

EXAMPLE 3. *Solve $u_{tt} = u_{xx}$ upto $t = 0.5$ with a spacing of 0.1 subject to $u(0, t) = 0$, $u(1, t) = 0$,*
$u_t(x, 0) = 0$ and $u(x, 0) = 10 + x(1 - x)$.   (Anna-2004)

SOLUTION.   Here, the given equation is

$$\frac{\partial u^2}{\partial t^2} = \frac{\partial^2 u}{\partial x^2}$$

Comparing with the standard equation, we get

$$a^2 = 1, h = 0.1, k = 0.1$$

From the initial conditions

$$u(0, t) = 0, u(1, t) = 0$$

which can be written in difference from as follows

$$u_{0, j} = 0, u_{i, j} = 0$$

Now, $u(x, 0) = 10 + x (1 - x)$

$$u_{i, 0} = 10 + i (1 - i) \qquad \qquad \dots(1)$$

Consider the recurrence relation

$$u_{i, j + 1} = u_{i + 1, j} + u_{i - 1, j} - u_{i, j - 1} \qquad \qquad \dots(2)$$

Putting $i = 0.1, 0.2, \dots, 0.9$ successively, we get

$$u_{0.1, 0} = 10.09, u_{0.2, 0} = 10.16, u_{0.3, 0} = 10.21$$

$$u_{0.4, 0} = 10.24, u_{0.5, 0} = 10.25, u_{0.6, 0} = 10.24$$

$$u_{0.7, 0} = 10.21, u_{0.8, 0} = 10.16, u_{0.9, 0} = 10.08$$

For second row, we have

$$u_t(x, 0) = 0 \Rightarrow \frac{1}{2}(u_{i, j+1}, - y_{i, j-1}) = 0 \qquad \qquad \dots(3)$$

If $j = 0$, then $u_{i, 1} = u_{i, -1}$

Now, putting $j = 0$ in equation (2), we get

$$u_{i,1} = u_{i-1,0} + u_{i+1,0} - u_{i,-1}$$

From (3), we get

$$u_{i,1} = \frac{1}{2}(u_{i-1,0} + u_{i+1,0})$$

Putting $i = 1, 2, 3, \dots 9$ successively, we get the values of second row.

In the similar way, we can fill the remaining rows as shown in following table :

i \ j	0	1	2	3	4	5	6	7	8	9	10
0	0	10.19	10.16	10.21	10.24	10.25	10.24	10.21	10.16	10.09	0
1	0	5.08	10.15	10.20	10.23	10.24	10.23	10.20	10.15	5.08	0
2	0	0.06	5.12	10.17	10.20	10.21	10.20	10.17	10.12	0.06	0
3	0	0.04	0.08	5.12	10.15	10.16	10.15	10.12	10.08	0.04	0
4	0	0.02	0.04	0.06	5.08	10.09	10.08	10.06	10.04	0.02	0
5	0	0.00	0	0	0	0	0	0	0	−0.02	0

10.13.2 Solution by the Method of Characteristics

In order to understand the method of characteristics, consider a quasi-linear PDE of the form

$$a\frac{\partial^2 u}{\partial x^2} + b\frac{\partial^2 u}{\partial x \partial y} + c\frac{\partial^2 u}{\partial y^2} + h = 0 \qquad \ldots(1)$$

where, the coefficients a, b, c and h may be functions of u, $\dfrac{\partial u}{\partial x}$ and $\dfrac{\partial u}{\partial y}$ only.

Here, we use the following notation

$$\frac{\partial u}{\partial x} = p, \frac{\partial u}{\partial y} = q, \frac{\partial^2 u}{\partial x^2} = r, \frac{\partial^2 u}{\partial x \partial y} = s, \frac{\partial^2 u}{\partial y^2} = t \qquad \ldots(2)$$

Then, (1) can be written as

$$ar + bs + ct + h = 0 \qquad \ldots(3)$$

Let C be the curve in the x-y plane such that the function representing C satisfies the above equation. The total differentials of p and q in the direction tangential to C are given by

$$dp = \frac{\partial p}{\partial x}dx + \frac{\partial p}{\partial y}dy \quad \Rightarrow \quad dp = rdx + sdy \qquad \ldots(4)$$

$$dq = \frac{\partial q}{\partial x}dx + \frac{\partial q}{\partial y}dy \quad \Rightarrow \quad dq = sdx + tdy \qquad \ldots(5)$$

Eliminate r and t from (3) using (4) and (5), we get

$$\frac{a}{dx}(dp - sdy) + bs + \frac{c}{dy}(dq - sdx) + h = 0$$

$$\Rightarrow \qquad s\left(-a\frac{dy}{dx} + b - c\Big/\frac{dy}{dx}\right) + a\frac{dp}{dx} + c\frac{dq}{dy} + h = 0$$

$$\Rightarrow \qquad s(am^2 - bm + c) - \left(am\frac{dp}{dx} + c\frac{dq}{dx} + hm\right) = 0 \qquad \ldots(6)$$

where, $m = \dfrac{dy}{dx}$.

Equation (6) is independent of r and t. In order to make (6) independent of s, also, we choose the curve C such that the slope of the tangent at every point on it satisfies the quadratic equation

$$am^2 - bm + c = 0 \qquad \ldots(7)$$

The two directions given by (7) are real and distinct when $b^2 - 4ac > 0$, i.e., the second order partial differential equation (1) is hyperbolic. The two curves characterised by the above two directions are called characteristic curves. The differential equation (3) along a characteristic curve, then, reduces to the simple form

$$am + \frac{dp}{dx} + c\frac{dq}{dx} + hm = 0 \qquad ...(8)$$

The two families of characteristic are termed as f- characteristic and g- characteristic, where

$$f = \frac{b + \sqrt{b^2 - 4ac}}{2a} = \text{constt.} \qquad ...(9)$$

$$g = \frac{b - \sqrt{b^2 - 4ac}}{2a} = \text{constt.} \qquad ...(10)$$

Let the value of u, p and q be prescribed on the initial curve C_1,

which does not belong to any characteristic family. Let E, F, G be any three neighbouring points on C_1.

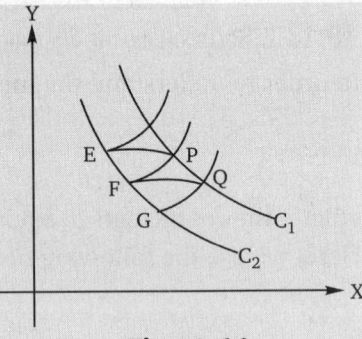

Fig. 10. 14

Let the f-characteristic through E intersect the g-characteristic through F at P. Similarly, the f-characteristic through F intersects the g-characteristic at Q and so on. The points P, Q would lie on a neighbouring curve, say C_2. Let the coordinate of P be (x_p, y_p) which are to be determined. Since E and F are close points. The curve EP and FP are approximated as straight line. If (x_E, y_E), (x_F, y_F) are the coordinates of E and F respectively, then we have

$$\frac{y_p - y_E}{x_p - x_E} = f_E \qquad ...(11)$$

$$\frac{y_p - y_F}{x_p - x_F} = g_f \qquad ...(12)$$

Since m ($= f_E$ or g_E) satisfies (8), we have

$$a_E f_E(p_P - P_E) + C_E(q_P - q_E) + h f_E = 0 \qquad ...(13)$$
$$a_F g_F(p_P - P_F) + C_F(q_P - q_F) + h g_F = 0 \qquad ...(14)$$

Using (11), (12) in (13) and (14) respectively, we have two equations for the two unknowns p_P and q_P. We can next determine u_P, the value p using

$$du = pdx + qdy \qquad ...(15)$$

This can be approximated as

$$u_P - u_E = \frac{P_P - P_E}{2}(x_P - x_E) + \frac{q_P - q_E}{2}(y_P - y_E)$$

$$\Rightarrow \qquad u_P = u_E + \frac{P_P - P_E}{2}(x_P - x_E) + \frac{q_P - q_E}{2}(y_P - y_E) \qquad ...(16)$$

The value of u at the point Q can be determined in a similar manner. Thus, we can obtain the solution along a neighbouring non-characteristic curve C_2 using the values u, p, q along C_2, we can proceed to determine the values along the next non- characteristic curve C_3 and so on.

▼ **REMARK**
▶ The method of characteristic cannot be applied to parabolic and elliptic partial differential equations, since the characteristic directions are coincident for parabolic and are imaginary for elliptic differential equations.

SOLVED EXAMPLE

EXAMPLE. *Use the method of characteristic to solve the boundary value problem*

$$\frac{\partial^2 u}{\partial x^2} + \frac{\partial^2 u}{\partial x \partial y} - 2\frac{\partial^2 u}{\partial y^2} + 1 = 0$$

$$u = x, \frac{\partial u}{\partial y} = x \ for \ y = 0, \ 0 \le x \le 1.$$

SOLUTION. Comparing the given equation, with the standard equation, we get $a = 1$, $b = 1$, $c = -2$, $h = -1$.

The characteristic directions are given by

$$m^2 - m - 2 = 0$$

$$\Rightarrow \qquad m = \frac{1 \pm 3}{2} \text{ which gives } f = 2, g = -1.$$

From the given condition, the initial curve C_1 is the straight line $y = 0$ and

$$u = x, \frac{\partial u}{\partial y} = x \text{ along } y = 0.$$

$$\therefore \qquad p = \frac{\partial u}{\partial x} = 1, q = \frac{\partial u}{\partial x} = x \text{ along } y = 0.$$

Let $\qquad x_E = 0.4$ and $\qquad x_F = 0.5$

$$\therefore \qquad q_E = 0.4 \qquad\qquad q_F = 0.5$$

$$u_E = x_E = 0.4, \qquad u_F = x_{\hat{F}} = 0.5.$$

Now, $\qquad y_P = y_E + f_E(x_P - x_E) = 0 + 2(x_P - 0.4) = 2x_P - 0.8,$

Also, $\qquad y_P = y_F + g_F(x_P - x_F) = 0 + (-1)(x_P - 0.5) = -x_P + 0.5$

Equating these two equations, we get

$$2x_P - 0.8 = -x_P + 0.5 \quad \Rightarrow \quad x_P = 0.433$$

Hence, $\qquad y_P = 0.5 - 0.433 = 0.067$

Now, $a_E f_E (p_P - p_E) + c_E(q_P - q_E) + h(y_P - y_E) = 0$

Therefore, $2(p_P - 1) - 2(q_P - 0.4) + 0.067 = 0$

$$\Rightarrow \qquad p_P - q_P = 0.4665$$

and $a_F q_F (p_P - p_F) + c_F(q_P - q_F) + h_F(y_P - y_F)(P_P - 1) = 0$

$$\Rightarrow -p_P - 2(q_P - 0.5) = -0.067$$

$$\Rightarrow \qquad p_P + 2q_P = 2.067$$

Solving $q_P = 0.5335$ and $p_P = 1$

$$\therefore \qquad u_P = u_E + \frac{p_P - p_E}{2}(x_P - x_E) + \frac{q_P - q_E}{2}(y_P - y_E)$$

$$\Rightarrow \qquad u_P = 0.4 + \frac{1+1}{2}(0.433 - 0.4) + \frac{0.5335 + 0.4}{2}(0.067) = 0.46425$$

Hence, the value of u at P is 0.46427 to a first approximation. The coordinates of p are $x_p = 0.433$, $y_p = 0.067$.

EXERCISE 10.1

1. Classify the following equations :
  (i) $3u_{xx} + u_{xy} - 4u_{yy} + 3u_y = 0$
  (ii) $u_{xx} - 6u_{xy} + 9u_{yy} - 17u_y = 0$

2. Solve the elliptic equation $u_{xx} + u_{yy} = 0$ for the following square mesh with boundary values as shown in the figure. Iterate until the maximum difference between two successive values at any point is less than 0.001.

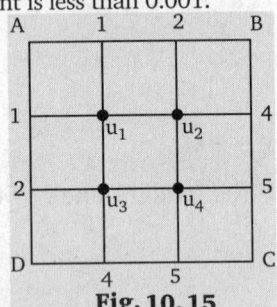

Fig. 10.15

3. Solve $u_{xx} - u_{yy} = 0$ over the square mesh of side four units satisfying the following boundary conditions :
  (i) $u(0, y) = 0$ for $0 \leq y \leq 4$
  (Cusat B.Tech-2008)
  (ii) $u(4, y) = 12 + y$ for $0 \leq y \leq 4$
  (iii) $u(x, 0) = 3x$ for $0 \leq x \leq 4$
  (iv) $u(x, 4) = x^2$ for $0 \leq x \leq 4$

4. Solve $\dfrac{\partial^2 u}{\partial x^2} + \dfrac{\partial^2 u}{\partial y^2} = 8x^2 y^2$ for square mesh, given $u = 0$ on the four boundaries dividing the square into 16-subsquares of length one unit. *(JNTU–2004S)*

5. Find the values of $u(x, y)$ satisfying the Laplace equation $\nabla^2 u = 0$ at the pivotal points of a square region with boundary values given below : *(VTU-2009)*

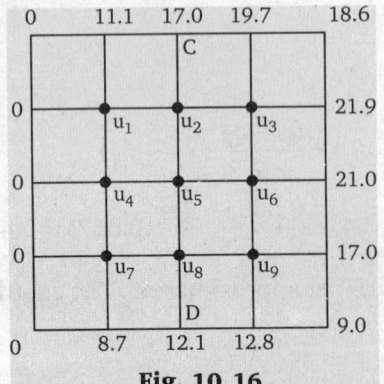

Fig. 10.16

6. Compute u for one time step by Crank-Nicholson method if $u_t = u_{xx}$; $0 < x < 5$, $t > 0$; $u(x, 0) = 20$, $u(0, t) = 0$ and $u(5, t) = 100$. *(Anna–2006)*

7. Find the numerical solution to solve $u_t = u_{xx}$, $0 \leq x \leq 1$, $t \geq 0$ under the conditions that $u(0, t) = u(1, t)$ and

$$u(x,0) = \begin{cases} 2x, & \text{for } 0 \leq x \leq 1/2 \\ 2(1-x), & \text{for } 1/2 \leq x \leq 1 \end{cases}$$

8. Solve the hyperbolic partial differential equation for one half period of oscillation taking $h = 1$.
  $u_{tt} = 25u_{xy}$, $u(0, t) = u(5, t) = 0$, $u_t(x, 0) = 0$

$$u(x,0) = \begin{cases} 2x, & \text{if } 0 \leq x \leq 2.5 \\ 10-x, & \text{if } 2.5 \leq x \leq 5 \end{cases}$$

9. Evaluate the pivotal values for the following equation taking $h = 1$ and upto one half of the period of vibration $16u_{xx} = u_{tt}$, given that $u(0, t) = u(5, t) = 0$, $u(x, 0) = x^2(x - 5)$ and $u_t(x, 0) = 0$. *(Madras–2006)*

10. Classify the following equations.
  (i) $u_{xx} + 2u_{xy} + u_{yy} = 0$ [Madras–2001]
  (ii) $y^2 u_{xx} - 2xy u_{xy} + x^2 u_{yy} + 2u_x - 3u = 0$
  (Madras–2003)
  (iii) $y^2 u_{xx} + u_{yy} + u^2_x + u^2 y + 7 = 0$
  (Madras–2003)

11. Solve the equation $u_{xx} + u_{yy} = 0$ for the square mesh with boundary values as shown in the fig. *(Delhi–2002)*

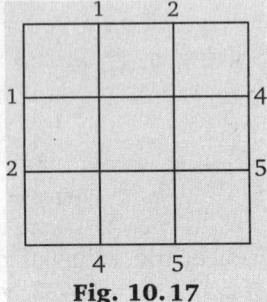

Fig. 10.17

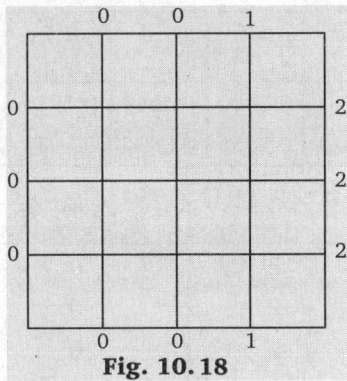

Fig. 10.18

12. Solve the elliptic equation $u_{xx} + u_{yy} = 0$ for the square mesh with boundary values as shown in fig. Iterate until the maximum difference between successive values at any point is less than 0.005.

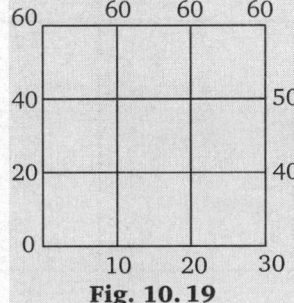

Fig. 10.19

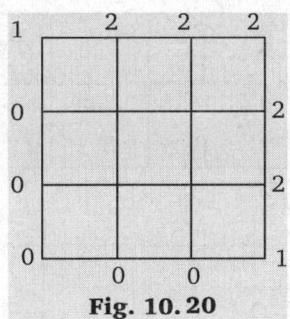

Fig. 10.20

13. Find the solution of the parabolic equation $u_{xx} = 2u$, when $u(0, t) = u(4, t) = 0$ and $4(x, 0) = x(4 - x)$, taking $h = 1$. Find the values upto $t = 5$. [Madras–2001]

14. Solve the equation $\dfrac{\partial^2 u}{\partial x^2} = \dfrac{\partial u}{\partial t}$ with the conditions $u(0, t)$, $u(x, 0) = x(1 - x)$ and $u(1, t) = 0$. Assume $h = 0.1$. Tabulate u for $t = k$, $2k$ and $3k$ choosing an appropriate value of k. [Anna–2004]

15. Solve the boundary value problem $u_{tt} = u_{xx}$ with the conditions $u(0, t) = u(1, t) = 0$, $u(x,0) = \dfrac{1}{2}x(1-x)$ and $u_i(x, 0) = 0$, taking $h = k = 0.1$ for $0 \le t \le 0.4$. Compare your solution with the exact solution at $x = 0.5$ and $t = 0.3$. [VTU–2003]

16. Solve $\dfrac{\partial^2 u}{\partial t^2} = \dfrac{\partial^2 u}{\partial x^2}$, $0 < x < 1$, $t > 0$, given $t > 0$, given $u(x, 0) = u_t(x, 0) = u(0,1) = 0$ and $u(1, t) = 100 \sin\pi t$ compute u for 4 times with $h = 0.25$. [Anna–2003]

Answers

1. (i) Hyperbolic; (ii) Parabolic

2. $u_1 = 1.999$, $u_2 = 2.999$, $u_4 = 3.999$

3. 2.37, 5.59, 9.87, 2.88, 6.13, 9.88, 3.01, 6.16, 9.51

4. –3, –2, –3, –2, –2, –2, –3, –2, –3

5. 7.9, 13.7, 17.9, 6.6, 11.9, 16.3, 6.6, 11.2, 14.3

6.

j \ i	0	0.25	0.5	0.75	1
0	0	0	0	0	0
1/16	0	0.00116	0.004464	0.01674	1/16
1/8	0	0.005899	0.019132	0.052771	1/8

7.

j \ i	0	1	2	3	4	5
0	0	20	20	20	20	100
1	0	9.80	20.19	30.72	59.92	100

8.

j \ i	0	0.2	0.4	0.6	0.8	1.0	0.8	0.6	0.4	0.2	0
0	0	0.1936	0.3689	0.5400	0.6461	0.6291	0.6461	0.5400	0.3689	0.1936	0
1	2	0.1989	0.3956	0.5834	0.7381	0.7691	0.7381	0.5834	0.3956	0.1989	0

9.

j \ i	0	1	2	3	4	5
0	0	2	4	4	2	0
0.2	0	2	4	4	2	0
0.4	0	2	2	2	2	0
0.6	0	0	0	0	0	0
0.8	0	−2	−2	−2	−2	0
1.0	0	−2	−4	−4	−2	0

10.

j \ i	0	1	2	3	4	5
0	0	4	12	18	16	0
1	0	4	12	18	16	0
2	0	8	10	10	2	0
3	0	6	6	−6	−6	0
4	0	−2	−10	−10	−8	0
5	0	−16	−18	−12	−4	0

11. (i) Parabolic; (ii) Parabolic (iii) Elliptic
12. $u_1 = 7.9, u_2 = 13.7, u_3 = 17.9, u_4 = 6.6, u_5 = 11.9, u_6 = 16.3, u_7 = 6.6, u_8 = 11.2, u_9 = 14.3$
13. $u_1 = 26.66, u_2 = 33.33, u_3 = 43.33, u_4 = 46.66$
.14. $u_1 = 0.99, u_2 = 1.49, u_3 = 0.49$

15.

j \ i	0	1	2	3	4
0	0	3	4	3	0
1	0	2	3	2	0
2	0	1.5	2	1.5	0
3	0	1	1.5	1	0
4	0	0.75	1	0.75	0
5	0	0.5	0.75	0.5	0

16.

j \ i	0	1	2	3	4	5	6	7	8	9	10
0	0	0.9	0.16	0.21	0.24	0.25	0.24	0.21	0.16	0.09	0
1	0	0.8	0.15	0.20	0.23	0.24	0.23	0.20	0.15	0.08	0
2	0	0.75	0.14	0.19	0.22	0.23	0.22	0.19	0.14	0.075	0
3	0	0.7	0.13	0.18	0.21	0.22	0.21	0.18	0.13	0.07	0

17.

$t = 0.3, x =$	0.1	0.2	0.3	0.4	0.5
Num. sol. u =	0.02	0.04	0.06	0.075	0.08
Exact. sol. u =	0.02	0.04	0.06	0.075	0.08

18.

i / j	0	1	2	3	4
0	0	0	0	0	0
1	0	0	0	0	70.7
2	0	0	0	70.7	100
3	0	0	70.7	100	70.7
4	0	70.7	100	70.7	0

MISCELLANEOUS EXERCISE

1. Solve the partial differential equation (viz. Laplace's equation) $\left(\dfrac{\partial^2 u}{\partial x^2}\right) + \left(\dfrac{\partial^2 u}{\partial y^2}\right) = 0$ for the figure given below :

Fig. 10.21

(PTU-2005, 09; Chennai-2001, 03, 04, 08; Raipur-2007, 09; UPTU-2008; Rohilkhand-2007; Bangluru-2005, 08)

2. Evaluate the pivotal values of the following equation taking $h = 1$ and upto one half of the period of vibration $16\dfrac{\partial^2 u}{\partial x^2} = \dfrac{\partial^2 u}{\partial t^2}$ given that $u(0, t) = u(5, t) = 0; u(x, 0) = x^2(5-x)$ and $u_t(x, 0) = 0$. (Meerut-2002; Chennai-2007; RGPV-2006)

3. Solve the equation $\dfrac{\partial^2 u}{\partial x^2} + \dfrac{\partial^2 u}{\partial y^2} = 0$ for the region boundary by the square $0 \le x \le 4$ and

$0 \le y \le 4$, subject to the bcs :
$u = 0$ at $x = 0$
$u + 8 + 2y$ at $x = 4$.
$u = x^2/2$ at $y = 0$
$u = x^2$ at $y = 4$.

Choose $h = k = 0.1$ and use Gauss-Seidel method to compute u at the interval mesh points.

4. Solve the heat equation $\dfrac{\partial u}{\partial t} = \dfrac{\partial^2 u}{\partial x^2}$ subject to the boundary and initial conditions as follows: $u(x, 0) = 0, u(0, t) = 0, u(1, t) = t$.

(Meerut-2003; Rohilkhand-2006, 09; Bhopal-2007; Bangluru-2004)

5. Solve the heat conduction equation $\dfrac{\partial u}{\partial t} = \dfrac{\partial^2 u}{\partial x^2}$ under the boundary and initial conditions :
(a) $u(x, t) = 0$ at $x = 0$ and $x = 1$ for $t > 0$.
(Kurukshetra-2002, 03, 06, 08)
(b) $u(x, t) = \sin \pi x$ at $t = 0$ for $0 \le x \le 1$, by Gauss-Seidel method.
(Madurai-2004; Bilaspur-2006, 09)

6. (a) Solve the elliptic equation $u_{xx} + u_{yy} = 0$ for the following square mesh with boundary values as shown.
(RGPV-2009; Bangaluru-2006; 07)
(b) Given the values of $u(x, y)$ on the boundary of the square in the figure and evaluate the function $u(x, y)$ satisfying the Laplace equation $\Delta^2 u = 0$ at the pivotal points of this figure by
(i) Jacobi's Method (Chennai-2004)
(ii) Gauss Seidel method.
(VTU-2007; Andhra-2000, 06)

Answers

1.

u_1	u_2	u_3	u_4	u_5	u_6	u_7	u_8	u_9
7.03	9.57	7.08	18.94	25.10	18.98	43.02	52.97	42.99
7.13	9.83	7.20	18.81	25.15	18.84	42.94	52.77	42.90
7.16	9.88	7.18	18.81	25.08	18.79	42.89	52.72	42.88
7.17	9.86	7.16	18.78	25.04	18.77	42.88	52.70	42.87
7.16	9.84	7.15	18.77	25.02	18.76	42.87	52.69	42.86

2.

i j	0	1	2	3	4	5
0	0	4	12	18	16	0
1	0	4	12	18	16	0
2	0	0+12−4 =8	4+18−12=10	12+16−18=10	18+0−16=2	0
3	0	0+10−4=6	8+10−12=6	10+2−18=−6	10+0−16=−6	0
4	0	−2	−10	−10	−8	0
5	0	−16	−18	−12	−4	0

3. $u_1 = 1208, u_2 = 782, u_3 = 1042, u_4 = 458$.

4. 0.53022.

5.

x $u(x)$	0.0 0.0	0.2 0.5878	0.4 0.9511	0.6 0.9511	0.8 0.5878	1.0 0.0
$n = 0$	0.0	0.5878	0.9511	0.9511	0.5878	0.0
$n = 1$	0.0	0.5129	0.8176	0.8078	0.4890	0.0
$n = 2$	0.0	0.4907	0.7900	0.7868	0.4855	0.0
$n = 3$	0.0	0.4861	0.7858	0.7855	0.4853	0.0
$n = 4$	0.0	0.4854	0.7854	0.7854	0.4853	0.0
$n = 5$	0.0	0.4853	0.7854	0.7854	0.4853	0.0
$n = 6$	0.0	0.4853	0.7854	0.7854	0.4853	0.0

6. (a) $u_1 = 939$, $u_2 = 1001$, $u_4 = 1251$ and 1126.

    (b) (i) $u_1 = 1208$, $u_2 = 792$, $u_3 = 1042$ and $u_4 = 458$.

       (ii) $u_1 = 1208$, $u_2 = 792$, $u_3 = 1042$ and $u_4 = 458$.

Objective Evaluations

FILL IN THE BLANKS

1. The Laplace equation is _____.

2. An equation of the type $\dfrac{\partial u}{\partial t} = \alpha^2 \dfrac{\partial^2 u}{\partial x^2}$ is known as _____ equation.

3. Parabolic equation is also known as _____ equation.

4. The equation _____ is known as hyperbolic equation.

5. The equation $3u_{xx} + u_{xy} - 4u_{yy} + 3u_y = 0$ is _____ equation.

6. If $u(x)$ is a function, then difference coefficient is _____.

TRUE/FALSE

Write 'T' for True and 'F' for False statement.

1. A difference quotient is the quotient obtained by dividing the difference between two values of a function by the difference between the two corresponding values of the independent variable. **(T/F)**

2. Laplace equation is also known as Heat equation. **(T/F)**

3. The equation
$$A\frac{\partial^2 u}{\partial x^2} + B\frac{\partial^2 u}{\partial x \partial y} + C\frac{\partial^2 u}{\partial y^2} + D\frac{\partial u}{\partial x} + E\frac{\partial u}{\partial y} A + Fu = 0$$
is said to be elliptic if $B^2 - 4AC < 0$. **(T/F)**

4. The Laplace equation $\dfrac{\partial^2 u}{\partial x^2} + \dfrac{\partial^2 u}{\partial y^2} = 0$ can be written as $u_{xx} + u_{yy} = 0$. **(T/F)**

5. Libermann iteration formula is
$$u_{i,j}^{(n+1)} = \frac{1}{4}\left[u_{i-1,j}^{(n+1)} + u_{i+1,j}^{(n)} + u_{i,j+1}^{(n+1)} + u_{i,j-1}^{(n)}\right]$$
(T/F)

6. Gauss-Seidal method is
$$u_{i,j}^{(n+1)} = \frac{1}{4}\left[u_{i-1,j}^{(n+1)} + u_{i+1,j}^{(n)} + u_{i,j+1}^{(n+1)} + u_{i,j-1}^{(n)}\right]$$
(T/F)

7. One dimentional heat equation is $\dfrac{\partial u}{\partial t} = \alpha^2 \dfrac{\partial^2 u}{\partial x^2}$ **(T/F)**

8. $\lambda\{u_{i+1,j+1} + u_{i+1,j+1}\} - 2(\lambda+1)u_{i,j+1} = 2(\lambda-1)u_i, -\lambda\{u_{i+1,j} + u_{i-1,j}\}$ is called Crank-Nicholson difference method. **(T/F)**

MULTIPLE CHOICE QUESTIONS

Choose the most appropriate one.

1. Parabolic equation is :

(a) $\dfrac{\partial^2 u}{\partial x^2} + \dfrac{\partial^2 u}{\partial y^2} = 0$ (b) $\dfrac{\partial u}{\partial t} = \alpha^2 \dfrac{\partial^2 u}{\partial x^2}$

(c) $a^2 \dfrac{\partial^2 u}{\partial x^2} = \dfrac{\partial^2 u}{\partial t^2}$ (d) none of these

2. Heat equation is also known as :

(a) Elliptic   (b) Hyperbolic

(c) Parabolic   (d) none of these

3. The formula
$$u_{i,j}^{(n+1)} = \frac{1}{4}\left[u_{i-1,j}^{(n+1)} + u_{i+1,j}^{(n)} + u_{i,j-1}^{(n)} + u_{i,j+1}^{(n+1)}\right]$$
is known by :

(a) Libermann interation formula

(b) Gauss-Seidel method

(c) Jacobi's method

(d) none of these

4. The formula
$$u_{i,j}^{(n+1)} = \frac{1}{4}\left[u_{i-1,j}^{(n)} + u_{i+1,j}^{(n)} + u_{i,j+1}^{(n+1)} + u_{i,j-1}^{(n)}\right]$$
is known by :

(a) Libermann iteration formula

(b) Gauss-Seidel method

(c) Jacobi's method

(d) none of these

5 $u_{i,j+1} = u_{i,j-1} + 2\alpha\left[u_{i-1,j} - (u_{i,j-1} + u_{i,j+1}) + u_{i+1,j}\right]$

    is known by :

(a) Iterative method

(b) Dufort and Frankel's method

(c) Crank Nicholson method

(d) none of these

6. $\lambda\{u_{i+1,j+1} + u_{i-1,j+1}\} - 2(\lambda+1)u_{i,j+1}$

    $= 2(\lambda-1)u_{i,j} - \lambda(u_{i+1,j} + u_{i-1,j})$

(a) Iterative method

(b) Dufort and Frankel's method

(c) Crank Nicholson method

(d) none of these

7. The equation $3u_{xx} + u_{xy} - 4u_{yy} + 3u_y = 0$ is :

(a) Hyperbolic   (b) Parabolic

(c) Elliptic   (d) none of these

8. The Crank-Nicholson formula is convergent for all values of :

(a) $(\lambda - 1)$

(b) λ

(c) $(\lambda + 1)$

(d) none of these

Answers

FILL IN THE BLANKS

   1. $\dfrac{\partial^2 u}{\partial x^2} + \dfrac{\partial^2 u}{\partial y^2} = 0$ **2.** Parabolic   **3.** Heat equation   **4.** $a^2 \dfrac{\partial^2 u}{\partial x^2} = \dfrac{\partial^2 u}{\partial t^2}$ **5.** Hyperbolic   **6.** $\dfrac{u(x+h) - u(x)}{h}$

TRUE/FALSE

   1. T   **2.** F   **3.** T   **4.** T   **5.** F   **6.** T   **7.** T   **8.** T

MULTIPLE CHOICE QUESTIONS

   1. (b)   **2.** (c)   **3.** (a)   **4.** (c)   **5.** (b)   **6.** (c)   **7.** (a)   **8.** (b)

COMPUTATIONAL TECHNIQUE LAB

1. SOLUTION OF P.D.E. (LAPLACE EQUATION) BY JACOBI METHOD

SYMBOLS USED

sm = Size of square mesh

ctr = It counts the number of iteration to achieve the desired level of accuracy.

flag = It controls the iteration.

t = It is used to store the values of $f[i][j]$ in each iteration temporarly.

ALGORITHM: SOLUTION OF PDE (LAPLACE EQUATION) BY JACOBI METHOD USING FIVE POINT FORMULA.

Step 1:	Start
Step 2:	Input sm
Step 3.	for $i = 1$ to sm
	for $j = 1$ to sm
	$f[i][j] = 0$
Step 4.	ctr = 0
Step 5.	$i = 1$, for $j = 1$ to sm
	input $f[i][j]$
Step 6.	$i = sm$; for $j = 1$ to sm
	input $f[i][j]$
Step 7.	$j = 1$, for $i = 2$ to (sm–1)
	input $f[i][j]$
Step 8.	$j = sm$, for $i = 2$ to (sm–1)
	input $f[i][j]$
Step 9.	Repeat steps 10 to 13 until flag $\neq$ 0
Step 10.	flag = 1
Step 11.	for $i = 2$ to (sm–1)
Step 12.	for $i = 2$ to (sm–1)
	$t = f[i-1][j] + f[i+1][j] + f[i][j+1] + f[i][j-1]/4$
	if $|f[i][j] - t| > 0.1$
	flag = 0
	$f[i][j] = t$
	End of step 12 loop\
Step 13.	ctr = ctr + 1
Step 14.	Print "Values of f at various mesh points are"
Step 15.	for i = 1 to sm
	for j = 1 to sm
	Print f[i][j]
	End of step 15 loop.
Step 16.	Stop

FLOW CHART: SOLUTION OF PDE BY JACOBI METHOD USING 5-POINT FORMULA.

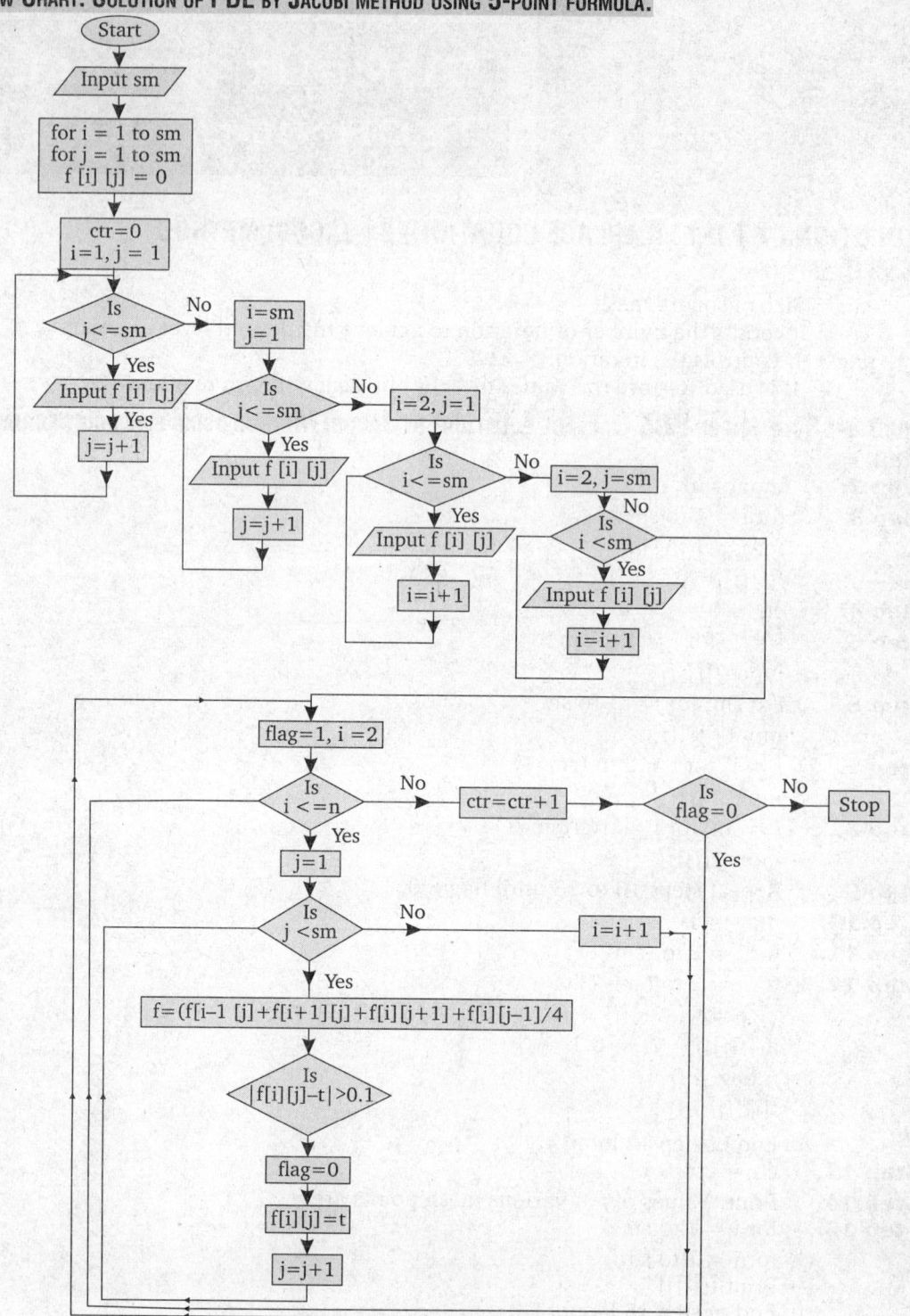

PROGRAM: *Following is a C program to solve the partial differential equation (Laplace Equation) by Jacobi using five point method.*

```c
/* program showing the solution of partial differential equation
(Laplace Equation)by Jacobi using five point method*/

#include<stdio.h>
#include<conio.h>
#include<math.h>
void main()
{
    int sm,ctr=0,i,j,k,flag;
    float f[20][20],t;
    clrscr();
    printf("\n enter the size of square mesh\n");
    scanf("%d",&sm);
    for(i=1;i<=sm;i++)
    for(j=1;j<=sm;j++)
        f[i][j]=0;
        printf("\n enter the boundary condition");
        for(i=1,j=1;j<=sm;j++)
        {
            printf("\n enter the values of f%d%d=",i,j);
            scanf("%f",&f[i][j]);
        }
        for(i=sm,j=1;j<=sm;j++)
        {
            printf("\n enter the values of f%d%d=",i,j);
            scanf("%f",&f[i][j]);
        }
        for(i=2,j=1;i<sm;i++)
        {
            printf("\n enter the values of f%d%d=",i,j);
            scanf("%f",&f[i][j]);
        }
        for(i=2,j=sm;i<sm;i++)
        {
            printf("\n enter the values of f%d%d=",i,j);
            scanf("%f",&f[i][j]);
        }
        do
        {
            flag=1;
            for(i=2;i<sm;i++)
```

```
        for(j=2;j<sm;j++)
        {
          t=(f[i-1][j]+f[i+1][j]+f[i][j+1]+f[i][j-1])/4;
          if(fabs(f[i][j]-t)>0.1)
                flag=0;
          f[i][j]=t;
        }
      ctr++;
      }while(flag==0);
      printf("\n the values of f at various mesh point");
      printf("\n after %d iteration are",ctr);
      for(i=1;i<=sm;i++)
        {
        printf("\n");
        for(j=1;j<=sm;j++)
        printf("  %5.2f",f[i][j]);
      }
      getch();
  }
```

Output: SOLUTION OF PARTIAL DIFFERENTIAL EQUATION LAPLACE EQUATION) BY JACOBI USING FIVE POINT METHOD

```
enter the size of square mesh 4

enter the values of f11=100

enter the values of f12=100

enter the values of f13=100

enter the values of f14=100

enter the values of f41=100

enter the values of f42=50

enter the values of f43=0

enter the values of f44=0

enter the values of f21=200
```

```
enter the values of f31=200

enter the values of f24=50

enter the values of f34=0

the values of f at various mesh point
after 7 iteration are

100.00   100.00   100.00   100.00
200.00   120.82    79.16    50.00
200.00   104.16    45.83     0.00
100.00    50.00     0.00     0.00
```

LAB ASSIGNMENT

1. Write a C program to find the solution of Laplace equation $f_{xx} + f_{yy} = 0$ at the pivotal points of the following figure by Jacobi method using 5 point formula.

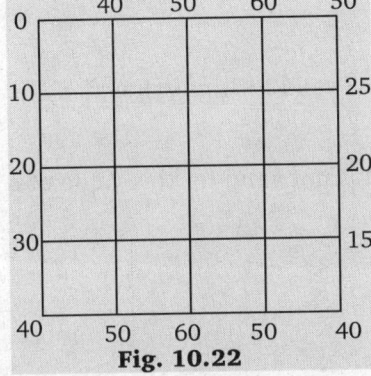

Fig. 10.22

Hint: Input sm = 5

$f_{11} = 0, f_{12} = 40, f_{13} = 50, f_{14} = 60, f_{15} = 50, f_{51} = 40, f_{52} = 50, f_{53} = 60,$
$f_{54} = 50, f_{55} = 40, f_{21} = 10, f_{31} = 20, f_{41} = 30, f_{52} = 25, f_{53} = 20 f_{54} = 15$

Output: Values of f at various mesh points are

0.00	40.00	50.00	60.00	50.00
10.00	29.93	38.77	39.07	25.00
20.00	31.09	36.18	32.56	20.00
30.00	38.36	42.38	34.98	15.00
40.00	50.00	60.00	50.00	40.00

2. SOLUTION OF ONE DIMENSIONAL HEAT EQUATION BY SCHMIDT METHOD

SYMBOLS USED

x = end values of x ; t = end values of t

$f_{xbt} = f(0, t)$

$f_{xt} = f(x, t)$

c = value of C^2 in one dimensional heat flow equation-

$$\frac{\partial f}{\partial t} = c^2 \frac{\partial^2 f}{\partial x^2}$$

i.e., $ft = C^2 f_{xx}$

$a = (c * k)/h^2$ *i.e.,* mesh ratio parameter

h = step size along x direction

k = step size along t direction

$f_x = f(x, 0)$

ALGORITHM : SCHMIDT METHOD

Step 1.	Start
Step 2.	Define fx()
Step 3.	Input x, t, fxbt, fxt, h, k, c
Step 4.	$a = (c * k)/h^2$
Step 5.	Print a
Step 6.	If $(0 < a < 0.5)$

Print "Schmid method is valid for this value of a"

goto step 7

else

Print "Schmidt method is not valid for this value of a"

Goto step 11

Step 7.   For $j = 0$ to t

         $f[0][j] = f_{xbt}$

         $f[x][j] = f_{xt}$

     End of step 7 loop

Step 8.   For $i = 1$ to $(x - 1)$

         $f[i][0] = fx[i]$

     End of step 8 loop

Step 9.   For $j = 0$ to $(t - 1)$

     For $i = 1$ to $(x - 1)$

$f[i][j + 1] = a * f[i - 1][j] + (1 - 2 * a) * f[i][j] + a * f[i + 1][j]$

     End of i loop

     End of step 9 loop

Step 10.   For $j = 0$ to t

     For $i = 0$ to x

     Print $f[i][j]$

     End of i loop

     End of step 10 loop

Step 11.   Stop

Flow Chart : Schmidt Method

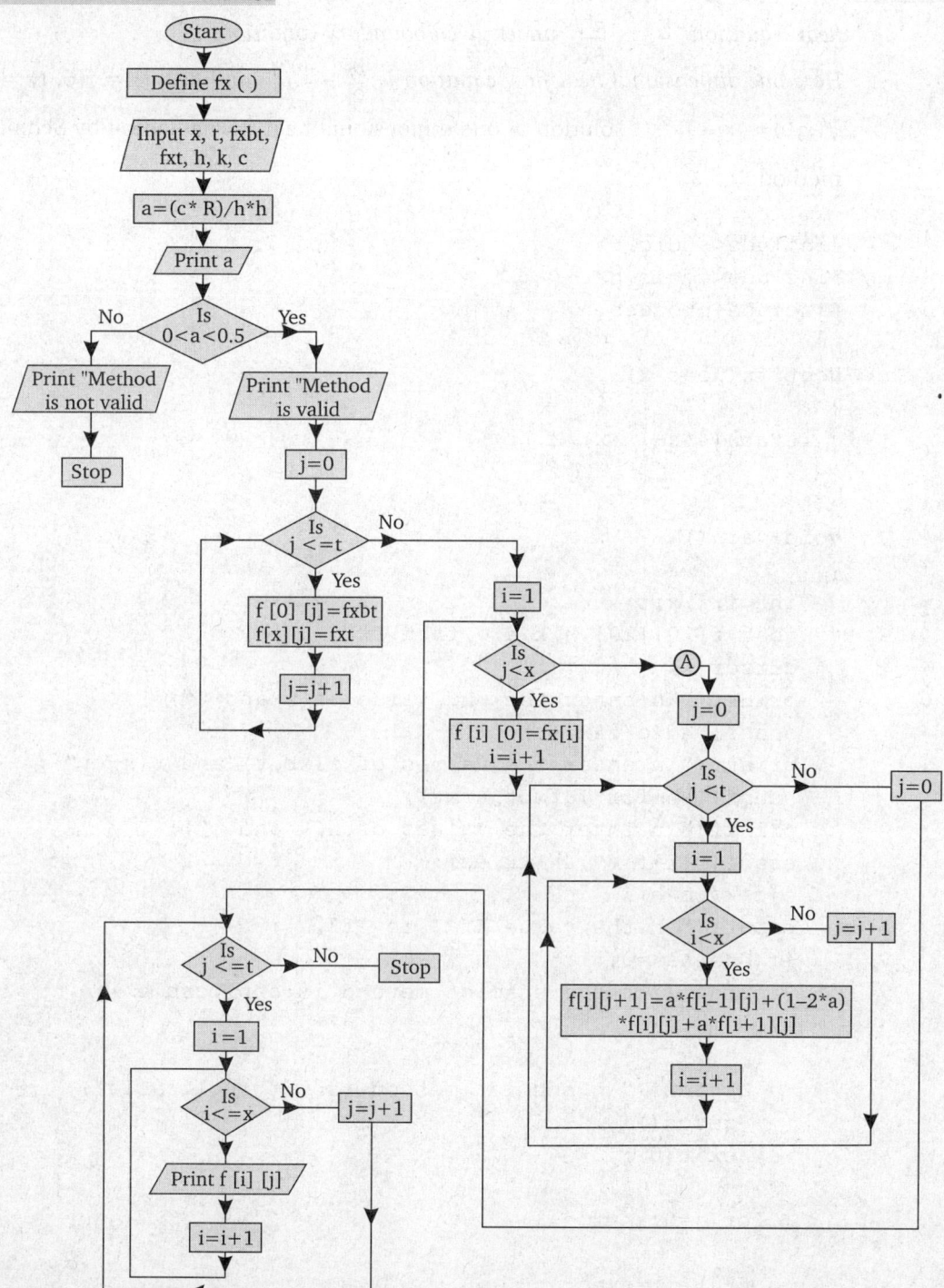

PROGRAM : *Following is a C program for Schmidt method to find the solution of one dimensional heat equation* $\frac{\partial f}{\partial t} = c^2 \frac{\partial^2 f}{\partial x^2}$ *under given boundary conditions.*

Here one dimensional heat flow equation is $\frac{\partial f}{\partial t} = \frac{4 \partial^2 f}{\partial x^2}$ *and* $F(0, t) = f(6, t) = 0$

$f(x, 0) = 4x - \frac{1}{2}x^2$ /* *solution of one dimensional heat flow equation by Schmidt method* */

```c
#include<stdio.h>
#include<conio.h>
#include<process.h>

float fx(float x)
{
   return(4*x-(x*x)/2 );
}

void main()
{
   int i,j,x,t;
   float f[20][20],h,k,a,c,fxbt,fxt;
   clrscr();
   printf("\n enter the end value of x and t");
   scanf("%d%d",&x,&t);
   printf("\n enter the values of f(xb,t) and f(x,t)");
   scanf("\n%f%f",&fxbt,&fxt);
   printf("\n enter the values of h,k and c");
   scanf("%f%f%f",&h,&k,&c);
   a=c*k/(h*h);
   printf("\n the value of a is %f",a);
   if(0<a&&a<=0.5)
     printf("\n the schmidt method is applicable");
   else
     {
       printf("\n schmidt method is not applicable");
       getch();
       exit(0);
     }
   for(j=0;j<=t;j++)
    {
     f[0][j]=fxbt;
     f[x][j]=fxt;
    }
```

```
        for (i=1;i<x;i++)
         f[i][0]=fx(i);
        for(j=0;j<t;j++)
        {
           for(i=1;i<x;i++)
               f[i][j+1]=a*f[i-1][j]+(1-2*a)*f[i][j]+a*f[i+1]
[j];
        }
        printf("\n the values of fij are");
        for(j=0;j<=t;j++)
          {
           printf("\n\n");
           for (i=0;i<=x;i++)
           printf("\t%.2f",f[i][j]);
          }
        getch();
          }
```

Output: SOLUTION OF ONE DIMENSIONAL HEAT FLOW EQUATION BY SCHMIDT METHOD

```
enter the end value of x and t    6    4

enter the values of f(xb,t) and f(x,t)    0 0

enter the values of h,k and c 1 .125    4

the value of a is 0.500000
the schmidt method is applicable
the values of fij are


        0.00    3.50    6.00    7.50    8.00    7.50    0.00


        0.00    3.00    5.50    7.00    7.50    4.00    0.00


        0.00    2.75    5.00    6.50    5.50    3.75    0.00


        0.00    2.50    4.62    5.25    5.12    2.75    0.00


        0.00    2.31    3.88    4.88    4.00    2.56    0.00
```

LAB ASSIGNMENT: SCHMIDT METHOD

1. Write a C language program for schmidt method to find the solution of the one dimensional heat flow equation $\dfrac{\partial f}{\partial t} = \dfrac{1}{2}\dfrac{\partial^2 f}{\partial x^2}$ when $f(0, t) = f(4, t) = 0$ and $f(x) = 4x - x^2$.

Taking $h = 1$ and find values of $f(x, t)$ for values upto $t = 5$.

Hint: Input: $x = 4, t = 5, h = 1, k = 1, c = 0.5$

Output: $a = 0.25$, Method is valid

Values of $f(x, t)$ are

0.00	3.00	4.00	3.00	0.00
0.00	2.00	3.00	2.00	0.00
0.00	1.50	2.00	1.50	0.00
0.00	1.00	1.50	1.00	0.00
0.00	0.75	1.00	0.75	0.00
0.00	0.50	0.75	0.50	0.00

⌘⌘⌘⌘⌘⌘

Method of Least Squares, Curve Fitting and Approximations

11.1 INTRODUCTION

If some values of a variate are collected in the arbitrary order, in which they occur, we cannot properly grasp the significance of the data.

For example: Consider the marks of 50 students in Mathematics, arranged according to their roll numbers, the maximum marks being 100.

9, 70, 75, 15, 0, 33, 69, 66, 37, 99, 81, 12, 31, 22, 60, 79, 46, 73, 46, 79, 75, 65, 85, 22, 8, 12, 41, 87, 82, 72, 50, 22, 87, 50, 89, 28, 29, 50, 40, 36, 40, 30, 28, 87, 81, 90, 22, 15, 30, 35.

The data given in the above form is called ungrouped data. If the data arranged in ascending or descending order of magnitude, it is said to be arranged in an array. If arranged the given data into class intervals 0-10, 10-20, ..., 90-100. Then, this method is known as tally method.

If the identity of the individuals about whom, a particular information is taken is not relevant, nor the order in which the observations arise, then we divide the observed range of variables into a suitable number of class intervals and record the number of observation in each class.

For example in the above case, the data may be expressed as in the following table :

Such a table showing the distribution of the frequencies in the different classes is called a frequency table and the way in which the class frequencies are distributed over the class intervals is called the grouped frequency distribution of the variable.

The following points may be considerable for classifications :

(i) The class should be clearly defined and should not lead to any ambiguity.

(ii) The number of class should never be less than 6 and not more than 30. Because with less number of classes, the accuracy may be lost, and with more number of classes, the computations become lengthy.

Marks	No. of Students
0–10	3
10–20	4
20–30	7
30–40	7
40–50	5
50–60	3
60–70	4
70–80	7
80–90	8
90–100	2

(iii) The observation corresponding to common point of two classes should always be put in the higher class. For example, a number corresponding to the values 20 is to be put up in the class 20-30 and not in 20-30.

(iv) The classes should be of equal width.

11.1.1 MAGNITUDE OF THE CLASS INTERVAL

Having fixed the number of classes, divide the range by it and nearest integer to this value gives the magnitude of the class interval.

11.1.2 CLASS LIMITS

The class limit should be chosen in such a way that the mid value of the class interval and actual average of the observations in that class intervals are as near to each other as possible.

11.2 CONTINUOUS FREQUENCY DISTRIBUTION

If we deal with a continuous variable, it is not possible to arrange the data in the class interval. Let us consider the distribution of age in years. If we take the intervals 15-19, 20-24, then the persons with ages between 19 and 20 years are not taken into consideration. In such a case, we form the class interval as follows.

Age in years

Below 5 :   5 or more but less than 10

                                      10 or more but less than 15

                                      15 or more but less than 20

                                      20 or more but less than 25

                                      ....

where all the persons with any fraction of age are included in one group or the other.

Practically, we re-write it as

                       0–5; 5–10; 10–15; 15–20; 20–25

This form of frequency distribution is known as continuous frequency distribution.

11.3 GRAPHICAL REPRESENTATION

It is often useful to represent a frequency distribution by means of a diagram which makes the unwidely data intelligible and conveys to the eye the general run of the observations. When data of two items is compared with one another, it is always easier to compare through graphs and diagrams. Here, we consider some important types of graphic representations.

11.3.1 HISTOGRAM

In drawing the histograms of a given grouped frequency distribution, first we mark off along a horizontal base line all the class-intervals using a suitable scale, then draw rectangles with the areas proportional to the frequencies of the respective class intervals. For equal class-intervals, the height of the rectangles will be proportional to the frequencies. For unequal class-intervals, the heights of the rectangles will be proportional to the ratios of the frequencies to the width of the corresponding class. Then the diagram of continuous rectangles so obtained is called histogram.

▶ **REMARKS**

▸ Histograms are useful, when the class intervals are not of the same width. They are appropriate to cases in which the frequency changes rapidly.

▸ To draw the histogram for an ungrouped frequency distribution of a variable, assume that the frequency corresponding to the variate value x is distributed over the interval $(x - h/2)$ to $(x + h/2)$, where h is the jump from one value to the next.

▸ If the grouped frequency distribution is not continuous, convert it into continuous distribution and then draw the histogram.

▸ The height of each rectangle is proportional to the frequency of the corresponding class, the height of a fraction of the rectangle is not proportional to the frequency of the corresponding fraction of the class, therefore, the histogram cannot be directly used to read frequency over a fraction of a class interval.

Consider the following example.

Marks	No. of Students	Marks	No. of Students
under 10	2	under 60	31
under 20	6	under 70	32
under 30	16	under 80	37
under 40	20	under 90	48
under 50	23	under 100	50

Then the histogram from the above data is :

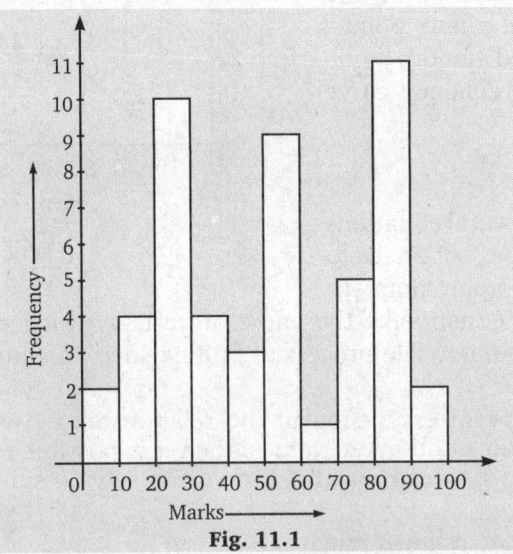

Fig. 11.1

11.3.2 FREQUENCY POLYGON

The frequency polygon is obtained by plotting points with abscissa as the variate values and the ordinates as the corresponding frequencies and joining the plotted points by a straight line taken in order. In a frequency polygon the variables or individuals of each class are assumed to be concentrated at the mid-points of the class-interval.

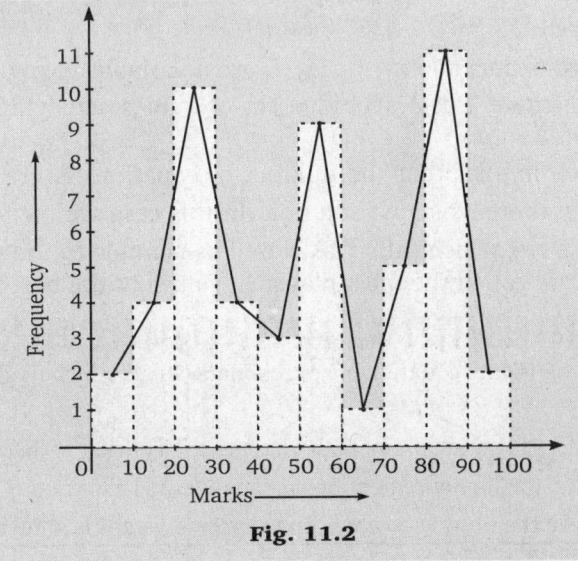

Fig. 11.2

11.3.3 FREQUENCY CURVE

(UPTU(MCA)–2006)

If through the vertices of a frequency polygon a smooth freehand curve is drawn we get the frequency curve.

11.3.4 COMMULATIVE FREQUENCY CURVE OR OGIVE

If the upper limits of the class taken as x-coordinate and the commulative frequencies as the y-coordinate and the points are plotted, then, these points, when joined by a freehand smooth curve given the commulative frequency curve or the ogive.

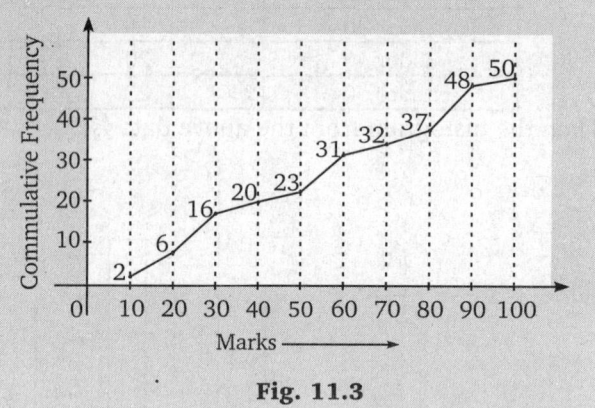

Fig. 11.3

11.4 CURVE FITTING

In applied mathematics, several equations of different types can be obtained to express the given data approximately. But we want to find the equation of the curve of best fit which may be most suitable for predicting the unknown values. The process of finding such an equation of best fit is known as curve fitting.

By curve fitting we means an expression of the relationship between two variables by an equation. If there are n pair of observed values, then it is possible to fit the given data to an equation that contains n arbitrary constants for we can solve n simultaneous equation for n unknowns.

Let us consider m independent linear equations

$$\left.\begin{array}{l} a_{11}x_1 + a_{12}x_2 + \ldots + a_{1n}x_n = b_1 \\ a_{21}x_1 + a_{22}x_2 + \ldots + a_{2n}x_n = b_2 \\ \ldots\ldots\ldots\ \ \ldots\ldots\ldots\ldots\ldots\ldots \\ \ldots\ldots\ldots\ \ \ldots\ldots\ldots\ldots\ldots\ldots \\ a_{m1}x_1 + a_{m2}x_2 + \ldots + a_{mn}x_n = b_m \end{array}\right\} \qquad \ldots(1)$$

where a's and b's are constants and $x_1, x_2, \ldots, x_n$ are n variables. Now, there are two cases :

(i) **If $m = n$**, then there exists a unique set of values satisfying the given system of equations.

(ii) **If $m > n$**, which implies that the number of equations is greater than the number of variables, then there exists no solution. In this case we try to find these values of variables $x_1, x_2, \ldots, x_n$ which satisfy as closed as possible to the given equation. These obtained values are called the most plausible values or the best fit values.

Recapitulations ||||||||||||||||||||||||||||||||||||||||||||||||||||||||||||||||||

- The general problem of finding equation of approximate curves which fit given set of data is called curve fitting.

- 'Curve fitting' is considered very important both from the point of view of theoretical and practical use.

- In theoretical statistics, the line of regression can be regarded as fitting of linear curves.

- The difference between the observed values and expected values is known as residual.

11.5 METHOD OF LEAST SQUARES

Consider m independent linear equations in n unknowns $x_1, x_2, ..., x_n$ where $m > n$ as

$$\left.\begin{aligned} a_{11}x_1 + a_{12}x_2 + ... + a_{1n}x_n &= b_1 \\ a_{21}x_1 + a_{22}x_2 + ... + a_{2n}x_n &= b_2 \\ \cdots\cdots \quad & \cdots\cdots\cdots \\ \cdots\cdots \quad & \cdots\cdots\cdots \\ a_{m1}x_1 + a_{m2}x_2 + ... + a_{mn}x_n &= b_m \end{aligned}\right\} \quad ...(1)$$

with constants a's and b's.

Let $x_1, x_2, ..., x_n$ be the most plausible values then, we have

$$E_i = (a_{i1}x_1 + a_{i2}x_2 + ... + a_{in}x_n - b_i) \text{ (known as residual or deviation,}$$

or the error E_i)

$$\Rightarrow \quad E_i = (a_{i1}x_1 + a_{i2}x_2 + ... + a_{in}x_n - b_i) \quad i = 1, 2, ... m. \quad ...(2)$$

Let us suppose S is the sum of the squares of E_i.

Then, we get
$$S = \sum_{i=1}^{n} (a_{i1}x_1 + a_{i2}x_2 + ... + a_{in}x_n - b_i)^2 = \sum_{i=1}^{n} E_i^2 \quad ...(3)$$

To find the maximum or minimum values of S, we must have

$$\frac{\partial S}{\partial x_1} = 0, \frac{\partial S}{\partial x_2} = 0, ... \frac{\partial S}{\partial x_n} = 0$$

Then (3) $\Rightarrow$
$$\sum_{i=1}^{m} a_{i1}E_i = 0, \sum_{i=1}^{m} a_{i2}E_i = 0, ... \sum_{i=1}^{m} a_{in}E_i = 0 \quad ...(4)$$

These n equations given by (4) are known as the normal equations and can be easily solved for the n variables $x_1, x_2, ..., x_n$. The values of $x_1, x_2, ..., x_n$ so obtained are the most plausible or best values.

REMARKS

- The principle of least squares, is first given by Gauss in 1795 but it was named and published for the first time in 1805 by Legendre.
- The method of least square does not help us to choose the degree of the curve to be fitted but helps us in finding the values of the constants when the form of the curve has already been given.

11.5.1 NORMAL EQUATIONS

Consider the curve of nth degree

$$y = a + bx + cx^2 + ... + kx^n \text{ with } (k \neq 0)$$

Then, the normal equation obtained from (4), are

$$\Sigma y = ma + b\Sigma x + ... + k\Sigma x^n$$

$$\Sigma xy = a\Sigma x + b\Sigma x^2 + ... + k\Sigma x^{n+1}$$

$$\Sigma x^2 y = a\Sigma x^2 + b\Sigma x^3 + ... + k\Sigma x^{n+2}$$

$$\Sigma x^n y = a\Sigma x^n + b\Sigma x^{n+1} + ... + k\Sigma x^{2n}$$

When second order partial derivatives are calculated and substituted these values, they gives a positive value of the function, which implies that S is minimum.

In particular, if we take $n = 1$ then we get the equation of straight line

$$y = a + bx$$

and the normal equations are

$$\Sigma y = ma + b\Sigma x + c\Sigma x^2 \; ; \quad \Sigma xy = a\Sigma x + b\Sigma x^2 + c\Sigma x^3 \; ; \quad \Sigma x^2 y = a\Sigma x^2 + b\Sigma x^3 + c\Sigma x^4$$

SOLVED EXAMPLES

EXAMPLE 1. *Find the normal equations and hence find the best fit values of x, y, z in the least square sense from the following equations*

$$x + 2y + z = 1; \; 2x + y + z = 4 \; ; \; -x + y + 2z = 4; \; 4x + 2y - 5z = -7$$

SOLUTION. The given equation can be written as

$$\left.\begin{array}{l} x + 2y + z - 1 = 0 \\ 2x + y + z - 4 = 0 \\ -x + y + 2z - 4 = 0 \\ 4x + 2y - 5z + 7 = 0 \end{array}\right] \qquad \text{...(1)}$$

Now, to obtain the normal equations of x, we multiply these equations by the coefficient of x in that equation and then add.

Then, the normal equations are

$$1(x+2y+z-1)+2(2x+y+z-4)+(-1)(-x+y+2z-4)+4(4x+2y-5z+7)=0$$

$$\Rightarrow \qquad\qquad\qquad 22x + 11y - 19z + 23 = 0 \qquad\qquad \text{...(2)}$$

Similarly, for the normal equation of y, multiply these equations by the coefficient of y in that equation and then add, we get

$$2(x+2y+z-1)+1(2x+y+z-4)+1(-x+y+2z-4)+2(4x+2y-5z+7) = 0$$

$$\Rightarrow \qquad\qquad\qquad 11x + 10y - 5z + 4 = 0 \qquad\qquad \text{...(3)}$$

Similarly, the normal equation for z is

$$1(x+2y+z-1)+1(2x+y+z-4)+2(-x+y+2z-4)-5(4x+2y-5z+7) = 0$$

$$\Rightarrow \qquad\qquad\qquad 19x + 5y - 31z + 48 = 0 \qquad\qquad \text{...(4)}$$

Solving (2), (3) and (4) for x, y and z we get

$$x = 0.910, y = -0.378 \text{ and } z = 2.045$$

EXAMPLE 2. *Find the normal equations for fitting the curve of type $y = ax + \dfrac{b}{x}$ by least square method.*

SOLUTION. The given equation can be written as

$$y = ax + bx^{-1}$$

Consider the n points $(x_1, y_1), (x_2, y_2), ..., (x_i, y_i), ..., (x_n, y_n)$.

Then the residual or deviation of the ith point (x_i, y_i) is

$$y_i = \left(ax_i + \frac{b}{x_i}\right)$$

The sum of the squares of the error is given by

$$S = \Sigma\left(y_i - ax_i - \frac{b}{x_i}\right)^2 \qquad\qquad \text{...(1)}$$

For the maxima and minima of S, we have

$$\frac{\partial S}{\partial a} = 0, \frac{\partial S}{\partial b} = 0,$$

$$\frac{\partial S}{\partial a} = 0 \Rightarrow \Sigma x_i \left(y_i - ax_i - \frac{b}{x_i} \right) = 0 \Rightarrow \Sigma (x_i y_i - ax_i^2 - b) = 0$$

$$\Rightarrow \qquad \Sigma x_i y_i = a\Sigma x_i^2 + nb \qquad \qquad ...(2)$$

$$\frac{\partial S}{\partial b} = 0 \quad \Rightarrow \quad \Sigma \left[\frac{1}{x_i} \left(y_i - ax_i - \frac{b}{x_i} \right) \right] = 0$$

$$\Rightarrow \qquad \Sigma \left[\frac{y_i}{x_i} - a - \frac{b}{x_i^2} \right] = 0 \qquad \Rightarrow \qquad \Sigma \frac{y_i}{x_i} = na + b\Sigma \frac{1}{x_i^2} \qquad \qquad ...(3)$$

Equation (2) and (3) gives the required normal equations.

EXAMPLE 3. *Find the most plausible values of x and y from the following equations :*
x + y = 3.00, 2x – y = 0.5, x + 3y = 7.25, 3x + y = 4.95.

SOLUTION. The given equations can be written as

$$\left. \begin{array}{l} x + y - 3.00 = 0 \\ 2x - y - 0.5 = 0 \\ x + 3y - 7.25 = 0 \\ 3x + y - 4.95 = 0 \end{array} \right] \qquad \qquad ...(1)$$

Now sum of the square 'S' of the errors is given by

$$S = (x+y-3.00)^2 + (2x-y-0.5)^2 + (x+3y-7.25)^2 + (3x+y-4.95)^2 = 0$$

For the maxima and minima of S, we must have

$$\frac{\partial S}{\partial a} = 0, \frac{\partial S}{\partial b} = 0,$$

Now, $\qquad \dfrac{\partial S}{\partial x} = 0$

$$\Rightarrow \quad (x+y-3.00) + 2(2x-y-0.50) + 1(x+3y-7.25) + 3(3x+y-4.95) = 0$$

$$\Rightarrow \qquad 15x + 5y - 26.10 = 0 \qquad \qquad ...(2)$$

and $\qquad \dfrac{\partial S}{\partial y} = 0$

$$\Rightarrow \quad (x+y-3.00) - (2x-y-0.50) + 3(x+3y-7.25) + 3(3x+y-4.95) = 0$$

$$\Rightarrow \qquad 13x + 14y - 39.20 = 0 \qquad \qquad ...(3)$$

Solving (2) and (3), we get

$$x = 1.234, \ y = 1.919.$$

EXAMPLE 4. *Find the most plausible values of x and y from the following equations*

$$3x + y = 4.95; \ x + y = 3.00; \ 2x - y = 0.5; \ x + 3y = 7.25 \qquad \text{(UPTU–2004)}$$

SOLUTION. Let $S = (3x+y-4.95)^2 + (x+y-3.00)^2 + (2x-y-0.5)^2 + (x+3y-7.25)^2 = 0$
$$\qquad \qquad ...(1)$$

$$\Rightarrow \quad \frac{\partial S}{\partial x} = 6(3x+y-4.95) + 2(x+y-3.00) + 4(2x-y-0.5) + 2(x+3y-7.25)$$

$$= 30x + 10y - 52.2$$

and $\dfrac{\partial S}{\partial y} = 10x + 24y - 58.4$

We obtained the normal equation by putting

$$\frac{\partial S}{\partial x} = 0 \text{ and } \frac{\partial S}{\partial y} = 0$$

Therefore, $3x + y = 5.22$; $x + 2.4y = 5.84$

On solving we get $x = 1.07871, y = 1.98387$

which is the required plausible solution.

EXERCISE 11.1

1. Find the values of x and y which satisfy the following equations most satisfactorily with the help of normal equation of x, y

   $x + 2.5y = 21, 3.2x - y = 28,$

   $4x + 1.2y = 42.04, 1.5x + 6.3y = 40.$

2. Find the most plausible values of x and y from the four equations

   $x - y + 2z = 3, 3x + 2y - 5z = 5,$

   $4x + y + 4z = 21$ and $-x + 3y + 3z = 14.$

   (Meerut-2002)

3. Use the method of least squares to find the most plausible values of x and y from the following equations

   $x + y = 3.01, 2x - y = 0.03, x + 3y = 7.03,$

   $3x + y = 4.97$

4. Find the most plausible values of x and y from the following equations

   $x + y = 301, 2x - y = 3, x + 3y = 703,$

   $3x + y = 497.$

Answers

1. $x = 9.620, y = 4.064$ **2.** $x = 2.47, y = 3.55, z = 1.92$ **3.** $x = 0.997, y = 2.0014.x = 100, y = 200$

11.6 METHOD OF CURVE-FITTING

Let us consider the rth degree curve

$$y = a + bx + cx^2 + \ldots + kx^r, \text{ with } k \neq 0 \qquad \ldots(i)$$

with the given values $(x_1, y_1), (x_2, y_2), \ldots, (x_m, y_m)$.

The curve given by (i) has $(r + 1)$ unknown $a, b, c, \ldots, k$ and so if $m = r + 1$, we get $(r + 1)$ equations when the values $(x_1, y_1), (x_2, y_2), \ldots, (x_m, y_m)$ are substituted for (x, y) in (i) and thus a unique solution of the values of the unknown $a, b, c, \ldots, k$ is possible.

Now let

$$y_i' = a + bx_i + cx_i^2 + \ldots + kx_i^r \qquad \ldots(ii)$$

and let y_i be the observed values y for x_i.

Then if u_i be the residual for this point, we have

$$u_i = y_i - y_i' = y_i - (a + bx_i + cx_i^2 + \ldots + kx_i^r), \qquad \text{from (ii)}$$

$$u_i = y_i - a - bx_i - cx_i^2 - \ldots - kx_i^r \qquad \ldots(iii)$$

Now in order to make the sum of the squares of the errors minimum, we define

$$S = \sum_{i=1}^{m} u_i^2 = \sum_{i=1}^{m} (y_i - a - bx_i - cx_i^2 - \ldots - kx_i^r) \qquad \ldots(iv)$$

∴ By the principle of maxima and minima, we must have

$$\frac{\partial S}{\partial a} = 0, \frac{\partial S}{\partial b} = 0, \ldots, \frac{\partial S}{\partial k} = 0 \qquad \ldots(v)$$

which gives the following $(r + 1)$ equations :

$$\frac{\partial S}{\partial a} = -2 \sum_{i=1}^{m} (y_i - a - bx_i - cx_i^2 - \ldots - kx_i^r) = 0$$

$\Rightarrow$

$$\sum_{i=1}^{m} (y_i - a - bx_i - cx_i^2 - \ldots - kx_i^r) = 0$$

$$\sum_{i=1}^{m} y_i - \sum_{i=1}^{m} a - \sum_{i=1}^{m} bx_i - \sum_{i=1}^{m} cx_i^2 - \ldots = 0$$

$\Rightarrow$

$$\Sigma y_i = ma + b\Sigma x_i + c\Sigma x_i^2 + \ldots \qquad \ldots(1)$$

Now

$$\frac{\partial S}{\partial b} = 0 = -2\Sigma(y_i - a - bx_i - cx_i^2 - \ldots - kx_i^r)x_i$$

or

$$\Sigma(y_i - a - bx_i - cx_i^2 - \ldots - kx_i^r)x_i = 0$$

$\therefore$

$$\Sigma x_i y_i = a\Sigma x_i + b\Sigma x_i^2 + c\Sigma x_i^3 + \ldots \qquad \ldots(2)$$

Similarly

$$\Sigma x_i^2 y_i = a\Sigma x_i^2 + b\Sigma x_i^3 + c\Sigma x_i^4 + \ldots \qquad \ldots(3)$$

The equations (1), (2) and (3) are known as normal equations and can be solved as simultaneous equations to evaluate $a, b, c, \ldots k$.

11.7 FITTING OF SOME SPECIAL CURVES

11.7.1 Fitting of a Straight line

Let

$$y = a + bx \qquad \ldots(1)$$

be the equation of the straight line, to be fitted.
Then the normal equations can be obtained as follows.

Let

$$u = (y_i - a - bx_i) \quad \text{and} \quad S = u^2 = \Sigma(y_i - a - bx_i)^2$$

For the maxima and minima of S, we must have $\dfrac{\partial S}{\partial a} = 0$

$\Rightarrow \qquad -2\Sigma(y_i - a - bx_i).1 = 0 \qquad \Rightarrow \qquad \Sigma(y_i - a - bx_i).1 = 0$

$\therefore \qquad \qquad \Sigma y_i = ma + b\Sigma x_i \qquad \ldots(2)$

Now

$$\frac{\partial S}{\partial a} = 0$$

$$-2\Sigma(y_i - a - bx_i)(-x) = 0 \qquad \Rightarrow \qquad \Sigma(y_i - a - bx_i)x = 0$$

$$\Sigma x_i y_i = a\Sigma x_i + b\Sigma x_i^2 \qquad \ldots(3)$$

Solving these two equations we can find the values of a and b.

Special Case : If the straight line to be fitted passes through the origin, then the equation of the straight line is given by

$$y = bx$$

$\therefore$ Its normal equation is $\Sigma xy = b\Sigma x^2$

we can find the value of constant b easily.

11.7.2 Fitting of a Parabolic Curve

Let

$$y = a + bx + cx^2 \qquad \ldots(1)$$

be the equations of the parabolic curve to be fitted.
Then the normal equations are

$$\left.\begin{array}{l} \Sigma y = ma + b\Sigma x + c\Sigma x^2 \\ \Sigma xy = a\Sigma x + b\Sigma x^2 + c\Sigma x^3 \\ \Sigma x^2 y = a\Sigma x^2 + b\Sigma x^3 + c\Sigma x^4 \end{array}\right\} \qquad \ldots(2)$$

Solving above three equations for a, b and c simultaneously we can find the values of a, b and c.

11.7.3 FITTING OF AN EXPONENTIAL CURVE

Let
$$y = ae^{bx} \qquad \qquad ...(1)$$
be the equation of the exponential curve to be fitted.

Taking logarithms of both sides of (1) to the base, 10.

We get
$$\log_{10} y = \log_{10} a + bx \log_{10} e \qquad ...(2)$$
This is of the form
$$Y = A + Bx \qquad ...(3)$$
where $Y = \log_{10} y, A = \log_{10} a$ and $B = b \log_{10} e$.

Now applying the method of fitting a straight lines we can find the values A and B and hence the values of a and b.

11.7.4 FITTING OF THE CURVE OF THE TYPE $y = ab^x$

Here the method of fitting is the same as given in part 1.7.3 above.

11.7.5 FITTING OF THE LOGARITHMIC CURVE OF THE FORM $y = ax^b$

We have
$$y = ax^b$$
Taking log of both the sides, we get
$$\log y = \log a + b \log x$$
Let us put $\log y = Y, \log a = A$ and $\log x = X$

Then, we have
$$Y = A + bX \qquad ...(1)$$
Normal equation for (1) are
$$\Sigma Y = Ax + b\Sigma X$$
$$\Sigma XY = A\Sigma X + b\Sigma X^2$$

11.7.6 FITTING OF THE CURVE $y = a + b/x$

Let put $\dfrac{1}{x} = X$. Then given equations becomes
$$y = a + bx \qquad ...(1)$$
The normal equation for (1) are
$$\Sigma y = an + b\Sigma x \qquad ...(2)$$
$$\Sigma Xy = a\Sigma X + b\Sigma X^2 \qquad ...(3)$$
Solving these equations we get the required value of a and b.

11.7.7 FITTING OF THE CURVE $y = ax + b/x$

Let the curve $y = ax + \dfrac{b}{x}$ passes through the point $(x_i, y_i), i = 1, 2, ..., n$. The error of estimate for i^{th} point (x_i, y_i) is

$$d_i = y_i - ax_i - \frac{b}{x_i}$$

By the principle of least squares, the sum of squares of the error should be minimum.

i.e.,
$$S = \sum_{i=1}^{n} d_i^2 \text{ should be minimum.}$$

$$\Rightarrow \qquad S = \sum_{i=1}^{n} \left(y_i - ax_i - \frac{b}{x_i} \right)^2 \text{ should be minimum.}$$

We obtained the normal equations by $\dfrac{\partial S}{\partial a} = 0$ and $\dfrac{\partial S}{\partial b} = 0$.

$\therefore$

$$\dfrac{\partial \Sigma}{\partial a} = 0 \implies -2 \sum_{i=1}^{n} x_i \left(y_i - ax_i - \dfrac{b}{x_i} \right) = 0$$

$\implies$

$$\sum_{i=1}^{n} x_i y_i = a \sum_{i=1}^{n} x_i^2 + bx \qquad \qquad \dots(1)$$

and

$$\dfrac{\partial \Sigma}{\partial b} = 0 \implies -2 \sum_{i=1}^{n} \dfrac{1}{x_i} \left(y_i - ax_i - \dfrac{b}{x_i} \right) = 0$$

$\implies$

$$\sum_{i=1}^{n} \dfrac{y_i}{x_i} = an + b \sum_{i=1}^{n} \dfrac{1}{x_i^2} \qquad \qquad \dots(2)$$

Solving (1) and (2) we get the required values of a and b.

11.7.8 FITTING OF THE CURVE $y = a + b/x + c/x^2$

We have $S = \displaystyle\sum_{i=1}^{n} \left(y_i - a - \dfrac{b}{x_i} - \dfrac{b}{x_i^2} \right)$ should be minimum.

Normal equations can be obtained by $\dfrac{\partial S}{\partial a} = 0, \dfrac{\partial S}{\partial b} = 0, \dfrac{\partial S}{\partial c} = 0$

Now $\dfrac{\partial S}{\partial a} = 0 \implies -2 \displaystyle\sum_{i=1}^{n} \left(y_i - a - \dfrac{b}{x_i} - \dfrac{c}{x_i^2} \right) = 0 \implies \Sigma y_i = an + b \displaystyle\sum_{i=1}^{n} \dfrac{1}{x_i} + c \displaystyle\sum_{i=1}^{n} \dfrac{1}{x_i^2}$

$\dfrac{\partial S}{\partial b} = 0 \implies -2 \displaystyle\sum_{i=1}^{n} \dfrac{1}{x_i} \left(y_i - a - \dfrac{b}{x_i} - \dfrac{c}{x_i^2} \right) = 0 \implies \displaystyle\sum_{i=1}^{n} \dfrac{y_i}{x_i} = a \displaystyle\sum_{i=1}^{n} \dfrac{1}{x_i} + b \displaystyle\sum_{i=1}^{n} \dfrac{1}{x_i^2} + c \displaystyle\sum_{i=1}^{n} \dfrac{1}{x_i^3}$

and $\dfrac{\partial S}{\partial c} = 0 \implies -2 \displaystyle\sum_{i=1}^{n} \dfrac{1}{x_i^2} \left(y_i - a - \dfrac{b}{x_i} - \dfrac{c}{x_i^2} \right) = 0 \implies \displaystyle\sum_{i=1}^{n} \dfrac{y_i}{x_i^2} = a \displaystyle\sum_{i=1}^{n} \dfrac{1}{x_i^2} + b \displaystyle\sum_{i=1}^{n} \dfrac{1}{x_i^3} + c \displaystyle\sum_{i=1}^{n} \dfrac{1}{x_i^4}$

$$\dots(3)$$

Solving (1), (2) and (3), we get the required values of a and b.

11.7.9 FITTING OF THE CURVE $y = ax^2 + b/x$

We have $\qquad S = \displaystyle\sum_{i=1}^{n} \left(y_i - ax_i^2 - \dfrac{b}{x_i} \right)^2$

Now $\qquad \dfrac{\partial S}{\partial a} = 0 \implies -2 \displaystyle\sum_{i=1}^{n} x_i^2 \left(y_i - ax_i^2 - \dfrac{b}{x_i} \right) = 0 \implies \displaystyle\sum_{i=1}^{n} x_i^2 y_i = a \displaystyle\sum_{i=1}^{n} x_i^4 + b \displaystyle\sum_{i=1}^{n} x_i$

$$\dots(1)$$

$\dfrac{\partial S}{\partial b} = 0 \implies -2 \displaystyle\sum_{i=1}^{n} \dfrac{1}{x_i} \left(y_i - ax_i^2 - \dfrac{b}{x_i} \right) = 0 \implies \displaystyle\sum_{i=1}^{n} \dfrac{y_i}{x_i} = a \displaystyle\sum_{i=1}^{n} x_i + b \displaystyle\sum_{i=1}^{n} \dfrac{1}{x_i^2}$

$$\dots(2)$$

Solving (1) and (2) we get the values of a and b.

11.7.10 FITTING OF THE CURVE $y = a/x + b\sqrt{x}$

Here we have
$$S = \sum_{i=1}^{n} \left(y_i - ax_i^2 - b\sqrt{x_i} \right)^2$$

Now $\dfrac{\partial S}{\partial a} = 0 \Rightarrow -2\sum_{i=1}^{n} \dfrac{1}{x_i}\left(y_i - \dfrac{a}{x_i} - b\sqrt{x_i} \right) = 0 \Rightarrow \sum_{i=1}^{n} \dfrac{y_i}{x_i} = a\sum_{i=1}^{n} \dfrac{1}{x_i^2} + b\sum_{i=1}^{n} \dfrac{1}{\sqrt{x_i}}$ ...(1)

$\dfrac{\partial S}{\partial b} = 0 \Rightarrow -2\sum_{i=1}^{n} \sqrt{x_i}\left(y_i - \dfrac{a}{x_i} - b\sqrt{x_i} \right) = 0 \Rightarrow \sum_{i=1}^{n} (\sqrt{x_i})y_i = a\sum_{i=1}^{n} \dfrac{1}{\sqrt{x_i}} + b\sum_{i=1}^{n} x_i$ ...(2)

On solving (1) and (2) we get the required values of a and b.

SOLVED EXAMPLES

EXAMPLE 1.   *Fit a straight line to the following data :*

x	1	2	3	4	5
y	5	7	9	10	11

SOLUTION.   Here, we have

x	1	2	3	4	5	$\Sigma x = 15$
y	5	7	9	10	11	$\Sigma y = 42$
xy	5	14	27	40	55	$\Sigma xy = 141$
x^2	1	4	9	16	25	$\Sigma x^2 = 55$

the equation of the line be $y = a + bx$ ...(1)

Normal equation are
$$\Sigma y = ma + b\Sigma x$$
and $\Sigma xy = a\Sigma x + b\Sigma x^2$

Using the given values,
$$42 = 5a + 15b \qquad\qquad ...(2)$$
$$141 = 15a + 55b \qquad\qquad ...(3)$$

Solving (2) and (3), we get $b = \dfrac{3}{2}, a = \dfrac{39}{10}$

or $b = 1.5, a = 3.9$

Hence, required straight line is given by
$$y = (3.9) + (1.5)x$$

▶ **REMARK**

▸ For the sake of convenience, it is sometimes easy to change the origin and scale with the substitution $x = \dfrac{x - A}{h}$ and $y = \dfrac{y - B}{h}$, where A and B are the middle values of x and y series respectively.

EXAMPLE 2.   *Fit a straight line to the following data :*

x	0	5	10	15	20	25
y	12	15	17	22	24	30

SOLUTION.   We observe that the number of values given is 6, *i.e.*, even and the difference in the values of x is 5. In such a case the calculations can be simplified if we take half the common distance which is 2.5 here as unit of measurement. Also the two mid-values are 10 and 15, so we take their mean, *i.e.*, $\frac{1}{2}(10+15)$, *i.e.*, 12.5 as the origin.

Therefore, introduce two new variables

$$u = \frac{x-12.5}{2.5} \text{ and } v = y - 20 \qquad \text{...(1)}$$

Then we have

$$u : \frac{0-12.5}{2.5}, \frac{5-12.5}{2.5}, \frac{10-12.5}{2.5}, \frac{15-12.5}{2.5}, \frac{20-12.5}{2.5}, \frac{25-12.5}{2.5}$$

							Total
u	-5	-3	-1	1	3	5	0
v	$12-20$	$15-20$	$17-20$	$22-20$	$24-20$	$30-20$	
v	-8	-5	-3	2	4	10	0
uv	40	15	3	2	12	50	122
u^2	25	9	1	1	9	25	70

Now let the equation of line be

$$v = a + bu \qquad \text{...(2)}$$

Then its normal equations are

$$\Sigma v = ma + b\Sigma u; \quad \Sigma vu = a\Sigma u + b\Sigma u^2$$

$$\Rightarrow \qquad 0 = 6.a + 0.b \qquad \text{...(3)}$$

$$122 = 0.a + 70.b \qquad \text{...(4)}$$

Solving (3) and (4) we get $a = 0$ and $b = 1.743$.

Substituting these values in (2), the equation of the line is

$$v = 0 + (1.743)u$$

or $\qquad y - 20 = (1.743)\left[\dfrac{x-12.5}{2.5}\right]$ \qquad [Using (1)]

or $\qquad y - 20 = \left(\dfrac{1.743}{2.5}\right)x - \dfrac{1.743 \times 12.5}{2.5}$

$$y - 20 = 0.7x - 8.715$$

$$\Rightarrow \qquad y = 0.7x + (20 - 8.715)$$

$$\Rightarrow \qquad y = 0.7x + 11.285$$

$$\Rightarrow \qquad y = 11.285 + 0.7x$$

▼ REMARK
▶ If in such a problem, the number of values given is odd then the calculations are simplified provided we take the common difference as the unit of measurement and the mid-values as the assumed origin.

EXAMPLE 3. *Fit a parabola of the second degree of the following data :*

x	1.0	1.5	2.0	2.5	3.0	3.5	4.0
y	1.1	1.3	1.6	2.6	2.7	3.4	4.1

(UPTU(MCA)–2006, VTU–2009, Bhopal–2008)

SOLUTION. Here we observe that the number of values given here is 7 which is odd, so we take the middle value 2.5, *i.e.*, 4th value as our assumed mean and let

$$u = \frac{x - 2.5}{0.5} \text{ and } v = y - 2.7$$

Let the second degree curve to be fitted is

$$v = a + bu + cu^2 \qquad \qquad ...(1)$$

Its normal equations are

$$\Sigma vu = a\Sigma u + b\Sigma u^2 + c\Sigma u^3 \qquad \qquad ...(2)$$

$$\Sigma vu^2 = a\Sigma u^2 + b\Sigma u^3 + c\Sigma u^4 \qquad \qquad ...(3)$$

Now from the given data, we have

Total

u	– 3	– 2	– 1	0	1	2	3	**0**
v	– 1.6	–1.4	–1.1	– 0.1	0	0.7	1.4	**– 14.1**
vu	4.8	2.8	1.1	0	0	1.4	4.2	**14.3**
u^2	9	4	1	0	1	4	9	**2.8**
vu^2	– 14.4	– 5.6	– 1.1	0	0	2.8	12.6	**– 5.7**
u^3	–27	–8	– 1	0	1	8	27	**0**
u^4	81	16	1	0	1	16	81	**196**

Substituting the values of $\Sigma u, \Sigma v, \Sigma uv, \Sigma vu^2, \Sigma u^2, \Sigma u^3, \Sigma u^4$ in (2), (3) and (4), we get

$$-2.1 = 7a + 0.b + 27.c \qquad \qquad ...(5)$$

$$14.3 = 0.a + 27b + 0.c \qquad \qquad ...(6)$$

$$-5.7 = 27a + 0.b + 196c \qquad \qquad ...(7)$$

Solving (5), (6) and (7), we get $a = -0.04, b = 0.53, c = 0.03$.

Substituting these values of a, b and c in (1), we get

$$v = -0.04 + 0.53u + 0.03u^2$$

$$\Rightarrow \quad y - 2.7 = -0.04 + 0.53\left(\frac{x - 2.5}{0.5}\right) + 0.03\left(\frac{x - 2.5}{0.5}\right)^2$$

$$\Rightarrow \quad y - 2.7 = -0.04 + \frac{0.53x}{0.5} - \frac{2.5 \times 0.53}{0.5} + 0.03\left(\frac{x - 2.5}{0.5}\right)^2$$

$$\Rightarrow \quad y - 2.7 = -0.04 + 1.06x - 2.65 + 0.03\left[\frac{x^2 + 6.25 - 5.0x}{0.5}\right]$$

$$\Rightarrow \qquad y = -0.04 + 2.7 - 2.65 + 1.06x + 0.06(x^2 - 5x + 6.25)$$

$$\Rightarrow \qquad y = 0.01 + 1.06x + 0.06x^2 - 0.30x + 0.375$$

$$\Rightarrow \qquad y = 0.385 + 0.76x + 0.66x^2$$

EXAMPLE 4. *Fit a parabolic curve of regression of y on x to the seven pairs of values*

x	1.0	1.5	2.0	2.5	3.0	3.5	4.0
y	1.1	1.3	1.6	2	2.7	3.4	4.1

(MDU(B.E.)–2007)

SOLUTION. Let $X = \dfrac{x - 2.5}{0.5} = 2x - 5$ and $Y = y$

Then we have the following table

x	y	X	Y	X^2	XY	X^2Y	X^3	X^4
1.0	1.1	–3	1.1	9	–3.3	9.9	–27	81
1.5	1.3	–2	1.3	4	–2.6	5.2	–8	16
2.0	1.6	–1	1.6	1	–1.6	1.6	–1	1
2.5	2.0	0	2.0	0	0	0	0	0
3.0	2.7	1	2.7	1	2.7	2.7	1	1
3.5	3.4	2	3.4	4	6.8	6.9	8	16
4.0	4.1	3	4.1	9	12.3	12.3	27	81
Total		**0**	**16.2**	**28**	**14.3**	**69.9**	**0**	**196**

Let the equation of the parabolic curve is

$$Y = a + bX + cX^2 \qquad \ldots(1)$$

The normal equation of (1) are given by

$$\left.\begin{array}{l} \Sigma Y = ma + b\Sigma X + c\Sigma X^2 \\ \Sigma YX = a\Sigma X + b\Sigma X^2 + c\Sigma X^3 \\ \Sigma YX^2 = a\Sigma X^2 + b\Sigma X^3 + c\Sigma X^4 \end{array}\right\} \qquad \ldots(2)$$

Putting the values in (2), from the table, we get

$$16.2 = 7a + 27c; \; 14.3 = 27b; \; 69.9 = 28a + 196c$$

$\Rightarrow \qquad a = 2.07, b = 0.511, c = 0.061$

Putting the values of a, b and c in (1), we get

$$Y = 2.07 + 0.511X + 0.061X^2$$

$\Rightarrow \qquad y = 2.07 + 0.511(2x - 5) + 0.061(2x - 5)^2$

$\Rightarrow \qquad y = 1.04 - 0.193x + 0.243x^2$

EXAMPLE 5. *Fit a second degree parabola to the following data*

x	1	2	3	4	5
y	1090	1220	1390	1625	1915

SOLUTION. Let us define u and v such that $u = x - 3, v = \dfrac{y - 1450}{5}$

and equation of the parabola is $v = a + bu + cu^2$ $\qquad \ldots(1)$

Therefore, we have the following table :

x	y	u	v	u^2	v^4	uv	u^2v
1	1090	−2	−72	4	16	144	−288
2	1220	−1	−46	1	1	46	−46
3	1390	0	−12	0	0	0	0
4	1625	1	35	1	1	35	35
5	1915	2	93	4	16	186	372
Total		**0**	**−2**	**10**	**34**	**411**	**73**

Then putting values in the normal equation

$\Sigma v = ma + b\Sigma u + c\Sigma u^2$; $\Sigma uv = a\Sigma u + b\Sigma u^2 + c\Sigma u^3$; $\Sigma u^2 v = a\Sigma u^2 + b\Sigma u^3 + c\Sigma u^4$

we get $-2 = 59 + 0 + 10c$; $411 = 0 + 106 + 0$; $73 = 10a + 3 + 34c$

Solving these equations for a, b and c we get

$$a = -11.4, b = 41.1, c = 5.5$$

Put in (1), we get $v = -11.4 + 41.1u + 5.5u^2$

Now changing the origin, we get the fit curve as

$\Rightarrow$ $y = 1024 + 40.5x + 27.5x^2$

EXAMPLE 6. *If P is a pull required to lift a load W by means of a pulley block, find a linear law of the form P = mW + c connecting P and W, using the following data :*

P	12	15	21	25
W	50	70	100	120

Compute P where W = 150 Kg.   (UPTU–2005, 07, VTU–2002)

SOLUTION. The given equation is $P = mW + c$ ...(1)

The normal equations are

$$\left.\begin{array}{l}\Sigma P = 4c + m\Sigma W \\ \Sigma WP = c\Sigma W + m\Sigma W^2\end{array}\right]$$ ...(2)

Then, we have

W	P	W^2	WP
50	12	2500	600
70	15	4900	1050
100	21	10000	2100
120	25	14400	3000
340	73	31800	6750

Putting these values in (2), we get

$$73 = 4c + 340m$$
$$6750 = 340c + 31800m$$

which gives $m = 0.1879, c = 2.2785$

Hence, the line of best fit is

$$P = 2.2759 + 0.1879W$$

Now, for $W = 1500$ Kg $P = 30.4635$ Kg

EXAMPLE 7. *The pressure and the volume of a gas are related by the equation $pV^r = K, r$ and K being constants. Fit this equation to the following set of observations.*

P(Kg/cm²)	0.5	1.0	1.5	2.0	2.5	3.0
V(litres)	1.62	1.00	0.75	0.62	0.52	0.46

(VTU–2011, Madras–2000)

SOLUTION. Take log of the given equation, we have

$$\log_{10} p + r \log_{10} V = \log_{10} K$$

$$\Rightarrow \qquad \log_{10} V = \frac{1}{r}\log_{10} K - \frac{1}{r}\log_{10} P \qquad \Rightarrow \qquad y = A + BX$$

where $\qquad X = \log_{10} p, Y = \log_{10} V, A = \dfrac{1}{r}\log_{10} K,$

$$A = \frac{1}{r}\log_{10} K, B = -\frac{1}{r}$$

Now

P	V	X	Y	XY	X²
0.5	1.62	–0.3010	0.2095	–0.0630	0.0906
1.0	1.00	0	0	0	0
1.5	0.75	0.1762	–0.1249	–0.0220	0.0310
2.0	0.62	0.3010	–0.2076	–0.0625	0.0906
2.5	0.52	0.3979	–0.2840	–0.1130	0.1583
3.0	0.46	0.4771	–0.3372	–0.1609	0.2276
Total		**1.0511**	**–0.7442**	**–0.4214**	**0.5981**

Now putting all these values in the normal equations and get the required fitted curve.

EXAMPLE 8. *Fit a second degree parabola to the following data taking as dependent variable.*

x	1	2	3	4	5	6	7	8	9
y	2	6	7	8	10	11	11	10	9

SOLUTION. Let us introduce two new variables X and Y such that $X = x - 5, Y = y - 7$

Also, let the curve to be fit be $Y = a + bX + cX^2$

Then, we have the following table.

x	y	X	Y	XY	X²	X²Y	X³	X⁴
1	2	–4	–5	20	16	–80	–64	256
2	6	–3	–1	3	9	–9	–27	81
3	7	–2	0	0	4	0	–8	16
4	8	–1	1	–1	1	1	–1	1
5	10	0	2	0	0	0	0	0
6	11	1	4	4	1	4	1	1
7	11	2	4	8	4	16	8	16
8	10	3	3	9	9	27	27	81
9	9	4	2	8	16	32	64	256
Total		**0**	**11**	**51**	**60**	**–9**	**0**	**708**

Therefore, the normal equations are

$$11 = 9a + 0 + 60c$$
$$51 = 0 + 60b$$
$$-9 = 60a + 0 + 708c$$

Solving these equations for a, b and c, we get $a = 3, b = 0.85$ and $c = -0.27$

Hence, the curve of fit is

$$Y = a + bX + cX^2 = 3 + 0.85X - 0.27X^2$$

$$\Rightarrow \qquad y - 7 = 3 + 0.85(x - 5) - 0.27(x - 5)^2$$

$$= 3 + 0.85x - 4.25 - 0.27x^2 + 2.7x - 6.75$$

$$\Rightarrow \qquad y = -1 + 3.55x - 0.27x^2$$

EXAMPLE 9.   *Fit a parabola* $y = ax^2 + bx + c$ *by the method of least square using following data*

x	1	2	3	4	5
y	5	12	26	60	97

SOLUTION.   The given equation of the parabola is $y = ax^2 + bx + c$

Therefore, the normal equations are

$$a\Sigma x^2 + b\Sigma x + n.c = \Sigma y$$
$$a\Sigma x^3 + b\Sigma x^2 + c\Sigma x = \Sigma xy \qquad \qquad \dots(1)$$
$$a\Sigma x^4 + b\Sigma x^3 + c\Sigma x^2 = \Sigma x^2 y$$

Here, we have $n = 5$

Now, we construct the following table.

x	y	x^2	x^3	x^4	xy	$x^2 y$
1	5	1	1	1	5	5
2	12	4	8	16	24	48
3	26	9	27	81	78	234
4	60	16	64	256	240	960
5	97	25	125	625	485	2425
$\Sigma x = 15$	$\Sigma y = 200$	$\Sigma x^2 = 55$	$\Sigma x^3 = 225$	$\Sigma x^4 = 979$	$\Sigma xy = 832$	$\Sigma x^2 y = 3672$

Putting all the above values in (1), we obtained

$$55a + 15b + 5c = 200$$
$$225a + 55b + 15c = 832 \qquad \qquad \dots(2)$$
$$979a + 225b + 55c = 3672$$

Solving (2), we get

$$a = 5.7143, b = -11.0858, c = 10.4001$$

Hence, the required parabola for best fit is

$$y = 5.7143x^2 - 11.0858x + 10.4001$$

EXAMPLE 10.   *By the method of least squares, find the curve* $y = ax + bx^2$ *that best fits the following data :*

x	1	2	3	4	5
y	1.8	5.1	8.9	14.1	19.8

SOLUTION.   Let error of estimate for i^{th} point (x_i, y_i) be $E_i = (y_i - ax_i - bx_i^2)$. By principle of least squares the values of a and b are such that

$$U = \sum_{i=1}^{5} E_i^2 = \sum_{i=1}^{5} (y_i - ax_i - bx_i^2) \text{ is minimum}$$

∴   Normal equation are given by $\dfrac{\partial u}{\partial a} = 0$.

$$\Rightarrow \qquad \sum_{i=1}^{5} x_i y_i = a \sum_{i=1}^{5} x_i^2 + b \sum_{i=1}^{5} x_i^3$$

and $\dfrac{\partial u}{\partial b} = 0$

Then, $$\sum_{i=1}^{5} x_i^2 y_i = a \sum_{i=1}^{5} x_i^3 + b \sum_{i=1}^{5} x_i^4$$

Delete the suffix i, then normal equations are

$$\Sigma xy = a\Sigma x^2 + b\Sigma x^3 \qquad\qquad …(A)$$

and $$\Sigma x^2 y = a\Sigma x^3 + b\Sigma x^4 \qquad\qquad …(B)$$

x	y	x^2	x^3	x^4	xy	$x^2 y$
1	1.8	1	1	1	1.8	1.8
2	5.1	4	8	16	10.2	20.4
3	8.9	9	27	81	26.7	80.1
4	14.1	16	64	256	56.4	225.6
5	19.8	25	125	625	99	49.5
		$\Sigma x^2 = 55$	$\Sigma x^3 = 255$	$\Sigma x^4 = 979$	$\Sigma xy = 194.1$	$\Sigma x^2 y = 822.9$

Putting all these values in equation (1) and (2), we get

$$1941 = 55a + 225b$$

and $$822.9 = 255a + 979b$$

On solving, we get

$$a = \frac{83.85}{55}; 1.52, b = \frac{317.4}{664}; 0.49$$

Hence, required parabolic curve is

$$y = 1.52x + 0.49x^2.$$

EXAMPLE 11.   *Find a relation of the form $y = AB^x$ for the following data by the method of least squares*

x	2	3	4	5	6
y	8.3	15.4	33.1	65.2	126.4

SOLUTION.   The curve to be fitted is $y = A(B)^x$.

or, $y = a + bX$ where

$$a = \log_{10} A, b = \log_{10} B \text{ and } h = \log_{10} y$$

Therefore, the normal equations are $\Sigma Y = 5a + b\Sigma X$ and $\Sigma XY = a\Sigma X + b\Sigma X^2$

X	Y	$Y = \log_{10}y$	X^2	XY
2	8.3	0.1191	4	1.8382
3	15.4	1.1872	9	3.5616
4	33.1	1.5198	16	6.0792
5	65.2	1.8142	25	9.0710
6	127.4	2.1052	36	12.6312
$\Sigma X = 20$		$\Sigma Y = 7.5455$	$\Sigma X^2 = 90$	$\Sigma XY = 33.1812$

Therefore, the normal equations, becomes $7.5455 = 5a + 20b$ and

$\qquad 33.1812 = 20a + 90b$

$\Rightarrow \qquad\qquad a = 0.31$ and $b = 0.3$

$\therefore \qquad\qquad A = $ Antilog $a = 2.04$

and $\qquad\quad B = $ Antilog $b = 1.995$

Hence, the required curve is $y = 2.04(1.995)^x$.

EXAMPLE 12. *Using the method of least square, fit the non-linear curve of the form $y = ae^{bx}$ to the following data :*

x	0	2	4
y	5.012	10	31.62

SOLUTION. The curve of the form $y = ae^{bx}$ is to fitted in the form

$\qquad\qquad Y = A + Bx$ ...(1)

where $\qquad Y = \log_{10} y, A = \log_{10} a$

and $\qquad\quad B = b \log_{10} e.$

The normal equations for (1) are

$\qquad\qquad \Sigma Y = An + b\Sigma x$ ...(2)

and $\qquad \Sigma xY = A\Sigma x + b\Sigma x^2$ ...(3)

The table is given below :

x	y	$Y = \log_{10}y$	xY	x^2
0	5.012	0.70001	0	0
2	10	1.00000	2.0000	4
4	31.62	1.49996	5.99984	16
6	46.632	3.19997	7.99984	20

Here

$n = 3, \Sigma Y = 3.19997, \Sigma xY = 7.99984, \Sigma x^2 = 20.$

Substituting these values in (2) and (3), we get

$\qquad\qquad 3.19997 = 3A + 6B$ ...(4)

and $\qquad\qquad 7.199984 = 6A + 20B$ ...(5)

Solving (4) and (5), we get

$\qquad\qquad A = 0.66667, B = 0.19999,$

since $A = \log_{10} a, B = b \log_{10} e = (0.43429)b$

Therefore $\qquad a = 10^A, b = \dfrac{B}{0.43429}$

$\Rightarrow \qquad a = (10)^{0.66667}, b = \dfrac{0.19999}{0.43429}$

i.e., $\qquad a = 4.64162, b = 0.6050$

Hence, the required curve of the best fit is given by $y = 4.64162e^{0.4605x}$

EXAMPLE 13. *Find the curve of the best fit of the type $y = ae^{bx}$ to the following data by the method of least square*

x	1	5	7	9	12
y	10	15	12	15	21

SOLUTION. The given equation of the curve can be written as

$\qquad Y = A + Bx$ \hfill ...(1)

where $\qquad Y = \log_{10} y, A = \log_{10} a$

and $\qquad B = b \log_{10} e.$

The normal equations for (1) are

$\qquad \Sigma Y = An + b\Sigma x$ \hfill ...(2)

and $\qquad \Sigma xY = A\Sigma x + b\Sigma x^2$ \hfill ...(3)

x	y	$Y = \log_{10}y$	xY	x^2
1	10	1	1	1
5	15	1.17609	5.88045	25
7	12	1.07918	7.55426	49
9	15	1.17609	10.58481	81
12	21	1.32222	15.86664	144
34	73	5.75358	40.88616	300

Putting the values from table in (2) and (3), we get

$\qquad 5.75358 = 5A + 34B$ \hfill ...(4)

$\qquad 40.88616 = 34A + 300B$ \hfill ...(5)

On solving (4) and (5), we get

$\qquad A = 0.97658, B = 0.02561$

$\Rightarrow \qquad a = 10^A = 10^{0.97658} = 9.47502,$

and $\qquad b = \dfrac{0.02561}{0.43429} = 0.05897$

Hence, the required curve of the best fit is

$\qquad y = 9.47502e^{0.05897x}.$

EXAMPLE 14. *Determine the constant a and b by the method of least squares such that $y = ae^{bx}$ fits the following data :*

x	2	4	6	8	10
y	4.077	11.084	30.128	81.897	222.62

(UPTU–2004)

SOLUTION. We have $\quad y = ae^{bx}$

Taking log, we get

$$\log y = \log a + bx$$

i.e., $\qquad Y = A + bx$; where $\quad Y = \log y$ ...(1)

and $\qquad A = \log a$

Normal equations of equation (1) are

$$\Sigma Y = An + b\Sigma x \qquad\qquad\qquad ...(2)$$

and $\qquad \Sigma xY = A\Sigma x + b\Sigma x^2 \qquad\qquad\qquad ...(3)$

x	y	$Y = \log y$	xY	x^2
2	4.077	1.4054	2.8107	4
4	11.084	2.4055	9.6220	16
6	30.128	3.4054	20.4327	36
8	81.897	4.4055	35.2437	64
10	222.62	5.4055	54.0547	100
30		17.0272	122.1638	220

Putting all these above values in equations (2) and (3), we get

$$17.0272 = 5A + 30b \qquad\qquad\qquad ...(4)$$
$$122.1638 = 30A + 220b \qquad\qquad\qquad ...(5)$$

From (4) and (5), we get $b = 0.5, A = 0.4054$, i.e., $\qquad A = \log a = 0.4054$

Taking antilog, we get $a = 1.4999 \implies a = 1.5$ (app.)

EXAMPLE 15. *Fit a second degree parabola to the following data :*

x	0	1	2	3	4
y	1	1.8	1.3	2.5	6.3

(PTU–2003, 06)

SOLUTION. Let $X = x - 2$ and $Y = y$ and the equation of parabolic curve is

$$Y = a + bx + cx^2 \qquad\qquad\qquad ...(1)$$

The normal equation of (1) are given by

$$\left.\begin{array}{l} \Sigma Y = ma + b\Sigma X + c\Sigma X^2 \\ \Sigma YX = a\Sigma X + b\Sigma X^2 + c\Sigma X^3 \\ \Sigma YX^2 = a\Sigma X^2 + b\Sigma X^3 + c\Sigma X^4 \end{array}\right\} \qquad ...(2)$$

Here $\qquad a = 1.48, b = 1.13$ and $c = 0.55$

Putting these value in eq^n (2), we get

$$Y = 1.48 + 1.13X + 0.55X^2$$

$\implies \qquad y = 1.48 + 1.13(x - 2) + 0.55(x - 2)^2$

Hence, $\qquad y = 1.42 - 1.07x + 0.55x^2$

EXAMPLE 16. *Fit a curve of the form $y = ae^{bx}$ to the following data :*

x	0	1	2	3
y	1.05	2.10	3.85	8.30

(JNTU–2009)

SOLUTION.   The given equation of curve can be written as

$$Y = A + Bx \qquad \qquad \qquad \text{...(1)}$$

Where $Y = \log_{10}y, A = \log_{10}a$ and $B = b\log_{10}e$

The normal equations for (1) are

$$\Sigma Y = An + b\Sigma x \qquad \qquad \text{...(2)}$$

and $\Sigma xY = A\Sigma x + b\Sigma x^2 \qquad \qquad \text{...(3)}$

x	y	$Y = \log_{10}y$	xy	x^2
0	1.05	0.0212	0	0
1	2.10	0.3222	0.3222	1
2	3.85	0.5855	1.1710	4
3	8.30	0.9191	2.7573	9
$\Sigma x = 6$		$\Sigma Y = 1.8480$	$\Sigma xy = 4.2505$	$\Sigma x^2 = 14$

Putting the values from table in (2) and (3), we get

$$1.848 = 4A + 6B$$
$$4.2505 = 6A + 14B$$

Solving these equations, we get

$$A = 0.0185, B = 0.2956$$

Now $a = \text{antilog } A = 1.0186$

$$b = B/\log_{10}e = 0.6806$$

$$y = 1.0186e^{0.6806x}$$

11.8 CURVE FITTING BY SUM OF EXPONENTIALS

Consider the equation

$$y = A_1e^{\lambda_1 x} + A_2e^{\lambda_2 x} \qquad \qquad \text{...(1)}$$

It can be seen that the function given by (1) satisfy a differential equation of the type

$$\frac{d^2y}{dx^2} = a_1\frac{dy}{dx} + a_2y \qquad \qquad \text{...(2)}$$

where a_1, a_2 are constants.

Assume that a is the initial value of x we get, by integrating (2) w.r.t. from a to x.

$$y'(x) - y'(a) = a_1y(x) - a_1y(a) + a_2\int_a^x y(x)dx$$

$\Rightarrow$ $y'(x) = \dfrac{dy}{dx}$

Again integrating (3) w.r.t. x from a to x, we get

$$y(x) - y(a) - (x - a)y'(a) = a_1\int_a^x y(x)dx - a_1(x - a)y(a) + a_2\int_a^x\int_a^x ydxdx \qquad \text{...(4)}$$

Now using the result

$$\underbrace{\int_a^x \cdots \int_a^x f(x)dx}_{n \text{ times}} = \frac{1}{(n-1)!}\int_a^x (x-t)^{n-1}f(t)dt \qquad \text{...(5)}$$

(By Convolution theorem of integration)

The equation (5) can be written as

$$y(x_1) - y(a) - (x - a)y'(a) = a_1\int_a^x ydx - a_1(x - a)y(a) + a_2\int_a^x (x-t)y(t)dt \qquad \text{...(6)}$$

Now, choosing two points x_1 and x_2 such that $a - x_1 = x_2 - a$, then (6) gives

$$y(x_1) - y(a) - (x_1 - a)y'(a) = a_1 \int_a^{x_1} y(x)dx - a_1(x_1 - a)y(a) + a_2 \int_a^{x_1} (x_1 - t)y(t)dt$$

and

$$y(x_2) - y(a) - (x_2 - a)y'(a) = a_1 \int_a^{x_2} y(x)dx - a_1(x_2 - a)y(a) + a_2 \int_a^{x_2} (x_2 - t)y(t)dt$$

Adding above two equations and simplifying by using $a - x_1 = x_2 - a$, we get

$$y(x_1) + y(x_2) - 2y(a) = a_1 \left[\int_a^{x_1} y(x)dx + \int_a^{x_2} y(x)dx \right] + a_2 \left[\int_a^{x_1} (x_1 - t)y(t)dt + \int_a^{x_2} (x_2 - t)y(t)dt \right]$$

$$\dots(7)$$

Now the equation (7) can be used to setup a linear system of equations for a_1 and a_2 and then obtain λ_1 and λ_2 from the following characteristic equation $\lambda^2 = a_1 \lambda + a_2$.
Finally, A_1 and A_2 can be obtained by the method of least square.

SOLVED EXAMPLE

EXAMPLE.   *Fit a function of the form* $y = A_1 e^{\lambda_1 x} + A_2 e^{\lambda_2 x}$ *to the following data :*

x	1.0	1.1	1.2	1.3	1.4	1.5	1.6	1.7	1.8
y	1.54	1.67	1.81	1.97	2.15	2.35	2.58	2.83	3.11

(UPTU–2003)

SOLUTION.   To fit the given curve we use the following steps :

Step 1 : Choose x_1 and x_2 such that

$$a - x_1 = x_2 - a \qquad \dots(1)$$

where a is the initial value of x taken from $x_i : i = 1, 2, \dots, n.$

Step 2 : Use

$$y(x_1) + y(x_2) - 2y(a)$$
$$= a_1 \left[\int_a^{x_1} y(x)dx + \int_a^{x_2} y(x)dx \right] + a_2 \left[\int_a^{x_1} (x_1 - t)y(t)dt + \int_a^{x_2} (x_2 - t)y(t)dt \right]$$

$$\dots(2)$$

Step 3 : Repeat the step I and II by choosing another set of x_1, x_2 and a such that (2) is true. Therefore, we obtain a linear system of equations for a_1 and a_2. Solve these equation to get a_1 and a_2.

Step 4 : Substitute the values of a_1 and a_2, obtained in step III in the characteristic equation $\lambda^2 = a_1 \lambda + a_2$

Step 5 : Finally, use the method of least square to obtain A_1 and A_2 choosing $x_1 = 1.0, a = 1.2$ and $x_2 = 1.4$, we have

Now equation (2) gives

$$1.54 + 2.15 - 3.62$$
$$= a_1 \left[\int_{1.2}^{1.0} ydx + \int_{1.2}^{1.4} ydx \right] + a_2 \left[\int_{1.2}^{1.0} (1.0 - t)y(t)dt + \int_{1.2}^{1.4} (1.4 - t)ydt \right]$$

$$\Rightarrow \quad 0.07 = a_1 \left[-\int_{1.0}^{1.2} ydx + \int_{1.2}^{1.4} ydx \right] + a_2 \left[-\int_{1.0}^{1.2} (1.0 - t)ydt + \int_{1.2}^{1.4} (1.4 - t)ydt \right]$$

On simplification, we get

$$1.81a_1 + 2.180a_2 = 2.10 \qquad \dots(4)$$

Again choosing $x_1 = 1.4, a = 1.6$ and $x_2 = 1.8$, so that we have $a - x_1 = 0.2 = x_2 - a$ and using equation (2), we get

$$2.88a_1 + 3.104a_2 = 3.00 \qquad \qquad ...(5)$$

Solving (4) and (5), we get

$$a_1 = 0.03204, a_2 = 0.9364$$

Put these values in (3), we get

$$\lambda^2 - 0.03204\lambda - 0.9364 = 0$$

Now, using the method of least squares, we get $A_1 = 0.499, A_2 = 0.491$
Hence, the required curve for best fit is

$$y = 0.499e^{0.99x} + 0.491e^{-0.96x}.$$

EXERCISE 11.2

1. Show that the line of fit to the following data :

x	6	7	7	8	8	8	9	9	10
y	5	5	4	5	4	3	4	3	3

is given by $y = -0.5x + 8$.   (JNTU–2008)

2. Show that the best-fitting linear function for the points $(x_1, y_1), (x_2, y_2)...(x_n, y_n)$ may be expressed in the form

$$\begin{vmatrix} x & y & 1 \\ \Sigma x & \Sigma y & n \\ \Sigma x^2 & \Sigma y^2 & \Sigma x \end{vmatrix} = 0$$

Show also that the line passes through the mean points $(\overline{x}, \overline{y})$.

3. Find the parabola of the form $a + bx + cx^2$ which fit most closely with the observations :

x	–3	–2	–1	0	1	2	3
y	4.63	2.11	0.67	0.09	0.63	2.15	4.58

(VTU–2006, JNTU–2000S)

4. Fit a second degree parabola to the following data

x	1.0	1.5	2.0	2.5	3.0	3.5	4.0
y	1.1	1.1	1.6	2.0	2.7	3.4	4.1

5. The following table gives the results of the measurement of train resistance; V is the velocity in miles per hour, R is the resistance in pounds per ton.

V	20	40	60	80	100	120
R	5.5	9.1	14.9	22.8	33.3	46.0

If R is related to V by the relation $R = a + bV + cV^2$, fit the curve. (UPTU–2002)

6. Fit the relation $R = a + bV^2$ from the following data :

V	10	20	30	40	50
R	8	10	15	21	30

(Indore–2008)

7. Fit a least square geometric curve $y = ax^b$ to the following data :

x	1	2	3	4	5
y	0.5	2	4.5	8	12.5

8. The voltage V across a capacitor at time t seconds is given by the following table :

t	0	2	4	6	8
V	150	63	28	12	5.6

Use the method of least square to fit a curve of the form $V = ae^{kt}$ to the above data.

(Meerut-2005; UPTU–2001)

9. Applying the method of least squares to fit the curves $y = ax^2 + \dfrac{b}{x}$ to the following data :

x	1	2	3	4
y	–1.51	0.99	3.88	7.66

(Meerut-2005; Madras–2003)

10. Use the method of least square to fit the curve $y = kx^m$ for the following data :

x	1	2	3	4	5
y	7.1	27.8	62.1	110	161

11. Use the method of least square to fit the straight line for the following data :

x	71	68	73	69	67	65	66	67
y	69	72	70	70	68	67	68	64

(UPTU–2004)

12. Fit an exponential curve of the form $y = ab^x$ to the following data :   (Tiruchirapalli–2001)

x	1	2	3	4	5	6	7	8
y	1	12	1.8	2.5	3.6	4.7	6.6	9.1

13. Fit the curve $y = ae^{bx}$ to the following data :

x	0	2	4
y	5.1	10	31.1

(Coimbatore–1997)

14. Fit a second degree parabola to the following data :

x	1989	1990	1991	1992	1993
y	352	356	357	358	360
x	1994	1995	1996	1997	
y	361	361	360	359	

(UPTU–2001, 09)

15. Find the best possible curve of the form $y = a + bx$, using method of least squares for the data :

x	1	3	4	6	8	9	11	14
y	1	2	4	4	5	7	8	9

(VTU–2011)

16. Fit a second degree parabola to the following data :

x	1	2	3	4	5
y	124	129	140	159	228
x	6	7	8	9	10
y	289	315	302	263	210

(UPTU–2009)

17. The velocity V of a liquid is known to vary with temperature according to a quadratic law $V = a + bT + cT^2$. Find the best values of a, b and c for the following table :

T	1	2	3	4	5	6	7
V	2.31	2.01	3.80	1.66	1.55	1.47	1.41

(UPTU(MCA)–2010)

18. Find the least squares fit of the form $y = a_0 + a_1 x^2$ to the following data :

x	–1	0	1	2
y	2	5	3	0

(UPTU–2008)

19. Predict the mean radiation dose at an altitude of 3000 feet by fitting an exponential curve to the given data :

Amplitude (x)	50	450	780	1200
Dose of radiation (y)	28	30	32	36
Amplitude (x)	4400	4800	5300	
Dose of radiation (y)	51	58	69	

(SVTU–2007, JNTU–2003)

20. Fit the curve $y = ax + b/x$ to the following data :

x	1	2	3	4
y	5.4	6.3	8.2	10.3
x	5	6	7	8
y	12.6	14.9	17.3	19.5

(UPTU–2010)

21. Predict y at $x = 3.75$, by fitting a power curve $y = ax^b$ to the given data :

x	1	2	3	4	5	6
y	2.98	4.26	5.21	6.10	6.80	7.50

(JNTU–2003)

22. Fit the curve of the form $y = ae^{bx}$ to the following data :

x	77	100	185	239	285
y	2.4	3.4	7.0	11.1	19.6

(VTU–2011S, JNTU–2006)

23. A chemical company, wishing to study the effect of extraction time (t) on the efficiency of an extraction operation (e) obtained the data shown in the following table:

t	27	45	41	19	3
e	57	64	80	46	62
t	39	19	49	15	31
e	72	52	77	57	68

Fit a straight line to the given data by the method of least square.   (JNTU–2004)

24. Find the parabola of the form $y = a + bx + cx^2$ which fits most closely with the observations:

x	–3	–2	–1	0	1	2	3
y	4.63	2.11	0.67	0.09	0.63	2.15	4.58

(VTU–2009)

25. By the method of least square, fit a parabola of the form $y = a + bx + cx^2$, to the following data:

x	2	4	6	8	10
y	6.07	12.85	31.47	57.38	91.29

(JNTU–2008)

26. If V(km/hr) and R(kg/ton) are related by a relation of the type $R = a + bV^2$, find by the method of least square a and b with the help of the following table :

V	10	20	30	40	50
R	8	10	15	21	30

(Indore–2008)

27. Estimate y at $x = 2.25$ by fitting the in difference curve of the form $xy = Ax + B$ to the following data:

x	1	2	3	4
y	3	1.5	6	7.5

(JNTU–2003)

28. Predict y at $x = 3.75$, by fitting a power curve $y = ax^b$ to the given data:

x	1	2	3	4	5	6
y	2.98	4.26	5.21	6.10	6.80	7.50

(JNTU–2003)

29. Fit the exponential curve $y = ae^{bx}$ to the following data:

x	2	4	6	8
y	25	38	56	84

(Madras–2003)

30. By the method of averages, fit a curve of the form $y = ae^{bx}$ to the following data:

x	5	15	20	30	35	40
y	10	14	25	40	50	62

(Madras–2002)

31. Fit a straight line to the following data, using the method of moments:

x	1	3	5	7	9
y	1.5	2.8	4.0	4.7	6.0

(Madras–2001)

32. By using the method of moments fit a parabola to the following data:

x	1	2	3	4
y	0.30	0.64	1.32	5.40

(Madras–2000)

Answers

3. $y = 1.243 - 0.004x + 0.22x^2$ **4.** $y = 10.4 - 0.198x + 0.244x^2$ **5.** $R = 3.48 - 0.002V + 0.003V^2$

6. $R = 6.32 + 0.0095V^2$ **7.** $a = 0.5012, b = 1.9977$ **8.** $a = 146.3, k = -0.412$

9. $a = 0.51, b = -2.04$ **10.** $k = 7.17, m = 1.95$ **12.** $y = 0.6823(1.384)^x$

13. $a = 4.1, b = 0.43$ **14.** $y = -1000106.41 + 1034.29x - 0.267x^2$ **15.** $a = 0.545, b = 0.636$

16. $y = 18.866 + 66.158x - 4.333$ **17.** $V = 2.593 - 0.326T + 0.023T^2$

18. $y = 4.167 - 1.111x^2$ **19.** $y = 44.9$ approx. **20.** $a = 3, b = 2$

21. $y = 2.978x^{0.5143}; 5.8769$ **22.** $a = 0.1839, b = 0.0221$ **23.** $y = 48.9 + 0.5067x$

24. $y = 1.243 - 0.004x + 0.22x^2$ **25.** $y = -0.703 - 0.858x + 0.992x^2$ **26.** $a = 6.32, b = 0.0095$

27. $y = 7.187 - 5.16/x; 4.894$ **28.** $y = 2.978x^{0.5143}, 5.8769$ **29.** $a = 9.484, b = 0.315$

30. $a = 1.459, b = 0.062$ **31.** $y = 1.184 + 0.523x$ **32.** $y = 0.485 + 0.397x + 0.124x^2$

11.9 SPLINE FUNCTION

Sometimes the problem of interpolation can be solved by dividing the entire range of points into subinterval and use low-order polynomials to interpolate each subinterval. Such types of polynomial are called piecewise polynomials. Here, we take these subinterval as the form of $[x_i, x_{i+1}]$ where i varies from 0 to 1.

Here, we observe that the piecewise polynomial shown in above figure exhibit discontinuity at some points, which connect these polynomials. Now it is possible to construct piecewise polynomials that prevent such type of discontinuities at the connecting points. Such piecewise polynomials are called Spline functions.

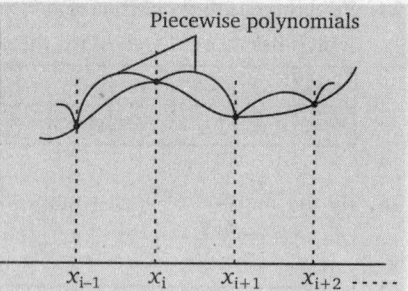

Fig. 11.4 : Piecewise polynomial interpolation

Let $s(x)$ denote the spline function of degree m. Then $s(x)$ must satisfying the following conditions :

    (i) $s(x)$ is a polynomial of degree atmost m in each of the subintervals (x_i, x_{i+1}), $i = 0, 1, 2, ..., n$.

    (ii) $s(x)$ and its derivatives of order 1, 2, ..., $m-1$ are continuous in (x_0, x_n).

The process of constructing such type of polynomials is called Spline interpolation.

Here, we shall discuss only cubic splines.

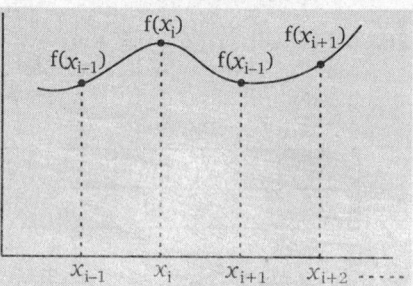

Fig. 11.5 : Spline polynomial

If the second order derivative or piecewise polynomials exist but may not be continuous at the knots. Then it is possible to determine $m_0, m_1, ..., m_n$ in such a way that the resulting piecewise cubic interpolation is twice continuously differentiable. Such type of interpolation is called cubic spline interpolation.

Definition. *A spline function of degree n with knots x_i, $i = 0, 1, ..., n$ is a function $s(x)$ with the properties.*

    (i) On each interval $[x_{i-1}, x_i]$, $1 \le i \le n$, $s(x)$ is a polynomial of degree n.

    (ii) $s(x)$ and $s'(x), s''(x), ..., s^{n-1}(x)$ are continuous on $[a, b]$.

Cubic splines are popular because of their ability to interpolate data with smooth curve.

To construct a cubic spline function which would interpolate the points $(x_0, f_0), (x_1, f_1), ..., (x_n, f_n)$. Then the cubic splines $s(x)$ consist of $(n-1)$ cubic corresponding to $(n-1)$ subintervals.

Let us denote such cubic by $s_i(x)$, then

$$s(x) = s_i(x), \quad i = 1, 2, ..., n$$

Then, these cubic must satisfy the following conditions :

    (1) $s(x)$ must interpolate f at all the points $x_0, x_1, ..., x_n$, i.e., for each i, $s(x_i) = f_i$.

    (2) The functions values must be equal at all the interior knots, i.e., $s_i(x_i) = s_{i+1}(x_i)$.

    (3) The first derivative at the interior knots must be equal.

    $\Rightarrow$ $s_i'(x_0) = s_{i+1}'(x_i)$

    (4) The second derivative at the interior knots must be equal.

    $\Rightarrow$ $s_i''(x_1) = s_{i+1}''(x_n)$

    (5) The second derivative at the end points are zero.

    $\Rightarrow$ $s_i''(x_0) = s_{i+1}''(x_n)$.

11.9.1 CONSTRUCTION OF A CUBIC SPLINE FUNCTION

Since $s(x)$ is a piecewise cubic polynomials, therefore, $s''(x)$ is a linear function of x in the interval (x_{i-1}, x_i) and hence, can be written as

$$s''(x) = \frac{x_i - x}{x_i - x_{i+1}} s''(x_{i-1}) + \frac{(x - x_{i-1})}{(x_i - x_{i-1})} s''(x_i) \qquad ...(1)$$

On integrating (1) twice, with respect to x, we get

$$s(x) = \frac{(x_i - x)^3}{6h_i} P_{i-1} + \frac{(x - x_{i-1})^3}{6h_i} P_i + C_1 x + C_2 \qquad \ldots(2)$$

where $P_i = s''(x_i)$ and C_1, C_2 are constant to be determined by choosing $s(x_{i-1}) = f(x_{i-1})$ and $s(x_i) = f(x_i)$.

Then, we have

$$C_1 = \frac{(f_i - f_{i-1})}{h_i} - \frac{1}{6}(P_i - P_{i-1})h_0$$

and

$$C_2 = \frac{(x_i f_{i-1} - x_{i-1} f_i)}{h_i} - \frac{1}{6}(x P_{i-1} - x_{i-1} P_i)h_i \qquad \ldots(3)$$

Putting the values of C_1 and C_2 in (2), we get

$$s(x) = \left(\frac{(x_i - x)[(x_i - x)^2 - h_i^2]}{6h_i} \right) P_{i-1} + \left(\frac{(x - x_{i-1})[(x - x_{i-1})^2 - h_i^2]}{6h} \right) P_i$$

$$+ \frac{1}{h_i}(x_i - x)f_{i-1} + \frac{1}{h_i}(x - x_{i-1})f_i \qquad \ldots(4)$$

and

$$s''(x) = \frac{(x_i - x)^2}{2h_i} P_{i-1} + \frac{(x - x_{i-1})^2}{2h_i} P_i - \frac{(P_i - P_{i-1})h_i}{6} + \frac{f_i - f_{i-1}}{h_i} \qquad \ldots(5)$$

We want that the derivative $f'(x)$ be continuous at $x = x_i \pm \varepsilon$ as $\varepsilon \to 0$. Letting $s'(x_i - \varepsilon) = s'(x_i + \varepsilon)$ as $\varepsilon \to 0$.

We get

$$\frac{h_i}{6} P_{i-1} + \frac{h_i}{3} P_i + \frac{1}{h_i}(f_i - f_{i-1}) = -\frac{h_{i+1}}{3} P_i - \frac{h_{i+1}}{6} P_{i+1} + \frac{1}{h_i + 1}(f_{i+1} - f_i)$$

$$\Rightarrow \qquad \frac{h_i}{6} P_{i-1} + \frac{h_i + h_{i+1}}{3} P_i + \frac{h_{i+1}}{6} P_{i+1} = \frac{1}{h_{i+1}}(f_{i+1} - f_i) - \frac{1}{h_i}(f_i - f_{i-1})$$

$$i = 1, 2, \ldots, n - 1 \qquad \ldots(6)$$

$$\begin{bmatrix} 2(h_1 + h_2) & h_2 & 0 & 0 & 0 & 0 \\ h_2 & 2(h_2 + h_3) & h_3 & 0 & 0 & 0 \\ \vdots & \vdots & \vdots & \vdots & \vdots & \vdots \\ \vdots & \vdots & \vdots & \vdots & \vdots & \vdots \\ 0 & 0 & 0 & h_{n-2} & 2(h_{n-2} + h_{n-1}) & h_{n-1} \\ 0 & 0 & 0 & 0 & h_{n-1} & 2(h_{n-2} + h_n) \end{bmatrix} \begin{bmatrix} P_1 \\ P_2 \\ \vdots \\ \vdots \\ \vdots \\ P_n \end{bmatrix} = \begin{bmatrix} M_1 \\ M_2 \\ \vdots \\ \vdots \\ \vdots \\ M_n \end{bmatrix}$$

which is the system of $(n-1)$ linear equations in $(n+1)$ unknowns $P_0, P_1, P_2, \ldots, P_n$ and can also be expressed as

where

$$M_i = 6\left[\frac{f_{i+1} - f_i}{h_{i+1}} - \frac{f_i - f_{i-1}}{h_i} \right]$$

$$h_i = x_i - x_{i-1} \qquad i = 1, 2, \ldots, n - 1$$

▶ **REMARKS**

▸ If $P_0 = P_n = 0$, then the spline is called the natural splines because the splines are assumed to take their natural straight line shape outside the intervals of approximations.

▸ If $P_0 = P_n, P_1 = P_{n+1}, f_0 = f_n, f_1 = f_{n+1}, h_1 = h_{n+1}$, then the spline is called periodic splines.

▸ For a non-periodic spline we use

$$s'(a) = f'(a) = f'(0) \quad \text{and} \quad s'(b) = f'(b) = f'_n$$

$$\Rightarrow \quad 2P_0 + P_1 = \frac{6}{h_1}\left(\frac{f_i - f_0}{h_1} - f'_0\right) \quad \Rightarrow \quad P_{n-1} + 2P_n = \frac{6}{h_n}\left(f'_n - f_n - \frac{f_{n-1}}{h_n}\right)$$

▸ For equidistant knots $h_i = h$, (4) and (6) gives

$$s(x) = \frac{1}{6h}[(x_i - x)^3 P_{i-1} + (x - x_{i-1})^3 P_i] + \frac{1}{h}(x_i - x)\left(f_{i-1} - \frac{h^2}{2}P_{i-1}\right) + \frac{1}{h}(x - x_{i-1})\left(f_i - \frac{h^2}{6}P_i\right)$$

and $\qquad P_{i-1} + 4P_i + P_{i+1} = \frac{6}{h^2}(f_{i+1} - 2f_i + 2f_{i-1})$

ALGORITHM

Step 1.	Give input data.
Step 2.	Obtain step lengths and form function differences
Step 3.	Obtain the coefficients of the tridiagonal matrix
Step 4.	Compute the RHS (M array) of the system of matrix
Step 5.	Use Gauss elimination method for P_i
Step 6.	Obtained the coefficients of natural cubic splines
Step 7.	Find the spline function at the given point
Step 8.	Print the obtained results.

SOLVED EXAMPLES

EXAMPLE 1. *Estimate the function value f at x = 7 using cubic splines from the following data :*

i	0	1	2
x_i	4	9	16
f_i	2	3	4

SOLUTION. Here, we get

$$h_1 = x_1 - x_0 = 9 - 4 = 5$$
$$h_2 = x_2 - x_1 = 16 - 9 = 7$$
$$f_0 = 2, f_1 = 3, f_2 = 4$$

Now, we have $\quad h_1 P_0 + 2(h_1 + h_2)a_1 + h_2 a_2 = 6\left[\frac{f_2 - f_1}{h_2} - \frac{f_1 - f_0}{h_1}\right]$

Using $P_2 = P_0 = 0$, Thus $2(5+7)P_1 = 6\left[\frac{1}{7} - \frac{1}{5}\right]$

Therefore, $\quad P_1 = \frac{6 \times (-2)}{35 \times 24} = -0.0143$

Now, since $n = 3$, therefore there are two cubic splines given by

$$s_1(x), x_0 \leq x \leq x_1$$

and $\qquad s_2(x) = x_1 \leq x \leq x_2$

Since we want to find the value of f at $x = 7$ which lie in the domain of $s_i(x)$, therefore we use only $s_i(x)$ for estimation

Now $\qquad s_1(x) = \dfrac{P_1(u_0^3 - h_1^2 u_0)}{6h_1} + \dfrac{1}{h_1}(f_1 u_0 - f_0 u_1)$

where $\qquad u_0 = x - x_0, u_1 = x - x_1$

Now substituting all the values, we get

$$s_1(7) = -\frac{0.0143}{6 \times 5}[(7-4)^3 - 5^2(7-4)] = 2.6229$$

EXAMPLE 2. *Obtain the cubic splines approximation for the function given by the following table*

x	0	1	2	3
f(x)	1	2	33	244

and $P(0) = 0, P(3) = 0.$

SOLUTION. Here, the points x_i are equispaced with $h = 1$, therefore, we have

$$P_{i-1} + 4P_i + P_{i+1} = 6(f_{i+1} - 2f_i + f_{i-1}) \qquad i = 1, 2$$

which implies that $\qquad P_0 + 4P_1 + P_2 = 6(f_2 - 2f_1 + f_0)$

$$P_1 + 4P_2 + P_3 = 6(f_3 - 2f_2 + f_1)$$

Now using $P_0 = 0, P_3 = 0$ and the given values of the function we get

$$4P_1 + P_2 = 180$$
$$P_1 + 4P_2 = 1080$$

$\Rightarrow \qquad\qquad\qquad P_1 = -24, P_2 = 276$

Thus, the cubic splines in the corresponding intervals become

$$-4x^3 + 5x + 1 \text{ in the interval } [0, 1]$$
$$50x^3 - 162x^2 + 167x - 53 \text{ in the interval } [1, 2]$$

and $\quad -46x^3 + 414x^2 - 985x + 715 \text{ in the interval } [2, 3]$

EXAMPLE 3. *Estimate the value of* $f(2.5)$ *using cubic spline functions with the following data :*

i	0	1	2	3
x_i	1	2	3	4
$f(x_i)$	0.5	0.3333	0.25	0.20

SOLUTION. Since the points are equally spaced.

Here $\qquad n = 4 \qquad \Rightarrow \quad h_1 = h_2 = h_3 = 1$

$\qquad\qquad\qquad\qquad \Rightarrow \quad$ we have three intervals and three cubic

Now, we obtained only P_0 and P_1.

We have $\qquad \begin{bmatrix} 4 & 1 \\ 1 & 4 \end{bmatrix}\begin{bmatrix} P_0 \\ P_1 \end{bmatrix} = \begin{bmatrix} M_1 \\ M_2 \end{bmatrix}$

$\Rightarrow \qquad\qquad M_1 = \dfrac{6}{h^2}(f_2 - 2f_1 + f_0) = 6(0.25 - 2 \times 0.3333 + 0.5) = 0.5004$

and $\qquad\qquad M_2 = \dfrac{6}{h^2}(f_3 - 2f_2 + f_1) = 6(0.2 - 2 \times 0.25 + 0.3333) = 0.1998$

Now $$P_1 = \frac{M_1 \times 4 - M_2 \times 1}{15} = \frac{0.5004 \times 4 - 1.998}{15} = 0.0199$$

and, $$P_2 = \frac{M_2 \times 4 - M_1 \times 1}{15} = \frac{0.1998 \times 4 - 0.5004}{15} = 0.0199$$

Now, we want to find the value of $f(x)$ at $x = 2.5$ which lies in the domain of $s_2(x)$.

Now, $$s_2(x) = \frac{P_1}{6}[(x - x_2) + (x - x_2)^3] + \frac{P_2}{6}[(x - x_1)^3 - (x - x_1)]$$

$$+ [f_2(x - x_1) - f_1(x - x_2)]$$

$$\Rightarrow \qquad s_3(2.5) = \frac{0.1201}{6}[(2.5 - 3) - (2.5 - 3)^3] + \frac{0.0119}{6}[(2.5 - 2)^3 - (2.5 - 2)]$$

$$+ (0.25)(2.5 - 2) - 0.3333(2.5 - 3)$$

$$= 0.2829$$

EXAMPLE 4.   *Evaluate* $I = \int_0^1 \frac{1}{1 + x} \, dx$, *using cubic spline method.*

SOLUTION.   Taking $n = 2$, *i.e.*, $h = 0.5$, then the tables of values of x and $y = \dfrac{1}{1 + x}$ is given below

x	0	0.5	1
y	1	2/3	1/2

Assuming $P_0 = P_2 = 0$,

Since we have $P_{i-1} + 4P_i + P_{i+1} = \dfrac{6}{h^2}(y_{i-1} - 2y_i + y_{i+1})$

Here $n = 2$ so we shall take $i = 1$ only.

$$\therefore \quad P_0 + 4P_1 + P_2 = \frac{6}{(0.5)^2}(y_0 - 2y_1 + y_2)$$

or $$4P_1 = 24\left[1 - 2\left(\frac{2}{3}\right) + \frac{1}{2}\right]$$

or $$4P_1 = 4 \quad \Rightarrow \quad P_1 = 1$$

Thus, $$I = \sum_{i=1}^{2}\left[\frac{h}{2}\left(y_{i-1} - \frac{h^3}{4}\right)(P_{i-1} + P_i)\right]$$

$$= \frac{h}{2}(y_0 + y_1) - \frac{h^3}{24}(P_0 + P_1) + \frac{h}{2}(y_1 + y_2) - \frac{h^3}{24}(P_1 + P_2)$$

$$\therefore \qquad I = \frac{0.5}{2}\left(1 + \frac{2}{3}\right) - \frac{(0.5)^3}{24}(0 + 1) + \frac{0.5}{2}\left(\frac{2}{3} + \frac{1}{21}\right) - \frac{(0.5)^3}{24}(1 + 0)$$

$$= \frac{5}{12} - \frac{1}{336} + \frac{7}{24} - \frac{1}{336} = \frac{17}{24} - \frac{2}{336} = \frac{118}{168} = 0.702380952.$$

EXERCISE 11.3

1. Fit the following four points by the cubic splines

i	0	1	2	3
x_i	1	2	3	4
y_i	1	5	11	8

Use the conditions $y_0'' = y_3'' = 0$.

Hence, find (i) $y(1.5)$ (ii) $y'(2)$.

2. Suppose $f_i = x_i^{-2}$ and $f_i' = -2x_i^{-3}$, where

$x = \dfrac{1}{2}, i = 1(1)4$ are given. Fit these values by

the piecewise cubic Hermite polynomials.

3. Show that $I = \int_0^1 \sin \pi x \, dx = 0.62500000$ using

cubic spline method.

4. Find the natural cubic spline which agrees with $y(x)$ of the set of data points

x	2	3	4
y	11	49	123

Hence, compute $y(2.5)$ and $y'(2)$.

5. Find the cubic spline fit for the following data :

x	0	1	2	3
$f(x)$	1	4	10	8

Using $f''(0) = f''(3) = 0$ and valid in (1, 2). Hence, obtain the estimate $f(1.5)$.

6. Show that the following function is a cubic spline

$$f(x) = \begin{cases} -x - 2x^3, & -1 \le x \le 0 \\ -x^2 + 2x^3, & 0 \le x \le 1 \end{cases}$$

7. Show that $f(x) = \begin{cases} -x - 2x^3, & -1 \le x \le 0 \\ -x^2 + 2x^3, & 0 \le x \le 1 \end{cases}$

is not cubic splines.

Answers

1. $f(x) = \begin{cases} \dfrac{1}{15}(17x^3 - 51x^2 + 94x - 45); & 1 \le x \le 2 \\[2mm] \dfrac{1}{15}(-55x^3 - 381x^2 - 770x + 53); & 2 \le x \le 3 \\[2mm] \dfrac{1}{15}(38x^3 - 456x^2 + 1741x - 1980); & 3 \le x \le 4 \end{cases}$; $y(1.5) = \dfrac{103}{40}, y'(2) = \dfrac{94}{15}$

2. $f(x) = \begin{cases} -24x^3 - 68x^2 - 66x + 23, & \dfrac{1}{2} \le x \le 1 \\[2mm] \dfrac{1}{27}(-40x^3 + 188x^2 - 310x + 189), & 1 \le x \le \dfrac{3}{2} \\[2mm] \dfrac{1}{108}(-28x^3 + 184x^2 - 427x + 369), & \dfrac{3}{2} \le x \le 2 \end{cases}$

4. For $n = 2, y(0.25) = \dfrac{213}{8}, y'(2) = 29$

5. $P_1 = 8, P_2 = -14, S(x) = \dfrac{1}{3}(-11x^3 + 45x^2 - 40x + 118)$

11.10 REGRESSION ANALYSIS

11.10.1 SCATTER OR DOT DIAGRAM

If we measure the heights and weights of a certain number of students, denote the quantity by x and y and plot them on a graph paper referring to two perpendicular axes. For each student, there shall be one point and thus we get scatter diagram.

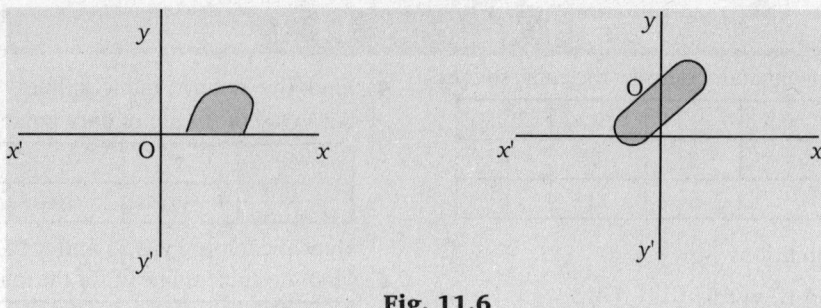

Fig. 11.6

If the origin of axes is taken as $(\bar{x}, \bar{y})$ where $\bar{x}, \bar{y}$ are the means of the values of x and y respectively, the points may be scattered all around the region. For points lying in I and III quadrants, the product $(x - \bar{x})(y - \bar{y})$ is positive and for those points which are lying in II and IV quadrants, it is negative.

11.10.2 REGRESSION

Let us suppose that the scatter diagram indicates some relationship between the two variables x and y, the dots of the scatter diagram will be more or less concentrated round a curve. This curve is called the curve regression.

The straight line about which the various points may be considered as scattered is called the regression line.

▼ REMARK

▶ It should be noted that one can predict exactly only if the two variables are perfectly related. In that case, there is no scatter in the data and the various points lie exactly on the regression line, but when the correlation is less than perfect, *i.e.*, there is a scatter of points on the scatter diagram, then the regression line is only a representation of the general trend.

11.10.3 EQUATION OF THE LINE OF REGRESSION

Let $y = ax + b$ is the given equation of straight line. The method of least square can be used to fit a straight line to the set of points given on the scatter diagram. Now, transfer the origin to the points $(\bar{x}, \bar{y})$, where $\bar{x}$ and $\bar{y}$ are the means of x-series and y-series, respectively.

Suppose that x, y be the deviation from the respective means $\bar{x}$ and $\bar{y}$.

Therefore, $x = X - \bar{x}$ and $y = Y - \bar{y}$

Let $Y = aX + b$ be the equation of the line of best fit of x. Changing the origin to $(\bar{x}, \bar{y})$, we get the form

$$y = ax + b; \text{ where } y = Y - \bar{y} \text{ and } x = X - \bar{x}.$$

Let $P(x_r, y_r)$ be any dot, then the difference between P and the line is

$$y_r - ax_r - b$$

Let I denote the sum of the squares of such distances given by

$$I = \Sigma(y - ax - b)^2 \text{ for all values of } r.$$

Now, using the principle of least squares, choose a and b such that I is minimum. For minima of I, we must have

$$\frac{\partial I}{\partial a} = -2\Sigma x\,(y - ax - b) = 0 \implies \Sigma xy - a\Sigma x^2 - b\Sigma x = 0$$

and $\qquad \dfrac{\partial I}{\partial b} = -2\Sigma(y - ax - b) = 0 \;\Rightarrow\; \Sigma y - a\Sigma x - nb = 0$

Since $\Sigma x = 0$, $\Sigma y = 0$, then we get

$$a = \dfrac{\Sigma xy}{\Sigma x^2} = \dfrac{r\sigma_y}{\sigma_x} \quad \text{and } b = 0 \qquad \left(\because \dfrac{r\sigma_y}{\sigma_x} = \dfrac{\Sigma xy}{\sqrt{\Sigma x^2 \, \Sigma y^2}} \times \sqrt{\left(\dfrac{\Sigma y^2}{n}\right)\left(\dfrac{n}{\Sigma x^2}\right)} \right)$$

Therefore, the line of fit is $\qquad y = r \dfrac{\sigma_y}{\sigma_x} \cdot x$

Now, rechanging the origin, we get $\quad Y - \bar{y} = r \dfrac{\sigma_y}{\sigma_x} (X - \bar{x})$

This is known as regression line of Y on X.

If X is taken to be dependent variable, then the regression line is $\quad X - \bar{x} = r \dfrac{\sigma_x}{\sigma_y} (Y - \bar{y})$

This is called the regression line of X on Y.

Recapitulations ||||||||||||||||||||||||||||||||||||||||||||||||||||

- If the straight line is chosen such that the sum of squares of deviation parallel to the axis of y is minimum, it is called the line of regression of y on x.

- The coefficients $r\dfrac{\sigma_y}{\sigma_x}$ and $r\dfrac{\sigma_x}{\sigma_y}$ are called the regression coefficients of y on x and of x on y, respectively.

- If $r = \pm 1$, the two regression lines will coincide.

11.10.4 LEAST SQUARE REGRESSION

The approach discussed above, is known as least square regression.

Here if $y = a + bx = f(x)$ is the given equation, then using the principle of least square and principle of maxima and minima, we may find the normal equations, given by

$$\Sigma x_i = na + b\Sigma y_i \quad \text{and} \quad \Sigma x_i y_i = a\Sigma x_i + b\Sigma x_i^2$$

Solving the above equations for a and b, we get

$$a = \dfrac{\Sigma y_i}{n} - b\dfrac{\Sigma x_i}{n} = \bar{y} - b\bar{x} \quad \text{and} \quad b = \dfrac{n\Sigma x_i y_i - \Sigma x_i \Sigma y_i}{\Sigma x_i^2 - (\Sigma x_i)^2}$$

where $\bar{x}, \bar{y}$ are the means of x-series and y-series respectively.

11.11 PROPERTIES OF REGRESSION COEFFICIENTS

THEOREM 1. *Correlation coefficient is the geometric mean between the regression coefficients.*

PROOF. The coefficient of regression are given by

$$\dfrac{r\sigma_y}{\sigma_x} \text{ and } \dfrac{r\sigma_x}{\sigma_y}$$

Geometric mean between these two coefficients is given by

$$\sqrt{\dfrac{r\sigma_y}{\sigma_x} \cdot \dfrac{r\sigma_x}{\sigma_y}} = \sqrt{r^2} = r = \text{coefficient of correlation.}$$

THEOREM 2. *If one of the regression coefficient is greater than unity, other must be less than unity.*

PROOF. We know that

Two regression coefficients are

$$b_{yx} = r\frac{\sigma_y}{\sigma_x} \text{ and } b_{xy} = r\frac{\sigma_x}{\sigma_y} \qquad \qquad ...(1)$$

Let us suppose $b_{yx} > 1$. Then $\dfrac{1}{b_{yx}} < 1$ ...(2)

From (1) we obtain

$$b_{yx} \cdot b_{xy} = r^2 \le 1 \qquad \Rightarrow \qquad b_{xy} \le \frac{1}{b_{xy}} < 1$$

Similarly if $b_{xy} > 1$ then $b_{yx} < 1$.

THEOREM 3. *Arithmetic mean of regression coefficients is greater than the correlation coefficient.*

PROOF. Here, we have to prove that

$$\frac{b_{yx} + b_{xy}}{2} > r \qquad \text{or} \qquad r\frac{\sigma_y}{\sigma_x} + r\frac{\sigma_x}{\sigma_y} > 2r$$

$$\Rightarrow \qquad \sigma_x^2 + \sigma_y^2 > 2\sigma_x\sigma_y$$

$$\Rightarrow \qquad \sigma_x^2 + \sigma_y^2 - 2\sigma_x\sigma_y > 0$$

$$\Rightarrow \qquad (\sigma_x - \sigma_y)^2 > 0 \text{ which is always true.}$$

THEOREM 4. *The regression coefficients are independent of the origin but not of scale.*

PROOF. Let $u = \dfrac{x-a}{h}, v = \dfrac{y-b}{k}$, where a, b, h and k are constants.

Now, $$b_{yx} = r\frac{\sigma_y}{\sigma_x} = r \cdot \frac{k\sigma_v}{\sigma_u} = \frac{k}{h}\left(\frac{r\sigma_v}{\sigma_u}\right) = \frac{k}{h}b_{vu}$$

Similarly $$b_{xy} = \frac{h}{k}b_{uv}$$

Hence, b_{yx} and b_{xy} are both independent of a and b but not of h and k.

THEOREM 5. *The correlation coefficient and the two regression coefficients have same sign.*

PROOF. The regression coefficient of y on $x = b_{yx} = r\dfrac{\sigma_y}{\sigma_x}$

and regression coefficient of x on $y = b_{xy} = r\dfrac{\sigma_x}{\sigma_y}$

Since, σ_x and σ_y are both positive. Hence, b_{yx}, b_{xy} and r having the same sign.

11.12 ANGLE BETWEEN TWO LINES OF REGRESSION

If θ is the acute angle between two regression lines in the case of two variables x and y, then

$$\tan\theta = \frac{1-r^2}{r} \cdot \frac{\sigma_x\sigma_y}{\sigma_x^2 + \sigma_y^2}$$

where, r, σ_x, σ_y have their usual meaning. Explain the significance of the formula, when $r = 0$ and $r = \pm 1$.

Proof. Equations of regression lines are given by

$$y - \bar{y} = r\frac{\sigma_y}{\sigma_x}(x - \bar{x}) \qquad \ldots (1)$$

and

$$x - \bar{x} = r\frac{\sigma_x}{\sigma_y}(y - \bar{y}) \qquad \ldots (2)$$

Slopes of (1) and (2) are given by $m_1 = r\dfrac{\sigma_y}{\sigma_x}$ and $m_2 = \dfrac{\sigma_y}{r\sigma_x}$

Now

$$\tan\theta = \pm\frac{m_2 - m_1}{1 + m_2 m_1} = \frac{\dfrac{\sigma_y}{r\sigma_x} - \dfrac{r\sigma_y}{\sigma_x}}{1 + \dfrac{\sigma_y^2}{\sigma_x^2}}$$

$$= \pm\frac{1 - r^2}{r}\cdot\frac{\sigma_y}{\sigma_x}\cdot\frac{\sigma_x^2}{\sigma_x^2 + \sigma_y^2} = \pm\frac{1 - r^2}{r}\cdot\frac{\sigma_x\,\sigma_y}{\sigma_x^2 + \sigma_y^2}$$

Now, since $r^2 \le 1$ and σ_x, σ_y are positive, therefore, positive sign gives the acute angle between the lines.

Hence,

$$\tan\theta = \frac{1 - r^2}{r}\cdot\frac{\sigma_x\,\sigma_y}{\sigma_x^2 + \sigma_y^2}$$

Also, when $r = 0$, $\theta = \dfrac{\pi}{2}$. Therefore, two lines of regression are perpendicular to each other.

Thus, the estimated value of y is the same for all values of x and vice-versa.

When, $r = \pm 1$, $\tan\theta = 0$ so that $\theta = 0$ or π.

Hence, the lines of regression coincide and there is a perfect correlation between two variates x and y.

SOLVED EXAMPLES

EXAMPLE 1. *Fit a straight line to the following set of data*

x	1	2	3	4	5
y	3	4	5	6	8

SOLUTION. We have

x_i	y_i	x_i^2	$x_i y_i$
1	3	1	3
2	4	4	8
3	5	9	15
4	6	16	24
5	8	25	40
15	**26**	**55**	**90**

Now,
$$b = \frac{n\Sigma x_i y_i - \Sigma x_i \Sigma y_i}{n\Sigma x_i^2 - (\Sigma x_i)^2}$$

$$= \frac{5 \times 90 - 15 \times 26}{5 \times 55 - 15 \times 15} = 1.20$$

and
$$a = \frac{\Sigma y_i}{n} - b\frac{\Sigma x_i}{n}$$

$$= \frac{26}{5} - 1.20 \times \frac{5}{5} = 1.60$$

Therefore, the linear equation is

$$y = 1.6 + 1.2x$$

which can be shown in the adjoining figure.

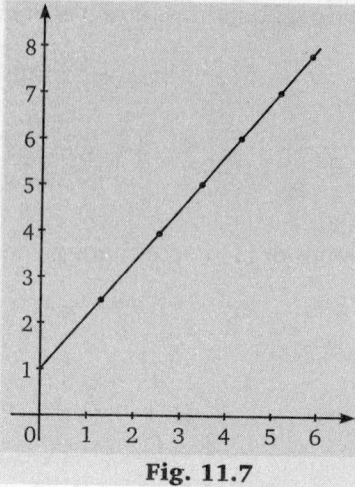

Fig. 11.7

EXAMPLE 2. *If the regression coefficients are 0.8 and 0.2, what would be the value of coefficient of correlation.*

[UPTU(B.Tech. (SUM))–2004]

SOLUTION. We have $r^2 = b_{yx} \cdot b_{xy} = 0.8 \times 0.2 = 0.16$

Since r, b_{xy} and b_{yx} having same sign as both the regression coefficients b_{yx} and b_{xy}.

Hence, $r = \sqrt{0.16} = 0.4$

EXAMPLE 3. *Find linear regression from the following data:*

x	1	2	3	4	5	6	7	8
y	3	7	10	12	14	17	20	24

[UPTU(MCA)–2001]

SOLUTION. The regression coefficients are given by

$$b_{yx} = \frac{n\Sigma xy - \Sigma x \Sigma y}{n\Sigma x^2 - (\Sigma x)^2} \qquad \text{... (1)}$$

$$b_{xy} = \frac{n\Sigma xy - \Sigma x \Sigma y}{n\Sigma y^2 - (\Sigma y)^2} \qquad \text{... (2)}$$

Here, we have the following table

x	y	x^2	y^2	xy
1	3	1	9	3
2	7	4	49	14
3	10	9	100	30
4	12	16	144	48
5	14	25	196	70
6	17	36	289	102
7	20	49	400	140
8	24	64	576	192
36	107	204	1763	599

Also, $n = 8$

Putting these values from above table in (1) and (2), we get

$$b_{yx} = \frac{(8 \times 599) - (36 \times 107)}{(8 \times 204) - (36)^2} = 2.7976$$

and

$$b_{xy} = \frac{(8 \times 599) - (36 \times 107)}{(8 \times 1763) - (107)^2} = 0.3540.$$

EXAMPLE 4. *The regression lines of y on x and x on y are respectively y = ax + b, x = cy + d. Show that*

$$\frac{\sigma_y}{\sigma_x} = \sqrt{\frac{a}{c}}, \quad \bar{x} = \frac{bc + d}{1 - ac} \quad \text{and} \quad \bar{y} = \frac{ad + b}{1 - ac}.$$

SOLUTION. The regression line of y on x is given by

$$y = ax + b \qquad \qquad \text{... (1)}$$

$\therefore \qquad \qquad b_{yx} = a$

Similarly, $\qquad \qquad b_{xy} = c \qquad \qquad \text{... (2)}$

Now, $\qquad \qquad b_{yx} = r \dfrac{\sigma_y}{\sigma_x} \qquad \qquad \text{... (3)}$

and $\qquad \qquad b_{xy} = r \dfrac{\sigma_x}{\sigma_y} \qquad \qquad \text{... (4)}$

Using (3) and (4), we get

$$\frac{b_{yx}}{b_{xy}} = \frac{\sigma_y^2}{\sigma_x^2} \Rightarrow \frac{a}{c} = \frac{\sigma_y^2}{\sigma_x^2}, \text{ i.e., } \frac{\sigma_y}{\sigma_x} = \sqrt{\frac{a}{c}}$$

Since, both the regression lines pass through the point $(\bar{x}, \bar{y})$, therefore

$$\bar{y} = a\bar{x} + b, \quad \bar{x} = c\bar{y} + d$$

$$\Rightarrow \qquad \qquad a\bar{x} - \bar{y} = -b \qquad \qquad \text{... (5)}$$

$$\bar{x} - c\bar{y} = d \qquad \qquad \text{... (6)}$$

Multiplying (6) by a and then subtracting from (5), we get

$$(ac - 1)\bar{y} = -ad - b$$

$$\Rightarrow \qquad \qquad \bar{y} = \frac{ad + b}{1 - ac}$$

Similarly, we get

$$\bar{x} = \frac{bc + d}{1 - ac}.$$

EXAMPLE 5. *For two random variables x and y with the same mean, the two regression equations are y = ax + b and x = αy + β*

Show that $\dfrac{b}{\beta} = \dfrac{1 - a}{1 - \alpha}$. Also, find the common ratio.

SOLUTION. We have $\qquad b_{yx} = a, b_{xy} = \alpha$

If m be the common mean, then regression lines are given by

$$y - m = a(x - m)$$

$$\Rightarrow \qquad \qquad y = ax + m(1 - \alpha) \qquad \qquad \text{...(1)}$$

and $\qquad \qquad x = ay + m(1 - \alpha) \qquad \qquad \text{...(2)}$

On comparing (1) and (2) with the given equation, we get
$$b = m(1-a), \beta = m(1-\alpha)$$

Therefore,
$$\frac{b}{\beta} = \frac{1-a}{1-\alpha}$$

Further
$$m = \frac{b}{1-a} = \frac{\beta}{1-\alpha}$$

Now since regression lines passes through $(\bar{x}, \bar{y})$

$\therefore$ $\quad\quad\quad\quad\quad \bar{x} = \alpha\bar{y} + \beta$

and $\quad\quad\quad\quad\quad \bar{y} = a\bar{x} + b$ will hold

$\Rightarrow$ $\quad\quad\quad\quad\quad m = am + b$

and $\quad\quad\quad\quad\quad m = \alpha m + \beta$

$\Rightarrow$ $\quad\quad\quad\quad\quad m = \dfrac{\beta - b}{a - \alpha}$.

EXAMPLE 6. *For 10 observations on price (x) and supply (y), the following data were obtained (in appropriate units) :*

$$\Sigma x = 130, \Sigma y = 220, \Sigma x^2 = 2288, \Sigma y^2 = 5506 \quad and \quad \Sigma xy = 3467$$

Find the two lines of regression and estimate the supply when the price is 16 units

SOLUTION. We have $n = 10$, $\bar{x} = \dfrac{\Sigma x}{n} = 13$, $\bar{y} = \dfrac{\Sigma y}{n} = 22$

Regression coefficient of y on x is
$$b_{yx} = \frac{n\Sigma xy - \Sigma x \Sigma y}{n\,\Sigma x^2 - (\Sigma x)^2} = \frac{(10 \times 3467) - (130 \times 220)}{(10 \times 2288) - (130)^2}$$

$$= \frac{34670 - 28600}{22880 - 16900} = \frac{6070}{5980} = 1.015$$

Therefore, regression line of y on x is
$$y - \bar{y} = b_{yx}(x - \bar{x})$$

$\Rightarrow$ $\quad\quad\quad y - 22 = 1.015(x - 13)$

$\Rightarrow$ $\quad\quad\quad\quad y = 1.015x + 8.805$

Similarly, regression coefficient of x on y is
$$b_{xy} = \frac{n\Sigma xy - \Sigma x\,\Sigma y}{n\Sigma y^2 - (\Sigma y)^2} = \frac{(10 \times 3467) - (130 \times 220)}{(10 \times 5506) - (220)^2}$$

$$= \frac{6070}{6600} = 0.9114$$

$\therefore$ Regression line of x on y is
$$x - \bar{x} = b_{xy}(y - \bar{y})$$

$\Rightarrow$ $\quad\quad\quad x - 13 = 0.9114\,(y - 22)$

$\Rightarrow$ $\quad\quad\quad\quad x = 0.9114y - 7.0508$

Now, since we are to estimate supply (y) when price (x) is given.

Therefore, we are to use regression line of y on x.

$\therefore$ When $x = 16$
$$y = 1.015(16) + 8.805 = 24.045$$

EXAMPLE 7. *Find the line of regression of y on x for the following data :*

x	1.53	1.78	2.60	2.95	3.42
y	33.50	36.30	40.00	45.80	53.50

SOLUTION. We know that regression line of y on x is given by

$$y - \bar{y} = b_{yx}(x - \bar{x}) \qquad \ldots (1)$$

where $\qquad b_{yx} = \dfrac{n\Sigma xy - \Sigma x\, \Sigma y}{n\,\Sigma x^2 - (\Sigma x)^2} \qquad \ldots (2)$

Here, we have the following table

x	y	x^2	xy
1.53	33.50	2.3409	51.255
1.78	36.30	2.1684	64.614
2.60	40.00	6.76	104.00
2.95	45.80	8.7025	135.11
3.42	53.50	11.6964	182.97
12.28	209.1	32.6682	537.949

Also, $n = 5$.

Then, $\qquad b_{yx} = \dfrac{(5 \times 537.949) - (12.28)(209.1)}{(5 \times 32.6682) - (12.28)^2} = 9.726$

Also, Mean, $\qquad \bar{x} = \dfrac{\Sigma x}{n} = \dfrac{12.28}{5} = 2.456$

and $\qquad \bar{y} = \dfrac{\Sigma y}{n} = \dfrac{20.91}{5} = 41.82$

Putting all these values in (1), we get

$$y - 41.82 = 9.726(x - 2.456) = 9.726x - 23.887$$

$\Rightarrow \qquad y = 17.932 + 9.726x$

which is the required equation of regression.

EXAMPLE 8. *The equations of two regression lines, obtained in a correlation analysis of 60 observations are :*

$$5x = 6y + 24 \quad and \quad 1000y = 768x - 3608$$

What is the correlation coefficient? Show that the ratio of the coefficients of variability of x to that of y is 5/24. What is the ratio of variances of x and y?

SOLUTION. As per given, the regression line of x on y is

$$5x = 6y + 24$$

$\Rightarrow \qquad x = \dfrac{6}{5}y + \dfrac{24}{5}$

$\Rightarrow \qquad b_{xy} = \dfrac{6}{5} \qquad \ldots (1)$

Also, regression line of y on x is

$$1000y = 768x - 3608$$

$\Rightarrow \qquad y = 0.768x - 3.608$

$\Rightarrow \qquad b_{yx} = 0.768$ ... (2)

Using (1), we get

$$r \frac{\sigma_x}{\sigma_y} = \frac{6}{5}$$ ... (3)

From (2) $\quad r \dfrac{\sigma_y}{\sigma_x} = 0.678$ ... (4)

Multiplying (3) and (4), we get

$$r^2 = 0.916 \ \Rightarrow \ r = 0.96$$ ... (5)

Now, dividing (4) by (3), we get

$$\frac{\sigma_x^2}{\sigma_y^2} = \frac{6}{5 \times 0.768} = 1.5625$$ ... (6)

$$\Rightarrow \qquad \frac{\sigma_x}{\sigma_y} = 1.25 = \frac{5}{4}$$

Now, since the regression line passes through the point $(\bar{x}, \bar{y})$, then

$$5\bar{x} = 6\bar{y} + 24 \quad \text{and} \quad 1000\bar{y} = 768\bar{x} - 3608$$

On solving, we get $\bar{x} = 6, \ \bar{y} = 1$

Also, coefficient of variability of $x = \dfrac{\sigma_x}{\bar{x}}$

and coefficient of variability of $\ y = \dfrac{\sigma_y}{\bar{y}}$

Hence, required ratio $= \dfrac{\sigma_x}{\bar{x}} \times \dfrac{\bar{y}}{\sigma_y} = \dfrac{\bar{y}}{\bar{x}} \left(\dfrac{\sigma_x}{\sigma_y} \right) = \dfrac{1}{6} \times \dfrac{5}{4} = \dfrac{5}{24}.$

EXAMPLE 9. *The following data regarding the heights (y) and weights (x) of 100 college students are given by*

$\Sigma x = 1500, \ \Sigma x^2 = 2272500, \ \Sigma y = 6800, \ \Sigma y^2 = 463025 \ and \ \Sigma xy = 1022250$

Find the equation of regression line of height on weight.

SOLUTION. We have

$$\bar{x} = \frac{\Sigma x}{n} = \frac{15000}{100} = 150$$

and $\qquad \bar{y} = \dfrac{\Sigma y}{n} = \dfrac{6800}{100} = 68$

Regression coefficients of y and x is given by

$$b_{yx} = \frac{n\Sigma xy - \Sigma x \Sigma y}{n\Sigma x^2 - (\Sigma x)^2}$$

$$= \frac{(100 \times 1022250) - (15000 \times 6800)}{(100 \times 2272500) - (15000)^2} = 0.1.$$

Hence, regression line of height (y) on weight (x) is given by

$$y - \bar{y} = b_{yx}(x - \bar{x})$$

$$\Rightarrow \qquad y - 68 = (0.1)(x - 150)$$

$$\Rightarrow \qquad y = 0.1x - 15 + 68$$

i.e., $\qquad y = 0.1x + 53$

EXAMPLE 10. *The following table shows the arithmetic mean and standard deviation of the advertising expenditure (x) and sales of company (y) for the year 2001-02.*

Statistical measure	Advertising expenditure (in lacs)	Sales (y) (in lacs)
Arithmetic Mean	20	100
Standard deviation	3	12

The coefficient of correlation between x and y is 0.8

(i) *Find the equation of two lines of regression.*

(ii) *What would be the expected sales of the company if the advertising expenditure is ₹ 32 lacs.*

SOLUTION. We have

$$\bar{x} = 20, \bar{y} = 100, \sigma_x = 3, \sigma_y = 12 \text{ and } r = 0.8$$

(i) Regression line of *y* and *x* is given by

$$y - \bar{y} = b_{yx}(x - \bar{x}) = r\frac{\sigma_y}{\sigma_x}(x - \bar{x})$$

$$\therefore \quad y - 100 = (0.8)\left(\frac{12}{3}\right)(x - 20) = 3.2(x - 20) = 3.2x - 64$$

$$\Rightarrow \qquad y = 3.2x + 36 \qquad\qquad\qquad ...(1)$$

Similarly regression line of *x* on *y* is given by

$$x - \bar{x} = r\frac{\sigma_x}{\sigma_y}(y - \bar{y})$$

$$\Rightarrow \qquad x - 20 = (0.8)\left(\frac{3}{12}\right)(y - 100)$$

$$= 0.2(y - 100) = 0.2y - 20$$

i.e., $\qquad x = 0.2y$

(ii) We have to determine expected sales of the company while advertising expenditure is given. So we will use the regression line of *y* on *x*.

Regression line of *y* on *x* is

$$y = 3.2x + 36$$

when *x* = 32,

$$y = 3.2(32) + 36$$

$$= 102.4 + 36 = 138.4 \text{ lacs}$$

EXAMPLE 11. *Find the coefficient of correlation when the two regression equations are x = –0.2y + 4.2 and y = –0.8x + 8.4.*

SOLUTION. Given that $\qquad x = -0.2y + 4.2 \qquad\qquad\qquad ...(1)$

$$y = -0.8x + 8.4 \qquad\qquad\qquad ...(2)$$

Assume that eq. (1) is the regression line of *x* on *y* and eq. (2) is the regression line of *y* on *x*. Then

Regression coefficients of *x* on *y* is

$$b_{xy} = -0.2$$

and Regression coefficient of y on x is

$$b_{yx} = -0.8$$

We know that b_{xy} and b_{yx} having the same sign and $b_{xy}b_{yx} = 0.16(< 1)$.
Therefore, our assumption is correct.

Also $b_{xy} \cdot b_{yx} = r^2$

$\Rightarrow$ $(-0.2)(-0.8) = r^2$

$\Rightarrow$ $r^2 = 0.16$

$\Rightarrow$ $r = -0.4$ ($\because$ r, σ_x and σ_y have the same sign)

11.13 FITTING A POLYNOMIAL FUNCTION

If a given set of data does not appear to satisfy a linear equation, then we try a suitable polynomial as a regression curve to fit the data. The least squares technique can be used to fit data to a polynomial.

Let $y = f(x)$ by a polynomial of degree $(m - 1)$ such that

$$y = a_1 + a_2x + a_3x^2 + ... + a_mx^{m-1} \qquad \text{(with } a_m \neq 0) \quad ...(1)$$
$$= f(x)$$

If the data contains n sets of x and y values, then the sum of squares of the errors is given by

$$I = \sum_{i=1}^{n} [y_i - f(x_i)]^2 \qquad\qquad ...(2)$$

Now, since $f(x)$ is a polynomial and contains coefficients $a_0, a_1, ..., a_m$. Then we want to find all the coefficients.

Using the principle of maxima and minima of I, we must have, from (2)

$$\frac{\partial I}{\partial a_1} = 0 \qquad\qquad \Rightarrow \qquad\qquad -2\sum_{i=1}^{n} [y_i - f(x_i)]\frac{\partial f(x_i)}{\partial a_1} = 0$$

$$\frac{\partial I}{\partial a_2} = 0 \qquad\qquad \Rightarrow \qquad\qquad -2\sum_{i=1}^{n} [y_i - f(x_i)]\frac{\partial f(x_i)}{\partial a_2} = 0$$

....

....

$$\frac{\partial I}{\partial a_m} = 0 \qquad\qquad \Rightarrow \qquad\qquad -2\sum_{i=1}^{n} [y_i - f(x_i)]\frac{\partial f(x_i)}{\partial a_m} = 0$$

which gives the general term

$$\frac{\partial I}{\partial a_j} = -2\sum_{i=1}^{N} [y_i - f(x_i)]\frac{\partial f(x_i)}{\partial a_j} = 0$$

$$\frac{\partial f(x_i)}{\partial a_j} = x_1^{j-1}$$

Therefore, we have $\sum_{i=1}^{n} [y_i - f(x_i)]x_i^{j-1} \qquad j = 1, 2, ..., m.$

$$\sum_{i=1}^{n} [y_ix_i^{j-1} - x_i^{j-1}f(x_i)] = 0$$

$\Rightarrow$ $\sum_{i=1}^{n} [x_i^{j-1}(a_1 + a_2x_i + a_3x_i^2 + ... + a_mx_i^{m-1})] = \sum_{i=1}^{n} y_ix_i^{j-1}$

[By substituting the values of $f(x)$]

Then, we get the following normal equations :

$$\Sigma y_i = na_1 + a_2\Sigma x_i + a_3\Sigma x_i^2 + ... + a_m\Sigma x_i^{m-1}$$

$$\Sigma x_i y_i = a_1\Sigma x_i + a_2\Sigma x_i^2 + a_3\Sigma x_i^3 + ... + a_m\Sigma x_i^m$$

$$...$$

$$...$$

$$\Sigma y_i x_i^{m-1} = a_1\Sigma x_i^{m-1} + a_2\Sigma x_i^m + a_3\Sigma x_i^{m+1} + ... + a_m\Sigma x_i^{2m-2}$$

The above set of normal equation can be written as in matrix form as follows :

$$CA = B$$

where

$$C = \begin{bmatrix} n & \Sigma x_i & \Sigma x_i^2 & \cdots & \Sigma x_i^{m-1} \\ \Sigma x_i & \Sigma x_i^2 & \Sigma x_i^3 & \cdots & \Sigma x_i^m \\ \cdots & \cdots & \cdots & \cdots & \cdots \\ \Sigma x_i^{m-1} & \Sigma x_i^m & \cdots & \cdots & \Sigma x_i^{2m-2} \end{bmatrix}$$

$$A = \begin{bmatrix} a_1 \\ a_2 \\ \vdots \\ \vdots \\ a_m \end{bmatrix} \quad \text{and } B = \begin{bmatrix} \Sigma y_i \\ \Sigma y_i x_i \\ \Sigma y_i x_i^2 \\ \vdots \\ \Sigma y_i x_i^{m-1} \end{bmatrix}$$

Here, the element of the matrix C is

$$C[j,k] = \sum_{i=1}^{n} x_i^{j+k-2} \qquad\qquad j = 1, 2, ..., m$$

Similarly,

$$B(j) = \sum_{j=1}^{n} y_i x_i^{j-1} \qquad\qquad k = 1, 2, ..., m$$

ALGORITHM

Step 1.	Read the number of data points n and the order of polynomial mp
Step 2.	Take data values
Step 3.	Check, If $n \leq mp$ then, the regression is not possible
Step 4.	Set $m = mp + 1$
Step 5.	Compute the coefficients of C and B matrices
Step 6.	Solve for $a_1, a_2, ..., a_m$
Step 7.	Write $a_1, a_2, ..., a_m$
Step 8.	Estimate the function value at the given value of independent variable.

SOLVED EXAMPLE

EXAMPLE . *Fit a second degree polynomial to the following data :*

x	1.0	2.0	3.0	4.0
y	6.0	11.0	18.0	27.0

SOLUTION. Since, we want to fit a second order polynomials therefore the order of polynomial is 2. So, we will have three simultaneous normal equation as given below :

$$a_1 x + a_2 \Sigma x_i + a_3 \Sigma x_i^2 = \Sigma y_i \qquad \ldots(1)$$

$$a_1 \Sigma x_i + a_2 \Sigma x_i^2 + a_3 \Sigma x_i^3 = \Sigma y_i x_i \qquad \ldots(2)$$

$$a_1 \Sigma x_i^2 + a_2 \Sigma x_i^3 + a_3 \Sigma x_i^4 = \Sigma y_i x_i^2 \qquad \ldots(3)$$

Then, we have the following table.

x	y	x^2	x^3	x^4	yx	yx^2
1	6	1	1	1	6	6
2	11	4	8	16	22	44
3	18	9	27	81	54	162
4	27	16	64	256	108	432
10	**62**	**30**	**100**	**354**	**190**	**644**

Putting all these values in (1), (2) and (3), we get

$$4a_1 + 10a_2 + 30a_3 = 62 \qquad \ldots(4)$$

$$10a_1 + 30a_2 + 100a_3 = 190 \qquad \ldots(5)$$

$$30a_1 + 100a_2 + 354a_3 = 644 \qquad \ldots(6)$$

Solving (4), (5) and (6) for a_1, a_2 and a_3, we get

$$a_1 = 3;\ a_2 = 2;\ a_3 = 1$$

Hence, the least square quadratic polynomial is

$$y = 3 + 2x + x^2.$$

11.14 NON-LINEAR REGRESSION

11.14.1 FITTING TRANSCENDENTAL EQUATIONS

Since the relationship between the dependent and independent variables is not always linear. The non-linear relationship between them may exist in the form of transcendental equations. For example (1), consider the equation for population growth given by

$$P = p_0 e^{kt} \qquad \ldots(1)$$

where p_0 is the initial value of the population P, k is the growth rate and 't' is the time.

(2) Consider the equation of gas law relating to the pressure and volume, given by

$$P = av^b$$

Let I be sum of the square of the derivation, then we have

$$I = \sum_{i=1}^{n} [p_i - av_i^b]^2$$

To minimize I, we must have

$$\frac{\partial I}{\partial a} = 0 \ \text{ and } \ \frac{\partial I}{\partial b} = 0$$

which gives

$$\Sigma p_1 v_i^b = a\Sigma (v_i^b)^2$$

$$\Sigma p_1 v_i^b \log v_i = a\Sigma (v_i^b)^2 \log v_i$$

Solve the above equation for a and b. Now, since b appears under the summation sign, use iterative technique to solve for a and b.

We get

$$b = \frac{n\Sigma \log v_i \log p_i - \Sigma \log v_i \Sigma \log p_i}{n\Sigma (\log v_i)^2 - (\Sigma \log p_i)^2}$$

$$a = \text{antilog}\left(\frac{1}{n}\Sigma \log p_i - b'\Sigma \log x_i\right)$$

SOLVED EXAMPLES

EXAMPLE 1. *Fit a power function model of the form* $y = ax^b$ *using the following data :*

x	1	2	3	4	5
y	0.5	2	4.5	8	12.5

SOLUTION. Since, given curve is

$$y = ax^b \qquad \Rightarrow \quad \log y = \log a + b \log x$$

Now, we have the following table :

x_i	y_i	$\log x_i$	$\log y_i$	$(\log x_i)^2$	$\log(x_i)\log(y_i)$
1	0.5	0	−0.6931	0	0
2	2	0.6391	0.6931	0.4805	0.4804
3	4.5	1.0986	1.5041	1.2069	1.6524
4	8	1.3863	2.0794	1.9218	2.8827
5	12.5	1.6094	2.5257	2.5903	4.0649
Total = 10	**27.5**	**4.7874**	**6.1092**	**6.1995**	**9.0804**

Therefore,

$$b = \frac{n\Sigma \log x_i \log y_i - \Sigma \log x_i \log y_i}{n\Sigma(\log x_i)^2 - (\Sigma \log x_i)^2}$$

$$= \frac{5 \times 9.0804 - 4.7874 \times 6.1092}{5 \times 6.1995 - (4.7874)^2}$$

$$= \frac{45.402 - 29.2472}{30.9975 - 22.9192} = 1.9998$$

Also

$$\log a = \frac{6.1092 - (1.9998)(4.7847)}{5} = -0.6929$$

$$\Rightarrow \qquad a = 0.5001$$

Hence, the fitted power function equation is given by

$$y = 0.5001x^{1.9998}$$

AN IMPORTANT DISCUSSION

If the data seems to fit a curve other than a straight line we can use same transformations to obtain a linear regression. We summarise in the table below some transformation which you can carry out to fit a given set of data of a straight line.

S. No.	Form of function	Transformation	Form of the linear regression equation
1.	Exponential : $y = ce^{bx}$	$y^* = \log y$	$y^* = a + bx$
2.	Power : $y = cx^d$	$y^* = \log y,\ x^* = \log x$	$y^* = a + bx^*$
3.	Reciprocal : $y = a + \dfrac{b}{x}$	$x^* = \dfrac{1}{x}$	$y = a + bx^*$
4.	Hyperbolic function : $\dfrac{x}{c + dx}$	$y^* = \dfrac{1}{y},\ x^* = \dfrac{1}{x}$	$y^* = a + bx^*$

EXAMPLE 2. *Given the following data :*

x	1	5	3	2	1	1	7	3
y	6	1	0	0	1	2	1	5

(i) Fit a regression line of y on x.   (ii) Fit a regression line of x on y.

SOLUTION.

x	y	x^2	y^2	xy
1	6	1	36	6
5	1	25	1	5
3	0	9	0	0
2	0	4	0	0
1	1	1	1	1
1	2	1	4	2
7	1	49	1	7
3	5	9	25	15
23	**16**	**99**	**68**	**36**

(i) Regression equation of y on x :

The regression equation of y on x is given by

$$y = a + bx$$

where the values of a and b are given by the normal equation

$$\Sigma y = na + b\Sigma x$$
$$\Sigma xy = a\Sigma x + b\Sigma x^2$$

$\Rightarrow \quad 16 = 8a + 23b$ ...(1)

$\quad\quad 36 = 23a + 99b$ ...(2)

Solving (1) and (2) for a and b, we get

$$a = 2.874$$
$$b = -0.304$$

Hence, the regression equation of y on x is

$$y = 2.874 - (0.304)x$$

(ii) Regression equation of x on y is given by

$\quad\quad x = c + by$ ...(1)

when c and d are given by normal equations

$$\Sigma x = nc + b\Sigma y$$
$$\Sigma xy = c\Sigma y + b\Sigma y^2$$

$\Rightarrow \quad 23 = 8c + 16b$ ...(2)

and $36 = 16c + 68b$ ...(3)

Now, solving (2) and (3) for b and c, we get

$$c = 3.431 \text{ and } \quad b = -0.278$$

Therefore, the required regression equation of x on y is given by

$$x = 3.431 - 0.278y$$

EXAMPLE 3. *Obtain regression line of x on y for the given data :*

x	1	2	3	4	5	6
y	5.0	8.1	10.6	13.1	16.2	20.0

SOLUTION.   Regression line of x on y for the data :

x	1	2	3	4	5	6
y	5.0	8.1	10.6	13.1	16.2	20.0

The normal for obtaining 'a' and 'b' are

$$\Sigma x_i = na + b\Sigma y_i$$
$$\Sigma x_i y_i = a\Sigma y_i + b\Sigma y_i^2$$

Substituting the value by given table

x	y	xy	y^2
1	5	5	25.00
2	8.1	16.2	65.61
3	10.6	31.8	112.36
4	13.1	52.4	171.61
5	16.2	66.0	262.44
6	20.0	60.0	400.00
21	**73.0**	**231.4**	**1037.02**

Substituting these values in required normal equations, we get

$$21 = 6a + 73b$$
$$231.4 = 73a + 1037b$$

11.15   SIMPLIFIED DETERMINATION OF REGRESSION ANALYSIS

Since we know that the regression line can be written as :

$$y - \bar{y} = b_{yx}(x - \bar{x}) \text{ ; for regression of } y \text{ on } x \text{ and} \qquad \text{... (1)}$$
$$x - \bar{x} = b_{xy}(y - \bar{y}) \text{ ; for regression of } x \text{ on } y. \qquad \text{... (2)}$$

This form leads to the following simplified equations :

$$b_{yx} = \dfrac{\dfrac{\Sigma(x - \bar{x})(y - \bar{y})}{n}}{\dfrac{\Sigma(x - \bar{x})^2}{n}}$$

SOLVED EXAMPLE

EXAMPLE .   *A panel of two judges P and Q graded seven dramatic performance by independently awarding marks as follows :*

Performance	1	2	3	4	5	6	7
Marks by P	46	42	44	40	43	41	45
Marks by Q	40	38	36	35	39	37	41

The eight performance, which judge Q would not attend, was awarded 37 marks by judge P. If judge Q had been present, how many marks would be expected to have been awarded by him to the eighth performance.

SOLUTION.   Here, we first calculate the means :

Mean marks by P(x variable) $= \bar{x} = \dfrac{301}{7} = 43$

Mean marks by Q(y variable) $= \bar{y} = \dfrac{266}{7} = 38$

To calculate b_{yx}, we have the following table

No.	x	$(x - \bar{x})$	$(x - \bar{x})^2$	y	$(y - \bar{y})$	$(y - \bar{y})(x - \bar{x})$
1	46	+3	9	40	+2	+6
2	42	−1	1	38	0	0
3	44	+1	1	36	−2	−2
4	40	−3	9	35	−3	+9
5	43	0	0	39	+1	0
6	41	−2	4	37	−1	+2
7	45	+2	4	41	+3	+6
Total	301		28	266		21

Now, we have

$$b_{yx} = \frac{\Sigma(x - \bar{x})(y - \bar{y})}{\Sigma(x - \bar{x})^2} = \frac{21}{28} = 0.75$$

Therefore, the regression line is

$$y - 38 = 0.75(x - 43)$$

$\Rightarrow$ $\qquad\qquad\qquad y = 0.75x + 5.75$

If $x = 37$, then we have

$$y = (0.75) \times (37) + 5.75 = 33.5$$

$\Rightarrow$ Judge Q was likely to award 33.5 marks.

11.16 REGRESSION ANALYSIS OF GROUPED DATA

The regression analysis of the grouped data 4 will follow the same procedure as above, except the fact that all items falling with in a specified group are approximated as having a value equal to the mid-point value of the group. Since the grouping may be on both variables x and y, the data is usually organized in a two way matrix. The following formula may be used for the value of b_{yx}.

$$b_{yx} = \left[\frac{\left(\dfrac{\Sigma fxy}{n} - \dfrac{\Sigma fx}{n} \cdot \dfrac{\Sigma fy}{n} \right)}{\left(\dfrac{\Sigma fx^2}{n} - \left(\dfrac{\Sigma fx}{n} \right)^2 \right)} \right]$$

11.17 MULTIPLE LINEAR REGRESSION

If the dependent variable is a function of two or more variables, then we cannot fit the regression line by using our usual methods.

For example, the salary of a salesman may be expressed as

$$y = 500 + 5x_1 + 8x_2$$

where x_1 and x_2 are the number of units sold of product 1 and 2 respectively.

Now, we shall discuss an approach to fit the experimental data where the variable under consideration is a linear function of two independent variables.

Consider a two-variable linear function

$$y = a_1 + a_2 x + a_3 z \qquad \text{... (1)}$$

Let I denote the sum of the squares of errors, then I is given by

$$I = \sum_{i=1}^{n} (y_i - a_1 - a_2 x_i - a_3 z_i)^2 \qquad \text{... (2)}$$

Differentiating (2) with respect to a_1, a_2 and a_3, we get

$$\frac{\partial I}{\partial a_1} = -2\Sigma(y_i - a_1 - a_2 x_i - a_3 z_i) \qquad \text{... (3)}$$

$$\frac{\partial I}{\partial a_2} = -2\Sigma(y_i - a_1 - a_2 x_i - a_3 z_i)\, x_i \qquad \text{... (4)}$$

and $$\frac{\partial I}{\partial a_3} = -2\Sigma(y_i - a_1 - a_2 x_i - a_3 z_i)\, y_i \qquad \text{... (5)}$$

By the principle of maxima and minima of I, we must have

$$\frac{\partial I}{\partial a_1} = 0, \quad \frac{\partial I}{\partial a_2} = 0 \text{ and } \frac{\partial I}{\partial a_3} = 0.$$

Thus, (3), (4) and (5) gives

$$\left.\begin{array}{l} \Sigma y_i = n a_1 + a_2 \Sigma x_i + a_3 \Sigma z_i \\ \Sigma y_i x_i = a_1 \Sigma x_i + a_2 \Sigma x_i^2 + a_3 \Sigma x_i z_i \\ \Sigma y_i z_i = a_i \Sigma z_i + a_2 \Sigma x_i z_i + a_3 \Sigma z_i^2 \end{array}\right\} \qquad \text{... (6)}$$

The system (6) can also be written in the matrix form as follows :

$$\begin{bmatrix} n & \Sigma x_i & \Sigma z_i \\ \Sigma x_i & \Sigma x_i^2 & \Sigma x_i z_i \\ \Sigma z_i & \Sigma x_i z_i & \Sigma z_i^2 \end{bmatrix} \begin{bmatrix} a_1 \\ a_2 \\ a_3 \end{bmatrix} = \begin{bmatrix} \Sigma y_i \\ \Sigma y_i x_i \\ \Sigma y_i z_i \end{bmatrix}$$

The above equation can be solved using any standard method.

▶ **REMARKS**

▸ Since, this is a two-dimensional case, therefore, we obtain a regression plane rather than regression line.

▸ This case can be easily extended to the more general case

$$y = a_1 + a_2 x_1 + a_3 x_2 + \dots + a_{n+1} x_m$$

SOLVED EXAMPLES

EXAMPLE 1. *Obtain a regression plane to fit the data, using the following table*

x	1	2	3	4
y	0	1	2	3
z	12	18	24	30

SOLUTION. Using the above table, we may get

x	z	y	x^2	z^2	xz	yx	yz
1	0	12	1	0	0	12	0
2	1	18	4	1	2	36	18
3	2	24	9	4	6	72	48
4	3	30	16	9	12	120	90
10	6	84	30	14	20	240	156

Now putting all these values in the following normal equation :

$$\begin{bmatrix} n & \Sigma x_i & \Sigma z_i \\ \Sigma x_i & \Sigma x_i^2 & \Sigma x_i z_i \\ \Sigma z_i & \Sigma x_i z_i & \Sigma z_i^2 \end{bmatrix} \begin{bmatrix} a_1 \\ a_2 \\ a_3 \end{bmatrix} = \begin{bmatrix} \Sigma y_i \\ \Sigma y_i x_i \\ \Sigma y_i z_i \end{bmatrix}$$

we get
$$4a_1 + 10a_2 + 5a_3 = 84$$
$$10a_1 + 30a_2 + 20a_3 = 240$$
$$6a_1 + 20a_2 + 14a_3 = 156$$

Solving these equations for a_1, a_2 and a_3 , we get
$$a_1 = 10, a_2 = 2, \ a_3 = 4$$
Hence, the required regression plane is
$$y = 10 + 2x + 4z .$$

EXAMPLE 2. *In a partially destroyed laboratory records of an analysis of correlation data, the following results only are legible*

Variance of x is 9. Regression equation: $8x - 10y + 66 = 0$ *and* $40x - 18y = 214$.

Calculate :
(a) The mean value of x and y   (b) The standard deviation of y
(c) The coefficient of correlation between x and y. (Meerut-2010, 14)
(UPTU(MCA)-2008, 11, 14)

SOLUTION. We know that two regression lines pass through the point $(\bar{x}, \bar{y})$ where $\bar{x}$ and $\bar{y}$ are mean of x and y respectively

(a) To find $\bar{x}$ and $\bar{y}$, solve the equation $\bar{x} = 13, \bar{y} = 17$

$$8\bar{x} - 10\bar{y} + 66 = 0 \qquad \text{... (1)}$$
and $\qquad 40\bar{x} - 18\bar{y} - 214 = 0 \qquad$ ... (2)

Solving (1) and (2), we get
Hence, the values of x and y are respectively 13 and 17.

(b) and (c) The equations of the two line regression lines can also be put in the following form
$$y = 0.80x + 6.60, \ x = 0.45y + 5.35$$
The regression coefficient of y on x is b_{yx} and is given by

$$b_{yx} = r\frac{\sigma_y}{\sigma_x} = 0.80 \qquad \text{... (3)}$$

and the regression coefficient of x on y is b_{xy} and is given by

$$b_{xy} = r \frac{\sigma_x}{\sigma_y} = 0.45 \qquad \text{... (4)}$$

From (3) and (4), we get

$$r^2 = 0.80 \times 0.45 = 0.36$$

$$\Rightarrow \qquad r = 0.60$$

∴ The coefficient of correlation between x and y is 0.60.

Also, since variance (σ_x^2) of x is given as 9. Therefore, S.D. = 3.

Putting these values in (4), we get $\sigma_y = 4$.

EXERCISE 11.4

1. The following table gives the various values of two variables :

x	42	44	58	55	89	98	66
y	56	49	53	58	64	76	58

Determine the regression equations which may be associated with these values.

2. Obtain the line of regression of the following data:

x	1	2	3	4	5	6	7
y	9	8	10	12	11	13	14

3. Find the regression lines using the following data:Mean height = 50.07, Mean age = 9.98, Standard deviation for height = 5.26, Standard deviation of age = 2.59 and $r = 0.898$.

4. Find the mean of the variables x and y and the coefficients of correlation, given the following regression equations $2y - x = 50$, $3y - 2x = 10$.

5. Two lines of regression are given by $x + 2y - 5 = 0$ and $2x + 3y - 8 = 0$ and $\sigma_x^2 = 12$. Calculate the mean value of x and y, variance of y and the coefficient of correlation between x and y.

6. Use multiple linear regression to fit :

x_1	1	2	3	4	5
x_2	4	3	2	1	0
y	18	16	16	12	10

Compute coefficients and the error of estimate.

7. Obtain a regression plane to fit the following data:

x_1	5	4	3	2	1
x_2	3	− 2	− 1	4	0
y	15	− 8	− 1	26	8

8. The mean of bivariate frequency distribution are at (3, 4) and $r = 0.4$. The line of regression of y on x is parallel to the line $y = x$. Find the two lines of regression and estimate value of x when $y = 1$.

9. The following table gives the results of the measurements of train resistance, V is the velocity in miles per hour, R is the resistance in pounds per ton

V	20	40	60	80	100	120
R	5.5	9.1	14.9	22.8	33.3	46

If R is related to V by the relation $R = a + bV + cV^2$, find a, b and c by using the method of least square.

10. Fit a second degree parabola to the following data, taking y as dependent variable

x	1	2	3	4	5	6	7	8	9
y	2	6	7	8	10	11	11	10	9

11. Find the multiple linear regression of x_1 on x_2 and x_3 using the following data:

x_1	4	6	7	9	23	15
x_2	15	12	8	6	4	3
x_3	30	24	20	14	10	4

■ ■ □   **Answers**   ■ □ □

1. $y = 0.372x + 35.27$, $x = 2.2y - 65.9$ **2.** $Y = 0.929x + 7.284$, $X = 0.929Y - 6.219$

3. $y = 0.422x - 12.15$, $x = 1.825y + 31.86$ **4.** Mean of $X = 130$, Mean of $Y = 90$, $r = 0.866$

5. $1, 2, \sigma_y^2 = 6, r = 0.86$ **8.** $y = x + 1$, $x = 0.16y + 2.36$, $x = 2.52$

9. $R = 4.35 + 0.00241V + 0.0028705V^2$ **10.** $y = -1 + 3.55x - 0.27x^2$

11.18 DATA APPROXIMATION OF FUNCTIONS

The problem of approximation of a function is an important problem in numerical analysis. The commonly used classes of functions are polynomials, trigonometric functions, exponential functions and rational functions. The polynomial functions are most widely used in many applications.

11.18.1 WEIRSTRESS THEOREM

Let a function $f(x)$ is continuous on an interval $[a, b]$, then for a given $\varepsilon > 0$, $\exists$ a $m > 0$ and a polynomial $P(x)$ of degree n such that $|f(x) - P(x)| < \varepsilon$, $\forall x \in [a, b]$

Let us assume an expression of the form

$$f(x) = P(x, c_0, c_1, ..., c_n)$$
$$= c_0g_0(x) + c_1g_1(x) + ... + c_ng_n(x) \quad\quad ...(1)$$

where $g_i(x)$, $i = 0, 1, 2, ..., n$ are linearly independent functions and c_i, $i = 0, 1, 2, ..., n$ are the parameters, which are to be determined.

Here the error of approximation is defined as

$$E(f, c) = \|f(x) - [c_0g_0(x) + c_1g_1(x) + ... + c_ng_n(x)]\| \quad\quad ...(2)$$

where $\|\cdot\|$ denote the norm of the function.

The problem of approximations is to determine c_i, $i = 0, 1, 2, ..., n$, such that this error is as small as possible.

▶ REMARKS

▸ The functions $g_i(x)$ are called coordinate functions and taken as $g_i(x) = x_i$, $i = 0, 1, 2, ..., n$ for polynomial approximation.

▸ The norm of the function $\|\cdot\|$, is the distance of the different points to the origin.

▸ The criterion or a norm which marks the error, smallest is called the best approximations.

▸ For the minimization of error norm (2) solve the problem of best approximation.

The Norm can be Classified as Follows :

(A) For Discrete data : Let $x = <x_i>$ be the sequence of real or complex numbers then we define

     (i) L^P Norm : as follow

$$\|x\| = \left[\sum_{i=1}^{n} |x|^P \right]^{1/p}$$

       $p \geq 1$, which can be written as $\|x\|_p$.

     (ii) Eucledian Norm : Take $p = 2$

$$\|x\| = \left[\sum_{i=1}^{n} |x_i|^2 \right]^{1/2}$$

The Eucledian norm is also known as square norm and written as $\|x\|_2$.

(iii) Uniform Norm : Take $p = 0$

$$\|x\| = \max_{1 \le j \le n} |x_j|$$

(B) For Continuous data : If the function $f(x)$ is continuous on $[a, b]$ and $|f(x)|^p$ is integrable on $[a, b]$, then

(i) L^P Norm

$$\|f\| = \left[\int_a^b |f(x)|^p \, dx \right]^{1/p}, p \ge 1.$$

(ii) Eucledian Norm : Take $p = 2$ we have the Eucledian norm.

$$\|f\| = \left[\int_a^b |f(x)|^2 \, dx \right]^{1/2}.$$

(iii) Uniform norm : Take $p = \infty$

We have, the uniform norm

$$\|f\| = \max_{a \le x \le b} |f(x)|.$$

▶ **REMARK**
▸ When we used the Eucledian norm, we obtained the least square approximation and when uniform norm is used, we obtained the uniform approximations.

11.19 TYPES OF APPROXIMATIONS

(i) Least Square Approximations : It is the most commonly used approximations for approximating of given function $f(x)$ in tabular form. To find the least square approximation, we always used the Eucledian norm. The approximation, for which the constant c_i, $i = 0, 1, 2, ..., n$ are determined in such a way that the aggregate of weight function $W(x)$ with E^2, i.e., $W(x)E^2$ is as small as possible, is known as best approximation in the least square sense.

Now, for discrete data, we have

$$I(c_0, c_1 ..., c_n) = \left[\sum_{k=0}^{n} W(x_k) - \sum_{i=0}^{n} c_i g_i(x_k) \right]^2$$

$$= \text{minimum} \qquad ...(1)$$

for continuous data, we have

$$I(c_0, c_1 ..., c_n) = \int_a^b W(x) \left[f(x) - \sum_{i=0}^{n} c_i g_i(x) \right]^2 dx$$

$$= \text{minimum} \qquad ...(2)$$

where, $\quad I(c_0, c_1, ..., c_n) = $ sum of the square of the errors or residuals.

For polynomial approximation, we choose

$$g_i(x) = x_i, i = 1, 2, ..., n \text{ and } W(x) = 1.$$

Now, the necessary conditions for (1) or (2) to have a minimum are that

$$\frac{\partial I}{\partial c_j} = 0, \quad j = 0, 1, 2, ..., n$$

which gives a system of $(n + 1)$ linear equations in $(n + 1)$ unknowns $c_0, c_1, ..., c_n$. These equations are called normal equations. The values of $c_0, c_1, ..., c_n$ can be obtained by solving these normal equations and substituted in the given equation.

SOLVED EXAMPLES

EXAMPLE 1.   *Obtain a linear polynomial approximations to the function* $f(x) = x^3$ *on the interval* [0, 1] *using the least square approximations with weight function* $W(x) = 1$. *Also find the linear polynomial approximation of the given function through the origin and draw the graph of the given function and the two linear polynomial approximation.*

SOLUTION.   Let us consider a linear polynomial approximation

$$P(x) = a_0 + a_1 x \qquad \qquad ...(1)$$

where a_0 and a_1 are the arbitrary parameters.

Now, the sum of the squares of the error is given by

$$I[a_0, a_1] = \int_0^1 [x^3 - (a_0 + a_1 x)]^2 dx$$

$$= \int_0^1 [x^6 - 2(a_0 x^3 + a_1 x^4) + a_0^2 + 2a_0 a_1 x + a_1^2 x^2] dx$$

$$= \left[\frac{x^7}{7} - 2\left(a_0 \cdot \frac{x^4}{4} + a_1 \frac{x^5}{5} \right) + a_0^2 x + a_0 a_1 x^2 + a_1^2 \frac{x^3}{3} \right]_0^1$$

$$= \frac{1}{7} - 2\left(\frac{a_0}{4} + \frac{a_1}{5} \right) + a_0^2 + a_0 a_1 + \frac{a_1^2}{3}$$

For the minima of $I(a_0, a_1)$, we must have

$$\frac{\partial I}{\partial a_0} = 0 \Rightarrow -\frac{1}{2} + 2a_0 + a_1 = 0$$

and

$$\frac{\partial I}{\partial a_1} = 0 \Rightarrow -\frac{2}{5} + a_0 + \frac{2}{3}a_1 = 0$$

$$\Rightarrow \quad a_0 = -\frac{1}{5}$$

$$a_1 = \frac{9}{10}$$

Put these values in (1), the required linear polynomial approximation is

$$P(x) = -\frac{1}{5} + \frac{9}{10} x = \left(\frac{9x - 2}{10} \right)$$

Also, the value of $I(a_0, a_1) = \frac{9}{700}$

Now, if we take the linear polynomial approximation through the origin, then

$$P(x) = ax.$$

We have $I(a)$ = sum of the squares of the deviations or errors

$$= \int_0^1 (x^3 - ax)^2 dx$$

$$= \int_0^1 (x^6 - 2ax^4 + a^2 x^2) dx = \left[\frac{x^7}{7} - \frac{2ax^5}{5} + \frac{a^2 x^3}{3} \right]_0^1$$

$$= \frac{1}{7} - \frac{2a}{5} + \frac{a^2}{3}$$

For the minima of I, we must have

$$\frac{dI}{da} = 0$$

$$\Rightarrow \qquad -\frac{2}{5} + \frac{2}{3}a = 0$$

$$\Rightarrow \qquad a = \frac{3}{5}.$$

Hence, the required linear polynomial approximation through the origin is given by

$$P(x) = \frac{3}{5}x \text{ and } I(a) = \frac{1}{7} - \frac{2}{5} \cdot \frac{3}{5} + \frac{1}{3} \cdot \frac{9}{25} = \frac{16}{700}.$$

Now, we plot the approximation as in fig. 11.10.

It may be noted that the approximating polynomials $P(x)$ may or may not have common values with $f(x)$.

EXAMPLE 2. *Obtain an approximation in the sense of the principle of least squares in the form of a polynomial of second degree to the function $f(x) = \dfrac{1}{1+x^2}$ in the range $-1 \le x \le 1$.*

SOLUTION. Let the required approximation be
$$P(x) = a_0 + a_1 x + a_2 x^2$$
where a_0, a_1, a_2 are arbitrary parameters.

Now, we have $I(a_0, a_1, a_2)$ = sum of the squares of the errors

$$= \int_{-1}^{1} \left[\frac{1}{1+x^2} - a_0 - a_1 x - a_2 x^2 \right]^2 dx$$

For the minima of I, we must have

$$\frac{\partial I}{\partial a_0} = 0$$

$$\Rightarrow \qquad \int_{-1}^{1} 2\left(\frac{1}{1+x^2} - a_0 - a_1 x_1 - a_2 x^2 \right)(-1)dx = 0$$

$$\Rightarrow \qquad \int_{-1}^{1} \left(\frac{1}{1+x^2} - a_0 - a_1 x - a_2 x^2 \right)dx = 0$$

$$\Rightarrow \qquad 2[\tan^{-1} x]_0^1 - a_0 (x)_{-1}^1 - 2a_2 \left[\frac{x^3}{3} \right]_0^1 = 0$$

$$\Rightarrow \qquad \frac{\pi}{2} - 2a_0 - \frac{2}{3}a_2 = 0 \qquad \qquad \text{...(1)}$$

Now, $\dfrac{\partial I}{\partial a_1} = 0 \Rightarrow \int_{-1}^{1} 2\left(\dfrac{1}{1+x^2} - a_0 - a_1 x - a_2 x^2 \right)(-x)dx = 0$

$$\Rightarrow \qquad -2a_1 \left[\frac{x^3}{3} \right]_0^1 = 0 \Rightarrow -\frac{2a_1}{3} = 0$$

$$\Rightarrow \qquad a_1 = 0 \qquad \qquad \text{...(2)}$$

Now, $\dfrac{\partial I}{\partial a_2} = 0 \Rightarrow \int_{-1}^{1} 2\left(\dfrac{1}{1+x^2} - a_0 - a_1 x - a_2 x^2 \right)(-x^2)dx = 0$

$$\Rightarrow \qquad \int_{-1}^{1} \left(\frac{x^2}{1+x^2} - a_0 x^2 - a_1 x^3 - a_2 x^4 \right)dx = 0$$

$$\Rightarrow \quad \int_{-1}^{1}\left(1-\frac{1}{1+x^2}-a_0 x^2 - a_1 x^3 - a_2 x^4\right)dx = 0$$

$$\Rightarrow \quad [x]_{-1}^{1} - 2[\tan^{-1} x]_0^1 - 2a_0\left[\frac{x^3}{3}\right]_0^1 - 2a_2\left[\frac{x^5}{5}\right]_0^1 = 0$$

$$\Rightarrow \quad 2 - \frac{\pi}{2} - \frac{2a_0}{3} - \frac{2a_2}{5} = 0 \qquad \qquad \text{...(3)}$$

Solving (1), (2), (3) for a_0, a_1 and a_2, we get

$$a_0 = \frac{3}{4}(2\pi - 5), \quad a_1 = 0, \quad a_2 = \frac{15}{4}(3 - \pi)$$

Hence, the required polynomial approximation is given by

$$P(x) = \frac{3}{4}(2\pi - 5) + \frac{15}{4}(3 - \pi)x^2$$

EXAMPLE 3. *Obtain the least squares polynomials of degree one and two for the function $f(x) = x^{1/2}$ on [0, 1].*

SOLUTION. Let us first suppose the least square polynomial approximation of degree 1 be
$$P(x) = a_0 + a_1 x$$
where a_0, a_1 and arbitrary parameters.
Now, we have

$$I(a_0, a_1) = \text{sum of the squares of the derivations or errors}$$
$$= \int_0^1 [x^{1/2} - a_0 - a_1 x]^2 dx$$
$$= \text{minimum}$$

For the minima of I, we must have

$$\frac{\partial I}{\partial a_0} = 0 \Rightarrow \int_0^1 2(x^{1/2} - a_0 - a_1 x)(-1)dx = 0 \quad \Rightarrow \int_0^1 (x^{1/2} - a_0 - a_1 x)(-1)dx = 0$$

$$\Rightarrow \left[\frac{2}{3}x^{3/2} - a_0 x - a_1 \frac{x^2}{2}\right]_0^1 = 0 \quad \Rightarrow \frac{2}{3} - a_0 - \frac{a_1}{2} = 0 \qquad \text{...(1)}$$

and $\frac{\partial I}{\partial a_1} = 0 \Rightarrow \int_0^1 2(x^{1/2} - a_0 - a_1 x)(-x)dx = 0 \quad \Rightarrow \int_0^1 (x^{1/2} - a_0 x - a_1 x^2)dx = 0$

$$\Rightarrow \left[\frac{2}{5}x^{5/2} - a_0 \frac{x^2}{2} - a_1 \frac{x^3}{3}\right]_0^1 = 0 \quad \Rightarrow \frac{2}{5} - \frac{a_0}{2} - \frac{a_1}{3} = 0 \qquad \text{...(2)}$$

Solving (1) and (2) for a_0 and a_1 we get

$$a_0 = \frac{4}{15} \quad \text{and} \quad a_1 = \frac{4}{5}$$

Hence, the required first degree least square approximations for $x^{1/2}$ on [0, 1] is

$$P(x) = \frac{4}{15} + \frac{4}{5}x = \frac{4}{15}(1 + 3x)$$

Now, we want to find the least square polynomial of approximation of degree 2.
Let us define

$$P(x) = a_0 + a_1 x + a_2 x^2 \qquad \text{(a polynomial of degree 2)}$$

where a_0, a_1, a_2 are arbitrary parameters.

Now, we have

$$I[a_0, a_1, a_2] = \text{sum of the squares of the deviation}$$
$$= \int_0^1 [x^{1/2} - a_0 - a_1 x - a_2 x^2]^2 dx$$
$$= \text{minimum}$$

For the minima of I, we must have

$$\frac{\partial I}{\partial a_0} = 0 \Rightarrow \int_0^1 [x^{1/2} - a_0 - a_1 x - a_2 x^2] dx = 0$$

$$\Rightarrow \frac{2}{3} - a_0 - \frac{a_1}{2} - \frac{a_2}{3} = 0$$

$$\Rightarrow a_0 + \frac{a_1}{2} + \frac{a_2}{3} = \frac{2}{3} \qquad \qquad \text{...(1)}$$

Also,

$$\frac{\partial I}{\partial a_1} = 0 \Rightarrow \int_0^1 [x^{1/2} - a_0 - a_1 x - a_2 x^2](-x) dx = 0$$

$$\Rightarrow \int_0^1 [x^{3/2} - a_0 x - a_1 x^2 - a_2 x^3] dx = 0$$

$$\Rightarrow \frac{2}{5} - \frac{a_0}{2} - \frac{a_1}{3} - \frac{a_2}{4} = 0$$

$$\Rightarrow \frac{a_0}{2} + \frac{a_1}{3} + \frac{a_2}{4} = \frac{2}{5} \qquad \qquad \text{...(2)}$$

and

$$\frac{\partial I}{\partial a_2} = 0 \Rightarrow \int_0^1 2[x^{1/2} - a_0 - a_1 x - a_2 x^2](-x^2) dx = 0$$

$$\Rightarrow \int_0^1 [x^{5/2} - a_0 x^2 - a_1 x^3 - a_2 x^4] dx = 0$$

$$\Rightarrow \frac{2}{7} - \frac{a_0}{3} - \frac{a_1}{4} - \frac{a_2}{5} = 0$$

$$\Rightarrow \frac{a_0}{3} + \frac{a_1}{4} + \frac{a_2}{3} = \frac{2}{7} \qquad \qquad \text{...(3)}$$

Solving (1), (2) and (3) for a_0, a_1 and a_2 we get

$$a_0 = \frac{6}{35}, a_1 = \frac{48}{35}, a_2 = -\frac{20}{35}$$

Hence, the required second degree least square approximation to $x^{1/2}$ on $(0, 1)$ is

$$P(x) = \frac{6}{35} + \frac{48}{35} x - \frac{20}{35} x^2$$

11.20 USE OF ORTHOGONAL FUNCTIONS

For the large n, the normal equation of the curve becomes ill-conditioned, which gives large amount of error in the parameters $a_1, a_2, ..., a_n$. This difficulties can be avoided if the coordinate function $g(x)$ be chosen such that they are orthogonal with respect to the weight function $W(x)$ on an interval (a, b).

11.20. 1 ORTHOGONAL FUNCTIONS

(i) **For Discrete Data:** A set of real functions $\{g_i(x)\}$ is said to be orthogonal over a set of points $x_1, x_2, ..., x_n$ with respect to the weight function $W(x)$ if

$$\sum_{k=1}^{m} W(x_k) g_i(x_k) g_j(x_k) = 0, \text{ whenever } i \neq j$$

(ii) For Continuous Data: A set of real function $\{g_i(x)\}$ is said to be orthogonal on an interval $[a, b]$ with respect to the weight function $W(x)$ if

$$\int_a^b W(x)g_i(x)g_j(x)dx = 0 \text{, whenever } i \neq j$$

If the functions $g_i(x)$, $i = 0, 1, ..., n$ is orthogonal, then we get

$$\sum_{k=1}^m W(x_i)f(x_k)g_i(x_k) = \sum_{k=1}^m a_i W(x_k)g_i^2(x_k)$$

$$\Rightarrow \qquad a_i = \frac{\displaystyle\sum_{k=1}^m W(x_k)g_i(x_k)f(x_k)}{\displaystyle\sum_{k=1}^m W(x_k)g_i^2(x_k)}, \; i = 0, 1, 2, ..., n$$

Similarly, for continuous data, we have

$$a_i = \frac{\int_a^b W(x_k)g_i(x_k)f(x_k)dx}{\int_a^b W(x_k)g_i^2(x_k)dx}, \; i = 0, 1, 2, ..., n$$

▶ **REMARKS**

▶ The use of orthogonal functions as coordinate functions not only avoids the problem of ill-conditioning in normal equations but also determines the constant a_i, $i = 0, 1, ..., n$ directly.

▶ Every set of linearly independent polynomials satisfies the condition of orthogonality. Sometimes it can be orthogonalized by the following methods.

11.21 GRAM-SCHMIDT ORTHOGONALIZING PROCESS

Let us suppose $g_i(x) = x^i$ be the given polynomial of degree i, $i = 0, 1, 2, ..., n$. Then the polynomial $g_i^*(x)$ of degree i which are orthogonal over $[a, b]$ with respect to the weight function $W(x)$ can be generated from the recursion relation

$$g_i^*(x) = x^i - \sum_{r=0}^{i-1} g_i^*(x), \; i = 0, 1, 2, ..., n$$

where $\qquad a_{ir} = \dfrac{\int_a^b W(x)^i g_r^*(x)dx}{\int_a^b W(x)x^i g_r^{*2}(x)dx}$ and $g_0^*(x) = 1$.

SOLVED EXAMPLE

EXAMPLE . *Using the Gram-Schmidt orthogonalization process, compute the first three orthogonal polynomials $P_0(x)$, $P_1(x)$, $P_2(x)$ which are orthogonal with respect to the weight function $W(x) = 1$, on $[0, 1]$. Use these polynomials to obtain the least square approximation of second degree for $f(x) = x^{1/2}$ on $[0, 1]$.*

SOLUTION. From the Gram-Schmidt orthogonalization process, we have

$$g_i^*(x) = x^i - \sum_{i=0}^{i-1} a_{ir}g_r^*(x), i = 1, 2, ..., n$$

where $\qquad a_{ir} = \dfrac{\int_a^b x^i g_r^*(x)dx}{\int_a^b g_r^{*2}(x)dx}$ and $\quad g_0^*(x) = 1$

which implies $\overset{*}{g_0}(x) = 1 = P_0(x)$ then $\overset{*}{g_i}(x) = x - a_{10}\overset{*}{g_0}(x)$

where
$$a_{10} = \frac{\int_0^1 x\,dx}{\int_0^1 dx} = \frac{1}{2}$$

$\therefore$
$$\overset{*}{g_1}(x) = x - \frac{1}{2} = P_1(x)$$

Also,
$$\overset{*}{g_2}(x) = x^2 - a_{20}\overset{*}{g_0}(x) - a_{21}\overset{*}{g_1}(x)$$

where $\quad a_{20} = \dfrac{\int_0^1 x^2\,dx}{\int_0^1 dx} = \dfrac{1}{3} \quad$ and $\quad a_{21} = \dfrac{\int_0^1 x^2\left(x - \dfrac{1}{2}\right)dx}{\int_0^1\left(x - \dfrac{1}{2}\right)dx} = 1$

$\therefore$
$$\overset{*}{g_2}(x) = x^2 - \frac{1}{3} - \left(x - \frac{1}{2}\right) = x^2 - x + \frac{1}{6} = P_6(x).$$

Using these polynomials, we have the sum of the square of the deviation
$$I[c_0, c_1, c_2] = \int_0^1 [x^{1/2} - c_0 P_0(x) - c_1 P_1(x) - c_2 P_2(x)]^2 dx$$
$$= \text{minimum}.$$

Now for the minima of I, we must have

$$\frac{\partial I}{\partial c_0} = 0 \Rightarrow \int_0^1 [x^{1/2} - c_0 P_0(x) - c_1 P_1(x) - c_2 P_2(x)]^2 P_0(x)dx = 0$$

$$\frac{\partial I}{\partial c_1} = 0 \Rightarrow \int_0^1 [x^{1/2} - c_0 P_0 - c_1 P_2]P_1 dx = 0$$

$$\frac{\partial I}{\partial c_2} = 0 \Rightarrow \int_0^1 [x^{1/2} - c_0 P_0 - c_1 P_1 - c_2 P_2]P_1 dx = 0$$

Now, using the orthogonality condition, we get

$$c_0 = \frac{\int_0^1 x^{1/2} P_0(x)dx}{\int_0^1 P_0^2(x)dx} = \frac{\int_0^1 x^{1/2}dx}{\int_0^1 dx} = \frac{2}{3}$$

and
$$c_1 = \frac{\int_0^1 x^{1/2} P_1(x)dx}{\int_0^1 P_1^2(x)dx} = \frac{\int_0^1 x^{1/2}\left(x - \dfrac{1}{2}\right)dx}{\int_0^1\left(x^2 - x + \dfrac{1}{6}\right)dx} = -\frac{4}{7}$$

Hence, the required least square approximation is given by

$$y = P(x) = \frac{2}{3}P_0(x) + \frac{4}{5}P_1(x) - \frac{4}{7}P_2(x)$$

$$= \frac{2}{3} + \frac{4}{5}\left(x - \frac{1}{2}\right) - \frac{4}{7}\left(x^2 - x + \frac{1}{6}\right)$$

$$y = \frac{1}{35}(6 + 48x - 20x^2)$$

11.22 LEGENDRE AND CHEBYSHEV POLYNOMIALS

In the theory of approximations, there are some orthogonal polynomials such as, Legendre and Chebyshev polynomial which can be used as the coordinate function, while applying the method of least squares.

11.22.1 LEGENDRE POLYNOMIALS

The Legendre polynomials $P_n(x)$ defined on an interval $(-1, 1)$ are given by

$$P_n(x) = \sum_{r=0}^{(n/2)} (-1)^r \frac{(2n-2r)!}{2^n r!(n-2r)!(n-r)!} \cdot x^{n-2r}$$

where

$$\left(\frac{n}{2}\right) = \begin{cases} n/2, & \text{if } n \text{ is even} \\ \dfrac{(n-1)}{2}, & \text{if } n \text{ is odd} \end{cases}$$

$P_n(x)$ is a polynomial in x of degree n and satisfies the Legendre differential equation

$$(1-x^2)\frac{d^2y}{dx^2} - 2x\frac{dy}{dx} + n(n+1)y = 0$$

The Legendre polynomials satisfies the recurrence relation.

$$(n+1)P_{n+1}(x) = (2n+1)xP_n(x) - nP_{n-1}(x).$$

11.22.2 PROPERTIES OF LEGENDRE POLYNOMIAL

(i) The polynomials $P_n(x)$ is an even polynomial if n is even and it is an odd polynomial if n is odd.

(ii) The Legendre polynomial $P_n(x)$ is an orthogonal polynomials and satisfy

$$\int_0^1 P_n(x)P_m(x)dx = \begin{cases} 0 & \text{if } m \neq n \\ \dfrac{2}{2n+1} & \text{if } m = n \end{cases}$$

(iii) $P_n(-x) = (-1)^n P_n(x)$.

11.22.3 CHEBYSHEV POLYNOMIALS

The Chebyshev polynomial of the first kind of degree n over the interval $[-1, 1]$ is denoted by $T_n(x)$, defined by the relation

$$T_n(x) = \cos(n \cos^{-1}x) = \cos n\theta$$

where

$$\theta = \cos^{-1}x \text{ or } x = \cos\theta$$

The Chebyshev polynomial $T_n(x)$ satisfy the recurrence relation

$$T_{n+1}(x) = 2xT_n(x) - T_{n-1}(x)$$

with

$$T_0(x) = 1, T_1(x) = x$$

The differential equation satisfying by the Chebyshew polynomial $T_n(x)$ is

$$(1-x^2)\frac{d^2y}{dx^2} - x\frac{dy}{dx} + n^2y = 0.$$

11.22.4 PROPERTIES OF CHEBYSHEV POLYNOMIAL

(i) $T_n(x)$ is a polynomial of degree n.

(ii) $T_n(x)$ is even function of x if x is even and it is an odd function of x if n is odd.

(iii) $T_n(x)$ has n simple zeroes.

(iv) $T_n(x)$ assumes extreme values at $(n + 1)$ points

$$x_k = \cos\left(\frac{k\pi}{n}\right),\ k = 0, 1, 2, ..., n$$

and the extreme value at x_k is $(-1)^k$.

(v) $|T_n(x)| \le 1,\ \forall\, x \in [-1, 1]$

(vi) $T_n(x)$ are orthogonal on the interval $[-1, -1]$ with respect to the weight function

$$W(x) = \frac{1}{\sqrt{1-x^2}}$$

(vii) $\int_0^1 T_m(x)T_n(x)\cos\theta = \begin{cases} 0 & \text{if} \quad m \ne n \\ \dfrac{\pi}{2} & \text{if} \quad m = n \ne 0 \\ \pi & \quad m = n = 0 \end{cases}$

(viii) If $P_n(x)$ is a monic polynomial of degree n and $\overline{T_n}(x) = \dfrac{T_n(x)}{2^{n-1}}$ is monic Chebyshew polynomial, then $\max\limits_{-1\le x\le 1} |T_n(x)| \le \max\limits_{-1\le x\le 1} |P_n(x)|$.

This important property of Chebyshew polynomial is known as minimax property and can be stated as follows.

"If all monic polynomial $P_n(x)$ of degree n, the polynomial $2^{1-n}T_n(x)$ has the least upper bound (Supremum) for its absolute value in the range $|x| \le 1$, *i.e.*, $-1 \le x \le 1$. Since $|T_n(x)| \le 1$, the upper bound referred to above is 2^{1-n}".

11.23 UNIFORM APPROXIMATION

In uniform approximation, we use the uniform norm to continuous function over a close interval. Here, it should be noted that Weierstrass approximation theorem provide the possibility of uniform approximation.

This theorem uses the Bernstein polynomial, as given below

$$B_n[f,x] = \sum_{k=0}^{n} f\left(\frac{k}{n}\right)\left(\frac{n}{k}\right)x^k(1-x)^{n-k} \quad ...(1)$$

defined on [0, 1]. It has been shown that

$$\lim_{n\to\infty} B_n[f,x] = f(x),\ \text{uniformly continuous on } [0, 1].$$

REMARK

- When we use Bernstein polynomial, then the convergence of the approximation is very slow. Chebyshev polynomials are the best known polynomials which give a good rapid uniform approximation to a continuous function $f(x)$.

11.23.1 UNIFORM POLYNOMIAL APPROXIMATION

Consider a continuous function $f(x)$ defined on a closed interval $[a, b]$, which is to be approximated by the polynomials $P_n(x) = a_0 + a_1x + ... + a_nx_n$ where $a_0, a_1, ..., a_n$ are constants.

In uniform polynomial approximation (also known as minimax approx.) we want to find the constant $a_0, a_1, ..., a_n$ is such a way that the error

$$E_n(f, a_0, a_1, ..., a_n) = f(x) - P_n(x)$$

satisfy the minimax principle.

$$\max_{a\le x\le b} |E_n(f,a_0,a_1,...,a_n)| = \min_{a\le x\le b} |E_n(f,a_0,a_1,...,a_n)|.$$

If $n = 0$, then we want to approximate the function $f(x)$ by a constant a_0, let

$$M = \max_{a\le x\le b} |f(x)| \text{ and } M = \min_{a\le x\le b} |f(x)|.$$

Now, by minimax principle, we require

$$\max_{a \le x \le b} | f(x) - a_0 | = \min_{a \le x \le b} | f(x) - a_0 |$$

$$\Rightarrow \quad M - a_0 = -(m - a_0) = -m + a_0$$

$$\Rightarrow \quad a_0 = \frac{1}{3}(m + M)$$

$$\Rightarrow \quad E_0[f, c_0] = M - \frac{1}{2}[M + m] = \frac{1}{2}(M - m)$$

The graph of the constant minimax approximation is given as follows :

Now, we find from figure, that when the error curve $\varepsilon(x) = f(x) - a_0$ is drawn, the value $\pm E_0(f, c_0)$ is assumed at least twice, once with plus sign and once with minus sign, but always of equal value.

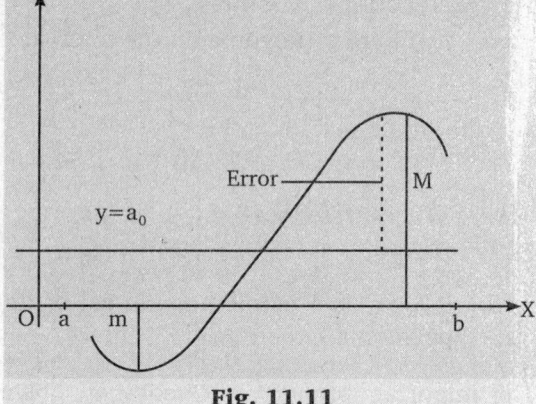

Fig. 11.11

If $n = 1$. Here, we want to approximate the function $f(x)$ by a first degree polynomial

$$P_1(x) = a_0 + a_1 x$$

The parameter a_0 and a_1 are to be so determined that the error

$$E_1(f, c_0, c_1) = f(x) - P_1(x)$$

satisfies the minimax principle

$$\max_{a \le x \le b} | E_1(f, c_0, c_1) | = \min_{a \le x \le b} | E_1(f, c_0, c_1) |.$$

Set $\varepsilon(x) = f(x) - c_0 - c_1 x$

$$\Rightarrow \quad \varepsilon'(x) = f'(x) - c_1.$$

Now take three points $x_1, x_2, x_3,\ a \le x_1 < x_2 < x_3 < b$ such that

$$\varepsilon(x_i) = \pm E_1,\ i = 1, 2, 3, \ldots$$

and the error must be alternate in signs at these points.

Now, assume that $x_1 = a, x_3 = b$ and x_2 is some interior point in the open interval (a, b)

i.e., $a < x_2 < b$

$$\Rightarrow \quad \varepsilon'(x_2) = 0$$

$$\Rightarrow \quad f'(x_2) - c_1 = 0$$

Therefore, we get the equation

$$f(a) - (a_0 + a_1 a) = -[f(x_2) - (a_0 + a_1 x_2)]$$
$$f(x_2) - (a_0 + a_1 x_2) = -[f(b) - (a_0 + a_1 b)]$$

and $f'(x_2) - a_1 = 0$

These equations give the values of a_0, a_1 and x_1. Then the graph of the linear minimax approximation is

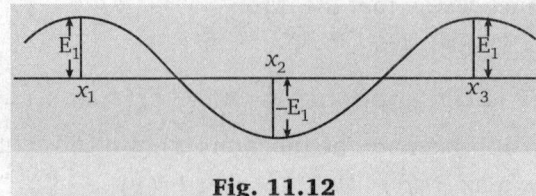

Fig. 11.12

11.24 CHEBYSHEV POLYNOMIAL APPROXIMATIONS

Let $f(x)$ be a continuous function defined on an interval $(-1, 1)$ and let $a_0 + a_1 x + ... + a_n x^n$ be the required minimax polynomial approximation for $f(x)$.

Let us suppose $f(x) = \dfrac{c_0}{2} + \sum\limits_{i=1}^{\infty} c_i T_i(x)$, is the Chebyshev series expansion for $f(x)$. Then, the series of the partial sum is

$$P_n(x) = \frac{c_0}{2} + \sum_{i=1}^{n} c_i T_i(x) \qquad \qquad ...(1)$$

is very nearly the solution to the problem

$$\max_{-1 \le x \le 1} \left| f(x) - \sum_{i=0}^{n} a_i x^i \right| = \max_{-1 \le x \le 1} \left| f(x) - \sum_{i=0}^{n} a_i x^i \right|$$

$\Rightarrow$ the partial sum (1) is nearly the best uniform approximation to $f(x)$. Because

$$f(x) = \frac{c_0}{2} + c_1 T_1(x) + c_2 T_2(x) + ... + c_n T_n(x) + c_{n+1} T_{n+1}(x) + \text{Remainder} \qquad ...(2)$$

Now, neglecting the remainder term, we get

$$f(x) - \left[\frac{c_0}{2} + \sum_{i=1}^{n} c_i T_i(x) \right] = c_{n+1} T_{n+1}(x) \qquad \qquad ...(3)$$

Now since $T_{n+1}(x)$ has $(n + 2)$ equal maxima and minima which is alternate is sign. Therefore the polynomial (1) of degree n is the best uniform approximation to $f(x)$.

ALGORITHM

To determine the best uniform approximation to a given continuous function defined on $(-1, 1)$, we find it easy to start with the truncated Chebyshev expansion of the function and then improve. The polynomial (1) is called the minimax polynomial. By this process we can obtain the best lower order approximation, called the minimax approximation, to a given polynomial.

11.24.1 CHEBYSHEV EQUIOSCILLATION THEOREM

Let $f(x)$ be a continuous function on a closed interval (a, b) and let $P_n(x)$ be the best uniform approximation according to the minimax principle

$$\max_{a \le x \le b} | f(x) - P_n(x) | = \min_{a \le x \le b} | f(x) - P_n(x) |$$

If we set

$$E_n[f, x] = \max_{a \le x \le b} | f(x) - P_n(x) | \text{ and } \varepsilon(x) = f(x) - P_n(x).$$

Then, there are at least $(n + 2)$ points

$$a = x_0 < x_1 < x_2 < ... < x_n < x_{n+1} = b$$

where (i) $\varepsilon(x_i) = \pm E_n$, $i = 0, 1, 2, ..., n + 1$
and (ii) $\varepsilon(x_i) = -\varepsilon(x_{i+1})$, $i = 0, 1, 2, ..., n.$
Here, the condition (ii), implies

$$\varepsilon'(x_i) = 0, i = 0, 1, 2, ..., n.$$

▼ REMARK
▸ The best approximation is completely and uniquely determined under the above condition (i) and (ii).

11.25 LANCZOS ECONOMIZATION OF POWER SERIES FOR A GENERAL FUNCTION

Let the given function $f(x)$ can be expressed as power series in x such that

$$f(x) = \sum_{i=1}^{\infty} a_i x^i, \quad -1 \le x \le 1 \qquad \qquad ...(1)$$

Here, we can change each power of x in (1) in the terms of Chebyshev polynomial and we get

$$f(x) = \sum_{i=1}^{\infty} c_i T_i(x) \qquad \qquad ...(2)$$

as the Chebyshev series expansion for $f(x)$ on $[-1, 1]$. Here, it should be noted that for a large number of functions $f(x)$, the series (2) converges more rapidly than the power series given by (1).

The partial sum of (2) is

$$P(x) = \sum_{i=1}^{\infty} c_i T_i(x) \qquad \qquad ...(3)$$

which is a good approximation to $f(x)$ in the sense

$$\max_{-1 \le x \le 1} |f(x) - P_n(x)| \le |c_{n+1}| + |c_{n+2}| +... \le \varepsilon \text{ (say)} | \text{ (say)}.$$

It for a given ε, it is possible to find the number of terms that should be retained in (3), then this process is known as Lanczos economization.

▶ **REMARK**

▷ To find the economized polynomial approximation for $f(x)$, replace each $T_i(x)$ in (3) by its polynomial form and then, rearranging the term.

11.26 RATIONAL APPROXIMATION

Since, the polynomial approximation is not always the best approximation. When the function behaves as a quotient of two polynomials, known as rational function, rational approximation can be used.

A rational approximation for a function $f(x)$ is written in the form

$$f(x) \approx R_{n,m}(x) = \frac{a_0 + a_1 x + ... + a_n x^n}{b_0 + b_1 x + ... + b_m x^m} = \frac{\sum_{i=0}^{n} a_i x^i}{\sum_{j=0}^{m} b_j x^j} \qquad \qquad ...(1)$$

The suffices n, m in $R_{n, m}(x)$ indicates that the numerator and denominator of the rational function (1) are polynomials of degree n and m respectively. Without any loss of generality we may take $b_0 = 1$, because we can always divide the numerator and the denominator in (1) by b_0. The constant b_0 will not be zero, otherwise the function is not defined.

Now, using the Maclaurin series expansion for $f(x)$ given by

$$f(x) = \sum_{j=1}^{\infty} c_j x^j \text{, in which } c_j\text{'s are known constant,}$$

we get from (1) $$f(x) - R_{n,m}(x) = \sum_{j=0}^{\infty} c_j x^j - \frac{\sum_{j=0}^{n} a_i x^i}{\sum_{j=0}^{m} b_j x^j}$$

$$= \frac{\left(\sum\limits_{j=0}^{\infty} c_j x^j\right)\left(\sum\limits_{j=0}^{m} b_j x^j\right)\left(\sum\limits_{j=0}^{n} a_j x^j\right)}{\sum\limits_{j=0}^{m} b_j x^j} \qquad \ldots(2)$$

Here, we determined $(n + m + 1)$ constants a_i, b_i by equating the coefficient of $x^j, j = 0, 1,$ $2, \ldots, (m + n)$ in the numerator of (1) to zero. The first non-zero term in the numerator of (2) gives the order of approximation. If we take $m = 0$, then we simply obtain Maclaurin's series expansion of $f(x)$, otherwise we get an approximation of order $M = n + m + 1$.

SOLVED EXAMPLES

EXAMPLE 1. *Determine the least square approximation of the type $ax^2 + bx + c$, to the function 2^x at the points $x_i = 0, 1, 2, 3, 4$.*

SOLUTION. Here, we want to determine a, b, c such that

$$I(a,b,c) = \sum_{i=0}^{4} [2^{x_i} - ax_i^2 - b_i - c]^2 \text{ is maximum.} \qquad \ldots(1)$$

The normal equation of (1) are given by

$$\left.\begin{array}{c} \sum\limits_{i=0}^{4} 2^x \cdot x_i^2 - a \sum\limits_{i=0}^{4} x_i^4 - b \sum\limits_{i=0}^{4} x_i^3 - c \sum\limits_{i=0}^{4} x_i^2 = 0 \\[2mm] \sum\limits_{i=0}^{4} 2^{x_i} x_i - a \sum\limits_{i=0}^{4} x_i^3 - b \sum\limits_{i=0}^{4} x_i^2 - c \sum\limits_{i=0}^{4} x_i = 0 \end{array}\right] \qquad \ldots(2)$$

and $\qquad \sum\limits_{i=0}^{4} 2^{x_i} - a \sum\limits_{i=0}^{4} x_i^2 - b \sum\limits_{i=0}^{4} x_i - 5c = 0$

Now, we have the following table

x	2^x	x^2	x^3	x^4	$x \cdot 2^x$	$x^2 \cdot 2^x$
0	1	0	0	0	0	0
1	2	1	1	1	2	2
2	4	4	8	16	8	16
3	8	9	27	81	24	72
4	16	16	64	256	64	256
Total	**10**	**31**	**30**	**100**	**354**	**98**	**346**

Substituting all the values from table in (2), we get

$$\left.\begin{array}{c} 354a + 100b + 30c = 346 \\ 100a + 30b + 10c = 94 \\ 3a + 10b + 5c = 31 \end{array}\right] \qquad \ldots(3)$$

Now, after solving (3) for a, b, c, we get
$$a = 1.143, b = -0.971, c = 1.286$$
Hence, the required least squares approximation to 2^x is
$$y = 1.143x^2 - 0.971x + 1.286$$

EXAMPLE 2. *The following measurements of a function f were made*

x	–2	–1	0	1	3
$f(x)$	7.0	4.8	2.3	2	13.8

Fit a third degree polynomial $P_3(x)$ to the data by the least squares method. As the value for $x = 1$ is known to be exact and $f'(1) = 1$. Also, we have $P_3(1) = 2$ and $P_3'(1) = 1$.

SOLUTION.   Let us take the polynomial of 3rd degree
$$P_3(x) = a_0 + a_1(x-1) + a^2(x-1)^2 + a_3(x-1)^3$$
since we have $P_3(1) = 2$ and $P_3'(1) = 1$.
$$\Rightarrow \qquad a_0 = 2, a_1 = 1$$
Therefore, by the method of least squares approximation, we want to find a_1, a_2 such that, the sum of the squares of deviation
$$I(a_2, P_3) = \sum_{i=1}^{5} [f(x_i) - 2 - (x_i - 1) - a_2(x_i - 1)^2 - a_3(x_i - 1)^3]^2 \text{ is minimum}$$

Then, the required normal equations are   ...(2)
$$\sum_{i=1}^{5}(x_i - 1)^2 f(x_i) - 2\sum_{i=1}^{5}(x_i - 1)^2 - \sum_{i=1}^{5}(x_i - 1)^3$$
$$-a_2\sum_{i=1}^{5}(x_i - 1)^4 - a_3\sum_{i=1}^{5}(x_i - 1)^5 = 0 \qquad\qquad ...(3)$$
and $\sum_{i=1}^{5}(x_i - 1)^2 f(x_i) - 2\sum_{i=1}^{5}(x_i - 1)^3 - \sum_{i=1}^{5}(x_i - 1)^4 - a_2\sum_{i=1}^{5}(x_i - 1)^5$
$$-a_3\sum_{i=1}^{5}(x_i - 1)^6 = 0 \qquad\qquad ...(4)$$

Now, using the given data in (3) and (4), we get
$$-114a_2 + 244a_3 = -131.7$$
and $\qquad 244a_2 - 858a_3 = 177.3$ ...(5)
Now solving (5) for a_2, a_3, we get
$$a_2 = 1.8220, a_3 = 0.3115$$
Hence, the required least square approximation is
$$P_3(x) = 2 + (x-1) + 1.822(x-1)^2 + 0.3115(x-1)^3$$

EXAMPLE 3.   *Using the Chebyshev polynomials, obtain the least square approximation of second degree for $f(x) = x^4$ on $[-1, 1]$.*

SOLUTION.   Defined the Chebyshev polynomial $P(x)$ as
$$f(x) = P(x) = c_0 T_0(x) + c_1 T_1(x) + c_2 T_2(x) \qquad\qquad ...(1)$$
Then, the sum of the square of the deviation
$$I(c_0, c_1, c_2) = \int_{-1}^{1} \frac{1}{\sqrt{1 - x^2}} [x^4 - c_0 T_0 - c_1 T_1 - c_2 T_2]^2 dx$$
which is to be minimum.
Then, the normal equations are given by

$$\left. \begin{array}{l} \dfrac{\partial I}{\partial c_0} = 0 \Rightarrow \int_{-1}^{1}[x^4 - c_0 T_0 - c_1 T_1 - c_2 T_2]\dfrac{T_0}{\sqrt{(1 - x^2)}} dx = 0 \\[4mm] \dfrac{\partial I}{\partial c_0} = 0 \Rightarrow \int_{-1}^{1}[x^4 - c_0 T_0 - c_1 T_1 - c_2 T_2]\dfrac{T_1}{\sqrt{(1 - x^2)}} dx = 0 \\[4mm] \text{and} \quad \dfrac{\partial I}{\partial c_0} = 0 \Rightarrow \int_{-1}^{1}[x^4 - c_0 T_0 - c_1 T_1 - c_2 T_2]\dfrac{T_2}{\sqrt{(1 - x^2)}} dx = 0 \end{array} \right\} \quad ...(2)$$

From (2), we obtained

$$c_0 = \frac{1}{\pi}\int_{-1}^{1}\frac{x^4 T_0}{\sqrt{1-x^2}}dx = \frac{3}{8}$$

$$c_1 = \frac{2}{\pi}\int_{-1}^{1}\frac{x^4 T_1}{\sqrt{1-x^2}}dx = 0$$

and $\qquad c_3 = \frac{2}{\pi}\int_{-1}^{1}\frac{x^4 T_2}{\sqrt{1-x^2}}dx = \frac{1}{2}$

Hence, the required approximation is given by

$$f(x) = \frac{3}{8}T_0 + \frac{1}{2}T_2$$

EXAMPLE 4. *Determine as accurately as possible a straight line $y = ax + b$ approximating $\dfrac{1}{x^2}$ in the Chebyshev sense on the interval $[1, 2]$. What is the maximum error? Calculate a and b to two correct decimals.*

SOLUTION. The error of approximation E is given by

$$E(x) = \frac{1}{x^2} - ax - b = \varepsilon(x) \text{ (say)} \qquad \qquad ...(1)$$

Let us choose the points $1, \alpha, 2$ and applying Chebyshev equioscillation theorem, we get

$$\varepsilon(1) = -\varepsilon(\alpha) \Rightarrow \varepsilon(1) + \varepsilon(\alpha) = 0 \qquad \qquad ...(2)$$
$$\varepsilon(\alpha) = -\varepsilon(2) \Rightarrow \varepsilon(2) + \varepsilon(\alpha) = 0 \qquad \qquad ...(3)$$

and $\qquad\qquad \varepsilon'(\alpha) = 0 \qquad \qquad ...(4)$

From (1), we have

$$\varepsilon'(x) = -\frac{2}{x^3} - a$$

Now from (2), (3) and (4), we get

$$\left. \begin{array}{r} 1 - a - b + \dfrac{1}{\alpha^2} - a\alpha - b = 0 \\[3mm] \dfrac{1}{4} - 2a - b + \dfrac{1}{\alpha^2} - a\alpha - b = 0 \\[3mm] \dfrac{2}{\alpha^3} + a = 0 \end{array} \right] \qquad ...(5)$$

and

Solving (5) for a, b and α, we get

$$a = -\frac{3}{4}, \quad b = 1.66, \quad \alpha^3 = \frac{8}{3}$$

Hence, the required approximating straight line is
$$y = 0.75x + 1.66$$
The maximum error in magnitude
$$= |\varepsilon(1)| = |1 - a - b| \approx 0.1$$

EXAMPLE 5. *Determine the values of a, b, c and d in the polynomial*
$$P(x) = ax^3 + bx^2 + cx + d$$
which minimize $\displaystyle\max_{-1\le x\le 1} |p(x) - |x||$.

SOLUTION.   Here, we want to find a, b and c such that

$$\max_{-1 \le x \le 1} \left\| p(x) - |x| \right\| \text{ is minimum.}$$

The error of approximation ε is given by $\varepsilon(x) = ax^3 + bx^2 + cx + d - |x|$

Take the points $x_0 = -1$, $x_1 = -\alpha$, $x_2 = 0$, $x_3 = \alpha$, $x_4 = 1$ and using Chebyshev equioscillation theorem, we get

$$\varepsilon(-1) + \varepsilon(-\alpha) = 0$$
$$\varepsilon(-\alpha) + \varepsilon(0) = 0$$
$$\varepsilon(0) + \varepsilon(\alpha) = 0$$
$$\varepsilon(\alpha) + \varepsilon(1) = 0$$

and $\varepsilon'(-\alpha) = 0 = \varepsilon'(0) = \varepsilon'(\alpha)$.

The solution of the system of equation is $b = \dfrac{1}{2\alpha}, d = \dfrac{\alpha}{4}, c = 0, a = 0$ and $\alpha = \dfrac{1}{2}$.

Hence, the values of a, b, c and d is given by $a = 0$, $b = 1$, $c = 0$, $d = \dfrac{1}{8}$ and the approximation is $x^2 + \dfrac{1}{8}$.

EXAMPLE 6.   *Find the polynomial of second degree, which is the best approximation in maximum norm to $\sqrt{x}$ on the point set $\left\{0, \dfrac{1}{9}, \dfrac{4}{9}, 1\right\}$.*

SOLUTION.   Let $P_2(x)$ is the required polynomial, which is given by

$$P_2(x) = ax^2 + bx + c$$

The error function $\varepsilon(x)$ is given by

$$\varepsilon(x) = ax^2 + bx + c - \sqrt{x}$$

Now, using Chebyshev equioscillation theorem, we have

$$\varepsilon(0) + \varepsilon\left(\frac{1}{9}\right) = 0$$

$$\varepsilon\left(\frac{1}{9}\right) + \varepsilon\left(\frac{4}{9}\right) = 0$$

$$\varepsilon\left(\frac{4}{9}\right) + \varepsilon(1) = 0$$

which gives $\dfrac{a}{81} + \dfrac{1}{9}b + 2c = \dfrac{1}{3}$; $\dfrac{17}{81}a + \dfrac{5}{9}b + 2c = 1$; $\dfrac{98}{81}a + \dfrac{13}{9}b + 2c = \dfrac{5}{3}$

Solving these equations for a, b, c, we get

$$a = -\frac{9}{8}, b = 2, c = \frac{1}{16}$$

Hence, the best polynomial approximation is

$$P_2(x) = \frac{1}{16} + 2x - \frac{9}{8}x^2$$

EXAMPLE 7.   *Find the best lower order approximation to the cubic polynomial $2x^3 + 3x^2$ in the interval $[-1, 1]$.*

SOLUTION.   Here, we have $x^3 = \dfrac{1}{4}[3T_1(x) + T_3(x)]$ therefore, $2x^3 + 3x^2$ in terms of Chebyshev polynomials, we get

$$2x^3 + 3x^2 = \frac{2}{4}[3T_1(x) + T_3(x)] + 3x^2$$

$$= 3x^2 + \frac{3}{2}T_1(x) + \frac{1}{2}T_3(x) = 3x^2 + \frac{3}{2}x + \frac{1}{2}T_3(x) \quad [\because T_1(x) = x]$$

Since $|T_3(x)| \leq 1, -1 \leq x \leq 1$ therefore the polynomial $3x^2 + \frac{3}{2}x$ is the required best lower order approximation to the given cubic with a maximum error $\pm\frac{1}{2}$ in the range $[-1, 1]$.

EXAMPLE 8. *Find a uniform polynomial approximation of degree 4 or less to e^x in $[-1, 1]$ using Lanczos economization with a tolerance of $\varepsilon = 0.02$.*

SOLUTION. Since, we have

$$f(x) = e^x = 1 + x + \frac{x^2}{2} + \frac{x^3}{6} + \frac{x^4}{24} + \frac{x^5}{120} + \dots$$

since $\quad \frac{1}{120} = 0.008\dots$

Therefore, we take $f(x)$ upto $\frac{x^4}{24}$ with a tolerance of $\varepsilon = 0.02$

s.t. $\quad f(x) = e^x = 1 + x + \frac{x^2}{2} + \frac{x^3}{6} + \frac{x^4}{24}$ ...(1)

Changing each power of x in (1) in terms of Chebyshev polynomials, we get

$$e^x = T_0 + T_1 + \frac{1}{4}(T_0 + T_1) + \frac{1}{24}(3T_1 + T_3) + \frac{1}{192}(3T_0 + 4T_2 + T_4)$$

$$: \frac{81}{64}T_0 + \frac{9}{8}T_1 + \frac{13}{48}T_2 + \frac{1}{24}T_3 + \frac{1}{192}T_4 \quad \text{...(2)}$$

Neglecting the last term because its magnitude 0.005 is less than 0.02.
Hence, the required economized polynomial approximation for e^x is given by

$$e^x \simeq \frac{81}{64}T_0 + \frac{9}{8}T_1 + \frac{13}{48}T_2 + \frac{1}{24}T_3 = \frac{x^3}{6} + \frac{13}{24}x^2 + x + \frac{191}{192}$$

EXAMPLE 9. *Find a polynomial $P(x)$ of degree as low as possible such that*

$$\max_{|x| \leq 1} |e^{x^2} - P(x)| \leq 0.05.$$

SOLUTION. We know that

$$e^{x^2} = 1 + x^2 + \frac{x^4}{2} + \frac{x^6}{6} + \frac{x^8}{24} + \frac{x^{10}}{120} + \dots.$$

$$= 1 + x^2 + \frac{x^4}{2} + \frac{x^6}{6} + \frac{x^8}{24} = P(x) \quad \text{...(1)}$$

with error in the leading term as $\frac{1}{120} \approx 0.0083$

Now expressing (1) in terms of Chebyshev polynomials, we get

$$P(x) = T_0 + \frac{1}{2}(T_0 + T_2) + \frac{1}{16}(3T_0 + 4T_2 + T_4) + \frac{1}{192}(10T_0 + 15T_2 + 6T_4 + T_6)$$

$$+ \frac{1}{3072}(35T_0 + 56T_2 + 28T_4 + 8T_6 + T_8)$$

$$= \frac{1}{3072}(5379T_0 + 2600T_2 + 316T_4 + 24T_6 + T_8)$$

Here, we have

$$\left| \frac{1}{3072}(24T_6 + T_8) \right| \leq 0.0082$$

                    < the required maximal error of approximation 0.05.
Therefore, the required approximation

$$e^{x^2} \approx \frac{1}{3072}(5379T_0 + 2600T_2 + 316T_4)$$

$$= \frac{1}{3072}[5379 + 2600(2x^2 - 1) + 316(8x^4 - 8x^2 + 1)]$$

$$= \frac{1}{3072}[3095 + 2672x^2 + 2528x^4]$$

$$= 1.0075 + 0.8698x^2 + 0.8229x^4.$$

EXAMPLE 10. *Find the lowest order polynomial which approximates the function* $f(x) = \sum\limits_{r=0}^{4} (-x)^r$ *in the range* $0 \leq x \leq 1$, *with an error less than 0.1.*

SOLUTION. Change the interval [0, 1] to [−1, 1] by using the transformation

$$x = \frac{1}{2}(1 + t)$$

Therefore, the function $f(x) = \sum\limits_{r=0}^{4} (-x)^r = 1 - x + x^2 - x^3 + x^4, 0 \leq x \leq 1$.

transform to $F(t) = 1 - \frac{1}{2}(1+t) + \frac{1}{4}(1+t)^2 - \frac{1}{8}(1+t)^3 + \frac{1}{16}(1+t)^4$

$$= \frac{11}{16} - \frac{1}{8}t + \frac{1}{4}t^2 + \frac{1}{8}t^3 + \frac{1}{16}t^4, -1 \leq t \leq 1.$$

Now changing the power of t in $F(t)$ in terms of Chebyshev polynomials, we get

$$F(t) = \frac{11}{16}T_0 - \frac{1}{8}T_1 + \frac{1}{8}(T_2 + T_0) + \frac{1}{32}(T_3 + 3T_1) + \frac{1}{128}(T_4 + 4T_2 + 3T_0)$$

$$= \frac{107}{128}T_0 - \frac{1}{32}T_1 + \frac{5}{32}T_2 + \frac{1}{32}T_3 + \frac{1}{128}T_4$$

Now, since $\left| \frac{1}{32}T_0 \right| + \left| \frac{1}{148}T_4 \right| < 0.1$, we get the approximation

$$F(t) = \frac{107}{128}T_0 - \frac{1}{32}T_1 + \frac{5}{32}T_2 = \frac{107}{128} - \frac{1}{32}t + \frac{5}{32}(2t^2 - 1)$$

$$= \frac{5}{16}t^2 - \frac{t}{32} + \frac{87}{128}$$

Replace t by $(2x - 1)$, we get the required polynomial approximation as

$$= \frac{1}{28}(160x^2 - 168x + 131)$$

EXAMPLE 11. *Obtain the rational approximation of the form* $\dfrac{a_0 + a_1 x}{1 + b_1 x + b_2 x^2}$ *to the function* e^x.

SOLUTION. Since, we know that

$$e^x = 1 + x + \frac{x^2}{2} + \frac{x^3}{6} + \frac{x^4}{24} + \dots$$

Let $e^x \simeq \dfrac{a_0 + a_1 x}{1 + b_1 x + b_2 x^2}$

Consider

$$e^x - \dfrac{a_0 + a_1 x}{1 + b_1 x + b_2 x^2} = \dfrac{\left(1 + x + \dfrac{x^2}{2} + \dfrac{x^3}{6} + \ldots\right)(a + b_1 x + b_2 x^2) - (a_0 + a_1 x)}{1 + b_1 x + b_2 x^2}$$

Set $\left(1 + x + \dfrac{x^2}{2} + \dfrac{x^3}{6} + \ldots\right)(1 + b_1 x + b_2 x^2) - (a_0 + a_1 x) = \sum\limits_{j=0}^{\infty} p_j x^j$.

To determine the four constants a_0, a_1, b_1 and b_2, we put $p_j = 0$, $j = 0, 1, 2, 3$ and we get

$$\left.\begin{aligned}
1 - a_0 &= 0 \\
b_1 + 1 - a_1 &= 0 \\
b_2 + b_1 + \dfrac{1}{2} &= 0 \\
b_2 + \dfrac{1}{2} b_1 + \dfrac{1}{6} &= 0
\end{aligned}\right] \qquad \ldots(1)$$

Solving (1) for a_0, a_1, b_1, b_2 we get

$$a_0 = 1, \ a_1 = \dfrac{1}{3}, \ b_1 = \dfrac{-2}{3}, \ b_2 = \dfrac{1}{6}$$

$$\therefore \qquad e^x \simeq \dfrac{1 + \dfrac{1}{3} x}{1 - \dfrac{2}{3} x + \dfrac{1}{6} x^4} + 0(x^4)$$

EXERCISE 11.5

1. Use the method of least squares to fit the curve $y = \dfrac{C_0}{2} + a \cdot \sqrt{x}$ to the following table

x	0.1	0.2	0.4	0.5	1	2
y	21	11	7	6	5	6

2. An experiment with a periodic process give the following data

t^0	0	50	100	150
y	0.754	1.762	2.041	1.412
t^0	200	250	300	350
y	0.303	−0.484	380	0.520

Estimate the parameters a and b in the model $y = b + a \sin t$, using the least square approximation.

3. Obtain the Chebyshev linear polynomial approximation to the function
$$f(x) = x^3 \text{ on } [0, 1].$$

4. Prove that the polynomial of best approximation of degree not exceeding 3 for $|x|$ in the interval $[-1, 1]$ is $x^2 + \dfrac{1}{8}$.

5. Obtain the Chebyshev polynomial approximation of second degree to the function $f(x) = x^3$ on $[0, 1]$. What is the maximal error?

6. Determine the best minimax approximation to $e^{|x|}$ with a polynomial of degree 0 and 1 for $|x| \le 1$.

7. Determine the polynomial of second degree, which is the best approximation in maximum norm to $\sqrt{x}$ on the point set $\left[0, \dfrac{1}{9}, \dfrac{4}{9}, 1\right]$.

8. Find the best uniform approximation of degree 3 or less to x^4 on $[-1, 1]$.

9. The function $P_3(x) = x^3 - 9x^2 - 20x + 5$ is given. Find a second degree polynomial $P_2(x)$ such that $\delta = \max\limits_{0 \le x \le 4} |P_3(x) - P_2(x)|$

becomes as small as possible. The value of δ and the value of x for which $|P_3(x) - P_2(x)| = \delta$ should also be given.

10. Economize the power series

$$\sin x \approx x - \frac{x^3}{6} + \frac{x^5}{120} - \frac{x^7}{5040}, \ -1 \le x \le 1.$$

11. The function f is defined by

$$f(x) = \frac{1}{2}\int_0^x \frac{1 - e^{-t^2}}{t^2}\, dt$$

Approximate f by a polynomial $P(x) = a + bx + cx^2$ such that

$$\max_{|x| \le 1} |f(x) - P(x)| \le 5 \times 10^{-3}$$

12. Approximate $F(x) = \frac{1}{x}\int_0^x \frac{e^t - 1}{t}\,dt$, by a third degree polynomial $P_3(x)$ so that

$$\max_{-1 \le x \le 1} |F(x) - P_3(x)| \le 3 \times 10^{-4}$$

13. If we want to approximate a continuous function $f(x)$ on $|x| \le 1$ by a polynomial $P_n(x)$ of degree n, suppose that we have found

$$f(x) - P_n(x) = \alpha_{n+1}(x)T_{n+1}(x) + r(x)$$

where T_{n+1} denotes the Chebyshev polynomial of degree $(n + 1)$ with

$$\frac{1}{2^{n+1}} \le |\alpha_{n+1}| \le \frac{1}{2^n}$$

and $\quad |r(x)| \le \frac{1}{10}\cdot|\alpha_{n+1}|, \ |x| \le 1$

Show that $\dfrac{0.4}{2^n} \le \max_{|x| \le 1}|f(x) - P_n^*(x)| \le \dfrac{1.1}{2^n}$

Where, $P_n^*(x)$ denotes the optimal polynomial of degree n for $f(x)$ on $|x| \le 1$.

14. Obtain the rational approximation for e^x of the form

(i) $\dfrac{a_0 + a_1 x}{1 + b_1 x}$ 　　(ii) $\dfrac{a_0 + a_1 x + a_2 x^2}{1 + b_1 x + b_2 x^2}$

(iii) $\dfrac{a_0 + a_1 x + a_2 x^2}{1 + b_1 x}$

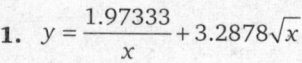**Answers**

1. $y = \dfrac{1.97333}{x} + 3.2878\sqrt{x}$ 　　**2.** $y = 0.752575 + 1.312810 \sin t$ 　**3.** $P_1(x) = x - \dfrac{\sqrt{3}}{9}$

5. $P_2(x) = \dfrac{1}{32}(48x^2 - 18x + 1)$ 　**6.** $P_1(x) = \dfrac{(e+1)}{2}$ 　　**7.** $P_2(x) = \dfrac{1}{16} + 2x - \dfrac{9}{8}x^2$

8. $P(x) = x^2 - \dfrac{1}{8}$ 　　　**9.** $P_2(x) = -3x^2 - 29x + 7$ 　　**10.** $\sin x = \dfrac{383}{384}x - \dfrac{5}{32}x^3$

11. $P(x) = 0.9966 - 0.1381x^2$ 　**12.** $P_3(x) = \dfrac{37}{3456}x^3 + \dfrac{103}{1800}x^2 + \dfrac{3455}{13824}x + \dfrac{4799}{4800}$

14. (i) $e^x = \dfrac{1 + \frac{1}{2}x}{1 - \frac{1}{2}x} + 0(x^3)$ 　(ii) $e^x = \dfrac{1 + \frac{1}{2}x + \frac{1}{12}x^2}{1 - \frac{1}{2}x + \frac{1}{12}x^2} + 0(x^5)$ 　(iii) $e^x = \dfrac{1 + \frac{2}{3}x + \frac{1}{6}x^2}{1 - \frac{1}{3}x} + 0(x^4)$

Objective Evaluations

MULTIPLE CHOICE QUESTIONS

Choose the most appropriate one.

1. If there are m independent linear equation in n unknowns then for the existence of unique solution, we must have
   (a) $m = n$ (b) $m > n$
   (c) $m < n$ (d) none of these

2. For no solution of m linear equation in n unknowns we must have:
   (a) $m = n$ (b) $m > n$
   (c) $m < n$ (d) none of these

3. The general problem of finding equation of approximate curves which fit given set of data is called :
   (a) data fitting   (b) curve fitting
   (c) both (a) and (b)   (d) none of these

4. The difference between the observed values and expected values is called :
   (a) residue   (b) residual
   (c) resistance   (d) none of these

5. For the line $y = a + bx$ the normal equations are given by :
   (a) $\Sigma y = ma + b\Sigma x + c\Sigma x^2$
   (b) $\Sigma xy = a\Sigma x + b\Sigma x^2 + c\Sigma x^3$
   (c) $\Sigma x^2 y = a\Sigma x^2 + b\Sigma x^3 + c\Sigma x^4$
   (d) all are true

6. Norm of x is defined by:
   (a) $\left[\sum_{i=1}^{n} |x_i|^p\right]^{1/p}$ (b) $\sum_{i=1}^{n} |x_i|^p$
   (c) both (a) and (b) are true
   (d) none of these

7. Euclidian norm is defined by :
   (a) $\left[\sum_{i=1}^{n} |x_i|^p\right]^{1/p}$ (b) $\left[\sum_{i=1}^{n} |x_i|^2\right]^{1/2}$
   (c) $\sum_{i=1}^{n} |x_i|^p$ (d) none of these

8. To find the least square approximation, we always use :
   (a) L^p-norm   (b) Euclidean
   (c) both (a) and (b) are true
   (d) none of these

9. To compute the values of the function like e^x, $\sin x$, $\cos x$ etc. for large values of the argument x we use :

   (a) least square approximation
   (b) rational approximation
   (c) polynomial approximation
   (d) none of these

10. The line of regression can be regarded as fitting of
    (a) linear curves   (b) quadratic curve
    (c) cubic curve   (d) none of these

11. The principle of least square was first given by :
    (a) Gauss   (b) Newton
    (c) Legendre   (d) none of these

12. The normal equations of the parabolic curve $y = a + bx + cx^2$ are given by :
    (a) $\Sigma y = ma + b\Sigma x + c\Sigma x^2$
    (b) $\Sigma xy = a\Sigma x + b\Sigma x^2 + c\Sigma x^3$
    (c) $\Sigma x^2 y = a\Sigma x^2 + b\Sigma x^3 + c\Sigma x^4$
    (d) all are true

13. The piecewise polynomial that prevent the discontinuities at the connecting points are called :
    (a) cubic function   (b) square function
    (c) spline function   (d) none of these

14. In spline function, the function values must be equal at :
    (a) all knots   (b) interior knots
    (c) exterior knots   (d) none of these

15. The second derivation of the cubic of line function at the interior knots must be :
    (a) zero   (b) equal
    (c) unequal   (d) none of these

16. The second derivative of the cubic of spline function at the end points are
    (a) zero   (b) equal
    (c) non-zero   (d) none of these

17. If $P_0 = P_n = 0$ then spline is called:
    (a) natural splines   (b) odd splines
    (c) even splines   (d) none of these

18. The criterion of a norm which makes the error smallest is called :
    (a) best approximation
    (b) rational approximation
    (c) both (a) and (b) are true
    (d) none of these

19. Which of the following is/are true
    (a) when we used Euclidian norm, we

obtained least square approximation

(b) when uniform norm is used, we obtained the uniform approximation

(c) both (a) and (b) are true

(d) none of these

20. The approximation for which the constant c_i, $i = 0, 1, 2, ..., n$ are determined in such a way that the aggregate of weight function is as smell as possible is called:

(a) best approximation in the least square sense

(b) best approximation in the uniform approximation sense

(c) both (a) and (b) are true

(d) none of these

21. A set of real functions $[g_i(x)]$ is said to be orthogonal over a set of points $x_1, x_2,..., x_n$ with respect of the weight function $W(x)$ if

$$\sum_{k=1}^{m} W(x_k)g_i(x_k)g_j(x_k) =$$

(a) 1

(b) 0

(c) 2

(d) none of these

22. The polynomial defined by

$$P_n(x) = \sum_{r=0}^{\left(\frac{n}{2}\right)} (-1)^r \frac{(2n-2r)!}{2^n r!(n-2r)!(n-r)!} x^{n-2r}$$

is called :

(a) Legendre polynomial

(b) Chebyshev's polynomial

(c) Bassel's polynomial

(d) none of these

23. If $T_n(x)$ is a Chobyshev's polynomial then which of the following is/are true

(a) $T_0(x) = 1$

(b) $T_1(x) = x$

(c) $T_{n+1}(x) = 2xT_n(x) - T_{n-1}(x)$

(d) all are true

24. The exponential curve $y = ab^x$ which can be fitted to the data :

x	1	2	3	4	5	6	7	8
y	1.0	1.2	1.8	2.5	3.6	4.7	6.6	9.1

is $y =$

(a) 2^x

(b) 3^x

(c) $(.68)(1.38)^x$

(d) none of these

25. The linear polynomial approximation for the function $f(x) = x^3$ through the origin is:

(a) $\dfrac{x}{5}$

(b) $\dfrac{3x}{5}$

(c) $\dfrac{4x}{5}$

(d) none of these

26. The polynomial of best approximation of degree not exceeding 3 for $|x|$ in the interval $(-1, 1)$ is :

(a) $x^2 + 8$

(b) $x^2 + \dfrac{1}{8}$

(c) $x^2 + \dfrac{3}{8}$

(d) none of these

27. The best fit values of x, y, z in the least square since from the following equations $x+2y+z=1$, $2x+y+z = 4, -x+y+2z = 4$ and $4x+2y-5z = -7$ are given by :

(a) $.910, -.378, 2.045$

(b) $.910, .378, .488$

(c) $1, 0, 0$

(d) none of these

28. The normal equation for fitting the curves of type $y = ax + \dfrac{b}{x}$ by least square method is :

(a) $\Sigma x_i y_i = a\Sigma x_i^2 + nb$ (b) $\dfrac{\Sigma y_i}{x_i} = na + b\Sigma \dfrac{1}{x_i^2}$

(c) both (a) and (b) are true

(d) none of these

29. The most plausible values of x and y from the following equations $x + y = 3, 2x - y = 0.5$, $x + 3y = 7.25, 3x + y = 4.95$ are given by:

(a) 1.234, 1.919

(b) 2.23, 1.9

(c) 2.23, 2.9

(d) none of these

30. If $x + 2.5y = 21, 3.2x - y = 28, 4x+1.2y = 42.04, 1.5x+6.3y = 40$ then most plausible values are given by:

(a) 9.620, 4.064

(b) 8.620, 5.064

(c) 7.620, 1.98

(d) none of these

Answers

MULTIPLE CHOICE QUESTIONS

1. (a)	2. (b)	3. (b)	4. (b)	5. (d)	6. (a)	7. (b)	8. (b)	9. (b)
10. (a)	11. (a)	12. (d)	13. (c)	14. (b)	15. (b)	16. (a)	17. (a)	18. (a)
19. (c)	20. (a)	21. (b)	22. (a)	23. (d)	24. (c)	25. (b)	26. (b)	27. (a)
28. (c)	29. (a)	30. (a)						

COMPUTATIONAL TECHNIQUE LAB

1. LEAST SQUARE METHOD

Fitting a straight line $y = a + bx$ using least square method.

Symbols Used

n = number of values of x and y

$x[i]$ = Different values of x

$y[i]$ = Different values of y

sx = sum of different values of $x[i]$'s *i.e.* $\Sigma x[i]$

sy = sum of different values of $y[i]$'s *i.e.* $\Sigma y[i]$

xx = sum of squares of values of $x[i]$'s *i.e.* $\Sigma x[i]2$

xy = sum of multiplication of different values of $x[i]$ and $y[i]$'s *i.e.* $\Sigma x[i].y[i]$

a = value of constant in the curve

b = coefficient of x in the curve

ALGORITHM : FITTING OF STRAIGHT LINE USING LEAST SQUARE METHOD

Step 1 :   Start

Step 2 :   Input n

Step 3 :   for i = 0 to (n – 1)

Step 4 :   input $x[i]$, y [i]

Step 5 : $sx = 0$, $sy = 0$, $xx = 0$, $xy = 0$

Step 6 :   For i = 0 to (n – 1)

Step 7 : $sx = sx + x[i]$

Step 8 : $sy = sy + y[i]$

Step 9 : $xx = xx + x[i] * x[i]$

Step 10 : $xy = xy + x[i] * y[i]$

Step 11 :   End of i loop

Step 12 : $b = (n * xy - sx * sy)/(n * xx - sx * sx)$

Step 13 : $a = (sy - b * sx)/n$

Step 14 :   if (b < 0)

Print "line is y =", a, b, 'x'

else

Print "line is y =", a, '+', b, 'x'

Step 15 :   Stop

FLOW CHART: FITTING OF STRAIGHT LINE

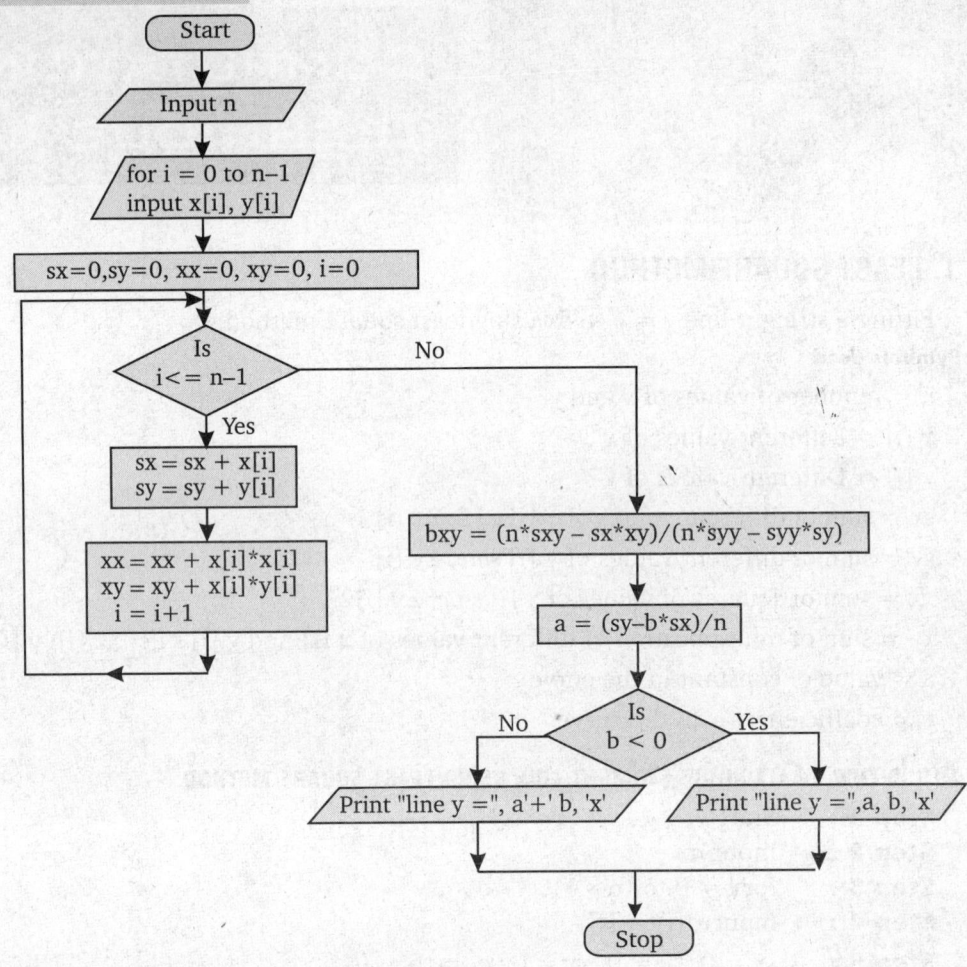

PROGRAM: *Following is a program in C language to fit a straight line by the method of least square.*

//FITTING OF STRAIGHT LINE y= a+bx

```
#include<stdio.h>
#include<conio.h>

void main()
    {
        clrscr();
        float  x[30],y[30],sx=0,sy=0,xx=0,xy=0,a,b;
        int n,i;
        printf("Enter number of values of x and y \n");
        scanf("%d", &n);
```

```
            printf("Enter the value of x and y\n ");
            for(i=0;i<n;i++)
            {
                scanf("%f %f",&x[i],&y[i]);
            }
            for(i=0;i<n;i++)
              {  sx+=x[i];
                 sy+=y[i];
                 xx+=x[i]*x[i];
                 xy+=x[i]*y[i];
              }
            b=(n*xy-sy*sx)/(n*xx-sx*sx);
            a=(sy-b*sx)/n;

            if(b<0)
             printf(" line y =%.2f%.2fx",a,b);
            else
               printf(" line y =%.2f%+.2fx",a,b);
            getch();

        }
```

Output: FITTING OF STRAIGHT LINE y = a+bx

```
Enter number of values of x and y
6
Enter the value of x and y
1       1200
2        900
3        600
4        200
5        110
6         50

line y = 1362.00-243.43x
```

Lab Assignment : Fitting of Straight Line

1. Write a C program to fit a straight line of the form y = a + bx from the data given below :

x	0	1	2	3	4	5
y	25	22	19	17	13	8

Hint : Input : n = 6

Output : Line y = 25.48 – 3.26 x

2. Write a C program to fit a straight line from the following data

x	1	2	3	4	5
y	2	4	6	8	10

Hint : Input : $n = 5$

Output : Line $y = 0 + 2x$

2. REGRESSION LINES

Regression line of y on x is $y - \overline{y} = b_{yx} (x - \overline{x})$ and regression line of x on y is $x - \overline{x} = b_{xy} (y - \overline{y})$.

SYMBOLS USED :

    n = number of different values of x and y

x [i] = different values of x

y [i] = different values of y

   sx = sum of different values of x, *i.e.*, Σxi

   sy = sum of different values of y, *i.e.*, Σyi

sxy = sum of product of different values of x and y

sxx = sum of square of x values

syy = sum of square of y values

byx = regression coefficient of y on x which is equal to $b_{yx} = \dfrac{n\Sigma xy - \Sigma x \Sigma y}{n\Sigma x^2 - (\Sigma x)^2}$

bxy = regression coefficient of x and y on $b_{xy} = \dfrac{n\Sigma xy - \Sigma x \cdot \Sigma y}{n\Sigma y^2 - (\Sigma y)^2}$

xm = mean of x series

ym = mean of y series

ALGORITHM : REGRESSION LINE Y ON X

Step 1 :	Start
Step 2 :	Input n
Step 3 :	for i = 0 to n – 1
Step 4 :	input x [i], y [i]
Step 5 :	End of i loop
Step 6 :	sx = 0, sy = 0, sxx = 0, sxy = 0
Step 7 :	For i = 0 to (n – 1)
Step 8 :	sx = sx + x [i]
Step 9 :	sy = sy + y [i]
Step 10 :	sxx = sxx + x [i] * x [i]
Step 11 :	sxy = sxy + x [i] * y [i]
Step 12 :	End of i loop
Step 13 :	xm = sx/n
Step 14 :	ym = sy/n

Step 15 :   byx = (n * sxy – sx * sy)/(n * sxx – sx * sx)

Step 16 :   t = ym – xm * byx

Step 17 :   If (by x < 0)

          Print "Regression line y on x is y =", t, byx, 'x'

      else

          Print "Regression line y on x is y =", t, '+', byx, 'x'

Step 18 :   Stop.

FLOW CHART : REGRESSION LINE Y ON X

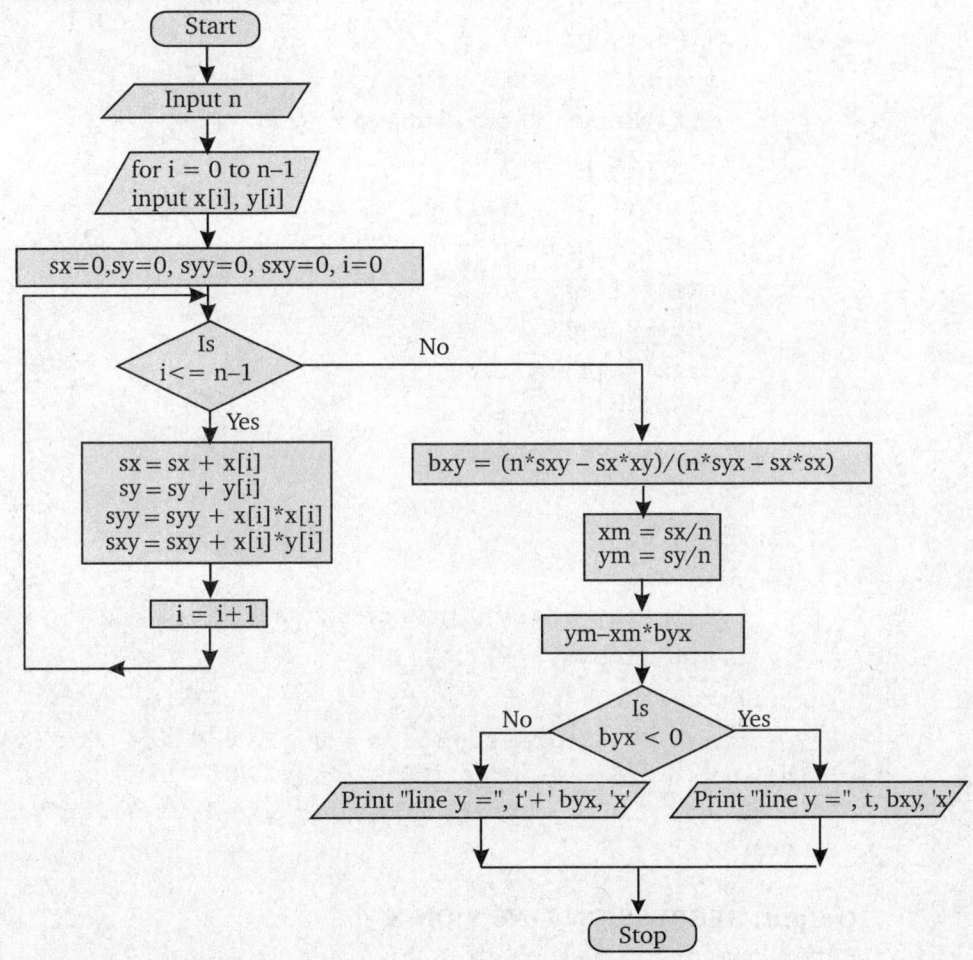

PROGRAM   *Following program in C language finds regression line y on x.*

//REGRESSION LINE Y ON X

```
#include<stdio.h>
#include<conio.h>
```

```
void main()
  {
    clrscr();
    float  x[30],y[30],sx=0,sy=0,sxx=0,syy=0,sxy=0;
    float xm,ym,byx,t1;
    int n,i;
    printf("Enter number of values of x and y \n");
    scanf("%d", &n);
    printf("Enter the values of x \n ");
    for(i=0;i< n;i++)
        scanf("%f",&x[i]);
    printf("Enter the values of  y\n ");
    for(i=0;i< n;i++)
        scanf("%f",&y[i]);
    for(i=0;i< n;i++)
      {  sx+=x[i];
         sy+=y[i];
         sxx+=x[i]*x[i];
         syy+=y[i]*y[i];
         sxy+=x[i]*y[i];
       }
     xm=sx/n;
     ym=sy/n;
     byx=(n*sxy-sx*sy)/(n*sxx-sx*sx);
     t1=ym-xm*byx;
    if(byx)
    printf("\n Regression line y on x is y =%.2f%.2fx",t1,byx);
  else
    printf("\nRegression line y on x is y=%.2f+%.2fx",t1,byx);
  getch();
  }
```

Output: REGRESSION LINE Y ON X

```
Enter number of values of x and y
5
Enter the values of x
1
2
3
4
```

```
5
Enter the values of  y
2
5
3
8
7


Regression line y on x is y =1.10+1.30x
```

3. REGRESSION LINE X ON Y

ALGORITHM : REGRESSION LINE X ON Y

Step 1 :	Start
Step 2 :	Input n
Step 3 :	for i = 0 to (n – 1)
Step 4 :	input x [i], y [i]
Step 5 :	End of i loop
Step 6 :	sx = 0, sy = 0, sxy = 0, syy = 0
Step 7 :	For i = 0 to (n – 1)
Step 8 :	sx = sx + x [i]
Step 9 :	sy = sy + y [i]
Step 10 :	syy = syy + y [i] * y [i]
Step 11 :	sxy = sxy + x [i] * y [i]
Step 12 :	End of i loop
Step 13 :	xm = sx/n
Step 14 :	ym = sy/n
Step 15 :	bxy = (n * sxy – sx * sy)/(n * syy – sy * sy)
Step 16 :	t = xm – ym * bxy
Step 17 :	If (bxy < 0)
	Print "Regression line x on y is x =", t, bxy, 'y'
	else
	Print "Regression line x on y is x =", t, '+', bxy, 'y'
Step 18 :	Stop.

FLOW CHART : REGRESSION LINE X ON Y

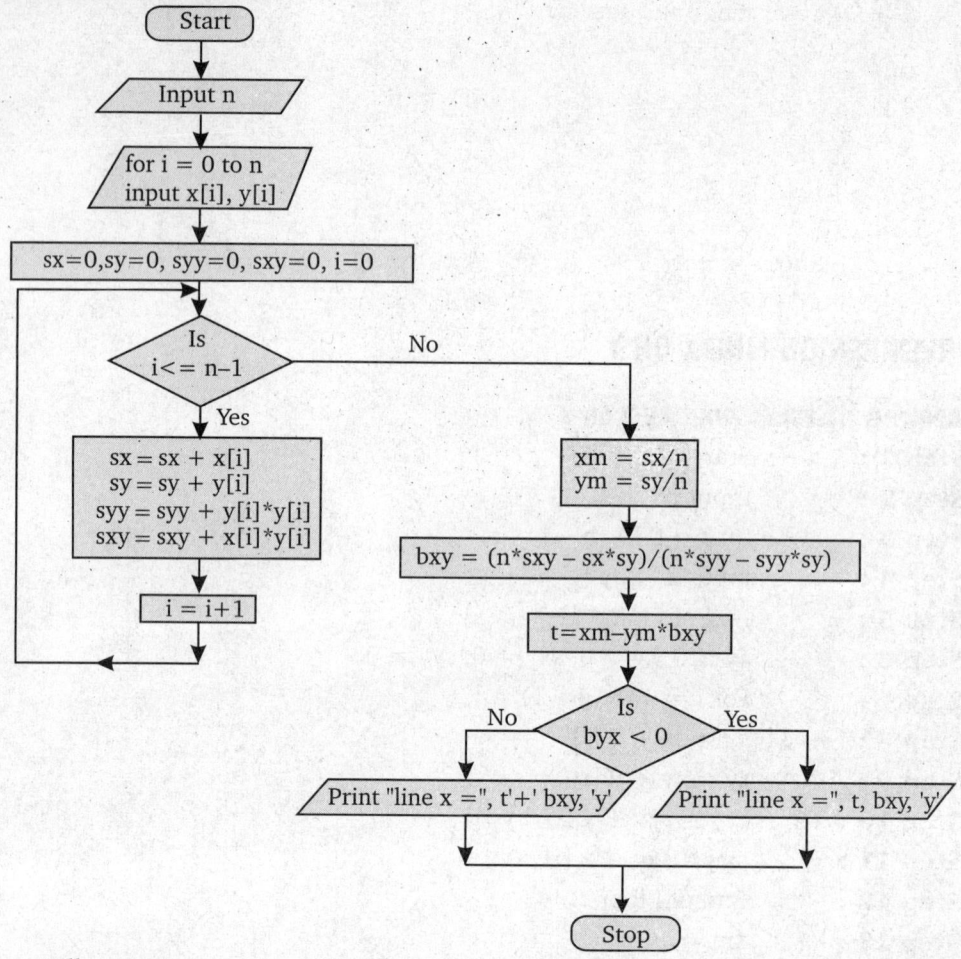

PROGRAM: *Following program in C language finds regression line x on y.*

```
//REGRESSION LINE X ON Y
#include<stdio.h>
#include<conio.h>
void main()
    {  clrscr();
       float  x[30],y[30],sx=0,sy=0,sxx=0,syy=0,sxy=0;
       float xm,ym,bxy,t;
       int n,i;
       printf("Enter number of values of x and y \n");
       scanf("%d", &n);
       printf("Enter the values of x \n ");
```

```
     for(i=0;i<n;i++)
        scanf("%f",&x[i]);
     printf("Enter the values of  y\n ");
     for(i=0;i<n;i++)
        scanf("%f",&y[i]);
     for(i=0;i<n;i++)
      {   sx+=x[i];
          sy+=y[i];
          sxx+=x[i]*x[i];
          syy+=y[i]*y[i];
          sxy+=x[i]*y[i];

      }
     xm=sx/n;
     ym=sy/n;
     bxy=(n*sxy-sx*sy)/(n*syy-sy*sy);
     t=xm-ym*bxy;
    if(bxy<0)
    printf("\n Regression line x on y is x =%.2f%.2fy",t,bxy);
   else
     printf("\n Regression line x on y is x =%.2f+%.2fy",t,bxy);
     getch();
  }
```

Output:REGRESSION LINE X ON Y

```
Enter number of values of x and y
8
Enter the values of x
1
2
3
4
5
6
7
8
Enter the values of  y
3
```

```
7
10
12
14
17
20
24
```

```
Regression line x on y is x =-0.24+0.35y
```

Lab Assignment : Regression Lines

1. Write a C program to find regression line y on x and x on y for the following data :

x	1.53	1.78	2.60	2.95	3.42
y	33.50	36.30	40.00	45.80	53.50

Hint : Input : n = 5

Output : Regression line y on x is y = 17.9315 + 9.726x

Regression line x on y is x = −1.54 + 0.10y

2. Write a C program to find both regression lines from the given data :

x	1	2	3	4	5
y	14	13	9	5	2

Hint : Input : n = 5

Output : Regression line y on x is x = 18.20 – 3.20x

Regression line x on y is x = 5.62 – 0.30y.

4. CORRELATION COEFFICIENT

To calculate correlation coefficient r, we use the following formula, known as Karl Pearson correlation coefficient

$$r = \frac{\Sigma(x_i - \bar{x})(y_i - \bar{y})}{\sqrt{\Sigma(x_i - \bar{x})^2 \Sigma(y_i - \bar{y})^2}}$$

where x = mean of x series ; y = mean of y series

Symbol Used :

n = number of different values of x and y ; x [i] = different values of x

y [i] = different values of y ; xm = mean of x series

ym = mean of y series ; sx = sum of x series

sy = sum of y series

dx [i] = deviation from xm of x series
dy [i] = deviation from ym of y series
   sdx = sum of deviations of x series;   sdy = sum of deviation of y series
  sdxy = sum of product of deviation of x and y series

ALGORITHM : CORRELATION COEFFICIENT r

Step 1 :	Start
Step 2 :	Input n
Step 3 :	For i = 0 to n – 1
Step 4 :	Input x [i], y [i]
Step 5 :	sx = 0, sy = 0, sdxy = 0, sdx = 0, sdy = 0
Step 6 :	For i = 0 to n – 1
Step 7 :	sx = sx + x [i]
Step 8 :	sy = sy + y [i]
Step 9 :	End of i loop
Step 10 :	sm = sx/n
Step 11 :	ym = sy/n
Step 12 :	For i = 0 to n – 1
Step 13 :	dx [i] = (x [i] – xm)
Step 14 :	dy [i] = (y [i] – ym)
Step 15 :	sdxy = sdxy + dx [i] * dy [i]
Step 16 :	sdx = sdx + dx [i]
Step 17 :	sdy = sdy + dy [i]
Step 18 :	End of i loop
Step 19 :	$r = sdxy/\sqrt{sdx * sdy}$
Step 20 :	Print "Correlation coefficient r =", r
Step 21 :	Stop

(1) FLOW CHART

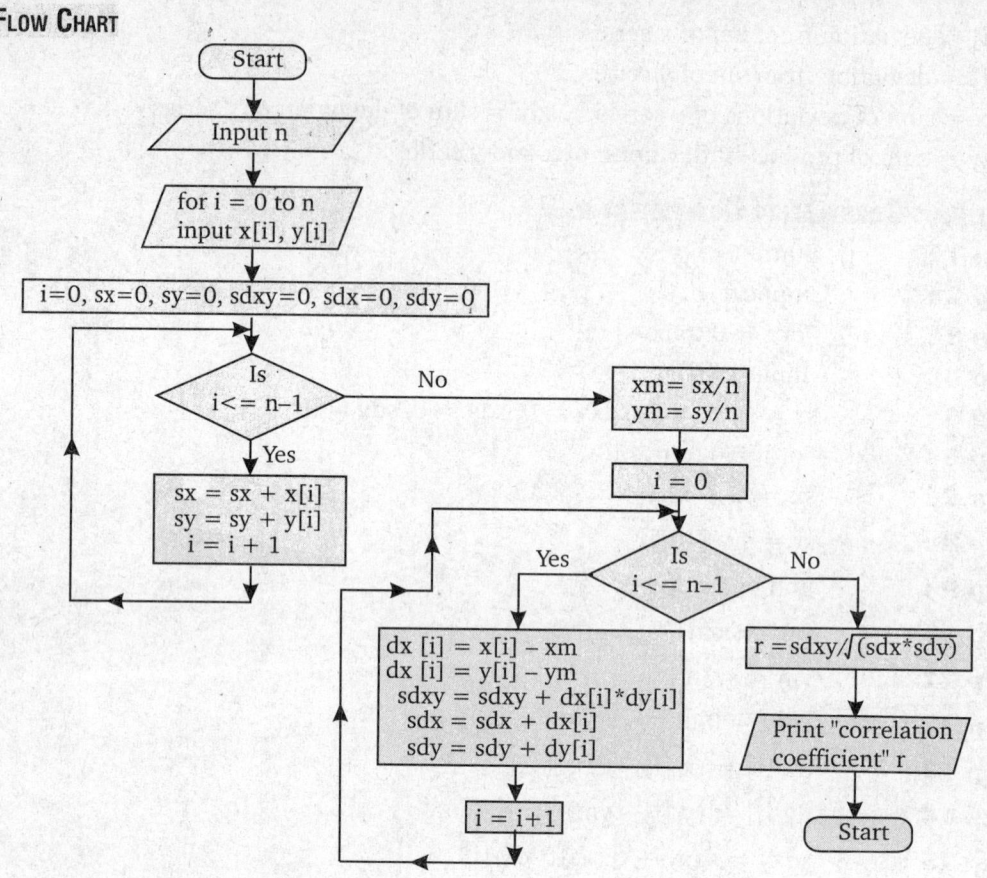

PROGRAM: *Following is a C program to calculate the Karl Pearson coefficient of correlation.*

```
//CORELATION COEFFICIENT
#include<stdio.h>
#include<conio.h>
#include<math.h>
void main()
    {
        clrscr();
        float  x[30],y[30],dx[30],dy[30],sx=0,sy=0,sdx=0;
        float sdxy=0,sdy=0,xm,ym,r;
        int n,i;
        printf("Enter number of values of x and y \n");
        scanf("%d", &n);
        printf("Enter the values of x \n ");
        for(i=0;i<n;i++)
```

```
            scanf("%f",&x[i]);
        printf("Enter the values of  y\n ");
        for(i=0;i<n;i++)
            scanf("%f",&y[i]);
        for(i=0;i<n;i++)
          {  sx+=x[i];
             sy+=y[i];
           }
         xm=sx/n;
         ym=sy/n;
         for (i=0;i<n;i++)
         {  dx[i]=x[i]-xm;
            dy[i]=y[i]-ym;
            sdxy+=dx[i]*dy[i];
            sdx+=dx[i]*dx[i];
            sdy+=dy[i]*dy[i];
          }
          r=sdxy/sqrt(sdx*sdy);
          printf("\n corelation coefficient  r=%f",r);
        getch();
      }
```

Output: CORELATION COEFFICIENT

Enter number of values of x and y

7

Enter the values of x

65

66

67

68

69

70

71

Enter the values of y

67

68

66

69

```
72
72
69
Corelation coefficient r =0.668153
```

LAB ASSIGNMENT : REGRESSION LINES

1. Write a C program to calculate the correlation coefficient from the given data :

x	1	2	3	4	5
y	14	13	9	5	2

Hint : Input : n = 5

Output : r = ...

Regression line x on y is x = −1.54 + 0.10y

2. Write a C program to calculate the correlation coefficient from the data given below:

x	0	1	2	3	4	5
y	25	22	19	17	13	8

Hint : Input : n = 5

Output : r = ...

❈❈❈❈❈❈

GLOSSARY

Absolute Error : If x^A is the approximate value of exact number x^T, then the absolute error denoted by E_a is defined by

$$E_a = \overline{\Delta x} = |x^T - x^A|$$
$$E_a = |x^T - x^A|$$

Angle between Two Lines of Regression : If θ is the angle between two regression lines in the case of two variables x and y, then

$$\tan\theta = \frac{1-r^2}{r} \cdot \frac{\sigma_x \sigma_y}{\sigma_x^2 + \sigma_y^2}$$

where, r, σ_x, σ_y have their usual meaning. Explain the significance of the formula, when $r = 0$ and $r = \pm 1$.

Approximate Numbers : These are numbers which are not exact.

Backward Difference : The differences $y_1 - y_0, y_2 - y_1, ..., y_n - y_{n-1}$ when denoted by $\nabla y_1, \nabla y_2, ..., \nabla y_n$ respectively are called the first backward differences, when ∇ is called backward difference operator.

Binary Arithmetic : Arithmetic operations additions, subtraction, multiplication and division on binary numbers constitute binary arithmetic.

Central Difference : If $y_1 - y_0 = \delta y_{1/2}, y_2 - y_1 = \delta y_{3/2}, y_n - y_{n-1} = \delta y_{n-2}$
Then these differences called central difference and δ is called central difference operator.

Degree of Difference Equation : The degree of difference equation is highest power of y and free from Δ.

Difference Equation : An equation that consists of an independent variable x, a dependent variable y_x and one or more of its difference $\Delta y_x, \Delta^2 y_x, ..., \Delta^n y_x$ is called a difference equation. It is of the form

$$F[x, y_x, \Delta y_x, \Delta^2 y_x, ..., \Delta^n y_x] = 0$$

Difference Quotients : A difference quotient is the quotient obtained by dividing the difference between two values of a function by the difference between the two corresponding values of the independent variable.

Divided Difference : Let $f(x_0), f(x_1), ..., f(x_n)$ be the values of the function corresponding to the values of $x_0, x_1, ..., x_n$ which are not equally spaced. We know that the difference of the function values with respect to the difference of the arguments are called divided differences.

Elliptic Equation : An equation of the type $\dfrac{\partial^2 u}{\partial x^2} + \dfrac{\partial^2 u}{\partial y^2} = 0$ is said to be elliptic equation.

Error : The quantity, True value – Approximate value is called the error.

Exact Numbers : The numbers in which, there is no uncertainty and no approximation, is said to be exact numbers.

Extrapolation : The process of computing the value of the function outside the given range is called extrapolation.

Factorial Notation : A product of the form $x(x-1)(x-2) \ldots (x-n+1)$ is denoted by $x^{(n)}$ and is called a factorial.

Forward Difference : The difference $[f(x_0 + h) - f(x_0)], [f(x + 2h) - f(x_0 + h)],$ $\ldots [f(x_0 + nh) - f(x_0 + \{n-1\}h)]$ are called first forward differences and are denoted by $\Delta f(x_0), \Delta f(x_0 + h) \ldots \Delta f(x_0 + (n-1)h)$. Here, Δ is known as forward difference operator.

Gram-Schmidt Orthogonalizing Process : Let $g_i(x) = x^i$ be the given polynomial of degree i, $i = 0, 1, 2, \ldots, n$. Then the polynomial $g_i^*(x)$ of degree i which are orthogonal over $[a, b]$ with respect to the weight function $W(x)$ can be generated from the recursion relation

$$g_i^*(x) = x^i - \sum_{r=0}^{i-1} g_i^*(x), \; i = 0, 1, 2, \ldots, n$$

where

$$a_{ir} = \frac{\int_a^b W(x)^i g_r^*(x)dx}{\int_a^b W(x) x^i g_r^{*2}(x)dx} \text{ and } g_0^*(x) = 1.$$

Hyperbolic Equation : An equation of the type $a^2 \dfrac{\partial^2 u}{\partial x^2} = \dfrac{\partial^2 u}{\partial t^2}$ or $a^2 u_{xx} - u_{tt} = 0$ is said to the hyperbolic equation.

Inherent Error : The errors which are already present in the statement of a problem before its solution are called inherent errors.

Interpolation : The method of obtaining the value of a function for any intermediate value of the argument from the given set of values of the function for certain values of arguments is know as interpolation.

Lagrange's Interpolation Formula : Let $y = f(x)$ be a function, which takes the values $f(x_0)$, $f(x_1), \ldots f(x_n)$, corresponding to the values of $x = x_0, x_1, \ldots, x_n$, not necessarily equally spaced. Then

$$f(x) = \frac{(x - x_1)(x - x_2)\ldots(x - x_n)}{(x_0 - x_1)(x_0 - x_2)\ldots(x_0 - x_n)} f(x_0) + \frac{(x - x_0)(x - x_2)\ldots(x - x_n)}{(x_1 - x_0)(x_1 - x_2)\ldots(x_1 - x_n)} f(x_1) + \ldots$$

$$+ \frac{(x - x_0)(x - x_1)\ldots(x - x_{n-1})}{(x_n - x_0)(x_n - x_1)\ldots(x_n - x_{n-1})} f(x_n)$$

Linearly Independent Solution or Fundamental Set of Solutions : The rth order homogeneous difference equation

$$y_{x+r} + a_1 y_{x+r-1} + a_2 y_{x+r-2} + \ldots + a_r y_x = 0 \qquad \ldots(1)$$

$y^{(1)}, y^{(2)}, \ldots, y^{(r)}$ the solutions of equation (1) are said to form a fundamental set of solutions of (1) if the rth order determinant

$$\begin{vmatrix} y_0^{(1)} & y_0^{(2)} & \ldots & y_0^{(r)} \\ y_1^{1} & y_1^{(2)} & \ldots & y_1^{(r)} \\ \vdots & \vdots & \vdots & \vdots \\ y_{r-1}^{(1)} & y_{r-1}^{(2)} & \ldots & y_{r-1}^{(r)} \end{vmatrix} \text{ is different from zero.}$$

LU Decomposition Method : This method is based on the fact that every square matrix A can be expressed as the form LU where L is a unit lower triangular matrix while U is upper triangular matrix and provided all the principal minors of A are non-singular.

Newton's Formula for Backward Interpolation : Newton's-Gregory's formula for backward interpolation formula with equal interval is

$$f(a+nh+uh) = f(a+nh) + u\Delta f(a+nh) + \frac{u(u+1)}{2!}\nabla^2 f(a+nh) + ...$$
$$+\frac{u(u+1)(u+2)...(u+n-1)}{n!}\nabla^n f(a+nh)$$

Newton's Formula for Forward Interpolation : Newton's-Gregory's formula for forward interpolation with equal interval is

$$f(x_0+hu) = f(x_0) + u\Delta f(x_0) + \frac{u(u-1)}{2!}\Delta^2 f(x_0) + \frac{u(u-1)(u-2)}{3!}\Delta^3 f(x_0) + ...$$
$$+\frac{u(u-1)(u-2)...(u-(n-1)h)}{n!}\Delta^n f(x_0)$$

where $u = \dfrac{x-x_0}{h}$.

Order of Difference Equation : The order of a difference equation is the difference between the highest and the lowest subscripts of the y, it is free from Δ.

Parabolic Equations : An equation of the type $\dfrac{\partial u}{\partial t} = \alpha^2 \dfrac{\partial^2 u}{\partial x^2}$ is known as parabolic or heat equation.

Percentage Error: The percentage error in x^A, which is the approximate value of x^T is

$$E_p = 100 \times E_r = 100 \times \left| \frac{x^T - x^A}{x^T} \right|$$

Reciprocal Factorial : The reciprocal factorial function $x^{(-n)}$ is defined by

$$x^{(-n)} = \frac{1}{(x+h)(x+2h)...(x+nh)} \quad \text{where } n \in \mathbf{Z}^+$$

Regression Line : The straight line about which the various points may be considered as scattered is called the regression line.

Relative Error : The relative error is the absolute error divided by the true value of the given quantity. It is denoted by E_r and defined as

$$E_r = \left| \frac{x^T - x^A}{x^T} \right| = \frac{\text{Absolute error}}{\text{True value}}$$

Rounding-off a number : The process of cutting-off superfluous digits and retaining as many digits as desired is known as rounding off a number.

Sensitivity Analysis : Investigation to see how small changes in input parameters influence the output are termed as sensitivity analysis, when problem is sensitive to small changes in its parameter, it is impossible to make a numerically stable method for its solution.

Solution of Difference Equation : A function y is a solution of a difference equation over a set A if the value of y make the difference equation a true statement for every point of A.

Spline Function : A spline function of degree n with knots x_i, $i = 0, 1, ..., n$ is a function $s(x)$ with the properties.

(i) On each interval $[x_{i-1}, x_i]$, $1 \leq i \leq n$, $s(x)$ is a polynomial of degree n.

(ii) $s(x)$ and $s'(x)$, $s''(x)$, ..., $s_{n-1}(x)$ are continuous on $[a, b]$.

Stirling's Difference Formula : The Stirling's formula can be obtained by taking the average of Gauss's forward difference formula and Gauss's backward difference formula.

The Cayley-Hamilton Theorem : Every square matrix satisfies its characteristic equation.

Triangularisation Method : This method is also known as 'Factorization' method. We factorize the given matrix A as LU.

i.e.,
$$A = LU$$

where L is a lower triangular matrix having diagonal elements unity and U is an upper triangular matrix.

Truncation Error : The truncation errors arises by using some approximations in place of an exact mathematical procedure.

Uniform Approximation : In uniform approximation, we use the uniform norm to continuous function over a close interval. Here, it should be noted that Weierstrass approximation theorem provide the possibility of uniform approximation.

□□□□□□

BIBLIOGRAPHY

1.	Bunchanan, JL and PR Turner	1992	*Numerical Methods and Analysis* McGraw-Hill book company
2.	Conte, SD and Carl De Boor	1981	*Elementary Numerical Analysis, 3rd Edition* McGraw-Hill Book Company,
3.	Mathews, JH	1944	*Numerical Methods for Mathematics, Science and Engineering, 2nd Edition* Prentice-Hall of India
4.	Anita, H.M	1991	*Numerical Methods for Scientists and Engineers,* TATA McGraw-Hill Publishing Company New Delhi.
5.	Balagurusamy, E.	1999	*Numerical Methods* Tata McGraw-Hill, New Delhi.
6.	Chapra, S.C and Canale, R.P	1989	*Numerical Methods for Engineers,* McGraw-Hill Book Company.
7.	Gerald, C.F. and Wheatly. P.O.	1994	*Applied Numerical Analysis,* Addison-Wesley Publishing Company.
8.	Jain, M.K Iyengar S.R.K and Jain R.K.	2003	*Numerical Methods for scientific and Engineering Computation* New Age International publisher, India.
9.	Pearson, C.E.	1986	*Numerical Methods for Engineering and Science* Van Nostrand Rainhold, New-York.
10.	Salvadori, M.G. and Baron, M.L.	1961	*Numerical Methods in Engineering* Prentice Hall, Englewood Cliffs, N.J.
11.	B.S. Grewal	2004	*Numerical Methods in Engineering and Science* Khanna Publishers.

❑❑❑❑❑❑

Index

A

Accuracy of numbers	93
Angle between two lines of regression	830
Application of error formula to the fundamental operations of arithmetics	117

B

Base conversion	134
Bessel's difference formula	382
Binary arithmetic	143
Bisection method	156
Blunders	129

C

Central differences interpolation formulae	379
Chebychev's formula	466
Chebyshev polynomial approximations	859
Choice to select the suitable interpolation formula	381
Choleski's method	541
Classification of partial differential equations	745
Complex roots	205
Computer software	133
Computer storage	109
Concept of normalized floating point	111
Contents of C-program	3
Continuous frequency distribution	796
Convergence of a method	724
Crank-Nicholson method	761
Crout's method	267
Curve fitting	798
Curve fitting by sum of exponentials	817

D

Data approximation of functions	848
Data input with scanf () function	18
Data output with printf () function	16
Decision control in the C-program	23
Deductions from Newton-Cote's formula	460
Degree of difference equation	632
Derivative using Newton's divided difference formula	426
Derivative using Newton's forward interpolation formula	423

E

Derivatives using Newton's backward difference formula	425
Derivatives using Stirling's formula	426
Difference equation as a relation among the value of y_x	631
Difference quotients	743
Difference schemes	299
Differences between divided difference and ordinary difference	323
Differences of zero	321
Divided difference	321
Dufort and Frankel's method	762

E

Eigen values and eigen vectors	553
Eigenvalues of special type of matrices	556
Elliptic equations	745
Enherent errors	103
Error analysis	722
Error analysis in numerical differentiation	442
Error in a series approximation	116
Error in determinants	117
Error in higher order derivatives	442
Errors and their analysis	96
Escalater method	547
Euler's method	671
Euler's modified method	674
Everett's difference formula	383
Existence and uniqueness of solution of differential equation	669
Existence and uniqueness theorem	635
Existence of solution	252

F

Factorial notation	303
Finite difference calculus	333
Fitting a polynomial function	838
Fitting of some special curves	803
Floating point arithmetic and errors	109
Functions	83
Fundamental theorem of difference calculus	307

G

Gauss elimination method 254
Gauss elimination method 527
Gauss's quadrature formula 463
Gauss-Jordan method 527
General solution of second order homogeneous difference equation 650
General solution of the homogeneous difference equation of order n 652
Given's method 608
Graeffe's root square method 221
Gram-Schmidt orthogonalizing process 854
Graphical representation 796

H

Header files 12
Hermite's interpolation formula 351
Hierarchy of operators 11
Higher order rules 490
House holder's method 613
Hyperbolic equation 771

I

Important features of C 1
Input-output manipulations 22
Instructions of a C-program 13
Interpolation 332
Inverse power method 597
Inversion of complex matrices 550
Iteration method 170
Iterative method 763
Iterative method for the system of non-linear equations 177
Iterative methods 273

J

Jacobi's method 602
Jordan's method 266

L

Lagrange's interpolation formula 363
Lanczos economization of power series for a general function 860
Legendre and Chebyshev polynomials 856
Lin-Bairstow's method 213
Linear difference equation 634
Linear equations 251
Linear homogeneous equation with constant coefficients 644

Linearly independent solution or fundamental set of solutions 646
LU decomposition method or method of factorization 260

M

Machine computations 130
Maxima and minima of tabulated function 427
Method of curve-fitting 802
Method of least squares 799
Methods of interpolation 332
Methods of solution 156
Milne's methods 708
Muller's method 208
Multiple linear regression 844

N

Newton's divided difference formula 355
Newton's formulae for interpolation 333
Newton's method for complex roots 206
Newton-Cote's formula 458
Newton-Raphson's method 191
Non-linear regression 840
Number system 133
Numerical evaluation of the singular integral 491
Numerical instability 130
Numerical quadrature 449

O

Operators in a C-program 8
Order of approximations 125
Order of difference equation 632

P

Parabolic equations 759
Particular solution of the complete difference equation 655
Picard's method of successive approximations 685
Pitfalls of floating point representation 111
Power method 588
Program writing in C 4
Propagation of error 126
Properties of Cote's numbers 459
Properties of regression coefficients 829
Properties of the equations and its roots 155

Q

Quadrature formula for equally spaced arguments 449

R

Rational approximation 860

Regression analysis 827

Regression analysis of grouped data 844

Regula-Falsi method (or method of false position) 179

Relation between eigenvalues and eigenvectors 554

Relation between the operators 302

Repetition control statements 41

Rounding off error 104

Runge-Kutta method 693

Rutishauser method 600

S

Secant method 167

Simplified determination of regression analysis 843

Simpson's 1/3 rule 451

Simpson's 3/8 rule 453

Simultaneous differential equations 703

Solution as sequences 641

Solution by Bender-Schmidt's method 760

Solution by forward difference method 759

Solution by Taylor series 680

Solution of difference equation 632

Solution of Laplace equations by Libermann's iteration process 746

Solution of second order differential equations 705

Solution of simultaneous difference equations 662

Solution of the equation $y_x + 1 = ay_x + b$ 636

Solution of two-dimensional heat equation: ADE method 769

Spline function 821

Stability analysis 724

Stirling's difference formula 380

String–the array of characters 74

Structured programming concept of modular programming 2

T

The Cayley-Hamilton theorem 576

The general formula for errors 106

The Trapezoidal rule 450

To express a given polynomial into factorial notation 304

Triangularisation method 533

Truncation error 104

Types of approximations 849

U

Uniform Approximation 857

Use of backslash characters 17

Use of orthogonal functions 853

W

Weddle's rule 455

Working with 'C' program 2

Working with arrays 60